THE PHYSICS TEXTBOOK 85% OF STUDENTS PREFER COMES WITH THE MOST EFFECTIVE AND VALUABLE MULTIMEDIA AVAILABLE TO HELP YOU SUCCEED IN YOUR COURSE.

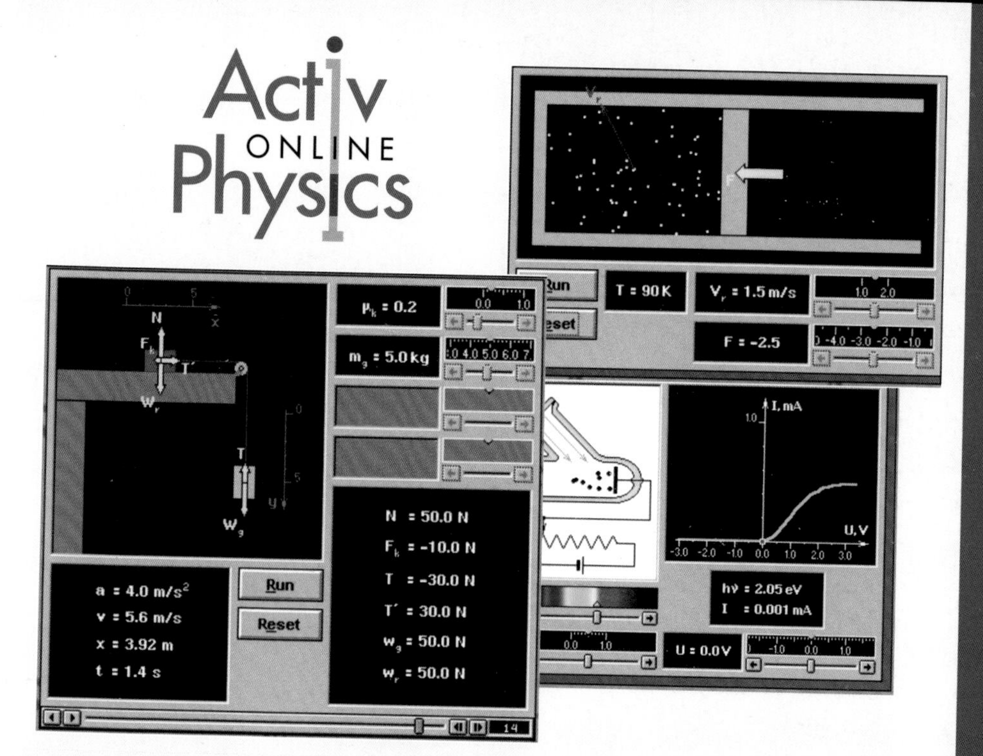

ActivPhysics™ OnLine utilizes visualization, simulation and multiple representations to help you better understand key physical processes, experiment quantitatively, and develop your critical-thinking skills. This library of online interactive simulations are coupled with thought-provoking questions and activities to guide your understanding of physics.

Website: www.aw-bc.com/knight

Minimum System Requirements:
Windows: 250 MHz; OS 98, NT, ME, 2000, XP
Macintosh: 233 MHz; OS 9.2, 10
Both:

- 64 RAM installed
- 1024 x 768 screen resolution
- Browsers: Internet Explorer 5.5, 6.0; Netscape 6.2.3, 7.0
- Plug Ins: Macromedia's Flash 6.2.3

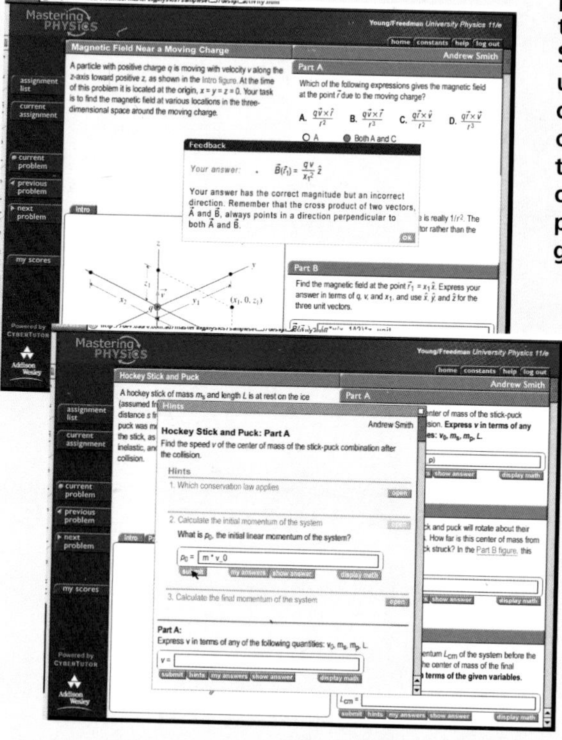

MasteringPhysics™ is the first Socratic tutoring system developed specifically for physics students like you. It is the result of years of detailed studies of how students work physics problems, and where they get stuck and need help. Studies show students who used MasteringPhysics significantly improved their scores on traditional final exams and the Force Concept Inventory (a conceptual test) when compared with traditional hand-graded homework.

With your purchase of a new copy of Knight's *Physics for Scientists and Engineers,* you should have received a Student Access Kit for MasteringPhysics™ if your professor required it as a component of your course. The kit contains instructions and a code for you to access MasteringPhysics.

If you did not purchase a new textbook and your professor requires you to enroll in the **MasteringPhysics** online homework and tutorial program, you may purchase an online subscription with a major credit card. Go to www.masteringphysics.com and follow the links to purchasing online.

Minimum System Requirements:
Windows: 250 MHz; OS 98, NT, ME, 2000, XP
Macintosh: 233 MHz; OS 9.2, 10
RedHat Linux 8.0
All:

- 64 RAM installed
- 1024 x 768 screen resolution
- Browsers: Internet Explorer 5.0, 5.5, 6.0; Netscape 6.2.3, 7.0; Mozilla 1.2, 1.3

MasteringPhysics™ is powered by MyCyberTutor by Effective Educational Technologies

Table of Problem-Solving Strategies

Note for users of the five-volume edition:
Volume 1 (pp. 1–481) includes chapters 1–15.
Volume 2 (pp. 482–607) includes chapters 16–19.
Volume 3 (pp. 608–779) includes chapters 20–24.
Volume 4 (pp. 780–1194) includes chapters 25–36.
Volume 5 (pp. 1148–1383) includes chapters 36–42.

Chapters 37–42 are not in the Standard Edition.

ActivPhysics™ OnLine Activities

www.aw-bc.com/knight

1.1 Analyzing Motion Using Diagrams
1.2 Analyzing Motion Using Graphs
1.3 Predicting Motion from Graphs
1.4 Predicting Motion from Equations
1.5 Problem-Solving Strategies for Kinematics
1.6 Skier Races Downhill
1.7 Balloonist Drops Lemonade
1.8 Seat Belts Save Lives
1.9 Screeching to a Halt
1.10 Pole-Vaulter Lands
1.11 Car Starts, Then Stops
1.12 Solving Two-Vehicle Problems
1.13 Car Catches Truck
1.14 Avoiding a Rear-End Collision
2.1.1 Force Magnitudes
2.1.2 Skydiver
2.1.3 Tension Change
2.1.4 Sliding on an Incline
2.1.5 Car Race
2.2 Lifting a Crate
2.3 Lowering a Crate
2.4 Rocket Blasts Off
2.5 Truck Pulls Crate
2.6 Pushing a Crate Up a Wall
2.7 Skier Goes Down a Slope
2.8 Skier and Rope Tow
2.9 Pole-Vaulter Vaults
2.10 Truck Pulls Two Crates
2.11 Modified Atwood Machine
3.1 Solving Projectile Motion Problems
3.2 Two Balls Falling
3.3 Changing the *x*-Velocity
3.4 Projectile *x*- and *y*-Accelerations
3.5 Initial Velocity Components
3.6 Target Practice I
3.7 Target Practice II
4.1 Magnitude of Centripetal Acceleration
4.2 Circular Motion Problem Solving
4.3 Cart Goes Over Circular Path
4.4 Ball Swings on a String
4.5 Car Circles a Track
4.6 Satellites Orbit
5.1 Work Calculations
5.2 Upward-Moving Elevator Stops
5.3 Stopping a Downward-Moving Elevator
5.4 Inverse Bungee Jumper
5.5 Spring-Launched Bowler
5.6 Skier Speed
5.7 Modified Atwood Machine
6.1 Momentum and Energy Change
6.2 Collisions and Elasticity
6.3 Momentum Conservation and Collisions
6.4 Collision Problems
6.5 Car Collision: Two Dimensions
6.6 Saving an Astronaut
6.7 Explosion Problems
6.8 Skier and Cart
6.9 Pendulum Bashes Box
6.10 Pendulum Person-Projectile Bowling
7.1 Calculating Torques
7.2 A Tilted Beam: Torques and Equilibrium
7.3 Arm Levers
7.4 Two Painters on a Beam
7.5 Lecturing from a Beam
7.6 Rotational Inertia
7.7 Rotational Kinematics
7.8 Rotoride: Dynamics Approach
7.9 Falling Ladder
7.10 Woman and Flywheel Elevator: Dynamics Approach
7.11 Race Between a Block and a Disk
7.12 Woman and Flywheel Elevator: Energy Approach
7.13 Rotoride: Energy Approach
7.14 Ball Hits Bat
8.1 Characteristics of a Gas
8.2 Maxwell-Boltzmann Distribution: Conceptual Analysis
8.3 Maxwell-Boltzmann Distribution: Quantitative Analysis
8.4 State Variables and Ideal Gas Law
8.5 Work Done by a Gas
8.6 Heat, Internal Energy, and First Law of Thermodynamics
8.7 Heat Capacity
8.8 Isochoric Process
8.9 Isobaric Process
8.10 Isothermal Process
8.11 Adiabatic Process
8.12 Cyclic Process: Strategies
8.13 Cyclic Process: Problems
8.14 Carnot Cycle
9.1 Position Graphs and Equations
9.2 Describing Vibrational Motion
9.3 Vibrational Energy
9.4 Two Ways to Weigh Young Tarzan
9.5 Ape Drops Tarzan
9.6 Releasing a Vibrating Skier I
9.7 Releasing a Vibrating Skier II
9.8 One- and Two-Spring Vibrating Systems
9.9 Vibro-Ride
9.10 Pendulum Frequency
9.12 Risky Pendulum Walk
9.13 Physical Pendulum
10.1 Properties of Mechanical Waves
10.2 Speed of Waves on a String
10.3 Speed of Sound in a Gas
10.4 Standing Waves on Strings
10.5 Tuning a Stringed Instrument: Standing Waves
10.6 String Mass and Standing Waves
10.7 Beats and Beat Frequency
10.8 Doppler Effect: Conceptual Introduction
10.9 Doppler Effect: Problems
10.10 Complex Waves: Fourier Analysis
11.1 Electric Force: Coulomb's Law
11.2 Electric Force: Superposition Principle
11.3 Electric Force Superposition Principle (Quantitative)
11.4 Electric Field: Point Charge
11.5 Electric Field Due to a Dipole
11.6 Electric Field: Problems
11.7 Electric Flux
11.8 Gauss's Law
11.9 Motion of a Charge in an Electric Field: Introduction
11.10 Motion in an Electric Field: Problems
11.11 Electric Potential: Qualitative Introduction
11.12 Electric Potential, Field, and Force
11.13 Electrical Potential Energy and Potential
12.1 DC Series Circuits (Qualitative)
12.2 DC Parallel Circuits
12.3 DC Circuit Puzzles
12.4 Using Ammeters and Voltmeters
12.5 Using Kirchhoff's Laws
12.6 Capacitance
12.7 Series and Parallel Capacitors
12.8 *RC* Circuit Time Constants
13.1 Magnetic Field of a Wire
13.2 Magnetic Field of a Loop
13.3 Magnetic Field of a Solenoid
13.4 Magnetic Force on a Particle
13.5 Magnetic Force on a Wire
13.6 Magnetic Torque on a Loop
13.7 Mass Spectrometer
13.8 Velocity Selector
13.9 Electromagnetic Induction
13.10 Motional emf
14.1 The *RL* Circuit
14.2 The *RLC* Oscillator
14.3 The Driven Oscillator
15.1 Reflection and Refraction
15.2 Total Internal Reflection
15.3 Refraction Applications
15.4 Plane Mirrors
15.5 Spherical Mirrors: Ray Diagrams
15.6 Spherical Mirror: The Mirror Equation
15.7 Spherical Mirror: Linear Magnification
15.8 Spherical Mirror: Problems
15.9 Thin-Lens Ray Diagrams
15.10 Converging Lens Problems
15.11 Diverging Lens Problems
15.12 Two-Lens Optical Systems
16.1 Two-Source Interference: Introduction
16.2 Two-Source Interference: Qualitative Questions
16.3 Two-Source Interference: Problems
16.4 The Grating: Introduction and Qualitative Questions
16.5 The Grating: Problems
16.6 Single-Slit Diffraction
16.7 Circular Hole Diffraction
16.8 Resolving Power
16.9 Polarization
17.1 Relativity of Time
17.2 Relativity of Length
17.3 Photoelectric Effect
17.4 Compton Scattering
17.5 Electron Interference
17.6 Uncertainty Principle
17.7 Wave Packets
18.1 The Bohr Model
18.2 Spectroscopy
18.3 The Laser
19.1 Particle Scattering
19.2 Nuclear Binding Energy
19.3 Fusion
19.4 Radioactivity
19.5 Particle Physics
20.1 Potential Energy Diagrams
20.2 Particle in a Box
20.3 Potential Wells
20.4 Potential Barriers

Physics for Scientists and Engineers Volume 1

A Strategic Approach

Randall D. Knight
California Polytechnic State University, San Luis Obispo

San Francisco Boston New York
Cape Town Hong Kong London Madrid Mexico City
Montreal Munich Paris Singapore Sydney Tokyo Toronto

Executive Editor:	Adam Black, Ph.D.
Development Editor:	Alice Houston, Ph.D.
Project Manager:	Laura Kenney Editorial & Production Services
Associate Editor:	Liana Allday
Media Producer:	Claire Masson
Marketing Manager:	Christy Lawrence
Market Development:	Susan Winslow
Manufacturing Supervisor:	Vivian McDougal
Art Director:	Blakely Kim
Production Service:	Thompson Steele, Inc.
Text Design:	Mark Ong, Side by Side Studios
Cover Design:	Yvo Riezebos Design
Illustrations:	Precision Graphics
Photo Research:	Cypress Integrated Systems
Cover Printer:	Phoenix Color Corporation
Printer and Binder:	R. R. Donnelley & Sons

Cover Image:	Rainbow/PictureQuest
Credits:	see page C–1

Library of Congress Cataloging-in-Publication Data
Knight, Randall Dewey.
Physics for scientists and engineers : a strategic approach / Randall D. Knight.
p. cm.
Includes index.
ISBN 0-8053-8960-1 (extended ed. with MasteringPhysics)
1. Physics I. Title.

QC23.2.K65 2004
530--dc22

2003062809

ISBN 0-8053-8964-4 Volume 1 with MasteringPhysics
ISBN 0-8053-9008-1 Volume 1 without MasteringPhysics

2 3 4 5 6 7 8 9 10—DOW—06 05 04
www.aw-bc.com

Brief Contents

About the Author

Randy Knight has taught introductory physics for over 20 years at Ohio State University and California Polytechnic University, where he is currently Professor of Physics. Professor Knight received a bachelor's degree in physics from Washington University in St. Louis and a Ph.D. in physics from the University of California, Berkeley. He was a post-doctoral fellow at the Harvard-Smithsonian Center for Astrophysics before joining the faculty at Ohio State University. It was at Ohio State that he began to learn about the research in physics education that, many years later, led to this book.

Professor Knight's research interests are in the field of lasers and spectroscopy, and he has published over 25 research papers. He recently led the effort to establish an environmental studies program at Cal Poly, where, in addition to teaching introductory physics, he also teaches classes on energy, oceanography, and environmental issues. When he's not in the classroom or in front of a computer, you can find Randy hiking, sea kayaking, playing the piano, or spending time with his wife Sally and their seven cats.

Preface to the Instructor

In 1997 we published *Physics: A Contemporary Perspective.* This was the first comprehensive, calculus-based textbook to make extensive use of results from physics education research. The development and testing that led to this book had been partially funded by the National Science Foundation. In the preface we noted that it was a "work in progress" and that we very much wanted to hear from users—both instructors and students—to help us shape the book into a final form.

And hear from you we did! We received feedback and reviews from roughly 150 professors and, especially important, 4500 of their students. This textbook, the newly titled *Physics for Scientists and Engineers: A Strategic Approach*, is the result of synthesizing that feedback and using it to produce a book that we hope is uniquely tuned to helping today's students succeed. It is the first introductory textbook built from the ground up on research into how students can more effectively learn physics.

Objectives

My primary goals in writing *Physics for Scientists and Engineers: A Strategic Approach* have been:

- To produce a textbook that is more focused and coherent, less encyclopedic.
- To move key results from physics education research into the classroom in a way that allows instructors to use a range of teaching styles.
- To provide a balance of quantitative reasoning and conceptual understanding, with special attention to concepts known to cause student difficulties.
- To develop students' problem-solving skills in a systematic manner.
- To support an active-learning environment.

These goals and the rationale behind them are discussed at length in my small paperback book, *Five Easy Lessons: Strategies for Successful Physics Teaching* (Addison Wesley, 2002). Please request a copy from your local Addison Wesley sales representative if it would be of interest to you (ISBN 0-8053-8702-1).

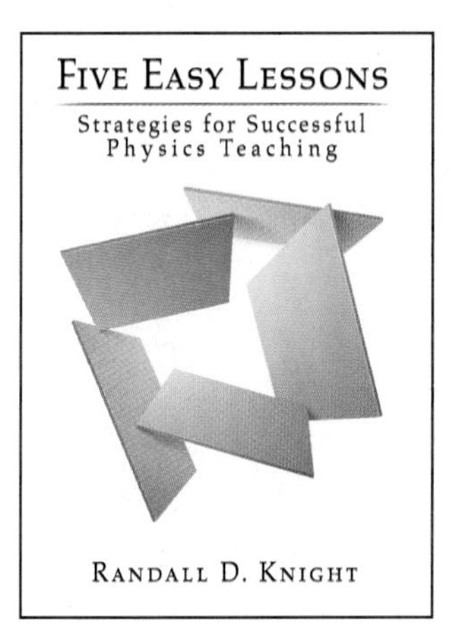

Textbook Organization

The 42-chapter extended edition (ISBN 0-8053-8685-8) of *Physics for Scientists and Engineers* is intended for use in a three-semester course. Most of the 36-chapter standard edition (ISBN 0-8053-8982-2), ending with relativity, can be covered in two semesters, but the judicious omission of a few chapters will avoid rushing through the material and give students more time to develop their knowledge and skills.

There's a growing sentiment that quantum physics is quickly becoming the province of engineers, not just scientists, and that even a two-semester course should include a reasonable introduction to quantum ideas. The *Instructor's Guide* outlines a couple of routes through the book that allow most of the quantum physics chapters to be reached in two semesters. I've written the book with the hope that an increasing number of instructors will choose one of these routes.

- **Extended edition,** with modern physics (ISBN 0-8053-8685-8): chapters 1–42.
- **Standard edition** (ISBN 0-8053-8982-2): chapters 1–36.
- **Volume 1** (ISBN 0-8053-8963-6) covers mechanics: chapters 1–15.
- **Volume 2** (ISBN 0-8053-8966-0) covers thermodynamics: chapters 16–19.
- **Volume 3** (ISBN 0-8053-8969-5) covers waves and optics: chapters 20–24.
- **Volume 4** (ISBN 0-8053-8972-5) covers electricity and magnetism, plus relativity: chapters 25–36.
- **Volume 5** (ISBN 0-8053-8975-X) covers relativity and quantum physics: chapters 36–42.
- **Volumes 1–5** boxed set (ISBN 0-8053-8978-4).

The full textbook is divided into seven parts: Part I: *Newton's Laws*, Part II: *Conservation Laws*, Part III: *Applications of Newtonian Mechanics*, Part IV: *Thermodynamics*, Part V: *Waves and Optics*, Part VI: *Electricity and Magnetism*, and Part VII: *Relativity and Quantum Mechanics*. Although I recommend covering the parts in this order (see below), doing so is by no means essential. Each topic is self-contained, and Parts III–VI can be rearranged to suit an instructor's needs. To facilitate a reordering of topics, the full text is available in the five individual volumes listed in the margin.

Organization Rationale: Thermodynamics is placed before waves because it is a continuation of ideas from mechanics. The key idea in thermodynamics is energy, and moving from mechanics into thermodynamics allows the uninterrupted development of this important idea. Further, waves introduce students to functions of two variables, and the mathematics of waves is more akin to electricity and magnetism than to mechanics. Thus moving from waves to fields to quantum physics provides a gradual transition of ideas and skills.

The purpose of placing optics with waves is to provide a coherent presentation of wave physics, one of the two pillars of classical physics. Optics as it is presented in introductory physics makes no use of the properties of electromagnetic fields. There's little reason other than historical tradition to delay optics until after E&M. The documented difficulties that students have with optics are difficulties with waves, not difficulties with electricity and magnetism. However, the optics chapters are easily deferred until the end of Part VI for instructors who prefer that ordering of topics.

More Effective Problem-Solving Instruction

Careful and systematic instruction is provided on all aspects of problem solving. Some of the features that support this approach are described here, and more details are provided in the *Instructor's Guide*.

- An emphasis on using *multiple representations*—descriptions in words, pictures, graphs, and mathematics—to look at a problem from many perspectives.
- The explicit use of *models*, such as the particle model, the wave model, and the field model, to help students recognize and isolate the essential features of a physical process.
- TACTICS BOXES for the development of particular skills, such as drawing a free-body diagram or using Lenz's law. Tactics Box steps are explicitly illustrated in subsequent worked examples, and these are often the starting point of a full problem-solving strategy.

TACTICS BOX 4.3 **Drawing a free-body diagram**

1. **Identify all forces acting on the object.** This step was described in Tactics Box 4.2.
2. **Draw a coordinate system.** Use the axes defined in your pictorial representation. If those axes are tilted, for motion along an incline, then the axes of the free-body diagram should be similarly tilted.
3. **Represent the object as a dot at the origin of the coordinate axes.** This is the particle model.
4. **Draw vectors representing each of the identified forces.** This was described in Tactics Box 4.1. Be sure to label each force vector.
5. **Draw and label the *net force* vector $\vec{F}_{net}$.** Draw this vector beside the diagram, not on the particle. Or, if appropriate, write $\vec{F}_{net} = \vec{0}$. Then check that $\vec{F}_{net}$ points in the same direction as the acceleration vector $\vec{a}$ on your motion diagram.

TACTICS BOX 32.2 **Evaluating line integrals**

1. If $\vec{B}$ is everywhere perpendicular to a line, the line integral of $\vec{B}$ is

$$\int_i^f \vec{B} \cdot d\vec{s} = 0$$

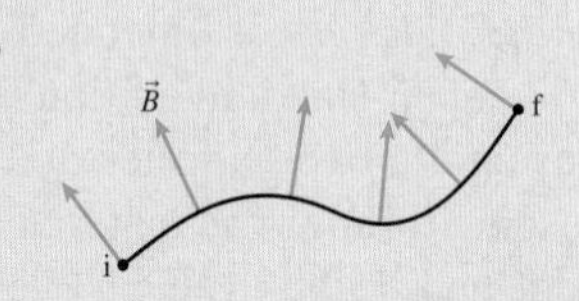

2. If $\vec{B}$ is everywhere tangent to a line of length L *and* has the same magnitude B at every point, the line integral of $\vec{B}$ is

$$\int_i^f \vec{B} \cdot d\vec{s} = BL$$

- **PROBLEM-SOLVING STRATEGIES** that help students develop confidence and more proficient problem-solving skills through the use of a consistent four-step approach: **MODEL, VISUALIZE, SOLVE, ASSESS.** Strategies are provided for each broad class of problems, such as dynamics problems or problems involving electromagnetic induction. The MP icon directs students to the specially developed *Skill Builder* tutorial problems in MasteringPhysics™ (see page xi), where they can interactively work through each of these strategies online.
- Worked **EXAMPLES** that illustrate good problem-solving practices through the consistent use of the four-step problem-solving approach and, where appropriate, the Tactics Box steps. The worked examples are often very detailed and carefully lead the student step by step through the *reasoning* behind the solution, not just through the numerical calculations. Steps that are often implicit or omitted in other textbooks, because they seem so obvious to experts, are explicitly discussed since research has shown these are often the points where students become confused.
- **NOTE ▶** Paragraphs within worked examples caution against common mistakes and point out useful tips for tackling problems.
- The *Student Workbook* (see page xi), a unique component of this text, bridges the gap between worked examples and end-of-chapter problems. It provides qualitative problems and exercises that focus on developing the skills and conceptual understanding necessary to solve problems with confidence.
- Approximately 3000 original and diverse *end-of-chapter problems* have been carefully crafted to exercise and test the full range of qualitative and quantitative problem-solving skills. *Exercises*, which are keyed to specific sections, allow students to practice basic skills and computations. *Problems* require a better understanding of the material and often draw upon multiple representations of knowledge. *Challenge Problems* are more likely to use calculus, utilize ideas from more than one chapter, and sometimes lead students to explore topics that weren't explicitly covered in the chapter.

MP **PROBLEM-SOLVING STRATEGY 5.2 Dynamics problems**

MODEL Make simplifying assumptions.

VISUALIZE

Pictorial representation. Show important points in the motion with a sketch, establish a coordinate system, define symbols, and identify what the problem is trying to find. This is the process of translating words to symbols.

Physical representation. Use a motion diagram to determine the object's acceleration vector $\vec{a}$. Then identify all forces acting on the object and show them on a free-body diagram.

It's OK to go back and forth between these two steps as you visualize the situation.

SOLVE The mathematical representation is based on Newton's second law

$$\vec{F}_{\text{net}} = \sum_i \vec{F}_i = m\vec{a}$$

The vector sum of the forces is found directly from the free-body diagram. Depending on the problem, either

- Solve for the acceleration, then use kinematics to find velocities and positions, or
- Use kinematics to determine the acceleration, then solve for unknown forces.

ASSESS Check that your result has the correct units, is reasonable, and answers the question.

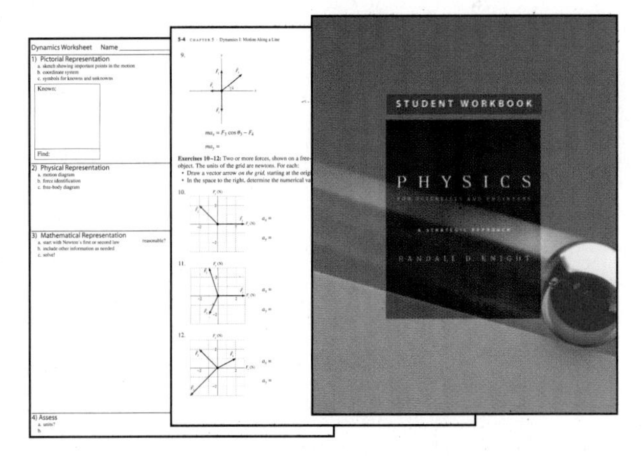

Proven Features to Promote Deeper Understanding

Research has shown that many students taking calculus-based physics arrive with a wealth of misconceptions and subsequently struggle to develop a coherent understanding of the subject. Using a number of unique, reinforcing techniques, this book tackles these issues head-on to enable students to build a solid foundation of understanding.

- A *concrete-to-abstract* approach introduces new concepts through observations about the real world and everyday experience. Step by step, the text then builds up the concepts and principles needed by a theory that will make sense of the observations and make new, testable predictions. This inductive approach better matches how students learn, and it reinforces how physics—and science in general—operates.

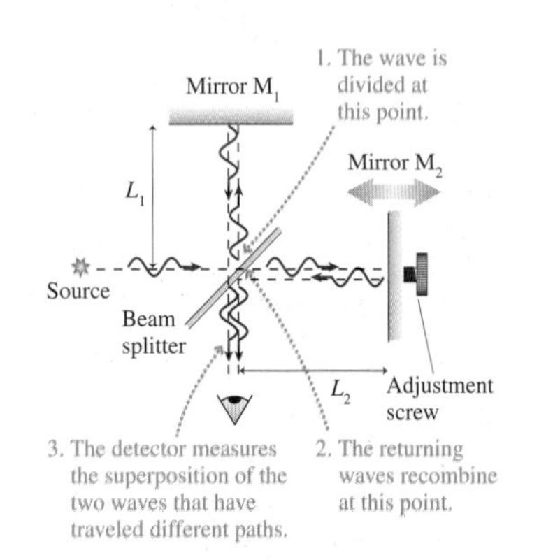
1. The wave is divided at this point.
Mirror M_1
Mirror M_2
L_1
Source
Beam splitter
L_2
Adjustment screw
3. The detector measures the superposition of the two waves that have traveled different paths.
2. The returning waves recombine at this point.

Annotated **FIGURE** showing the operation of the Michelson interferometer.

- **STOP TO THINK** questions embedded in each chapter allow students to assess whether they've understood the main idea of a section. The *Stop to Think* questions, which include concept questions, ratio reasoning, and ranking tasks, are primarily derived from physics education research.
- **NOTE ▶** paragraphs draw attention to common misconceptions, clarify possible confusions in terminology and notation, and provide important links to previous topics.
- Unique *annotated figures*, based on research into visual learning modes, make the artwork a teaching tool on a par with the written text. Commentary in blue—the "instructor's voice"—helps students "read" the figure. Students "learn by viewing" how to interpret a graph, how to translate between multiple representations, how to grasp a difficult concept through a visual analogy, and many other important skills.
- The learning goals and links that begin each chapter outline what the student needs to remember from previous chapters and what to focus on in the chapter ahead.
  - ▶ **Looking Ahead** lists key concepts and skills the student will learn in the coming chapter.
  - ◀ **Looking Back** suggests important topics students should review from previous chapters.
- Unique schematic *Chapter Summaries* help students organize their knowledge in an expert-like hierarchy, from general principles (top) to applications (bottom). Side-by-side pictorial, graphical, textual, and mathematical representations are used to help students with different learning styles and enable them to better translate between these key representations.
- *Part Overviews and Summaries* provide a global framework for the student's learning. Each part begins with an overview of the chapters ahead. It then concludes with a broad summary to help students draw connections between the concepts presented in that set of chapters. **KNOWLEDGE STRUCTURE** tables in the part summaries, similar to the chapter summaries, help students see a forest rather than dozens of individual trees.

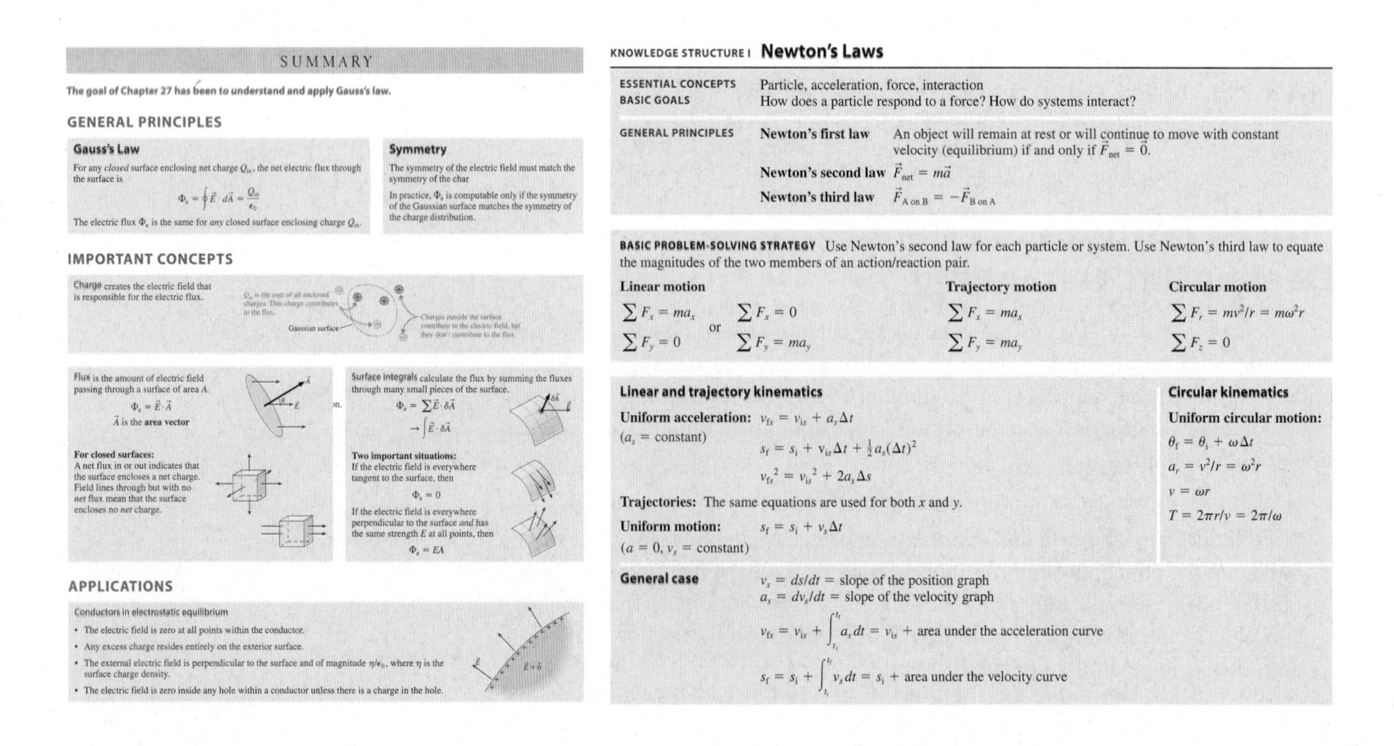

SUMMARY

The goal of Chapter 27 has been to understand and apply Gauss's law.

GENERAL PRINCIPLES

Gauss's Law

For any *closed* surface enclosing net charge Q_{in}, the net electric flux through the surface is

$$\Phi_e = \oint \vec{E} \cdot d\vec{A} = \frac{Q_{in}}{\epsilon_0}$$

The electric flux Φ_e is the same for *any* closed surface enclosing charge Q_{in}.

Symmetry

The symmetry of the electric field must match the symmetry of the char

In practice, Φ_e is computable only if the symmetry of the Gaussian surface matches the symmetry of the charge distribution.

IMPORTANT CONCEPTS

Charge creates the electric field that is responsible for the electric flux.

Gaussian surface

Flux is the amount of electric field passing through a surface of area A.

$$\Phi_e = \vec{E} \cdot \vec{A}$$

$\vec{A}$ is the **area vector**

For closed surfaces: A net flux in or out indicates that the surface encloses a net charge. Field lines through but with no *net* flux mean that the surface encloses no *net* charge.

Surface integrals calculate the flux by summing the fluxes through many small pieces of the surface.

$$\Phi_e = \sum \vec{E} \cdot \delta\vec{A} \rightarrow \int \vec{E} \cdot \delta\vec{A}$$

Two important situations:
If the electric field is everywhere tangent to the surface, then

$$\Phi_e = 0$$

If the electric field is everywhere perpendicular to the surface *and* has the same strength E at all points, then

$$\Phi_e = EA$$

APPLICATIONS

Conductors in electrostatic equilibrium

- The electric field is zero at all points within the conductor.
- Any excess charge resides entirely on the exterior surface.
- The external electric field is perpendicular to the surface and of magnitude η/ϵ_0, where η is the surface charge density.
- The electric field is zero inside any hole within a conductor unless there is a charge in the hole.

KNOWLEDGE STRUCTURE I **Newton's Laws**

ESSENTIAL CONCEPTS Particle, acceleration, force, interaction
BASIC GOALS How does a particle respond to a force? How do systems interact?

GENERAL PRINCIPLES

Newton's first law An object will remain at rest or will continue to move with constant velocity (equilibrium) if and only if $\vec{F}_{net} = \vec{0}$.

Newton's second law $\vec{F}_{net} = m\vec{a}$

Newton's third law $\vec{F}_{A\ on\ B} = -\vec{F}_{B\ on\ A}$

BASIC PROBLEM-SOLVING STRATEGY Use Newton's second law for each particle or system. Use Newton's third law to equate the magnitudes of the two members of an action/reaction pair.

Linear motion
$\sum F_x = ma_x$, $\sum F_y = 0$ or $\sum F_x = 0$, $\sum F_y = ma_y$

Trajectory motion
$\sum F_x = ma_x$, $\sum F_y = ma_y$

Circular motion
$\sum F_r = mv^2/r = m\omega^2 r$, $\sum F_z = 0$

Linear and trajectory kinematics

Uniform acceleration: (a_s = constant)

$$v_{fs} = v_{is} + a_s \Delta t$$
$$s_f = s_i + v_{is}\Delta t + \tfrac{1}{2}a_s(\Delta t)^2$$
$$v_{fs}^2 = v_{is}^2 + 2a_s \Delta s$$

Trajectories: The same equations are used for both x and y.

Uniform motion: ($a = 0$, v_s = constant) $s_f = s_i + v_s \Delta t$

Circular kinematics

Uniform circular motion:

$$\theta_f = \theta_i + \omega \Delta t$$
$$a_r = v^2/r = \omega^2 r$$
$$v = \omega r$$
$$T = 2\pi r/v = 2\pi/\omega$$

General case

$v_s = ds/dt$ = slope of the position graph
$a_s = dv_s/dt$ = slope of the velocity graph

$$v_{fs} = v_{is} + \int_{t_i}^{t_f} a_s\, dt = v_{is} + \text{area under the acceleration curve}$$

$$s_f = s_i + \int_{t_i}^{t_f} v_s\, dt = s_i + \text{area under the velocity curve}$$

The Student Workbook

A key component of *Physics for Scientists and Engineers: A Strategic Approach* is the accompanying *Student Workbook*. The workbook bridges the gap between textbook and homework problems by providing students the opportunity to learn and practice skills prior to using those skills in quantitative end-of-chapter problems, much as a musician practices technique separately from performance pieces. The workbook exercises, which are keyed to each section of the textbook, focus on developing specific skills, ranging from identifying forces and drawing free-body diagrams to interpreting wave functions.

The workbook exercises, which are generally qualitative and/or graphical, draw heavily upon the physics education research literature. The exercises deal with issues known to cause student difficulties and employ techniques that have proven to be effective at overcoming those difficulties. The workbook exercises can be used in-class as part of an active-learning teaching strategy, in recitation sections, or as assigned homework. More information about effective use of the *Student Workbook* can be found in the *Instructor's Guide*.

Available versions: Extended (ISBN 0-8053-8961-X), Standard (ISBN 0-8053-8984-9), Volume 1 (ISBN 0-8053-8965-2), Volume 2 (ISBN 0-8053-8968-7), Volume 3 (ISBN 0-8053-8971-7), Volume 4 (ISBN 0-8053-8974-1), and Volume 5 (ISBN 0-8053-8977-6).

Force and Motion · CHAPTER 4 **4-3**

4.4 What Do Forces Do? A Virtual Experiment

9. The figure shows an acceleration-versus-force graph for an object of mass m. Data have been plotted as individual points, and a line has been drawn through the points.
Draw and label, directly on the figure, the acceleration-versus-force graphs for objects of mass
a. $2m$ b. $0.5m$
Use triangles ▲ to show four points for the object of mass $2m$, then draw a line through the points. Use squares ■ for the object of mass $0.5m$.

10. A constant force applied to object A causes A to accelerate at 5 m/s^2. The same force applied to object B causes an acceleration of 3 m/s^2. Applied to object C, it causes an acceleration of 8 m/s^2.
a. Which object has the largest mass? ______
b. Which object has the smallest mass? ______
c. What is the ratio of mass A to mass B? (m_A/m_B) = ______

11. A constant force applied to an object causes the object to accelerate at 10 m/s^2. What will the acceleration of this object be if
a. The force is doubled? ______ b. The mass is doubled? ______
c. The force is doubled *and* the mass is doubled? ______
d. The force is doubled *and* the mass is halved? ______

12. A constant force applied to an object causes the object to accelerate at 8 m/s^2. What will the acceleration of this object be if
a. The force is halved? b. The mass is halved?
c. The force is halved *and* the mass is halved?
d. The force is halved *and* the mass is doubled?

4.5 Newton's Second Law

13. Forces are shown on two objects. For each:
a. Draw and label the net force vector. Do this right on the figure.
b. Below the figure, draw and label the object's acceleration vector.

Instructor Supplements

- The **Instructor's Guide for Physics for Scientists and Engineers** (ISBN 0-8053-8985-7) offers detailed comments and suggested teaching ideas for every chapter, an extensive review of what has been learned from physics education research, and guidelines for using active-learning techniques in your classroom.
- The **Instructor's Solutions Manuals, Chapters 1–19** (ISBN 0-8053-8986-5), and **Chapters 20–42** (ISBN 0-8053-8989-X), written by Professors Pawan Kahol and Donald Foster, at Wichita State University, provide *complete* solutions to all the end-of-chapter problems. The solutions follow the four-step Model/Visualize/Solve/Assess procedure used in the *Problem-Solving Strategies* and all worked examples. Emphasis is placed on the reasoning behind the solution, rather than just the numerical manipulations. The full text of each solution is available as an editable Word document and as a pdf file on the *Instructor's Supplement CD-ROM* for your own use or for posting on your course website.
- The cross-platform **Instructor's Resource CD-ROMs** (ISBN 0-8053-8996-2) consists of the **Simulation and Image Presentation CD-ROM** and the **Instructor's Supplement CD-ROM**. The *Simulation and Image Presentation CD-ROM* provides a comprehensive library of more than 220 applets from *ActivPhysics OnLine*, as well as all the figures from the textbook (excluding photographs) in JPEG format. In addition, all the tables, chapter summaries, and knowledge structures are provided as JPEGs, and the Tactics Boxes, Problem-Solving Strategies, and key (boxed) equations are provided in editable Word format. The *Instructor's Supplement CD-ROM* provides editable Word versions and pdf files of the *Instructor's Guide* and the *Instructor's Solutions Manuals*. Complete *Student Workbook* solutions are also provided as pdf files.
- **MasteringPhysics**™ (www.masteringphysics.com) is a sophisticated, research-proven online tutorial and homework assignment system that provides students with individualized feedback and hints based on their input. It provides a comprehensive library of conceptual tutorials (including one for each

Problem-Solving Strategy in this textbook), multistep self-tutoring problems, and end-of-chapter problems from *Physics for Scientists and Engineers*. *MasteringPhysics*™ provides instructors with a fast and effective way to assign online homework assignments that comprise a range of problem types. The powerful post-assignment diagnostics allow instructors to assess the progress of their class as a whole or to quickly identify individual students' areas of difficulty.

- **ActivPhysics**™ **OnLine** (www.aw-bc.com/knight) provides a comprehensive library of more than 420 tried and tested *ActivPhysics* applets updated for web delivery using the latest online technologies. In addition, it provides a suite of highly regarded applet-based tutorials developed by education pioneers Professors Alan Van Heuvelen and Paul D'Alessandris. The *ActivPhysics* margin icon directs students to specific exercises that complement the textbook discussion.

  The online exercises are designed to encourage students to confront misconceptions, reason qualitatively about physical processes, experiment quantitatively, and learn to think critically. They cover all topics from mechanics to electricity and magnetism and from optics to modern physics. The highly acclaimed *ActivPhysics OnLine* companion workbooks help students work through complex concepts and understand them more clearly. More than 220 applets from the *ActivPhysics OnLine* library are also available on the *Simulation and Image Presentation CD-ROM*.
- The **Printed Test Bank** (ISBN 0-8053-8994-6) and cross-platform **Computerized Test Bank** (ISBN 0-8053-8995-4), prepared by Professor Benjamin Grinstein, at the University of California, San Diego, contain more than 1500 high-quality problems, with a range of multiple-choice, true/false, short-answer, and regular homework-type questions. In the computerized version, more than half of the questions have numerical values that can be randomly assigned for each student.
- The **Transparency Acetates** (ISBN 0-8053-8993-8) provide more than 200 key figures from *Physics for Scientists and Engineers* for classroom presentation.

Student Supplements

- The **Student Solutions Manuals Chapters 1–19** (ISBN 0-8053-8708-0) and **Chapters 20–42** (ISBN 0-8053-8998-9), written by Professors Pawan Kahol and Donald Foster at Wichita State University, provides *detailed* solutions to more than half of the odd-numbered end-of-chapter problems. The solutions follow the four-step Model/Visualize/Solve/Assess procedure used in the *Problem-Solving Strategies* and all worked examples.

- **MasteringPhysics**™ (www.masteringphysics.com) provides students with individualized online tutoring by responding to their wrong answers and providing hints for solving multistep problems. It gives them immediate and up-to-date assessment of their progress, and shows where they need to practice more.

- **ActivPhysics**™ **OnLine** (www.aw-bc.com/knight) provides students with a suite of highly regarded applet-based tutorials (see above). The accompanying workbooks help students work though complex concepts and understand them more clearly. The *ActivPhysics* margin icon directs students to specific exercises that complement the textbook discussion.
- **ActivPhysics OnLine Workbook Volume 1: Mechanics • Thermal Physics • Oscillations & Waves** (ISBN 0-8053-9060-X)
- **ActivPhysics OnLine Workbook Volume 2: Electricity & Magnetism • Optics • Modern Physics** (ISBN 0-8053-9061-8)

- The **Addison-Wesley Tutor Center** (www.aw.com/tutorcenter) provides one-on-one tutoring via telephone, fax, email, or interactive website during evening hours and on weekends. Qualified college instructors answer questions and provide instruction for *Mastering Physics*™ and for the examples, exercises, and problems in *Physics for Scientists and Engineers*.

Acknowledgments

I have relied upon conversations with and, especially, the written publications of many members of the physics education community. Those who may recognize their influence include Arnold Arons, Uri Ganiel, Ibrahim Halloun, Richard Hake, David Hestenes, Leonard Jossem, Jill Larkin, Priscilla Laws, John Mallinckrodt, Lillian McDermott, Edward "Joe" Redish, Fred Reif, Rachel Scherr, Bruce Sherwood, David Sokoloff, Ronald Thornton, Sheila Tobias, and Alan Van Heuleven. John Rigden, founder and director of the Introductory University Physics Project, provided the impetus that got me started down this path. Early development of the materials was supported by the National Science Foundation as the *Physics for the Year 2000* project; their support is gratefully acknowledged.

I am grateful to Pawan Kahol and Don Foster for the difficult task of writing the *Instructor's Solutions Manuals*; to Jim Andrews and Susan Cable for writing the workbook answers; to Wayne Anderson, Jim Andrews, Dave Ettestad, Stuart Field, Robert Glosser, and Charlie Hibbard for their contributions to the end-of-chapter problems; and to my colleague Matt Moelter for many valuable contributions and suggestions.

I especially want to thank my editor Adam Black, development editor Alice Houston, editorial assistant Liana Allday, and all the other staff at Addison Wesley for their enthusiasm and hard work on this project. Project manager Laura Kenney, Carolyn Field and the team at Thompson Steele, Inc., copy editor Kevin Gleason, photo researcher Brian Donnelly, and page-layout artist Judy Maenle get much of the credit for making this complex project all come together. In addition to the reviewers and classroom testers listed below, who gave invaluable feedback, I am particularly grateful to Wendell Potter and Susan Cable for their close scrutiny of every word and figure.

Finally, I am endlessly grateful to my wife Sally for her love, encouragement, and patience, and to our many cats for their innate abilities to hold down piles of papers and to type qqqqqqqq whenever it was needed.

Randy Knight, September 2003
rknight@calpoly.edu

Reviewers and Classroom Testers

Gary B. Adams, *Arizona State University*
Wayne R. Anderson, *Sacramento City College*
James H. Andrews, *Youngstown State University*
David Balogh, *Fresno City College*
Dewayne Beery, *Buffalo State College*
Joseph Bellina, *Saint Mary's College*
James R. Benbrook, *University of Houston*
David Besson, *University of Kansas*
Randy Bohn, *University of Toledo*
Art Braundmeier, *University of Southern Illinois, Edwardsville*
Carl Bromberg, *Michigan State University*
Douglas Brown, *Cabrillo College*
Ronald Brown, *California Polytechnic State University, San Luis Obispo*
Mike Broyles, *Collin County Community College*

James Carolan, *University of British Columbia*
Michael Crescimanno, *Youngstown State University*
Wei Cui, *Purdue University*
Robert J. Culbertson, *Arizona State University*
Purna C. Das, *Purdue University North Central*
Dwain Desbien, *Estrella Mountain Community College*
John F. Devlin, *University of Michigan, Dearborn*
Alex Dickison, *Seminole Community College*
Chaden Djalali, *University of South Carolina*
Sandra Doty, *Denison University*
Miles J. Dresser, *Washington State University*
Charlotte Elster, *Ohio University*
Robert J. Endorf, *University of Cincinnati*
Tilahun Eneyew, *Embry-Riddle Aeronautical University*
F. Paul Esposito, *University of Cincinnati*
John Evans, *Lee University*
Michael R. Falvo, *University of North Carolina*
Abbas Faridi, *Orange Coast College*
Stuart Field, *Colorado State University*
Daniel Finley, *University of New Mexico*
Jane D. Flood, *Muhlenberg College*
Thomas Furtak, *Colorado School of Mines*
Richard Gass, *University of Cincinnati*
J. David Gavenda, *University of Texas, Austin*
Stuart Gazes, *University of Chicago*
Katherine M. Gietzen, *Southwest Missouri State University*
Robert Glosser, *University of Texas, Dallas*
William Golightly, *University of California, Berkeley*
Paul Gresser, *University of Maryland*
C. Frank Griffin, *University of Akron*
John B. Gruber, *San Jose State University*
Randy Harris, *University of California, Davis*
Stephen Haas, *University of Southern California*
Nicole Herbots, *Arizona State University*
Scott Hildreth, *Chabot College*
David Hobbs, *South Plains College*
Laurent Hodges, *Iowa State University*
John L. Hubisz, *North Carolina State University*
George Igo, *University of California, Los Angeles*
Bob Jacobsen, *University of California, Berkeley*
Rong-Sheng Jin, *Florida Institute of Technology*
Marty Johnston, *University of St. Thomas*
Stanley T. Jones, *University of Alabama*
Darrell Judge, *University of Southern California*
Pawan Kahol, *Wichita State University*
Teruki Kamon, *Texas A&M University*
Richard Karas, *California State University, San Marcos*
Deborah Katz, *U.S. Naval Academy*
Miron Kaufman, *Cleveland State University*
M. Kotlarchyk, *Rochester Institute of Technology*
Cagliyan Kurdak, *University of Michigan*
Fred Krauss, *Delta College*
H. Sarma Lakkaraju, *San Jose State University*
Darrell R. Lamm, *Georgia Institute of Technology*
Robert LaMontagne, *Providence College*
Alessandra Lanzara, *University of California, Berkeley*
Sen-Ben Liao, *Massachusetts Institute of Technology*
Dean Livelybrooks, *University of Oregon*
Chun-Min Lo, *University of South Florida*
Richard McCorkle, *University of Rhode Island*
James McGuire, *Tulane University*
Theresa Moreau, *Amherst College*
Gary Morris, *Rice University*
Michael A. Morrison, *University of Oklahoma*
Richard Mowat, *North Carolina State University*
Taha Mzoughi, *Mississippi State University*
Vaman M. Naik, *University of Michigan, Dearborn*
Craig Ogilvie, *Iowa State University*
Martin Okafor, *Georgia Perimeter College*
Benedict Y. Oh, *University of Wisconsin*
Georgia Papaefthymiou, *Villanova University*
Peggy Perozzo, *Mary Baldwin College*
Brian K. Pickett, *Purdue University, Calumet*
Joe Pifer, *Rutgers University*
Dale Pleticha, *Gordon College*
Robert Pompi, *SUNY-Binghamton*
David Potter, *Austin Community College*
Chandra Prayaga, *University of West Florida*
Didarul Qadir, *Central Michigan University*
Michael Read, *College of the Siskiyous*
Michael Rodman, *Spokane Falls Community College*
Sharon Rosell, *Central Washington University*
Anthony Russo, *Okaloosa-Walton Community College*
Otto F. Sankey, *Arizona State University*
Rachel E. Scherr, *University of Maryland*
Bruce Schumm, *University of California, Santa Cruz*
Douglas Sherman, *San Jose State University*
Elizabeth H. Simmons, *Boston University*
Alan Slavin, *Trent College*
William Smith, *Boise State University*
Paul Sokol, *Pennsylvania State University*
Chris Sorensen, *Kansas State University*
Anna and Ivan Stern, *AW Tutor Center*
Michael Strauss, *University of Oklahoma*
Arthur Viescas, *Pennsylvania State University*
Chris Vuille, *Embry-Riddle Aeronautical University*
Ernst D. Von Meerwall, *University of Akron*
Robert Webb, *Texas A&M University*
Zodiac Webster, *California State University, San Bernardino*
Robert Weidman, *Michigan Technical University*
Jeff Allen Winger, *Mississippi State University*
Ronald Zammit, *California Polytechnic State University, San Luis Obispo*
Darin T. Zimmerman, *Pennsylvania State University, Altoona*

Preface to the Student

From Me to You

The most incomprehensible thing about the universe is that it is comprehensible.
—Albert Einstein

The day I went into physics class it was death.
—Sylvia Plath, *The Bell Jar*

Let's have a little chat before we start. A rather one-sided chat, admittedly, because you can't respond, but that's OK. I've heard from many of your fellow students over the years, so I have a pretty good idea of what's on your mind.

What's your reaction to taking physics? Fear and loathing? Uncertainty? Excitement? All of the above? Let's face it, physics has a bit of an image problem on campus. You've probably heard that it's difficult, maybe downright impossible unless you're an Einstein. Things that you've heard, your experiences in other science courses, and many other factors all color your *expectations* about what this course is going to be like.

It's true that there are many new ideas to be learned in physics and that the course, like college courses in general, is going to be much faster paced than science courses you had in high school. I think it's fair to say that it will be an *intense* course. But we can avoid many potential problems and difficulties if we can establish, here at the beginning, what this course is about and what is expected of you—and of me!

Just what is physics, anyway? Physics is a way of thinking about the physical aspects of nature. Physics is not better than art or biology or poetry or religion, which are also ways to think about nature; it's simply different. One of the things this course will emphasize is that physics is a human endeavor. The information content of this book was not found in a cave or conveyed to us by aliens; it was discovered by real people engaged in a struggle with real issues. I hope to convey to you something of the history and the process by which we have come to accept the principles that form the foundation of today's science and engineering.

You might be surprised to hear that physics is not about "facts." Oh, not that facts are unimportant, but physics is far more focused on discovering *relationships* that exist between facts and *patterns* that exist in nature than on learning facts for their own sake. As a consequence, there's not a lot of memorization when you study physics. Some—there are still definitions and equations to learn—but less than in many other courses. Our emphasis, instead, will be on thinking and reasoning. This is important to factor into your expectations for the course.

Perhaps most important of all, *physics is not math!* Physics is much broader. We're going to look for patterns and relationships in nature, develop the logic that relates different ideas, and search for the reasons *why* things happen as they do. In doing so, we're going to stress qualitative reasoning, pictorial and graphical reasoning, and reasoning by analogy. And yes, we will use math, but it's just one tool among many.

It will save you much frustration if you're aware of this physics–math distinction up front. Many of you, I know, want to find a formula and plug numbers into it—that is, to do a math problem. Maybe that's what you learned in high school science courses, but it is *not* what this course expects of you. We'll certainly do

(a) X-ray diffraction pattern

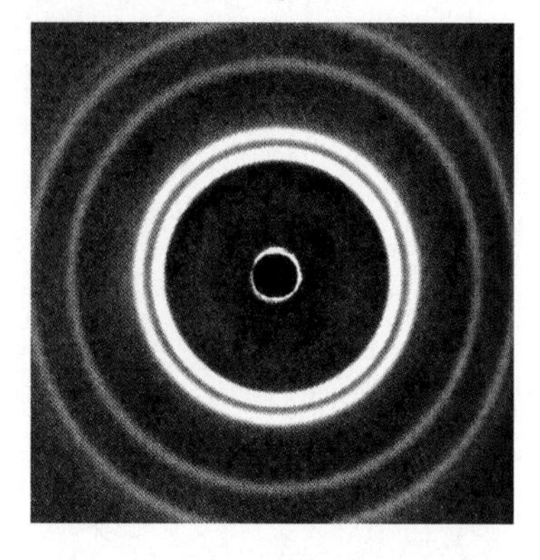

(b) Electron diffraction pattern

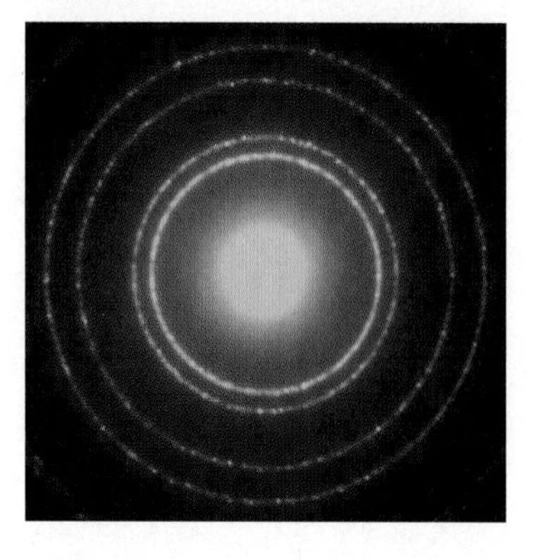

many calculations, but the specific numbers are usually the last and least important step in the analysis.

Physics is about recognizing patterns. The top photograph is an x-ray diffraction pattern that shows how a collimated beam of x rays spreads out after passing through a crystal. The bottom photograph shows what happens when a collimated beam of electrons is shot through the same crystal. What does the obvious similarity in these two photographs tell us about the nature of light and about the nature of matter?

As you study, you'll sometimes be baffled, puzzled, and confused. That's perfectly normal and to be expected. Making mistakes is OK too *if* you're willing to learn from the experience. No one is born knowing how to do physics any more than he or she is born knowing how to play the piano or shoot basketballs. The ability to do physics comes from practice, repetition, and struggling with the ideas until you "own" them and can apply them yourself in new situations. There's no way to make learning effortless, at least for anything worth learning, so expect to have some difficult moments ahead.

But also expect to have some moments of excitement at the joy of discovery. There will be instants at which the pieces suddenly click into place and you *know* that you understand a difficult idea. There will be times when you'll surprise yourself by successfully working a difficult problem that you didn't think you could solve. My hope, as an author, is that the excitement and sense of adventure will far outweigh the difficulties and frustrations.

Many of you, I suspect, would like to know the "best" way to study for this course. There is no best way. People are too different, and what works for one student works less effectively for another. But I do want to stress that *reading the text* is vitally important. Class time will be used to clarify difficulties and to develop tools for using the knowledge, but your instructor will *not* use class time simply to repeat information in the text. The basic knowledge for this course is written down within these pages, and the *number one expectation* is that you will read carefully and thoroughly to find and learn that knowledge.

Despite there being no best way to study, I will suggest *one* way that is successful for many students. It consists of the following four steps:

1. **Read each chapter *before* it is discussed in class.** I cannot stress too highly how important this step is. Class attendance is largely ineffective if you have not prepared. When you first read a chapter, focus on learning new vocabulary, definitions, and notation. There's a list of terms and notations at the end of each chapter. Learn them! You won't understand what's being discussed or how the ideas are being used if you don't know what the terms and symbols mean.
2. **Participate actively in class.** Take notes, ask and answer questions, take part in discussion groups. There is ample scientific evidence that *active participation* is far more effective for learning science than is passive listening.
3. **After class, go back for a *careful* rereading of the chapter.** In your second reading, pay closer attention to the details and the worked examples. Look for the *logic* behind each example (and I've tried to help make this clear), not just at what formula is being used. Do the *Student Workbook* exercises for each section as you finish your reading of it.
4. **Finally, apply what you have learned to the homework problems at the end of each chapter.** I strongly encourage you to form a study group with two or three classmates. There's good evidence that students who study regularly with a group do better than the rugged individualists who try to go it alone.

Did someone mention a workbook? The companion *Student Workbook* is a vital part of this course. It contains questions and exercises that ask you to reason *qualitatively*, to use graphical information, and to give explanations. It is through these exercises that you will learn what the concepts mean and will practice the reasoning skills appropriate to the chapter. You will then have acquired the baseline knowledge that you need *before* turning to the end-of-chapter homework problems. In sports or in music, you would never think of performing before you practice, so why would you want to do so in physics? The workbook is where you practice and work on basic skills.

Many of you, I know, would like to go straight to the homework problems and then thumb through the text looking for a formula that seems like it will work. That approach will not succeed in this course, and it's guaranteed to make you frustrated and discouraged. Very few homework problems are "plug and chug" problems where you simply put numbers into a formula. To work the homework problems successfully, you need a better study strategy—either that outlined above or your own—that helps you learn the concepts and the relationships between the ideas. Many of the chapters in this book have Problem-Solving Strategies to help you develop effective problem-solving skills.

A traditional guideline in college is to study two hours outside of class for every hour spent in class, and this text is designed with that expectation. Of course, two hours is an average. Some chapters are fairly straightforward and will go quickly. Others likely will require much more than two study hours per class hour.

Now that you know more about what is expected of you, what can you expect of me? That's a little trickier, because the book is already written! Nonetheless, it was prepared on the basis of what I think my students throughout the years have expected—and wanted—from their physics textbook.

You should know that these course materials—the text and the workbook—are based upon extensive research about how students learn physics and the challenges they face. The effectiveness of many of the exercises has been demonstrated through extensive class testing. I've written the book in an informal style that I hope you will find appealing and that will encourage you to do the reading. And finally, I have endeavored to make clear not only that physics, as a technical body of knowledge, is relevant to your profession but also that physics is an exciting adventure of the human mind.

I hope you'll enjoy the time we're going to spend together.

Detailed Contents

Volume 1 contains chapters 1–15; Volume 2 contains chapters 16-19; Volume 3 contains chapters 20–24; Volume 4 contains chapters 25–36; Volume 5 contains chapters 36–42.

Part VI Electricity and Magnetism

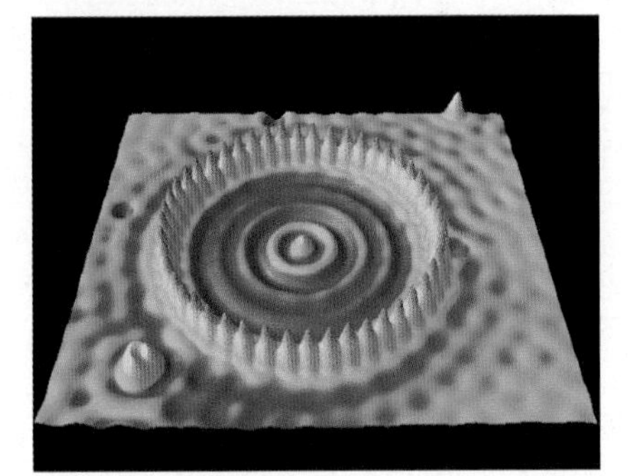

Introduction

Journey into Physics

Said Alice to the Cheshire cat,

"Cheshire-Puss, would you tell me, please, which way I ought to go from here?"
"That depends a good deal on where you want to go," said the Cat.
"I don't much care where—" said Alice.
"Then it doesn't matter which way you go," said the Cat.
—Lewis Carroll, *Alice in Wonderland*

Have you ever wondered about questions such as

Why is the sky blue?

Why is glass an insulator but metal a conductor?

What, really, is an atom?

These are the questions of which physics is made. Physicists try to understand the universe in which we live by observing the phenomena of nature—such as the sky being blue—and by looking for patterns and principles to explain these phenomena. Many of the discoveries made by physicists, from electromagnetic waves to nuclear energy, have forever altered the ways in which we live and think.

You are about to embark on a journey into the realm of physics. It is a journey in which you will learn about many physical phenomena and find the answers to questions such as the ones posed above. Along the way, you will also learn how to use physics to analyze and solve many practical problems.

As you proceed, you are going to see the methods by which physicists have come to understand the laws of nature. The ideas and theories of physics are not arbitrary; they are firmly grounded in experiments and measurements. By the time you finish this text, you will be able to recognize the *evidence* upon which our present knowledge of the universe is based.

Which Way Should We Go?

We are rather like Alice in Wonderland, here at the start of the journey, in that we must decide which way to go. Physics is an immense body of knowledge, and without specific goals it would not much matter which topics we study. But unlike Alice, we *do* have some particular destinations that we would like to visit.

The physics that provides the foundation for all of modern science and engineering can be divided into three broad categories:

- Particles and energy.
- Fields and waves.
- The atomic structure of matter.

A particle, in the sense that we'll use the term, is an idealization of a physical object. We will use particles to understand how objects move and how they interact with each other. One of the most important properties of a particle or a collection of particles is *energy*. We will study energy both for its value in understanding physical processes and because of its practical importance in a technological society.

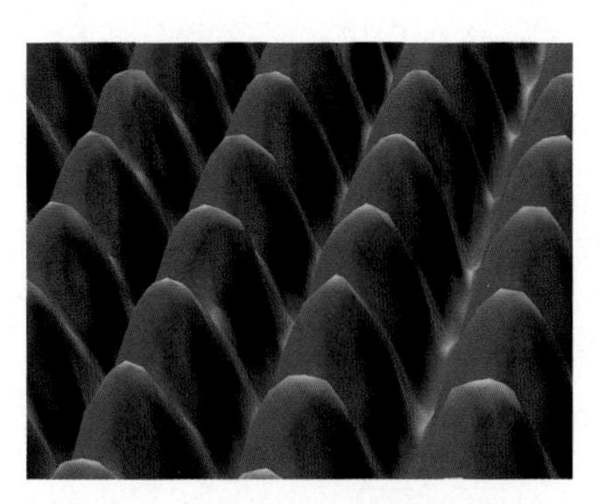

A scanning tunneling microscope allows us to "see" the individual atoms on a surface. One of our goals is to understand how an image such as this is made.

Particles are discrete, localized objects. Although many phenomena can be understood in terms of particles and their interactions, the long-range interactions of gravity, electricity, and magnetism are best understood in terms of *fields*, such as the gravitational field and the electric field. Rather than being discrete, fields spread continuously through space. Much of the second half of this book will be focused on understanding fields and the interactions between fields and particles.

Certainly one of the most significant discoveries of the past 500 years is that matter consists of atoms. Atoms and their properties are described by quantum physics, but we cannot leap directly into that subject and expect that it would make any sense. To reach our destination, we are going to have to study many other topics along the way—rather like having to visit the Rocky Mountains if you want to drive from New York to San Francisco. All our knowledge of particles and fields will come into play as we end our journey by studying the atomic structure of matter.

The Route Ahead

Here at the beginning, we can survey the route ahead. Where will our journey take us? What scenic vistas will we view along the way?

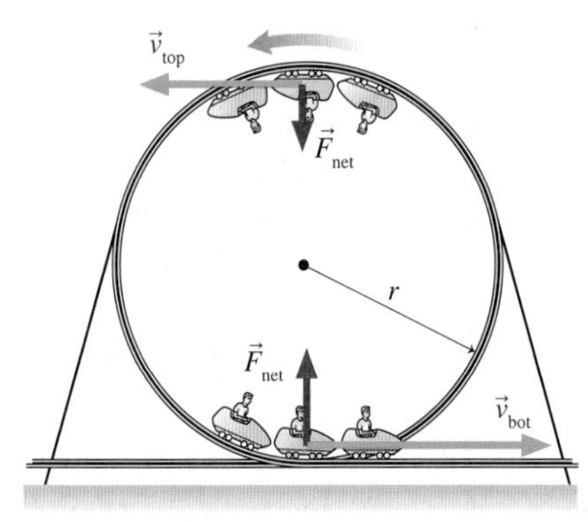

Parts I and II, *Newton's Laws* and *Conservation Laws*, form the basis of what is called *classical mechanics*. Classical mechanics is the study of motion. (It is called *classical* to distinguish it from the modern theory of motion at the atomic level, which is called *quantum mechanics*.) The first two parts of this textbook establish the basic language and concepts of motion. Part I will look at motion in terms of *particles* and *forces*. We will use these concepts to study the motion of everything from accelerating sprinters to orbiting satellites. Then, in Part II, we will introduce the ideas of *momentum* and *energy*. These concepts—especially energy—will give us a new perspective on motion and extend our ability to analyze motion.

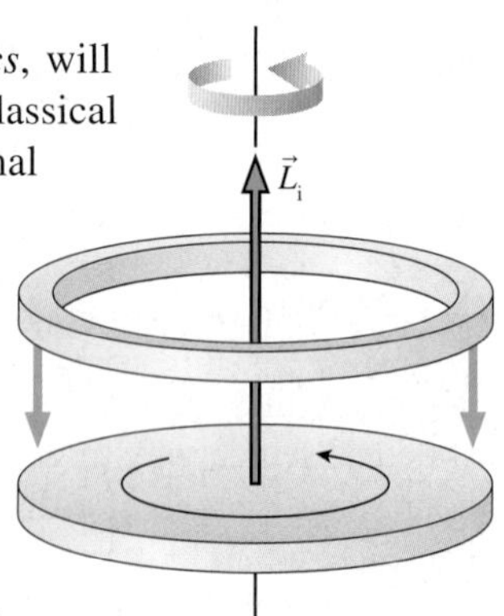

Part III, *Applications of Newtonian Mechanics*, will pause to look at four important applications of classical mechanics: Newton's theory of gravity, rotational motion, oscillatory motion, and the motion of fluids. Only oscillatory motion is a prerequisite for later chapters. Your instructor may choose to cover some or all of the other chapters, depending upon the time available, but your study of Parts IV–VII will not be hampered if these chapters are omitted.

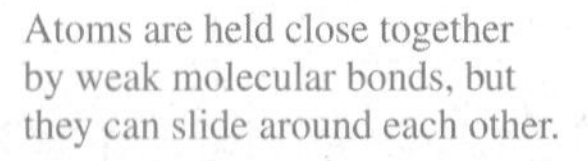

Atoms are held close together
by weak molecular bonds, but
they can slide around each other.

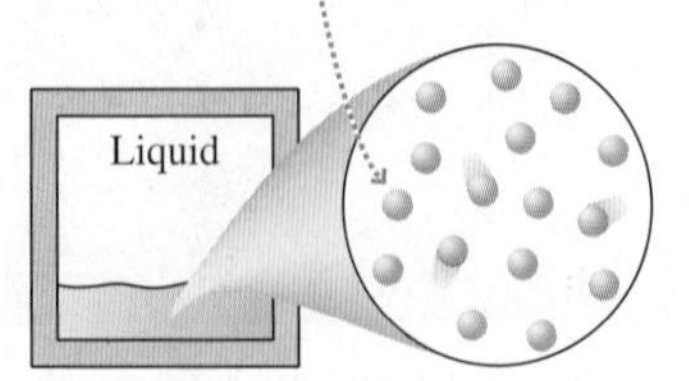

Part IV, *Thermodynamics*, extends the ideas of particles and energy to systems such as liquids and gases that contain vast numbers of particles. Here we will look for connections between the *microscopic* behavior of large numbers of atoms and the *macroscopic* properties of bulk matter. You will find that some of the properties of gases that you know from chemistry, such as the ideal gas law, turn out to be direct consequences of the underlying atomic structure of the gas. We will also expand the concept of energy and study how energy is transferred and utilized.

Waves are ubiquitous in nature, whether they be large-scale oscillations like ocean waves, the less obvious motions of sound waves, or the subtle undulations of light waves and matter waves that go to the heart of the atomic structure of matter. In **Part V,** *Waves and Optics*, we will emphasize the unity of wave physics and find that many diverse wave phenomena can be analyzed with the same concepts and mathematical language. It is here we will begin to accumulate evidence that the theory of classical mechanics is inadequate to explain the observed behavior of atoms, and we will end this section with some atomic puzzles that seem to defy understanding.

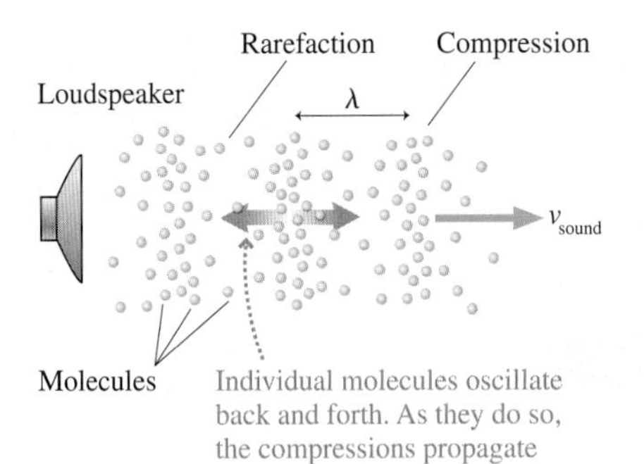
Rarefaction
Compression
Loudspeaker
λ
v_{sound}
Molecules
Individual molecules oscillate back and forth. As they do so, the compressions propagate forward at speed v_{sound}.

Part VI, *Electricity and Magnetism*, is devoted to the *electromagnetic force*, one of the most important forces in nature. In essence, the electromagnetic force is the "glue" that holds atoms together. It is also the force that makes this the "electronic age." We'll begin this part of the journey with simple observations of static electricity. Bit by bit, we'll be led to the basic ideas behind electrical circuits, to magnetism, and eventually to the discovery of electromagnetic waves.

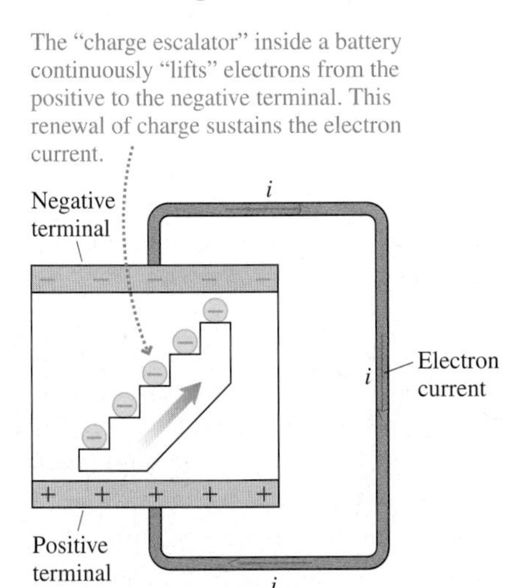
The "charge escalator" inside a battery continuously "lifts" electrons from the positive to the negative terminal. This renewal of charge sustains the electron current.
Negative terminal
i
Electron current
Positive terminal

Part VII is *Relativity and Quantum Physics*. We'll start by exploring the strange world of Einstein's theory of *relativity*, a world in which space and time aren't quite what they appear to be. Then we will enter the microscopic domain of *atoms*, where the behaviors of light and matter are at complete odds with what our common sense tells us is possible. Although the mathematics of quantum theory quickly gets beyond the level of this text, and time will be running out, you will see that the quantum theory of atoms and nuclei explains many of the things that you learned simply as rules in chemistry.

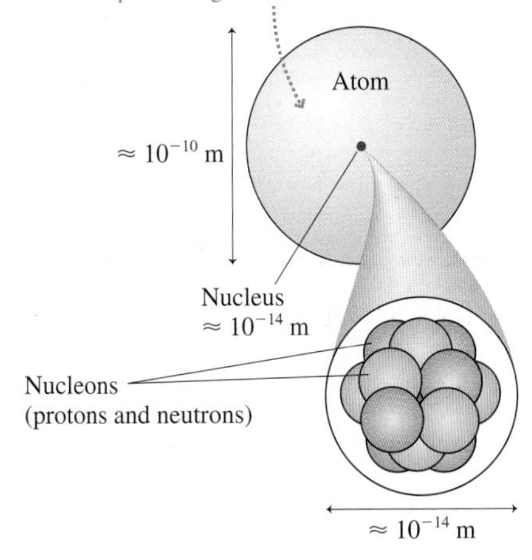
This picture of an atom would need to be 10 m in diameter if it were drawn to the same scale as the dot representing the nucleus.
Atom
$\approx 10^{-10}$ m
Nucleus
$\approx 10^{-14}$ m
Nucleons
(protons and neutrons)
$\approx 10^{-14}$ m

We will not have visited all of physics on our travels. There just isn't time. Many exciting topics, ranging from quarks to black holes, will have to remain unexplored. But this particular journey need not be the last. As you finish this text, you will have the background and the experience to explore new topics further in more advanced courses or for yourself.

With that said, let us take the first step.

PART I

If the acceleration of a Podracer can reach 50 m/s^2, what are the maximum tensions in the two large cables? To find out, what quantities do you need to estimate?

Newton's Laws

Why Things Change

Each of the seven parts of this book opens with an overview. The overview gives you a look ahead, a glimpse of where your journey will take you in the next few chapters. It's easy to lose sight of the big picture while you're busy negotiating the terrain of each chapter. You may find it helpful to look back at the overviews several times as your travels take you through the different parts of the book.

Change

Simple observations of the world around you show that most things change, few things remain the same. Some changes, such as aging, are biological. Others, such as sugar dissolving in your coffee, are chemical. We're going to be concerned about changes that involve *motion* of one form or another. For example, you were standing, now you're sitting. The leaf that was on the tree has fallen. The air molecules in the room have moved to new positions.

Part I of this textbook, *Newton's Laws,* is about motion. The "laws of motion" were discovered by Isaac Newton roughly 350 years ago, so our study of motion is hardly cutting edge science. Nonetheless, it is still extremely important. Mechanics—the science of motion—is the basis for much of engineering and applied science. And many of the ideas introduced in mechanics will be needed later to understand things like the motion of waves or the motion of electrons in a semiconductor. Newton's mechanics is the foundation of much of contemporary science, so it is important that we start at the beginning.

Part I is going to focus on the motion of "things" such as balls, cars, rockets, and satellites. These are *macroscopic* objects, as opposed to microscopic atoms and molecules. They are also objects with well-defined boundaries, in contrast to the spread-out motion of a wave. We'll get to the motion of waves and atoms at later stages of our journey.

There are two big questions we must tackle to study how things change by moving:

- **How do we describe motion?** It is easy to say that an object moves, but it's not obvious how we should measure or characterize the motion if we want to analyze it mathematically. The mathematical description of motion is called *kinematics,* and it is the subject matter of Chapters 1 and 2.
- **How do we explain motion?** Why do objects have the particular motion they do? Why, when you toss a ball upward, does it go up and then come back down rather than keep going up? Are there "laws of nature" that allow us to predict

an object's motion? The explanation of motion in terms of its causes is called *dynamics,* and it is the topic of Chapters 4 through 8.

Chapter 3, wedged between kinematics and dynamics, is about *vectors.* Our universe is three dimensional, and vectors are a mathematical tool for describing motion in three dimensions. Much of physics and engineering is written in the language of vectors, hence it is important that you become fluent in this language at an early stage of your journey.

Two key ideas that will help answer these questions are *force* (the "cause") and *acceleration* (the "effect"). One of our goals is to develop a "Newtonian intuition" for the connection between force and acceleration. Much of Chapters 1 through 4 is devoted to helping you develop your intuition about the nature of motion. Pictures and graphs will play a big role in this development. You'll then put this knowledge to use in Chapters 5 through 8 as you analyze motion of increasing complexity.

Models: A Vehicle for Understanding

Reality is extremely complicated. We would never be able to develop a science if we had to keep track of every little detail of every situation. Consider a simple example: throwing a ball. To understand the motion of the ball, is it necessary to keep track of every atom inside? Of every quark inside every proton inside every nucleus inside every atom inside the ball? Do we need to analyze what you ate for breakfast and the biochemistry of how that was translated into muscle power? In principle, the answer to all these questions is "Yes." But in practice these are all details that have no influence at all on the measurements you might make on the ball.

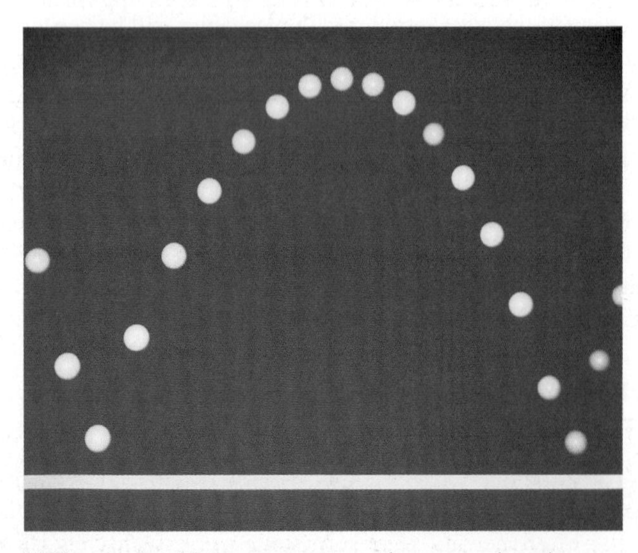

We'll make simplifying assumptions when we analyze the motion of a ball in Chapter 5. This simplified description of reality is a *model* of the situation.

We can do a perfectly fine analysis if we treat the ball as a round solid and your hand as another solid that exerts a force on the ball. This is now what we call a *model* of the situation. A model is a simplified description of reality—much as a model airplane is a simplified version of a real airplane—that is used to reduce the complexity of a problem to the point where it can be analyzed and understood.

Much of physics and engineering is a matter of model building—simplifying the situation, isolating the essential features, and developing a set of equations that provides an adequate, although not perfect, description of reality. Physics, in particular, attempts to strip a phenomenon down to its barest essentials in order to illustrate the physical principles involved. Many of the features neglected by the physicist as "unnecessary details" would be considered absolutely essential by an engineer or a chemist. Both are right. The model each investigator uses has to match the needs each of them is trying to meet.

In building a model, you need to isolate and keep just those features that are essential to the problem. In the case of the ball, for example, you can ignore the ball's atomic structure because it doesn't affect the motion of the ball as a whole. But you cannot ignore gravity. It is an essential feature of the model because it directly influences the motion of the ball. What about air resistance? If you are throwing a very hard, dense ball a very short distance, then ignoring air resistance is probably acceptable. But if you are throwing a ping-pong ball from the Empire State Building, then certainly air resistance is an essential feature of the model. The more details you omit, the simpler the model and the easier it will be to solve the model's equations—but at the expense of less accuracy and poorer agreement with reality.

Model building is a major part of the strategy that we will develop for solving problems. It is, however, a skill that takes some practice and experience to acquire. We will pay close attention, especially in the earlier chapters, to where simplifying assumptions are being made, and why. Learning *how* to simplify a situation is the essence of successful modeling. As you begin to apply these ideas in your own homework, you will be gaining experience with model building and learning how to be a sophisticated problem solver.

1 Concepts of Motion

This snowboarder is demonstrating a fairly extreme form of motion.

▶ Looking Ahead
The goal of Chapter 1 is to introduce the fundamental concepts of motion. In this chapter you will learn to:

- Draw and interpret motion diagrams.
- Describe motion with vectors.
- Use the concepts of position, velocity, and acceleration.
- Use multiple representations of motion.
- Analyze and interpret motion problems.

The universe in which we live is one of change and motion. This snowboarder was clearly in motion when the photograph was taken. In the course of a day you probably walk, run, bicycle, or drive your car, all forms of motion. The clock hands are moving inexorably forward as you read this text. The pages of this book may look quite still, but a microscopic view would reveal jostling atoms and whirling electrons. The stars look as permanent as anything, yet the astronomer's telescope reveals them to be ceaselessly moving within galaxies that rotate and orbit yet other galaxies.

Motion is a theme that will appear in one form or another throughout this entire book. Although we all have intuition about motion, based on our experiences, some of the important aspects of motion turn out to be rather subtle. So rather than jumping immediately into a lot of mathematics and calculations, this first chapter focuses on *visualizing* motion and becoming familiar with the *concepts* needed to describe a moving object. We will use mathematical ideas when needed, because they increase the precision of our thoughts, but we will defer actual calculations until Chapter 2. Our goal is to lay the foundations for understanding motion.

1.1 Motion Diagrams

Socrates: *The nature of motion appears to be the question with which we begin.*
Plato, 375 BCE

The quest to understand motion dates to antiquity. The ancient Babylonians, Chinese, and Greeks were especially interested in the celestial motions of the night sky. The Greek philosopher and scientist Aristotle wrote extensively about the nature of moving objects. However, our modern understanding of motion did not begin until Galileo (1564–1642) first formulated the concepts of motion in

Translational motion

Circular motion

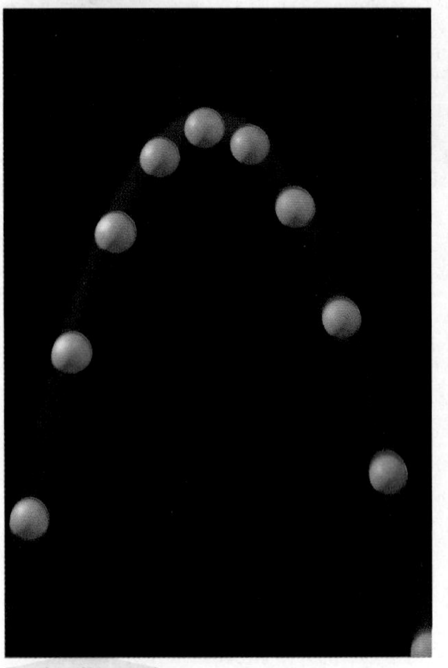
Projectile motion

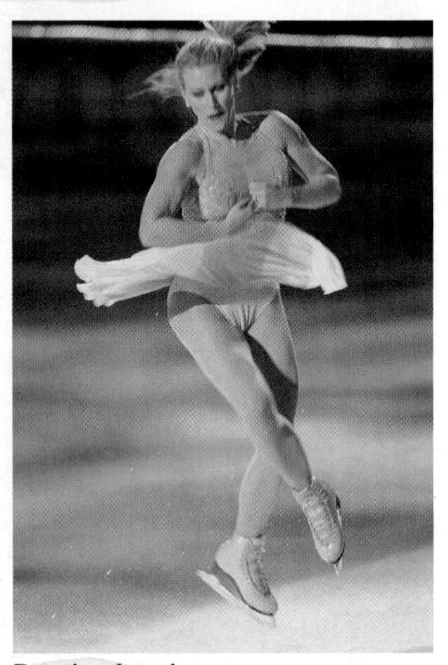
Rotational motion

FIGURE 1.1 Four basic types of motion.

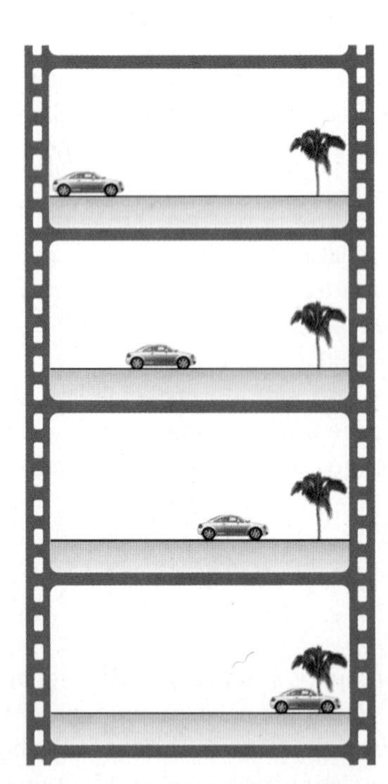
FIGURE 1.2 Several frames from the movie of a car.

mathematical terms. And it took Newton (1642–1727) and the invention of calculus to put the concepts of motion on a firm and rigorous footing.

As a starting point, let's define **motion** as the change of an object's position with time. Examples of motion are easy to list. Bicycles, baseballs, cars, airplanes, and rockets are all objects that move. The path along which an object moves, which might be a straight line or might be curved, is called the object's **trajectory.**

Figure 1.1 shows four basic types of motion that we will study in this book. Rotational motion is somewhat different from the other three in that rotation is a change of the object's *angular* position. We'll defer rotational motion until later and, for now, focus on motion along a line, circular motion, and projectile motion.

The fundamental question we want to ask is: What *concepts* are needed to give a full and accurate description of motion?

Making a Motion Diagram

An easy way to study motion is to make a movie of a moving object. A movie camera, as you probably know, takes photographs at a fixed rate, typically 30 photographs every second. Each separate photo is called a *frame,* and the frames are all lined up one after the other in a *filmstrip.* As an example, Figure 1.2 shows a few frames from the movie of a car going past. Not surprisingly, the car is in a somewhat different position in each frame.

Suppose we now cut the individual frames of the film strip apart, stack them on top of each other, and then project the entire stack at once onto a screen for viewing. The result is shown in Figure 1.3. This composite photo, showing an object's position at several *equally spaced instants of time,* is called a **motion diagram.** As simple as motion diagrams seem, they will turn out to be a powerful tool for analyzing motion.

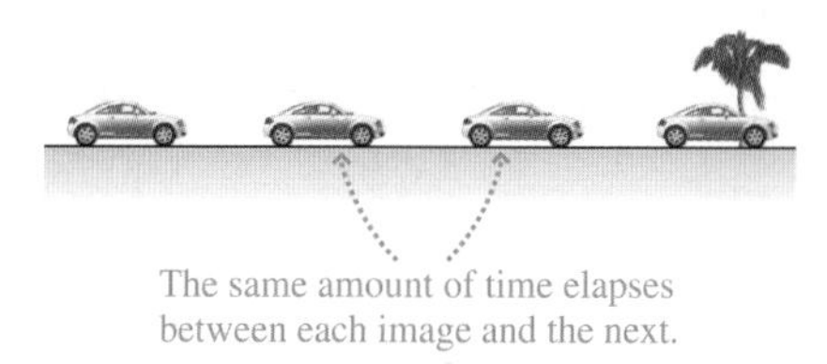

FIGURE 1.3 A motion diagram of the car shows all the frames simultaneously.

NOTE ► It's important to keep the camera in a *fixed position* as the object moves by. Don't "pan" it to track the moving object. ◄

Now let's take our camera out into the world and make a few motion diagrams. The following table shows how we can see important aspects of the motion in a motion diagram.

Examples of motion diagrams

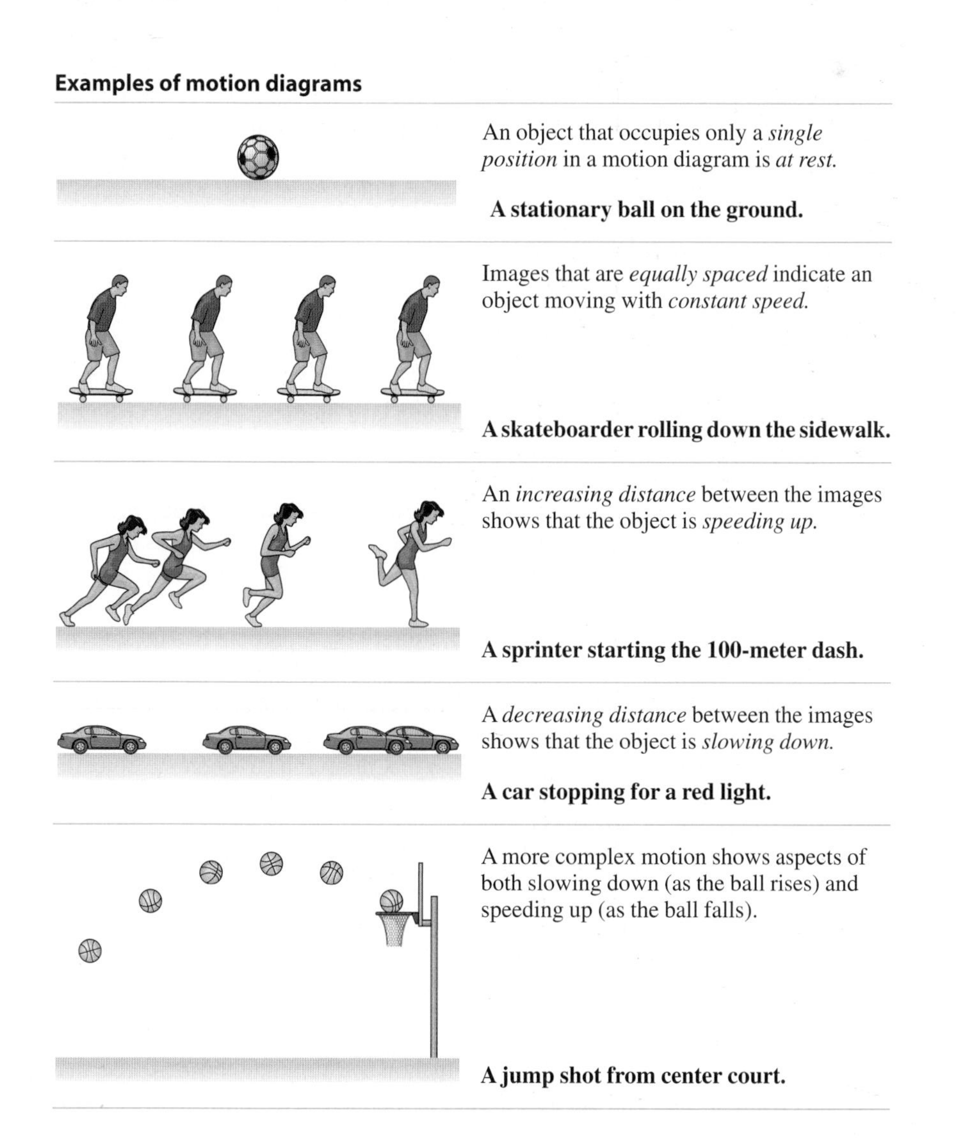

An object that occupies only a *single position* in a motion diagram is *at rest.*
A stationary ball on the ground.
Images that are *equally spaced* indicate an object moving with *constant speed.*
A skateboarder rolling down the sidewalk.
An *increasing distance* between the images shows that the object is *speeding up.*
A sprinter starting the 100-meter dash.
A *decreasing distance* between the images shows that the object is *slowing down.*
A car stopping for a red light.
A more complex motion shows aspects of both slowing down (as the ball rises) and speeding up (as the ball falls).
A jump shot from center court.

We have defined several concepts (at rest, constant speed, speeding up, and slowing down) in terms of how the moving object appears in a motion diagram. These are called **operational definitions,** meaning that the concepts are defined in terms of a particular procedure or operation performed by the investigator. For example, we could answer the question "Is the airplane speeding up?" by checking whether or not the images in the plane's motion diagram are getting farther apart. Many of the concepts in physics will be introduced as operational definitions. This reminds us that physics is an experimental science.

STOP TO THINK 1.1 Which car is going faster, A or B? Assume there are equal intervals of time between the frames of both movies.

NOTE ► Each chapter in this textbook will have several *Stop to Think* questions. These questions are designed to see if you've understood the basic ideas that have been presented. The answers are given at the end of the chapter, but you should make a serious effort to think about these questions before turning to the answers. If you answer correctly, and are sure of your answer rather than just guessing, you can proceed to the next section with confidence. But if you answer incorrectly, it would be wise to reread the preceding sections carefully before proceeding onward. ◄

1.2 The Particle Model

For many objects, such as cars and rockets, the motion of the object *as a whole* is not influenced by the "details" of the object's size and shape. To describe the object's motion, all we really need to keep track of is the motion of a single point, such as a white dot painted on the side of the object.

If we restrict our attention to objects undergoing **translational motion,** which is the motion of an object along a trajectory, we can consider the object *as if* it were just a single point, without size or shape. We can also treat the object *as if* all of its mass were concentrated into this single point. An object that can be represented as a mass at a single point in space is called a **particle.** A particle has no size, no shape, and no distinction between top and bottom or between front and back.

If we treat an object as a particle, we can represent the object in each frame of a motion diagram as a simple dot rather than having to draw a full picture. Figure 1.4 shows how much simpler motion diagrams appear when the object is represented as a particle. Note that the dots have been numbered 0, 1, 2, . . . to tell the sequence in which the frames were exposed. These diagrams are more abstract than the pictures, but they are easier to draw and they still convey our full understanding of the object's motion.

(a) Motion diagram of a rocket launch

4 ●

Numbers show order in which frames were exposed.

3 ●

2 ●

1 ●
0 ●

(b) Motion diagram of a car stopping

● 0 ● 1 ● 2 ● 3 ● 4

The same amount of time elapses between each image and the next.

FIGURE 1.4 Motion diagrams in which the object is represented as a particle.

Using the Particle Model

Treating an object as a particle is, of course, a simplification of reality. As we noted in the overview, such a simplification is called a *model.* Models allow us to focus on the important aspects of a phenomenon by excluding those aspects that play only a minor role. The **particle model** of motion is a simplification in which we treat a moving object as if all of its mass were concentrated at a single point.

The particle model is an excellent approximation of reality for the motion of cars, planes, rockets, and similar objects. People are somewhat more complex, because of moving arms and legs, but the motion of a person's body as a whole is still described reasonably well within the particle model. A ball rolling on a surface is a combination of translational and rotational motion, but we can treat the ball as a particle if its translational motion is all we care about. In fact, so successful has the particle model been that the motion of more complex objects, which cannot be treated as a single particle, is often analyzed as if the object were a collection of particles.

Not all motions can be reduced to the motion of a single point. Consider a rotating gear. The center of the gear doesn't move at all, and each tooth on the gear is moving in a different direction. Rotational motion is qualitatively different than translational motion, and we'll need to go beyond the particle model later when we study rotational motion.

STOP TO THINK 1.2 Three motion diagrams are shown. Which is a dust particle settling to the floor at constant speed, which is a ball dropped from the roof of a building, and which is a descending rocket slowing to make a soft landing on Mars?

(a) 0 ● 1 ● 2 ● 3 ● 4 ● 5 ●

(b) 0 ● 1 ● 2 ● 3 ● 4 ● 5 ●

(c) 0 ● 1 ● 2 ● 3 ● 4 ● 5 ●

1.3 Position and Time

To develop motion diagrams further, we need to be able to make measurements. As we look at a motion diagram, it would be useful to know *where* the object is (i.e., its *position*) and *when* the object was at that position (i.e., the *time*). These are easy measurements to make.

Position measurements can be made by laying a coordinate system grid over a motion diagram. You can then measure the (x, y) coordinates of each point in the motion diagram. Of course, the world does not come with a coordinate system attached. A coordinate system is an artificial grid that *you* place over a problem in order to analyze the motion. You place the origin of your coordinate system wherever you wish, and different observers of a moving object might all choose to use different origins. Likewise, you can choose the orientation of the x-axis and y-axis to be helpful for that particular problem. The conventional choice is for the x-axis to point to the right and the y-axis to point upward, but there is nothing sacred about this choice. We will soon have many occasions to tilt the axes at an angle.

Time, in a sense, is also a coordinate system, although you may never have thought of time this way. You can pick an arbitrary point in the motion and label it "$t = 0$ seconds." This is simply the instant you decide to start your clock or stopwatch, so it is the origin of your time coordinate. Different observers might choose to start their clocks at different moments. A movie frame labeled "$t = 4$ seconds" means it was taken 4 seconds after you started your clock.

We typically choose $t = 0$ to represent the "beginning" of a problem, but the object may have been moving before then. Those earlier instants would be measured as negative times, just as objects on the x-axis to the left of the origin have negative values of position. Negative numbers are not to be avoided; they simply locate an event in space or time *relative to an origin.*

To illustrate, Figure 1.5a on the next page shows an xy-coordinate system and time information superimposed over the motion diagram of a basketball. You can see that ball's position is $(x_4, y_4) = (4\text{ m}, 3\text{ m})$ at time $t_4 = 2.0\text{ s}$. Notice how we've used subscripts to indicate the time and the object's position in a specific frame of the motion diagram.

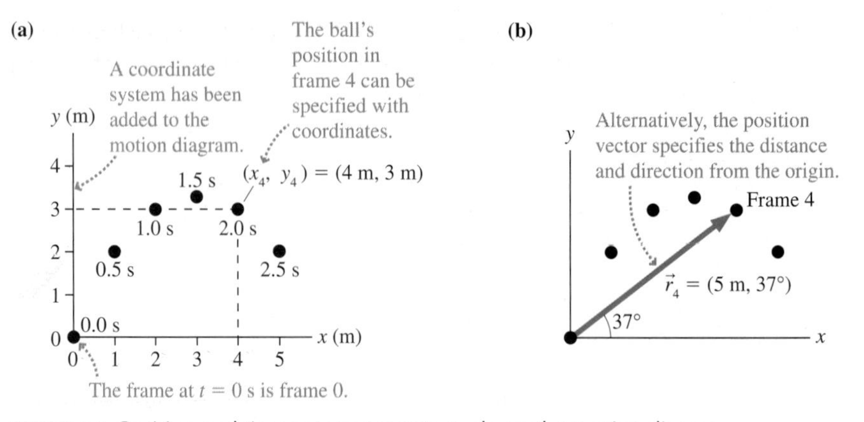

FIGURE 1.5 Position and time measurements made on the motion diagram of a basketball.

NOTE ► The first frame is labeled 0 to correspond with time $t = 0$. That is why the fifth frame is labeled 4. ◄

Another way to locate the ball is to draw an arrow from the origin to the point representing the ball. You can then specify the length and direction of the arrow. An arrow drawn from the origin to an object's position is called the **position vector** of the object, and it is given the symbol $\vec{r}$. Figure 1.5b shows the position vector $\vec{r}_4 = (5\text{ m}, 37°)$.

The position vector $\vec{r}$ does not tell us anything different than the coordinates (x, y). It simply provides the information in an alternative form. Although you're probably more familiar with (x, y) coordinates than with vectors, you will find that vectors are a useful and powerful way to describe many concepts in physics.

A Word About Vectors and Notation

Before we continue our discussion, we should take a closer look at what a vector is. Vectors will be studied thoroughly in Chapter 3, so all we need for now is a little basic information. Some physical quantities, such as time, mass, and temperature, can be described completely by a single number with a unit. For example, the mass of an object is 6 kg and its temperature is 30°C. When a physical quantity is described by a single number (with a unit), we call it a **scalar quantity.** A scalar can be positive, negative, or zero.

Many other quantities, however, have a directional quality and cannot be described by a single number. To describe the motion of a car, for example, you must specify not only how fast it is moving, but also the *direction* in which it is moving. A **vector quantity** is a quantity that has both a *size* (the "How far?" or "How fast?") and a *direction* (the "Which way?"). The size or length of a vector is called its *magnitude*. The magnitude of a vector can be positive or zero, but it cannot be negative.

When we want to represent a vector quantity with a symbol, we need somehow to indicate that the symbol is for a vector rather than for a scalar. We do this by drawing an arrow over the letter that represents the quantity. Thus $\vec{r}$ and $\vec{A}$ are symbols for vectors, whereas r and A, without the arrows, are symbols for scalars. In handwritten work you must draw arrows over all symbols that represent vectors. This may seem strange until you get used to it, but it is very important because we will often use both r and $\vec{r}$, or both A and $\vec{A}$, in the same problem, and they mean different things! Without the arrow, you will be using the same symbol with two different meanings and will likely end up making a mistake. Note that the arrow over the symbol always points to the right, regardless of which direction the actual vector points. Thus we write $\vec{r}$ or $\vec{A}$, never $\overleftarrow{r}$ or $\overleftarrow{A}$.

NOTE ▶ Some textbooks represent vectors with boldface type, such as **r** or **A**. This book will consistently display the vector arrow over vector symbols, just as you should do in handwritten work. ◀

Change in Position

Now that you've seen how to measure position and time, let's return to the problem of motion, where we need to measure *changes* in position that occur over time. Consider the following:

> Sam is standing 50 feet (ft) east of the corner of 12th Street and Vine. He then walks northeast for 100 ft to a second point. What is Sam's change of position?

Figure 1.6 shows Sam's motion in terms of position vectors. Because we're free to place the origin of our coordinate system wherever we wish, we've placed it at the intersection. Sam's initial position is the vector $\vec{r}_0$ drawn from the origin to the point where he starts walking. Vector $\vec{r}_1$ is his position after he finishes walking. You can see that Sam has changed position, and a *change* of position is called a **displacement.** His displacement is the vector labeled $\Delta\vec{r}$. The Greek letter delta (Δ) is used in math and science to indicate the *change* in a quantity. Here it indicates a change in the position $\vec{r}$.

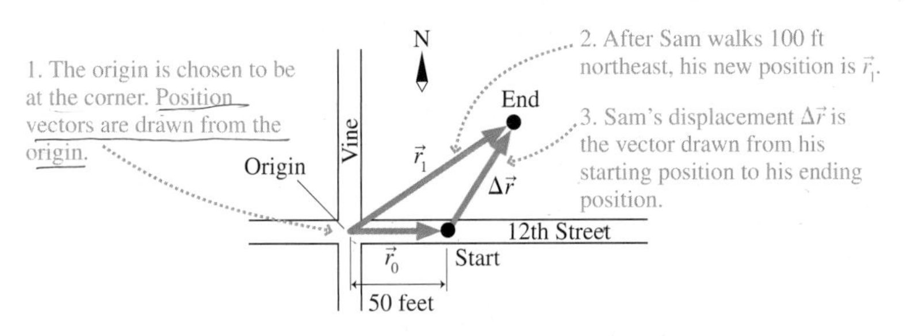

FIGURE 1.6 Sam undergoes a displacement $\Delta\vec{r}$ from position $\vec{r}_0$ to position $\vec{r}_1$.

NOTE ▶ $\Delta\vec{r}$ is a *single* symbol. You cannot cancel out or remove the Δ in algebraic operations. ◀

Displacement is a vector quantity; it requires both a length and a direction to describe it. Specifically, the displacement $\Delta\vec{r}$ is a vector drawn *from* a starting position *to* an ending position. Sam's displacement vector is written

$$\Delta\vec{r} = (100 \text{ ft, northeast})$$

where we've given both the length and the direction. The length, or magnitude, of a displacement vector is simply the straight-line distance between the starting and ending positions.

Suppose you start 10 ft from a door and walk directly away from the door for 5 ft. Although it's clear that you end up 15 ft from the door, the *procedure* by which you learn this is to *add* your change in position (5 ft) to your initial position (10 ft).

Similarly, we can answer the question "Where does Sam end up?" if we *add* his change in position (his displacement $\Delta\vec{r}$) to his initial position, the vector $\vec{r}_0$. Sam's final position in Figure 1.6, vector $\vec{r}_1$, can be seen as a combination of vector $\vec{r}_0$ *plus* vector $\Delta\vec{r}$. In fact, $\vec{r}_1$ is the *vector sum* of vectors $\vec{r}_0$ and $\Delta\vec{r}$. This is written

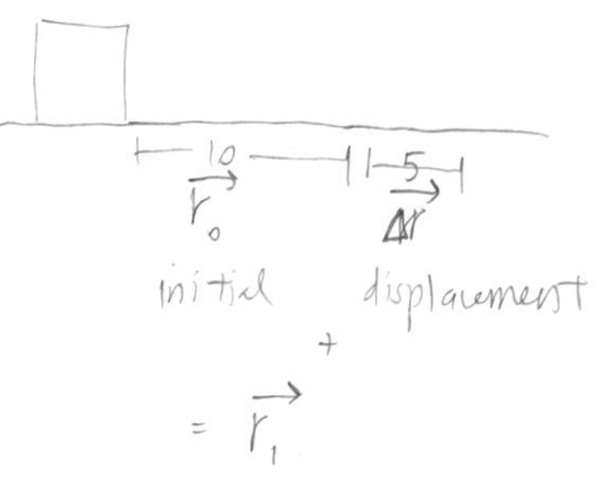

$$\vec{r}_1 = \vec{r}_0 + \Delta\vec{r} \tag{1.1}$$

Notice, however, that we are now adding vector quantities, not scalar quantities. Vector addition is a different process from "regular" addition. You can add two vectors $\vec{A}$ and $\vec{B}$ with the following three-step procedure.

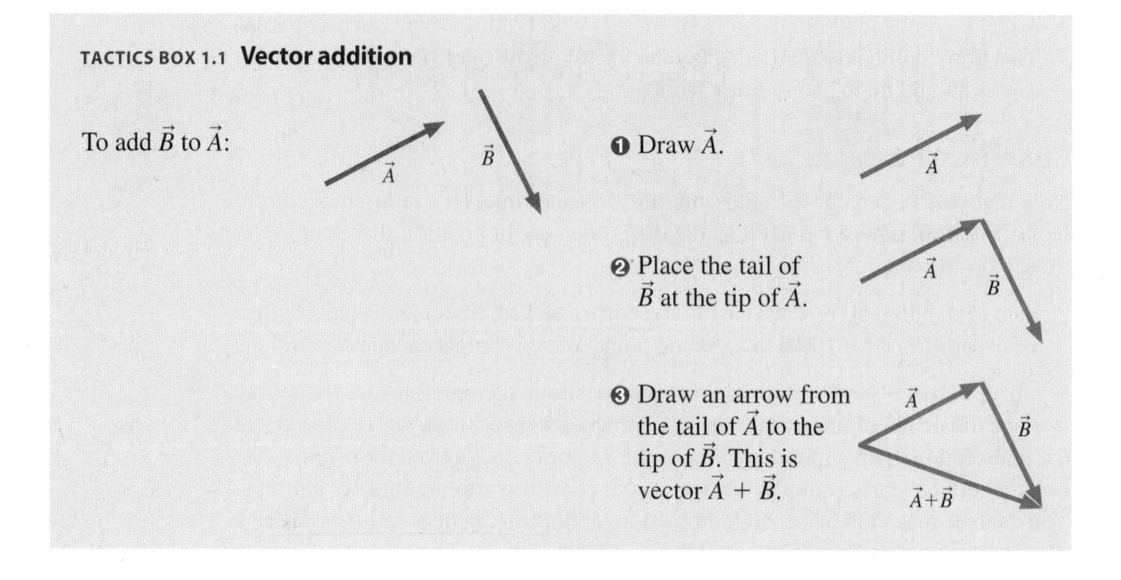

TACTICS BOX 1.1 Vector addition
To add $\vec{B}$ to $\vec{A}$:
$\vec{A}$
$\vec{B}$
❶ Draw $\vec{A}$.
❷ Place the tail of $\vec{B}$ at the tip of $\vec{A}$.
❸ Draw an arrow from the tail of $\vec{A}$ to the tip of $\vec{B}$. This is vector $\vec{A} + \vec{B}$.
$\vec{A}+\vec{B}$

This is exactly how $\vec{r}_0$ and $\Delta\vec{r}$ are added in Figure 1.6 to give $\vec{r}_1$.

NOTE ► A vector is not tied to a particular location on the page. You can move a vector around on a figure as long as you don't change its length or the direction it points. Vector $\vec{B}$ is not changed by sliding it over to where its tail is at the tip of $\vec{A}$. ◄

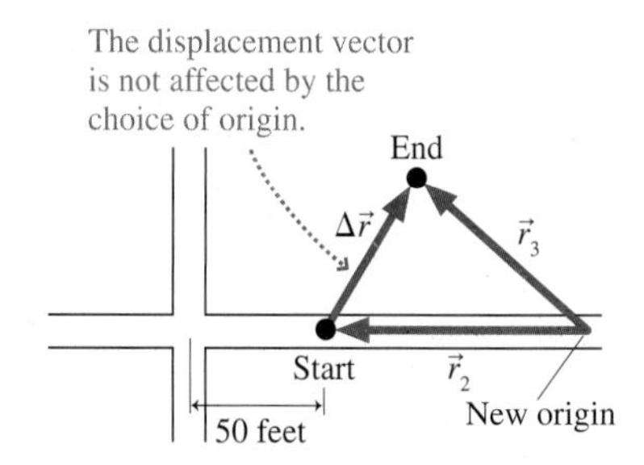

FIGURE 1.7 Sam's displacement $\Delta\vec{r}$ is unchanged by using a different coordinate system.

In Figure 1.6, we chose *arbitrarily* to put the origin of the coordinate system at the corner. While this might be convenient, it certainly is not mandatory. Figure 1.7 shows a different choice of where to place the origin. Notice something interesting. Vectors $\vec{r}_0$ and $\vec{r}_1$ have become new vectors $\vec{r}_2$ and $\vec{r}_3$, but the displacement vector $\Delta\vec{r}$ has not changed! **The displacement is a quantity that is independent of the coordinate system.** In other words, the arrow drawn from the one position of an object to the next is the same no matter what coordinate system you choose. This independence gives the displacement $\Delta\vec{r}$ more *physical significance* than the position vectors themselves have.

This observation suggests that the displacement, rather than the actual position, is what we want to focus on as we analyze the motion of an object. Equation 1.1 told us that $\vec{r}_1 = \vec{r}_0 + \Delta\vec{r}$. This is easily rearranged to give a more precise definition of displacement: **The displacement $\Delta\vec{r}$ of an object as it moves from an initial position $\vec{r}_i$ to a final position $\vec{r}_f$ is**

$$\Delta\vec{r} = \vec{r}_f - \vec{r}_i \tag{1.2}$$

Graphically, $\Delta\vec{r}$ is a vector arrow drawn *from* position $\vec{r}_i$ *to* position $\vec{r}_f$. The displacement vector is independent of the coordinate system.

NOTE ► To be more general, we've written Equation 1.2 in terms of an *initial position* and a *final position,* indicated by subscripts i and f. We'll frequently use i and f when writing general equations, then use specific numbers or values, such as 0 and 1, when working a problem. ◄

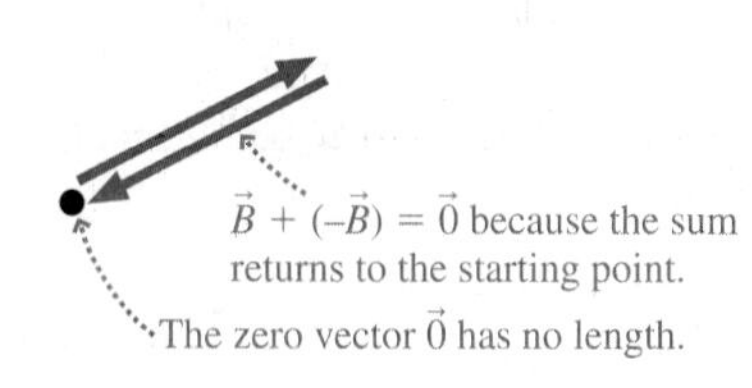

FIGURE 1.8 The negative of a vector.

This definition of $\Delta\vec{r}$ involves *vector subtraction.* With numbers, subtraction is the same as the addition of a negative number. That is, $5 - 3$ is the same as $5 + (-3)$. Similarly, we can use the rules for vector addition to find $\vec{A} - \vec{B} = \vec{A} + (-\vec{B})$ if we first define what we mean by $-\vec{B}$. As Figure 1.8 shows, the negative of vector $\vec{B}$ is a vector with the same length but pointing in the opposite direction. This makes sense because $\vec{B} - \vec{B} = \vec{B} + (-\vec{B}) = \vec{0}$, where $\vec{0}$, a vector with zero length, is called the **zero vector.**

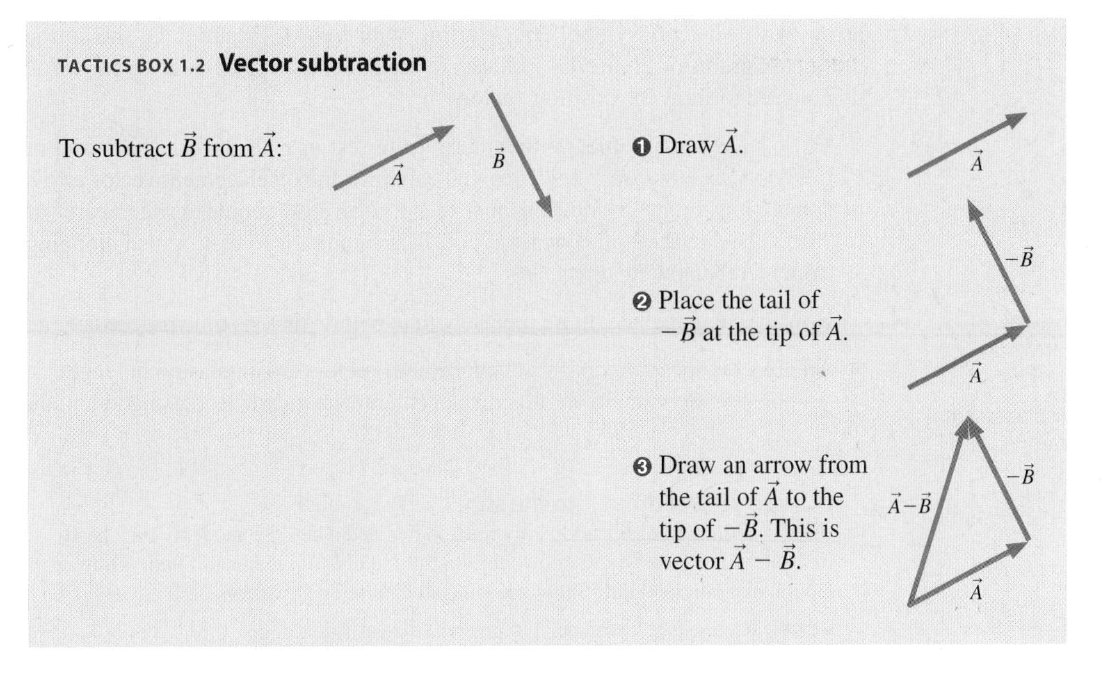
TACTICS BOX 1.2 Vector subtraction
To subtract $\vec{B}$ from $\vec{A}$:
$\vec{A}$
$\vec{B}$
❶ Draw $\vec{A}$.
❷ Place the tail of $-\vec{B}$ at the tip of $\vec{A}$.
$-\vec{B}$
❸ Draw an arrow from the tail of $\vec{A}$ to the tip of $-\vec{B}$. This is vector $\vec{A} - \vec{B}$.
$\vec{A}-\vec{B}$

Figure 1.9 shows how to use the vector subtraction rules to find the displacement $\Delta\vec{r}$ in going from position $\vec{r}_i$ to position $\vec{r}_f$.

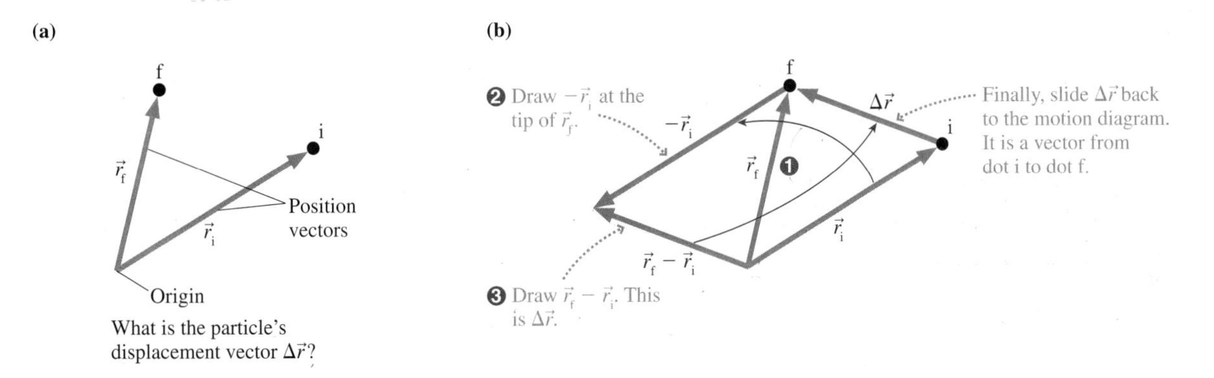
(a)
f
i
$\vec{r}_f$
$\vec{r}_i$
Position vectors
Origin
What is the particle's displacement vector $\Delta\vec{r}$?
(b)
❷ Draw $-\vec{r}_i$ at the tip of $\vec{r}_f$.
$-\vec{r}_i$
$\Delta\vec{r}$
Finally, slide $\Delta\vec{r}$ back to the motion diagram. It is a vector from dot i to dot f.
❶
$\vec{r}_f - \vec{r}_i$
❸ Draw $\vec{r}_f - \vec{r}_i$. This is $\Delta\vec{r}$.

FIGURE 1.9 Using vector subtraction to find $\Delta\vec{r}$.

Change in Time

It's also useful to consider a *change* in time. For example, the clock readings of two frames of film might be t_1 and t_2. The specific values are arbitrary because they are timed relative to an arbitrary instant that you chose to call $t = 0$. But the **time interval** $\Delta t = t_2 - t_1$ is *not* arbitrary. It represents the elapsed time for the object to move from one position to the next. All observers will measure the same value for Δt, regardless of when they choose to start their clocks.

The time interval $\Delta t = t_f - t_i$ measures the elapsed time as an object moves from an initial position $\vec{r}_i$ at time t_i to a final position $\vec{r}_f$ at time t_f. The value of Δt is independent of the specific clock used to measure the times.

A stopwatch is used to measure a time interval.

Application to Motion Diagrams

A motion diagram is a series of position measurements. Suppose an object moves from position $\vec{r}_n$ in frame n of a motion diagram to position $\vec{r}_{n+1}$ in frame $n + 1$. As Figure 1.10 on the next page shows, the displacement vector $\Delta\vec{r}$ is simply the vector drawn from dot n to dot $n + 1$.

The first step in analyzing a motion diagram is to determine all of the displacement vectors by drawing arrows connecting each dot to the next. Label each

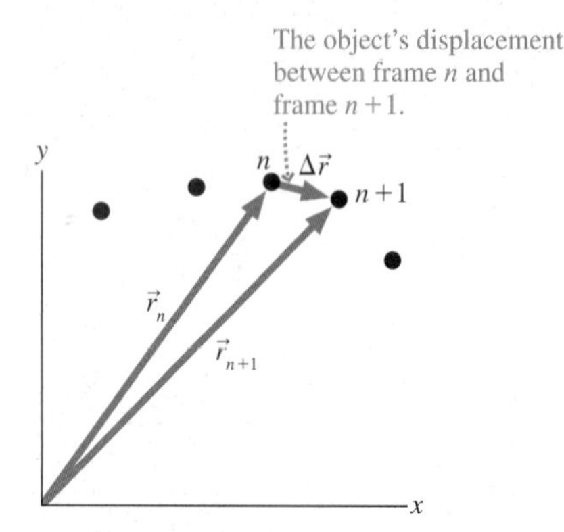
The object's displacement between frame n and frame $n+1$.
y
n
$\Delta\vec{r}$
$n+1$
$\vec{r}_n$
$\vec{r}_{n+1}$
x

FIGURE 1.10 Drawing a displacement vector on a motion diagram.

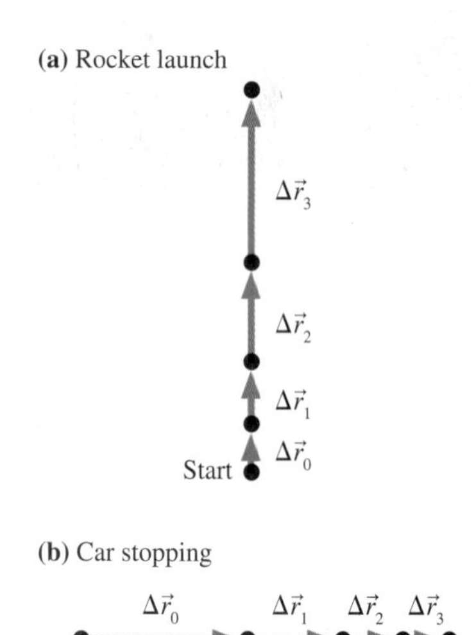
(a) Rocket launch
$\Delta\vec{r}_3$
$\Delta\vec{r}_2$
$\Delta\vec{r}_1$
$\Delta\vec{r}_0$
Start
(b) Car stopping
$\Delta\vec{r}_0$
$\Delta\vec{r}_1$ $\Delta\vec{r}_2$ $\Delta\vec{r}_3$
Stop

FIGURE 1.11 Motion diagrams with the displacement vectors.

arrow with a *vector* symbol $\Delta\vec{r}_n$, starting with $n = 0$. Figure 1.11 shows the motion diagrams of Figure 1.4 redrawn to include the displacement vectors. You do not need to show the position vectors.

NOTE ▶ When an object either starts from rest or ends at rest, the initial or final dots are *as close together* as you can draw the displacement vector arrow connecting them. In addition, just to be clear, you should write "Start" or "Stop" beside the initial or final dot. It is important to distinguish stopping from merely slowing down. ◀

Now we can conclude, more precisely than before, that, as time proceeds:

- An object is speeding up if its displacement vectors are increasing in length.
- An object is slowing down if its displacement vectors are decreasing in length.

EXAMPLE 1.1 Headfirst into the snow

Alice is sliding along a smooth, icy road on her sled when she suddenly runs headfirst into a large, very soft snowbank that gradually brings her to a halt. Draw a motion diagram for Alice. Show and label all displacement vectors.

MODEL Use the particle model to represent Alice as a dot.

VISUALIZE Figure 1.12 shows Alice's motion diagram. The problem statement suggests that Alice's speed is very nearly constant until she hits the snowbank. Thus her displacement vectors are of equal length as she slides along the icy road. She begins slowing when she hits the snowbank, so the displacement vectors then get shorter until she stops. It is reasonable to assume that her stopping distance in the snow is less than the distance she had slid along the road, but we do not want to make her stop *too* quickly.

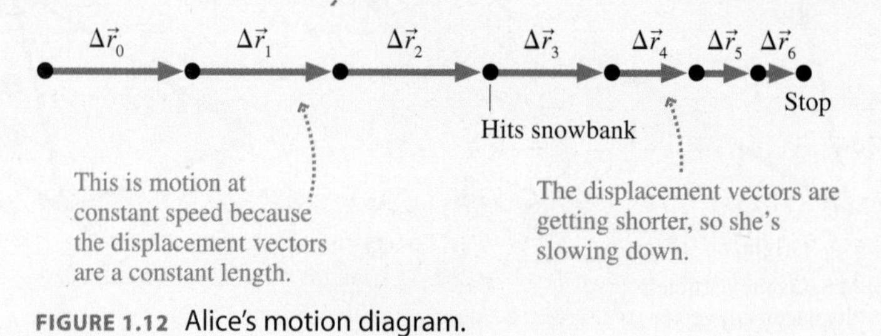
$\Delta\vec{r}_0$ $\Delta\vec{r}_1$ $\Delta\vec{r}_2$ $\Delta\vec{r}_3$ $\Delta\vec{r}_4$ $\Delta\vec{r}_5$ $\Delta\vec{r}_6$
Stop
Hits snowbank
This is motion at constant speed because the displacement vectors are a constant length.
The displacement vectors are getting shorter, so she's slowing down.

FIGURE 1.12 Alice's motion diagram.

To summarize the main idea of this section, we have added coordinate systems and clocks to our motion diagrams in order to measure *when* each frame was exposed and *where* the object was located at that time. Different observers of the motion may choose different coordinate systems and different clocks. Thus a particular value of the position $\vec{r}$ or the time t is arbitrary because each is measured relative to an arbitrarily chosen origin. However, all observers find the *same* values for the displacements $\Delta\vec{r}$ and the time intervals Δt because these are independent of the specific coordinate system used to measure them.

1.4 Velocity

It's no surprise that, during a given time interval, a speeding bullet travels farther than a speeding snail. To extend our study of motion so that we can compare the bullet to the snail, we need a way to measure how fast or how slowly an object moves.

One quantity that measures an object's fastness or slowness is its **average speed,** defined as the ratio

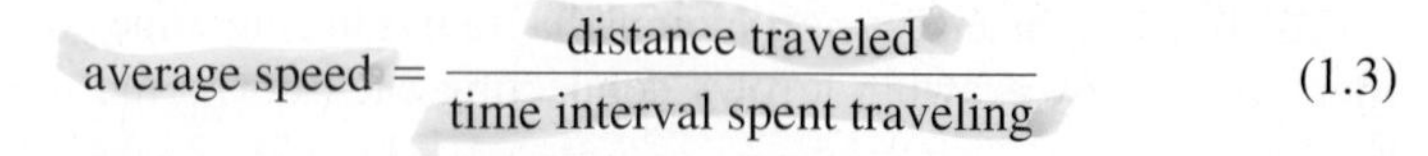

$$\text{average speed} = \frac{\text{distance traveled}}{\text{time interval spent traveling}} \tag{1.3}$$

If you drive 15 miles (mi) in 30 minutes ($\frac{1}{2}$ hour), your average speed is

$$\text{average speed} = \frac{15 \text{ mi}}{\frac{1}{2}\text{hour}} = 30 \text{ mph} \tag{1.4}$$

Although the concept of speed is widely used in our day-to-day lives, it is not a sufficient basis for a science of motion. To see why, imagine you're trying to land a jet plane on an aircraft carrier. It matters a great deal to you whether the aircraft carrier is moving at 20 mph (miles per hour) to the north or 20 mph to the east. Simply knowing that the boat's speed is 20 mph is not enough information! The difficulty with speed is that it tells us nothing about the *direction* in which an object is moving.

A more useful question than "How fast does an object move?" is the question "How quickly (or slowly) does an object move from one position to another?" These may seem to be the same question, but they're not. An object's change of position is measured by its displacement $\Delta\vec{r}$, a vector quantity. The displacement tells us not only the distance traveled by an object, but also the *direction* of motion.

Consider the ratio $\Delta\vec{r}/\Delta t$ for an object that undergoes a displacement $\Delta\vec{r}$ during the time interval Δt. This ratio is a vector, because $\Delta\vec{r}$ is a vector, so it has both a magnitude and a direction. The size, or magnitude, of this ratio is very similar to the definition of speed: The ratio will be larger for a fast object than for a slow object. But in addition to measuring how fast an object moves, this ratio is a vector that points in the same direction as $\Delta\vec{r}$. That is, it points in the direction of motion.

It is convenient to give this ratio a name. We call it the **average velocity,** and it has the symbol $\vec{v}_{avg}$. **The average velocity of an object during the time interval Δt, in which the object undergoes a displacement $\Delta\vec{r}$, is the vector**

$$\vec{v}_{avg} = \frac{\Delta\vec{r}}{\Delta t} \tag{1.5}$$

An object's average-velocity vector points in the same direction as the displacement vector $\Delta\vec{r}$. This is the direction of motion.

NOTE ▶ In everyday language we do not make a distinction between speed and velocity, but in physics *the distinction is very important.* In particular, speed is simply "How fast," whereas velocity is "How fast, and in which direction." As we go along we will be giving other words more precise meaning in physics than they have in everyday language. ◀

As an example, Figure 1.13a shows two ships that start from the same position and move 5 miles in 15 minutes. Both ships have a speed of 20 mph, but their velocities are different. Because their displacements during Δt are $\Delta\vec{r}_A$ = (5 mi, north) and $\Delta\vec{r}_B$ = (5 mi, east), we can write their velocities as

$$\begin{aligned}\vec{v}_{avg\,A} &= (20 \text{ mph, north})\\ \vec{v}_{avg\,B} &= (20 \text{ mph, east})\end{aligned} \tag{1.6}$$

Notice how the velocity *vectors* in Figure 1.13b point in the direction of motion.

NOTE ▶ Our goal in this chapter is to *visualize* motion with motion diagrams. Strictly speaking, the vector we have defined in Equation 1.5, and the vector we will show on motion diagrams, is the *average* velocity $\vec{v}_{avg}$. But to allow the motion diagram to be a useful tool, we will drop the subscript and refer to the average velocity as simply $\vec{v}$. Our definitions and symbols, which somewhat blur the distinction between average and instantaneous quantities, are adequate for visualization purposes, but they're not the final word on the subject. We will refine these definitions in Chapter 2, where our goal will be to develop the mathematics of motion. ◀

The victory goes to the runner with the highest average speed.

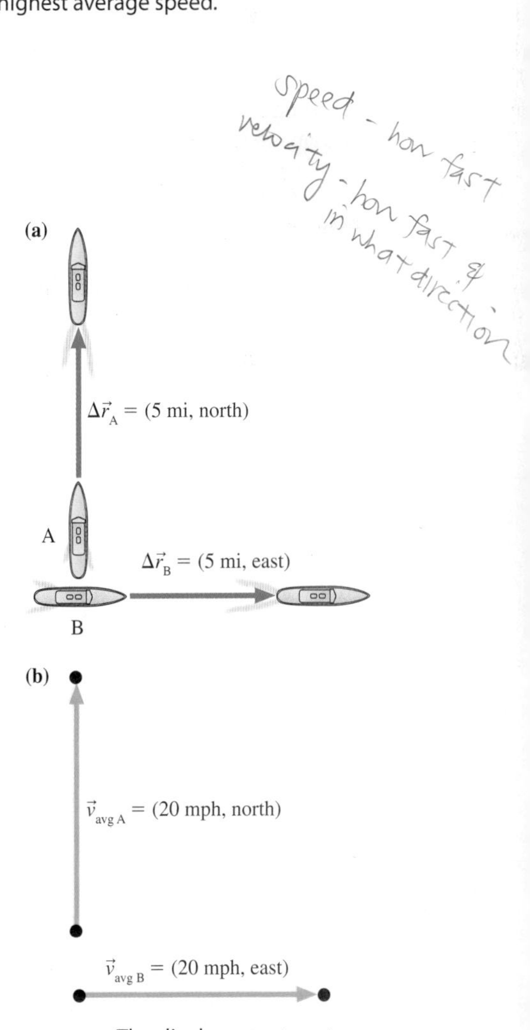

FIGURE 1.13 The displacement vectors and velocities of ships A and B.

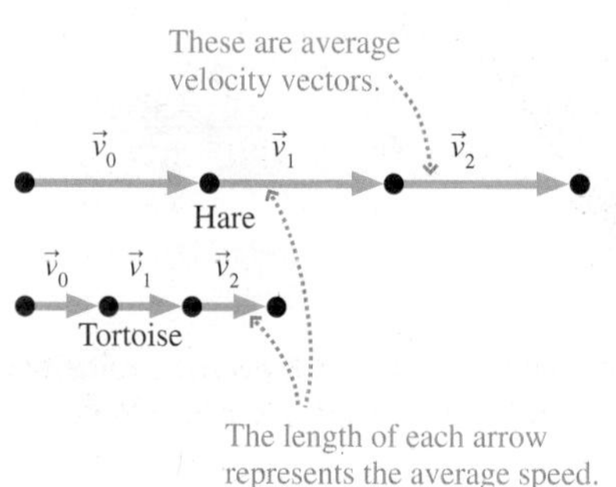
These are average velocity vectors.
$\vec{v}_0$
$\vec{v}_1$
$\vec{v}_2$
Hare
$\vec{v}_0$
$\vec{v}_1$
$\vec{v}_2$
Tortoise
The length of each arrow represents the average speed. The hare moves faster than the tortoise.

FIGURE 1.14 Motion diagram of the tortoise racing the hare.

Motion Diagrams with Velocity Vectors

The velocity vector, as we've defined it, points in the same direction as the displacement $\Delta\vec{r}$, and the length of $\vec{v}$ is directly proportional to the length of $\Delta\vec{r}$. Consequently, the vectors connecting each dot of a motion diagram to the next, which we previously labeled as displacement vectors, could equally well be identified as velocity vectors.

This idea is illustrated in Figure 1.14, which shows four frames from the motion diagram of a tortoise racing a hare. The vectors connecting the dots are now labeled as velocity vectors $\vec{v}$. The length of each velocity vector is no longer a distance. Instead, **the length of a velocity vector represents the average speed with which the object moves between the two points.** Longer velocity vectors indicate faster motion. You can see from the diagram that the hare moves faster than the tortoise.

Notice that the hare's velocity vectors do not change; each has the same length and direction. We say the hare is moving with *constant velocity.* The tortoise is also moving with its own constant velocity.

EXAMPLE 1.2 Accelerating up a hill

The light turns green and a car accelerates, starting from rest, up a 20° hill. Draw a motion diagram that shows the car's velocity.

MODEL Use the particle model to represent the car as a dot.

VISUALIZE The car's motion takes place along a straight line, but the line is neither horizontal nor vertical. Because a motion diagram is made from frames of a movie, it will show the object moving with the correct orientation—in this case, at an angle of 20°. Figure 1.15 shows several frames of the motion diagram, where we see the car speeding up. The car starts from rest, so the first arrow is drawn as short as possible and the first dot is labeled "Start." The displacement vectors have been drawn from each dot to the next, but then they have been identified and labeled as average velocity vectors $\vec{v}$.

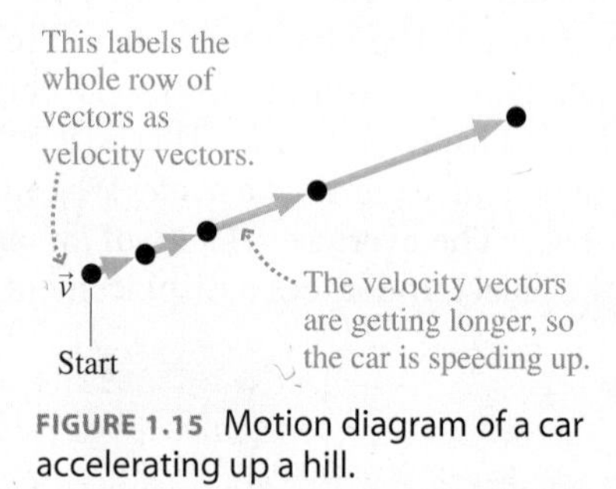
This labels the whole row of vectors as velocity vectors.
$\vec{v}$
Start
The velocity vectors are getting longer, so the car is speeding up.

FIGURE 1.15 Motion diagram of a car accelerating up a hill.

NOTE ► Rather than label every single vector, it's easier to give one label to the entire row of velocity vectors. You can see this in Figure 1.15. ◄

EXAMPLE 1.3 It's a hit!

Jake hits a ball at a 60° angle. It is caught by Jim. Draw a motion diagram of the ball.

MODEL This example is typical of how many problems in science and engineering are worded. The problem does not give a clear statement of where the motion begins or ends. Are we interested in the motion of the ball just during the time it is in the air between Jake and Jim? What about the motion *as* Jake hits it (ball rapidly speeding up) or *as* Jim catches it (ball rapidly slowing down)? Should we include Jim dropping the ball after he catches it? The point is that *you* will often be called on to make a *reasonable interpretation* of a problem statement. In this problem, the details of hitting and catching the ball are complex. The motion of the ball through the air is easier to describe, and it's a motion you might expect to learn about in a physics class. So our *interpretation* is that the motion diagram should start as the ball leaves Jake's bat (ball already moving) and should end the instant it touches Jim's hand (ball still moving). We will model the ball as a particle.

VISUALIZE With this interpretation in mind, Figure 1.16 shows the motion diagram of the ball. Notice how, in contrast to the car of Figure 1.15, the ball is already moving as the motion diagram movie begins. As before, the average velocity vectors are found by connecting the dots with arrows. You can see that the average velocity vectors get shorter (ball slowing down), get longer (ball speeding up), and change direction. Each $\vec{v}$ is different, so this is *not* constant velocity motion.

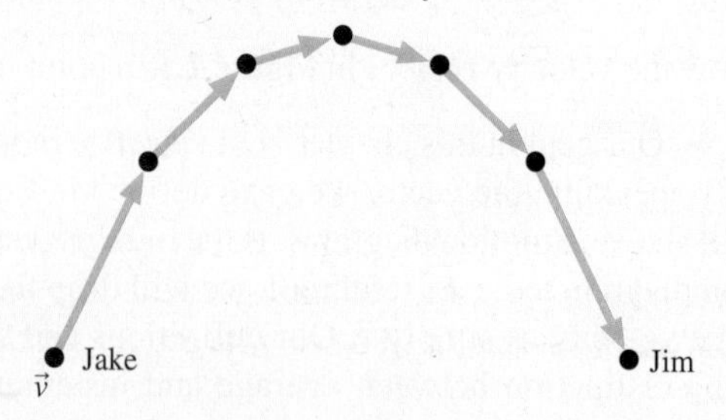
Jake
$\vec{v}$
Jim

FIGURE 1.16 Motion diagram of a ball traveling from Jake to Jim.

Relating Position to Velocity

We defined the average velocity $\vec{v}$ in terms of the displacement $\Delta\vec{r}$, but let's turn that around. Suppose a ball at position $\vec{r}_1$ has velocity $\vec{v}$. As you can see in Figure 1.17, the ball is displaced from its starting position $\vec{r}_1$ to a new position $\vec{r}_2$ during the time interval Δt. We can combine $\vec{v} = \Delta\vec{r}/\Delta t$ and $\Delta\vec{r} = \vec{r}_2 - \vec{r}_1$ to write

$$\vec{r}_2 = \vec{r}_1 + \vec{v}\,\Delta t \tag{1.7}$$

Knowing an object's velocity is the key to finding its position at later instants of time.

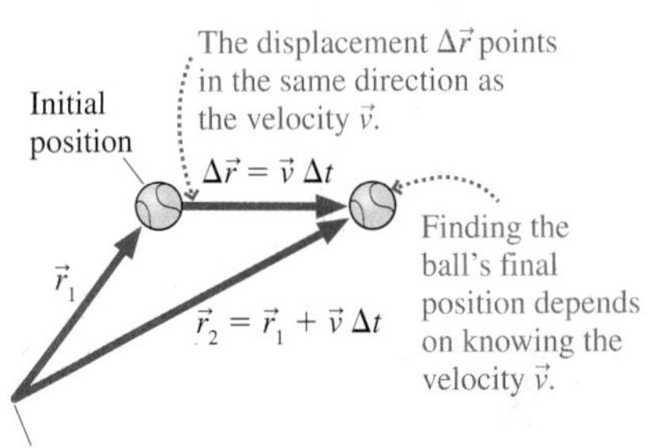

FIGURE 1.17 A ball with average velocity $\vec{v}$ moves from position $\vec{r}_1$ to position $\vec{r}_2$.

STOP TO THINK 1.3 A particle moves from position 1 to position 2 during the interval Δt. Which vector shows the particle's average velocity?

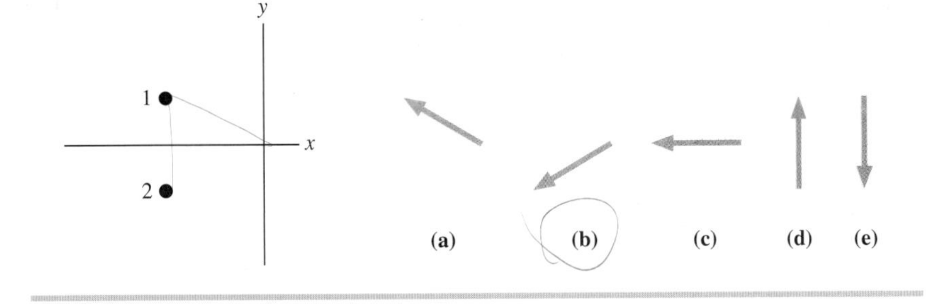

1.5 Acceleration

The goal of this chapter is to find a set of concepts with which to describe motion. Position, time, and velocity are important concepts, and at first glance they might appear to be sufficient. But that is not the case. Sometimes an object's velocity is constant, as it was in Figure 1.14. More often, an object's velocity changes as it moves, as in Figures 1.15 and 1.16. We need one more motion concept, one that will describe a *change* in the velocity.

Because velocity is a vector, it can change in two possible ways:

1. The magnitude can change, indicating a change in speed, or
2. The direction can change, indicating that the object has changed direction.

Figure 1.15 showed the motion diagram of a car speeding up. That was an example in which the magnitude of the velocity vector changed but not the direction.

As an example where only the direction changes, but not the magnitude, Figure 1.18 is the motion diagram of a runner going at constant speed around a circular track. The lengths of all the average velocity vectors are the same, showing that the runner's speed is constant, but the direction of each velocity vector is different. Thus Figure 1.18 also represents a changing velocity.

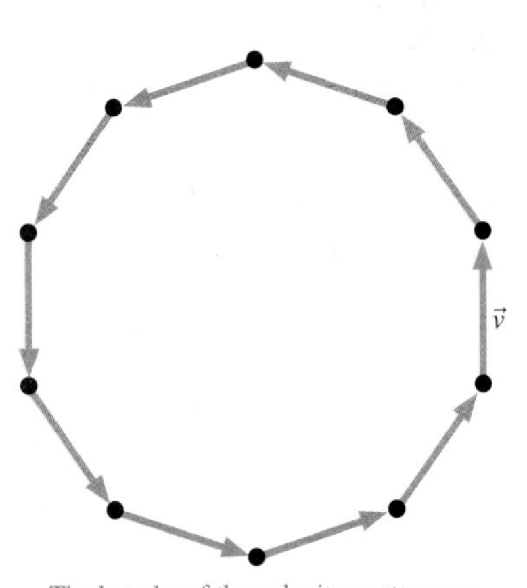

$\vec{v}$
The lengths of the velocity vectors are the same, indicating constant speed, but the direction of each vector is different. This is a changing velocity.

FIGURE 1.18 Motion diagram of a runner on a circular track.

How can we measure the change of velocity in a meaningful way? When we wanted to measure changes in position, the ratio $\Delta\vec{r}/\Delta t$ was useful. This ratio is the *rate of change of position*. By analogy, consider an object whose velocity changes from $\vec{v}_1$ to $\vec{v}_2$ during the time interval Δt. The ratio $\Delta\vec{v}/\Delta t$, where $\Delta\vec{v} = \vec{v}_2 - \vec{v}_1$, is the *rate of change of velocity*. But what does it measure?

Consider two cars, a Volkswagen Beetle and a fancy Porsche. Let them start from rest, and measure their velocities after an elapsed time of 10 seconds. The Porsche, we can assume, will have a larger $\Delta\vec{v}$. Consequently, it will have the larger value of the ratio $\Delta\vec{v}/\Delta t$.

Restart the two cars, but now measure the time interval Δt it takes for each of them to reach a speed of 60 mph. The Porsche will have a smaller value of Δt, which again gives it a larger value of the ratio $\Delta\vec{v}/\Delta t$. Thus this ratio appears to measure how quickly the car speeds up. It has a large magnitude for objects that speed up quickly and a small magnitude for objects that speed up slowly. The ratio $\Delta\vec{v}/\Delta t$ is called the **average acceleration,** and its symbol is $\vec{a}_{avg}$. **The average acceleration of an object during the time interval Δt, in which the object's velocity changes by $\Delta\vec{v}$, is the vector**

$$\vec{a}_{avg} = \frac{\Delta\vec{v}}{\Delta t} \tag{1.8}$$

An object's average-acceleration vector points in the same direction as the vector $\Delta\vec{v}$. Note that acceleration, like position and velocity, is a vector. Both its magnitude and its direction are important pieces of information.

Acceleration is a fairly abstract concept. Position and time are our real hands-on measurements of an object, and they are easy to understand. You can "see" where the object is located and the time on the clock. Velocity is a bit more abstract, being a relationship between the change of position and the change of time. Motion diagrams help us visualize velocity as the vector arrows connecting one position of the object to the next. Acceleration is an even more abstract idea about changes in the velocity. Yet it is essential to develop a good intuition about acceleration because it will be a key concept for understanding why objects move as they do.

NOTE ▶ As we did with velocity, we will drop the subscript and refer to the average acceleration as simply $\vec{a}$. This is adequate for visualization purposes, but not the final word on the subject. We will refine the definition of acceleration in Chapter 2. ◀

Finding the Acceleration Vectors on a Motion Diagram

Let's look at how we can determine the average acceleration vector $\vec{a}$ from a motion diagram. From its definition, we see that $\vec{a}$ points in the same direction as $\Delta\vec{v}$, the change of velocity. This critically important idea is the basis for a technique to find $\vec{a}$.

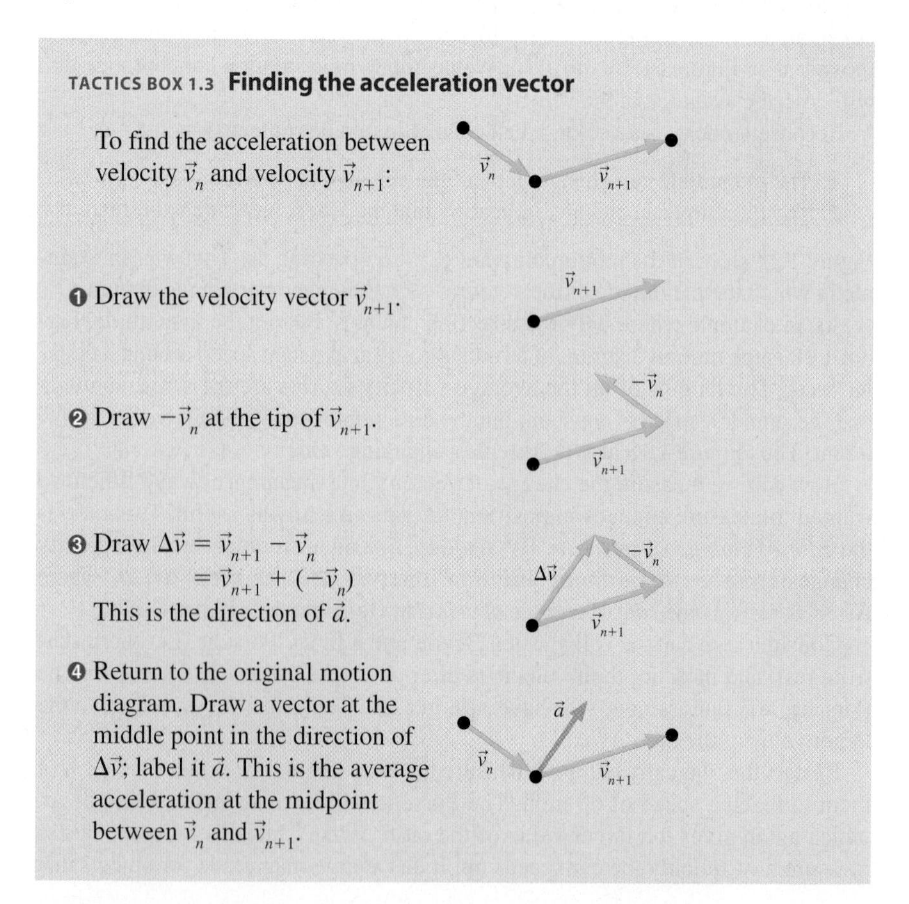

TACTICS BOX 1.3 Finding the acceleration vector
To find the acceleration between velocity $\vec{v}_n$ and velocity $\vec{v}_{n+1}$:
$\vec{v}_n$
$\vec{v}_{n+1}$
❶ Draw the velocity vector $\vec{v}_{n+1}$.
$\vec{v}_{n+1}$
❷ Draw $-\vec{v}_n$ at the tip of $\vec{v}_{n+1}$.
$-\vec{v}_n$
$\vec{v}_{n+1}$
❸ Draw $\Delta\vec{v} = \vec{v}_{n+1} - \vec{v}_n = \vec{v}_{n+1} + (-\vec{v}_n)$
This is the direction of $\vec{a}$.
$\Delta\vec{v}$
$-\vec{v}_n$
$\vec{v}_{n+1}$
❹ Return to the original motion diagram. Draw a vector at the middle point in the direction of $\Delta\vec{v}$; label it $\vec{a}$. This is the average acceleration at the midpoint between $\vec{v}_n$ and $\vec{v}_{n+1}$.
$\vec{a}$
$\vec{v}_n$
$\vec{v}_{n+1}$

Notice that the acceleration vector goes beside the dot, not beside the velocity vectors. This is because each acceleration vector is determined as the *difference* between the two velocity vectors on either side of a dot. The length of $\vec{a}$ does not have to be the exact length of $\Delta\vec{v}$; it is the direction of $\vec{a}$ that is most important.

The procedure of Tactics Box 1.3 can be repeated to find $\vec{a}$ at each point in the motion diagram. Note that we cannot determine $\vec{a}$ at the first and last points because we have only one velocity vector and can't find $\Delta\vec{v}$.

We can turn the Equation 1.8 definition of average acceleration around. Suppose a ball moving with velocity $\vec{v}_1$ has acceleration $\vec{a}$. Because of the acceleration, the ball's velocity changes during the time interval Δt to velocity $\vec{v}_2$. We can combine $\vec{a} = \Delta\vec{v}/\Delta t$ and $\Delta\vec{v} = \vec{v}_2 - \vec{v}_1$ to write

$$\vec{v}_2 = \vec{v}_1 + \vec{a}\,\Delta t \tag{1.9}$$

Knowing an object's acceleration is the key to finding its velocity at later instants of time.

The Complete Motion Diagram

You've now seen several *Tactics Boxes* that help you achieve specific tasks. Tactics Boxes will appear in nearly every chapter in this book. We'll also, where appropriate, provide *Problem-Solving Strategies.* Problem solving will be discussed in more detail later in the chapter, but this is a good place for the first problem-solving strategy.

MP PROBLEM-SOLVING STRATEGY 1.1 Motion diagrams

MODEL Represent the moving object as a particle. Make simplifying assumptions when interpreting the problem statement.

VISUALIZE A complete motion diagram consists of:

- The position of the object in each frame of the film, shown as a dot. Use five or six dots to make the motion clear but without overcrowding the picture. More complex motions may need more dots.
- The average velocity vectors, found by connecting each dot in the motion diagram to the next with a vector arrow. There is *one* velocity vector linking each *two* position dots. Label the row of velocity vectors $\vec{v}$.
- The average acceleration vectors, found using Tactics Box 1.3. There is *one* acceleration vector linking each *two* velocity vectors. Each acceleration vector is drawn at the dot between the two velocity vectors it links. Use $\vec{0}$ to indicate a point at which the acceleration is zero. Label the row of acceleration vectors $\vec{a}$.

STOP TO THINK 1.4 A particle undergoes acceleration $\vec{a}$ while moving from point 1 to point 2. Which of the choices shows the velocity vector $\vec{v}_2$ as the particle moves away from point 2?

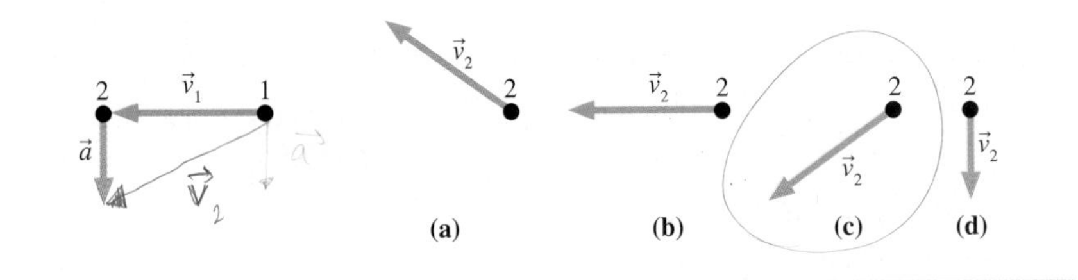

1.6 Examples of Motion Diagrams

This section will look at examples of the full strategy for drawing motion diagrams.

EXAMPLE 1.4 Skiing through the woods

A skier glides along smooth, horizontal snow at constant speed, then speeds up going down a hill. Draw the skier's motion diagram.

MODEL Represent the skier as a particle. It's reasonable to assume that the downhill slope is a straight line.

VISUALIZE Figure 1.19 shows a complete motion diagram of the skier. The dots are equally spaced for the horizontal motion, indicating constant speed, then the dots get further apart as the skier speeds up down the hill. The insets show how the average acceleration vector $\vec{a}$ is determined. All the other acceleration vectors along the slope will be similar to the one shown because each velocity vector is longer than the preceding one. The acceleration at the point where the slope changes is a little tricky because the velocity changes in both length and direction, but the procedure in Tactics Box 1.3 is designed to handle situations exactly like this. Notice that we've explicitly written $\vec{0}$ for the acceleration beside the dots where the velocity is constant.

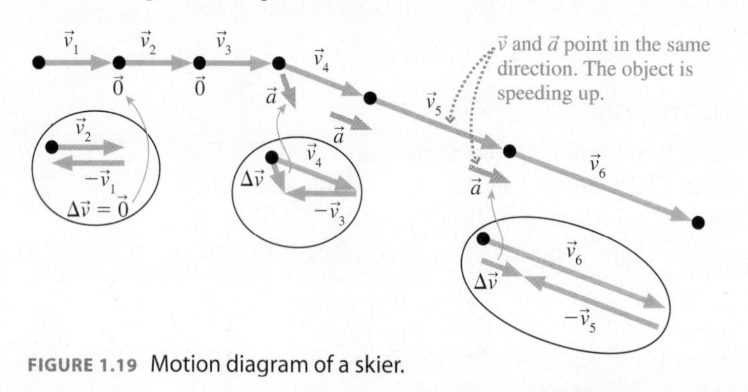
$\vec{v}$ and $\vec{a}$ point in the same direction. The object is speeding up.

FIGURE 1.19 Motion diagram of a skier.

EXAMPLE 1.5 The first astronauts land on Mars

A spaceship carrying the first astronauts to Mars descends safely to the surface. Draw a motion diagram for the last few seconds of the descent.

MODEL Represent the spaceship as a particle. It's reasonable to assume that its motion in the last few seconds is straight down. The problem ends as the spacecraft touches the surface.

VISUALIZE Figure 1.20 shows a complete motion diagram as the spaceship descends and slows, using its rockets, until it comes to rest on the surface. Notice how the dots get closer together as it slows. The inset shows how the acceleration vector $\vec{a}$ is determined at one point. All the other acceleration vectors will be similar, because for each pair of velocity vectors the earlier one is longer than the later one.

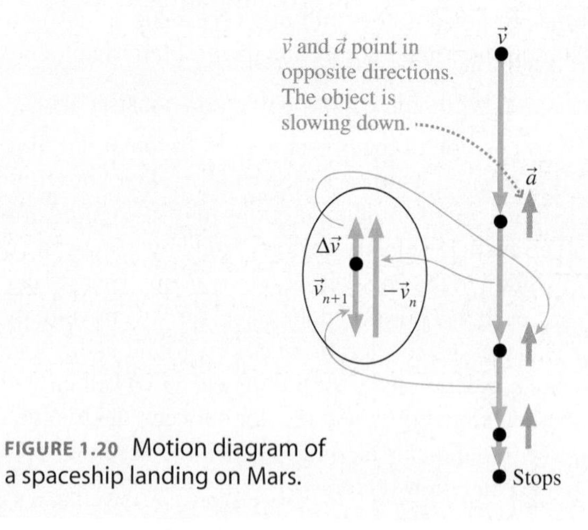
$\vec{v}$ and $\vec{a}$ point in opposite directions. The object is slowing down.
Stops

FIGURE 1.20 Motion diagram of a spaceship landing on Mars.

Speeding Up and Slowing Down

Notice something interesting in Figures 1.19 and 1.20. Where the object is speeding up, the acceleration and velocity vectors point in the *same direction.* Where the object is slowing down, the acceleration and velocity vectors point in *opposite directions.* These results, which are consistent with Equation 1.9, are always true for motion in a straight line. **For motion along a line:**

- **An object is speeding up if and only if $\vec{v}$ and $\vec{a}$ point in the same direction.**
- **An object is slowing down if and only if $\vec{v}$ and $\vec{a}$ point in opposite directions.**
- **An object's velocity is constant if and only if $\vec{a} = \vec{0}$.**

NOTE ▶ In everyday language, we use the word *accelerate* to mean "speed up" and the word *decelerate* to mean "slow down." But speeding up and slowing down are both changes in the velocity and consequently, by our definition, *both* are accelerations. In physics, *acceleration* refers to changing the velocity, no matter what the change is, and not just to speeding up. ◄

More Examples of Motion Diagrams

EXAMPLE 1.6 At the amusement park

Anne rides the Ferris wheel at an amusement park. Draw Anne's motion diagram.

MODEL Represent Anne as a particle. Assume that the Ferris wheel turns at constant speed.

VISUALIZE The motion diagram of Figure 1.21 uses 10 frames of film to show one complete revolution of the Ferris wheel. The seat on the Ferris wheel moves in a circle at a constant speed, so we've shown equal distances between each dot and the next. As before, the velocity vectors are found by connecting each dot to the next. Note that the velocity vectors are *straight lines,* not curves. This is because the displacement vectors $\Delta\vec{r}$ are *not* the same as the actual path followed.

Although Anne moves at constant speed, she does *not* move with constant velocity. All the velocity vectors have the same length, but each has a different *direction,* and that means Anne is accelerating. This is not a "speeding up" or "slowing down" acceleration, but it is still a change of the velocity with time. The inset to Figure 1.21 shows how to find the acceleration at the bottom of the circle. Vector $\vec{v}_n$ is the velocity vector that leads into this dot, while $\vec{v}_{n+1}$ moves away from it. From the circular geometry of the main figure, the two angles marked α are equal. Thus we see that $\vec{v}_{n+1}$ and $-\vec{v}_n$ form an isosceles triangle and vector $\Delta\vec{v}$ is exactly vertical. When $\vec{a}$ is drawn on the

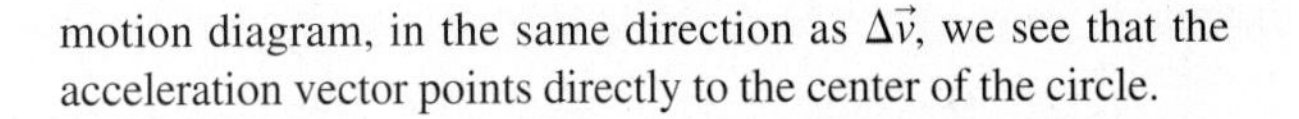
motion diagram, in the same direction as $\Delta\vec{v}$, we see that the acceleration vector points directly to the center of the circle.

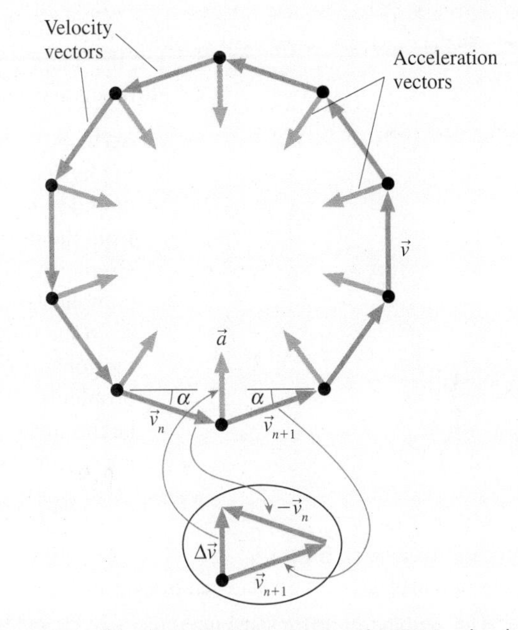

FIGURE 1.21 Anne's motion diagram on the Ferris wheel.

No matter which dot you select on the motion diagram in Figure 1.21, the velocities leading to and away from that dot change in such a way as to cause the acceleration to point directly to the center of the circle. You should convince yourself of this by finding $\vec{a}$ at several other points around the circle. An acceleration that always points directly toward the center of a circle is called a *centripetal acceleration.* The word "centripetal" comes from a Greek root meaning "center seeking." We will have a lot to say about centripetal acceleration in later chapters.

EXAMPLE 1.7 Throwing the shot

The Hulk is an Olympic shot putter. Draw a motion diagram for the shot from the moment it leaves the Hulk's hand until it hits the ground.

MODEL Represent the shot as a particle.

VISUALIZE The shot is an example of projectile motion. You probably know, from watching baseballs, basketballs, rocks, and so on, that projectiles move in some sort of an arc. Projectiles also slow down as they rise and speed up again as they fall. The top of the arc is where the object has the slowest motion, which is why a ball seems to hesitate there before falling, but it is not at rest. Figure 1.22a shows a photo of a projectile moving through the air. The motion diagram of the shot is shown in Figure 1.22b. Notice that the shot does *not* start from rest; it leaves the Hulk's hand with an initial velocity. The velocity vectors along the ascent get

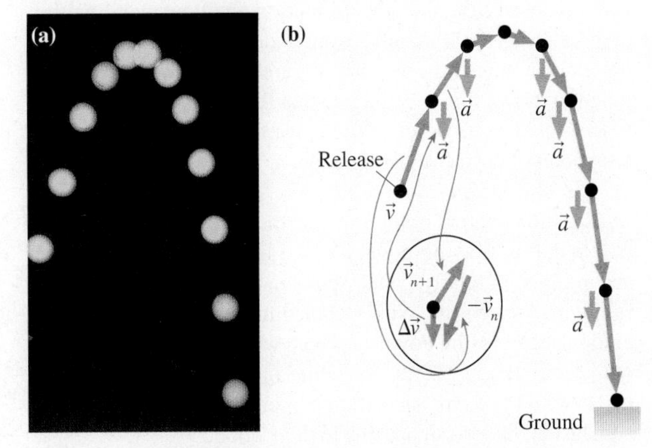

FIGURE 1.22 Photograph and motion diagram of a projectile.

shorter, because the shot slows while rising, then lengthen again as the shot falls. The last position is at the instant of impact.

Both the length and the direction of the velocity vector are changing, so the shot must have an acceleration. The inset shows the determination of $\vec{a}$ at one point along the trajectory. The acceleration must be pointed more or less downward, but we cannot determine the exact direction of $\vec{a}$ without further information about the exact shape of the arc. It turns out, for projectile motion, that the acceleration vector at every point is *straight down.* This is a conclusion that we will justify in Chapter 6 but will simply assert, without proof, for now. This very specific acceleration is called the *acceleration due to gravity.*

It is a good idea to compare the Example 1.7 projectile motion to the constant-speed circular motion in Example 1.6. Both objects follow curved trajectories, but the shapes of the curves are different. Consequently, the motions have very different acceleration vectors.

NOTE ► Example 1.7 revisits the important issue of how we interpret problems in physics. The Hulk clearly had to flex his muscles and move various parts of his body to get the shot moving. That part of the motion is complicated. But from the moment the shot leaves his hand until it hits the ground, it has a fairly simple motion. The impact and any subsequent motion again become very complicated. In this case, you were explicitly asked to consider the motion only from the time the Hulk releases the shot until it hits the ground; the complex motions of the throw and the impact are not part of the problem. But many problems will not be this specific. *You* have to interpret when the problem begins and ends. The key is to focus on isolating those parts of the problem that are essential while disregarding the nonessential aspects. ◄

EXAMPLE 1.8 Tossing a ball

Draw the motion diagram of a ball tossed straight up in the air.

MODEL This problem calls for some interpretation. Should we include the toss itself, or only the motion after the tosser releases the ball? Should we include the ball hitting the ground? It appears that this problem is really concerned with the ball's motion through the air. Consequently, we begin the motion diagram at the moment that the tosser releases the ball and end the diagram at the moment the ball hits the ground. We will consider neither the toss nor the impact. And, of course, we will represent the ball as a particle.

VISUALIZE We have a slight difficulty here because the ball retraces its route as it falls. A literal motion diagram would show the upward motion and downward motion on top of each other, leading to confusion. We can avoid this difficulty by horizontally separating the upward motion and downward motion diagrams. This will not affect our conclusions because it does not change any of the vectors. Figure 1.23 shows the motion diagram drawn this way. Notice that the very top dot is shown twice—as the end point of the upward motion and the beginning point of the downward motion.

The ball slows down as it rises. You've learned that the acceleration vectors point opposite the velocity vectors for an object that is slowing down along a line, and they are shown accordingly. Similarly, $\vec{a}$ and $\vec{v}$ point in the same direction as the falling ball speeds up. Notice something interesting: the acceleration vectors point downward both while the ball is rising *and* while it is falling. Both "speeding up" and "slowing down" occur with the *same* acceleration vector. This is an important conclusion, one worth pausing to think about.

Now let's look at the top point on the ball's trajectory. The velocity vectors are pointing upward but getting shorter as the ball approaches the top. As it starts to fall, the velocity vectors are pointing downward and getting longer. There must be a moment—just an instant as $\vec{v}$ switches from pointing up to pointing down—when the velocity is zero. Indeed, the ball's velocity *is* zero for an instant at the precise top of the motion!

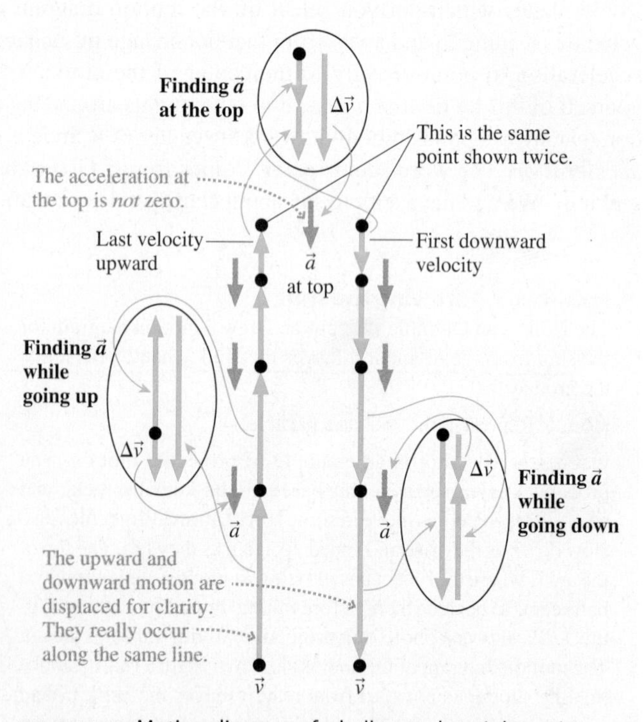

FIGURE 1.23 Motion diagram of a ball tossed straight up in the air.

But what about the acceleration at the top? The inset shows how the average acceleration is determined from the last upward velocity before the top point and the first downward velocity. We find that the acceleration at the top is pointing downward, just as it does elsewhere in the motion.

Many people expect the acceleration to be zero at the highest point. But recall that the velocity at the top point *is* changing—from up to down. If the velocity is changing, there *must* be an acceleration. A downward-pointing acceleration vector is needed to turn the velocity vector from up to down. Another way to think about this is to note that zero acceleration would mean no change of velocity. When the ball reached zero velocity at the top, it would hang there and not fall if the acceleration were also zero!

The motion of the ball in Example 1.8 is really a special case of projectile motion. The difference between the tossed ball and the shot thrown by the Hulk is that the ball happens to start out moving exactly vertically. We saw in Example 1.7 that the acceleration is a constant downward vector for projectile motion. That is exactly what we have found again for the ball.

1.7 From Words to Symbols

Physics is not mathematics. Math problems are clearly stated, such as "What is 2 + 2?" Physics is about the world around us, and to describe that world we must use language. Now, language is wonderful—we couldn't communicate without it—but it can sometimes be imprecise or ambiguous.

The challenge when reading a physics problem is to translate the words into symbols that can be manipulated, calculated, and graphed. This translation from words to symbols is the heart of problem solving in physics. This is the point where ambiguous words and phrases must be clarified, where the imprecise must be made precise, and where you arrive at an understanding of exactly what the question is asking.

You may have been told that the first step in solving a physics problem is to "draw a picture," but perhaps you didn't know why or what to draw. The purpose of drawing a picture is to aid you in the words-to-symbols translation. And there really is a *method* for drawing pictures, one that will help you be a better problem solver. It is called the **pictorial representation** of the problem.

TACTICS BOX 1.4 **Drawing a pictorial representation**

1. **Sketch the situation.** Not just any sketch. Show the object at the *beginning* of the motion, at the *end,* and at any point where the character of the motion changes. Very simple drawings are adequate.
2. **Establish a coordinate system.** Select your axes and origin to match the motion.
3. **Define symbols.** Use the sketch to define symbols representing quantities such as position, velocity, acceleration, and time. *Every* variable used later in the mathematical solution should be defined on the sketch. Some will have known values, others are initially unknown, but all should be given symbolic names.
4. **List known information.** Make a table of the quantities whose values you can determine from the problem statement or that can be found quickly with simple geometry or unit conversions. Some quantities are implied by the problem, rather than explicitly given. Others are determined by your choice of coordinate system.
5. **Identify the desired unknowns.** What quantity or quantities will allow you to answer the question? These should have been defined as symbols in step 3. Don't list every unknown; only the one or two needed to answer the question.

It's not an overstatement to say that a well-done pictorial representation of the problem will take you halfway to the solution. The following example illustrates how to construct a pictorial representation for a problem that is typical of problems you will see in the next few chapters.

EXAMPLE 1.9 Drawing a pictorial representation

Draw a pictorial representation for the following problem:

> A rocket sled accelerates at 50 m/s^2 for 5 seconds, then coasts for 3 seconds. What is the total distance traveled?

VISUALIZE The motion has a beginning, an end, and a point where the nature of the motion changes from accelerating to coasting. These are the three points sketched in Figure 1.24. A coordinate system has been chosen with the origin at the starting point. The quantities x, v, and t are needed at each of three *points*, so these have been defined on the sketch and distinguished by subscripts. Accelerations are associated with *intervals* between the points, so only two accelerations are defined. Values for three quantities are given in the problem statement. Others, such as $x_0 = 0$ m and $t_0 = 0$ s, are inferred from our choice of coordinate system. The value $v_{0x} = 0$ m/s is part of our *interpretation* of the problem. Finally, we identify x_2 as the quantity that will answer the question. We now understand quite a bit about the problem and would be ready to start a quantitative analysis.

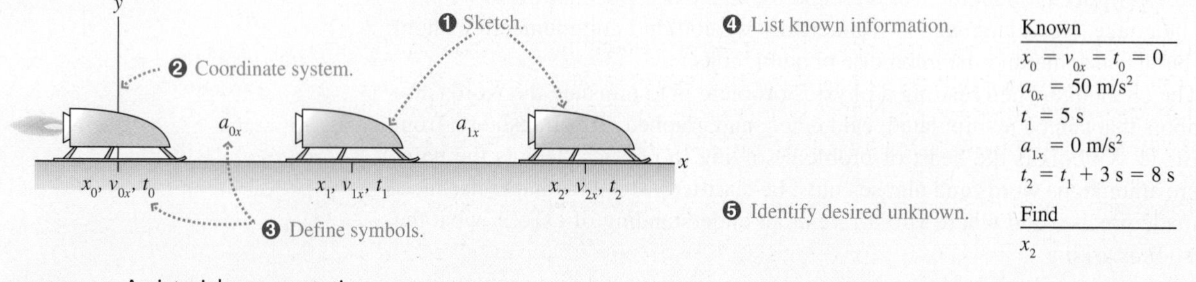

FIGURE 1.24 A pictorial representation.

We didn't *solve* the problem; that is not the purpose of the pictorial representation. The pictorial representation is a systematic way to go about interpreting a problem and getting ready for a mathematical solution. Although this is a simple problem, and you probably know how to solve it if you've taken physics before, you will soon be faced with much more challenging problems. Learning good problem-solving skills at the beginning, while the problems are easy, will make them second nature later when you really need them.

Using Symbols

Symbols are a language that allows us to talk with precision about the relationships in a problem. As with any language, we all need to agree to use words or symbols in the same way if we want to be able to communicate with each other. Many of the ways we use symbols in science and engineering are somewhat arbitrary, often reflecting historical roots. Nonetheless, practicing scientists and engineers have come to agree on how to use the language of symbols. Learning this language is part of learning physics.

We use subscripts in physics to designate a particular point in the problem. Scientists usually label the starting point of the problem with the subscript "0," not the subscript "1" that you might expect. When using subscripts, make sure that all symbols referring to the same point in the problem have the *same numerical subscript.* To have the one point in a problem characterized by position x_1 but velocity v_{2x} is guaranteed to lead to confusion!

Notation for Motion Along a Line

Our goal in this section is simply to learn how to *analyze* a problem statement, not to solve it. We will limit ourselves to motion along a straight line. Although the position, average velocity, and average acceleration of an object are vectors, the full vector notation is not needed for motion along a line. Instead, we can measure the one-dimensional position, velocity, and acceleration of an object with the quantities x, v_x, and a_x (or y, v_y, and a_y if the motion is vertical). The vector nature of these quantities appears through their *signs:*

- v_x (or v_y) is positive if the velocity vector $\vec{v}$ points to the right (or up). It is negative if the velocity vector $\vec{v}$ points to the left (or down).
- a_x (or a_y) is positive if the acceleration vector $\vec{a}$ points to the right (or up). It is negative if the acceleration vector $\vec{a}$ points to the left (or down).

ActivPhysics ONLINE 1.1

The appropriate sign for v is usually clear. Determining the sign of a is more difficult, and this is where a motion diagram can help.

We will examine one-dimensional motion in detail in Chapter 2, but these simple rules are enough to understand the examples and homework problems of this chapter.

Representations

A picture is one way to *represent* your knowledge of a situation. You could also represent your knowledge using words, graphs, or equations. Each **representation of knowledge** gives us a different perspective on the problem. The more tools you have for thinking about a complex problem, the more likely you are to solve it.

There are five representations of knowledge that we will use over and over:

1. The *verbal* representation. A problem statement, in words, is a verbal representation of knowledge. So is an explanation that you write.
2. The *pictorial* representation. The pictorial representation, which we've just presented, is the most literal depiction of the situation.
3. The *physical* representation. We will develop many special techniques and diagrams that allow us to analyze aspects of the physics. The motion diagram, which allows us to find the acceleration vectors, is part of the physical representation.
4. The *graphical* representation. We will make extensive use of graphs to portray information.
5. The *mathematical* representation. Equations that can be used to find the numerical values of specific quantities are the mathematical representation.

NOTE ► The mathematical representation is only one of many. Much of physics is more about thinking and reasoning than it is about solving equations. ◄

1.8 A Problem-Solving Strategy

One of the goals of this textbook is to help you learn a *strategy* for solving physics problems. The purpose of a strategy is to guide you in the right direction with minimal wasted effort. The problem-solving strategy shown on the next page is based on using different representations of knowledge.

Throughout this textbook we will emphasize the first two steps. They are the *physics* of the problem, as opposed to the mathematics of solving the resulting equations. This is not to say that those mathematical operations are always easy—in many cases they are not. But our primary goal is to understand the physics.

Building a house requires careful planning. The architect's visualization and drawings have to be complete before the detailed procedures of construction get underway. The same is true for solving problems in physics.

Problem-Solving Strategy

MODEL It's impossible to treat every detail of a situation. Simplify the situation with a model that captures the essential features. For example, the object in a mechanics problem is usually represented as a particle.

VISUALIZE This is where expert problem solvers put most of their effort.

- Draw a *pictorial representation.* This helps you assess the information you are given and starts the process of translating the problem into symbols.
- Draw a *physical representation.* This helps you visualize important aspects of the physics. Motion diagrams are part of the physical representation. Chapter 4 will introduce free-body diagrams to display information about forces.
- Use a *graphical representation* if it is appropriate for the problem.
- Go back and forth between these three representations; they need not be done in any particular order.

SOLVE Only after modeling and visualizing are complete is it time to develop a *mathematical representation* with specific equations that must be solved. All symbols used here should have been defined in the pictorial representation.

ASSESS Is your result believable? Does it have proper units? Does it make sense?

Using the Problem-Solving Strategy

A couple of final examples will summarize the ideas of this chapter. Our task, at this point, is not to *solve* the problem, but to focus on what is happening in the problem—in other words, to make the translation from words to symbols in preparation for subsequent mathematical analysis. We will do so, in each case, by developing the pictorial representation and drawing the motion diagram.

EXAMPLE 1.10 Launching a weather rocket

Use the first two steps of the problem-solving strategy to analyze the following problem:

> A small rocket, such as those used for meteorological measurements of the atmosphere, is launched vertically with an acceleration of 30 m/s². It runs out of fuel after 30 seconds. What is its maximum altitude?

MODEL We need to do some interpretation. Common sense tells us that the rocket does not stop the instant it runs out of fuel. Instead, it continues upward, while slowing, until reaching its maximum altitude. This second half of the motion, after running out of fuel, is like the ball that was tossed upward in the first half of Example 1.8. Because the problem does not ask about the rocket's descent, we conclude that the problem ends at the point of maximum altitude. We'll represent the rocket as a particle.

VISUALIZE Figure 1.25 shows the physical represention (i.e., the motion diagram) and the pictorial representation. The rocket is speeding up during the first half of the motion, so $\vec{a}_0$ is parallel to $\vec{v}$. The initial acceleration $a_{0y} = 30\text{ m/s}^2$ is given in the problem. During the second half, as the rocket slows, $\vec{a}_1$ points downward. Thus a_{1y} is a negative number.

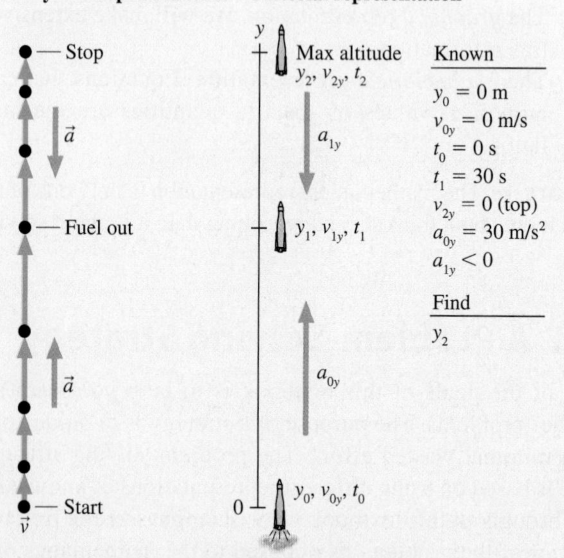

FIGURE 1.25 Physical and pictorial representations for the rocket.

This information is then transferred to the pictorial representation. Although the velocity v_{2y} wasn't given in the problem statement, we know it must be zero at the very top of the trajectory. Lastly, we have identified y_2 as the desired unknown. This, of course, is not the only unknown in the problem, but it is the one we are specifically asked to find.

ASSESS If you've had a previous physics class, you may be tempted to assign a_{1y} the value $-9.8\ \text{m/s}^2$, the acceleration due to gravity. However, that would be true only if there is no air resistance on the rocket. We will need to consider the *forces* acting on the rocket during the second half of its motion before we can determine a value for a_{1y}. For now, all that we can safely conclude is that a_{1y} is negative.

EXAMPLE 1.11 Making the tackle

Use the first two steps of the problem-solving strategy to analyze the following problem:

> Fred catches the football while standing directly on the goal line. He immediately starts running forward with an acceleration of 6 ft/s². At the moment the catch is made, Tommy is 20 yards away and heading directly toward Fred with a steady speed of 15 ft/s. If neither deviates from a straight-ahead path, where will Tommy tackle Fred?

MODEL Both Fred and Tommy will be represented as particles.

VISUALIZE Here is a problem with two moving objects. Figure 1.26 first shows the physical representation, with Fred accelerating toward the right and Tommy moving to the left. This is our choice—the problem didn't say which way Fred runs. Our choice gives positive numbers for Fred's velocity and acceleration, but you could just as well set up the problem with Fred running toward the left. Tommy moves toward the left, so he has a *negative* value for $(v_{0x})_T$. The ending point of the problem is fairly clear: when Fred and Tommy collide (ouch!).

Notice how we have used subscripts F and T to distinguish symbols for Fred's motion from similar symbols describing Tommy's motion. Using the same symbol to mean more than one thing is guaranteed to lead to confusion and errors. The time, however, does not have an extra subscript because only one clock is needed to time both Fred and Tommy. Thus *both* starting positions are described with the same time symbol t_0.

We know intuitively how the problem ends—they collide—but how do we state that more precisely? To collide, they must both reach the same position at the same time (t_1), and so the relation we are looking for is $(x_1)_F = (x_1)_T$. Keep in mind that $(x_1)_T$ is Tommy's *position,* measured with respect to the coordinate system, not the distance that Tommy has traveled.

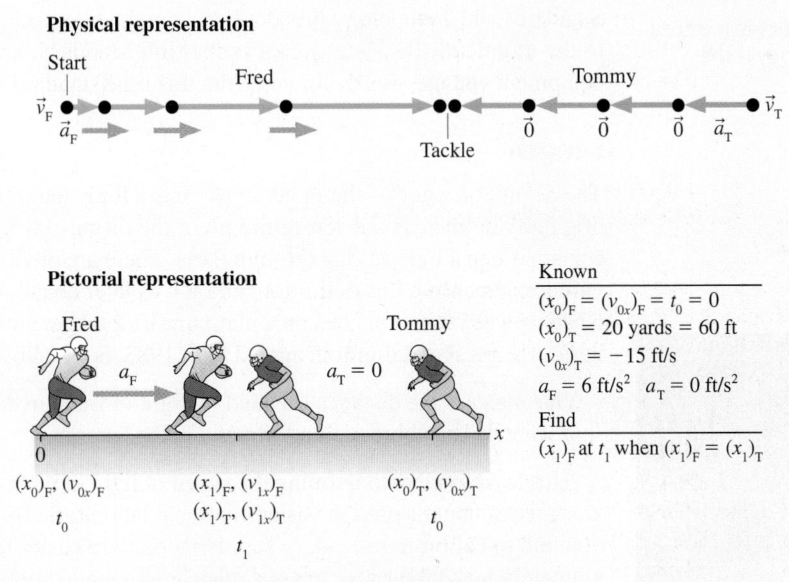

FIGURE 1.26 Physical and pictorial representations.

As you finish this chapter, you now have some powerful tools for thinking about and analyzing problems of motion. Continued practice will make these second nature to you, and coming chapters will give you plenty of opportunity for that practice.

1.9 Units and Significant Figures

Science is based upon experimental measurements, and measurements require *units*. The system of units currently used in science is called *le Système Internationale d'Unités*. These are commonly referred to as **SI units.** Older books often referred to *mks units,* which stands for "meter-kilogram-second," or *cgs units,* which is "centimeter-gram-second." For practical purposes, SI units are the same as mks units. In casual speaking we often refer to *metric units,* although this could mean either mks or cgs units. To be precise, we will follow the internationally accepted custom of calling our system of units *SI units.*

TABLE 1.1 Some basic SI units

Quantity	Unit	Abbreviation
time	second	s
length	meter	m
mass	kilogram	kg

All of the quantities needed to understand motion can be expressed in terms of the three basic SI units shown in Table 1.1. Other quantities can be expressed as a combination of these basic units. Velocity, expressed in meters per second or m/s, is a ratio of the length unit to the time unit.

Time

The standard of time prior to 1960 was based on the *mean solar day.* As timekeeping accuracy and astronomical observations improved, it became apparent that the earth's rotation is not perfectly steady. Meanwhile, physicists had been developing a device called an *atomic clock.* This instrument is able to measure, with incredibly high precision, the frequency of radio waves absorbed by atoms as they move between two closely spaced energy levels. This frequency can be reproduced with great accuracy at many laboratories around the world. Consequently, the SI unit of time—the second—was redefined in 1967 as follows:

> One *second* is the time required for 9,192,631,770 oscillations of the radio wave absorbed by the cesium-133 atom. The abbreviation for second is the letter s.

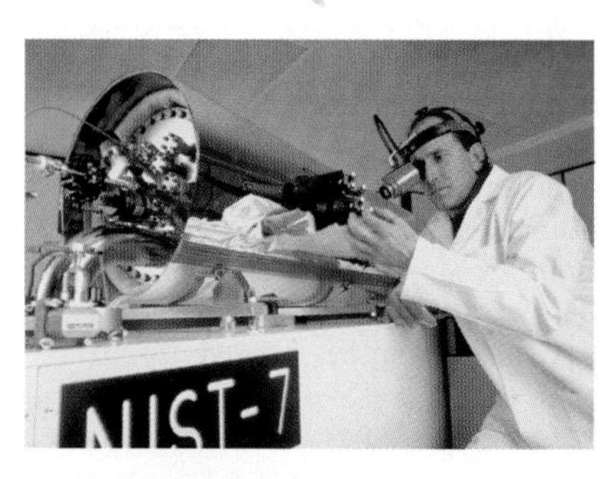

An atomic clock at the National Institute of Standards and Technology is the primary standard of time.

A radio station operated (in the United States) by the National Institute of Standards and Technology broadcasts a signal whose frequency is linked directly to the atomic clocks. This signal is the time standard, and any time-measuring equipment you use was calibrated from this time standard.

Length

The SI unit of length—the meter—also has a long and interesting history. It was originally defined as one ten-millionth of the distance from the North Pole to the equator along a line passing through Paris. There are obvious practical difficulties with implementing this definition, and it was later abandoned in favor of the distance between two scratches on a platinum-iridium bar stored in a special vault in Paris. The present definition, agreed to in 1983, is as follows:

> One *meter* is the distance traveled by light in vacuum during 1/299,792,458 of a second. The abbreviation for meter is the letter m.

This is equivalent to defining the speed of light to be exactly 299,792,458 m/s. Laser technology is used in various national laboratories to implement this definition and to calibrate secondary standards that are easier to use. These standards ultimately make their way to your ruler or to a meter stick. It is worth keeping in mind that any measuring device you use is only as accurate as the care with which it was calibrated.

By international agreement, this metal cylinder, stored in Paris, is the definition of the kilogram.

Mass

The original unit of mass, the gram, was defined as the mass of 1 cubic centimeter of water. That is why you know the density of water as 1 g/cm^3. This definition proved to be impractical when scientists needed to make very accurate measurements. The SI unit of mass—the kilogram—was redefined in 1889 as:

> One *kilogram* is the mass of the international standard kilogram, a polished platinum-iridium cylinder stored in Paris. The abbreviation for kilogram is the symbol kg.

The kilogram is the only SI unit still defined by a manufactured object. Despite the prefix *kilo,* it is the kilogram, not the gram, that is the proper SI unit.

Using Prefixes

We will have many occasions to use lengths, times, and masses that are either much less or much greater than the standards of 1 meter, 1 second, and 1 kilogram. We will do so by using *prefixes* to denote various powers of ten. Table 1.2 lists the common prefixes that will be used frequently throughout this book. Memorize it! Few things in science are learned by rote memory, but this list is one of them. A more extensive list of prefixes is shown inside the cover of the book.

Although prefixes make it easier to talk about quantities, the proper SI units are meters, seconds, and kilograms. Quantities given with prefixed units must be converted to SI units before any calculations are done. Unit conversions are best done at the very beginning of a problem, as part of the pictorial representation.

TABLE 1.2 Common prefixes

Prefix	Power of 10	Abbreviation
mega-	10^6	M
kilo-	10^3	k
centi-	10^{-2}	c
milli-	10^{-3}	m
micro-	10^{-6}	μ
nano-	10^{-9}	n

Unit Conversions

Although SI units are our standard, we cannot entirely forget that the United States still uses English units. Many engineering calculations are done in English units. And even after repeated exposure to metric units in classes, most of us "think" in the English units we grew up with. Thus it remains important to be able to convert back and forth between SI units and English units. Table 1.3 shows a few frequently used conversions, and these are worth memorizing if you do not already know them. While the English system was originally based on the length of the king's foot, it is interesting to note that today the conversion 1 in = 2.54 cm is the *definition* of the inch. In other words, the English system for lengths is now based on the meter!

TABLE 1.3 Useful unit conversions

1 in = 2.54 cm
1 mi = 1.609 km
1 mph = 0.447 m/s
1 m = 39.37 in
1 km = 0.621 mi
1 m/s = 2.24 mph

There are various techniques for doing unit conversions. One effective method is to write the conversion factor as a ratio equal to one. For example, using information in Tables 1.2 and 1.3,

$$\frac{10^{-6}\ \text{m}}{1\ \mu\text{m}} = 1 \quad \text{and} \quad \frac{2.54\ \text{cm}}{1\ \text{in}} = 1$$

Because multiplying any expression by 1 does not change its value, these ratios are easily used for conversions. To convert 3.5 μm to meters we would compute

$$3.5\ \cancel{\mu\text{m}} \times \frac{10^{-6}\ \text{m}}{1\ \cancel{\mu\text{m}}} = 3.5 \times 10^{-6}\ \text{m}.$$

Similarly, the conversion of 2 feet to meters would be

$$2\ \cancel{\text{ft}} \times \frac{12\ \cancel{\text{in}}}{1\ \cancel{\text{ft}}} \times \frac{2.54\ \cancel{\text{cm}}}{1\ \cancel{\text{in}}} \times \frac{10^{-2}\ \text{m}}{1\ \cancel{\text{cm}}} = 0.610\ \text{m}.$$

Notice how units in the numerator and in the denominator cancel until just the desired units remain at the end. You can continue this process of multiplying by 1 as many times as necessary to complete all the conversions.

Assessment

As we get further into problem solving, we will need to decide whether or not the answer to a problem "makes sense." To determine this, at least until you have more experience with SI units, you may need to convert from SI units back to the English units in which you think. But this conversion does not need to be very accurate. For example, if you are working a problem about automobile speeds and reach an answer of 35 m/s, all you really want to know is whether or not this is a realistic speed for a car. That requires a "quick and dirty" conversion, not a conversion of great accuracy.

TABLE 1.4 Approximate conversion factors

1 cm $\approx \frac{1}{2}$ in
10 cm $\approx$ 4 in
1 m $\approx$ 1 yard
1 m $\approx$ 3 feet
1 km $\approx$ 0.6 mile
1 m/s $\approx$ 2 mph

Table 1.4 shows a number of approximate conversion factors that can be used to assess the answer to a problem. Using 1 m/s $\approx$ 2 mph, you find that 35 m/s is roughly 70 mph, a reasonable speed for a car. But an answer of 350 m/s, which you might get after making a calculation error, would be an unreasonable 700 mph. Practice with these will allow you to develop intuition for metric units.

NOTE ► These approximate conversion factors are accurate only to one significant figure. This is sufficient to assess the answer to a problem, but do *not* use the conversion factors from Table 1.4 for converting English units to SI units at the start of a problem. Use Table 1.3. ◄

Significant Figures

It is necessary to say a few words about a perennial source of difficulty: significant figures. Mathematics is a subject where numbers and relationships can be as precise as desired, but physics deals with a real world of ambiguity and imprecision. It is important in all areas of science and engineering to state clearly what you know about a situation—no less and, especially, no more. Numbers provide one way to specify your knowledge.

If you report that a length has a value of 6.2 m, the implication is that the actual value falls between 6.15 m and 6.25 m and thus rounds to 6.2 m. If that is the case, then reporting a value of simply 6 m is saying less than you know; you are withholding information. On the other hand, to report the number as 6.213 m is wrong. Any person reviewing your work—perhaps a client who hired you—would interpret the number 6.213 m as meaning that the actual length falls between 6.2125 m and 6.2135 m, thus rounding to 6.213 m. In this case, you are claiming to have knowledge and information that you do not really possess.

The way to state your knowledge precisely is through the proper use of **significant figures.** You can think of a significant figure as being a digit that is reliably known. A number such as 6.2 m has *two* significant figures because the next decimal place—the one-hundredths—is not reliably known. As Figure 1.27 shows, the best way to determine how many significant figures a number has is to write it in scientific notation.

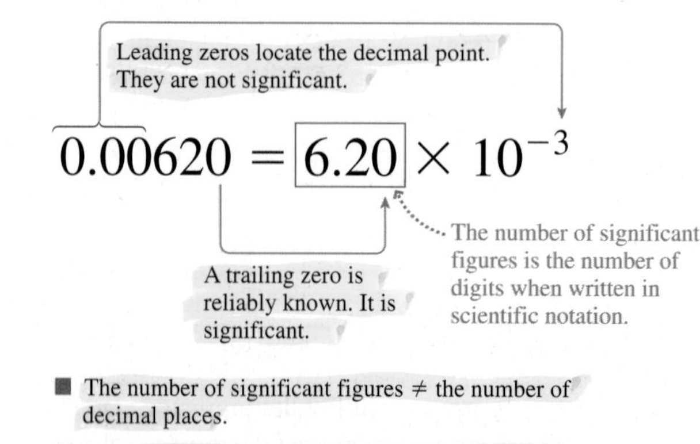

Leading zeros locate the decimal point. They are not significant.
0.00620 = 6.20 × 10⁻³
A trailing zero is reliably known. It is significant.
The number of significant figures is the number of digits when written in scientific notation.

■ The number of significant figures ≠ the number of decimal places.

■ Changing units shifts the decimal point but does not change the number of significant figures.

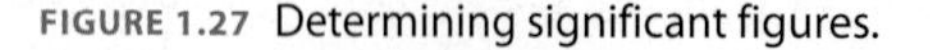

FIGURE 1.27 Determining significant figures.

Calculations with numbers follow the "weakest link" rule. The saying, which you probably know, is that "a chain is only as strong as its weakest link." If nine out of ten links in a chain can support a 1000 pound weight, that strength is meaningless if the tenth link can support only 200 pounds. Nine out of the ten numbers used in a calculation might be known with a precision of 0.01%; but if the tenth number is poorly known, with a precision of only 10%, then the result of the calculation cannot possibly be more precise than 10%. The weak link rules!

TACTICS BOX 1.5 Using significant figures

1. When multiplying or dividing several numbers, or taking roots, the number of significant figures in the answer should match the number of significant figures of the *least* precisely known number used in the calculation.
2. When adding or subtracting several numbers, the number of decimal places in the answer should match the *smallest* number of decimal places of any number used in the calculation.

EXAMPLE 1.12 Using significant figures

An object consists of two pieces. The mass of one piece has been measured to be 6.47 kg. The volume of the second piece, which is made of aluminum, has been measured to be $4.44 \times 10^{-4}\ \text{m}^3$. A handbook lists the density of aluminum as $2.7 \times 10^3\ \text{kg/m}^3$. What is the total mass of the object?

SOLVE First, calculate the mass of the second piece:

$$m = (4.44 \times 10^{-4}\ \text{m}^3)(2.7 \times 10^3\ \text{kg/m}^3)$$
$$= 1.199\ \text{kg} = 1.2\ \text{kg}$$

The number of significant figures of a product must match that of the *least* precisely known number, which is the two-significant-figure density of aluminum. Now add the two masses:

$$\begin{array}{r} 6.47\ \text{kg} \\ +\ 1.2\ \text{kg} \\ \hline 7.7\ \text{kg} \end{array}$$

The sum is 7.67 kg, but the hundredths place is not reliable because the second mass has no reliable information about this digit. Thus we must round to the one decimal place of the 1.2 kg. The best we can say, with reliability, is that the total mass is 7.7 kg.

There are two notable exceptions to these rules:

1. It is customary to keep one extra significant figure if (and only if) the number starts with a 1. For example, 10.43 could be used in a calculation with 8.91. The rationale for this exception is that four significant figures for numbers starting with 1 has roughly the same percentage accuracy as three significant figures for numbers starting with 2–9.
2. It is acceptable to keep one or two extra digits during intermediate steps of a calculation, as long as the final answer is reported with the proper number of significant figures. The goal is to minimize round-off errors in the calculation. But only one or two extra digits, not the seven or eight shown in your calculator display.

In laboratory work, the proper number of significant figures is determined by the experiment. The least precisely measured quantity sets the proper number. Textbook problems in science and engineering use an accepted standard of *three* significant figures for nearly all calculations. Two significant figures is too imprecise for most problems, while four is unnecessary *unless* the problem happens to give very accurate data with which to work. **Three significant figures will be the standard in this textbook, unless a problem provides data with more than three significant figures.** All data supplied with a problem can be assumed to have three-significant-figure accuracy even when, to avoid being unduly pedantic, integer values are given as "20 kg" rather than "2.00×10^1 kg."

NOTE ▶ Be careful! Many calculators have a default setting that shows two decimal places, such as 5.23. This is dangerous. If you need to calculate

TABLE 1.5 Some approximate lengths

	Length (m)
Circumference of the earth	4×10^7
New York to Los Angeles	5×10^6
Distance you can drive in 1 hour	1×10^5
Altitude of jet planes	1×10^4
Distance across a college campus	1000
Length of a football field	100
Length of a classroom	10
Length of your arm	1
Width of a textbook	0.1
Length of your little fingernail	0.01
Diameter of a pencil lead	1×10^{-3}
Thickness of a sheet of paper	1×10^{-4}
Diameter of a dust particle	1×10^{-5}

TABLE 1.6 Some approximate masses

	Mass (kg)
Large airliner	1×10^5
Small car	1000
Large human	100
Medium-size dog	10
Science textbook	1
Apple	0.1
Pencil	0.01
Raisin	1×10^{-3}
Fly	1×10^{-4}

5.23/58.5, your calculator will show a result of 0.09 and it is all too easy to write that down as an answer. But by doing so, you have reduced a calculation of two numbers having three significant figures to an answer with only one significant figure. The proper result of this division is 0.0894 or 8.94×10^{-2}. You will avoid this error if you keep your calculator set to display numbers in *scientific notation* with two decimal places. ◀

Proper use of significant figures is part of the "culture" of science and engineering. We will frequently emphasize these "cultural issues" because you must learn to speak the same language as the natives if you wish to communicate effectively. Most students "know" the rules of significant figures, having learned them in high school, but many fail to apply them. It is important that you understand the reasons for significant figures and that you get in the habit of using them properly.

Orders of Magnitude and Estimating

Precise calculations are appropriate when we have precise data, but there are many times when a very rough estimate is sufficient. Suppose you see a rock fall off a cliff and would like to know how fast it was going when it hit the ground. By doing a mental comparison with the speeds of familiar objects, such as cars and bicycles, you might judge that the rock was traveling at "about" 20 mph.

This is a one-significant-figure estimate. With some luck, you can probably distinguish 20 mph from either 10 mph or 30 mph, but you certainly, just from a visual appearance, cannot distinguish 20 mph from 21 mph. A one-significant-figure estimate or calculation, such as this, is called an **order-of-magnitude estimate.** An order-of-magnitude estimate is indicated by the symbol $\sim$, which indicates even less precision than the "approximately equal" symbol $\approx$. You would say that the speed of the falling rock is $v \sim 20$ mph.

A useful skill is to make reliable order-of-magnitude estimates on the basis of known information, simple reasoning, and common sense. This is a skill that is acquired by practice. Most chapters in this book will have homework problems that ask you to make order-of-magnitude estimates. The following example is typical of an estimation problem.

Tables 1.5 and 1.6 have information that will be useful for doing estimates.

EXAMPLE 1.13 Estimating a sprinter's speed

Estimate the speed with which an Olympic sprinter crosses the finish line of the 100 m dash.

SOLVE We do need one piece of information, but it is a widely known piece of sports trivia. That is, world-class sprinters run the 100 m dash in about 10 s. Their *average* speed is $v_{avg} \approx (100\text{ m})/(10\text{ s}) \approx 10$ m/s. But that's only average. They go slower than average at the beginning, and they cross the finish line at a speed faster than average. How much faster? Twice as fast, 20 m/s, would be $\approx$40 mph. Sprinters don't seem like they're running as fast as a 40 mph car, so this probably is too fast. Let's *estimate* that their final speed is 50% faster than the average. Thus they cross the finish line at $v \sim 15$ m/s.

STOP TO THINK 1.5 Rank in order, from the most to the least, the number of significant figures in the following numbers. For example, if b has more than c, c has the same number as a, and a has more than d, you could give your answer as b > c = a > d.

a. 8200 b. 0.0052 c. 0.430 d. 4.321×10^{-10}

SUMMARY

The goal of Chapter 1 has been to introduce the fundamental concepts of motion.

GENERAL STRATEGY

Motion Diagrams

- Help visualize motion.
- Provide a tool for finding acceleration vectors.

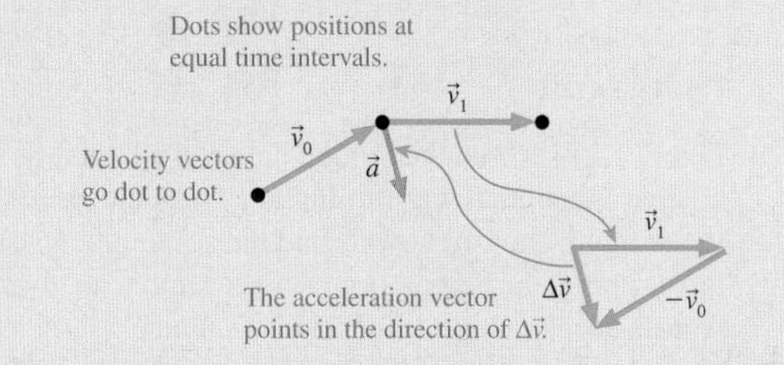

Dots show positions at equal time intervals.
Velocity vectors go dot to dot.
$\vec{v}_0$
$\vec{v}_1$
$\vec{a}$
$\vec{v}_1$
$\Delta\vec{v}$
$-\vec{v}_0$
The acceleration vector points in the direction of $\Delta\vec{v}$.

► These are the average velocity and the average acceleration vectors.

Problem Solving

MODEL Make simplifying assumptions.

VISUALIZE Use:

- **Pictorial representation**
- **Physical representation**
- **Graphical representation**

SOLVE Use a **mathematical representation** to find numerical answers.

ASSESS Does the answer have the proper units? Does it make sense?

IMPORTANT CONCEPTS

The particle model represents a moving object as if all its mass were concentrated at a single point.

Position locates an object with respect to a chosen coordinate system. Change in position is called displacement.

Velocity is the rate of change of the position vector $\vec{r}$.

Acceleration is the rate of change of the velocity vector $\vec{v}$. An object has an acceleration if it

- Changes speed and/or
- Changes direction.

Pictorial Representation

❶ Sketch the situation.

❷ Establish coordinates.

❸ Define symbols.

❹ List knowns.

❺ Identify desired unknown.

a_x

x

0

$x_0,\ v_{0x},\ t_0$ $\quad$ $x_1,\ v_{1x},\ t_1$

Known

$x_0 = v_{0x} = t_0 = 0$

$a_x = 2\text{ m/s}^2 \quad t_1 = 2\text{ s}$

Find

x_1

APPLICATIONS

For **motion along a line:**

- Speeding up: $\vec{v}$ and $\vec{a}$ point in the same direction.
- Slowing down: $\vec{v}$ and $\vec{a}$ point in opposite directions.
- Constant speed: $\vec{a} = \vec{0}$.

Significant figures are reliably known digits. Three significant figures is the standard for this book. The number of significant figures for:

- **Multiplication, division, powers** is set by the value with the fewest significant figures.
- **Addition, subtraction** is set by the value with the smallest number of decimal places.

TERMS AND NOTATION

motion
trajectory
motion diagram
operational definition
translational motion
particle
particle model
position vector, $\vec{r}$
scalar quantity
vector quantity
displacement, $\Delta\vec{r}$
zero vector, $\vec{0}$
time interval, Δt
average speed
average velocity, $\vec{v}$
average acceleration, $\vec{a}$
pictorial representation
representation of knowledge
SI units
significant figures
order-of-magnitude estimate

EXERCISES AND PROBLEMS

Exercises

Section 1.1 Motion Diagrams

1. A car skids to a halt to avoid hitting an object in the road. Draw a basic motion diagram, using the images from the movie, from the time the skid begins until the car is stopped.
2. You drop a soccer ball from your third-story balcony. Draw a basic motion diagram, using the images from the movie, from the time you release the ball until it touches the ground.

Section 1.2 The Particle Model

3. a. Write a paragraph describing the *particle model.* What is it, and why is it important?
   b. Give two examples of situations, different from those described in the text, for which the particle model is appropriate.
   c. Give an example of a situation, different from those described in the text, for which it would be inappropriate.

Section 1.3 Position and Time

4. Write a sentence or two describing the difference between position and displacement. Give one example of each.

Section 1.4 Velocity

5. a. What is an *operational definition?*
   b. Give operational definitions of displacement and velocity. Your definition should be given mostly in words and pictures, with a minimum of symbols or mathematics.
6. A softball player hits the ball and starts running toward first base. Draw a motion diagram, using the particle model, showing her position and her average velocity vectors during the first few seconds of her run.
7. A softball player slides into second base. Draw a basic motion diagram, using the particle model, showing his position and his average velocity vectors from the time he begins to slide until he reaches the base.

Section 1.5 Acceleration

8. Give an operational definition of acceleration. Your definition should be given mostly in words and pictures, with a minimum of symbols or mathematics.
9. a. Find the average acceleration vector at point 1 of this three-point motion diagram.
   b. Is the object's average speed between points 1 and 2 greater than, less than, or equal to its average speed between points 0 and 1? Explain how you can tell.

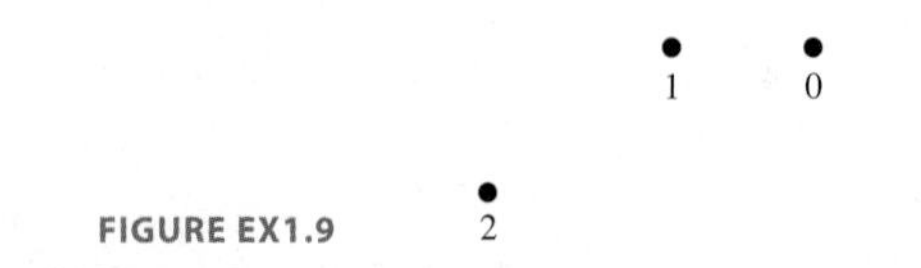

FIGURE EX1.9

10. a. Find the average acceleration vector at point 1 of this three-point motion diagram.
    b. Is the object's average speed between points 1 and 2 greater than, less than, or equal to its average speed between points 0 and 1? Explain how you can tell.

2
1
0

FIGURE EX1.10

11. Figure 1.21 showed the motion diagram for Anne as she rode a Ferris wheel that was turning at a constant speed. The inset to the figure showed how to find the acceleration vector at the lowest point in her motion. Use a similar analysis to find Anne's acceleration vector at the 12 o'clock, 4 o'clock, and 8 o'clock positions of the motion diagram. Use a ruler so that your analysis is accurate.
12. Figure 1.18 showed the motion diagram of a runner on a circular track. Find the acceleration vector when the runner is at the top of the diagram and when the runner is at the bottom of the diagram.

Section 1.6 Examples of Motion Diagrams

13. A car travels to the left at a steady speed for a few seconds, then brakes for a stop sign. Draw a complete motion diagram of the car.
14. A child is sledding on a smooth, level patch of snow. She encounters a rocky patch and slows to a stop. Draw a complete motion diagram of the child and her sled.
15. A roof tile falls straight down from a two-story building. It lands in a swimming pool and settles gently to the bottom. Draw a complete motion diagram of the tile.
16. Your roommate drops a tennis ball from a third story balcony. It hits the sidewalk and bounces as high as the second story. Draw a complete motion diagram of the tennis ball from the time it is released until it reaches the maximum height on its bounce. Be sure to determine and show the acceleration at the lowest point.
17. A car is driving north at steady speed. It makes a gradual 90° left turn without losing speed, then continues driving to the west. Draw a complete motion diagram as seen from a helicopter hovering over the highway.
18. A toy car rolls down a ramp, then across a smooth, horizontal floor. Draw a complete motion diagram of the toy car.

Section 1.7 From Words to Symbols

Section 1.8 A Problem-Solving Strategy

19. Draw a pictorial representation for the following problem. Do *not* solve the problem. The light turns green, and a bicyclist starts forward with an acceleration of 1.5 m/s^2. How far must she travel to reach a speed of 7.5 m/s?
20. Draw a pictorial representation for the following problem. Do *not* solve the problem. What acceleration does a rocket need to reach a speed of 200 m/s at a height of 1.0 km?

Section 1.9 Units and Significant Figures

21. Convert the following to SI units:
    a. 9.12 μs b. 3.42 km
    c. 44 cm/ms d. 80 km/hour
22. Convert the following to SI units:
    a. 8 in b. 66 ft/s
    c. 60 mph d. 14 in^2
23. Convert the following to SI units:
    a. 1 hour b. 1 day
    c. 1 year d. 32 ft/s^2
24. Using the approximate conversion factors in Table 1.4, convert the following to SI units *without* using your calculator.
    a. 20 ft b. 60 mi
    c. 60 mph d. 8 in
25. A regulation soccer field for international play is a rectangle with a length between 100 m and 110 m and a width between 64 m and 75 m. What are the smallest and largest areas that the field could be?
26. The quantity called *mass density* is the mass per unit volume of a substance. Express the following mass densities in SI units.
    a. Aluminum, 2.7×10^{-3} kg/cm^3
    b. Alcohol, 0.81 g/cm^3
27. How many significant figures does each of the following numbers have?
    a. 6.21 b. 62.1
    c. 0.620 d. 0.062
28. How many significant figures does each of the following numbers have?
    a. 6200 b. 0.006200
    c. 1.0621 d. 6.21×10^3
29. Compute the following numbers, applying the significant figure rule adopted in this textbook.
    a. 33.3×25.4 b. $33.3 - 25.4$
    c. $\sqrt{33.3}$ d. $333.3 \div 25.4$
30. Compute the following numbers, applying the significant figure rule adopted in this textbook.
    a. 33.3^2 b. 33.3×45.1
    c. $\sqrt{22.2 - 1.2}$ d. 44.4^{-1}
31. Estimate (don't measure!) the length of a typical car. Give your answer in both feet and meters. Briefly describe how you arrived at this estimate.
32. Estimate the height of a telephone pole. Give your answer in both feet and meters. Briefly describe how you arrived at this estimate.
33. Estimate the average speed with which you go from home to campus via whatever mode of transportation you use most commonly. Give your answer in both mph and m/s. Briefly describe how you arrived at this estimate.
34. Estimate the average speed with which the hair on your head grows. Give your answer in both m/s and μm/hour. Briefly describe how you arrived at this estimate.

Problems

For Problems 35 through 44, draw a complete motion diagram *and* a pictorial representation. Do *not* solve these problems or do any mathematics.

35. A Porsche accelerates from a stoplight at 5.0 m/s^2 for five seconds, then coasts for three more seconds. How far has it traveled?
36. Billy drops a watermelon from the top of a three-story building, 10 m above the sidewalk. How fast is the watermelon going when it hits?
37. Sam is recklessly driving 60 mph in a 30 mph speed zone when he suddenly sees the police. He steps on the brakes and slows to 30 mph in three seconds, looking nonchalant as he passes the officer. How far does he travel while braking?
38. A speed skater moving across frictionless ice at 8.0 m/s hits a 5.0-m-wide patch of rough ice. She slows steadily, then continues on at 6.0 m/s. What is her acceleration on the rough ice?
39. You would like to stick a wet spit wad on the ceiling, so you toss it straight up with a speed of 10 m/s. How long does it take to reach the ceiling, 3.0 m above?
40. A student standing on the ground throws a ball straight up. The ball leaves the student's hand with a speed of 15 m/s when the hand is 2.0 m above the ground. How long is the ball in the air before it hits the ground? (The student moves her hand out of the way.)
41. A ball rolls along a smooth horizontal floor at 10 m/s, then starts up a 20° ramp. How high does it go before rolling back down?
42. A motorist is traveling at 20 m/s. He is 60 m from a stop light when he sees it turn yellow. His reaction time, before stepping on the brake, is 0.50 s. What steady deceleration while braking will bring him to a stop right at the light?
43. Ice hockey star Bruce Blades is 5.0 m from the blue line and gliding toward it at a speed of 4.0 m/s. You are 20 m from the blue line, directly behind Bruce. You want to pass the puck to Bruce. With what speed should you shoot the puck down the ice so that it reaches Bruce exactly as he crosses the blue line?
44. You are standing still as Fred runs past you with the football at a speed of 6.0 yards per second. He has only 30 yards left to go before reaching the goal line to score the winning touchdown. If you begin running at the exact instant he passes you, what acceleration must you maintain to catch him 5.0 yards in front of the goal line?

Problems 45 through 50 show a motion diagram. For each of these problems, write a one or two sentence "story" about a *real object* that has this motion diagram. Your stories should talk about people or objects by name and say what they are doing. Problems 35–44 are examples of motion short stories.

45.

46.

47.

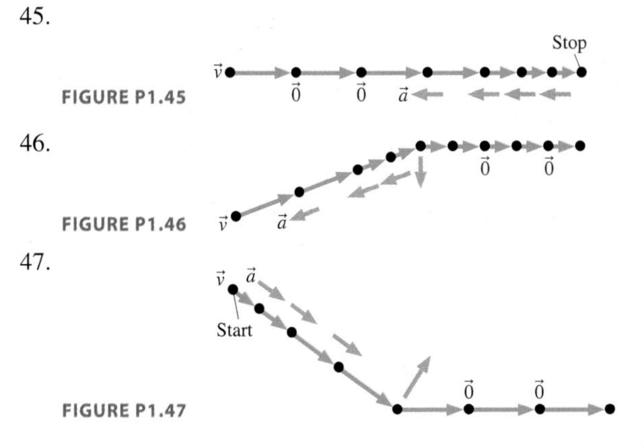

FIGURE P1.45

FIGURE P1.46

FIGURE P1.47

48.

49.

50.

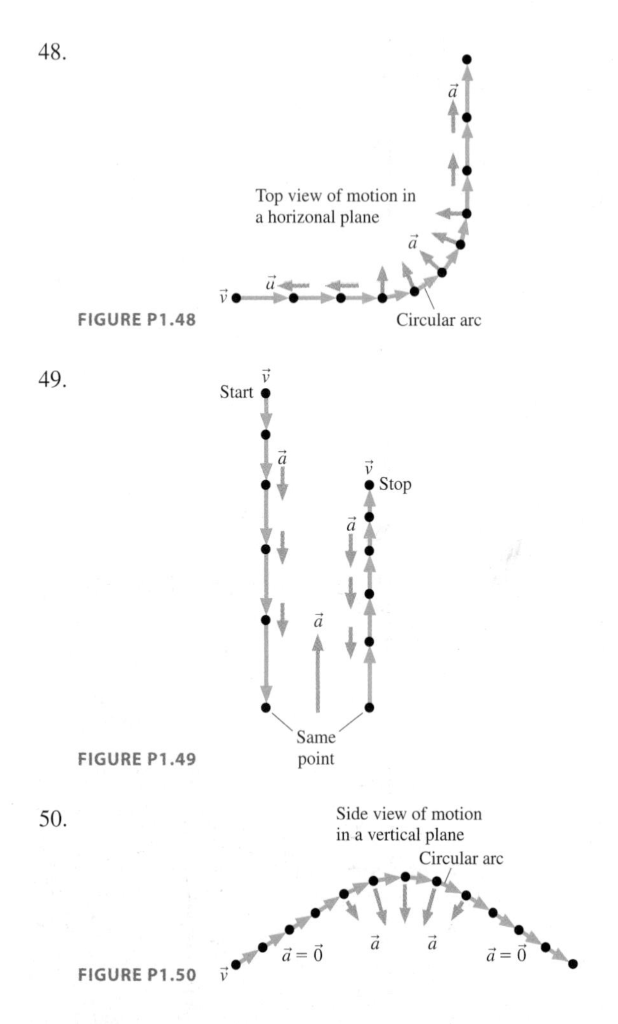

FIGURE P1.48

FIGURE P1.49

FIGURE P1.50

Problems 51 through 56 show a partial motion diagram. For each:

a. Complete the motion diagram by adding acceleration vectors.
b. Write a physics *problem* for which this is the correct motion diagram. Be imaginative! Don't forget to include enough information to make the problem complete and to state clearly what is to be found.
c. Draw a full pictorial representation for your problem.

51.

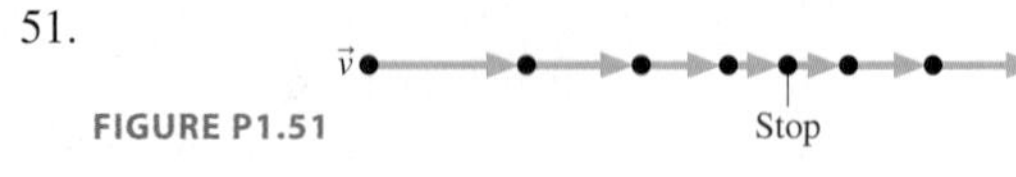

FIGURE P1.51

52.

53.

54.

55.

56.

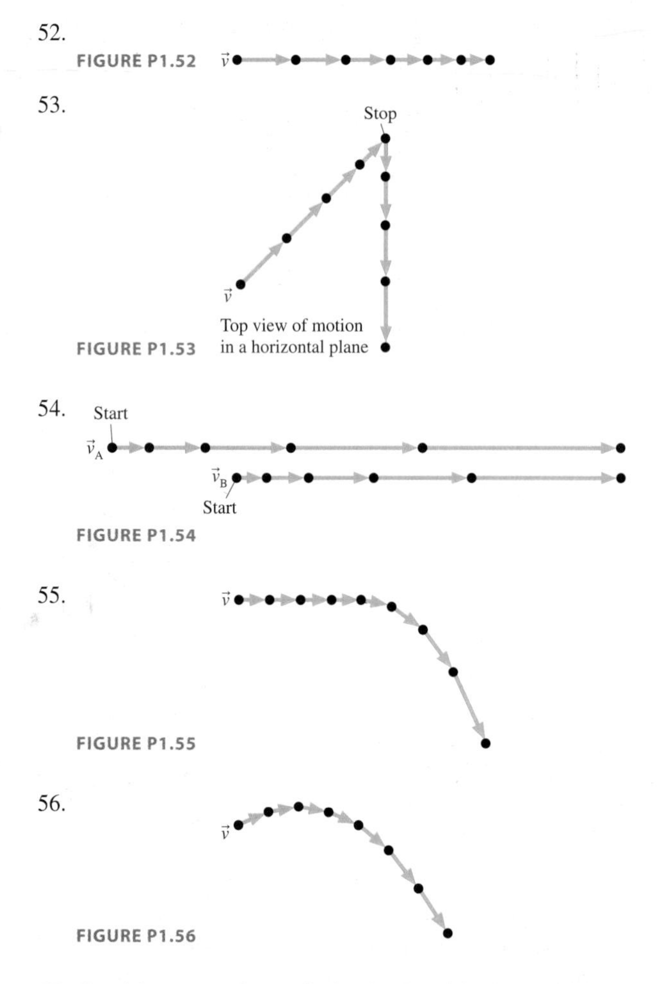

FIGURE P1.52

FIGURE P1.53

FIGURE P1.54

FIGURE P1.55

FIGURE P1.56

57. Consider a pendulum swinging back and forth on a string. Use a motion diagram analysis and a written explanation to answer the following questions.
   a. At the lowest point in the motion, is the velocity zero or nonzero? Is the acceleration zero or nonzero? If these vectors aren't zero, which way do they point?
   b. At the end of its arc, when the pendulum is at the highest point on the right or left side, is the velocity zero or nonzero? Is the acceleration zero or nonzero? If these vectors aren't zero, which way do they point?

STOP TO THINK ANSWERS

Stop to Think 1.1: B. The images of B are farther apart, so it travels a larger distance than does A during the same intervals of time.

Stop to Think 1.2: a. Dropped ball. **b.** Dust particle. **c.** Descending rocket.

Stop to Think 1.3: e. The average velocity vector is found by connecting one dot in the motion diagram to the next.

Stop to Think 1.4: c. The velocity leading away from this point is $\vec{v}_2 = \vec{v}_1 + \vec{a}\Delta t$.

$\vec{v}_1$

$\Delta\vec{v} = \vec{a}\,\Delta t$ $\quad \vec{v}_2 = \vec{v}_1 + \Delta\vec{v}$

Stop to Think 1.5: d > c > b = a.

2 Kinematics: The Mathematics of Motion

World-class sprinters have a tremendous acceleration at the start of a race.

► **Looking Ahead**

The goal of Chapter 2 is to learn how to solve problems about motion in a straight line. In this chapter you will learn to:

- Understand the mathematics of position, velocity, and acceleration for motion along a straight line.
- Use a graphical representation of motion.
- Use an explicit problem-solving strategy for kinematics problems.
- Understand free-fall motion and motion along inclined planes.

◄ **Looking Back**

Each chapter in this textbook builds on ideas and techniques from previous chapters. The Looking Back feature calls your attention to specific sections that are of major significance to the present chapter. A brief review of these sections will improve your study of this chapter. Please review:

- Sections 1.4–1.5 Velocity and acceleration.
- Sections 1.7–1.8 Problem solving in physics.

A race, whether between runners, bicyclists, or drag racers, exemplifies the idea of motion. Today, we use electronic stopwatches, video recorders, and other sophisticated instruments to analyze motion, but it hasn't always been so. Galileo, who in the early 1600s was the first scientist to study motion experimentally, used his pulse to measure time!

Galileo made a useful distinction between the *cause* of motion and the *description* of motion. **Kinematics** is the modern name for the mathematical description of motion without regard to causes. The term comes from the Greek word *kinema,* meaning "movement." You know this word through its English variation *cinema*—motion pictures! In this chapter on kinematics we'll develop the mathematical tools for describing motion. Then, in Chapter 4, we'll turn our attention to the *cause* of motion.

We will begin our study of kinematics with motion in one dimension; that is, motion along a straight line. Runners, drag racers, and skiers are just a few examples of motion in one dimension. The kinematics of two-dimensional motion—projectile motion and circular motion—will be considered in later chapters.

2.1 Motion in One Dimension

You learned in Chapter 1 that an object's motion can be described in terms of three fundamental quantities: its position $\vec{r}$, velocity $\vec{v}$, and acceleration $\vec{a}$. These quantities are vectors, having a direction as well as a magnitude. But for motion in one dimension, the vectors are restricted to point only "forward" or "backward." Consequently, we can describe one-dimensional motion with the simpler quantities x, v_x, and a_x (or y, v_y, and a_y). However, we need to give each of these quantities an explicit *sign,* positive or negative, to indicate whether the position, velocity, or acceleration vector points forward or backward. Learning to use the signs correctly is an important goal of this chapter.

Determining the Signs of Position, Velocity, and Acceleration

1.1 ActivPhysics Online

Position, velocity, and acceleration are measured with respect to a coordinate system. You will recall that a coordinate system is a grid or axis that *you* impose on a problem to analyze the motion. We will find it convenient to use an x-axis to describe both horizontal motion and motion along an inclined plane. A y-axis will be used for vertical motion. A coordinate axis has two essential features:

1. An origin, to define zero, and
2. An x or y label to indicate the positive end of the axis.

We will adopt the convention that the positive end of an x-axis is to the right and the positive end of a y-axis is up. The signs of position, velocity, and acceleration are based on this convention.

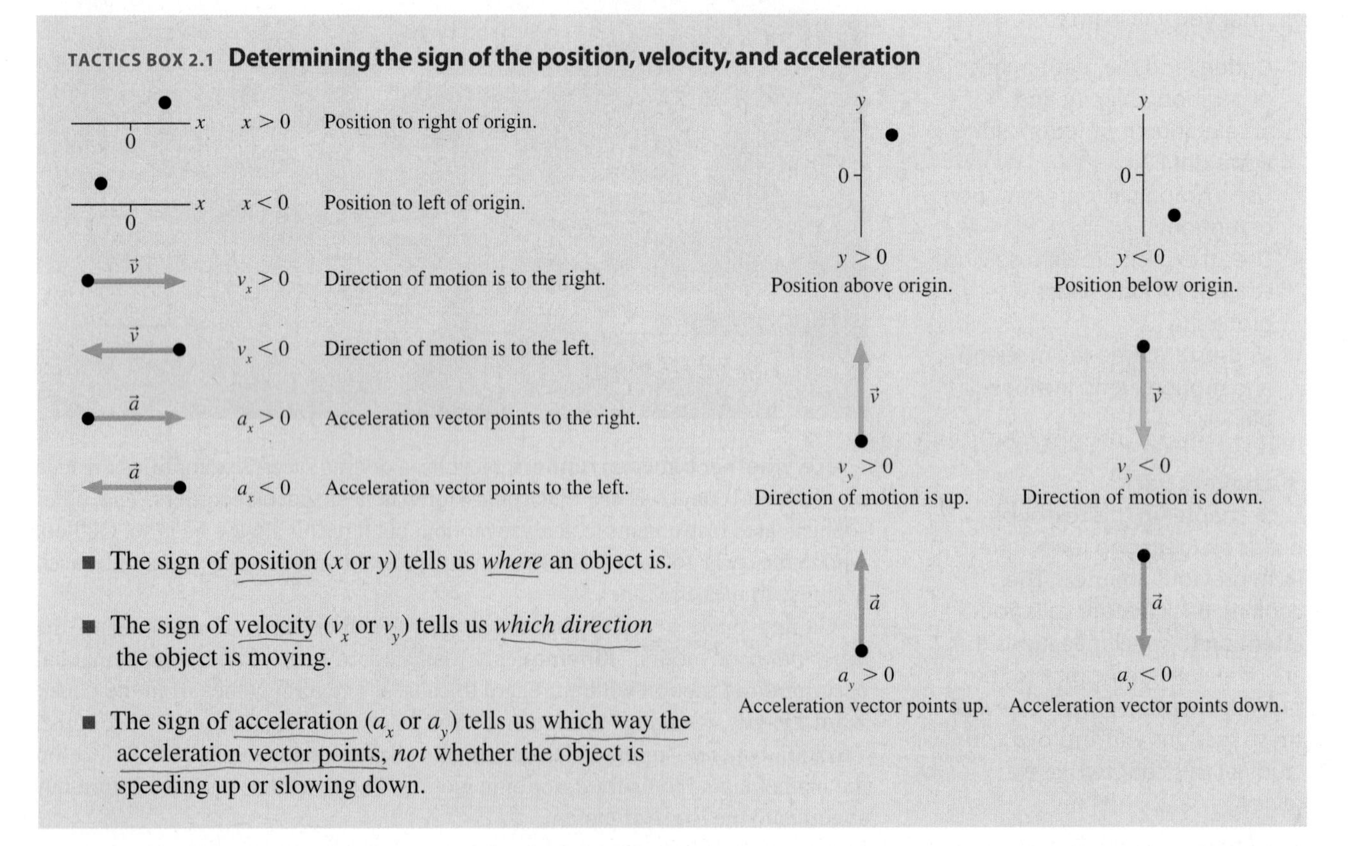
TACTICS BOX 2.1 Determining the sign of the position, velocity, and acceleration
x > 0 Position to right of origin.
x < 0 Position to left of origin.
v_x > 0 Direction of motion is to the right.
v_x < 0 Direction of motion is to the left.
a_x > 0 Acceleration vector points to the right.
a_x < 0 Acceleration vector points to the left.
y > 0 Position above origin.
y < 0 Position below origin.
v_y > 0 Direction of motion is up.
v_y < 0 Direction of motion is down.
a_y > 0 Acceleration vector points up.
a_y < 0 Acceleration vector points down.
The sign of position (x or y) tells us where an object is.
The sign of velocity (v_x or v_y) tells us which direction the object is moving.
The sign of acceleration (a_x or a_y) tells us which way the acceleration vector points, not whether the object is speeding up or slowing down.

Acceleration is where things get a bit tricky. A natural tendency is to think that a positive value of a_x or a_y describes an object that is speeding up while a negative value describes an object that is slowing down (decelerating). However, this interpretation *does not work.*

Acceleration was defined as $\vec{a}_{avg} = \Delta\vec{v}/\Delta t$. The direction of $\vec{a}$ can be determined by using a motion diagram to find the direction of $\Delta\vec{v}$. The one-dimensional acceleration a_x (or a_y) is then positive if the vector $\vec{a}$ points to the right (or up), negative if $\vec{a}$ points to the left (or down).

Figure 2.1 shows that this method for determining the sign of a does not conform to the simple idea of speeding up and slowing down. The object in Figure 2.1a has a positive acceleration ($a_x > 0$) not because it is speeding up but because the vector $\vec{a}$ points to the right. Compare this with the motion diagram of Figure 2.1b. Here the object is slowing down, but it still has a positive acceleration ($a_x > 0$) because $\vec{a}$ points to the right.

In Chapter 1 we found that an object is speeding up if $\vec{v}$ and $\vec{a}$ point in the same direction, slowing down if they point in opposite directions. For one-dimensional motion this rule becomes:

- **An object is speeding up if and only if v_x and a_x have the same sign.**
- **An object is slowing down if and only if v_x and a_x have opposite signs.**
- **An object's velocity is constant if and only if $a_x = 0$.**

Notice how the first two of these rules are at work in Figure 2.1

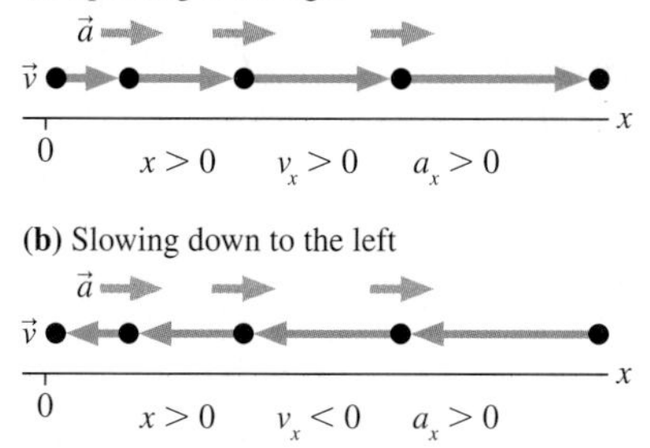

FIGURE 2.1 One of these objects is speeding up, the other slowing down, but they both have a positive acceleration a_x.

Position-versus-Time Graphs

Figure 2.2 is a motion diagram, made at 1 frame per minute, of a student walking to school. You can see that she leaves home at a time we choose to call $t = 0$ min and makes steady progress for a while. Beginning at $t = 3$ min there is a period where the distance traveled during each time interval becomes less—perhaps she slowed down to speak with a friend. Then she picks up the pace and the distances within each interval are longer.

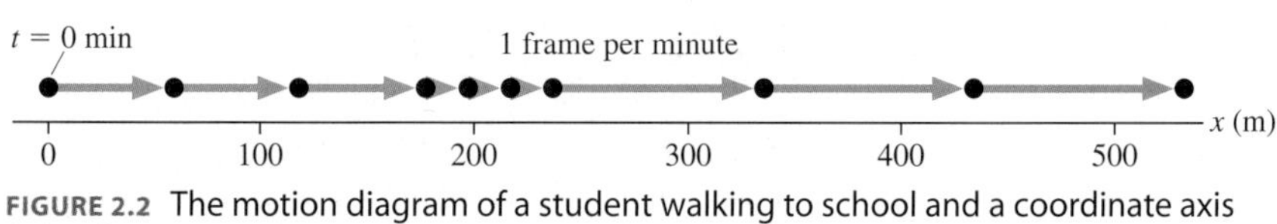

FIGURE 2.2 The motion diagram of a student walking to school and a coordinate axis for making measurements.

TABLE 2.1 Measured positions of a student walking to school

Time t (min)	Position x (m)
0	0
1	60
2	120
3	180
4	200
5	220
6	240
7	340
8	440
9	540

Figure 2.2 includes a coordinate axis, and you can see that every dot in a motion diagram occurs at a specific position. Table 2.1 shows the student's positions at different times as measured along this axis. For example, she is at position $x = 120$ m at $t = 2$ min.

The motion diagram is one way to represent the student's motion. Another is to make a graph of the measurements in Table 2.1. Figure 2.3a is a graph of x versus t for the student. The motion diagram tells us only where the student is at a few discrete points of time, so this graph of the data shows only points, no lines.

NOTE ► A graph of "a versus b" means that a is graphed on the vertical axis and b on the horizontal axis. Saying "graph a versus b" is really a shorthand way of saying "graph a as a function of b." ◄

However, common sense tells us the following. First, the student was *somewhere specific* at all times. That is, there was never a time when she failed to have a well-defined position, nor could she occupy two positions at one time. (As reasonable as this belief appears to be, it will be severely questioned and found not entirely accurate when we get to quantum physics!) Second, the student moved *continuously* through all intervening points of space. She could not go from $x = 100$ m to $x = 200$ m without passing through every point in between. It is thus quite reasonable to believe that her motion can be shown as a continuous

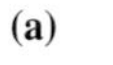
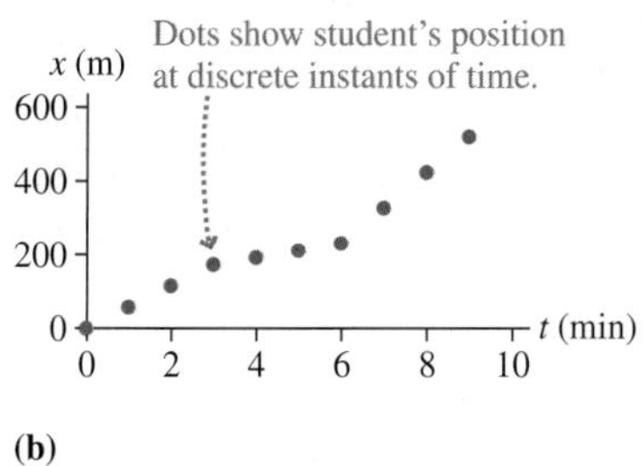

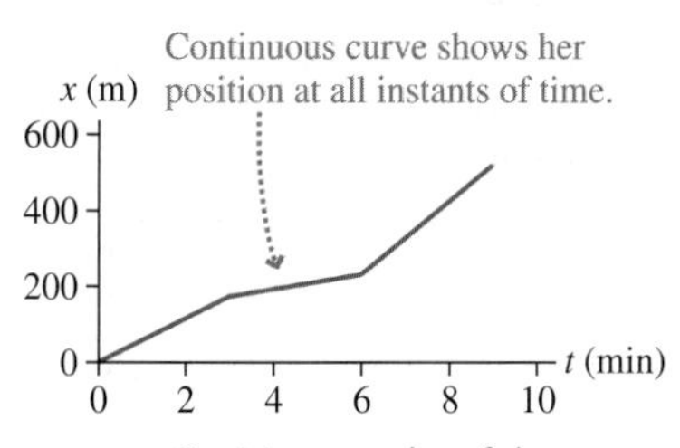

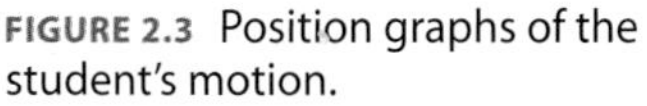

FIGURE 2.3 Position graphs of the student's motion.

curve passing through the measured points, as shown in Figure 2.3b. A continuous curve that shows an object's position as a function of time is called a **position-versus-time graph** or, sometimes, just a *position graph.*

NOTE ▶ A graph is *not* a "picture" of the motion. The student is walking along a straight line, but the graph itself is not a straight line. Further, we've graphed her position on the vertical axis even though her motion is horizontal. Graphs are *abstract representations* of motion. We will place significant emphasis on the process of interpreting graphs, and many of the exercises and problems will give you a chance to practice these skills. ◀

EXAMPLE 2.1 Interpreting a position graph

The graph in Figure 2.4a represents the motion of a car along a straight road. Describe the motion of the car.

MODEL Consider the car to be a particle that occupies a single point in space.

VISUALIZE As Figure 2.4b shows, the graph represents a car that travels to the left for 30 minutes, stops for 10 minutes, then travels back to the right for 40 minutes.

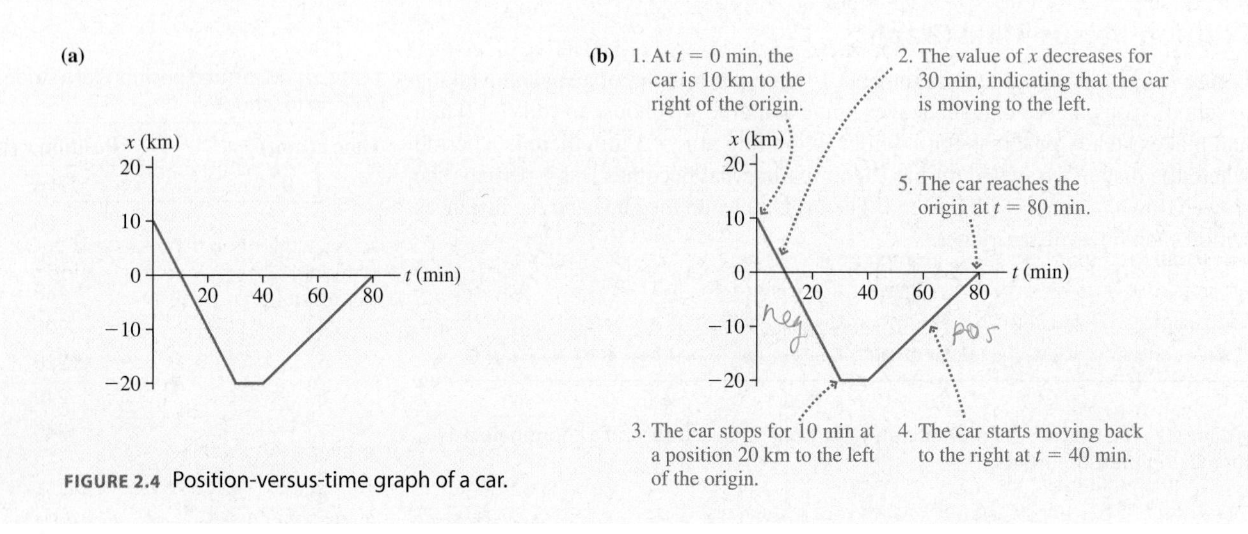

FIGURE 2.4 Position-versus-time graph of a car.

2.2 Uniform Motion

If you drive your car at a perfectly steady 60 miles per hour (mph), you will cover 60 mi during the first hour, another 60 mi during the second hour, yet another 60 mi during the third hour, and so on. This is an example of what we call *uniform motion.* In this case, 60 mi is not your position, but rather the *change* in your position during each hour; that is, your displacement Δx. Similarly, 1 hour is a time interval Δt rather than a specific instant of time. This suggests the following definition: **Straight-line motion in which equal displacements occur during *any* successive equal-time intervals is called uniform motion.**

The qualifier "any" is important. If during each hour you drive 120 mph for 30 minutes and stop for 30 minutes, you will cover 60 mi during each successive 1-hour interval. But you would *not* have equal displacements during successive 30-minute intervals, so this motion is not uniform. Your constant 60 mph driving is uniform motion because you will find equal displacements no matter how you choose your successive time intervals.

Figure 2.5 shows how uniform and nonuniform motion appear in motion diagrams and position-versus-time graphs. Notice that the position-versus-time graph for uniform motion is a straight line. This follows from the requirement that all Δx corresponding to the same Δt be equal. In fact, an alternative definition of uniform motion is: **An object's motion is uniform if and only if its position-versus-time graph is a straight line.**

The slope of a straight-line graph is defined as "rise over run." Because position is graphed on the vertical axis, the "rise" of a position-versus-time graph is the object's displacement Δx. The "run" is the time interval Δt. Consequently, the slope is $\Delta x/\Delta t$. The slope of a straight-line graph is constant, so an object in uniform motion has the *same* value of $\Delta x/\Delta t$ during *any* time interval Δt.

Chapter 1 defined the *average velocity* as $\Delta\vec{r}/\Delta t$. For one-dimensional motion this is simply

$$v_{avg} \equiv \frac{\Delta x}{\Delta t} \text{ or } \frac{\Delta y}{\Delta t} = \text{slope of the position-versus-time graph} \qquad (2.1)$$

That is, **the average velocity is the slope of the position-versus-time graph.** Velocity has units of "length per time," such as "miles per hour." The SI units of velocity are meters per second, abbreviated m/s.

NOTE ► The symbol $\equiv$ in Equation 2.1 stands for "is defined as" or "is equivalent to." This is a stronger statement than the two sides simply being equal. ◄

Equation 2.1 allows us to associate the slope of the position-versus-time graph, a *geometrical* quantity, with the *physical* quantity that we call the average velocity v_{avg}. This is an extremely important idea. In the case of uniform motion, where the slope $\Delta x/\Delta t$ is the same at all times, it appears that the average velocity is constant and unchanging. Consequently, a final definition of uniform motion is: **An object's motion is uniform if and only if its velocity v_x or v_y is constant and unchanging.** There's no real need to specify "average" for a velocity that doesn't change, so we will drop the subscript and refer to the average velocity as v_x or v_y.

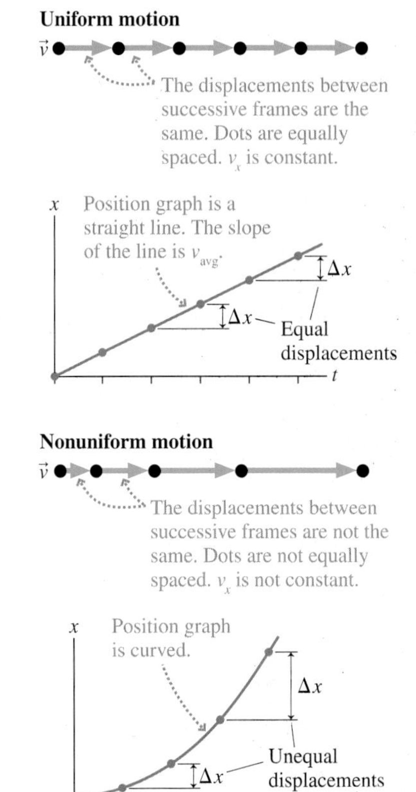
Uniform motion
The displacements between successive frames are the same. Dots are equally spaced. v_x is constant.
Position graph is a straight line. The slope of the line is v_{avg}.
Δx
Equal displacements
x
t
Nonuniform motion
The displacements between successive frames are not the same. Dots are not equally spaced. v_x is not constant.
Position graph is curved.
Unequal displacements

FIGURE 2.5 Motion diagrams and position graphs for uniform motion and nonuniform motion.

EXAMPLE 2.2 Skating with constant velocity

The position-versus-time graph of Figure 2.6a represents the motion of two students on roller blades. Determine their velocities and describe their motion.

MODEL Represent the two students as particles.

VISUALIZE Figure 2.6a is a graphical representation of the students' motion. Both graphs are straight lines, telling us that both skaters are moving uniformly with constant velocities.

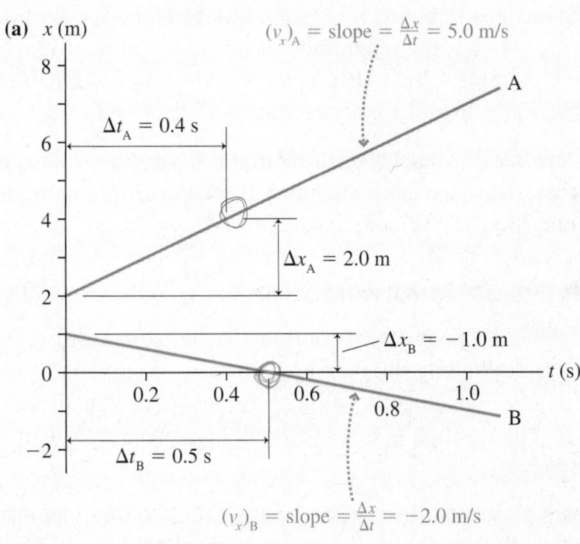
(a) x (m)
$(v_x)_A$ = slope = Δx/Δt = 5.0 m/s
Δt_A = 0.4 s
Δx_A = 2.0 m
Δx_B = −1.0 m
Δt_B = 0.5 s
A
B
t (s)

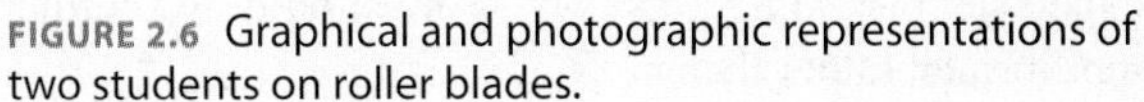
$(v_x)_B$ = slope = Δx/Δt = −2.0 m/s

(b)

FIGURE 2.6 Graphical and photographic representations of two students on roller blades.

SOLVE We can determine the students' velocities by measuring the slopes of the graphs. Skater A undergoes a displacement $\Delta x_A = 2.0$ m during the time interval $\Delta t_A = 0.40$ s. Thus his velocity is

$$(v_x)_A = \frac{\Delta x_A}{\Delta t_A} = \frac{2.0\text{ m}}{0.40\text{ s}} = 5.0\text{ m/s}$$

We need to be more careful with skater B. Although he moves a distance of 1.0 m in 0.50 s, his *displacement* Δx has a very precise definition:

$$\Delta x_B = x_{\text{at }0.5\text{ s}} - x_{\text{at }0.0\text{ s}} = 0.0\text{ m} - 1.0\text{ m} = -1.0\text{ m}$$

Careful attention to the signs is very important! This leads to

$$(v_x)_B = \frac{\Delta x_B}{\Delta t_B} = \frac{-1.0\text{ m}}{0.50\text{ s}} = -2.0\text{ m/s}$$

ASSESS The minus sign indicates that skater B is moving to the left. Our interpretation of this graph is that two students on roller blades are moving with constant velocities in opposite directions, as the photograph of Figure 2.6b on page 39 shows. Skater A starts at $x = 2.0$ m and moves to the right with a velocity of 5.0 m/s. Skater B starts at $x = 1.0$ m and moves to the left with a velocity of -2.0 m/s. Their speeds, of $\approx$10 mph and $\approx$4 mph, are reasonable for skaters on roller blades.

Example 2.2 brought out several points that are worth emphasizing.

TACTICS BOX 2.2 Interpreting position-versus-time graphs

1. Steeper slopes correspond to faster speeds.
2. Negative slopes correspond to negative velocities and, hence, to motion to the left (or down).
3. The slope is a ratio of intervals, $\Delta x/\Delta t$, not a ratio of coordinates. That is, the slope is *not* simply x/t.
4. We are distinguishing between the *actual* slope and the *physically meaningful* slope. If you were to use a ruler to measure the rise and the run of the graph, you could compute the actual slope of the line as drawn on the page. That is not the slope to which we are referring when we equate the velocity with the slope of the line. Instead, we find the *physically meaningful* slope by measuring the rise and run using the scales along the axes. The "rise" Δx is some number of meters; the "run" Δt is some number of seconds. The physically meaningful rise and run include units, and the ratio of these units gives the units of the slope.

An object's **speed** v is how fast it's going, independent of direction. This is simply $v = |v_x|$ or $v = |v_y|$, the magnitude or absolute value of its velocity. In Example 2.2, for example, skater B's *velocity* is -2.0 m/s but his *speed* is 2.0 m/s. Speed is a scalar quantity, not a vector.

NOTE ▶ Our mathematical analysis of motion is based on velocity, not speed. The subscript in v_x or v_y is an essential part of the notation, reminding us that, even in one dimension, the velocity is a vector. ◀

The Mathematics of Uniform Motion

We need a mathematical analysis of motion that will be valid regardless of whether an object moves along the x-axis, the y-axis, or any other straight line. Consequently, it will be convenient to write equations for a "generic axis" that we will call the s-axis. The position of an object will be represented by the symbol s and its velocity by v_s.

NOTE ▶ Equations written in terms of s are valid for any one-dimensional motion. In a specific problem, however, you should use either x or y, whichever is appropriate, rather than s. ◀

Consider an object in uniform motion along the s-axis with the linear position-versus-time graph shown in Figure 2.7. The object's **initial position** is s_i at time t_i. The term *initial position* refers to the starting point of our analysis or the starting point in a problem; the object may or may not have been in motion prior to t_i. At a later time t_f, the ending point of our analysis or the ending point of a problem, the object's **final position** is s_f.

The object's velocity v_s along the s-axis can be determined by finding the slope of the graph:

$$v_s = \frac{\text{rise}}{\text{run}} = \frac{\Delta s}{\Delta t} = \frac{s_f - s_i}{t_f - t_i} \tag{2.2}$$

Equation 2.2 is easily rearranged to give

$$s_f = s_i + v_s \Delta t \text{ (uniform motion)} \tag{2.3}$$

Equation 2.3 applies to any time interval Δt during which the velocity is constant.

The velocity of a uniformly moving object tells us the amount by which its position changes during each second. A particle with a velocity of 20 m/s *changes* its position by 20 m during every second of motion: by 20 m during the first second of its motion, by another 20 m during the next second, and so on. If the object starts at $s_i = 10$ m, it will be at $s = 30$ m after 1 second of motion and at $s = 50$ m after 2 seconds of motion. Thinking of velocity like this will help you develop an intuitive understanding of the connection between velocity and position.

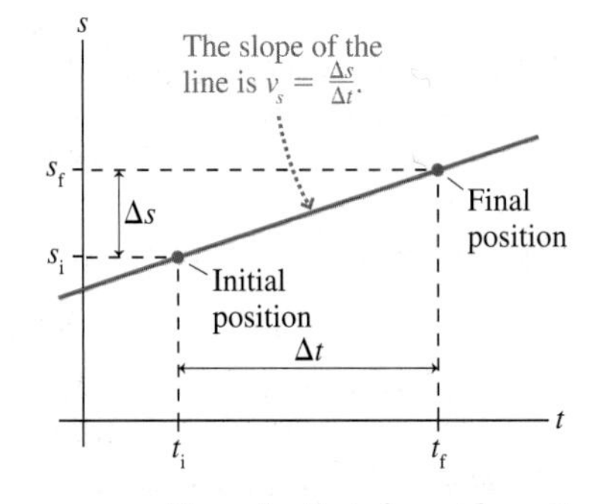

FIGURE 2.7 The velocity is found from the slope of the position-versus-time graph.

EXAMPLE 2.3 Lunch in Cleveland?

Bob leaves home in Chicago at 9:00 A.M. and travels east at a steady 60 mph. Susan, 400 miles to the east in Pittsburgh, leaves at the same time and travels west at a steady 40 mph. Where will they meet for lunch?

MODEL Here is a problem where, for the first time, we can really put all four aspects of our problem-solving strategy into play. To begin, represent Bob and Susan as particles.

VISUALIZE Figure 2.8 shows the physical representation (the motion diagram) and the pictorial representation. The equal spacings of the dots in the motion diagram indicate that the motion is uniform. In evaluating the given information, we recognize that the starting time of 9:00 A.M. is not relevant to the problem. Consequently, the initial time is chosen as simply $t_0 = 0$ hr. Bob and Susan are traveling in opposite directions, hence one of the velocities must be a negative number. We have

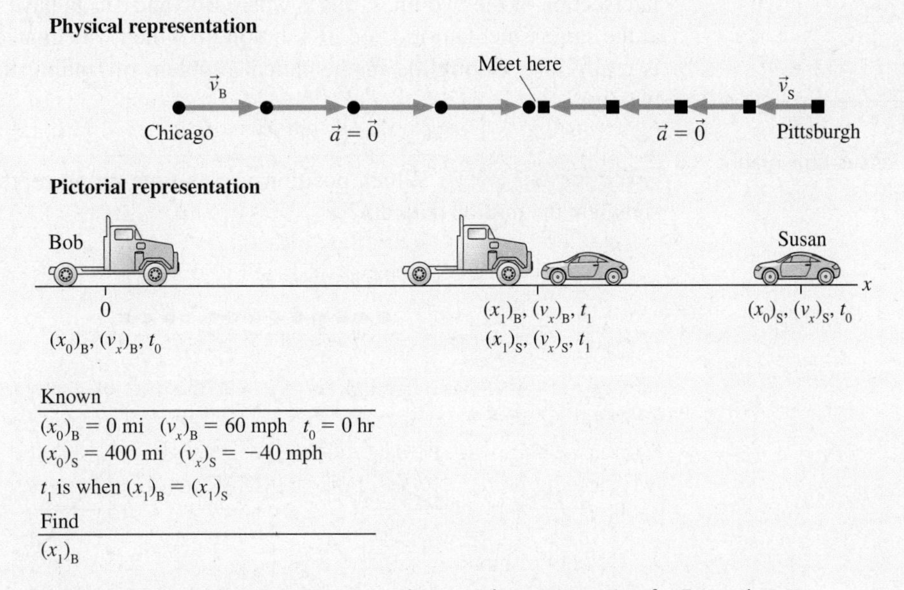

FIGURE 2.8 Physical representation and pictorial representation for Example 2.3.

chosen a coordinate system in which Bob starts at the origin and moves to the right (east) while Susan is moving to the left (west). Thus Susan has the negative velocity. Notice how we've assigned position, velocity, and time symbols to each point in the motion. Pay special attention to how subscripts are used to distinguish different points in the problem and to distinguish Bob's symbols from Susan's.

One purpose of the pictorial representation is to establish what we need to find. Bob and Susan meet when they have the same position at the same time t_1. Thus we want to find $(x_1)_B$ at the time when $(x_1)_B = (x_1)_S$. Notice that $(x_1)_B$ and $(x_1)_S$ are their *positions,* which are equal when they meet, not the distances they have traveled.

SOLVE The goal of the mathematical representation is to proceed from the pictorial representation to a mathematical solution of the problem. We can begin by using Equation 2.3 to find Bob's and Susan's positions at time t_1 when they meet:

$$(x_1)_B = (x_0)_B + (v_x)_B(t_1 - t_0) = (v_x)_B t_1$$

$$(x_1)_S = (x_0)_S + (v_x)_S(t_1 - t_0) = (x_0)_S + (v_x)_S t_1$$

Notice two things. First, we started by writing the *full* statement of Equation 2.3. Only then did we simplify by dropping those terms known to be zero. You're less likely to make accidental errors if you follow this procedure. Second, we replaced the generic symbol *s* with the specific horizontal-position symbol *x*, and we replaced the generic subscripts i and f with the specific symbols 0 and 1 that we defined in the pictorial representation. This is also good problem-solving technique.

The condition that Bob and Susan meet is

$$(x_1)_B = (x_1)_S$$

By equating the right-hand sides of the above equations, we get

$$(v_x)_B t_1 = (x_0)_S + (v_x)_S t_1$$

Solving for t_1, we find that they meet at time

$$t_1 = \frac{(x_0)_S}{(v_x)_B - (v_x)_S} = \frac{400 \text{ miles}}{60 \text{ mph} - (-40) \text{ mph}} = 4.0 \text{ hours}$$

Finally, inserting this time back into the equation for x_{B1} gives

$$(x_1)_B = \left(60 \frac{\text{miles}}{\text{hour}}\right) \times (4.0 \text{ hours}) = 240 \text{ miles}$$

While this is a number, it is not yet the answer to the question. The phrase "240 miles" by itself does not say anything meaningful. Because this is the value of Bob's *position,* and Bob was driving east, the answer to the question is, "They meet 240 miles east of Chicago."

ASSESS Before stopping, we should check whether or not this answer seems reasonable. We certainly expected an answer between 0 miles and 400 miles. We also know that Bob is driving faster than Susan, so we expect that their meeting point will be *more* than halfway from Chicago to Pittsburgh. Our assessment tells us that 240 miles is a reasonable answer.

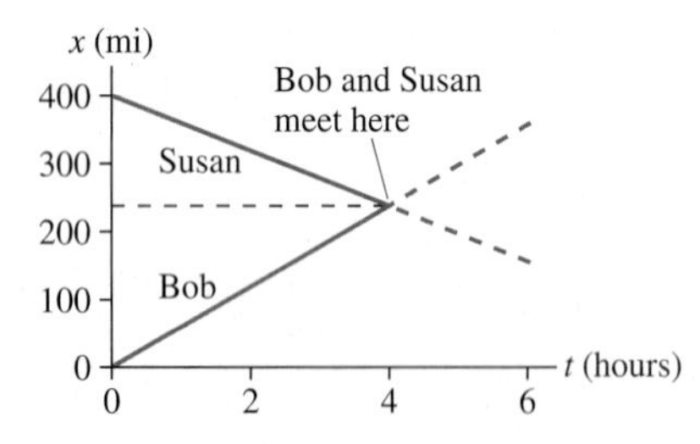

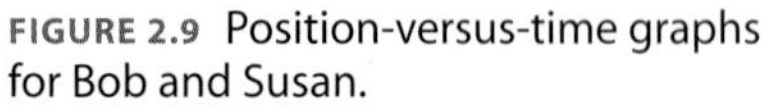
FIGURE 2.9 Position-versus-time graphs for Bob and Susan.

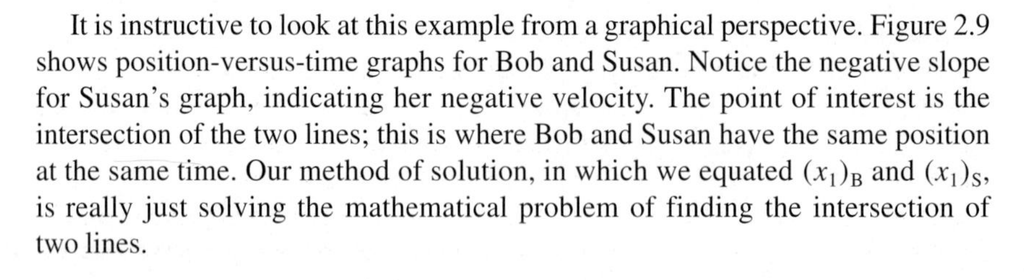
It is instructive to look at this example from a graphical perspective. Figure 2.9 shows position-versus-time graphs for Bob and Susan. Notice the negative slope for Susan's graph, indicating her negative velocity. The point of interest is the intersection of the two lines; this is where Bob and Susan have the same position at the same time. Our method of solution, in which we equated $(x_1)_B$ and $(x_1)_S$, is really just solving the mathematical problem of finding the intersection of two lines.

STOP TO THINK 2.1 Which position-versus-time graph represents the motion shown in the motion diagram?

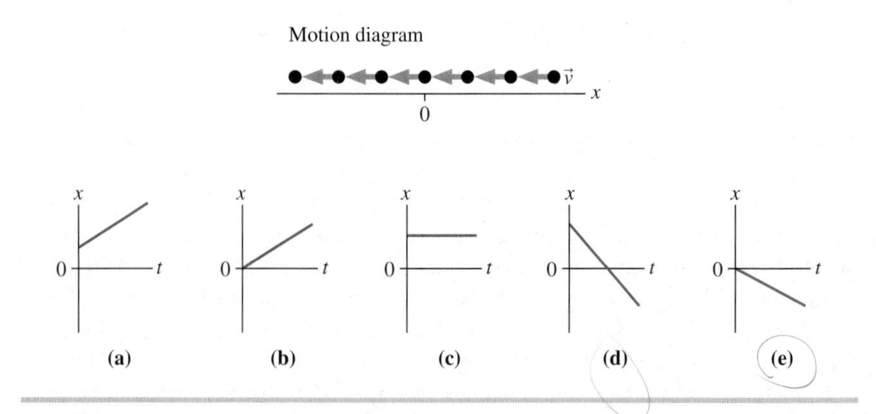

2.3 Instantaneous Velocity

Not many objects in the universe move with constant velocity. Consider, for example, what happens when you drive a car. You start from rest, accelerate, perhaps drive at steady velocity for a while (constant v), decelerate, and ultimately stop. This is clearly a more complex problem than we considered in the last section. If we were to graph your car's position at various times, the graph would not be a straight line.

Figure 2.10a shows the motion diagram of a jet as it takes off. Figure 2.10b is the corresponding position-versus-time graph. The graph curves upward as the spacing between the motion-diagram dots increases. We can determine the jet's average speed v_{avg} between any two times t_i and t_f by selecting those two points on the graph, drawing the straight-line connection between them, measuring Δx and Δt, and finally using these to compute $v_{avg} = \Delta x/\Delta t$. Graphically, v_{avg} is simply the slope of the straight-line connection between the two points.

This jet plane most definitely does not move with a constant velocity.

Average velocity has only limited usefulness for an object whose velocity isn't constant. Suppose, for example, that you and Frank and Karen all leave at precisely 8:00 A.M. in your cars. At 9:00 A.M. you have each traveled exactly 60 mi. All you discern from this information is that each of you had the same average velocity $v_{avg} = 60$ mph. Suppose, however, that Frank started out going faster than 60 mph and later slowed down; Karen got a slow start, but later sped up; and you drove at a steady but boring 60 mph the entire hour. What would each of you see if you read your car's speedometer at 8:10? You would see 60 mph, but Frank would see a speed greater than 60 mph while Karen would see a speed less than 60 mph. Later, at 8:50, Frank's speedometer reads less than 60 mph, Karen's reads more than 60 mph, while yours has not changed.

The speedometer reading tells you how fast you're going *at that instant,* rather than averaged over the entire hour. We can define an object's **instantaneous velocity** to be its velocity—a speed *and* a direction—at a single *instant* of time t.

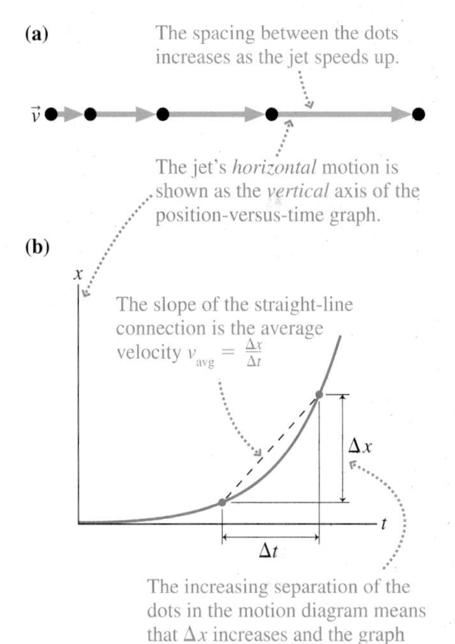
(a)
The spacing between the dots increases as the jet speeds up.
$\vec{v}$
The jet's *horizontal* motion is shown as the *vertical* axis of the position-versus-time graph.
(b)
x
The slope of the straight-line connection is the average velocity $v_{avg} = \frac{\Delta x}{\Delta t}$
Δx
t
Δt
The increasing separation of the dots in the motion diagram means that Δx increases and the graph curves upward.

FIGURE 2.10 Motion diagram and position graph of a jet during take off.

Such a definition, though, raises some difficult issues. Just what does it mean to have a velocity "at an instant"? Suppose a police officer pulls you over and says, "I just clocked you going 80 miles per hour." You might respond, "But that's impossible. I've only been driving for 20 minutes, so I can't possibly have gone 80 miles." Unfortunately for you, the police officer was a physics major. He replies, "I mean that at the instant I measured your velocity, you were moving at a rate such that you *would* cover a distance of 80 miles *if* you were to continue at that velocity without change for 1 hour. That will be a $200 fine."

Here, again, is the idea that velocity is the *rate* at which an object changes its position. Rates tell us how quickly or how slowly things change, and that idea is conveyed by the word "per." An instantaneous velocity of 80 mph means that the rate at which your car's position is changing—at that exact instant—is such that it would travel a distance of 80 miles in 1 hour *if* it continued at that rate without change. Whether or not it actually does travel at that velocity for another hour, or even for another millisecond, is not relevant.

Using Motion Diagrams and Graphs

Let's use motion diagrams to analyze an accelerating jet plane. Figure 2.11a on the next page shows a motion diagram made using a normal 30-frames-per-second camera. The third velocity vector is the *average* velocity during the time interval Δt it took the object to move the distance Δs from point 2 to point 3. We would like to determine the *instantaneous* velocity v_s at the position marked with the red ×, slightly before point 3. Because the jet is accelerating, its velocity at this point—like Karen's at 8:50—is larger than the average velocity between 2 and 3. How can we measure it?

60
70
80
50
90
40
100
30
110
20
120
10
miles
km/h
MPH

The speedometer reading tells you how fast you're going *at that instant.*

Suppose we use a high-speed camera, one that takes 100 frames per second, to film just the segment of motion between points 2 and 3. This "magnified" motion

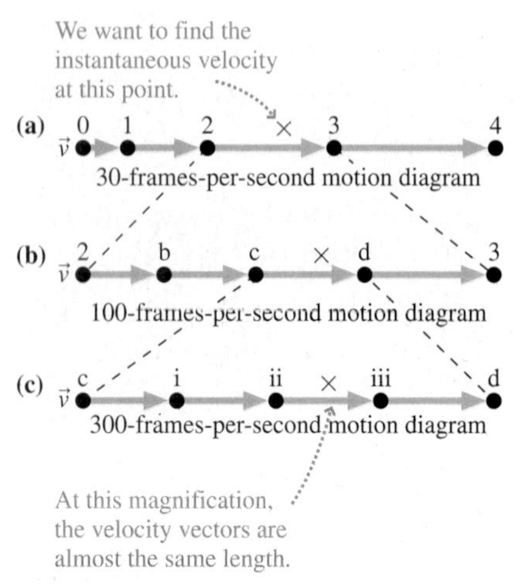

FIGURE 2.11 Motion diagram of an accelerating jet made with increasingly fast movie cameras.

diagram is shown in Figure 2.11b. While the velocity changed a lot between points 0 and 4, causing the length of the arrows in Figure 2.11a to change greatly, the velocity change between 2 and 3 is much less. The velocity arrow from c to d is the *average* velocity between these two points. This average is a much closer approximation to the instantaneous velocity at × because the lengths of the velocity arrows on either side are nearly the same. Even so, the velocity is still changing. The instantaneous velocity we seek is not the same as the average velocity.

Finally we get out our really-high-speed 300-frames-per-second camera and film just the interval between points c and d. The result is shown in Figure 2.11c. Now, at this level of magnification, each velocity vector is *almost* the same length. On this time scale, the motion near × appears very nearly uniform! *If* the motion were to continue at a constant velocity after ×, then the velocity it would have is the velocity of Figure 2.11c.

The point of Figure 2.11 is that the average velocity $v_{avg} = \Delta s/\Delta t$ becomes a better and better approximation to the instantaneous velocity v_s as the time interval Δt over which the displacement is measured gets smaller and smaller. By magnifying the motion diagram, we are using smaller and smaller time intervals Δt. But even 300 frames per second isn't fast enough. We need to let $\Delta t \rightarrow 0$.

We can state this idea mathematically in terms of a limit:

$$v_s \equiv \lim_{\Delta t \to 0} \frac{\Delta s}{\Delta t} = \frac{ds}{dt} \quad \text{(instantaneous velocity)} \tag{2.4}$$

As Δt gets smaller and smaller, the average velocity $v_{avg} = \Delta s/\Delta t$ reaches a constant value—the limit—and no longer changes. This limit is called *the derivative of s with respect to t,* and it is denoted ds/dt. We'll look at derivatives in the next section.

Now let's analyze the same accelerating jet with position-versus-time graphs. Figure 2.12a shows two points that are separated by $\Delta t = 1/30$ s. These are points 2 and 3 in the 30-frames-per-second motion diagram of Figure 2.11a. The slope of the straight line connecting these points is the average velocity over this time interval. Figures 2.12b and 2.12c show the effect of decreasing the time interval Δt.

As Δt gets smaller, the straight line becomes a better and better approximation of the curve between the two points and the average velocity becomes a better and better approximation of the instantaneous velocity at point ×. Notice that the slope in Figure 2.12c is a little larger than the slope in Figure 2.12a. This is because, as we had noted, the instantaneous velocity at × is larger than the average velocity between points 2 and 3.

Finally, in Figure 2.12d, we reach the limit $\Delta t \rightarrow 0$. In this limit, the straight line is tangent to the curve at point ×. **The instantaneous velocity at time t is the slope of the line that is tangent to the position-versus-time graph at time t.** The practical issue will be how to determine the slope of the tangent line.

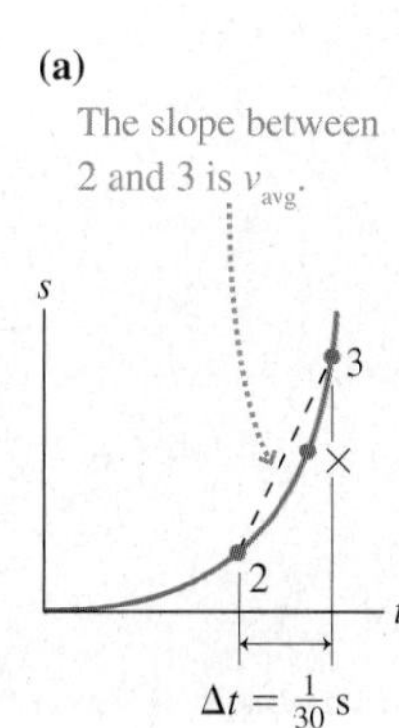

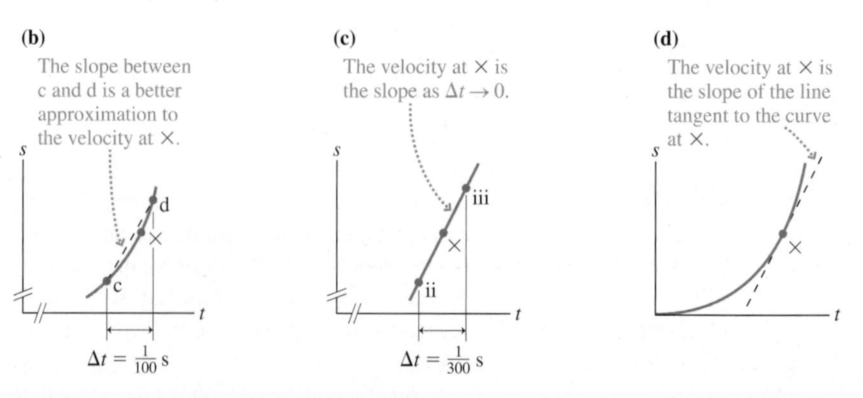

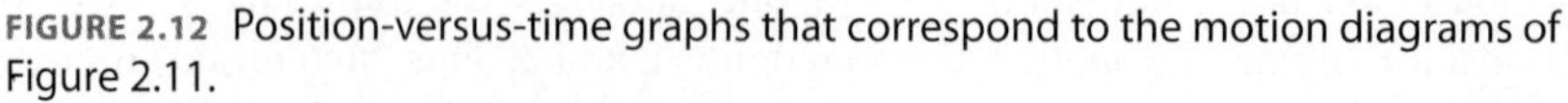

FIGURE 2.12 Position-versus-time graphs that correspond to the motion diagrams of Figure 2.11.

EXAMPLE 2.4 Relating a velocity graph to a position graph

Figure 2.13a is the position-versus-time graph of a car.

a. Draw the car's velocity-versus-time graph.
b. Describe the car's motion.

MODEL Represent the car as a particle, with a well-defined position at each instant of time.

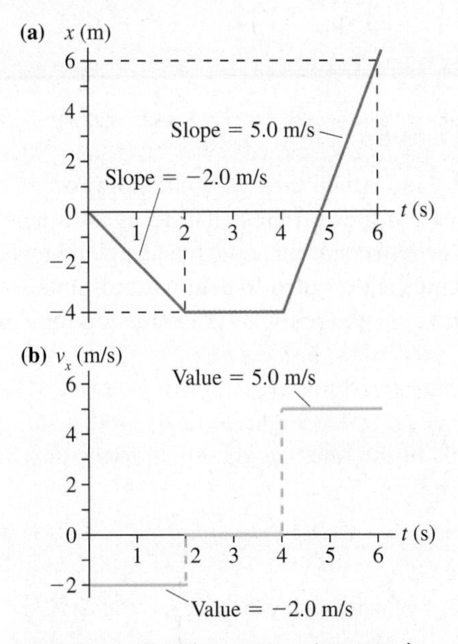

FIGURE 2.13 Position-versus-time graph and the corresponding velocity-versus-time graph.

VISUALIZE Figure 2.13a is a graphical representation of the motion.

SOLVE

a. The car's position-versus-time graph is a sequence of three straight lines. Each of these straight lines represents uniform motion at a constant velocity. We can determine the car's velocity during each interval of time by measuring the slope of the line. From $t = 0$ s to $t = 2$ s ($\Delta t = 2$ s) the car's displacement is $\Delta x = -4\text{ m} - 0\text{ m} = -4$ m. The velocity during this interval is

$$v_x = \frac{\Delta x}{\Delta t} = \frac{-4.0\text{ m}}{2.0\text{ s}} = -2.0\text{ m/s}$$

The car's position does not change from $t = 2$ s to $t = 4$ s ($\Delta x = 0$), so $v_x = 0$. Finally, the displacement between $t = 4$ s and $t = 6$ s ($\Delta t = 2$ s) is $\Delta x = 10$ m. Thus the velocity during this interval is

$$v_x = \frac{10\text{ m}}{2.0\text{ s}} = 5.0\text{ m/s}$$

These velocities are shown on the velocity-versus-time graph of Figure 2.13b.

b. The car backs up for 2 s at 2 m/s, sits at rest for 2 s, then drives forward at 5 m/s for at least 2 s. We can't tell from the graph what happens for $t > 6$ s.

ASSESS The velocity graph and the position graph look completely different. The *value* of the velocity graph at any instant of time equals the *slope* of the position graph.

EXAMPLE 2.5 Finding velocity from position graphically

Figure 2.14 shows the position-versus-time graph of a particle that moves along the y-axis.

a. At which labeled point or points is the particle moving the slowest?
b. At which point or points is the particle moving the fastest?
c. Sketch an approximate velocity-versus-time graph for the particle.

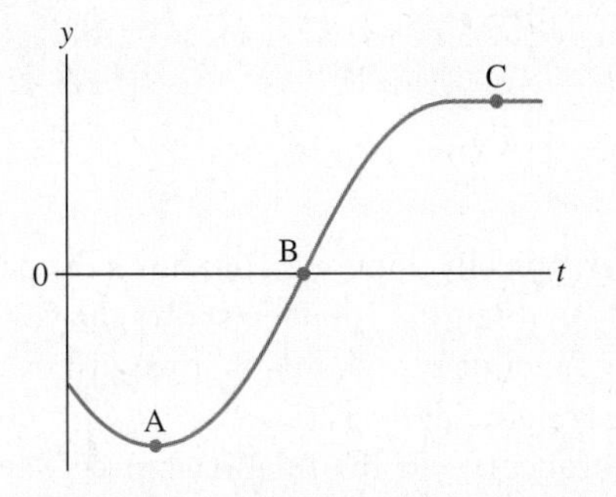

FIGURE 2.14 Position-versus-time graph.

MODEL The problem statement tells us the object is a particle.

VISUALIZE Figure 2.15 is a graphical representation of the motion.

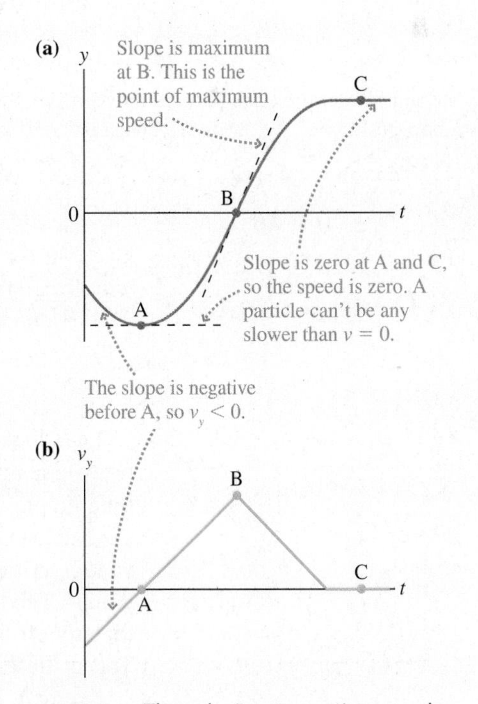

FIGURE 2.15 The velocity-versus-time graph is found from the position graph.

SOLVE

a. Figure 2.15a shows that the particle is slowest—no speed at all!—at points A and C. At point A, the speed is only instantaneously zero. At point C, the particle has actually stopped and remains at rest.
b. The particle moves the fastest at point B.
c. Although we cannot find an exact velocity-versus-time graph, we can see that the slope, and hence v_y, is initially negative, becomes zero at point A, rises to a maximum value at point B, decreases back to zero a little before point C, then remains at zero thereafter. Thus Figure 2.15b shows, at least approximately, the particle's velocity-versus-time graph.

ASSESS Once again, the shape of the velocity graph bears no resemblance to the shape of the position graph. You must transfer *slope* information from the position graph to *value* information on the velocity graph.

A Little Calculus: Derivatives

We have reached the point beyond which Galileo could not proceed because he lacked the mathematical tools. Further progress had to await a new branch of mathematics called *calculus,* invented simultaneously in England by Newton and in Germany by Leibniz. Calculus is designed to deal with instantaneous quantities. In other words, it provides us with the tools for evaluating limits such as the one in Equation 2.4.

The notation ds/dt is called *the derivative of s with respect to t*, and Equation 2.4 defines it as the limiting value of a ratio. As Figure 2.12 showed, ds/dt can be interpreted graphically as the slope of the line that is tangent to the position-versus-time graph at time t.

EXAMPLE 2.6 Finding velocity from position as a derivative

The position of a particle as a function of time is $s = 2t^2$ m, where t is in s. What is the velocity v_s as a function of time?

SOLVE To solve this problem we need to "take the derivative" of s.

$$v_s = \frac{ds}{dt} = \lim_{\Delta t \to 0} \frac{\Delta s}{\Delta t}$$

During the time interval Δt, the particle moves from position $s_{\text{at } t}$ to the new position $s_{\text{at } t+\Delta t}$. Its displacement is

$$\begin{aligned}\Delta s &= s_{\text{at } t+\Delta t} - s_{\text{at } t} \\ &= 2(t + \Delta t)^2 - 2t^2 \\ &= 2(t^2 + 2t\Delta t + (\Delta t)^2) - 2t^2 \\ &= 4t\Delta t + 2(\Delta t)^2\end{aligned}$$

The average velocity during the time interval Δt is

$$v_{\text{avg}} = \frac{\Delta s}{\Delta t} = \frac{4t\Delta t + 2(\Delta t)^2}{\Delta t} = 4t + 2\Delta t$$

We can finish by taking the limit $\Delta t \to 0$ to find

$$v_s = \frac{ds}{dt} = \lim_{\Delta t \to 0} (4t + 2\Delta t) = 4t \text{ m/s}$$

In other words, the function for calculating the velocity at any instant of time is $v_s = 4t$ m/s, where t is in s. At $t = 3$ s, for example, the particle is located at position $s = 18$ m and its instantaneous velocity, at just that instant, is $v_s = 12$ m/s.

Let's look at this example graphically. Figure 2.16a shows the particle's position-versus-time graph $s = 2t^2$ m. Figure 2.16b then shows the velocity-versus-time graph, using the velocity function $v = 4t$ m/s that we just calculated. You see that the velocity graph is a straight line.

It is critically important to understand the relationship between these two graphs. The *value* of the velocity graph at any instant of time, which we can read directly off the vertical axis, is the *slope* of the position graph at that same time. This is illustrated at $t = 1$ s and $t = 3$ s.

Example 2.6 showed how the limit of $\Delta s/\Delta t$ can be evaluated to find a derivative, but the procedure is clearly rather tedious. It would hinder us significantly if

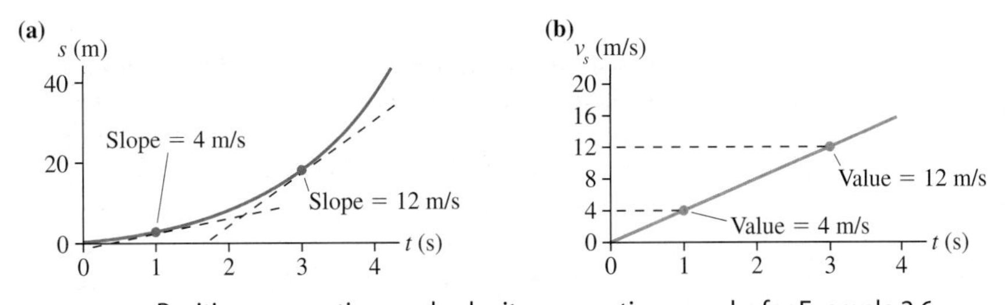

FIGURE 2.16 Position-versus-time and velocity-versus-time graphs for Example 2.6.

we had to do this for every new situation. Fortunately, we need only a few basic derivatives in this text. Learn these, and you do not have to go all the way back to the definition in terms of limits.

The only functions we will use in Parts I and II of this book are powers and polynomials. Consider the function $u = ct^n$, where c and n are constants. The following result is proven in calculus:

$$\text{The derivative of } u = ct^n \text{ is } \frac{du}{dt} = nct^{n-1} \tag{2.5}$$

NOTE ▶ The symbol u is a "dummy name." Equation 2.5 can be used to take the derivative of *any* function of the form ct^n. ◀

Example 2.6 needed to find the derivative of the function $s = 2t^2$. Using Equation 2.5 with $c = 2$ and $n = 2$, the derivative of $s = 2t^2$ with respect to t is

$$v_s = \frac{ds}{dt} = 2 \cdot 2t^{2-1} = 4t$$

Similarly, the derivative of the function $x = 3/t^2 = 3t^{-2}$ is

$$\frac{dx}{dt} = (-2) \cdot 3t^{-2-1} = -6t^{-3} = -\frac{6}{t^3}$$

A value that doesn't change with time, such as the position of an object at rest, can be represented by the function $u = c =$ constant. That is, the exponent of t^n is $n = 0$. You can see from Equation 2.5 that the derivative of a constant is zero. That is,

$$\frac{du}{dt} = 0 \text{ if } u = c = \text{constant} \tag{2.6}$$

This makes sense. The graph of the function $u = c$ is simply a horizontal line at height c. The slope of a horizontal line—which is what the derivative du/dt measures—is zero.

The only other information we need about derivatives for now is how to evaluate the derivative of the sum of two or more functions. Let u and w be two separate functions of time. You will learn in calculus that

$$\frac{d}{dt}(u + w) = \frac{du}{dt} + \frac{dw}{dt} \tag{2.7}$$

That is, the derivative of a sum is the sum of the derivatives.

NOTE ▶ You may have learned in calculus to take the derivative dy/dx, where y is a function of x. The derivatives we use in physics are the same; only the notation is different. We're interested in how quantities change with time, so our derivatives are with respect to t instead of x. ◀

EXAMPLE 2.7 Using calculus to find the velocity

A particle's position is given by the function $x = (-t^3 + 3t)$ m, where t is in s.

a. What is the particle's position and velocity at $t = 2$ s?
b. Draw graphs of x and v_x during the interval $-3 \text{ s} \le t \le 3$ s.
c. Draw a motion diagram to illustrate this motion.

SOLVE

a. We can compute the position at $t = 2$ s directly from the function x:

$$x(\text{at } t = 2 \text{ s}) = -(2)^3 + (3)(2) = -8 + 6 = -2 \text{ m}$$

The velocity is then $v_x = dx/dt$. The function for x is the sum of two polynomials, so

$$v_x = \frac{dx}{dt} = \frac{d}{dt}(-t^3 + 3t) = \frac{d}{dt}(-t^3) + \frac{d}{dt}(3t)$$

The first derivative is a power with $c = -1$ and $n = 3$; the second has $c = 3$ and $n = 1$. Using Equation 2.5,

$$v_x = (-3t^2 + 3) \text{ m/s}$$

where t is in s. Evaluating the velocity at $t = 2$ s gives

$$v_x(\text{at } t = 2 \text{ s}) = -3(2)^2 + 3 = -9 \text{ m/s}$$

The negative sign indicates that the particle, at this instant of time, is moving to the *left* at a speed of 9 m/s.

b. Figure 2.17 shows the position graph and the velocity graph. These were created by computing, and then graphing, the values of x and v_x at several points between -3 s and 3 s. The slope of the position-versus-time graph at $t = 2$ s is -9 m/s; this becomes the *value* that is graphed for the velocity at $t = 2$ s. Similar measurements are shown at $t = -1$ s, where the velocity is instantaneously zero.

c. Finally, we can interpret the graphs of part b to draw the motion diagram shown in Figure 2.18.

- The particle is initially to the right of the origin ($x > 0$ at $t = -3$ s) but moving to the left ($v_x < 0$). Its *speed* is slowing ($v = |v_x|$ is decreasing), so the velocity vector arrows are getting shorter.
- The particle passes the origin at $t \approx -1.5$ s, but it is still moving to the left.
- The position reaches a minimum at $t = -1$ s; the particle is as far left as it is going. The velocity is *instantaneously* $v_x = 0$ m/s as the particle reverses direction.
- The particle moves back to the right between $t = -1$ s and $t = 1$ s ($v_x > 0$).
- The particle turns around again at $t = 1$ s and begins moving back to the left ($v_x < 0$). It keeps speeding up, then disappears off to the left.

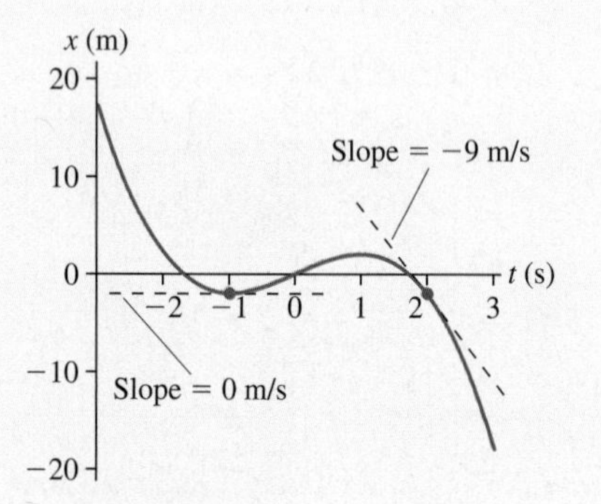

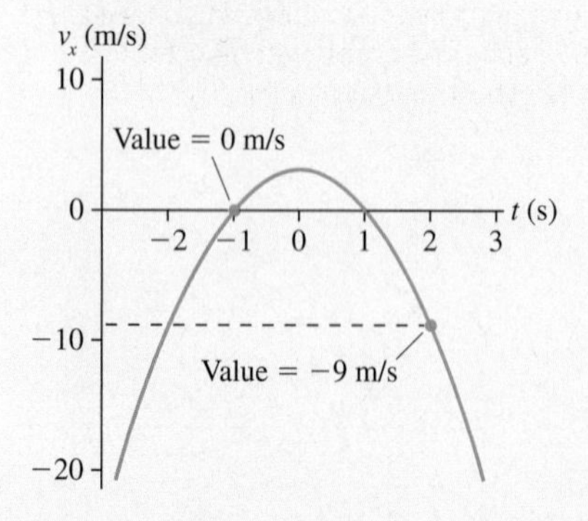

FIGURE 2.17 Position and velocity graphs.

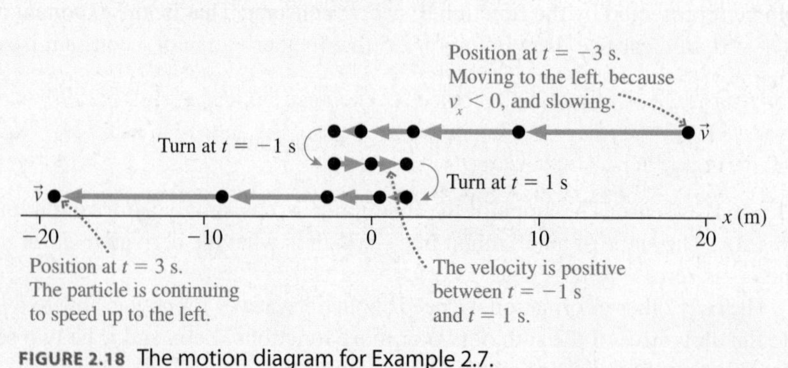

FIGURE 2.18 The motion diagram for Example 2.7.

The particle in this example moved out to $x = -2$ m at $t = -1$ s, then returned. The point in its motion where it reversed direction is called a *turning point.* Because the velocity was negative just before reaching the turning point and positive just after, it had to pass through $v_x = 0$ m/s. Thus, a **turning point** is a point where the velocity is instantaneously zero as the particle reverses direction. A second turning point occurs at $t = 1$ s as the particle reaches $x = 2$ m. We will see many future examples of turning points.

STOP TO THINK 2.2 Which velocity-versus-time graph goes with the position-versus-time graph on the left?

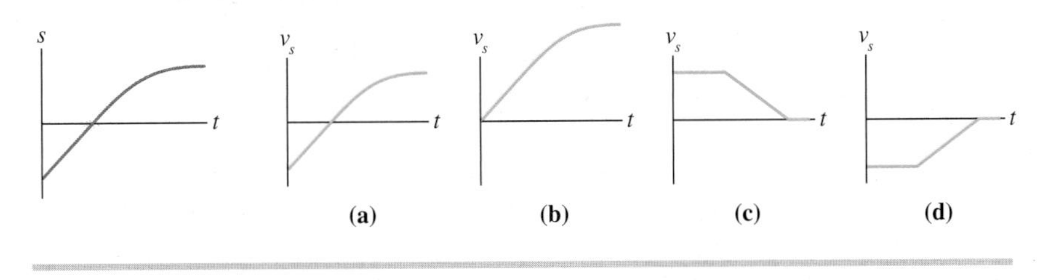

2.4 Finding Position from Velocity

Equation 2.4 provides a means of finding the instantaneous velocity v_s if we know the position s as a function of time. In mathematical terms, the velocity is the derivative of the position function. Graphically, the velocity is the slope of the position-versus-time graph.

But what about the reverse problem? Can we use the object's velocity to predict its position at some future time t? Equation 2.3, $s_f = s_i + v_s\Delta t$, does this for the case of uniform motion with a constant velocity. We need to find a more general expression that is valid when v_s is not constant.

Figure 2.19a is a velocity-versus-time graph for a particle whose velocity varies with time. Suppose we know the object's position to be s_i at an initial time t_i. Our goal is to find its position s_f at a later time t_f.

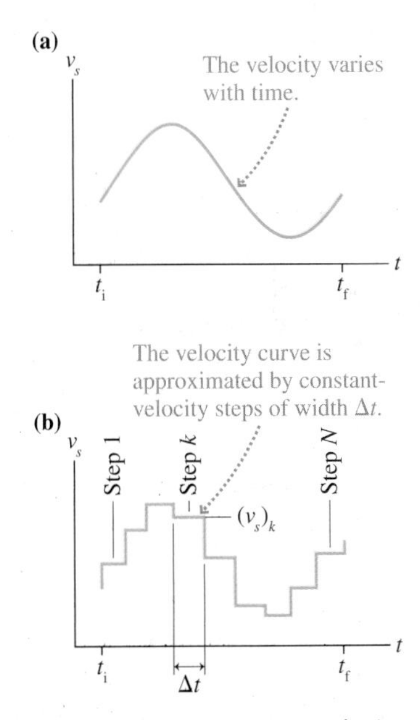

FIGURE 2.19 Approximating a velocity-versus-time graph with a series of constant-velocity steps.

Because we know how to handle constant velocities, using Equation 2.3, let's *approximate* the velocity function of Figure 2.19a as a series of constant-velocity steps of width Δt. This is illustrated in Figure 2.19b. During the first step, from time t_i to time $t_i + \Delta t$, the velocity has the constant value $(v_s)_1$. The velocity is a constant $(v_s)_2$ during the second step from $t_i + \Delta t$ to $t_i + 2\Delta t$, and so on. The velocity during step k has the constant value $(v_s)_k$. Altogether the velocity-versus-time curve has been divided into N constant-velocity steps of equal width Δt. Although the approximation shown in the figure is rather rough, with only nine steps, we can easily imagine that it could be made as accurate as desired by having more and more ever-narrower steps .

The velocity during each step is constant (uniform motion), so we can apply Equation 2.3 to each step. The object's displacement Δs_1 during the first step is simply $\Delta s_1 = (v_s)_1\Delta t$. The displacement during the second step $\Delta s_2 = (v_s)_2\Delta t$, and during step k the displacement is $\Delta s_k = (v_s)_k\Delta t$.

The total displacement of the object between t_i and t_f can be approximated as the sum of the all the individual displacements during each of the N constant-velocity steps. That is,

$$\Delta s = s_f - s_i \approx \Delta s_1 + \Delta s_2 + \cdots + \Delta s_N = \sum_{k=1}^{N} (v_s)_k \Delta t \tag{2.8}$$

where Σ (Greek sigma) is the symbol for summation. With a simple rearrangement, the particle's final position is

$$s_f \approx s_i + \sum_{k=1}^{N} (v_s)_k \Delta t \tag{2.9}$$

Our goal was to use the velocity to find the final position s_f. Equation 2.9 nearly reaches that goal, but Equation 2.9 is only approximate because the constant-velocity steps are only an approximation of the true velocity graph. But if we now let $\Delta t \to 0$, each step's width approaches zero while the total number of steps N approaches infinity. In this limit, the series of steps becomes a perfect replica of the velocity-versus-time graph and Equation 2.9 becomes exact. Thus

$$s_f = s_i + \lim_{\Delta t \to 0} \sum_{k=1}^{N} (v_s)_k \Delta t = s_i + \int_{t_i}^{t_f} v_s \, dt \tag{2.10}$$

The curlicue symbol is called an *integral*. The expression on the right is read, "the integral of $v_s\,dt$ from t_i to t_f." Equation 2.10 is the result that we were seeking. It allows us to predict an object's position s_f at a future time t_f.

We can give Equation 2.10 an important geometric interpretation. Figure 2.20 shows step k in the approximation of the velocity graph as a long, thin rectangle of height $(v_s)_k$ and width Δt. The product $\Delta s_k = (v_s)_k \Delta t$ is the area (base × height) of this small rectangle. The sum in Equation 2.10 adds up all of these rectangular areas to give the total area enclosed between the t-axis and the tops of the steps. The limit of this sum as $\Delta t \to 0$ is the total area enclosed between the t-axis and the velocity curve. This is called the "area under the curve." Thus a graphical interpretation of Equation 2.10 is:

During step k, the product $\Delta s_k = (v_s)_k \Delta t$ is the area of the shaded rectangle.

v_s

Δt

t

t_i

t_f

During the interval t_i to t_f, the total displacement Δs is the "area under the curve."

FIGURE 2.20 The total displacement Δs is the "area under the curve."

$$s_f = s_i + \text{area under the velocity curve } v_s \text{ between } t_i \text{ and } t_f \tag{2.11}$$

NOTE ▶ Wait a minute! The displacement $\Delta s = s_f - s_i$ is a length. How can a length equal an area? Recall earlier, when we found that the velocity is the slope of the position graph, we made a distinction between the *actual* slope and the *physically meaningful* slope? The same distinction applies here. The velocity graph does indeed bound a certain area on the page. That is the actual area, but it is *not* the area to which we are referring. Once again, we need to measure the quantities we are using, v_s and Δt, by referring to the scales on the axes. Δt is some number of seconds while v_s is some number of meters per second. When these are multiplied together, the *physically meaningful* area has units of meters, appropriate for a displacement. The following examples will help make this clear. ◀

EXAMPLE 2.8 The displacement during a drag race

Figure 2.21 shows the velocity-versus-time graph of a drag racer. How far does the racer move during the first 3.0 s?

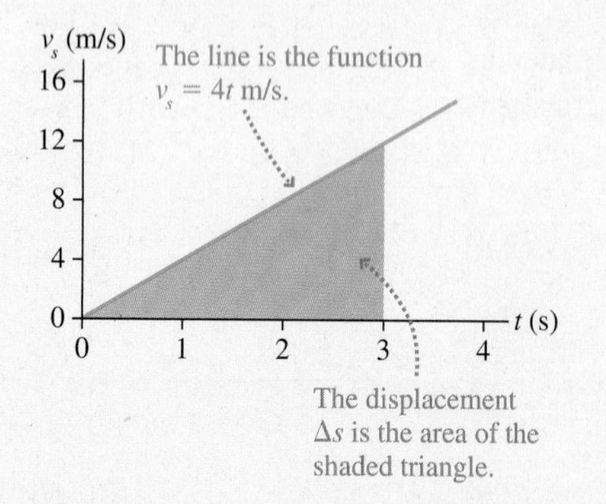

v_s (m/s)
16
12
8
4
0
0
1
2
3
4
t (s)
The line is the function $v_s = 4t$ m/s.
The displacement Δs is the area of the shaded triangle.

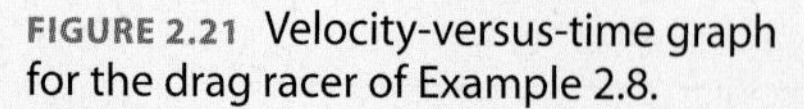

FIGURE 2.21 Velocity-versus-time graph for the drag racer of Example 2.8.

MODEL Represent the drag racer as a particle with a well-defined position at all times.

VISUALIZE Figure 2.21 is a graphical representation of the motion.

SOLVE The question "how far" indicates that we need to find a displacement Δs rather than a position s. According to Equation 2.11, the car's displacement $\Delta s = s_f - s_i$ between $t = 0$ s and $t = 3$ s is the area under the curve from $t = 0$ s to $t = 3$ s. The curve in this case is an angled line, so the area is that of a triangle:

$$\begin{aligned}\Delta s &= \text{area of triangle between } t = 0 \text{ s and } t = 3 \text{ s} \\ &= \tfrac{1}{2} \times \text{base} \times \text{height} \\ &= \tfrac{1}{2} \times 3 \text{ s} \times 12 \text{ m/s} \\ &= 18 \text{ m}\end{aligned}$$

The drag racer moves 18 m during the first 3 seconds.

ASSESS The "area" is a product of s with m/s, so Δs has the proper units of m.

EXAMPLE 2.9 Finding an expression for the racer's position

a. Find an algebraic expression for the position s as a function of time t for the drag racer whose velocity-versus-time graph was shown in Figure 2.21. Assume the car's initial position is $s_i = 0$ m at $t_i = 0$ s.
b. Draw the car's position-versus-time graph.

SOLVE

a. Let $s_i = 0$ at $t_i = 0$ and let s be the position at later time t. The straight line for v_s in Figure 2.21 is described by the linear function $v_s = 4t$ m/s, where t is in s. Then

$$s = s_i + \int_0^t v_s dt = 0 + \text{area under the triangle between 0 and } t$$

$$= 0 + \tfrac{1}{2}(t - 0)(4t - 0)$$

$$= 2t^2 \text{ m, where } t \text{ is in s}$$

b. Figure 2.22 shows the drag racer's position-versus-time graph. It's simply a graph of the function $s = 2t^2$ m, where t is in s. Notice that the *linear* velocity graph of Figure 2.21 is associated with a *parabolic* position graph. This is a general result that we will see again.

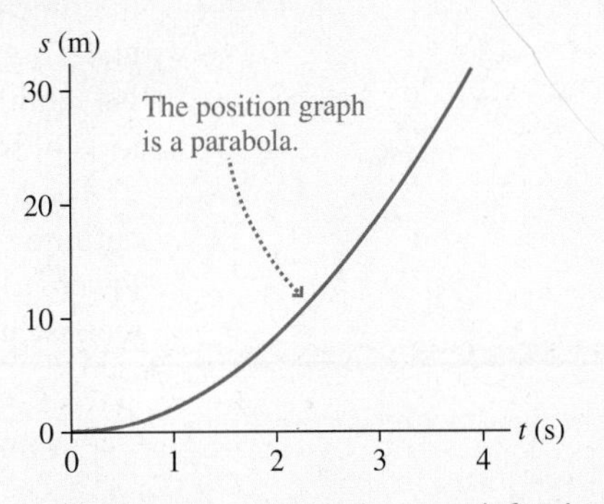

FIGURE 2.22 The position-versus-time graph for the drag racer whose velocity graph was shown in Figure 2.21.

ASSESS This is exactly Example 2.6 in reverse! There we found, by taking the derivative, that a particle whose position is $s = 2t^2$ m has a velocity described by $v_s = 4t$ m/s. Here we have found, by integration, that a drag racer whose velocity is $v_s = 4t$ m/s has a position described by $s = 2t^2$ m.

EXAMPLE 2.10 Finding the turning point

Figure 2.23 is the velocity graph for a particle that starts at $x_i = 30$ m at time $t_i = 0$ s.

a. Draw a motion diagram for the particle.
b. Where is the particle's turning point?
c. At what time does the particle reach the origin?

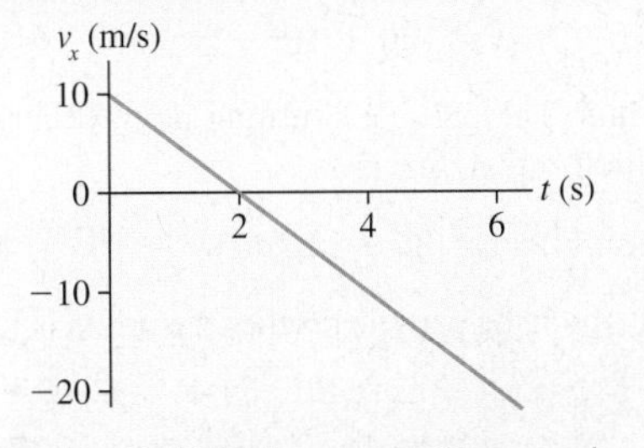

FIGURE 2.23 Velocity-versus-time graph for the particle of Example 2.10.

VISUALIZE The particle is initially 30 m to the right of the origin and moving *to the right* ($v_x > 0$) with a speed of 10 m/s. But v_x is decreasing, so the particle is slowing down. At $t = 2$ s the velocity, just for an instant, is zero before becoming negative. This is the turning point. The velocity is negative for $t > 2$ s, so the particle has reversed direction and moves back toward the origin. At some later time, which we want to find, the particle will pass $x = 0$ m.

SOLVE

a. Figure 2.24 shows the motion diagram. The distance scale will be established in parts b and c but is shown here for convenience.

b. The particle reaches the turning point at $t = 2$ s. To learn *where* it is at that time we need to find the displacement during the first two seconds. We can do this by finding the area under the curve between $t = 0$ s and $t = 2$ s:

$$x(\text{at } t = 2\text{ s}) = x_i + \int_{0\text{ s}}^{2\text{ s}} v_x dt$$

$$= x_i + \text{area under the curve between 0 s and 2 s}$$

$$= 30\text{ m} + \tfrac{1}{2}(2\text{ s} - 0\text{ s})(10\text{ m/s} - 0\text{ m/s})$$

$$= 40\text{ m}$$

The turning point is at $x = 40$ m.

c. The particle needs to move $\Delta x = -40$ m to get from the turning point to the origin. That is, the area under the curve from $t = 2$ s to the desired time t needs to be -40 m. Because the curve is below the axis, with negative values of v_x, the area to the right of $t = 2$ s is a *negative* area. With a bit of geometry, you will find that the triangle with a base extending from $t = 2$ s to $t = 6$ s has an area of -40 m. Thus the particle reaches the origin at $t = 6$ s.

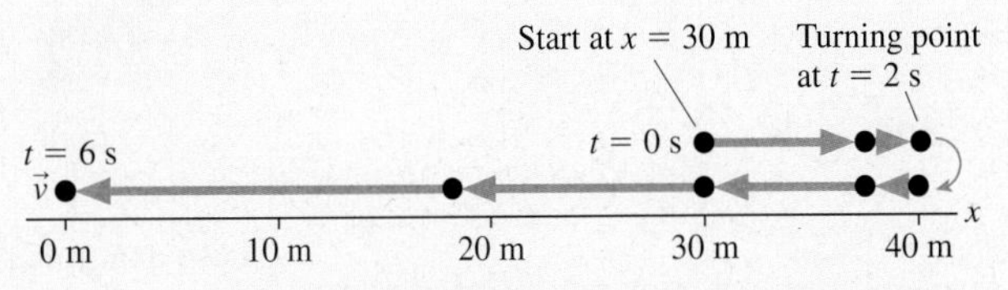

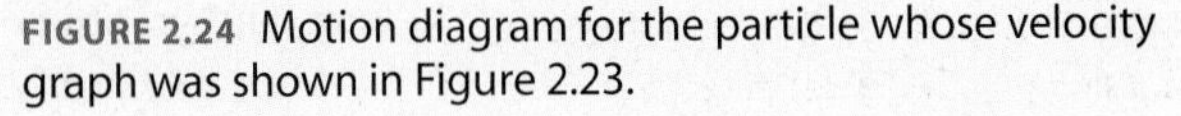

FIGURE 2.24 Motion diagram for the particle whose velocity graph was shown in Figure 2.23.

A Little More Calculus: Integrals

Taking the derivative of a function is equivalent to finding the slope of a graph of the function. Similarly, evaluating an integral is equivalent to finding the area under a graph of the function. The graphical method is very important for building intuition about motion but is limited in its practical application. Just as derivatives of standard functions can be evaluated and tabulated, so can integrals.

The integral in Equation 2.10 is called a *definite integral* because there are two definite boundaries to the area we want to find. These boundaries are called the lower (t_i) and upper (t_f) *limits of integration.* For the important function $u = ct^n$, the essential result from calculus is that

$$\int_{t_i}^{t_f} u\,dt = \int_{t_i}^{t_f} ct^n\,dt = \left.\frac{ct^{n+1}}{n+1}\right|_{t_i}^{t_f} = \frac{ct_f^{\,n+1}}{n+1} - \frac{ct_i^{\,n+1}}{n+1} \quad (n \neq -1) \tag{2.12}$$

The vertical bar in the third step with subscript t_i and superscript t_f is a shorthand notation from calculus that means—as seen in the last step—the integral evaluated at the upper limit t_f *minus* the integral evaluated at the lower limit t_i. You also need to know that for two functions u and w,

$$\int_{t_i}^{t_f} (u + w)\,dt = \int_{t_i}^{t_f} u\,dt + \int_{t_i}^{t_f} w\,dt \tag{2.13}$$

That is, the integral of a sum is equal to the sum of the integrals.

EXAMPLE 2.11 Using calculus to find the position

Use calculus to solve Example 2.10.

SOLVE Figure 2.23 is a linear graph. Its "y-intercept" is seen to be 10 m/s and its slope is −5 (m/s)/s. Thus the velocity graphed here can be described by the equation

$$v_x = (10 - 5t)\text{ m/s}$$

where t is in s. We can find the position x at time t by using Equation 2.10:

$$x = x_i + \int_0^t v_x\,dt = 30\text{ m} + \int_0^t (10 - 5t)\,dt$$

$$= 30\text{ m} + \int_0^t 10\,dt - \int_0^t 5t\,dt$$

We used Equation 2.13 for the integral of a sum to get the final expression. The first integral is a function of the form $u = ct^n$ with $c = 10$ and $n = 0$; the second is of the form $u = ct^n$ with $c = 5$ and $n = 1$. Using Equation 2.12,

$$\int_0^t 10\,dt = 10t\Big|_0^t = 10 \cdot t - 10 \cdot 0 = 10t\text{ m}$$

and

$$\int_0^t 5t\,dt = \tfrac{5}{2}t^2\Big|_0^t = \tfrac{5}{2} \cdot t^2 - \tfrac{5}{2} \cdot 0^2 = \tfrac{5}{2}t^2\text{ m}$$

Combining the pieces gives

$$x = (30 + 10t - \tfrac{5}{2}t^2)\text{ m}$$

where t is in s. The particle's turning point occurs at $t = 2$ s, and its position at that time is

$$x(\text{at } t = 2\text{ s}) = 30 + (10)(2) - \tfrac{5}{2}(2)^2 = 40\text{ m}$$

The time at which the particle reaches the origin is found by setting $x = 0$ m:

$$30 + 10t - \tfrac{5}{2}t^2 = 0$$

This quadratic equation has two solutions: $t = -2$ s or $t = 6$ s.

When we solve a quadratic equation, we cannot just arbitrarily select the root we want. Instead, we must decide which is the *meaningful* root. Here the negative root refers to a time before the problem began, so the meaningful one is the positive root, $t = 6$ s.

ASSESS The results agree with the answers we found previously from a graphical solution.

These examples make the point that there are often many ways to solve a problem. The graphical procedures for finding derivatives and integrals are simple, but they work only for a limited range of problems—those where the geometry is simple. The techniques of calculus are more demanding, but these techniques allow us to deal with functions whose graphs are quite complex.

Summing Up

As you work on building intuition about motion, you need to be able to move back and forth between four different representations of the motion:

- The motion diagram;
- The position-versus-time graph;
- The velocity-versus-time graph;
- The description in words.

Given a description of a certain motion, you should be able to sketch the motion diagram and the position and velocity graphs. Given one graph, you should be able to generate the other. And given position and velocity graphs, you should be able to "interpret" them by describing the motion in words or in a motion diagram.

STOP TO THINK 2.3 Which position-versus-time graph goes with the velocity-versus-time graph on the left? The particle's position at $t_i = 0$ s is $x_i = -10$ m.

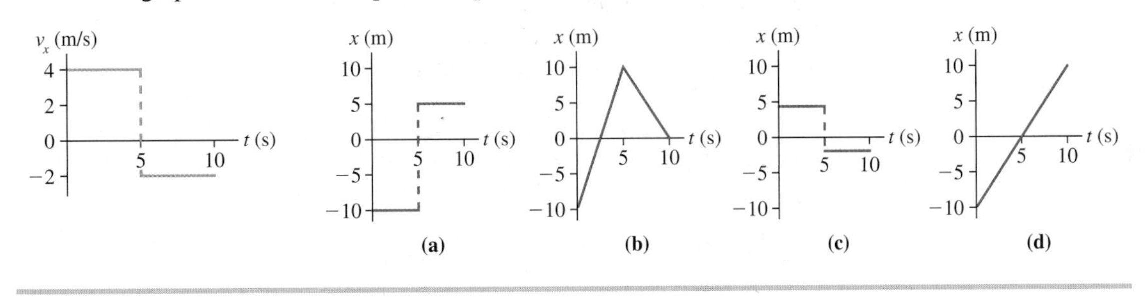

2.5 Motion with Constant Acceleration

We need one more major concept to describe one-dimensional motion: acceleration. Acceleration, as we noted in Chapter 1, is a rather abstract concept. You cannot "see" the value of acceleration, as you can that of position, nor can you judge it by looking to see if an object is moving quickly or slowly. Nonetheless, acceleration is the linchpin of mechanics. We will see very shortly that Newton's laws relate the acceleration of an object to the forces that are exerted on it.

Let's conduct a race between a Volkswagen Beetle and a Porsche to see which can achieve a velocity of 30 m/s ($\approx$60 mph) in the shortest time. Both cars are equipped with computers that will record the speedometer reading 10 times each second. This gives a nearly continuous record of the *instantaneous* velocity of each car. Table 2.2 shows some of the data. The velocity-versus-time graphs, based on these data, are shown in Figure 2.25.

TABLE 2.2 Velocities of a Porsche and a Volkswagen Beetle

t(s)	$v_{Porsche}$ (m/s)	v_{VW} (m/s)
0.0	0.0	0.0
0.1	0.5	0.2
0.2	1.0	0.4
0.3	1.5	0.6
0.4	2.0	0.8
⋮	⋮	⋮

How can we describe the difference in performance of the two cars? It is not that one has a different velocity from the other; both achieve every velocity between 0 and 30 m/s. The distinction is how long it took each to *change* its velocity from 0 to 30 m/s. The Porsche changed velocity quickly, in 6 s, while the VW needed 15 s to make the same velocity change. This suggests that the distinction is, once again, a *rate*.

In this case, as we compare the two cars, we are looking at the rate at which their velocities change. Because the Porsche had a velocity change $\Delta v_s = 30$ m/s during a time interval $\Delta t = 6$ s, the *rate* at which its velocity changed was

$$\text{rate of velocity change} = \frac{\Delta v_s}{\Delta t} = \frac{30 \text{ m/s}}{6.0 \text{ s}} = 5.0 \text{ (m/s)/s} \tag{2.14}$$

Notice the units. They are units of "velocity per second." A rate of velocity change of 5.0 "meters per second per second" means that the velocity increases

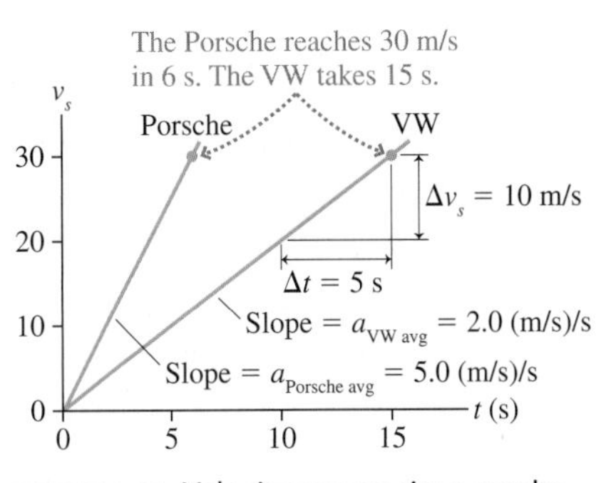

FIGURE 2.25 Velocity-versus-time graphs for the Porsche and the VW Beetle.

by 5.0 m/s during the first second, by another 5.0 m/s during the next second, and so on. In fact, the velocity will increase by 5.0 m/s during any second in which it is changing at the rate of 5.0 (m/s)/s.

Chapter 1 introduced *acceleration* as "the rate of change of velocity." That is, acceleration measures how quickly or slowly an object's velocity changes. The Porsche's velocity changed quickly, so it had a large acceleration. The VW's velocity changed more slowly, so its acceleration was less. In parallel with our treatment of velocity, let's call this the **average acceleration** a_{avg} during the time interval Δt:

$$a_{avg} \equiv \frac{\Delta v_s}{\Delta t} \text{ (average acceleration)} \tag{2.15}$$

Because Δv_s and Δt are the "rise" and "run" of a velocity-versus-time graph, we see that a_{avg} can be interpreted graphically as the *slope* of a straight-line velocity-versus-time graph. Figure 2.25 uses this idea to show that the VW's average acceleration is

$$a_{VW\,avg} = \frac{\Delta v_s}{\Delta t} = \frac{10 \text{ m/s}}{5.0 \text{ s}} = 2.0 \text{ (m/s)/s} \tag{2.16}$$

This is less than the acceleration of the Porsche, as expected.

1.2, 1.3 ActivPhysics ONLINE

An object whose velocity-versus-time graph is a straight-line graph has a steady and unchanging acceleration. Such a graph represents motion with *constant acceleration,* which we call **uniformly accelerated motion: An object has uniformly accelerated motion if and only if its acceleration a_s is constant and unchanging. The object's velocity-versus-time graph is a straight line, and a_s is the slope of the line.** There's no need to specify "average" if the acceleration is constant, so we'll use the symbol a_s as we discuss motion along the s-axis with constant acceleration.

NOTE ► An important aspect of acceleration is its *sign.* Acceleration $\vec{a}$, like position $\vec{r}$ and velocity $\vec{v}$, is a vector. For motion in one dimension the sign of a_x (or a_y) is positive if the vector $\vec{a}$ points to the right (or up), negative if it points to the left (or down). This was illustrated in Figure 2.1, which you may wish to review. It's particularly important to emphasize that positive and negative values of a_s do *not* correspond to "speeding up" and "slowing down." ◄

EXAMPLE 2.12 Relating acceleration to velocity

a. A particle has a velocity of 10 m/s and a constant acceleration of 2 (m/s)/s. What is its velocity 1 s later? 2 s later?
b. A particle has a velocity of −10 m/s and a constant acceleration of 2 (m/s)/s. What is its velocity 1 s later? 2 s later?

SOLVE

a. An acceleration of 2 (m/s)/s *means* that the velocity increases by 2 m/s every 1 s. If the particle's initial velocity is 10 m/s, then 1 s later its velocity will be 12 m/s. After 2 s, which is 1 additional second later, it will increase by another 2 m/s to 14 m/s. After 3 s it will be 16 m/s. Here a positive a_s is causing the particle to speed up.
b. If the particle's initial velocity is a *negative* −10 m/s but the acceleration is a positive +2 (m/s)/s, then 1 s later the velocity will be −8 m/s. After 2 s it will be −6 m/s, and so on. In this case, a positive a_s is causing the object to *slow down* (decreasing speed v). This agrees with our rule from Section 2.1: An object is slowing down if and only if v_s and a_s have opposite signs.

NOTE ► It is customary to abbreviate the acceleration units (m/s)/s as m/s^2. For example, the particles in Example 2.12 had an acceleration of 2 m/s^2. We will use this notation, but keep in mind the *meaning* of the notation as "(meters per second) per second." ◄

EXAMPLE 2.13 Running the court

A basketball player starts at the left end of the court and moves with the velocity shown in Figure 2.26. Draw a motion diagram and an acceleration-versus-time graph for the basketball player.

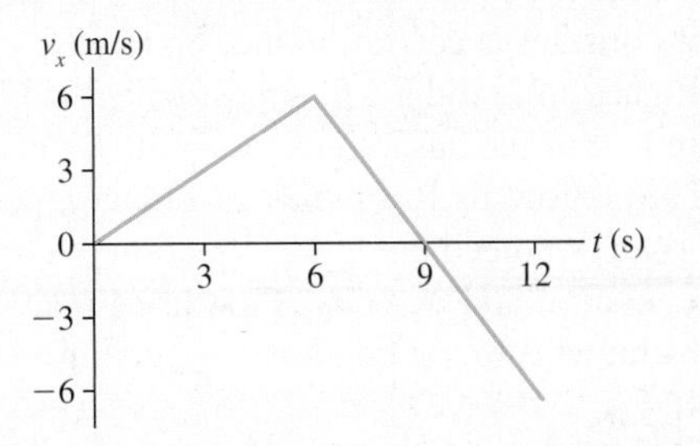

FIGURE 2.26 Velocity-versus-time graph for the basketball player of Example 2.13.

VISUALIZE The velocity is positive (motion to the right) and increasing for the first 6 seconds, so the velocity arrows in the motion diagram are to the right and getting longer. From $t = 6$ s to 9 s the motion is still to the right (v_x is still positive), but the arrows are getting shorter because v_x is decreasing. There's a turning point at $t = 9$ s, when $v_x = 0$, and after that the motion is to the left (v_x is negative) and getting faster. The motion diagram of Figure 2.27a shows the velocity vectors and the acceleration vectors.

SOLVE Acceleration is the slope of the velocity graph. For the first 6 s, the slope has the constant value

$$a_x = \frac{\Delta v_x}{\Delta t} = \frac{6.0 \text{ m/s}}{6.0 \text{ s}} = 1.0 \text{ m/s}^2$$

The velocity decreases by 12 m/s during the 6-s interval from $t = 6$ s to $t = 12$ s, so

$$a_x = \frac{\Delta v_x}{\Delta t} = \frac{-12 \text{ m/s}}{6.0 \text{ s}} = -2.0 \text{ m/s}^2$$

The acceleration graph for these 12 s is shown in Figure 2.27b. Although there are two segments of the motion, each segment is uniformly accelerated motion with constant acceleration. Notice that there is no change in the acceleration at $t = 9$ s, the turning point.

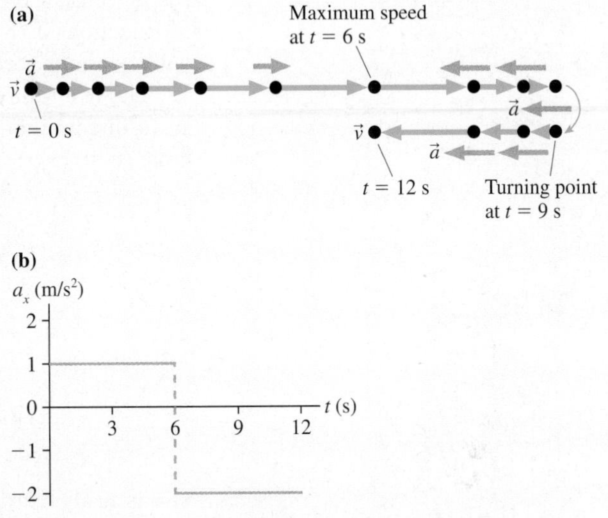

FIGURE 2.27 Motion diagram and acceleration graph for Example 2.13.

ASSESS The *sign* of a_x does *not* tell us whether or not the object is speeding up or slowing down. The basketball player is slowing down from $t = 6$ s to $t = 9$ s, then speeding up from $t = 9$ s to $t = 12$ s. Nonetheless, his acceleration is negative during this entire interval because his acceleration vector, as seen in the motion diagram, always points to the left.

The Kinematic Equations of Constant Acceleration

Consider an object whose acceleration a_s remains constant during the time interval $\Delta t = t_f - t_i$. At the beginning of this interval, at time t_i, the object has initial velocity v_{is} and initial position s_i. Note that t_i is often zero, but it does not have to be. Figure 2.28a shows the acceleration-versus-time graph. It is a horizontal line between t_i and t_f, indicating a *constant* acceleration.

The object's velocity is changing because the object is accelerating. It is not hard to find the object's velocity v_{fs} at a later time t_f. By definition,

$$a_s = \frac{\Delta v_s}{\Delta t} = \frac{v_{fs} - v_{is}}{\Delta t} \tag{2.17}$$

which is easily rearranged to give

$$v_{fs} = v_{is} + a_s \Delta t \tag{2.18}$$

The velocity-versus-time graph, shown in Figure 2.28b, is a straight line that starts at v_{is} and has slope a_s.

We would also like to know the object's position s_f at time t_f. As you learned in the last section,

$$s_f = s_i + \text{area under the velocity curve } v_s \text{ between } t_i \text{ and } t_f \tag{2.19}$$

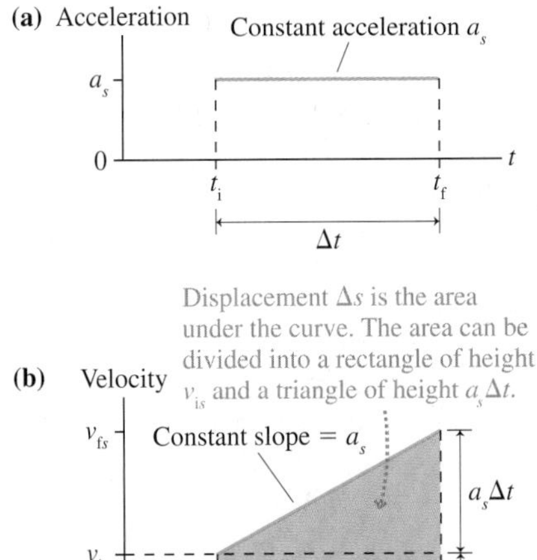

FIGURE 2.28 Acceleration and velocity graphs for motion with constant acceleration.

The shaded area in Figure 2.28b can be subdivided into a rectangle of area $v_{is}\Delta t$ and a triangle of area $\frac{1}{2}(a_s\Delta t)(\Delta t) = \frac{1}{2}a_s(\Delta t)^2$. Adding these gives

$$s_f = s_i + v_{is}\Delta t + \tfrac{1}{2}a_s(\Delta t)^2 \tag{2.20}$$

where $\Delta t = t_f - t_i$ is the elapsed time. The quadratic dependence on Δt causes the position-versus-time graph for constant-acceleration motion to have a parabolic shape. You saw this earlier in Figure 2.22, and it will appear below in Figure 2.29.

Equations 2.18 and 2.20 are two of the basic kinematic equations for motion with *constant* acceleration. They allow us to predict an object's position and velocity at a future instant of time. We need one more equation to complete our set, a direct relation between position and velocity. First use Equation 2.18 to write $\Delta t = (v_{fs} - v_{is})/a_s$. Substitute this into Equation 2.20, giving

$$\begin{aligned} s_f &= s_i + v_{is}\left(\frac{v_{fs} - v_{is}}{a_s}\right) + \tfrac{1}{2}a_s\left(\frac{v_{fs} - v_{is}}{a_s}\right)^2 \\ &= s_i + \left(\frac{v_{is}v_{fs}}{a_s} - \frac{v_{is}^2}{a_s}\right) + \left(\frac{v_{fs}^2}{2a_s} - \frac{v_{is}v_{fs}}{a_s} + \frac{v_{is}^2}{2a_s}\right) \\ &= s_i + \frac{v_{fs}^2 - v_{is}^2}{2a_s} \end{aligned} \tag{2.21}$$

This is easily rearranged to read

$$v_{fs}^2 = v_{is}^2 + 2a_s\,\Delta s \tag{2.22}$$

where $\Delta s = s_f - s_i$ is the *displacement* (not the distance!).

Equations 2.18, 2.20, and 2.22, which are summarized in Table 2.3, are the key results for motion with constant acceleration.

TABLE 2.3 The kinematic equations for motion with constant acceleration

$v_{fs} = v_{is} + a_s\Delta t$
$s_f = s_i + v_{is}\,\Delta t + \tfrac{1}{2}a_s(\Delta t)^2$
$v_{fs}^2 = v_{is}^2 + 2a_s\,\Delta s$

Figure 2.29 is a comparison of motion with constant velocity (uniform motion) and motion with constant acceleration (uniformly accelerated motion). Notice that uniform motion is really a special case of uniformly accelerated motion in which the constant acceleration happens to be zero. The graphs for a negative acceleration are left as an exercise.

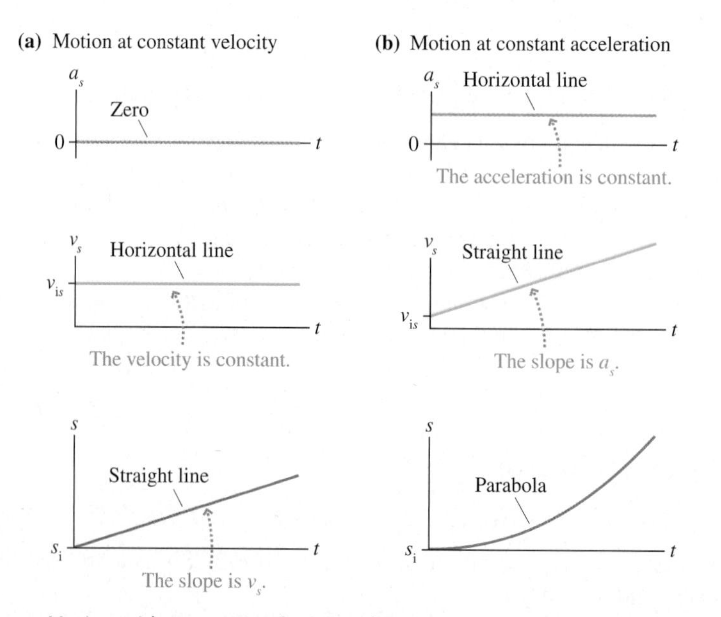

FIGURE 2.29 Motion with constant velocity and constant acceleration. These graphs assume $s_i = 0$, $v_{is} > 0$, and (for constant acceleration) $a_s > 0$.

A Problem-Solving Strategy

This information can be assembled into a problem-solving strategy for kinematics with constant acceleration.

PROBLEM-SOLVING STRATEGY 2.1 Kinematics with constant acceleration

MODEL Use the particle model. Make simplifying assumptions.

VISUALIZE Use different representations of the information in the problem.

- Draw a *motion diagram.* Motion diagrams are part of the physical representation.
- Draw a *pictorial representation.* This helps you assess the information you are given and starts the process of translating the problem into symbols.
- Use a *graphical representation* if it is appropriate for the problem.
- Go back and forth between these three representations as needed.

SOLVE The mathematical representation is based on the three kinematic equations

$$v_{fs} = v_{is} + a_s \Delta t$$

$$s_f = s_i + v_{is} \Delta t + \tfrac{1}{2} a_s (\Delta t)^2$$

$$v_{fs}^2 = v_{is}^2 + 2a_s \Delta s$$

- Use x or y, as appropriate to the problem, rather than the generic s.
- Replace i and f with numerical subscripts defined in the pictorial representation.
- Uniform motion with constant velocity has $a_s = 0$.

ASSESS Is your result believable? Does it have proper units? Does it make sense?

EXAMPLE 2.14 The motion of a rocket sled

A rocket sled accelerates at 50 m/s^2 for 5.0 s, coasts for 3.0 s, then deploys a braking parachute and decelerates at 3.0 m/s^2 until coming to a halt.

a. What is the maximum velocity of the rocket sled?
b. What is the total distance traveled?

MODEL Represent the rocket sled as a particle.

VISUALIZE Figure 2.30 shows the physical and pictorial representations. Recall that we discussed the first two-thirds of this problem as Example 1.9 in Chapter 1.

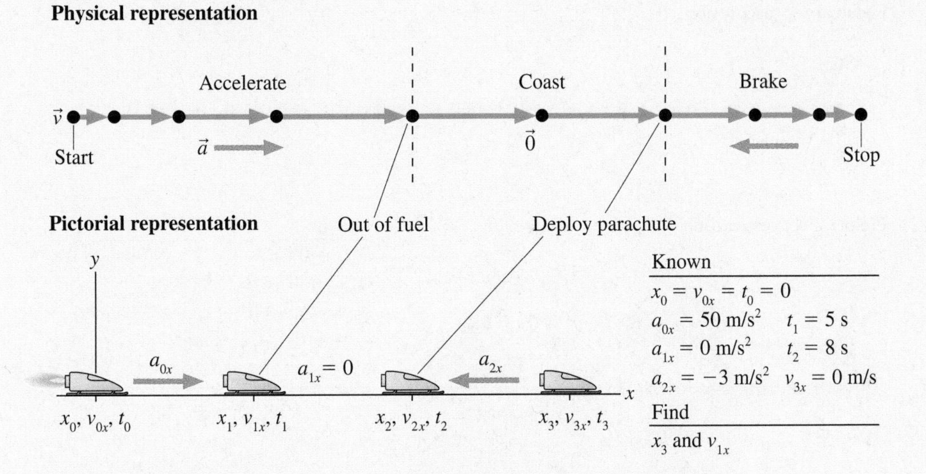

FIGURE 2.30 The physical and pictorial representations of the rocket sled.

SOLVE

a. The maximum velocity is identified in the pictorial representation as v_{1x}, the velocity at time t_1 when the acceleration phase ends. The first kinematic equation in Table 2.3 gives

$$v_{1x} = v_{0x} + a_{0x}(t_1 - t_0) = a_{0x}t_1$$
$$= (50 \text{ m/s}^2)(5.0 \text{ s}) = 250 \text{ m/s}$$

We started with the complete equation, then simplified by noting which terms were zero.

b. Finding the total distance requires several steps. First, the sled's position when the acceleration ends at t_1 is found from the second equation in Table 2.3:

$$x_1 = x_0 + v_{0x}(t_1 - t_0) + \tfrac{1}{2}a_{0x}(t_1 - t_0)^2 = \tfrac{1}{2}a_{0x}t_1^2$$
$$= \tfrac{1}{2}(50 \text{ m/s}^2)(5.0 \text{ s})^2 = 625 \text{ m}$$

During the coasting phase, which is uniform motion with no acceleration ($a_{1x} = 0$),

$$x_2 = x_1 + v_{1x}\Delta t = x_1 + v_{1x}(t_2 - t_1)$$
$$= 625 \text{ m} + (250 \text{ m/s})(3.0 \text{ s}) = 1375 \text{ m}$$

Notice that, in this case, Δt is not simply t. The braking phase is a little different because we don't know how long it lasts. But we do know that the sled ends with $v_{3x} = 0$ m/s, so we can use the third equation in Table 2.3:

$$v_{3x}^2 = v_{2x}^2 + 2a_{2x}\Delta x = v_{2x}^2 + 2a_{2x}(x_3 - x_2)$$

This can be solved for x_3:

$$x_3 = x_2 + \frac{v_{3x}^2 - v_{2x}^2}{2a_{2x}}$$
$$= 1375 \text{ m} + \frac{0 - (250 \text{ m/s})^2}{2(-3.0 \text{ m/s}^2)} = 11{,}800 \text{ m}$$

ASSESS Using the approximate conversion factor 1 m/s $\approx$ 2 mph from Table 1.4, we see that the top speed is $\approx$ 500 mph. The total distance traveled is $\approx$ 12 km $\approx$ 7 mi. This is reasonable because it takes a very long distance to stop from a top speed of 500 mph!

NOTE ► We used explicit numerical subscripts throughout the mathematical representation, each referring to a symbol that was defined in the pictorial representation. The subscripts i and f in the Table 2.3 equations are just generic "place holders" and don't have unique values. During the acceleration phase we had i = 0 and f = 1. Later, during the coasting phase, these became i = 1 and f = 2. The numerical subscripts have a clear meaning and are less likely to lead to confusion. ◄

EXAMPLE 2.15 Friday night football

Fred catches the football while standing directly on the goal line. He immediately starts running forward with an acceleration of 6 ft/s². At the moment the catch is made, Tommy is 20 yards away and heading directly toward Fred with a steady speed of 15 ft/s. If neither deviates from a straight-ahead path, where will Tommy tackle Fred?

MODEL Represent Fred and Tommy as particles.

VISUALIZE This problem statement was analyzed in Example 1.11. The physical and pictorial representations are shown again in Figure 2.31.

SOLVE We want to find *where* Fred and Tommy have the same position. The pictorial representation designates time t_1 as *when* they meet. The axes have been chosen so that Fred starts at

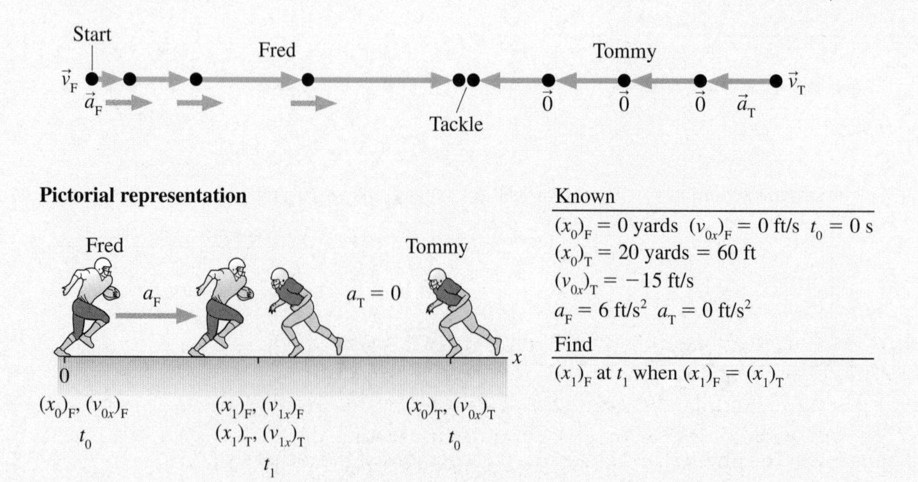

FIGURE 2.31 The physical and pictorial representations for Example 2.15.

$(x_0)_F = 0$ ft and moves to the right while Tommy starts at $(x_0)_T = 60$ ft and runs to the left with a *negative* velocity. The first equation of Table 2.3 allows us to find their positions at time t_1. These are:

$$(x_1)_F = (x_0)_F + (v_{0x})_F(t_1 - t_0) + \tfrac{1}{2}(a_x)_F(t_1 - t_0)^2$$
$$= \tfrac{1}{2}(a_x)_F t_1^2$$
$$(x_1)_T = (x_0)_T + (v_{0x})_T(t_1 - t_0) + \tfrac{1}{2}(a_x)_T(t_1 - t_0)^2$$
$$= (x_0)_T + (v_{0x})_T t_1$$

Notice that Tommy's position equation contains the term $(v_{0x})_T t_1$, not $-(v_{0x})_T t_1$. The fact that he is moving to the left has already been considered in assigning a *negative value* to $(v_{0x})_T$, hence we don't want to add any additional negative signs in the equation. If we now set $(x_1)_F$ and $(x_1)_T$ equal to each other, indicating the point of the tackle, we can solve for t_1:

$$\tfrac{1}{2}(a_x)_F t_1^2 = (x_0)_T + (v_{0x})_T t_1$$
$$\tfrac{1}{2}(a_x)_F t_1^2 - (v_{0x})_T t_1 - (x_0)_T = 0$$
$$3t_1^2 + 15t_1 - 60 = 0$$

The solutions of this quadratic equation for t_1 are $t_1 = (-7.62\text{ s}, +2.62\text{ s})$. The negative time is not meaningful in this problem, so the time of the tackle is $t_1 = 2.62$ s. Using this to compute $(x_1)_F$ gives

$$(x_1)_F = \tfrac{1}{2}(a_x)_F t_1^2 = 20.6\text{ feet} = 6.9\text{ yards}$$

Tommy makes the tackle at just about the 7-yard line!

ASSESS The answer had to be between 0 yards and 20 yards. Because Tommy was already running, whereas Fred started from rest, it is reasonable that Fred will cover less than half the 20-yard separation before meeting Tommy. Thus 6.9 yards is a reasonable answer.

NOTE ▶ The purpose of the assessment step is not to prove that an answer must be right but to rule out answers that, with a little thought, are clearly wrong. ◀

It is worth exploring Example 2.15 graphically. Figure 2.32 shows position-versus-time graphs for Fred and Tommy. The curves intersect at $t = 2.62$ s, and that is where the tackle occurs. You should compare this problem to Example 2.3 and Figure 2.9 for Bob and Susan to notice the similarities and the differences.

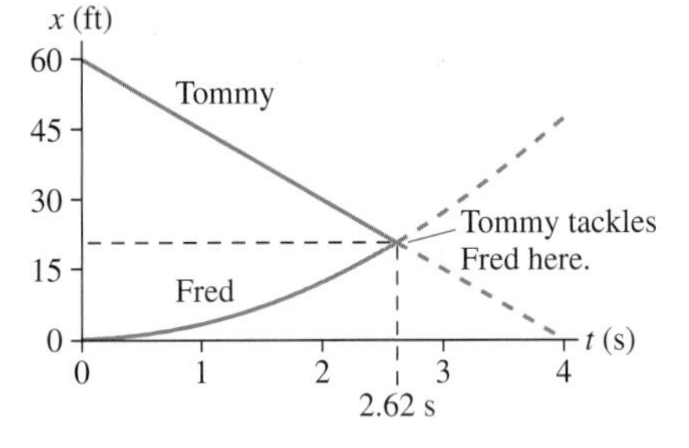

FIGURE 2.32 Position-versus-time graphs for Fred and Tommy.

STOP TO THINK 2.4 Which velocity-versus-time graph or graphs goes with this acceleration-versus-time graph? The particle is initially moving to the right.

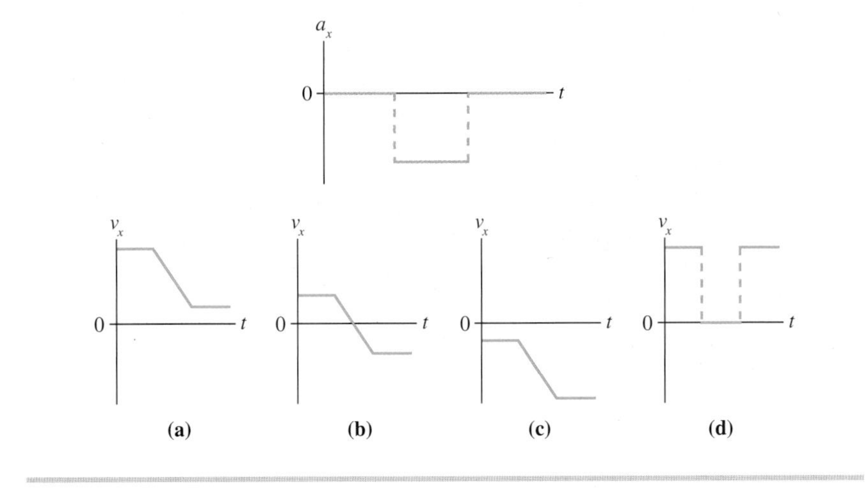

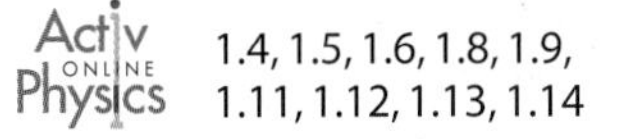

2.6 Free Fall

The motion of an object moving under the influence of gravity only, and no other forces, is called **free fall.** Strictly speaking, free fall occurs only in a vacuum, where there is no air resistance. Fortunately, the effect of air resistance is small for "heavy objects," so we'll make only a very slight error in treating these objects *as if* they were in free fall. For very light objects, such as a feather, or for objects that fall through very large distances and gain very high speeds, the effect

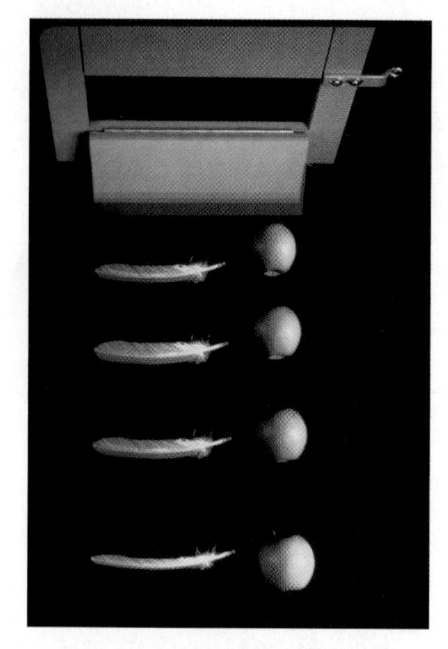

In the absence of air resistance, any two objects fall at the same rate and hit the ground at the same time. The apple and feather in this photograph are falling in a vacuum.

of air resistance is *not* negligible. Motion with air resistance is a problem we will study in Chapter 5. Until then, we will restrict our attention to "heavy objects" and will make the reasonable assumption that falling objects are in free fall.

The motion of falling objects has interested scientists since antiquity. The ancient Greek scientist and philosopher Aristotle asserted that heavier objects "fall faster" than do light objects. After all, a rock falls to the ground much more quickly than a feather. Aristotle's claim was based on casual observations and on what he thought "should" happen rather than on actual experiments.

Galileo, in the 17th century, was the first to challenge Aristotle and to put Aristotle's assertion to a rigorous experimental test. The story of Galileo dropping different weights from the leaning bell tower at the cathedral in Pisa is well known, although historians cannot confirm its truth. But bell towers were common in the Italy of Galileo's day, so he had ample opportunity to make the measurements and observations that he describes in his writings.

Careful observations show that falling objects *don't* "hit the ground" at the same time. There are slight differences in the arrival times, but Galileo correctly identified these differences as due to air resistance. He then formulated a general conclusion for an idealized situation of motion in a vacuum. In doing so, Galileo developed a *model* of motion—motion in the absence of air resistance—that could only be approximated by any real object. It was Galileo's innovative use of experiments, models, and mathematics that made him the first "modern" scientist.

Galileo's discovery can be summarized as follows:

- Two objects dropped from the same height will, if air resistance can be neglected, hit the ground at the same time and with the same speed.
- Consequently, **any two objects in free fall, regardless of their mass, have the same acceleration $\vec{a}_{\text{free fall}}$.** This is an especially important conclusion.

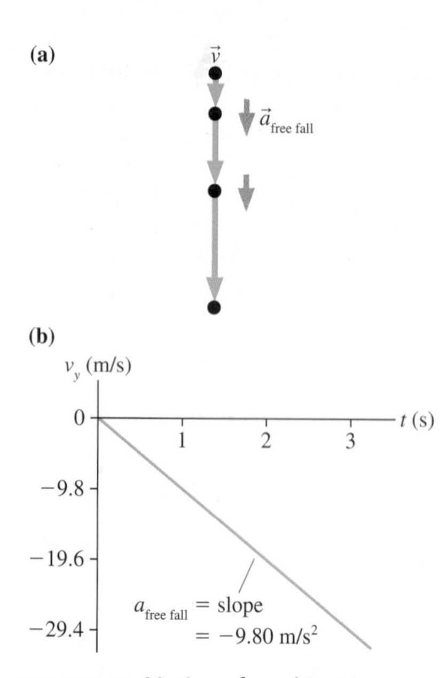

FIGURE 2.33 Motion of an object in free fall.

Figure 2.33a shows the motion diagram of an object that was released from rest and falls freely. Figure 2.33b shows the object's velocity graph. The motion diagram and graph are identical for a falling pea and a falling boulder. The acceleration $a_{\text{free fall}}$ is easily found from the slope of the velocity graph. Careful measurements show that the value of the free-fall acceleration varies ever-so-slightly at different places on the earth, due to the slightly nonspherical shape of the earth and to the fact that the earth is rotating. A global average, at sea level, is

$$\vec{a}_{\text{free fall}} = (9.80 \text{ m/s}^2, \text{vertically downward}) \tag{2.23}$$

where *vertically downward* means along a line toward the center of the earth.

The length, or magnitude, of the free-fall acceleration is known as the **acceleration due to gravity,** and has the special symbol g:

$$g = 9.80 \text{ m/s}^2 \quad \text{(acceleration due to gravity)}$$

Several points about free fall are worthy of note:

- g, by definition, is *always* positive. **There will never be a problem that will use a negative value for g.** But, you say, objects fall when you release them rather than rise, so how can g be positive?
- g is *not* the acceleration $a_{\text{free fall}}$, but simply its magnitude. Because we've chosen the y-axis to point vertically up, the downward acceleration vector $\vec{a}_{\text{free fall}}$ has the one-dimensional acceleration

$$a_y = a_{\text{free fall}} = -g \tag{2.24}$$

  It is a_y that is negative, not g.
- Because free fall is motion with constant acceleration, we can use the kinematic equations of Table 2.3 with the acceleration being due to gravity, $a_y = -g$.

- g is not called "gravity." Gravity is a force, not an acceleration. g is *the acceleration due to gravity.*
- $g = 9.80\text{ m/s}^2$ only on earth. Other planets have different values of g. You will learn in Chapter 12 how to determine g for other planets.

NOTE ▶ Despite the name, free fall is not restricted to objects that are literally falling. Any object moving under the influence of gravity only, and no other forces, is in free fall. This includes objects falling straight down, objects that have been tossed or shot straight up, and projectile motion. This chapter considers only objects that move up and down along a vertical line; projectile motion will be studied in Chapter 6. ◀

ActivPhysics ONLINE 1.7, 1.10

EXAMPLE 2.16 A falling rock

A rock is released from rest at the top of a 100-m-tall building. How long does the rock take to fall to the ground, and what is its impact velocity?

MODEL Represent the rock as a particle. Assume air resistance is negligible.

VISUALIZE Figure 2.34 shows the physical and pictorial representations. We have placed the origin at the ground, which makes $y_0 = 100\text{ m}$. Although the rock falls 100 m, it is important to notice that the *displacement* is $\Delta y = y_1 - y_0 = -100\text{ m}$.

SOLVE Free fall is motion with the specific constant acceleration $a_y = -g$. The first question involves a relation between time and distance, so only the second equation in Table 2.3 is relevant. Using $v_{0y} = 0\text{ m/s}$ and $t_0 = 0\text{ s}$, we find

$$y_1 = y_0 + v_{0y}\Delta t + \tfrac{1}{2}a_y\Delta t^2 = y_0 + v_{0y}\Delta t - \tfrac{1}{2}g\Delta t^2 = y_0 - \tfrac{1}{2}gt_1^2$$

We can now solve for t_1, finding:

$$t_1 = \sqrt{\frac{2(y_0 - y_1)}{g}} = \sqrt{\frac{2(100\text{ m} - 0\text{ m})}{9.80\text{ m/s}^2}} = \pm 4.52\text{ s}$$

The $\pm$ sign indicates that there are two mathematical solutions; therefore we have to use physical reasoning to choose between them. A negative t_1 would refer to a time before we dropped the rock, so we select the positive root: $t_1 = 4.52\text{ s}$.

Now that we know the fall time, we can use the first kinematic equation to find v_{1y}:

$$v_{1y} = v_{0y} - g\Delta t = -gt_1 = -(9.80\text{ m/s}^2)(4.52\text{ s}) = -44.3\text{ m/s}.$$

Alternatively, we could work directly from the third kinematic equation:

$$v_{1y} = \sqrt{v_{0y}^2 - 2g\Delta y} = \sqrt{-2g(y_1 - y_0)} = \sqrt{-2(9.80\text{ m/s}^2)(0\text{ m} - 100\text{ m})} = \pm 44.3\text{ m/s}$$

This method is useful if you don't know Δt. However, we must again choose the correct sign of the square root. Because the velocity vector points downward, the sign of v_y has to be negative. Thus $v_{1y} = -44.3\text{ m/s}$. The importance of careful attention to the signs cannot be overemphasized!

A common error would be to say "The rock fell 100 m, so $\Delta y = 100\text{ m}$." This would have you trying to take the square root of a negative number. As noted above, Δy is not a distance. It is a *displacement,* with a carefully defined meaning of $y_f - y_i$. In this case, $\Delta y = y_1 - y_0 = -100\text{ m}$.

ASSESS Are the answers reasonable? Well, 100 m is about 300 feet, which is about the height of a 30-floor building. How long does it take something to fall 30 floors? Four or five seconds seems pretty reasonable. How fast would it be going at the bottom? Using $1\text{ m/s} \approx 2\text{ mph}$, we find that $44.3\text{ m/s} \approx 90\text{ mph}$. That also seems pretty reasonable after falling 30 floors. Had we misplaced a decimal point, though, and found 443 m/s, we would be suspicious when we converted this to ≈ 900 mph! The answers all seem reasonable.

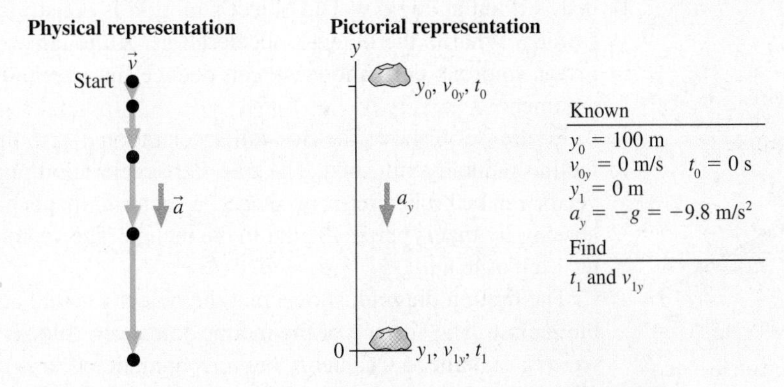

FIGURE 2.34 Physical and pictorial representations of a falling rock.

EXAMPLE 2.17 A vertical cannonball

A cannonball is shot straight up with an initial speed of 100 m/s. How high does it go?

MODEL Represent the cannonball as a particle. Assume air resistance is negligible.

VISUALIZE Figure 2.35 shows the physical and pictorial representations for the cannonball's motion. Even though the ball was shot upward, this is a free-fall problem because the ball (after being launched) is moving under the influence of gravity *only*. A critical aspect of the problem is knowing where it ends. How do we put "how high" into symbols? The clue is that the very top point of the trajectory is a *turning point*. Recall that the instantaneous velocity at a turning point is $v = 0$. Thus we can characterize the "top" of the trajectory as the point where $v_{1y} = 0$ m/s. This was not explicitly stated but is part of our interpretation of the problem.

SOLVE We are looking for a relationship between distance and velocity, without knowing the time interval. This relationship is described mathematically by the third kinematic equation in Table 2.3. Using $y_0 = 0$ m and $v_{1y} = 0$ m/s, we have

$$v_{1y}{}^2 = 0 = v_{0y}{}^2 - 2g\Delta y = v_{0y}{}^2 - 2gy_1$$

Solving for y_1, we find that the cannonball reaches a height

$$y_1 = \frac{v_{0y}{}^2}{2g} = \frac{(100 \text{ m/s})^2}{2(9.80 \text{ m/s}^2)} = 510 \text{ m}$$

ASSESS Is this answer reasonable? A speed of 100 m/s is ≈200 mph—that's pretty fast! The calculated height is 510 m ≈ 1500 ft. In Example 2.16 we found that an object dropped from 100 m is going 44 m/s when it hits the ground, so it seems reasonable that an object shot upward at 100 m/s will go significantly higher than 100 m. While we cannot say that 510 m is necessarily better than 400 m or 600 m, we can say that it is not unreasonable. The point of the assessment is not to prove that the answer *has* to be right, but to find answers that are obviously wrong.

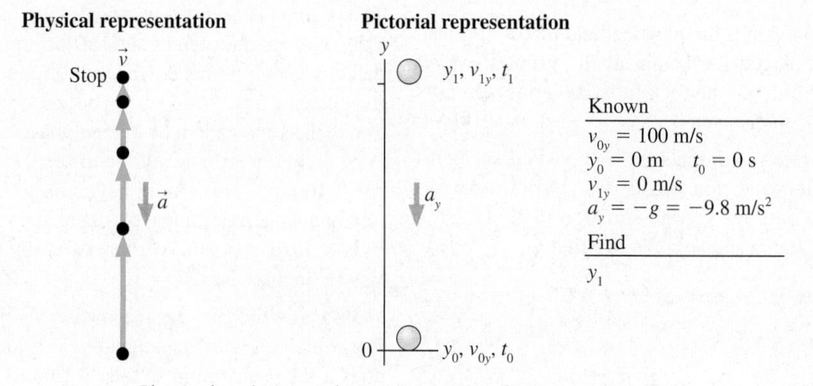

FIGURE 2.35 Physical and pictorial representations for Example 2.17.

2.7 Motion on an Inclined Plane

A problem closely related to free fall is that of an object moving down a straight, but frictionless, inclined plane, such as a skier going down a slope on frictionless snow. In practice, we can come very close to this ideal by having a ball rolling on a very smooth surface.

Figure 2.36a shows an object accelerating down a frictionless, inclined plane that is tilted at angle θ. The object's motion is constrained to be parallel to the surface. What is the object's acceleration? Although we're not yet prepared to give a rigorous derivation, we can deduce the acceleration with a plausibility argument.

Figure 2.36b shows the free-fall acceleration $\vec{a}_{\text{free fall}}$ the ball would have if the incline suddenly vanished. The free-fall acceleration points straight down. This vector can be broken into two pieces: a vector $\vec{a}_\parallel$ that is parallel to the incline and a vector $\vec{a}_\perp$ that is perpendicular to the incline. The vector addition rules of Chapter 1 tell us that $\vec{a}_{\text{free fall}} = \vec{a}_\parallel + \vec{a}_\perp$.

The motion diagram shows that the object's actual acceleration is parallel to the incline. The surface of the incline somehow "blocks" $\vec{a}_\perp$, through a process we will examine in Chapter 5, but $\vec{a}_\parallel$ is unhindered. It is this piece of $\vec{a}_{\text{free fall}}$, parallel to the incline, that accelerates the object.

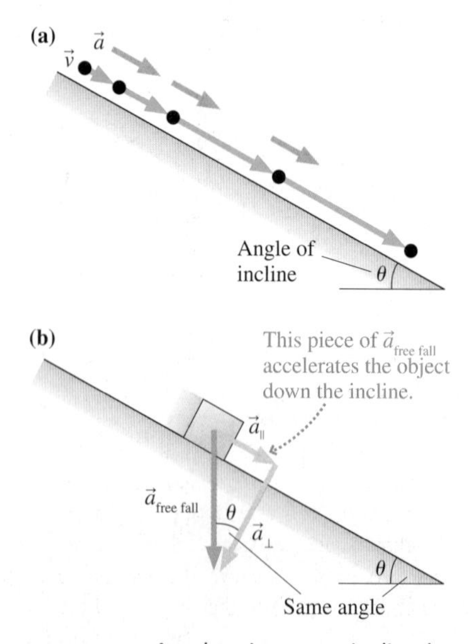

FIGURE 2.36 Acceleration on an inclined plane.

Consider an inclined plane tilted at angle θ. Figure 2.36b shows that the three vectors form a right triangle with angle θ at the bottom. By definition, the length, or magnitude, of $\vec{a}_{\text{free fall}}$ is g. Vector $\vec{a}_{\parallel}$ is opposite angle θ, so the length, or magnitude, of $\vec{a}_{\parallel}$ must be $g\sin\theta$. Consequently, the one-dimensional acceleration along the incline is

$$a_s = \pm g \sin\theta \tag{2.25}$$

The correct sign depends on the direction in which the ramp is tilted, as the following examples will illustrate.

Skiing is an example of motion on an inclined plane.

Equation 2.25 makes sense. Suppose the plane is perfectly horizontal. If you place an object on a horizontal surface, you expect it to stay at rest with no acceleration. Equation 2.25 gives $a_s = 0$ when $\theta = 0°$, in agreement with our expectations. Now suppose you tilt the plane until it becomes vertical, at $\theta = 90°$. Without friction, an object would simply fall, in free fall, parallel to the vertical surface. Equation 2.25 gives $a_s = -g = a_{\text{free fall}}$ when $\theta = 90°$, again in agreement with our expectations. We'll use Newton's laws of motion in Chapter 5 to verify Equation 2.25, but we can have confidence in it now because we see that it gives the correct result in these *limiting cases.*

EXAMPLE 2.18 Skiing down an incline

A skier's speed at the bottom of a 100-m-long, frictionless, snow-covered slope is 20 m/s. What is the angle of the slope?

MODEL Represent the skier as a particle. Assume that air resistance is negligible. Assume that the slope is a straight line.

VISUALIZE Figure 2.37 shows the physical and pictorial representations of the skier.

SOLVE The motion diagram shows that the acceleration vector $\vec{a}_{\parallel}$ points in the positive s-direction. Thus the one-dimensional acceleration is $a_s = +g\sin\theta$. This is constant-acceleration motion. The third kinematic equation from Table 2.3 is

$$v_{1s}^2 = v_{0s}^2 + 2a_s\Delta x = 2g\sin\theta\Delta x$$

where we used $v_{0s} = 0$ m/s. Solving for $\sin\theta$, we find

$$\sin\theta = \frac{v_{1s}^2}{2g\Delta x} = \frac{(20\text{ m/s})^2}{2(9.80\text{ m/s}^2)(100\text{ m})} = 0.204$$

Thus

$$\theta = \sin^{-1}(0.204) = 11.8°$$

ASSESS A 100-m-long slope and a speed of 20 m/s $\approx$ 40 mph are fairly typical parameters for skiing. A 1° angle or an 80° angle would be unrealistic, but 12° seems plausible.

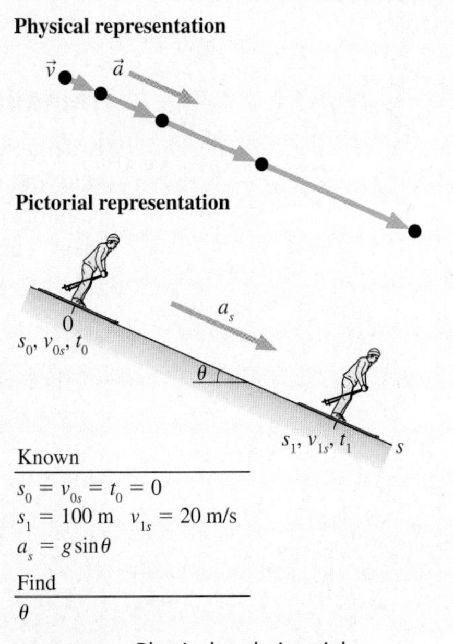

FIGURE 2.37 Physical and pictorial representations for the skier of Example 2.18.

EXAMPLE 2.19 At the amusement park

An amusement park ride shoots a car up a frictionless track inclined at 30°. The car rolls up, then rolls back down. If the height of the track is 20 m, what is the maximum allowable speed with which the car can start?

MODEL Represent the car as a particle. Assume air resistance is negligible.

VISUALIZE Figure 2.38 on the next page shows the physical and pictorial representations of the car. The problem starts as the car is shot up the incline, and it ends when the car reaches its highest point. The highest point is a turning point, so $v_{1s} = 0$ m/s. The motion diagram shows that the acceleration vector $\vec{a}_{\parallel}$ points in the negative s-direction, so $a_s = -g\sin\theta$. The *maximum* starting speed is that at which the car goes to the very top of the ramp, a height of 20 m.

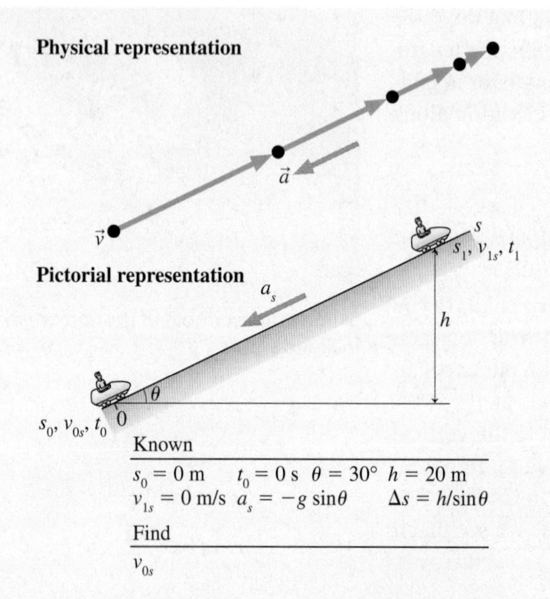

FIGURE 2.38 Physical and pictorial representations for the car of Example 2.19.

SOLVE The maximum possible displacement Δs_{max} is related to the height h by

$$\Delta s_{max} = s_1 - s_0 = \frac{h}{\sin 30°} = \frac{20 \text{ m}}{\sin 30°} = 40 \text{ m}$$

The initial speed v_{0s} that allows the car to travel this distance is found from

$$v_{1s}{}^2 = 0 = v_{0s}{}^2 + 2a_s\Delta x = v_{0s}{}^2 - 2g \sin\theta\Delta x$$

$$v_{0s} = \sqrt{2g \sin 30°\Delta x} = \sqrt{2(9.8 \text{ m/s}^2)(0.500)(40 \text{ m})}$$

$$= 19.8 \text{ m/s}$$

This is the maximum speed, because a car starting any faster will run off the top.

ASSESS 20 m ≈ 60 feet and 19.8 m/s ≈ 40 mph. It seems plausible that a car would need to be going this fast to gain 60 feet of elevation rolling up a ramp. Be sure you understand why the sign of a_s is negative here but positive in Example 2.18.

Thinking Graphically

Kinematics is the language of motion. We will spend the entire rest of this course studying moving objects, from baseballs to electrons, and the concepts we have developed in this chapter will be used extensively. One of the most important ideas, summarized in Tactics Box 2.3, has been that the relationships between position, velocity, and acceleration can be expressed graphically.

TACTICS BOX 2.3 Interpreting graphical representations of motion

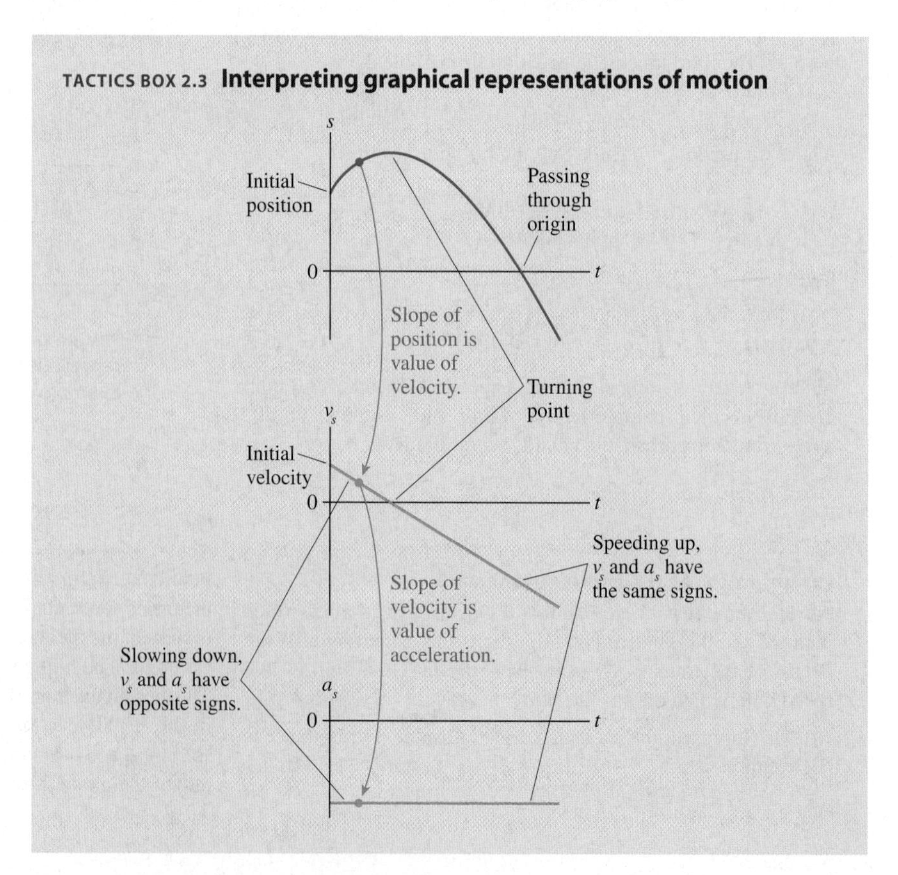

A good way to solidify your understanding of motion graphs is to consider the problem of a hard, smooth ball rolling on a smooth (i.e., frictionless) track. The track is made up of several straight segments connected together. Each segment may be either horizontal or inclined. Your task will be to analyze the ball's motion graphically. This will require you to reason about, rather than calculate, the relationships between s, v_s, and a_s.

There are two variations to this type of problem. In the first, you are given a picture of a track and the initial condition of the ball. The problem is then to draw graphs of s, v_s, and a_s. In the second, you are given the graphs, and the problem is to deduce the shape of the track on which the ball is rolling.

There are a small number of rules to follow in each of these problems:

1. Assume that the ball passes smoothly from one segment of the track to the next, with no loss of speed and without ever leaving the track.
2. The position, velocity, and acceleration graphs should be stacked vertically. They should each have the same horizontal scale so that a vertical line drawn through all three connects points describing the same instant of time.
3. Although the graphs have no numbers, they should show the correct *relationships*. For example, if the velocity is greater during the first part of the motion than during the second part, then the position graph should be steeper in the first part than in the second. Similarly, longer-lasting motions should span a greater horizontal range than shorter-lasting motions.
4. The position s is the position measured *along* the track. Similarly, v_s and a_s are the velocity and acceleration parallel to the track.

EXAMPLE 2.20 From track to graphs I

Draw position, velocity, and acceleration graphs for the ball on the track of Figure 2.39.

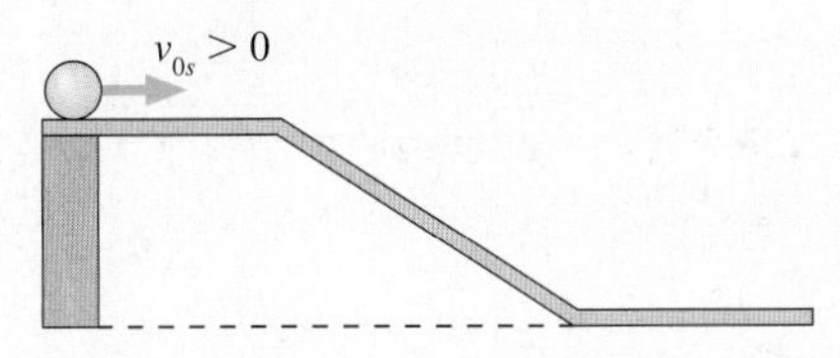

FIGURE 2.39 A ball rolling along a frictionless track.

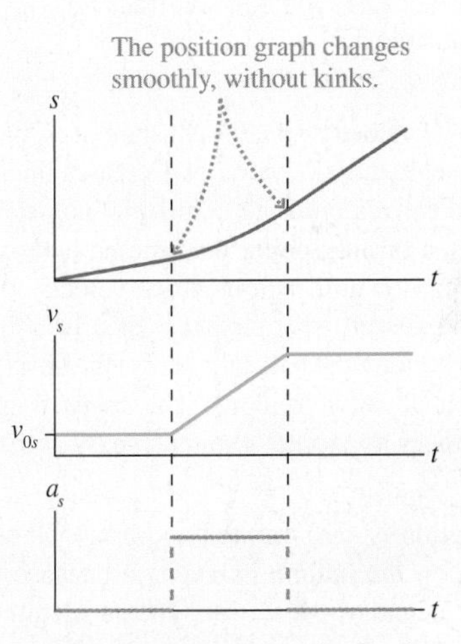

FIGURE 2.40 Motion graphs for the ball in Example 2.20.

VISUALIZE It is often easiest to begin with the velocity. Here the ball starts with an initial velocity v_{0s}. There is no acceleration on the horizontal surface ($a_s = 0$ if $\theta = 0°$), so the velocity remains constant until the ball reaches the slope. The slope is an inclined plane that, as we have learned, has constant acceleration. The velocity increases linearly with time during constant-acceleration motion. The ball returns to constant-velocity motion after reaching the bottom horizontal segment. The middle graph of Figure 2.40 shows the velocity.

We have enough information to draw the acceleration graph. We noted that the acceleration is zero while the ball is on the horizontal segments, and a_s has a constant positive value on the slope. These accelerations are consistent with the slope of the velocity graph: zero slope, then positive slope, then a return to zero slope. The bottom part of Figure 2.40 shows the acceleration graph. The acceleration cannot *really* change instantly from zero to a nonzero value, but the change can be so quick that we do not see it on the time scale of the graph. That is what the vertical dotted lines imply.

Finally, we need to find the position-versus-time graph. You might want to refer back to Figure 2.29 to review how the position graph looks for constant-velocity and constant-acceleration motion. The position increases linearly with time during the first segment at constant velocity. It also does so during the third segment of motion, but with a steeper slope to indicate a faster velocity. In between, while the acceleration is nonzero but constant, the position graph has a *parabolic* shape. The top part of Figure 2.40 shows the position-versus-time graph.

Two points are worth noting:

1. The dotted vertical lines through the graphs show the instants when the ball moves from one segment of the track to the next. Because of Rule 1, the speed does not change abruptly at these points; it changes gradually.
2. The parabolic section of the position-versus-time graph blends *smoothly* into the straight lines on either side. This is a consequence of Rule 1. An abrupt change of slope (a "kink") would indicate an abrupt change in velocity and would violate Rule 1.

EXAMPLE 2.21 From track to graphs II

Figure 2.41 shows a track with a "switch." A ball moving left-to-right passes through and heads up the incline, but a ball rolling down the incline goes straight through and continues downhill. Draw position, velocity, and acceleration graphs of the ball's motion.

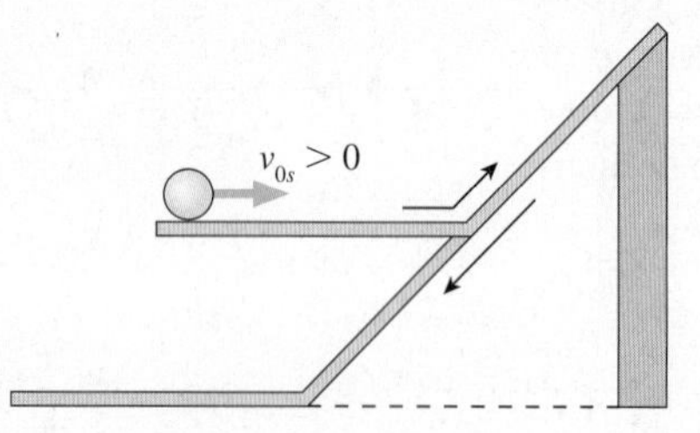

FIGURE 2.41 The ball and track for Example 2.21.

VISUALIZE The velocity remains constant at v_{0s} while the ball is on the level segment. The velocity decreases linearly with time after the ball moves onto the uphill incline. At some point v_s reaches zero, a turning point, then the ball starts rolling back down. Rolling downhill is a *negative* velocity (motion to the left), but the velocity still changes linearly with time. Upon reaching the bottom, the ball rolls across the horizontal segment with constant negative velocity. The *speed* is greater than it was on the upper horizontal segment, so $|v_s|$ at the end is larger than v_{0s}.

The acceleration is zero on the two horizontal segments. The acceleration on the incline is *negative* because the vector $\vec{a}_{\parallel}$ points in the negative s-direction. The acceleration is constant the whole time that the ball is on the incline, regardless of whether it is moving up or down. "Moving uphill" or "moving downhill" is determined by the sign of the velocity, not by the acceleration. This constant negative acceleration is consistent with the constant negative slope of the velocity graph. The velocity changes sign at the turning point, where $v = 0$, but **the acceleration does not change at this point.**

The position changes linearly while the velocity is constant. The position changes parabolically while the acceleration is constant and reaches a *maximum value* at the turning point, which is the top of the parabola. The constant negative velocity during the last segment implies a straight-line position graph with negative slope. All three motion graphs are shown in Figure 2.42.

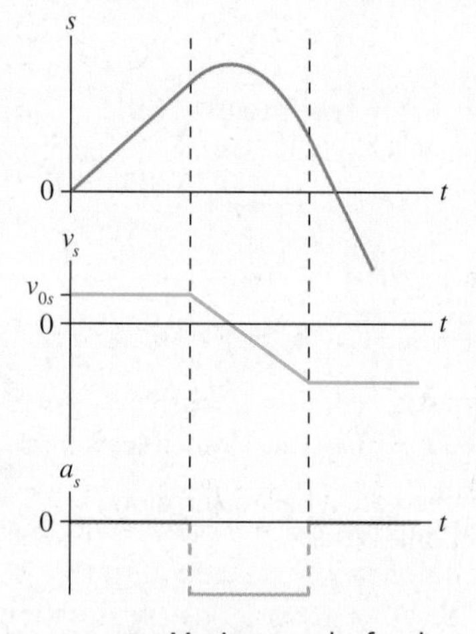

FIGURE 2.42 Motion graphs for the ball in Example 2.21.

EXAMPLE 2.22 From graphs to track

Figure 2.43 shows a set of motion graphs for a ball moving on a track. Draw a picture of the track and describe the ball's initial condition. Each segment of the track is *straight,* but the segments may be tilted.

VISUALIZE As with the last two examples, let's start by examining the velocity graph. The ball starts with initial velocity $v_{0s} = 0$, then the velocity increases linearly with time. This indicates that the ball is released from rest while on a slope, then starts rolling downhill in the positive s-direction. The acceleration graph confirms this: The initial acceleration is positive. That the motion is to the right (positive s-direction) is also seen from the position graph, where s becomes increasingly positive. After a while, the acceleration drops to zero and the velocity holds constant—a horizontal segment! Then the acceleration becomes negative while the velocity is *positive* but decreasing. This is motion to the right while slowing down, implying that the ball is going uphill. The acceleration has a smaller magnitude in the third segment than it had during the first segment, and the velocity graph is less steep. This indicates that the uphill tilt is less than the initial downhill tilt.

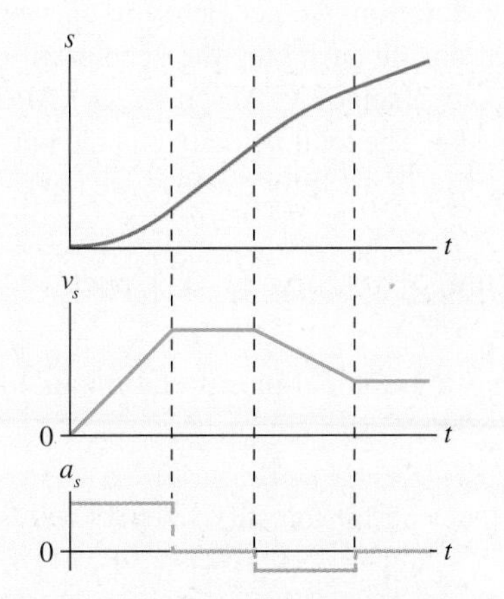

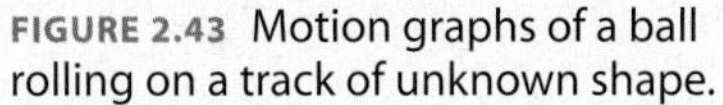

FIGURE 2.43 Motion graphs of a ball rolling on a track of unknown shape.

Lastly, the acceleration is again zero and the velocity is constant, so the final segment is horizontal. The position has increased throughout and v_s is never negative, so the motion is purely left-to-right with no turning points. Figure 2.44 shows the track and the initial conditions that are responsible for the graphs of Figure 2.43.

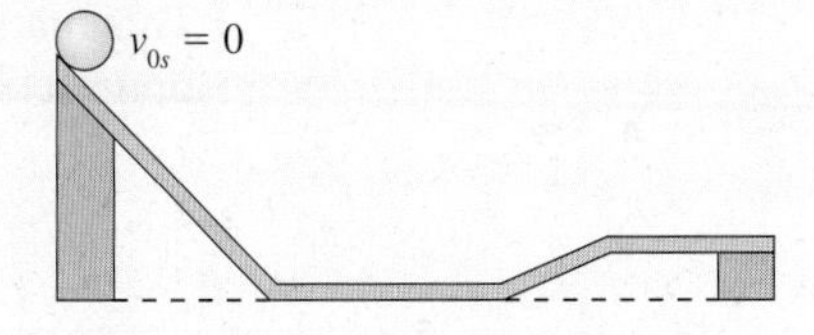

FIGURE 2.44 Track responsible for the motion graphs of Figure 2.43.

STOP TO THINK 2.5 The ball rolls up the ramp, then back down. Which is the correct acceleration graph?

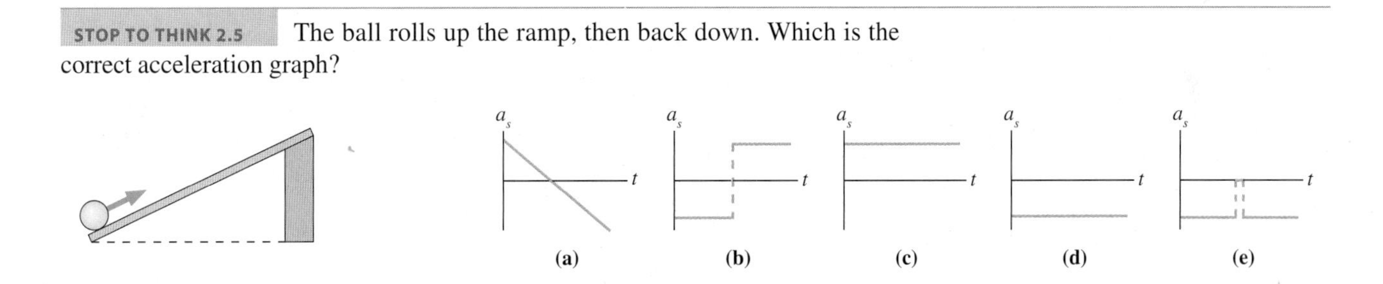

2.8 Instantaneous Acceleration

Figure 2.45 shows a velocity that increases with time, reaches a maximum, then decreases. This is *not* uniformly accelerated motion. We can still define an average acceleration, as we did in Section 2.5, but the average acceleration does not give a complete description of motion with nonuniform acceleration. Instead, we need the acceleration at each *instant* of time.

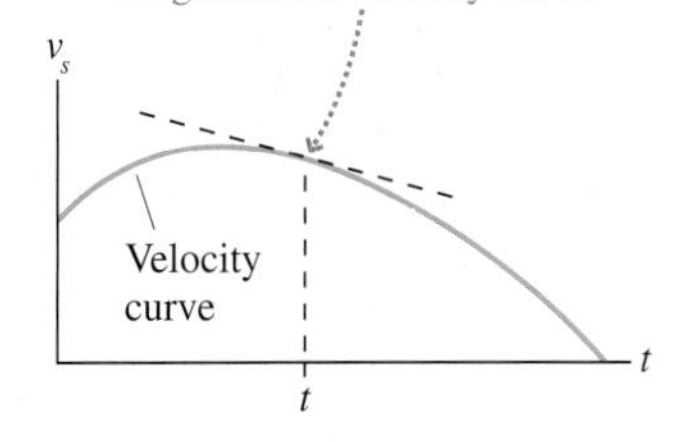

FIGURE 2.45 Motion with nonuniform acceleration.

We can define an instantaneous acceleration in much the same way that we defined the instantaneous velocity. The instantaneous velocity was found to be the limit of the average velocity as the time interval $\Delta t \to 0$. Graphically, the instantaneous velocity at time t is the slope of the position-versus-time graph at that time. By analogy: **The instantaneous acceleration a_s at a specific instant of time t is the slope of the line that is tangent to the velocity-versus-time curve at time t.** Mathematically, this is

$$a_s \equiv \lim_{\Delta t \to 0} \frac{\Delta v_s}{\Delta t} = \frac{dv_s}{dt} \quad \text{(instantaneous acceleration)} \tag{2.26}$$

The instantaneous acceleration is the derivative (i.e., the rate of change) of the velocity.

The reverse problem—to find the velocity v_s if we know the acceleration a_s at all instants of time—is also important. When we wanted to find the position from the velocity, we took a velocity curve, divided it into N steps, found that the displacement Δs_k during step k was the area $(v_s)_k \Delta t$ of a small rectangle, then we added all the steps (i.e., integrated) to find s_f.

We can do the same with acceleration. An acceleration curve can be divided into N very narrow steps so that during each step the acceleration is essentially constant. During step k, the velocity changes by $\Delta(v_s)_k = (a_s)_k\Delta t$. This is the area of the small rectangle under the step. The total velocity change between t_i and t_f is found by adding all the small $\Delta(v_s)_k$. In the limit $\Delta t \to 0$, we have

$$v_{fs} = v_{is} + \lim_{\Delta t \to 0} \sum_{k=1}^{N} (a_s)_k \Delta t = v_{is} + \int_{t_i}^{t_f} a_s\,dt \tag{2.27}$$

This mathematical statement has a graphical interpretation analogous to Equation 2.11. In this case:

$$v_{fs} = v_{is} + \text{area under the acceleration curve } a_s \text{ between } t_i \text{ and } t_f \tag{2.28}$$

The constant acceleration equation $v_{fs} = v_{is} + a_s\Delta t$ is a special example of Equation 2.28. If you look back at Figure 2.28a you will see that the quantity $a_s\Delta t$ is the rectangular area under the horizontal acceleration curve.

EXAMPLE 2.23 Finding velocity from acceleration

Figure 2.46 shows the acceleration graph for a particle with an initial velocity of 10 m/s. What is the particle's velocity at $t = 8$ s?

MODEL We're told this is the motion of a particle.

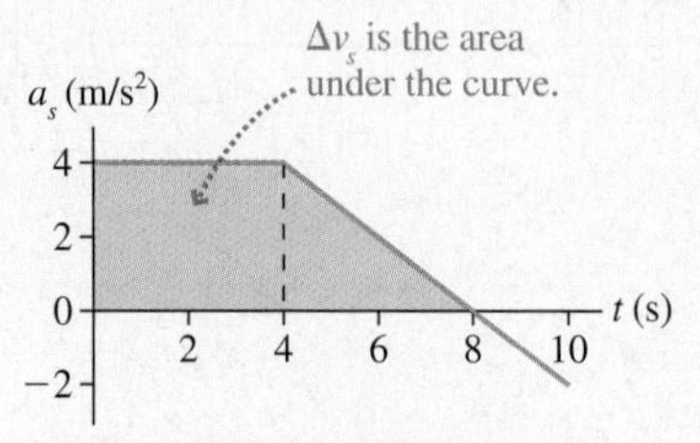

FIGURE 2.46 Acceleration graph for Example 2.23.

VISUALIZE Figure 2.46 is a graphical representation of the motion.

SOLVE The change in velocity is found as the area under the acceleration curve:

$$v_{fs} = v_{is} + \text{area under the acceleration curve } a_s \text{ between } t_i \text{ and } t_f$$

The area under the curve between $t_i = 0$ s and $t_f = 8$ s can be subdivided into a rectangle ($0\text{ s} \le t \le 4\text{ s}$) and a triangle ($4\text{ s} \le t \le 8\text{ s}$). These areas are easily computed. Thus

$$v_s(\text{at } t = 8\text{ s}) = 10\text{ m/s} + (4\text{ (m/s)/s})(4\text{ s}) + \tfrac{1}{2}(4\text{ (m/s)/s})(4\text{ s}) = 34\text{ m/s}$$

EXAMPLE 2.24 A nonuniform acceleration

Figure 2.47a shows the velocity-versus-time graph for a particle whose velocity is given by $v_s = [10 - (t-5)^2]$ m/s, where t is in s.

a. Find an expression for the particle's acceleration a_s and draw the acceleration-versus-time graph.
b. Describe the motion.

MODEL We're told that this is a particle.

VISUALIZE The figure shows the velocity graph. It is a parabola centered at $t = 5$ s with an apex $v_{max} = 10$ m/s. The slope of v_s is positive but decreasing in magnitude for $t < 5$ s. The slope is zero at $t = 5$ s, and it is negative and increasing in magnitude for $t > 5$ s. Thus the acceleration graph should start positive, decrease steadily, pass through zero at $t = 5$ s, then become increasingly negative.

SOLVE

a. We can find an expression for a_s by taking the derivative of v_s. First, expand the square to give

$$v_s = (-t^2 + 10t - 15)\text{ m/s}$$

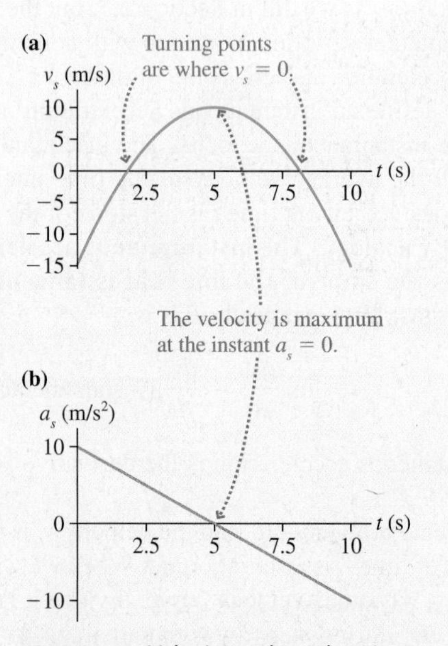

FIGURE 2.47 Velocity and acceleration graphs for Example 2.24.

Then use the derivative rule (Equation 2.5) to find

$$a_s = \frac{dv_s}{dt} = (-2t + 10)\ \mathrm{m/s^2}$$

where t is in s. This is a linear equation that is graphed in Figure 2.47b. The graph meets our expectations.

b. This is a complex motion. The particle starts out moving to the left ($v_s < 0$) at 15 m/s. The positive acceleration causes the speed to decrease (slowing down because v_s and a_s have opposite signs) until the particle reaches a turning point ($v_s = 0$) just before $t = 2$ s. The particle then moves to the right ($v_s > 0$) and speeds up until reaching maximum speed at $t = 5$ s. From $t = 5$ s to just after $t = 8$ s, the particle is still moving to the right ($v_s > 0$) but slowing down. Another turning point occurs just after $t = 8$ s. Then the particle moves back to the left and gains speed as the negative a_s makes the velocity ever more negative.

STOP TO THINK 2.6 Rank in order, from largest to smallest, the accelerations at points A to C.

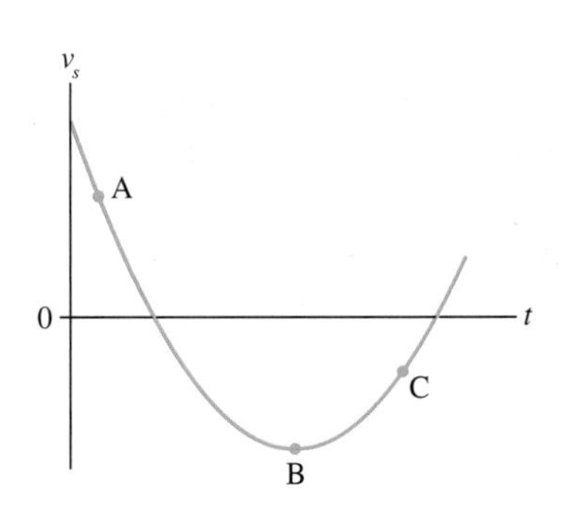

a. $a_A > a_B > a_C$.
b. $a_C > a_A > a_B$.
c. $a_C > a_B > a_A$.
d. $a_B > a_A > a_C$.

SUMMARY

The goal of Chapter 2 has been to learn how to solve problems about motion in a straight line.

GENERAL PRINCIPLES

Kinematics describes motion in terms of position, velocity, and acceleration.

General kinematic relationships are given **mathematically** by:

Instantaneous velocity $v_s = ds/dt$ = slope of position graph

Instantaneous acceleration $a_s = dv_s/dt$ = slope of velocity graph

Final position $s_f = s_i + \int_{t_i}^{t_f} v_s\,dt = s_i + \begin{cases}\text{area under the velocity curve}\\ \text{from } t_i \text{ to } t_f\end{cases}$

Final velocity $v_{fs} = v_{is} + \int_{t_i}^{t_f} a_s\,dt = v_{is} + \begin{cases}\text{area under the acceleration}\\ \text{curve from } t_i \text{ to } t_f\end{cases}$

The kinematic equations for **motion with constant acceleration:**

$$v_{fs} = v_{is} + a_s\Delta t$$

$$s_f = s_i + v_{is}\Delta t + \tfrac{1}{2}a_s(\Delta t)^2$$

$$v_{fs}^2 = v_{is}^2 + 2a_s\Delta s$$

IMPORTANT CONCEPTS

Position, velocity, and acceleration are related **graphically.**

- The slope of the position-versus-time graph is the value on the velocity graph.
- The slope of the velocity graph is the value on the acceleration graph.
- s is a maximum or minimum at a turning point, and $v_s = 0$.

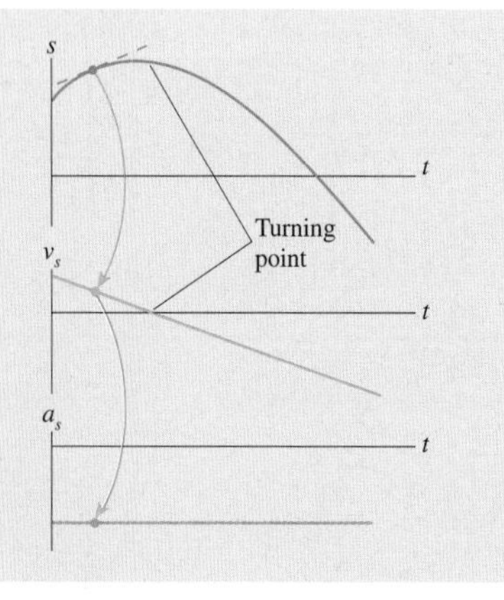

Motion with constant acceleration is uniformly accelerated motion.

Uniform motion is motion with constant velocity and zero acceleration.

$$s_f = s_i + v_s\Delta t$$

APPLICATIONS

The **sign of v_s** indicates the direction of motion.

- $v_s > 0$ is motion to the right or up.
- $v_s < 0$ is motion to the left or down.

The **sign of a_s** indicates which way $\vec{a}$ points, *not* whether the object is speeding up or slowing down.

- $a_s > 0$ if $\vec{a}$ points to the right or up.
- $a_s < 0$ if $\vec{a}$ points to the left or down.
- The direction of $\vec{a}$ is found with a motion diagram.

An object is **speeding up** if and only if v_s and a_s have the same sign. An object is **slowing down** if and only if v_s and a_s have opposite signs.

Free fall is constant-acceleration motion with $a_y = -g = -9.80\ \text{m/s}^2$.

Motion on an inclined plane has $a_s = \pm g\sin\theta$. The sign depends on the direction of the tilt.

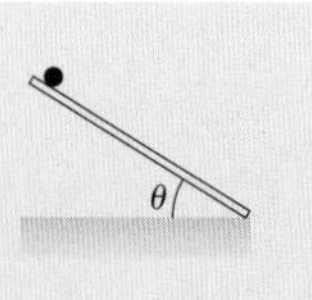

TERMS AND NOTATION

kinematics
position-versus-time graph
uniform motion
speed, v
initial position, s_i
final position, s_f
instantaneous velocity, v_s
turning point
average acceleration, a_{avg}
uniformly accelerated motion
free fall
acceleration due to gravity, g
instantaneous acceleration, a_s

EXERCISES AND PROBLEMS

The icon in front of a problem indicates that the problem can be done on a Dynamics Worksheet. Dynamics Worksheets are found at the back of the *Student Workbook*. If you use a worksheet, draw a motion diagram in the Physical Representation section, establish your coordinate system and symbols in the Pictorial Representation section, then solve the problem in the Mathematical Representation section.

Exercises

Section 2.1 Motion in One Dimension

1. Figure Ex2.1 shows a motion diagram of a car traveling down a street. The camera took one frame every second. A distance scale is provided.
   a. Measure the x-value of the car at each dot. Place your data in a table, similar to Table 2.1, showing each position and the instant of time at which it occurred.
   b. Make a position-versus-time graph for the ball. Because you have data only at certain instants of time, your graph should consist of dots that are not connected together.

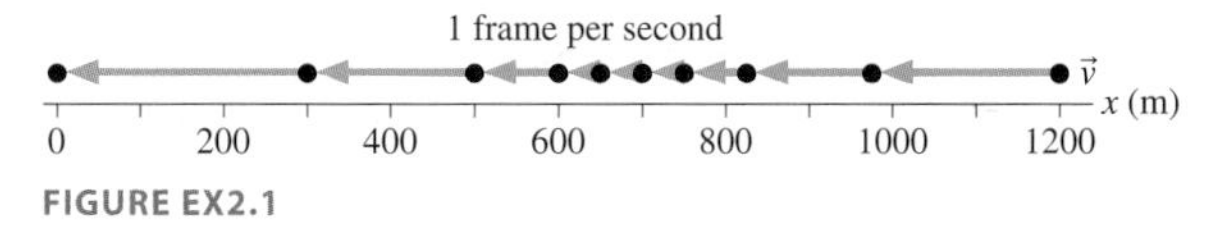

FIGURE EX2.1

2. For each motion diagram, determine the sign (positive or negative) of the position, the velocity, and the acceleration.

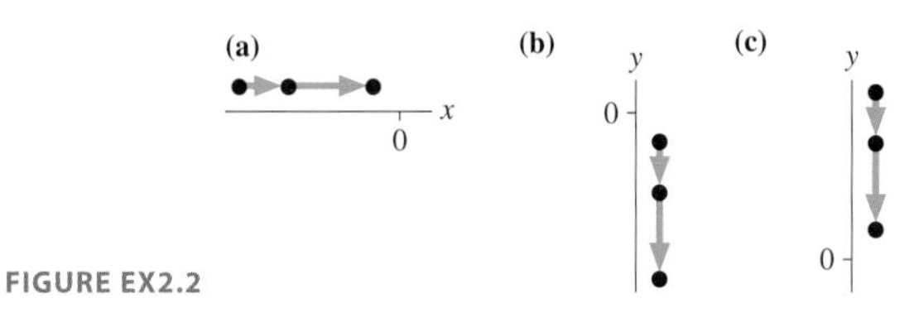

FIGURE EX2.2

3. Write a short description of the motion of a real object for which this would be a realistic position-versus-time graph.

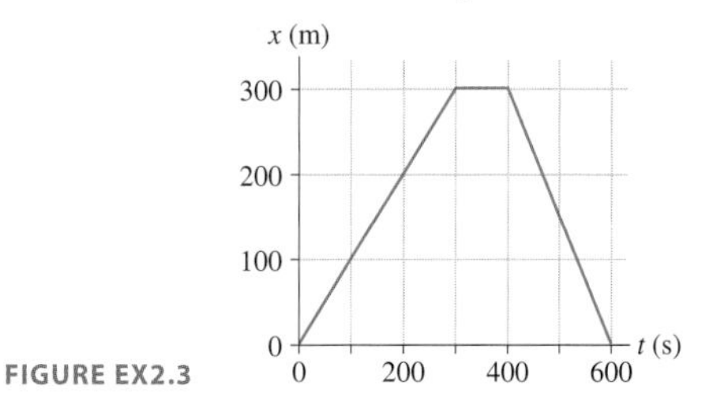

FIGURE EX2.3

4. Write a short description of the motion of a real object for which this would be a realistic position-versus-time graph.

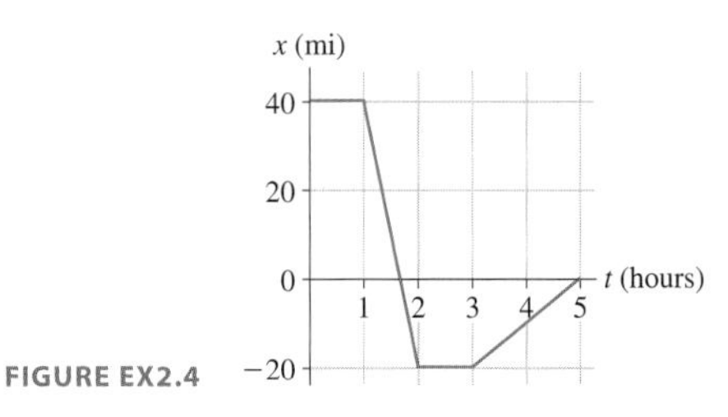

FIGURE EX2.4

Section 2.2 Uniform Motion

5. A car starts at the origin and moves with velocity $\vec{v}$ = (10 m/s, northeast). How far from the origin will the car be after traveling for 45 s?
6. Larry leaves home at 9:05 and runs at constant speed to the lamppost. He reaches the lamppost at 9:07, immediately turns, and runs to the tree. Larry arrives at the tree at 9:10.
   a. What is Larry 's average velocity during each of these two intervals?
   b. What is the average velocity for Larry's entire run?

FIGURE EX2.6

7. Alan leaves Los Angeles at 8:00 A.M. to drive to San Francisco, 400 mi away. He travels at a steady 50 mph. Beth leaves Los Angeles at 9:00 A.M. and drives a steady 60 mph.
   a. Who gets to San Francisco first?
   b. How long does the first to arrive have to wait for the second?
8. A bicyclist has the position-versus-time graph shown. What is the bicyclist's velocity at t = 10 s, at t = 25 s, and at t = 35 s?

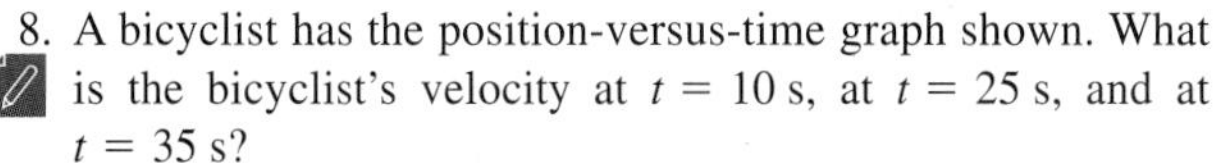

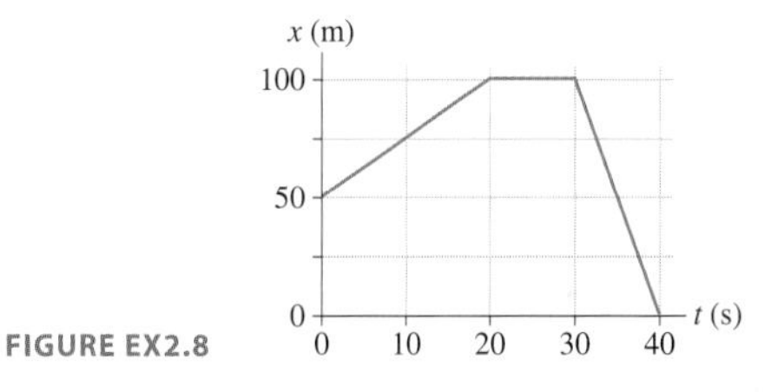

FIGURE EX2.8

9. Julie drives 100 mi to Grandmother's house. On the way to Grandmother's, Julie drives half the *distance* at 40 mph and half the distance at 60 mph. On her return trip, she drives half the *time* at 40 mph and half the time at 60 mph.
   a. What is Julie's average speed on the way to Grandmother's house?
   b. What is her average speed on the return trip?

Section 2.3 Instantaneous Velocity

Section 2.4 Finding Position from Velocity

10. Figure Ex2.10 shows the position graph of a particle.
    a. Draw the particle's velocity graph for the interval $0 \text{ s} \le t \le 4 \text{ s}$.
    b. Does this particle have a turning point or points? If so, at what time or times?

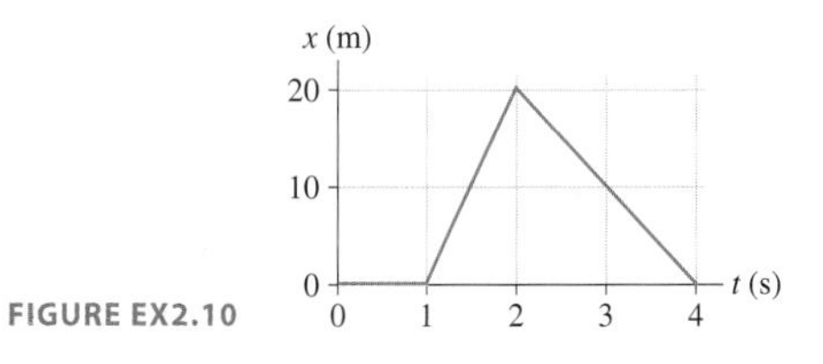

FIGURE EX2.10

11. A particle starts from $x_0 = 10$ m at $t_0 = 0$ and moves with the velocity graph shown in Figure Ex2.11.
    a. What is the object's position at $t = 2$ s, 3 s, and 4 s?
    b. Does this particle have a turning point? If so, at what time?

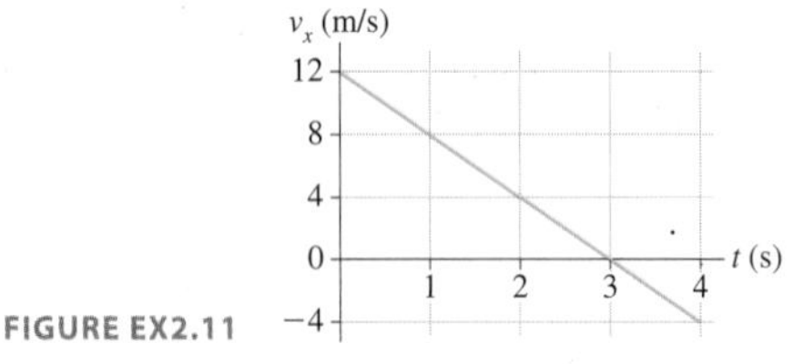

FIGURE EX2.11

Section 2.5 Motion with Constant Acceleration

12. Figure Ex2.12 shows the velocity graph of a particle. Draw the particle's acceleration graph for the interval $0 \text{ s} \leq t \leq 4 \text{ s}$. Give both axes an appropriate numerical scale.

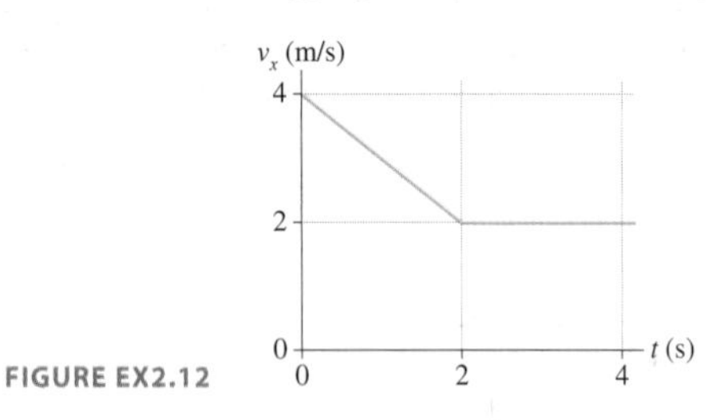

FIGURE EX2.12

13. Figure Ex2.13 shows the velocity graph of a train that starts from the origin at $t = 0$ s.
    a. Draw position and acceleration graphs for the train.
    b. Find the acceleration of the train at $t = 3.0$ s.

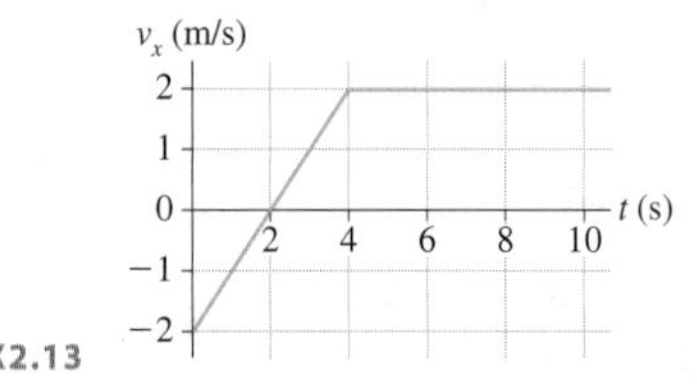

FIGURE EX2.13

14. Figure Ex2.14 shows the velocity graph of a particle moving along the x-axis. Its initial position is $x_0 = 2$ m at $t_0 = 0$ s. At $t = 2$ s, what are the particle's (a) position, (b) velocity, and (c) acceleration?

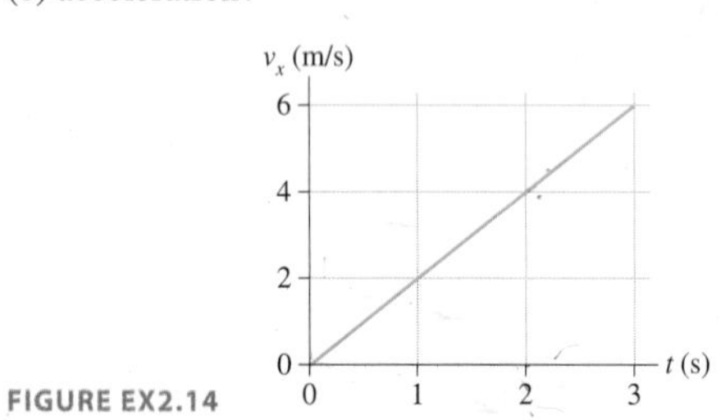

FIGURE EX2.14

15. a. What constant acceleration, in SI units, must a car have to go from zero to 60 mph in 10 s?
    b. What fraction of g is this?
    c. How far has the car traveled when it reaches 60 mph? Give your answer both in SI units and in feet.

16.  A jet plane is cruising at 300 m/s when suddenly the pilot turns the engines up to full throttle. After traveling 4.0 km, the jet is moving with a speed of 400 m/s.
    a. What is the jet's acceleration, assuming it to be a constant acceleration?
    b. Is your answer reasonable? Explain.

17. A speed skater moving across frictionless ice at 8.0 m/s hits a 5.0-m-wide patch of rough ice. She slows steadily, then continues on at 6.0 m/s. What is her acceleration on the rough ice?

18. a. How many days will it take a spaceship to accelerate to the speed of light (3.0×10^8 m/s) with the acceleration g?
    b. How far will it travel during this interval?
    c. What fraction of a light year is your answer to part b? A *light year* is the distance light travels in one year.

NOTE ▶ We know, from Einstein's theory of relativity, that no object can travel at the speed of light. So this problem, while interesting and instructive, is not realistic. ◀

Section 2.6 Free Fall

19. Ball bearings are made by letting spherical drops of molten metal fall inside a tall tower—called a *shot tower*—and solidify as they fall.
    a. If a bearing needs 4.0 s to solidify enough for impact, how high must the tower be?
    b. What is the bearing's impact velocity?

20. A ball is thrown vertically upward with a speed of 19.6 m/s.
    a. What is the ball's velocity and its height after 1, 2, 3, and 4 s?
    b. Draw the ball's velocity-versus-time graph. Give both axes an appropriate numerical scale.

21. A student standing on the ground throws a ball straight up. The ball leaves the student's hand with a speed of 15 m/s when the hand is 2.0 m above the ground. How long is the ball in the air before it hits the ground? (The student moves her hand out of the way.)

22. A rock is tossed straight up with a speed of 20 m/s. When it returns, it falls into a hole 10 m deep.
    a. What is the rock's velocity as it hits the bottom of the hole?
    b. How long is the rock in the air, from the instant it is released until it hits the bottom of the hole?

Section 2.7 Motion on an Inclined Plane

23. A car traveling at 30 m/s runs out of gas while traveling up a 20° slope. How far up the hill will it coast before starting to roll back down?

24. A skier is gliding along at 3.0 m/s on horizontal, frictionless snow. He suddenly starts down a 10° incline. His speed at the bottom is 15 m/s.
    a. What is the length of the incline?
    b. How long does it take him to reach the bottom?

Section 2.8 Instantaneous Acceleration

25. A particle moving along the x-axis has its position described by the function $x = (2t^3 - t + 1)$ m, where t is in s. At $t = 2$ s, what are the particle's (a) position, (b) velocity, and (c) acceleration?

26. A particle moving along the x-axis has its velocity described by the function $v_x = 2t^2$ m/s, where t is in s. Its initial position is $x_0 = 1$ m at $t_0 = 0$ s. At $t = 1$ s, what are the particle's (a) position, (b) velocity, and (c) acceleration?

27. Figure Ex2.27 shows the acceleration-versus-time graph of a particle moving along the x-axis. Its initial velocity is $v_{0x} = 8.0$ m/s at $t_0 = 0$ s. What is the particle's velocity at $t = 4.0$ s?

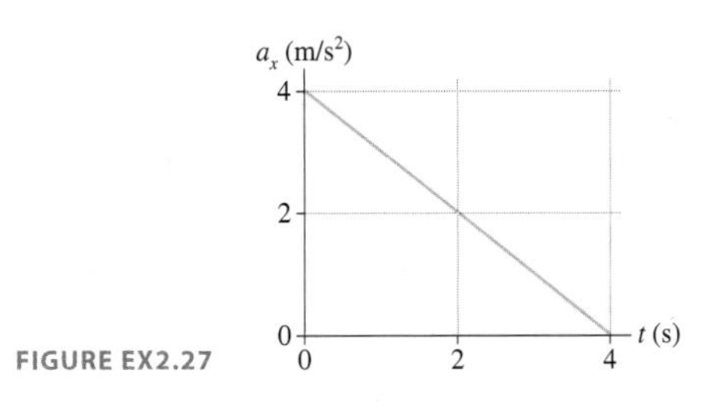

FIGURE EX2.27

Problems

28. Figure P2.28 shows the motion diagram, made at two frames of film per second, of a ball rolling along a track. The track has a 3.0-m-long sticky section.
    a. Use the meter stick to measure the positions of the center of the ball. Place your data in a table, similar to Table 2.1, showing each position and the instant of time at which it occurred.
    b. Make a position-versus-time graph for the ball. Because you have data only at certain instants of time, your graph should consist of dots that are not connected together.
    c. What is the *change* in the ball's position from $t = 0$ s to $t = 1.0$ s?
    d. What is the *change* in the ball's position from $t = 2.0$ s to $t = 4.0$ s?
    e. What is the ball's velocity before reaching the sticky section?
    f. What is the ball's velocity after passing the sticky section?
    g. Determine the ball's acceleration on the sticky section of the track.

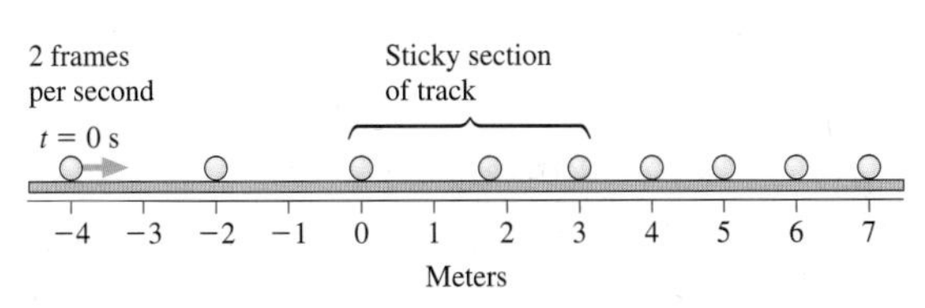

FIGURE P2.28

29. A particle's position on the x-axis is given by the function $x = (t^2 - 4t + 2)$ m, where t is in s.
    a. Make a position-versus-time graph for the interval $0 \text{ s} \le t \le 5$ s. Do this by calculating and plotting x every 0.5 s from 0 s to 5 s, then drawing a smooth curve through the points.
    b. Determine the particle's velocity at $t = 1.0$ s by drawing the tangent line on your graph and measuring its slope.
    c. Determine the particle's velocity at $t = 1.0$ s by evaluating the derivative at that instant. Compare this to your result from part b.
    d. Are there any turning points in the particle's motion? If so, at what position or positions?
    e. Where is the particle when $v_x = 4.0$ m/s?
    f. Draw a motion diagram for the particle.

30. The velocity-versus-time graph is shown for a particle moving along the x-axis. Its initial position is $x_0 = 2.0$ m at $t_0 = 0$ s.
    a. What are the particle's position, velocity, and acceleration at $t = 1.0$ s?
    b. What are the particle's position, velocity, and acceleration at $t = 3.0$ s?

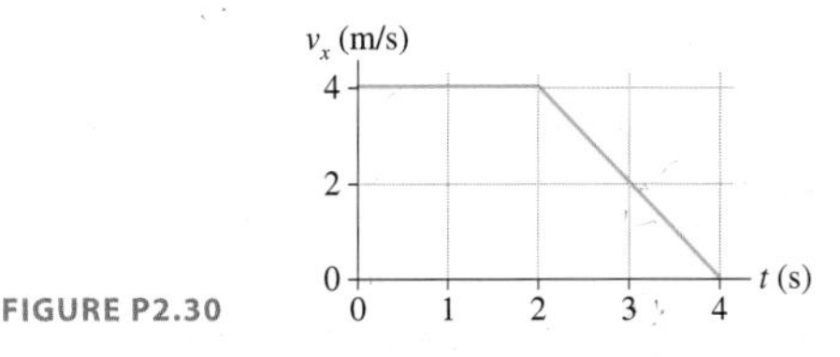

FIGURE P2.30

31. Three particles move along the x-axis, each starting with $v_{0x} = 10$ m/s at $t_0 = 0$ s. The graph for A is a position-versus-time graph; the graph for B is a velocity-versus-time graph; the graph for C is an acceleration-versus-time graph. Find each particle's velocity at $t = 7.0$ s. Work with the geometry of the graphs, not with kinematic equations.

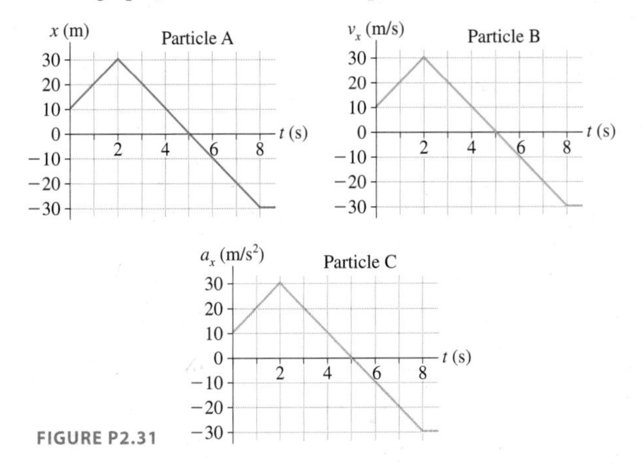

FIGURE P2.31

32. The velocity graph is shown for a particle having initial position $x_0 = 0$ m at $t_0 = 0$ s.
    a. At what time or times is the particle found at $x = 35$ m? Work with the geometry of the graph, not with kinematic equations.
    b. Draw a motion diagram for the particle.

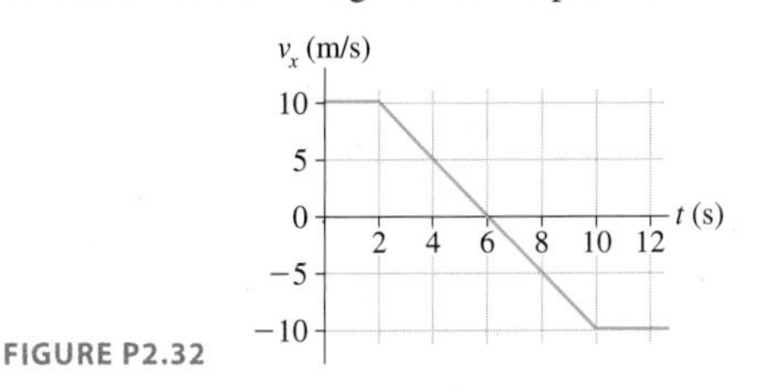

FIGURE P2.32

33. The acceleration graph is shown for a particle that starts from rest at $t = 0$ s. Determine the object's velocity at times $t =$ 0 s, 2 s, 4 s, 6 s, and 8 s.

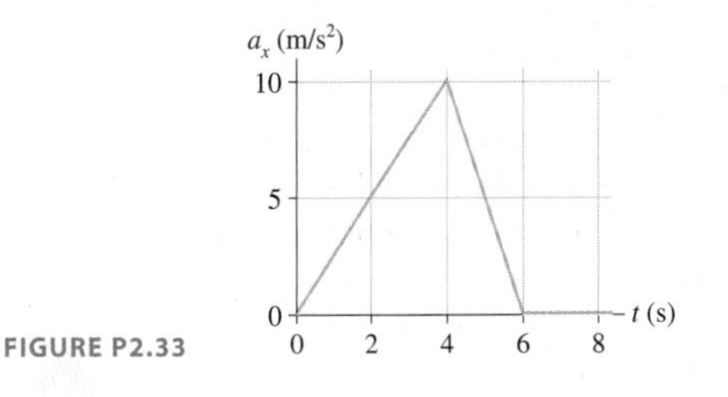

FIGURE P2.33

34. A block is suspended from a spring, pulled down, and released. The block's position-versus-time graph is shown in Figure P2.34.
    a. At what times is the velocity zero? At what times is the velocity most positive? Most negative?
    b. Draw a reasonable velocity-versus-time graph.

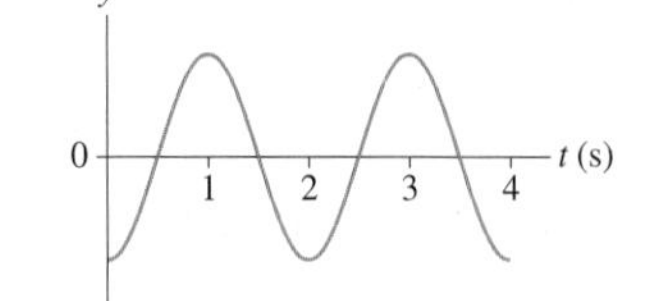

FIGURE P2.34

35. The acceleration graph is shown for a particle that starts from rest at $t = 0$ s.
    a. Draw the particle's velocity graph over the interval $0 \text{ s} \leq t \leq 10$ s. Include an appropriate numerical scale on both axes.
    b. Describe, in words, how the velocity graph would differ if the particle had an initial velocity of 2.0 m/s.

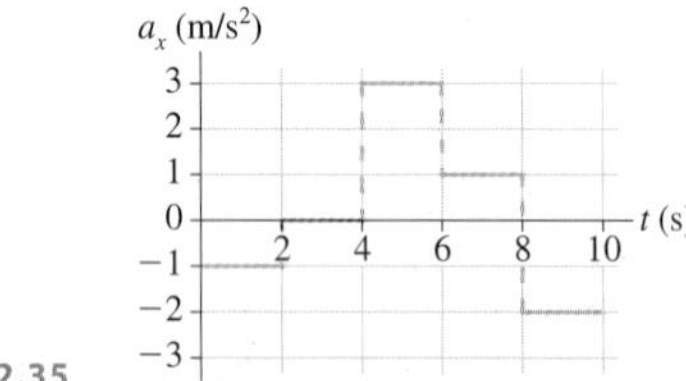

FIGURE P2.35

36. Drop a rubber ball or a tennis ball from a height of about 25 cm ($\approx$1 ft) and watch carefully as it bounces. Then draw a position graph, a velocity graph, and an acceleration graph from the instant you drop the ball until it returns to its maximum height. Stack your three graphs vertically so that the time axes are aligned with each other. Pay particular attention to the time when the ball is in contact with the ground. This is a short interval of time, but it's not zero.
37. The position of a particle is given by the function $x = (2t^3 - 9t^2 + 12)$ m, where t is in s.
    a. At what time or times is $v_x = 0$ m/s?
    b. What are the particle's position and its acceleration at this time(s)?
38. An object starts from rest at $x = 0$ m at time $t = 0$ s. Five seconds later, at $t = 5.0$ s, the object is observed to be at $x = 40.0$ m and to have velocity $v_x = 11$ m/s.
    a. Was the object's acceleration uniform or nonuniform? Explain your reasoning.
    b. Sketch the velocity-versus-time graph implied by these data. Is the graph a straight line or curved? If curved, is it concave upward or downward?
39. A particle's velocity is described by the function $v_x = kt^2$ m/s, where k is a constant and t is in s. The particle's position at $t_0 = 0$ s is $x_0 = -9.0$ m. At $t_1 = 3.0$ s, the particle is at $x_1 = 9.0$ m. Determine the value of the constant k. Be sure to include the proper units.
40. A particle's acceleration is described by the function $a_x = (10 - t)$ m/s², where t is in s. Its initial conditions are $x_0 = 0$ m and $v_{0x} = 0$ m/s at $t = 0$ s.
    a. At what time is the velocity again zero?
    b. What is the particle's position at that time?
41. A ball rolls along the frictionless track shown. Each segment of the track is straight, and the ball passes smoothly from one segment to the next without changing speed or leaving the track. Draw three vertically stacked graphs showing position, velocity, and acceleration versus time. Each graph should have the same time axis, and the proportions of the graph should be qualitatively correct. Assume that the ball has enough speed to reach the top.

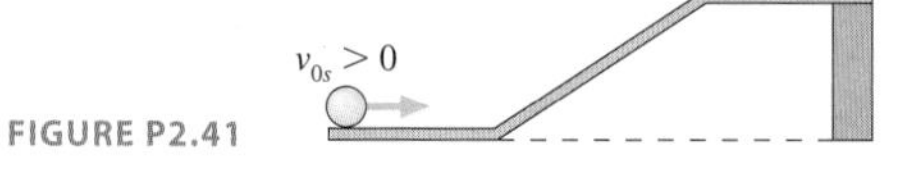

FIGURE P2.41

42. Draw position, velocity, and acceleration graphs for the ball shown here. See Problem 41 for more information.

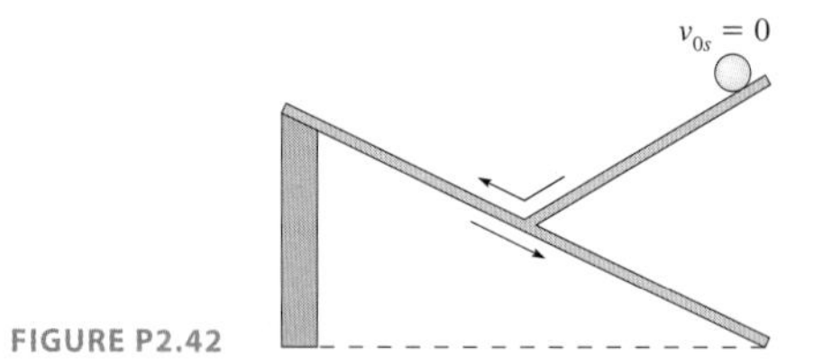

FIGURE P2.42

43. Draw position, velocity, and acceleration graphs for the ball shown here. See Problem 41 for more information. The ball changes direction but not speed as it bounces from the reflecting wall.

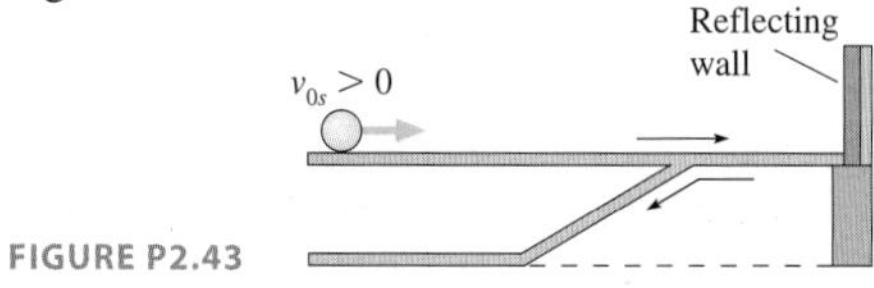

FIGURE P2.43

44. Figure P2.44 shows a set of kinematic graphs for a ball rolling on a track. All segments of the track are straight lines, but some may be tilted. Draw a picture of the track and also indicate the ball's initial condition.

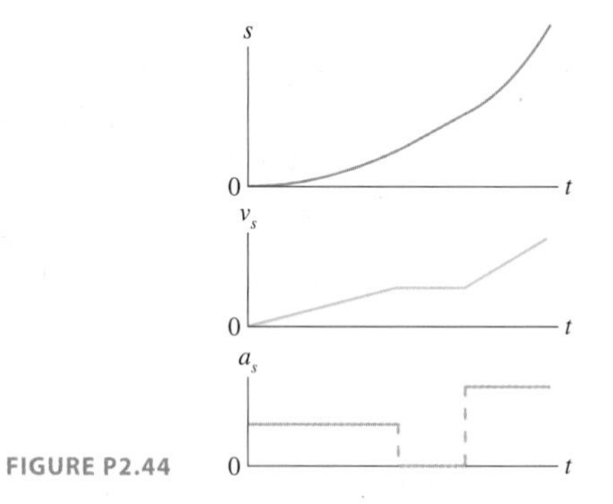

FIGURE P2.44

45. Figure P2.45 shows a set of kinematic graphs for a ball rolling on a track. All segments of the track are straight lines, but some may be tilted. Draw a picture of the track and also indicate the ball's initial condition.

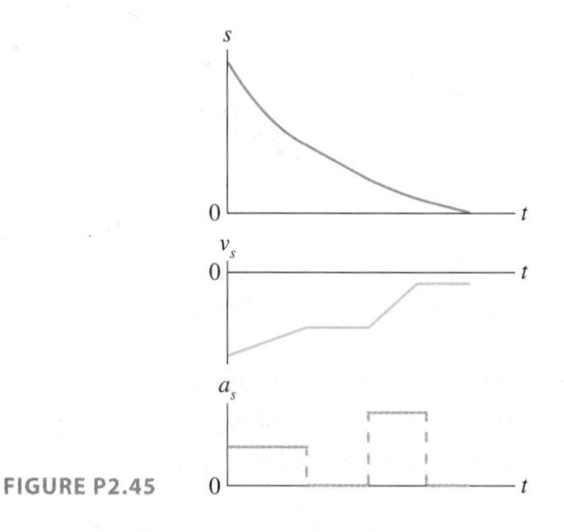

FIGURE P2.45

46. Figure P2.46 shows a set of kinematic graphs for a ball rolling on a track. All segments of the track are straight lines, but some may be tilted. Draw a picture of the track and also indicate the ball's initial condition.

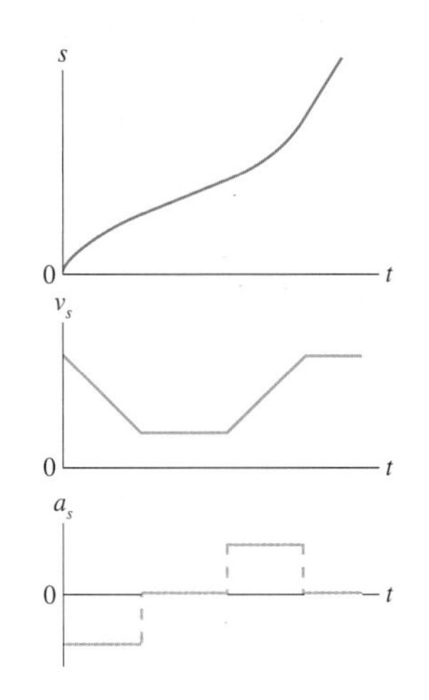

FIGURE P2.46

47. The takeoff speed for an Airbus A320 jetliner is 80 m/s. Velocity data measured during takeoff are as follows:

t(s)	v_s (m/s)
0	0
10	23
20	46
30	69

a. What is the takeoff speed in miles per hour?
b. Is the jetliner's acceleration constant during takeoff? Explain.
c. At what time do the wheels leave the ground?
d. For safety reasons, in case of an aborted takeoff, the runway must be three times the takeoff distance. Can an A320 take off safely on a 2.5-mi-long runway?

48. Does a real automobile have constant acceleration? Measured data for a Porsche 944 Turbo at maximum acceleration is as follows:

t(s)	v_s (mph)
0	0
2	28
4	46
6	60
8	70
10	78

a. Make a graph of velocity versus time. Based on your graph, is the acceleration constant? Explain.
b. Draw a smooth curve through the points on your graph, then use your graph to estimate the car's acceleration at 2.0 s and 8.0 s. Give your answer in SI units.
c. Use your graph to estimate the distance traveled in the first 10 s.

Hint: Approximate the curve with five steps, each of width $\Delta t = 2$ s. Let the height of each step be the average of the velocities at the beginning and end of the step.

49. A driver has a reaction time of 0.50 s, and the maximum deceleration of her car is 6.0 m/s^2. She is driving at 20 m/s when suddenly she sees an obstacle in the road 50 m in front of her. Can she stop the car in time to avoid a collision?

50. You are driving to the grocery store at 20 m/s. You are 110 m from an intersection when the traffic light turns red. Assume that your reaction time is 0.50 s and that your car brakes with constant acceleration.
a. How far are you from the intersection when you begin to apply the brakes?
b. What acceleration will bring you to rest right at the intersection?
c. How long does it take you to stop?

51. The minimum stopping distance for a car traveling at a speed of 30 m/s is 60 m, including the distance traveled during the driver's reaction time of 0.50 s.
a. What is the minimum stopping distance for the same car traveling at a speed of 40 m/s?
b. Draw a position-versus-time graph for the motion of the car in part a. Assume the car is at $x_0 = 0$ m when the driver first sees the emergency situation ahead that calls for a rapid halt.

52. You're driving down the highway late one night at 20 m/s when a deer steps onto the road 35 m in front of you. Your reaction time before stepping on the brakes is 0.5 s, and the maximum deceleration of your car is 10 m/s^2.
a. How much distance is between you and the deer when you come to a stop?
b. What is the maximum speed you could have and still not hit the deer?

53. A 200kg weather rocket is loaded with 100 kg of fuel and fired straight up. It accelerates upward at 30 m/s^2 for 30 s, then runs out of fuel. Ignore any air resistance effects.
a. What is the rocket's maximum altitude?
b. How long is the rocket in the air?
c. Draw a velocity-versus-time graph for the rocket from liftoff until it hits the ground.

54. A 1000kg weather rocket is launched straight up. The rocket motor provides a constant acceleration for 16 s, then the motor stops. The rocket altitude 20 s after launch is 5100 m. You can ignore any effects of air resistance.
a. What was the rocket's acceleration during the first 16 s?
b. What is the rocket's speed as it passes through a cloud 5100 m above the ground?

55. A lead ball is dropped into a lake from a diving board 5.0 m above the water. After entering the water, it sinks to the bottom with a constant velocity equal to the velocity with which it hit the water. The ball reaches the bottom 3.0 s after it is released. How deep is the lake?

56. A hotel elevator ascends 200 m with a maximum speed of 5.0 m/s. Its acceleration and deceleration both have a magnitude of 1.0 m/s^2.
a. How far does the elevator move while accelerating to full speed from rest?
b. How long does it take to make the complete trip from bottom to top?

57. A car starts from rest at a stop sign. It accelerates at 4.0 m/s^2 for 6 seconds, coasts for 2 s, and then slows down at a rate of 3.0 m/s^2 for the next stop sign. How far apart are the stop signs?

58. A car accelerates at $2.0\ m/s^2$ along a straight road. It passes two marks that are 30 m apart at times $t = 4.0$ s and $t = 5.0$ s. What was the car's initial velocity?
59. Santa loses his footing and slides down a frictionless, snowy roof that is tilted at an angle of 30°. If Santa slides 10 m before reaching the edge, what is his speed as he leaves the roof?
60. Ann and Carol are driving their cars along the same straight road. Carol is located at $x = 2.4$ mi at $t = 0$ hours and drives at a steady 36 mph. Ann, who is traveling in the same direction, is located at $x = 0.0$ mi at $t = 0.50$ hours and drives at a steady 50 mph.
    a. At what time does Ann overtake Carol?
    b. What is their position at this instant?
    c. Draw a position-versus-time graph showing the motion of both Ann and Carol.
61. A ball rolls along the smooth track shown in the figure with an initial speed of 5.0 m/s. Assume that the ball turns all the corners smoothly, with no loss of speed.
    a. What is the ball's speed as it goes over the top?
    b. What is its speed when it reaches the level track on the right side?
    c. By what percentage does the ball's final speed differ from its initial speed? Is this surprising?

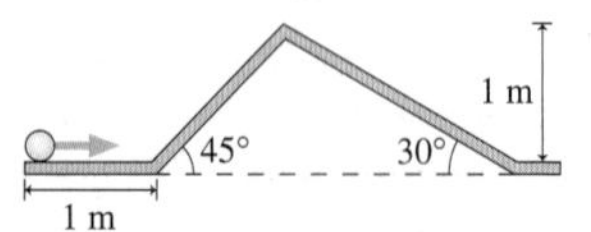

FIGURE P2.61

62. A toy train is pushed forward and released at $x_0 = 2.0$ m with a speed of 2.0 m/s. It rolls at a steady speed for 2.0 s, then one wheel begins to stick. The train comes to a stop 6.0 m from its starting point. What is the train's acceleration after its wheel begins to stick?
63. Bob is driving the getaway car after the big bank robbery. He's going 50 m/s when his headlights suddenly reveal a nail strip that the cops have placed across the road 150 m in front of him. If Bob can stop in time, he can throw the car into reverse and escape. But if he crosses the nail strip, all his tires will go flat and he will be caught. Bob's reaction time before he can hit the brakes is 0.60 s, and his car's maximum deceleration is $10\ m/s^2$. Is Bob in jail?
64. One game at the amusement park has you push a puck up a long, frictionless ramp. You win a stuffed animal if the puck, at its highest point, comes to within 10 cm of the end of the ramp without going off. You give the puck a push, releasing it with a speed of 5.0 m/s when it is 8.50 m from the end of the ramp. The puck's speed after traveling 3.0 m is 4.0 m/s. Are you a winner?
65. A professional skier's *initial* acceleration on fresh snow is 90% of the acceleration expected on a frictionless, inclined plane, the loss being due to friction. Due to air resistance, his acceleration slowly decreases as he picks up speed. The speed record on a mountain in Oregon is 180 kilometers per hour at the bottom of a 25° slope that drops 200 m.
    a. What exit speed could a skier reach in the absence of air resistance?
    b. What percentage of this ideal speed is lost to air resistance?
66. Heather and Jerry are standing on a bridge 50 m above a river. Heather throws a rock straight down with a speed of 20 m/s. Jerry, at exactly the same instant of time, throws a rock straight up with the same speed. Ignore air resistance.
    a. How much time elapses between the first splash and the second splash?
    b. Which rock has the faster speed as it hits the water?
67. Nicole throws a ball straight up. Chad watches the ball from a window 5.0 m above the point where Nicole released it. The ball passes Chad on the way up, and it has a speed of 10 m/s as it passes him on the way back down. How fast did Nicole throw the ball?
68. A motorist is driving at 20 m/s when she sees that a traffic light 200 m ahead has just turned red. She knows that this light stays red for 15 s, and she wants to reach the light just as it turns green again. It takes her 1.0 s to step on the brakes and begin slowing. What is her speed as she reaches the light at the instant it turns green?
69. David is driving a steady 30 m/s when he passes Tina, who is sitting in her car at rest. Tina begins to accelerate at a steady $2.0\ m/s^2$ at the instant when David passes.
    a. How far does Tina drive before passing David?
    b. What is her speed as she passes him?
70. A Porsche challenges a Honda to a 400-m race. Because the Porsche's acceleration of $3.5\ m/s^2$ is larger than the Honda's $3.0\ m/s^2$, the Honda gets a 50-m head start. Both cars start accelerating at the same instant. Who wins?
71. A cat is sleeping on the floor in the middle of a 3.0-m-wide room when a barking dog enters with a speed of 1.50 m/s. As the dog enters, the cat (as only cats can do) immediately accelerates at $0.85\ m/s^2$ toward an open window on the opposite side of the room. The dog (all bark and no bite) is a bit startled by the cat and begins to slow down at $0.10\ m/s^2$ as soon as it enters the room. Does the dog catch the cat before the cat is able to leap through the window?
72. You want to visit your friend in Seattle during spring break. To save money, you decide to travel there by train. Unfortunately, your physics final exam took the full 3 hours, so you are late in arriving at the train station. You run as fast as you can, but just as you reach the platform you see your train, 30 m ahead of you down the platform, begin to accelerate at $1.0\ m/s^2$. You chase after the train at your maximum speed of 8.0 m/s, but there's a barrier 50 m ahead. Will you be able to leap onto the back step of the train before you crash into the barrier?
73. A rocket is launched straight up with constant acceleration. Four seconds after liftoff, a bolt falls off the side of the rocket. The bolt hits the ground 6.0 s later. What was the rocket's acceleration?

In Problems 74 through 77 you are given the kinematic equation or equations that are used to solve a problem. For each of these, you are to

a. Write a *realistic* problem for which this is the correct equation(s). Be sure that the answer your problem requests is consistent with the equation(s) given.
b. Draw the motion diagram and the pictorial representation for your problem.
c. Finish the solution of the problem.

74. $64\text{ m} = 0\text{ m} + (32\text{ m/s})(4\text{ s} - 0\text{ s}) + \frac{1}{2}a_x(4\text{ s} - 0\text{ s})^2$

75. $(10\text{ m/s})^2 = v_{0y}{}^2 - 2(9.8\text{ m/s}^2)(10\text{ m} - 0\text{ m})$

76. $(0\text{ m/s})^2 = (5\text{ m/s})^2 - 2(9.8\text{ m/s}^2)(\sin 10°)(x_1 - 0\text{ m})$

77. $v_{1x} = 0\text{ m/s} + (20\text{ m/s}^2)(5\text{ s} - 0\text{ s})$
$x_1 = 0\text{ m} + (0\text{ m/s})(5\text{ s} - 0\text{ s}) + \frac{1}{2}(20\text{ m/s}^2)(5\text{ s} - 0\text{ s})^2$
$x_2 = x_1 + v_{1x}(10\text{ s} - 5\text{ s})$

Challenge Problems

78. Jill has just gotten out of her car in the grocery store parking lot. The parking lot is on a hill and is tilted 3°. Fifty meters downhill from Jill, a little old lady lets go of a fully loaded shopping cart. The cart, with frictionless wheels, starts to roll straight downhill. Jill immediately starts to sprint after the cart with her top acceleration of 2.0 m/s^2. How far has the cart rolled before Jill catches it?

79. As a science project, you drop a watermelon off the top of the Empire State Building, 320 m above the sidewalk. It so happens that Superman flies by at the instant you release the watermelon. Superman is headed straight down with a speed of 35 m/s. How fast is the watermelon going when it passes Superman?

80. Your school science club has devised a special event for homecoming. You've attached a rocket to the rear of a small car that has been decorated in the blue-and-gold school colors. The rocket provides a constant acceleration for 9.0 s. As the rocket shuts off, a parachute opens and slows the car at a rate of 5.0 m/s^2. The car passes the judges' box in the center of the grandstand, 990 m from the starting line, exactly 12 s after you fire the rocket. What is the car's speed as it passes the judges?

81. Careful measurements have been made of Olympic sprinters in the 100-meter dash. A simple but reasonably accurate model is that a sprinter accelerates at 3.6 m/s^2 for $3\frac{1}{3}$ s, then runs at constant velocity to the finish line.
    a. What is the race time for a sprinter who follows this model?
    b. A sprinter could run a faster race by accelerating faster at the beginning, thus reaching top speed sooner. If a sprinter's top speed is the same as in part a, what acceleration would he need to run the 100-meter dash in 9.9 s?
    c. By what percent did the sprinter need to increase his acceleration in order to decrease his time by 1%?

82. A sprinter can accelerate with constant acceleration for 4.0 s before reaching top speed. He can run the 100-meter dash in 10 s. What is his speed as he crosses the finish line?

83. The Starship Enterprise returns from warp drive to ordinary space with a forward speed of 50 km/s. To the crew's great surprise, a Klingon ship is 100 km directly ahead, traveling in the same direction at a mere 20 km/s. Without evasive action, the Enterprise will overtake and collide with the Klingons in just slightly over 3.0 s. The Enterprise's computers react instantly to brake the ship. What acceleration does the Enterprise need to just barely avoid a collision with the Klingon ship? Assume the acceleration is constant.
    Hint: Draw a position-versus-time graph showing the motions of both the Enterprise and the Klingon ship. Let $x_0 = 0$ km be the location of the Enterprise as it returns from warp drive. How do you show graphically the situation in which the collision is "barely avoided"? Once you decide what it looks like graphically, express that situation mathematically.

STOP TO THINK ANSWERS

Stop to Think 2.1: d. The particle starts with positive x and moves to negative x.

Stop to Think 2.2: c. The velocity is the slope of the position graph. The slope is positive and constant until the position graph crosses the axis, then positive but decreasing, and finally zero when the position graph is horizontal.

Stop to Think 2.3: b. A constant positive v_x corresponds to a linearly increasing x, starting from $x_i = -10$ m. The constant negative v_x then corresponds to a linearly decreasing x.

Stop to Think 2.4: a or **b.** The velocity is constant while $a = 0$, it decreases linearly while a is negative. Graphs a, b, and c all have the same acceleration, but only graphs a and b have a positive initial velocity that represents a particle moving to the right.

Stop to Think 2.5: d. The acceleration vector $\vec{a}_{\|}$ points downhill (negative s-direction) and has the constant value $-g\sin\theta$ throughout the motion.

Stop to Think 2.6: c. Acceleration is the slope of the graph. The slope is zero at B. Although the graph is steepest at A, the slope at that point is negative, and so $a_A < a_B$. Only C has a positive slope, so $a_C > a_B$.

3 Vectors and Coordinate Systems

Wind has both a speed and a direction, hence the motion of the wind is described by a vector.

► **Looking Ahead**
The goals of Chapter 3 are to learn how vectors are represented and used. In this chapter you will learn to:

- Understand and use the basic properties of vectors.
- Decompose a vector into its components and reassemble vector components into a magnitude and direction.
- Add and subtract vectors both graphically and using components.

◄ **Looking Back**
This chapter continues the development of vectors that was begun in Chapter 1. Please review:

- Section 1.3 Vector addition and subtraction.

Many of the quantities that we use to describe the physical world are simply numbers. For example, the mass of an object is 2 kg, its temperature is 21°C, and it occupies a volume of 250 cm^3. A quantity that is fully described by a single number (with units) is called a **scalar quantity.** Mass, temperature, and volume are all scalars. Other scalar quantities include pressure, density, energy, charge, and voltage. We will often use an algebraic symbol to represent a scalar quantity. Thus m will represent mass, T temperature, V volume, E energy, and so on. Notice that scalars, in printed text, are shown in italics.

Our universe has three dimensions, so some quantities also need a direction for a full description. If you ask someone for directions to the post office, the reply "Go three blocks" will not be very helpful. A full description might be, "Go three blocks south." A quantity having both a size and a direction is called a **vector quantity.**

You met examples of vector quantities in Chapter 1: position, displacement, velocity, and acceleration. You will soon make the acquaintance of others, such as force, momentum, and the electric field. Now, before we begin a study of forces, it's worth spending a little time to look more closely at vectors.

3.1 Scalars and Vectors

Suppose you are assigned the task of measuring the temperature at various points throughout a building and then showing the information on a building floor plan. To do this, you could put little dots on the floor plan, to show the points at which you made measurements, then write the temperature at that point beside the dot.

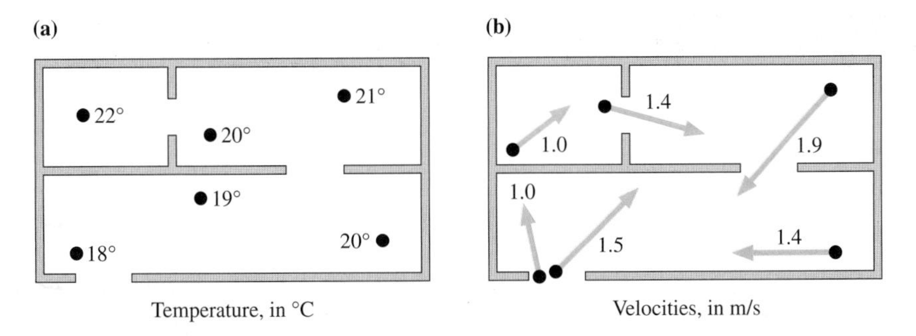

FIGURE 3.1 Measurements of scalar and vector quantities.

In other words, as Figure 3.1a shows, you can represent the temperature at each point with a simple number (with units). Temperature is a scalar quantity.

Having done such a good job on your first assignment, you are next assigned the task of measuring the velocities of several employees as they move about in their work. Recall from Chapter 1 that velocity is a vector; it has both a size and a direction. Simply writing each employee's speed is not sufficient because speed doesn't take into account the direction in which the person moved. After some thought, you conclude that a good way to represent the velocity is by drawing an arrow whose length is proportional to the speed and that points in the direction of motion. Further, as Figure 3.1b shows, you decide to place the *tail* of an arrow at the point where you measured the velocity.

As this example illustrates, the *geometric representation* of a vector is an arrow, with the tail of the arrow (not its tip!) placed at the point where the measurement is made. The vector then seems to radiate outward from the point to which it is attached. An arrow makes a natural representation of a vector because it inherently has both a length and a direction.

The mathematical term for the length, or size, of a vector is **magnitude,** so we can say that **a vector is a quantity having a magnitude and a direction.** As an example, Figure 3.2 shows the geometric representation of a particle's velocity vector $\vec{v}$. The particle's speed at this point is 5 m/s, *and* it is moving in the direction indicated by the arrow. The arrow is drawn with its tail at the point where the velocity was measured.

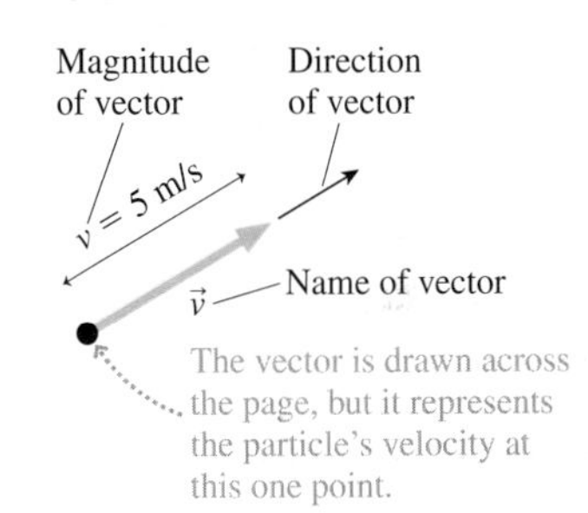

FIGURE 3.2 The velocity vector $\vec{v}$ has both a magnitude and a direction.

NOTE ► Although the vector arrow is drawn across the page, from its tail to its tip, this does *not* indicate that the vector "stretches" across this distance. Instead, the vector arrow tells us the value of the vector quantity only at the one point where the tail of the vector is placed. ◄

Arrows are good for pictures, but we also need an *algebraic representation* of vectors to use in labels and in equations. We do this by drawing a small arrow over the letter that represents the vector: $\vec{r}$ for position, $\vec{v}$ for velocity, $\vec{a}$ for acceleration, and so on.

The *magnitude* of a vector is indicated by the letter without the arrow. For example, the magnitude of the velocity vector in Figure 3.2 is $v = 5$ m/s. This is the object's *speed.* The magnitude of the acceleration vector $\vec{a}$ is written a. **The magnitude of a vector is a scalar quantity.**

NOTE ► The magnitude of a vector cannot be a negative number; it must be positive or zero, with appropriate units. ◄

It is important to get in the habit of using the arrow symbol for vectors. If you omit the vector arrow from the velocity vector $\vec{v}$ and write only v, then you're referring only to the object's speed, not its velocity. The symbols $\vec{r}$ and r, or $\vec{v}$ and v, do *not* represent the same thing, so if you omit the vector arrow from vector symbols you will soon have confusion and mistakes.

The boat's displacement is the straight-line connection from its initial to its final position.

3.2 Properties of Vectors

Recall from Chapter 1 that the *displacement* is a vector drawn from an object's initial position to its position at some later time. Because displacement is an easy concept to think about, we can use it to introduce some of the properties of vectors. However, these properties apply to *all* vectors, not just to displacement.

Suppose that Sam starts from his front door, walks across the street, and ends up 200 ft to the northeast of where he started. Sam's displacement, which we will label $\vec{S}$, is shown in Figure 3.3a. The displacement vector is a *straight-line connection* from his initial to his final position, not necessarily his actual path. The dotted line indicates a possible route Sam might have taken, but his displacement is the vector $\vec{S}$.

To describe a vector we must specify both its magnitude and its direction. We can write Sam's displacement as

$$\vec{S} = (200 \text{ ft, northeast})$$

where the first number specifies the magnitude and the second number is the direction. The magnitude of Sam's displacement is $S = 200$ ft, the distance between his initial and final points.

Sam's next-door neighbor Bill also walks 200 ft to the northeast, starting from his own front door. Bill's displacement $\vec{B} = (200 \text{ ft, northeast})$ has the same magnitude and direction as Sam's displacement $\vec{S}$. Because vectors are defined by their magnitude and direction, **two vectors are equal if they have the same magnitude and direction.** This is true regardless of the starting points of the vectors. Thus the two displacements in Figure 3.3b are equal to each other, and we can write $\vec{B} = \vec{S}$.

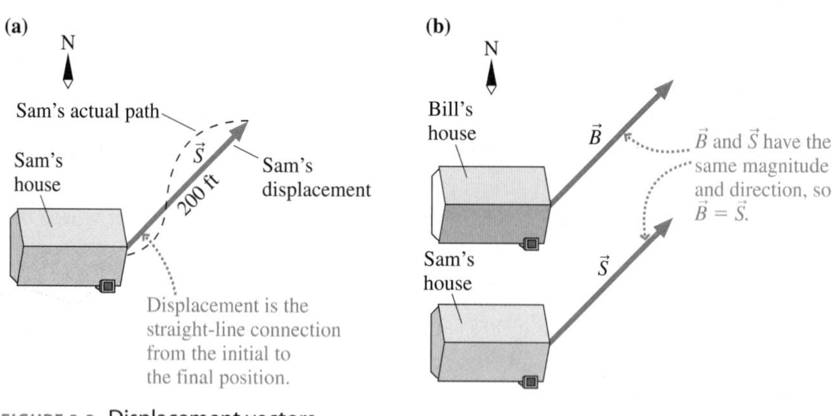

FIGURE 3.3 Displacement vectors.

NOTE ► A vector is unchanged if you move it to a different point on the page as long you don't change its length or the direction it points. We used this idea in Chapter 1 when we moved velocity vectors around in order to find the average acceleration vector $\vec{a}$. ◄

Vector Addition

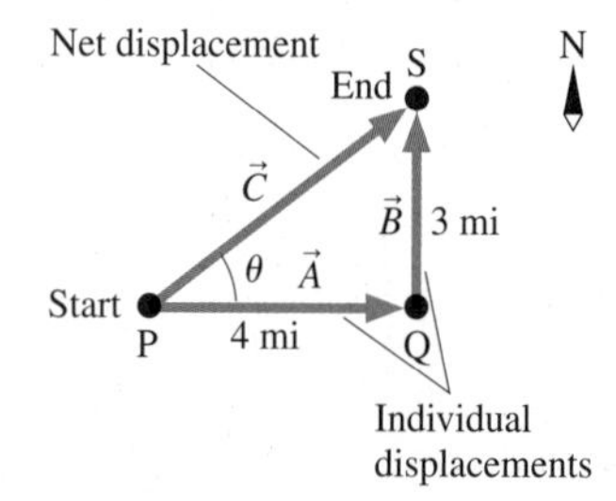

FIGURE 3.4 The net displacement $\vec{C}$ resulting from two displacements $\vec{A}$ and $\vec{B}$.

Figure 3.4 shows the displacement of a hiker who starts at point P and ends at point S. She first hikes 4 miles to the east, then 3 miles to the north. The first leg of the hike is described by the displacement $\vec{A} = (4 \text{ mi, east})$. The second leg of the hike has displacement $\vec{B} = (3 \text{ mi, north})$. Now, by definition, a vector from the initial position P to the final position S is also a displacement. This is vector $\vec{C}$ on the figure. $\vec{C}$ is the *net displacement* because it describes the net result of the hiker's first having displacement $\vec{A}$, then displacement $\vec{B}$.

If you earn \$50 on Saturday and \$60 on Sunday, your *net* income for the weekend is the sum of \$50 and \$60. With scalars, the word *net* implies addition. The same is true with vectors. The net displacement $\vec{C}$ is an initial displacement $\vec{A}$ *plus* a second displacement $\vec{B}$, or

$$\vec{C} = \vec{A} + \vec{B} \tag{3.1}$$

The sum of two vectors is called the **resultant vector.** It's not hard to show that vector addition is commutative: $\vec{A} + \vec{B} = \vec{B} + \vec{A}$. That is, you can add vectors in any order you wish.

Look back at Tactics Box 1.1 on page 10 to see the three-step procedure for adding two vectors. This tip-to-tail method for adding vectors, which is used to find $\vec{C} = \vec{A} + \vec{B}$ in Figure 3.4, is called **graphical addition.** Any two vectors of the same type—two velocity vectors or two force vectors—can be added in exactly the same way.

The graphical method for adding vectors is straightforward, but we need to do a little geometry to come up with a complete description of the resultant vector $\vec{C}$. Vector $\vec{C}$ of Figure 3.4 is defined by its magnitude C and by its direction. Because the three vectors $\vec{A}$, $\vec{B}$, and $\vec{C}$ form a right triangle, the magnitude, or length, of $\vec{C}$ is given by the Pythagorean theorem:

$$C = \sqrt{A^2 + B^2} = \sqrt{(4\text{ mi})^2 + (3\text{ mi})^2} = 5\text{ mi} \tag{3.2}$$

Notice that Equation 3.2 uses the magnitudes A and B of the vectors $\vec{A}$ and $\vec{B}$. The angle θ, which is used in Figure 3.4 to describe the direction of $\vec{C}$, is easily found for a right triangle:

$$\theta = \tan^{-1}\left(\frac{B}{A}\right) = \tan^{-1}\left(\frac{3\text{ mi}}{4\text{ mi}}\right) = 37° \tag{3.3}$$

Altogether, the hiker's net displacement is

$$\vec{C} = \vec{A} + \vec{B} = (5\text{ mi, } 37° \text{ north of east}) \tag{3.4}$$

NOTE ► Vector mathematics makes extensive use of geometry and trigonometry. Appendix A, at the end of this book, contains a brief review of these topics. ◄

EXAMPLE 3.1 Using graphical addition to find a displacement

A bird flies 100 m due east from a tree, then 200 m northwest (that is, 45° north of west). What is the bird's net displacement?

VISUALIZE Figure 3.5 shows the two individual displacements, which we've called $\vec{A}$ and $\vec{B}$. The net displacement is the vector sum $\vec{C} = \vec{A} + \vec{B}$, which is found graphically.

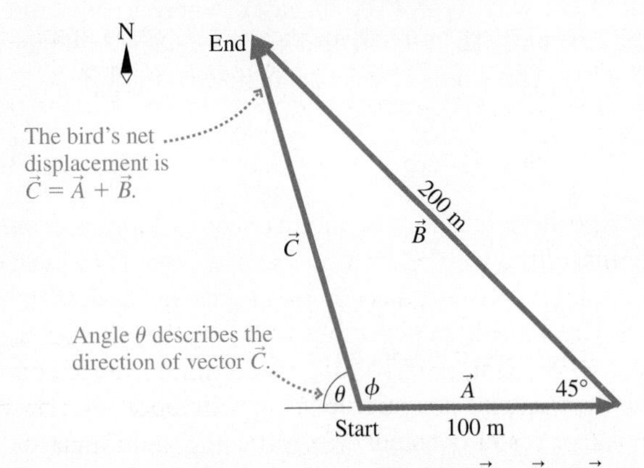

FIGURE 3.5 The bird's net displacement is $\vec{C} = \vec{A} + \vec{B}$.

SOLVE The two displacements are $\vec{A} = (100\text{ m, east})$ and $\vec{B} = (200\text{ m, northwest})$. The net displacement $\vec{C} = \vec{A} + \vec{B}$ is found by drawing a vector from the initial to the final position. But describing $\vec{C}$ is a bit trickier than the example of the hiker because $\vec{A}$ and $\vec{B}$ are not at right angles. First, we can find the magnitude of $\vec{C}$ by using the law of cosines from trigonometry:

$$\begin{aligned} C^2 &= A^2 + B^2 - 2AB\cos(45°) \\ &= (100\text{ m})^2 + (200\text{ m})^2 - 2(100\text{ m})(200\text{ m})\cos(45°) \\ &= 21{,}720\text{ m}^2 \end{aligned}$$

Thus $C = \sqrt{21{,}720\text{ m}^2} = 147$ m. Then a second use of the law of cosines can determine angle ϕ (the Greek letter phi):

$$B^2 = A^2 + C^2 - 2AC\cos\phi$$

$$\phi = \cos^{-1}\left[\frac{A^2 + C^2 - B^2}{2AC}\right] = 106°$$

It is easier to describe $\vec{C}$ with the angle $\theta = 180° - \phi = 74°$. The bird's net displacement is

$$\vec{C} = (147\text{ m, } 74° \text{ north of west})$$

When two vectors are to be added, it is often convenient to draw them with their tails together, as shown in Figure 3.6a. To evaluate $\vec{D} + \vec{E}$, you could move vector $\vec{E}$ over to where its tail is on the tip of $\vec{D}$, then use the tip-to-tail rule of graphical addition. The gives vector $\vec{F} = \vec{D} + \vec{E}$ in Figure 3.6b. Alternatively, Figure 3.6c shows that the vector sum $\vec{D} + \vec{E}$ can be found as the diagonal of the parallelogram defined by $\vec{D}$ and $\vec{E}$. This method for vector addition, which some of you may have learned, is called the *parallelogram rule* of vector addition.

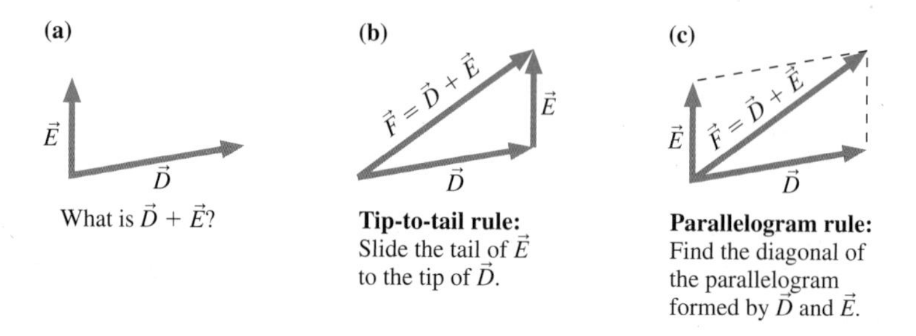

FIGURE 3.6 Two vectors can be added using the tip-to-tail rule or the parallelogram rule.

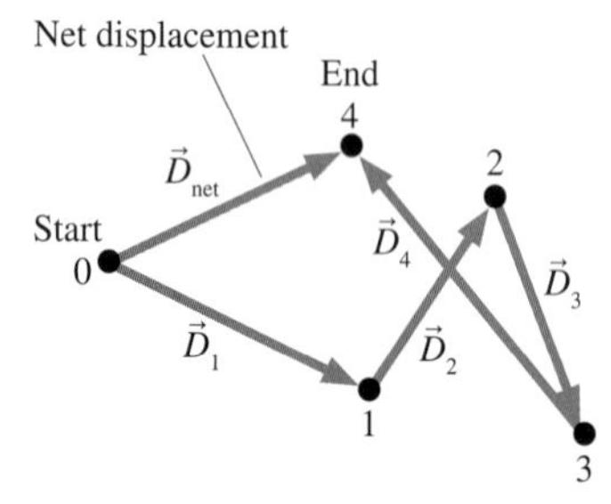

FIGURE 3.7 The net displacement after four individual displacements.

Vector addition is easily extended to more than two vectors. Figure 3.7 shows a hiker moving from initial position 0 to position 1, then position 2, then position 3, and finally arriving at position 4. These four segments are described by displacement vectors $\vec{D}_1$, $\vec{D}_2$, $\vec{D}_3$, and $\vec{D}_4$. The hiker's *net* displacement, an arrow from position 0 to position 4, is the vector $\vec{D}_{net}$. In this case,

$$\vec{D}_{net} = \vec{D}_1 + \vec{D}_2 + \vec{D}_3 + \vec{D}_4 \tag{3.5}$$

The vector sum is found by using the tip-to-tail method three times in succession.

STOP TO THINK 3.1 Which figure shows $\vec{A}_1 + \vec{A}_2 + \vec{A}_3$?

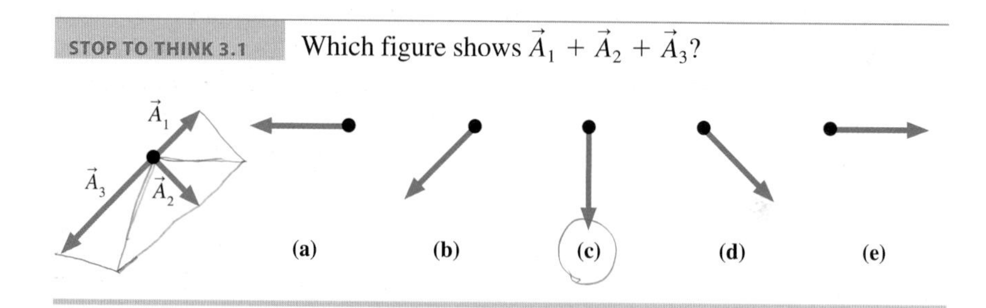

Multiplication by a Scalar

Suppose a second bird flies twice as far to the east as the bird in Example 3.1. The first bird's displacement was $\vec{A}_1 = (100 \text{ m, east})$, where a subscript has been added to denote the first bird. The second bird's displacement will then certainly be $\vec{A}_2 = (200 \text{ m, east})$. The words "twice as" indicate a multiplication, so we can say

$$\vec{A}_2 = 2\vec{A}_1$$

Multiplying a vector by a positive scalar gives another vector of *different magnitude* but pointing in the *same direction.*

Let the vector $\vec{A}$ be

$$\vec{A} = (A, \theta_A) \tag{3.6}$$

where we've specified the vector's magnitude A and direction θ_A. Now let $\vec{B} = c\vec{A}$, where c is a positive scalar constant. We define the multiplication of a vector by a scalar such that

$$\vec{B} = c\vec{A} \text{ means that } (B, \theta_B) = (cA, \theta_A) \tag{3.7}$$

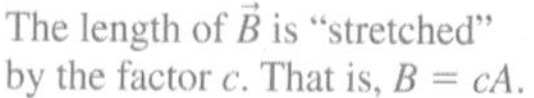

The length of $\vec{B}$ is "stretched" by the factor c. That is, $B = cA$.

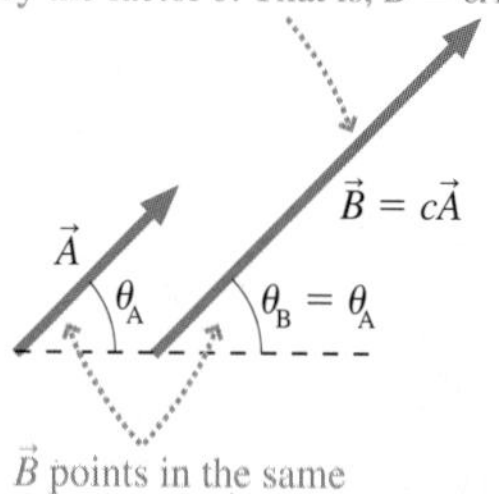

FIGURE 3.8 Multiplication of a vector by a scalar.

In other words, the vector is stretched or compressed by the factor c (i.e., vector $\vec{B}$ has magnitude $B = cA$), but $\vec{B}$ points in the same direction as $\vec{A}$. This is illustrated in Figure 3.8 on the previous page.

We used this property of vectors in Chapter 1 when we asserted that vector $\vec{a}$ points in the same direction as $\Delta\vec{v}$. From the definition

$$\vec{a} = \frac{\Delta\vec{v}}{\Delta t} = \left(\frac{1}{\Delta t}\right)\Delta\vec{v} \tag{3.8}$$

where $(1/\Delta t)$ is a scalar constant, we see that $\vec{a}$ points in the same direction as $\Delta\vec{v}$ but differs in length by the factor $(1/\Delta t)$.

Suppose we multiply $\vec{A}$ by zero. Using Equation 3.7,

$$0 \cdot \vec{A} = \vec{0} = (0 \text{ m, direction undefined}) \tag{3.9}$$

The product is a vector having zero length or magnitude. This vector is known as the **zero vector,** denoted $\vec{0}$. The direction of the zero vector is irrelevant; you cannot describe the direction of an arrow of zero length!

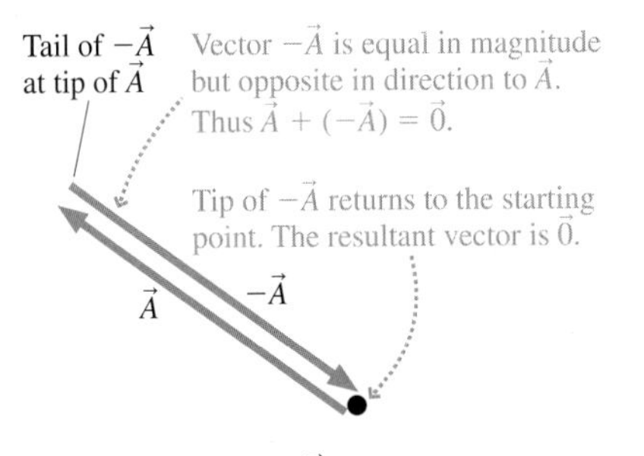

FIGURE 3.9 Vector $-\vec{A}$.

What happens if we multiply a vector by a negative number? Equation 3.7 does not apply if $c < 0$ because vector $\vec{B}$ cannot have a negative magnitude. Consider the vector $-\vec{A}$, which is equivalent to multiplying $\vec{A}$ by -1. Because

$$\vec{A} + (-\vec{A}) = \vec{0} \tag{3.10}$$

the vector $-\vec{A}$ must be such that, when it is added to $\vec{A}$, the resultant is the zero vector $\vec{0}$. In other words, the *tip* of $-\vec{A}$ must return to the *tail* of $\vec{A}$, as shown in Figure 3.9. This will be true only if $-\vec{A}$ is equal in magnitude to $\vec{A}$, but opposite in direction. Thus we can conclude that

$$-\vec{A} = (A, \text{direction opposite } \vec{A}) \tag{3.11}$$

That is, **multiplying a vector by -1 reverses its direction without changing its length.**

As an example, Figure 3.10 shows vectors $\vec{A}$, $2\vec{A}$, and $-3\vec{A}$. Multiplication by 2 doubles the length of the vector but does not change its direction. Multiplication by -3 stretches the length by a factor of 3 *and* reverses the direction.

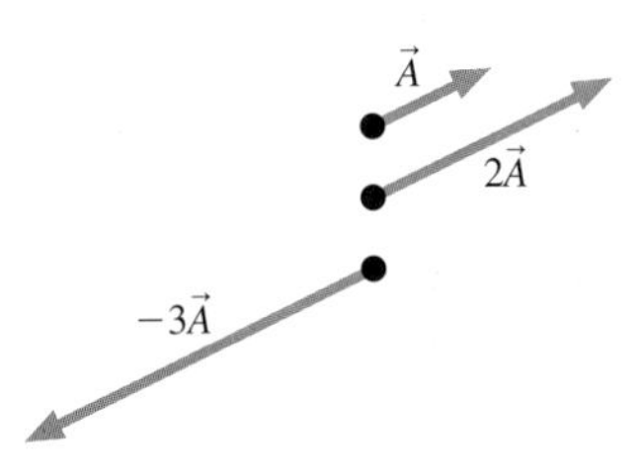

FIGURE 3.10 Vectors $\vec{A}$, $2\vec{A}$, and $-3\vec{A}$.

EXAMPLE 3.2 Velocity and displacement

Carolyn drives her car north at 30 km/hr for 1 hour, east at 60 km/hr for 2 hours, then north at 50 km/hr for 1 hour. What is Carolyn's net displacement?

SOLVE Chapter 1 defined velocity as

$$\vec{v} = \frac{\Delta\vec{r}}{\Delta t}$$

so the displacement $\Delta\vec{r}$ during the time interval Δt is $\Delta\vec{r} = (\Delta t)\vec{v}$. This is multiplication of the vector $\vec{v}$ by the scalar Δt. Carolyn's velocity during the first hour is $\vec{v}_1 = (30 \text{ km/hr, north})$, so her displacement during this interval is

$$\Delta\vec{r}_1 = (1 \text{ hour})(30 \text{ km/hr, north}) = (30 \text{ km, north})$$

Similarly,

$$\Delta\vec{r}_2 = (2 \text{ hours})(60 \text{ km/hr, east}) = (120 \text{ km, east})$$

$$\Delta\vec{r}_3 = (1 \text{ hour})(50 \text{ km/hr, north}) = (50 \text{ km, north})$$

In this case, multiplication by a scalar changes not only the length of the vector but also its units, from km/hr to km. The direction, however, is unchanged. Carolyn's net displacement is

$$\Delta\vec{r}_{\text{net}} = \Delta\vec{r}_1 + \Delta\vec{r}_2 + \Delta\vec{r}_3$$

This addition of the three vectors is shown in Figure 3.11, using the tip-to-tail method. $\Delta\vec{r}_{\text{net}}$ stretches from Carolyn's initial position to her final position. The magnitude of her net displacement is found using the Pythagorean theorem:

$$r_{\text{net}} = \sqrt{(120 \text{ km})^2 + (80 \text{ km})^2} = 144 \text{ km}$$

The direction of $\Delta\vec{r}_{\text{net}}$ is described by angle θ, which is

$$\theta = \tan^{-1}\left(\frac{80 \text{ km}}{120 \text{ km}}\right) = 33.7°$$

Thus Carolyn's net displacement is $\Delta\vec{r}_{\text{net}} = (144 \text{ km}, 33.7° \text{ north of east})$.

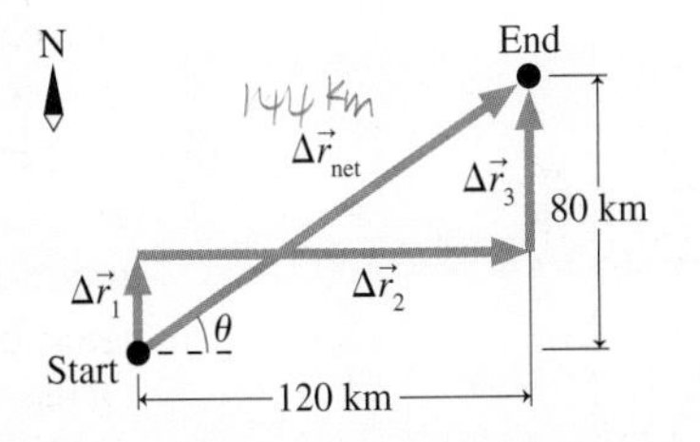

FIGURE 3.11 The net displacement is the vector sum $\Delta\vec{r}_{\text{net}} = \Delta\vec{r}_1 + \Delta\vec{r}_2 + \Delta\vec{r}_3$.

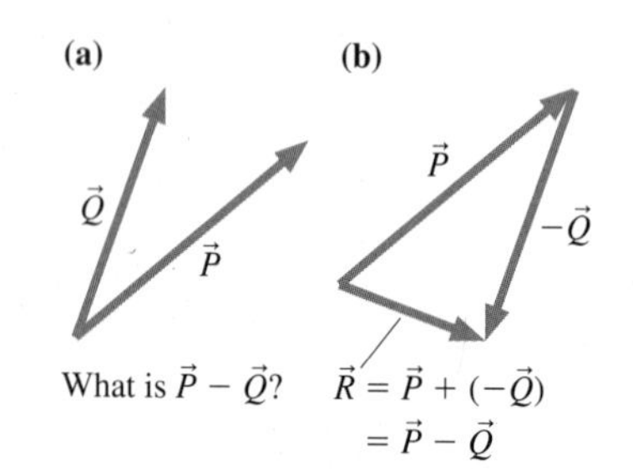

FIGURE 3.12 Vector subtraction.

Vector Subtraction

Figure 3.12a shows two vectors, $\vec{P}$ and $\vec{Q}$. What is $\vec{R} = \vec{P} - \vec{Q}$? Look back at Tactics Box 1.2 on page 11, which showed how to perform vector subtraction graphically. Figure 3.12b finds $\vec{P} - \vec{Q}$ by writing $\vec{R} = \vec{P} + (-\vec{Q})$, then using the rules of vector addition.

STOP TO THINK 3.2 Which figure shows $2\vec{A} - \vec{B}$?

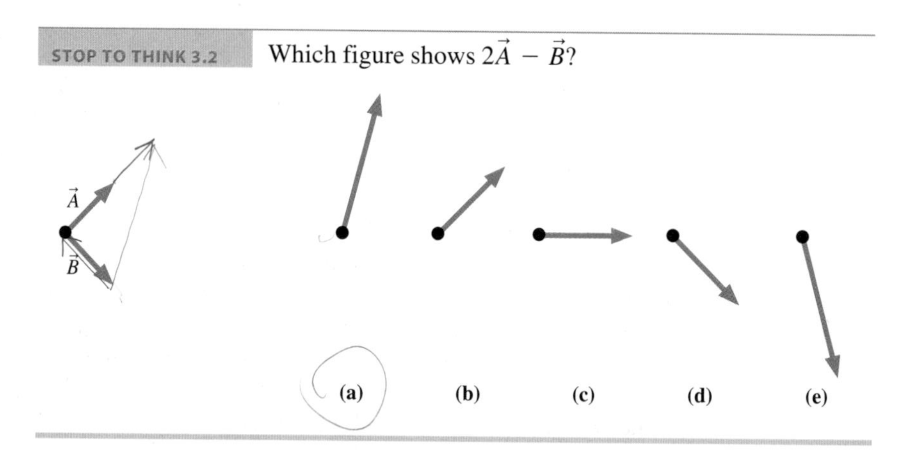

3.3 Coordinate Systems and Vector Components

Thus far, our discussion of vectors and their properties has not used a coordinate system at all. Vectors do not require a coordinate system. We can add and subtract vectors graphically, and we will do so frequently to clarify our understanding of a situation. But the graphical addition of vectors is not an especially good way to find quantitative results. In this section we will introduce a *coordinate description* of vectors that will be the basis of an easier method for doing vector calculations.

Coordinate Systems

As we noted in the first chapter, the world does not come with a coordinate system attached to it. A coordinate system is an artificially imposed grid that you place on a problem in order to make quantitative measurements. It may be helpful to think of drawing a grid on a piece of transparent plastic that you can then overlay on top of the problem. This conveys the idea that *you* choose:

- Where to place the origin, and
- How to orient the axes.

Different problem solvers may choose to use different coordinate systems; that is perfectly acceptable. However, some coordinate systems will make a problem easier to solve. Part of our goal is to learn how to choose an appropriate coordinate system for each problem.

We will generally use **Cartesian coordinates.** This is a coordinate system with the axes perpendicular to each other, forming a rectangular grid. The standard xy-coordinate system with which you are familiar is a Cartesian coordinate system. An xyz-coordinate system would be a Cartesian coordinate system in three dimensions. There are other possible coordinate systems, such as polar coordinates, but we will not be concerned with those for now.

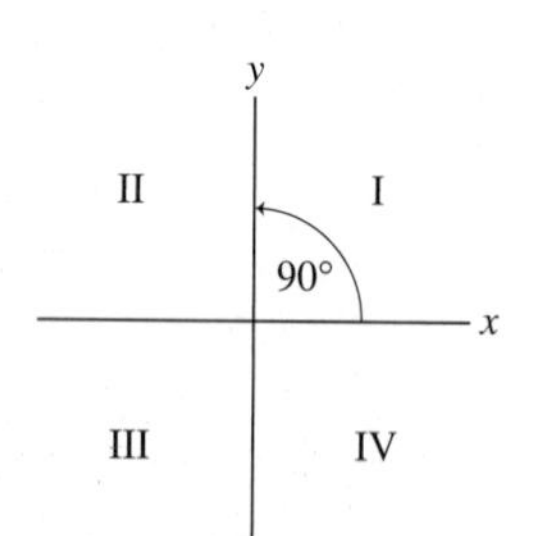

FIGURE 3.13 A conventional Cartesian coordinate system and the quadrants of the xy-plane.

The placement of the axes is not entirely arbitrary. By convention, the positive y-axis is located 90° *counterclockwise* (ccw) from the positive x-axis, as illustrated in Figure 3.13. Figure 3.13 also identifies the four **quadrants** of the coordinate system, I through IV. Notice that the quadrants are counted ccw from the positive x-axis.

Coordinate axes have a positive end and a negative end, separated by zero at the origin where the two axes cross. When you draw a coordinate system, it is important to label the axes. This is done by placing x and y labels at the *positive* ends of the axes, as in Figure 3.13. The purpose of the labels is twofold:

- To identify which axis is which, and
- To identify the positive ends of the axes.

This will be important when you need to determine whether the quantities in a problem should be assigned positive or negative values.

Component Vectors

Let's see how we can use a coordinate system to describe a vector. Figure 3.14 shows a vector $\vec{A}$ and an xy-coordinate system that we've chosen. Once the directions of the axes are known, we can define two new vectors *parallel to the axes* that we call the **component vectors** of $\vec{A}$. Vector $\vec{A}_x$, called the *x-component vector,* is the projection of $\vec{A}$ along the x-axis. Vector $\vec{A}_y$, the *y-component vector*, is the projection of $\vec{A}$ along the y-axis. Notice that the component vectors are perpendicular to each other.

You can see, using the parallelogram rule, that $\vec{A}$ is the vector sum of the two component vectors:

$$\vec{A} = \vec{A}_x + \vec{A}_y \tag{3.12}$$

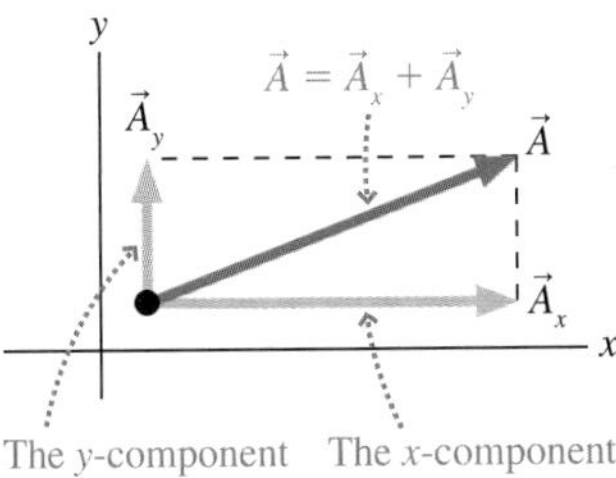
y
$\vec{A} = \vec{A}_x + \vec{A}_y$
$\vec{A}_y$
$\vec{A}$
$\vec{A}_x$
x
The y-component vector is parallel to the y-axis.
The x-component vector is parallel to the x-axis.

FIGURE 3.14 Component vectors $\vec{A}_x$ and $\vec{A}_y$ are drawn parallel to the coordinate axes such that $\vec{A} = \vec{A}_x + \vec{A}_y$.

In essence, we have broken vector $\vec{A}$ into two perpendicular vectors that are parallel to the coordinate axes. This process is called the **decomposition** of vector $\vec{A}$ into its component vectors.

NOTE ▶ It is not necessary for the tail of $\vec{A}$ to be at the origin. All we need to know is the *orientation* of the coordinate system so that we can draw $\vec{A}_x$ and $\vec{A}_y$ parallel to the axes. ◀

Components

You learned in Chapter 2 to give the one-dimensional kinematic variable v_x a positive sign if the velocity vector $\vec{v}$ points toward the positive end of the x-axis, a negative sign if $\vec{v}$ points in the negative x-direction. The basis of that rule is that v_x is what we call the *x-component* of the velocity vector. We need to extend this idea to vectors in general.

Suppose vector $\vec{A}$ has been decomposed into component vectors $\vec{A}_x$ and $\vec{A}_y$ parallel to the coordinate axes. We can describe each component vector with a single number (a scalar) called the **component.** The *x-component* and *y-component* of vector $\vec{A}$, denoted A_x and A_y, are determined as follows:

TACTICS BOX 3.1 Determining the components of a vector

1. The absolute value $|A_x|$ of the x-component A_x is the magnitude of the component vector $\vec{A}_x$.
2. The *sign* of A_x is positive if $\vec{A}_x$ points in the positive x-direction, negative if $\vec{A}_x$ points in the negative x-direction.
3. The y-component A_y is determined similarly.

In other words, the component A_x tells us two things: how big $\vec{A}_x$ is and, with its sign, which end of the axis $\vec{A}_x$ points toward. Figure 3.15 on the next page shows three examples of determining the components of a vector.

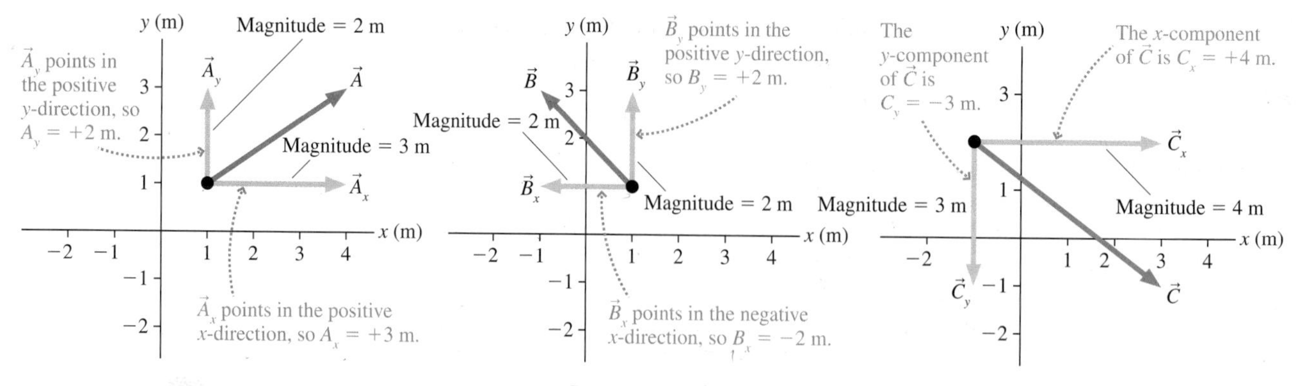

FIGURE 3.15 Determining the components of a vector.

NOTE ▶ Beware of the somewhat confusing terminology. $\vec{A}_x$ and $\vec{A}_y$ are called *component vectors,* whereas A_x and A_y are simply called *components.* The components A_x and A_y are scalars—just numbers (with units)—so make sure you do *not* put arrow symbols over the components. ◀

Much of physics is expressed in the language of vectors. We will frequently need to decompose a vector into its components. We will also need to "reassemble" a vector from its components. In other words, we need to move back and forth between the graphical and the component representations of a vector. To do so we apply geometry and trigonometry.

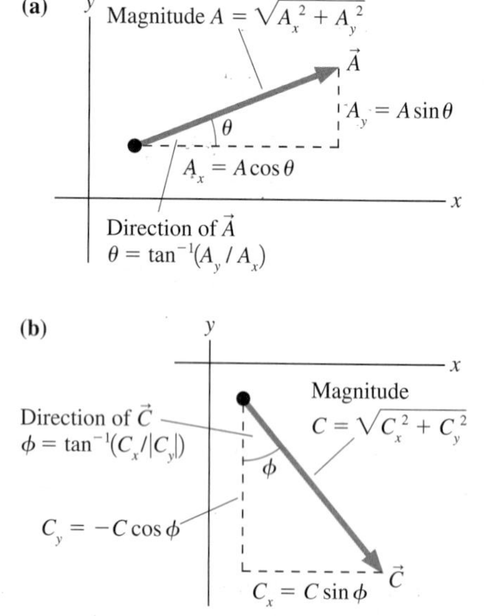

FIGURE 3.16 Moving between the graphical representation and the component representation.

Consider first the problem of decomposing a vector into its x- and y-components. Figure 3.16a shows a vector $\vec{A}$ at angle θ from the x-axis. It is *essential* to use a picture or diagram such as this to define the angle you are using to describe the vector's direction.

$\vec{A}$ points to the right and up, so Tactics Box 3.1 tells us that the components A_x and A_y are both positive. We can use trigonometry to find

$$A_x = A\cos\theta \qquad A_y = A\sin\theta \tag{3.13}$$

where A is the magnitude, or length, of $\vec{A}$. These equations convert the length and angle description of vector $\vec{A}$ into the vector's components, but they are correct *only* if $\vec{A}$ is in the first quadrant.

Figure 3.16b shows vector $\vec{C}$ in the fourth quadrant. In this case, where the component vector $\vec{A}_y$ is pointing *down,* in the negative y-direction, the y-component C_y is a *negative* number. The angle ϕ is measured from the y-axis, so the components of $\vec{C}$ are

$$C_x = C\sin\phi \qquad C_y = -C\cos\phi \tag{3.14}$$

The role of sine and cosine is reversed from that in Equations 3.13 because we are using a different angle.

NOTE ▶ Each decomposition requires that you pay close attention to the direction in which the vector points and the angles that are defined. The minus sign, when needed, must be inserted manually. ◀

We can also go in the opposite direction and determine the length and angle of a vector from its x- and y-components. Because A in Figure 3.16a is the hypotenuse of a right triangle, its length is given by the Pythagorean theorem:

$$A = \sqrt{A_x^2 + A_y^2} \tag{3.15}$$

Similarly, the tangent of angle θ is the ratio of the far side to the adjacent side, so

$$\theta = \tan^{-1}\left(\frac{A_y}{A_x}\right) \tag{3.16}$$

where $\tan^{-1}$ is the inverse tangent function. Equations 3.15 and 3.16 can be thought of as the "reverse" of Equations 3.13.

Equation 3.15 always works for finding the length or magnitude of a vector because the squares eliminate any concerns over the signs of the components. But finding the angle, just like finding the components, requires close attention to how the angle is defined and to the signs of the components. For example, finding the angle of vector $\vec{C}$ in Figure 3.16b requires the length of C_y *without* the minus sign. Thus vector $\vec{C}$ has magnitude and direction

$$C = \sqrt{C_x^2 + C_y^2}$$
$$\phi = \tan^{-1}\left(\frac{C_x}{|C_y|}\right) \tag{3.17}$$

Notice that the roles of x and y differ from those in Equation 3.16.

EXAMPLE 3.3 Finding the components of an acceleration vector

Find the x- and y-components of the acceleration vector $\vec{a}$ shown in Figure 3.17a.

VISUALIZE It's important to *draw* vectors. Figure 3.17b shows the original vector $\vec{a}$ decomposed into components parallel to the axes.

SOLVE The acceleration vector $\vec{a} = (6\ \text{m/s}^2,\ 30°$ below the negative x-axis) points to the left (negative x-direction) and down (negative y-direction), so the components a_x and a_y are both negative:

$$a_x = -a\cos 30° = -(6\ \text{m/s}^2)\cos 30° = -5.2\ \text{m/s}^2$$
$$a_y = -a\sin 30° = -(6\ \text{m/s}^2)\sin 30° = -3.0\ \text{m/s}^2$$

ASSESS The units of a_x and a_y are the same as the units of vector $\vec{a}$. Notice that we had to insert the minus signs manually by observing that the vector is in the third quadrant.

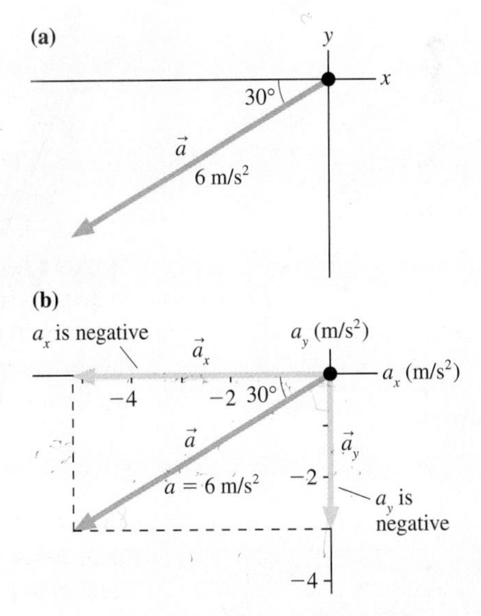

FIGURE 3.17 The acceleration vector $\vec{a}$ of Example 3.3.

EXAMPLE 3.4 Finding the direction of motion

Figure 3.18a shows a particle's velocity vector $\vec{v}$. Determine the particle's speed and direction of motion.

VISUALIZE Figure 3.18b shows the components v_x and v_y and defines an angle θ with which we can specify the direction of motion.

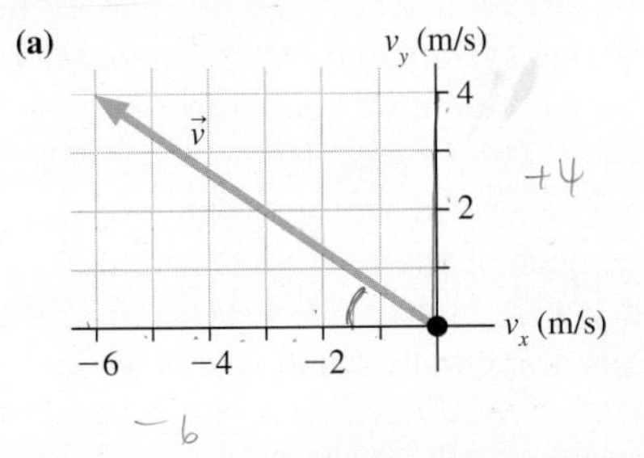

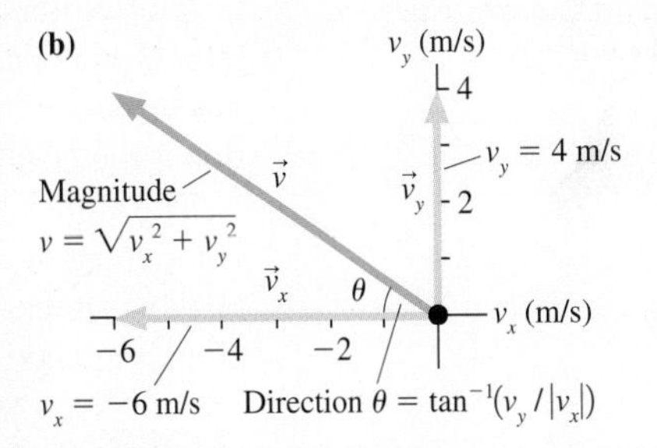

FIGURE 3.18 The velocity vector $\vec{v}$ of Example 3.4.

SOLVE We can read the components of $\vec{v}$ directly from the axes: $v_x = -6$ m/s and $v_y = 4$ m/s. Notice that v_x is negative. This is enough information to find the particle's speed v, which is the magnitude of $\vec{v}$:

$$v = \sqrt{v_x^2 + v_y^2} = \sqrt{(-6\text{ m/s})^2 + (4\text{ m/s})^2} = 7.2\text{ m/s}$$

From trigonometry, angle θ is

$$\theta = \tan^{-1}\left(\frac{v_y}{|v_x|}\right) = \tan^{-1}\left(\frac{4\text{ m/s}}{6\text{ m/s}}\right) = 33.7°$$

The absolute value signs are necessary because v_x is a negative number. The velocity vector $\vec{v}$ can be written in terms of the speed and the direction of motion as

$$\vec{v} = (7.2\text{ m/s}, 33.7° \text{ above the negative } x\text{-axis})$$

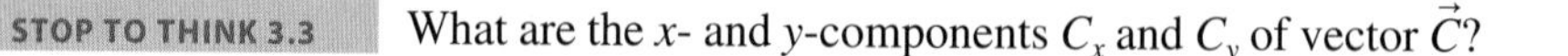

STOP TO THINK 3.3 What are the x- and y-components C_x and C_y of vector $\vec{C}$?

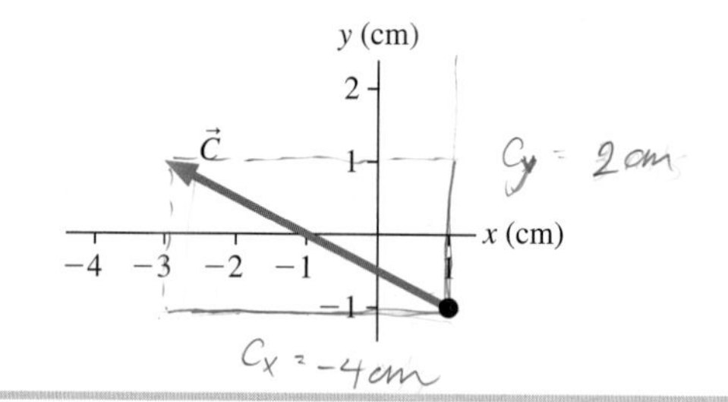

3.4 Vector Algebra

Vector components are a powerful tool for doing mathematics with vectors. In this section you'll learn how to use components to add and subtract vectors. First, we'll introduce an efficient way to write a vector in terms of its components.

Unit Vectors

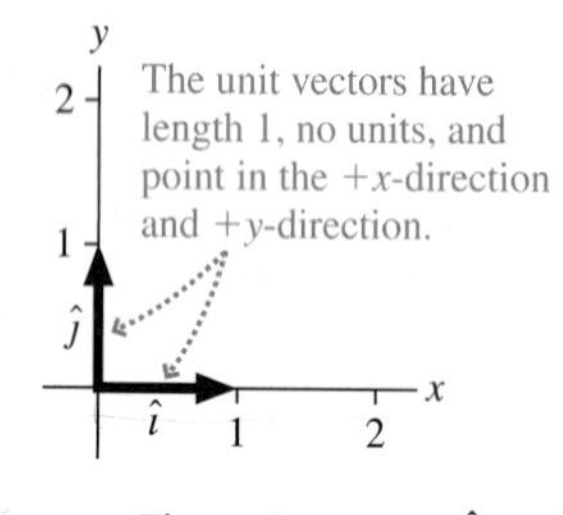

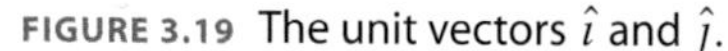

FIGURE 3.19 The unit vectors $\hat{\imath}$ and $\hat{j}$.

The vectors (1, +x-direction) and (1, +y-direction), shown in Figure 3.19, have some interesting and useful properties. Each has a magnitude of 1, no units, and is parallel to a coordinate axis. A vector with these properties is called a **unit vector.** These unit vectors have the special symbols

$$\hat{\imath} \equiv (1, +x\text{-direction})$$
$$\hat{j} \equiv (1, +y\text{-direction})$$

The notation $\hat{\imath}$ (read "i hat") and $\hat{j}$ (read "j hat") indicates a unit vector with a magnitude of 1.

Unit vectors establish the directions of the positive axes of the coordinate system. Our choice of a coordinate system may be arbitrary, but once we decide to place a coordinate system on a problem we need something to tell us "That direction is the positive x-direction." This is what the unit vectors do.

The unit vectors provide a useful way to write component vectors. The component vector $\vec{A}_x$ is the piece of vector $\vec{A}$ that is parallel to the x-axis. Similarly, $\vec{A}_y$ is parallel to the y-axis. Because, by definition, the vector $\hat{\imath}$ points along the x-axis and $\hat{j}$ points along the y-axis, we can write

$$\vec{A}_x = A_x\hat{\imath}$$
$$\vec{A}_y = A_y\hat{j} \tag{3.18}$$

Equations 3.18 separate each component vector into a scalar piece of length A_x (or A_y) and a directional piece $\hat{\imath}$ (or $\hat{\jmath}$). The full decomposition of vector $\vec{A}$ can then be written

$$\vec{A} = \vec{A}_x + \vec{A}_y = A_x\hat{\imath} + A_y\hat{\jmath} \tag{3.19}$$

Figure 3.20 shows how the unit vectors and the components fit together to form vector $\vec{A}$.

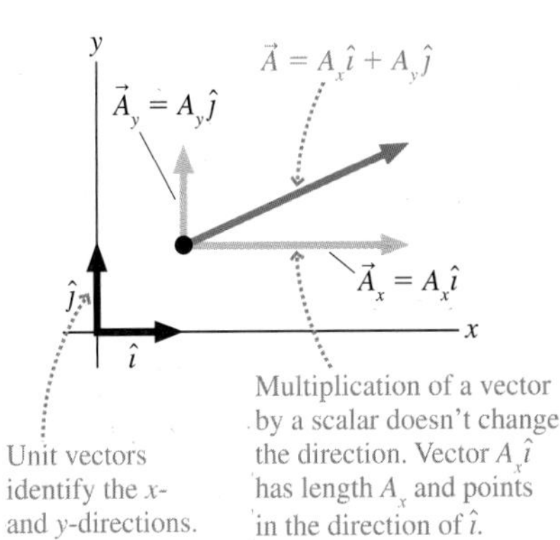

FIGURE 3.20 The decomposition of vector $\vec{A}$ is $A_x\hat{\imath} + A_y\hat{\jmath}$.

NOTE ► In three dimensions, the unit vector along the $+z$-direction is called $\hat{k}$, and to describe vector $\vec{A}$ we would include an additional component vector $\vec{A}_z = A_z\hat{k}$. ◄

You may have learned in a math class to think of vectors as pairs or triplets of numbers, such as $(4, -2, 5)$. This is another, and completely equivalent, way to write the components of a vector. Thus we could write, for a vector in three dimensions,

$$\vec{B} = 4\hat{\imath} - 2\hat{\jmath} + 5\hat{k} = (4,-2,5)$$

You will find the notation using unit vectors to be more convenient for the equations we will use in physics, but rest assured that you already know a lot about vectors if you learned about them as pairs or triplets of numbers.

EXAMPLE 3.5 Run rabbit run!

A rabbit, escaping a fox, runs 40° north of west at 10 m/s. A coordinate system is established with the positive x-axis to the east and the positive y-axis to the north. Write the rabbit's velocity in terms of components and unit vectors.

VISUALIZE Figure 3.21 shows the rabbit's velocity vector and the coordinate axes. We're showing a velocity vector, so the axes are labeled v_x and v_y rather than x and y.

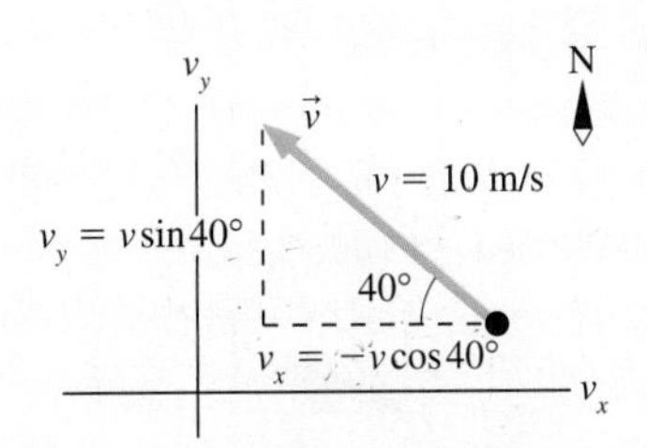

FIGURE 3.21 The velocity vector $\vec{v}$ is decomposed into components v_x and v_y.

SOLVE 10 m/s is the rabbit's *speed,* not its velocity. The velocity, which includes directional information, is

$$\vec{v} = (10 \text{ m/s, } 40° \text{ north of west})$$

Vector $\vec{v}$ points to the left and up, so the components v_x and v_y are negative and positive, respectively. The components are

$$v_x = -(10 \text{ m/s})\cos 40° = -7.66 \text{ m/s}$$
$$v_y = +(10 \text{ m/s})\sin 40° = 6.43 \text{ m/s}$$

With v_x and v_y now known, the rabbit's velocity vector is

$$\vec{v} = v_x\hat{\imath} + v_y\hat{\jmath} = (-7.66\hat{\imath} + 6.43\hat{\jmath}) \text{ m/s}$$

Notice that we've pulled the units to the end, rather than writing them with each component.

ASSESS Notice that the minus sign for v_x was inserted manually. Signs don't occur automatically; you have to set them after checking the vector's direction.

Working with Vectors

You learned in Section 3.2 how to add vectors graphically, but it is a tedious problem in geometry and trigonometry to find precise values for the magnitude and direction of the resultant. The addition and subtraction of vectors becomes much easier if we use components and unit vectors.

To see this, let's evaluate the vector sum $\vec{D} = \vec{A} + \vec{B} + \vec{C}$. To begin, write this sum in terms of the components of each vector:

$$\vec{D} = D_x\hat{\imath} + D_y\hat{\jmath} = \vec{A} + \vec{B} + \vec{C} = (A_x\hat{\imath} + A_y\hat{\jmath}) + (B_x\hat{\imath} + B_y\hat{\jmath}) + (C_x\hat{\imath} + C_y\hat{\jmath}) \tag{3.20}$$

We can group together all the x-components and all the y-components on the right side, in which case Equation 3.20 is

$$(D_x)\hat{\imath} + (D_y)\hat{\jmath} = (A_x + B_x + C_x)\hat{\imath} + (A_y + B_y + C_y)\hat{\jmath} \qquad (3.21)$$

Comparing the x- and y-components on the left and right sides of Equation 3.21, we find:

$$\begin{aligned} D_x &= A_x + B_x + C_x \\ D_y &= A_y + B_y + C_y \end{aligned} \qquad (3.22)$$

Stated in words, Equation 3.22 says that we can perform vector addition by adding the x-components of the individual vectors to give the x-component of the resultant and by adding the y-components of the individual vectors to give the y-component of the resultant. This method of vector addition is called **algebraic addition.**

EXAMPLE 3.6 Using algebraic addition to find a displacement

Example 3.1 was about a bird that flew 100 m to the east, then 200 m to the northwest. Use the algebraic addition of vectors to find the bird's net displacement. Compare the result to Example 3.1.

VISUALIZE Figure 3.22 shows displacement vectors $\vec{A} = (100 \text{ m, east})$ and $\vec{B} = (200 \text{ m, northwest})$. We draw vectors tip-to-tail if we are going to add them graphically, but it's usually easier to draw them all from the origin if we are going to use algebraic addition.

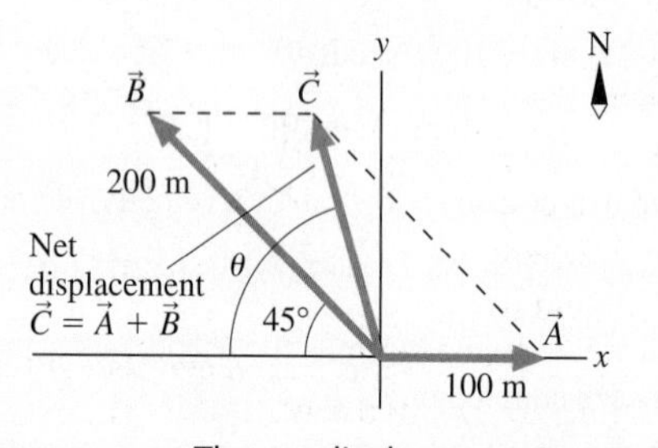

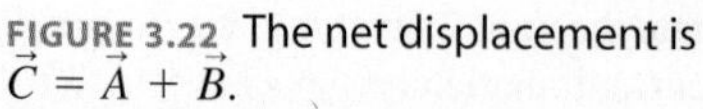
FIGURE 3.22 The net displacement is $\vec{C} = \vec{A} + \vec{B}$.

SOLVE To add the vectors algebraically we must know their components. From the figure these are seen to be

$$\begin{aligned} \vec{A} &= 100\,\hat{\imath}\text{ m} \\ \vec{B} &= (-200\cos 45^\circ\,\hat{\imath} + 200\sin 45^\circ\,\hat{\jmath})\text{ m} \\ &= (-141\hat{\imath} + 141\hat{\jmath})\text{ m} \end{aligned}$$

Notice that vector quantities must include units. Also notice, as you would expect from the figure, that $\vec{B}$ has a negative x-component. Adding $\vec{A}$ and $\vec{B}$ by components gives

$$\begin{aligned} \vec{C} = \vec{A} + \vec{B} &= 100\hat{\imath}\text{ m} + (-141\hat{\imath} + 141\hat{\jmath})\text{ m} \\ &= (100\text{ m} - 141\text{ m})\hat{\imath} + (141\text{ m})\hat{\jmath} \\ &= (-41\hat{\imath} + 141\hat{\jmath})\text{ m} \end{aligned}$$

This would be a perfectly acceptable answer for many purposes. However, we need to calculate the magnitude and direction of $\vec{C}$ if we want to compare this result to our earlier answer. The magnitude of $\vec{C}$ is

$$C = \sqrt{C_x^2 + C_y^2} = \sqrt{(-41\text{ m})^2 + (141\text{ m})^2} = 147\text{ m}$$

The angle θ, as defined in Figure 3.22, is

$$\theta = \tan^{-1}\left(\frac{C_y}{|C_x|}\right) = \tan^{-1}\left(\frac{141\text{ m}}{41\text{ m}}\right) = 74^\circ$$

Thus $\vec{C} = (147\text{ m}, 74^\circ \text{ north of west})$, in perfect agreement with Example 3.1.

Vector subtraction and the multiplication of a vector by a scalar, using components, are very much like vector addition. To find $\vec{R} = \vec{P} - \vec{Q}$ we would compute

$$\begin{aligned} R_x &= P_x - Q_x \\ R_y &= P_y - Q_y \end{aligned} \qquad (3.23)$$

Similarly, $\vec{T} = c\vec{S}$ would be

$$\begin{aligned} T_x &= cS_x \\ T_y &= cS_y \end{aligned} \qquad (3.24)$$

The next few chapters will make frequent use of *vector equations*. For example, you will learn that the equation to calculate the force on a car skidding to a stop is

$$\vec{F} = \vec{n} + \vec{w} + \mu\vec{f} \tag{3.25}$$

The following general rule is used to evaluate such an equation:

The x-component of the left-hand side of a vector equation is found by doing scalar calculations (addition, subtraction, multiplication) with just the x-components of all the vectors on the right-hand side. A separate set of calculations uses just the y-components and, if needed, the z-components.

Thus Equation 3.25 is really just a shorthand way of writing three simultaneous equations:

$$\begin{aligned} F_x &= n_x + w_x + \mu f_x \\ F_y &= n_y + w_y + \mu f_y \\ F_z &= n_z + w_z + \mu f_z \end{aligned} \tag{3.26}$$

In other words, a vector equation is interpreted as meaning: Equate the x-components on both sides of the equals sign, then equate the y-components, and then the z-components. Vector notation allows us to write these three equations in a much more compact form.

Tilted Axes and Arbitrary Directions

As we've noted, the coordinate system is entirely your choice. It is a grid that you impose on the problem in a manner that will make the problem easiest to solve. We will soon meet problems where it will be convenient to tilt the axes of the coordinate system, such as those shown in Figure 3.23. Although you may not have seen such a coordinate system before, it is perfectly legitimate. The axes are perpendicular, and the y-axis is oriented correctly with respect to the x-axis. While we are used to having the x-axis horizontal, there is no requirement that it has to be that way.

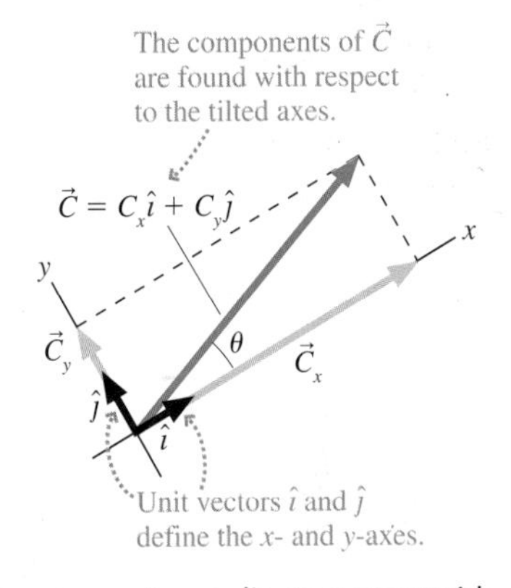

FIGURE 3.23 A coordinate system with tilted axes.

Finding components with tilted axes is no harder than what we have done so far. Vector $\vec{C}$ in Figure 3.23 can be decomposed $\vec{C} = C_x\hat{\imath} + C_y\hat{\jmath}$, where $C_x = C\cos\theta$ and $C_y = C\sin\theta$. Note that the unit vectors $\hat{\imath}$ and $\hat{\jmath}$ correspond to the *axes,* not to "horizontal" and "vertical," so they are also tilted.

Tilted axes are useful if you need to determine component vectors "parallel to" and "perpendicular to" an arbitrary line or surface. For example, we will soon need to decompose a force vector into component vectors parallel to and perpendicular to a surface.

Figure 3.24a shows a vector $\vec{A}$ and a tilted line. Suppose we would like to find the component vectors of $\vec{A}$ parallel and perpendicular to the line. To do so, establish a tilted coordinate system with the x-axis parallel to the line and the y-axis perpendicular to the line, as shown in Figure 3.24b. Then $\vec{A}_x$ is equivalent to vector $\vec{A}_\parallel$, the component of $\vec{A}$ parallel to the line, and $\vec{A}_y$ is equivalent to the perpendicular component vector $\vec{A}_\perp$. Notice that $\vec{A} = \vec{A}_\parallel + \vec{A}_\perp$.

If ϕ is the angle between $\vec{A}$ and the line, we can easily calculate the parallel and perpendicular components of $\vec{A}$:

$$\begin{aligned} A_\parallel &= A_x = A\cos\phi \\ A_\perp &= A_y = A\sin\phi \end{aligned} \tag{3.27}$$

It was not necessary to have the tail of $\vec{A}$ on the line in order to find a component of $\vec{A}$ parallel to the line. The line simply indicates a direction, and the component vector $\vec{A}_\parallel$ points in that direction.

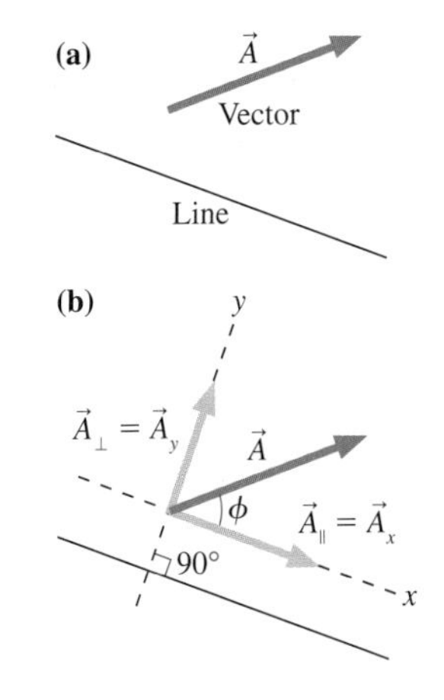

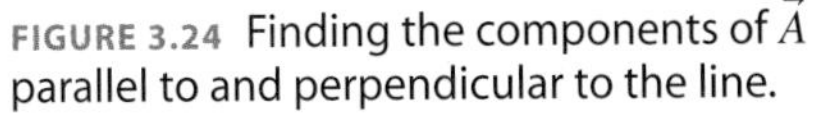

FIGURE 3.24 Finding the components of $\vec{A}$ parallel to and perpendicular to the line.

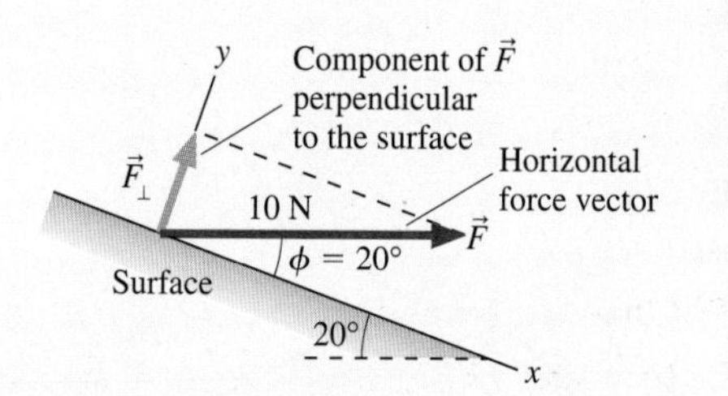

FIGURE 3.25 Finding the component of a force vector perpendicular to a surface.

EXAMPLE 3.7 Finding the force perpendicular to a surface

A horizontal force $\vec{F}$ with a strength of 10 N is applied to a surface. (You'll learn in Chapter 4 that force is a vector quantity measured in units of *newtons,* abbreviated N.) The surface is tilted at a 20° angle. Find the component of the force vector perpendicular to the surface.

VISUALIZE Figure 3.25 shows a horizontal force $\vec{F}$ applied to the surface. A tilted coordinate system has its y-axis perpendicular to the surface, so the perpendicular component is $F_{\perp} = F_y$.

SOLVE From geometry, the force vector $\vec{F}$ makes an angle $\phi = 20°$ with the tilted x-axis. The perpendicular component of $\vec{F}$ is thus

$$F_{\perp} = F \sin 20° = (10\text{ N}) \sin 20° = 3.42\text{ N}$$

STOP TO THINK 3.4 Angle ϕ that specifies the direction of $\vec{C}$ is given by

a. $\tan^{-1}(C_x/C_y)$.
b. $\tan^{-1}(C_x/|C_y|)$.
c. $\tan^{-1}(|C_x|/|C_y|)$.
d. $\tan^{-1}(C_y/C_x)$.
e. $\tan^{-1}(C_y/|C_x|)$.
f. $\tan^{-1}(|C_y|/|C_x|)$.

SUMMARY

The goal of Chapter 3 has been to learn how vectors are represented and used.

GENERAL PRINCIPLES

A vector is a quantity described by both a magnitude and a direction.

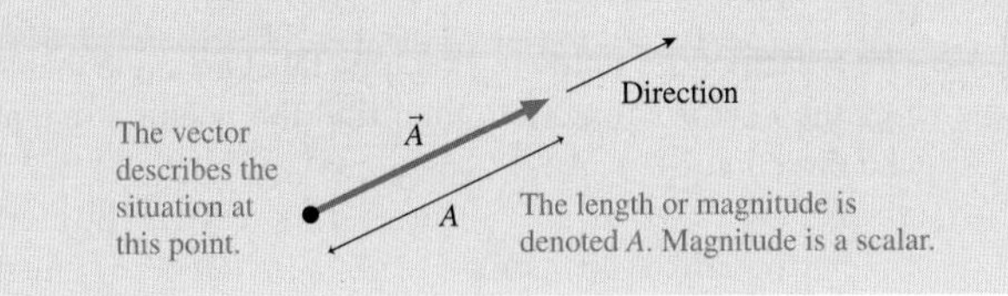

Unit Vectors

Unit vectors have magnitude 1 and no units. Unit vectors $\hat{\imath}$ and $\hat{\jmath}$ define the directions of the x- and y-axes.

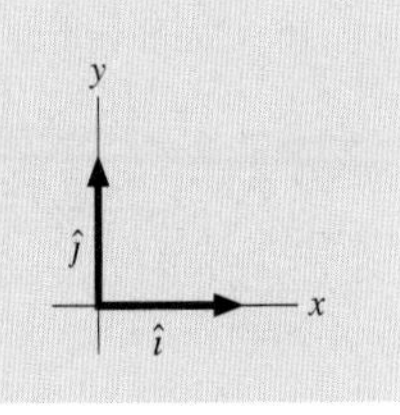

USING VECTORS

Components

The component vectors are parallel to the x- and y-axes.

$$\vec{A} = \vec{A}_x + \vec{A}_y = A_x\hat{\imath} + A_y\hat{\jmath}$$

In the figure at the right, for example:

$$A_x = A\cos\theta \qquad A = \sqrt{A_x^2 + A_y^2}$$

$$A_y = A\sin\theta \qquad \theta = \tan^{-1}(A_y/A_x)$$

▶ Minus signs need to be included if the vector points down or left.

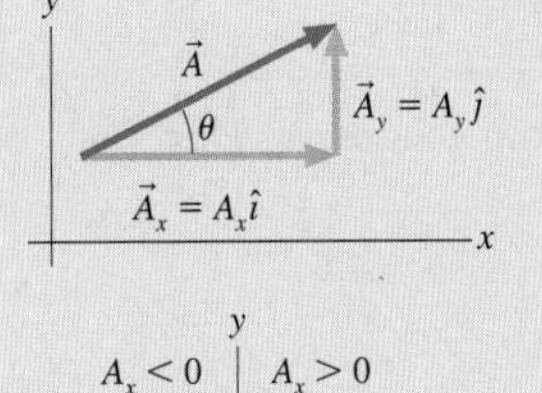

$A_x < 0$, $A_y > 0$	$A_x > 0$, $A_y > 0$
$A_x < 0$, $A_y < 0$	$A_x > 0$, $A_y < 0$

The components A_x and A_y are the magnitudes of the component vectors $\vec{A}_x$ and $\vec{A}_y$ *and* a plus or minus sign to show whether the component vector points toward the positive end or the negative end of the axis.

Working Graphically

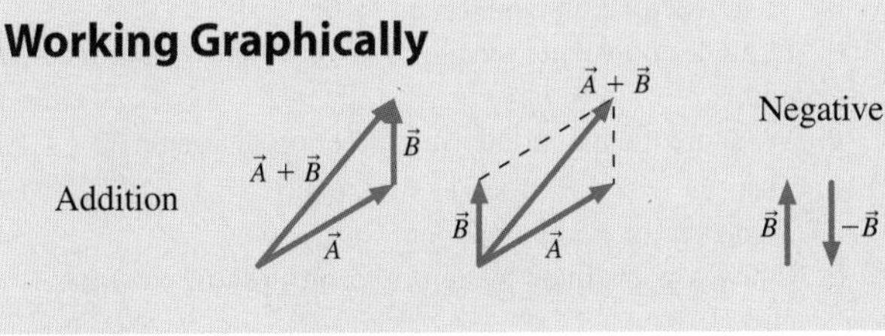

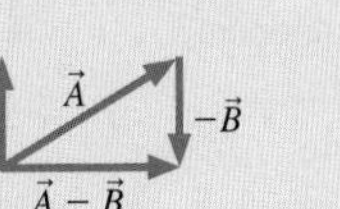

Working Algebraically

Vector calculations are done component by component.

$$\vec{C} = 2\vec{A} + \vec{B} \quad \text{means} \quad \begin{cases} C_x = 2A_x + B_x \\ C_y = 2A_y + B_y \end{cases}$$

The magnitude of $\vec{C}$ is then $C = \sqrt{C_x^2 + C_y^2}$ and its direction is found using $\tan^{-1}$.

TERMS AND NOTATION

scalar quantity
vector quantity
magnitude
resultant vector
graphical addition
zero vector, $\vec{0}$
Cartesian coordinates
quadrants
component vector
decomposition
component
unit vector, $\hat{\imath}$ or $\hat{\jmath}$
algebraic addition

EXERCISES AND PROBLEMS

Exercises

Section 3.2 Properties of Vectors

1. a. Can a vector have nonzero magnitude if a component is zero? If no, why not? If yes, give an example.
   b. Can a vector have zero magnitude and a nonzero component? If no, why not? If yes, give an example.
2. Suppose $\vec{C} = \vec{A} + \vec{B}$.
   a. Under what circumstances does $C = A + B$?
   b. Could $C = A - B$? If so, how? If not, why not?
3. Suppose $\vec{C} = \vec{A} - \vec{B}$.
   a. Under what circumstances does $C = A - B$?
   b. Could $C = A + B$? If so, how? If not, why not?
4. Trace the vectors in Figure Ex3.4 onto your paper. Then find (a) $\vec{A} + \vec{B}$ and (b) $\vec{A} - \vec{B}$.

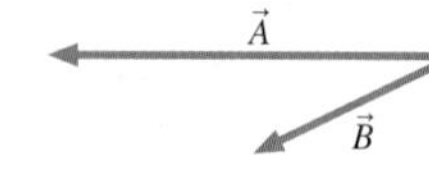

FIGURE EX3.4

5. Trace the vectors in Figure Ex3.5 onto your paper. Then find (a) $\vec{A} + \vec{B}$ and (b) $\vec{A} - \vec{B}$.

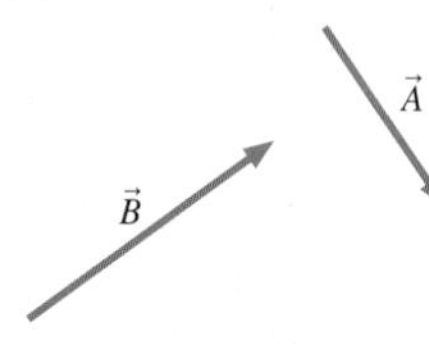

FIGURE EX3.5

Section 3.3 Coordinate Systems and Vector Components

6. A position vector in the first quadrant has an x-component of 6 m and a magnitude of 10 m. What is the value of its y-component?
7. A velocity vector 40° below the positive x-axis has a y-component of 10 m/s. What is the value of its x-component?
8. a. What are the x- and y-components of vector $\vec{E}$ in terms of the angle θ and the magnitude E shown in Figure Ex3.8?
   b. For the same vector, what are the x- and y-components in terms of the angle ϕ and the magnitude E?

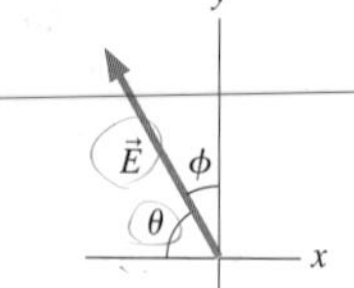

FIGURE EX3.8

9. Draw each of the following vectors, then find its x- and y-components.
   a. $\vec{r} = (100\text{ m}, 45°\text{ below }+x\text{-axis})$
   b. $\vec{v} = (300\text{ m/s}, 20°\text{ above }+x\text{-axis})$
   c. $\vec{a} = (5.0\text{ m/s}^2, -y\text{-direction})$
   d. $\vec{F} = (50\text{ N}, 36.9°\text{ above }-x\text{-axis})$
10. Draw each of the following vectors, then find its x- and y-components.
   a. $\vec{r} = (2\text{ km}, 30°\text{ left of }+y\text{-axis})$
   b. $\vec{v} = (5\text{ cm/s}, -x\text{-direction})$
   c. $\vec{a} = (10\text{ m/s}^2, 40°\text{ left of }-y\text{-axis})$
   d. $\vec{F} = (50\text{ N}, 36.9°\text{ right of }+y\text{-axis})$
11. Let $\vec{C} = (3.15\text{ m}, 15°\text{ above the negative }x\text{-axis})$ and $\vec{D} = (25.67, 30°\text{ to the right of the negative }y\text{-axis})$. Find the magnitude, the x-component, and the y-component of each vector.
12. The quantity called the *electric field* is a vector. The electric field inside a scientific instrument is $\vec{E} = (125\hat{\imath} - 250\hat{\jmath})$ V/m, where V/m stands for volts per meter. What are the magnitude and direction of the electric field?

Section 3.4 Vector Algebra

13. Draw each of the following vectors, label an angle that specifies the vector's direction, then find the vector's magnitude and direction.
   a. $\vec{A} = 4\hat{\imath} - 6\hat{\jmath}$
   b. $\vec{r} = (50\hat{\imath} + 80\hat{\jmath})$ m
   c. $\vec{v} = (-20\hat{\imath} + 40\hat{\jmath})$ m/s
   d. $\vec{a} = (2\hat{\imath} - 6\hat{\jmath})$ m/s^2
14. Draw each of the following vectors, label an angle that specifies the vector's direction, then find its magnitude and direction.
   a. $\vec{B} = -4\hat{\imath} + 4\hat{\jmath}$
   b. $\vec{r} = (-2\hat{\imath} - \hat{\jmath})$ cm
   c. $\vec{v} = (-10\hat{\imath} - 100\hat{\jmath})$ mph
   d. $\vec{a} = (20\hat{\imath} + 10\hat{\jmath})$ m/s^2
15. Let $\vec{A} = 2\hat{\imath} + 3\hat{\jmath}$ and $\vec{B} = 4\hat{\imath} - 2\hat{\jmath}$.
   a. Draw a coordinate system and on it show vectors $\vec{A}$ and $\vec{B}$.
   b. Use graphical vector subtraction to find $\vec{C} = \vec{A} - \vec{B}$.
16. Let $\vec{A} = 5\hat{\imath} + 2\hat{\jmath}$, $\vec{B} = -3\hat{\imath} - 5\hat{\jmath}$, and $\vec{C} = \vec{A} + \vec{B}$.
   a. Write vector $\vec{C}$ in component form.
   b. Draw a coordinate system and on it show vectors $\vec{A}$, $\vec{B}$, and $\vec{C}$.
   c. What are the magnitude and direction of vector $\vec{C}$?
17. Let $\vec{A} = 5\hat{\imath} + 2\hat{\jmath}$, $\vec{B} = -3\hat{\imath} - 5\hat{\jmath}$, and $\vec{D} = \vec{A} - \vec{B}$.
   a. Write vector $\vec{D}$ in component form.
   b. Draw a coordinate system and on it show vectors $\vec{A}$, $\vec{B}$, and $\vec{D}$.
   c. What are the magnitude and direction of vector $\vec{D}$?
18. Let $\vec{A} = 5\hat{\imath} + 2\hat{\jmath}$, $\vec{B} = -3\hat{\imath} - 5\hat{\jmath}$, and $\vec{E} = 2\vec{A} + 3\vec{B}$.
   a. Write vector $\vec{E}$ in component form.
   b. Draw a coordinate system and on it show vectors $\vec{A}$, $\vec{B}$, and $\vec{E}$.
   c. What are the magnitude and direction of vector $\vec{E}$?
19. Let $\vec{A} = 5\hat{\imath} + 2\hat{\jmath}$, $\vec{B} = -3\hat{\imath} - 5\hat{\jmath}$, and $\vec{F} = \vec{A} - 4\vec{B}$.
   a. Write vector $\vec{F}$ in component form.
   b. Draw a coordinate system and on it show vectors $\vec{A}$, $\vec{B}$, and $\vec{F}$.
   c. What are the magnitude and direction of vector $\vec{F}$?
20. Are the following statements true or false? Explain your answer.
   a. The magnitude of a vector can be different in different coordinate systems.
   b. The direction of a vector can be different in different coordinate systems.
   c. The components of a vector can be different in different coordinate systems.
21. Let $\vec{A} = (4.0\text{ m, vertically downward})$ and $\vec{B} = (5.0\text{ m}, 120°\text{ clockwise from }\vec{A})$. Find the x- and y-components of $\vec{A}$ and $\vec{B}$ in each of the two coordinate systems shown in Figure Ex3.21.

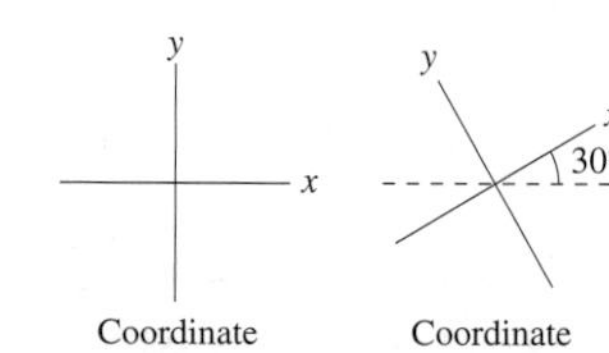

FIGURE EX3.21

22. What are the x- and y-components of the velocity vector shown in Figure Ex3.22?

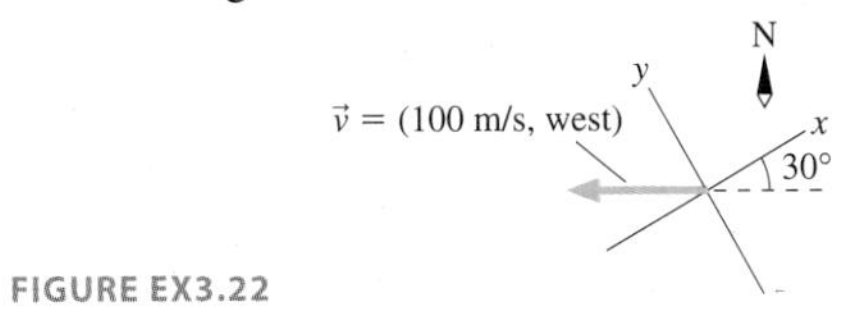

FIGURE EX3.22

Problems

23. Figure P3.23 shows vectors $\vec{A}$ and $\vec{B}$. Let $\vec{C} = \vec{A} + \vec{B}$.
   a. Reproduce the figure on your page as accurately as possible, using a ruler and protractor. Draw vector $\vec{C}$ on your figure, using the graphical addition of $\vec{A}$ and $\vec{B}$. Then determine the magnitude and direction of $\vec{C}$ by *measuring* it with a ruler and protractor.
   b. Based on your figure of part a, use geometry and trigonometry to *calculate* the magnitude and direction of $\vec{C}$.
   c. Decompose vectors $\vec{A}$ and $\vec{B}$ into components, then use these to calculate algebraically the magnitude and direction of $\vec{C}$.

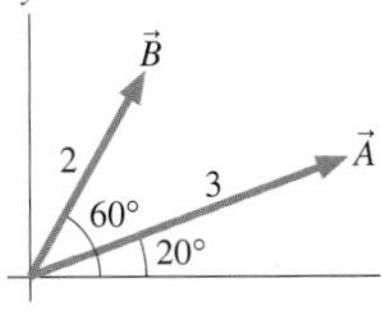

FIGURE P3.23

24. a. What is the angle ϕ between vectors $\vec{E}$ and $\vec{F}$ in Figure P3.24?
   b. Use geometry and trigonometry to determine the magnitude and direction of $\vec{G} = \vec{E} + \vec{F}$.
   c. Use components to determine the magnitude and direction of $\vec{G} = \vec{E} + \vec{F}$.

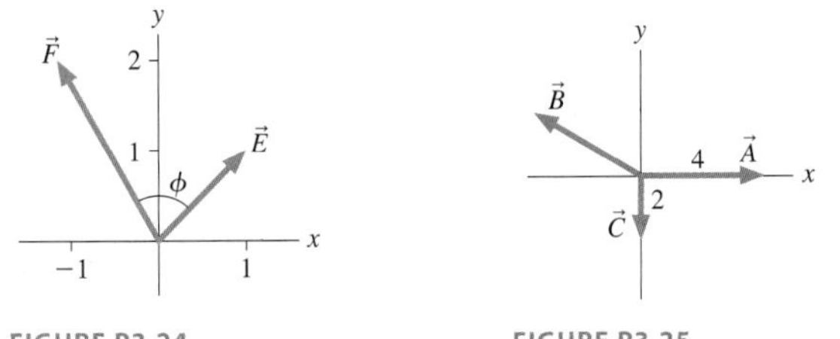

FIGURE P3.24

FIGURE P3.25

25. For the three vectors shown above in Figure P3.25, $\vec{A} + \vec{B} + \vec{C} = -2\hat{\imath}$. What is vector $\vec{B}$?
   a. Write $\vec{B}$ in component form.
   b. Write $\vec{B}$ as a magnitude and a direction.
26. Figure P3.26 shows vectors $\vec{A}$ and $\vec{B}$. Find vector $\vec{C}$ such that $\vec{A} + \vec{B} + \vec{C} = \vec{0}$. Write your answer in component form.

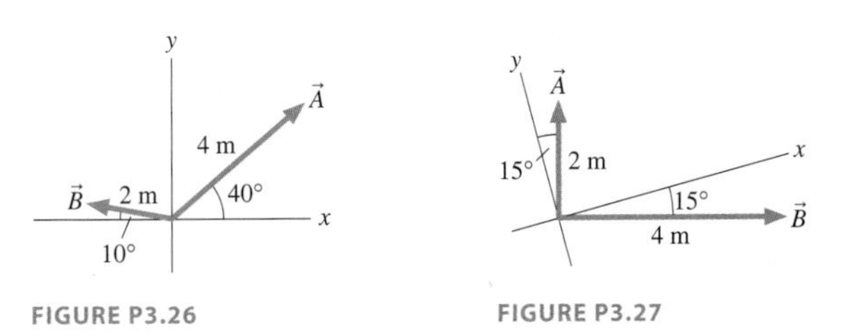

FIGURE P3.26

FIGURE P3.27

27. Figure P3.27 shows vectors $\vec{A}$ and $\vec{B}$. Find $\vec{D} = 2\vec{A} + \vec{B}$. Write your answer in component form.
28. Let $\vec{A}$ = (3.0 m, 20° south of east), $\vec{B}$ = (2.0 m, north), and $\vec{C}$ = (5.0 m, 70° south of west).
   a. Draw and label $\vec{A}$, $\vec{B}$, and $\vec{C}$ with their tails at the origin. Use a coordinate system with the x-axis to the east.
   b. Write $\vec{A}$, $\vec{B}$, and $\vec{C}$ in component form, using unit vectors.
   c. Find the magnitude and the direction of $\vec{D} = \vec{A} + \vec{B} + \vec{C}$.
29. Trace the vectors in Figure P3.29 onto your paper. Use the graphical method of vector addition and subtraction to find the following.
   a. $\vec{D} + \vec{E} + \vec{F}$
   b. $\vec{D} + 2\vec{E}$
   c. $\vec{D} - 2\vec{E} + \vec{F}$

FIGURE P3.29

30. Let $\vec{E} = 2\hat{\imath} + 3\hat{\jmath}$ and $\vec{F} = 2\hat{\imath} - 2\hat{\jmath}$. Find the magnitude of
   a. $\vec{E}$ and $\vec{F}$ b. $\vec{E} + \vec{F}$ c. $-\vec{E} - 2\vec{F}$
31. Find a vector that points in the same direction as the vector $(\hat{\imath} + \hat{\jmath})$ and whose magnitude is 1.
32. The position of a particle as a function of time is given by $\vec{r} = (5\hat{\imath} + 4\hat{\jmath})t^2$ m, where t is in seconds.
   a. What is the particle's distance from the origin at $t = 0$, 2, and 5 s?
   b. Find an expression for the particle's velocity $\vec{v}$ as a function of time.
   c. What is the particle's speed at $t = 0$, 2, and 5 s?
33. While vacationing in the mountains you do some hiking. In the morning, your displacement is $\vec{S}_{morning}$ = (2000 m, east) + (3000 m, north) + (200 m, vertical). After lunch, your displacement is $\vec{S}_{afternoon}$ = (1500 m, west) + (2000 m, north) − (300 m, vertical).
   a. At the end of the hike, how much higher or lower are you compared to your starting point?
   b. What is your total displacement?
34. The minute hand on a watch is 2.0 cm in length. What is the displacement vector of the tip of the minute hand
   a. From 8:00 to 8:20 A.M.?
   b. From 8:00 to 9:00 A.M.?
35. Bob walks 200 m south, then jogs 400 m southwest, then walks 200 m in a direction 30° east of north.
   a. Draw an accurate graphical representation of Bob's motion. Use a ruler and a protractor!
   b. Use either trigonometry or components to find the displacement that will return Bob to his starting point by the most direct route. Give your answer as a distance and a direction.
   c. Does your answer to part b agree with what you can measure on your diagram of part a?
36. Jim's dog Sparky runs 50 m northeast to a tree, then 70 m west to a second tree, and finally 20 m south to a third tree.
   a. Draw a picture and establish a coordinate system.
   b. Calculate Sparky's net displacement in component form.
   c. Calculate Sparky's net displacement as a magnitude and an angle.
37. A field mouse trying to escape a hawk runs east for 5.0 m, darts southeast for 3.0 m, then drops 1.0 m down a hole into its burrow. What is the magnitude of the net displacement of the mouse?
38. Carlos runs with velocity $\vec{v}$ = (5 m/s, 25° north of east) for 10 minutes. How far to the north of his starting position does Carlos end up?
39. A cannon tilted upward at 30° fires a cannonball with a speed of 100 m/s. What is the component of the cannonball's velocity parallel to the ground?

40. Jack and Jill ran up the hill at 3.0 m/s. The horizontal component of Jill's velocity vector was 2.5 m/s.
    a. What was the angle of the hill?
    b. What was the vertical component of Jill's velocity?
41. The treasure map in Figure P3.41 gives the following directions to the buried treasure: "Start at the old oak tree, walk due north for 500 paces, then due east for 100 paces. Dig." But when you arrive, you find an angry dragon just north of the tree. To avoid the dragon, you set off along the yellow brick road at an angle 60° east of north. After walking 300 paces you see an opening through the woods. Which direction should you go, and how far, to reach the treasure?

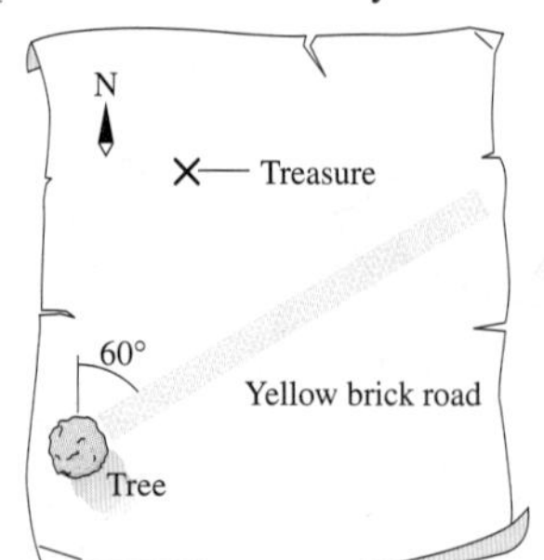

FIGURE P3.41

42. Mary needs to row her boat across a 100-m-wide river that is flowing to the east at a speed of 1.0 m/s. Mary can row the boat with a speed of 2.0 m/s relative to the water.
    a. If Mary rows straight north, how far downstream will she land?
    b. Draw a picture showing Mary's displacement due to rowing, her displacement due to the river's motion, and her net displacement.
43. A jet plane is flying horizontally with a speed of 500 m/s over a hill that slopes upward with a 3% grade (i.e., the "rise" is 3% of the "run"). What is the component of the plane's velocity perpendicular to the ground?
44. A flock of ducks is trying to migrate south for the winter, but they keep being blown off course by a wind blowing from the west at 6.0 m/s. A wise elder duck finally realizes that the solution is to fly at an angle to the wind. If the ducks can fly at 8.0 m/s relative to the air, what direction should they head in order to move directly south?
45. A pine cone falls straight down from a pine tree growing on a 20° slope. The pine cone hits the ground with a speed of 10 m/s. What is the component of the pine cone's impact velocity (a) parallel to the ground and (b) perpendicular to the ground?
46. The car in Figure P3.46 speeds up as it turns a quarter-circle curve from north to east. When exactly halfway around the curve, the car's acceleration is $\vec{a} = (2\text{ m/s}^2, 15°$ south of east). At this point, what is the component of $\vec{a}$ (a) tangent to the circle and (b) perpendicular to the circle?

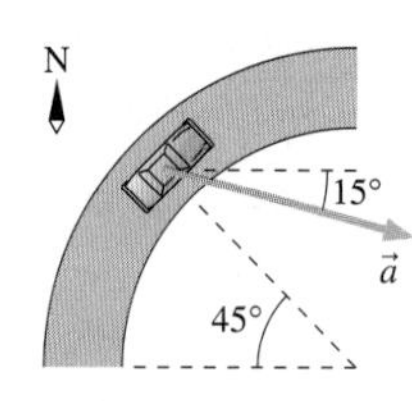

FIGURE P3.46

47. Figure P3.47 shows three ropes tied together in a knot. One of your friends pulls on a rope with 3 units of force and another pulls on a second rope with 5 units of force. How hard and in what direction must you pull on the third rope to keep the knot from moving?

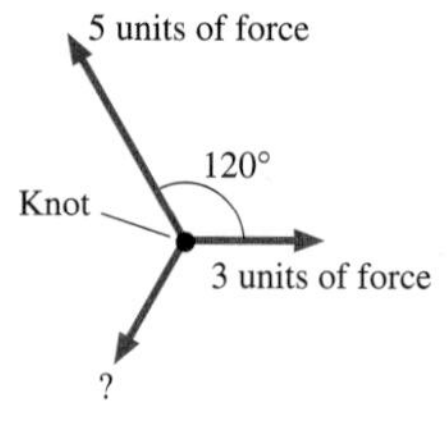

FIGURE P3.47

48. Three forces are exerted on an object placed on a tilted floor in Figure P3.48. The forces are measured in newtons (N). Assuming that forces are vectors,
    a. What is the component of the *net force* $\vec{F}_{net} = \vec{F}_1 + \vec{F}_2 + \vec{F}_3$ parallel to the floor?
    b. What is the component of $\vec{F}_{net}$ perpendicular to the floor?
    c. What are the magnitude and direction of $\vec{F}_{net}$?

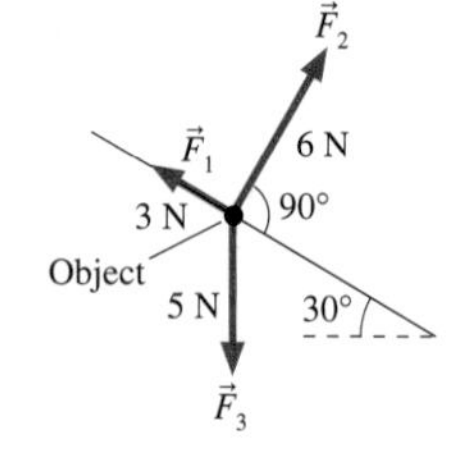

FIGURE P3.48

49. Figure P3.49 shows four electrical charges located at the corners of a rectangle. Like charges, you will recall, repel each other while opposite charges attract. Charge B exerts a repulsive force (directly *away from* B) on charge A of 3 N. Charge C exerts an attractive force (directly *toward* C) on charge A of 6 N. Finally, charge D exerts an attractive force of 2 N on charge A. Assuming that forces are vectors, what is the magnitude and direction of the net force $\vec{F}_{net}$ exerted on charge A?

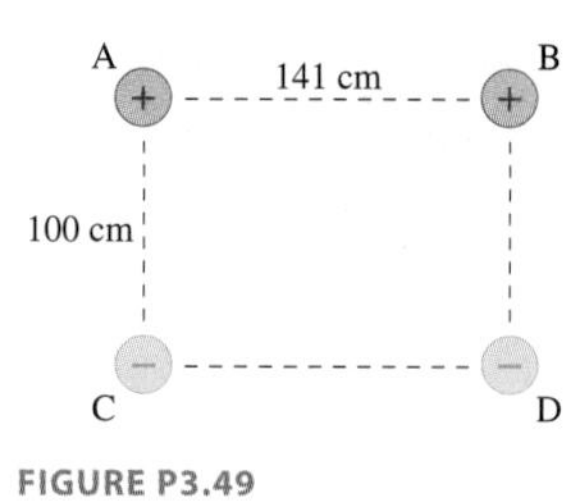

FIGURE P3.49

STOP TO THINK ANSWERS

Stop to Think 3.1: c. The graphical construction of $\vec{A}_1 + \vec{A}_2 + \vec{A}_3$ is shown below.

Stop to Think 3.2: a. The graphical construction of $2\vec{A} - \vec{B}$ is shown below.

Stop to Think 3.3: $C_x = -4$ **cm,** $C_y = 2$ **cm.**

Stop to Think 3.4: c. Vector $\vec{C}$ points to the left and down, so both C_x and C_y are negative. C_x is in the numerator because it is the side opposite ϕ.

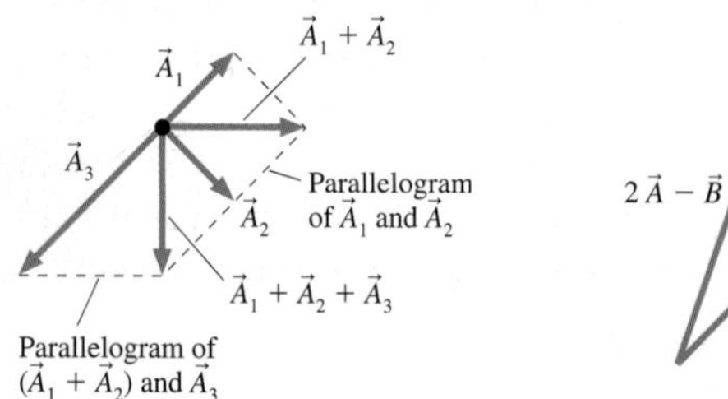

STOP TO THINK 3.1

STOP TO THINK 3.2

4 Force and Motion

A drag race is a memorable example of the connection between force and motion.

▶ Looking Ahead
The goal of Chapter 4 is to establish a connection between force and motion. In this chapter you will learn to:

- Recognize what a force is and is not.
- Identify the specific forces acting on an object.
- Draw free-body diagrams.
- Understand the connection between force and motion.

◀ Looking Back
To master the material introduced in this chapter, you must understand how acceleration is determined and how vectors are used. Please review:

- Section 1.5 Acceleration.
- Section 3.2 Properties of vectors.

This drag racer can cover a quarter mile from a standing start in less than 5 seconds. Not bad! We could use kinematics to describe the car's motion with pictures, graphs, and equations. By defining position, velocity, and acceleration and dressing them in mathematical clothing, kinematics provides a language to describe *how* something moves. But kinematics would tell us nothing about *why* the car accelerates so quickly. For the more fundamental task of understanding the *cause* of motion, we turn our attention to **dynamics.** Dynamics joins with kinematics to form **mechanics,** the general science of motion. We study dynamics qualitatively in this chapter, then develop it quantitatively in the next four chapters.

The theory of mechanics originated in the mid-1600s when Sir Isaac Newton formulated his laws of motion. These fundamental principles of mechanics explain how motion occurs as a consequence of forces. Newton's laws are more than 300 years old, but they still form the basis for our contemporary understanding of motion.

A challenge in learning physics is that a textbook is not an experiment. The book can assert that an experiment will have a certain outcome, but you may not be convinced unless you see or do the experiment yourself. Newton's laws are frequently contrary to our intuition, and a lack of familiarity with the evidence for Newton's laws is a source of difficulty for many people. You will have an opportunity through lecture demonstrations and in the laboratory to see for yourself the evidence supporting Newton's laws. Physics is not an arbitrary collection of definitions and formulas, but a consistent theory as to how the universe really works. It is only with experience and evidence that we learn to separate physical fact from fantasy.

4.1 Force

If you kick a ball, it rolls across the floor. If you pull on a door handle, the door opens. You know, from many years of experience, that some sort of *force* is required to move these objects. Our goal is to understand *why* motion occurs, and the observation that force and motion are related is a good place to start.

The two major issues that this chapter will examine are:

- What is a force?
- What is the connection between force and motion?

We begin with the first of these questions in the table below.

What is a force?

A force is a push or a pull.

Our commonsense idea of a **force** is that it is a *push* or a *pull*. We will refine this idea as we go along, but it is an adequate starting point. Notice our careful choice of words: We refer to "*a* force," rather than simply "force." We want to think of a force as a very specific *action,* so that we can talk about a single force or perhaps about two or three individual forces that we can clearly distinguish. Hence the concrete idea of "a force" acting on an object.

Object

A force acts on an object.

Implicit in our concept of force is that **a force acts on an object.** In other words, pushes and pulls are applied *to* something—an object. From the object's perspective, it has a force *exerted* on it. Forces do not exist in isolation from the object that experiences them.

Agent

A force requires an agent.

Every force has an **agent,** something that acts or exerts power. That is, a force has a specific, identifiable *cause*. As you throw a ball, it is your hand, while in contact with the ball, that is the agent or the cause of the force exerted on the ball. *If* a force is being exerted on an object, you must be able to identify a specific cause (i.e., the agent) of that force. Conversely, a force is not exerted on an object *unless* you can identify a specific cause or agent. Although this idea may seem to be stating the obvious, you will find it to be a powerful tool for avoiding some common misconceptions about what is and is not a force.

A force is a vector.

If you push an object, you can push either gently or very hard. Similarly, you can push either left or right, up or down. To quantify a push, we need to specify both a magnitude *and* a direction. It should thus come as no surprise that a force is a vector quantity. The symbol for a force is the vector symbol $\vec{F}$. The size or strength of a force is its magnitude F.

A force can be either a contact force . . .

There are two basic classes of forces, depending on whether the agent touches the object or not. **Contact forces** are forces that act on an object by touching it at a point of contact. The bat must touch the ball to hit it. A string must be tied to an object to pull it. The majority of forces that we will examine are contact forces.

. . . or a long-range force.

Long-range forces are forces that act on an object without physical contact. Magnetism is an example of a long-range force. You have undoubtedly held a magnet over a paper clip and seen the paper clip leap up to the magnet. A coffee cup released from your hand is pulled to the earth by the long-range force of gravity.

Let's summarize these ideas as our definition of force:

- A force is a push or a pull on an object.
- A force is a vector. It has both a magnitude and a direction.
- A force requires an agent. Something does the pushing or pulling.
- A force is either a contact force or a long-range force. Gravity is the only long-range force we will deal with until much later in the book.

There's one more important aspect of forces. If you push against a door (the object) to close it, the door pushes back against your hand (the agent). If a tow rope pulls on a car (the object), the car pulls back on the rope (the agent). Thus, in general, if an agent exerts a force on an object, the object exerts a force on the agent. We really need to think of a force as an *interaction* between two objects. Although the interaction perspective is a more exact way to view forces, it adds complications that we would like to avoid for now. Our approach will be to start by focusing on how a single object responds to forces exerted on it. Then, in Chapter 8, we'll return to the larger issue of how two or more objects interact with each other.

NOTE ▶ In the particle model, objects cannot exert forces on themselves. A force on an object will always have an agent or cause external to the object. Now, there are certainly objects that have internal forces (think of all the forces inside the engine of your car!), but the particle model is not valid if you need to consider those internal forces. If you are going to treat your car as a particle and look only at the overall motion of the car as a whole, that motion will be a consequence of external forces acting on the car. ◀

Force Vectors

We can use a simple diagram to visualize how forces are exerted on objects. Because we are using the particle model, in which objects are treated as points, the process of drawing a force vector is straightforward. Here is how it goes:

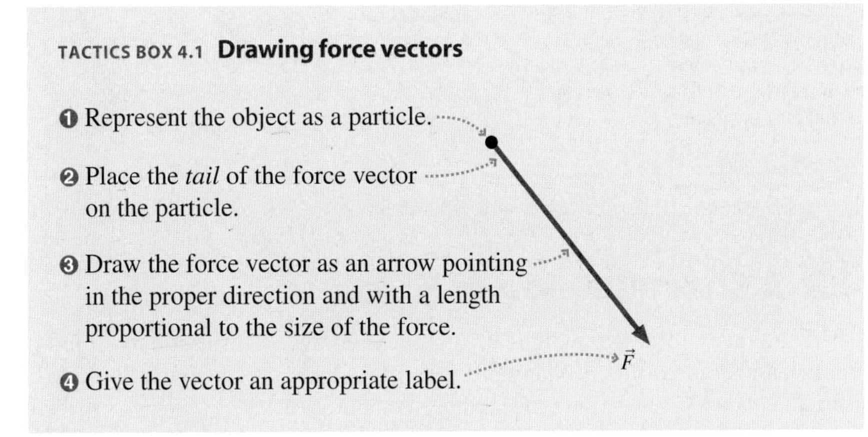

TACTICS BOX 4.1 Drawing force vectors
❶ Represent the object as a particle.
❷ Place the *tail* of the force vector on the particle.
❸ Draw the force vector as an arrow pointing in the proper direction and with a length proportional to the size of the force.
❹ Give the vector an appropriate label.
$\vec{F}$

Step 2 may seem contrary to what a "push" should do, but recall that moving a vector does not change it as long as the length and angle do not change. The vector $\vec{F}$ is the same regardless of whether the tail or the tip is placed on the particle. Our reason for using the tail will become clear when we consider how to combine several forces.

Figure 4.1 on the next page shows three examples of force vectors. One is a push, one a pull, and one a long-range force, but in all three the *tail* of the force vector is placed on the particle representing the object.

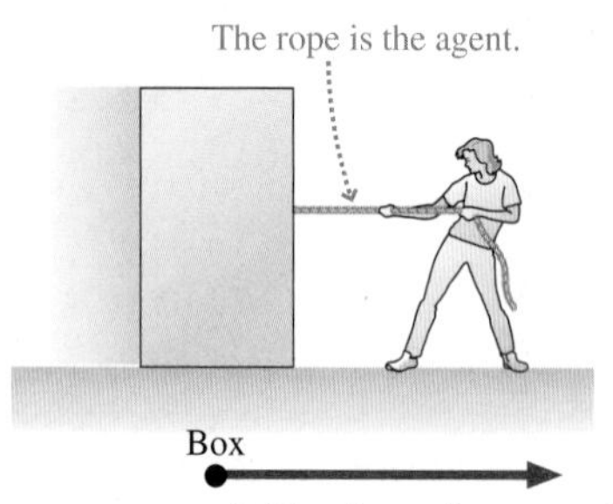

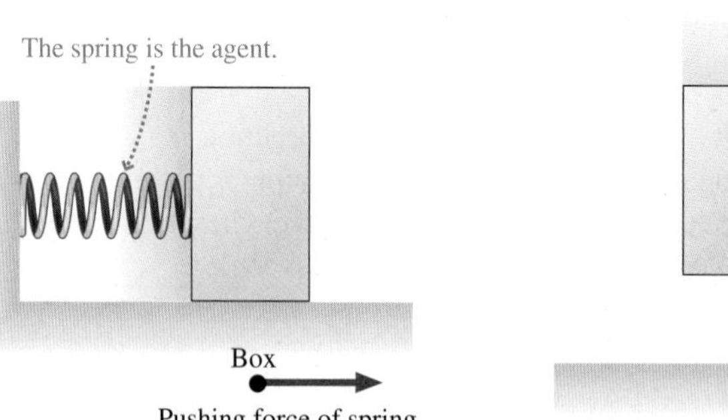

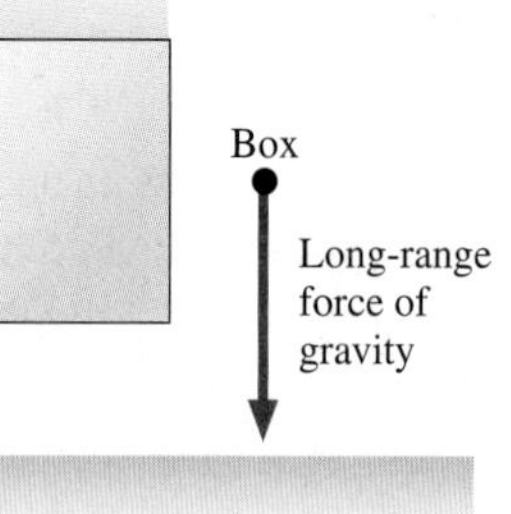

FIGURE 4.1 Three force vectors.

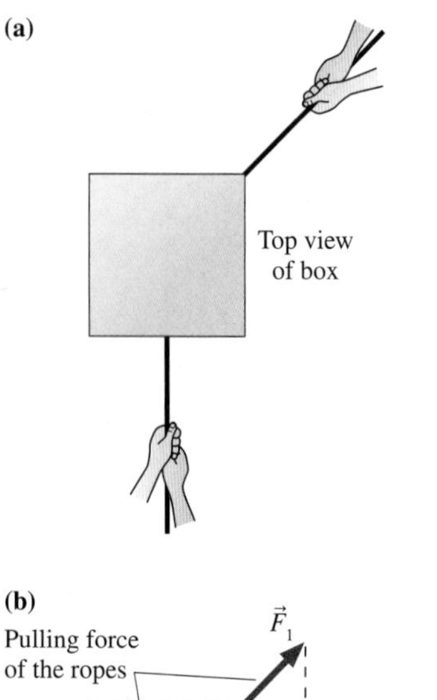

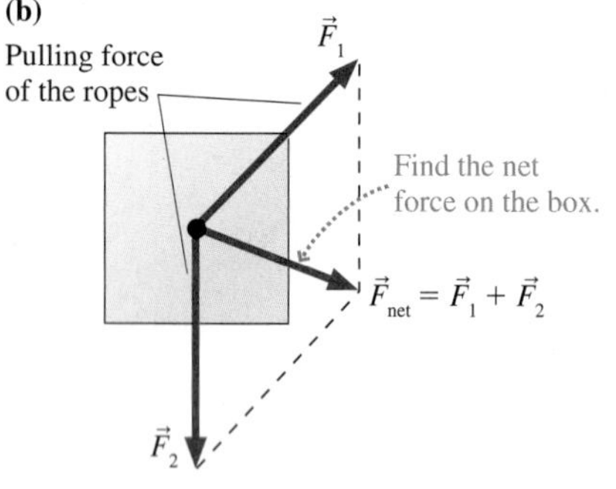

FIGURE 4.2 Two forces applied to a box.

Combining Forces

Figure 4.2a shows a box being pulled by two ropes, each exerting a force on the box. How will the box respond? Experimentally, we find that when several individual forces $\vec{F}_1, \vec{F}_2, \vec{F}_3, \ldots$ are exerted on an object, they combine to form a **net force** given by the *vector* sum of the individual forces:

$$\vec{F}_{net} \equiv \sum_{i=1}^{N} \vec{F}_i = \vec{F}_1 + \vec{F}_2 + \cdots + \vec{F}_N \tag{4.1}$$

Recall that $\equiv$ is the symbol meaning "is defined as." Mathematically, this summation is called a **superposition of forces.** The net force is sometimes called the *resultant force.* Figure 4.2b shows the net force on the box.

STOP TO THINK 4.1 Two forces are exerted on an object. What third force would make the net force point to the left?

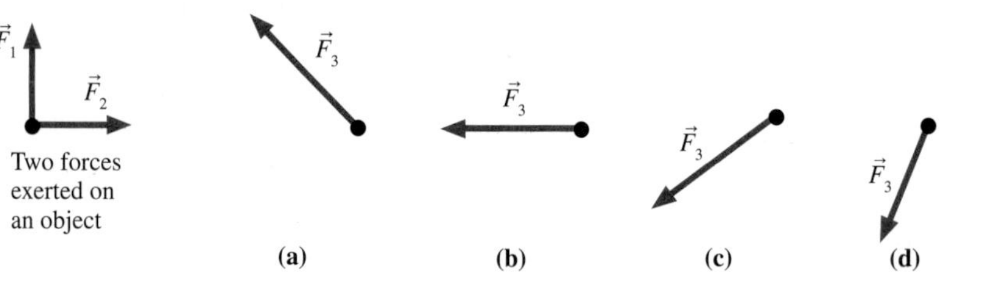

4.2 A Short Catalog of Forces

There are many forces we will deal with over and over. This section will introduce you to some of them. Many of these forces have special symbols. As you learn the major forces, be sure to learn the symbol for each.

Weight

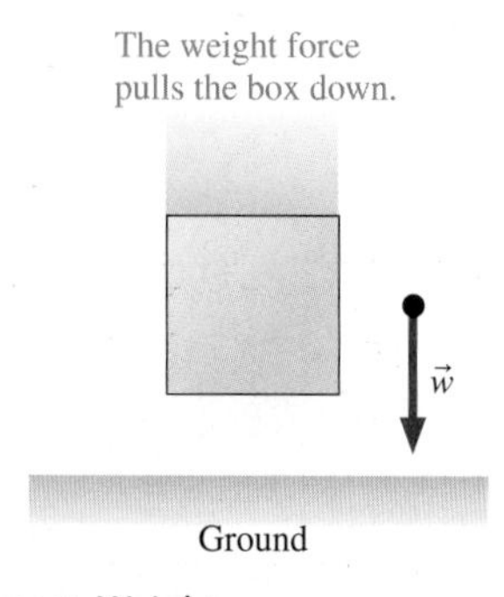

FIGURE 4.3 Weight.

A falling rock is pulled toward the earth by the long-range force of gravity. Gravity is what keeps you in your chair, keeps the planets in their orbits around the sun, and shapes the large-scale structure of the universe. We'll have a thorough look at gravity in Chapter 12. For now we'll concentrate on objects on or near the surface of the earth (or other planet).

The gravitational pull of the earth on an object on or near the surface of the earth is called **weight.** The symbol for weight is $\vec{w}$. Weight is the only long-range force we will encounter in the next few chapters. The agent for the weight force is the *entire earth* pulling on an object. Weight acts on an object whether the object is moving or at rest. The weight vector always points vertically downward, as shown in Figure 4.3 on the previous page.

NOTE ► We often refer to "the weight" of an object. This is an informal expression for w, the magnitude of the weight force exerted on the object. Note that **weight is not the same thing as mass.** We will briefly examine mass later in the chapter and explore the connection between weight and mass in Chapter 5. ◄

A stretched spring exerts a force on an object.

Spring Force

Springs exert one of the most common contact forces. A spring can either push (when compressed) or pull (when stretched). Figure 4.4 shows the spring force. In both cases, pushing and pulling, the tail of the force vector is placed on the particle in the force diagram. There is no special symbol for a spring force, so we simply use a subscript label: $\vec{F}_{sp}$.

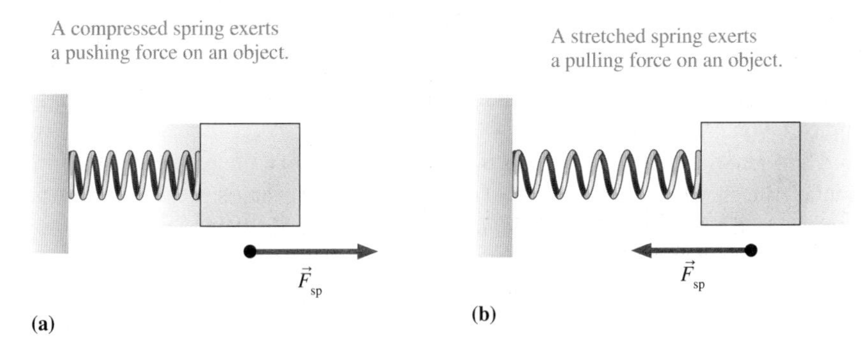

FIGURE 4.4 The spring force.

Although you may think of a spring as a metal coil that can be stretched or compressed, this is only one type of spring. Hold a ruler, or any other thin piece of wood or metal, by the ends and bend it slightly. It flexes. When you let go, it "springs" back to its original shape. This is just as much a spring as is a metal coil.

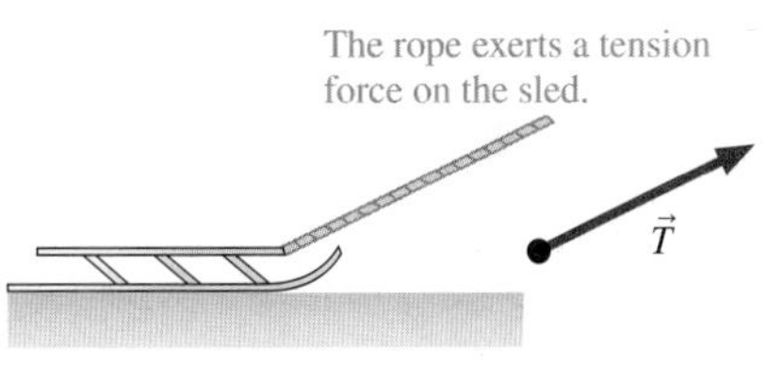

FIGURE 4.5 Tension.

Tension Force

When a string or rope or wire pulls on an object, it exerts a contact force that we call the **tension force,** represented by a capital $\vec{T}$. The direction of the tension force is always in the direction of the string or rope, as you can see in Figure 4.5. The commonplace reference to "the tension" in a string is an informal expression for T, the size or magnitude of the tension force.

If you were to use a very powerful microscope to look inside a rope, you would "see" that it is made of *atoms* joined together by *molecular bonds*. Molecular bonds are not rigid connections between the atoms. They are more accurately thought of as tiny *springs* holding the atoms together, as in Figure 4.6. These are very stiff springs, to be sure, but pulling on the ends of a string or rope stretches the molecular springs ever so slightly. The tension within a rope and the tension experienced by an object at the end of the rope are really the net spring force being exerted by billions and billions of microscopic springs.

This atomic-level view of tension introduces a new idea: a microscopic **atomic model** for understanding the behavior and properties of macroscopic objects. We will frequently use an atomic model to obtain a deeper understanding of our observations.

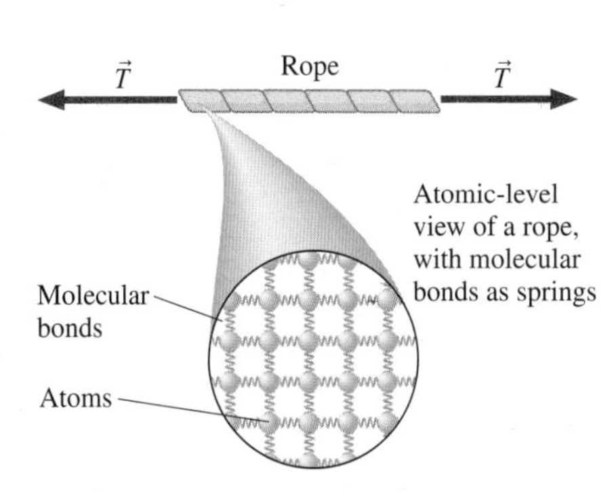

FIGURE 4.6 An atomic-level view of tension.

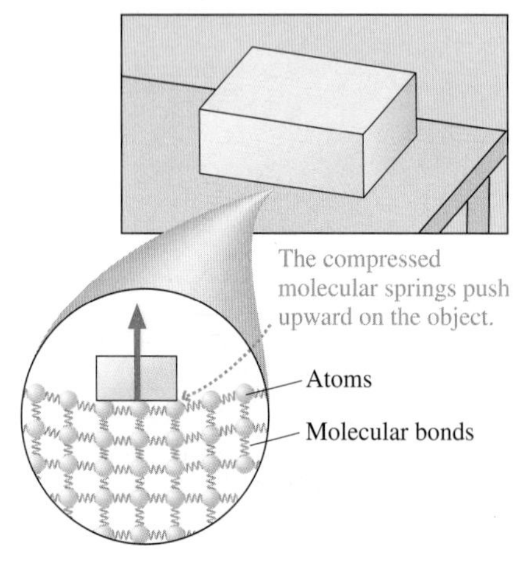
The compressed molecular springs push upward on the object.
Atoms
Molecular bonds

FIGURE 4.7 Atomic-level view of the force exerted by a table.

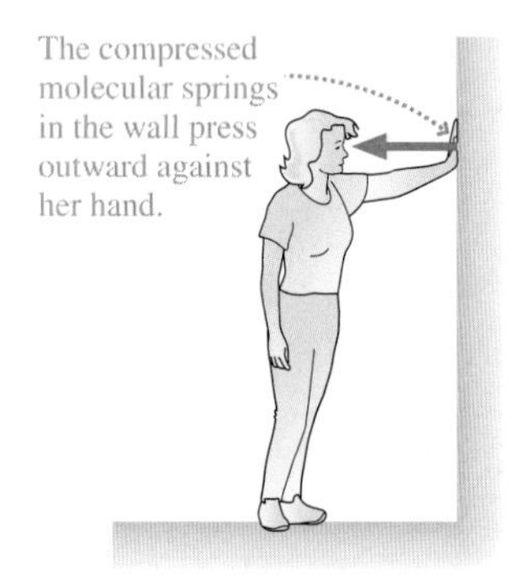
The compressed molecular springs in the wall press outward against her hand.

FIGURE 4.8 The wall pushes outward against your hand.

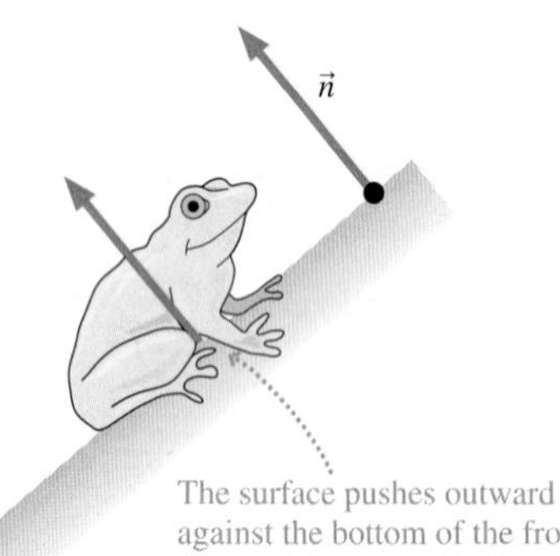
$\vec{n}$
The surface pushes outward against the bottom of the frog. The push is perpendicular to the surface.

FIGURE 4.9 The normal force.

Normal Force

If you sit on a bed, the springs in the mattress compress and, as a consequence of the compression, exert an upward force on you. Stiffer springs would show less compression but still exert an upward force. The compression of extremely stiff springs might be measurable only by sensitive instruments. Nonetheless, the springs would compress ever so slightly and exert an upward spring force on you.

Figure 4.7 shows an object resting on top of a sturdy table. The table may not visibly flex or sag, but—just as you do to the bed—the object compresses the molecular springs in the table. The size of the compression is very small because molecular springs are so stiff, but it is not zero. As a consequence, the compressed molecular springs *push upward* on the object. We say that "the table" exerts the upward force, but it is important to understand that the pushing is *really* done by molecular springs. Similarly, an object resting on the ground compresses the molecular springs holding the ground together and, as a consequence, the ground pushes up on the object.

We can extend this idea. Suppose you place your hand on a wall and lean against it, as shown in Figure 4.8. Does the wall exert a force on your hand? As you lean, you compress the molecular springs in the wall and, as a consequence, they push outward against your hand. So the answer is "yes," the wall does exert a force on you.

The force the table surface exerts is vertical, the force the wall exerts is horizontal. But in all cases, the force exerted on an object that is pressing against a surface is in a direction *perpendicular* to the surface. Mathematicians refer to a line that is perpendicular to a surface as being *normal* to the surface. In keeping with this terminology, we define the **normal force** as the force exerted by a surface (the agent) against an object that is pressing against the surface. The symbol for the normal force is $\vec{n}$.

We're not using the word *normal* to imply that the force is an "ordinary" force or to distinguish it from an "abnormal force." A surface exerts a force *perpendicular* (i.e., normal) to itself as the molecular springs press *outward.* Figure 4.9 shows an object on an inclined surface, a common situation. Notice how the normal force $\vec{n}$ is perpendicular to the surface.

We have spent a lot of time describing the normal force because many people have a difficult time understanding it. The normal force is a very real force arising from the very real compression of molecular bonds. It is in essence just a spring force, but one exerted by a vast number of microscopic springs acting at once. The normal force is responsible for the "solidness" of solids. It is what prevents you from passing right through the chair you are sitting in and what causes the pain and the lump if you bang your head into a door. Your head can then tell you that the force exerted on it by the door was very real!

Friction

You've certainly observed that a rolling or sliding object, if not pushed or propelled, slows down and eventually stops. You've probably discovered that you can slide better across a sheet of ice than across asphalt. And you also know that most objects stay in place on a table without sliding off even if the table isn't absolutely level. The force responsible for these sorts of behavior is **friction.** The symbol for friction is a lower case $\vec{f}$.

Friction, like the normal force, is exerted by a surface. On a microscopic level, friction arises as atoms from the object and atoms on the surface run into each other. The rougher the surface is, the more these atoms are forced into close proximity and, as a result, the larger the friction force. We will develop a simple model of friction in the next chapter that will be sufficient for our needs. For now, it is useful to distinguish between two kinds of friction:

- *Kinetic friction,* denoted $\vec{f}_k$, appears as an object slides across a surface. This is a force that "opposes the motion," meaning that the friction force vector $\vec{f}_k$ points in a direction opposite the velocity vector $\vec{v}$ (i.e., "the motion").
- *Static friction,* denoted $\vec{f}_s$, is the force that keeps an object "stuck" on a surface and prevents its motion. Finding the direction of $\vec{f}_s$ is a little trickier than finding it for $\vec{f}_k$. Static friction points opposite the direction in which the object *would* move if there were no friction. That is, it points in the direction necessary to *prevent* motion.

Figure 4.10 shows examples of kinetic and static friction.

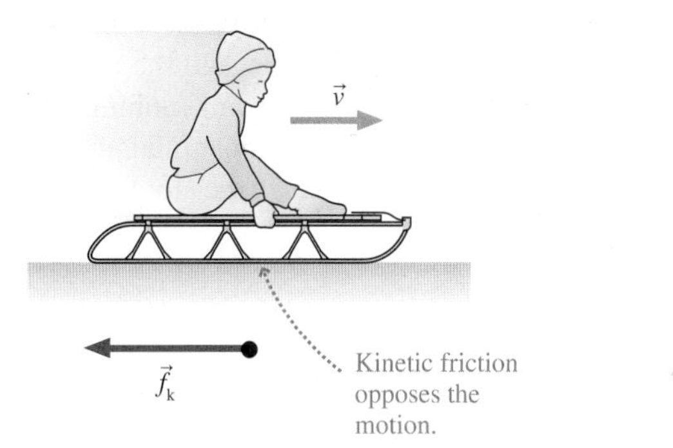
$\vec{v}$
$\vec{f}_k$
Kinetic friction opposes the motion.
$\vec{f}_s$
Static friction acts in the direction that prevents slipping.

FIGURE 4.10 Kinetic and static friction.

> **NOTE ▶** A surface exerts a kinetic friction force when an object moves *relative to* the surface. A package on a conveyor belt is in motion, but it does not experience a kinetic friction force because it is not moving relative to the belt. So to be precise, we should say that the kinetic friction force points opposite to an object's motion *relative to* a surface. ◀

Drag

Friction at a surface is one example of a *resistive force,* a force that opposes or resists motion. Resistive forces are also experienced by objects moving through fluids—gases and liquids. The resistive force of a fluid is called **drag** and is symbolized as $\vec{D}$. Drag, like kinetic friction, points opposite the direction of motion. Figure 4.11 shows an example of drag.

Drag can be a large force for objects moving at high speeds or in dense fluids. Hold your arm out the window as you ride in a car and feel how the air resistance against it increases rapidly as the car's speed increases. Drop a lightweight object into a beaker of water and watch how slowly it settles to the bottom. In both cases the drag force is very significant.

For objects that are heavy and compact, that move in air, and whose speed is not too great, the drag force of air resistance is fairly small. To keep things as simple as possible, **you can neglect air resistance in all problems unless a problem explicitly asks you to include it.** The error introduced into calculations by this approximation is generally pretty small. This textbook will not consider objects moving in liquids.

Air resistance is a significant force on falling leaves. It points opposite the direction of motion.

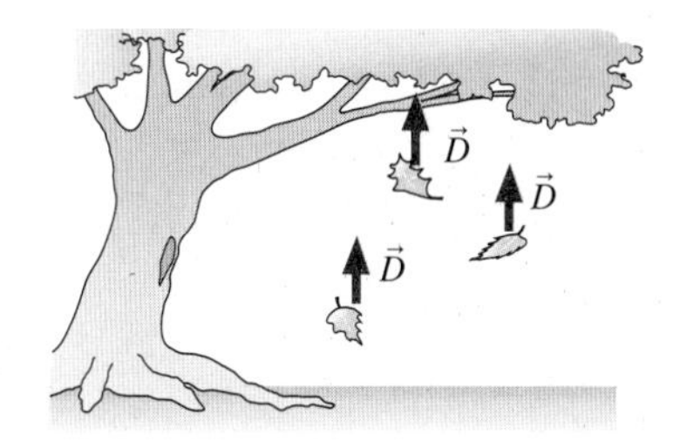
$\vec{D}$
$\vec{D}$
$\vec{D}$

FIGURE 4.11 Air resistance is an example of drag.

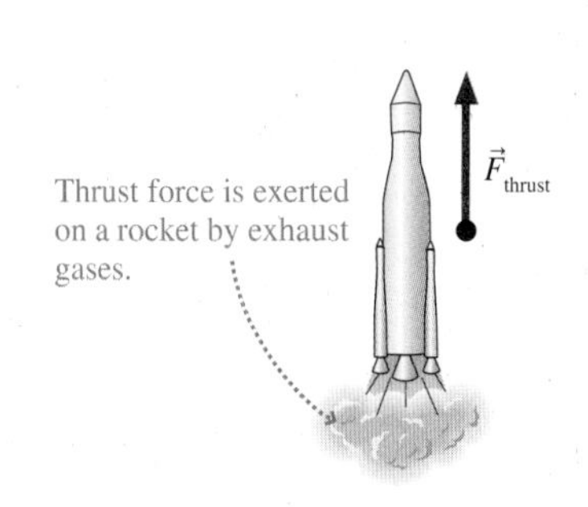
Thrust force is exerted on a rocket by exhaust gases.
$\vec{F}_{thrust}$

FIGURE 4.12 Thrust force on a rocket.

Thrust

A jet airplane obviously has a force that propels it forward during takeoff. Likewise for the rocket being launched in Figure 4.12. This force, called **thrust,** occurs when a jet or rocket engine expels gas molecules at high speed. Thrust is a contact force, with the exhaust gas being the agent that pushes on the engine. The process by which thrust is generated is rather subtle, and we will postpone a full discussion until we introduce Newton's third law in Chapter 8. For now, we will treat thrust as a force opposite the direction in which the exhaust gas is expelled. There's no special symbol for thrust, so we will call it $\vec{F}_{\text{thrust}}$.

Force	Notation
General force	$\vec{F}$
Weight	$\vec{w}$
Spring force	$\vec{F}_{\text{sp}}$
Tension	$\vec{T}$
Normal force	$\vec{n}$
Static friction	$\vec{f}_{\text{s}}$
Kinetic friction	$\vec{f}_{\text{k}}$
Drag	$\vec{D}$
Thrust	$\vec{F}_{\text{thrust}}$

Electric and Magnetic Forces

Electricity and magnetism, like gravity, exert long-range forces. The forces of electricity and magnetism act on charged particles. We will study electric and magnetic forces in detail in Part VI of this textbook. For now, it is worth noting that the forces holding molecules together—the molecular bonds—are not actually tiny springs. Atoms and molecules are made of charged particles—electrons and protons—and what we call a molecular bond is really an attractive electric force between these particles. So when we say that the normal force and the tension force are due to "molecular springs," or that friction is due to atoms running into each other, what we're really saying is that these forces, at the most fundamental level, are actually electric forces between the charged particles in the atoms.

4.3 Identifying Forces

Force and motion problems generally have two basic steps:

1. Identify all of the forces acting on an object.
2. Use Newton's laws and kinematics to determine the motion.

Understanding the first step is the primary goal of this chapter. We'll turn our attention to step 2 in the next chapter.

A typical physics problem describes an object that is being pushed and pulled in various directions. Some forces are given explicitly, others are only implied. In order to proceed, it is necessary to determine all the forces that act on the object. It is also necessary to avoid including forces that do not really exist. Now that you have learned the properties of forces and seen a catalog of typical forces, we can develop a step-by-step method for identifying each force in a problem. This procedure for identifying forces is part of the *physical representation* of the problem.

TACTICS BOX 4.2 Identifying forces

1. **Identify "the system" and "the environment."** The system is the object whose motion you wish to study; the environment is everything else.
2. **Draw a picture of the situation.** Show the object—the system—and everything in the environment that touches the system. Ropes, springs, and surfaces are all parts of the environment.
3. **Draw a closed curve around the system.** Only the object is inside the curve; everything else is outside.
4. **Locate every point on the boundary of this curve where the environment touches the system.** These are the points where the environment exerts *contact forces* on the object.
5. **Name and label each contact force acting on the object.** There is at least one force at each point of contact; there may be more than one. When necessary, use subscripts to distinguish forces of the same type.
6. **Name and label each long-range force acting on the object.** For now, the only long-range force is weight.

EXAMPLE 4.1 Forces on a bungee jumper

A bungee jumper has leapt off a bridge and is nearing the bottom of her fall. What forces are being exerted on the bungee jumper?

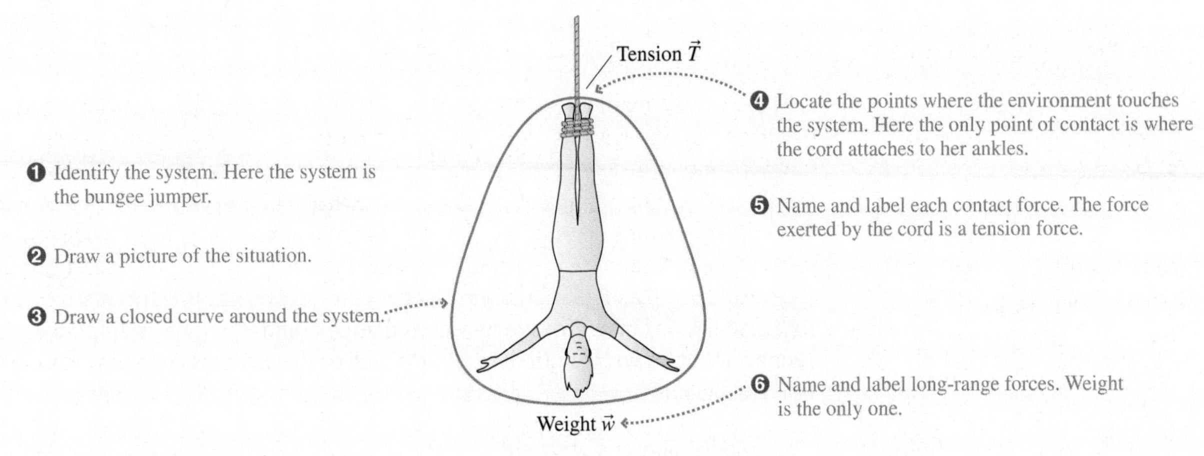

Tension $\vec{T}$
❹ Locate the points where the environment touches the system. Here the only point of contact is where the cord attaches to her ankles.
❶ Identify the system. Here the system is the bungee jumper.
❺ Name and label each contact force. The force exerted by the cord is a tension force.
❷ Draw a picture of the situation.
❸ Draw a closed curve around the system.
❻ Name and label long-range forces. Weight is the only one.
Weight $\vec{w}$

FIGURE 4.13 Forces on a bungee jumper.

EXAMPLE 4.2 Forces on a skier

A skier is being towed up a snow-covered hill by a tow rope. What forces are being exerted on the skier?

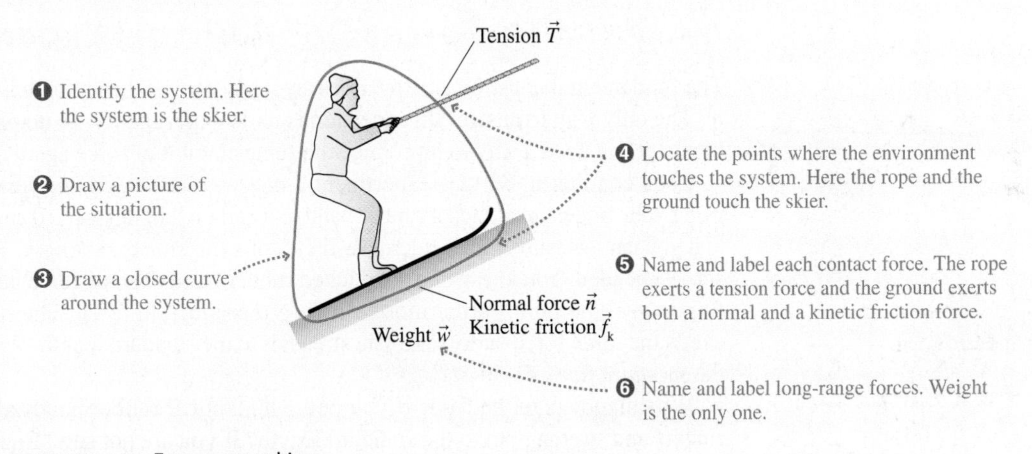

Tension $\vec{T}$
❶ Identify the system. Here the system is the skier.
❹ Locate the points where the environment touches the system. Here the rope and the ground touch the skier.
❷ Draw a picture of the situation.
❸ Draw a closed curve around the system.
❺ Name and label each contact force. The rope exerts a tension force and the ground exerts both a normal and a kinetic friction force.
Normal force $\vec{n}$
Kinetic friction $\vec{f}_k$
Weight $\vec{w}$
❻ Name and label long-range forces. Weight is the only one.

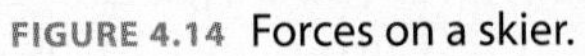

FIGURE 4.14 Forces on a skier.

NOTE ▶ You might have expected two friction forces and two normal forces in Example 4.2, one on each ski. Keep in mind, however, that we're working within the particle model, which represents the skier by a single point. A particle has only one contact with the ground, so there is a single normal force and a single friction force. The particle model is valid if we want to analyze the translational motion of the skier as a whole, but we would have to go beyond the particle model to find out what happens to each ski. ◀

Now that you're getting the hang of this, the next example is meant to look much more like a sketch you should make when asked to identify forces in a homework problem.

EXAMPLE 4.3 Forces on a rocket

A rocket is being launched to place a new satellite in orbit. Air resistance is not negligible. What forces are being exerted on the rocket?

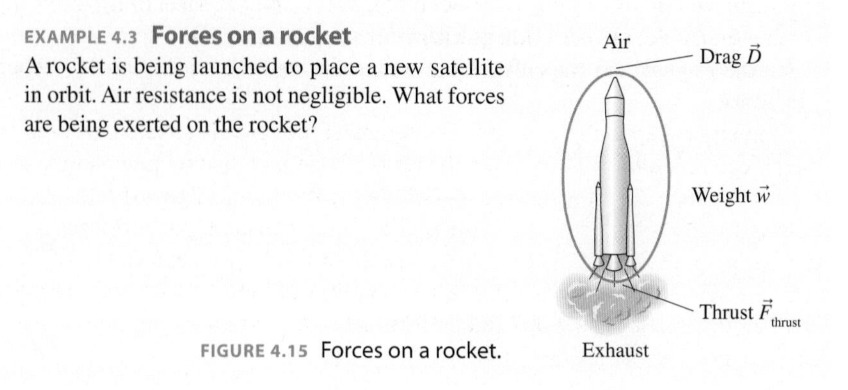

FIGURE 4.15 Forces on a rocket.

STOP TO THINK 4.2 You've just kicked a rock, and it is now sliding across the ground about 2 meters in front of you. Which of these forces act on the rock? List all that apply.

a. Gravity, acting downward.
b. The normal force, acting upward.
c. The force of the kick, acting in the direction of motion.
d. Friction, acting opposite the direction of motion.
e. Air resistance, acting opposite the direction of motion.

4.4 What Do Forces Do? A Virtual Experiment

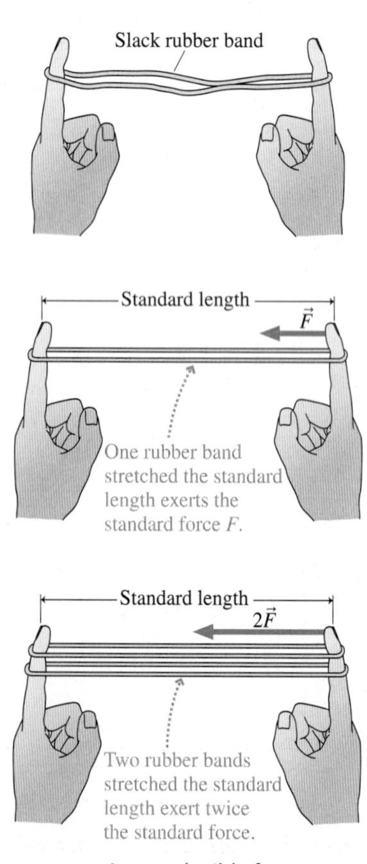

FIGURE 4.16 A reproducible force.

The fundamental question is: How does an object move when a force is exerted on it? The only way to answer this question is to do experiments. To do experiments, however, we need a way to reproduce the same amount of force again and again.

Let's conduct a "virtual experiment," one you can easily visualize. Imagine using your fingers to stretch a rubber band to a certain length—say 10 centimeters—that you can measure with a ruler. We'll call this the *standard length.* Figure 4.16 shows the idea. You know that a stretched rubber band exerts a force because your fingers *feel* the pull. Furthermore, this is a reproducible force. The rubber band exerts the same force every time you stretch it to the standard length. We'll call this the *standard force F*.

What happens to the force if you put *two* identical rubber bands around your fingers and stretch both to the standard length? If you are not sure, find a few rubber bands and try this. You will discover that two rubber bands exert a larger pulling force than one rubber band. This is not surprising. If two rubber bands are each pulling equally hard, the net pull is twice that of one rubber band: $F_{net} = 2F$. N side-by-side rubber bands, each pulled to the standard length, will exert N times the standard force: $F_{net} = NF$.

Now we're ready to start the virtual experiment. Imagine an object to which you can attach rubber bands, such as a block of wood with a hook. If you attach a rubber band and stretch it to the standard length, the object experiences the same force F as did your finger. N rubber bands attached to the object will exert N times the force of one rubber band. The rubber bands give us a way of applying a known and reproducible force to an object.

Our next task is to measure the object's motion in response to these forces. Imagine using the rubber bands to pull the object across a horizontal table. Friction between the object and the surface might affect our results, so let's just eliminate friction. This is, after all, a virtual experiment! (In practice you could nearly eliminate friction by pulling a smooth block over a smooth sheet of ice or by supporting the object on a cushion of air.) To make measurements, lay a meter stick along the edge of the table and hang a movie camera over the table to record the motion.

Our experiment will be easiest to interpret if the force is *constant* throughout the object's motion. If you stretch the rubber band and then release the object, it moves toward your hand. But as it does so, the rubber band gets shorter and the pulling force decreases. To keep the pulling force constant, you must *move your hand* at just the right speed to keep the length of the rubber band from changing! Figure 4.17 shows the experiment being carried out. Once the motion is complete, you can use motion diagrams (made from the movie frames) and kinematics to analyze the object's motion.

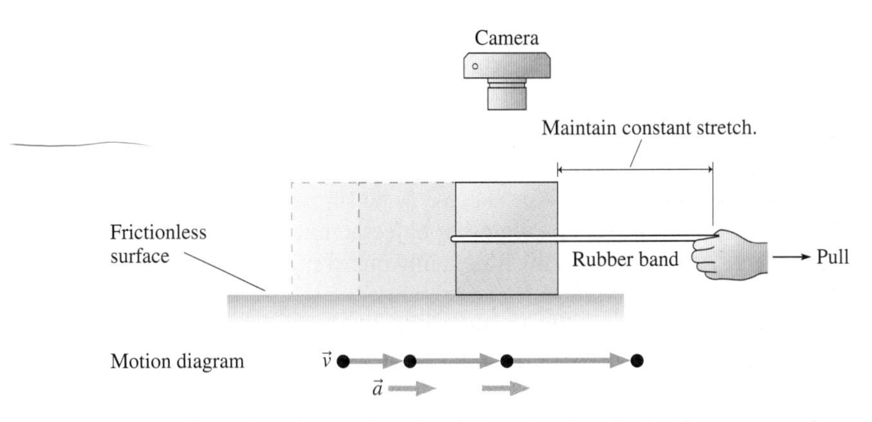

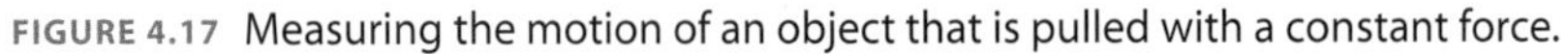

FIGURE 4.17 Measuring the motion of an object that is pulled with a constant force.

The first important finding of this experiment is that **an object pulled with a constant force moves with a constant acceleration.** This finding could not have been anticipated in advance. It's conceivable that the object would speed up for a while, then move with a steady speed. Or that it would continue to speed up, but that the *rate* of increase, the acceleration, would steadily decline. These are conceivable motions, but they're not what happens. Instead, the object continues to accelerate *with a constant acceleration* for as long as you pull it with a constant force.

The next question is: What happens if you increase the force by using several rubber bands? To find out, use 2 rubber bands. Stretch both to the standard length to double the force, then measure the acceleration. Then measure the acceleration due to 3 rubber bands, then 4, and so on. Table 4.1 shows the results of this experiment. You can see that doubling the force causes twice the acceleration, tripling the force causes three times the acceleration, and so on.

Figure 4.18 is a graph of the data. Force is the independent variable, the one you can control, so we've placed force on the horizontal axis to make an acceleration-versus-force graph. The graph shows that **the acceleration is directly proportional to the force.** This is our second important finding. Recall that proportionality indicates a linear relationship whose graph passes through the origin (y-intercept of zero). This result can be written

$$a = cF \tag{4.2}$$

where c is called the *proportionality constant.* The proportionality constant c is the slope of the graph.

The final question for our virtual experiment is: How does the acceleration depend on the size of the object? (The "size" of an object is somewhat ambiguous. We'll be more precise below.) To find out, glue the original object and an identical copy together, and then, applying the *same force* as you applied to the

TABLE 4.1 Acceleration due to an increasing force

Rubber bands	Force	Acceleration
1	F	a_1
2	$2F$	$a_2 = 2a_1$
3	$3F$	$a_3 = 3a_1$
⋮	⋮	⋮
N	NF	$a_N = Na_1$

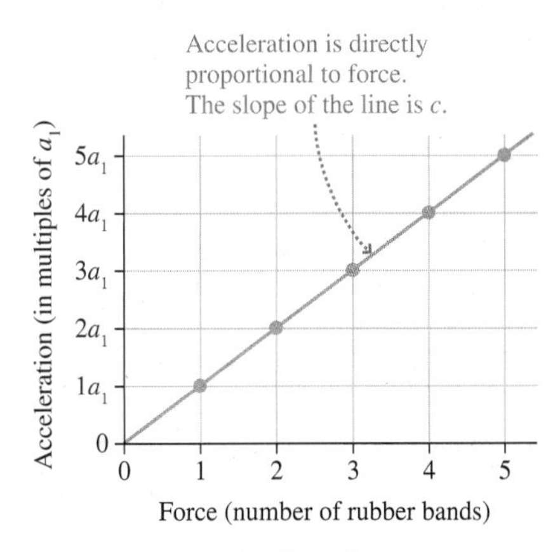

FIGURE 4.18 Graph of acceleration versus force.

TABLE 4.2 Acceleration with different numbers of objects

Number of objects	Acceleration
1	a_1
2	$a_2 = \frac{1}{2}a_1$
3	$a_3 = \frac{1}{3}a_1$
⋮	⋮
N	$a_N = \frac{1}{N}a_1$

original, single object, measure the acceleration of this new object. Three objects glued together make an object three times the size of the original. Doing several such experiments, applying the same force to each object, would give you the results shown in Table 4.2. An object twice the size of the original has only half the acceleration of the original object when both are subjected to the same force. An object three times the size of the original has one-third the acceleration.

Figure 4.19 shows these results added to the graph of Figure 4.18. You can see that the proportionality constant c between acceleration and force—the slope of the line—changes with the size of the object. The graph for an object twice the size of the original is a line with half the slope. It may seem surprising that larger objects have smaller slopes, so you'll want to think about this carefully.

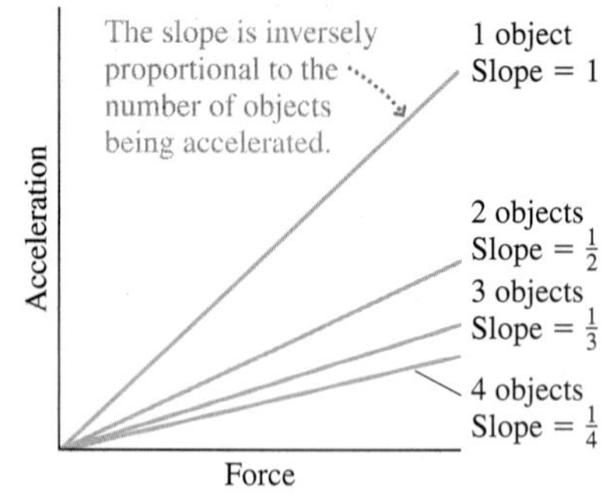

FIGURE 4.19 Acceleration-versus-force graphs for objects of different size.

Mass

Now, "twice the size" is a little vague; we could mean the object's external dimensions or some other measure. Although *mass* is a common word, we've avoided the term so far because we first need to define what mass is. Because we made the larger objects in our experiment from the same material as the original object, an object twice the size has twice as many atoms—twice the amount of matter—as the original. Thus it should come as no surprise that it has twice the mass as the original. Loosely speaking, **an object's mass is a measure of the amount of matter it contains.** This is certainly our everyday meaning of *mass,* but it is not yet a precise definition.

Figure 4.19 showed that an object with twice the amount of matter as the original accelerates only half as quickly if both experience the same force. An object with N times as much matter has only $\frac{1}{N}$ of the original acceleration. The more matter an object has, the more it *resists* accelerating in response to a force. You're familiar with this idea: Your car is much harder to push than your bicycle. The tendency of an object to resist a *change* in its velocity (i.e., to resist acceleration) is called **inertia.** Figure 4.19 tells us that larger objects have more inertia than smaller objects of the same material.

We can make this idea precise by defining the **inertial mass** m of an object to be

$$m \equiv \frac{1}{\text{slope of the acceleration-versus-force graph}} = \frac{F}{a}$$

We usually refer to the inertial mass as simply "the mass." Mass is an *intrinsic* property of an object. It is the property that determines how an object accelerates in response to an applied force.

STOP TO THINK 4.3 Two rubber bands stretched to the standard length cause an object to accelerate at 2 m/s^2. Suppose another object with twice the mass is pulled by four rubber bands stretched to the standard length. The acceleration of this second object is

a. 1 m/s^2. b. 2 m/s^2. c. 4 m/s^2. d. 8 m/s^2. e. 16 m/s^2.

A = F/M

4.5 Newton's Second Law

We can now summarize the results of our experiment. Figure 4.18 showed that the acceleration is directly proportional to the force, a conclusion that we wrote in Equation 4.2 with the unspecified proportionality constant c. Now we see that c, the slope of the acceleration-versus-force graph, is the inverse of the inertial mass m.

Thus we've found that a force of magnitude F causes an object of mass m to accelerate with

$$a = \frac{F}{m}$$

This simple equation answers the question with which we started: How does an object move when a force is exerted on it? A force causes an object to *accelerate!* Furthermore, the size of the acceleration is directly proportional to the size of the force and inversely proportional to the object's mass.

This is an important finding, but our experiment was limited to looking at an object's response to a single applied force. Realistically, an object is likely to be subjected to several distinct forces $\vec{F}_1, \vec{F}_2, \vec{F}_3, \ldots$ that may point in different directions. What happens then? In that case, it is found experimentally that the acceleration is determined by the *net* force.

Newton was the first to recognize the connection between force and motion. This relationship is known today as Newton's second law.

Newton's second law An object of mass m subjected to forces $\vec{F}_1, \vec{F}_2, \vec{F}_3, \ldots$ will undergo an acceleration $\vec{a}$ given by

$$\vec{a} = \frac{\vec{F}_{net}}{m} \qquad (4.3)$$

where the net force $\vec{F}_{net} = \vec{F}_1 + \vec{F}_2 + \vec{F}_3 + \cdots$ is the vector sum of the individual forces. The acceleration vector $\vec{a}$ points in the same direction as the net force vector $\vec{F}_{net}$.

It may seem puzzling that we've skipped over Newton's first law. The reasons for this will become clear as we continue our discussion of dynamics. For now, the critical idea is that **an object's acceleration vector $\vec{a}$ points in the same direction as the net force vector $\vec{F}_{net}$.**

The significance of Newton's second law cannot be overstated. There was no reason to suspect that there should be any simple relationship between force and acceleration. Yet there it is, a simple but exceedingly powerful equation relating the two. Newton's work, preceded to some extent by Galileo's, marks the beginning of a highly successful period in the history of science during which it was learned that the behavior of physical objects can often be described and predicted by mathematical relationships. While some relationships are found to apply only in special circumstances, others seem to have universal applicability. Those equations that appear to apply at all times and under all conditions have come to be called "laws of nature." Newton's second law is a law of nature; you will meet others as we go through this book.

We can rewrite Newton's second law in the form

$$\vec{F}_{net} = m\vec{a} \qquad (4.4)$$

which is how you'll see it presented in many textbooks. Equations 4.3 and 4.4 are mathematically equivalent, but Equation 4.3 better describes the central idea of Newtonian mechanics: A force applied to an object causes the object to accelerate.

Be careful not to think that one force "overcomes" the others to determine the motion. Forces are not in competition with each other! It is $\vec{F}_{net}$, the sum of *all* the forces, that determines the acceleration $\vec{a}$.

As an example, Figure 4.20a shows a box being pulled by two ropes. The ropes exert tension forces $\vec{T}_1$ and $\vec{T}_2$ on the box. Figure 4.20b represents the box as a particle, shows the forces acting on the box, and adds them graphically to

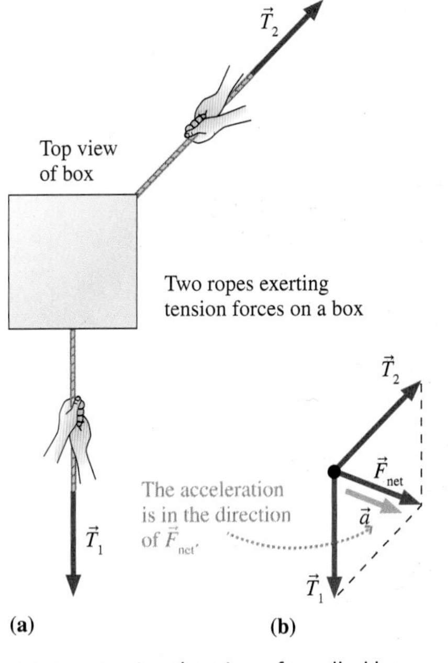

FIGURE 4.20 Acceleration of a pulled box.

find the net force $\vec{F}_{net}$. The box will accelerate in the direction of $\vec{F}_{net}$ with an acceleration of magnitude

$$\vec{a} = \frac{\vec{F}_{net}}{m} = \frac{\vec{T}_1 + \vec{T}_2}{m}$$

NOTE ▶ The acceleration is *not* $(T_1 + T_2)/m$. You must add the forces as *vectors,* not merely add their magnitudes as scalars. ◀

Units of Force

Because $\vec{F}_{net} = m\vec{a}$, the units of force must be mass units multiplied by acceleration units. We've previously specified the SI unit of mass as the kilogram. We can now define the basic unit of force as "the force that causes a 1 kg mass to accelerate at 1 m/s^2." From the second law, this force is

$$1 \text{ basic unit of force} \equiv 1 \text{ kg} \times 1\frac{\text{m}}{\text{s}^2} = 1\frac{\text{kg m}}{\text{s}^2}$$

This basic unit of force is called a newton:

> One **newton** is the force that causes a 1 kg mass to accelerate at 1 m/s^2. The abbreviation for newton is N. Mathematically, $1 \text{ N} = 1 \text{ kg m/s}^2$.

The newton is a *secondary unit,* meaning that it is defined in terms of the *primary units* of kilograms, meters, and seconds. We will introduce other secondary units as needed.

It is important to develop a feeling for what the size of forces should be. Table 4.3 shows some typical forces. As you can see, "typical" forces on "typical" objects are likely to be in the range 0.01–10,000 N. Forces less than 0.01 N are too small to consider unless you are dealing with very small objects. Forces greater than 10,000 N would make sense only if applied to very massive objects.

TABLE 4.3 Approximate magnitude of some typical forces

Force	Approximate magnitude (newtons)
Weight of a U.S. quarter	0.05
Weight of a 1-pound object	5
Weight of a 110-pound person	500
Propulsion force of a car	5,000
Thrust force of a rocket motor	5,000,000

The unit of force in the English system is the *pound* (abbreviated lb). Although the definition of the pound has varied throughout history, it is now defined in terms of the newton:

$$1 \text{ pound} = 1 \text{ lb} \equiv 4.45 \text{ N}$$

You very likely associate pounds with kilograms rather than with newtons. Everyday language often confuses the ideas of mass and weight, but we're going to need to make a clear distinction between them. More on this in the next chapter.

STOP TO THINK 4.4 Three forces act on an object. In which direction does the object accelerate?

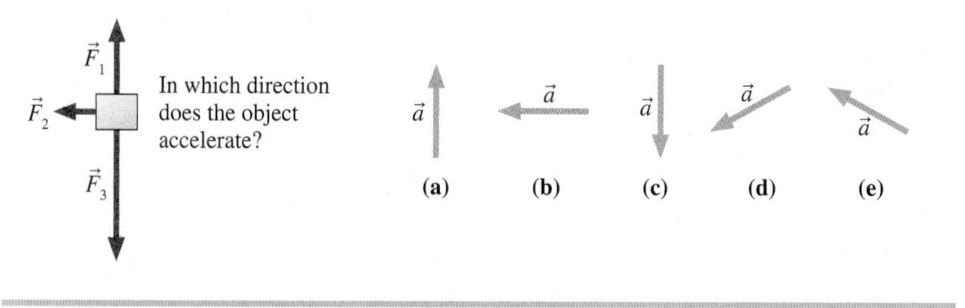

4.6 Newton's First Law

As we remarked earlier, Aristotle and his contemporaries in the world of ancient Greece were very interested in motion. One question they asked was: What is the "natural state" of an object if left to itself? It does not take an expensive research program to see that every moving object on earth, if left to itself, eventually comes to rest. Aristotle concluded that the natural state of an earthly object is to be at rest. An object at rest requires no explanation; it is doing precisely what comes naturally to it. A moving object, though, is not in its natural state and thus requires an explanation: Why is this object moving? What keeps it going and prevents it from being in its natural state?

Galileo reopened the question of the "natural state" of objects. He suggested focusing on the *limiting case* in which resistance to the motion (e.g., friction or air resistance) is zero. This is an idealization that may not be realizable in practice, but Galileo had asserted previously, with great success, that the idealized case can establish a *general principle.* Many careful experiments in which he minimized the influence of friction led Galileo to a conclusion that was in sharp contrast to Aristotle's belief that rest is an object's natural state.

Galileo found that an external influence (i.e., a force) is needed to make an object accelerate—to *change* its velocity. In particular, a force is needed to put an object in motion. But, in the absence of friction or air resistance, a moving object continues to move along a straight line forever with no loss of speed. In other words, the natural state of an object—its behavior if free of external influences—is *uniform motion* with constant velocity! This does not happen in practice because friction or air resistance prevents the object from being left alone. "At rest" has no special significance in Galileo's view of motion; it is simply uniform motion that happens to have $\vec{v} = \vec{0}$.

Galileo's experiments were limited to motion along horizontal surfaces. It was left to Newton to generalize this result, and today we call it Newton's first law of motion.

Newton's first law An object that is at rest will remain at rest, or an object that is moving will continue to move in a straight line with constant velocity, if and only if the net force acting on the object is zero.

Newton's first law is also known as the *law of inertia.* If an object is at rest, it has a tendency to stay at rest. If it is moving, it has a tendency to continue moving with the *same velocity.*

NOTE ► The first law refers to *net* force. An object can remain at rest, or can move in a straight line with constant velocity, even though forces are exerted on it as long as the *net* force is zero. ◄

Notice the "if and only if" aspect of Newton's first law. If an object is at rest or moves with constant velocity, then we can conclude that there is no net force acting on it. Conversely, if no net force is acting on it, we can conclude that the object will have constant velocity, not just constant speed. The direction remains constant, too!

An object on which the net force is zero, $\vec{F}_{net} = \vec{0}$ is said to be in **mechanical equilibrium.** According to Newton's first law, there are two distinct forms of mechanical equilibrium:

1. The object is at rest. This is **static equilibrium.**
2. The object is moving in a straight line with constant velocity. This is **dynamic equilibrium.**

Two examples of mechanical equilibrium are shown in Figure 4.21. Both share the common feature that the acceleration is zero: $\vec{a} = \vec{0}$.

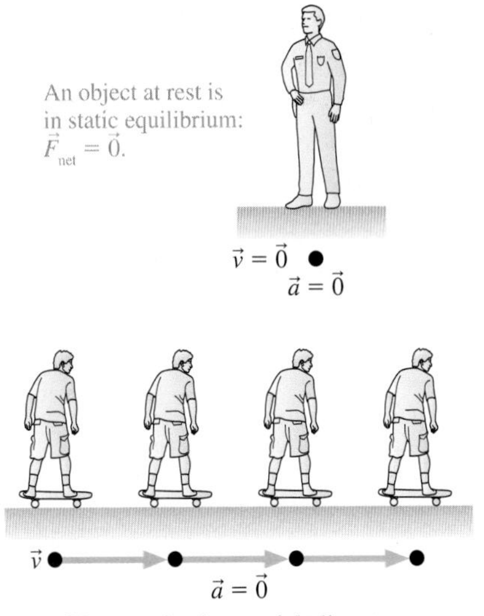
An object at rest is in static equilibrium: $\vec{F}_{net} = \vec{0}$.
$\vec{v} = \vec{0}$
$\vec{a} = \vec{0}$

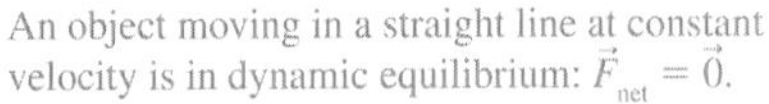
$\vec{v}$
$\vec{a} = \vec{0}$
An object moving in a straight line at constant velocity is in dynamic equilibrium: $\vec{F}_{net} = \vec{0}$.

FIGURE 4.21 Two examples of mechanical equilibrium.

What Good Is Newton's First Law?

The first law completes our definition of force. It answers the question: What is a force? If an "influence" on an object causes the object's velocity to change, the influence is a force.

Newton's first law changes the question the ancient Greeks were trying to answer: What causes an object to move? Newton's first law says **no cause is needed for an object to move!** Uniform motion is the object's natural state. Nothing at all is required for it to remain in that state. The proper question, according to Newton, is: What causes an object to *change* its velocity? Newton, with Galileo's help, also gave us the answer. **A *force* is what causes an object to change its velocity.**

The preceding paragraph contains the essence of Newtonian mechanics. This new perspective on motion, however, is often contrary to our common experience. We all know perfectly well that you must keep pushing an object—exerting a force on it—to keep it moving. Newton is asking us to change our point of view and to consider motion *from the object's perspective* rather than from our personal perspective. As far as the object is concerned, our push is just one of several forces acting on it. Others might include friction, air resistance, or gravity. Only by knowing the *net* force can we determine the object's motion.

Newton's first law may seem to be merely a special case of Newton's second law. After all, the equation $\vec{F}_{\text{net}} = m\vec{a}$ tells us that an object moving with constant velocity ($\vec{a} = \vec{0}$) has $\vec{F}_{\text{net}} = \vec{0}$. The difficulty is that the second law assumes that we already know what force is. The purpose of the first law is to *identify* a force as something that disturbs a state of equilibrium. The second law then describes how the object responds to this force. Thus from a *logical* perspective, the first law really is a separate statement that must precede the second law. But this is a rather formal distinction. From a pedagogical perspective it is better—as we have done—to use a commonsense understanding of force and start with Newton's second law.

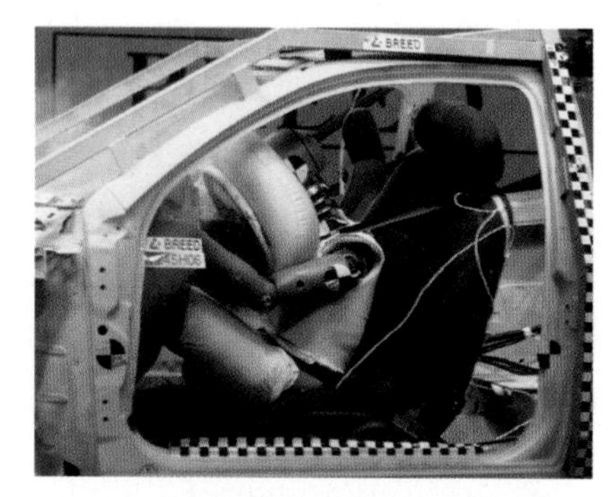

This guy thinks there's a force hurling him into the windshield. What a dummy!

Inertial Reference Frames

If a car stops suddenly, you may be "thrown" into the windshield if you're not wearing your seat belt. You have a very real forward acceleration *relative to the car*, but is there a force pushing you forward? A force is a push or a pull caused by an identifiable agent in contact with the object. Although you *seem* to be pushed forward, there's no agent to do the pushing.

The difficulty—an acceleration without an apparent force—comes from using an inappropriate coordinate system. Your acceleration measured in a coordinate system attached to the car is not the same as your acceleration measured in a coordinate system attached to the ground. Newton's second law says $\vec{F}_{\text{net}} = m\vec{a}$. But which $\vec{a}$? Measured in which coordinate system?

We define an **inertial reference frame** as a coordinate system in which Newton's laws are valid. The first law provides a convenient way to test whether a coordinate system is an inertial reference frame. If $\vec{a} = \vec{0}$ (an object is at rest or moving with constant velocity) only when $\vec{F}_{\text{net}} = \vec{0}$, then the coordinate system in which $\vec{a}$ is measured is an inertial reference frame.

Not all coordinate systems are inertial reference frames. Figure 4.22a shows a physics student cruising at constant velocity in an airplane. If the student places a ball on the floor, it stays there. There are no horizontal forces, and the ball remains at rest relative to the airplane. That is, $\vec{a} = \vec{0}$ in the airplane's coordinate system when $\vec{F}_{\text{net}} = \vec{0}$. Newton's first law is satisfied, so this airplane is an inertial reference frame.

The physics student in Figure 4.22b conducts the same experiment during takeoff. She carefully places the ball on the floor just as the airplane starts to accelerate down the runway. You can imagine what happens. The ball rolls to the back of the plane as the passengers are being pressed back into their seats. Nothing exerts a horizontal contact force on the ball, yet the ball accelerates *in the plane's coordinate system*. This violates Newton's first law, so the plane is *not* an inertial reference frame during takeoff.

(a)

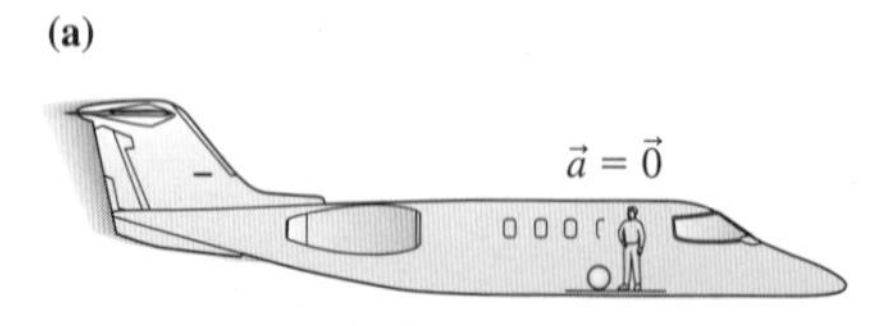
$\vec{a} = \vec{0}$
The ball stays in place.

A ball with no horizontal forces stays at rest in an airplane cruising at constant velocity. The airplane is an inertial reference frame.

(b)

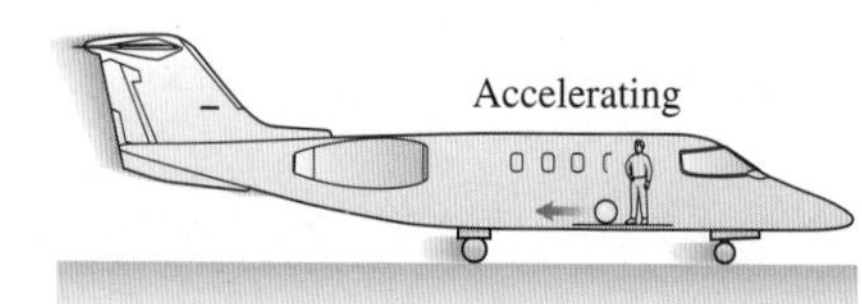
Accelerating
The ball rolls to the back.

The ball rolls to the back of the plane during takeoff. An accelerating plane is not an inertial reference frame.

FIGURE 4.22 Reference frames.

In the first example, the plane is traveling with constant velocity. In the second, the plane is accelerating. **Accelerating reference frames are not inertial reference frames.** Consequently, Newton's laws are not valid in a coordinate system attached to an accelerating object.

But accelerating with respect to what? The plane accelerated with respect to the earth. But the earth is accelerating as it rotates on its axis and revolves around the sun. The entire solar system is accelerating as our Milky Way galaxy rotates.

This is a subtle and difficult question, one that Einstein grappled with as he developed his theory of relativity. We cannot give a complete answer here, but suffice it to say that an inertial reference frame is a reference frame that is not accelerating *with respect to the distant stars.* Thus the earth is not exactly an inertial reference frame. However, the earth's acceleration with respect to the distant stars is so small that violations of Newton's laws can be measured only in extremely high-precision experiments. We will treat the earth and laboratories attached to the earth as inertial reference frames, an approximation that is exceedingly well justified.

We will prove in Chapter 6 that a coordinate system moving with constant velocity relative to an inertial reference frame is also an inertial reference frame, a reference frame in which Newton's laws are valid. Because the earth is an inertial reference frame, the airplane of Figure 4.22a is also an inertial reference frame. But a car braking to a stop is not, so you *cannot* use Newton's laws in the car's reference frame.

To understand the motion of objects in the car, such as the passengers, you need to measure velocities and accelerations *relative to the ground.* From the perspective of an observer on the ground, the body of a passenger in a braking car tries to continue moving forward with constant velocity, exactly as we would expect on the basis of Newton's first law, while his immediate surroundings are decelerating. The passenger is not "thrown" into the windshield. Instead, the windshield runs into the passenger!

Common Misconceptions About Force

It is important to identify correctly all the forces acting on an object. It is equally important not to include forces that do not really exist. We have established a number of criteria for identifying forces; the two critical ones are:

- A force has an agent. Something tangible and identifiable causes the force.
- Forces exist at the point of contact between the agent and the object experiencing the force (except for the few special cases of long-range forces).

We all have had many experiences suggesting that a force is necessary to keep something moving. Consider a bowling ball rolling along on a smooth floor. It is very tempting to think that a horizontal "force of motion" keeps it moving in the forward direction. But if we draw a closed curve around the ball, *nothing contacts it* except the floor. No agent is giving the ball a forward push. According to our definition, then, there is *no* forward "force of motion" acting on the ball. So what keeps it going? Recall our discussion of the first law: *no* cause is needed to keep an object moving at constant velocity. It continues to move forward simply because of its inertia.

One reason for wanting to include a "force of motion" is that we tend to view the problem from our perspective as one of the agents of force. You certainly have to keep pushing to shove a box across the floor at constant velocity. If you stop, it stops. Newton's laws, though, require that we adopt the object's perspective. The box experiences your pushing force in one direction *and* a friction force in the opposite direction. The box moves at constant velocity if the *net* force is zero. This will be true as long as your pushing force exactly balances the friction force.

When you stop pushing, the friction force causes an acceleration that slows and stops the box.

A related problem occurs if you throw a ball. A pushing force was indeed required to accelerate the ball *as it was thrown.* But that force disappears the instant the ball loses contact with your hand. The force does not stick with the ball as the ball travels through the air. Once the ball has acquired a velocity, *nothing* is needed to keep it moving with that velocity.

A final difficulty worth noting is the force due to air pressure. You may have learned in an earlier science class that air, like any fluid, exerts forces on objects. Perhaps you learned this idea as "the air presses down with a weight of 15 pounds on every square inch." There is only one error here, but it is a serious one: the word *down.* Air pressure, at sea level, does indeed exert a force of 15 pounds per square inch, but in *all* directions. It presses down on the top of an object, but also inward on the sides and upward on the bottom. For most purposes, the *net* force due to air pressure is zero! The only way to experience an air pressure force is to form a seal around one side of the object and then remove the air, creating a *vacuum.* When you press a suction cup against the wall, you press the air out and the rubber forms a seal that prevents the air from returning. Now the air pressure does hold the suction cup in place! We do not need to be concerned with air pressure until Part III of this book.

4.7 Free-Body Diagrams

Having discussed at length what is and is not a force, we are ready to assemble our knowledge about force and motion into a single diagram called a *free-body diagram.* You will learn in the next chapter how to write the equations of motion directly from the free-body diagram. Solution of the equations is a mathematical exercise—possibly a difficult one, but nonetheless an exercise that could be done by a computer. The *physics* of the problem, as distinct from the purely calculational aspects, are the steps that lead to the free-body diagram.

A **free-body diagram** represents the object as a particle and shows *all* of the forces acting on the object. The free-body diagram joins with the motion diagram and force identification to form the full *physical representation* of a problem.

TACTICS BOX 4.3 Drawing a free-body diagram

1. **Identify all forces acting on the object.** This step was described in Tactics Box 4.2.
2. **Draw a coordinate system.** Use the axes defined in your pictorial representation. If those axes are tilted, for motion along an incline, then the axes of the free-body diagram should be similarly tilted.
3. **Represent the object as a dot at the origin of the coordinate axes.** This is the particle model.
4. **Draw vectors representing each of the identified forces.** This was described in Tactics Box 4.1. Be sure to label each force vector.
5. **Draw and label the *net force* vector $\vec{F}_{net}$.** Draw this vector beside the diagram, not on the particle. Or, if appropriate, write $\vec{F}_{net} = \vec{0}$. Then check that $\vec{F}_{net}$ points in the same direction as the acceleration vector $\vec{a}$ on your motion diagram.

EXAMPLE 4.4 An elevator accelerates upward

An elevator, suspended by a cable, speeds up as it moves upward from the ground floor. Draw a free-body diagram of the elevator.

MODEL Treat the elevator as a particle.

VISUALIZE

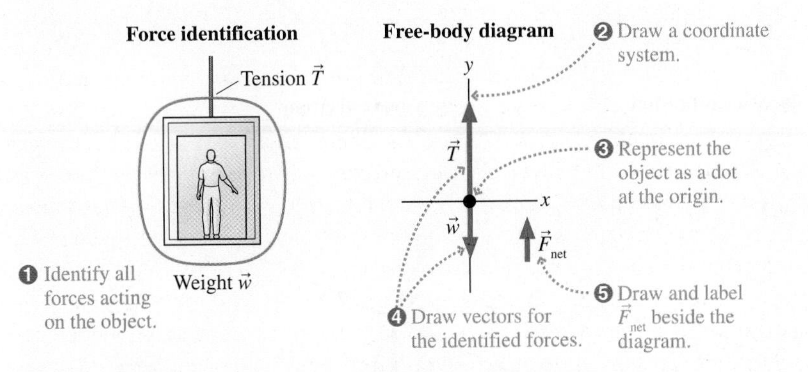

FIGURE 4.23 Free-body diagram of an elevator accelerating upward.

ASSESS The coordinate axes, with a vertical y-axis, are the ones we would use in a pictorial representation of the motion. The elevator is accelerating upward, so $\vec{F}_{net}$ must point upward. For this to be true, the magnitude of $\vec{T}$ must be larger than the magnitude of $\vec{w}$. The diagram has been drawn accordingly.

EXAMPLE 4.5 An ice block shoots across a frozen lake

Bobby straps a small model rocket to a block of ice and shoots it across the smooth surface of a frozen lake. Friction is negligible. Draw a full physical representation of the block of ice.

MODEL Treat the block of ice as a particle. The full physical representation consists of a motion diagram to determine $\vec{a}$, a force identification picture, and a free-body diagram. The statement of the situation implies that friction is negligible.

VISUALIZE

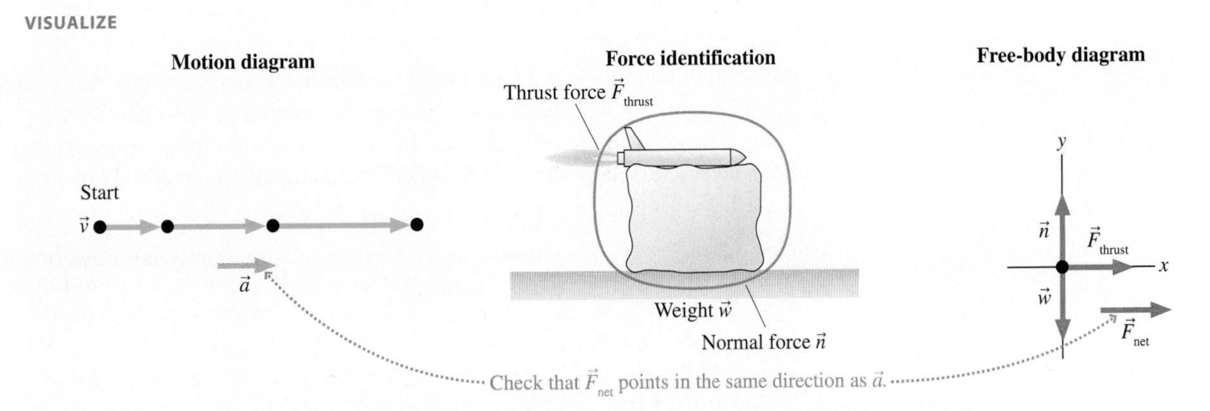

FIGURE 4.24 Physical representation for a block of ice shooting across a frictionless frozen lake.

ASSESS The motion diagram tells us that the acceleration is in the $+x$-direction. According to the rules of vector addition, this can be true only if the upward-pointing $\vec{n}$ and the downward-pointing $\vec{w}$ are equal in magnitude and thus cancel each other $(w_y = -n_y)$. The vectors have been drawn accordingly, and this leaves the net force vector pointing toward the right, in agreement with $\vec{a}$ from the motion diagram.

EXAMPLE 4.6 A skier is pulled up a hill

A tow rope pulls a skier up a snow-covered hill at a constant speed. Draw a full physical representation of the skier.

MODEL This is Example 4.2 again with the additional information that the skier is moving at constant speed. The skier will be treated as a particle in *dynamic equilibrium*. If we were doing a kinematics problem, the pictorial representation would use a tilted coordinate system with the x-axis parallel to the slope, so we use these same tilted coordinate axes for the free-body diagram.

VISUALIZE

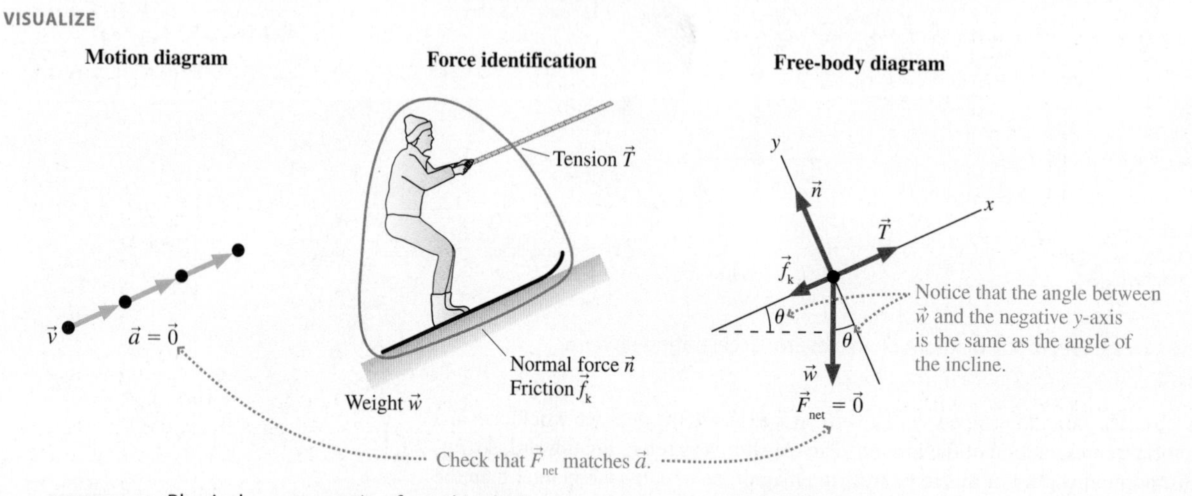

FIGURE 4.25 Physical representation for a skier being towed at a constant speed.

ASSESS We have shown $\vec{T}$ pulling parallel to the slope and $\vec{f}_k$, which opposes the direction of motion, pointing down the slope. $\vec{n}$ is perpendicular to the surface and thus along the y-axis. Finally, and this is important, the weight $\vec{w}$ is *vertically* downward, *not* along the negative y-axis. In fact, you should convince yourself from the geometry that the angle θ between the $\vec{w}$ vector and the negative y-axis is the same as the angle θ of the incline above the horizontal. The skier moves in a straight line with constant speed, so $\vec{a} = \vec{0}$ and, from Newton's first law, $\vec{F}_{net} = \vec{0}$. Thus we have drawn the vectors such that the y-component of $\vec{w}$ is equal in magnitude to $\vec{n}$. Similarly, $\vec{T}$ must be large enough to match the negative x-components of both $\vec{f}_k$ and $\vec{w}$.

Free-body diagrams will be our major tool for the next several chapters. Careful practice with the workbook exercises and homework in this chapter will pay immediate benefits in the next chapter. Indeed, it is not too much to assert that a problem is half solved, or even more, when you complete the free-body diagram.

STOP TO THINK 4.5 An elevator suspended by a cable is moving upward and slowing to a stop. Which free-body diagram is correct?

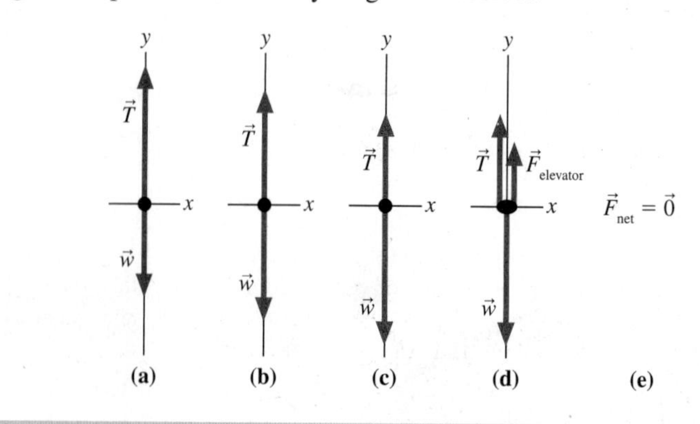

SUMMARY

The goal of Chapter 4 has been to learn how force and motion are connected.

GENERAL PRINCIPLES

Newton's First Law

An object at rest will remain at rest, or an object that is moving will continue to move in a straight line with constant velocity, if and only if the net force on the object is zero.

$\vec{F}_{net} = \vec{0}$

$\vec{v}$ $\vec{v}$ $\vec{v}$ $\vec{v}$ $\vec{v}$

$\vec{a} = \vec{0}$

The first law tells us that no "cause" is needed for motion. Uniform motion is the "natural state" of an object.

Newton's laws are valid only in inertial reference frames.

Newton's Second Law

An object with mass m will undergo acceleration

$$\vec{a} = \frac{1}{m}\vec{F}_{net}$$

where $\vec{F}_{net} = \vec{F}_1 + \vec{F}_2 + \vec{F}_3 + \cdots$ is the vector sum of all the individual forces acting on the object.

$\vec{F}$

$\vec{v}$ $\vec{v}$ $\vec{v}$ $\vec{v}$ $\vec{v}$

$\vec{a}$

The second law tells us that a net force causes an object to accelerate. This is the connection between force and motion that we are seeking.

IMPORTANT CONCEPTS

Acceleration is the link to kinematics.

From a, find v and x.
From v and x, find a.

$\vec{a} = \vec{0}$ is the condition for equilibrium.

Static equilibrium if $\vec{v} = \vec{0}$.
Dynamic equilibrium if $\vec{v} =$ constant.

Equilibrium occurs if and only if $\vec{F}_{net} = \vec{0}$.

Mass is the resistance of an object to acceleration. It is an intrinsic property of an object.

Force is a push or a pull on an object.

- Force is a vector, with a magnitude and a direction.
- Force requires an agent.
- Force is either a contact force or a long-range force.

KEY SKILLS

Identifying Forces

Forces are identified by locating the points where the environment touches the system. These are points where contact forces are exerted. In addition, objects with mass feel a long-range weight force.

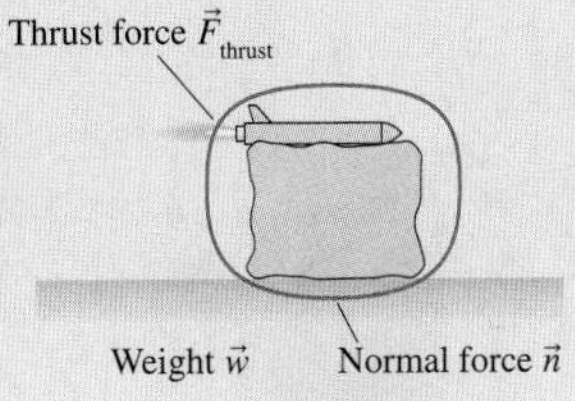

Free-Body Diagrams

A free-body diagram represents the object as a particle at the origin of a coordinate system. Force vectors are drawn with their tails on the particle. The net force vector is drawn beside the diagram.

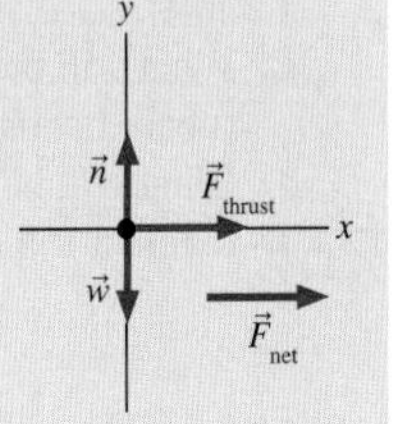

TERMS AND NOTATION

dynamics
mechanics
force, $\vec{F}$
agent
contact force
long-range force
net force, $\vec{F}_{net}$
superposition of forces
weight, $\vec{w}$
tension force, $\vec{T}$
atomic model
normal force, $\vec{n}$
friction, $\vec{f}_k$ or $\vec{f}_s$
drag, $\vec{D}$
thrust, $\vec{F}_{thrust}$
inertia
inertial mass, m
Newton's second law
newton, N
Newton's first law
mechanical equilibrium
static equilibrium
dynamic equilibrium
inertial reference frame
free-body diagram

EXERCISES AND PROBLEMS

Exercises

Section 4.1 Force

1. Write a one-paragraph essay on the topic "What Is a Force?" Explain in your own words how you recognize a force and what the properties of forces are.

Section 4.3 Identifying Forces

2. A mountain climber is hanging from a rope in the middle of a crevasse. The rope is vertical. Identify the forces on the mountain climber.
3. A baseball player is sliding into second base. Identify the forces on the baseball player.
4. A jet plane is speeding down the runway during takeoff. Air resistance is not negligible. Identify the forces on the jet.

Section 4.4 What Do Forces Do?

5. Figure Ex4.5 shows an acceleration-versus-force graph for three objects pulled by rubber bands. The mass of object 2 is 0.20 kg. What are the masses of objects 1 and 3? Explain your reasoning.

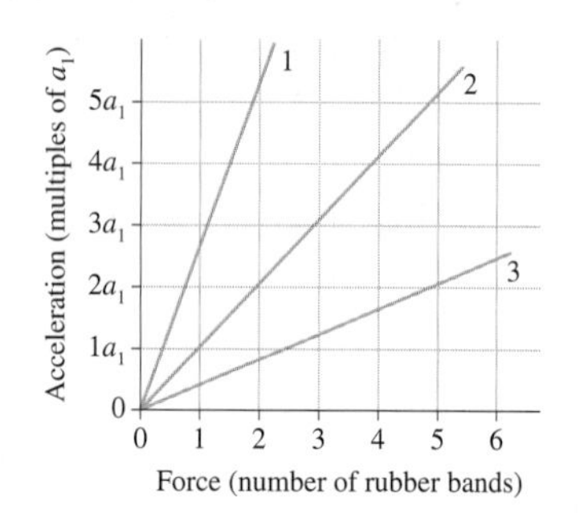

FIGURE EX4.5

6. Two rubber bands pulling on an object cause it to accelerate at 1.2 m/s^2.
   a. What will be the object's acceleration if it is pulled by four rubber bands?
   b. What will be the acceleration of two of these objects glued together if they are pulled by two rubber bands?

Section 4.5 Newton's Second Law

7. Write a one-paragraph essay on the topic "Force and Motion." Explain in your own words the connection between force and motion. Where possible, cite *evidence* supporting your statements.
8. Figure Ex4.8 shows an acceleration-versus-force graph for a 500 g object. Redraw this graph and add appropriate acceleration values on the vertical scale.

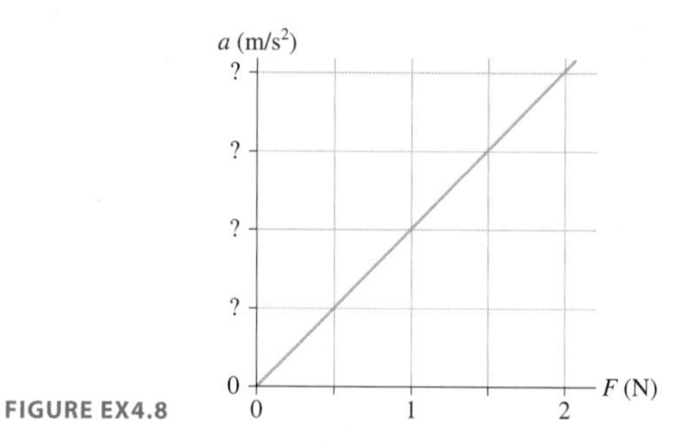

FIGURE EX4.8

9. Figure Ex4.9 shows an object's acceleration-versus-force graph. What is the object's mass?

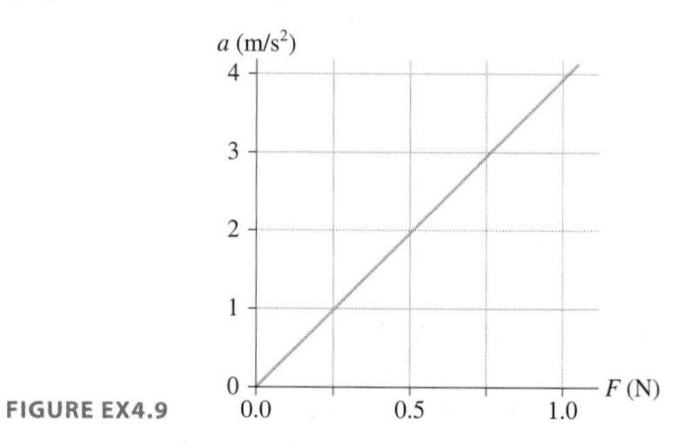

FIGURE EX4.9

10. Based on the information in Table 4.3, estimate
    a. The weight of this textbook.
    b. The propulsion force of a bicycle.
11. Based on the information in Table 4.3, estimate
    a. The weight of a pencil.
    b. The propulsion force of a sprinter.

Section 4.6 Newton's First Law

Exercises 12 through 14 show two forces acting on an object. Redraw the diagram, then add a third force that will cause the object to be in equilibrium. Label the new force $\vec{F}_3$.

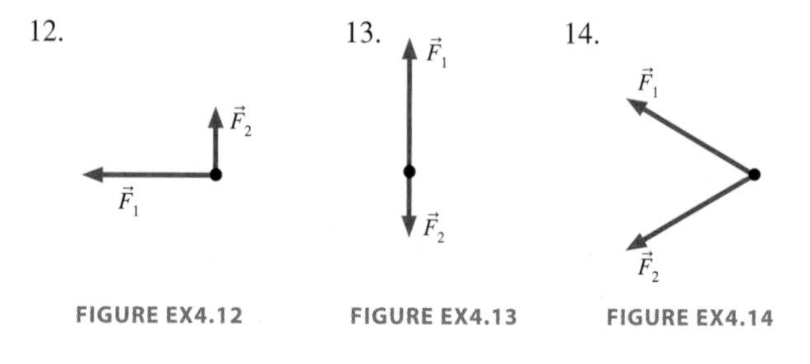

FIGURE EX4.12 FIGURE EX4.13 FIGURE EX4.14

Section 4.7 Free-Body Diagrams

Exercises 15 through 17 show a free-body diagram. For each:

a. Redraw the free-body diagram.
b. Write a short description of a real object for which this is the correct free-body diagram. Use Examples 4.4, 4.5, and 4.6 as models of what a description should be like.

15. 16. 17.

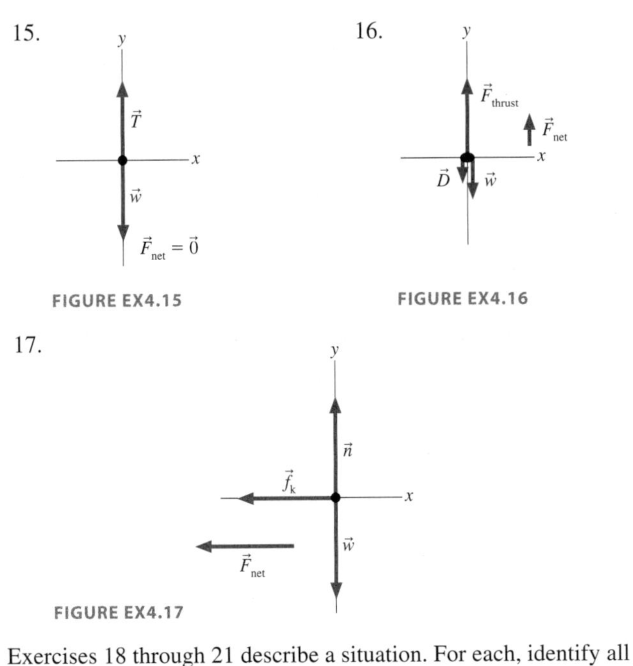

FIGURE EX4.15

FIGURE EX4.16

FIGURE EX4.17

Exercises 18 through 21 describe a situation. For each, identify all forces acting on the object and draw a free-body diagram of the object.

18. Your car is sitting in the parking lot.
19. An ice hockey puck glides across frictionless ice.
20. An elevator, hanging from a cable, descends at steady speed.
21. Your physics textbook is sliding across the table.

Problems

22. Redraw the two motion diagrams shown in Figure P4.22, then draw a vector beside each one to show the direction of the net force acting on the object. Explain your reasoning.

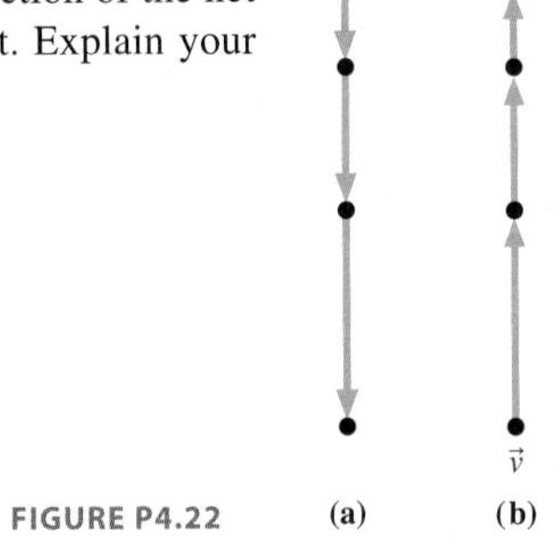

FIGURE P4.22

23. Redraw the two motion diagrams shown in Figure P4.23, then draw a vector beside each one to show the direction of the net force acting on the object. Explain your reasoning.

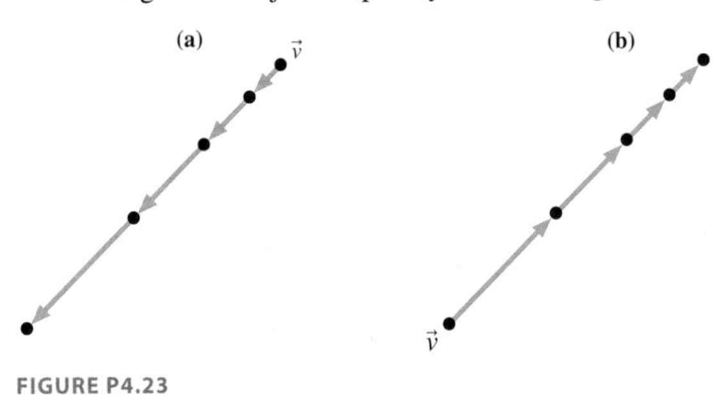

FIGURE P4.23

24. A force with x-component F_x acts on a 2.0 kg object as it moves along the x-axis. The object's acceleration graph (a_x versus t) is shown in Figure P4.24. Draw a graph of F_x versus t.

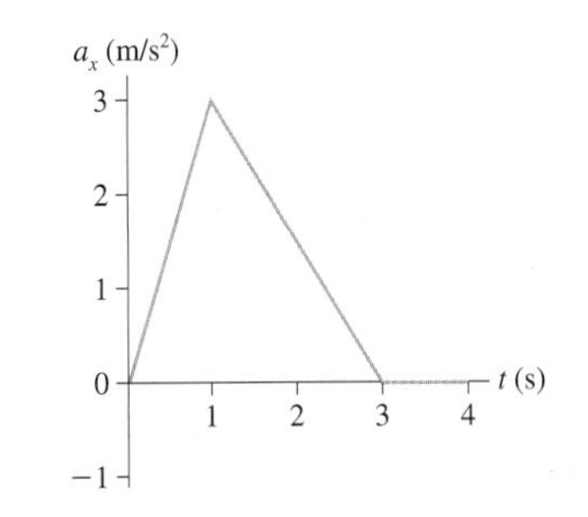

FIGURE P4.24

25. A force with x-component F_x acts on a 500 g object as it moves along the x-axis. The object's acceleration graph (a_x versus t) is shown in Figure P4.25. Draw a graph of F_x versus t.

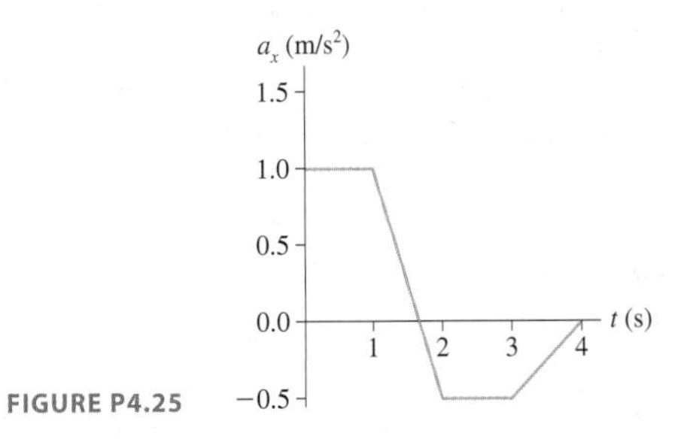

FIGURE P4.25

26. A force with x-component F_x acts on a 2.0 kg object as it moves along the x-axis. A graph of F_x versus t is shown in Figure P4.26. Draw an acceleration graph (a_x versus t) for this object.

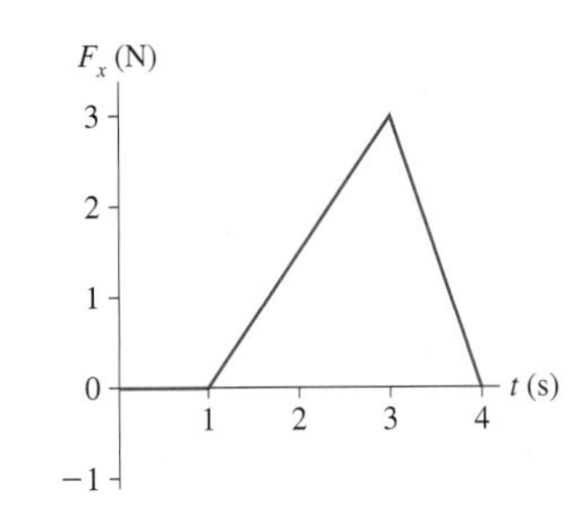

FIGURE P4.26

27. A force with x-component F_x acts on a 500 g object as it moves along the x-axis. A graph of F_x versus t is shown in Figure P4.27. Draw an acceleration graph (a_x versus t) for this object.

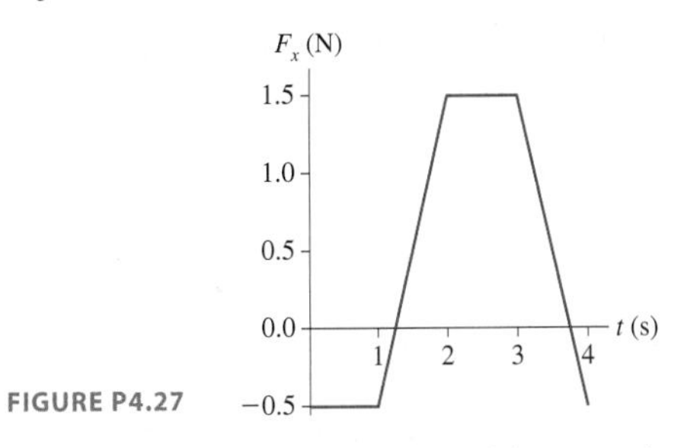

FIGURE P4.27

28. A constant force is applied to an object, causing the object to accelerate at 10 m/s^2. What will the acceleration be if
    a. The force is halved?
    b. The object's mass is halved?
    c. The force and the object's mass are both halved?
    d. The force is halved and the object's mass is doubled?

29. A constant force is applied to an object, causing the object to accelerate at 8.0 m/s^2. What will the acceleration be if
    a. The force is doubled?
    b. The object's mass is doubled?
    c. The force and the object's mass are both doubled?
    d. The force is doubled and the object's mass is halved?

Problems 30 through 36 show a free-body diagram. For each:
  a. Redraw the diagram.
  b. Identify the direction of the acceleration vector $\vec{a}$ and show it as a vector next to your diagram. Or, if appropriate, write $\vec{a} = \vec{0}$.
  c. If possible, identify the direction of the velocity vector $\vec{v}$ and show it as a labeled vector.
  d. Write a short description of a real object for which this is the correct free-body diagram. Use Examples 4.4, 4.5, and 4.6 as models of what a description should be like.

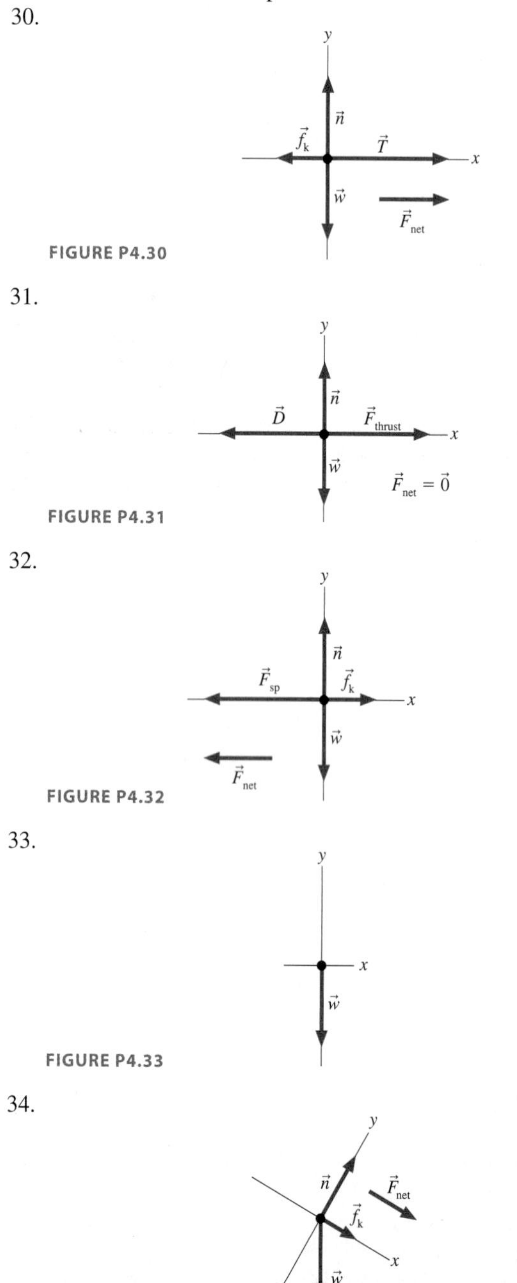

FIGURE P4.30

FIGURE P4.31

FIGURE P4.32

FIGURE P4.33

FIGURE P4.34

35.

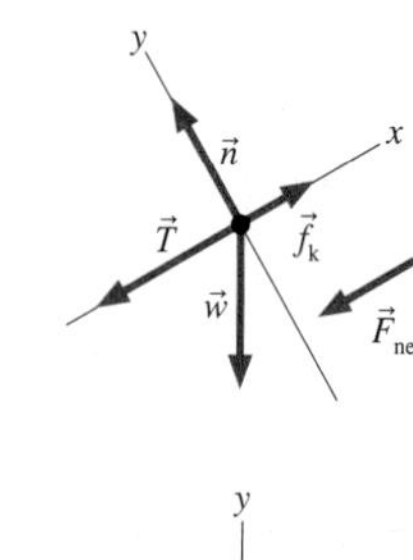

FIGURE P4.35

36.

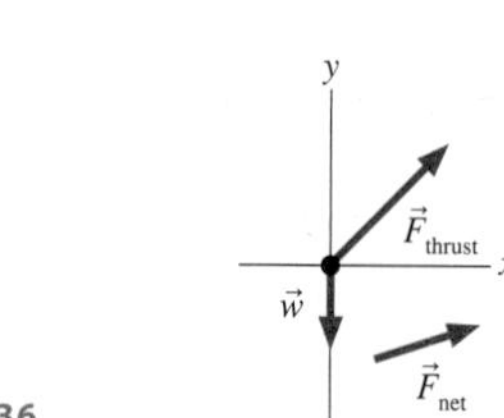

FIGURE P4.36

Problems 37 through 46 describe a situation. For each, draw a full physical representation. A full physical representation is a motion diagram, force identification, and a free-body diagram.

37. An elevator, suspended by a single cable, has just left the tenth floor and is speeding up as it descends toward the ground floor.
38. A rocket is being launched straight up. Air resistance is not negligible.
39. A jet plane is speeding down the runway during takeoff. Air resistance is not negligible.
40. You've slammed on the brakes and your car is skidding to a stop while going down a 20° hill.
41. A skier is going down a 20° slope. A *horizontal* headwind is blowing in the skier's face. Friction is small, but not zero.
42. You've just kicked a soccer ball and it is now rolling across the grass.
43. A styrofoam ball has just been shot straight up. Air resistance is not negligible.
44. A spring-loaded gun shoots a plastic ball. The trigger has just been pulled and the ball is starting to move down the barrel. The barrel is horizontal.
45. A person on a bridge throws a rock straight down toward the water. The rock has just been released.
46. A gymnast has just landed on a trampoline. She's still moving downward as the trampoline stretches.

Challenge Problems

47. A heavy box is in the back of a truck. The truck is accelerating to the right. Draw a full physical representation of the box.
48. A bag of groceries is on the back seat of your car as you stop for a stop light. The bag does not slide. Draw a full physical representation of the bag.
49. A rubber ball bounces. We'd like to understand *how* the ball bounces.
    a. A rubber ball has been dropped and is bouncing off the floor. Draw a motion diagram of the ball during the brief time interval that it is in contact with the floor. Show 4 or 5 frames as the ball compresses, then another 4 or 5 frames as it expands. What is the direction of $\vec{a}$ during each of these parts of the motion?
    b. Draw a picture of the ball in contact with the floor and identify all forces acting on the ball.

c. Draw a free-body diagram of the ball during its contact with the ground. Is there a net force acting on the ball? If so, in which direction?

d. Write a paragraph in which you describe what you learned from parts a to c and in which you answer the question: How does a ball bounce?

50. If a car stops suddenly, you feel "thrown forward." We'd like to understand what happens to the passengers as a car stops. Imagine yourself sitting on a *very* slippery bench inside a car. This bench has no friction, no seat back, and there's nothing for you to hold to.

a. Draw a picture and identify all of the forces acting on you as the car travels at a perfectly steady speed on level ground.

b. Draw your free-body diagram. Is there a net force on you? If so, in which direction?

c. Repeat parts a and b with the car slowing down.

d. Describe what happens to you as the car slows down.

e. Use Newton's laws to explain why you seem to be "thrown forward" as the car stops. Is there really a force pushing you forward?

f. Suppose now that the bench is not slippery. As the car slows down, you stay on the bench and don't slide off. What force is responsible for your deceleration? In which direction does this force point? Include a free-body diagram as part of your answer.

STOP TO THINK ANSWERS

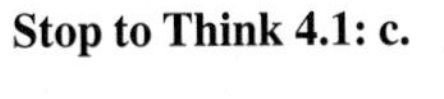

Stop to Think 4.1: c.

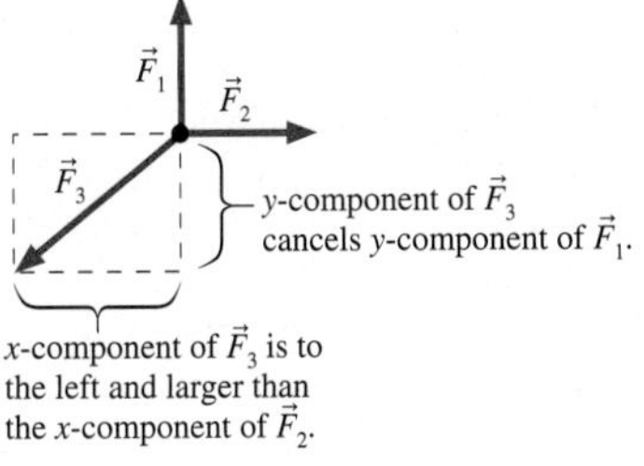
$\vec{F}_1$
$\vec{F}_2$
$\vec{F}_3$
y-component of $\vec{F}_3$ cancels y-component of $\vec{F}_1$.
x-component of $\vec{F}_3$ is to the left and larger than the x-component of $\vec{F}_2$.

Stop to Think 4.2: a, b, and d. Friction and the normal force are the only contact forces. Nothing is touching the rock to provide a "force of the kick." We've agreed to ignore air resistance unless a problem specifically calls for it.

Stop to Think 4.3: b. Acceleration is proportional to force, so doubling the number of rubber bands doubles the acceleration of the original object from 2 m/s^2 to 4 m/s^2. But acceleration is also inversely proportional to mass. Doubling the mass cuts the acceleration in half, back to 2 m/s^2.

Stop to Think 4.4: d.

$\vec{F}_1$ $\vec{F}_2$ $\vec{F}_3$ $\vec{F}_1 + \vec{F}_2$ $\vec{F}_{net}$ $\vec{a}$

First add $\vec{F}_1$ and $\vec{F}_2$. Then add $\vec{F}_3$. This is $\vec{F}_{net}$. $\vec{a}$ is in the same direction as $\vec{F}_{net}$.

Stop to Think 4.5: c. The acceleration vector points downward as the elevator slows. $\vec{F}_{net}$ points in the same direction as $\vec{a}$, so $\vec{F}_{net}$ also points down. This will be true if the tension is less than the weight: $T < w$.

5 Dynamics I: Motion Along a Line

This skydiver may not know it, but he is testing Newton's second law as he plunges toward the ground below.

▶ Looking Ahead

The goal of Chapter 5 is to learn how to solve problems about motion in a straight line. In this chapter you will learn to:

- Solve static and dynamic equilibrium problems by applying a Newton's-first-law strategy.
- Solve dynamics problems by applying a Newton's-second-law strategy.
- Understand how mass, weight, and apparent weight differ.
- Use simple models of friction and drag.

◀ Looking Back

This chapter pulls together many strands of thought from Chapters 1–4. Please review:

- Sections 2.5–2.7 Constant acceleration kinematics, including free fall.
- Sections 3.3–3.4 Working with vectors and vector components.
- Sections 4.2, 4.3, and 4.7 Identifying forces and drawing free-body diagrams.

A skydiver accelerates until reaching a *terminal speed* of about 140 mph. To understand the skydiver's motion, we need to look closely at the forces exerted on him. We also need to understand how those forces determine his motion.

In Chapter 4 we learned what a force is and is not. We also discovered the fundamental relationship between force and motion: Newton's second law. Chapter 5 begins to develop a *strategy* for solving force and motion problems. Our strategy is to learn a set of *procedures,* not to memorize a set of equations.

This chapter focuses on objects that move in a straight line, such as runners, bicycles, cars, planes, and rockets. Weight, tension, thrust, friction, and drag forces will be essential to our understanding. Projectile motion and motion in a circle are the topics of the next two chapters.

5.1 Equilibrium

The concept of equilibrium is essential for the engineering analysis of stationary objects such as bridges.

An object on which the net force is zero is said to be in *equilibrium.* The object might be at rest in *static equilibrium,* or it might be moving along a straight line with constant velocity in *dynamic equilibrium.* Both are identical from a Newtonian perspective because $\vec{F}_{net} = \vec{0}$ and $\vec{a} = \vec{0}$.

Newton's first law is the basis for a four-step *strategy* for solving equilibrium problems.

MP PROBLEM-SOLVING STRATEGY 5.1 Equilibrium problems

MODEL Make simplifying assumptions.

VISUALIZE

Physical representation. Identify all forces acting on the object and show them on a free-body diagram.

Pictorial representation. The free-body diagram is usually sufficient as a picture for equilibrium problems, but you still must translate words to symbols and identify what the problem is trying to find.

It's OK to go back and forth between these two steps as you visualize the situation.

SOLVE The mathematical representation is based on Newton's first law

$$\vec{F}_{net} = \sum_i \vec{F}_i = \vec{0}$$

The vector sum of the forces is found directly from the free-body diagram.

ASSESS Check that your result has the correct units, is reasonable, and answers the question.

Newton's laws are *vector equations.* Recall from Chapter 3 that the vector equation in the step labeled Solve is a shorthand way of writing two simultaneous equations:

$$\begin{aligned}(F_{net})_x &= \sum_i (F_i)_x = 0 \\ (F_{net})_y &= \sum_i (F_i)_y = 0\end{aligned} \qquad (5.1)$$

In other words, each component of $\vec{F}_{net}$ must simultaneously be zero. Although real-world situations often have forces pointing in three dimensions, thus requiring a third equation for the z-component of $\vec{F}_{net}$, we will restrict ourselves for now to problems that can be analyzed in two dimensions.

Equilibrium problems occur frequently, especially in engineering applications. Let's look at a couple of examples.

Static Equilibrium

EXAMPLE 5.1 Three-way tug-of-war

You and two friends find three ropes tied together with a single knot and decide to have a three-way tug-of-war. Alice pulls to the west with 100 N of force while Bob pulls to the south with 200 N. How hard, and in which direction, should you pull to keep the knot from moving?

MODEL We'll treat the *knot* in the rope as a particle in static equilibrium.

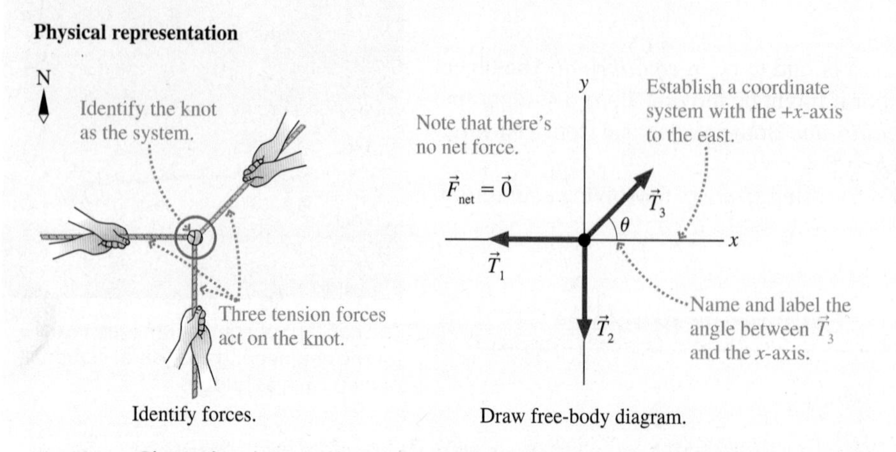
Physical representation
N
Identify the knot as the system.
Three tension forces act on the knot.
Identify forces.
Note that there's no net force.
$\vec{F}_{net} = \vec{0}$
Establish a coordinate system with the +x-axis to the east.
y
x
θ
$\vec{T}_1$
$\vec{T}_2$
$\vec{T}_3$
Name and label the angle between $\vec{T}_3$ and the x-axis.
Draw free-body diagram.

Pictorial representation
A free-body diagram usually suffices in equilibrium problems. All you need to do is add the known information and identify what you are trying to find.

Known
$T_1 = 100$ N
$T_2 = 200$ N

Find
T_3 and θ

List knowns and unknowns.

FIGURE 5.1 Physical and pictorial representations for a knot in static equilibrium.

VISUALIZE Figure 5.1 shows how to carry out the physical and pictorial representations. Notice that we've *defined* angle θ to indicate the direction of your pull.

SOLVE The free-body diagram shows tension forces $\vec{T}_1$, $\vec{T}_2$, and $\vec{T}_3$ acting on the knot. Newton's first law, written in component form, is

$$(F_{net})_x = \sum_i (F_i)_x = T_{1x} + T_{2x} + T_{3x} = 0$$

$$(F_{net})_y = \sum_i (F_i)_y = T_{1y} + T_{2y} + T_{3y} = 0$$

NOTE ▶ You might have been tempted to write $-T_{1x}$ in the first equation because $\vec{T}_1$ points in the negative x-direction. But the net force, by definition, is the *sum* of all the individual forces. The fact that $\vec{T}_1$ points to the left will be taken into account when we *evaluate* the components. ◀

The components of the force vectors can be evaluated directly from the free-body diagram:

$$T_{1x} = -T_1 \qquad T_{1y} = 0$$
$$T_{2x} = 0 \qquad T_{2y} = -T_2$$
$$T_{3x} = +T_3\cos\theta \qquad T_{3y} = +T_3\sin\theta$$

This is where the signs enter, with T_{1x} being assigned a negative value because $\vec{T}_1$ points to the left. Similarly, $T_{2y} = -T_2$. With these components, Newton's first law becomes

$$-T_1 + T_3\cos\theta = 0$$
$$-T_2 + T_3\sin\theta = 0$$

These are two simultaneous equations for the two unknowns T_3 and θ. We will encounter equations of this form on many occasions, so make a note of the method of solution. First, rewrite the two equations as

$$T_1 = T_3\cos\theta$$
$$T_2 = T_3\sin\theta$$

Next, divide the second equation by the first to eliminate T_3:

$$\frac{T_2}{T_1} = \frac{T_3\sin\theta}{T_3\cos\theta} = \tan\theta$$

Then solve for θ:

$$\theta = \tan^{-1}\left(\frac{T_2}{T_1}\right) = \tan^{-1}\left(\frac{200\text{ N}}{100\text{ N}}\right) = 63.4°$$

Finally, use θ to find T_3:

$$T_3 = \frac{T_1}{\cos\theta} = \frac{100\text{ N}}{\cos 63.4°} = 224\text{ N}$$

The force that maintains equilibrium and prevents the knot from moving is thus

$$\vec{T}_3 = (224\text{ N, }63.4°\text{ north of east})$$

ASSESS Is this result reasonable? Because your friends pulled west and south, you expected to pull in a generally northeast direction. You also expected to pull harder than either of them but, because they didn't pull in the same direction, less than the sum of their pulls. The result for $\vec{T}_3$ meets these expectations.

Dynamic Equilibrium

EXAMPLE 5.2 Towing a car up a hill
A car with a weight of 15,000 N is being towed up a 20° slope at constant velocity. Friction is negligible. The tow rope is rated at 6000 N maximum tension. Will it break?

MODEL We'll treat the car as a particle in dynamic equilibrium. We'll ignore friction.

VISUALIZE This problem asks for a yes or no answer, not a number, but we still need a quantitative analysis. Part of our analysis

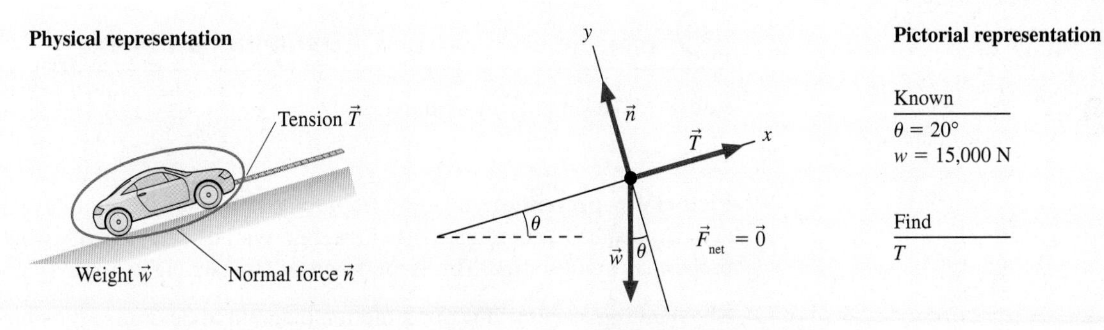

FIGURE 5.2 The physical and pictorial representations of a car being towed up a hill.

of the problem statement is to determine which quantity or quantities allow us to answer the question. In this case the answer is clear: We need to calculate the tension in the rope. Figure 5.2 shows the physical and pictorial representations. Note the similarities to Examples 4.2 and 4.6 in Chapter 4, which you may want to review.

SOLVE The free-body diagram shows forces $\vec{T}$, $\vec{n}$, and $\vec{w}$ acting on the car. Newton's first law is

$$(F_{net})_x = \sum F_x = T_x + n_x + w_x = 0$$
$$(F_{net})_y = \sum F_y = T_y + n_y + w_y = 0$$

Notice that we dropped the label *i* from the sum. From here on, we'll use ΣF_x and ΣF_y as a simple shorthand notation to indicate that we're adding all the *x*-components and all the *y*-components of the forces.

We can deduce the components directly from the free-body diagram:

$$T_x = T \qquad T_y = 0$$
$$n_x = 0 \qquad n_y = n$$
$$w_x = -w\sin\theta \qquad w_y = -w\cos\theta$$

NOTE ▶ The weight has both *x*- and *y*-components in this coordinate system, both of which are negative due to the direction of the vector $\vec{w}$. You'll see this situation often, so be sure you understand where w_x and w_y come from. ◀

With these components, the first law becomes

$$T - w\sin\theta = 0$$
$$n - w\cos\theta = 0$$

The first of these can be rewritten as

$$T = w\sin\theta$$
$$= (15{,}000\ \text{N})\sin 20° = 5130\ \text{N}$$

Because $T < 6000$ N, we conclude that the rope will *not* break. It turned out that we did not need the *y*-component equation in this problem.

ASSESS Because there's no friction, it would not take *any* tension force to keep the car rolling along a horizontal surface ($\theta = 0°$). At the other extreme, $\theta = 90°$, the tension force would need to equal the car's weight ($T = w = 15{,}000$ N) to lift the car straight up at constant velocity. The tension force for a 20° slope should be somewhere in between, and 5130 N is a little less than half the weight of the car. That our result is reasonable doesn't prove it's right, but we have at least ruled out careless errors that give unreasonable results.

5.2 Using Newton's Second Law

Equilibrium is important, but it is a special case of motion. Newton's second law is a more general link between force and motion. We now need a strategy for using Newton's second law to solve dynamics problems.

The essence of Newtonian mechanics can be expressed in two steps:

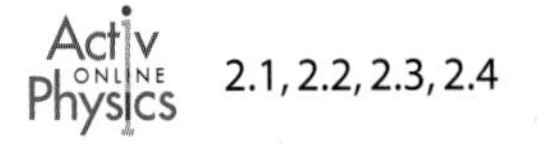

- The forces on an object determine its acceleration $\vec{a} = \vec{F}_{net}/m$.
- The object's trajectory can be determined by using $\vec{a}$ in the equations of kinematics.

These two ideas are the basis of a strategy for solving dynamics problems.

MP PROBLEM-SOLVING STRATEGY 5.2 **Dynamics problems**

MODEL Make simplifying assumptions.

VISUALIZE

Pictorial representation. Show important points in the motion with a sketch, establish a coordinate system, define symbols, and identify what the problem is trying to find. This is the process of translating words to symbols.

Physical representation. Use a motion diagram to determine the object's acceleration vector $\vec{a}$. Then identify all forces acting on the object and show them on a free-body diagram.

It's OK to go back and forth between these two steps as you visualize the situation.

SOLVE The mathematical representation is based on Newton's second law

$$\vec{F}_{net} = \sum_i \vec{F}_i = m\vec{a}$$

The vector sum of the forces is found directly from the free-body diagram. Depending on the problem, either

- Solve for the acceleration, then use kinematics to find velocities and positions, or
- Use kinematics to determine the acceleration, then solve for unknown forces.

ASSESS Check that your result has the correct units, is reasonable, and answers the question.

Newton's second law is a vector equation. To apply the step labeled Solve, you must write the second law as two simultaneous equations:

$$(F_{net})_x = \sum F_x = ma_x$$
$$(F_{net})_y = \sum F_y = ma_y \tag{5.2}$$

The primary goal of this chapter is to illustrate the use of this strategy. Let's start with some examples.

EXAMPLE 5.3 **Speed of a towed car**

A 1500 kg car is pulled by a tow truck. The tension in the tow rope is 2500 N, and a 200 N friction force opposes the motion. If the car starts from rest, what is its speed after 5.0 seconds?

MODEL We'll treat the car as an accelerating particle. We'll assume, as part of our *interpretation* of the problem, that the road is horizontal and that the direction of motion is to the right.

VISUALIZE Figure 5.3 shows the pictorial and physical representations. We've established a coordinate system and defined symbols to represent kinematic quantities. We've identified the speed v_1, rather than the velocity v_{1x}, as what we're trying to find.

SOLVE We begin the mathematical representation with Newton's second law:

$$(F_{net})_x = \sum F_x = T_x + f_x + n_x + w_x = ma_x$$
$$(F_{net})_y = \sum F_y = T_y + f_y + n_y + w_y = ma_y$$

All four forces acting on the car have been included in the vector sum. The equations are perfectly general, with + signs everywhere, because the four vectors are *added* to give $\vec{F}_{net}$. We can now "read" the vector components directly from the free-body diagram:

$$T_x = +T \qquad T_y = 0$$
$$n_x = 0 \qquad n_y = +n$$
$$f_x = -f \qquad f_y = 0$$
$$w_x = 0 \qquad w_y = -w$$

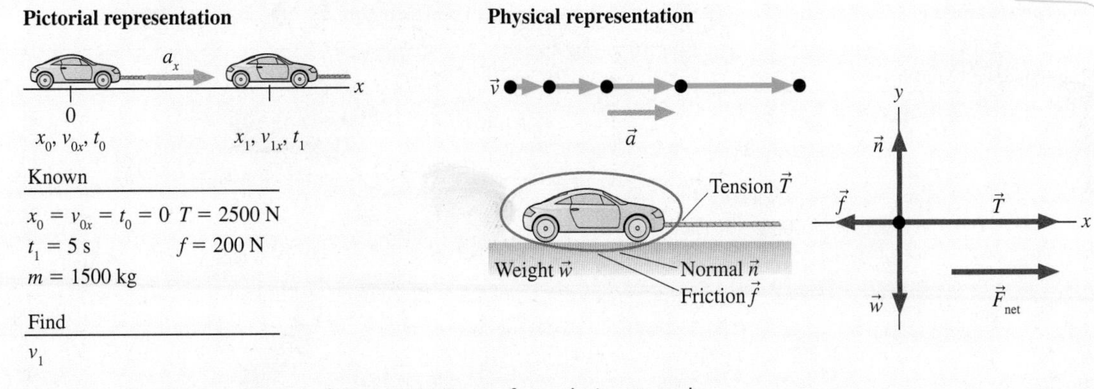

FIGURE 5.3 Pictorial and physical representations of a car being towed.

The signs of the components depend on which way the vectors point. Substituting these into the second-law equations and dividing by m gives

$$a_x = \frac{1}{m}(T - f)$$

$$= \frac{1}{1500 \text{ kg}}(2500 \text{ N} - 200 \text{ N}) = 1.53 \text{ m/s}^2$$

$$a_y = \frac{1}{m}(n - w)$$

NOTE ▶ Newton's second law has allowed us to determine a_x exactly but has given only an algebraic expression for a_y. However, we know ***from the motion diagram*** that $a_y = 0$! That is, the motion is purely along the x-axis, so there is *no* acceleration along the y-axis. The requirement $a_y = 0$ allows us to conclude that $n = w$. Although we do not need n for this problem, it will be important in many future problems. ◀

We can finish by using constant-acceleration kinematics to find the velocity:

$$v_{1x} = v_{0x} + a_x \Delta t$$

$$= 0 + (1.53 \text{ m/s}^2)(5.0 \text{ s})$$

$$= 7.65 \text{ m/s}$$

The problem asked for the *speed* after 5.0 s, which is $v_1 = 7.65$ m/s.

ASSESS 7.65 m/s ≈ 15 mph, a reasonable speed after 5 s of acceleration.

EXAMPLE 5.4 Altitude of a rocket

A 500 g model rocket with a weight of 4.9 N is launched straight up. The small rocket motor burns for 5.0 s and has a steady thrust of 20 N. What maximum altitude does the rocket reach? Assume that the mass loss of the burned fuel is negligible.

MODEL We'll treat the rocket as an accelerating particle. Air resistance will be neglected.

VISUALIZE The pictorial representation of Figure 5.4 on the next page finds that this is a two-part problem. First, the rocket accelerates straight up. Second, the rocket continues going up as it slows down, a free-fall situation. The maximum altitude is at the end of the second part of the motion.

SOLVE We now know what the problem is asking, have established relevant symbols and coordinates, and know what the forces are. We begin the mathematical representation by writing Newton's second law, in component form, as the rocket accelerates upward. The free-body diagram shows two forces, so

$$(F_{net})_x = \sum F_x = (F_{thrust})_x + w_x = ma_{0x}$$

$$(F_{net})_y = \sum F_y = (F_{thrust})_y + w_y = ma_{0y}$$

The fact that vector $\vec{w}$ points downward—and which might have tempted you to use a minus sign in the y-equation—will be taken into account when we *evaluate* the components. None of the vectors in this problem has an x-component, so only the y-component of the second law is needed. We can use the free-body diagram to see that

$$(F_{thrust})_y = +F_{thrust}$$

$$w_y = -w$$

This is the point at which the directional information about the force vectors enters. The y-component of the second law is then

$$a_{0y} = \frac{1}{m}(F_{thrust} - w)$$

$$= \frac{20.0 \text{ N} - 4.9 \text{ N}}{0.500 \text{ kg}} = 30.2 \text{ m/s}^2$$

Notice that we converted the mass to SI units of kilograms before doing any calculations and that, because of the definition of the newton, the division of newtons by kilograms automatically gives the correct SI units of acceleration.

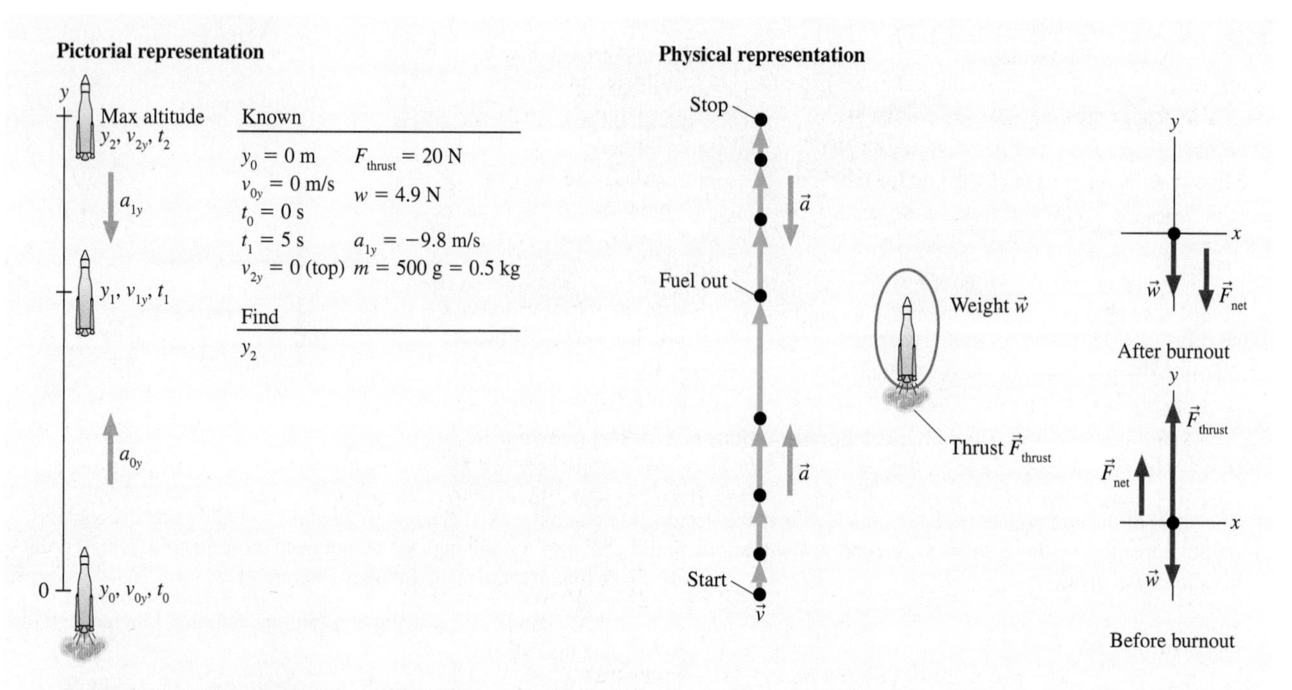

FIGURE 5.4 Pictorial and physical representations of a rocket launch.

The acceleration of the rocket is constant until it runs out of fuel, so we can use constant-acceleration kinematics to find the altitude and velocity at burnout ($\Delta t = t_1 = 5.0$ s):

$$y_1 = y_0 + v_{0y}\,\Delta t + \tfrac{1}{2}a_{0y}(\Delta t)^2$$
$$= \tfrac{1}{2}a_{0y}(\Delta t)^2 = 377 \text{ m}$$
$$v_{1y} = v_{0y} + a_{0y}\,\Delta t = a_{0y}\,\Delta t = 151 \text{ m/s}$$

The only force on the rocket after burnout is gravity, so the second part of the motion is free-fall with $a_{1y} = -g$. We do not know how long it takes to reach the top, but we do know that the final velocity is $v_{2y} = 0$.

We can use free-fall kinematics to find the maximum altitude:

$$v_{2y}^2 = 0 = v_{1y}^2 - 2g\,\Delta y = v_{1y}^2 - 2g(y_2 - y_1)$$

which we can solve to find

$$y_2 = y_1 + \frac{v_{1y}^2}{2g} = 377 \text{ m} + \frac{(151 \text{ m/s})^2}{2(9.80 \text{ m/s}^2)}$$
$$= 1540 \text{ m} = 1.54 \text{ km}$$

ASSESS The maximum altitude reached by this rocket is 1.54 km, or just slightly under one mile. While this is fairly high, it does not seem unreasonable for a high-acceleration rocket.

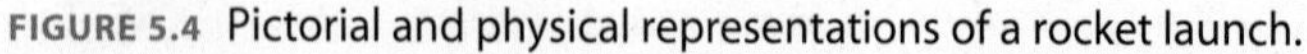

These first examples have shown all the details. Our purpose has been to show how the problem-solving strategy is put into practice. Future examples will be briefer, but the basic *procedure* will remain the same.

STOP TO THINK 5.1 A Martian lander is approaching the surface. It is slowing its descent by firing its rocket motor. Which is the correct free-body diagram for the lander?

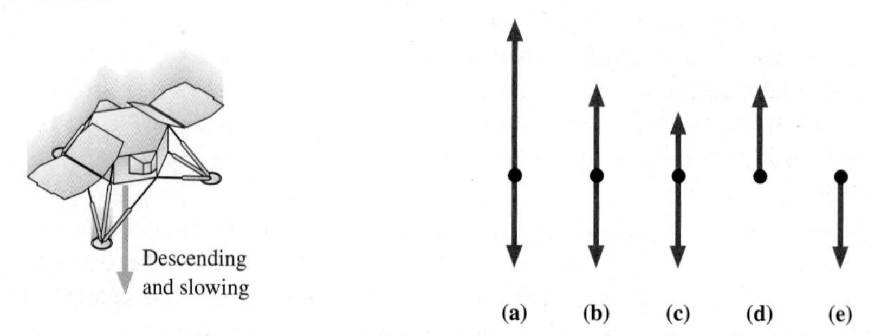

5.3 Mass and Weight

When the doctor asks what you weigh, what does she really mean? We do not make a large distinction in our ordinary use of language between the terms *weight* and *mass*, but in physics their distinction is of critical importance.

Mass, you'll recall from Chapter 4, is a scalar quantity that describes an object's inertia. Loosely speaking, it also describes the amount of matter in an object. Mass, measured in kilograms, is an intrinsic property of an object; it has the same value wherever the object may be and whatever forces might be acting on it.

Weight, on the other hand, is a *force*. Specifically, it is the gravitational force exerted on an object by a planet. Weight is a vector, not a scalar, and the vector's direction is always straight down. Weight is measured in newtons.

Mass and weight are not the same thing, but they are related. We can use Galileo's discovery about free fall to make the connection. Figure 5.5 shows the free-body diagram of an object in free fall. The *only* force acting on this object is its weight, the downward pull of gravity, so $\vec{F}_{\text{net}} = \vec{w}$. Newton's second law is thus

$$\vec{F}_{\text{net}} = \vec{w} = m\vec{a} \tag{5.3}$$

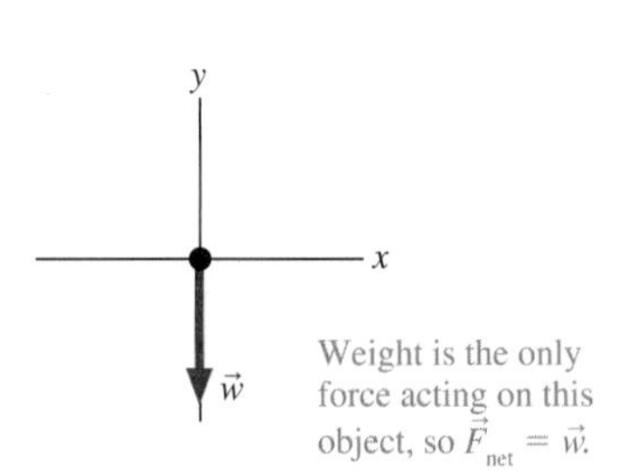

FIGURE 5.5 The free-body diagram of an object in free fall.

Recall Galileo's discovery that *any* object in free fall, regardless of its mass, has the same acceleration:

$$\vec{a}_{\text{free fall}} = (9.80\ \text{m/s}^2, \text{downward}) = (g, \text{downward}) \tag{5.4}$$

where $g = 9.80\ \text{m/s}^2$ is the acceleration due to gravity at the Earth's surface. If we use the free-fall acceleration in Equation 5.3, the object's weight is

$$\vec{w} = (mg, \text{downward}) \tag{5.5}$$

The magnitude of the weight force, which we call simply "the weight," is directly proportional to the mass, with g as the constant of proportionality:

$$w = mg \tag{5.6}$$

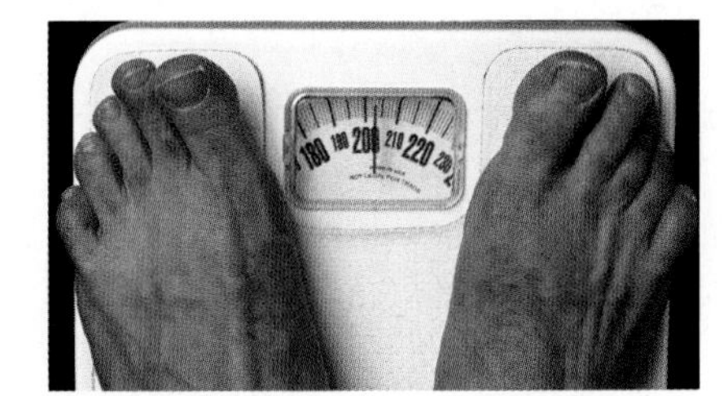

A spring scale, such as the familiar bathroom scale, measures weight, not mass.

Because an object's weight depends on g, and the value of g varies from planet to planet, weight is not a fixed, constant property of an object. The value of g at the surface of the moon is about one-sixth its earthly value, so an object on the moon would have only one-sixth its weight on Earth. The object's weight on Jupiter would be larger than its weight on Earth. Its mass, however, would be the same. The amount of matter has not changed, only the gravitational force exerted on that matter.

So when the doctor asks what you weigh, she really wants to know your *mass.* That's the amount of matter in your body. You can't really "lose weight" by going to the moon, even though you would weigh less there!

Measuring Mass and Weight

A *pan balance*, shown in Figure 5.6, is a device for measuring *mass*. You may have used a pan balance to "weigh" chemicals in a chemistry lab. An unknown mass is placed in one pan, then known masses are added to the other until the pans balance. Gravity pulls down on both sides, effectively *comparing* the masses, and the unknown mass equals the sum of the known masses that balance it. Although a pan balance requires gravity in order to function, it does not depend on the value of g. Consequently, the pan balance would give the same result on another planet.

Spring scales, such as the two shown in Figure 5.7 on the next page, measure weight, not mass. Hanging an item on the scale in Figure 5.7a, which might be used to weigh items in the grocery store, stretches the spring. The spring in the "bathroom scale" in Figure 5.7b is compressed when you stand on it.

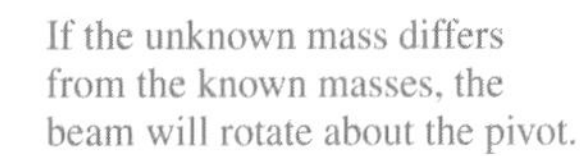

If the unknown mass differs from the known masses, the beam will rotate about the pivot.

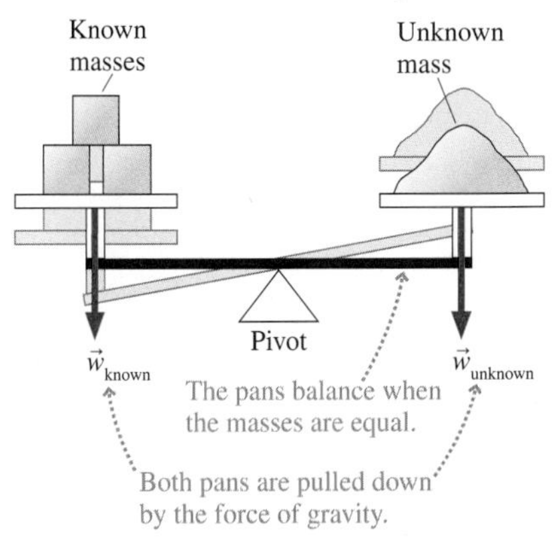

FIGURE 5.6 A pan balance measures mass.

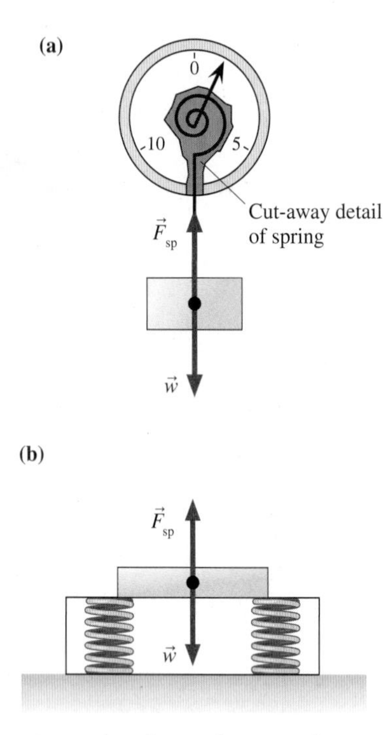

FIGURE 5.7 A spring scale measures weight.

A spring scale can be understood on the basis of Newton's first law. The object being weighed is at rest, in static equilibrium, so the net force on it must be zero. The stretched spring in Figure 5.7a *pulls* up; the compressed spring in Figure 5.7b *pushes* up. But in both cases, in order to have $\vec{F}_{net} = \vec{0}$, the upward spring force must exactly balance the downward weight force:

$$F_{sp} = w = mg \tag{5.7}$$

The *reading* of a spring scale is F_{sp}, the magnitude of the force that the spring is exerting. If the object is in equilibrium, then F_{sp} is exactly equal to the object's weight w. The scale does not "know" the weight of the object. All it can do is to measure how much the spring is stretched or compressed. On a different planet, with a different value for g, the expansion or compression of the spring would be different and the scale's reading would be different.

The unit of force in the English system is the *pound.* We noted in Chapter 4 that the pound is defined as 1 lb $\equiv$ 4.45 N. An object whose weight $w = mg$ is 4.45 N has a mass

$$m = \frac{w}{g} = \frac{4.45 \text{ N}}{9.80 \text{ m/s}^2} = 0.454 \text{ kg} = 454 \text{ g}$$

You may have learned in previous science classes that "1 pound = 454 grams" or, equivalently, that "1 kg = 2.2 lb." Strictly speaking, these well-known "conversion factors" are not true. They are comparing a weight (pounds) to a mass (kilograms). The correct statement would be, "A mass of 1 kg has a weight on *earth* of 2.2 pounds." On another planet, the weight of a 1 kg mass would be something other than 2.2 pounds.

EXAMPLE 5.5 Mass and weight on Jupiter

What is the kilograms-to-pounds conversion factor on Jupiter, where the acceleration due to gravity is 25.9 m/s^2?

SOLVE Consider an object with a mass of 1 kg. Its weight on Jupiter is

$$w_{Jupiter} = mg_{Jupiter} = (1 \text{ kg})(25.9 \text{ m/s}^2)$$
$$= 25.9 \text{ N} \times \frac{1 \text{ lb}}{4.45 \text{ N}} = 5.82 \text{ lb}$$

If you had gone to school on Jupiter, you would have learned that 1 kg = 5.82 lb.

Apparent Weight

The weight of an object is the force of gravity on that object. You may never have thought about it, but gravity is not a force that you can feel or sense directly. Your *sensation* of weight—how heavy you feel—is due to *contact forces* pressing against you. Surfaces touch you and activate nerve endings in your skin. As you read this, your sensation of weight is due to the normal force exerted on you by the chair in which you are sitting. When you stand, you feel the contact force of the floor pushing against your feet. If you hang from a rope, your sensation of weight is due to the tension force pulling up on you.

If you stand at rest, with $\vec{a} = \vec{0}$, then the force you *feel,* the normal force pressing against your feet, is exactly equal in magnitude to your weight. This would be a trivial conclusion if objects were always at rest, but what happens if $\vec{a} \neq \vec{0}$?

Recall the sensations you feel while being accelerated. You feel "heavy" when an elevator suddenly accelerates upward or when an airplane accelerates for take-off. This sensation vanishes as soon as the elevator or airplane reaches a steady cruising speed. Your stomach seems to rise a little and you feel lighter than normal as the upward-moving elevator brakes to a halt or a roller coaster goes over the top. Your true weight $w = mg$ has not changed during these events, but your *apparent weight* has.

To investigate this, imagine a man weighing himself by standing on a spring scale in an elevator as it accelerates upward. What does the scale read? How does the scale reading correspond to the man's *sensation* of weight?

As Figure 5.8 shows, the only forces acting on the man are the upward spring force of the scale and the downward weight force. This seems to be the same situation as Figure 5.7b, but there's one big difference. The man is accelerating; he's not in equilibrium. Thus, according to Newton's laws, there must be a net force acting on the man in the direction of $\vec{a}$.

For the net force $\vec{F}_{\text{net}}$ to point upward, the magnitude of the spring force must be *greater* than the magnitude of the weight force. That is, $F_{\text{sp}} > w$. This conclusion has major implications. Looking at the free-body diagram in Figure 5.8, we see that the y-component of Newton's second law is

$$(F_{\text{net}})_y = (F_{\text{sp}})_y + w_y = F_{\text{sp}} - w = F_{\text{sp}} - mg = ma_y \tag{5.8}$$

where m is the man's mass.

The scale reading is the value of F_{sp}, the magnitude of the force that the scale exerts on the man. Solving Equation 5.8 for F_{sp} gives

$$F_{\text{sp}} = \text{scale reading} = mg + ma_y = mg\left(1 + \frac{a_y}{g}\right) = w\left(1 + \frac{a_y}{g}\right) \tag{5.9}$$

If the elevator is either at rest or moving with constant velocity, then $a_y = 0$ and the man is in equilibrium. In that case, $F_{\text{sp}} = w$ and the scale correctly reads his weight. But if $a_y \neq 0$, the scale's reading is *not* the man's true weight.

Let's define an object's **apparent weight** w_{app} as the magnitude of the contact force that supports the object. From Equation 5.9, this is

$$w_{\text{app}} = w\left(1 + \frac{a_y}{g}\right) \tag{5.10}$$

We've found w_{app} for a situation where a contact force supports the object from below. A homework problem will let you show that the expression is also valid when the object is supported from above by the tension in a rope or cable.

An object *appears* to weigh whatever the scale reads, although, as the saying goes, appearances can be deceiving. As the elevator accelerates upward, $a_y > 0$ and $w_{\text{app}} > w$. You *feel* heavier than normal, and a scale would read more than your true weight. The acceleration vector $\vec{a}$ points downward and $a_y < 0$ when the elevator brakes, so $w_{\text{app}} < w$. You feel lighter and, if you were standing on a scale, it would read less than your true weight.

An object doesn't have to be on a scale for its apparent weight to differ from its true weight. An object's apparent weight is the magnitude of the contact force supporting it. It makes no difference whether this is the spring force of the scale or simply the normal force of the floor.

The idea of apparent weight has important applications. Astronauts are nearly crushed by their apparent weight during a rocket launch when $a_y \gg g$. Much of the thrill of amusement park rides, such as roller coasters, comes from rapid changes in your apparent weight. In Chapter 7, the concept of apparent weight will help us understand how you can swing a bucket of water over your head without the water falling out!

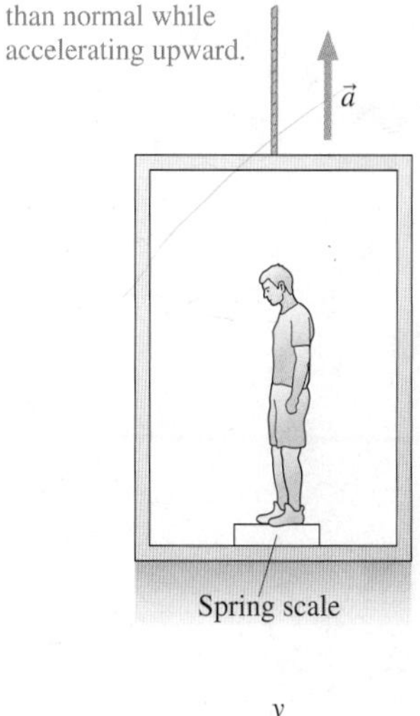

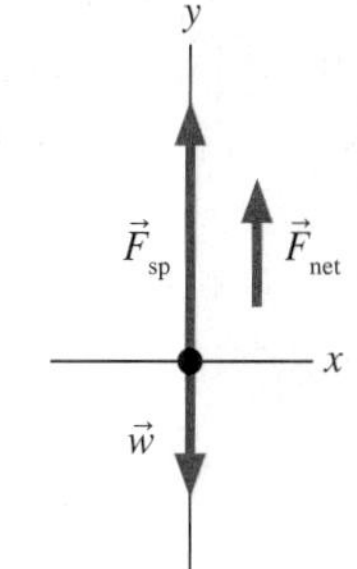

FIGURE 5.8 A man weighing himself in an accelerating elevator.

Weightlessness

One last issue before leaving this topic: Suppose the elevator cable breaks and the elevator, along with the man and his scale, plunges straight down in free fall! What will the scale read? The acceleration in free fall is $a_y = -g$. When this

Astronauts are weightless as they orbit the earth.

acceleration is used in Equation 5.10, we find that $w_{app} = 0$! In other words, the man has *no sensation* of weight.

Think about this carefully. Suppose, as the elevator falls, the man inside releases a ball from his hand. In the absence of air resistance, as Galileo discovered, both the man and the ball would fall at the same rate. From the man's perspective, the ball would appear to "float" beside him. Similarly, the scale would float beneath him and not press against his feet. He is what we call *weightless.*

Surprisingly, "weightless" does *not* mean "no weight." An object that is **weightless** has no *apparent* weight. The distinction is significant. The man's weight is still mg, because gravity is still pulling down on him, but he has no *sensation* of weight as he free falls. The term "weightless" is a very poor one, likely to cause confusion because it implies that objects have no weight. As we see, that is not the case.

But isn't this exactly what happens to astronauts orbiting the earth? You've seen films of astronauts and various objects floating inside the Space Shuttle. If an astronaut tries to stand on a scale, it does not exert any force against her feet and reads zero. She is said to be weightless. But if the criterion to be weightless is to be in free fall, and if astronauts orbiting the earth are weightless, does this mean that they are in free fall? This is a very interesting question to which we shall return in Chapter 7.

STOP TO THINK 5.2 An elevator that has descended from the 50th floor is coming to a halt at the 1st floor. As it does, your apparent weight is

a. More than your true weight.
b. Less than your true weight.
c. Equal to your true weight.
d. Zero.

5.4 Friction

In everyday life, friction is everywhere. Friction is absolutely essential for many things we do. Without friction you could not walk, drive, or even sit down (you would slide right off the chair!). It is sometimes useful to think about idealized frictionless situations, but it is equally necessary to understand a real world where friction is present. Although friction is a complicated force, many aspects of friction can be described with a simple model.

Static Friction

Chapter 4 defined *static friction* $\vec{f}_s$ as the force on an object that keeps it from slipping. Figure 5.9 shows a person pushing on a box with horizontal force $\vec{F}_{push}$. If the box remains at rest, "stuck" to the floor, it must be because of a static friction force pushing back to the left. The box is in static equilibrium, so the static friction must exactly balance the pushing force:

$$f_s = F_{push} \tag{5.11}$$

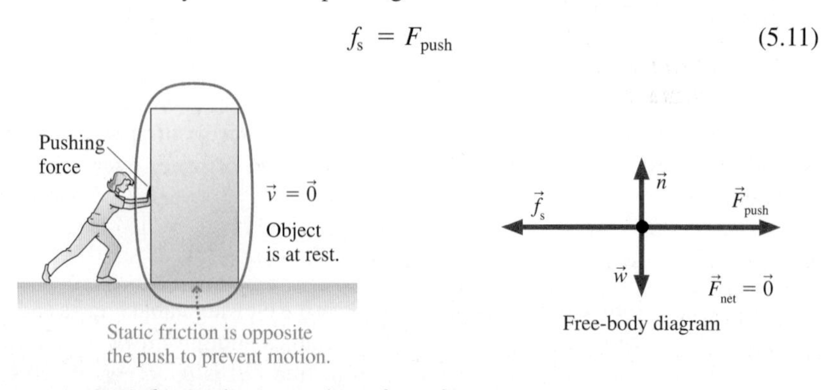

FIGURE 5.9 Static friction keeps an object from slipping.

To determine the direction of $\vec{f}_s$, decide which way the object would move if there were no friction. The static friction force $\vec{f}_s$ points in the *opposite* direction to prevent the motion.

Unlike weight, which has the precise and unambiguous magnitude $w = mg$, the size of the static friction force depends on how hard you push. The harder the person in Figure 5.9 pushes, the harder the floor pushes back. Reduce the pushing force, and the static friction force will automatically be reduced to match. Static friction acts in *response* to an applied force. Figure 5.10 illustrates this idea.

But there's clearly a limit to how big f_s can get. If you push hard enough, the object slips and starts to move. In other words, the static friction force has a *maximum* possible size $f_{s\,max}$.

- An object remains at rest as long as $f_s < f_{s\,max}$.
- The object slips when $f_s = f_{s\,max}$.
- A static friction force $f_s > f_{s\,max}$ is not physically possible.

Experiments with friction (first done by Leonardo da Vinci) show that $f_{s\,max}$ is proportional to the magnitude of the normal force. That is,

$$f_{s\,max} = \mu_s n \tag{5.12}$$

where μ_s is called the **coefficient of static friction.** The coefficient is a dimensionless number that depends on the materials of which the object and the surface are made. Table 5.1 shows some typical values of coefficients of friction. It is to be emphasized that these are only approximate. The exact value of the coefficient depends on the roughness, cleanliness, and dryness of the surfaces.

NOTE ▶ Equation 5.12 does *not* say $f_s = \mu_s n$. The value of f_s depends on the force or forces that static friction has to balance to keep the object from moving. It can have any value from 0 up to, but not exceeding, $\mu_s n$. ◀

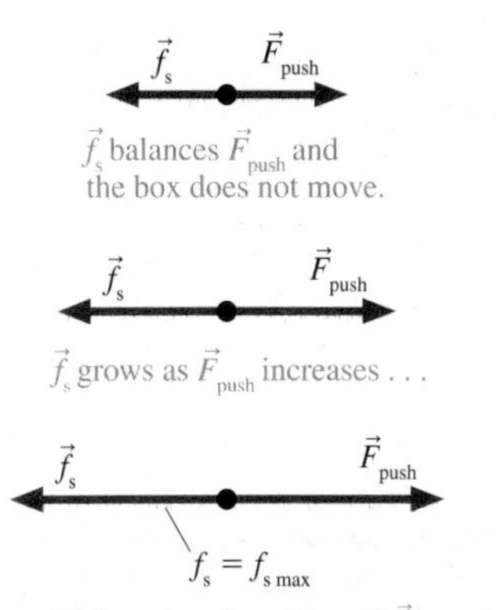
$\vec{f}_s$
$\vec{F}_{push}$
$\vec{f}_s$ balances $\vec{F}_{push}$ and the box does not move.
$\vec{f}_s$
$\vec{F}_{push}$
$\vec{f}_s$ grows as $\vec{F}_{push}$ increases . . .
$\vec{f}_s$
$\vec{F}_{push}$
$f_s = f_{s\,max}$
. . . until f_s reaches $f_{s\,max}$. Now, if $\vec{F}_{push}$ gets any bigger, the object will start to move.

FIGURE 5.10 Static friction acts in *response* to an applied force.

TABLE 5.1 Coefficients of friction

Materials	Static μ_s	Kinetic μ_k	Rolling μ_r
Rubber on concrete	1.00	0.80	0.02
Steel on steel (dry)	0.80	0.60	0.002
Steel on steel (lubricated)	0.10	0.05	
Wood on wood	0.50	0.20	
Wood on snow	0.12	0.06	
Ice on ice	0.10	0.03	

Kinetic Friction

Once the box starts to slide, in Figure 5.11, the static friction force is replaced by a kinetic friction force $\vec{f}_k$. Experiments show that kinetic friction, unlike static friction, has a nearly *constant* magnitude. Furthermore, the size of the kinetic friction force is *less* than the maximum static friction, $f_k < f_{s\,max}$, which explains why it is easier to keep the box moving than it was to start it moving. The direction of $\vec{f}_k$ is always opposite to the direction in which an object slides across the surface.

The kinetic friction force is also proportional to the magnitude of the normal force:

$$f_k = \mu_k n \tag{5.13}$$

where μ_k is called the **coefficient of kinetic friction.** Table 5.1 includes typical values of μ_k. You can see that $\mu_k < \mu_s$, causing the kinetic friction to be less than the maximum static friction.

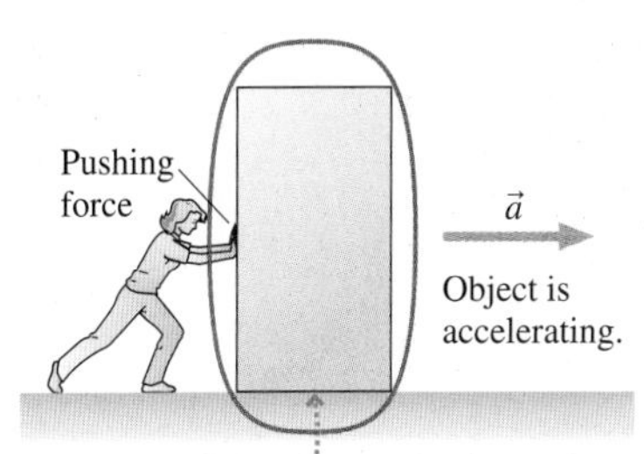
Pushing force
$\vec{a}$
Object is accelerating.
Kinetic friction is opposite the motion.

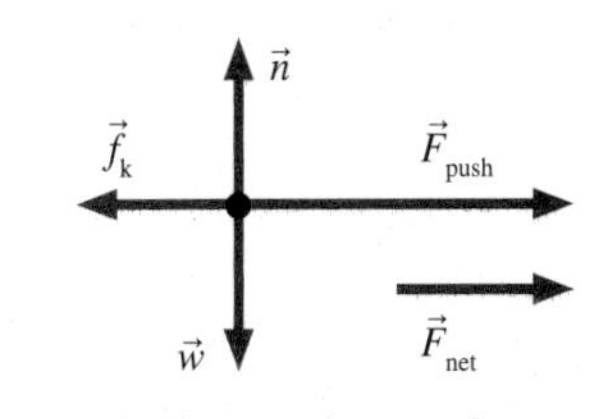
$\vec{n}$
$\vec{f}_k$
$\vec{F}_{push}$
$\vec{w}$
$\vec{F}_{net}$

FIGURE 5.11 The kinetic friction force is opposite the direction of motion.

Rolling Friction

If you slam on the brakes hard enough, your car tires slide against the road surface and leave skid marks. This is kinetic friction. A wheel *rolling* on a surface also experiences friction, but not kinetic friction. The portion of the wheel that contacts the surface is stationary with respect to the surface, not sliding. To see this, roll a wheel slowly and watch how it touches the ground.

Textbooks draw wheels as circles, but no wheel is perfectly round. The weight of the wheel, and of any object supported by the wheel, causes the bottom of the wheel to flatten where it touches the surface, as Figure 5.12 on the next page

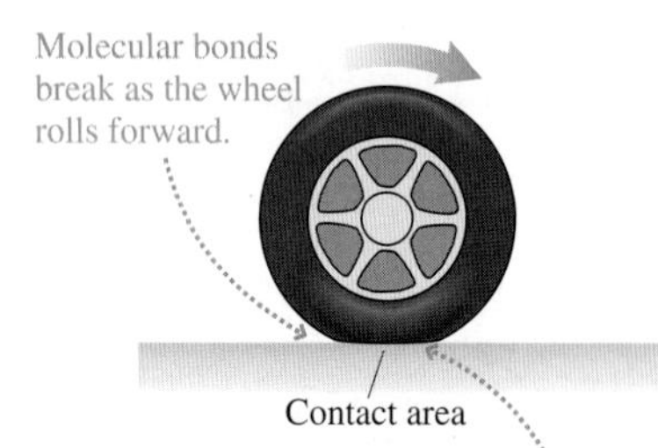

FIGURE 5.12 Rolling friction is due to the contact area between a wheel and the surface.

shows. The contact area between a car tire and the road is fairly large. The contact area between a steel locomotive wheel and a steel rail is much less, but it's not zero.

Molecular bonds are quickly established where the wheel presses against the surface. These bonds have to be broken as the wheel rolls forward, and the effort needed to break them causes **rolling friction.** (Think how it is to walk with a wad of chewing gum stuck to the sole of your shoe!) The force of rolling friction can be calculated in terms of a **coefficient of rolling friction** μ_r:

$$\vec{f}_r = (\mu_r n, \text{direction opposite the motion}) \tag{5.14}$$

Rolling friction acts very much like kinetic friction, but values of μ_r (see Table 5.1) are much less than values of μ_k. This is why it is easier to roll an object on wheels than to slide it.

A Model of Friction

These ideas can be summarized in a *model* of friction:

2.5, 2.6 ActivPhysics ONLINE

$$\begin{aligned} &\text{Static: } \vec{f}_s \leq (\mu_s n, \text{direction as necessary to prevent motion}) \\ &\text{Kinetic: } \vec{f}_k = (\mu_k n, \text{direction opposite the motion}) \\ &\text{Rolling: } \vec{f}_r = (\mu_r n, \text{direction opposite the motion}) \end{aligned} \tag{5.15}$$

Here "motion" means "motion relative to the surface." The maximum value of static friction $f_{s\,max} = \mu_s n$ occurs at the point where the object slips and begins to move.

NOTE ▶ Equations 5.15 are a "model" of friction, not a "law" of friction. These equations provide a reasonably accurate, but not perfect, description of how friction forces act. For example, we've ignored the surface area of the object because surface area has little effect. Likewise, our model assumes that the kinetic friction force is independent of the object's speed. This is a fairly good, but not perfect, approximation. Equations 5.15 are a simplification of reality that works reasonably well, which is what we mean by a "model." They are not a "law of nature" on a level with Newton's laws. ◀

Figure 5.13 summarizes these ideas graphically by showing how the friction force changes as the magnitude of an applied force $\vec{F}_{push}$ increases.

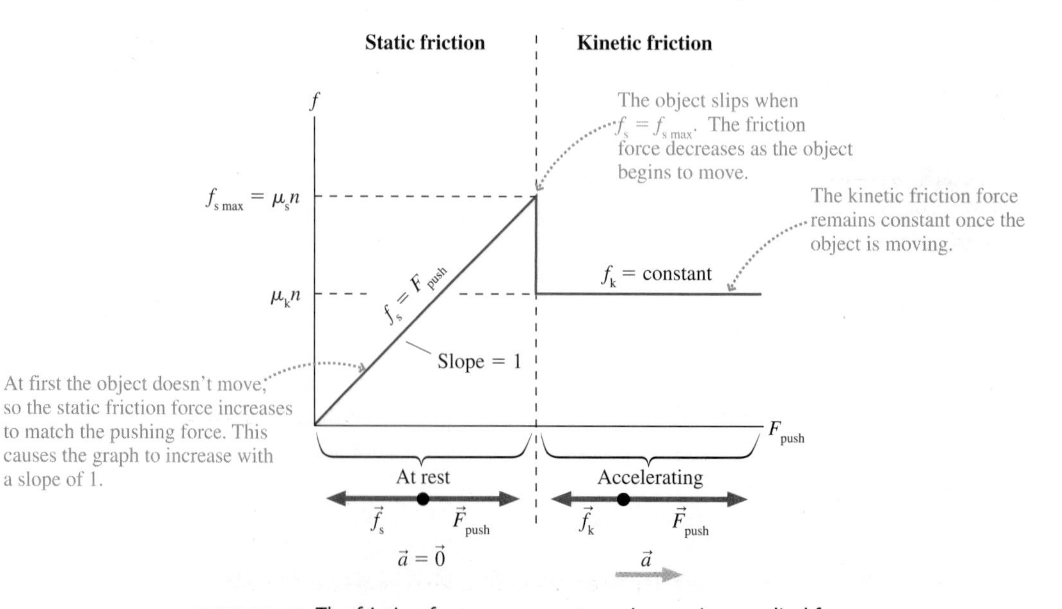

FIGURE 5.13 The friction force response to an increasing applied force.

STOP TO THINK 5.3 Rank in order, from largest to smallest, the size of the friction forces $\vec{f}_a$ to $\vec{f}_e$ in these 5 different situations. The box and the floor are made of the same materials in all situations.

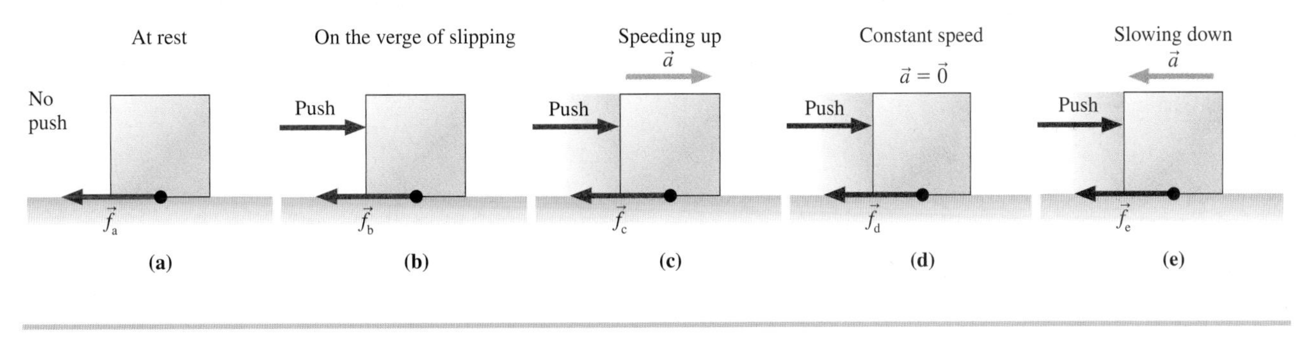

EXAMPLE 5.6 How far does a box slide?

Carol pushes a 50.0 kg wood box across a wood floor at a steady speed of 2.0 m/s. How much force does Carol exert on the box? If she stops pushing, how far will the box slide before coming to rest?

MODEL We model the box as a particle and we describe the friction forces with the model of static and kinetic friction. This is a two-part problem: first while Carol is pushing the box, then as it slides after she releases it.

VISUALIZE This is a fairly complex situation, one that calls for careful visualization. Figure 5.14 shows the pictorial and physical representations both while Carol pushes, when $\vec{a} = \vec{0}$, and after she stops. We've placed $x = 0$ at the point where she stops pushing because this is the point where the kinematics calculation for "How far?" will begin. Notice that each part of the motion needs its own free-body diagram. The box is moving until the very instant that the problem ends, so only kinetic friction is relevant.

SOLVE We'll start by finding how hard Carol has to push to keep the box moving at a steady speed. The box is in dynamic equilibrium ($\vec{a} = \vec{0}$), and Newton's first law is

$$\sum F_x = F_{\text{push}} - f_k = 0$$

$$\sum F_y = n - w = n - mg = 0$$

where we've used $w = mg$ for the weight. The negative sign occurs in the first equation because $\vec{f}_k$ points to the left and thus the *component* is negative: $(f_k)_x = -f_k$. Similarly, $w_y = -w$ because the weight vector points down. In addition to Newton's laws, we also have our model of kinetic friction:

$$f_k = \mu_k n$$

Altogether we have three simultaneous equations in the three unknowns F_{push}, f_k, and n. Fortunately, these equations are easy to solve. The y-component of Newton's law tells us that $n = mg$. We can then find the friction force to be

$$f_k = \mu_k mg$$

Substitute this into the x-component of the first law, giving

$$F_{\text{push}} = f_k = \mu_k mg$$

$$= (0.20)(50.0\text{ kg})(9.80\text{ m/s}^2) = 98.0\text{ N}$$

where μ_k for wood on wood was taken from Table 5.1. This is how hard Carol pushes to keep the box moving at a steady speed.

The box is not in equilibrium after Carol stops pushing it. Our strategy for the second half of the problem is to use Newton's second law to find the acceleration, then use kinematics to find how far the box moves before stopping. We see from the

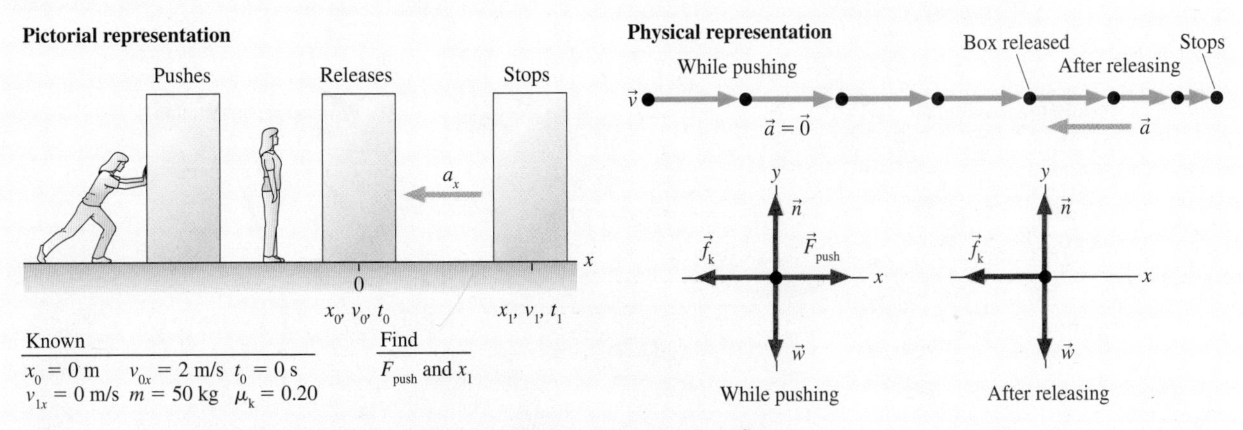

FIGURE 5.14 Pictorial and physical representations of a box sliding across a floor.

motion diagram that $a_y = 0$. Newton's second law, applied to the second free-body diagram of Figure 5.14, is

$$\sum F_x = -f_k = ma_x$$
$$\sum F_y = n - mg = ma_y = 0$$

We also have our model of friction,

$$f_k = \mu_k n$$

We see from the y-component equation that $n = mg$, and thus $f_k = \mu_k mg$. Using this in the x-component equation gives

$$ma_x = -f_k = -\mu_k mg$$

This is easily solved to find the box's acceleration:

$$a_x = -\mu_k g = -(0.20)(9.80 \text{ m/s}^2) = -1.96 \text{ m/s}^2$$

The acceleration component a_x is negative because the acceleration vector $\vec{a}$ points to the left, as we see from the motion diagram.

Now we are left with a problem of constant-acceleration kinematics. We are interested in a distance, rather than a time interval, so the easiest way to proceed is

$$v_{1x}^2 = 0 = v_{0x}^2 + 2a_x\Delta x = v_{0x}^2 + 2a_x x_1$$

from which the distance that the box slides is

$$x_1 = \frac{-v_{0x}^2}{2a_x} = \frac{-(2.0 \text{ m/s})^2}{2(-1.96 \text{ m/s}^2)} = 1.02 \text{ m}$$

We get a positive answer because the two negative signs cancel.

ASSESS Carol was pushing at 2 m/s $\approx$ 4 mph, which is fairly fast. The box slides 1.02 m, which is slightly over 3 feet. That sounds reasonable.

NOTE ► We needed both the horizontal and the vertical components of the second law even though the motion was entirely horizontal. This need is typical when friction is involved because we must find the normal force before we can evaluate the friction force. ◄

EXAMPLE 5.7 Dumping a file cabinet

A 50.0 kg steel file cabinet is in the back of a dump truck. The truck's bed, also made of steel, is slowly tilted. What is the size of the static friction force on the cabinet when the bed is tilted 20°? At what angle will the file cabinet begin to slide?

MODEL We'll model the file cabinet as a particle. We'll also use the model of static friction. The file cabinet will slip when the static friction force reaches its maximum value $f_{s\,max}$.

VISUALIZE Figure 5.15 shows the pictorial and physical representations when the truck bed is tilted at angle θ. This is a static equilibrium problem, so the free-body diagram suffices for a pictorial representation. We can make the analysis easier if we tilt the coordinate system to match the bed of the truck. To prevent the file cabinet from slipping, the static friction force must point *up* the slope.

SOLVE The file cabinet is in static equilibrium. Newton's first law is

$$(F_{net})_x = \sum F_x = n_x + w_x + (f_s)_x = 0$$
$$(F_{net})_y = \sum F_y = n_y + w_y + (f_s)_y = 0$$

From the free-body diagram we see that f_s has only a *negative* x-component and that n has only a positive y-component. The weight vector can be written $\vec{w} = +w\sin\theta\,\hat{\imath} - w\cos\theta\,\hat{\jmath}$, so $\vec{w}$ has both x- and y-components in this coordinate system. Thus the first law becomes

$$\sum F_x = w\sin\theta - f_s = mg\sin\theta - f_s = 0$$
$$\sum F_y = n - w\cos\theta = n - mg\cos\theta = 0$$

where we've used $w = mg$. The x-component equation allows us to determine the size of the static friction force when $\theta = 20°$:

$$f_s = mg\sin\theta = (50.0 \text{ kg})(9.80 \text{ m/s}^2)\sin 20° = 168 \text{ N}$$

This value does not require knowing μ_s. We simply have to find the size of the friction force that will balance the component of $\vec{w}$ that points down the slope. The coefficient of static friction only enters when we want to find the angle at which the file

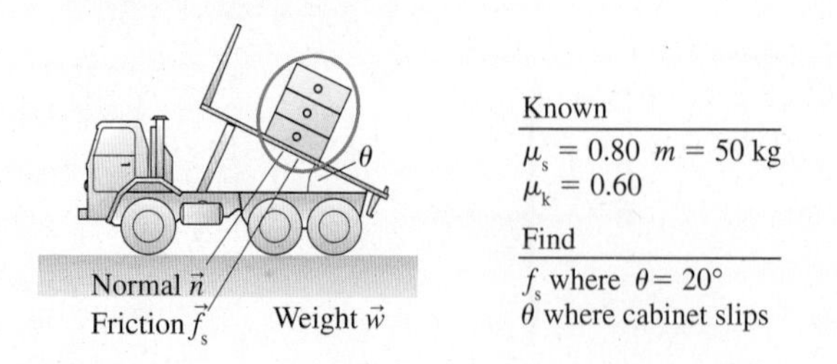

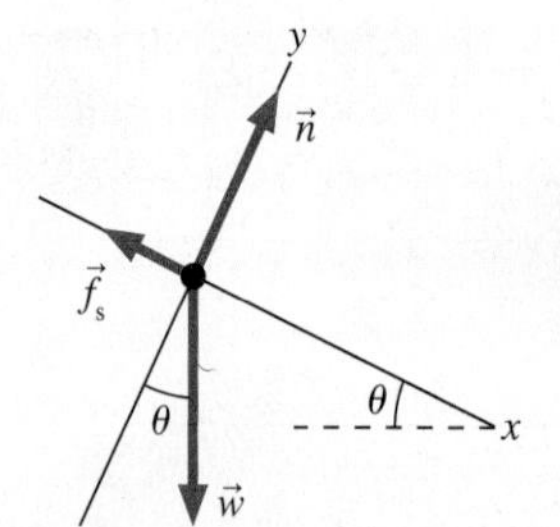

FIGURE 5.15 The pictorial and physical representations of a file cabinet in a tilted dump truck.

cabinet slips. Slipping occurs when the static friction reaches its maximum value

$$f_s = f_{s\,max} = \mu_s n$$

From the y-component of Newton's law we see that $n = mg\cos\theta$. Consequently,

$$f_{s\,max} = \mu_s mg\cos\theta$$

Substituting this into the x-component of the first law gives

$$mg\sin\theta - \mu_s mg\cos\theta = 0$$

The mg in both terms cancels, and we find

$$\frac{\sin\theta}{\cos\theta} = \tan\theta = \mu_s$$

$$\theta = \tan^{-1}\mu_s = \tan^{-1}(0.80) = 38.7°$$

ASSESS Steel doesn't slide all that well on unlubricated steel, so a fairly large angle is not surprising. The answer seems reasonable. It is worth noting that $n = mg\cos\theta$ in this example. A common error is to use simply $n = mg$. Be sure to evaluate the normal force within the context of each specific problem.

The angle at which slipping begins is called the *angle of repose*. Figure 5.16 shows that knowing the angle of repose can be very important because it is the angle at which loose materials (gravel, sand, snow, etc.) begin to slide on a mountainside, leading to landslides and avalanches.

FIGURE 5.16 The angle of repose is the angle at which loose materials, such as gravel or snow, begin to slide.

Causes of Friction

It is worth a brief pause to look at the *causes* of friction. All surfaces, even those quite smooth to the touch, are very rough on a microscopic scale. When two objects are placed in contact, they do not make a smooth fit. Instead, as Figure 5.17 shows, the high points on one surface become jammed against the high points on the other surface while the low points are not in contact at all. Only a very small fraction (typically 10^{-4}) of the surface area is in actual contact. The amount of contact depends on how hard the surfaces are pushed together, which is why friction forces are proportional to n.

At the points of actual contact, the atoms in the two materials are pressed closely together and molecular bonds are established between them. These bonds are the "cause" of the static friction force. For an object to slip, you must push it hard enough to break these molecular bonds between the surfaces. Once they are broken, and the two surfaces are sliding against each other, there are still attractive forces between the atoms on the opposing surfaces as the high points of the materials push past each other. However, the atoms move past each other so quickly that they do not have time to establish the tight bonds of static friction. That is why the kinetic friction force is smaller.

Occasionally, in the course of sliding, two high points will be forced together so closely that they do form a tight bond. As the motion continues, it is not this surface bond that breaks but weaker bonds at the *base* of one of the high points. When this happens, a small piece of the object is left behind "embedded" in the surface. This is what we call *abrasion*. Abrasion causes materials to wear out as a result of friction, be they the piston rings in your car or the seat of your pants. In machines, abrasion is minimized with lubrication, a very thin film of liquid between the surfaces that allows them to "float" past each other with many fewer points in actual contact.

Friction, as seen at the atomic level, is a very complex phenomenon. A detailed understanding of friction is at the forefront of engineering research today, where it is especially important for designing highly miniaturized machines and nanostructures.

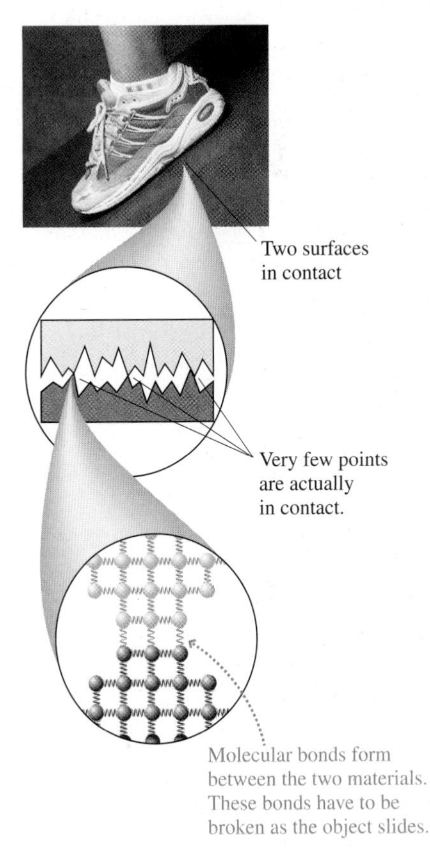

Two surfaces in contact
Very few points are actually in contact.
Molecular bonds form between the two materials. These bonds have to be broken as the object slides.

FIGURE 5.17 An atomic-level view of friction.

5.5 Drag

The air exerts a drag force on objects as they move through the air. You experience drag forces every day as you jog, bicycle, ski, or drive your car. The drag force is especially important for the skydiver at the beginning of the chapter.

FIGURE 5.18 The drag force on a high-speed motorcyclist is significant.

The drag force $\vec{D}$

- Is opposite in direction to $\vec{v}$.
- Increases in magnitude as the object's speed increases.

Figure 5.18 illustrates the drag force.

Drag is a more complex force than ordinary friction because drag depends on the object's speed. At relatively low speeds the drag force is small and can usually be neglected, but drag plays an important role as speeds increase.

Experimental studies have found that the drag force depends on an object's speed in a complicated way. Fortunately, we can use a fairly simple *model* of drag if the following three conditions are met:

- The object's size (diameter) is between a few millimeters and a few meters.
- The object's speed is less than a few hundred meters per second.
- The object is moving through the air near the earth's surface.

These conditions are usually satisfied for balls, people, cars, and many other objects of the everyday world. Under these conditions, the drag force can be written

$$\vec{D} \approx \left(\tfrac{1}{4}Av^2, \text{ direction opposite the motion}\right) \tag{5.16}$$

where A is the cross-section area of the object. The size of the drag force is proportional to the *square* of the object's speed. This model of drag fails for objects that are very small (such as dust particles), very fast (such as jet planes), or that move in other media (such as water). We'll leave those situations to more advanced textbooks.

NOTE ▶ Let's look at this model more closely. You may have noticed that an area multiplied by a speed squared does not give units of force. Unlike the $\frac{1}{2}$ in $\Delta x = \frac{1}{2}a(\Delta t)^2$, which is a "pure" number, the $\frac{1}{4}$ in the expression for $\vec{D}$ has units. This number depends on the air's density and viscosity, and it's actually $\frac{1}{4}$ kg/m^3. We've suppressed the units in Equation 5.16, but doing so gives us an expression that works *only* if A is in m^2 and v is in m/s. Equation 5.16 cannot be converted to other units. And the number is not exactly $\frac{1}{4}$, which is why Equation 5.16 has an $\approx$ sign rather than an $=$ sign, but it's close enough to allow Equation 5.16 to be a reasonable yet simple model of drag. ◀

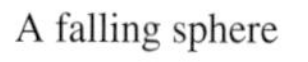

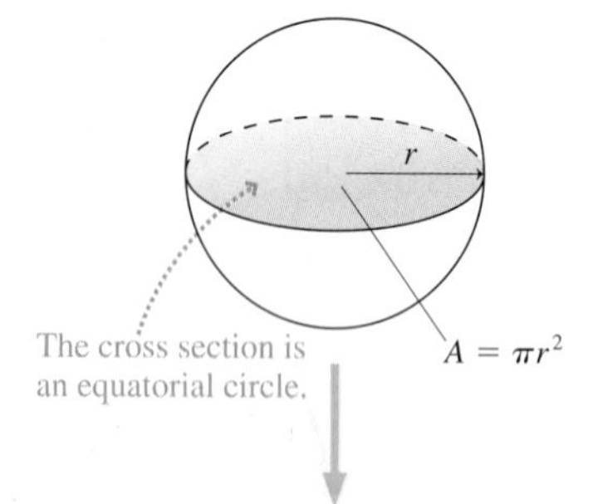

The area in Equation 5.16 is the cross section of the object as it "faces into the wind." Figure 5.19 shows how to calculate the cross-section area for objects of different shape. It's interesting to note that the magnitude of the drag force, $\frac{1}{4}Av^2$, depends on the object's *size and shape* but not on its *mass*. We will see shortly that the irrelevance of the mass has important consequences.

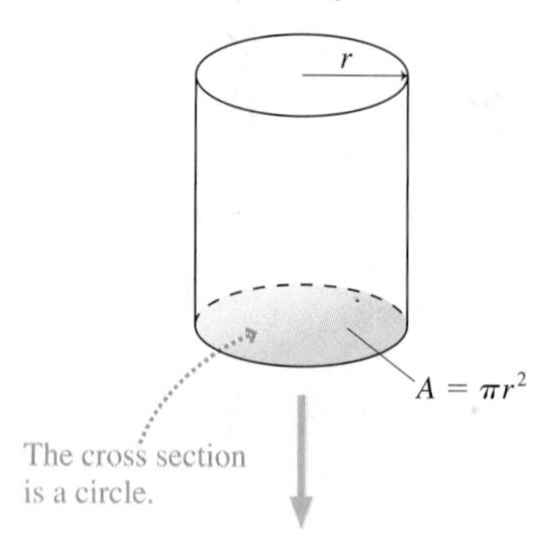

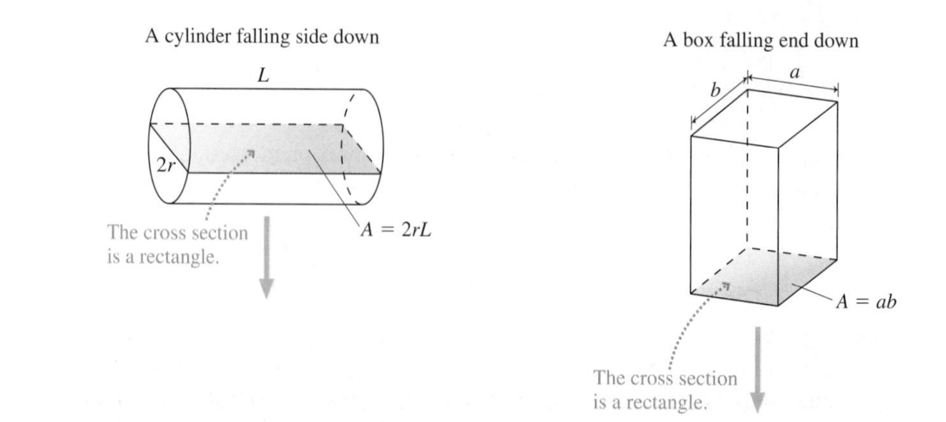

FIGURE 5.19 Cross-section areas for objects of different shape.

EXAMPLE 5.8 Air resistance compared to rolling friction

The profile of a typical 1500 kg passenger car, as seen from the front, is 1.6 m wide and 1.4 m high. At what speed does the magnitude of the drag equal the magnitude of the rolling friction?

MODEL Treat the car as a particle. Use the models of rolling friction and drag.

VISUALIZE Figure 5.20 shows the car and a free-body diagram. A full pictorial representation is not needed because we won't be doing any kinematics calculations.

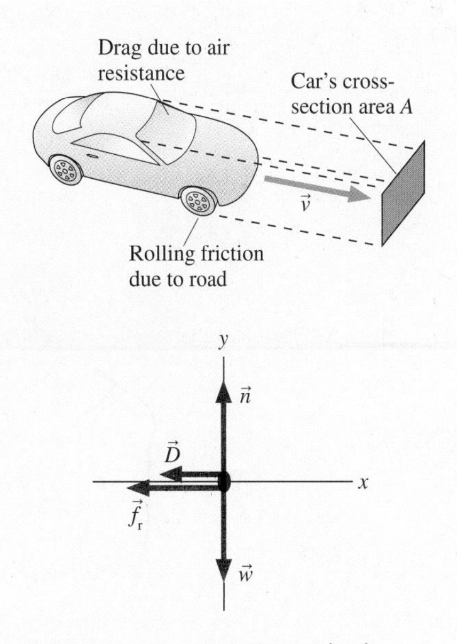

FIGURE 5.20 A car experiences both rolling friction and drag.

SOLVE Drag is less than friction at low speeds, where air resistance is negligible. But drag increases as v increases, so there will be a speed at which the two forces are equal in size. Above this speed, drag is more important than rolling friction.

The magnitudes of the forces are $D \approx \frac{1}{4}Av^2$ and $f_r = \mu_r n$. There's no motion and no acceleration in the vertical direction, so we can see from the free-body diagram that $n = w = mg$. Thus $f_r = \mu_r mg$. Equating friction and drag, we have

$$\frac{1}{4}Av^2 = \mu_r mg$$

Solving for v, we find

$$v = \sqrt{\frac{4\mu_r mg}{A}} = \sqrt{\frac{4(0.02)(1500\ \text{kg})(9.8\ \text{m/s}^2)}{(1.4\ \text{m})(1.6\ \text{m})}} = 23\ \text{m/s}$$

where the value of μ_r for rubber on concrete was taken from Table 5.1.

ASSESS 23 m/s is approximately 50 mph, a reasonable result. This calculation shows that our assumption that we can ignore air resistance is really quite good for car speeds less than 30 or 40 mph. Calculations that neglect drag will be increasingly inaccurate as speeds go above 50 mph.

Figure 5.21 shows a ball moving up and down vertically. If there were no air resistance, the ball would be in free fall with $a_{\text{free fall}} = -g$ throughout its flight. Let's see how drag changes this.

Referring to Figure 5.21:

1. The drag force $\vec{D}$ points down as the ball rises. This *increases* the net force on the ball and causes the ball to slow down *more quickly* than it would in a vacuum. The magnitude of the acceleration, which we'll calculate below, is $a > g$.
2. The drag force decreases as the ball slows.
3. $\vec{v} = \vec{0}$ at the highest point in the ball's motion, so there's no drag and the acceleration is simply $a_{\text{free fall}} = -g$.
4. The drag force increases as the ball speeds up.
5. The drag force $\vec{D}$ points up as the ball falls. This *decreases* the net force on the ball and causes the ball to speed up *less quickly* than it would in a vacuum. The magnitude of the acceleration is $a < g$.

We can use Newton's second law to find the ball's acceleration $a_\uparrow$ as it rises. You can see from the forces in Figure 5.21 that

$$a_\uparrow = \frac{(F_{\text{net}})_y}{m} = \frac{-mg - D}{m} = -\left(g + \frac{D}{m}\right) \tag{5.17}$$

The magnitude of $a_\uparrow$, which is the ball's deceleration as it rises, is $g + D/m$. Air resistance causes the ball to slow down *more quickly* than it would in a vacuum. But Equation 5.17 tells us more. Because D depends on the object's size but not on its mass, drag has a larger *effect* (larger acceleration) on a less massive ball than on a more massive ball of the same size.

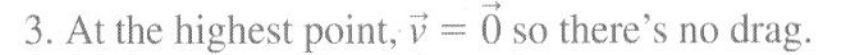

3. At the highest point, $\vec{v} = \vec{0}$ so there's no drag.

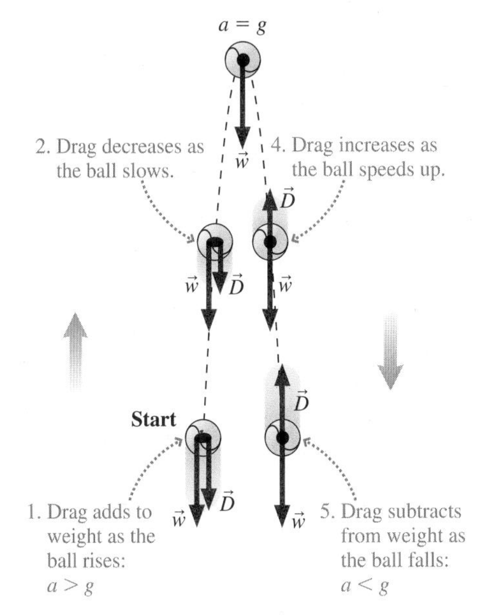

5. Drag subtracts from weight as the ball falls: $a < g$

FIGURE 5.21 Drag force on a ball moving vertically.

A Ping-Pong ball and a golf ball are about the same size, but it's harder to throw the Ping-Pong ball than the golf ball. We can now give an *explanation:*

- The drag force has the same magnitude for two objects of equal size.
- According to Newton's second law, the acceleration (the *effect* of the force) depends inversely on the mass.
- Therefore, the effect of the drag force is larger on a less massive ball than on a more massive ball of equal size.

As the ball in Figure 5.21 falls, its acceleration $a_\downarrow$ is

$$a_\downarrow = \frac{(F_{\text{net}})_y}{m} = \frac{-mg + D}{m} = -\left(g - \frac{D}{m}\right) \tag{5.18}$$

The magnitude of $a_\downarrow$ is $g - D/m$, so the ball speeds up *less quickly* than it would in a vacuum. Once again, the effect is larger for a less massive ball than for a more massive ball of equal size.

Terminal Speed

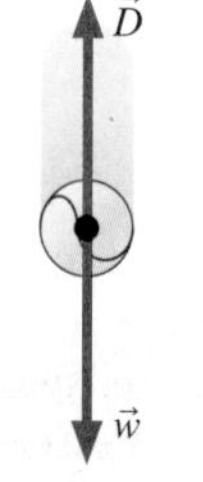
$\vec{D}$
Terminal speed is reached when the drag exactly balances the weight: $\vec{a} = \vec{0}$.
$\vec{w}$

FIGURE 5.22 An object falling at terminal speed.

The drag force increases as an object falls and gains speed. If the object falls far enough, it will eventually reach a speed, shown in Figure 5.22, at which $D = w$. That is, the drag force will be equal and opposite to the weight force. The net force at this speed is $\vec{F}_{\text{net}} = \vec{0}$, so there is no further acceleration and the object falls with a *constant* speed. The speed at which the exact balance between the upward drag force and the downward weight force causes an object to fall without acceleration is called the **terminal speed** v_{term}. Once an object has reached terminal speed, it will continue falling at that speed until it hits the ground.

It's not hard to compute the terminal speed. It is the speed, by definition, at which $D = w$ or, equivalently, $\frac{1}{4}Av^2 \approx mg$. This speed is

$$v_{\text{term}} \approx \sqrt{\frac{4mg}{A}} \tag{5.19}$$

A more massive object has a larger terminal speed than a less massive object of equal size. A 10-cm-diameter lead ball, with a mass of 6 kg, has a terminal speed of 170 m/s while a 10-cm-diameter Styrofoam ball, with a mass of 50 g, has a terminal speed of only 15 m/s.

A popular use of Equation 5.19 is to find the terminal speed of a skydiver. A skydiver is rather like the cylinder of Figure 5.19 falling "side down." A typical skydiver is 1.8 m long and 0.40 m wide ($A = 0.72\text{ m}^2$) and has a mass of 75 kg. His terminal speed is

$$v_{\text{term}} \approx \sqrt{\frac{4mg}{A}} = \sqrt{\frac{4(75\text{ kg})(9.8\text{ m/s}^2)}{0.72\text{ m}^2}} = 64\text{ m/s}$$

This is roughly 140 mph. A higher speed can be reached by falling feet first or head first, which reduces the area A.

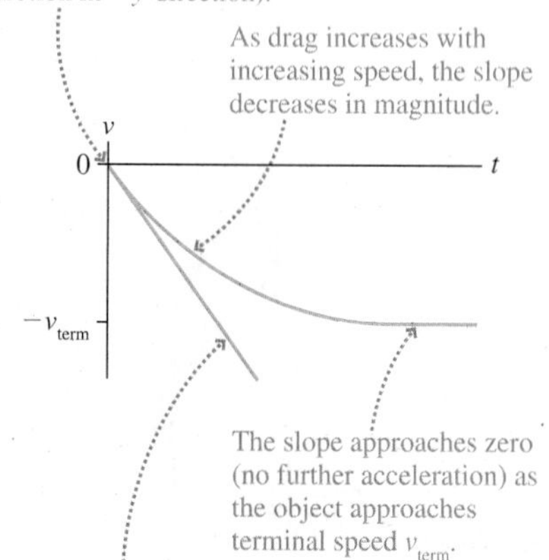
The velocity starts at zero, then becomes increasingly negative (motion in $-y$-direction).
As drag increases with increasing speed, the slope decreases in magnitude.
v
0
t
$-v_{\text{term}}$
The slope approaches zero (no further acceleration) as the object approaches terminal speed v_{term}.
Without drag, the graph is a straight line with slope $a_y = -g$.

FIGURE 5.23 The velocity-versus-time graph of a falling object with and without drag.

Figure 5.23 shows the results of a more detailed calculation for a falling object. Without drag, the velocity graph is a straight line with slope $= a_y = -g$. When drag is included, the slope steadily decreases in magnitude and approaches zero (no further acceleration) as the object reaches terminal speed.

Although we've focused our analysis on objects moving vertically, the same ideas apply to objects moving horizontally. If an object is thrown or shot horizontally, $\vec{D}$ causes the object to slow down. An airplane reaches its maximum speed, which is analogous to the terminal speed, when the drag is equal and opposite to the thrust: $D = F_{\text{thrust}}$. The net force is then zero and the plane cannot go any faster. The maximum speed of a passenger jet is about 550 mph.

We will continue to neglect drag unless a problem specifically calls for drag to be considered.

STOP TO THINK 5.4 The terminal speed of a Styrofoam ball is 15 m/s. Suppose a Styrofoam ball is shot straight down with an initial speed of 30 m/s. Which velocity graph is correct?

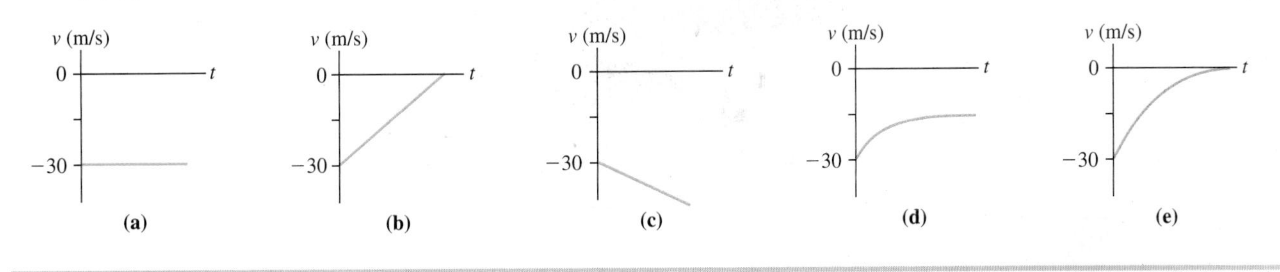

5.6 More Examples of Newton's Second Law

We will finish this chapter with several additional examples in which we use the problem-solving strategy in more complex scenarios.

ActivPhysics ONLINE 2.7, 2.8, 2.9

EXAMPLE 5.9 Stopping distances

A 1500 kg car is traveling at a speed of 30 m/s when the driver slams on the brakes and skids to a halt. Determine the stopping distance if the car is traveling up a 10° slope, down a 10° slope, or on a level road.

MODEL We'll represent the car as a particle and we'll use the model of kinetic friction. We want to solve the problem only once, not three separate times, so we'll leave the slope angle θ unspecified until the end.

VISUALIZE Figure 5.24 shows pictorial and physical representations. We've shown the car sliding uphill, but these representations work equally well for a level or downhill slide if we let θ be zero or negative, respectively. We've used a tilted coordinate system so that the motion is along one of the axes. We've *assumed* that the car is traveling to the right, although the problem didn't state this. You could equally well make the opposite assumption, but you would have to be careful with negative values of x. The car *skids* to a halt, so we've taken the coefficient of *kinetic* friction for rubber on concrete from Table 5.1.

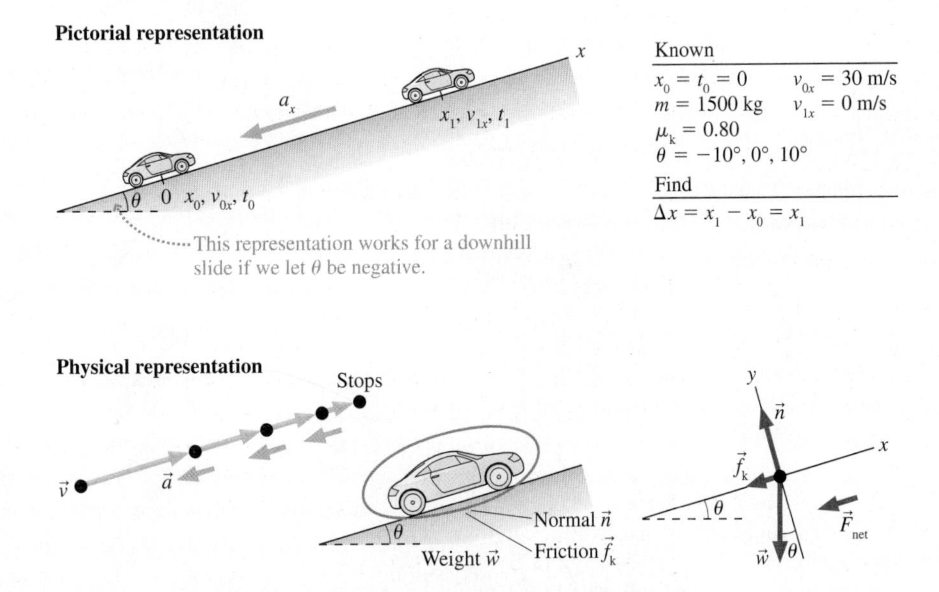

FIGURE 5.24 Pictorial and physical representations of a skidding car.

SOLVE Newton's second law and the model of kinetic friction are

$$\sum F_x = n_x + w_x + (f_k)_x$$
$$= -mg\sin\theta - f_k = ma_x$$
$$\sum F_y = n_y + w_y + (f_k)_y$$
$$= n - mg\cos\theta = ma_y = 0$$
$$f_k = \mu_k n$$

We've written these equations by "reading" the motion diagram and the free-body diagram. Notice that both components of the weight vector $\vec{w}$ are negative. $a_y = 0$, because the motion is entirely along the x-axis.

The second equation gives $n = mg\cos\theta$. Using this in the friction model, we find $f_k = \mu_k mg\cos\theta$. Inserting this result back into the first equation then gives

$$ma_x = -mg\sin\theta - \mu_k mg\cos\theta$$
$$= -mg(\sin\theta + \mu_k\cos\theta)$$
$$a_x = -g(\sin\theta + \mu_k\cos\theta)$$

This is a constant acceleration. Constant-acceleration kinematics gives

$$v_{1x}^2 = 0 = v_{0x}^2 + 2a_x(x_1 - x_0) = v_{0x}^2 + 2a_x x_1$$

which we can solve for the stopping distance x_1:

$$x_1 = -\frac{v_{0x}^2}{2a_x} = \frac{v_{0x}^2}{2g(\sin\theta + \mu_k\cos\theta)}$$

Notice how the minus sign in the expression for a canceled the minus sign in the expression for x_1. Evaluating our result at the three different angles gives the stopping distances:

$$x_1 = \begin{cases} 48\text{ m} & \theta = 10° & \text{uphill} \\ 57\text{ m} & \theta = 0° & \text{level} \\ 75\text{ m} & \theta = -10° & \text{downhill} \end{cases}$$

The implications are clear about the danger of driving downhill too fast!

ASSESS 30 m/s $\approx$ 60 mph and 57 m $\approx$ 180 feet on a level surface. This is similar to the stopping distances you learned when you got your driver's license, so the results seem reasonable. Additional confirmation comes from noting that the expression for a_x becomes $-g\sin\theta$ if $\mu_k = 0$. This is what you learned in Chapter 3 for the acceleration on a frictionless inclined plane.

This is a good example for pointing out the advantages of working problems *algebraically*. If you had started plugging in numbers early, you would not have found that the mass eventually cancels out and you would have done several needless calculations. In addition, it is now easy to calculate the stopping distance for different angles. Had you been computing numbers, rather than algebraic expressions, you would have had to go all the way back to the beginning for each angle.

EXAMPLE 5.10 A dog sled race

It's dog sled race day in Alaska! A wooden sled, with rider and supplies, has a mass of 200 kg. When the starting gun sounds, it takes the dogs 15 meters to reach their "cruising speed" of 5.0 m/s across the snow. Two ropes are attached to the sled, one on each side of the dogs. The ropes pull upward at 10°. What are the tensions in the ropes at the start of the race?

MODEL We'll represent the sled as a particle and we'll use the model of kinetic friction. We interpret the question as asking for the *magnitude* T of the tension forces. We'll assume that the tensions in the two ropes are equal.

VISUALIZE Figure 5.25 shows the pictorial and physical representations. Notice that the tensions $\vec{T}_1$ and $\vec{T}_2$ are tilted up, but the net force is directly to the right in order to match the acceleration $\vec{a}$ of the motion diagram.

SOLVE We have enough information to calculate the acceleration. We can then use $\vec{a}$ to find the tension. We'll assume that the acceleration is constant during the first 15 m. Then

$$v_{1x}^2 = v_{0x}^2 + 2a_x(x_1 - x_0) = 2a_x x_1$$
$$a_x = \frac{v_{1x}^2}{2x_1} = \frac{(5.0\text{ m/s})^2}{2(15\text{ m})} = 0.833\text{ m/s}^2$$

Newton's second law can be written by "reading" the free-body diagram:

$$\sum F_x = n_x + T_{1x} + T_{2x} + w_x + (f_k)_x$$
$$= 2T\cos\theta - f_k = ma_x$$
$$\sum F_y = n_y + T_{1y} + T_{2y} + w_y + (f_k)_y$$
$$= n + 2T\sin\theta - mg = ma_y = 0$$

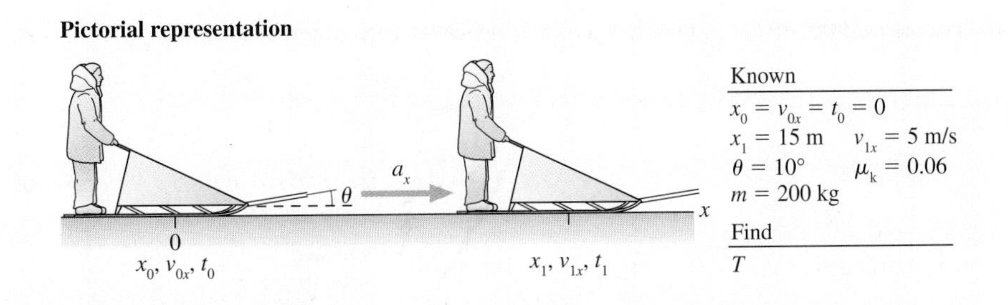

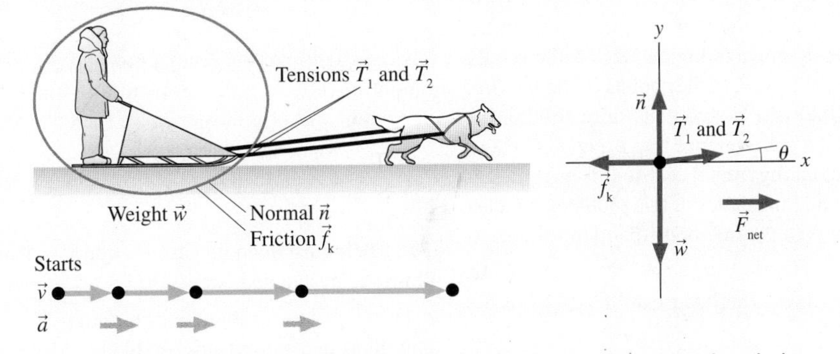

FIGURE 5.25 Pictorial and physical representations of an accelerating dog sled.

Make sure you understand where all the terms come from, including their signs. We've used $w = mg$ and our knowledge that $\vec{a}$ has only an x-component. The tensions $\vec{T}_1$ and $\vec{T}_2$ have both x- and y-components. The assumption of equal tensions allows us to write $T_1 = T_2 = T$, and this introduces the factors of 2.

In addition, we have the model of kinetic friction

$$f_k = \mu_k n$$

From the y-equation and the friction equation,

$$n = mg - 2T\sin\theta$$
$$f_k = \mu_k n = \mu_k mg - 2\mu_k T\sin\theta$$

Notice that n is *not* equal to mg. The y-components of the tension forces support part of the weight, so the ground does not press against the bottom of the sled as hard as it would otherwise.

Substituting the friction back into the x-equation gives

$$2T\cos\theta - (\mu_k mg - 2\mu_k T\sin\theta)$$
$$= 2T(\cos\theta + \mu_k\sin\theta) - \mu_k mg = ma_x$$
$$T = \frac{1}{2}\frac{m(a_x + \mu_k g)}{\cos\theta + \mu_k\sin\theta}$$

Using $a_x = 0.833$ from above with $m = 200$ kg and $\theta = 10°$, we find that the tension is

$$T = 143 \text{ N}$$

ASSESS It's a bit hard to assess this result. We do know that the weight of the sled is $mg \approx 2000$ N. We also know that the dogs can drag a sled over snow (small μ_k) but probably can't lift the sled straight up, so we anticipate that $T \ll 2000$ N. Our calculation agrees.

EXAMPLE 5.11 Make sure the cargo doesn't slide

A 100 kg box of dimensions 50 cm × 50 cm × 50 cm is in the back of a flatbed truck. The coefficients of friction between the box and the bed of the truck are $\mu_s = 0.4$ and $\mu_k = 0.2$. What is the maximum acceleration the truck can have without the box slipping?

MODEL This is a somewhat different problem from any we have looked at thus far. Let the box, which we'll model as a particle, be the system. It contacts its environment only where it touches the truck bed, so only the truck can exert contact forces on the box. If the box does *not* slip, then there is no motion of the box *relative to the truck* and the box must accelerate *with the truck:* $a_{box} = a_{truck}$. As the box accelerates, it must, according to Newton's second law, have a net force acting on it. But from what?

Imagine, for a moment, that the truck bed is frictionless. The box would slide backwards (as seen in the truck's reference frame) as the truck accelerates. The force that prevents sliding is *static friction,* so the truck must exert a static friction force on the box to "pull" the box along with it and prevent the box from sliding *relative to the truck.*

Pictorial representation

Known

$m = 100$ kg
Box dimensions 50 cm × 50 cm × 50 cm
$\mu_s = 0.4 \quad \mu_k = 0.2$

Find

Acceleration at which box slips

Physical representation

Weight $\vec{w}$ Normal $\vec{n}$ Static friction $\vec{f}_s$

y $\vec{n}$ $\vec{f}_s$ x $\vec{w}$ $\vec{F}_{net}$

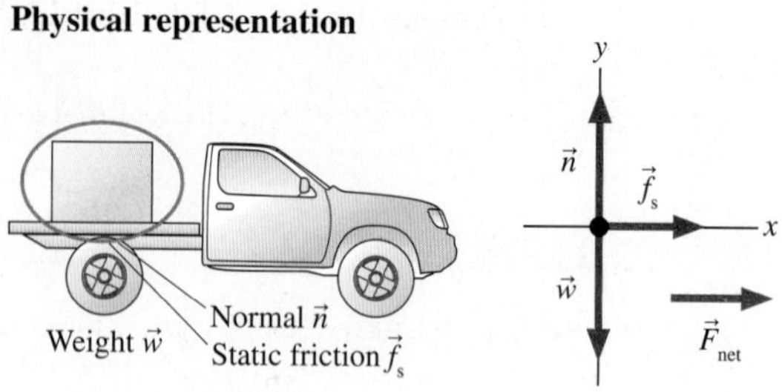

FIGURE 5.26 Pictorial and physical representations for the box in a flatbed truck.

VISUALIZE This situation is shown in Figure 5.26. There is only one horizontal force on the box, $\vec{f}_s$, and it points in the *forward* direction to accelerate the box. Notice that we're solving the problem with the ground as our reference frame. Newton's laws are not valid in the accelerating truck because it is not an inertial reference. There are no kinematics in this problem, so a list of known information next to the free-body diagram suffices as a pictorial representation.

SOLVE Newton's second law, which we can "read" from the free-body diagram, is

$$\sum F_x = f_s = ma_x$$

$$\sum F_y = n - w = n - mg = ma_y = 0$$

Now, static friction, you will recall, can be *any* value between 0 and $f_{s\,max}$. If the truck accelerates slowly, so that the box doesn't slip, then $f_s < f_{s\,max}$. However, we're interested in the acceleration a_{max} at which the box begins to slip. This is the acceleration at which f_s reaches its maximum possible value

$$f_s = f_{s\,max} = \mu_s n$$

The y-equation of the second law and the friction model combine to give $f_s = \mu_s mg$. Substituting this into the x-equation, and noting that a_x is now a_{max}, we find

$$a_{max} = \frac{f_s}{m} = \mu_s g = 3.9 \text{ m/s}^2$$

The truck must keep its acceleration less than 3.9 m/s^2 if slipping is to be avoided.

ASSESS 3.9 m/s^2 is about one-third of g. You may have noticed that items in a car or truck are likely to *tip over* when you start or stop, but they slide only if you really floor it and accelerate very quickly. So this answer seems reasonable. Notice that the dimensions of the crate were not needed. Real-world situations rarely have exactly the information you need, no more and no less. Many problems in this textbook will require you to assess the information in the problem statement in order to learn which is relevant to the solution.

The mathematical representation of this last example was quite straightforward. The challenge was in the analysis that preceded the mathematics—that is, in the *physics* of the problem rather than the mathematics. It is here that our analysis tools—motion diagrams, force identification, and free-body diagrams—prove their value.

SUMMARY

The goal of Chapter 5 has been to learn how to solve problems about motion in a straight line.

GENERAL STRATEGY

All examples in this chapter follow a four-part strategy. You'll become a better problem solver if you adhere to it as you do the homework problems. The *Dynamics Worksheets* will help you structure your work in this way.

Equilibrium Problems

Object at rest or moving with constant velocity.

MODEL Make simplifying assumptions.

VISUALIZE

- **Physical representation:** Forces and free-body diagram
- **Pictorial representation:** Translate words to symbols.

SOLVE Use Newton's first law

$$\vec{F}_{net} = \sum_i \vec{F}_i = \vec{0}$$

"Read" the vectors from the free-body diagram.

ASSESS Is the result reasonable?

Go back and forth between representations as needed.

Dynamics Problems

Object accelerating.

MODEL Make simplifying assumptions.

VISUALIZE

- **Pictorial representation:** Sketch to define situation. Translate words to symbols.
- **Physical representation:** Forces and free-body diagram

SOLVE Use Newton's second law

$$\vec{F}_{net} = \sum_i \vec{F}_i = m\vec{a}$$

"Read" the vectors from the free-body diagram. Use kinematics to find velocities and positions.

ASSESS Is the result reasonable?

IMPORTANT CONCEPTS

Specific information about three important forces:

Weight $\vec{w} = (mg, \text{downwards})$

Friction $\vec{f}_s = (0 \text{ to } \mu_s n, \text{direction as necessary to prevent motion})$

$\vec{f}_k = (\mu_k n, \text{direction opposite the motion})$

$\vec{f}_r = (\mu_r n, \text{direction opposite the motion})$

Drag $\vec{D} \approx (\frac{1}{4}Av^2, \text{direction opposite the motion})$

Newton's laws are vector expressions. You must write them out by components:

$$(F_{net})_x = \sum F_x = ma_x \text{ or } 0$$

$$(F_{net})_y = \sum F_y = ma_y \text{ or } 0$$

APPLICATIONS

Apparent weight is the magnitude of the contact force supporting an object. It is what a scale would read, and it is your sensation of weight. It equals your true weight $w = mg$ only when $a = 0$.

$$w_{app} = w\left(1 + \frac{a_y}{g}\right)$$

Terminal speed is $v_{term} \approx \sqrt{\dfrac{4mg}{A}}$

TERMS AND NOTATION

apparent weight, w_{app}
weightless
coefficient of static friction, μ_s
coefficient of kinetic friction, μ_k
rolling friction
coefficient of rolling friction, μ_r
terminal speed, v_{term}

EXERCISES AND PROBLEMS

The ✎ icon indicates that the problem can be done on a Dynamics Worksheet.

Exercises

Section 5.1 Equilibrium

1. The three ropes in the figure are tied to a small, very light ring. Two of the ropes are anchored to walls at right angles, and the third rope pulls as shown. What are T_1 and T_2, the magnitudes of the tension forces in the first two ropes?

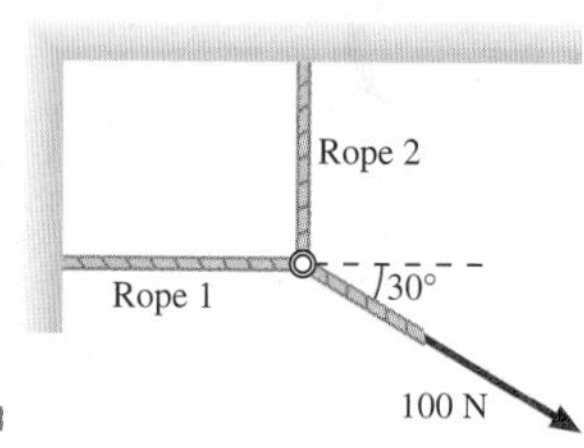

FIGURE EX5.1

2. The three ropes in the figure are tied to a small, very light ring. Two of these ropes are anchored to walls at right angles with the tensions shown in the figure. What are the magnitude and direction of the tension $\vec{T}_3$ in the third rope?

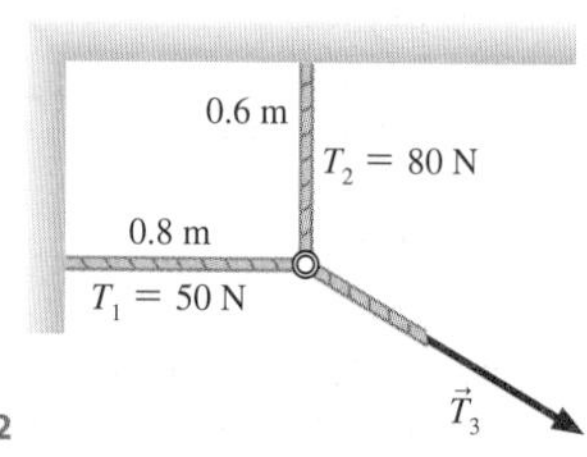

FIGURE EX5.2

3. A 20 kg loudspeaker is suspended 2.0 m below the ceiling by two 3.0-m-long cables that angle outward at equal angles. What is the tension in the cables?
4. A football coach sits on a sled while two of his players build their strength by dragging the sled across the field with ropes. The friction force on the sled is 1000 N and the angle between the two ropes is 20°. How hard must each player pull to drag the coach at a steady 2.0 m/s?

Section 5.2 Using Newton's Second Law

5. In each of the two free-body diagrams, the forces are acting on a 2.0 kg object. For each diagram, find the values of a_x and a_y, the x- and y-components of the acceleration.

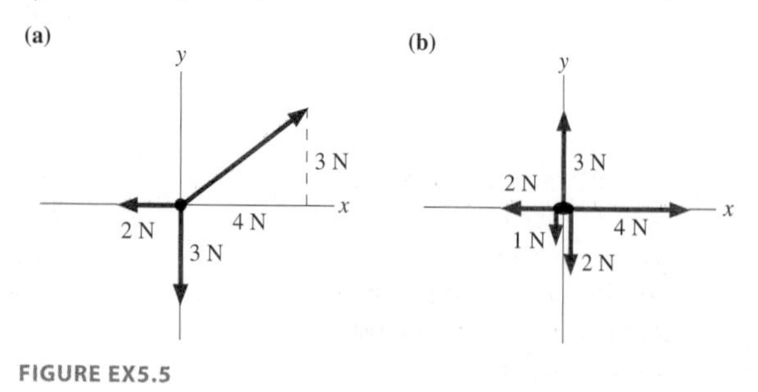

FIGURE EX5.5

6. In each of the two free-body diagrams, the forces are acting on a 2.0 kg object. For each diagram, find the values of a_x and a_y, the x- and y-components of the acceleration.

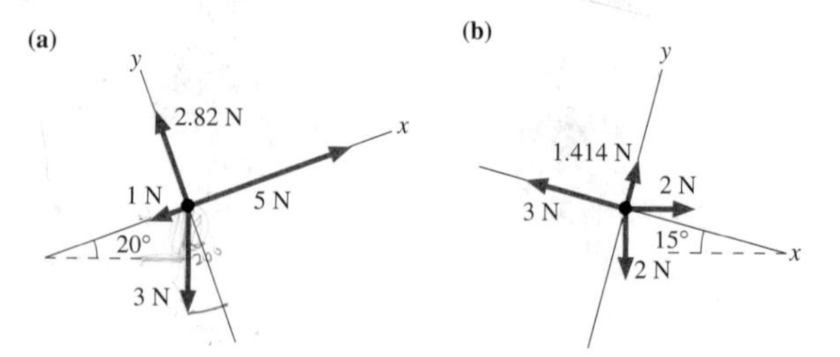

FIGURE EX5.6

7. Figure Ex5.7 shows the velocity graph of a 2.0 kg object as it moves along the x-axis. What is the net force acting on this object at $t = 1$ s? At 4 s? At 7 s?

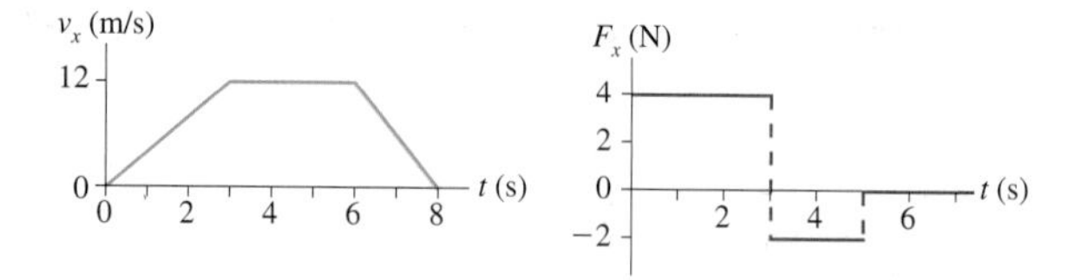

FIGURE EX5.7 FIGURE EX. 5.8

8. Figure Ex5.8 shows the force acting on a 2.0 kg object as it moves along the x-axis. The object is at rest at the origin at $t = 0$ s. What are its acceleration and velocity at $t = 6$ s?
9. A horizontal rope is tied to a 50 kg box on frictionless ice. What is the tension in the rope if:
   a. The box is at rest?
   b. The box moves at a steady 5.0 m/s?
   c. The box has $v_x = 5.0$ m/s and $a_x = 5.0$ m/s^2?
10. A 50 kg box hangs from a rope. What is the tension in the rope if:
    a. The box is at rest?
    b. The box moves up a steady 5.0 m/s?
    c. The box has $v_y = 5.0$ m/s and is speeding up at 5.0 m/s^2?
    d. The box has $v_y = 5.0$ m/s and is slowing down at 5.0 m/s^2?

Section 5.3 Mass and Weight

11. An astronaut's weight on earth is 800 N. What is his weight on Mars, where $g = 3.76$ m/s^2?
12. A woman has a mass of 55 kg.
    a. What is her weight on earth?
    b. What are her mass and her weight on the moon, where $g = 1.62$ m/s^2?
13. It takes the elevator in a skyscraper 4.0 s to reach its cruising speed of 10 m/s. A 60 kg passenger gets aboard on the ground floor. What is the passenger's apparent weight
    a. Before the elevator starts moving?
    b. While the elevator is speeding up?
    c. After the elevator reaches its cruising speed?
14. Figure Ex5.14 shows the velocity graph of a 75 kg passenger in an elevator. What is the passenger's apparent weight at $t = 1$ s? At 5 s? At 9 s?

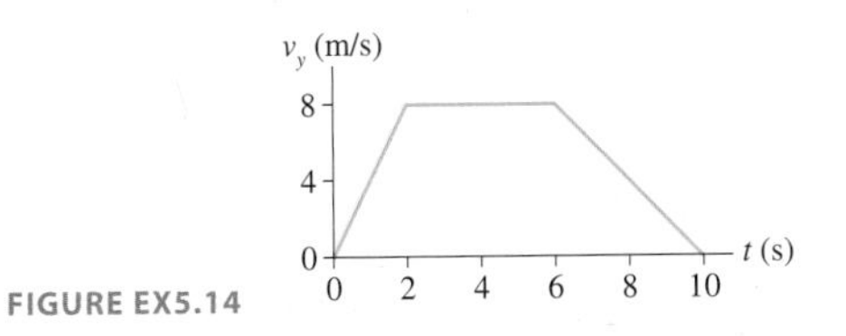

FIGURE EX5.14

Section 5.4 Friction

15. Bonnie and Clyde are sliding a 300 kg bank safe across the floor to their getaway car. The safe slides with a constant speed if Clyde pushes from behind with 385 N of force while Bonnie pulls forward on a rope with 350 N of force. What is the safe's coefficient of kinetic friction on the bank floor?
16. A 4000 kg truck is parked on a 15° slope. How big is the friction force on the truck?
17. A 1000 kg car traveling at a speed of 40 m/s skids to a halt on wet concrete where $\mu_k = 0.6$. How long are the skid marks?
18. An Airbus A320 jetliner has a takeoff mass of 75,000 kg. It reaches its takeoff speed of 82 m/s (180 mph) in 35 s. What is the thrust of the engines? You can neglect air resistance but not rolling friction.
19. A 50,000 kg locomotive is traveling at 10 m/s when its engine and brakes both fail. How far will the locomotive roll before it comes to a stop?
20. A stubborn, 120 kg mule sits down and refuses to move. To drag the mule to the barn, the exasperated farmer ties a rope around the mule and pulls with his maximum force of 800 N. The coefficients of friction between the mule and the ground are $\mu_s = 0.8$ and $\mu_k = 0.5$. Is the farmer able to move the mule?

Section 5.5 Drag

21. A 75 kg skydiver can be modeled as a rectangular "box" with dimensions 20 cm × 40 cm × 180 cm. What is his terminal speed if he falls feet first?
22. A 22-cm-diameter bowling ball has a terminal speed of 85 m/s. What is the ball's mass?

Problems

23. Use information that you can find in Chapters 4 and 5 to *estimate* the acceleration of a car.
24. Use information that you can find in Chapters 4 and 5 to *estimate* the size of the friction force on a baseball player sliding into second base.
25. A 5.0 kg object initially at rest at the origin is subjected to the time-varying force shown in Figure P5.25. What is the object's velocity at $t = 6$ s?

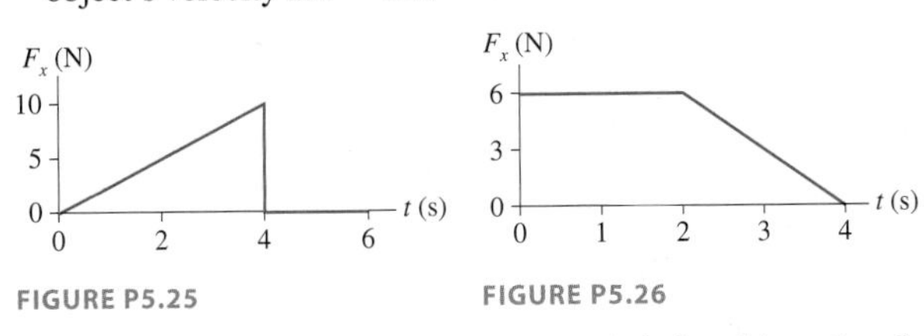

FIGURE P5.25 FIGURE P5.26

26. A 2.0 kg object initially at rest at the origin is subjected to the time-varying force shown in Figure P5.26. What is the object's velocity at $t = 4$ s?
27. A 1000 kg steel beam is supported by two ropes. Each rope has a maximum sustained tension of 6000 N. Does either rope break? If so, which one(s)?

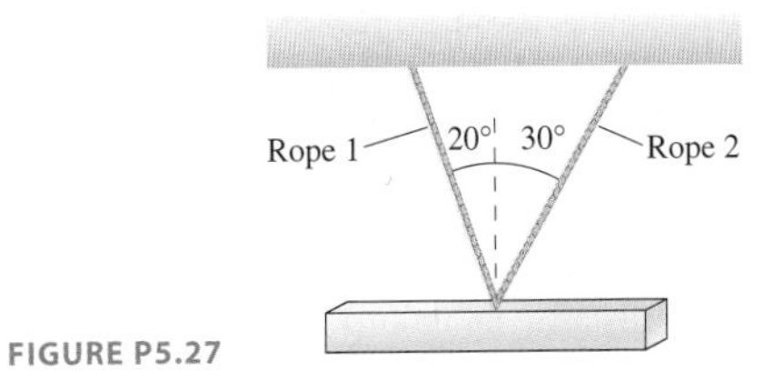

FIGURE P5.27

28. A 500 kg piano is being lowered into position by a crane while two people steady it with ropes pulling to the sides. Bob's rope pulls to the left, 15° below horizontal, with 500 N of tension. Ellen's rope pulls toward the right, 25° below horizontal.
    a. What tension must Ellen maintain in her rope to keep the piano descending at a steady speed?
    b. What is the tension in the main cable supporting the piano?
29. In an electricity experiment, a 1.0 g plastic ball is suspended on a 60-cm-long string and given an electric charge. A charged rod brought near the ball exerts a horizontal electrical force $\vec{F}_{elec}$ on it, causing the ball to swing out to a 20° angle and remain there.
    a. What is the magnitude of $\vec{F}_{elec}$?
    b. What is the tension in the string?
30. Henry gets into an elevator on the 50th floor of a building and it begins moving at $t = 0$ s. The figure shows his apparent weight over the next 12 s.
    a. Is the elevator's initial direction up or down? Explain how you can tell.
    b. What is Henry's mass?
    c. How far has Henry traveled at $t = 12$ s?

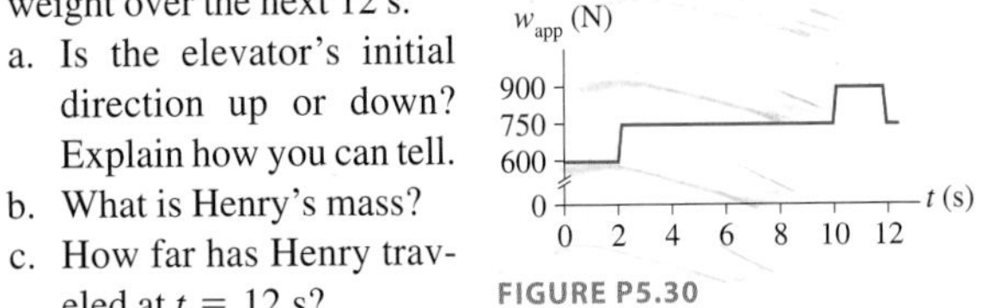

FIGURE P5.30

31. Zach, whose mass is 80 kg, is in an elevator descending at 10 m/s. The elevator takes 3.0 s to brake to a stop at the first floor.
    a. What is Zach's apparent weight before the elevator starts braking?
    b. What is Zach's apparent weight while the elevator is braking?
32. You've always wondered about the acceleration of the elevators in the 101-story-tall Empire State Building. One day, while visiting New York, you take your bathroom scale into the elevator and stand on them. The scales read 150 lb as the door closes. The reading varies between 120 lb and 170 lb as the elevator travels 101 floors. What conclusions can you draw?
33. An accident victim with a broken leg is being placed in traction. The patient wears a special boot with a pulley attached to the sole. The foot and boot together have a mass of 4.0 kg, and the doctor has decided to hang a 6.0 kg mass from the rope. The boot is held suspended by the ropes and does not touch the bed.
    a. Determine the amount of tension in the rope by using Newton's laws to analyze the hanging mass.

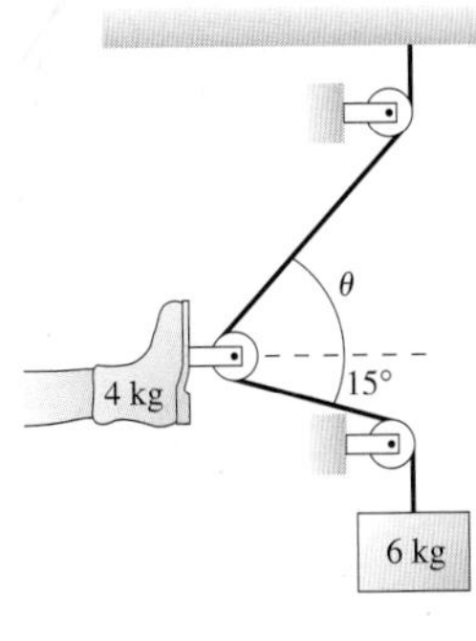

FIGURE P5.33

b. The net traction force needs to pull straight out on the leg. What is the proper angle θ for the upper rope?
c. What is the net traction force pulling on the leg?

Hint: If the pulleys are frictionless, which we will assume, the tension in the rope is constant from one end to the other.

34. Seat belts and air bags save lives by reducing the forces exerted on the driver and passengers in an automobile collision. Cars are designed with a "crumple zone" in the front of the car. In the event of an impact, the passenger compartment decelerates over a distance of about 1 m as the front of the car crumples. An occupant restrained by seat belts and air bags decelerates with the car. By contrast, an unrestrained occupant keeps moving forward with no loss of speed (Newton's first law!) until hitting the dashboard or windshield. These are unyielding surfaces, and the unfortunate occupant then decelerates over a distance of only about 5 mm.
   a. A 60 kg person is in a head-on collision. The car's speed at impact is 15 m/s. Estimate the net force on the person if he or she is wearing a seat belt and if the air bag deploys.
   b. Estimate the net force that ultimately stops the person if he or she is not restrained by a seat belt or air bag.
   c. How do these two forces compare to the person's weight?
35. A rifle with a barrel length of 60 cm fires a 10 g bullet with a horizontal speed of 400 m/s. The bullet strikes a block of wood and penetrates to a depth of 12 cm.
   a. What frictional force (assumed to be constant) does the wood exert on the bullet?
   b. How long does it take the bullet to come to rest?
   c. Draw a velocity-versus-time graph for the bullet in the wood.
36. Compressed air is used to fire a 50 g ball vertically upward from a 1.0-m-tall tube. The air exerts an upward force of 2.0 N on the ball as long as it is in the tube. How high does the ball go above the top of the tube?
37. What thrust does a 200 g model rocket need in order to have a vertical acceleration of 10 m/s^2
   a. On Earth?
   b. On the moon, where $g = 1.62\ m/s^2$?
38. A 20,000 kg rocket has a rocket motor that generates 3.0×10^5 N of thrust.
   a. What is the rocket's initial upward acceleration?
   b. At an altitude of 5000 m the rocket's acceleration has increased to 6.0 m/s^2. What mass of fuel has it burned?
39. A 2.0 kg steel block is at rest on a steel table. A horizontal string pulls on the block.
   a. What is the minimum string tension needed to move the block?
   b. If the string tension is 20 N, what is the block's speed after moving 1.0 m?
   c. If the string tension is 20 N and the table is coated with oil, what is the block's speed after moving 1.0 m?
40. Sam, whose mass is 75 kg, takes off across level snow on his jet-powered skis. The skis have a thrust of 200 N and a coefficient of kinetic friction on snow of 0.1. Unfortunately, the skis run out of fuel after only 10 s.
   a. What is Sam's top speed?
   b. How far has Sam traveled when he finally coasts to a stop?
41. Sam, whose mass is 75 kg, takes off down a 50-m-high, 10° slope on his jet-powered skis. The skis have a thrust of 200 N. Sam's speed at the bottom is 40 m/s. What is the coefficient of kinetic friction of his skis on snow?
42. A 10 kg crate is placed on a horizontal conveyor belt. The materials are such that $\mu_s = 0.5$ and $\mu_k = 0.3$.
   a. Draw a free-body diagram showing all the forces on the crate if the conveyer belt runs at constant speed.
   b. Draw a free-body diagram showing all the forces on the crate if the conveyer belt is speeding up.
   c. What is the maximum acceleration the belt can have without the crate slipping?
43. A baggage handler drops your 10 kg suitcase onto a conveyor belt running at 2.0 m/s. The materials are such that $\mu_s = 0.5$ and $\mu_k = 0.3$. How far is your suitcase dragged before it is riding smoothly on the belt?
44. Johnny jumps off a swing, lands sitting down on a grassy 20° slope, and slides 3.5 m down the slope before stopping. The coefficient of kinetic friction between grass and the seat of Johnny's pants is 0.5. What was his initial speed on the grass?
45. A 2.0 kg wood block is launched up a wooden ramp that is inclined at a 30° angle. The block's initial speed is 10 m/s.
   a. What vertical height does the block reach above its starting point?
   b. What speed does it have when it slides back down to its starting point?
46. It's moving day, and you need to push a 100 kg box up a 20° ramp into the truck. The coefficients of friction for the box on the ramp are $\mu_s = 0.9$ and $\mu_k = 0.6$. Your largest pushing force is 1000 N. Can you get the box into the truck without assistance if you get a running start at the ramp? If you stop on the ramp, will you be able to get the box moving again?
47. It's a snowy day and you're pulling a friend along a level road on a sled. You've both been taking physics, so she asks what you think the coefficient of friction between the sled and the snow is. You've been walking at a steady 1.5 m/s, and the rope pulls up on the sled at a 30° angle. You estimate that the mass of the sled, with your friend on it, is 60 kg and that you're pulling with a force of 75 N. What answer will you give?
48. A horizontal rope pulls a 10 kg wood sled across frictionless snow. A 5.0 kg wood box rides on the sled. What is the largest tension force for which the box doesn't slip?
49. A pickup truck with a steel bed is carrying a steel file cabinet. If the truck's speed is 15 m/s, what is shortest distance in which it can stop without the file cabinet sliding?
50. You're driving along at 25 m/s with your aunt's valuable antiques in the back of your pickup truck when suddenly you see a giant hole in the road 55 m ahead of you. Fortunately, your foot is right beside the brake and your reaction time is zero! Will the antiques be as fortunate?
   a. Can you stop the truck before it falls into the hole?
   b. If your answer to part a is yes, can you stop without the antiques sliding and being damaged? Their coefficients of friction are $\mu_s = 0.6$ and $\mu_k = 0.3$.

   Hint: You're not trying to stop in the shortest possible distance. What's your best strategy for avoiding damage to the antiques?
51. A 2.0 kg wood box slides down a vertical wood wall while you push on it at a 45° angle. What magnitude of force should you apply to cause the box to slide down at a constant speed?

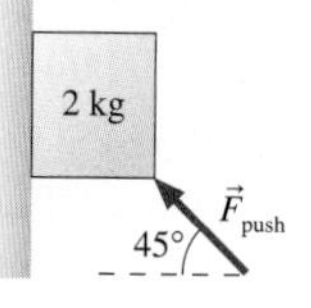

FIGURE P5.51

52. A 1.0 kg wood block is pressed against a vertical wood wall by the 12 N force shown. If the block is initially at rest, will it move upward, move downward, or stay at rest?

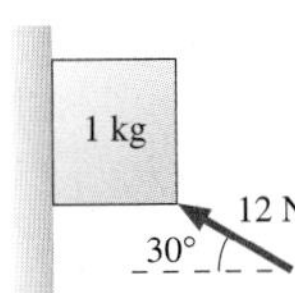

FIGURE P5.52

53. What is the terminal speed for an 80 kg skier going down a 40° snow-covered slope on wooden skis? Assume that the skier is 1.8 m tall and 0.40 m wide.

54. A 10 g Ping-Pong ball has a diameter of 3.5 cm.
   a. The ball is shot straight up at twice its terminal speed. What is its initial acceleration?
   b. The ball is shot straight down at twice its terminal speed. What is its initial acceleration?
   c. The ball is shot straight down at twice its terminal speed. Draw a plausible velocity-versus-time graph.

55. Try this! Hold your right hand out with your palm perpendicular to the ground, as if you were getting ready to shake hands. You can't hold anything in your palm this way because it would fall straight down. Use your left hand to hold a small object, such as a ball or a coin, against your outstretched palm, then let go as you quickly swing your hand to the left across your body, parallel to the ground. You'll find that the object stays against your palm; it doesn't slip or fall.
   a. Is the condition for keeping the object against your palm one of maintaining a certain minimum velocity v_{min}? Or one of maintaining a certain minimum acceleration a_{min}? Explain.
   b. Suppose the object's mass is 50 g, with $\mu_s = 0.8$ and $\mu_k = 0.4$. Determine either v_{min} or a_{min}, whichever you answered in part a.

56. Suppose you use your hand to *push* a ball straight down toward the floor. The ball's weight is 1.0 N.
   a. Draw the ball's free-body diagram while you're pushing it.
   b. Is F_{net} larger than, smaller than, or equal to the ball's weight w? Explain.
   c. Find an expression for the ball's apparent weight when its acceleration is a_y. Evaluate w_{app} for a_y equal to $-g$, $-1.5g$, and $-2g$.

57. You've been called in to investigate a construction accident in which the cable broke while a crane was lifting a 4500 kg container. The steel cable is 2.0 cm in diameter and has a safety rating of 50,000 N. The crane is designed not to exceed speeds of 3.0 m/s or accelerations of 1.0 m/s^2, and your tests find that the crane is not defective. What is your conclusion? Did the crane operator recklessly lift too heavy a load? Or was the cable defective?

58. An artist friend of yours needs help hanging a 500 lb sculpture from the ceiling. For artistic reasons, she wants to use just two ropes. One will be 30° from vertical, the other 60°. She needs you to determine the smallest diameter rope that can safely support this expensive piece of art. On a visit to the hardware store you find that rope is sold in increments of $\frac{1}{8}$-inch diameter and that the safety rating is 4000 pounds per square inch of cross section. What diameter rope should you buy?

59. A machine has an 800 g steel shuttle that is pulled along a square steel rail by an elastic cord. The shuttle is released when the elastic cord has 20 N tension at a 45° angle. What is the initial acceleration of the shuttle?

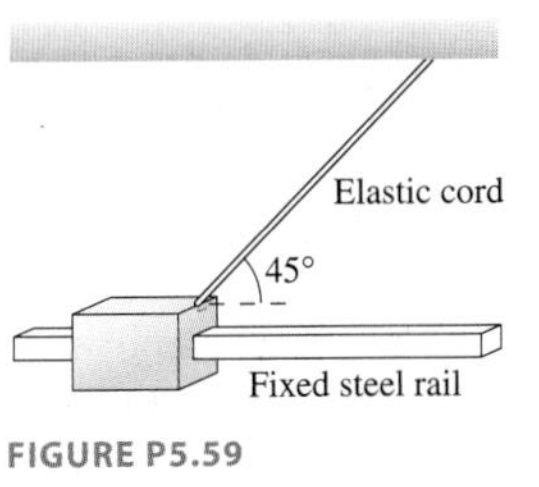

FIGURE P5.59

60. A 1.0 kg ball hangs from the ceiling of a truck by a 1.0-m-long string. The back of the truck, where you are riding with the ball, has no windows and has been completely soundproofed. The truck travels along an exceedingly smooth test track, and you feel no bumps or bounces as it moves. Your only instruments are a meter stick, a protractor, and a stopwatch.
   a. The driver tells you, over a loudspeaker, that the truck is either at rest, or it is moving forward at a steady speed of 5 m/s. Can you determine which it is? If so, how? If not, why not?
   b. Next, the driver tells you that the truck is either moving forward with a steady speed of 5 m/s, or it is accelerating at 5 m/s^2. Can you determine which it is? If so, how? If not, why not?
   c. Suppose the truck has been accelerating forward at 5 m/s^2 long enough for the ball to achieve a steady position. Does the ball have an acceleration? If so, what are the magnitude and direction of the ball's acceleration?
   d. Draw a free-body diagram that shows all forces acting on the ball as the truck accelerates.
   e. Suppose the ball makes a 10° angle with the vertical. If possible, determine the truck's velocity. If possible, determine the truck's acceleration.

61. Imagine *hanging* from a big spring scale as it moves vertically with acceleration a. Show that Equation 5.10 is the correct expression for your apparent weight.

Problems 62 through 65 show a free-body diagram. For each:
   a. Write a realistic dynamics problem for which this is the correct free-body diagram. Your problem should ask a question that can be answered with a value of position or velocity (such as "How far?" or "How fast?"), and should give sufficient information to allow a solution.
   b. Solve your problem!

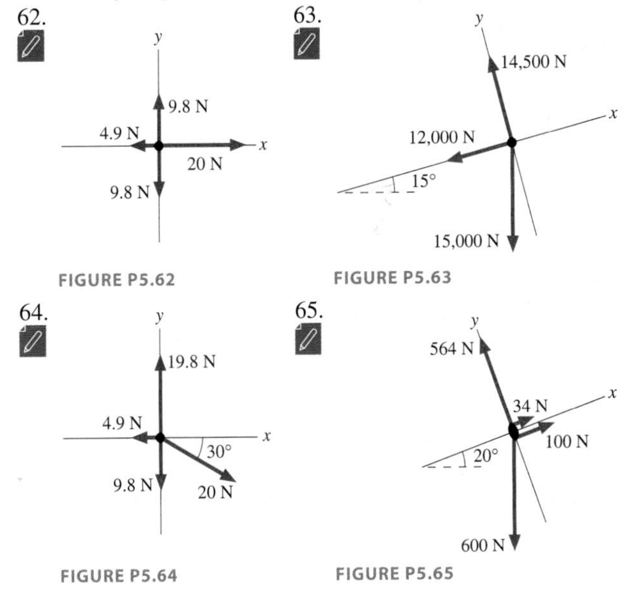

FIGURE P5.62

FIGURE P5.63

FIGURE P5.64

FIGURE P5.65

In Problems 66 through 69 you are given the dynamics equations that are used to solve a problem. For each of these, you are to

a. Write a realistic problem for which these are the correct equations.
b. Draw the free-body diagram and the pictorial representation for your problem.
c. Finish the solution of the problem.

66. $-0.8n = (1500\text{ kg})a_x$
$n - (1500\text{ kg})(9.80\text{ m/s}^2) = 0$

67. $T - 0.2n - (20\text{ kg})(9.80\text{ m/s}^2)\sin 20° = (20\text{ kg})(2\text{ m/s}^2)$
$n - (20\text{ kg})(9.80\text{ m/s}^2)\cos 20° = 0$

68. $(100\text{ N})\cos 30° - f_k = (20\text{ kg})a_x$
$n + (100\text{ N})\sin 30° - (20\text{ kg})(9.80\text{ m/s}^2) = 0$
$f_k = 0.2n$

69. $-f_k + (20\text{ kg})(9.80\text{ m/s}^2)\sin\theta = 0$
$n - (20\text{ kg})(9.80\text{ m/s}^2)\cos\theta = 0$
$f_k = 0.2n$

Challenge Problems

70. The figure shows an *accelerometer,* a device for measuring the horizontal acceleration of cars and airplanes. A ball is free to roll on a parabolic track described by the equation $y = x^2$, where both x and y are in meters. A scale along the bottom is used to measure the ball's horizontal position x.

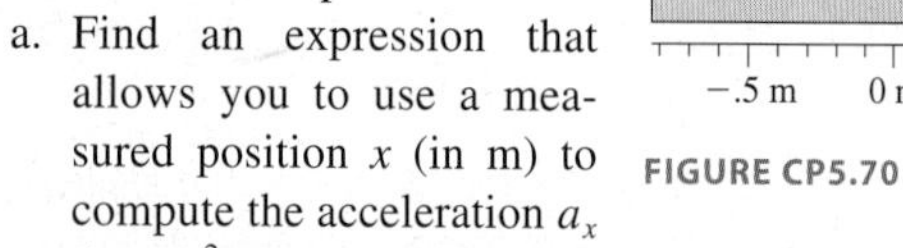

FIGURE CP5.70

a. Find an expression that allows you to use a measured position x (in m) to compute the acceleration a_x (in m/s²). (For example, $a_x = 3x$ is a possible expression.)
b. What is the acceleration if $x = 20$ cm?

71. You've entered a "slow ski race" where the winner is the skier who takes the *longest* time to go down a 15° slope without ever stopping. You need to choose the best wax to apply to your skis. Red wax has a coefficient of kinetic friction 0.25, yellow is 0.20, green is 0.15, and blue is 0.10. Having just finished taking physics, you realize that a wax too slippery will cause you to accelerate down the slope and lose the race. But a wax that's too sticky will cause you to stop and be disqualified. You know that a strong headwind will apply a 50 N horizontal force against you as you ski, and you know that your mass is 75 kg. Which wax do you choose?

72. A testing laboratory wants to determine if a new widget can withstand large accelerations and decelerations. To find out, they glue a 5.0 kg widget to a test stand that will drive it vertically up and down. The graph shows its acceleration during the first second, starting from rest.

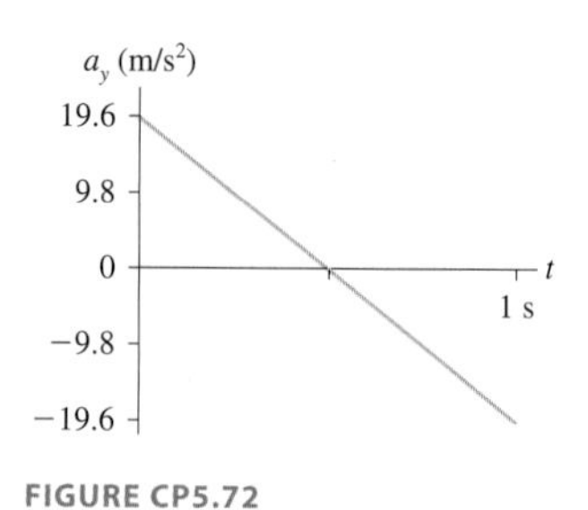

FIGURE CP5.72

a. Identify the forces acting on the widget and draw a free-body diagram.
b. Determine the value of n_y, the y-component of the normal force acting on the widget, during the first second of motion. Give your answer as a graph of n_y-versus-t.
c. Your answer to part b should show an interval of time during which n_y is negative. How can this be? Explain what it means physically for n_y to be negative.
d. At what time is the apparent weight of the widget a maximum? What is the acceleration at this time?
e. Is the apparent weight of the widget ever zero? If so, at what instant of time does this happen? What is the acceleration at that time?
f. Suppose the technician forgets to glue the widget to the test stand. Will the widget remain on the test stand throughout the first second, or will it fly off the stand at some instant of time? If so, at what time will this occur?

73. An object with cross section A is shot horizontally across frictionless ice. Its initial velocity is v_{0x} at $t_0 = 0$ s. Air resistance is not negligible.

a. Show that the velocity at time t is given by the expression

$$v_x = \frac{v_{0x}}{1 + Av_{0x}t/4m}$$

b. A 1.6 m wide, 1.4 m high, 1500 kg car hits a very slick patch of ice while going 20 m/s. If friction is neglected, how long will it take until the car's speed drops to 10 m/s? To 5 m/s?
c. Assess whether or not it is reasonable to neglect kinetic friction.

STOP TO THINK ANSWERS

Stop to Think 5.1: a. The lander is descending and slowing. The acceleration vector points upward, and so $\vec{F}_{net}$ points upward. This can be true only if the thrust has a larger magnitude than the weight.

Stop to Think 5.2: a. You are descending and slowing, so your acceleration vector points upward and there is a net upward force on you. The floor pushes up against your feet harder than gravity pulls down.

Stop to Think 5.3: $f_b > f_c = f_d = f_e > f_a$. Situations c, d, and e are all kinetic friction, which does not depend on either velocity or acceleration. Kinetic friction is smaller than the maximum static friction that is exerted in b. $f_a = 0$ because no friction is needed to keep the object at rest.

Stop to Think 5.4: d. The ball is shot *down* at 30 m/s, so $v_{0y} = -30$ m/s. This exceeds the terminal speed, so the upward drag force is *larger* than the downward weight force. Thus the ball *slows down* even though it is "falling." It will slow until $v_y = -15$ m/s, the terminal velocity, then maintain that velocity.

6 Dynamics II: Motion in a Plane

This diver is a spinning projectile following a parabolic trajectory.

▶ **Looking Ahead**

The goal of Chapter 6 is to learn to solve problems about motion in a plane. In this chapter you will learn to:

- Understand kinematics and dynamics in two dimensions.
- Understand projectile motion.
- Explore the issues of relative motion.

◀ **Looking Back**

This chapter is an extension of several ideas introduced in Chapters 1–5. Please review:

- Section 1.5 Finding acceleration vectors on a motion diagram.
- Sections 2.5–2.6 Constant-acceleration kinematics and free fall.
- Section 4.6 Inertial reference frames.
- Section 5.2 Solving problems with Newton's second law.

We have limited ourselves thus far to motion along a straight line. One-dimensional motion includes a lot of interesting physics and applications, but motion in the real world is often in two or more dimensions. A car turning a corner, a planet orbiting the sun, and the diver in the photograph are examples of two-dimensional motion. Restricting ourselves to one dimension has allowed us to concentrate on basic physics principles, but the time has come to broaden our horizons and consider a wider variety of motion.

Newton's laws are "laws of nature," meaning that they describe all motion, not just motion along a straight line. This chapter and the next will extend the application of Newton's laws to new situations. Chapter 6 will focus on motion in which we can treat the x- and y-components of the acceleration independently of each other. Projectile motion is an important example. Chapter 7 will cover circular motion, where the components are *not* independent.

Motion in a plane will provide us with the tools to look at an important question. Suppose you and I are moving relative to each other. Perhaps I'm standing still while you drive past in your car. How do physical measurements that I make in my coordinate system compare to measurements you make in your coordinate system? Are "my physics" and "your physics" the same? Questions such as these led Einstein to his theory of relativity. We will begin to answer these questions in the context of what is called Galilean relativity.

6.1 Kinematics in Two Dimensions

To begin, let's look at how motion in two dimensions differs from motion in one dimension. As an example, Figure 6.1 shows the motion diagram of a roller coaster car. Here the object moves in a vertical plane, but we'll also look at situations where the plane of motion is horizontal. We'll call it the *xy*-plane regardless of its orientation.

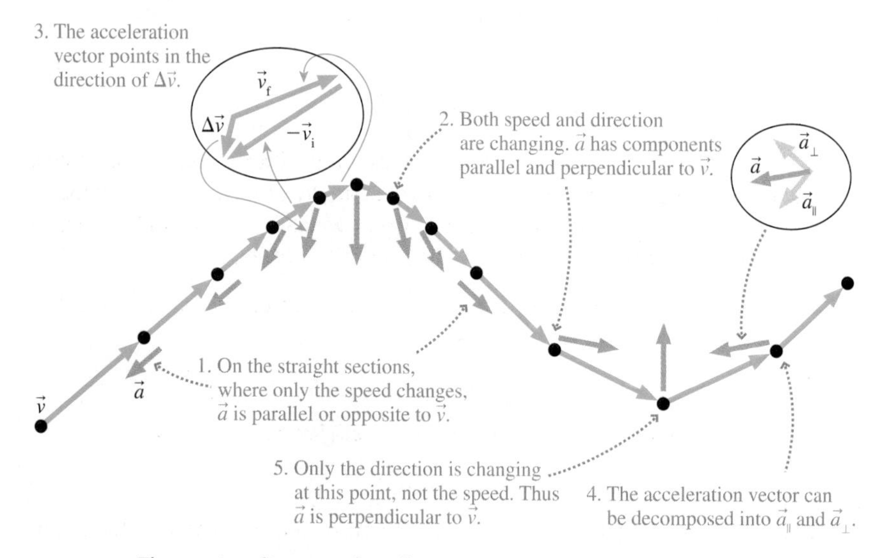

3. The acceleration vector points in the direction of $\Delta\vec{v}$.
$\vec{v}_f$
$\Delta\vec{v}$
$-\vec{v}_i$
2. Both speed and direction are changing. $\vec{a}$ has components parallel and perpendicular to $\vec{v}$.
$\vec{a}_\perp$
$\vec{a}$
$\vec{a}_\parallel$
$\vec{v}$
$\vec{a}$
1. On the straight sections, where only the speed changes, $\vec{a}$ is parallel or opposite to $\vec{v}$.

5. Only the direction is changing at this point, not the speed. Thus $\vec{a}$ is perpendicular to $\vec{v}$.
4. The acceleration vector can be decomposed into $\vec{a}_\parallel$ and $\vec{a}_\perp$.

FIGURE 6.1 The motion diagram of a roller coaster car.

What we most need to understand is how velocity and acceleration are related in two dimensions. The average acceleration was defined in Chapter 1 as $\vec{a}_{avg} = \Delta\vec{v}/\Delta t$. That is, the acceleration vector $\vec{a}$ points in the direction of $\Delta\vec{v}$, the change in velocity. Because velocity is a *vector*, a change can be either a change in length (i.e., a change of speed) or a change in direction.

On the straight sections of the roller coaster, you can see that a particle moving along a *straight line* speeds up if $\vec{a}$ and $\vec{v}$ point in the same direction and slows down if $\vec{a}$ and $\vec{v}$ point in opposite directions. This idea was the basis for the one-dimensional kinematics we developed in Chapter 2. For linear motion, acceleration is a change of speed.

When the direction of $\vec{v}$ changes, as it does when the roller coaster car goes over the hill or through the valley, the acceleration gets a bit tricky. You learned in Section 1.5 how to use vector subtraction to find the direction of $\vec{a}$, and a review is well worthwhile. The procedure is shown at one point in the motion diagram.

Chapter 3 showed how to decompose a vector into two perpendicular components. At point 4 in Figure 6.1 the acceleration vector $\vec{a}$ has been decomposed into a piece $\vec{a}_\parallel$ that is parallel to $\vec{v}$ and a piece $\vec{a}_\perp$ that is perpendicular to $\vec{v}$. **$\vec{a}_\parallel$ is the piece of the acceleration vector that changes the speed.** In this case it is slowing the car because $\vec{a}_\parallel$ is opposite the motion. **The component $\vec{a}_\perp$ is the piece of the acceleration that causes the velocity to change direction.**

Notice that $\vec{a}$ *always* has a perpendicular component at points where the direction of $\vec{v}$ is changing. At point 5, where only the direction is changing, not the speed, the parallel component vanishes and $\vec{a}$ is perpendicular to $\vec{v}$.

Position and Velocity

Motion diagrams are an important tool for visualizing motion, but we also need to develop a mathematical description of motion in two dimensions. It will be easiest to use x- and y-components of vectors, rather than components parallel and perpendicular to the motion. We'll point out, as we go along, the connection between these two points of view.

Figure 6.2 shows a particle moving along a curved path—its *trajectory*—in the xy-plane. At time t_1, the particle is at point 1. We can locate the particle in terms of its position vector

$$\vec{r}_1 = r_{1x}\hat{\imath} + r_{1y}\hat{\jmath}$$

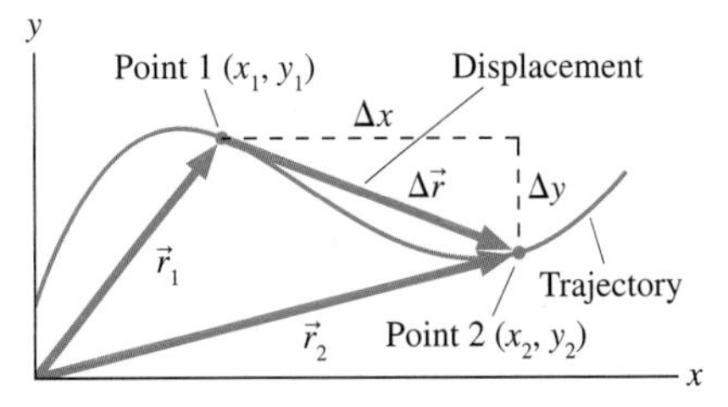

FIGURE 6.2 A particle moving along a trajectory in the xy-plane.

But r_{1x}, the x-component of $\vec{r}_1$, is simply x_1, the x-coordinate of the point. Similarly, r_{1y} is the y-coordinate y_1. Hence the position vector is

$$\vec{r}_1 = x_1\hat{\imath} + y_1\hat{\jmath} \tag{6.1}$$

A short time later, at t_2, the position vector is $\vec{r}_2 = x_2\hat{\imath} + y_2\hat{\jmath}$.

NOTE ▶ In Chapter 2 we made extensive use of position-versus-time graphs, either x-versus-t or y-versus-t. Figure 6.2, like many of the graphs we'll use in this chapter, is a graph of y-versus-x. In other words, it's an actual *picture* of the trajectory, not an abstract representation of the motion. ◀

The vector connecting these two points on the trajectory is the *displacement vector*

$$\Delta\vec{r} = \vec{r}_2 - \vec{r}_1$$

We can write the displacement vector in component form as

$$\Delta\vec{r} = \Delta x\hat{\imath} + \Delta y\hat{\jmath} \tag{6.2}$$

where $\Delta x = x_2 - x_1$ and $\Delta y = y_2 - y_1$ are the horizontal and vertical changes of position.

Chapter 1 defined the *average velocity* of a particle moving through a displacement $\Delta\vec{r}$ in a time interval Δt as

$$\vec{v}_{avg} = \frac{\Delta\vec{r}}{\Delta t} = \frac{\Delta x}{\Delta t}\hat{\imath} + \frac{\Delta y}{\Delta t}\hat{\jmath} \tag{6.3}$$

You learned in Chapter 2 that the *instantaneous velocity* is the limit of $\vec{v}_{avg}$ as $\Delta t \to 0$. Taking the limit of Equation 6.3 gives the instantaneous velocity in two dimensions:

$$\vec{v} = \lim_{\Delta t\to 0}\frac{\Delta\vec{r}}{\Delta t} = \frac{d\vec{r}}{dt} = \frac{dx}{dt}\hat{\imath} + \frac{dy}{dt}\hat{\jmath} \tag{6.4}$$

But we can also write the velocity vector in terms of its x- and y-components as

$$\vec{v} = v_x\hat{\imath} + v_y\hat{\jmath} \tag{6.5}$$

Comparing Equations 6.4 and 6.5, you can see that the velocity vector $\vec{v}$ has x- and y-components

$$v_x = \frac{dx}{dt} \quad \text{and} \quad v_y = \frac{dy}{dt} \tag{6.6}$$

That is, the x-component v_x of the velocity vector is the rate dx/dt at which the particle's x-coordinate is changing. The y-component is similar.

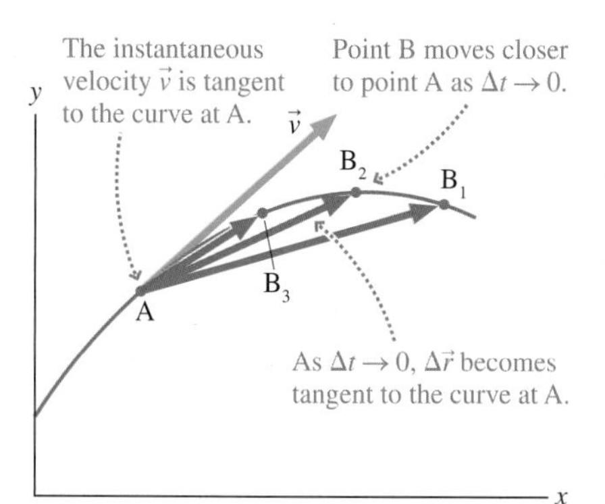

FIGURE 6.3 The instantaneous velocity vector $\vec{v}$ is tangent to the trajectory.

The average velocity $\vec{v}_{avg}$ points in the direction of $\Delta\vec{r}$, a fact we used in Chapter 1 to draw the velocity vectors on motion diagrams. Figure 6.3 shows that $\Delta\vec{r}$ becomes tangent to the trajectory as $\Delta t \to 0$. Consequently, **the instantaneous velocity vector $\vec{v}$ is tangent to the trajectory.**

Figure 6.4 illustrates another important feature of the velocity vector. If the vector's angle θ is measured from the positive x-axis, the velocity vector components are

$$v_x = \frac{dx}{dt} = v\cos\theta$$
$$v_y = \frac{dy}{dt} = v\sin\theta \tag{6.7}$$

where

$$v = \sqrt{v_x^2 + v_y^2} \tag{6.8}$$

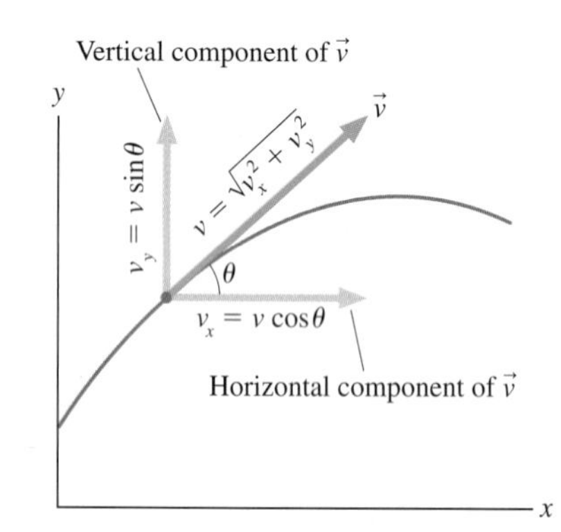

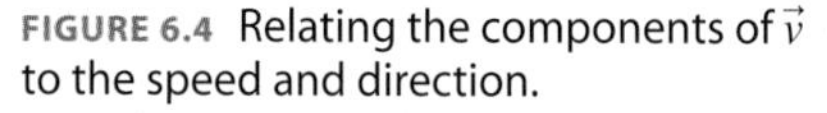
FIGURE 6.4 Relating the components of $\vec{v}$ to the speed and direction.

is the particle's *speed* at that point. Speed is always a positive number (or zero), whereas the components are *signed* quantities to convey information about the direction of the velocity vector. Conversely, we can use the two velocity components to determine the direction of motion:

$$\tan\theta = \frac{v_y}{v_x} \tag{6.9}$$

NOTE ▶ In Chapter 2, you learned that the *value* of the velocity component v_s at time t is given by the *slope* of the position-versus-time graph at time t. Now we see that the *direction* of the velocity vector $\vec{v}$ is given by the *tangent* to the y-versus-x graph of the trajectory. Figure 6.5 reminds you that these two graphs use different interpretations of the tangent lines. The tangent to the trajectory does not tell us anything about how fast the particle is moving, only its direction. ◀

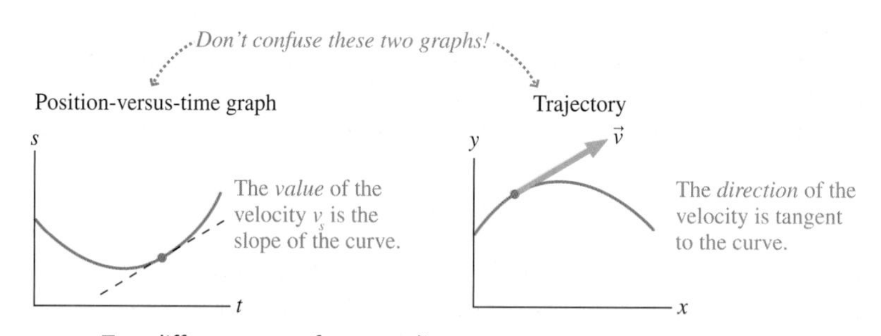

FIGURE 6.5 Two different uses of tangent lines.

EXAMPLE 6.1 Describing the motion with graphs

A particle's motion is described by the two equations

$$x = 2t^2 \text{ m}$$
$$y = (5t + 5) \text{ m}$$

where the time t is in s.

a. Draw a graph of the particle's trajectory.
b. Draw a graph of the particle's speed as a function of time.

MODEL These are *parametric equations* that give the particle's coordinates x and y separately in terms of the parameter t.

SOLVE

a. The trajectory is a curve in the xy-plane. The easiest way to proceed is to calculate x and y at several instants of time.

t (s)	x (m)	y (m)	v (m/s)
0	0	5	5.0
1	2	10	6.4
2	8	15	9.4
3	18	20	13.0
4	32	25	16.8

These points are plotted in Figure 6.6a, then a smooth curve is drawn through them to show the trajectory.

b. The particle's speed is given by Equation 6.8. We first need to use Equation 6.6 to find the components of the velocity vector:

$$v_x = \frac{dx}{dt} = 4t \text{ m/s} \quad \text{and} \quad v_y = \frac{dy}{dt} = 5 \text{ m/s}$$

Using these gives the particle's speed at time t:

$$v = \sqrt{v_x^2 + v_y^2} = \sqrt{16t^2 + 25} \text{ m/s}$$

The speed was computed in the table above and is graphed in Figure 6.6b.

ASSESS The y-versus-x graph of Figure 6.6a is a trajectory, not a position-versus-time graph. Thus the slope is *not* the particle's speed. The particle is speeding up, as you can see in the second graph, even though the slope of the trajectory is decreasing.

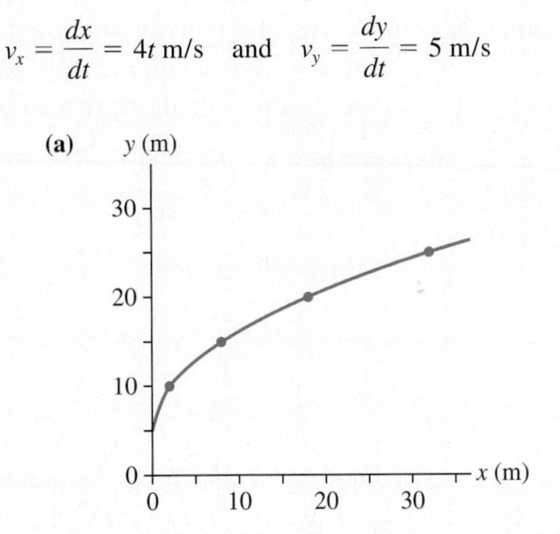

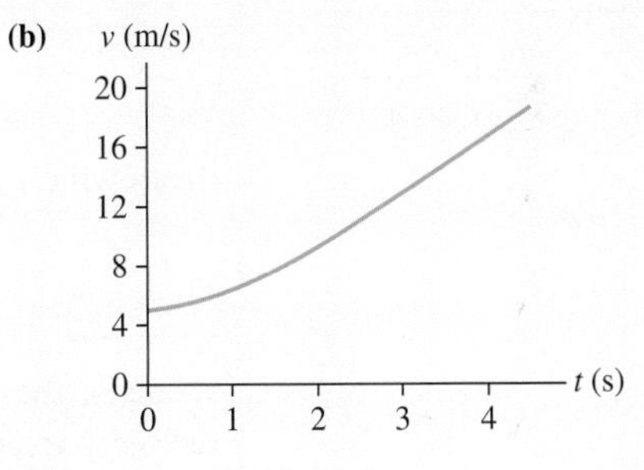

FIGURE 6.6 Two motion graphs for the particle of Example 6.1.

Acceleration

Let's return to the particle moving along a trajectory in the xy-plane. Figure 6.7 shows the instantaneous velocity $\vec{v}_1$ at point 1 and, a short time later, velocity $\vec{v}_2$ at point 2. These two vectors are tangent to the trajectory.

In Chapter 1 we defined the particle's *average acceleration* as

$$\vec{a}_{avg} = \frac{\Delta\vec{v}}{\Delta t} \tag{6.10}$$

where $\Delta\vec{v} = \vec{v}_2 - \vec{v}_1$ is the change in velocity during the interval Δt. We can use the vector-subtraction technique of Chapter 1 to find $\vec{a}_{avg}$ on this segment of the trajectory. This is shown in the insert to Figure 6.7.

If we now take the limit $\Delta t \to 0$, the *instantaneous acceleration* is

$$\vec{a} = \lim_{\Delta t \to 0} \frac{\Delta\vec{v}}{\Delta t} = \frac{d\vec{v}}{dt} \tag{6.11}$$

As $\Delta t \to 0$, points 1 and 2 in Figure 6.7 merge, and the instantaneous acceleration $\vec{a}$ is found at the same point on the trajectory (and the same instant of time) as the instantaneous velocity $\vec{v}$. This is shown in Figure 6.8.

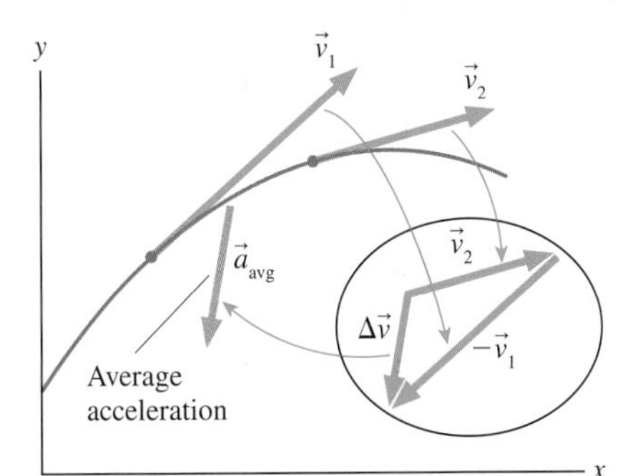

FIGURE 6.7 The average acceleration vector at a point on a curved trajectory.

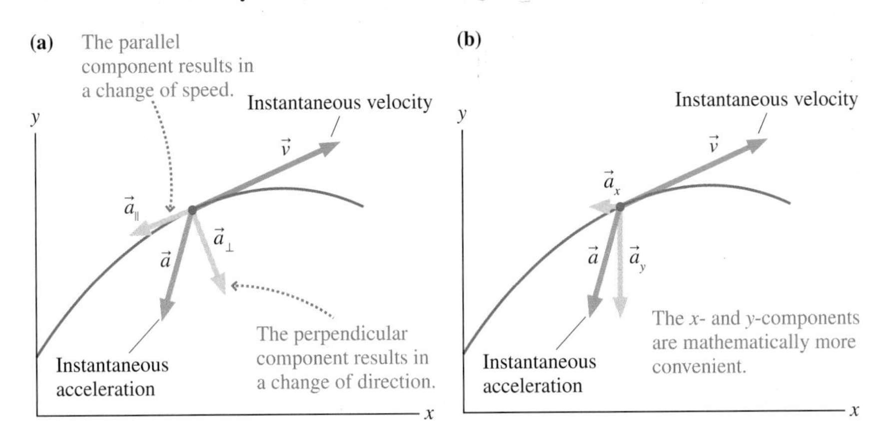

FIGURE 6.8 The instantaneous acceleration $\vec{a}$ can be decomposed into parallel and perpendicular components or into x- and y-components.

The acceleration vector $\vec{a}$ is the rate at which $\vec{v}$ is changing at that instant. To show this, Figure 6.8a decomposes $\vec{a}$ into components $\vec{a}_{\parallel}$ and $\vec{a}_{\perp}$ that are parallel and perpendicular to the trajectory. $\vec{a}_{\parallel}$ is associated with a change of speed and $\vec{a}_{\perp}$ is associated with a change in direction. Both kinds of changes are accelerations. Notice that $\vec{a}_{\perp}$ always points toward the "inside" of the curve because that is the direction in which $\vec{v}$ is changing.

The parallel and perpendicular components of $\vec{a}$ convey important ideas about acceleration, but the directions of $\vec{a}_{\parallel}$ and $\vec{a}_{\perp}$ keep changing. It's usually more practical to write $\vec{a}$ in terms of the x- and y-components shown in Figure 6.8b. Because $\vec{v} = v_x\hat{\imath} + v_y\hat{\jmath}$, we find

$$\vec{a} = a_x\hat{\imath} + a_y\hat{\jmath} = \frac{d\vec{v}}{dt} = \frac{dv_x}{dt}\hat{\imath} + \frac{dv_y}{dt}\hat{\jmath} \tag{6.12}$$

from which we see that

$$a_x = \frac{dv_x}{dt} \quad \text{and} \quad a_y = \frac{dv_y}{dt} \tag{6.13}$$

That is, the x-component of $\vec{a}$ is the rate dv_x/dt at which the x-component of velocity is changing.

Constant Acceleration

We're going to restrict our study of motion in a plane to situations where the acceleration $\vec{a} = a_x\hat{\imath} + a_y\hat{\jmath}$ is constant. This implies that the two components a_x and a_y are both constant (including, perhaps, zero). In this case, everything you learned about constant-acceleration kinematics in Chapter 2 carries over to the x- and y-components of two-dimensional motion.

Consider a particle that moves with constant acceleration from an initial position $\vec{r}_i = x_i\hat{\imath} + y_i\hat{\jmath}$, starting with initial velocity $\vec{v}_i = v_{ix}\hat{\imath} + v_{iy}\hat{\jmath}$. Its position and velocity at a final point f are

$$\begin{aligned} x_f &= x_i + v_{ix}\Delta t + \tfrac{1}{2}a_x(\Delta t)^2 \qquad & y_f &= y_i + v_{iy}\Delta t + \tfrac{1}{2}a_y(\Delta t)^2 \\ v_{fx} &= v_{ix} + a_x\Delta t & v_{fy} &= v_{iy} + a_y\Delta t \end{aligned} \tag{6.14}$$

There are *many* quantities to keep track of in two-dimensional kinematics, making the pictorial representation all the more important.

NOTE ▶ For constant acceleration, the x-component of the motion and the y-component of the motion are independent of each other. However, they remain connected through the fact that Δt must be the same for both. ◀

EXAMPLE 6.2 Plotting the trajectory of the shuttlecraft

The up thrusters on the shuttlecraft of the starship *Enterprise* give it an upward acceleration of 5.0 m/s^2. Its forward thrusters provide a forward acceleration of 20 m/s^2. As it leaves the *Enterprise*, the shuttlecraft turns on only the upthrusters. After clearing the flight deck, 3.0 s later, it adds the forward thrusters. Plot a trajectory of the shuttlecraft for its first 6 s.

MODEL Represent the shuttlecraft as a particle. There are two segments of constant-acceleration motion.

VISUALIZE Figure 6.9 shows a pictorial representation. The coordinate system has been chosen so that the shuttlecraft starts at the origin and initially moves along the y-axis. The craft moves vertically for 3 s, then begins to acquire a forward motion. There are three points in the motion: the beginning, the end, and the point at which forward thrusters are turned on. These points are labeled (x_0, y_0), (x_1, y_1), and (x_2, y_2). The velocities are (v_{0x}, v_{0y}), (v_{1x}, v_{1y}), and (v_{2x}, v_{2y}). This will be our standard labeling scheme for trajectories, where it is essential to keep the x-components and y-components separate.

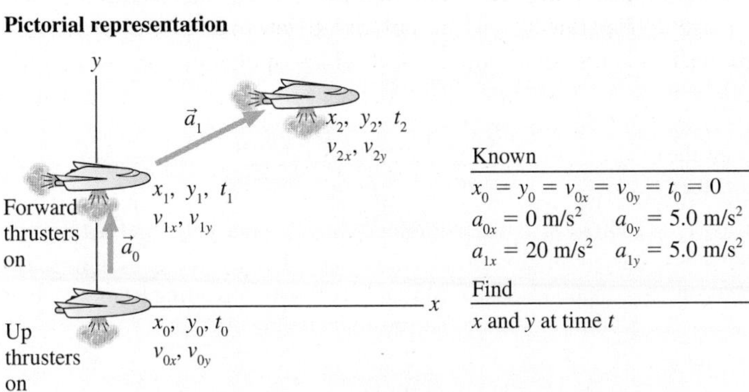

FIGURE 6.9 Pictorial representation of the motion of the shuttlecraft.

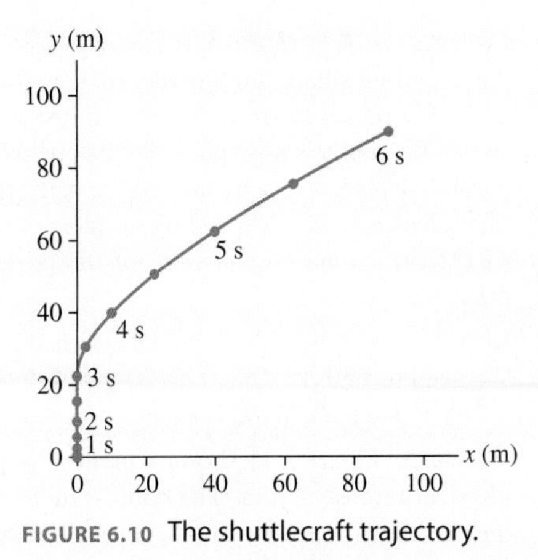

FIGURE 6.10 The shuttlecraft trajectory.

SOLVE During the first phase of the acceleration, when $a_{0x} = 0\text{ m/s}^2$ and $a_{0y} = 5.0\text{ m/s}^2$, the motion is described by

$$y = y_0 + v_{0y}(t - t_0) + \frac{1}{2}a_{0y}(t - t_0)^2 = 2.5t^2\text{ m}$$

$$v_y = v_{0y} + a_{0y}(t - t_0) = 5.0t\text{ m/s}$$

where the time t is in s. These equations allow us to calculate the position and velocity at any time t. At $t_1 = 3.0$ s, when the first phase of the motion ends, we find that

$$x_1 = 0\text{ m} \qquad v_{1x} = 0\text{ m/s}$$

$$y_1 = 22.5\text{ m} \qquad v_{1y} = 15\text{ m/s}$$

During the next 3 s, when $a_{1x} = 20\text{ m/s}^2$ and $a_{1y} = 5.0\text{ m/s}^2$, the x- and y-coordinates are

$$x = x_1 + v_{1x}(t - t_1) + \tfrac{1}{2}a_{1x}(t - t_1)^2$$
$$= 10(t - 3.0)^2\text{ m}$$

$$y = y_1 + v_{1y}(t - t_1) + \tfrac{1}{2}a_{1y}(t - t_1)^2$$
$$= (22.5 + 15(t - 3.0) + 2.5(t - 3.0)^2)\text{ m}$$

where, again, t is in s. To show the trajectory, we've calculated x and y every 0.5 s, plotted the points in Figure 6.10, and drawn a smooth curve through the points. You can see the shuttlecraft "lift off" during the first 3 s, then begin to accelerate forward.

STOP TO THINK 6.1 This acceleration will cause the particle to

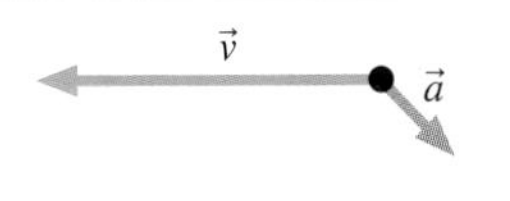

a. Speed up and curve upward.
b. Speed up and curve downward.
c. Slow down and curve upward.
d. Slow down and curve downward.
e. Move to the right and down.
f. Reverse direction.

6.2 Dynamics in Two Dimensions

Newton's second law $\vec{a} = \vec{F}_{net}/m$ determines an object's acceleration. It makes no distinction between linear motion and nonlinear motion. The x- and y-components of the acceleration vector are given by

$$a_x = \frac{(F_{net})_x}{m} \quad \text{and} \quad a_y = \frac{(F_{net})_y}{m} \tag{6.15}$$

Problem-Solving Strategy 5.2 for dynamics problems, on page 126, is still valid. As a quick review, you should

1. Draw a pictorial representation and a physical representation (motion diagram and free-body diagram).
2. Use Newton's second law in component form:

$$(F_{net})_x = \sum F_x = ma_x \quad \text{and} \quad (F_{net})_y = \sum F_y = ma_y$$

The force components (including proper signs) are found from the free-body diagram. Solve for the acceleration, then use the Equation 6.14 kinematic equations to find velocities and positions.

EXAMPLE 6.3 A rocketing hockey puck

Alice tapes a small 200 g model rocket to a 400 g ice hockey puck. The rocket generates 8.0 N of thrust. She orients the puck so that the rocket's nose points in the positive y-direction, then pushes the puck across frictionless ice in the positive x-direction. She releases it with a speed of 2.0 m/s at the exact instant the rocket fires. Find an equation for the puck's trajectory, then graph it.

MODEL Model the puck with the attached rocket as a particle. We need to find the *function* $y(x)$ that describes the curve followed by the puck in the xy-plane.

VISUALIZE Figure 6.11 shows a pictorial representation and a physical representation. The coordinate axes were defined in the problem statement, and we'll specify that the motion starts at the origin. We can identify three forces acting on the puck, but only $\vec{F}_{\text{thrust}}$ acts in the plane of motion. The normal force $\vec{n}$ and the weight $\vec{w}$ are perpendicular to the plane of motion (along the z-axis) and cancel each other to give $(F_{\text{net}})_z = 0$.

SOLVE The net force in the xy-plane is $\vec{F}_{\text{net}} = F_{\text{thrust}}\hat{j}$. The x-component of $\vec{F}_{\text{net}}$ is zero, so $a_x = 0$ and the motion along the x-axis is uniform motion with constant velocity. That is, the rocket thrust along the y-axis causes no change in the x-component of the puck's velocity. It will remain at $v_x = 2.0$ m/s. Simultaneously, $\vec{F}_{\text{thrust}}$ will cause the puck to accelerate in the positive y-direction. Hence y and v_y will steadily increase from their initial values of zero. We can use the free-body diagram of Figure 6.11 and Newton's second law to find the acceleration:

$$a_x = \frac{(F_{\text{net}})_x}{m} = 0$$

$$a_y = \frac{(F_{\text{net}})_y}{m} = \frac{F_{\text{thrust}}}{m} = \frac{8.0\ \text{N}}{0.60\ \text{kg}} = 13.33\ \text{m/s}^2$$

where m is the combined mass of the puck and the rocket. The x-motion is one of constant velocity at $v_{0y} = 2.0$ m/s. The y-motion is one of constant acceleration, starting from $y_0 = 0$ m with $v_{0y} = 0$ m/s. The position at time t is

$$x = x_0 + v_{0x}(t - t_0) + \frac{1}{2}a_x(t - t_0)^2$$

$$= v_{0x}t = 2.0t\ \text{m}$$

$$y = y_0 + v_{0y}(t - t_0) + \frac{1}{2}a_y(t - t_0)^2$$

$$= \frac{1}{2}a_y t^2 = 6.67t^2\ \text{m}$$

where t is in s. This solution gives x and y as explicit functions of time, but it is not the $y(x)$ solution we were looking for. To express y in terms of x, we need to eliminate the time variable from these two equations. Using the x-equation, we find that $t = x/2.0$. Substituting this into the y-equation, the function that describes the trajectory is

$$y(x) = 6.67\left(\frac{x}{2.0}\right)^2 = 1.67x^2\ \text{m}$$

where x is in m. The trajectory equation is of the form $y(x) = cx^2$, which is the equation of a parabola. Figure 6.12 shows the trajectory calculated from this equation.

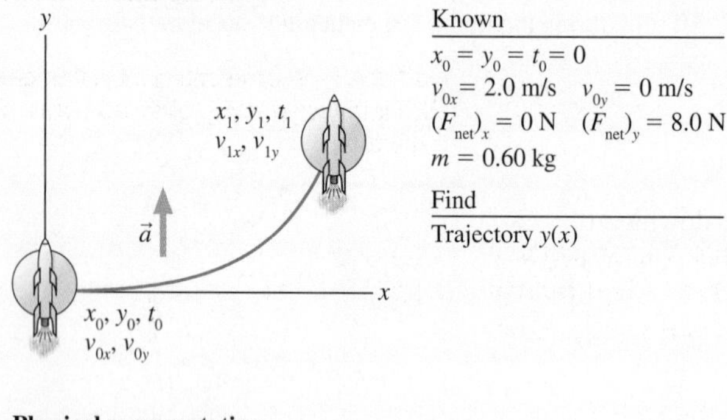

FIGURE 6.11 Pictorial and physical representations of the hockey puck.

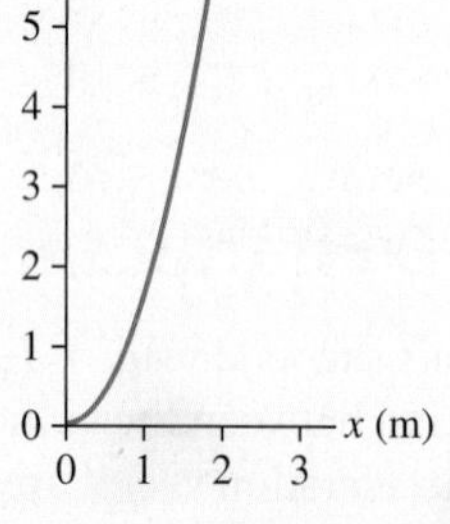

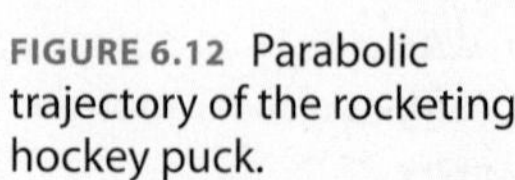

FIGURE 6.12 Parabolic trajectory of the rocketing hockey puck.

ASSESS The solution depended on the fact that the time parameter t is the *same* for both components of the motion.

Although Example 6.3 was about a rocketing hockey puck, our conclusion is quite general. **Any object for which one component of the acceleration is zero while the other has a constant, nonzero value follows a parabolic trajectory.** Important examples of motion along a parabolic trajectory include projectiles, which we will study in the next section, and electrons moving through the "deflection plates" that steer them to the face of your computer display terminal.

STOP TO THINK 6.2 The components of this particle's acceleration are

a. $a_x > 0, a_y > 0.$ b. $a_x = 0, a_y > 0.$ c. $a_x < 0, a_y > 0.$
d. $a_x > 0, a_y < 0.$ e. $a_x = 0, a_y < 0.$ f. $a_x < 0, a_y < 0.$

6.3 Projectile Motion

Baseballs and tennis balls flying through the air, Olympic divers, daredevils shot from cannons, and bullets shot from guns all exhibit what we call *projectile motion.* **A projectile is an object that moves in two dimensions under the influence of only the gravitational force.** Projectile motion is an extension of the free-fall motion we studied in Chapter 2. We will continue to neglect the influence of air resistance, leading to results that are a good approximation of reality for objects moving relatively slowly over relatively short distances.

If the only force acting on an object is its weight, pointing in the negative y-direction, then $a_x = 0$ while a_y has a constant, nonzero value. As we just noted, these are the conditions for a *parabolic trajectory* and, indeed, projectiles do move along parabolic paths. Figure 6.13 shows the parabolic trajectory of a bouncing ball. You should also look back at Example 1.7, where we examined the motion diagram of a projectile.

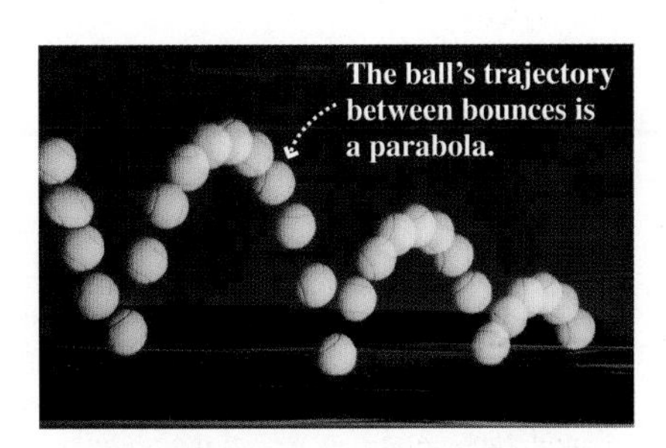

FIGURE 6.13 The parabolic trajectory of a bouncing ball.

The start of a projectile's motion, be it thrown by hand or shot from a gun, is called the *launch*, and the angle θ of the initial velocity $\vec{v}_i$ above the horizontal (i.e., above the x-axis) is called the **launch angle.** Figure 6.14 illustrates the relationship between the initial velocity vector $\vec{v}_i$ and the initial values of the components v_{ix} and v_{iy}. You can see that

$$v_{ix} = v_i \cos\theta$$
$$v_{iy} = v_i \sin\theta \qquad (6.16)$$

where v_i is the initial speed.

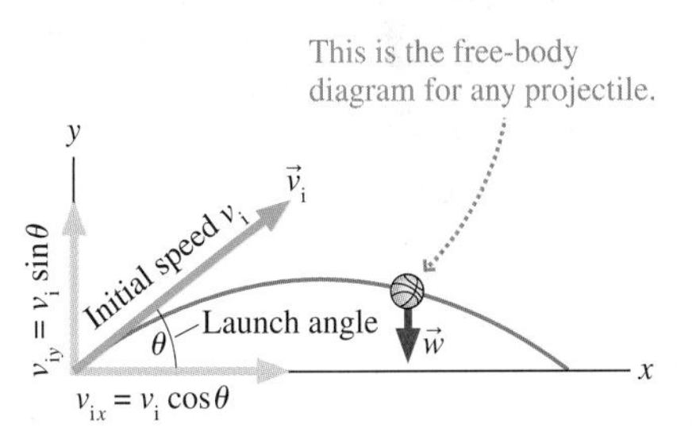

FIGURE 6.14 A projectile that is launched with initial velocity $\vec{v}_i$ follows a parabolic trajectory.

NOTE ► The components v_{ix} and v_{iy} are not always positive. In particular, a projectile launched at an angle *below* the horizontal (such as a ball thrown downward from the roof of a building) has *negative* values for θ and v_{iy}. However, the *speed* v_i is always positive. ◄

ActivPhysics ONLINE 3.1–3.7

The net force on a projectile is simply its weight: $\vec{F}_{net} = \vec{w} = -mg\hat{j}$, as shown in Figure 6.14. We can use Newton's second law to find the acceleration:

$$a_x = \frac{(F_{net})_x}{m} = 0$$
$$a_y = \frac{(F_{net})_y}{m} = \frac{-mg}{m} = -g \qquad (6.17)$$

In other words, **the vertical component of acceleration a_y is just the familiar $-g$ of free fall while the horizontal component a_x is zero.**

To see how these conditions influence the motion, Figure 6.15 shows a projectile launched from $(x_i, y_i) = (0 \text{ m}, 0 \text{ m})$ with an initial velocity $\vec{v}_i = (9.8\hat{\imath} + 19.6\hat{\jmath})$ m/s. The velocity and acceleration vectors are then shown every 1.0 s. The value of v_x never changes, but v_y decreases by 9.8 m/s^2 every second. This is what it *means* to accelerate at $a_y = -9.8 \text{ m/s}^2 = (-9.8 \text{ m/s})$ per second. Be sure to notice that nothing *pushes* the projectile along the curve. Instead, the downward weight force causes a downward acceleration that changes the velocity vector as shown.

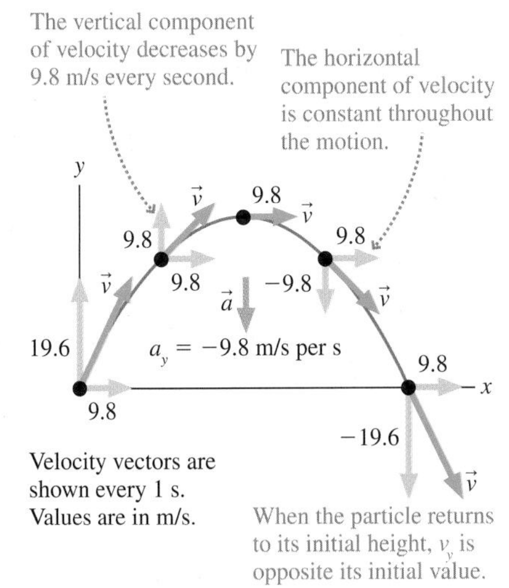

FIGURE 6.15 The velocity and acceleration vectors of a projectile moving along a parabolic trajectory.

You can see from Figure 6.15 that **projectile motion is made up of two independent motions:** uniform motion at constant velocity in the horizontal direction and free-fall motion in the vertical direction. The kinematic equations that describe these two motions are

$$x_f = x_i + v_{ix}\Delta t \qquad y_f = y_i + v_{iy}\Delta t - \tfrac{1}{2}g(\Delta t)^2$$
$$v_{fx} = v_{ix} = \text{constant} \qquad v_{fy} = v_{iy} - g\Delta t \tag{6.18}$$

These are parametric equations for the parabolic trajectory of a projectile.

EXAMPLE 6.4 Don't try this at home!

A stunt man drives a car off a 10-m-high cliff at a speed of 20 m/s. How far does the car land from the base of the cliff?

MODEL Represent the car as a particle in free fall. Assume that the car is moving horizontally as it leaves the cliff.

VISUALIZE The pictorial representation, shown in Figure 6.16, is *very* important because the number of symbols in projectile motion problems can be quite large. We have chosen to put the origin at the base of the cliff. The assumption that the car is moving horizontally as it leaves the cliff leads to $v_{0x} = v_0$ and $v_{0y} = 0$ m/s. A physical representation is not needed in projectile motion problems because we already know that a free-body diagram consists of a single force vector—the downward weight force—and that $\vec{a} = -g\hat{j}$.

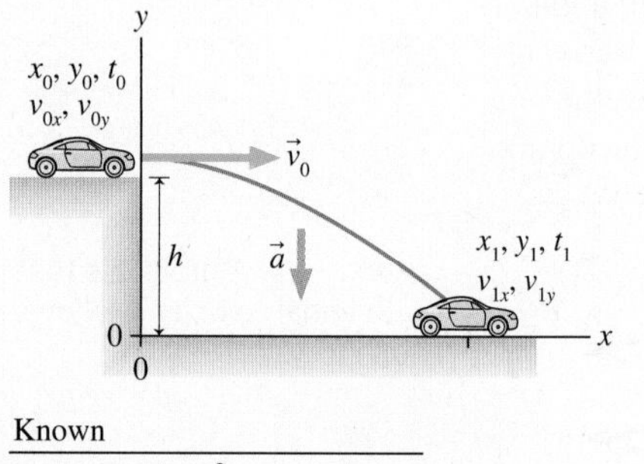

Known
$x_0 = v_{0y} = t_0 = 0$
$y_0 = 10$ m $\quad v_{0x} = v_0 = 20$ m/s
$a_x = 0$ m/s^2 $\quad a_y = -g \quad y_1 = 0$ m

Find
x_1

FIGURE 6.16 Pictorial representation for the car of Example 6.4.

SOLVE Each point on the trajectory has x- and y-components of position, velocity, and acceleration but only *one* value of time. The time needed to move horizontally to x_1 is the *same* time needed to fall vertically through distance h. **Although the horizontal and vertical motions are independent, they are connected through the time t.** This is a critical observation for solving projectile motion problems. The kinematics equations are

$$x_1 = x_0 + v_{0x}(t_1 - t_0) = v_0 t_1$$
$$y_1 = 0 = y_0 + v_{0y}(t_1 - t_0) - \frac{1}{2}g(t_1 - t_0)^2 = h - \frac{1}{2}g t_1^2$$

We can use the vertical equation to determine the time t_1 needed to fall distance h:

$$t_1 = \sqrt{\frac{2h}{g}} = \sqrt{\frac{2(10\text{ m})}{9.80\text{ m/s}^2}} = 1.43\text{ s}$$

We then insert this expression for t into the horizontal equation to find the distance traveled:

$$x_1 = v_0 t_1 = (20\text{ m/s})(1.43\text{ s}) = 28.6\text{ m}$$

ASSESS The cliff height is $h \approx 33$ ft and the initial speed is $v_0 \approx 40$ mph. Traveling $x_1 = 29$ m ≈ 95 ft before hitting the ground seems reasonable.

Reasoning About Projectile Motion

Think about the following question:

> A rifle fires a bullet exactly horizontally at height h above a horizontal field. At the exact instant that the bullet is fired, a second bullet is simply dropped from height h. Which bullet hits the ground first?

It may seem hard to believe, but they hit the ground *simultaneously*. They do so because the horizontal and vertical components of projectile motion are independent of each other. The initial horizontal velocity of the first bullet has *no* influence over its vertical motion. Neither bullet has any initial motion in the vertical direction, so both fall distance h in the same amount of time. You can see this in Figure 6.17, where one ball is shot horizontally and the other released from rest at the same instant. The *vertical* motions of the two balls are identical, and they hit the floor simultaneously.

Figure 6.18a shows a useful way to think about the trajectory of a projectile. Without gravity, a projectile would follow a straight line. Because of gravity, the particle at time t has "fallen" a distance $\frac{1}{2}gt^2$ below this line. The separation grows as $\frac{1}{2}gt^2$, giving the trajectory its parabolic shape.

Figure 6.18a can help you understand a "classic" problem in physics.

> A hungry hunter in the jungle wants to shoot down a coconut that is hanging from the branch of a tree. He aims the gun directly at the coconut, but as luck would have it the coconut falls from the branch at the *exact* instant the hunter pulls the trigger. Does the bullet hit the coconut?

You might think that the bullet will miss, but it doesn't. Although the bullet travels very fast, it follows a slightly curved trajectory, not a straight line. Had the coconut stayed on the tree, the bullet would have curved under its target as gravity causes it to fall a distance $\frac{1}{2}gt^2$ below the straight line. But $\frac{1}{2}gt^2$ is also the distance the coconut falls while the bullet is in flight. Thus, as Figure 6.18b shows, the bullet and the coconut fall the same distance and meet at the same point!

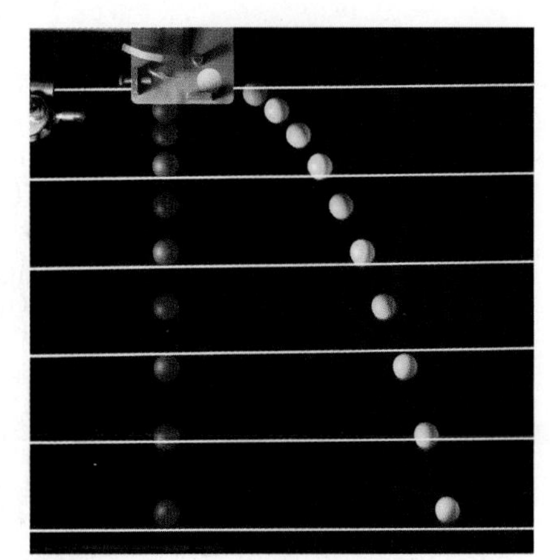

FIGURE 6.17 A projectile launched horizontally falls in the same time as a projectile that is released from rest.

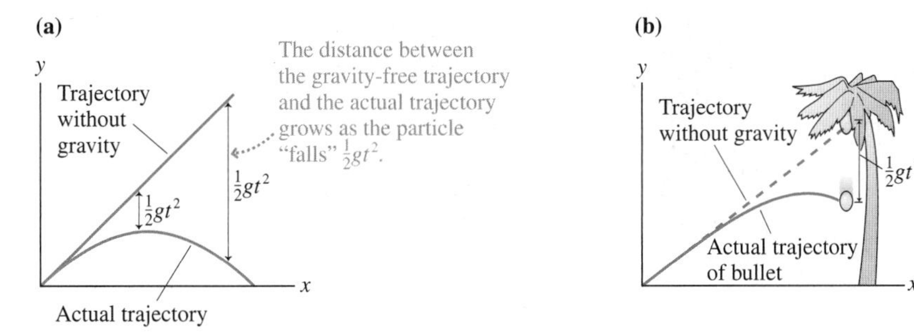

FIGURE 6.18 A projectile follows a parabolic trajectory because it "falls" a distance $\frac{1}{2}gt^2$ below a straight-line trajectory.

Solving Projectile Motion Problems

Let's summarize this information with a problem-solving strategy.

MP PROBLEM-SOLVING STRATEGY 6.1 Projectile motion problems

MODEL Make simplifying assumptions.

VISUALIZE Use a pictorial representation. Establish a coordinate system with the x-axis horizontal and the y-axis vertical. Show important points in the motion on a sketch. Define symbols and identify what the problem is trying to find.

SOLVE The acceleration is known: $a_x = 0$ and $a_y = -g$. Thus the problem becomes one of kinematics. The kinematic equations are

$$x_f = x_i + v_{ix}\Delta t \qquad y_f = y_i + v_{iy}\Delta t - \tfrac{1}{2}g(\Delta t)^2$$

$$v_{fx} = v_{ix} = \text{constant} \qquad v_{fy} = v_{iy} - g\Delta t$$

Δt is the same for the horizontal and vertical components of the motion. Find Δt from one component, then use that value for the other component.

ASSESS Check that your result has the correct units, is reasonable, and answers the question.

EXAMPLE 6.5 The distance of a fly ball

A baseball is hit at angle θ and is caught at the height from which it was hit.

a. If the ball is hit at a 30° angle, with what speed must it leave the bat to travel 100 m?

b. What angle causes the ball to go the maximum distance?

MODEL Represent the ball as a particle. Ignore air resistance.

VISUALIZE Figure 6.19 shows the pictorial representation. The height above the ground at which the ball was hit and caught is not relevant, because they are the same, so we have placed the origin at the point where the ball is hit. The ball travels distance x_1.

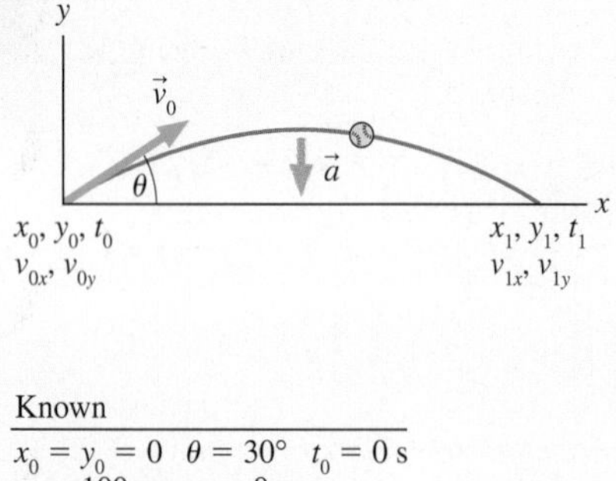

Known
$x_0 = y_0 = 0 \quad \theta = 30° \quad t_0 = 0$ s
$x_1 = 100$ m $\quad y_1 = 0$ m

Find
v_0

FIGURE 6.19 Pictorial representation for the baseball of Example 6.5.

SOLVE

a. The initial x- and y-components of the ball's velocity are

$$v_{0x} = v_0 \cos\theta$$
$$v_{0y} = v_0 \sin\theta$$

where v_0 is the initial speed that we need to find. The kinematic equations of projectile motion are

$$x_1 = x_0 + v_{0x}(t_1 - t_0) = (v_0\cos\theta)t_1$$

$$y_1 = 0 = y_0 + v_{0y}(t_1 - t_0) - \frac{1}{2}g(t_1 - t_0)^2 = (v_0\sin\theta)t_1 - \frac{1}{2}gt_1^2$$

We can use the vertical equation to find the time of flight:

$$0 = (v_0\sin\theta)t_1 - \frac{1}{2}gt_1^2 = \left(v_0\sin\theta - \frac{1}{2}gt_1\right)t_1$$

$$t_1 = 0 \quad \text{or} \quad \frac{2v_0\sin\theta}{g}$$

Both values are legitimate solutions. The first corresponds to the instant when $y = 0$ at the beginning of the trajectory and the second to when $y = 0$ at the end. Clearly, though, we want the second solution. Substituting this expression for t_1 into the equation for x_1 gives

$$x_1 = (v_0\cos\theta)\frac{2v_0\sin\theta}{g} = \frac{2v_0^2\sin\theta\cos\theta}{g}$$

We can simplify this result by using the trigonometric identity $2\sin\theta\cos\theta = \sin(2\theta)$. The distance traveled by the ball when hit at angle θ is

$$x_1 = \frac{v_0^2\sin(2\theta)}{g}$$

Setting $x_1 = 100$ m and solving for the speed v_0 gives

$$v_0 = \sqrt{\frac{gx_1}{\sin(2\theta)}} = \sqrt{\frac{(9.80\text{ m/s}^2)(100\text{ m})}{\sin 60°}} = 33.6\text{ m/s}$$

b. What value of θ will maximize x_1? As you know, the sine function has a maximum value of 1 at an angle of 90°. Because the equation for x_1 contains the expression $\sin(2\theta)$, it will reach a maximum for $\theta_{max} = 45°$. If the batter hitting the ball with a speed of 33.6 m/s had hit it at a 45° angle, the ball would have traveled 115 m and just cleared the left field wall (at 110 m)—winning the World Series and bringing fame and fortune to our hero. Instead, he hit at a mere 30° angle, the ball was caught, and history has forgotten his name.

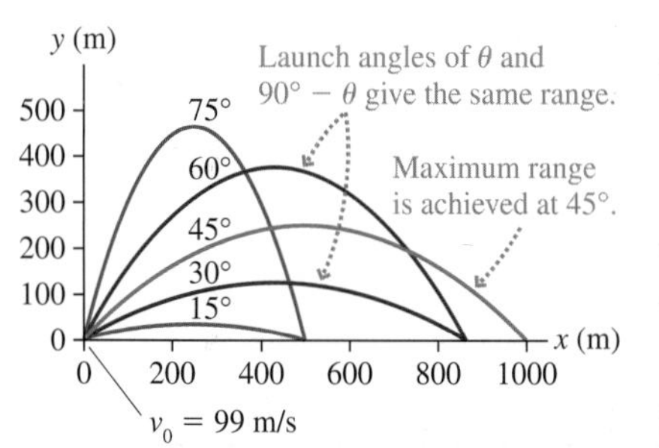

FIGURE 6.20 Trajectories of a projectile launched at different angles with a speed of 99 m/s.

As Example 6.5 found, a projectile that lands at the same elevation from which it was fired travels distance

$$\text{distance} = \frac{v_0^2\sin(2\theta)}{g} \tag{6.19}$$

The maximum distance occurs for $\theta = 45°$, where $\sin(2\theta) = 1$. But there's more that we can learn from this equation. Because $\sin(180° - x) = \sin x$, it follows that $\sin(2(90° - \theta)) = \sin(2\theta)$. Consequently, a projectile launched either at angle θ *or* at angle $(90° - \theta)$ will travel the same distance. Figure 6.20 shows the trajectories of projectiles launched with the same initial speed in 15° increments of angle.

NOTE ► Equation 6.19 is *not* a general result. It applies *only* in situations where the projectile lands at the same elevation from which it was fired. ◄

EXAMPLE 6.6 Santa's sleigh ride

Santa parked his sleigh 5.0 m from the edge of a 20° roof. Unfortunately, the parking brake failed. Santa's sleigh slid down the roof and landed on the ground 6.0 m from the wall of the house. The sleigh's coefficient of kinetic friction on the snowy roof is 0.08. How high was the wall of the house?

MODEL Represent the sleigh as a particle. This is a two-part problem. First, the sleigh slides down the roof—motion in a straight line. The sleigh then becomes a projectile until it hits the ground. These two problems are connected through the fact that the *final* speed of the linear motion problem is the *initial* speed of the projectile problem.

VISUALIZE Figure 6.21 shows separate pictorial representations for the two parts of the motion. Notice that we're using different coordinate systems for the two parts of the problem; each is chosen to serve a specific need. Point 2 is the same as point 1, but we've given it a new number in order to write its coordinates in the new coordinate system without causing confusion. We've explicitly noted that $v_2 = v_1$, which connects the linear motion problem to the projectile motion problem. The linear motion includes a free-body diagram, but none is needed for the projectile motion.

SOLVE The first part of the motion is similar to problems you solved in Chapter 5. Newton's second law and the model of kinetic friction are

$$\sum F_x = w_x + (f_k)_x = mg\sin\theta - f_k = ma_x$$

$$\sum F_y = w_y + n_y = n - mg\cos\theta = ma_y = 0$$

$$f_k = \mu_k n$$

We've written these equations, as we did in Chapter 5, by "reading" the free-body diagram. We know that $a_y = 0$ because the motion is entirely along the x-axis. The y-equation gives $n = mg\cos\theta$, from which we find that the friction is $f_k = \mu_k mg\cos\theta$. Substituting this into the x-equation gives

$$a = \frac{mg\sin\theta - \mu_k mg\cos\theta}{m}$$

$$= g(\sin\theta - \mu_k\cos\theta) = 2.62 \text{ m/s}^2$$

Notice that the mass canceled out, which is fortunate because we weren't given a value. The velocity after sliding 5.0 m is found from one-dimensional kinematics:

$$v_{1x}^2 = v_{0x}^2 + 2a_x\Delta x = 2a_x x_1$$

$$v_{1x} = \sqrt{2a_x x_1} = 5.11 \text{ m/s}$$

Now we have to solve a projectile problem. Notice that in the pictorial model we've restarted the clock: $t_2 = 0$ s. The initial velocity $\vec{v}_2$ of the projectile motion is the *same* as the final velocity $\vec{v}_1$ of the linear motion, which we can write

$$\vec{v}_1 = (5.11 \text{ m/s}, 20° \text{ below horizontal})$$

The initial velocity components are

$$v_{2x} = v_2\cos\theta = (5.11 \text{ m/s})\cos(-20°) = 4.80 \text{ m/s}$$

$$v_{2y} = v_2\sin\theta = (5.11 \text{ m/s})\sin(-20°) = -1.75 \text{ m/s}$$

This is an example where the initial value of v_y is negative. The kinematic equations are

$$x_3 = x_2 + v_{2x}(t_3 - t_2) = v_{2x}t_3$$

$$y_3 = 0 = y_2 + v_{2y}(t_3 - t_2) - \frac{1}{2}g(t_3 - t_2)^2$$

$$= h + v_{2y}t_3 - \frac{1}{2}gt_3^2$$

We can use the horizontal motion to find that the sleigh hits the ground at time

$$t_3 = \frac{x_3}{v_{2x}} = \frac{6.0 \text{ m}}{4.80 \text{ m/s}} = 1.25 \text{ s}$$

We can now use t_3 in the vertical-motion equation to solve for the wall height:

$$h = \frac{1}{2}gt_3^2 - v_{2y}t_3$$

$$= \frac{1}{2}(9.80 \text{ m/s}^2)(1.25 \text{ s})^2 - (-1.75 \text{ m/s})(1.25 \text{ s})$$

$$= 9.84 \text{ m}$$

The wall of the house is 9.84 m high, or roughly 30 ft.

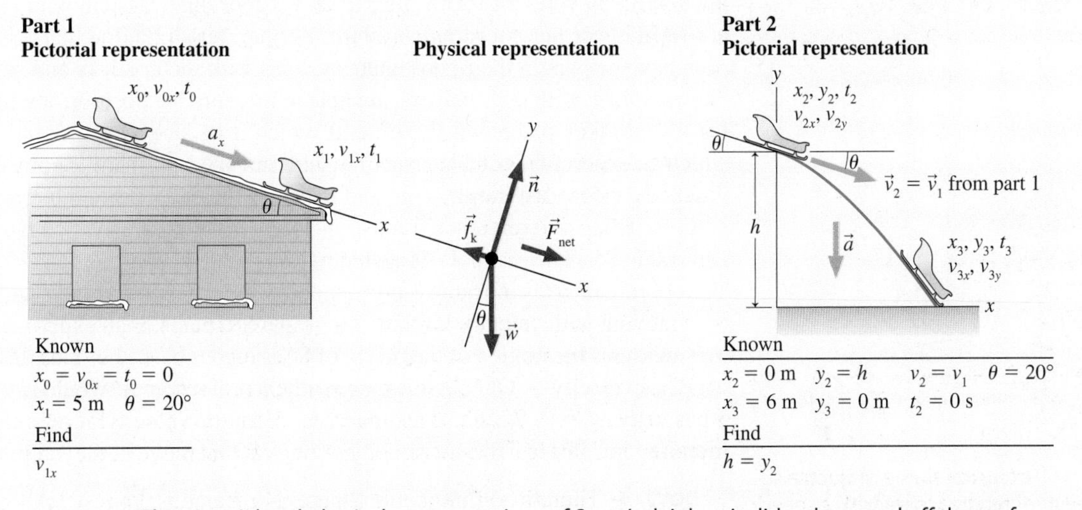

FIGURE 6.21 The pictorial and physical representations of Santa's sleigh as it slides down and off the roof.

STOP TO THINK 6.3 A 100 g ball rolls off a table and lands 2 m from the base of the table. A 200 g ball rolls off the same table with the same speed. It lands at distance

a. < 1 m. b. 1 m. c. Between 1 m and 2 m.
d. 2 m. e. Between 2 m and 4 m. f. 4 m.

6.4 Relative Motion

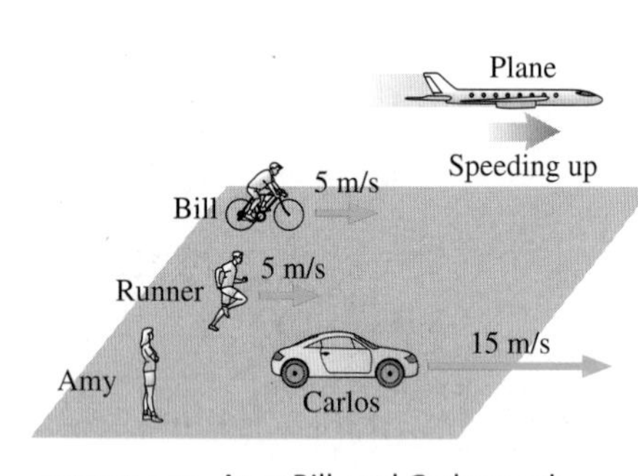

FIGURE 6.22 Amy, Bill, and Carlos each measure the velocity of the runner and the acceleration of the jet plane. The velocities are shown in Amy's reference frame.

You've now dealt many times with problems that say something like "A car travels at 30 m/s" or "A plane travels at 300 m/s." But just what do these statements really mean?

In Figure 6.22, Amy, Bill, and Carlos are watching a runner. According to Amy, the runner's velocity is $v_x = 5$ m/s. But to Bill, who's riding alongside, the runner is lifting his legs up and down but going neither forward nor backward relative to Bill. As far as Bill is concerned, the runner's velocity is $v_x = 0$ m/s. Carlos sees the runner receding in his rearview mirror, in the *negative* x-direction, getting 10 m further away from him every second. According to Carlos, the runner's velocity is $v_x = -10$ m/s. Which is the runner's *true* velocity?

Velocity is not a concept that can be true or false. The runner's velocity *relative to Amy* is 5 m/s. That is, his velocity is 5 m/s in a coordinate system attached to Amy and in which Amy is at rest. The runner's velocity relative to Bill is 0 m/s, and the velocity relative to Carlos is -10 m/s. These are all valid descriptions of the runner's motion.

What about the jet plane, which is speeding up? Suppose Amy, Bill, and Carlos each uses his or her coordinate system to measure the plane's acceleration. Do their values agree or disagree? Acceleration is important because the question we ultimately want to address in this section is "In which coordinate systems are Newton's laws valid?" If Newton's laws are to be the foundation of mechanics, we certainly should know the coordinate systems in which we can use these laws. And because Newton's second law is about acceleration, not velocity, we need to explore how acceleration is measured in different coordinate systems. First, however, we need to look at relative position and relative velocity.

Relative Position

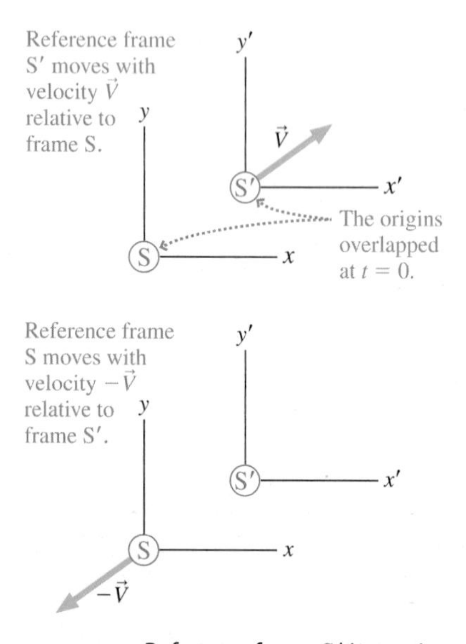

FIGURE 6.23 Reference frame S′ is moving with velocity $\vec{V}$ relative to reference frame S.

Suppose that Amy and Bill each have a coordinate system attached to their bodies. As Bill bicycles past Amy, he carries his coordinate system with him. Each is at rest in his or her coordinate system. Further, let's imagine that Amy and Bill each have helpers in their coordinate systems with meter sticks and stopwatches. Amy and Bill, with their helpers, are able to measure the position at which a physical event takes place and the time at which it occurs. A coordinate system in which an experimenter makes position and time measurements of physical events is called a **reference frame.** Amy and Bill each have their own reference frame.

Let's define two reference frames, shown in Figure 6.23, that we'll call frame S and frame S′. (The symbol ′ is called a *prime*, and S′ is pronounced "S prime.") The coordinate axes in frame S are x and y, while those in S′ are x' and y'. Frame S′ is moving with velocity $\vec{V}$ relative to frame S. That is, if an experimenter at rest in S measures the motion of the origin of S′ as it goes past, she finds that the origin of S′ has velocity $\vec{V}$. Of course, an experimenter at rest in S′ would say that frame S has velocity $-\vec{V}$. We'll use an uppercase V for the velocity of reference frames, reserving lowercase v for the velocity of objects that move in the reference frames.

NOTE ► There's no implication that either frame is "at rest." All we know is that the two frames are moving *relative* to each other with velocity $\vec{V}$. ◄

We will stipulate four conditions for reference frames:

1. The frames are oriented the same, with the x- and x'-axes parallel to each other.
2. The origins of frame S and frame S′ coincide at $t = 0$.
3. All motion is in the xy-plane, so we don't need to consider the z-axis.
4. The relative velocity $\vec{V}$ is *constant*.

The first three are a matter of how we define the coordinate systems. Item 4, by contrast, is a choice with consequences. It says that we will consider only reference frames that move with constant speed in a straight line. These are called **inertial reference frames,** a term introduced in Chapter 4. We'll see that these are the reference frames in which Newton's laws are valid.

Suppose a light bulb flashes at time t. Experimenters in both reference frames see the flash and measure its position. Observers in S place the flash at position $\vec{r}$, as measured with respect to the coordinate system of frame S. Similarly, experimenters in S′ determine that the flash occurred at position $\vec{r}\,'$, relative to the origin of S′. (We'll use primes to indicate positions and velocities measured in frame S′.)

What is the relationship between the position vectors $\vec{r}$ and $\vec{r}\,'$? It's not hard to see, from Figure 6.24, that

$$\vec{r} = \vec{r}\,' + \vec{R} \tag{6.20}$$

where $\vec{R}$ is the position vector of the origin of frame S′ as measured in frame S.

Frame S′ is traveling with velocity $\vec{V}$ relative to frame S, and their origins coincide at $t = 0$. At time t, when the light flashes, the origin of S′ has moved to position $\vec{R} = t\vec{V}$. (We've written $t\vec{V}$ rather than $\vec{V}t$ because it is customary to write the scalar first.) Thus

$$\vec{r} = \vec{r}\,' + t\vec{V} \quad \text{or} \quad \vec{r}\,' = \vec{r} - t\vec{V} \tag{6.21}$$

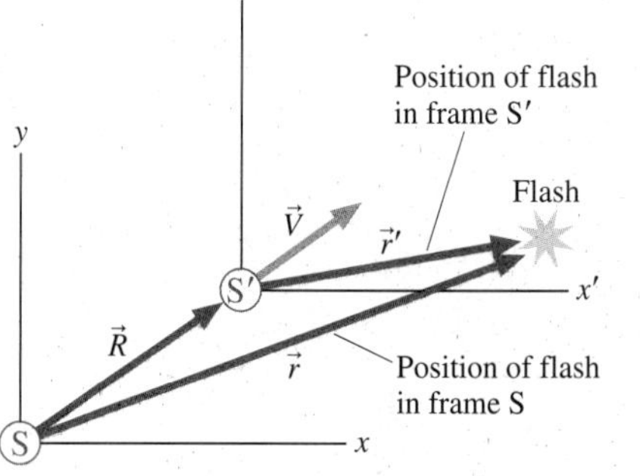

FIGURE 6.24 Measurements in frame S find that a light flash occurred at position $\vec{r}$. The same flash occurred at position $\vec{r}\,'$ in frame S′.

Equation 6.21 is called the **Galilean transformation of position.** It will be easiest for most purposes to write this in terms of components:

$$\begin{aligned} x &= x' + V_x t \\ y &= y' + V_y t \end{aligned} \quad \text{or} \quad \begin{aligned} x' &= x - V_x t \\ y' &= y - V_y t \end{aligned} \tag{6.22}$$

If we know *where* and *when* an event occurred in one reference frame, we can *transform* that position into any other reference frame that moves relative to the first with constant velocity $\vec{V}$.

EXAMPLE 6.7 Watching a ball toss

Mike throws a ball upward at a 63° angle with a speed of 22 m/s. Nancy rides past Mike on her bicycle at 10 m/s at the instant he releases the ball.

a. Find and graph the ball's trajectory as seen by Mike.
b. Find and graph the ball's trajectory as seen by Nancy.

SOLVE

a. For Mike, the ball is a projectile that follows a parabolic trajectory. This problem is almost exactly the same as Example 6.5. The components of the initial velocity are

$$v_{0x} = v_0 \cos\theta = (22 \text{ m/s})\cos 63° = 10.0 \text{ m/s}$$
$$v_{0y} = v_0 \sin\theta = (22 \text{ m/s})\sin 63° = 19.6 \text{ m/s}$$

The x- and y-equations of motion for the ball's position at time t are

$$x = x_0 + v_{0x}(t - t_0) = 10.0t \text{ m}$$
$$y = y_0 + v_{0y}(t - t_0) - \frac{1}{2}g(t - t_0)^2 = (19.6t - 4.9t^2) \text{ m}$$

where t is in s. It's not hard to show that the ball reaches height $y_{max} = 19.6$ m at $t = 2$ s and hits the ground at $t = 4$ s. The trajectory is shown in Figure 6.25a on the next page.

b. We can determine the trajectory Nancy sees by using Equations 6.22 to transform the ball's position from Mike's reference frame into Nancy's reference frame. Let Mike be in frame S and Nancy in frame S′. Nancy moves with velocity $\vec{V} = 10.0\hat{\imath}$ m/s relative to S. In terms of components,

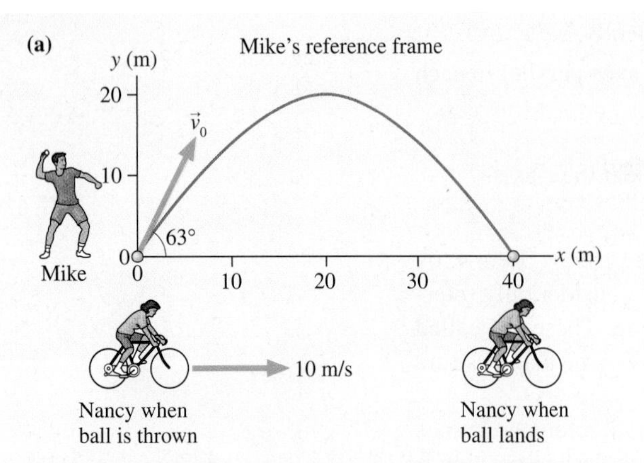

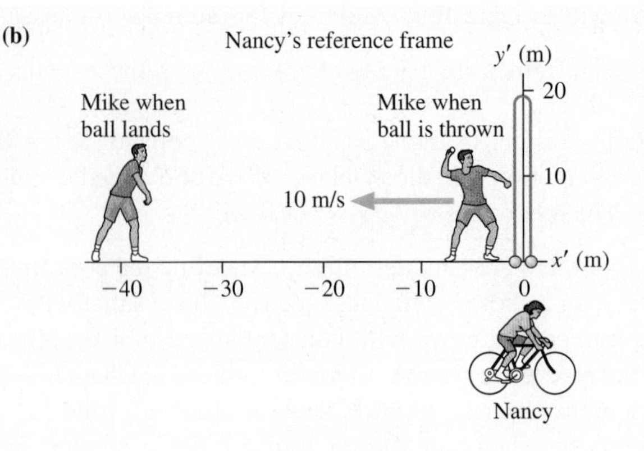

FIGURE 6.25 A ball's trajectory as seen by Mike and Nancy.

$V_x = 10$ m/s and $V_y = 0$ m/s. When Mike, at time t, measures the ball at position (x, y) in frame S, Nancy finds the ball at

$$x' = x - V_x t = 10.0t - 10.0t = 0$$
$$y' = y - V_y t = y$$

Because Nancy's horizontal motion is the same as the ball's ($V_x = v_x = 10$ m/s in Mike's frame), she doesn't see the ball moving either right or left. Nancy's experience is like that of Bill riding beside the runner in Figure 6.22. The ball moves *vertically* up and down in frame S′. Further, the vertical position y' in S′ is the same as the vertical position y in S. According to Nancy, the ball goes straight up, reaches a height of 19.6 m, and falls straight back down. It hits the ground right beside her bicycle at $t = 4$ s. This is seen in Figure 6.25b.

In Chapter 2 we studied *free fall*, vertical motion straight up and down. In Section 6.3 we studied the parabolic trajectories of *projectile motion*. Now, from Example 6.7 we see that **free-fall motion and projectile motion are really the same motion, simply seen from two different reference frames.** The motion is vertical in the *one* reference frame whose horizontal motion is the same as the ball's. The trajectory is a parabola in any other reference frame.

Relative Velocity

Let's think a bit more about Example 6.7. According to an observer in Mike's reference frame, Mike throws the ball with velocity $\vec{v}_0 = (10.0\hat{\imath} + 19.6\hat{\jmath})$ m/s. The ball's initial speed is $v_0 = 22.0$ m/s. But in frame S′, where Nancy sees the ball go straight up and down, Mike throws the ball with velocity $\vec{v}_0' = 19.6\hat{\jmath}$ m/s. An object's velocity measured in frame S is *not* the same as its velocity measured in frame S′. Our goal is to find a general relationship between an object's velocity $\vec{v}$ as measured in reference frame S and its velocity $\vec{v}\,'$ as measured in a different frame S′.

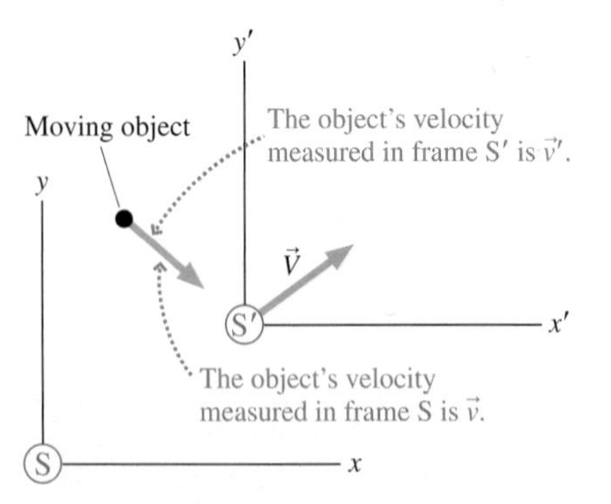

FIGURE 6.26 A velocity of a moving object is measured by experimenters in two different reference frames.

Figure 6.26 shows a *moving object* that is observed from reference frames S and S′. Experimenters in frame S locate the object at position $\vec{r}$ and measure its velocity to be $\vec{v}$. Simultaneously, experimenters in S′ measure position $\vec{r}\,'$ and velocity $\vec{v}\,'$. The position vectors, which are related by $\vec{r} = \vec{r}\,' + \vec{R}$, change as the object moves. In addition, $\vec{R}$ changes as the reference frames move relative to each other. The *rate* of change is

$$\frac{d\vec{r}}{dt} = \frac{d\vec{r}\,'}{dt} + \frac{d\vec{R}}{dt} \tag{6.23}$$

The derivative $d\vec{r}/dt$, by definition, is the object's velocity $\vec{v}$ measured in frame S. Similarly, $d\vec{r}\,'/dt$ is the object's velocity $\vec{v}\,'$ measured in frame S′. And

$d\vec{R}/dt$ is the velocity $\vec{V}$ of frame S′ relative to frame S. Consequently, Equation 6.23 tells us that

$$\vec{v} = \vec{v}\,' + \vec{V} \quad \text{or} \quad \vec{v}\,' = \vec{v} - \vec{V} \tag{6.24}$$

Equation 6.24 is the **Galilean transformation of velocity.** If we know an object's velocity measured in one reference frame, we can transform it into the velocity that would be measured by an experimenter in a different reference frame. As Figure 6.27 shows, doing so is an exercise in vector addition.

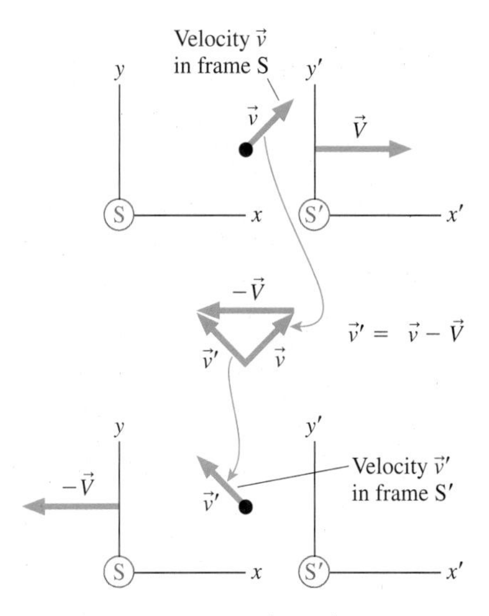

FIGURE 6.27 Velocities $\vec{v}$ and $\vec{v}\,'$, as measured in frames S and S′, are related by vector addition.

We will often find it convenient, as we did with position, to write Equation 6.24 in terms of components:

$$\begin{aligned} v_x &= v_x' + V_x \\ v_y &= v_y' + V_y \end{aligned} \quad \text{or} \quad \begin{aligned} v_x' &= v_x - V_x \\ v_y' &= v_y - V_y \end{aligned} \tag{6.25}$$

This relationship between velocities measured by experimenters in different frames of reference was recognized by Galileo in his pioneering studies of motion, hence its name.

Let's apply Equation 6.25 to Mike and Nancy. We've already noted that the ball's initial velocity in Mike's frame, frame S, is $\vec{v}_0 = (10.0\hat{\imath} + 19.6\hat{\jmath})$ m/s. Nancy was moving relative to Mike at velocity $\vec{V} = 10.0\hat{\imath}$ m/s. We can use Equation 6.25 to transform the velocity to Nancy's frame, frame S′, finding

$$v_x' = v_x - V_x = 10.0 \text{ m/s} - 10.0 \text{ m/s} = 0 \text{ m/s}$$

$$v_y' = v_y - V_y = 19.6 \text{ m/s} - 0 \text{ m/s} = 19.6 \text{ m/s}$$

Thus $\vec{v}_0\,' = 19.6\,\hat{\jmath}$ m/s. This agrees with our conclusion from Example 6.7.

It's important to understand the distinction between the three velocities $\vec{v}$, $\vec{v}\,'$, and $\vec{V}$. $\vec{v}$ and $\vec{v}\,'$ are the velocities of an *object* that is observed from both reference frames. Experimenters in S use their meter sticks and stopwatches to measure the object's velocity $\vec{v}$ in their reference frame. At the same time, experimenters in S′ measure the velocity of the same object to be $\vec{v}\,'$. $\vec{V}$ is the relative velocity between two *reference frames*; the velocity of S′ as measured by an experimenter in S. $\vec{V}$ has nothing to do with the object. It may happen that either $\vec{v}$ or $\vec{v}\,'$ is zero, meaning that the object is at rest in one reference frame, but we still must distinguish between the object and the reference frame.

EXAMPLE 6.8 A speeding bullet

The police are chasing a bank robber. While driving at 50 m/s, they fire a bullet to shoot out a tire of his car. The police gun shoots bullets at 300 m/s. What is the bullet's speed as measured by a TV camera crew parked beside the road?

MODEL Assume that all motion is along the x-axis. Let the earth be frame S and a frame attached to the police car be S′. Frame S′ moves relative to frame S with $V_x = 50$ m/s.

SOLVE The bullet is the moving object that will be observed from both frames. The gun is in frame S′, so the bullet travels in this frame with $v_x' = 300$ m/s. We can use Equation 6.25 to transform the bullet's velocity into the earth reference frame:

$$v_x = v_x' + V_x = 300 \text{ m/s} + 50 \text{ m/s} = 350 \text{ m/s}$$

The Galilean velocity transformations are pretty much common sense for one-dimensional motion. Their real usefulness appears when an object travels in a *medium* that moves with respect to the earth. For example, a boat moves relative to the water. What is the boat's net motion if the water is a flowing river? Airplanes fly relative to the air, but the air at high altitudes often flows at high speed. Navigation of boats and planes requires knowing both the motion of the vessel in the medium and the motion of the medium relative to the earth.

EXAMPLE 6.9 Flying to Cleveland I

Cleveland is 300 miles east of Chicago. A plane leaves Chicago flying due east at 500 mph. The pilot forgot to check the weather and doesn't know that the wind is blowing to the south at 50 mph. What is the plane's ground speed? Where is the plane 0.60 hours later, when the pilot expects to land in Cleveland?

MODEL Let the earth be reference frame S. Chicago and Cleveland are at rest in the earth's frame. Let the air be frame S′. If the x-axis points east and the y-axis north, then the air is moving with respect to the earth at $\vec{V} = -50\hat{j}$ mph. The plane flies in the air, so its velocity in frame S′ is $\vec{v}' = 500\hat{i}$ mph.

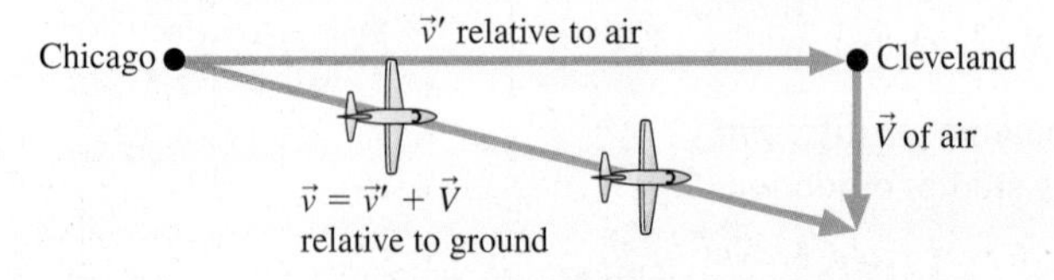

FIGURE 6.28 The wind causes a plane flying due east in the air to move to the southeast relative to the earth.

SOLVE The velocity transformation equation $\vec{v} = \vec{v}' + \vec{V}$ is a vector addition equation. Figure 6.28 shows graphically what happens. Although the nose of the plane points east, the wind carries the plane in a direction somewhat south of east. The plane's velocity relative to the ground is

$$\vec{v} = \vec{v}' + \vec{V} = (500\hat{i} - 50\hat{j})\text{ mph}$$

The plane's ground speed, its speed in frame S, is

$$v = \sqrt{v_x^2 + v_y^2} = 502\text{ mph}$$

After flying for 0.6 hours at this velocity, the plane's location (relative to Chicago) is

$$x = v_x t = (500\text{ mph})(0.6\text{ hr}) = 300\text{ mi}$$

$$y = v_y t = (-50\text{ mph})(0.6\text{ hr}) = -30\text{ mi}$$

The plane is 30 mi due south of Cleveland! Although the pilot thought he was flying to the east, his actual heading has been $\tan^{-1}(V/v) = \tan^{-1}(0.10) = 5.71°$ south of east.

EXAMPLE 6.10 Flying to Cleveland II

A wiser pilot flying from Chicago to Cleveland on the same day plots a course that will take her directly to Cleveland. In which direction does she fly the plane? How long does it take to reach Cleveland?

MODEL Let the earth be reference frame S. Let the air be frame S′. If the x-axis points east and the y-axis north, then the air is moving with respect to the earth at $\vec{V} = -50\hat{j}$ mph.

SOLVE The objective of navigation is to move between two points on the earth's surface, in frame S. The wiser pilot, who knows that the wind will affect her plane, draws the vector picture of Figure 6.29. The plane's velocity in frame S is

$$v_x = v_x' + V_x = (500\text{ mph})\cos\theta$$

$$v_y = v_y' + V_y = (500\text{ mph})\sin\theta - 50\text{ mph}$$

In plotting her course, the pilot knows that she wants $v_y = 0$ in order to fly due east to Cleveland in the earth's frame. To achieve this, she'll actually have to point the nose of the plane somewhat north of east. The proper heading is

$$\theta = \sin^{-1}\left(\frac{50\text{ mph}}{500\text{ mph}}\right) = 5.74°$$

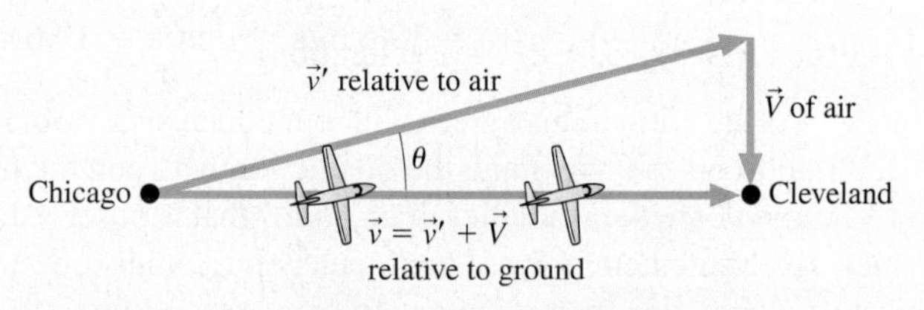

FIGURE 6.29 To travel due east in a south wind, a pilot has to point the plane somewhat to the northeast.

The plane's velocity in frame S is then $\vec{v} = (500\text{ mph})\cos 5.74\hat{i} = 497\hat{i}$ mph. You can see from Figure 6.29 that the plane's speed v in the earth's frame is slower than its speed v' in the air's reference frame. The time needed to fly to Cleveland at this speed is

$$t = \frac{300\text{ mi}}{497\text{ mph}} = 0.604\text{ hr}$$

It takes 0.004 hr = 14 s more time to reach Cleveland than it would on a day without wind.

ASSESS A boat crossing a river or an ocean current faces the same difficulties. These are exactly the kinds of calculations performed by pilots of boats and planes as part of navigation.

The Galilean Principle of Relativity

The most important question we raised at the beginning of this section was "In which reference frames are Newton's laws valid?" That is, in which reference frames does $\vec{F}_{net} = m\vec{a}$?

Suppose a net force $\vec{F}_{net}$ acts on an object in frame S. Further, suppose that Newton's laws have been tested and found valid in frame S. Then experimenters in S will find that the object has acceleration $\vec{a}$ such that $\vec{F}_{net} = m\vec{a}$.

What is the situation in frame S′ that moves relative to frame S with velocity $\vec{V}$? Does $\vec{F}'_{\text{net}} = m\vec{a}'$? To answer this question we must transform the force and acceleration measurements of frame S to frame S′.

Recall that a force is a push or pull, the strength of which can be measured with a spring scale. The strength doesn't depend on a coordinate system. If experimenters in frame S see the scale reading 5 N, experimenters in frame S′ will see the same reading. In other words, the size and direction of a force are the same in all reference frames. Thus $\vec{F}'_{\text{net}} = \vec{F}_{\text{net}}$.

So what about acceleration? The velocity transformation equation, Equation 6.24, is $\vec{v}' = \vec{v} - \vec{V}$. If we take the time derivative of this equation, we find

$$\frac{d\vec{v}'}{dt} = \frac{d\vec{v}}{dt} - \frac{d\vec{V}}{dt}$$
$$\vec{a}' = \vec{a} - \frac{d\vec{V}}{dt} \tag{6.26}$$

where we've used the definitions of $\vec{a}$ and $\vec{a}'$. Now, one of the four conditions we placed on reference frames S and S′ was the requirement that $\vec{V}$ be a *constant* velocity. Thus $d\vec{V}/dt = 0$, and the **Galilean transformation of acceleration** is

$$\vec{a}' = \vec{a} \tag{6.27}$$

Observers in frames S and S′ may measure different positions and velocities for an object, but they *agree* on its acceleration.

Reference frames that move with constant velocity are inertial reference frames. Equation 6.27 tells us that experimenters in two inertial reference frames measure the *same acceleration* for an object. Those experimenters also measure the *same force* exerted on the object. Thus if Newton's laws are tested and found to be valid in any one particular inertial reference—and they have been—we can conclude that Newton's laws are valid in *all* inertial reference frames. This idea is known as the Galilean principle of relativity:

Galilean principle of relativity Newton's laws of motion are valid in all inertial reference frames.

A reference frame that is speeding up, slowing down, or turning is not an inertial reference frame. Equation 6.27 is *not* true if S′ is a noninertial frame because $d\vec{V}/dt \neq 0$. Thus $\vec{F}_{\text{net}} \neq m\vec{a}$ in a noninertial reference frame and Newton's laws cannot be used.

While the Galilean transformations and the Galilean principle of relativity seem to be almost obvious, they're actually based on quite fundamental assumptions about the nature of space and time. Those assumptions were unquestioned—in fact, not even recognized—until the beginning of the 20th century. To see where problems arose, consider the situation in Figure 6.30. Tom is shooting his laser pointer in the positive x-direction. The laser beam is moving away from Tom at the speed of light, $v_x = 3.0 \times 10^8$ m/s.

Sue flies by in her spaceship, traveling in the positive x-direction with velocity $V_x = 2.0 \times 10^8$ m/s. This is the velocity of her reference frame relative to Tom's. Sue sees the laser beam traveling past the window of her spaceship and uses instruments on board her spaceship to measure the velocity of the light. According to Equation 6.25, the Galilean transformation of velocity, Sue *should* find that the light travels with velocity

$$v'_x = v_x - V_x = 1.0 \times 10^8 \text{ m/s}$$

After all, this is really no different from the police and the bullet of Example 6.8.

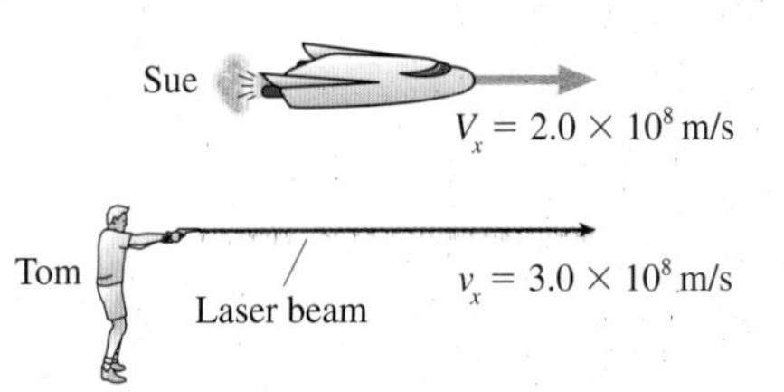

FIGURE 6.30 The laser beam moves away from Tom at the speed of light, $v_x = 3.0 \times 10^8$ m/s. How fast does the laser beam go past Sue's window?

But when the experiment is done, Sue finds that the laser beam travels in her reference frame with $v'_x = 3.0 \times 10^8$ m/s, exactly the same velocity that Tom measured. That is, $v'_x = v_x$! This is the same as if the TV crew and the police in Example 6.8 *both* observed the bullet to have $v_x = 300$ m/s.

Now, it's true that we can't do the experiment exactly as described here, but physicists did figure out over 100 years ago how to measure the speed of light in moving reference frames. Many experiments that are the equivalent of Tom and Sue's have been done, and they all find the same result: **The speed of light is the same in all inertial reference frames, no matter how fast the reference frames are moving with respect to each other.**

This finding is totally at odds with the Galilean transformation of velocity. It is also totally at odds with common sense. It turns out that the universal constancy of the speed of light can be understood only if *time* is different for experimenters moving relative to each other. It was Albert Einstein who, in 1905, first suggested that space and time are more complex and more subtle than is assumed in Newtonian mechanics. This is the basis of his theory of relativity, a topic that we'll return to in Chapter 36.

Fortunately for us, Galilean relativity works extremely well for objects whose speed is much less than the speed of light. Einstein's relativity becomes important only when speeds exceed about 10% of the speed of light. The fastest human spacecraft doesn't come anywhere close to this, so relativity is rarely an issue in engineering or everyday life. But atomic particles such as electrons and protons routinely travel at more than 99% of the speed of light in particle accelerators. Relativity, which is one of the cornerstones of modern physics, is essential for understanding their behavior.

STOP TO THINK 6.4 A plane traveling horizontally to the right at 100 m/s flies past a helicopter that is going straight up at 20 m/s. From the helicopter's perspective, the plane's direction and speed are

a. Right and up, less than 100 m/s.
b. Right and up, 100 m/s.
c. Right and up, more than 100 m/s.
d. Right and down, less than 100 m/s.
e. Right and down, 100 m/s.
f. Right and down, more than 100 m/s.

SUMMARY

The goal of Chapter 6 has been to learn to solve problems about motion in a plane.

GENERAL PRINCIPLES

Galilean Principle of Relativity

Newton's laws of motion are valid in all inertial reference frames.

Newton's Second Law

Expressed in x- and y-component form:

$$(F_{net})_x = \sum F_x = ma_x$$

$$(F_{net})_y = \sum F_y = ma_y$$

IMPORTANT CONCEPTS

Relative motion

Inertial reference frames move relative to each other with constant velocity $\vec{V}$. Measurements of position and velocity measured in frame S are related to measurements in frame S′ by the Galilean transformations

$$x' = x - V_x t \qquad v_x' = v_x - V_x$$

$$y' = y - V_y t \qquad v_y' = v_y - V_y$$

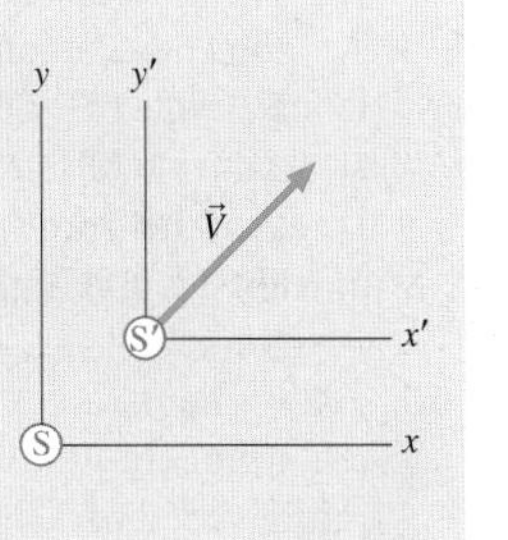

The instantaneous velocity

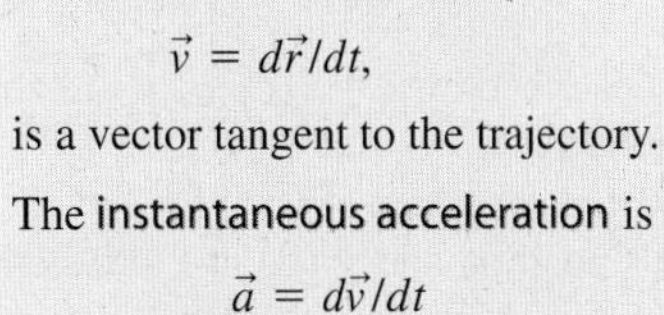

$$\vec{v} = d\vec{r}/dt,$$

is a vector tangent to the trajectory.

The instantaneous acceleration is

$$\vec{a} = d\vec{v}/dt$$

$\vec{a}_\parallel$, the component of $\vec{a}$ parallel to $\vec{v}$, is responsible for change of *speed.* $\vec{a}_\perp$, the component of $\vec{a}$ perpendicular to $\vec{v}$, is responsible for change of *direction.*

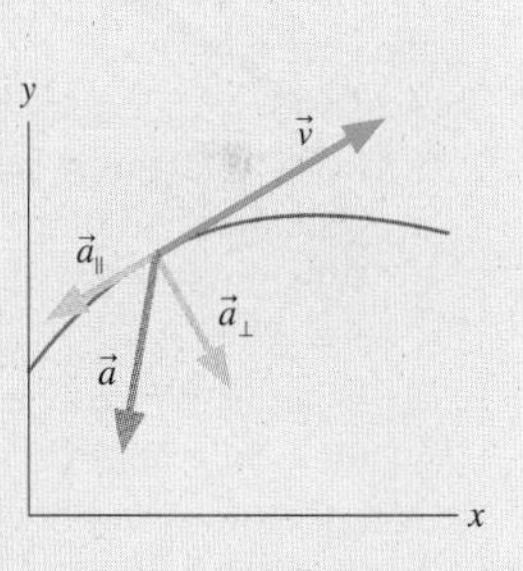

APPLICATIONS

Kinematics in two dimensions

If $\vec{a}$ is constant, then the x- and y-components of motion are independent of each other. For a particle that starts from initial position $\vec{r}_i$ and velocity $\vec{v}_i$, its position and velocity at a final point f are

$$x_f = x_i + v_{ix}\Delta t + \tfrac{1}{2}a_x(\Delta t)^2$$

$$y_f = y_i + v_{iy}\Delta t + \tfrac{1}{2}a_y(\Delta t)^2$$

$$v_{fx} = v_{ix} + a_x\Delta t$$

$$v_{fy} = v_{iy} + a_y\Delta t$$

Projectile motion occurs if the only force on the object is its weight.

- Uniform motion in the horizontal direction with $v_{0x} = v_0 \cos\theta$.
- Free-fall motion in the vertical direction with $a_y = -g$ and $v_{0y} = v_0 \sin\theta$.
- The combined motion is a parabola.
- The x and y kinematic equations have the *same* value for Δt.

TERMS AND NOTATION

projectile
launch angle, θ
reference frame
inertial reference frame
Galilean transformation of position
Galilean transformation of velocity
Galilean transformation of acceleration
Galilean principle of relativity

EXERCISES AND PROBLEMS

The icon in front of a problem indicates that the problem can be done on a Dynamics Worksheet.

Exercises

Section 6.1 Kinematics in Two Dimensions

1. A particle moves in the xy-plane with constant acceleration. The particle is located at $\vec{r} = (2\hat{\imath} + 4\hat{\jmath})$ m at $t = 0$ s. At $t = 3$ s it is at $\vec{r} = (8\hat{\imath} - 2\hat{\jmath})$ m and has velocity $\vec{v} = (5\hat{\imath} - 5\hat{\jmath})$ m/s.
   a. What is the particle's acceleration vector $\vec{a}$?
   b. What are its position, velocity, and speed at $t = 5$ s?
2. A sailboat is traveling east at 5.0 m/s. A sudden gust of wind gives the boat an acceleration $\vec{a} =$ (0.80 m/s^2, 40° north of east). What are the boat's speed and direction 6.0 s later when the gust subsides?
3. A particle's trajectory is described by $x = (\frac{1}{2}t^3 - 2t^2)$ m and $y = (\frac{1}{2}t^2 - 2t)$ m, where t is in s.
   a. Calculate and plot the trajectory from $t = -2$ s to $t = 5$ s.
   b. What are the particle's position and speed at $t = 0$ s and $t = 4$ s?
   c. What is the particle's direction of motion, measured from the x-axis, at $t = 0$ s and $t = 4$ s?
4. A rocket-powered hockey puck moves on a horizontal frictionless table. Figure Ex6.4 shows graphs of v_x and v_y, the x- and y-components of the puck's velocity. The puck starts from the origin.
   a. In which direction is the puck moving at $t = 2.0$ s?
   b. How far from the origin is the puck at $t = 5.0$ s?

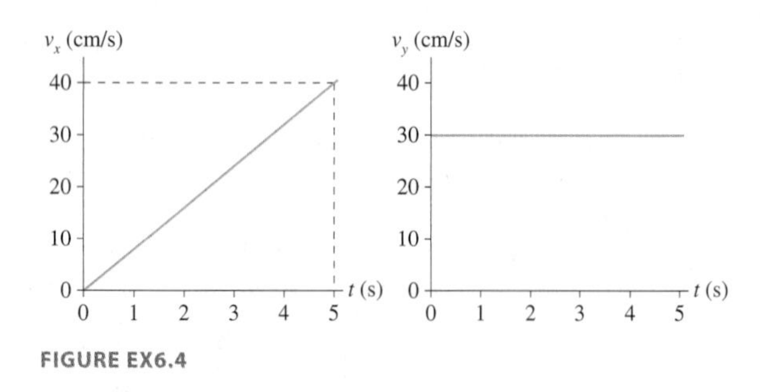

FIGURE EX6.4

Section 6.2 Dynamics in Two Dimensions

5. As a science fair project, you want to launch an 800 g model rocket straight up and hit a horizontally moving target as it passes 30 m above the launch point. The rocket engine provides a constant thrust of 15.0 N. The target is approaching at a speed of 15 m/s. At what horizontal distance between the target and the rocket should you launch?
6. A 500 g model rocket is on a cart that is rolling to the right at a speed of 3.0 m/s. The rocket engine, when it is fired, exerts an 8.0 N thrust on the rocket. Your goal is to have the rocket pass through a small horizontal hoop that is 20 m above the launch point. At what horizontal distance left of the hoop should you launch?

Section 6.3 Projectile Motion

7. For a projectile, which of the following quantities are constant during the flight: x, y, r, v_x, v_y, v, a_x, a_y, F_x, F_y? Which of the quantities are zero throughout the flight?
8. A physics student on Planet Exidor throws a ball, and it follows the parabolic trajectory shown in Figure Ex6.8. The ball's position is shown at 1 s intervals until $t = 3$ s. At $t = 1$ s, the ball's velocity is $\vec{v} = (2.0\hat{\imath} + 2.0\hat{\jmath})$ m/s.

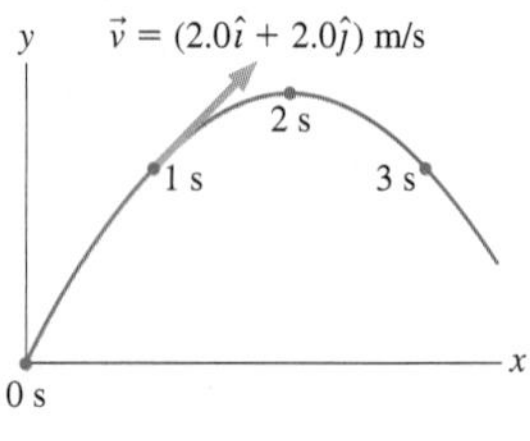

FIGURE EX6.8

   a. Determine the ball's velocity at $t = 0$ s, 2 s, and 3 s.
   b. What is the value of g on Planet Exidor?
   c. What was the ball's launch angle?
9. An object is launched with an initial velocity of 50 m/s at a launch angle of 36.9° above the horizontal.
   a. Make a table showing values of x, y, v_x, v_y, and the speed v every 1 s from $t = 0$ s to $t = 6$ s.
   b. Plot a graph of the object's trajectory during the first 6 s of motion.
10. Two spheres are launched horizontally from a 1.0-m-high table. Sphere A has a mass of 1.0 kg and is launched with an initial speed of 5.0 m/s. Sphere B has a mass of 0.4 kg and is launched with an initial speed of 2.5 m/s.
   a. Compare the times for each sphere to hit the floor.
   b. Compare the distances that each travels from the edge of the table.
11. A rifle is aimed horizontally at a target 50 m away. The bullet hits the target 2.0 cm below the aim point.
   a. What was the bullet's flight time?
   b. What was the bullet's speed as it left the barrel?
12. A ball thrown horizontally at 25 m/s travels a horizontal distance of 50 m before hitting the ground. From what height was the ball thrown?

Section 6.4 Relative Motion

13. A boat takes 3.0 hours to travel 30 km down a river, then 5.0 hours to return. How fast is the river flowing?
14. When the moving sidewalk at the airport is broken, as it often seems to be, it takes you 50 s to walk from your gate to baggage claim. When it is working and you stand on the moving sidewalk the entire way, without walking, it takes 75 s to travel the same distance. How long will it take you to travel from the gate to baggage claim if you walk while riding on the moving sidewalk?
15. An assembly line has a staple gun that rolls to the left at 1.0 m/s while parts to be stapled roll past it to the right at 3.0 m/s. The staple gun fires 10 staples per second. How far apart are the staples in the finished part?
16. Ted is sitting in his lawn chair when Stella flies directly overhead, going southeast at 100 m/s. Five seconds later, a firecracker explodes 200 m east of Ted. What are the coordinates of the explosion in Stella's reference frame? Let Stella be at the origin, with her x-axis pointing to the east.

17. Ships A and B leave port together. For the next two hours, ship A travels at 20 mph in a direction 30° west of north while the ship B travels 20° east of north at 25 mph.
    a. What is the distance between the two ships two hours after they depart?
    b. What is the speed of ship A as seen by ship B?

Problems

18. A particle starts from rest at $\vec{r}_0 = 9.0\hat{j}$ m and moves in the xy-plane with the velocity shown in Figure P6.18. The particle passes through a wire hoop located at $\vec{r}_1 = 20\hat{i}$ m, then continues onward.
    a. At what time does the particle pass through the hoop?
    b. What is the value of v_{4y}, the y-component of the particle's velocity at $t = 4$ s?
    c. Calculate and plot the particle's trajectory from $t = 0$ s to $t = 4$ s.

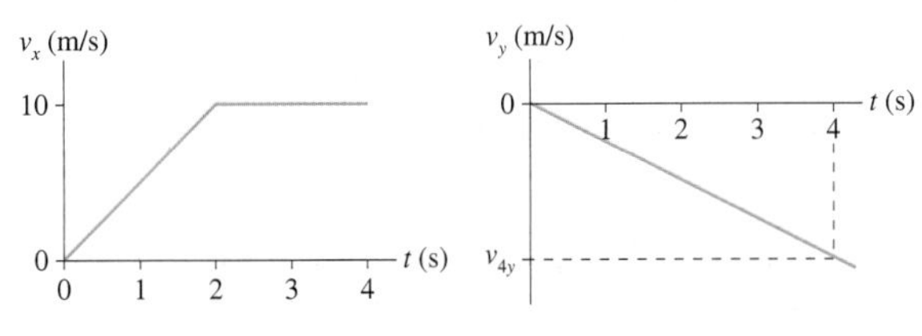

FIGURE P6.18

19. A 4.0×10^{10} kg asteroid is heading directly toward the center of the earth at a steady 20 km/s. To save the planet, astronauts strap a giant rocket to the asteroid perpendicular to its direction of travel. The rocket generates 5.0×10^9 N of thrust. The rocket is fired when the asteroid is 4.0×10^6 km away from earth. You can ignore the rotational motion of the earth and asteroid around the sun.
    a. If the mission fails, how many hours is it until the asteroid impacts the earth?
    b. The radius of the earth is 6400 km. By what minimum angle must the asteroid be deflected to just miss the earth?
    c. The rocket fires at full thrust for 300 s before running out of fuel. Is the earth saved?
20. A rocket-powered hockey puck on frictionless ice, such as in Example 6.3, slides along the y-axis with speed v_0. The front of the rocket is tilted at angle θ from the x-axis. The rocket motor ignites as the puck crosses the origin, exerting force $\vec{F}_{thrust}$ on the puck.
    a. Find an algebraic expression $y(x)$ for the puck's trajectory.
    b. Suppose the thrust is 2.0 N, the combined mass of the puck and rocket is 1.0 kg, and the initial speed is 2.0 m/s. Make a graph of your function $y(x)$ from $x = 0$ to $x = 20$ m for the two cases $\theta = 45°$ and $\theta = -45°$.
21. You are asked to consult for the city's research hospital, where a group of doctors is investigating the bombardment of cancer tumors with high-energy ions. The ions are fired directly toward the center of the tumor at speeds of 5.0×10^6 m/s. To cover the entire tumor area, the ions are deflected sideways by passing them between two charged metal plates that accelerate the ions perpendicular to the direction of their initial motion. The acceleration region is 5.0 cm long, and the ends of the acceleration plates are 1.5 m from the patient. What acceleration is required to move an ion 2.0 cm across the tumor?

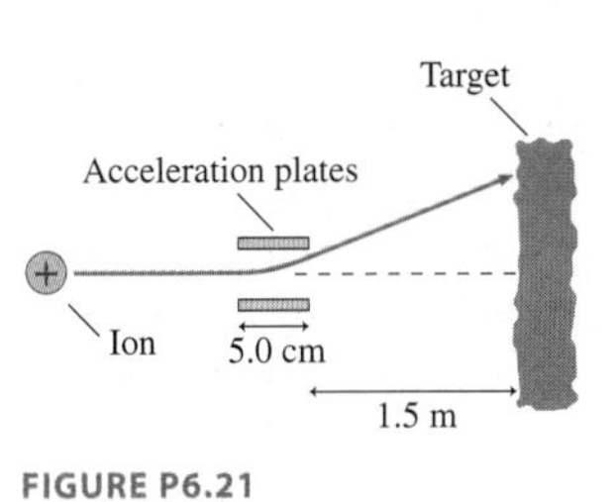

FIGURE P6.21

22. A projectile's horizontal range on level ground is $R = v_0^2 \sin 2\theta / g$. At what launch angle or angles will the projectile land at half of its maximum possible range?
23. a. A projectile is launched with speed v_0 and angle θ. Derive an expression for the projectile's maximum height h.
    b. A baseball is hit with a speed of 33.6 m/s. Calculate its height and the distance traveled if it is hit at angles of 30°, 45°, and 60°.
24. A sailor climbs to the top of the mast, 15 m above the deck, to look for land while his ship moves steadily forward through calm waters at 4.0 m/s. Unfortunately, he drops his spyglass to the deck below.
    a. Where does it land with respect to the base of the mast below him?
    b. Where does it land with respect to a fisherman sitting at rest in his dinghy as the ship goes past? Assume that the fisherman is even with the mast at the instant the spyglass is dropped.
25. A projectile is fired with an initial speed of 30 m/s at an angle of 60° above the horizontal. The object hits the ground 7.5 s later.
    a. How much higher or lower is the launch point relative to the point where the projectile hits the ground?
    b. To what maximum height above the launch point does the projectile rise?
    c. What are the magnitude and direction of the projectile's velocity at the instant it hits the ground?
26. In the Olympic shotput event, an athlete throws the shot with an initial speed of 12 m/s at a 40.0° angle from the horizontal. The shot leaves her hand at a height of 1.8 m above the ground.
    a. How far does the shot travel?
    b. Repeat the calculation of part (a) for angles 42.5°, 45.0°, and 47.5°. Put all your results, including 40.0°, in a table. At what angle of release does she throw the farthest?
27. On the Apollo 14 mission to the moon, astronaut Alan Shepard hit a golf ball with a 6 iron. The acceleration due to gravity on the moon is 1/6 of its value on earth. Suppose he hits the ball with a speed of 25 m/s at an angle 30° above the horizontal.
    a. How much farther did the ball travel on the moon than it would have on earth?
    b. For how much more time was the ball in flight?
28. A ball is thrown toward a cliff of height h with a speed of 30 m/s and an angle of 60° above horizontal. It lands on the edge of the cliff 4.0 s later.
    a. How high is the cliff?
    b. What was the maximum height of the ball?
    c. What is the ball's impact speed?
29. A tennis player hits a ball 2.0 m above the ground. The ball leaves his racquet with a speed of 20.0 m/s at an angle 5° above the horizontal. The horizontal distance to the net is 7.0 m, and the net is 1.0 m high. Does the ball clear the net? If so, by how much? If not, by how much does it miss?
30. A baseball player friend of yours wants to determine his pitching speed. You have him stand on a ledge and throw the ball

horizontally from an elevation 4.0 m above the ground. The ball lands 25 m away.

a. What is his pitching speed?
b. As you think about it, you're not sure he threw the ball exactly horizontally. As you watch him throw, the pitches seem to vary from 5° below horizontal to 5° above horizontal. What is the *range* of speeds with which the ball might have left his hand?

31. You are playing right field for the baseball team. Your team is up by one run in the bottom of the last inning of the game when a ground ball slips through the infield and comes straight toward you. As you pick up the ball 65 m from home plate, you see a runner rounding third base and heading for home with the tying run. You throw the ball at an angle of 30° above the horizontal with just the right speed so that the ball is caught by the catcher, standing on home plate, at the same height as you threw it. As you release the ball, the runner is 20.0 m from home plate and running full speed at 8.0 m/s. Will the ball arrive in time for your team's catcher to make the tag and win the game?

32. A stunt man drives a car at a speed of 20 m/s off a 30-m-high cliff. The road leading to the cliff is inclined upward at an angle of 20°.
a. How far from the base of the cliff does the car land?
b. What is the car's impact speed?

33. In one contest at the county fair, a spring-loaded plunger launches a ball at a speed of 3.0 m/s from one corner of a smooth, flat board that is tilted up at a 20° angle. To win, you must make the ball hit a small target at the adjacent corner, 2.50 m away. At what angle θ should you tilt the ball launcher?

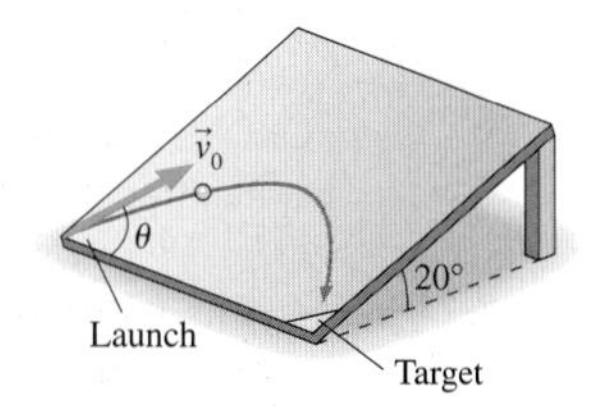

FIGURE P6.33

34. You're 6.0 m from one wall of a house. You want to toss a ball to your friend who is 6.0 m from the opposite wall. The throw and catch each occur 1.0 m above the ground.
a. What minimum speed will allow the ball to clear the roof?
b. At what angle should you toss the ball?

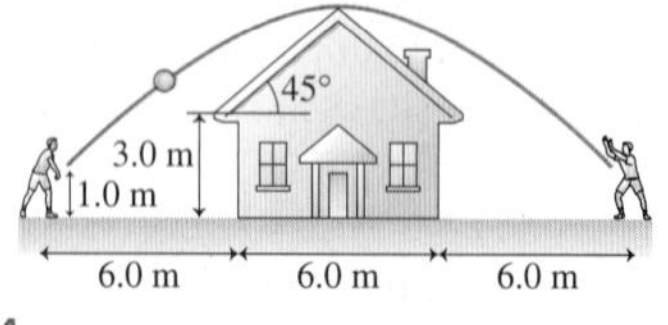

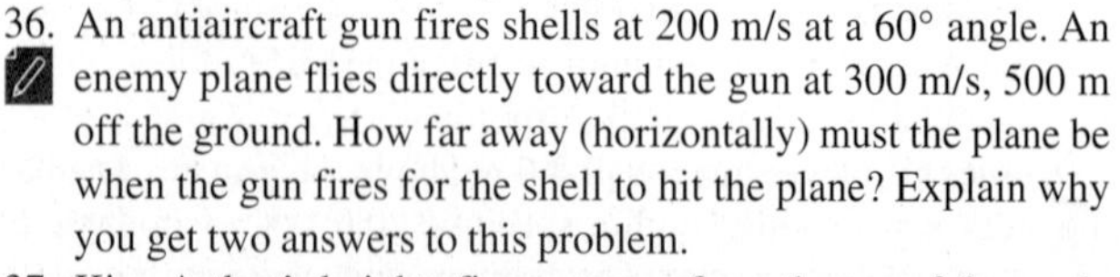
FIGURE P6.34

35. A supply plane needs to drop a package of food to scientists working on a glacier in Greenland. The plane flies 100 m above the glacier at a speed of 150 m/s. How far short of the target should it drop the package?

36. An antiaircraft gun fires shells at 200 m/s at a 60° angle. An enemy plane flies directly toward the gun at 300 m/s, 500 m off the ground. How far away (horizontally) must the plane be when the gun fires for the shell to hit the plane? Explain why you get two answers to this problem.

37. King Arthur's knights fire a cannon from the top of the castle wall. The cannonball is fired at a speed of 50 m/s and an angle of 30°. A cannonball that was accidentally dropped hits the moat below in 1.5 s.
a. How far from the castle wall does the cannonball hit the ground?
b. What is the ball's maximum height above the ground?

38. Quarterback Fred is going to throw a pass to tight end Doug. Doug is 20 m in front of Fred and running straight away at 6.0 m/s when Fred throws the 500 g football at a 40° angle. Doug catches the ball without having to alter his speed and runs for the game-winning touchdown. How fast did Fred throw the ball?

39. You are watching an archery tournament when you start wondering how fast an arrow is shot from the bow. Remembering your physics, you ask one of the archers to shoot an arrow parallel to the ground. You find the arrow stuck in the ground 60 m away, making a 3° angle with the ground. How fast was the arrow shot?

40. An archer standing on a 15° slope shoots an arrow 20° above the horizontal, as shown in Figure P6.40. How far down the slope does the arrow hit if it is shot with a speed of 50 m/s from 1.75 m above the ground?

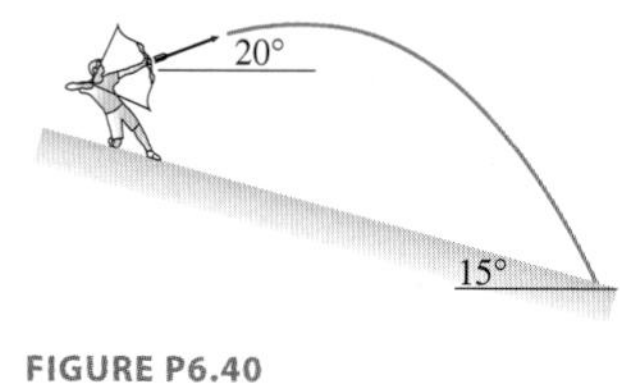

FIGURE P6.40

41. A popular pastime is to see who can push an object closest to the edge of a table without its going off. You push the 100 g object and release it 2.0 m from the table edge. Unfortunately, you push a little too hard. The object slides across, sails off the edge, falls 1.0 m to the floor, and lands 30 cm from the edge of the table. If the coefficient of kinetic friction is 0.5, what was the object's speed as you released it?

42. Sand moves without slipping at 6.0 m/s down a conveyer that is tilted at 15°. The sand enters a pipe 3.0 m below the end of the conveyer belt, as shown in Figure P6.42. What is the horizontal distance d between the conveyer belt and the pipe?

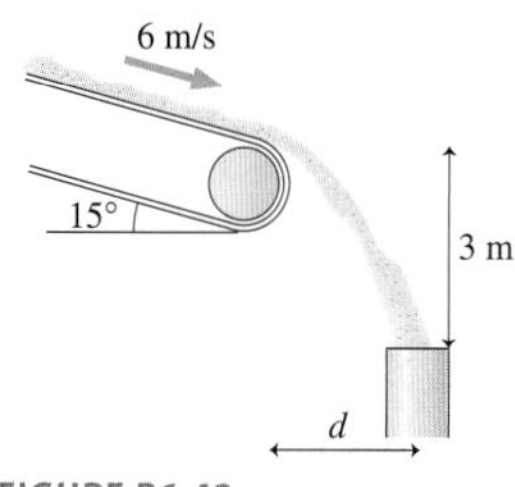

FIGURE P6.42

43. Sam (75 kg) takes off up a 50-m-high, 10° frictionless slope on his jet-powered skis. The skis have a thrust of 200 N. He keeps his skis tilted at 10° after becoming airborne, as shown in Figure P6.43. How far does Sam land from the base of the cliff?

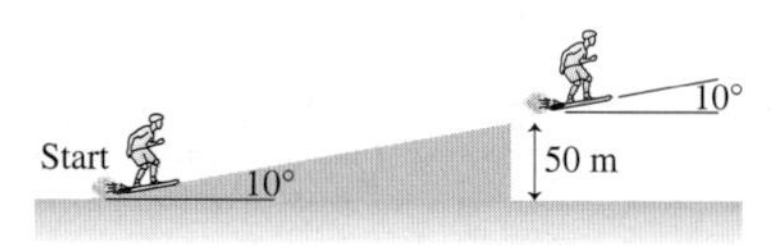

FIGURE P6.43

44. A skateboarder starts up a 1.0-m-high, 30° ramp at a speed of 7.0 m/s. The skateboard wheels roll without friction. How far from the end of the ramp does the skateboarder touch down?

45. A motorcycle daredevil plans to ride up a 2.0-m-high, 20° ramp, sail across a 10-m-wide pool filled with hungry crocodiles, and land at ground level on the other side. He has done this stunt many times and approaches it with confidence. Unfortunately, the motorcycle engine dies just as he starts up the ramp. He is going 11 m/s at that instant, and the rolling

friction of his rubber tires is not negligible. Does he survive, or does he become crocodile food?

46. A 5000 kg interceptor rocket is launched at an angle of 44.7°. The thrust of the rocket motor is 140,700 N.
    a. Find an equation $y(x)$ that describes the rocket's trajectory.
    b. What is the shape of the trajectory?
    c. At what elevation does the rocket reach the speed of sound, 330 m/s?
47. A rocket-powered hockey puck has a thrust of 2.0 N and a total mass of 1.0 kg. It is released from rest on a frictionless table, 4.0 m from the edge of a 2.0 m drop. The front of the rocket is pointed directly toward the edge. How far does the puck land from the base of the table?
48. A 500 g model rocket is resting horizontally at the top edge of a 40-m-high wall when it is accidentally bumped. The bump pushes it off the edge with a horizontal speed of 0.5 m/s and at the same time causes the engine to ignite. When the engine fires, it exerts a constant 20 N horizontal thrust away from the wall.
    a. How far from the base of the wall does the rocket land?
    b. Describe the trajectory of the rocket while it travels to the ground.

In Problems 49 through 51 you are given the equations that are used to solve a problem. For each of these, you are to

a. Write a realistic problem for which these are the correct equations. Be sure that the answer your problem requests is consistent with the equations given.
b. Finish the solution of the problem.

49. $100 \text{ m} = 0 \text{ m} + (50\cos\theta \text{ m/s})t_1$

    $0 \text{ m} = 0 \text{ m} + (50\sin\theta \text{ m/s})t_1 - \frac{1}{2}(9.80 \text{ m/s}^2)t_1^2$
50. $x_1 = 0 \text{ m} + (30 \text{ m/s})t_1$

    $0 \text{ m} = 300 \text{ m} - \frac{1}{2}(9.80 \text{ m/s}^2)t_1^2$
51. $v_x = -(6.0\cos 45°) \text{ m/s} + 3.0 \text{ m/s}$

    $v_y = (6.0\sin 45°) \text{ m/s} + 0 \text{ m/s}$

    $100 \text{ m} = v_y t_1$

    $x_1 = v_x t_1$
52. Write a realistic problem for which the x-versus-t and y-versus-t graphs shown in Figure P6.52 represent the motion of an object. Be sure the answer your problem requests is consistent with the graphs. Then finish the solution of the problem.

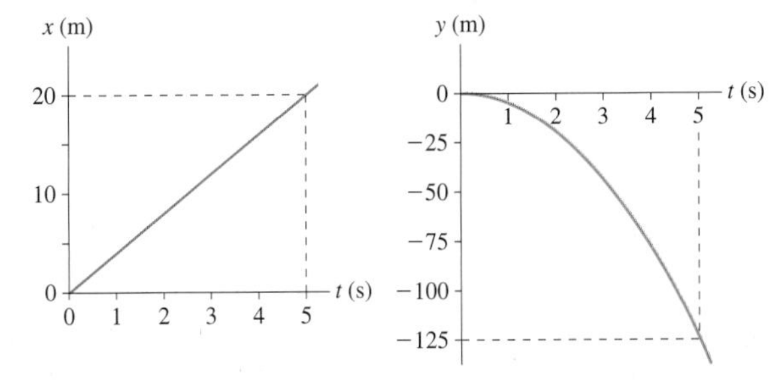

FIGURE P6.52

53. Mary needs to row her boat across a 100-m-wide river that is flowing to the east at a speed of 3.0 m/s. Mary can row with a speed of 2.0 m/s.
    a. If Mary rows straight north, where will she land?
    b. Draw a picture showing her displacement due to rowing, her displacement due to the river's motion, and her net displacement.
54. A kayaker needs to paddle north across a 100-m-wide harbor. The tide is going out, creating a tidal current that flows to the east at 2.0 m/s. The kayaker can paddle with a speed of 3.0 m/s.
    a. In which direction should he paddle in order to travel straight across the harbor?
    b. How long will it take him to cross?
55. Mike throws a ball upward at a 63° angle with a speed of 22 m/s. Nancy drives past Mike at 30 m/s at the instant he releases the ball.
    a. What is the ball's initial angle in Nancy's reference frame?
    b. Find and graph the ball's trajectory as seen by Nancy.
56. A sailboat is sailing due east at 8.0 mph. The wind appears to blow from the southwest at 12.0 mph.
    a. What are the true wind speed and direction?
    b. What are the true wind speed and direction if the wind appears to blow from the northeast at 12.0 mph?
57. A child in danger of drowning in a river is being carried downstream by a current that flows uniformly with a speed of 2.0 m/s. The child is 200 m from the shore and 1500 m upstream of the boat dock from which the rescue team sets out. If their boat speed is 8.0 m/s with respect to the water, at what angle should the pilot leave the shore to go directly to the child?
58. Quarterback Fred is going to throw a pass to tight end Alberto. Fred is being chased, however, and he throws the ball at 20 m/s while running directly toward the nearest sideline at 4.0 m/s. Alberto is standing still directly upfield from Fred at the moment the ball is released. In what direction should Fred throw so that Alberto can catch the ball without moving?
59. The paper delivery boy tries to throw the paper into your narrow driveway without slowing down. His pickup truck travels at 10 mph, and he throws the paper at 20 mph just as the truck passes the driveway.
    a. In what direction should he throw the paper in order for it to land in the driveway?
    b. What is the paper's speed relative to the ground?
60. While driving north at 25 m/s during a rainstorm you notice that the rain makes an angle of 38° with the vertical. While driving back home moments later at the same speed but in the opposite direction, you see that the rain is falling straight down. From these observations, determine the speed and angle of the raindrops relative to the ground.
61. A plane has an airspeed of 200 mph. The pilot wishes to reach a destination 600 mi due east, but a wind is blowing at 50 mph in the direction 30° north of east.
    a. In what direction must the pilot head the plane in order to reach her destination?
    b. How long will the trip take?
62. Susan, driving north at 60 mph, and Shawn, driving east at 45 mph, are approaching an intersection. What is Shawn's speed relative to Susan's reference frame?
63. As is discussed more fully in Chapter 42, one of the processes by which a radioactive nucleus can decay to a more stable state is by the emission of a very energetic particle called a gamma-ray photon. Like the laser beam discussed in this chapter, the gamma-ray photon is a form of light and travels at the speed of light, 3.0×10^8 m/s. Suppose a radioactive nucleus emits a gamma-ray photon while moving through the laboratory toward the east at 1.5×10^8 m/s. If the photon is emitted toward the east, parallel to the motion of the nucleus, what is the speed of the gamma-ray photon as determined by a scientist in the laboratory reference frame?

Challenge Problems

64. In the absence of air resistance, a projectile that lands at the elevation from which it was launched achieves maximum range when launched at a 45° angle. Suppose a projectile of mass m is launched with speed v_0 into a headwind that exerts a constant, horizontal retarding force $\vec{F}_{\text{wind}} = -F_{\text{wind}}\hat{\imath}$.
   a. Find an expression for the angle at which the range is maximum.
   b. By what percentage is the maximum range of a 0.50 kg ball reduced if $F_{\text{wind}} = 0.60$ N?
65. You have been hired to assist with stunts for a new action movie. In one scene, the writers want to drop a package from a small plane into a moving convertible sports car. The car will drive along a horizontal road at 30 m/s. The plane will approach the car from behind at an altitude of 60 m and a speed of 50 m/s relative to the ground. The copilot will view the car through a sighting tube that measures the angle below horizontal. At what angle should the package be released?
66. A cat is chasing a mouse. The mouse runs in a straight line at a speed of 1.5 m/s. If the cat leaps off the floor at a 30° angle and a speed of 4.0 m/s, at what distance behind the mouse should the cat leap in order to land on the poor mouse?
67. A rubber ball is dropped onto a ramp that is tilted at 20°, as shown in Figure CP6.67. A bouncing ball obeys the "law of reflection," which says that the ball leaves the surface at the same angle it approached the surface. The ball's next bounce is 3 m to the right of its first bounce. What is the ball's rebound speed on its first bounce?

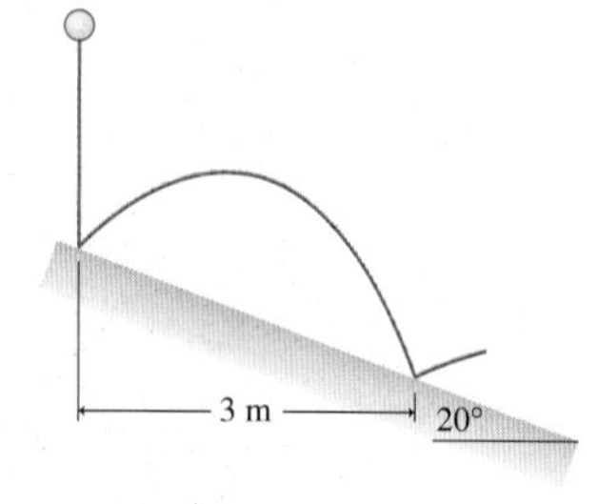

FIGURE CP6.67

68. A motorcycle daredevil wants to set a record for jumping over burning school buses. He has hired you to help with the design. He intends to ride off a horizontal platform at 40 m/s, cross the burning buses in a pit below him, then land on a ramp sloping down at 20°. It's very important that he not bounce when he hits the landing ramp because that could cause him to lose control and crash. You immediately recognize that he won't bounce if his velocity is parallel to the ramp as he touches down. This can be accomplished if the ramp is tangent to his trajectory *and* if he lands right on the front edge of the ramp. There's no room for error! Your task is to determine where to place the landing ramp. That is, how far from the edge of the launching platform should the front edge of the landing ramp be horizontally and how far below it? There's a clause in your contract that requires you to test your design before the hero goes on national television to set the record.
69. Driving a spaceship isn't as easy as it looks in the movies. Imagine you're a physics student in the 31st century. You live in a remote space colony where the gravitational force from any stars or planets is negligible. You're on your way home from school, coasting along in your 20,000 kg personal spacecraft at 2.0 km/s, when the computer alerts you to the fact that the entrance to your pod is 500 km away along a line 30° from your present heading, as shown in Figure CP6.69. You need to make a left turn so that you can enter the pod going straight ahead at 1.0 km/s. You could do this with a series of small rocket burns, but you want to impress the girls in the spacecraft behind you by getting through the entrance with a single rocket burn. You can use small thrusters to quickly rotate your spacecraft to a different orientation before and after the main rocket burn.

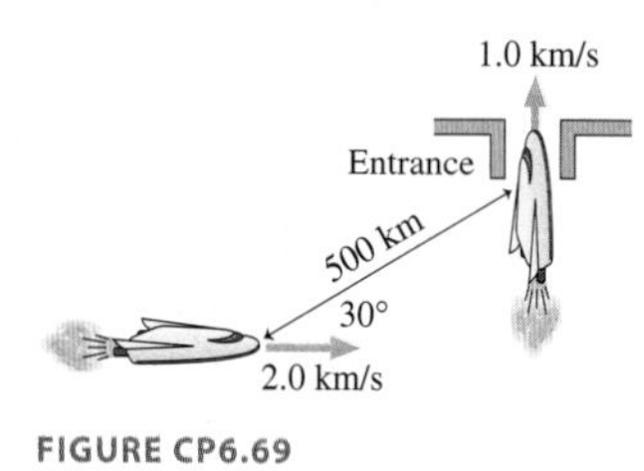

FIGURE CP6.69

   a. You need to determine three things: How to orient your spacecraft for the main rocket burn, the magnitude F_{thrust} of the rocket burn, and the length of the burn. Use a coordinate system in which you start at the origin and are initially moving along the x-axis. Measure the orientation of your spacecraft by the angle it makes with the positive x-axis. Your initial orientation is 0°. You can end the burn before you reach the entrance, but you're not allowed to have the engine on as you pass through the entrance. Mass loss during the burn is negligible.
   b. Calculate your position coordinates every 50 s until you reach the entrance, then plot a graph of your trajectory. Be sure to label the position of the entrance.
70. Uri is on a flight from Boston to Los Angeles. His plane is traveling 20° south of west at 500 mph. Val is on a flight from Miami to Seattle. Her plane is traveling 30° north of west at 500 mph. Somewhere over Kansas, Uri's plane passes 1000 ft directly over Val's plane. Uri is sitting on the right side and can see Val's plane below him after they pass. Uri notices that the fuselage of Val's plane doesn't point in the direction that her plane is moving. What is the angle between the fuselage and the direction of motion?

STOP TO THINK ANSWERS

Stop to Think 6.1: d. The parallel component of $\vec{a}$ is opposite $\vec{v}$ and will cause the particle to slow down. The perpendicular component of $\vec{a}$ will cause the particle to change directions in a downward direction.

Stop to Think 6.2: b. A parabola requires acceleration in one direction but not the other. The horizontal motion is constant velocity, with $a_x = 0$.

Stop to Think 6.3: d. A projectile's acceleration $\vec{a} = -g\hat{\jmath}$ does not depend on its mass. The second ball has the same initial velocity and same acceleration, so it follows the same trajectory and lands at the same position.

Stop to Think 6.4: f. The helicopter frame S′ moves with $\vec{V} = 20\hat{\jmath}$ m/s relative to the earth frame S. The plane moves with $\vec{v} = 100\hat{\imath}$ m/s in the earth's frame. The vector addition in the figure shows that $\vec{v}\,'$ is longer than $\vec{v}$.

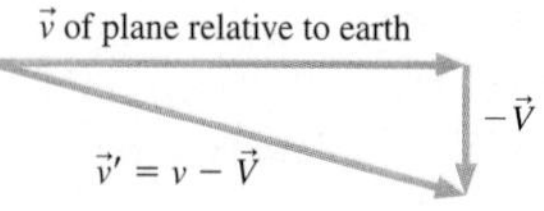

7 Dynamics III: Motion in a Circle

Why doesn't the roller coaster fall off the track at the top of the loop?

► **Looking Ahead**

The goal of Chapter 7 is to learn to solve problems about motion in a circle. In this chapter you will learn to:

- Understand the mathematics of circular kinematics.
- Use Newton's laws to analyze the dynamics of circular motion.
- Understand circular orbits of satellites and planets.
- Think about apparent weight and fictitious forces for objects in circular motion.

◄ **Looking Back**

This chapter, like Chapter 6, extends ideas of one-dimensional kinematics and dynamics into two dimensions. Please review:

- Sections 1.5–1.6 Finding acceleration vectors on a motion diagram.
- Sections 2.2 and 2.5 Uniform motion and constant-acceleration kinematics.
- Sections 5.2 and 5.3 Newton's second law and apparent weight.
- Section 6.1 Kinematics in two dimensions.

A roller coaster doing a loop-the-loop is a dramatic example of circular motion. But why doesn't it fall off the track when it's upside-down at the top of the loop? How is it that you can swing a bucket of water over your head without the water falling out? Why does your car "spin out" of a curve on an wet or icy road but not when the pavement is dry?

To answer these questions, we must study how objects move in circles. We can understand circular motion in terms of forces and Newton's laws, but first we'll need to develop some new concepts and mathematical tools for describing motion in a circle. We'll be able to apply these new ideas to a wide variety of interesting and important problems, from a car rounding a curve to the orbit of the earth around the sun.

Although this chapter will continue to focus on the motion of particles, circular motion is a prelude to studying the rotational motion of solid objects. The ideas introduced in this chapter will be the basis for Chapter 13, "Rotation of a Rigid Body."

7.1 Uniform Circular Motion

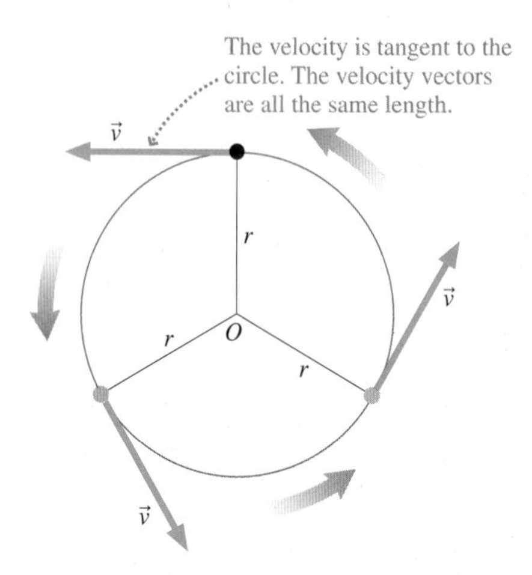
The velocity is tangent to the circle. The velocity vectors are all the same length.
$\vec{v}$
r
O
r
r
$\vec{v}$
$\vec{v}$

FIGURE 7.1 A particle in uniform circular motion.

To begin, consider a particle that moves at *constant speed* around a circle of radius r. This is called **uniform circular motion.** Our first task is to *describe* the motion, and we will do so in this section and the next by developing the kinematics of uniform circular motion. We'll need some new mathematical ideas because the xy-coordinate system we've been using is not the best coordinate system for circular motion. Our second task, in Section 7.3, will be to *explain* the motion in terms of forces.

Figure 7.1 shows a particle moving around a circle of radius r. The particle might be a satellite moving in an orbit, a ball on the end of a string, or even just a dot painted on the side of a wheel. Regardless of what the particle represents, its velocity vector $\vec{v}$ is always tangent to the circle. The particle's speed v is constant, so the vector $\vec{v}$ stays the same length as the particle moves around the circle.

Period

The time interval it takes the particle to go around the circle once, completing one revolution (abbreviated rev), is called the **period** of the motion. Period is represented by the symbol T. This is a logical symbol, because period is an interval of time, but there's a risk of confusing the period T with the symbol $\vec{T}$ for tension or, even worse, the identical symbol T for the magnitude of the tension force.

NOTE ► The number of symbols used in science and engineering far exceeds the number of letters in the English alphabet. Even after we've borrowed from the Greek alphabet, it's inevitable that some letters are used several times to represent entirely different quantities. The use of T is the first time we've run into this problem, but it won't be the last. You must be alert to the *context* of a symbol's use in order to deduce its meaning. Always remembering to use vector arrows over vector symbols, such as $\vec{T}$, helps to clarify the meaning of symbols. ◄

It's easy to relate the particle's period T to its speed v. For a particle moving with constant speed, speed is simply distance/time. In one period, the particle moves once around a circle of radius r and travels the circumference $2\pi r$. Thus

$$v = \frac{1 \text{ circumference}}{1 \text{ period}} = \frac{2\pi r}{T} \tag{7.1}$$

EXAMPLE 7.1 A rotating crankshaft

A 4.0-cm-diameter crankshaft turns at 2400 rpm (revolutions per minute). What is the speed of a point on the surface of the crankshaft?

SOLVE We need to determine the time it takes the crankshaft to make 1 rev. First, convert 2400 rpm to revolutions per second:

$$\frac{2400 \text{ rev}}{1 \text{ min}} \times \frac{1 \text{ min}}{60 \text{ s}} = 40 \text{ rev/s}$$

If the crankshaft turns 40 times in 1 s, the time for 1 rev is

$$T = \frac{1}{40} \text{ s} = 0.025 \text{ s}$$

Thus the speed of a point on the surface, where $r = 2.0 \text{ cm} = 0.020 \text{ m}$, is

$$v = \frac{2\pi r}{T} = \frac{2\pi(0.020 \text{ m})}{0.025 \text{ s}} = 5.03 \text{ m/s}$$

Angular Position

Rather than using xy-coordinates, it will be more convenient to describe the position of the particle by its distance r from the center of the circle (labeled O) and its angle θ from the positive x-axis. This is shown in Figure 7.2. The angle θ is the **angular position** of the particle.

We can distinguish a position above the x-axis from a position that is an equal angle below the x-axis by *defining* θ to be positive when measured *counterclockwise* (ccw) from the positive x-axis. An angle measured clockwise (cw) from the positive x-axis has a negative value. "Clockwise" and "counterclockwise" in circular motion are analogous, respectively, to "left of the origin" and "right of the origin" in linear motion, which we associated with negative and positive values of x. A particle 30° below the positive x-axis is equally well described by either $\theta = -30°$ or $\theta = +330°$. We could also describe this particle by $\theta = \frac{11}{12}$ rev, where *revolutions* are another way to measure the angle.

Although degrees and revolutions are widely used measures of angle, mathematicians and scientists usually find it more convenient to measure the angle θ in Figure 7.2 by using the **arc length** s that the particle travels along the edge of a circle of radius r. We define the angular unit of **radians** such that

$$\theta \text{ (radians)} \equiv \frac{s}{r} \tag{7.2}$$

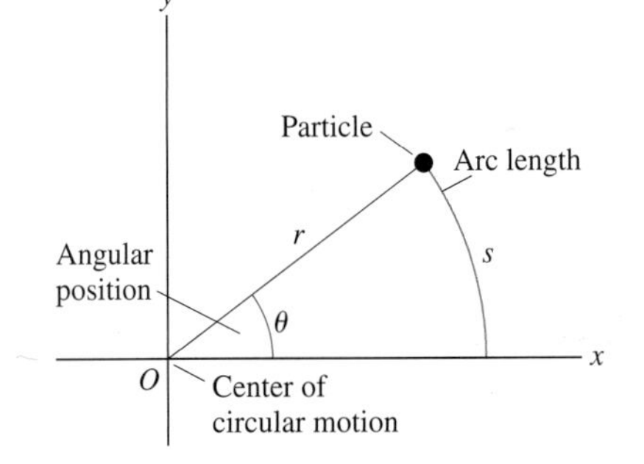

FIGURE 7.2 A particle's position is described by distance r and angle θ.

The radian, which is abbreviated rad, is the SI unit of an angle. An angle of 1 rad has an arc length s exactly equal to the radius r.

The arc length completely around a circle is the circle's circumference $2\pi r$. Thus the angle of a full circle is

$$\theta_{\text{full circle}} = \frac{2\pi r}{r} = 2\pi \text{ rad}$$

This relationship is the basis for the well-known conversion factors

$$1 \text{ rev} = 360° = 2\pi \text{ rad}$$

As a simple example of converting between radians and degrees, let's convert an angle of 1 rad to degrees:

$$1 \text{ rad} = 1 \text{ rad} \times \frac{360°}{2\pi \text{ rad}} = 57.3°$$

Thus a reasonable approximation is $1 \text{ rad} \approx 60°$. We will often specify angles in degrees, but keep in mind that the SI unit is the radian.

An important consequence of Equation 7.2 is that the arc length spanning angle θ is

$$s = r\theta \tag{7.3}$$

This is a result that we will use often, but it is valid *only* if θ is measured in radians and not in degrees. This very simple relationship between angle and arc length is one of the primary motivations for using radians.

NOTE ▶ While the concept of measuring an angle is simple, units of angle are often troublesome. Unlike the kilogram or the second, for which we have standards, the radian is a *defined* unit. Further, its definition as a ratio of two lengths makes it a *pure number* without dimensions. Thus the unit of angle, be it radians or degrees or revolutions, is really just a *name* to remind us that we're dealing with an angle. The practical implication is that the radian unit sometimes appears or disappears without warning. This seems rather mysterious until you get used to it. This textbook will call your attention to such behavior the first few times it occurs. With a little practice, you'll soon learn when the rad unit is needed and when it's not. ◀

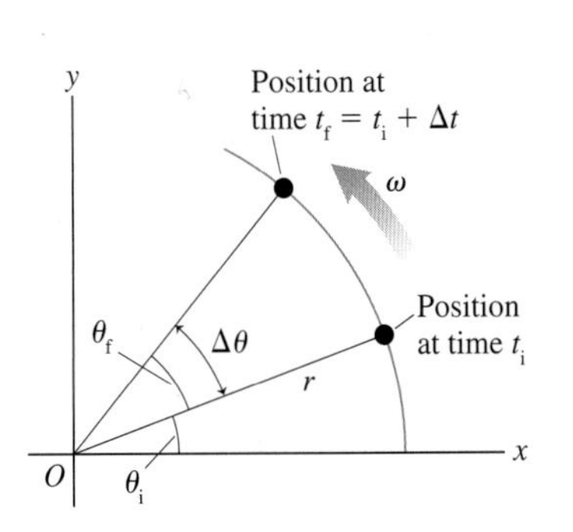

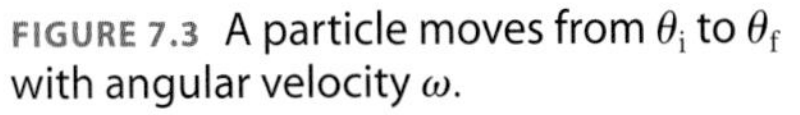
FIGURE 7.3 A particle moves from θ_i to θ_f with angular velocity ω.

Angular Velocity

Figure 7.3 shows a particle moving in a circle from an initial angular position θ_i at time t_i to a final angular position θ_f at a later time t_f. The change $\Delta\theta = \theta_f - \theta_i$ is called the **angular displacement.** We can measure the particle's circular motion in terms of the rate of change of θ, just as we measured the particle's linear motion in terms of the rate of change of its position x.

In analogy with linear motion, let's define the *average angular velocity* to be

$$\text{average angular velocity} \equiv \frac{\Delta\theta}{\Delta t} \tag{7.4}$$

As the time interval Δt becomes very small, $\Delta t \to 0$, we arrive at the definition of the instantaneous **angular velocity**

$$\omega \equiv \lim_{\Delta t \to 0} \frac{\Delta\theta}{\Delta t} = \frac{d\theta}{dt} \quad \text{(angular velocity)} \tag{7.5}$$

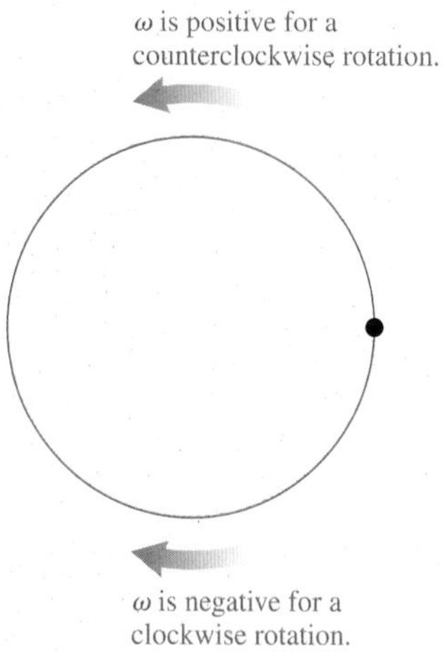

ω is positive for a counterclockwise rotation.

ω is negative for a clockwise rotation.

FIGURE 7.4 Positive and negative angular velocities.

The symbol ω is a lowercase Greek omega, *not* an ordinary w. The SI unit of angular velocity is rad/s, but °/s, rev/s, and rev/min are also common units. Revolutions per minute is often abbreviated rpm.

Angular velocity is the *rate* at which a particle's angular position is changing. A particle that starts from $\theta = 0$ rad with an angular velocity of 0.5 rad/s will be at angle $\theta = 0.5$ rad after 1 s, at $\theta = 1.0$ rad after 2 s, at $\theta = 1.5$ rad after 3 s, and so on. Its angular position is increasing at the *rate* of 0.5 radians per second. In analogy with uniform linear motion, which you studied in Chapter 2, uniform circular motion is motion in which the angle increases at a *constant* rate: **A particle moves with uniform circular motion if and only if its angular velocity ω is constant and unchanging.**

Angular velocity, like the velocity v_s of one-dimensional motion, can be positive or negative. The signs shown in Figure 7.4 are based on the fact that θ was defined to be positive for a counterclockwise rotation. Because the definition $\omega = d\theta/dt$ for circular motion parallels the definition $v_s = ds/dt$ for linear motion, the graphical relationships we found between v_s and s in Chapter 2 apply equally well to ω and θ:

- $\omega =$ slope of the θ-versus-t graph at time t
- $\theta_f = \theta_i +$ area under the ω-versus-t graph between t_i and t_f

EXAMPLE 7.2 A graphical representation of circular motion

Figure 7.5 shows the angular position of a particle moving around a circle of radius r. Describe the particle's motion and draw an ω-versus-t graph.

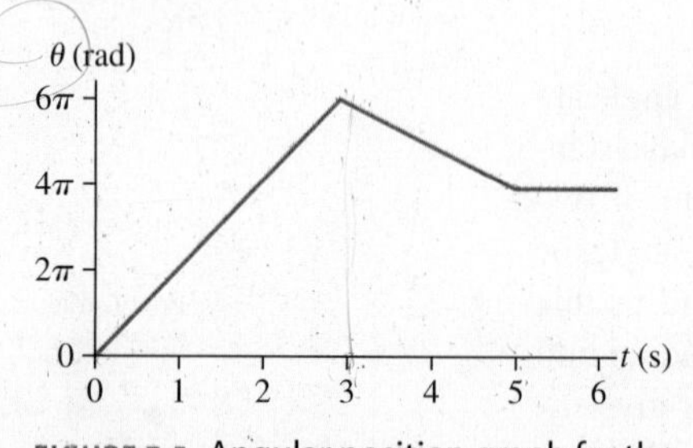

FIGURE 7.5 Angular position graph for the particle of Example 7.2.

SOLVE Although circular motion seems to "start over" every revolution (every 2π rad), the angular position θ continues to increase. $\theta = 6\pi$ rad corresponds to three revolutions. This particle makes 3 ccw (because θ is getting more positive) rev in 3 s, immediately reverses direction and makes 1 cw rev in 2 s, then stops at $t = 5$ s and holds the position $\theta = 4\pi$ rad. The angular velocity is found by measuring the slope of the graph:

$t = 0 - 3$ s	slope $= \Delta\theta/\Delta t = 6\pi$ rad/3 s $= 2\pi$ rad/s
$t = 3 - 5$ s	slope $= \Delta\theta/\Delta t = -2\pi$ rad/2 s $= -\pi$ rad/s
$t > 5$ s	slope $= \Delta\theta/\Delta t = 0$ rad/s

These results are shown as an ω-versus-t graph in Figure 7.6. For the first 3 s, the motion is uniform circular motion with $\omega = 2\pi$ rad/s. The particle then changes to a different uniform circular motion with $\omega = -\pi$ rad/s for 2 s, then stops.

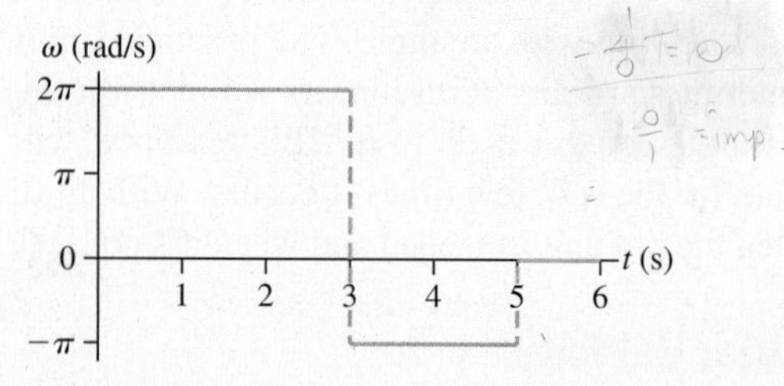

FIGURE 7.6 ω-versus-t graph for the particle of Example 7.2.

NOTE ▶ In physics, we nearly always want to give results as numerical values. Example 7.1 had a π in the equation, but we used its numerical value to compute $v = 5.03$ m/s. However, angles in radians are an exception to this rule. It's okay to leave a π in the value of θ or ω, and we have done so in Example 7.2. ◀

The angular velocity is constant during uniform circular motion, so the ω-versus-t graph is a horizontal line. It's easy to see from Figure 7.7 that the area under the curve from t_i to t_f is simply $\omega \Delta t$. Consequently,

$$\theta_f = \theta_i + \omega \Delta t \qquad \text{(uniform circular motion)} \tag{7.6}$$

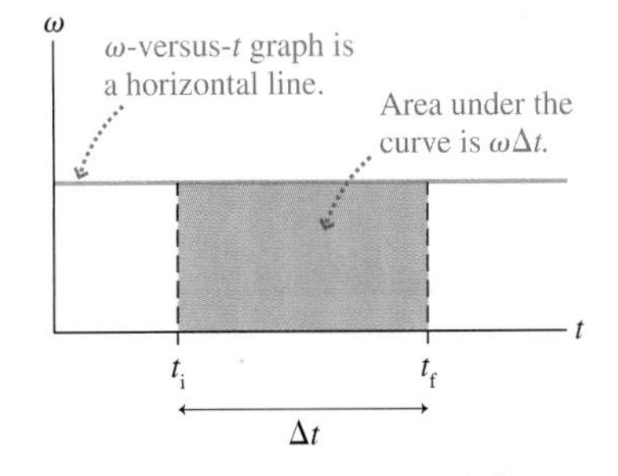

FIGURE 7.7 The ω-versus-t graph for uniform circular motion is a horizontal line.

Equation 7.6 is equivalent, with different variables, to the result $s_f = s_i + v_s \Delta t$ for uniform linear motion. You will see many more instances where circular motion is analogous to linear motion with angular variables replacing linear variables. Thus much of what you learned about linear kinematics and dynamics carries over to circular motion.

Not surprisingly, the angular velocity ω is closely related to the *period T* of the motion. As a particle goes around a circle one time, its angular displacement is $\Delta\theta = 2\pi$ rad during the interval $\Delta t = T$. Thus, using the definition of angular velocity, we find

$$|\omega| = \frac{2\pi \text{ rad}}{T} \quad \text{or} \quad T = \frac{2\pi \text{ rad}}{|\omega|} \tag{7.7}$$

The period alone gives only the absolute value of $|\omega|$. To determine the sign of ω you need to know the direction of motion.

EXAMPLE 7.3 At the roulette wheel

A small, steel roulette ball rolls around the inside of a 30-cm-diameter roulette wheel. The ball completes 2 rev in 1.20 s.

a. What is the ball's angular velocity?
b. What is the ball's position at $t = 2.0$ s? Assume $\theta_i = 0$.

MODEL Model the ball as a particle in uniform circular motion.

SOLVE

a. The period of the ball's motion, the time for 1 rev, is $T = 0.60$ s. Thus

$$\omega = \frac{2\pi \text{ rad}}{T} = \frac{2\pi \text{ rad}}{0.60 \text{ s}} = 10.47 \text{ rad/s}$$

b. The ball starts at $\theta_i = 0$ rad. After $\Delta t = 2.0$ s, its position is given by Equation 7.6:

$$\theta_f = 0 \text{ rad} + (10.47 \text{ rad/s})(2.0 \text{ s}) = 20.94 \text{ rad}$$

Although this is a mathematically acceptable answer, an observer would say that the ball is always located somewhere between 0° and 360°. Thus it is common practice to subtract off an integer number of 2π, representing the completed revolutions. Because $20.94/2\pi = 3.333$, we can write

$$\begin{aligned}\theta_f &= 20.94 \text{ rad} = 3.333 \times 2\pi \text{ rad} \\ &= 3 \times 2\pi \text{ rad} + 0.333 \times 2\pi \text{ rad} \\ &= 3 \times 2\pi \text{ rad} + 2.09 \text{ rad}\end{aligned}$$

In other words, at $t = 2.0$ s the ball has completed 3 rev and is 2.09 rad = 120° into its fourth revolution. An observer would say that the ball's position is $\theta = 120°$.

STOP TO THINK 7.1 A particle moves cw around a circle at constant speed for 2.0 s. It then reverses direction and moves ccw at half the original speed until it has traveled through the same angle. Which is the particle's angle-versus-time graph?

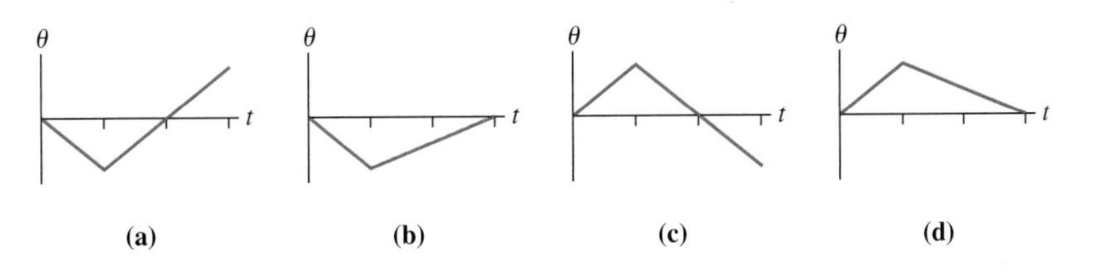

7.2 Velocity and Acceleration in Uniform Circular Motion

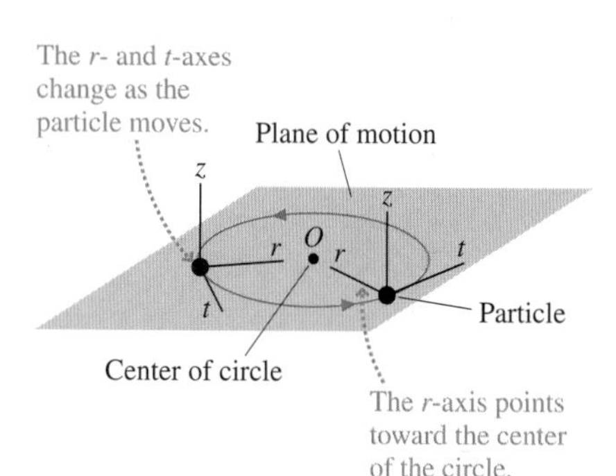

FIGURE 7.8 The axes of the rtz-coordinate system.

The xy-coordinate system we've been using for linear motion and projectile motion is not the best coordinate system for circular motion. Figure 7.8 shows a circular trajectory and the plane in which the circle lies. Let's establish a coordinate system with its origin at the point where the particle is located. The axes are defined as follows:

- The r-axis (radial axis) points *from* the particle *toward* the center of the circle.
- The t-axis (tangential axis) is tangent to the circle, pointing in the ccw direction.
- The z-axis is perpendicular to the plane of motion.

The three axes of this rtz-coordinate system are mutually perpendicular, just like the axes of the familiar xyz-coordinate system. Notice how the axes move with the particle so that the r-axis always points to the center of the circle. It will take a little getting used to, but you will soon see that circular motion problems are most easily described in these coordinates.

Figure 7.9 shows a vector $\vec{A}$ in the plane of motion. We can decompose $\vec{A}$ into its radial and tangential components:

$$A_r = A\cos\phi$$
$$A_t = A\sin\phi$$

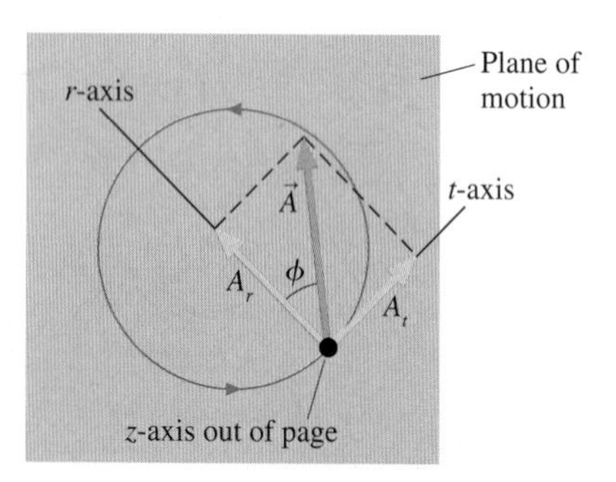

FIGURE 7.9 Vector $\vec{A}$ can be decomposed into radial and tangential components.

where ϕ is the angle with the r-axis. The positive r-direction, by definition, is toward the center of the circle, so the radial component A_r has a positive value. $\vec{A}$ lies in the plane of motion, so its perpendicular component is $A_z = 0$.

NOTE ▶ In Chapter 6, we noted that the acceleration vector $\vec{a}$ can be decomposed into a component $\vec{a}_\parallel$ parallel to the motion and a component $\vec{a}_\perp$ perpendicular to the motion. That idea is the basis for the rtz-coordinate system. Because the velocity vector $\vec{v}$ is tangent to the circle, the tangential component A_t of vector $\vec{A}$ is the component of $\vec{A}$ parallel to the motion. The radial component A_r is the component perpendicular to the motion. ◀

Velocity

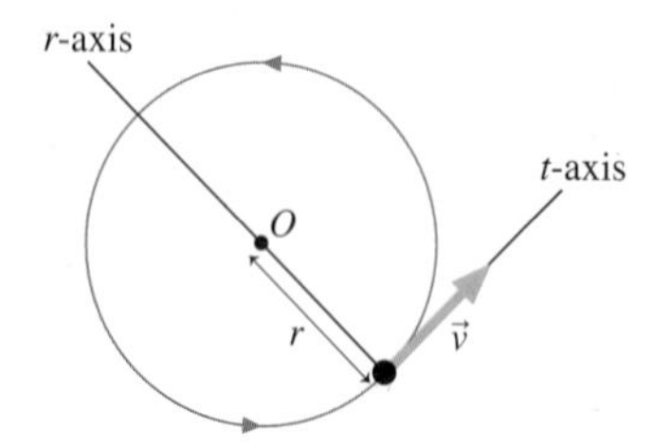

FIGURE 7.10 The velocity vector $\vec{v}$ has only a tangential component v_t.

For a particle in circular motion, such as the one seen in Figure 7.10, the velocity vector $\vec{v}$ is tangent to the circle. In other words, the velocity vector has only a tangential component v_t. The radial and perpendicular components of $\vec{v}$ are always zero.

The tangential velocity component v_t is the rate ds/dt at which the particle moves *around* the circle, where s is the arc length measured from the positive x-axis. From Equation 7.3, the arc length is $s = r\theta$. Taking the derivative, we find

$$v_t = \frac{ds}{dt} = r\frac{d\theta}{dt}$$

But $d\theta/dt$ is the angular velocity ω. Thus the tangential velocity and the angular velocity are related by

$$v_t = \omega r \qquad \text{(with } \omega \text{ in rad/s)} \tag{7.8}$$

NOTE ▶ ω is restricted to rad/s because the relationship $s = r\theta$ is the definition of radians. While it may be convenient in some problems to measure ω in rev/s or rpm, you must convert to SI units of rad/s before using Equation 7.8. ◀

The tangential velocity v_t is positive for ccw motion, negative for cw motion. Because v_t is the only nonzero component of $\vec{v}$, the particle's speed is $v = |v_t| = |\omega|r$. We'll sometimes write this as $v = \omega r$ if there's no ambiguity about the sign of ω.

As a simple example, a particle moving cw at 2.0 m/s in a circle of radius 40 cm has an angular velocity

$$\omega = \frac{v_t}{r} = \frac{-2.0 \text{ m/s}}{0.40 \text{ m}} = -5.0 \text{ rad/s}$$

where v_t and ω are negative because the motion is clockwise. Notice the units. Velocity divided by distance has units of s^{-1}. But because the division, in this case, gives us an angular quantity, we've inserted the *dimensionless* unit rad to give ω the appropriate units of rad/s.

To summarize, the velocity in the *rtz*-coordinates is

$$\begin{aligned} v_r &= 0 \\ v_t &= \omega r \\ v_z &= 0 \end{aligned} \tag{7.9}$$

Acceleration

Example 1.6 in Chapter 1 looked at the uniform circular motion of a Ferris wheel. The motion diagram from that example is shown again as Figure 7.11a. Although the particle moves with constant speed, it has an acceleration because the *direction* of the velocity vector $\vec{v}$ is always changing. A motion diagram analysis shows that the **acceleration $\vec{a}$ points toward the center of the circle.** The instantaneous velocity is tangent to the circle, so $\vec{v}$ and $\vec{a}$ are perpendicular to each other at all points on the circle, as Figure 7.11b shows.

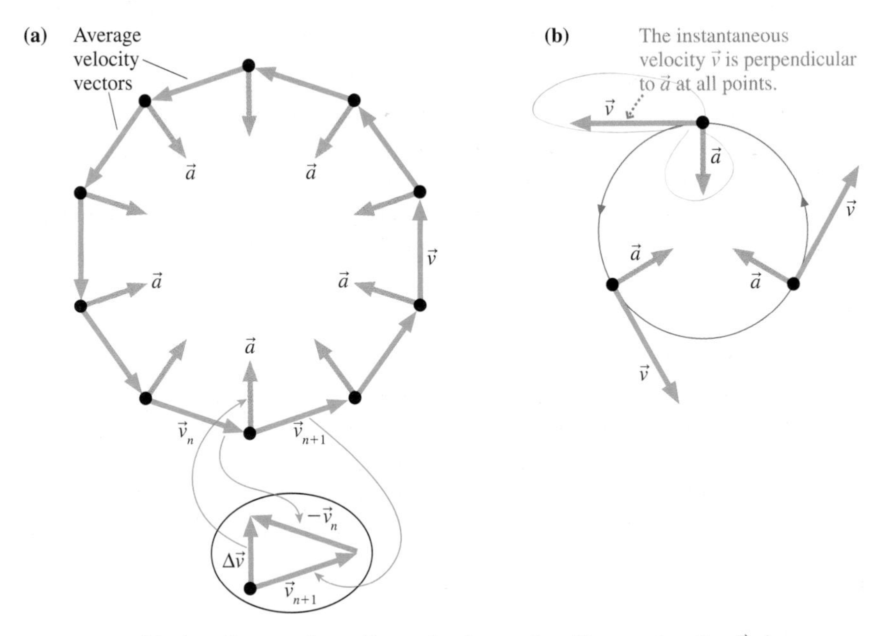

FIGURE 7.11 Motion diagram for uniform circular motion. The acceleration $\vec{a}$ always points to the center.

The acceleration of uniform circular motion is called **centripetal acceleration,** a term from a Greek root meaning "center seeking." Centripetal acceleration is not a new type of acceleration; all we are doing is *naming* an acceleration that corresponds to a particular type of motion. The magnitude of the centripetal acceleration is constant because each successive $\Delta\vec{v}$ in the motion diagram has the same length.

The motion diagram tells us the direction that $\vec{a}$ points in, but it doesn't give us a value for a. To complete our description of circular motion, we need to find a quantitative relationship between a and the tangential velocity v_t.

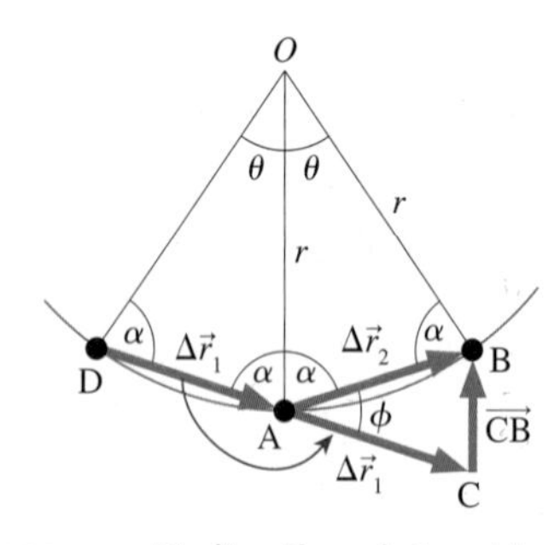

FIGURE 7.12 Finding the relationship between velocity and acceleration for circular motion.

Figure 7.12 shows in more detail the three points at the bottom of the motion diagram of Figure 7.11. Vectors $\Delta\vec{r}_1$ and $\Delta\vec{r}_2$ have the same length, namely $v\Delta t$. Vector $\Delta\vec{r}_1$ has been redrawn with the tails of the two vectors together at point A. The vector that points straight up from C to B is

$$\overrightarrow{CB} = \Delta\vec{r}_2 - \Delta\vec{r}_1 = \vec{v}_2\Delta t - \vec{v}_1\Delta t = (\vec{v}_2 - \vec{v}_1)\Delta t = \Delta\vec{v}\,\Delta t$$

Triangle ABC is an isosceles triangle because the motion is at constant speed. Triangle ABO is also isosceles because sides AO and BO are both of length r. Furthermore, because $\theta + \alpha + \alpha = 180°$ in triangle ABO and $\phi + \alpha + \alpha = 180°$ along line DAC, you can see that $\phi = \theta$ and hence ABC and ABO are *similar triangles* with side AB in common.

Recall, from geometry, that the ratios of the sides of similar triangles are equal. Thus

$$\frac{\text{CB}}{\text{AB}} = \frac{\text{AB}}{\text{AO}} \quad \text{or} \quad \frac{|\Delta\vec{v}|\Delta t}{v\Delta t} = \frac{v\Delta t}{r} \tag{7.10}$$

where $|\Delta\vec{v}|$ indicates the magnitude of vector $\Delta\vec{v}$. Rearranging this equation gives

$$\frac{|\Delta\vec{v}|}{\Delta t} = \frac{v^2}{r} \tag{7.11}$$

But $|\Delta\vec{v}|/\Delta t$ is the magnitude of the average acceleration vector: $\vec{a}_{avg} = \Delta\vec{v}/\Delta t$. Thus

$$a_{avg} = \frac{v^2}{r} \tag{7.12}$$

This analysis has been in terms of the *average* velocities and the *average* acceleration. If we now let $\Delta t \to 0$, the three points in Figure 7.12 move together but the geometrical relationships do not change. Thus Equation 7.12 remains valid when we take the limit, and it applies equally well to instantaneous velocities and accelerations.

4.1 

Thus the acceleration of uniform circular motion points to the center of the circle, because $\Delta\vec{v}$ does, and has magnitude v^2/r. In vector notation,

$$\vec{a} = \left(\frac{v^2}{r}, \text{ toward center of circle}\right) \qquad \text{(centripetal acceleration)} \tag{7.13}$$

This acceleration is conveniently written in the *rtz*-coordinate system as

$$\begin{aligned} a_r &= \frac{v^2}{r} = \omega^2 r \\ a_t &= 0 \\ a_z &= 0 \end{aligned} \tag{7.14}$$

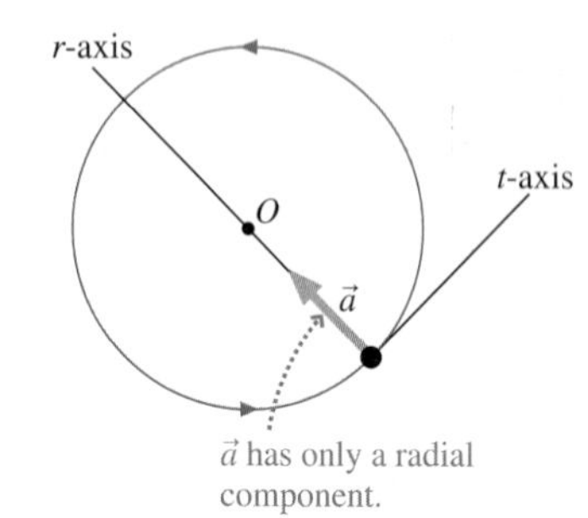

FIGURE 7.13 The acceleration vector in the *rtz*-coordinate system.

We used Equation 7.8, $v = \omega r$, to write a_r in terms of the angular velocity ω. For convenience, we'll often refer to the component a_r as "the centripetal acceleration."

Figure 7.13 shows the acceleration vector in the *rtz*-coordinate system. Compare this to the velocity vector in Figure 7.10. You can begin to see the advantages of the *rtz*-coordinate system. Only one component of $\vec{v}$ and one component of $\vec{a}$ are nonzero.

EXAMPLE 7.4 The acceleration of an atomic electron

We will later study the Bohr atom. This is a simple model of the hydrogen atom in which an electron orbits a proton at a radius of 5.29×10^{-11} m with a period of 1.52×10^{-16} s. What is the electron's centripetal acceleration?

SOLVE From Equation 7.1, the electron's speed is

$$v = \frac{2\pi r}{T} = \frac{2\pi(5.29 \times 10^{-11}\text{ m})}{1.52 \times 10^{-16}\text{ s}} = 2.19 \times 10^6 \text{ m/s}$$

r, v
given   need to calculate
v = 2πr

Then from Equation 7.14,

$$a_r = \frac{v^2}{r} = \frac{(2.19 \times 10^6 \text{ m/s})^2}{5.29 \times 10^{-11} \text{ m}} = 9.07 \times 10^{22} \text{ m/s}^2$$

ASSESS This was not intended as a profound problem, merely to illustrate how a centripetal acceleration is computed. In addition, it demonstrates the unbelievably enormous accelerations that take place at the atomic level. It should then come as no surprise that atomic particles may behave in ways that our intuition, trained by accelerations of only a few m/s^2, cannot easily grasp.

STOP TO THINK 7.2 Rank in order, from largest to smallest, the centripetal accelerations $(a_r)_a$ to $(a_r)_e$ of particles a to e.

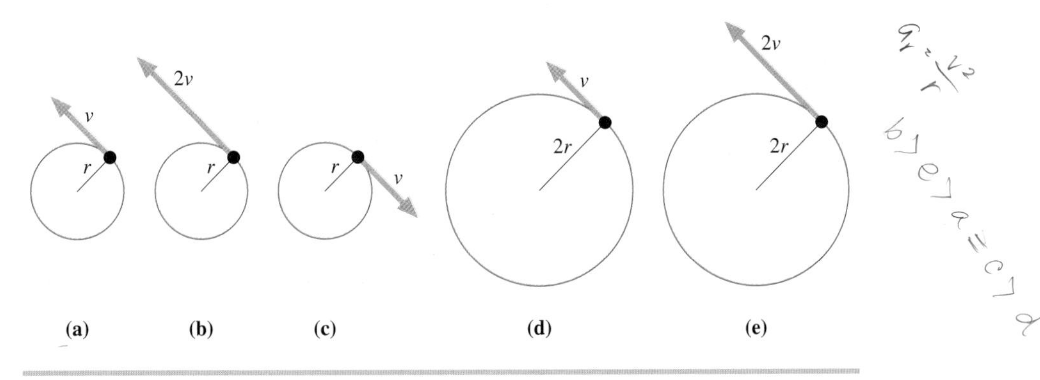

7.3 Dynamics of Uniform Circular Motion

Highway and racetrack curves are banked to allow the normal force of the road to provide the centripetal acceleration of the turn.

A particle in uniform circular motion is clearly not traveling at constant velocity in a straight line. Consequently, according to Newton's first law, the particle *must* have a net force acting on it. We've already determined the acceleration of a particle in uniform circular motion—the centripetal acceleration of Equation 7.13. Newton's second law tells us exactly how much net force is needed to cause this acceleration:

$$\vec{F}_{net} = m\vec{a} = \left(\frac{mv^2}{r}, \text{ toward center of circle}\right) \tag{7.15}$$

In other words, a particle of mass m moving at constant speed v around a circle of radius r must have a net force of magnitude mv^2/r pointing toward the center of the circle. Without such a force, the particle would move off in a straight line tangent to the circle.

Figure 7.14 shows the net force $\vec{F}_{net}$ acting on a particle as it undergoes uniform circular motion. You can see that **$\vec{F}_{net}$ points along the radial axis of the *rtz*-coordinate system, toward the center of the circle.** The tangential and perpendicular components of $\vec{F}_{net}$ are zero.

NOTE ▶ The force described by Equation 7.15 is not a *new* force. Our rules for identifying forces have not changed. What we are saying is that a particle moves with uniform circular motion *if and only if* a net force always points toward the center of the circle. The force itself must have an identifiable agent and will be one of our familiar forces, such as tension, friction, or the normal force. Equation 7.15 simply tells us how the force needs to act—how strong and in which direction—to cause the particle to move with speed v in a circle of radius r. ◀

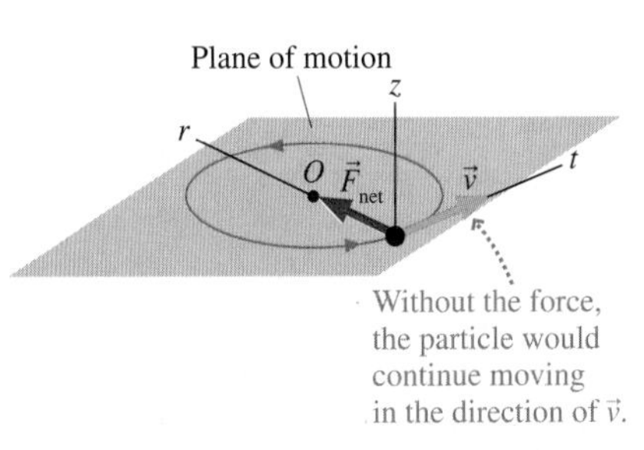

FIGURE 7.14 The net force points in the radial direction, toward the center of the circle.

The usefulness of the *rtz*-coordinate system becomes apparent when we write Newton's second law, Equation 7.15, in terms of the *r*-, *t*-, and *z*-components:

$$(F_{net})_r = \sum F_r = ma_r = \frac{mv^2}{r} = m\omega^2 r$$
$$(F_{net})_t = \sum F_t = ma_t = 0 \qquad (7.16)$$
$$(F_{net})_z = \sum F_z = ma_z = 0$$

Notice that we've used our explicit knowledge of the acceleration, as given in Equation 7.14, to write the right-hand side of these equations. **For uniform circular motion, the sum of the forces along the *t*-axis and along the *z*-axis *must* equal zero, and the sum of the forces along the *r*-axis *must* equal ma_r where a_r is the centripetal acceleration.**

It is time for some examples to clarify these ideas and to see how some of the forces you've come to know can be involved in circular motion.

EXAMPLE 7.5 Spinning in a circle

An energetic father places his 20 kg child on a 5.0 kg cart to which a 2.0-m-long rope is attached. He then holds the end of the rope and spins the cart and child around in a circle, keeping the rope parallel to the ground. If the tension in the rope is 100 N, how many revolutions per minute (rpm) does the cart make? Rolling friction between the cart's wheels and the ground is negligible.

MODEL Model the child in the cart as a particle in uniform circular motion.

VISUALIZE Figure 7.15 shows the pictorial and physical representations. The main idea of the pictorial representation is to illustrate the relevant geometry and to define the symbols that will be used. A circular-motion problem usually does not have starting and ending points like a projectile problem, so numerical subscripts such as x_1 or y_2 are usually not needed. Here we need to define the cart's speed v and the radius r of the circle.

A motion diagram is not needed for uniform circular motion because we know the acceleration $\vec{a}$ points to the center of the circle. Motion diagrams have been very helpful, but they should be second nature to you now. We'll continue to use motion diagrams when they help clarify a situation, but we won't include them in every solution. The essential part of the physical representation is the free-body diagram. **For uniform circular motion we'll draw the free-body diagram in the *rz*-plane, looking at the edge of the circle, because this is the plane of the forces.** The contact forces acting on the cart are the normal force of the ground and the tension force of the rope. The normal force is perpendicular to the plane of the motion and thus in the *z*-direction. The direction of $\vec{T}$ is determined by the statement that the rope is parallel to the ground. In addition, there is the long-range weight force $\vec{w}$.

SOLVE We defined the *r*-axis to point toward the center of the circle, so $\vec{T}$ points in the positive *r*-direction and has *r*-component $T_r = T$. Newton's second law, using the *rtz*-components of Equation 7.16, is

$$\sum F_r = T = \frac{mv^2}{r}$$
$$\sum F_z = n - w = 0$$

We've taken the *r*- and *z*-components of the forces directly from the free-body diagram, as you learned to do in Chapter 5. Then we've *explicitly* equated the sums to $a_r = v^2/r$ and $a_z = 0$. This is the basic strategy for all uniform circular-motion problems. From the *z*-equation we can find that $n = w$. This would be useful if we needed to determine a friction force, but it's not

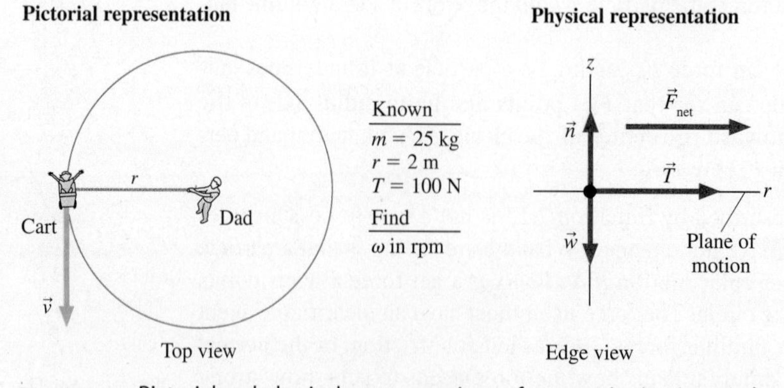

FIGURE 7.15 Pictorial and physical representations of a cart spinning in a circle.

needed in this problem. From the r-equation, the speed of the cart is

$$v = \sqrt{\frac{rT}{m}} = \sqrt{\frac{(2.0\text{ m})(100\text{ N})}{25\text{ kg}}} = 2.83\text{ m/s}$$

The cart's angular velocity is then found from Equation 7.8:

$$\omega = \frac{v_t}{r} = \frac{v}{r} = \frac{2.83\text{ m/s}}{2.0\text{ m}} = 1.41\text{ rad/s}$$

This is another case where we inserted the radian unit because ω is specifically an *angular* velocity. Finally, we need to convert ω to rpm:

$$\omega = \frac{1.41\text{ rad}}{1\text{ s}} \times \frac{1\text{ rev}}{2\pi\text{ rad}} \times \frac{60\text{ s}}{1\text{ min}} = 13.5\text{ rpm}$$

ASSESS 13.5 rpm corresponds to a period $T \approx 4.5$ s. This result is reasonable.

This has been a fairly typical circular-motion problem. You might want to think about how the solution would change if the rope is *not* parallel to the ground.

EXAMPLE 7.6 Turning the corner I

What is the maximum speed with which a 1500 kg car can make a left turn around a curve of radius 50 m on a level (unbanked) road without sliding?

MODEL Although the car turns only a quarter of a circle, we can model the car as a particle in uniform circular motion as it goes around the turn. Assume that rolling friction is negligible.

VISUALIZE Figure 7.16 shows the pictorial and physical representations. The car moves along a circular arc at constant speed for the quarter-circle necessary to complete the turn. The motion before and after the turn is not relevant to the problem. The more interesting issue is *how* a car turns a corner. What force or forces can we identify that cause the direction of the velocity vector to change? Imagine you are driving a car on a completely frictionless road, such as a very icy road. You would not be able to turn a corner. Turning the steering wheel would be of no use; the car would slide straight ahead, in accordance with both Newton's first law and the experience of anyone who has ever driven on ice! So it must be *friction* that somehow allows the car to turn.

The force-identification section of Figure 7.16 shows the top view of a tire as it turns a corner. If the road surface were frictionless, the tire would slide straight ahead. The force that prevents an object from sliding across a surface is *static friction.* Static friction $\vec{f}_s$ pushes *sideways* on the tire, toward the center of the circle. How do we know the direction is sideways? If $\vec{f}_s$ had a component either parallel to $\vec{v}$ or opposite to $\vec{v}$, it would cause the car to speed up or slow down. Because the car changes direction but not speed, static friction must be perpendicular to $\vec{v}$. Thus $\vec{f}_s$ causes the centripetal acceleration of circular motion around the curve. With this in mind, the free-body diagram, drawn from behind the car, shows the static friction force pointing toward the center of the circle.

SOLVE Because the static friction force has a maximum value, there will be a maximum speed with which a car can turn without sliding. The maximum speed is reached when the static friction force reaches its maximum $f_{s\,max} = \mu_s n$. If the car enters the curve at a speed higher than the maximum, static friction will not be large enough to provide the necessary centripetal acceleration and the car will slide.

The static friction force points in the positive r-direction, so its radial component is simply the magnitude of the vector: $(f_s)_r = f_s$. Newton's second law in the rtz-coordinate system is

$$\sum F_r = f_s = \frac{mv^2}{r}$$

$$\sum F_z = n - w = 0$$

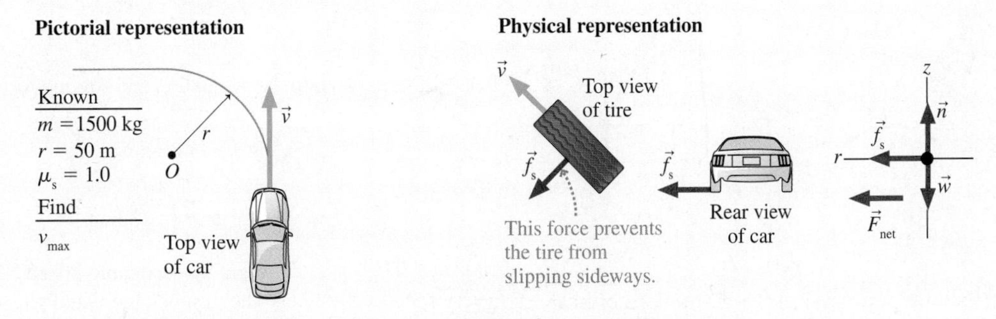

FIGURE 7.16 Pictorial and physical representations of a car turning a corner.

The only difference from Example 7.5 is that the tension force toward the center has been replaced by a static friction force toward the center. From the radial equation, the speed is

$$v = \sqrt{\frac{rf_s}{m}}$$

The speed will be a maximum when f_s reaches its maximum value

$$f_s = f_{s\,max} = \mu_s n = \mu_s w = \mu_s mg$$

where we used $n = w$ from the z-equation. At that point,

$$v_{max} = \sqrt{\frac{rf_{s\,max}}{m}} = \sqrt{\mu_s rg}$$

$$= \sqrt{(1.00)(50\text{ m})(9.80\text{ m/s}^2)} = 22.1\text{ m/s}$$

where we found μ_s in Table 5.1.

ASSESS 22.1 m/s ≈ 45 mph, a reasonable answer for how fast a car can take an unbanked curve. Notice that the car's mass canceled out and that the final equation for v_{max} is quite simple. This is another example of why it pays to work algebraically until the very end.

4.5 ActivPhysics ONLINE

Because μ_s depends on road conditions, the maximum safe speed through turns can vary dramatically. Wet roads, in particular, lower the value of μ_s and thus lower the speed of turns. A car that handles normally while driving straight ahead on a wet road can suddenly slide out of control when turning a corner. Icy conditions are even worse. The corner you turn every day at 45 mph will require a speed of no more than 15 mph if the coefficient of static friction drops to 0.1.

EXAMPLE 7.7 Turning the corner II

A highway curve of radius 70 m is banked at a 15° angle. At what speed v_0 can a car take this curve without assistance from friction?

MODEL The car is a particle in uniform circular motion.

VISUALIZE Having just discussed the role of friction in turning corners, it is perhaps surprising to suggest that the same turn can also be accomplished without friction. Example 7.6 considered a level roadway, but real highway curves are *banked* by being tilted up at the outside edge of the curve. The angle is modest on ordinary highways, but it can be quite large on high-speed racetracks. The purpose of banking becomes clear if you look at the free-body diagram in Figure 7.17. The normal force $\vec{n}$ is perpendicular to the road, so tilting the road causes $\vec{n}$ to have a component toward the center of the circle. The radial component n_r is the inward force that causes the centripetal acceleration needed to turn the car. Notice that we are *not* using a tilted coordinate system, although this looks rather like an inclined-plane problem. The center of the circle is in the same horizontal plane as the car, and for circular motion problems we need the r-axis to pass through the center. Tilted axes are for *linear* motion along an incline.

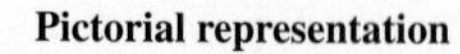

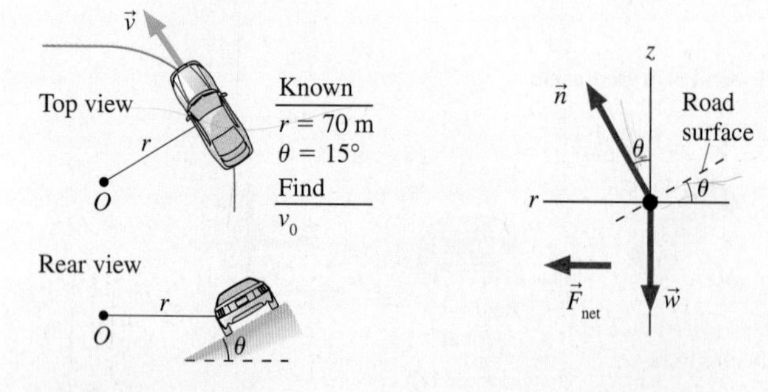

FIGURE 7.17 Pictorial and physical representations of a car on a banked curve.

SOLVE Without friction, $n_r = n\sin\theta$ is the only component of force in the radial direction. It is this inward component of the normal force on the car that causes it to turn the corner. Newton's second law is

$$\sum F_r = n\sin\theta = \frac{mv_0^2}{r}$$

$$\sum F_z = n\cos\theta - w = 0$$

where θ is the angle at which the road is banked and we've assumed that the car is traveling at the correct speed v_0. From the z-equation,

$$n = \frac{w}{\cos\theta} = \frac{mg}{\cos\theta}$$

Substituting this into the r-equation and solving for v_0 gives

$$\frac{mg}{\cos\theta}\sin\theta = mg\tan\theta = \frac{mv_0^2}{r}$$

$$v_0 = \sqrt{rg\tan\theta} = 13.6\text{ m/s}$$

ASSESS This is ≈ 27 mph, a reasonable speed. Only at this very specific speed can the turn be negotiated without reliance on friction forces.

It's interesting to explore what happens at other speeds. The car will need to rely on both the banking *and* friction if it takes the curve at a speed higher or lower than v_0. Figure 7.18a has modified the free-body diagram to include a static friction force. Remember that $\vec{f}_s$ must be parallel to the surface, so it is tilted downward at angle θ. Because $\vec{f}_s$ has a component in the positive r-direction, the *net* radial force is larger than that provided by $\vec{n}$ alone. This will allow the car to take the curve at $v > v_0$. We could use a quantitative analysis similar to Example 7.6 to determine the maximum speed on a banked curve by analyzing Figure 7.18a when $f_s = f_{s\,max}$.

But what about taking the curve at a speed $v < v_0$? In this situation, the r-component of the normal force is too big; not that much center-directed force is needed. As Figure 7.18b shows, the net force can be reduced by having $\vec{f}_s$ point *up* the slope! This seems very strange at first, but consider the limiting case in which the car is parked on the banked curve, with $v = 0$. Were it not for a static friction force pointing *up* the slope, the car would slide sideways down the incline. In fact, for any speed less than v_0 the car will slip to the inside of the curve unless it is prevented from doing so by a static friction force pointing up the slope.

Our analysis thus finds three divisions of speed. At v_0, the car turns the corner with no assistance from friction. At greater speeds, the car will slide out of the curve unless an inward-directed friction force increases the size of the net force. And lastly, at lesser speeds the car will slip down the incline unless an outward-directed friction force prevents it from doing so.

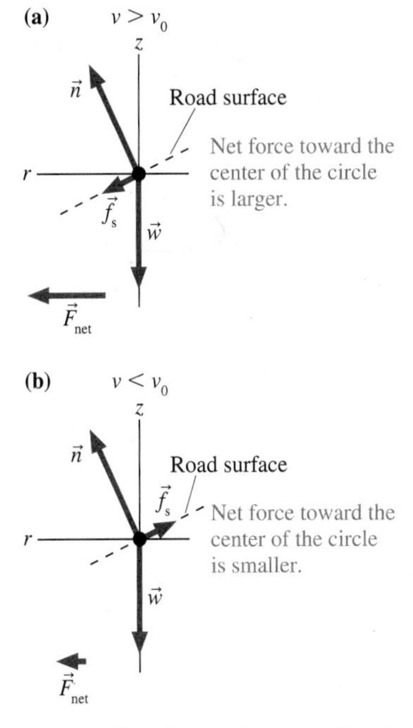

FIGURE 7.18 Free-body diagrams showing the static friction force when $v > v_0$ and when $v < v_0$.

EXAMPLE 7.8 A rock in a sling

A Stone Age hunter places a 1.0 kg rock in a sling and swings it in a horizontal circle around his head on a 1.0-m-long vine. If the vine breaks at a tension of 200 N, what is the maximum angular velocity, in rpm, with which he can swing the rock?

MODEL Model the rock as a particle in uniform circular motion.

VISUALIZE This problem appears, at first, to be essentially the same as Example 7.5, where the father spun his child around on a rope. However, the lack of a normal force from a supporting surface makes a *big* difference. In this case, the *only* contact force on the rock is the tension in the vine. Because the rock moves in a horizontal circle, you may be tempted to draw a free-body diagram like Figure 7.19a where $\vec{T}$ is directed along the r-axis. You will quickly run into trouble, however, because this diagram has a net force in the z-direction and it is impossible to satisfy $\sum F_z = 0$. The weight $\vec{w}$ certainly points vertically downward, so the difficulty must be with $\vec{T}$.

As an experiment, tie a small weight to a string, swing it over your head, and check the *angle* of the string. You will quickly discover that the string is *not* horizontal but, instead, is angled downward. The pictorial model of Figure 7.19b labels the angle θ. Notice that the rock moves in a *horizontal* circle, so the center of the circle is *not* at his hand. The r-axis point to the center of the circle, but the tension force is directed along the vine. Thus the correct free-body diagram is the one in Figure 7.19b.

SOLVE The free-body diagram shows that the downward weight force is balanced by an upward component of the tension, leaving the radial component of the tension to cause the centripetal acceleration. Newton's second law is

$$\sum F_r = T\cos\theta = \frac{mv^2}{r}$$

$$\sum F_z = T\sin\theta - w = T\sin\theta - mg = 0$$

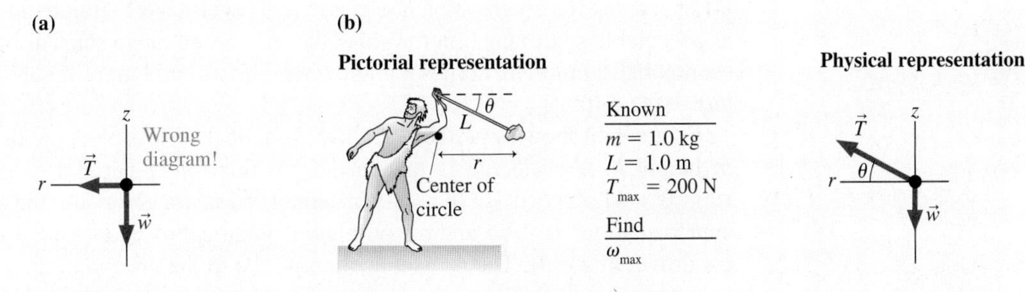

FIGURE 7.19 Pictorial and physical representations of a rock in a sling.

where θ is the angle of the vine below horizontal. From the z-equation we find

$$\sin\theta = \frac{mg}{T}$$

$$\theta = \sin^{-1}\left(\frac{(1.0\text{ kg})(9.8\text{ m/s}^2)}{200\text{ N}}\right) = 2.81°$$

where we've evaluated the angle at the maximum tension of 200 N. The vine's angle of inclination is small but not zero. Turning now to the r-equation, we find the rock's speed is

$$v = \sqrt{\frac{rT\cos\theta}{m}}$$

Careful! The radius r of the circle is *not* the length L of the vine. You can see in Figure 7.19b that $r = L\cos\theta$. Thus

$$v = \sqrt{\frac{LT\cos^2\theta}{m}} = \sqrt{\frac{(1.0\text{ m})(200\text{ N})(\cos 2.81°)^2}{1.0\text{ kg}}} = 14.1\text{ m/s}$$

We can now find the maximum angular velocity, the value of ω that brings the tension to the breaking point, to be

$$\omega_{\max} = \frac{v}{r} = \frac{v}{L\cos\theta} = \frac{14.1\text{ rad}}{1\text{ s}} \times \frac{1\text{ rev}}{2\pi\text{ rad}} \times \frac{60\text{ s}}{1\text{ min}} = 135\text{ rpm}$$

STOP TO THINK 7.3 A block on a string spins in a horizontal circle on a frictionless table. Rank order, from largest to smallest, the tensions T_a to T_e acting on blocks a to e.

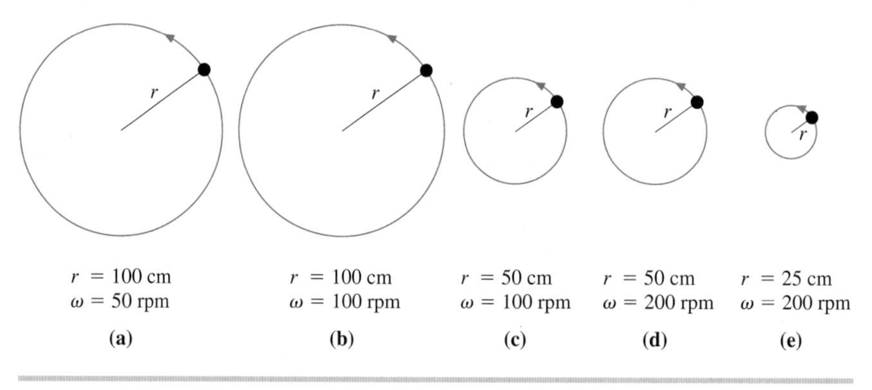

7.4 Circular Orbits

Satellites orbit the earth, the earth orbits the sun, and our entire solar system orbits the center of the Milky Way galaxy. Not all orbits are circular, but in this section we'll limit our analysis to circular orbits. We'll look at the elliptical orbits of satellites and planets in Chapter 12.

How does a satellite orbit the earth? What forces act on it? Why does it move in a circle? To answer these important questions, let's return, for a moment, to projectile motion. Projectile motion occurs when the only force on an object is gravity. Our analysis of projectiles made an implicit assumption that the earth is flat and that the acceleration due to gravity is everywhere straight down. This is an acceptable approximation for projectiles of limited range, such as baseballs or cannon balls, but there comes a point where we can no longer ignore the curvature of the earth.

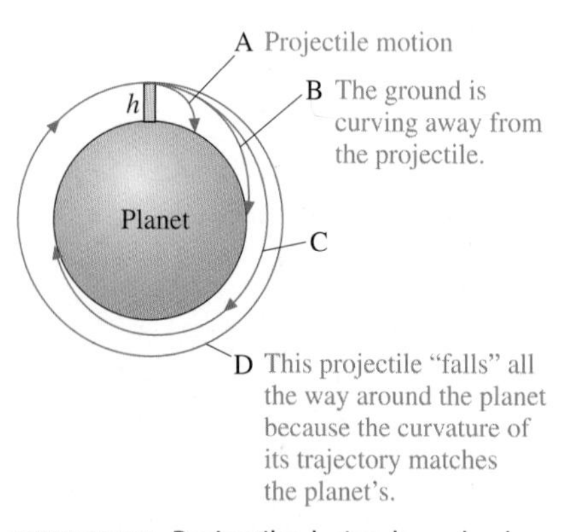

FIGURE 7.20 Projectiles being launched at increasing speeds from height h on a smooth, airless planet.

Figure 7.20 shows a perfectly smooth, spherical, airless planet with one tower of height h. A projectile is launched from this tower parallel to the ground ($\theta = 0°$) with speed v_0. If v_0 is very small, as in trajectory A, the "flat-earth approximation" is valid and the problem is identical to Example 6.4 in which a car drove off a cliff. The projectile simply falls to the ground along a parabolic trajectory.

As the initial speed v_0 is increased, the projectile begins to notice that the ground is curving out from beneath it. It is falling the entire time, always getting closer to the ground, but the distance that the projectile travels before finally reaching the ground—that is, its range—increases because the projectile must "catch up" with the ground that is curving away from it. Trajectories B and C are of this type. The actual calculation of these trajectories is beyond the scope of this textbook, but you should be able to understand the factors that influence the trajectory.

If the launch speed v_0 is sufficiently large, there comes a point where the curve of the trajectory and the curve of the earth are parallel. In this case, the projectile "falls" but it never gets any closer to the ground! This is the situation for trajectory D. A closed trajectory around a planet or star, such as trajectory D, is called an **orbit.**

The most important point of this qualitative analysis is that **an orbiting projectile is in free fall.** This is, admittedly, a strange idea, but one worth careful thought. An orbiting projectile is really no different from a thrown baseball or a car driving off a cliff. The only force acting on it is gravity, but its tangential velocity is so large that the curvature of its trajectory matches the curvature of the earth. When this happens, the projectile "falls" under the influence of gravity but never gets any closer to the surface, which curves away beneath it.

In the flat-earth approximation, shown in Figure 7.21a, the weight force acting on an object of mass m is

$$\vec{w} = (mg, \text{vertically downward}) \qquad \text{(flat-earth approximation)} \qquad (7.17)$$

But since stars and planets are actually spherical (or very close to it), the "real" force of gravity acting on an object is directed toward the *center* of the planet, as shown in Figure 7.21b. In this case the weight is

$$\vec{w} = (mg, \text{toward center}) \qquad \text{(spherical planet)} \qquad (7.18)$$

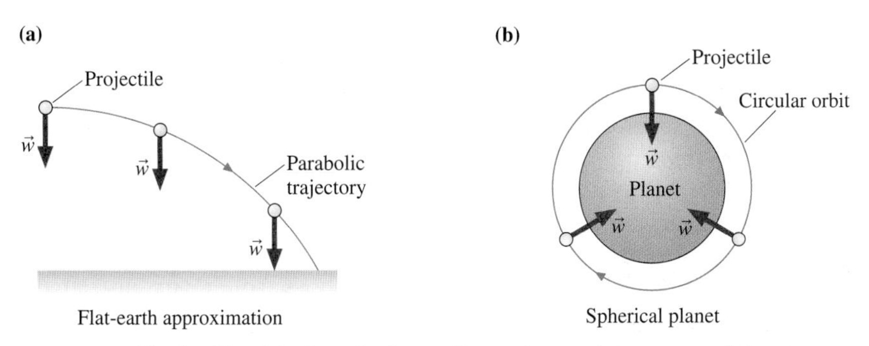

FIGURE 7.21 The "real" weight force is always directed toward the center of the planet.

As you have learned, a force of constant magnitude that always points toward the center of a circle causes the centripetal acceleration of uniform circular motion. Thus the weight force of Equation 7.18 on the object in Figure 7.21b causes it to have acceleration

$$\vec{a} = \frac{\vec{F}_{\text{net}}}{m} = (g, \text{toward center}) \qquad (7.19)$$

An object moving in a circle of radius r at speed v_{orbit} will have this centripetal acceleration if

$$a_r = \frac{(v_{\text{orbit}})^2}{r} = g \qquad (7.20)$$

That is, if an object moves parallel to the surface with the speed

$$v_{\text{orbit}} = \sqrt{rg} \qquad (7.21)$$

then the acceleration due to gravity provides exactly the centripetal acceleration needed for a circular orbit of radius r. An object with any other speed will not follow a circular orbit.

The earth's radius is $r = R_e = 6.37 \times 10^6$ m. (A table of useful astronomical data is inside the cover of this book.) The orbital speed of a projectile just skimming the surface of an airless, bald earth is

$$v_{orbit} = \sqrt{rg} = \sqrt{(6.37 \times 10^6 \text{ m})(9.80 \text{ m/s}^2)} = 7900 \text{ m/s} \approx 16{,}000 \text{ mph}$$

Even if there were no trees and mountains, a real projectile moving at this speed would burn up from the friction of air resistance.

Suppose, however, that we launched the projectile from a tower of height $h = 200$ mi $\approx 3.2 \times 10^5$ m, just above the earth's atmosphere. This is approximately the height of low-earth-orbit satellites, such as the Space Shuttle. Note that $h \ll R_e$, so the radius of the orbit $r = R_e + h = 6.69 \times 10^6$ m is only 5% greater than the earth's radius. Many people have a mental image that satellites orbit far above the earth, but in fact many satellites come pretty close to skimming the surface. Our calculation of v_{orbit} thus turns out to be quite a good estimate of the speed of a satellite in low earth orbit. (You will see later that the value of g differs slightly at a satellite's altitude, but not by much. Our calculation remains a good estimate.)

We can use v_{orbit} to calculate the period of a satellite orbit:

$$T = \frac{2\pi r}{v_{orbit}} = 2\pi\sqrt{\frac{r}{g}} \tag{7.22}$$

For a low earth orbit, with $r = R_e + 200$ miles, we find $T = 5192$ s $= 87$ min. The period of the space shuttle at an altitude of 200 mi is, indeed, just about 87 minutes.

When we discussed *weightlessness* in Chapter 5, we discovered that it occurs during free fall. We asked the question, at the end of Section 5.3, whether astronauts and their spacecraft were in free fall. We can now give an affirmative answer: They are, indeed, in free fall. They are falling continuously around the earth, under the influence of only the gravitational force, but never getting any closer to the ground because the earth's surface curves beneath them. Weightlessness in space is no different from the weightlessness in a free-falling elevator. It does *not* occur from an absence of weight or an absence of gravity. Instead, the astronaut, the spacecraft, and everything in it are "weightless" because they are all falling together.

Gravity

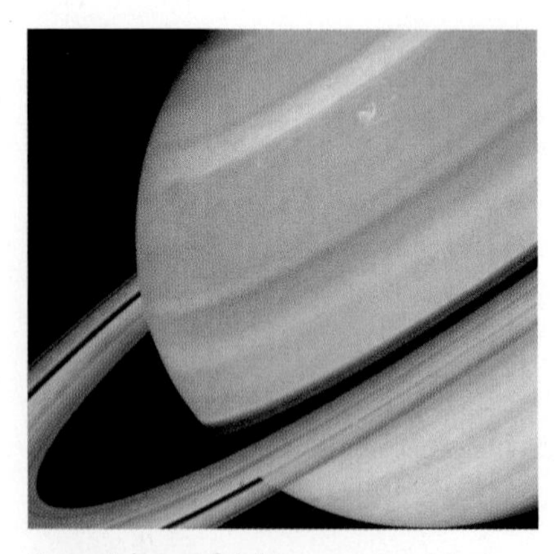

Saturn's beautiful rings consist of dust particles and small rocks orbiting the planet.

We can leave this section with a glance ahead, where we will look at the gravitational force more closely. If a satellite is simply "falling" around the earth, with the gravitational force causing a centripetal acceleration, then what about the moon? Is it obeying the same laws of physics? Or do celestial objects obey laws that we cannot discover by experiments here on earth?

The radius of the moon's orbit around the earth is $r = R_m = 3.84 \times 10^8$ m. If we use Equation 7.22 to calculate the period of the moon's orbit, the time it takes the moon to circle the earth once, we get

$$T = 2\pi\sqrt{\frac{r}{g}} = 2\pi\sqrt{\frac{3.84 \times 10^8 \text{ m}}{9.80 \text{ m/s}^2}} = 655 \text{ min} \approx 11 \text{ hours}$$

This is clearly wrong. As you probably know, the full moon occurs roughly once a month. More exactly, we know from astronomical measurements that the period of the moon's orbit is $T = 27.3$ days $= 2.36 \times 10^6$ s, a factor of 60 longer than we calculated it to be.

Newton believed that the laws of motion he had discovered were *universal*. That is, they should apply to the motion of the moon as well as to the motion of objects in the laboratory. But why should we assume that the acceleration due to gravity g is the same at the distance of the moon as it is on or near the earth's surface? If gravity is the force of the earth pulling on an object, it seems plausible that the size of that force, and thus the size of g, should diminish with increasing distance from the earth.

If the moon orbits the earth because of the earth's gravitational pull, what value of g would be needed to explain the moon's period? We can calculate $g_{\text{at moon}}$ from Equation 7.22 and the observed value of the moon's period:

$$g_{\text{at moon}} = \frac{4\pi^2 R_{\text{m}}}{T_{\text{moon}}{}^2} = 0.00272 \text{ m/s}^2$$

This is much less than the earth-bound value of 9.8 m/s^2.

Newton proposed the idea that the earth's force of gravity decreases inversely with the square of the distance from the earth. This is the basis of *Newton's law of gravity*, a topic you will study in Chapter 12. There we will be able to use the mass of the earth and the distance to the moon to *predict* that $g_{\text{at moon}} = 0.00272 \text{ m/s}^2$, exactly as expected. The moon, just like the space shuttle, is simply "falling" around the earth!

7.5 Fictitious Forces and Apparent Weight

If you are riding in a car that makes a sudden stop, you may feel as if a force "throws" you forward toward the windshield. But there really is no such force. You can not identify any agent that does the throwing. An observer watching from beside the road would simply see you continuing forward as the car stops.

The decelerating car is not an inertial reference frame. You learned in Chapter 4, and in more detail in Chapter 6, that Newton's laws are valid only in inertial reference frames. The roadside observer is in the earth's inertial reference frame. His observations of the car decelerating relative to the earth while you continue forward with constant velocity are in accord with Newton's laws.

Nonetheless, the fact that you *seem* to be hurled forward relative to the car is a very real experience. You can describe your experience in terms of what are called **fictitious forces**. These are not real forces, because no agent is exerting them, but they describe your motion *relative to a noninertial reference frame*. Figure 7.22 shows the situation from both reference frames.

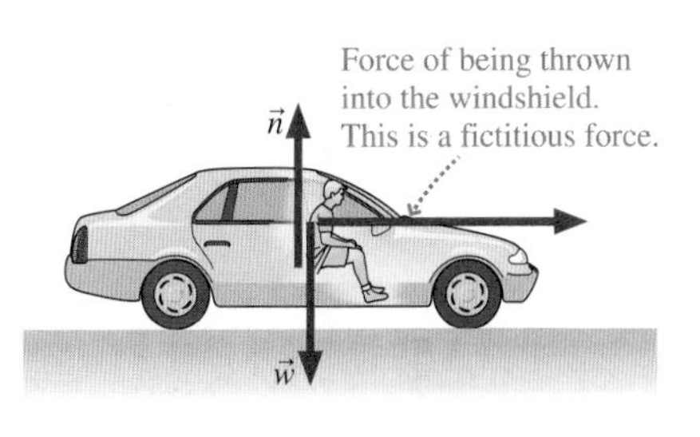

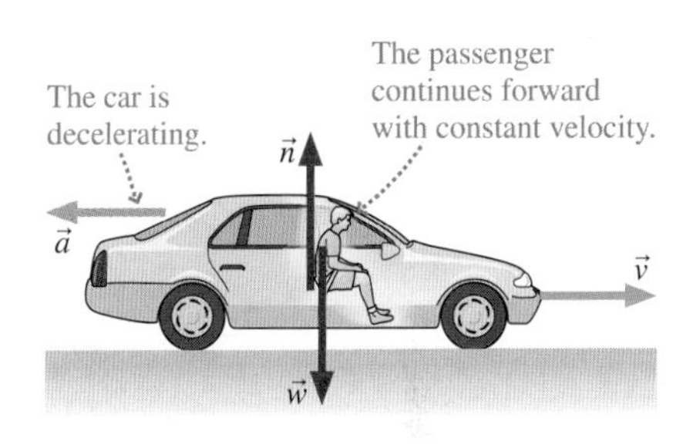

FIGURE 7.22 The forces are properly identified only in an inertial reference frame.

Centrifugal Force?

If the car turns a corner quickly, you feel "thrown" against the door. But is there really such a force? Figure 7.23 shows a bird's-eye view of you riding in a car as it makes a left turn. You try to continue moving in a straight line, obeying Newton's first law, when—without having been provoked—the door suddenly turns in front of you and runs into you! You do, indeed, then feel the force of the door because it is now the normal force of the door, pointing *inward* toward the center of the curve, that is causing you to turn the corner. But you were not "thrown" into the door; the door ran into you. The bird's-eye view, from an inertial reference frame, gives the proper perspective of what happens.

The "force" that seems to push an object to the outside of a circle is called the *centrifugal force*. Despite having a name, the centrifugal force is a fictitious force. It describes your experience *relative to a noninertial reference frame*, but there really is no such force. You must always use Newton's laws in an inertial reference frame, such as the reference frame of the ground. **A centrifugal force will never appear on a free-body diagram and never be included in Newton's laws.**

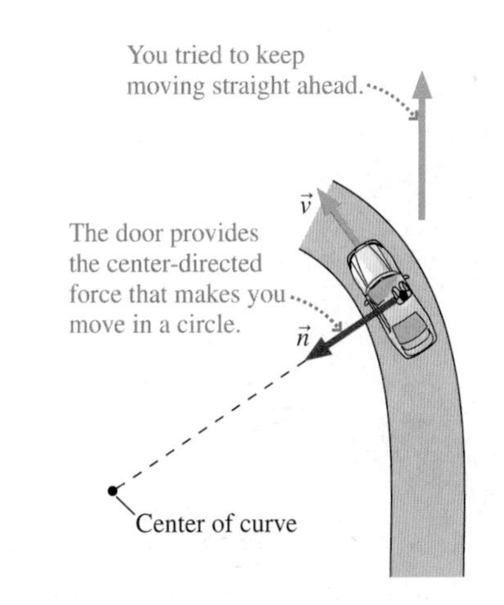

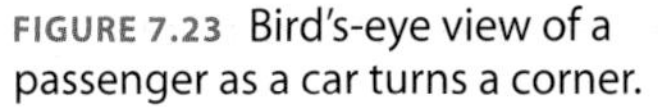
FIGURE 7.23 Bird's-eye view of a passenger as a car turns a corner.

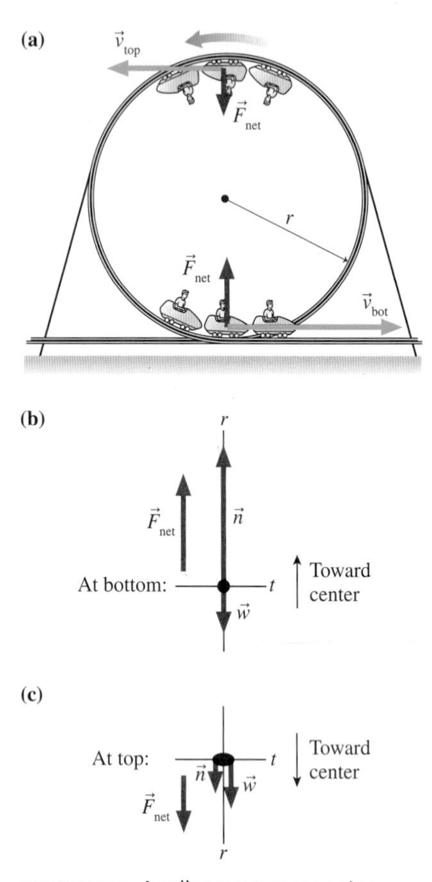

FIGURE 7.24 A roller coaster car going around a loop-the-loop.

NOTE ▶ You might wonder if the *rtz*-coordinate system is an inertial reference frame. It is, and Newton's laws apply, although the reason is rather subtle. We're using the *rtz*-coordinates to establish directions for decomposing vectors, but we're not making measurements in the *rtz*-system. That is, velocities and accelerations are measured in the laboratory reference frame. The particle would always be at rest ($\vec{v} = \vec{0}$) if we measured velocities in a reference frame attached to the particle. Thus the analysis of this chapter really is in the laboratory's inertial reference frame. ◀

Why Does the Water Stay in the Bucket?

Imagine swinging a bucket of water over your head. If you swing the bucket quickly, the water stays in. But you'll get a shower if you swing too slowly. Why does the water stay in the bucket? You might have thought there was a centrifugal force holding the water in, but we see that there isn't a centrifugal force. Analyzing this question will tell us a lot about forces in general and circular motion in particular. We'll begin by looking at a similar situation—a roller coaster—then return to the water in the bucket.

Figure 7.24a shows a roller coaster car going around a vertical loop-the-loop of radius r. We'll assume that the motion makes a complete circle and not worry about the entrance to and exit from the loop. Now, motion in a vertical circle is *not* uniform circular motion. The car slows down as it goes up one side and speeds up as it comes back down the other, so there is a component of the acceleration $\vec{a}$ that is *tangent* to the circle. But when the car is at the very top and very bottom points, it is changing only direction, not speed, so at those points the acceleration is purely centripetal and the circular-motion version of Newton's second law from Section 7.3 applies.

If you've ever ridden a roller coaster, you know that your sensation of weight changes as you go over the crests and through the dips. To understand why, Figure 7.24b shows a passenger's free-body diagram at the *bottom* of the loop. The only forces acting on the passenger are her weight $\vec{w}$ and the normal force $\vec{n}$ of the seat pushing up on her. Recall, from Chapter 5, that the passenger's apparent weight, her sensation of weight, is the magnitude of the force supporting her. Here the seat is supporting her with the normal force $\vec{n}$, so her apparent weight is $w_{app} = n$.

From this information we can deduce the following:

- She's moving in a circle, so there *must* be a net force toward the center of the circle—above her head—to provide the centripetal acceleration.
- The net force points *upward*, so it must be the case that $n > w$.
- Her apparent weight is $w_{app} = n$, so her apparent weight is larger than her true weight ($w_{app} > w$). Thus she "feels heavy" at the bottom of the circle.

In short, the normal force has to *exceed* the weight force to provide the net force she needs to "turn the corner" at the bottom of the circle. The logic of this analysis is especially important.

To analyze the situation quantitatively, notice that the r-axis, which must point toward the center of the circle, points *upward*. Thus the r-component of Newton's second law for circular motion is

$$\sum F_r = n_r + w_r = n - w = ma_r = \frac{m(v_{bot})^2}{r} \tag{7.23}$$

From Equation 7.23 we find

$$w_{app} = n = w + \frac{m(v_{bot})^2}{r} \tag{7.24}$$

The passenger's apparent weight at the bottom is *larger* than her true weight w, which agrees with your experience when you go through a dip or a valley.

Many popular amusement park rides are based on circular motion.

Now let's look at the roller coaster car as it crosses the top of the loop. This looks more like the water in the bucket, but things are a little trickier here. Whereas the normal force of the track pushes up when the car is at the bottom of the circle, it *presses down* when the car is at the top and the track is above the car. Figure 7.24c shows the car's free-body diagram at the top of the loop. Think about this diagram carefully to make sure you agree.

The car is still moving in a circle, so there *must* be a net force toward the center of the circle to provide the centripetal acceleration. The r-axis, which must point toward the center of the circle, now points *downward*. Consequently, both forces have *positive* components. Newton's second law at the top of the circle is

$$\sum F_r = n_r + w_r = n + w = \frac{m(v_{\text{top}})^2}{r} \tag{7.25}$$

Be sure you understand why this equation differs from Equation 7.23, describing what happens at the bottom.

From Equation 7.25, the normal force that the track exerts on the car is

$$n = \frac{m(v_{\text{top}})^2}{r} - w \tag{7.26}$$

If v_{top} is sufficiently large, the apparent weight $w_{\text{app}} = n$ of the car (and of the passengers in the car) can be larger than the true weight.

Our interest, however, is in what happens as the car gets slower and slower. Notice from Equation 7.26 that, as v_{top} decreases, there comes a point when n reaches zero. At that point, the track is *not* pushing against the car. Instead, the car is able to complete the circle because the weight force alone provides sufficient centripetal acceleration.

The speed at which $n = 0$ is called the *critical speed* v_c:

$$v_c = \sqrt{\frac{rw}{m}} = \sqrt{\frac{rmg}{m}} = \sqrt{rg} \tag{7.27}$$

The critical speed is the slowest speed at which the car can complete the circle. To understand why, notice that Equation 7.26 gives a negative value for n if $v < v_c$. But that is physically impossible. The track can push against the wheels of the car ($n > 0$), but it can't pull on them. When a solution becomes physically impossible, it usually indicates that we've made an incorrect assumption about the situation. In this case, we *assumed* that the motion was circular. But if we find that $n < 0$, our assumption is no longer valid. If $v < v_c$, the car cannot turn the full loop but, instead, comes off the track and becomes a projectile!

If you look back at the free-body diagram, **the critical speed v_c is the speed at which the weight force alone is sufficient to cause circular motion at the top.** The normal force has shrunk to zero. Circular motion with a speed less than v_c isn't possible because there's *too much* downward force. If the car attempts to go around at a lower speed, the normal force drops to zero *before* the car reaches the top. "No normal force" means "no contact." The car leaves the track when n reaches zero, becoming a projectile moving under the influence of only the weight force. Figure 7.25 summarizes this reasoning for the car on the loop-the-loop.

Returning now to the water in the bucket, Figure 7.26 on the next page shows a water-filled bucket at the top of a circle of radius r. Notice that the water has a tangential velocity. If the bucket suddenly disappeared, the water wouldn't fall straight down. Instead, it would fall along a parabolic trajectory like a ball that is thrown horizontally. This is the motion of an object acted on only by gravity.

If the water is to follow a circular trajectory with *more curvature* than the parabola, it needs *more force* than just its weight. The extra force is provided by

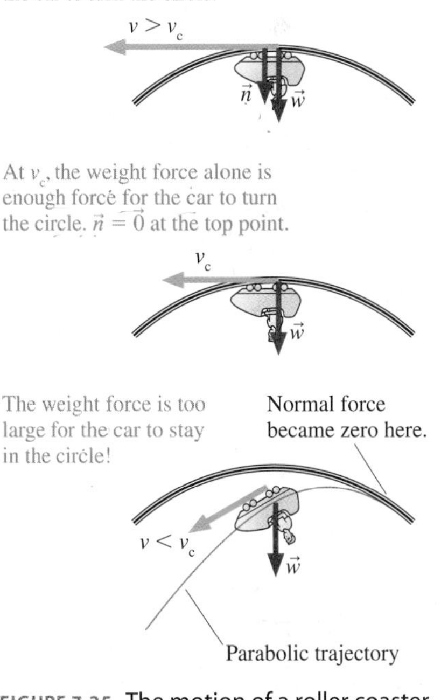
The normal force adds to the weight to make a large enough force for the car to turn the circle.
$v > v_c$
$\vec{n}$
$\vec{w}$
At v_c, the weight force alone is enough force for the car to turn the circle. $\vec{n} = \vec{0}$ at the top point.
v_c
$\vec{w}$
The weight force is too large for the car to stay in the circle!
Normal force became zero here.
$v < v_c$
$\vec{w}$
Parabolic trajectory

FIGURE 7.25 The motion of a roller coaster car for speeds above and below the critical speed.

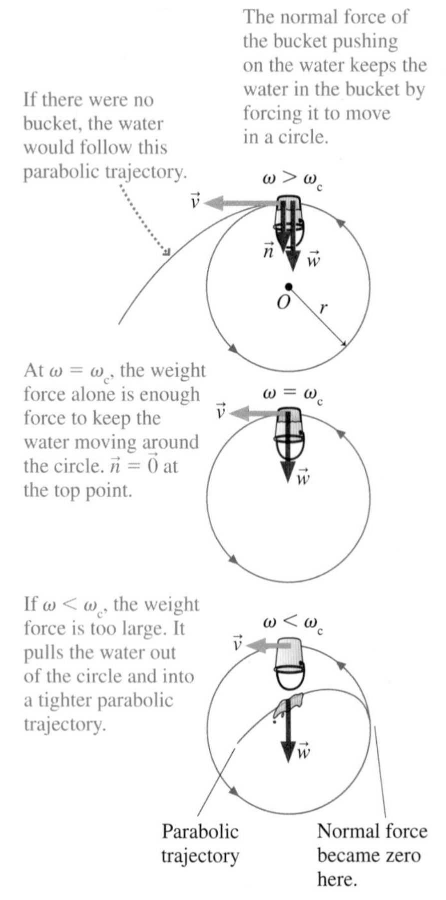

The normal force of the bucket pushing on the water keeps the water in the bucket by forcing it to move in a circle.
If there were no bucket, the water would follow this parabolic trajectory.
$\omega > \omega_c$
$\vec{v}$
$\vec{n}$
$\vec{w}$
O
r
At $\omega = \omega_c$, the weight force alone is enough force to keep the water moving around the circle. $\vec{n} = \vec{0}$ at the top point.
$\omega = \omega_c$
$\vec{v}$
$\vec{w}$
If $\omega < \omega_c$, the weight force is too large. It pulls the water out of the circle and into a tighter parabolic trajectory.
$\omega < \omega_c$
$\vec{v}$
$\vec{w}$
Parabolic trajectory
Normal force became zero here.

FIGURE 7.26 The forces on the water in a bucket.

the bottom of the bucket pushing on the water. As long as the bucket is pushing on the water, the water and the bucket are in contact and thus the water is "in" the bucket. (Water is a deformable substance, so the sides of the bucket are needed to keep the water together but otherwise aren't relevant to the motion.)

Notice the similarity to the car making the left turn in Figure 7.23. The passenger feels like he's being "hurled" into the door by a centrifugal force, but it's actually the pushing force from the door, pushing inward toward the center of the circle, that causes the passenger to turn the corner instead of moving straight ahead. Here it seems like the water is being "pinned" against the bottom of the bucket by a centrifugal force, but it's really the pushing force from the bottom of the bucket that causes the water to move in a circle instead of following a free-fall parabola.

As you gradually slow the speed of the bucket, the normal force of the bucket on the water gets smaller and smaller. There comes a point, as the angular velocity ω decreases, when n reaches zero. At that point, the bucket is *not* pushing against the water. Instead, the water is able to complete the circle because the weight force alone provides sufficient centripetal acceleration.

The critical angular velocity ω_c is the angular velocity at which the weight force alone is sufficient to cause circular motion at the top. Circular motion with an angular velocity less than ω_c isn't possible because there's *too much* downward force. If you attempt to swing the bucket with a smaller angular velocity, the normal force drops to zero *before* the water reaches the top. "No normal force" means "no contact." The water leaves the bucket when n becomes zero, becoming a projectile moving under the influence of only the weight force. That's when you get wet!

The analysis is exactly the same as for the roller coaster car. The only difference is that you're likely to swing a bucket with constant angular velocity (the roller coaster car does *not* have constant angular velocity), so it's more useful to calculate the critical angular velocity rather than the critical speed. We can find the critical angular velocity from Equation 7.27 by using $v_c = \omega_c r$. This gives

$$\omega_c = \sqrt{\frac{g}{r}} \tag{7.28}$$

It's not easy to understand why the water stays in the bucket. Careful thought about the *reasoning* presented in this section will greatly increase your understanding of forces and circular motion.

STOP TO THINK 7.4 A car is rolling over the top of a hill at speed v. At this instant,

a. $n > w$.
b. $n < w$.
c. $n = w$.
d. We can't tell about n without knowing v.

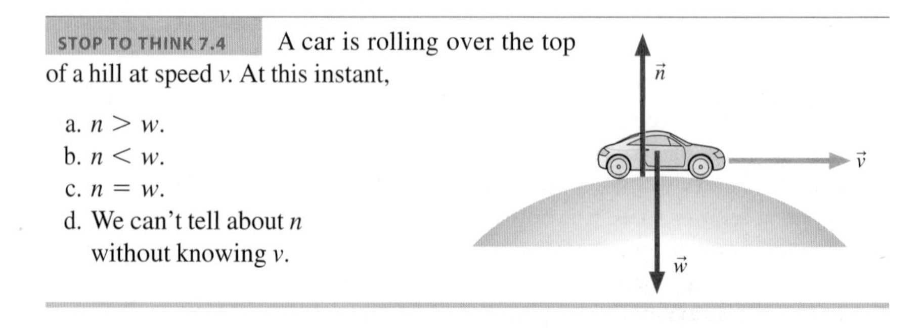

7.6 Nonuniform Circular Motion

Many interesting examples of circular motion involve objects whose speed changes. A roller coaster car doing a loop-the-loop slows down as it goes up one side, speeds up as it comes back down the other. The same is true if a ball on a string is swung in a vertical circle. The ball in a roulette wheel gradually slows until it stops. Circular motion with a changing speed is called **nonuniform circular motion.**

Figure 7.27a reminds you of uniform circular motion. The nonuniform circular motion of Figure 7.27b still has $v_r = 0$, because the particle is constrained to follow the circle, but v_t is no longer constant. The tangential component of velocity v_t changes if the particle speeds up or slows down. A changing tangential velocity implies that the particle has a **tangential acceleration**

$$a_t = \frac{dv_t}{dt} \tag{7.29}$$

The tangential acceleration is what causes the particle to change the speed with which it goes *around* the circle.

The tangential acceleration that changes the particle's speed is in addition to the radial, or centripetal, acceleration that changes the particle's direction of motion. The full acceleration vector $\vec{a}$, shown in Figure 7.28, can be written in terms of its radial and tangential component vectors as

$$\vec{a} = \vec{a}_r + \vec{a}_t \tag{7.30}$$

The magnitude of the acceleration is

$$a = \sqrt{a_r^2 + a_t^2} \tag{7.31}$$

and its angle from the r-axis, which we'll call ϕ, is

$$\phi = \tan^{-1}\left(\frac{a_t}{a_r}\right) \tag{7.32}$$

The acceleration vector is a pure centripetal acceleration, pointing toward the center of the circle ($\phi = 0$), only if $a_t = 0$.

If a_t is constant, then the arc length s traveled by the particle around the circle and the tangential velocity v_t are found from constant-acceleration kinematics:

$$\begin{aligned} s_f &= s_i + v_{it}\Delta t + \tfrac{1}{2}a_t(\Delta t)^2 \\ v_{ft} &= v_{it} + a_t\Delta t \end{aligned} \tag{7.33}$$

We can express these kinematic equations in terms of angular quantities by dividing both sides of Equation 7.33 by the radius r:

$$\begin{aligned} \frac{s_f}{r} &= \frac{s_i}{r} + \frac{v_{it}}{r}\Delta t + \frac{a_t}{2r}(\Delta t)^2 \\ \frac{v_{ft}}{r} &= \frac{v_{it}}{r} + \frac{a_t}{r}\Delta t \end{aligned} \tag{7.34}$$

You'll recognize that s/r is the angular position θ and v_t/r is the angular velocity ω. Thus the angular position and velocity after undergoing tangential acceleration a_t are

$$\begin{aligned} \theta_f &= \theta_i + \omega_i\Delta t + \frac{a_t}{2r}(\Delta t)^2 \\ \omega_f &= \omega_i + \frac{a_t}{r}(\Delta t) \end{aligned} \quad \text{(nonuniform circular motion)} \tag{7.35}$$

Equations 7.35 are an extension of Equation 7.6 to the case of nonuniform circular motion. In addition, the centripetal acceleration equation $a_r = v^2/r = \omega^2 r$ is still valid.

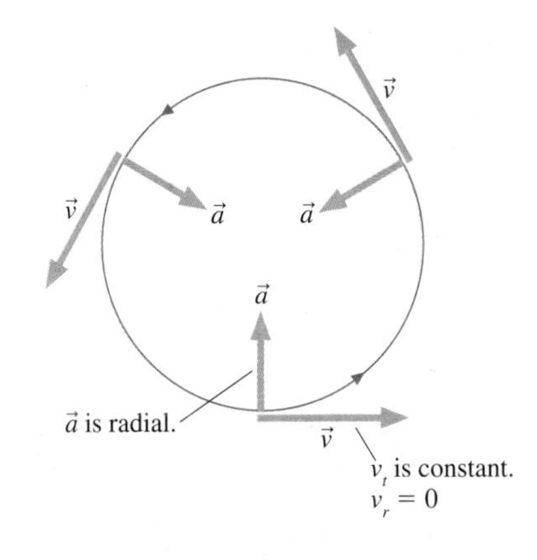
(a) Uniform circular motion
$\vec{v}$
$\vec{a}$
$\vec{a}$ is radial.
v_t is constant.
$v_r = 0$

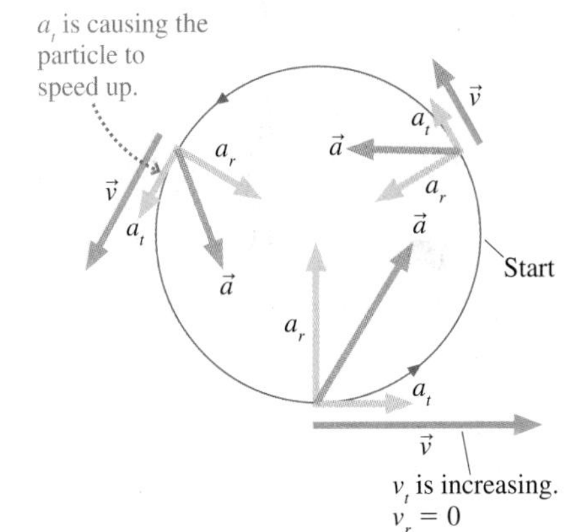
(b) Nonuniform circular motion
a_t is causing the particle to speed up.
$\vec{v}$
$\vec{a}$
a_r
a_t
Start
v_t is increasing.
$v_r = 0$

FIGURE 7.27 Uniform and nonuniform circular motion.

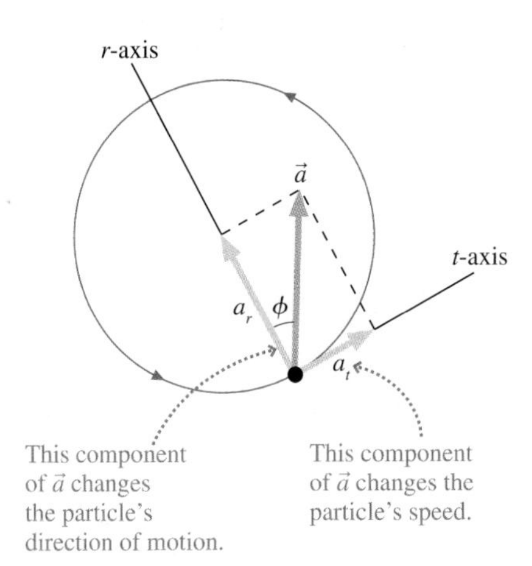
r-axis
$\vec{a}$
t-axis
a_r
ϕ
a_t
This component of $\vec{a}$ changes the particle's direction of motion.
This component of $\vec{a}$ changes the particle's speed.

FIGURE 7.28 The acceleration vector has a radial and tangential component.

EXAMPLE 7.9 Circular rocket motion

A model rocket is attached to the end of a 2.0-m-long massless, rigid rod. The other end of the rod rotates on a frictionless pivot, causing the rocket to move in a horizontal circle. The rocket accelerates at 1.0 m/s^2 for 10 s, then runs out of fuel.

a. What is the rocket's angular velocity, in rpm, when it runs out of fuel?
b. How many revolutions has the rocket made at that time?
c. What is the magnitude of $\vec{a}$ and its angle from the r-axis at $t = 2$ s?

MODEL Model the rocket as a particle in nonuniform circular motion. Assume that the rocket starts from rest.

VISUALIZE Figure 7.29 is a pictorial representation of the situation. The acceleration caused by the rocket motor is the tangential acceleration, $a_t = 1.0$ m/s^2.

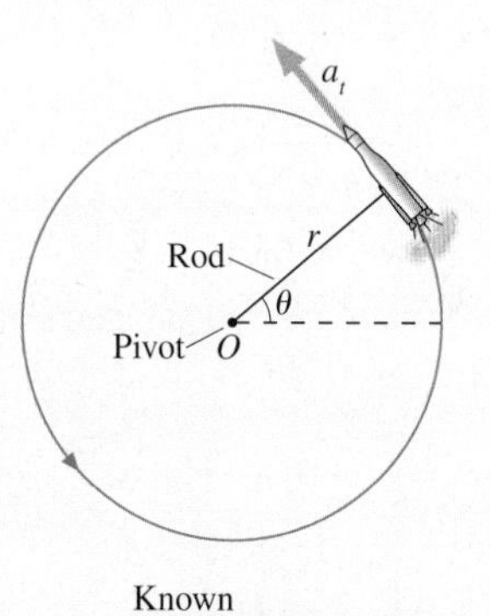

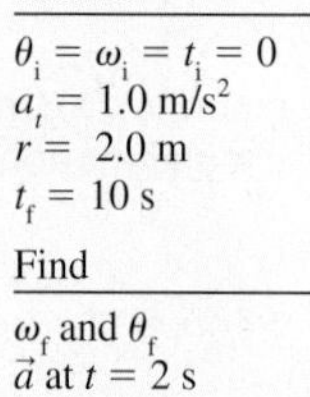

FIGURE 7.29 The nonuniform circular motion of a model rocket.

SOLVE

a. The angular velocity after 10 s is

$$\omega_f = \omega_i + \frac{a_t}{r}\Delta t = 0 + \frac{1.0 \text{ m/s}^2}{2.0 \text{ m}} 10 \text{ s} = 5.0 \text{ rad/s}$$

This is another situation in which we explicitly inserted the rad unit. Converting to rpm:

$$\omega_f = \frac{5.0 \text{ rad}}{1 \text{ s}} \times \frac{1 \text{ rev}}{2\pi \text{ rad}} \times \frac{60 \text{ s}}{1 \text{ min}} = 48 \text{ rpm}$$

b. The angular position at 10 s is

$$\theta_f = \theta_i + \omega_i \Delta t + \frac{a_t}{2r}(\Delta t)^2$$
$$= 0 + 0 + \frac{1.0 \text{ m/s}^2}{2(2.0 \text{ m})}(10 \text{ s})^2 = 25 \text{ rad}$$

Converting to revolutions:

$$\theta_f = 25 \text{ rad} \times \frac{1 \text{ rev}}{2\pi \text{ rad}} = 3.98 \text{ rev} \approx 4 \text{ rev}$$

c. The rocket motor creates the tangential acceleration $a_t = 1.0$ m/s^2. As the rocket speeds up, tension in the rod causes a radial acceleration $a_r = \omega^2 r$. At $t = 2$ s,

$$\omega_{2\text{ s}} = \omega_i + \frac{a_t}{r}\Delta t = 0 + \frac{1.0 \text{ m/s}^2}{2.0 \text{ m}} 2.0 \text{ s} = 1.0 \text{ rad/s}$$

$$a_r = (\omega_{2\text{ s}})^2 r = (1.0 \text{ rad/s})^2(2.0 \text{ m}) = 2.0 \text{ m/s}^2$$

This is a situation where we explicitly *dropped* rad^2 from the units of the answer because the final result is a linear quantity, not an angular quantity. Now we can use Equations 7.31 and 7.32 to find

$$a = \sqrt{a_r^2 + a_t^2} = \sqrt{(2.0 \text{ m/s}^2)^2 + (1.0 \text{ m/s}^2)^2} = 2.24 \text{ m/s}^2$$

$$\phi = \tan^{-1}\left(\frac{a_t}{a_r}\right) = \tan^{-1}\left(\frac{1.0 \text{ m/s}^2}{2.00 \text{ m/s}^2}\right) = 27°$$

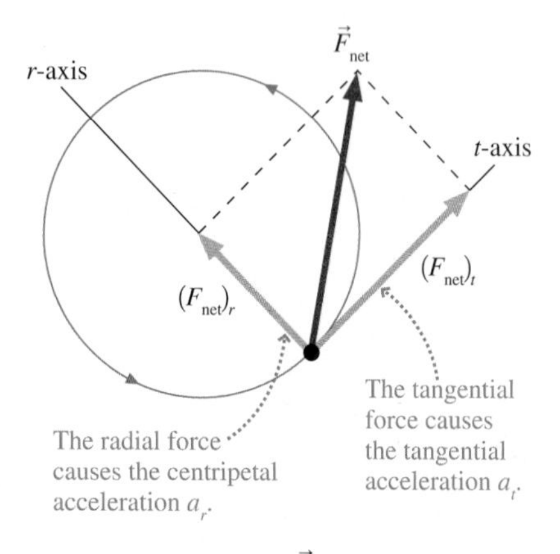

FIGURE 7.30 Net force $\vec{F}_{net}$ is applied to a particle moving in a circle.

Dynamics of Nonuniform Circular Motion

Figure 7.30 shows a net force $\vec{F}_{net}$ acting on a particle as it moves around a circle of radius r. $\vec{F}_{net}$ is likely to be a superposition of several forces, such as a tension force in a string, a thrust force, a friction force, and so on.

We can decompose the force vector $\vec{F}_{net}$ into a *tangential* component $(F_{net})_t$ and a radial component $(F_{net})_r$. The component $(F_{net})_t$ is positive for a tangential force in the ccw direction, negative for a tangential force in the cw direction. Because of our definition of the r-axis, the component $(F_{net})_r$ is positive for a radial force *toward* the center, negative for a radial force away from the center. For example, the particular force illustrated in Figure 7.30 has positive values for both $(F_{net})_t$ and $(F_{net})_r$.

The force component $(F_{net})_r$ perpendicular to the trajectory creates a centripetal acceleration and causes the particle to change directions. It is the component $(F_{net})_t$ parallel to the trajectory that creates a tangential acceleration and causes the particle to change speed. Force and acceleration are related to each other through Newton's second law:

$$\begin{aligned}(F_{\text{net}})_r &= \sum F_r = ma_r = \frac{mv^2}{r} = m\omega^2 r\\(F_{\text{net}})_t &= \sum F_t = ma_t\\(F_{\text{net}})_z &= \sum F_z = 0\end{aligned} \tag{7.36}$$

NOTE ▶ Equations 7.36 differ from Equations 7.16 for uniform circular motion only in the fact that a_t is no longer constrained to be zero. ◀

EXAMPLE 7.10 Slowing circular motion

A motor spins a 2.0 kg steel block around on an 80-cm-long arm at 200 rpm. The block is supported by a steel table. After the motor stops, how long does the block take to come to rest? How many revolutions does the block make during this time? Assume that the axle is frictionless.

MODEL Model the steel block as a particle in nonuniform circular motion.

VISUALIZE Figure 7.31 shows the pictorial and physical representations. Notice that, for the first time, we need a free-body diagram showing forces in three dimensions.

SOLVE If the table were frictionless, the block would spin around forever because of the frictionless axle. However, friction between the block and table exerts a retarding force $\vec{f}_k$ on the block. Kinetic friction is always opposite the direction of motion $\vec{v}$, so $\vec{f}_k$ is *tangent* to the circle.

The magnitude of the friction force is $f_k = \mu_k n$. The vertical forces, perpendicular to the plane of the motion, are the normal force $\vec{n}$ and the weight $\vec{w}$. There's no net force in the vertical direction, so the z-component of the second law is

$$\sum F_z = n - w = 0$$

from which we can conclude that $n = w = mg$ and thus $f_k = \mu_k mg$. The friction force is the only tangential component of force, so the t-component of Newton's second law is

$$\sum F_t = (f_k)_t = -f_k = ma_t$$

$$a_t = \frac{-f_k}{m} = \frac{-\mu_k mg}{m} = -\mu_k g = -5.88 \text{ m/s}^2$$

The coefficient of friction for steel on steel was taken from Table 5.1. The component $(f_k)_t$ is negative because the friction force vector points in the clockwise direction. The initial angular velocity needs to be converted to rad/s:

$$\omega_i = \frac{200 \text{ rev}}{1 \text{ min}} \times \frac{2\pi \text{ rad}}{1 \text{ rev}} \times \frac{1 \text{ min}}{60 \text{ s}} = 20.9 \text{ rad/s}$$

We can now use Equation 7.35 for circular kinematics to find the time it takes the block to come to rest:

$$\omega_f = 0 \text{ rad/s} = \omega_i + \frac{a_t}{r}(\Delta t) = \omega_i + \frac{a_t}{r} t_f$$

$$t_f = -\frac{r\omega_i}{a_t} = -\frac{(0.80 \text{ m})(20.9 \text{ rad/s})}{-5.88 \text{ m/s}^2} = 2.84 \text{ s}$$

The angular displacement while the block slows to a stop is then

$$\begin{aligned}\Delta\theta = \theta_f - \theta_i &= \omega_i t_f + \frac{a_t}{2r} t_f^2\\&= (20.9 \text{ rad/s})(2.84 \text{ s}) + \frac{(-5.88 \text{ m/s})}{2(0.80 \text{ m})}(2.84 \text{ s})^2\\&= 29.7 \text{ rad} \times \frac{1 \text{ rev}}{2\pi \text{ rad}} = 4.73 \text{ rev}\end{aligned}$$

ASSESS Is this answer reasonable? The block was moving pretty fast—200 rpm on an arm about 30 in long. Even though friction of steel on steel is fairly large, it's reasonable that the block would make several revolutions before stopping. The purpose of the assessment, as always, is not to prove that the answer is right but to rule out obviously unreasonable answers that have been reached by mistake.

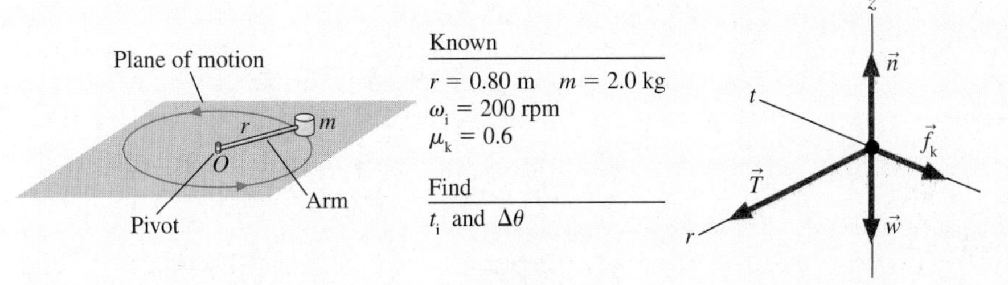

FIGURE 7.31 Pictorial and physical representations of the block of Example 7.10.

4.2, 4.3, 4.4 ActivPhysics ONLINE

We've come a long way since our first dynamics problems in Chapter 5, but our basic strategy has not changed.

PROBLEM-SOLVING STRATEGY 7.1 Circular motion problems

MODEL Make simplifying assumptions.

VISUALIZE **Pictorial representation.** Establish a coordinate system with the r-axis pointing toward the center of the circle. Show important points in the motion on a sketch. Define symbols and identify what the problem is trying to find.

Physical representation. Identify the forces and show them on a free-body diagram.

SOLVE Newton's second law is

$$(F_{\text{net}})_r = \sum F_r = ma_r = \frac{mv^2}{r} = m\omega^2 r$$
$$(F_{\text{net}})_t = \sum F_t = ma_t$$
$$(F_{\text{net}})_z = \sum F_z = 0$$

- Determine the force components from the free-body diagram. Be careful with signs.
- Solve for the acceleration, then use kinematics to find velocities and positions.

ASSESS Check that your result has the correct units, is reasonable, and answers the question.

STOP TO THINK 7.5 A ball on a string is swung in a vertical circle. The string happens to break when it is parallel to the ground and the ball is moving up. Which trajectory does the ball follow?

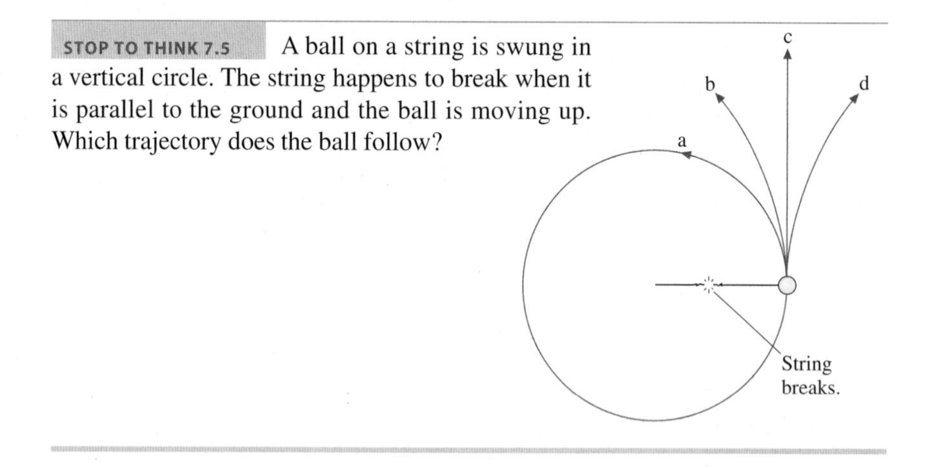

SUMMARY

The goal of Chapter 7 has been to learn to solve problems about motion in a circle.

GENERAL PRINCIPLES

Newton's Second Law

Expressed in *rtz*-component form:

$$(F_{\text{net}})_r = \sum F_r = ma_r = \frac{mv^2}{r} = m\omega^2 r \qquad (F_{\text{net}})_t = \sum F_t = \begin{cases} 0 & \text{uniform motion} \\ ma_t & \text{nonuniform motion} \end{cases} \qquad (F_{\text{net}})_z = \sum F_z = 0$$

Uniform Circular Motion

- v is constant.
- $\vec{F}_{\text{net}}$ points toward the center of the circle.
- The **centripetal acceleration** $\vec{a}$ points toward the center of the circle. It changes the particle's direction but not its speed.

Nonuniform Circular Motion

- v changes.
- $\vec{a}$ is parallel to $\vec{F}_{\text{net}}$.
- The radial component a_r changes the particle's direction.
- The tangential component a_t changes the particle's speed.

IMPORTANT CONCEPTS

rtz-coordinates

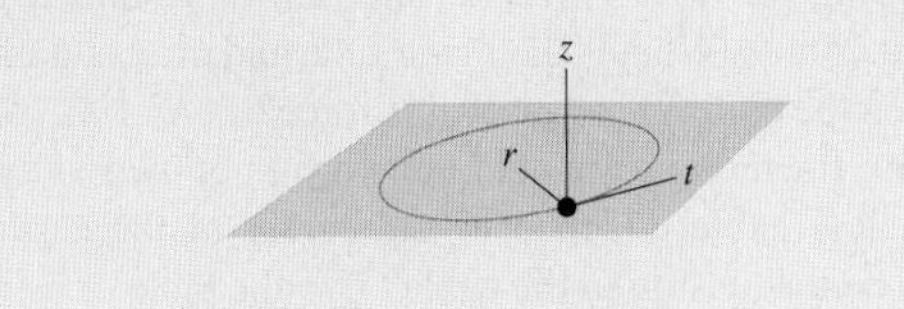

Angular position

$\theta = s/r$

Angular velocity

$\omega = d\theta/dt$

$v_t = \omega r$

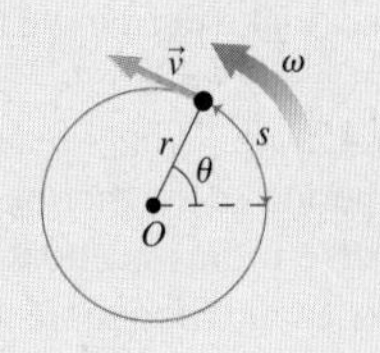

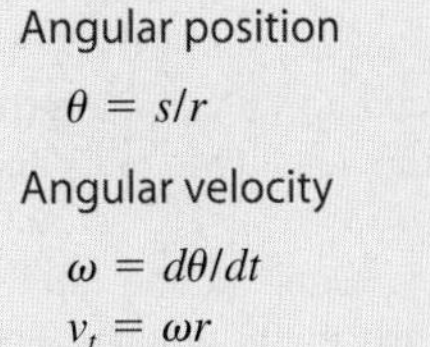

APPLICATIONS

Circular motion kinematics

Period $T = \dfrac{2\pi r}{v} = \dfrac{2\pi}{\omega}$

Uniform circular motion

$v_t = \text{constant} \qquad \omega = \text{constant}$

$\theta_f = \theta_i + \omega\Delta t$

Nonuniform circular motion

$\theta_f = \theta_i + \omega_i\Delta t + \dfrac{a_t}{2r}(\Delta t)^2$

$\omega_f = \omega_i + \dfrac{a_t}{r}\Delta t$

Orbits

A circular orbit has radius r if

$v = \sqrt{rg}$

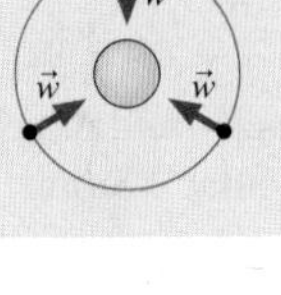

Apparent weight

Circular motion requires a net force pointing to the center. The apparent weight $w_{\text{app}} = n$ is usually not the same as the true weight w. n must be > 0 for the object to be in contact with a surface.

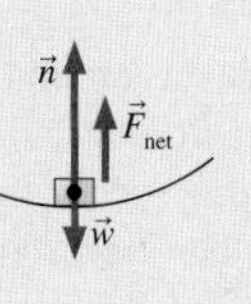

TERMS AND NOTATION

uniform circular motion
period, T
angular position, θ
arc length, s
radians
angular displacement, $\Delta\theta$
angular velocity, ω
centripetal acceleration, a_r
orbit
fictitious force
nonuniform circular motion
tangential acceleration, a_t

EXERCISES AND PROBLEMS

The icon indicates that the problem can be done on a Dynamics Worksheet.

Exercises

Section 7.1 Uniform Circular Motion

1. Figure Ex7.1 shows the angular-position-versus-time graph for a particle moving in a circle.
   a. Write a description of the particle's motion.
   b. Draw the angular-velocity-versus-time graph.

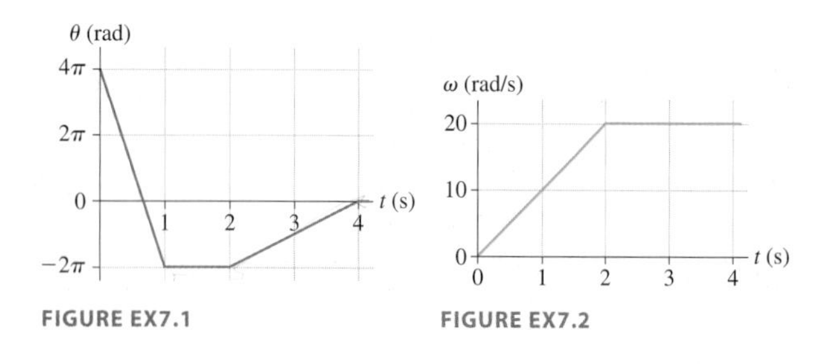

FIGURE EX7.1

FIGURE EX7.2

2. Figure Ex7.2 shows the angular-velocity-versus-time graph for a particle moving in a circle. How many revolutions does the object make during the first 4 s?
3. An old-fashioned single-play vinyl record rotates on a turntable at 45 rpm. What are (a) the angular velocity in rad/s and (b) the period of the motion?
4. The earth's radius is about 4000 miles. Kampala, the capital of Uganda, and Singapore are both nearly on the equator. The distance between them is 5000 miles.
   a. Through what angle do you turn, relative to the earth, if you fly from Kampala to Singapore? Give your answer in both radians and degrees.
   b. The flight from Kampala to Singapore takes 9 hours. What is the plane's angular velocity?

Section 7.2 Velocity and Acceleration in Uniform Circular Motion

5. The radius of the earth's very nearly circular orbit around the sun is 1.5×10^{11} m. Find the magnitude of the earth's (a) velocity, (b) angular velocity, and (c) centripetal acceleration as it travels around the sun. Assume a year of 365 days.
6. In uniform circular motion, which of the following quantities are constant: speed, instantaneous velocity, radial velocity, radial acceleration, tangential acceleration, the magnitude of the net force? Which of the quantities are zero throughout the motion?
7. Your roommate is working on his bicycle and has the bike upside down. He spins the 60-cm-diameter wheel, and you notice that a pebble stuck in the tread goes by three times every second. What are the pebble's speed and acceleration?
8. A 300-m-tall tower is built on the equator. How much faster does a point at the top of the tower move than a point at the bottom?
9. To withstand "g-forces" of up to 10 g's, caused by suddenly pulling out of a steep dive, fighter jet pilots train on a "human centrifuge." 10 g's is an acceleration of 98 m/s^2. If the length of the centrifuge arm is 12 m, at what speed is the rider moving when she experiences 10 g's?
10. How fast must a plane fly along the earth's equator so that the sun stands still relative to the passengers? In which direction must the plane fly, east to west or west to east? Give your answer in both km/hr and mph. The radius of the earth is 6400 km.

Section 7.3 Dynamics of Uniform Circular Motion

11. A 200 g block on a 50-cm-long string swings in a circle on a horizontal, frictionless table at 75 rpm.
    a. What is the speed of the block?
    b. What is the tension in the string?
12. In the Bohr model of the hydrogen atom, an electron (mass $m = 9.1\times 10^{-31}$ kg) orbits a proton at a distance of 5.3×10^{-11} m. The proton pulls on the electron with an electric force of 9.2×10^{-8} N. How many revolutions per second does the electron make?
13. A highway curve of radius 500 m is designed for traffic moving at a speed of 90 km/hr. What is the correct banking angle of the road?
14. A 1500 kg car drives around a flat 200-m-diameter circular track at 25 m/s. What is the magnitude and direction of the net force on the car? What causes this force?
15. Suppose the moon were held in its orbit not by gravity but by a massless cable attached to the center of the earth. What would be the tension in the cable? Use the table of astronomical data inside the cover of the book.
16. A 30 g ball rolls around a 40-cm-diameter L-shaped track, shown in Figure Ex7.16, at 60 rpm. What is the magnitude of the net force that the track exerts on the ball? Rolling friction can be neglected.

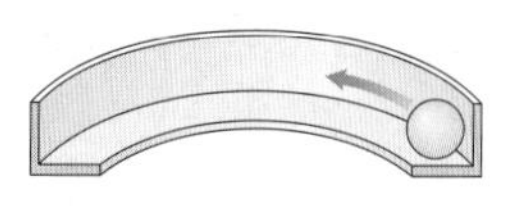

FIGURE EX7.16

Section 7.4 Circular Orbits

17. A satellite orbiting the moon very near the surface has a period of 110 min. What is the moon's acceleration due to gravity?
18. What is the acceleration due to gravity of the sun at the distance of the earth's orbit?

Section 7.5 Fictitious Forces and Apparent Weight

19. The passengers in a roller coaster car feel 50% heavier than their true weight as the car goes through a dip with a 30 m radius of curvature. What is the car's speed at the bottom of the dip?
20. A roller coaster car crosses the top of a circular loop-the-loop at twice the critical speed. What is the ratio of the car's apparent weight to its true weight?

21. As a roller coaster car crosses the top of a 40-m-diameter loop-the-loop, its apparent weight is the same as its true weight. What is the car's speed at the top?
22. Measure the length of your arms. Then estimate the minimum angular velocity (in rpm) for swinging a bucket of water in a vertical circle without spilling any.

Section 7.6 Nonuniform Circular Motion

23. A car speeds up as it turns from traveling due south to heading due east. When exactly halfway around the curve, the car's acceleration is 3.0 m/s^2, 20° north of east. What are the radial and tangential components of the acceleration at that point?
24. A 5.0-m-diameter merry-go-round is initially turning with a 4.0 s period. It slows down and stops in 20 s.
    a. Before slowing, what is the speed of a child on the rim?
    b. How many revolutions does the merry-go-round make as it stops?
25. A 3.0-cm-diameter crankshaft that is rotating at 2500 rpm comes to a halt in 1.5 s.
    a. What is the tangential acceleration of a point on the surface of the crankshaft?
    b. How many revolutions does the crankshaft make as it stops?
26. A computer disk is 8.0 cm in diameter. A reference dot on the edge of the disk is initially located at $\theta = 45°$. The disk accelerates steadily for $\frac{1}{2}$ second, reaching 2000 rpm, then coasts at steady angular velocity for another $\frac{1}{2}$ second. What are the location and speed of the reference dot at $t = 1$ s?

Problems

27. A car starts from rest on a curve with a radius of 120 m and accelerates at 1.0 m/s^2. Through what angle will the car have traveled when the magnitude of its total acceleration is 2.0 m/s^2?
28. A typical laboratory centrifuge rotates at 4000 rpm. Test tubes have to be placed into a centrifuge very carefully because of the very large accelerations.
    a. What is the acceleration at the end of a test tube that is 10 cm from the axis of rotation?
    b. For comparison, what is the magnitude of the acceleration a test tube would experience if dropped from a height of 1.0 m and stopped in a 1.0-ms-long encounter with a hard floor?
29. Astronauts use a centrifuge to simulate the acceleration of a rocket launch. The centrifuge takes 30 s to speed up from rest to its top speed of 1 rotation every 1.3 s. The astronaut is strapped into a seat 6.0 m from the axis.
    a. What is the astronaut's tangential acceleration during the first 30 s?
    b. How many g's of acceleration does the astronaut experience when the device is rotating at top speed? Each 9.8 m/s^2 of acceleration is 1 g.
30. Communications satellites are placed in a circular orbit where they stay directly over a fixed point on the equator as the earth rotates. These are called *geosynchronous orbits*. The altitude of a geosynchronous orbit is 3.58×10^7 m ($\approx$22,000 miles).
    a. What is the period of a satellite in a geosynchronous orbit?
    b. Find the value of g at this altitude.
    c. What is the apparent weight of a 2000 kg satellite in a geosynchronous orbit?
31. A 75 kg man weighs himself at the north pole and at the equator. Which scale reading is higher? By how much?
32. A 1500 kg car takes a 50-m-radius unbanked curve at 15 m/s. What is the size of the friction force on the car?
33. The father of Example 7.5 stands at the summit of a conical hill as he spins his 20 kg child around on a 5.0 kg cart with a 2.0 m long rope. The sides of the hill are inclined at 20°. He again keeps the rope parallel to the ground, and friction is negligible. What rope tension will allow the cart to spin with the same 13.5 rpm it had in the example?
34. A 500 g ball swings in a vertical circle at the end of a 1.5-m-long string. When the ball is at the bottom of the circle, the tension in the string is 15 N. What is the speed of the ball at that point?
35. A concrete highway curve of radius 70 m is banked at a 15° angle. What is the maximum speed with which a 1500 kg rubber-tired car can take this curve without sliding?
36. A student ties a 500 g rock to a 1.0-m-long string and swings it around her head in a horizontal circle. At what angular velocity, in rpm, does the string tilt down at a 10° angle?
37. A 5.0 g coin is placed 15 cm from the center of a turntable. The coin has static and kinetic coefficients of friction with the turntable surface of $\mu_s = 0.80$ and $\mu_k = 0.50$. The turntable very slowly speeds up to 60 rpm. Does the coin slide off?
38. You've taken your neighbor's young child to the carnival to ride the rides. She wants to ride The Rocket. Eight rocket-shaped cars hang by chains from the outside edge of a large steel disk. A vertical axle through the center of the ride turns the disk, causing the cars to revolve in a circle. You've just finished taking physics, so you decide to figure out the speed of the cars while you wait. You estimate that the disk is 5 m in diameter and the chains are 6 m long. The ride takes 10 s to reach full speed, then the cars swing out until the chains are 20° from vertical. What is the car's speed?
39. A *conical pendulum* is formed by attaching a 500 g ball to a 1.0-m-long string, then allowing the mass to move in a horizontal circle of radius 20 cm. Figure P7.39 shows that the string traces out the surface of a cone, hence the name.
    a. What is the tension in the string?
    b. What is the ball's angular velocity, in rpm?

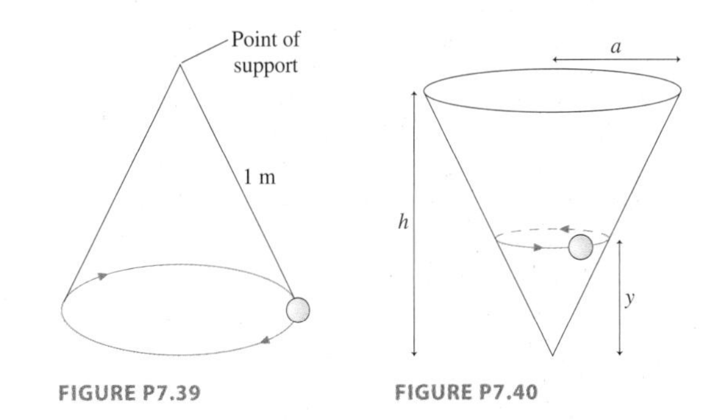

FIGURE P7.39 FIGURE P7.40

40. A small ball rolls around a horizontal circle at height y inside the cone shown in Figure P7.40. Find an expression of the ball's speed in terms of a, h, y, and g.
41. In an old-fashioned amusement park ride, passengers stand inside a 5.0-m-diameter hollow steel cylinder with their backs against the wall. The cylinder begins to rotate about a vertical axis. Then the floor on which the passengers are standing

suddenly drops away! If all goes well, the passengers will "stick" to the wall and not slide. Clothing has a static coefficient of friction against steel in the range 0.6 to 1.0 and a kinetic coefficient in the range 0.4 to 0.7. A sign next to the entrance says "No children under 30 kg allowed." What is the minimum angular velocity, in rpm, for which the ride is safe?

42. A 10 g steel marble is spun so that it rolls at 150 rpm around the *inside* of a vertically oriented steel tube. The tube, shown in Figure P7.42, is 12 cm in diameter. Assume that the rolling resistance is small enough for the marble to maintain 150 rpm for several seconds. During this time, will the marble spin in a horizontal circle, at constant height, or will it spiral down the inside of the tube?

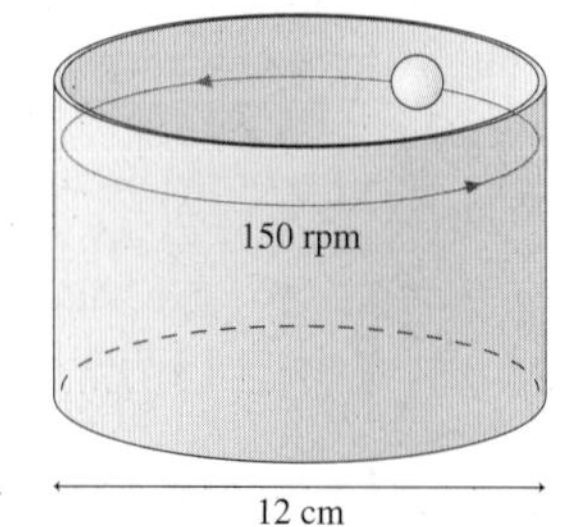

FIGURE P7.42

43. Three cars are driving at 25 m/s along the road shown in Figure P7.43. Car B is at the bottom of the hill and car C is at the top. Suppose each car suddenly brakes hard and starts to skid. What is the tangential acceleration (i.e., the acceleration parallel to the road) of each car? Assume $\mu_k = 1.0$.

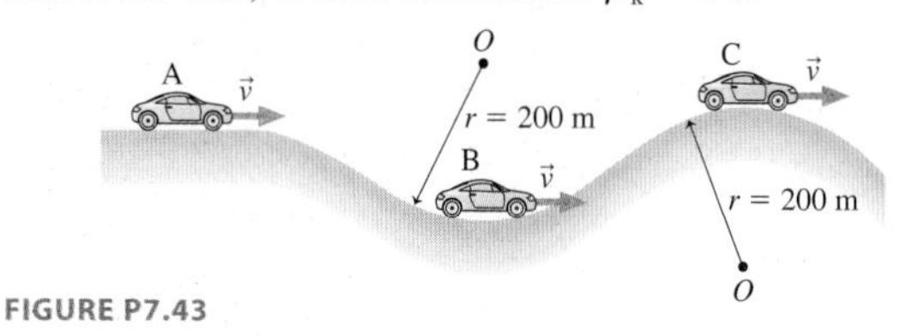

FIGURE P7.43

44. A car drives over the top of a hill that has a radius of 50 m. What maximum speed can the car have without flying off the road at the top of the hill?

45. A 500 g ball moves in a vertical circle on a 102-cm-long string. If the speed at the top is 4.0 m/s, then the speed at the bottom will be 7.5 m/s. (You'll learn how to show this in Chapter 10.)
    a. What is the ball's weight?
    b. What is the tension in the string when the ball is at the top?
    c. What is the tension in the string when the ball is at the bottom?

46. While at the county fair, you decide to ride the Ferris wheel. Having eaten too many candy apples and elephant ears, you find the motion somewhat unpleasant. To take your mind off your stomach, you wonder about the motion of the ride. You estimate the radius of the big wheel to be 15 m, and you use your watch to find that each loop around takes 25 s.
    a. What are your speed and magnitude of your acceleration?
    b. What is the ratio of your apparent weight to your true weight at the top of the ride?
    c. What is the ratio of your apparent weight to your true weight at the botom?

47. In an amusement park ride called The Roundup, passengers stand inside a 16-m-diameter rotating ring. After the ring has acquired sufficient speed, it tilts into a vertical plane, as shown in Figure P7.47.
    a. Suppose the ring rotates once every 4.5 s. If a rider's mass is 55 kg, with how much force does the ring push on her at the top of the ride? At the bottom?
    b. What is the longest rotation period of the wheel that will prevent the riders from falling off at the top?

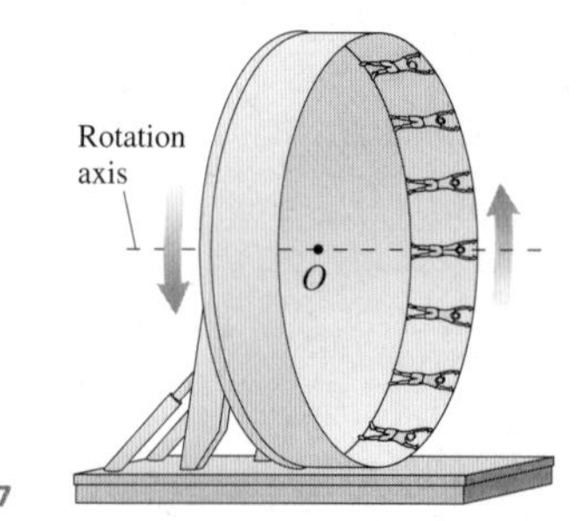

FIGURE P7.47

48. You have a new job designing rides for an amusement park. In one ride, the rider's chair is attached by a 9.0-m-long chain to the top of a tall rotating tower. The tower spins the chair and rider around at the rate of 1 rev every 4.0 s. In your design, you've assumed that the maximum possible combined weight of the chair and rider is 150 kg. You've found a great price for chain at the local discount store, but your supervisor wonders if the chain is strong enough. You contact the manufacturer and learn that the chain is rated to withstand a tension of 3000 N. Will this chain be strong enough for the ride?

49. Suppose you swing a ball in a vertical circle on a 1.0-m-long string. As you probably know from experience, there is a *minimum* angular velocity ω_{min} you must maintain if you want the ball to complete the full circle. If you swing the ball at $\omega < \omega_{min}$, then the string goes slack before the ball reaches the top of the circle. What is ω_{min}? Give your answer in rpm.

50. It is proposed that future space stations create an artificial gravity by rotating. Suppose a space station is constructed as a 1000-m-diameter cylinder that rotates about its axis. The inside surface is the deck of the space station. What rotation period will provide "normal" gravity?

51. A 100 g ball on a 60-cm-long string is swung in a vertical circle about a point 200 cm above the floor. The tension in the string when the ball is at the very bottom of the circle is 5.0 N. A very sharp knife is suddenly inserted, as shown in Figure P7.51, to cut the string directly below the point of support. Where does the ball hit the floor?

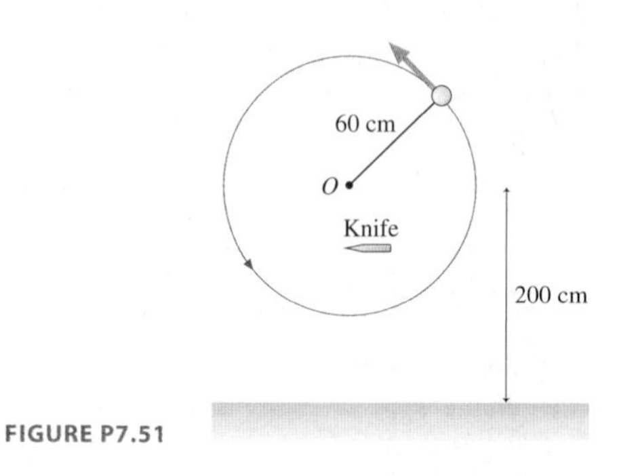

FIGURE P7.51

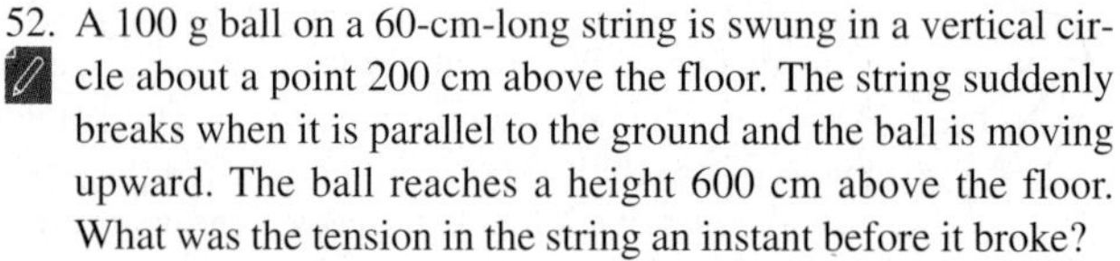

52. A 100 g ball on a 60-cm-long string is swung in a vertical circle about a point 200 cm above the floor. The string suddenly breaks when it is parallel to the ground and the ball is moving upward. The ball reaches a height 600 cm above the floor. What was the tension in the string an instant before it broke?

53. A 1500 kg car starts from rest and drives around a flat 50-m-diameter circular track. The forward force provided by the car's drive wheels is a constant 1000 N.
 a. What are the magnitude and direction of the car's acceleration at $t = 10$ s?
 b. If the car has rubber tires and the track is concrete, at what time does the car begin to slide out of the circle?

54. A 500 g steel block rotates on a steel table while attached to a 2.0-m-long massless rod. Compressed air fed through the rod is ejected from a nozzle on the back of the block, exerting a thrust force of 3.5 N. The nozzle is 70° from the radial line, as shown in Figure P7.54. The block starts from rest.
 a. What is the block's angular velocity after 10 rev?
 b. What is the tension in the rod after 10 rev?

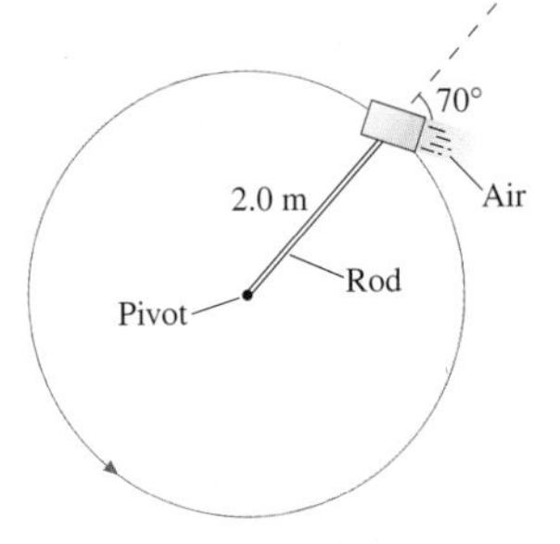

FIGURE P7.54

55. A 500 g steel block rotates on a steel table while attached to a 1.2-m-long hollow tube. Compressed air fed through the tube and ejected from a nozzle on the back of the block exerts a thrust force of 4.0 N perpendicular to the tube. The maximum tension the tube can withstand without breaking is 50 N. If the block starts from rest, how many revolutions does it make before the tube breaks?

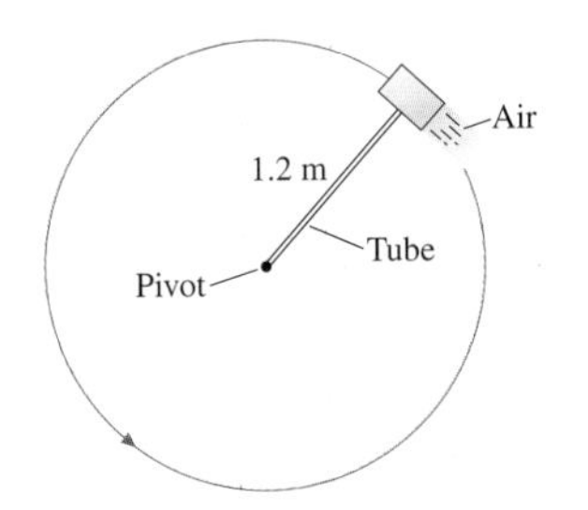

FIGURE P7.55

56. A 2.0 kg ball swings in a vertical circle on the end of an 80-cm-long string. The tension in the string is 20 N when its angle from the highest point on the circle is $\theta = 30°$.
 a. What is the ball's speed when $\theta = 30°$?
 b. What are the magnitude and direction of the ball's acceleration when $\theta = 30°$?

In Problems 57 through 59 you are given the equation (or equations) used to solve a problem. For each of these, you are to
 a. Write a realistic problem for which this is the correct equation. Be sure that the answer your problem requests is consistent with the equation given.
 b. Finish the solution of the problem.

57. $60\text{ N} = (0.30\text{ kg})\omega^2(0.50\text{ m})$

58. $(1500\text{ kg})(9.8\text{ m/s}^2) - 11{,}760\text{ N} = (1500\text{ kg})\,v^2/(200\text{ m})$

59. $2.5\text{ rad} = 0\text{ rad} + \omega_i(10\text{ s}) + ((1.5\text{ m/s}^2)/2(50\text{ m}))(10\text{ s})^2$
 $\omega_f = \omega_i + ((1.5\text{ m/s}^2)/(50\text{ m}))(10\text{ s})$

Challenge Problems

60. Two wires are tied to the 2.0 kg sphere shown in Figure CP7.60. The sphere revolves in a horizontal circle at constant speed.
 a. For what speed is the tension the same in both wires?
 b. What is the tension?

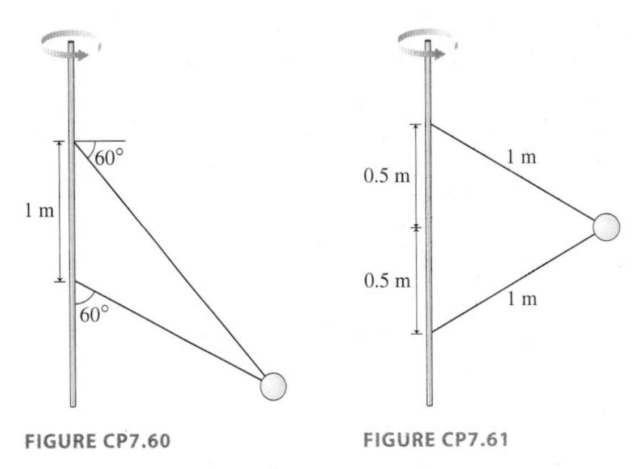

FIGURE CP7.60 FIGURE CP7.61

61. Two wires are tied to the 300 g sphere shown in Figure CP7.61. The sphere revolves in a horizontal circle at a constant speed of 7.5 m/s. What is the tension in each of the wires?

62. A 60 g ball is tied to the end of a 50-cm-long string and swung in a vertical circle. The center of the circle, as shown in Figure CP7.62, is 150 cm above the floor. The ball is swung at the minimum speed necessary to make it over the top without the string going slack. If the string is released at the instant the ball is at the top of the loop, where does the ball hit the ground?

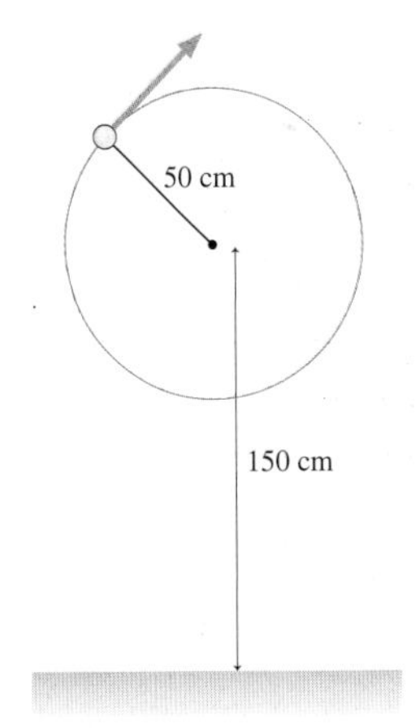

FIGURE P7.62

63. A small ball rolls around a horizontal circle at height y inside a frictionless hemispherical bowl of radius R, as shown in Figure CP7.63.
 a. Find an expression for the ball's angular velocity in terms of R, y, and g.
 b. What is the minimum value of ω for which the ball can move in a circle?
 c. What is ω in rpm if $R = 20$ cm and the ball is halfway up?

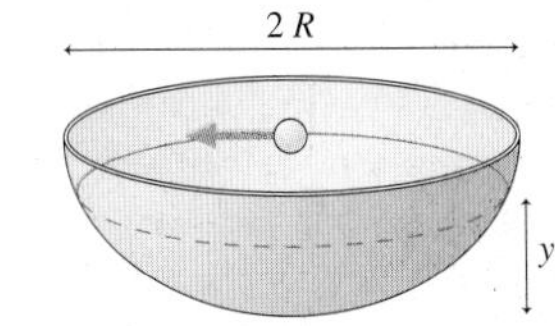

FIGURE CP7.63

64. You are flying to New York. You've been reading the in-flight magazine, which has an article about the physics of flying. You learned that the airflow over the wings creates a *lift force* that is always perpendicular to the wings. In level flight, the upward lift force exactly balances the downward weight force. The pilot comes on to say that, because of heavy traffic, the plane is going to circle the airport for a while. She says that you'll maintain a speed of 400 mph at an altitude of 20,000 ft. You start to wonder what the diameter of the plane's circle around the airport is. You notice that the pilot has banked the plane so that the wings are 10° from horizontal. The safety card in the seatback pocket informs you that the plane's wing span is 250 ft. What can you learn about the diameter?

65. If a vertical cylinder of water (or any other liquid) rotates about its axis, as shown in Figure CP7.65, the surface forms a smooth curve. Assuming that the water rotates as a unit (i.e., all the water rotates with the same angular velocity), show that the shape of the surface is a parabola described by the equation $z = (\omega^2/2g)r^2$.
Hint: Each particle of water on the surface is subject to only two forces: gravity and the normal force due to the water underneath it. The normal force, as always, acts perpendicular to the surface.

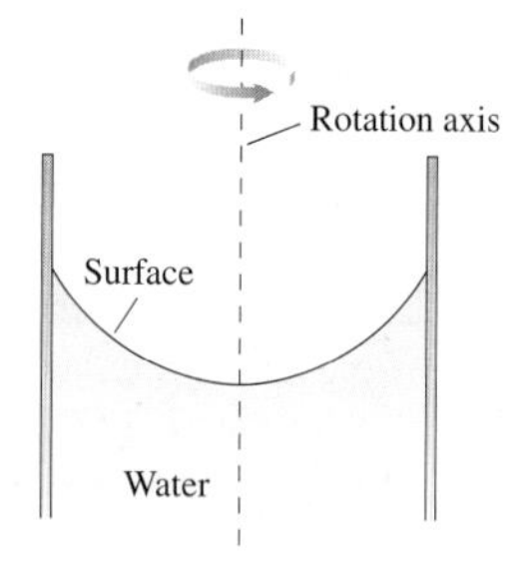

FIGURE CP7.65

STOP TO THINK ANSWERS

Stop to Think 7.1: b. An initial cw rotation causes the particle's angular position to become increasingly negative. The speed drops to half after reversing direction, so the slope becomes positive and is half as steep as the initial slope. Turning through the same angle returns the particle to $\theta = 0°$.

Stop to Think 7.2: $(a_r)_b > (a_r)_e > (a_r)_a = (a_r)_c > (a_r)_d$. Centripetal acceleration is v^2/r. Doubling r decreases a_r by a factor of 2. Doubling v increases a_r by a factor of 4. Reversing direction doesn't change a_r.

Stop to Think 7.3: $T_d > T_b = T_e > T_c > T_a$. The center-directed force is $m\omega^2 r$. Changing r by a factor of 2 changes the tension by a factor of 2, but changing ω by a factor of 2 changes the tension by a factor of 4.

Stop to Think 7.4: b. The car is moving in a circle, so there must be a net force toward the center of the circle. The circle is below the car, so the net force must point downward. This can be true only if $w > n$.

Stop to Think 7.5: c. The ball does not have a "memory" of its previous motion. The velocity $\vec{v}$ is straight up at the instant the string breaks. The only force on the ball after the string breaks is the weight force, straight down. This is just like tossing a ball straight up.

8 Newton's Third Law

These two sumo wrestlers are *interacting* with each other.

► **Looking Ahead**

The goal of Chapter 8 is to use Newton's third law to understand interacting systems. In this chapter you will learn to:

- Identify action/reaction pairs of forces in interacting systems.
- Understand and use Newton's third law.
- Use an expanded problem-solving strategy for dynamics problems.
- Understand the role of strings, ropes, and pulleys.

◄ **Looking Back**

This chapter further develops the concept of force. Please review:

- Sections 4.1–4.3 The basic concept of force and the atomic-level view of tension.
- Section 5.2 The basic problem-solving strategy for dynamics.

Rather than a single particle responding to a well-defined force, such as you've learned to deal with in the last few chapters, these sumo wrestlers are two systems *interacting* with each other. The harder one sumo wrestler pushes, the harder the other pushes back. A hammer and a nail, your foot and a soccer ball, and the earth-moon system are other examples of interacting systems.

Newton's second law is not sufficient to explain what happens when two or more objects interact. Newton's second law, the essence of single-particle dynamics, treats a system as an isolated entity being acting upon by external forces. Chapter 8 will introduce a new law of physics, Newton's *third* law, that describes how two systems interact with each other. We will then expand our problem-solving strategy to make simultaneous use of Newton's second and third laws when solving problems of interacting systems.

Newton's third law brings us to the pinnacle of Newton's theory of forces and motion. The tools you will have learned when you finish this chapter can be used to solve complex but realistic dynamics problems.

8.1 Interacting Systems

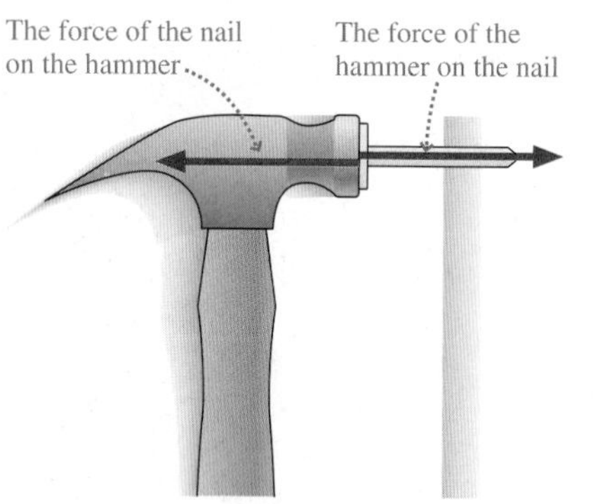

FIGURE 8.1 The hammer and nail are interacting with each other.

Our goal is to understand how two systems interact. Think about the hammer and nail in Figure 8.1. The hammer certainly exerts a force on the nail as it drives the nail forward. At the same time, the nail exerts a force on the hammer. If you are not sure that it does, imagine hitting the nail with a glass hammer. It's the force of the nail on the hammer that causes the glass to shatter.

If you stop to think about it, any time that object A pushes or pulls on object B, object B pushes or pulls back on object A. As sumo wrestler A pushes on sumo wrestler B, B pushes back on A. (If A pushed forward without B pushing back, A would fall over in the same way you do if someone suddenly opens a door you're leaning against.) Your chair pushes upward on you (a normal force) while, at the same time, you push down on the chair. These are examples of what we call an *interaction*. An **interaction** is the mutual influence of two systems on each other.

To be more specific, if object A exerts a force $\vec{F}_{\text{A on B}}$ on object B, then object B exerts a force $\vec{F}_{\text{B on A}}$ on object A. This pair of forces, shown in Figure 8.2, is called an **action/reaction pair.** Two systems interact by exerting an action/reaction pair of forces on each other. Notice the very explicit subscripts on the force vectors. The first letter is the *agent*, the second letter is the object on which the force acts. $\vec{F}_{\text{A on B}}$ is a force exerted *by* A *on* B. The distinction is important, and we will use this explicit notation for much of this chapter.

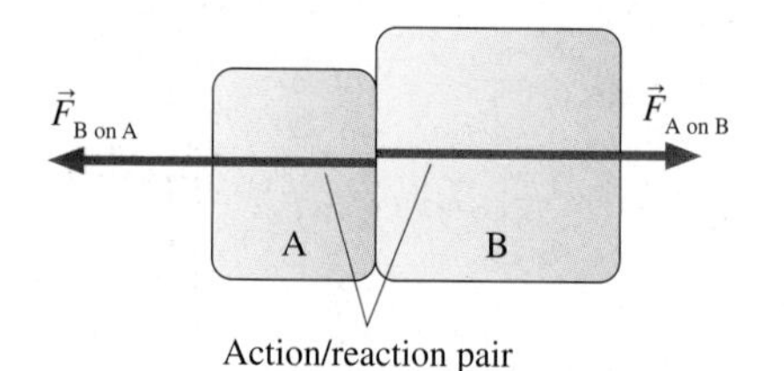

FIGURE 8.2 An action/reaction pair of forces.

NOTE ▶ The name "action/reaction pair" is somewhat misleading. The forces occur simultaneously, and we cannot say which is the "action" and which the "reaction." Neither is there any implication about cause and effect; the action does not *cause* the reaction. **An action/reaction pair of forces exists as a pair, or not at all.** In identifying action/reaction pairs, the labels are the key. Force $\vec{F}_{\text{A on B}}$ is paired with force $\vec{F}_{\text{B on A}}$. ◀

The sumo wrestlers and the hammer and nail interact through contact forces. The same idea holds true for long-range forces. You probably have played with kitchen magnets or bar magnets. As you hold two magnets, you can feel with your fingertips that *both* have forces pulling on them.

But what about gravity? If you release a ball, it falls because the earth's gravity exerts a downward force $\vec{F}_{\text{earth on ball}}$ on it. This force is what we've been calling *weight*. But does the ball also pull upward on the earth? That is, is there a force $\vec{F}_{\text{ball on earth}}$?

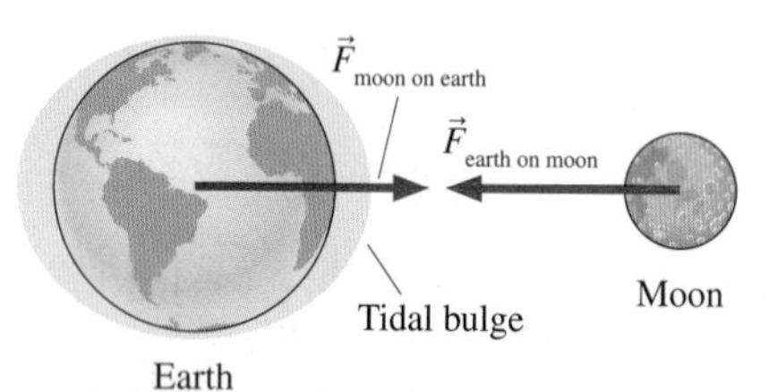

FIGURE 8.3 The ocean tides are an indication of the long-range gravitational interaction of the earth and the moon.

Newton was the first to recognize that, indeed, the ball *does* pull upward on the earth. Likewise, the moon pulls on the earth in response to the earth's gravity pulling on the moon. Newton's evidence was the tides. Scientists and astronomers have studied and timed the ocean's tides since antiquity. It was known that the tides depend on the phase of the moon, but Newton was the first to understand that the tides are the ocean's response to the gravitational pull of the moon on the earth. As Figure 8.3 shows, the flexible water bulges toward the moon while the relatively inflexible crust of the earth remains stationary.

Systems and the Environment

In earlier chapters we considered forces acting on a single object that we called the *system*. The forces originated from agents in the *environment*. Figure 8.4a shows a diagrammatic representation of single-particle dynamics. If all the forces acting on the particle are known, we can use Newton's second law $\vec{F}_{\text{net}} = m\vec{a}$ to determine the particle's acceleration.

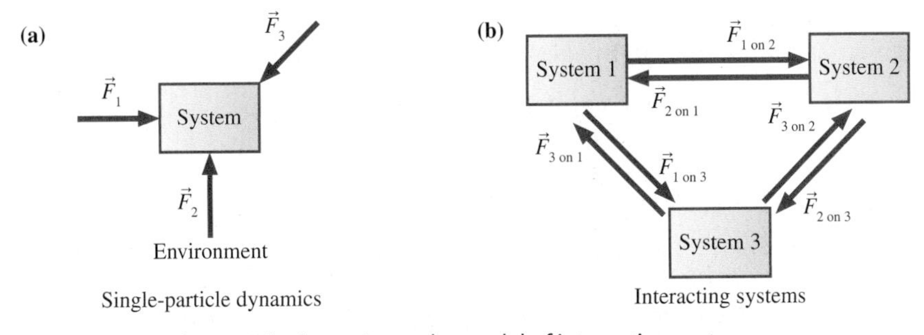

FIGURE 8.4 Single-particle dynamics and a model of interacting systems.

We now want to extend the particle model to situations in which two or more objects, each represented as a particle, interact with each other. For example, Figure 8.4b shows three systems interacting via action/reaction pairs of forces. The forces can be given labels such as $\vec{F}_{1\text{ on }2}$ and $\vec{F}_{2\text{ on }1}$.

We will often be interested in the motion of Systems 1 and 2 but not of System 3. For example, Systems 1 and 2 might be the hammer and the nail while System 3 is the earth. The earth interacts with both the hammer and the nail, but in a practical sense the earth remains "at rest" while the hammer and nail move. It will be convenient to separate "systems of interest" from "systems in the environment," as shown in Figure 8.5a. Forces originating in the environment are called **external forces.**

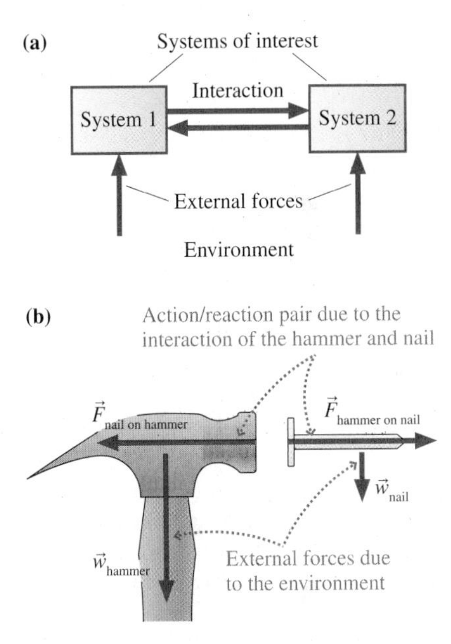

FIGURE 8.5 As a practical matter, an interaction with a system whose motion is not of interest can be called an external force.

NOTE ► This is a practical distinction, not a fundamental distinction. If object B pushes or pulls on object A whenever A pushes or pulls on B, then *every* force is one member of an action/reaction pair. There is no such thing as a true "external force." What we call an external force is an interaction between a system of interest and a system whose motion is not of interest. ◄

Figure 8.5b illustrates this idea for the hammer and the nail. $\vec{F}_{\text{hammer on nail}}$ and $\vec{F}_{\text{nail on hammer}}$ are an action/reaction pair that describe the interaction between the hammer and the nail. The force labeled $\vec{w}_{\text{hammer}}$ is really force $\vec{F}_{\text{earth on hammer}}$. That is, the hammer interacts with the earth through the action/reaction pair of forces $\vec{F}_{\text{earth on hammer}}$ and $\vec{F}_{\text{hammer on earth}}$. But we only want to know how the hammer moves, not how the earth moves, so the force $\vec{F}_{\text{earth on hammer}}$ can be treated as an external force.

Newton's second law $\vec{a} = \vec{F}_{\text{net}}/m$ applies *separately* to Systems 1 and 2 in Figure 8.5a:

$$\begin{aligned} \text{System 1:} \quad & \vec{a}_1 = \frac{\vec{F}_{1\text{ net}}}{m_1} = \frac{1}{m_1}\sum \vec{F}_{\text{on }1} \\ \text{System 2:} \quad & \vec{a}_2 = \frac{\vec{F}_{2\text{ net}}}{m_2} = \frac{1}{m_2}\sum \vec{F}_{\text{on }2} \end{aligned} \tag{8.1}$$

The net force on System 1, denoted $\sum \vec{F}_{\text{on }1}$, is the sum of *all* forces acting *on* System 1. The sum includes both forces due to System 2 ($\vec{F}_{2\text{ on }1}$) and any external forces originating in the environment.

NOTE ► Forces exerted *by* System 1, such as $\vec{F}_{1\text{ on }2}$, do *not* appear in the equation for System 1. Objects change their motion in response to forces exerted *on* them, not by forces exerted *by* them. ◄

The bat and the ball are interacting with each other.

8.2 Identifying Action/Reaction Pairs

The key step for analyzing interacting systems is the identification of all the action/reaction pairs of forces.

TACTICS BOX 8.1 Identifying action/reaction pairs

1. **Draw each object separately.** Place them in the correct position relative to other objects. Don't forget to include objects like the earth that may not be mentioned in the problem.
2. **Identify every force.** Draw the force vector on the object on which it acts. Label each with a subscripted label such as $\vec{F}_{A \text{ on } B}$. The usual force symbols, such as $\vec{n}$ and $\vec{w}$, can be used.
3. **Identify the action/reaction pairs.** Force $\vec{F}_{A \text{ on } B}$ goes with force $\vec{F}_{B \text{ on } A}$. Connect the two force vectors of each action/reaction pair with a dotted line. When you're done, there should be no unpaired forces.
4. **Identify the objects that are systems of interest.** Other objects whose motion you don't care about are part of the environment.
5. **Draw a free-body diagram for each system of interest.** Include only the forces acting *on* the system, not forces that the system exerts on other objects.

We'll illustrate these ideas with two concrete examples. The first example will be much longer than usual because we'll go carefully through all the steps in the reasoning.

EXAMPLE 8.1 The forces involved in pushing a crate

Figure 8.6a shows a person pushing a large crate across a rough surface. Identify all action/reaction pairs, show them on a figure, then draw free-body diagrams of the person and the crate.

VISUALIZE Figure 8.6b redraws the figure with every object in the correct position but separated from all other objects. The person and the crate are obvious objects. The earth is also an object that both exerts and experiences forces. We've represented the earth rather abstractly as a long rectangle so that it can be "under" both the person and the crate. It will be useful to distinguish between the surface, which exerts contact forces, and the earth as a whole, which exerts the long-range force of gravity. The letters P (person), C (crate), S (surface), and E (the earth as a whole) will be used for labels.

Figure 8.6b also identifies the various interactions. Some, like the pushing interaction between the person and the crate, are fairly obvious. The interactions with the earth are a little trickier. Gravity, which is a long-range force, is an interaction between each object and the earth as a whole. Friction forces and normal forces are a contact interaction between each object and the earth's surface. Altogether, there are seven interactions.

NOTE ► Interactions are between two *different* objects. None of the interactions are between an object and itself. ◄

We can now start identifying and labeling forces, as you learned to do in Chapter 4. These are shown in Figure 8.7. We'll begin with the crate: It is pushed with force $\vec{F}_{P \text{ on } C}$, it experiences an upward normal force $\vec{n}_{S \text{ on } C}$, and it has a kinetic friction force $\vec{f}_{S \text{ on } C}$ in the direction opposite the motion. The crate also has a downward weight force $\vec{w}_{E \text{ on } C}$ exerted on it by the earth as a whole. These are all forces $\vec{F}_{\text{something on } C}$, and all the force vectors are drawn *on* crate C.

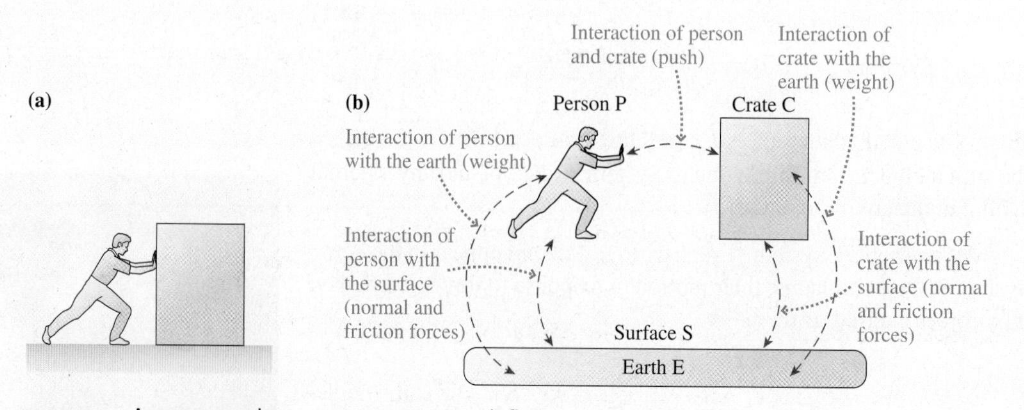

FIGURE 8.6 A person pushes a crate across a rough floor.

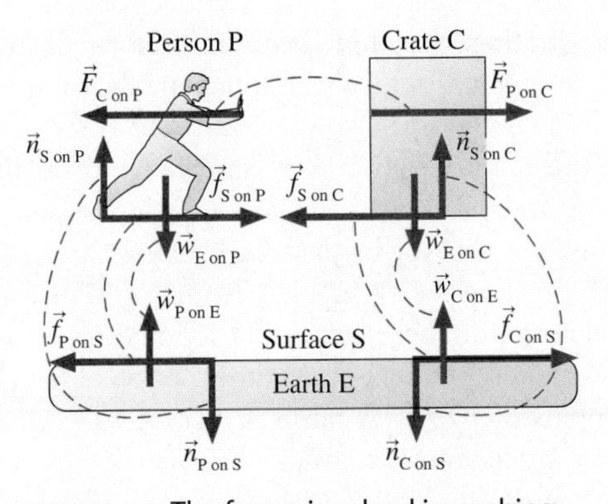

FIGURE 8.7 The forces involved in pushing a crate.

If A pushes or pulls on B, then B pushes or pulls back on A. The reaction to force $\vec{F}_{A\ on\ B}$ is $\vec{F}_{B\ on\ A}$. We can use this reasoning to deduce the reactions to the four forces exerted on crate C. The person pushes the crate, so the crate pushes against the person with force $\vec{F}_{C\ on\ P}$. This is a force exerted *on the person*, so it is drawn on person P. Forces $\vec{F}_{P\ on\ C}$ and $\vec{F}_{C\ on\ P}$ are then connected with a dotted line to show that they are an action/reaction pair. Similarly, force $\vec{n}_{C\ on\ S}$ is the force of the crate pushing down on the surface. This is a contact force, so it is drawn on the earth's surface. The weight force $\vec{w}_{E\ on\ C}$ is the gravitational force *by* the earth *on* the crate. By "earth" we mean the *entire* earth, not just the surface. To identify the reaction force, simply reverse the letters in the labels. The reaction to $\vec{w}_{E\ on\ C}$ is $\vec{w}_{C\ on\ E}$. In other words, the crate pulls up on the entire earth.

We have to be careful with friction. Force $\vec{f}_{S\ on\ C}$ is a kinetic friction force retarding the crate's motion, so it points to the left. The crate exerts force $\vec{f}_{C\ on\ S}$ on the surface. But in which direction? Imagine the floor is covered with sand. As the crate slides, it tries to push the sand to the right. Thus force $\vec{f}_{C\ on\ S}$, which forms an action/reaction pair with $\vec{f}_{S\ on\ C}$, points to the right.

The person experiences similar forces. Force $\vec{F}_{C\ on\ P}$ pushes against the person's hands while the normal force $\vec{n}_{S\ on\ P}$ pushes up and the weight force $\vec{w}_{E\ on\ P}$ pulls down. These are forces exerted *on* the person, so the force vectors are drawn on P.

The hardest interaction to understand is the friction between the person and the surface. It is tempting to draw force $\vec{f}_{S\ on\ P}$ pointing to the left. After all, friction forces are supposed to be in the direction opposite the motion. But if we did so, the person would have two forces to the left ($\vec{F}_{C\ on\ P}$ and $\vec{f}_{S\ on\ P}$) and none to the right, causing the person to accelerate *backward*! That is clearly not what happens, so what is wrong?

Imagine pushing a crate to the right across loose sand. Each time you take a step, you tend to kick the sand to the *left*, behind you. Thus friction force $\vec{f}_{P\ on\ S}$, the force of the person pushing against the earth, is to the *left*. In reaction, the force of the earth's surface against the person is a friction force to the *right*. This force, $\vec{f}_{S\ on\ P}$, causes the person to accelerate in the forward direction. Forces $\vec{f}_{S\ on\ P}$ to the right and $\vec{f}_{P\ on\ S}$ to the left form another action/reaction pair.

Notice that *every* force is one member of an action/reaction pair. Further, each member of a pair is attached to a different object. **The two forces of an action/reaction pair never occur on the same object.**

But surely 14 forces is an excessive number for such a simple situation! While these forces all exist, we don't need all of them to understand how the crate moves. This is the point where we can distinguish between systems of interest and the environment. The *crate* and the *person* are systems of interest because they move. The surface and the earth do not move, so we can locate them in the environment.

Figure 8.8 shows a free body diagram of just the two systems of interest. Forces $\vec{F}_{P\ on\ C}$ and $\vec{F}_{C\ on\ P}$ are an action/reaction pair between the systems of interest, so we've kept their full labels and connected them with a dotted line. We can now treat all other forces as external forces, and can simplify their labels. Even so, it is important to use subscript labels such as $\vec{w}_P$ and $\vec{w}_C$, or $\vec{n}_P$ and $\vec{n}_C$, to distinguish external forces on the crate from external forces on the person.

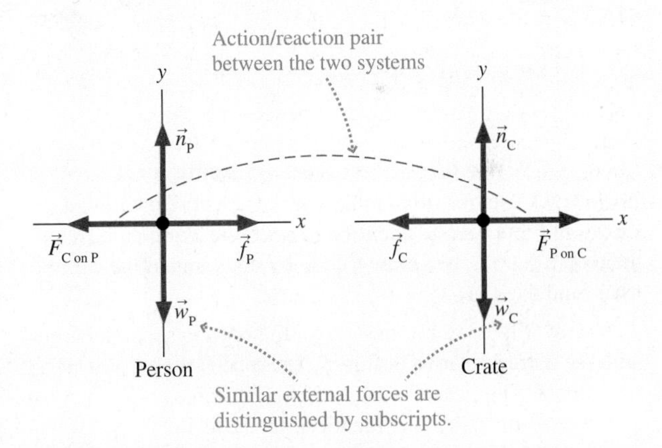

FIGURE 8.8 Free-body diagrams of the person and the crate.

Propulsion

The friction force $\vec{f}_{S\ on\ P}$ is an example of **propulsion.** It is the force that a system with an internal source of energy uses to drive itself forward. Propulsion is an important feature not only of walking or running but also of the forward motion of cars, jets, and rockets. Propulsion is somewhat counterintuitive, so it is worth a closer look.

If you try to walk across a frictionless floor, your foot slips and slides *backward*. In order for you to walk, the floor needs to have friction so that your foot *sticks* to the floor as you straighten your leg, moving your body forward. The

What force causes this sprinter to accelerate?

friction that prevents slipping is *static* friction. Static friction, you will recall, acts in the direction that prevents slipping. The static friction force $\vec{f}_{\text{S on P}}$ has to point in the *forward* direction to prevent your foot from slipping backward. It is this forward-directed static friction force that propels you forward! The force of your foot on the floor, the other half of the action/reaction pair, is in the opposite direction.

The distinction between you and the crate is that you have an *internal source of energy* that allows you to straighten your leg by pushing backward against the surface. In essence, you walk by pushing the earth away from you. The earth's surface responds by pushing you forward. These are static friction forces. In contrast, all the crate can do is slide, so *kinetic* friction opposes the motion of the crate.

Figure 8.9 shows how propulsion works. A car uses its motor to spin the tires, causing the tires to push backward against the ground. This is why dirt and gravel are kicked backward, not forward. The earth's surface responds by pushing the car forward. These are also *static* friction forces. The tire is rolling, but the bottom of the tire, where it contacts the road, is instantaneously at rest. If it weren't, you would leave one giant skid mark as you drove and would burn off the tread within a few miles.

Rocket motors are somewhat different because they are not pushing *against* anything. That's why rocket propulsion works in the vacuum of space. Instead, the rocket engine pushes hot, expanding gases out of the back of the rocket. In response, the exhaust gases push the rocket forward with the force we've called *thrust*.

The person pushes backward against the earth. The earth pushes forward on the person. Static friction.

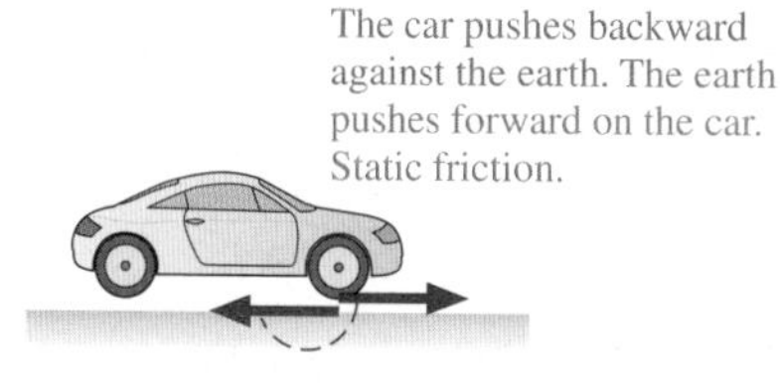

The car pushes backward against the earth. The earth pushes forward on the car. Static friction.

The rocket pushes the hot gases backward. The gases push the rocket forward. Thrust force.

FIGURE 8.9 Examples of propulsion.

EXAMPLE 8.2 The forces involved in towing a car

A tow truck uses a rope to pull a car along a horizontal road, as shown in Figure 8.10a. Identify all action/reaction pairs, show them on a figure, then draw free-body diagrams of the car, the truck, and the rope.

VISUALIZE Figure 8.10b has drawn the objects separately, but with the correct relative positions. The rope is shown as a separate object. The normal force and weight force action/reaction pairs are identical to those in Example 8.1. We've assumed that the weight of the rope is negligible in comparison with all the other forces.

NOTE ► Make sure you avoid the common error of considering $\vec{n}$ and $\vec{w}$ to be an action/reaction pair. These are both forces on the *same* object, whereas the two forces of an action/reaction pair are always on two *different* objects that are interacting with each other. ◄

What about the friction forces? The car is an inert object rolling along. It would slow and stop if the rope were cut, so the

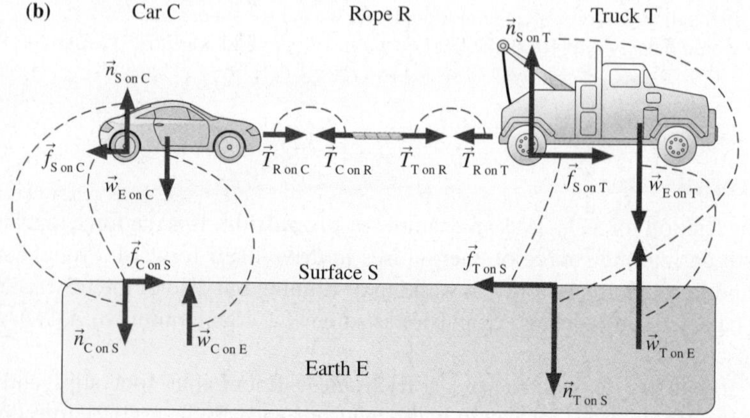

FIGURE 8.10 Identifying the forces as a truck tows a car.

surface must exert a rolling friction force $\vec{f}_{\text{S on C}}$ to the left. The reaction of the car trying to drag the ground along with it is the force $\vec{f}_{\text{C on S}}$ to the right. The truck, however, has an internal source of energy. The truck's drive wheels push the ground to the left with force $\vec{f}_{\text{T on S}}$. In reaction, the ground propels the truck forward, to the right, with force $\vec{f}_{\text{S on T}}$.

Finally, we need to identify the forces between the car, the truck, and the rope. What pulls on what in the horizontal direction? The rope pulls on the car with a tension force $\vec{T}_{\text{R on C}}$. You might be tempted to put the reaction force on the truck, because we say that "the truck pulls the car," but the truck is not in contact with the car. The truck pulls on the rope, then the rope pulls on the car. Thus the reaction to $\vec{T}_{\text{R on C}}$ is a force on the *rope*: $\vec{T}_{\text{C on R}}$. At the other end, $\vec{T}_{\text{T on R}}$ and $\vec{T}_{\text{R on T}}$ are an action/reaction pair.

We have *three* systems of interest: the truck, the rope, and the car. The surface and the earth are in the environment, so normal forces, weight forces, and friction forces can be treated as external forces. This information is shown on the free-body diagrams of Figure 8.11.

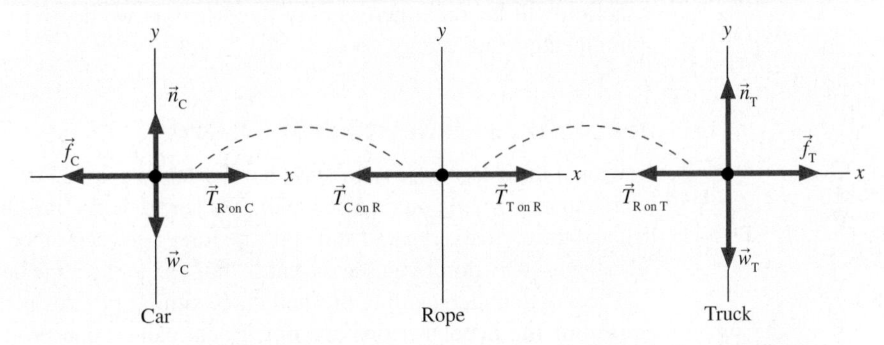

FIGURE 8.11 Free-body diagrams of the three systems of interest in Example 8.2.

STOP TO THINK 8.1 What, if anything, is wrong with this force diagram for a bicycle that is accelerating toward the right?

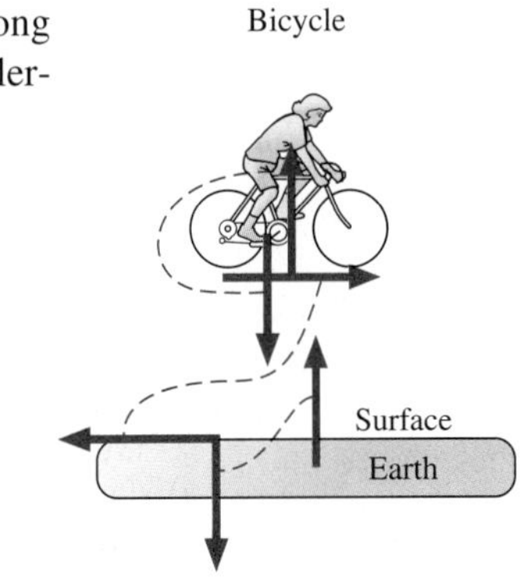

8.3 Newton's Third Law

Newton was the first to recognize how the two members of an action/reaction pair of forces are related to each other. Today we know this as Newton's third law:

> **Newton's third law** Every force occurs as one member of an action/reaction pair of forces.
>
> - The two members of an action/reaction pair act on two *different* objects.
> - The two members of an action/reaction pair are equal in magnitude but opposite in direction: $\vec{F}_{\text{A on B}} = -\vec{F}_{\text{B on A}}$.

We deduced most of the third law in Section 8.2. There we found that the two members of an action/reaction pair are always opposite in direction (see Figures 8.7 and 8.10). According to the third law, this will always be true. But the most significant portion of the third law, which is by no means obvious, is that the two members of an action/reaction pair have *equal* magnitudes. That is, $F_{\text{A on B}} = F_{\text{B on A}}$. This is the quantitative relationship that will allow you to solve problems of interacting systems.

Newton's third law is frequently stated as "For every action there is an equal but opposite reaction." While this is indeed a catchy phrase, it lacks the preciseness of our preferred version. In particular, it fails to capture an essential feature of action/reaction pairs—that they each act on a *different* object.

NOTE ▶ Newton's third law extends and completes our concept of *force*. We can now recognize force as an *interaction* between objects rather than as some "thing" with an independent existence of its own. The concept of an interaction will become increasingly important as we begin to study the laws of momentum and energy. ◀

Reasoning with Newton's Third Law

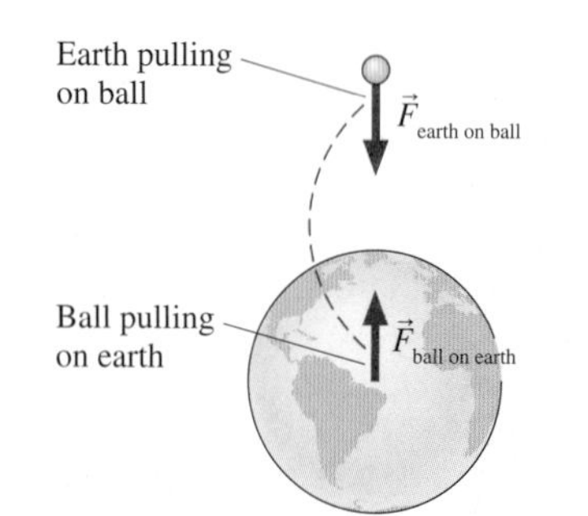

FIGURE 8.12 The action/reaction forces of a ball and the earth are equal in magnitude.

Newton's third law is easy to state but harder to grasp. For example, consider what happens when you release a ball. Not surprisingly, it falls down. But if the ball and the earth exert equal and opposite forces on each other, as Newton's third law alleges, why don't you see the earth "fall up" to meet the ball?

The key to understanding this and many similar puzzles is that **the forces are equal but the accelerations are not.** Equal causes can produce very unequal effects. Figure 8.12 shows equal-magnitude forces on the ball and the earth. The force on ball B is simply the weight force of Chapter 5:

$$\vec{F}_{\text{earth on ball}} = \vec{w}_{\text{B}} = -m_{\text{B}}g\hat{j} \tag{8.2}$$

where m_{B} is the mass of the ball. According to Newton's second law, this force gives the ball an acceleration

$$\vec{a}_{\text{B}} = \frac{\vec{w}_{\text{B}}}{m_{\text{B}}} = -g\hat{j} \tag{8.3}$$

This is just the familiar free-fall acceleration due to gravity.

According to Newton's third law, the ball pulls up on the earth with force $\vec{F}_{\text{ball on earth}}$. As the ball accelerates down, the earth as a whole has an upward acceleration

$$\vec{a}_{\text{E}} = \frac{\vec{F}_{\text{ball on earth}}}{m_{\text{E}}} \tag{8.4}$$

where m_{E} is the mass of the earth. Because $\vec{F}_{\text{earth on ball}}$ and $\vec{F}_{\text{ball on earth}}$ are an action/reaction pair, $\vec{F}_{\text{ball on earth}}$ must be equal in magnitude and opposite in direction to $\vec{F}_{\text{earth on ball}}$. That is,

$$\vec{F}_{\text{ball on earth}} = -\vec{F}_{\text{earth on ball}} = -\vec{w}_{\text{B}} = +m_{\text{B}}g\hat{j} \tag{8.5}$$

Using this result in Equation 8.4, the upward acceleration of the earth as a whole is

$$\vec{a}_{\text{E}} = \frac{\vec{F}_{\text{ball on earth}}}{m_{\text{E}}} = \frac{m_{\text{B}}g\hat{j}}{m_{\text{E}}} = \left(\frac{m_{\text{B}}}{m_{\text{E}}}\right)g\hat{j} \tag{8.6}$$

The upward acceleration of the earth is less than the downward acceleration of the ball by the factor $m_{\text{B}}/m_{\text{E}}$. If we assume a 1 kg ball, we can estimate the magnitude of $\vec{a}_{\text{E}}$:

$$a_{\text{E}} = \frac{1\text{ kg}}{6 \times 10^{24}\text{ kg}}g \approx 2 \times 10^{-24}\text{ m/s}^2$$

With this incredibly small acceleration, it would take the earth 8×10^{15} years, approximately 500,000 times the age of the universe, to reach a speed of 1 mph! So we certainly would not expect to see or feel the earth "fall up" after dropping a ball.

NOTE ▶ Newton's third law equates the size of two forces, not two accelerations. The acceleration continues to depend on the mass, as Newton's second law states. **In an interaction between two objects of different mass, the lighter mass will do essentially all of the accelerating even though the forces exerted on the two objects are equal.** ◀

EXAMPLE 8.3 The forces on accelerating boxes

The hand shown in Figure 8.13 pushes boxes A and B to the right across a frictionless table. The mass of B is larger than the mass of A.

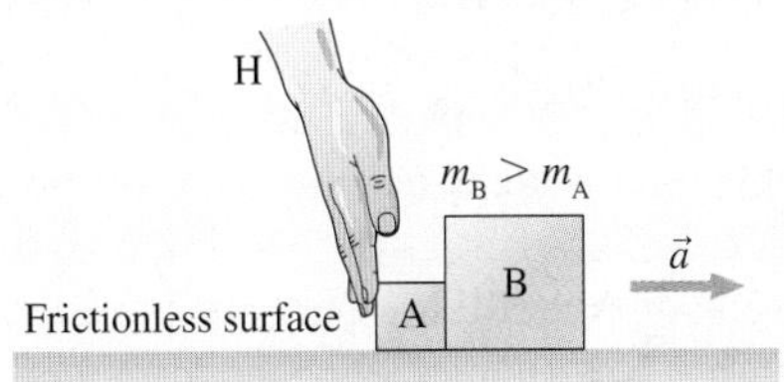

FIGURE 8.13 Hand H pushes boxes A and B across a frictionless table.

a. Draw free-body diagrams of A, B, and the hand H, showing only the *horizontal* forces. Connect action/reaction pairs with dotted lines.
b. Rank in order, from largest to smallest, the horizontal forces shown on your free-body diagrams.

VISUALIZE

a. The hand H pushes on box A, and A pushes back on H. Thus $\vec{F}_{H\text{ on }A}$ and $\vec{F}_{A\text{ on }H}$ are an action/reaction pair. Similarly, A pushes on B and B pushes back on A. The hand H does not touch box B, so there is no interaction between them. There is no friction. Figure 8.14 shows the four horizontal forces and identifies two action/reaction pairs. (We've chosen to ignore forces of the wrist or arm on the hand because our systems of interest are the boxes A and B.) Notice that each force is shown on the free-body diagram of the object that it acts *on*.
b. According to Newton's third law, $F_{A\text{ on }H} = F_{H\text{ on }A}$ and $F_{A\text{ on }B} = F_{B\text{ on }A}$. But the third law is not our only tool. Because the boxes are accelerating to the right, Newton's *second* law tells us that box A must have a net force to the right. Consequently, $F_{H\text{ on }A} > F_{B\text{ on }A}$. Thus

$$F_{A\text{ on }H} = F_{H\text{ on }A} > F_{A\text{ on }B} = F_{B\text{ on }A}$$

ASSESS You might have expected $F_{A\text{ on }B}$ to be larger than $F_{H\text{ on }A}$ because $m_B > m_A$. It's true that the *net* force on B is larger than the *net* force on A, but we have to reason more closely to judge the individual forces. Notice how we used both the second and the third laws to answer this question.

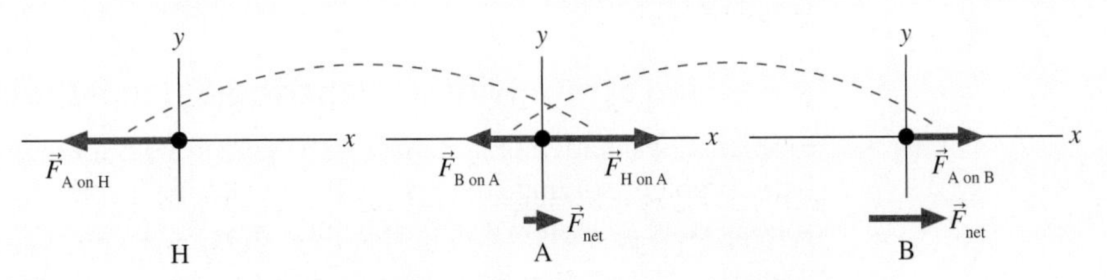

FIGURE 8.14 The free-body diagrams, showing only the horizontal forces.

STOP TO THINK 8.2 Car B is stopped for a red light. Car A, which has the same mass as car B, doesn't see the red light and runs into the back of B. Which of the following statements is true?

a. B exerts a force on A but A doesn't exert a force on B.
b. B exerts a larger force on A than A exerts on B.
c. B exerts the same amount of force on A as A exerts on B.
d. A exerts a larger force on B than B exerts on A.
e. A exerts a force on B but B doesn't exert a force on A.

Acceleration Constraints

Newton's third law is one quantitative relationship you can use to solve problems of interacting systems. In addition, we frequently have other information about the motion in a problem. For example, think about the two boxes in Example 8.3. As long as they're touching, box A *has* to have exactly the same acceleration as box B. If they were to accelerate differently, either box B would take off on its own or it would suddenly slow down and box A would run over it! Our problem implicitly assumes that neither of these is happening. Thus the two accelerations are *constrained* to be equal: $\vec{a}_A = \vec{a}_B$. A well-defined relationship between the accelerations of two or more systems is called an **acceleration constraint.** It is an independent piece of information that can help solve a problem.

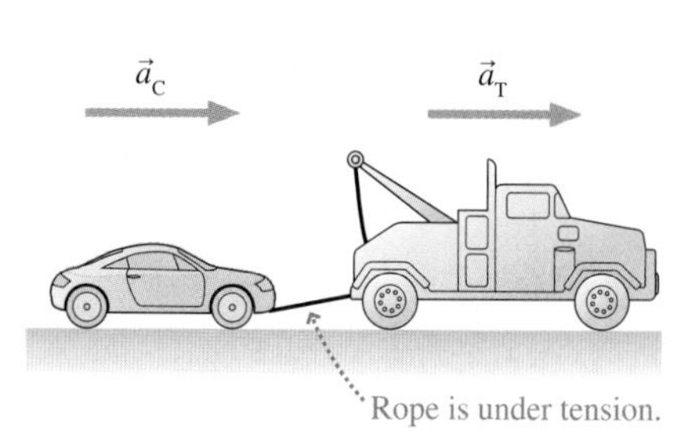

FIGURE 8.15 The car and the truck have the same acceleration.

In practice, we'll express acceleration constraints in terms of the x- and y-components of $\vec{a}$. Consider the car being towed in Figure 8.15. As long as the rope is under tension, the accelerations are constrained to be equal: $\vec{a}_C = \vec{a}_T$. This is one-dimensional motion, so for problem solving we would use just the x-components a_{Cx} and a_{Tx}. In terms of these components, the acceleration constraint is

$$a_{Cx} = a_{Tx} = a_x$$

Because the accelerations of both systems are equal, we can drop the subscripts C and T and call both of them a_x.

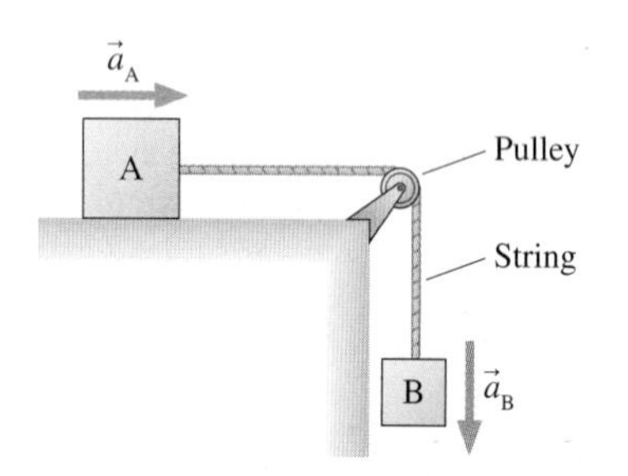

FIGURE 8.16 The string constrains the two systems to accelerate together.

Don't assume the accelerations of A and B will always have the same sign. Consider blocks A and B in Figure 8.16. The blocks are connected by a string, so they are constrained to move together and their accelerations have equal magnitudes. But A has a positive acceleration (to the right) in the x-direction while B has a negative acceleration (downward) in the y-direction. Thus the acceleration constraint is

$$a_{Ax} = -a_{By}$$

This relationship does *not* say that a_{Ax} is a negative number. It is simply a relational statement, saying that a_{Ax} is (-1) times whatever a_{By} happens to be. The acceleration a_{By} in Figure 8.16 is a negative number, so a_{Ax} is positive. In some problems, the signs of a_{Ax} and a_{By} may not be known until the problem is solved, but the *relationship* is known from the beginning.

A Revised Strategy for Interacting-System Problems

Problems of interacting systems can be solved with a few modifications to the basic problem-solving strategy we developed in Chapter 5. A revised problem-solving strategy is shown at the top of the next page.

NOTE ► We have dropped the motion diagram from the physical representation. Motion diagrams served a useful function in the early chapters, but by now you should be able to determine the directions of the acceleration vectors without the need for an explicit diagram. But if you are uncertain—use one! ◄

You might be puzzled that the Solve step calls for the use of the third law to equate just the *magnitudes* of action/reaction forces. What about the "opposite in direction" part of the third law? You have already used it! Your free-body diagrams should show the two members of an action/reaction pair to be opposite in direction, and that information will have been utilized in writing the second-law equations. Because the directional information has already been used, all that is left is the magnitude information.

PROBLEM-SOLVING STRATEGY 8.1 Interacting-system problems

MODEL Identify which objects are systems and which are part of the environment. Make simplifying assumptions.

VISUALIZE **Pictorial representation.** Show important points in the motion with a sketch. You may want to give each system a separate coordinate system. Define symbols and identify what the problem is trying to find. Include acceleration constraints as part of the pictorial model.
Physical representation. Identify all forces acting on each system and all action/reaction pairs. Draw a *separate* free-body diagram for each system. Connect the force vectors of action/reaction pairs with dotted lines. Use subscript labels to distinguish forces, such as $\vec{n}$ and $\vec{w}$, that act independently on more than one system.

SOLVE Use Newton's second and third laws:

- Write the equations of Newton's second law for each system, using the force information from the free-body diagrams.
- Equate the magnitudes of action/reaction pairs.
- Include the acceleration constraints, the friction model, and other quantitative information relevant to the problem.
- Solve for the acceleration, then use kinematics to find velocities and positions.

ASSESS Check that your result has the correct units, is reasonable, and answers the question.

NOTE ▶ Two steps are especially important when drawing the free-body diagrams. First, draw a *separate* diagram for each system. They need not have the same coordinate system. Second, show only the forces acting *on* that system. The force $\vec{F}_{\text{A on B}}$ goes on the free-body diagram of System B, but $\vec{F}_{\text{B on A}}$ goes on the diagram of System A. The two members of an action/reaction pair *always* appear on two different free-body diagrams—*never* on the same diagram. ◀

EXAMPLE 8.4 Keep the crate from sliding

You and a friend have just loaded a 200 kg crate filled with priceless art objects into the back of a 2000 kg truck. As you press down on the accelerator, force $\vec{F}_{\text{surface on truck}}$ propels the truck forward. To keep things simple, call this just $\vec{F}_{\text{T}}$. What is the maximum magnitude $\vec{F}_{\text{T}}$ can have without the crate sliding? The static and kinetic coefficients of friction between the crate and the bed of the truck are 0.8 and 0.3. Rolling friction of the truck is negligible.

MODEL The crate and the truck are separate systems that we'll call C and T. We'll model them as particles. The earth and the road surface are part of the environment.

VISUALIZE We're not doing any kinematics in this problem, so the pictorial representation of Figure 8.17 on the next page is minimal. We need a coordinate system, the known information, and—new to problems of interacting systems—the acceleration constraint. As long as the crate doesn't slip, it must accelerate *with* the truck. Both accelerations are in the positive x-direction, so the acceleration constraint in this problem is

$$a_{\text{C}x} = a_{\text{T}x} = a_x$$

Figure 8.17 also shows the free-body diagrams. The crate's diagram is drawn above the truck's diagram to reflect their relative positions. Force $\vec{n}_{\text{T on C}}$ is the normal force of the *truck* pushing up on the crate. The crate does not contact the ground, so the ground cannot exert forces on the crate. Although the crate moves, there's no motion of the crate *relative to* the truck. The force that prevents slipping is static friction. To prevent the crate from sliding out the back of the truck, a *static* friction force $\vec{f}_{\text{T on C}}$ points in the forward direction. The value of μ_{k} isn't relevant, so only μ_{s} is included as known information.

The truck has a weight force $\vec{w}_{\text{T}}$ and a normal force $\vec{n}_{\text{T}}$ of the *ground* pushing up on the truck. The truck tires try to push the ground backward as they spin, so the ground reacts with the forward force $\vec{F}_{\text{T}}$ that propels the truck forward. The truck experiences two other forces. Because $\vec{n}_{\text{T on C}}$ is a force of the truck on the crate, the truck experiences the reaction force $\vec{n}_{\text{C on T}}$ of the crate pressing down on the truck. Similarly, $\vec{f}_{\text{C on T}}$ is the reaction to friction force $\vec{f}_{\text{T on C}}$ of the truck on the crate. The truck tries to drag the crate forward; the crate tries to hold the truck back.

Notice that $\vec{n}_{\text{T}}$ and $\vec{w}_{\text{T}}$ are *not* an action/reaction pair. They are really the forces $\vec{n}_{\text{surface on T}}$ and $\vec{w}_{\text{earth on T}}$, so their reactions

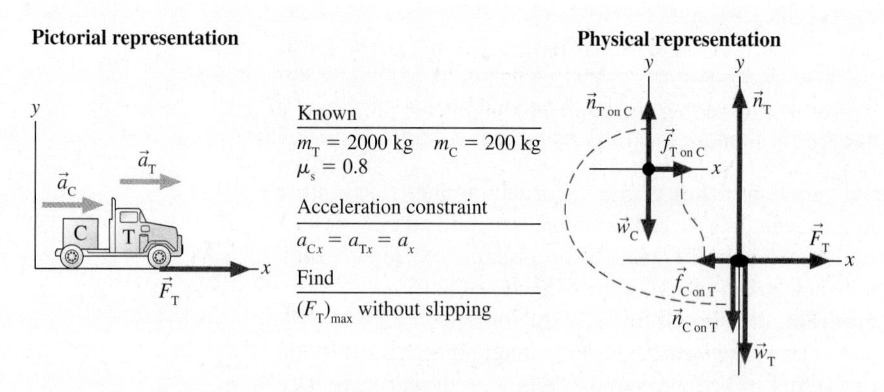

FIGURE 8.17 Pictorial and physical representations of the crate and truck in Example 8.4.

act on the surface and the earth as a whole. The two members of an action/reaction pair *never* appear on the same free-body diagram.

SOLVE Now we're ready to write Newton's second law. For the crate:

$$\sum (F_{\text{on crate}})_x = f_{\text{T on C}} = m_C a_{Cx} = m_C a_x$$

$$\sum (F_{\text{on crate}})_y = n_{\text{T on C}} - w_C = n_{\text{T on C}} - m_C g = 0$$

For the truck:

$$\sum (F_{\text{on truck}})_x = F_T - f_{\text{C on T}} = m_T a_{Tx} = m_T a_x$$

$$\sum (F_{\text{on truck}})_y = n_T - w_T - n_{\text{C on T}} = n_T - m_T g - n_{\text{C on T}} = 0$$

Be sure you agree with all the signs, which are based on the free-body diagrams. The net force in the y-direction is zero because there's no motion in the y-direction. It may seem like a lot of effort to write all the subscripts, but it is very important in problems with more than one system.

Notice that we've already used the acceleration constraint $a_{Cx} = a_{Tx} = a_x$. Another important piece of information is Newton's third law, which tells us that $f_{\text{C on T}} = f_{\text{T on C}}$ and $n_{\text{C on T}} = n_{\text{T on C}}$. Finally, we know that the maximum value of F_T will occur when the static friction on the crate reaches its maximum value

$$f_{\text{T on C}} = f_{\text{s max}} = \mu_s n_{\text{T on C}}$$

The friction depends on the normal force on the crate, not the normal force on the truck.

Now we can assemble all the pieces. From the y-equation of the crate, $n_{\text{T on C}} = m_C g$. Thus

$$f_{\text{T on C}} = \mu_s n_{\text{T on C}} = \mu_s m_C g$$

Using this in the x-equation of the crate, we find that the acceleration is

$$a_x = \frac{f_{\text{T on C}}}{m_C} = \mu_s g$$

This is the crate's maximum acceleration without slipping. Now use this acceleration *and* the fact that $f_{\text{C on T}} = f_{\text{T on C}} = \mu_s m_C g$ in the x-equation of the truck to find

$$F_T - f_{\text{C on T}} = F_T - \mu_s m_C g = m_T a_x = m_T \mu_s g$$

Solving for F_T, the maximum propulsion without the crate sliding is

$$(F_T)_{\max} = \mu_s(m_T + m_C)g$$
$$= 0.8(2200\ \text{kg})(9.80\ \text{m/s}^2) = 17{,}200\ \text{N}$$

ASSESS This is a hard result to assess. Few of us have any intuition about the size of forces that propel cars and trucks. Even so, the fact that the forward force on the truck is a significant fraction (80%) of the combined weight of the truck and the crate seems plausible. We might have been suspicious if F_T had been only a tiny fraction of the weight or much greater than the weight.

As you can see, there are many equations and many pieces of information to keep track of when solving a problem of interacting systems. These problems are not inherently harder than the problems you learned to solve in Chapters 5–7, but they do require a high level of organization. Using the systematic approach of the problem-solving strategy will help you solve similar problems successfully.

STOP TO THINK 8.3 Boxes A and B are sliding to the right across a frictionless table. The hand H is slowing them down. The mass of A is larger than the mass of B. Rank in order, from largest to smallest, the *horizontal* forces on A, B, and H.

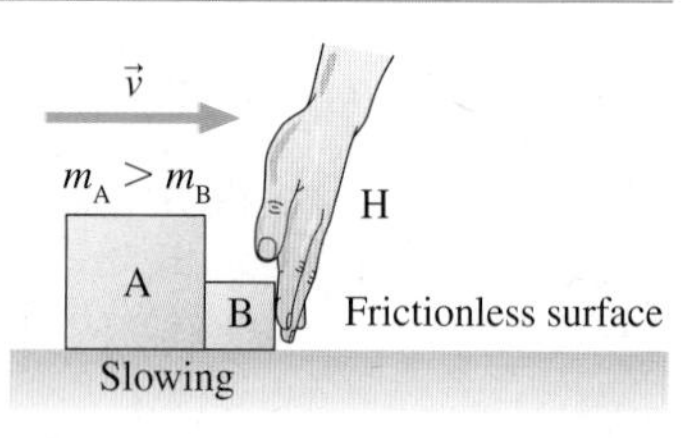

a. $F_{\text{B on H}} = F_{\text{H on B}} = F_{\text{A on B}} = F_{\text{B on A}}$
b. $F_{\text{B on H}} = F_{\text{H on B}} > F_{\text{A on B}} = F_{\text{B on A}}$
c. $F_{\text{B on H}} = F_{\text{H on B}} < F_{\text{A on B}} = F_{\text{B on A}}$
d. $F_{\text{H on B}} = F_{\text{H on A}} > F_{\text{A on B}}$

8.4 Ropes and Pulleys

Many systems are connected by strings, ropes, cables, and so on. In single-particle dynamics, we defined *tension* as the force exerted on an object by a rope or string. Now we need to think more carefully about the string itself. Just what do we mean when we talk about the tension "in" a string?

Tension Revisited

Figure 8.18a shows a heavy safe hanging from a rope, placing the rope under tension. If you cut the rope, the safe and the lower portion of the rope will fall. Thus there must be a force *within* the rope by which the upper portion of the rope pulls upward on the lower portion to prevent it from falling.

Chapter 4 introduced an atomic-level model in which tension is due to the stretching of spring-like molecular bonds within the rope. Stretched springs exert pulling forces, and the combined pulling force of billions of stretched molecular springs in a string or rope is what we call *tension*.

An important aspect of tension is that it pulls equally *in both directions*. Figure 8.18b is a very thin cross section through the rope. This small piece of rope is in equilibrium, so it must be pulled equally from both sides. To gain a mental picture, imagine holding your arms outstretched and having two friends pull on them. You'll remain at rest—but "in tension"—as long as they pull with equal strength in opposite directions. But if one lets go, analogous to the breaking of molecular bonds if a rope breaks or is cut, you'll fly off in the other direction!

The bottom layer of molecules in the rope is in contact with the safe. Here the combined pulling force of the stretched molecular springs pulls up on the safe with force $\vec{T}_{\text{R on S}}$. This is the force that we've been calling simply $\vec{T}$, the tension force of the rope pulling on the safe. Because the tension pulls equally in both directions, the tension *on the safe* at the end of the rope is the same strength as the tension *in the rope* near the end of the rope.

Figure 8.19 shows forces on the safe, the rope, and the earth. Newton's third law tells us that $T_{\text{R on S}} = T_{\text{S on R}}$ and $w_{\text{E on S}} = w_{\text{S on E}}$. These relationships are true whether the safe is at rest or accelerating. We need Newton's second law to compare $T_{\text{R on S}}$ to $w_{\text{E on S}}$.

- If the safe is in equilibrium, either at rest or moving with constant velocity, then $\vec{F}_{\text{net}} = \vec{0}$. Thus $T_{\text{R on S}} = w_{\text{E on S}}$. The tension in the rope is equal to the weight of the safe.
- If the safe is accelerating, it must have a net force acting on it. Thus $T_{\text{R on S}} \neq w_{\text{E on S}}$. The tension in the rope is *not* equal to the weight of the safe.

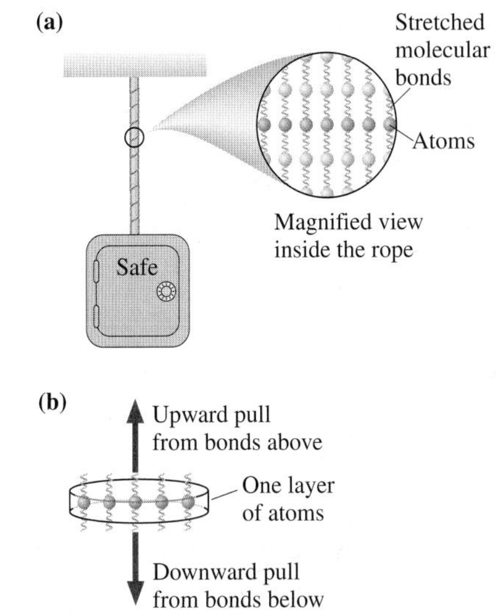

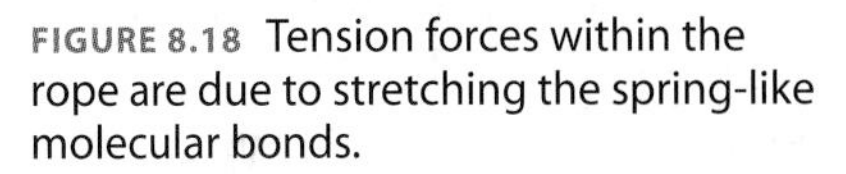

FIGURE 8.18 Tension forces within the rope are due to stretching the spring-like molecular bonds.

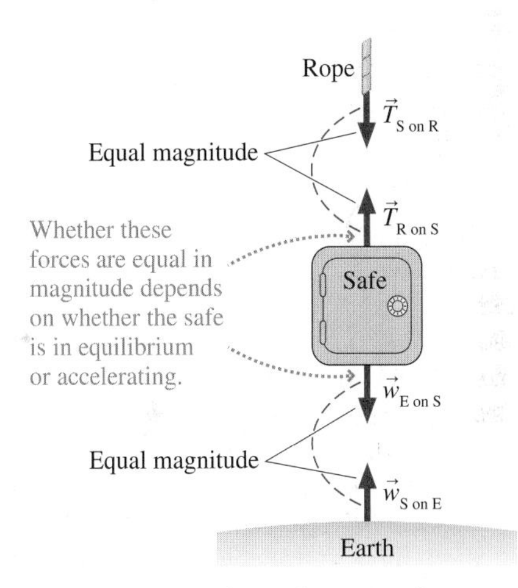

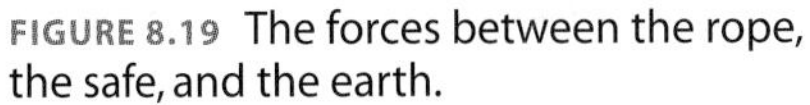

FIGURE 8.19 The forces between the rope, the safe, and the earth.

EXAMPLE 8.5 Pulling a rope

Figure 8.20a shows a student pulling horizontally with a 100 N force on a rope that is attached to a wall. In Figure 8.20b, two students in a tug-of-war pull on opposite ends of a rope with 100 N each. Is the tension in the second rope larger, smaller, or the same as that in the first?

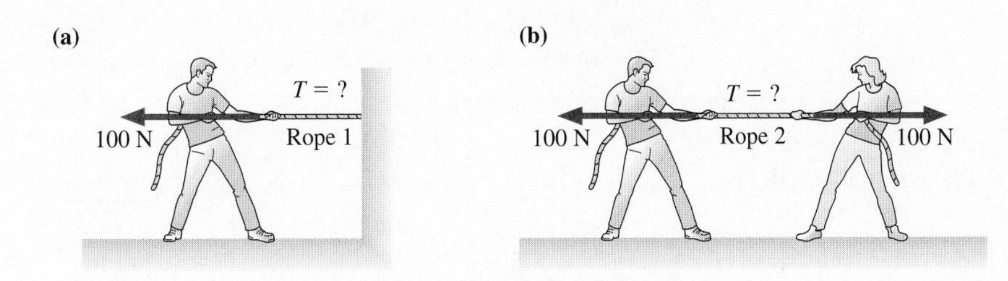

FIGURE 8.20 Pulling on a rope. Which produces a larger tension?

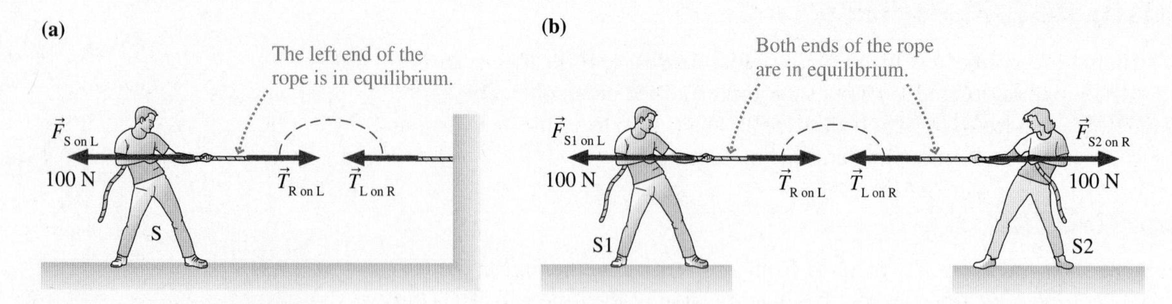

FIGURE 8.21 Analysis of tension forces.

SOLVE Surely pulling on a rope from both ends causes more tension than pulling on one end. Right? Before jumping to conclusions, let's analyze the situation carefully.

Suppose we make an imaginary slice through the rope, as shown in Figure 8.21. The right half of the rope pulls on the left half with $\vec{T}_{\text{R on L}}$ while the left half pulls back on the right half with $\vec{T}_{\text{L on R}}$. These two forces are an action/reaction pair, and their magnitude is what we *mean* by "the tension in the rope." The left half of the rope is in equilibrium, so force $\vec{T}_{\text{R on L}}$ has to balance exactly the 100 N force with which the student is pulling. That is, $T_{\text{R on L}} = F_{\text{S on L}}$. Thus

$$T_{\text{L on R}} = T_{\text{R on L}} = F_{\text{S on L}} = 100 \text{ N}$$

The first equality is based on Newton's third law (action/reaction pair). The second equality follows from Newton's second law (left half is in equilibrium). This reasoning leads us to the conclusion that the tension in the first rope is 100 N.

Now make an imaginary slice through the second rope. The left half of the rope is pulled by forces $\vec{T}_{\text{R on L}}$ and $\vec{F}_{\text{S1 on L}}$. This half of the rope is again in equilibrium, because the rope is at rest, so from Newton's second law

$$T_{\text{R on L}} = F_{\text{S1 on L}} = 100 \text{ N}$$

Similarly, the right half of the rope is pulled by forces $\vec{T}_{\text{L on R}}$ and $\vec{F}_{\text{S2 on R}}$. This piece of the rope is also in equilibrium, so

$$T_{\text{L on R}} = F_{\text{S2 on R}} = 100 \text{ N}$$

And $\vec{T}_{\text{L on R}}$ and $\vec{T}_{\text{R on L}}$ are again an action/reaction pair, so from Newton's third law

$$T_{\text{L on R}} = T_{\text{R on L}} = 100 \text{ N}$$

The tension in the rope has not changed! It is still 100 N.

You may have *assumed* that the student on the right in Figure 8.20b is doing something to the rope that the wall in Figure 8.20a does not do. But let's look more closely. Figure 8.22 shows a detailed view of the point at which the rope of Figure 8.20a is tied to the wall. Because the rope pulls on the wall with force $\vec{F}_{\text{R on W}}$, the wall must pull back on the rope (action/reaction pair) with force $\vec{F}_{\text{W on R}}$. And because the rope as a whole is in equilibrium, the wall's pull to the right must balance the student's pull to the left: $F_{\text{W on R}} = F_{\text{S on R}} = 100$ N.

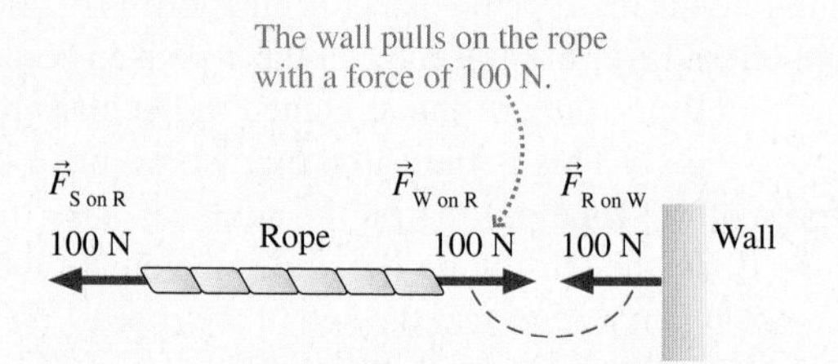

FIGURE 8.22 A closer look at the forces on the rope.

In other words, the wall in Figure 8.20a pulls the right end of the rope with a force of 100 N. The student in Figure 8.20b pulls the right end of the rope with a force of 100 N. The forces are the same in both situations! The rope does not care whether it is pulled by a wall or by a hand. It experiences the same forces in both cases, so the rope's tension is the same 100 N in both.

STOP TO THINK 8.4 All three 50 kg blocks are at rest. Is the tension in rope 2 greater than, less than, or equal to the tension in rope 1?

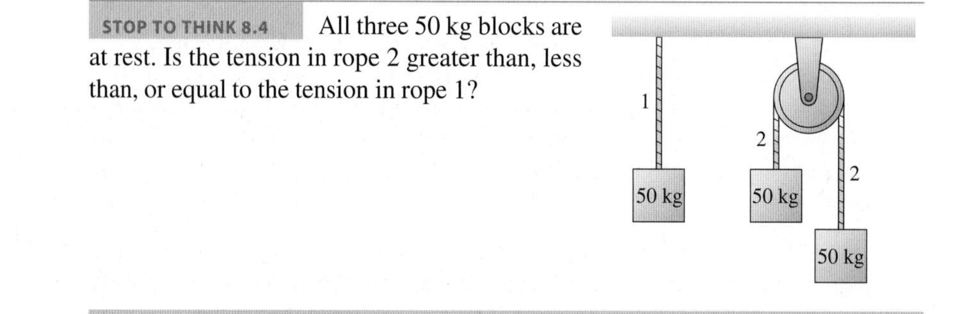

The Massless String Approximation

Example 8.5 showed that the tension is constant throughout a rope that is in equilibrium. But what happens if the rope is accelerating? For example, Figure 8.23a shows two blocks connected by a string and being pulled by force $\vec{F}$. Is the string's tension at the right end, where it pulls back on B, the same as the tension at the left end, where it pulls on A?

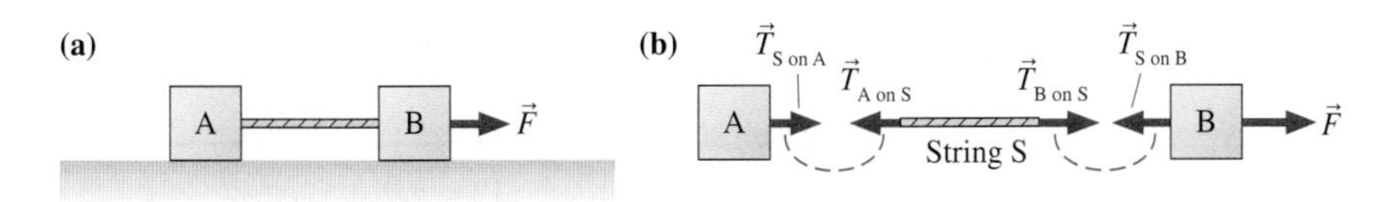

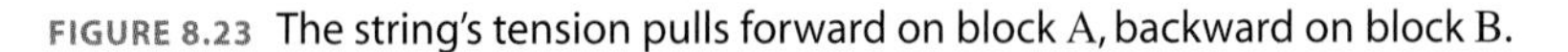

FIGURE 8.23 The string's tension pulls forward on block A, backward on block B.

Figure 8.23b shows the horizontal forces acting on the blocks and the string. If the string is accelerating, then it must have a net force applied to it. The only forces acting on the string are $\vec{T}_{\text{A on S}}$ and $\vec{T}_{\text{B on S}}$, so Newton's second law *for the string* is

$$(F_{\text{net}})_x = T_{\text{B on S}} - T_{\text{A on S}} = m_s a_x \qquad (8.7)$$

where m_s is the mass of the string.

If the string is accelerating, then the tensions at the two ends can *not* be the same. In fact, you can see that

$$T_{\text{B on S}} = T_{\text{A on S}} + m_s a_x \qquad (8.8)$$

The tension at the "front" of the string is higher than the tension at the "back." This difference in the tensions is necessary to accelerate the string! On the other hand, the tension is constant throughout a string in equilibrium ($a_x = 0$). This was the situation in Example 8.5.

Often in physics and engineering problems the mass of the string or rope is much less than the masses of the objects that it connects. In such cases, we can adopt the **massless string approximation.** In the limit $m_s \to 0$, Equation 8.8 becomes

$$T_{\text{B on S}} = T_{\text{A on S}} \quad \text{(massless string approximation)} \qquad (8.9)$$

In other words, **the tension in a massless string is constant.** This is nice, but it isn't the primary justification for the massless string approximation.

Look again at Figure 8.23b. If $T_{\text{B on S}} = T_{\text{A on S}}$, then

$$\vec{T}_{\text{S on A}} = -\vec{T}_{\text{S on B}} \qquad (8.10)$$

That is, the force on block A is equal and opposite to the force on block B. Forces $\vec{T}_{\text{S on A}}$ and $\vec{T}_{\text{S on B}}$ act *as if* they are an action/reaction pair of forces. Thus we can draw the simplified diagram of Figure 8.24 in which the string is missing and blocks A and B interact directly with each other through forces that we can call $\vec{T}_{\text{A on B}}$ and $\vec{T}_{\text{B on A}}$.

In other words, **if systems A and B interact with each other through a massless string, we can omit the string and treat forces $\vec{F}_{\text{A on B}}$ and $\vec{F}_{\text{B on A}}$ *as if* they are an action/reaction pair.** This is not literally true, because A and B are not in contact. Nonetheless, all a massless string does is transmit a force from A to B without changing the magnitude of that force. This is the real significance of the massless string approximation.

NOTE ▶ For problems in this book, you can assume that any strings or ropes are massless unless the problem explicitly states otherwise. The simplified view of Figure 8.24 is appropriate under these conditions. But if the string has a mass, it must be treated as a separate system. ◀

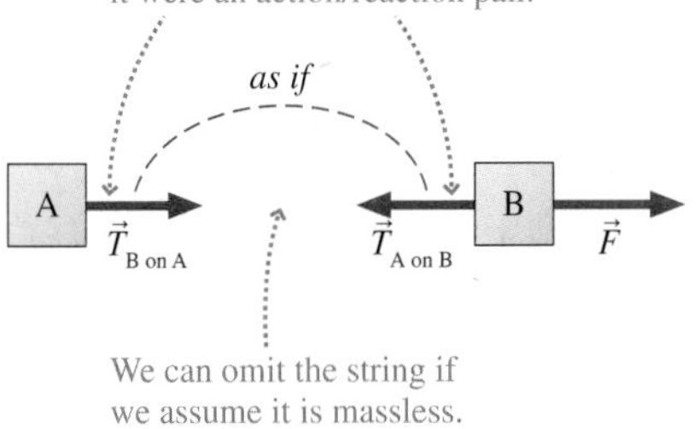

FIGURE 8.24 The massless string approximation allows systems A and B to act *as if* they are directly interacting.

EXAMPLE 8.6 Comparing two tensions

Blocks A and B in Figure 8.25 are connected by massless string 2 and pulled across a frictionless table by massless string 1. B has a larger mass than A. Is the tension in string 2 larger, smaller, or equal to the tension in string 1?

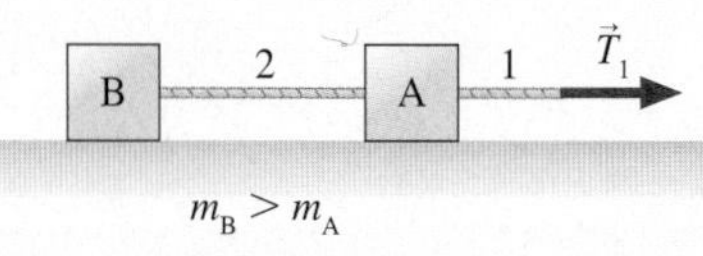

FIGURE 8.25 Blocks A and B are pulled across a frictionless table by massless strings.

MODEL The massless string approximation allows us to treat A and B *as if* they interact directly with each other. The blocks and strings must be accelerating because there's a force to the right and no friction.

SOLVE Both blocks have the same acceleration ($a_{Ax} = a_{Bx}$). B has a larger mass, so it may be tempting to conclude that string 2, which pulls B, has a greater tension than string 1, which pulls A. The flaw in this reasoning is that Newton's second law tells us only about the *net* force. The net force on B *is* larger than the net force on A, but the net force on A is *not* just the tension $\vec{T}_1$ in the forward direction. The tension in string 2 also pulls *backward* on A!

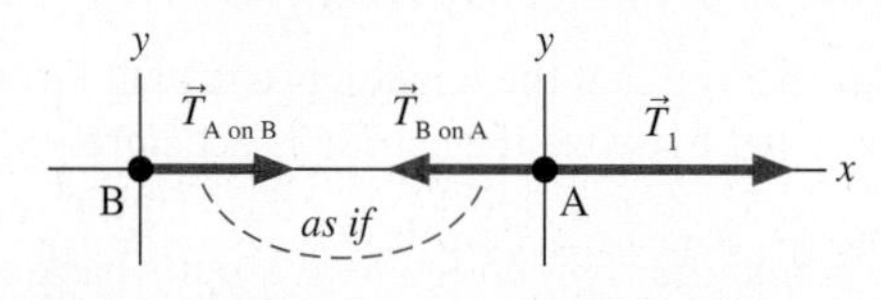

FIGURE 8.26 The horizontal forces on blocks A and B.

Figure 8.26 shows the horizontal forces in this frictionless situation. Forces $\vec{T}_{A\text{ on }B}$ and $\vec{T}_{B\text{ on }A}$ act *as if* they are an action/reaction pair. From Newton's third law,

$$T_{A\text{ on }B} = T_{B\text{ on }A} = T_2$$

where T_2 is the tension in string 2. From Newton's second law, the net force on A is

$$(F_{A\text{ net}})_x = T_1 - T_{B\text{ on }A} = T_1 - T_2 = m_A a_{Ax}$$

The net force on A is the *difference* in tension between string 1 pulling forward and string 2 pulling backward. The blocks are accelerating to the right, making $a_{Ax} > 0$, so

$$T_1 > T_2$$

The tension in string 2 is *smaller* than the tension in string 1.

ASSESS This is not an intuitively obvious result. A careful study of the reasoning in this example is worthwhile. An alternative analysis would note that $\vec{T}_1$ pulls *both* blocks, of combined mass $(m_A + m_B)$, whereas $\vec{T}_2$ pulls only block B. Thus string 1 must have the larger tension.

Pulleys

Strings and ropes often pass over pulleys. The application might be as simple as lifting a heavy weight or as complex as the internal cable-and-pulley arrangement that precisely moves a robot arm.

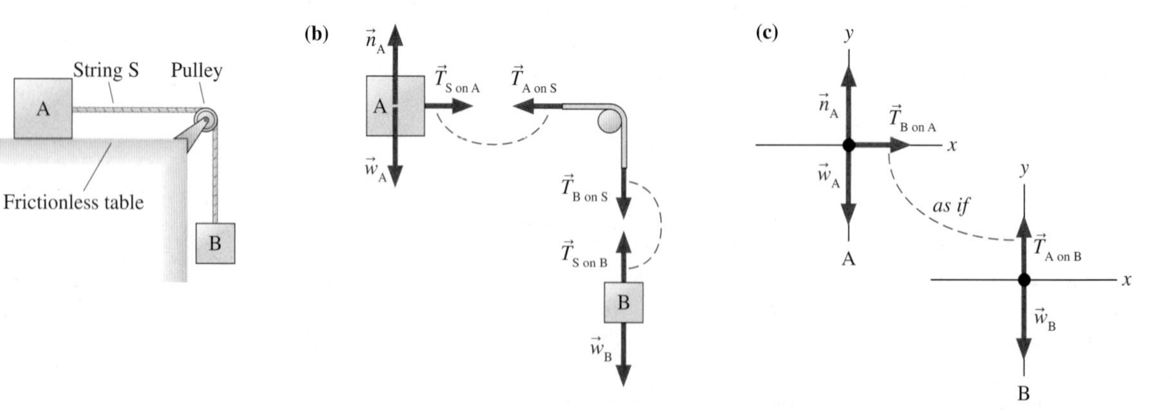

FIGURE 8.27 Blocks A and B are connected by a string that passes over a pulley.

Figure 8.27a shows a simple situation in which block B drags block A across a frictionless table as it falls. Figure 8.27b has drawn the objects separately and shown the forces. As the string moves, static friction between the string and pulley causes the pulley to turn. If we assume that

- The string *and* the pulley are both massless, and
- There is no friction where the pulley turns on its axle,

then no net force is needed to accelerate the string or turn the pulley. In this case,

$$T_{\text{A on S}} = T_{\text{B on S}}$$

In other words, the **tension in a massless string remains constant as it passes over a massless, frictionless pulley.**

Tension forces $\vec{T}_{\text{A on S}}$ and $\vec{T}_{\text{S on A}}$ are a true action/reaction pair, as are $\vec{T}_{\text{B on S}}$ and $T_{\text{S on B}}$. If we have a massless string and a massless, frictionless pulley, so that $T_{\text{A on S}} = T_{\text{B on S}}$, then it must be the case that

$$T_{\text{S on A}} = T_{\text{S on B}}$$

That is, the force of the string pulling on A equals the force of the string pulling on B.

Because of this, we can draw the simplified free-body diagram of Figure 8.27c in which the string and pulley are omitted. Forces $\vec{T}_{\text{A on B}}$ and $\vec{T}_{\text{B on A}}$ act *as if* they are an action/reaction pair, even though they are not opposite in direction. We can again say that A and B are systems that interact with each other *through the string*, and thus the force of A on B is paired with the force of B on A. The tension force gets "turned" by the pulley, which is why the two forces are not opposite each other but we can still equate their magnitudes.

STOP TO THINK 8.5 In Figure 8.27, is the tension in the string greater than, less than, or equal to the weight of block B?

8.5 Examples of Interacting-System Problems

We will conclude this chapter with several extended examples. Although the mathematics will be more involved than in any of our work up to this point, we will continue to emphasize the *reasoning* one uses in approaching problems such as these. The solutions will be based on Problem-Solving Strategy 8.1. In fact, these problems are now reaching a level of complexity that, for all practical purposes, it becomes impossible to work them unless you are following a well-planned strategy. Our earlier emphasis upon identifying forces and using free-body diagrams will now really begin to pay off!

ActivPhysics ONLINE 2.10, 2.11

EXAMPLE 8.7 Mountain climbing

A 90 kg mountain climber is suspended from the ropes shown in Figure 8.28a. The maximum tension that rope 3 can withstand before breaking is 1500 N. What is the smallest that angle θ can become before the rope breaks and the climber falls into the gorge?

MODEL Climber C, who can be modeled as a particle, is one system. The other point where forces are exerted is the knot, where the three ropes are tied together. We'll consider knot K to be a second system. Both systems are in static equilibrium. We'll assume massless ropes.

VISUALIZE Figure 8.28b shows two free-body diagrams. Forces $\vec{T}_{\text{C on K}}$ and $\vec{T}_{\text{K on C}}$ are not, strictly speaking, an action/reaction pair because the climber is not in contact with the knot. But if the ropes are massless, $\vec{T}_{\text{C on K}}$ and $\vec{T}_{\text{K on C}}$ act *as if* they are an action/reaction pair.

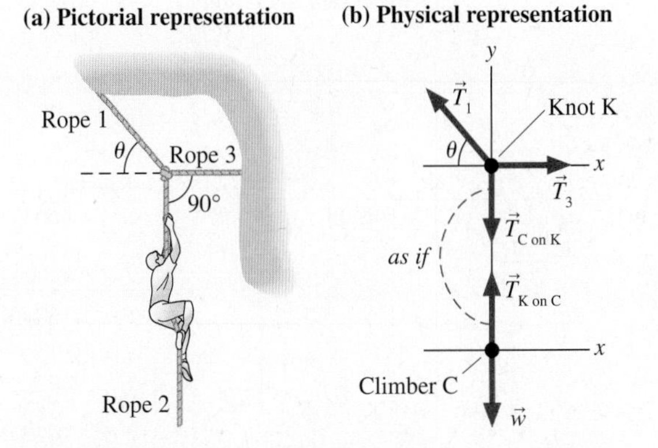

FIGURE 8.28 A mountain climber hanging from ropes and the free-body diagrams of the knot and of the climber.

SOLVE This is static equilibrium, so the net force on the climber and on the knot are zero. For the climber:

$$\sum (F_{\text{on C}})_y = T_{\text{K on C}} - w = T_{\text{K on C}} - mg = 0$$

And for the knot:

$$\sum (F_{\text{on K}})_x = T_3 - T_1 \cos\theta = 0$$
$$\sum (F_{\text{on K}})_y = T_1 \sin\theta - T_{\text{C on K}} = 0$$

From Newton's third law,

$$T_{\text{C on K}} = T_{\text{K on C}}$$

But $T_{\text{K on C}} = mg$, from the climber's equation, so $T_{\text{C on K}} = mg$. Using this gives us the knot's equations:

$$T_1 \cos\theta = T_3$$
$$T_1 \sin\theta = mg$$

Dividing the second of these by the first gives

$$\frac{T_1 \sin\theta}{T_1 \cos\theta} = \tan\theta = \frac{mg}{T_3}$$

If angle θ is too small, tension T_3 will exceed 1500 N. The smallest possible θ, at which T_3 reaches its maximum value of 1500 N, is

$$\theta_{\min} = \tan^{-1}\left(\frac{mg}{T_{3\,\max}}\right) = \tan^{-1}\left(\frac{(90\text{ kg})(9.80\text{ m/s}^2)}{1500\text{ N}}\right) = 30.5°$$

EXAMPLE 8.8 The show must go on!

A 200 kg set used in a play is stored in the loft above the stage. The rope holding the set passes up and over a pulley, then is tied backstage. The director tells a 100 kg stagehand to lower the set. When he unties the rope, the set falls and the unfortunate man is hoisted into the loft. What is the stagehand's acceleration?

MODEL The systems of interest are the stagehand M and the set S, which we will model as particles. Assume a massless rope and a massless, frictionless pulley.

VISUALIZE Figure 8.29 shows the pictorial and physical representations. The man's acceleration a_{My} is positive while the set's acceleration a_{Sy} is negative. These two accelerations have the same magnitude, because the two systems are connected by a rope, but they have opposite signs. Thus the acceleration constraint is $a_{Sy} = -a_{My}$. Forces $\vec{T}_{\text{M on S}}$ and $\vec{T}_{\text{S on M}}$ are not literally an action/reaction pair, but they act *as if* they are because the rope is massless and the pulley is massless and frictionless. Notice that the pulley has "turned" the tension force so that $\vec{T}_{\text{M on S}}$ and $\vec{T}_{\text{S on M}}$ are *parallel* to each other rather than opposite, as members of a true action/reaction pair would have to be.

SOLVE Newton's second law for the man and the set are

$$\sum (F_{\text{on M}})_y = T_{\text{S on M}} - w_{\text{M}} = T_{\text{S on M}} - m_{\text{M}}g = m_{\text{M}}a_{\text{M}y}$$
$$\sum (F_{\text{on S}})_y = T_{\text{M on S}} - w_{\text{S}}$$
$$= T_{\text{M on S}} - m_{\text{S}}g = m_{\text{S}}a_{\text{S}y} = -m_{\text{S}}a_{\text{M}y}$$

Only the y-equations are needed. Notice that we used the acceleration constraint in the last step. Newton's third law is

$$T_{\text{M on S}} = T_{\text{S on M}} = T$$

where we can drop the subscripts and call the tension simply T. With this substitution, the two second-law equations can be written

$$T - m_{\text{M}}g = m_{\text{M}}a_{\text{M}y}$$
$$T - m_{\text{S}}g = -m_{\text{S}}a_{\text{M}y}$$

Pictorial representation

Known
$m_M = 100$ kg $m_S = 200$ kg
Acceleration constraint
$a_{Sy} = -a_{My}$
Find
a_{My}

Physical representation

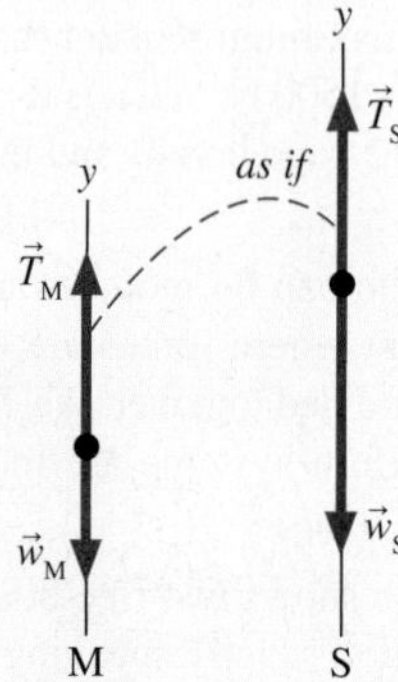

FIGURE 8.29 The pictorial and physical representations for Example 8.8.

These are simultaneous equations in the two unknowns T and a_{My}. We can eliminate T by subtracting the second equation from the first to give

$$(m_S - m_M)g = (m_S + m_M)a_{My}$$

Finally, we can solve for the hapless stagehand's acceleration:

$$a_{My} = \frac{m_S - m_M}{m_S + m_M}g = \frac{100 \text{ kg}}{300 \text{ kg}}9.80 \text{ m/s}^2 = 3.27 \text{ m/s}^2$$

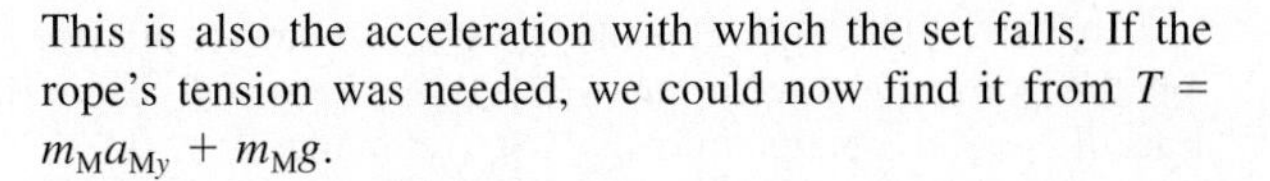

This is also the acceleration with which the set falls. If the rope's tension was needed, we could now find it from $T = m_M a_{My} + m_M g$.

ASSESS If the stagehand weren't holding on, the set would fall with free-fall acceleration g. The stagehand acts as a *counterweight* to reduce the acceleration.

EXAMPLE 8.9 A not-so-clever bank robbery

Bank robbers have pushed a 1000 kg safe to a second-story floor-to-ceiling window. They plan to break the window, then lower the safe 3.0 m to their truck. Not being too clever, they stack up 500 kg of furniture, tie a rope between the safe and the furniture, and place the rope over a pulley. Then they push the safe out the window. What is the safe's speed when it hits the truck? The coefficient of kinetic friction between the furniture and the floor is 0.5.

MODEL This is a continuation of the situation that we analyzed in Figures 8.16 and 8.27. The systems of interest are the safe S and the furniture F, which we will model as particles. We will assume a massless rope and a massless, frictionless pulley.

VISUALIZE The pictorial representation in Figure 8.30 establishes a coordinate system and defines the symbols that will be needed to calculate the safe's motion. The safe and the furniture are tied together, so their accelerations have the same magnitude. The safe has a y-component of acceleration a_{Sy} that is negative because the safe accelerates in the $-y$-direction. The furniture has an x-component a_F that is positive. Thus the acceleration constraint is

$$a_{Fx} = -a_{Sy}$$

The free-body diagrams of Figure 8.30 are modeled after Figure 8.27 but now include a kinetic friction force on the furniture. Forces $\vec{T}_{F \text{ on } S}$ and $\vec{T}_{S \text{ on } F}$ act *as if* they are an action/reaction pair, so they have been connected with a dotted line.

SOLVE We can write Newton's second law directly from the free-body diagrams. For the furniture,

$$\sum (F_{\text{on F}})_x = T_{S \text{ on } F} - f_k = T - f_k = m_F a_{Fx} = -m_F a_{Sy}$$
$$\sum (F_{\text{on F}})_y = n - w_F = n - m_F g = 0$$

And for the safe,

$$\sum (F_{\text{on S}})_y = T_{F \text{ on } S} - w_S = T - m_S g = m_S a_{Sy}$$

Notice how we used the acceleration constraint in the first equation. We also went ahead and made use of Newton's third law: $T_{F \text{ on } S} = T_{S \text{ on } F} = T$. We have one additional piece of information, the model of kinetic friction:

$$f_k = \mu_k n = \mu_k m_F g$$

where we used the y-equation of the furniture to deduce that $n = m_F g$. Substitute this result for f_k into the x-equation of the furniture, then rewrite the furniture's x-equation and the safe's y-equation:

$$T - \mu_k m_F g = -m_F a_{Sy}$$
$$T - m_S g = m_S a_{Sy}$$

Pictorial representation

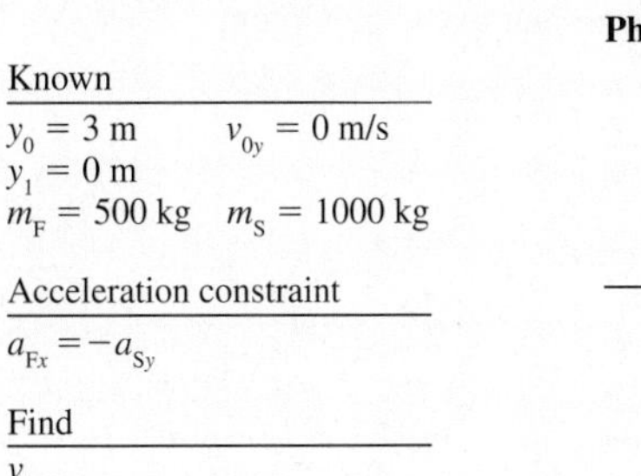

Physical representation

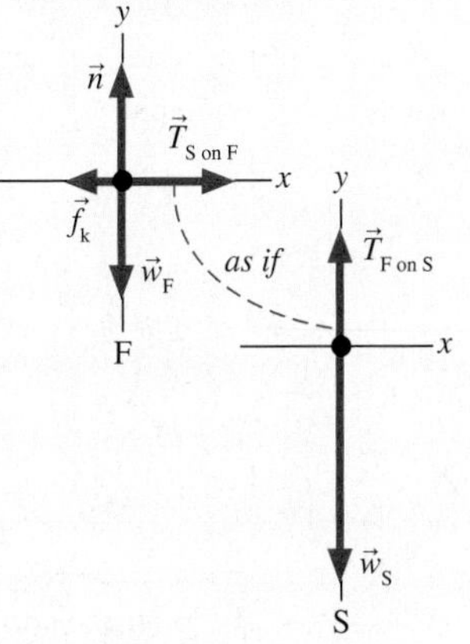

FIGURE 8.30 The pictorial and physical representations for Example 8.9.

We have succeeded in reducing our knowledge to two simultaneous equations in the two unknowns a_{Sy} and T. Subtract the second equation from the first to eliminate T:

$$(m_S - \mu_k m_F)g = -(m_S + m_F)a_{Sy}$$

Finally, solve for the safe's acceleration:

$$a_{Sy} = -\left(\frac{m_S - \mu_k m_F}{m_S + m_F}\right)g$$

$$= -\frac{1000\text{ kg} - 0.5(500\text{ kg})}{1000\text{ kg} + 500\text{ kg}}9.80\text{ m/s}^2 = -4.9\text{ m/s}^2$$

Now we need to calculate the kinematics of the falling safe. Because the time of the fall is not known or needed, we can use

$$v_{1y}{}^2 = v_{0y}{}^2 + 2a_{Sy}\Delta y = 0 + 2a_{Sy}(y_1 - y_0) = -2a_{Sy}y_0$$

$$v_1 = \sqrt{-2a_{Sy}y_0} = \sqrt{-2(-4.9\text{ m/s}^2)(3.0\text{ m})} = 5.42\text{ m/s}$$

The value of v_{1y} is negative, but we only needed to find the speed so we took the absolute value. It seems unlikely that the truck will survive the impact of the 1000 kg safe!

EXAMPLE 8.10 Pushing a package

A 40 kg boy works at his dad's hardware store. One of the boy's jobs is to unload the delivery truck. He places each package on a 30° ramp and shoves it up the ramp into the storeroom. He needs to shove the package with an acceleration of at least 1.0 m/s^2 in order for the package to make it to the top of the ramp. One day the ground is wet with rain and he's wearing slick leather-soled shoes. The coefficient of static friction between his shoes and the ground is only 0.25. The largest package of the day is 15 kg, and its coefficient of kinetic friction on the ramp is 0.40. Can he give the package a big enough shove to reach the top of the ramp without his feet slipping?

MODEL The systems of interest are the boy B and the package P, which we will model as particles.

VISUALIZE There's a lot of information in this problem, so the pictorial representation of Figure 8.31 is essential. The package is moving up an incline whereas the boy, if he slips, will move horizontally. Consequently, it is useful to give them different coordinate systems. The free-body diagrams show that the boy pushes the package with force $\vec{F}_{\text{B on P}}$ and the package pushes back with force $\vec{F}_{\text{P on B}}$. If static friction does its job, it must point *forward* to prevent the boy's feet from slipping backward. To answer the question, we'll first calculate how much static friction is needed for the boy to push the package with an acceleration of 1.0 m/s^2. Then we'll compare that to the maximum possible static friction $f_{s\,max}$.

SOLVE Now we're ready to write Newton's second law. The boy is in static equilibrium, with $\vec{F}_{net} = \vec{0}$, so his equations are

$$\sum(F_{\text{on B}})_x = f_s - F_{\text{P on B}}\cos\theta = f_s - F\cos\theta = 0$$

$$\sum(F_{\text{on B}})_y = n_B - w_B - F_{\text{P on B}}\sin\theta$$

$$= n_B - m_B g - F\sin\theta = 0$$

The package is accelerating up the ramp as he pushes it, so the package's equations are

$$\sum(F_{\text{on P}})_x = F_{\text{B on P}} - f_k - w_P\sin\theta$$

$$= F - f_k - m_P g\sin\theta = m_P a_x$$

$$\sum(F_{\text{on P}})_y = n_P - w_P\cos\theta = n_P - m_P g\cos\theta = 0$$

We went ahead and made use of Newton's third law: $F_{\text{P on B}} = F_{\text{B on P}} = F$. The package's y-equation tells us that $n_P = m_P g\cos\theta$, so the kinetic friction on the package is

$$f_k = \mu_k n_P = \mu_k m_P g\cos\theta$$

If we substitute this into the package's x-equation, we can solve for force F:

$$F - \mu_k m_P g\cos\theta - m_P g\sin\theta = m_P a_x$$

$$F = m_P(a_x + g\sin\theta + \mu_k g\cos\theta) = 139\text{ N}$$

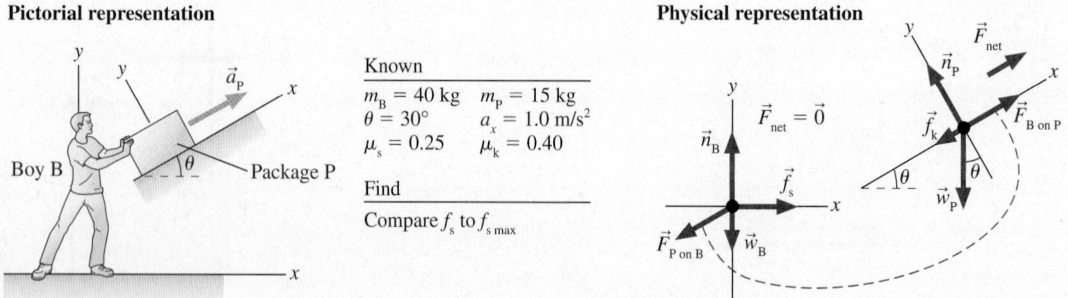

FIGURE 8.31 The pictorial and physical representations for Example 8.10.

This is the size of the force that will accelerate the package up the ramp at 1.0 m/s^2. If we now use this in the boy's x-equation, we find

$$f_s = F\cos\theta = 120 \text{ N}$$

The boy *needs* this much static friction to push the package without slipping. But needing 120 N of friction doesn't mean that 120 N is available. The maximum possible static friction is $f_{s\,max} = \mu_s n_B$. The normal force acting on the boy is not simply his weight. The normal force is affected by the vertical component of $\vec{F}_{P\text{ on }B}$. From his y-equation we find

$$n_B = m_B g + F\sin\theta = 462 \text{ N}$$

Thus the maximum static friction without slipping is

$$f_{s\,max} = \mu_s n_B = 115 \text{ N}$$

Consequently, the boy *cannot* shove hard enough without slipping.

ASSESS This is an excellent illustration of how crucial it is to focus on clarifying information, identifying forces, and drawing free-body diagrams. The rest of the problem was not trivial, but we could work our way through it with confidence after having found the free-body diagrams. It would be hopeless, even for an experienced physicist, to try to go directly to Newton's laws without this analysis.

Newton's third law brings us to the end of Part I on Newtonian mechanics. Rather than pursue ever more complex dynamics problems, our road will turn in a new direction as we begin to look at the new and important concepts of momentum and energy. Forces will remain with us, always forming the most basic level of our understanding of dynamics, but they need to begin to share the stage with newer ideas.

STOP TO THINK 8.6 A small car is pushing a larger truck that has a dead battery. The mass of the truck is larger than the mass of the car. Which of the following statements is true?

a. The car exerts a force on the truck, but the truck doesn't exert a force on the car.
b. The car exerts a larger force on the truck than the truck exerts on the car.
c. The car exerts the same amount of force on the truck as the truck exerts on the car.
d. The truck exerts a larger force on the car than the car exerts on the truck.
e. The truck exerts a force on the car, but the car doesn't exert a force on the truck.

SUMMARY

The goal of Chapter 8 has been to learn to use Newton's third law to understand interacting systems.

GENERAL PRINCIPLES

Newton's Third Law

Every force occurs as one member of an **action/reaction pair** of forces. The two members of an action/reaction pair:

- Act on two *different* objects.
- Are equal in magnitude but opposite in direction:

$$\vec{F}_{\text{A on B}} = -\vec{F}_{\text{B on A}}$$

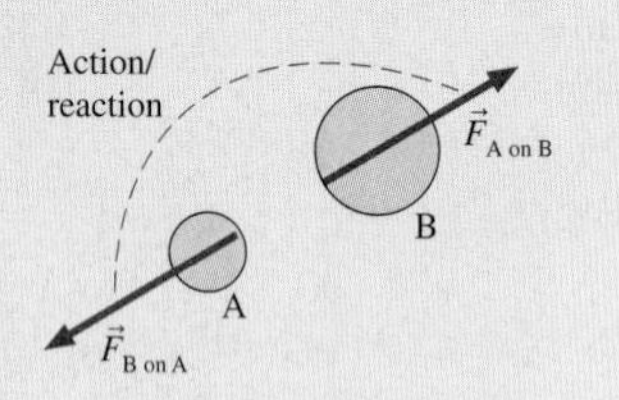

Solving Interacting-System Problems

MODEL Choose the systems of interest.

VISUALIZE

Pictorial representation:
Sketch and define coordinates.
Identify acceleration constants.

Physical representation:
Draw a separate free-body diagram for each system.
Connect action/reaction pairs with dotted lines.

SOLVE Write Newton's second law for each system.
Include *all* forces acting *on* each system.
Use Newton's third law to equate the magnitudes of action/reaction pairs.
Include acceleration constraints and friction.

ASSESS Is the result reasonable?

IMPORTANT CONCEPTS

Interacting systems and the environment

Two systems interact by exerting forces on each other.
Systems whose motion is not of interest form the environment.
The systems of interest interact with the environment, but those interactions can be considered external forces.

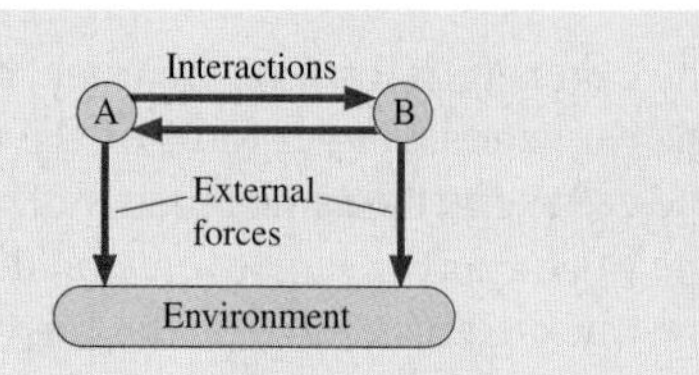

APPLICATIONS

Acceleration constraints

Objects that are constrained to move together must have accelerations of equal magnitude: $a_A = a_B$.
This must be expressed in terms of components, such as $a_{Ax} = -a_{By}$.

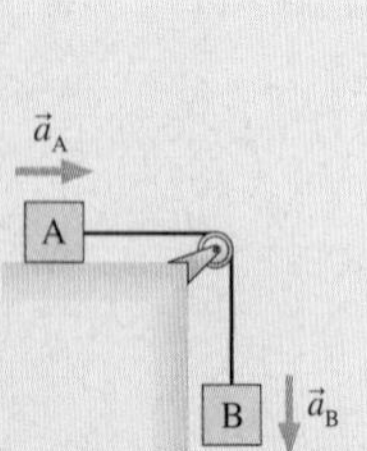

Strings and pulleys

The tension in a string or rope pulls in both directions. The tension is constant in a string if the string is:

- Massless, or
- In equilibrium

Systems connected by massless strings passing over massless, frictionless pulleys act *as if* they interact via an action/reaction pair of forces.

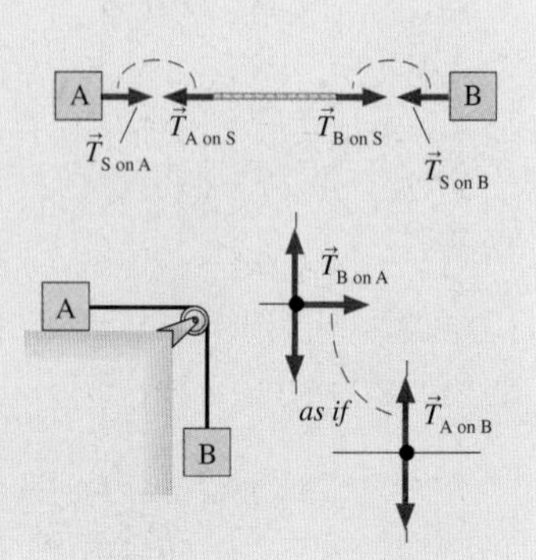

TERMS AND NOTATION

interaction
action/reaction pair
external force
propulsion
Newton's third law
acceleration constraint
massless string approximation

EXERCISES AND PROBLEMS

The icon indicates that the problem can be done on a Dynamics Worksheet.

Exercises

Section 8.2 Identifying Action/Reaction Pairs

1. A weight lifter stands up from a squatting position while holding a heavy barbell across his shoulders. Identify all action/reaction pairs, show them on a figure, then draw free-body diagrams for the weight lifter and the barbell. Use dotted lines to connect the members of an action/reaction pair.
2. A softball player is throwing the ball. Her arm has come forward to where it is beside her head, but she hasn't yet released the ball. Identify all action/reaction pairs, show them on a figure, then draw free-body diagrams for the ball player and the ball. Use dotted lines to connect the members of an action/reaction pair.
3. A soccer ball and a bowling ball roll across a hard floor and collide head on. Identify all action/reaction pairs during the time that the balls are in contact, show them on a figure, then draw free-body diagrams for each ball. Use dotted lines to connect the members of an action/reaction pair. Rolling friction is negligible.
4. A mountain climber is using a massless rope to pull a bag of supplies up a 45° slope. Identify all action/reaction pairs, show them on a figure, then draw free-body diagrams for the mountain climber, the rope, and the bag. Use dotted lines to connect the members of an action/reaction pair.
5. Block A in Figure Ex8.5 is heavier than block B and is sliding down the incline. All surfaces have kinetic friction. Identify all action/reaction pairs, show them on a figure, then draw free-body diagrams for both blocks, the rope, and the pulley. The pulley holder can be considered part of the earth. Use dotted lines to connect the members of an action/reaction pair.

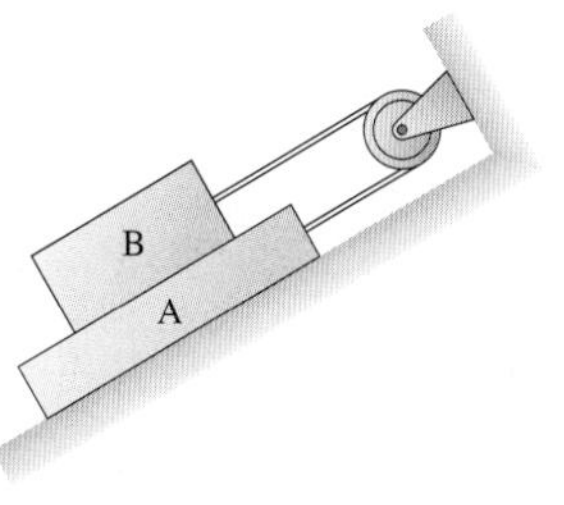

FIGURE EX8.5

Section 8.3 Newton's Third Law

6. Figure Ex8.6 shows two strong magnets on opposite sides of a small table. The long-range attractive force between the magnets keeps the lower magnet in place.
   a. Identify all action/reaction pairs, show them on a figure, then draw free-body diagrams for both magnets and the table. Use dotted lines to connect the members of an action/reaction pair.
   b. Suppose the weight of the table is 20 N, the weight of each magnet is 2.0 N, and the magnetic force on the lower magnet is three times its weight. Find the magnitude of each of the forces shown on your free-body diagrams.

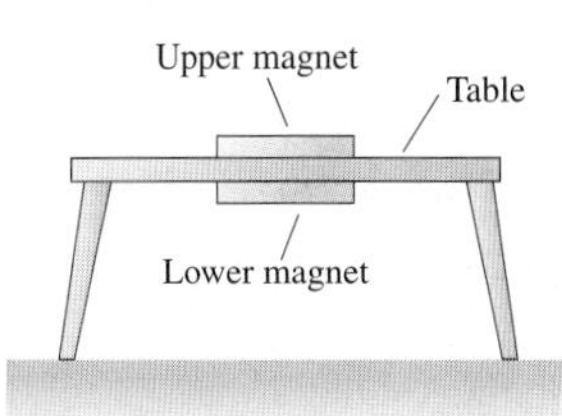

FIGURE EX8.6

7. a. How much force does an 80 kg astronaut exert on his chair while sitting at rest on the launch pad?
   b. How much force does the astronaut exert on his chair while accelerating straight up at 10 m/s^2?
8. A 1000 kg car pushes a 2000 kg truck that has a dead battery. When the driver steps on the accelerator, the drive wheels of the car push against the ground with a force of 4500 N.
   a. What is the magnitude of the force of the car on the truck?
   b. What is the magnitude of the force of the truck on the car?
9. Blocks with masses of 1 kg, 2 kg, and 3 kg are lined up in a row on a frictionless table. All three are pushed forward by a 12 N force applied to the 1 kg block. How much force does the 2 kg block exert on the 3 kg block? How much force does the 2 kg block exert on the 1 kg block?
10. An 80 kg spacewalking astronaut pushes off a 640 kg satellite, exerting a 100 N force for the 0.50 s it takes him to straighten his arms. How far apart are the astronaut and the satellite after 1.0 min?

Section 8.4 Ropes and Pulleys

11. What is the tension in the rope of Figure Ex8.11?

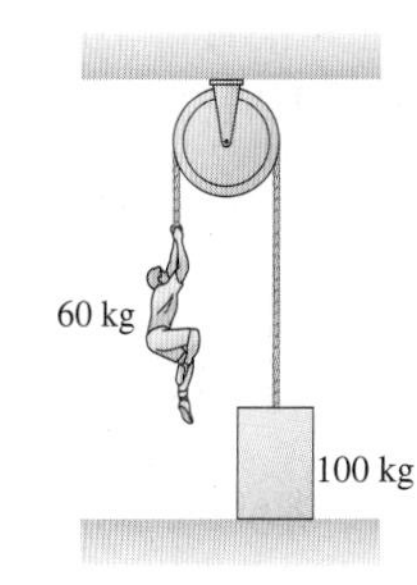

FIGURE EX8.11

12. Jimmy has caught two fish in Yellow Creek. He has tied the line holding the 3.0 kg steelhead trout to the tail of the 1.5 kg carp. To show the fish to a friend, he lifts upward on the carp with a force of 60 N.
   a. Draw separate free-body diagrams for the trout and the carp. Label all forces, then use dotted lines to connect action/reaction pairs or forces that act as if they are a pair.
   b. Rank in order, from largest to smallest, the magnitudes of all the forces shown on your free-body diagrams. Explain your reasoning.

13. A 2-m-long, 500 g rope pulls a 10 kg block of ice across a horizontal, frictionless surface. The block accelerates at 2.0 m/s^2. How much force pulls forward on (a) the ice, (b) the rope?
14. The cable cars in San Francisco are pulled along their tracks by an underground steel cable that moves along at 9.5 mph. The cable is driven by large motors at a central power station and extends, via an intricate pulley arrangement, for several miles beneath the city streets. The length of a cable stretches by up to 100 ft during its lifetime. To keep the tension constant, the cable passes around a 1.5-m-diameter "tensioning pulley" that rolls back and forth on rails, as shown in Figure Ex8.14. A 2000 kg block is attached to the tensioning pulley's cart, via a rope and pulley, and is suspended in a deep hole. What is the tension in the cable car's cable?

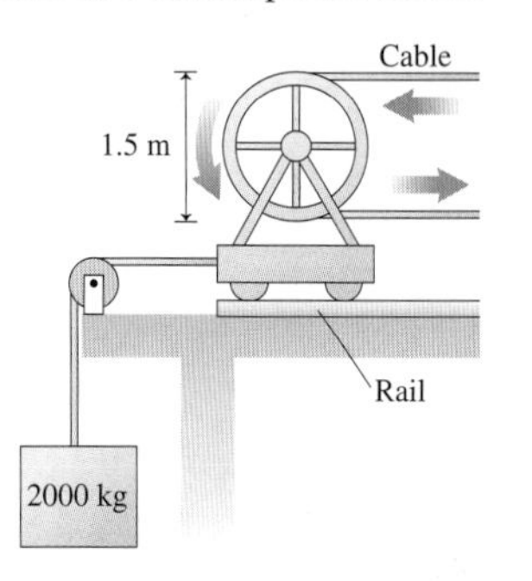

FIGURE EX8.14

15. A mobile at the art museum has a 2.0 kg steel cat and 4.0 kg steel dog suspended from a lightweight cable, as shown in Figure Ex8.15. It is found that $\theta_1 = 20°$ when the center rope is adjusted to be perfectly horizontal. What are the tension and the angle of rope 3?

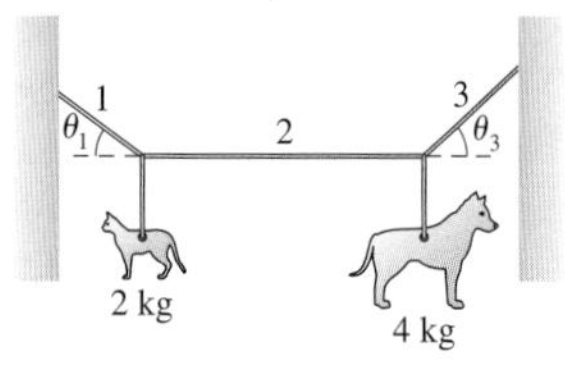

FIGURE EX8.15

Problems

16. A massive steel cable drags a 20 kg block across a horizontal, frictionless surface. A 100 N force applied to the cable causes the block to reach a speed of 4.0 m/s in a distance of 2.0 m. What is the mass of the cable?
17. A massive steel cable drags a 20 kg block across a horizontal, frictionless surface. A 100 N force applied to the cable causes the block to reach a speed of 4.0 m/s in 2.0 s. What is the difference in tension between the two ends of the cable?
18. A 1.0-m-long massive steel cable drags a 20 kg block across a horizontal, frictionless surface. A 100 N force applied to the cable causes the block to travel 4.0 m in 2.0 s. Graph the tension in the cable as a function of position along the cable, starting at the point where the cable is attached to the block.
19. A 3.0-m-long, 2.2 kg rope is suspended from the ceiling. Graph the tension in the rope as a function of position along the rope, starting from the bottom.
20. The sled dog in Figure P8.20 drags sleds A and B across the snow. The coefficient of friction between the sleds and the snow is 0.10. If the tension in rope 1 is 150 N, what is the tension in rope 2?

FIGURE P8.20

21. While driving to work last year, I was holding my coffee mug in my left hand while changing the CD with my right hand. Then the cell phone rang, so I placed the mug on the flat part of my dashboard. Then, believe it or not, a deer ran out of the woods and on to the road right in front of me. Fortunately, my reaction time was zero, and I was able to stop from a speed of 20 m/s in a mere 50 m, just barely avoiding the deer. Later tests revealed that the static and kinetic coefficients of friction of the coffee mug on the dash are 0.50 and 0.30, respectively; the coffee and mug had a mass of 0.50 kg; and the mass of the deer was 120 kg. Did my coffee mug slide?
22. a. Describe how a car accelerates from rest. Your explanation should be in terms of forces and physical laws. You should include and use a free-body diagram.
    b. Why can a car accelerate but a house cannot? Again, your explanation should be in terms of forces and their properties.
    c. Two-thirds of the weight of a 1500 kg car rests on the drive wheels. What is the maximum acceleration of this car on a concrete surface?
23. A Federation starship (2.0×10^6 kg) uses its tractor beam to pull a shuttlecraft (2.0×10^4 kg) aboard from a distance of 10 km away. The tractor beam exerts a constant force of 4.0×10^4 N on the shuttlecraft. Both spacecraft are initially at rest. How far does the starship move as it pulls the shuttlecraft aboard?
24. Bob, who has a mass of 75 kg, can throw a 500 g rock with a speed of 30 m/s. The distance through which his hand moves as he accelerates the rock forward from rest until he releases it is 1.0 m.
    a. What constant force must Bob exert on the rock to throw it with this speed?
    b. If Bob is standing on frictionless ice, what is his recoil speed after releasing the rock?
25. Two packages at UPS start sliding down the 20° ramp shown in Figure P8.25. Package A has a mass of 5.0 kg and a coefficient of friction of 0.20. Package B has a mass of 10 kg and a coefficient of friction of 0.15. How long does it take package A to reach the bottom?

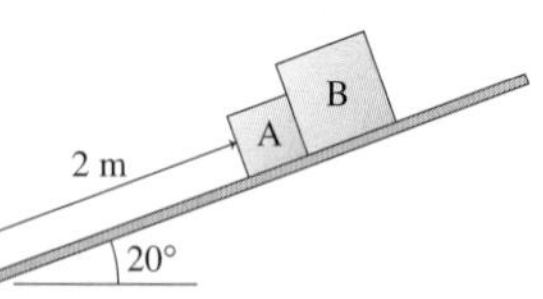

FIGURE P8.25

26. Figure P8.26 shows two 1.0 kg blocks connected by a rope. A second rope hangs beneath the lower block. Both ropes have a mass of 250 g. The entire assembly is accelerated upward at 3.0 m/s^2 by force $\vec{F}$.
    a. What is F?
    b. What is the tension at the top end of rope 1?
    c. What is the tension at the bottom end of rope 1?
    d. What is the tension at the top end of rope 2?

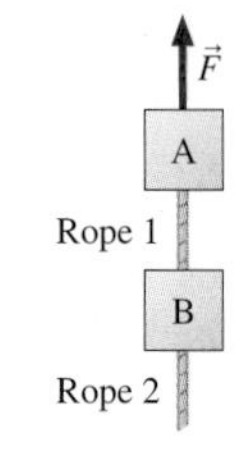

FIGURE P8.26

27. The 1.0 kg block in Figure P8.27 is tied to the wall with a rope. It sits on top of the 2.0 kg block. The lower block is pulled to the right with a tension force of 20 N. The coefficient of kinetic friction at both the lower and upper surfaces of the 2.0 kg block is $\mu_k = 0.40$.
   a. What is the tension in the rope holding the 1.0 kg block to the wall?
   b. What is the acceleration of the 2.0 kg block?

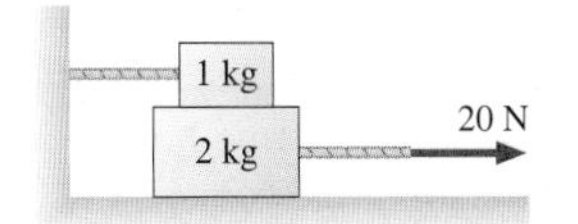

FIGURE P8.27

28. The lower block in Figure P8.28 is pulled on by a rope with a tension force of 20 N. The coefficient of kinetic friction between the lower block and the surface is 0.30. The coefficient of kinetic friction between the lower block and the upper block is also 0.30. What is the acceleration of the 2.0 kg block?

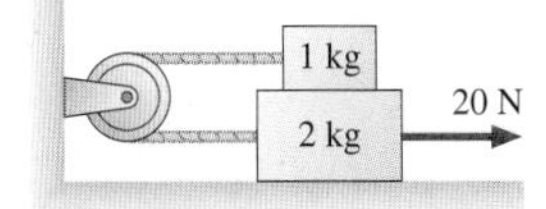

FIGURE P8.28

29. A rope attached to a 20 kg wood sled pulls the sled up a 20° snow-covered hill. A 10 kg wood box rides on top of the sled. If the tension in the rope steadily increases, at what value of the tension does the box slip?
30. Mass m_1 on the frictionless table of Figure P8.30 is connected by a string through a hole in the table to a hanging mass m_2. With what speed must m_1 rotate in a circle of radius r if m_2 is to remain hanging at rest?

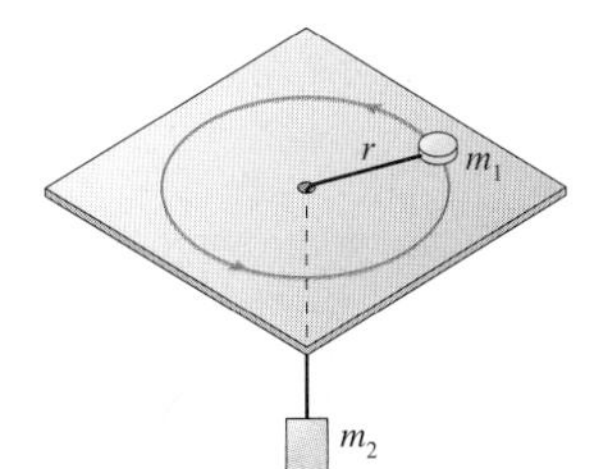

FIGURE P8.30

31. You see the boy next door trying to push a crate down the sidewalk. He can barely keep it moving, and his feet occasionally slip. You start to wonder how heavy the crate is. You call to ask the boy his mass, and he replies "50 kg." From your recent physics class you estimate that the static and kinetic coefficients of friction are 0.8 and 0.4 for the boy's shoes, and 0.5 and 0.2 for the crate. Estimate the mass of the crate.
32. The coefficient of static friction is 0.60 between the two blocks in Figure P8.32. The coefficient of kinetic friction between the lower block and the floor is 0.20. Force $\vec{F}$ causes both blocks to cross a distance of 5.0 m, starting from rest. What is the least amount of time in which this motion can be completed without the top block sliding on the lower block?

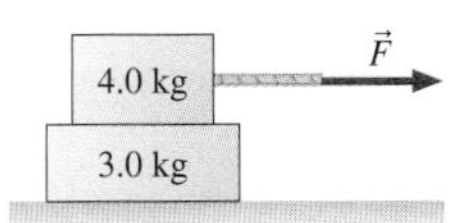

FIGURE P8.32

33. The 100 kg block in Figure P8.33 takes 6.0 s to reach the floor after being released from rest. What is the mass of the block on the left?

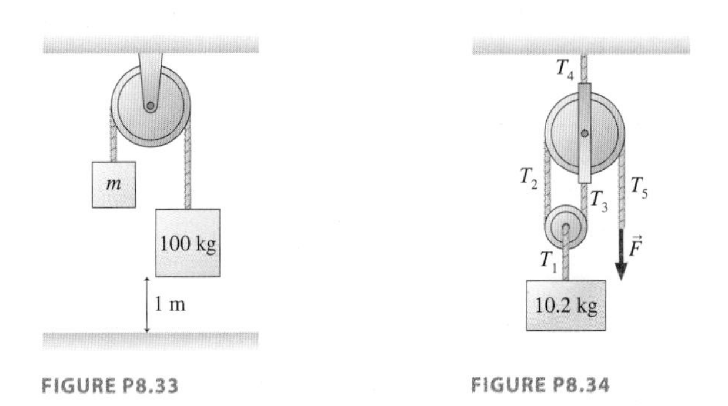

FIGURE P8.33

FIGURE P8.34

34. The 10.2 kg block in Figure P8.34 is held in place by the massless rope passing over two massless, frictionless pulleys. Find the tensions T_1 to T_5 and the magnitude of force $\vec{F}$.
35. The coefficient of kinetic friction between the 2.0 kg block in Figure P8.35 and the table is 0.30. What is the acceleration of the 2.0 kg block?

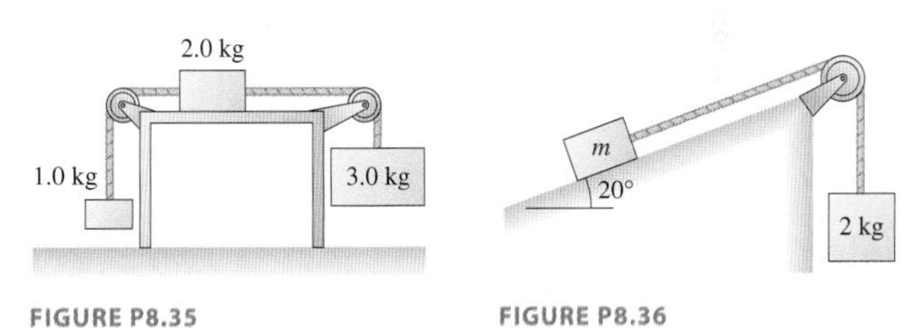

FIGURE P8.35

FIGURE P8.36

36. Figure P8.36 shows a block of mass m resting on a 20° slope. The block has coefficients of friction $\mu_s = 0.80$ and $\mu_k = 0.50$ with the surface. It is connected via a massless string over a massless, frictionless pulley to a hanging block of mass 2.0 kg.
   a. What is the minimum mass m that will stick and not slip?
   b. If this minimum mass is nudged ever so slightly, it will start being pulled up the incline. What acceleration will it have?
37. A 4.0 kg box is on a frictionless 35° slope and is connected via a massless string over a massless, frictionless pulley to a hanging 2.0 kg weight. The picture for this situation is similar to Figure P8.36.
   a. What is the tension in the string if the 4.0 kg box is *held* in place, so that it cannot move?
   b. If the box is then released, which way will it move on the slope?
   c. What is the tension in the string once the box begins to move?
38. The 1.0 kg physics book in Figure P8.38 is connected by a string to a 500 g coffee cup. The book is given a push up the slope and released with a speed of 3.0 m/s. The coefficients of friction are $\mu_s = 0.50$ and $\mu_k = 0.20$.
   a. How far does the book slide?
   b. At the highest point, does the book stick to the slope, or does it slide back down?

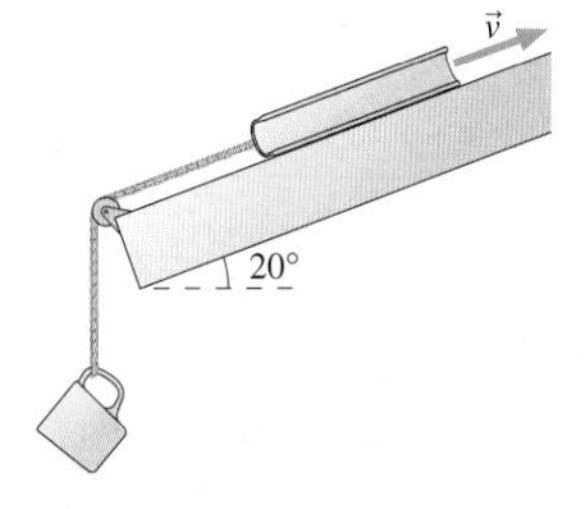

FIGURE P8.38

39. The 2000 kg cable car shown in Figure P8.39 descends a 200-m-high hill. In addition to its brakes, the cable car controls its speed by pulling an 1800 kg counterweight up the other side of the hill. The rolling friction of both the cable car and the counterweight are negligible.
    a. How much braking force does the cable car need to descend at constant speed?
    b. One day the brakes fail just as the cable car leaves the top on its downward journey. What is the runaway car's speed at the bottom of the hill?

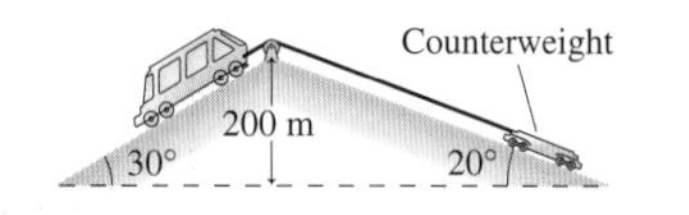

FIGURE P8.39

40. In Figure P8.40, find an expression for the acceleration of m_1. **Hint:** Think carefully about the acceleration constraint.

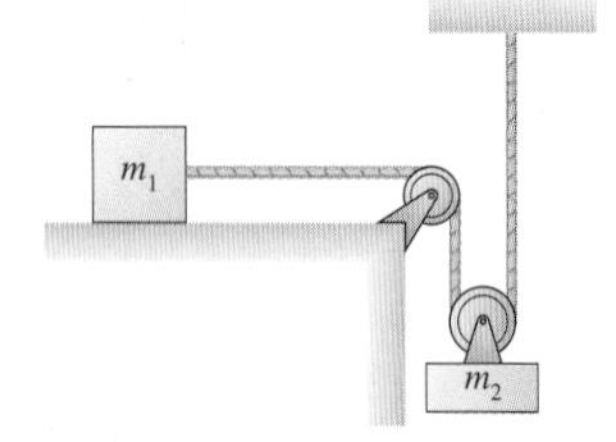

FIGURE P8.40

41. What is the acceleration of the 2.0 kg block in Figure P8.41 across the frictionless table?
Hint: Think carefully about the acceleration constraint.

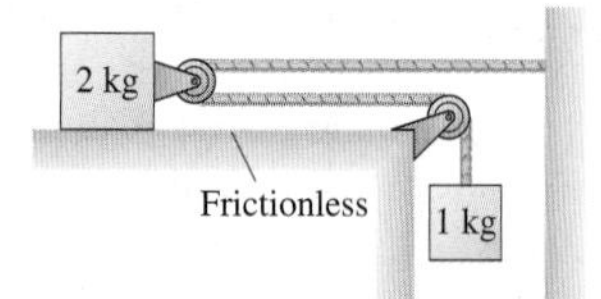

FIGURE P8.41

42. A house painter uses the chair and pulley arrangement of Figure P8.42 to lift himself up the side of a house. The painter's mass is 70 kg and the chair's mass is 10 kg. With what force must he pull down on the rope in order to accelerate upward at 0.20 m/s^2?

FIGURE P8.42

43. A 70 kg tightrope walker stands at the center of a rope. The rope supports are 10 m apart and the rope sags 10° at each end. The tightrope walker crouches down, then leaps straight up with an acceleration of 8.0 m/s^2 to catch a passing trapeze. What is the tension in the rope as he jumps?

44. Find an expression for the magnitude of the horizontal force F in Figure P8.44 for which m_1 does not slip either up or down along the wedge. All surfaces are frictionless.

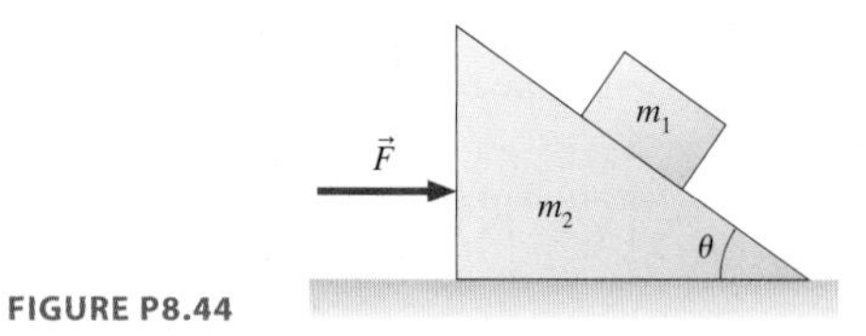

FIGURE P8.44

Problems 45 and 46 show the free-body diagrams of two interacting systems. For each of these, you are to
    a. Write a realistic problem for which these are the correct free-body diagrams. Be sure that the answer your problem requests is consistent with the diagrams shown.
    b. Finish the solution of the problem.

45.
46.

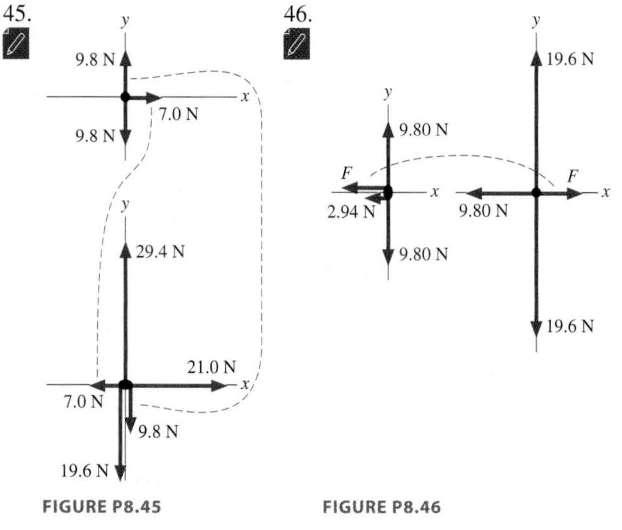

FIGURE P8.45
FIGURE P8.46

Challenge Problems

47. A 100 g ball of clay is thrown horizontally with a speed of 10 m/s toward a 900 g block resting on a frictionless surface. It hits the block and sticks. The clay exerts a constant force on the block during the 10 ms it takes the clay to come to rest relative to the block. After 10 ms, the block and the clay are sliding along the surface as a single system.
    a. What is their speed after the collision?
    b. What is the force of the clay on the block during the collision?
    c. What is the force of the block on the clay?

NOTE ► This problem can be worked using the conservation laws you will be learning in the next few chapters. However, here you're asked to solve the problem using Newton's laws. ◄

48. A 100 kg basketball player can leap straight up in the air to a height of 80 cm, as shown in Figure CP8.48. You can understand how by analyzing the situation as follows:
    a. The player bends his legs until the upper part of his body has dropped by 60 cm, then he begins his jump. Draw separate free-body diagrams for the player and for the floor *as* he is jumping, but before his feet leave the ground.

b. Is there a net force on the player as he jumps (before his feet leave the ground)? How can that be? Explain.
c. With what speed must the player leave the ground to reach a height of 80 cm?
d. What was his acceleration, assumed to be constant, as he jumped?
e. Suppose the player jumps while standing on a bathroom scale that reads in newtons. What does the scale read before he jumps, as he is jumping, and after his feet leave the ground?

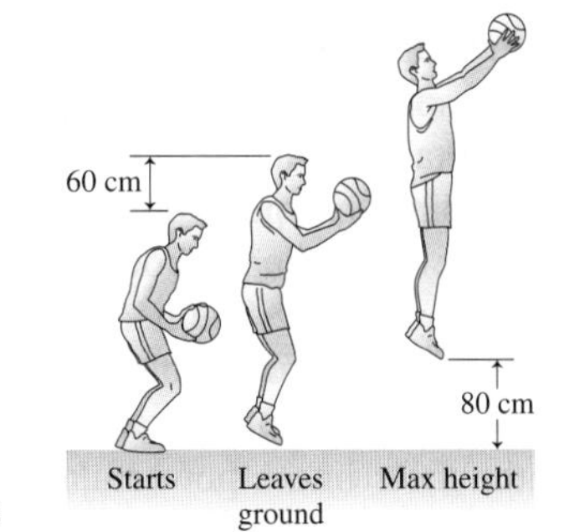

FIGURE CP8.48

49. Figure CP8.49 shows a 200 g hamster sitting on an 800 g wedge-shaped block. The block, in turn, rests on a spring scale.
a. Initially, static friction is sufficient to keep the hamster from moving. In this case, the hamster and the block are effectively a single 1000 g mass and the scale should read 9.8 N. Show that this is the case by treating the hamster and the block as *separate* systems and analyzing the forces.

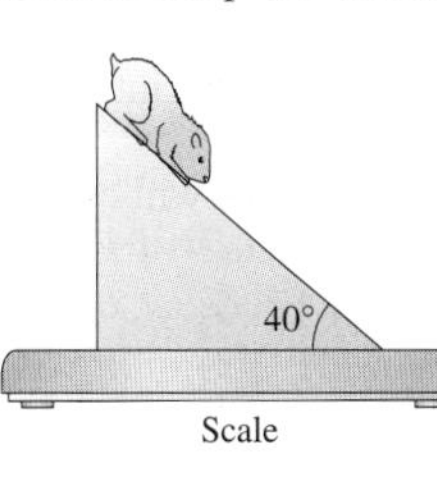

FIGURE CP8.49

b. An extra-fine lubricating oil having $\mu_s = \mu_k = 0$ is sprayed on the top surface of the block, causing the hamster to slide down. Friction between the block and the scale is large enough that the block does *not* slip on the scale. What does the scale read as the hamster slides down?

50. Figure CP8.50 shows three hanging masses connected by massless strings over two massless, frictionless pulleys.
a. Find the acceleration constraint for this system. It is a single equation relating a_{1y}, a_{2y}, and a_{3y}.
Hint: y_A isn't constant.
b. Find an expression for the tension in string A.
Hint: You should be able to write four second-law equations. These, plus the acceleration constraint, are five equations in five unknowns.
c. Suppose: $m_1 = 2.5$ kg, $m_2 = 1.5$ kg, and $m_3 = 4.0$ kg. Find the acceleration of each.
d. The 4.0 kg mass would appear to be in equilibrium. Explain why it accelerates.

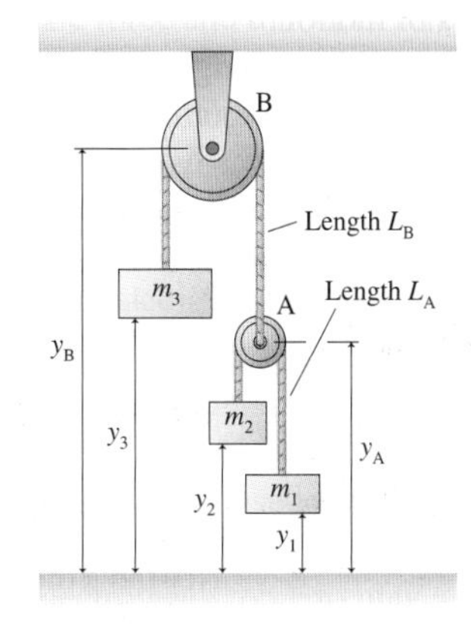

FIGURE CP8.50

STOP TO THINK ANSWERS

Stop to Think 8.1: The weight force and the normal force are incorrectly identified as an action/reaction pair. The normal force is $\vec{n}_{S\text{ on }B}$, so it is paired with force $\vec{n}_{B\text{ on }S}$ that acts *on the surface*. The weight force is $\vec{w}_{E\text{ on }B}$, so it is paired with force $\vec{w}_{B\text{ on }E}$ that acts *on the earth*.

Stop to Think 8.2: c. Newton's third law says that the force of A on B is *equal* and opposite to the force of B on A. This is always true. The speed of the objects isn't relevant.

Stop to Think 8.3: b. $F_{B\text{ on }H} = F_{H\text{ on }B}$ and $F_{A\text{ on }B} = F_{B\text{ on }A}$ because these are action/reaction pairs. Box B is slowing down and therefore must have a net force to the left. So from Newton's second law we also know that $F_{H\text{ on }B} > F_{A\text{ on }B}$.

Stop to Think 8.4: Equal to. Each block is hanging in equilibrium, with no net force, so the upward tension force is mg.

Stop to Think 8.5: Less. Block B is *accelerating* downward, so the net force on B must point down. The only forces acting on B are the tension and the weight, so $T_{S\text{ on }B} < w_{E\text{ on }B}$.

Stop to Think 8.6: c. Newton's third law says that the force of A on B is *equal* and opposite to the force of B on A. This is always true. The mass of the objects isn't relevant.

profound statement about nature and, at the same time, a very practical tool for problem solving.

In the case of chemical reactions, the final mass M_f after the reaction is complete is the same as the initial mass M_i *if the system is closed*. That is, once the box is closed you're not allowed to add or remove any matter. Our knowledge about mass can be stated as a *conservation law*.

Law of conservation of mass The total mass in a closed system is constant. Mathematically, $M_f = M_i$.

The qualification "in a closed system" is important. The final mass certainly won't equal the initial mass if you open the box halfway through and remove some of the matter. Other conservation laws that we discover will also have qualifications stating the circumstances under which they apply. Knowing when a law can be used is just as important as knowing the law! The fact that some quantity Q is conserved under some circumstances doesn't mean that it will always be conserved.

The law of conservation of mass is a critical underpinning of all of chemistry. In fact, conservation of mass is tacitly assumed in writing a reaction equation such as $2H_2 + O_2 \rightarrow 2H_2O$. Although our common sense and everyday experience suggests that mass is conserved, it was only after many long and precise experiments that conservation of mass was accepted as a fundamental law of nature.

Surprisingly, just as the laws of chemistry were becoming well established at the beginning of the twentieth century, Einstein's 1905 theory of relativity showed that there are circumstances in which mass actually is *not* conserved but can be converted to energy in accordance with his famous formula $E = mc^2$. Nonetheless, conservation of mass is an exceedingly good approximation in nearly all applications of science and engineering.

Conservation Laws

A system of interacting particles has another curious property. Each system is characterized by a certain number, and no matter how complex the interactions, the value of this number never changes. This number is called the *energy* of the system, and the fact that it never changes is called the *law of conservation of energy*. It is, perhaps, the single most important physical law ever discovered.

The law of conservation of energy is more fundamental than Newton's laws of motion. Newton's laws are a very good description of motion under many circumstances, but they fail for objects moving at extremely high speeds or for objects that are the size of atoms. But as far as we know, the law of conservation of energy is valid under all circumstances. No violation of this law has ever been observed. Energy will be *the* most important concept throughout the remainder of this textbook.

But what is energy? How do you determine the energy number for a system? These are not easy questions. Energy is an abstract idea, not as tangible or easy to picture as mass or force. Our modern concept of energy wasn't fully formulated until the middle of the nineteenth century, nearly two hundred years after Newton. Conservation of energy was not recognized as a law until the relationship between *energy* and *heat* was understood. That is a topic we will take up in Part IV, where the full concept of energy will be found to be the basis of thermodynamics. But all that in due time. In Part II we will be content to introduce the concept of energy and show how energy gives us an important new perspective on interacting particles.

Another important quantity we will study is called *momentum*. The law of conservation of mass is perhaps not surprising because of our everyday experience with mass. But we don't have everyday experience with momentum, so we have no expectation that it should be conserved. But under the proper circumstances, analogous to having the box tightly sealed, only those interactions can occur that do not change the momentum of a system.

Conservation laws tell us that nature is not free to act in arbitrary ways. Many conceivable interactions simply don't happen. We could *imagine* a situation in which 10 g of A and 10 g of B react to give 19 g of product, but such reactions simply don't happen. The only possible reactions are ones that conserve mass. In other situations, the only possible outcomes are ones that conserve momentum or energy. This is a very powerful statement about nature.

Conservation laws will give us a new and different *perspective* on motion. This is not insignificant. You probably have seen optical illusions where a figure appears first one way, then another, even though the basic information in the figure has not changed. Likewise with motion. We will soon see that there are some situations most easily analyzed from the perspective of Newton's laws, but others that make much more sense when analyzed from a conservation-law perspective. An important goal of Part II is to learn which perspective is best for a given problem.

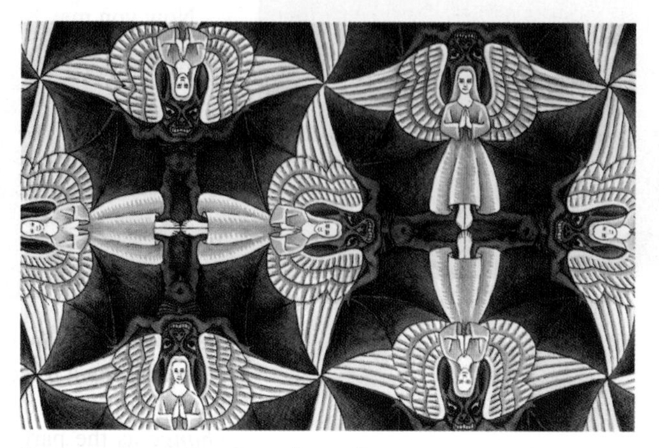

You can see angels or bats, depending on your perspective, but not both at once.

9 Impulse and Momentum

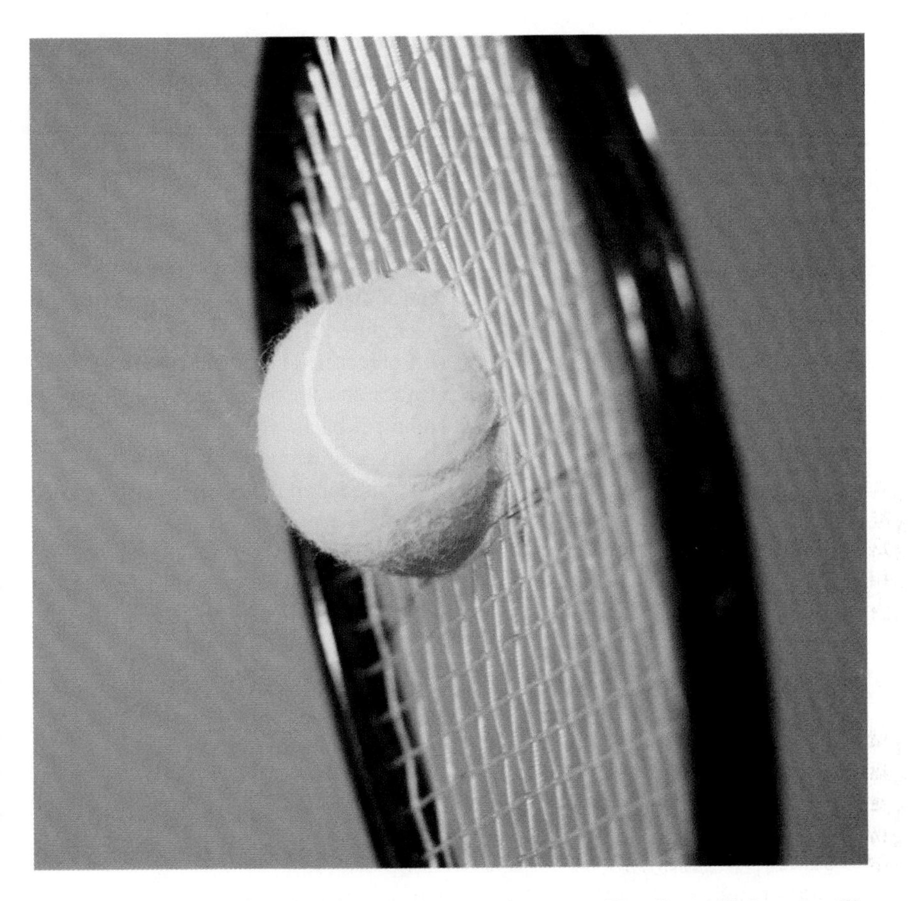

A tennis ball collides with a racket. Notice that the right side of the ball is flattened.

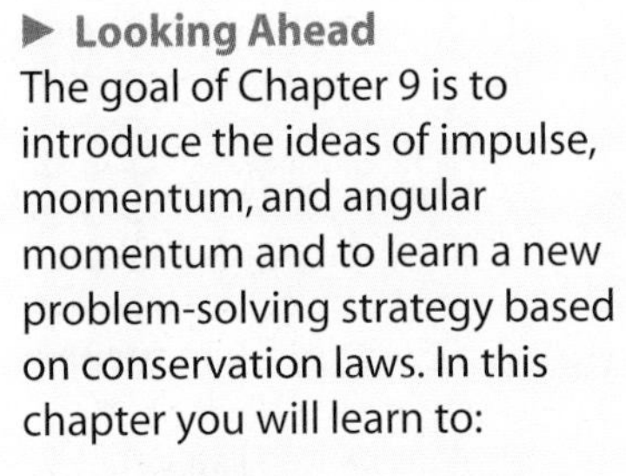

▶ Looking Ahead

The goal of Chapter 9 is to introduce the ideas of impulse, momentum, and angular momentum and to learn a new problem-solving strategy based on conservation laws. In this chapter you will learn to:

- Understand and use the concepts of impulse and momentum.
- Use a new before-and-after pictorial representation.
- Solve problems using the law of conservation of momentum.
- Apply these ideas to explosions and collisions.
- Use the law of conservation of angular momentum in simple situations.

◀ Looking Back

The law of conservation of momentum is based on Newton's third law. Please review:

- Sections 8.2–8.3 Action/reaction force pairs and Newton's third law.

A racket hitting a tennis ball is an example of what we'll call a *collision*. A collision is a complex interaction between two objects, and using Newton's second law to predict the outcome of a collision would be a daunting challenge. Nevertheless, some collisions have very simple outcomes. For example, consider a train car rolling along the tracks toward an identical car at rest. The two cars couple together upon impact and then roll down the tracks together. The forces between the train cars during the collision are unimaginably complex. Yet if you were to measure their speeds, you would find that the two coupled cars have exactly half the speed of the single car before impact. How can such a complex interaction give rise to such a simple outcome?

The opposite of a collision is an interaction that forces two objects apart. These interactions are called *explosions*, even though they may lack a flash or a pop. As an example, imagine a 75 kg archer on ice skates. If the archer shoots a 75 g arrow forward, the archer recoils backward. The interaction between the archer, the bow, and the arrow is very complex, yet the archer's recoil speed is always 1/1000 of the speed of the arrow. Another simple outcome.

Our goal in this chapter is to learn how to predict these simple outcomes without having to know all the details of the interaction forces. The new idea that will make this possible is *momentum*, a concept we will use to relate the situation "before" an interaction to the situation "after" the interaction. This before-and-after perspective will be a powerful new problem-solving tool.

9.1 Momentum and Impulse

Suppose that two or more objects have an intense and perhaps complex interaction, such as a collision or an explosion. Our goal is to find a relationship between the velocities of the objects before the interaction and their velocities after the interaction. We'll start by looking at collisions.

A **collision** is a short-duration interaction between two objects. The collision between a tennis ball and a racket, or a baseball and a bat, may seem instantaneous to your eye, but that is a limitation of your perception. A careful look at the photograph that opens this chapter reveals that the right side of the ball is flattened and pressed up against the strings of the racket. It takes time to compress the ball, and more time for the ball to re-expand as it leaves the racket.

The duration of a collision depends on the materials from which the objects are made, but 1 to 10 ms (0.001 to 0.010 s) is typical. This is the time during which the two objects are in contact with each other. The harder the objects, the shorter the contact time. A collision between two steel balls lasts less than 1 ms.

Figure 9.1 shows a microscopic view of a collision in which object A bounces off object B. The spring-like molecular bonds—the same bonds that cause normal forces and tension forces—compress during the collision, then re-expand as A bounces back. The molecular springs of a hard material, such as steel, are stiffer than the molecular springs of a rubber ball. Thus the deformation of steel is less than rubber and the duration of the collision is shorter, but neither is zero.

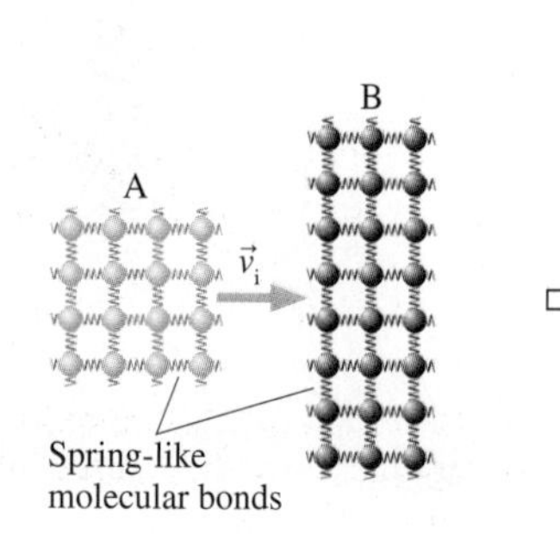

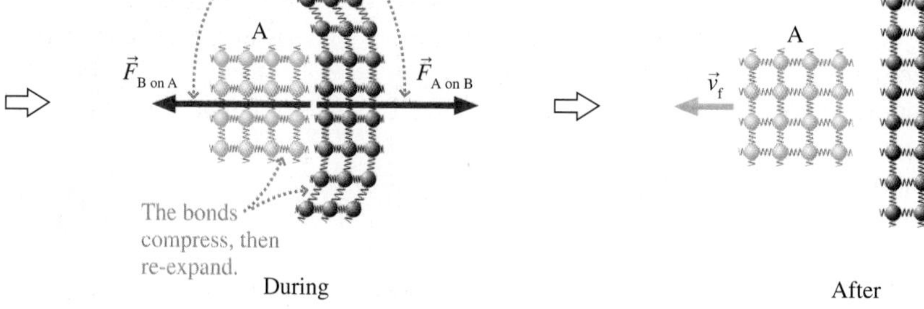

FIGURE 9.1 A microscopic view of a collision.

As A and B come into contact, the molecular springs begin to compress. The forces $\vec{F}_{A\text{ on }B}$ and $\vec{F}_{B\text{ on }A}$ are an action/reaction pair and, according to Newton's third law, have equal magnitudes: $F_{A\text{ on }B} = F_{B\text{ on }A}$. The force increases rapidly as the bonds compress, reaches a maximum at the instant A is at rest (point of maximum compression), then decreases as the bonds re-expand. This idea is shown graphically in Figure 9.2.

A large force exerted during a small interval of time is called an **impulsive force.** The force of a tennis racket on a ball, which would look much like Figure 9.2, is a good example of an impulsive force. Notice that an impulsive force has a well-defined duration.

NOTE ► Until now, we have not dealt with forces that change with time. Because an impulsive force is a function of time, we will write it as $F(t)$. ◄

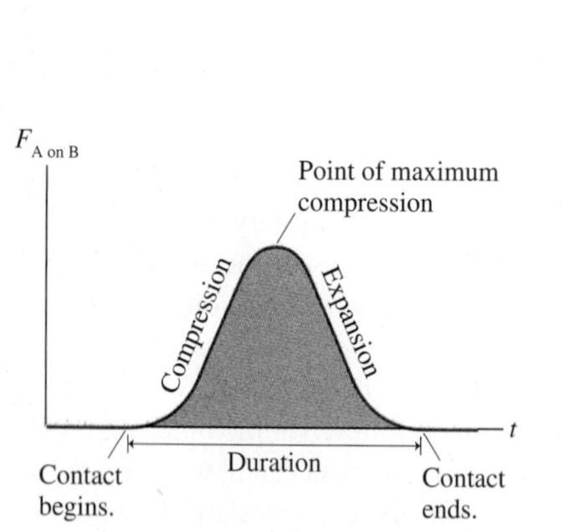

FIGURE 9.2 The rapidly changing magnitude of the force during a collision.

Consider a particle traveling in a straight line along the x-axis with initial velocity v_{ix}. The particle suddenly collides with another object and experiences an impulsive force $F_x(t)$ that begins at time t_i and ends at time t_f. After the collision, the particle has final velocity v_{fx}.

NOTE ▶ Both v_x and F_x are components of vectors and thus have *signs* indicating which way the vectors point. ◀

We can analyze the collision with Newton's second law to find the final velocity. Acceleration in one dimension is $a_x = dv_x/dt$, so the second law is

$$ma_x = m\frac{dv_x}{dt} = F_x(t)$$

After multiplying both sides by dt, we can write the second law as

$$m\,dv_x = F_x(t)\,dt \tag{9.1}$$

The force is nonzero only during the interval of time from t_i to t_f, so let's integrate Equation 9.1 over this interval. The velocity changes from v_{ix} to v_{fx} during the collision, thus

$$m\int_{v_i}^{v_f} dv_x = mv_{fx} - mv_{ix} = \int_{t_i}^{t_f} F_x(t)\,dt \tag{9.2}$$

We need some new tools to help us make sense of Equation 9.2.

Momentum

The product of the particle's mass and velocity is called the *momentum* of the particle:

$$\textbf{momentum} = \vec{p} = m\vec{v} \tag{9.3}$$

Momentum, like velocity, is a vector. The units of momentum are kg m/s.

Figure 9.3 shows that the momentum vector $\vec{p}$ is parallel to the velocity vector $\vec{v}$. Like any vector, $\vec{p}$ can be decomposed into x- and y-components. Equation 9.3, which is a vector equation, is a shorthand way to write the simultaneous equations

$$p_x = mv_x$$
$$p_y = mv_y$$

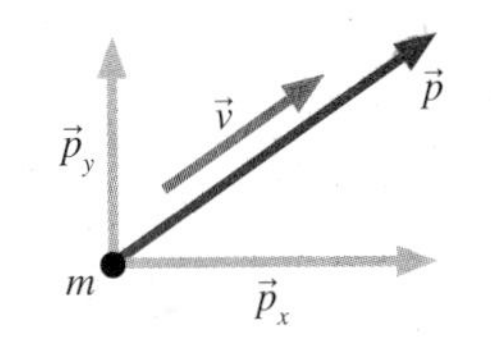

FIGURE 9.3 A particle's momentum vector $\vec{p}$ can be decomposed into x- and y-components.

NOTE ▶ One of the most common errors in momentum problems is a failure to use the appropriate signs. The momentum component p_x has the same sign as v_x. Momentum is *negative* for a particle moving to the left (on the x-axis) or down (on the y-axis). ◀

Momentum is another term that we use in everyday speech without a precise definition. In physics and engineering, momentum is a technical term whose meaning is defined in Equation 9.3. An object can have a large momentum either by having a small mass but a large velocity (a bullet fired from a rifle) or a small velocity but a large mass (a large truck rolling at a slow 1 mph).

Newton actually formulated his second law in terms of momentum rather than acceleration:

$$\vec{F} = m\vec{a} = m\frac{d\vec{v}}{dt} = \frac{d(m\vec{v})}{dt} = \frac{d\vec{p}}{dt} \tag{9.4}$$

This statement of the second law, saying that **force is the rate of change of momentum,** is more general than our earlier version $\vec{F} = m\vec{a}$. It allows for the possibility that the mass of the object might change, such as a rocket that is losing mass as it burns fuel.

NOTE ▶ The plural of *momentum* is *momenta*, from its Latin origin. ◀

Returning to Equation 9.2, you can see that mv_{ix} and mv_{fx} are p_{ix} and p_{fx}, the x-component of the particle's momentum before and after the collision. In terms of momentum, Equation 9.2 is

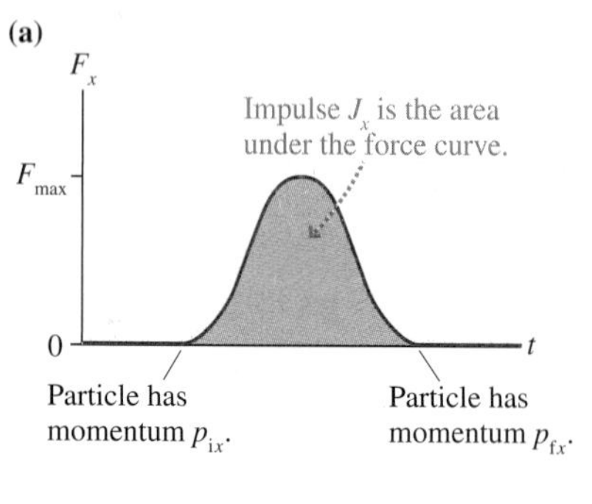

$$\Delta p_x = p_{fx} - p_{ix} = \int_{t_i}^{t_f} F_x(t)\,dt \tag{9.5}$$

Now we need to examine the right-hand side of Equation 9.5.

Impulse

Equation 9.5 tells us that the particle's change in momentum is related to the time integral of the force. Let's define a quantity J_x called the *impulse* to be

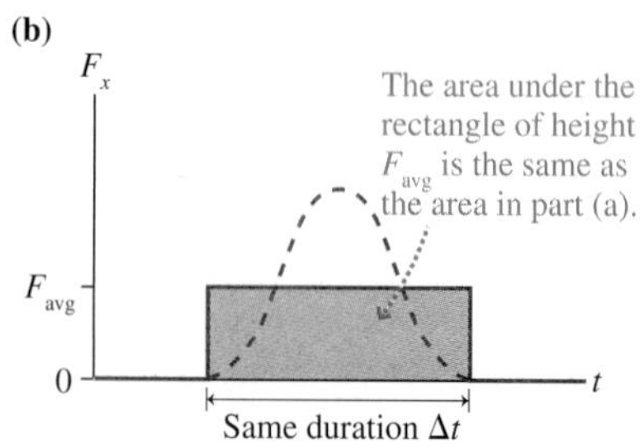

FIGURE 9.4 Looking at the impulse graphically.

$$\textbf{impulse} = J_x = \int_{t_i}^{t_f} F_x(t)\,dt$$
$$= \text{area under the } F_x(t) \text{ curve between } t_i \text{ and } t_f \tag{9.6}$$

Strictly speaking, impulse has units of N s, but you should be able to show that N s are equivalent to kg m/s, the units of momentum.

Figure 9.4a portrays the impulse graphically as the area under the force curve. Because the force changes in a complicated way during a collision, it is often useful to describe the collision in terms of an *average* force F_{avg}. As Figure 9.4b shows, F_{avg} is the height of a rectangle that has the same area, and thus the same impulse, as the real force curve. The impulse exerted during the collision is

$$J_x = F_{avg}\Delta t \tag{9.7}$$

Equation 9.2, which we found by integrating Newton's second law, can now be rewritten in terms of impulse and momentum as

$$\Delta p_x = J_x \quad \text{(impulse-momentum theorem)} \tag{9.8}$$

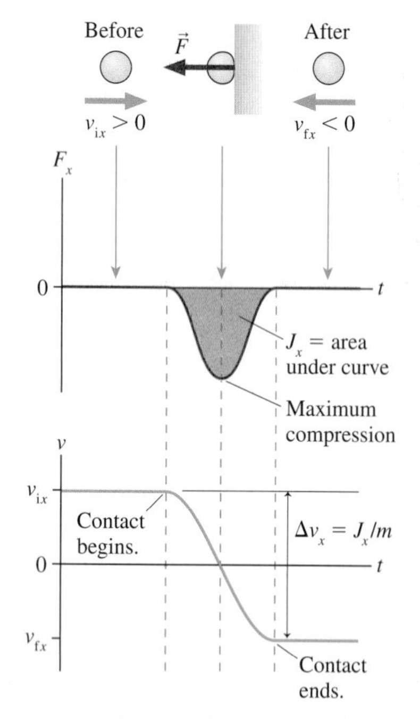

FIGURE 9.5 The impulse-momentum theorem helps us understand a rubber ball bouncing off a wall.

This result, called the **impulse-momentum theorem,** tells us that **an impulse delivered to a particle changes the particle's momentum.** The momentum p_{fx} "after" an interaction, such as a collision or an explosion, is equal to the momentum p_{ix} "before" the interaction *plus* the impulse that arises from the interaction:

$$p_f = p_i + J_x \tag{9.9}$$

The impulse-momentum theorem tells us that we do *not* need to know all the details of the force function $F_x(t)$. No matter how complicated the force, only the integral of the force—the area under the force curve—is needed to find p_{fx}. We opened this chapter with examples in which complex interactions led to very simple outcomes. The impulse-momentum theorem is an important step toward understanding these outcomes.

Figure 9.5 illustrates the impulse-momentum theorem for a rubber ball bouncing off a wall. Notice the signs; they are very important. The ball is initially traveling toward the right, so v_{ix} and p_{ix} are positive. After the bounce, v_{fx} and p_{fx} are negative. The force *on the ball* is toward the left, so F_x is also negative. The graphs show how the force and the velocity change with time.

Although the interaction is very complex, the impulse—the area under the force graph—is all we need to know to find the ball's velocity as it rebounds from the wall. The final momentum is

$$p_{fx} = p_{ix} + J_x = p_{ix} + \text{area under the force curve}$$

Thus the final velocity is

$$v_{fx} = \frac{p_{fx}}{m} = v_{ix} + \frac{\text{area under the force curve}}{m}$$

In this example, the area has a negative value.

STOP TO THINK 9.1 The cart's change of momentum is

a. -30 kg m/s.
b. -20 kg m/s.
c. 0 kg m/s.
d. 10 kg m/s.
e. 20 kg m/s.
f. 30 kg m/s.

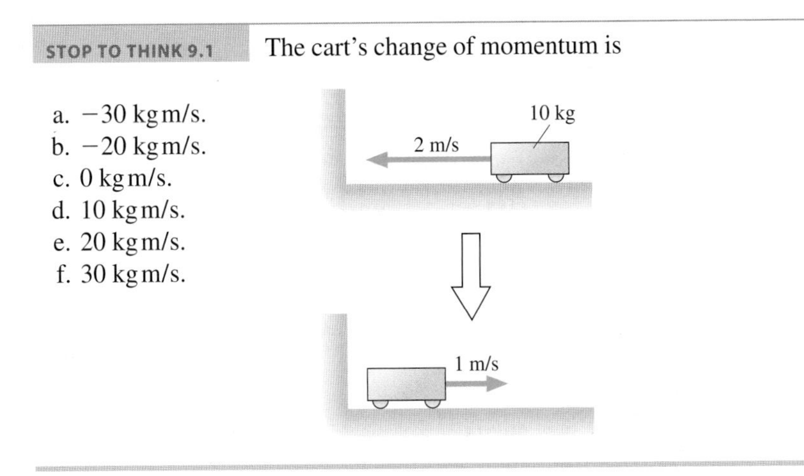

9.2 Solving Impulse and Momentum Problems

Pictorial representations have become an important problem-solving tool. The pictorial representations and free-body diagrams that you learned to draw in Part I were oriented toward the use of Newton's laws and a subsequent kinematical analysis. Now we are interested in making a connection between "before" and "after."

TACTICS BOX 9.1 Drawing a before-and-after pictorial representation

1. **Sketch the situation.** Use two drawings, labeled "Before" and "After," to show the objects *before* they interact and again *after* they interact.
2. **Establish a coordinate system.** Select your axes to match the motion.
3. **Define symbols.** Define symbols for the masses and for the velocities before and after the interaction. Position and time are not needed.
4. **List known information.** Give the values of quantities known from the problem statement or that can be found quickly with simple geometry or unit conversions. Before-and-after pictures are usually simpler than the pictures you used for dynamics problems, so listing known information on the sketch is adequate.
5. **Identify the desired unknowns.** What quantity or quantities will allow you to answer the question? These should have been defined as symbols in step 3.

NOTE ► The generic subscripts i and f, for *initial* and *final* are adequate in equations for a simple problem, but in more complex problems using numerical subscripts, such as v_{1x} and v_{2x}, will help keep all the symbols straight. ◄

EXAMPLE 9.1 Hitting a baseball

A 150 g baseball is thrown with a speed of 20 m/s. It is hit straight back toward the pitcher at a speed of 40 m/s. The interaction force between the ball and the bat has the shape shown in Figure 9.6. What is the *maximum* force F_{max} that the bat exerts on the ball? What is the *average* force that the bat exerts on the ball?

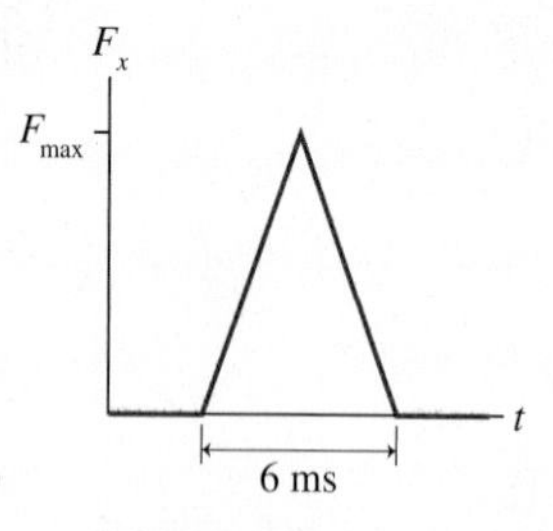

FIGURE 9.6 The interaction force between the baseball and the bat.

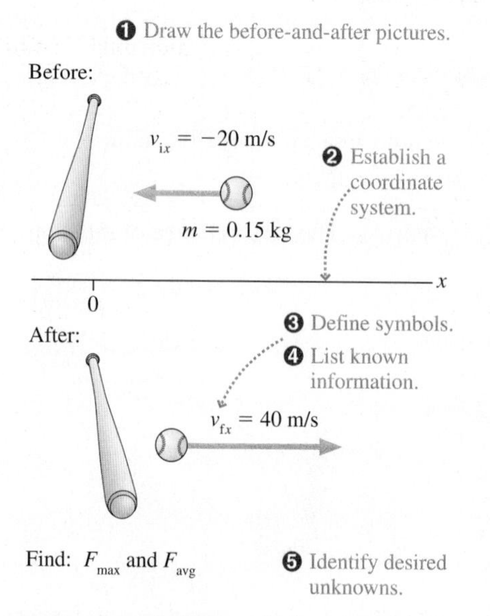

FIGURE 9.7 A before-and-after pictorial representation.

MODEL Model the baseball as a particle and the interaction as a collision.

VISUALIZE Figure 9.7 is a before-and-after pictorial representation. The steps from Tactics Box 9.1 are explicitly noted. Because F_x is positive (a force to the right), we know the ball was initially moving toward the left and is hit back toward the right. Thus we converted the statements about *speeds* into information about *velocities*, with v_{ix} negative.

SOLVE So far we've consistently started the mathematical representation with Newton's second law. Now we want to use the impulse-momentum theorem:

$$\Delta p_x = J_x = \text{area under the force curve}$$

We know the velocities before and after the collision, so we can find the change in the ball's momentum:

$$\Delta p_x = mv_{fx} - mv_{ix} = (0.15\text{ kg})(40\text{ m/s} - (-20\text{ m/s}))$$
$$= 9.0\text{ kg m/s}$$

The force curve is a triangle with height F_{max} and width 6.0 ms. The area under the curve is

$$J_x = \text{area} = \frac{1}{2} \times F_{max} \times (0.0060\text{ s}) = (F_{max})(0.0030\text{ s})$$

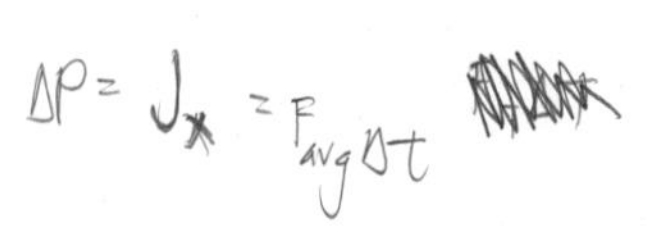

According to the impulse-momentum theorem,

$$9.0\text{ kg m/s} = (F_{max})(0.0030\text{ s})$$

Thus the *maximum* force is

$$F_{max} = \frac{9.0\text{ kg m/s}}{0.0030\text{ s}} = 3000\text{ N}$$

The *average* force, which depends on the collision duration $\Delta t = 0.0060$ s, has the smaller value

$$F_{avg} = \frac{J_x}{\Delta t} = \frac{\Delta p_x}{\Delta t} = \frac{9.0\text{ kg m/s}}{0.0060\text{ s}} = 1500\text{ N}$$

ASSESS F_{max} is a large force, but quite typical of the impulsive forces during collisions. The main thing to focus on is our new perspective: an impulse changes the momentum of an object.

Other forces often act on an object during a collision or other brief interaction. In Example 9.1, for instance, the baseball also has a weight force acting on it. Usually these other forces are *much* smaller than the interaction forces. The 1.5 N weight of the ball is vastly less than the 3000 N force of the bat on the ball. We can reasonably neglect these small forces *during* the brief time of the impulsive force by using what is called the **impulse approximation.**

When we use the impulse approximation, p_{ix} and p_{fx} (and v_{ix} and v_{fx}) are then the momenta (and velocities) *immediately* before and *immediately* after the collision. For example, the velocities in Example 9.1 are those of the ball just before and after it collides with the bat. We could then do a follow-up problem, including weight and drag, to find the ball's speed a second later as the second baseman catches it. We'll look at some two-part examples later in the chapter.

Momentum Bar Charts

The impulse-momentum theorem tells us that **impulse transfers momentum to an object.** If an object has 2 kg m/s of momentum, a 1 kg m/s impulse exerted on the object increases its momentum to 3 kg m/s. That is, $p_{fx} = p_{ix} + J_x$.

We can represent this "momentum accounting" with a **momentum bar chart.** Figure 9.8a shows a bar chart in which one unit of impulse adds to an initial two units of momentum to give three units of momentum. The bar chart of Figure 9.8b represents the baseball of Example 9.1, which started with negative momentum because it was moving to the left. Momentum bar charts, like before-and-after pictorial representations, are a tool for visualizing an interaction.

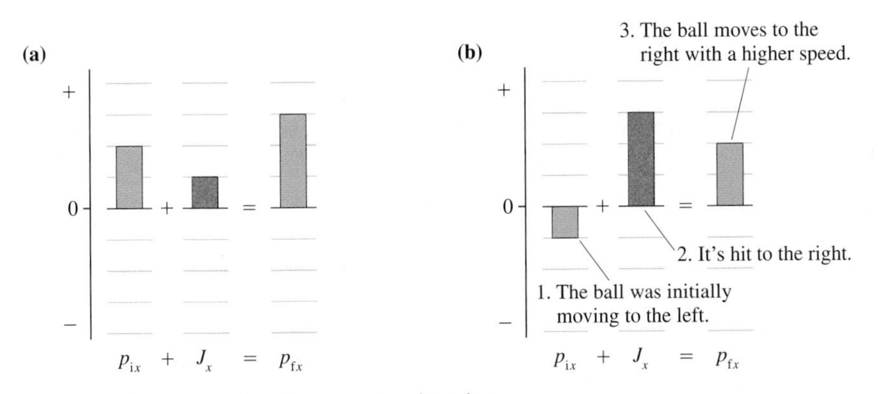

FIGURE 9.8 Two examples of momentum bar charts.

NOTE ▶ The vertical scale of a momentum bar chart has no numbers; it can be adjusted to match any problem. However, be sure that all bars in a given problem use a consistent scale. ◀

EXAMPLE 9.2 A bouncing ball

A 100 g rubber ball is dropped from a height of 2.0 m onto a hard floor. Figure 9.9 shows the force that the floor exerts on the ball. How high does the ball bounce?

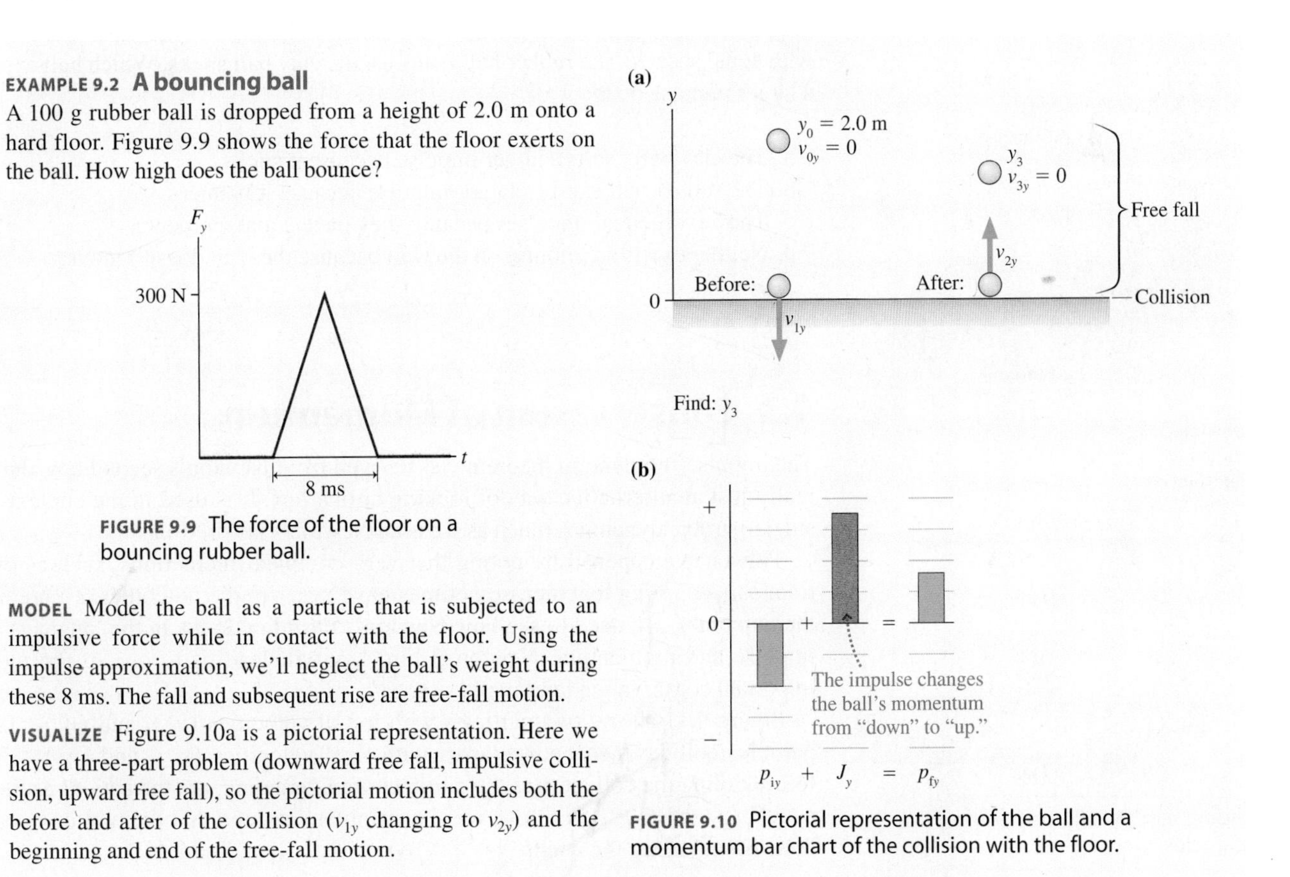

FIGURE 9.9 The force of the floor on a bouncing rubber ball.

FIGURE 9.10 Pictorial representation of the ball and a momentum bar chart of the collision with the floor.

MODEL Model the ball as a particle that is subjected to an impulsive force while in contact with the floor. Using the impulse approximation, we'll neglect the ball's weight during these 8 ms. The fall and subsequent rise are free-fall motion.

VISUALIZE Figure 9.10a is a pictorial representation. Here we have a three-part problem (downward free fall, impulsive collision, upward free fall), so the pictorial motion includes both the before and after of the collision (v_{1y} changing to v_{2y}) and the beginning and end of the free-fall motion.

SOLVE Velocity v_{1y}, the ball's velocity *immediately* before the collision, is found using free-fall kinematics with $\Delta y = -2.0$ m:

$$v_{1y}^2 = v_{0y}^2 - 2g\Delta y = 0 - 2g\Delta y$$

$$v_{1y} = \sqrt{-2g\Delta y} = \sqrt{-2(9.80 \text{ m/s}^2)(-2.0 \text{ m})} = -6.26 \text{ m/s}$$

We've chosen the negative root because the ball is moving in the negative y-direction.

The impulse-momentum theorem is $p_{2y} = p_{1y} + J_y$. The initial momentum, just before the collision, is $p_{1y} = mv_{1y} = -0.626$ kg m/s. The force of the floor is upward, so J_y is positive. The final momentum p_{2y}, as the ball leaves the floor, is positive but probably smaller in magnitude than p_{1y} because we know that rubber balls don't bounce back to their initial height. This relationship between p_{1y}, p_{2y}, and J_y is shown in the momentum bar chart of Figure 9.10b.

From Figure 9.8, the impulse J_y is

$$J_y = \text{area under the force curve} = \frac{1}{2} \times (300 \text{ N}) \times (0.0080 \text{ s})$$
$$= 1.200 \text{ N s}$$

Thus

$$p_{2y} = p_{1y} + J_y = (-0.626 \text{ kg m/s}) + 1.200 \text{ N s} = 0.574 \text{ kg m/s}$$

and the post-collision velocity is

$$v_{2y} = \frac{p_{2y}}{m} = \frac{0.574 \text{ kg m/s}}{0.10 \text{ kg}} = 5.74 \text{ m/s}$$

The rebound speed is less than the impact speed, as expected. Finally a second use of free-fall kinematics yields

$$v_{3y}^2 = 0 = v_{2y}^2 - 2g\Delta y = v_{2y}^2 - 2gy_3$$

$$y_3 = \frac{v_{2y}^2}{2g} = \frac{(5.74 \text{ m/s})^2}{2(9.80 \text{ m/s}^2)} = 1.68 \text{ m}$$

The ball bounces back to a height of 1.68 m.

ASSESS The ball bounces back to less than its initial height, which is realistic.

NOTE ▶ Example 9.2 illustrates an important point: The impulse-momentum theorem applies *only* during the brief interval in which an impulsive force is applied. Many problems will have segments of the motion that must be analyzed with kinematics or Newton's laws. The impulse-momentum theorem is a new and useful tool, but it doesn't replace all that you've learned up until now. ◀

STOP TO THINK 9.2 A 10 g rubber ball and a 10 g clay ball are thrown at a wall with equal speeds. The rubber ball bounces, the clay ball sticks. Which ball exerts a larger impulse on the wall?

a. The clay ball exerts a larger impulse because it sticks.
b. The rubber ball exerts a larger impulse because it bounces.
c. They exert equal impulses because they have equal momenta.
d. Neither exerts an impulse on the wall because the wall doesn't move.

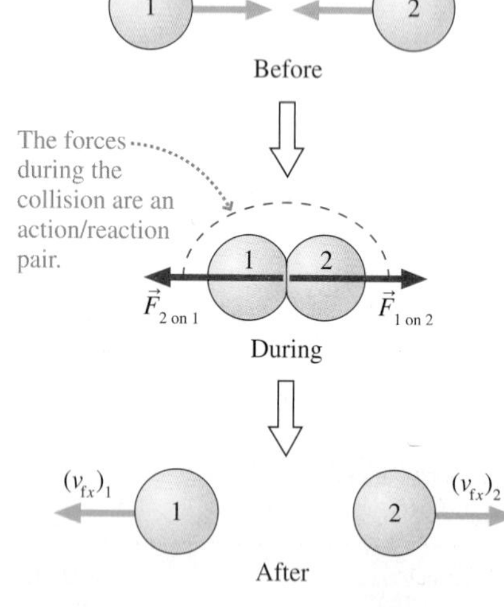

FIGURE 9.11 A collision between two particles.

9.3 Conservation of Momentum

The impulse-momentum theorem was derived from Newton's second law and is really just an alternative way of looking at that law. It is used in the context of single-particle dynamics, much as we used Newton's law in Chapters 4–7.

This chapter opened by noting that very complex interactions, such as two train cars coupling together, sometimes have very simple outcomes. To predict the outcomes, we need to see how Newton's *third* law looks in the language of impulse and momentum. Newton's third law will lead us to one of the most important conservation laws in physics.

Figure 9.11 shows two particles with initial velocities $(v_{ix})_1$ and $(v_{ix})_2$. The particles collide, then bounce apart with final velocities $(v_{fx})_1$ and $(v_{fx})_2$. The forces during the collision, as the particles are interacting, are the action/reaction pair $\vec{F}_{1 \text{ on } 2}$ and $\vec{F}_{2 \text{ on } 1}$. For now, we'll continue to assume that the motion is one dimensional along the x-axis.

NOTE ▶ The notation, with all the subscripts, may seem excessive. But there are two particles, and each has an initial and a final velocity, so we need to distinguish among four different velocities. ◀

Newton's second law for each particle *during* the collision is

$$\begin{aligned} \frac{d(p_x)_1}{dt} &= (F_x)_{2\text{ on }1} \\ \frac{d(p_x)_2}{dt} &= (F_x)_{1\text{ on }2} = -(F_x)_{2\text{ on }1} \end{aligned} \tag{9.10}$$

We made explicit use of Newton's third law in the second equation.

Although Equations 9.10 are for two different particles, suppose—just to see what happens—we were to *add* these two equations. If we do, we find that

$$\frac{d(p_x)_1}{dt} + \frac{d(p_x)_2}{dt} = \frac{d}{dt}((p_x)_1 + (p_x)_2) = (F_x)_{2\text{ on }1} + (-(F_x)_{2\text{ on }1}) = 0 \tag{9.11}$$

If the time derivative of the quantity $(p_x)_1 + (p_x)_2$ is zero, it must be the case that

$$(p_x)_1 + (p_x)_2 = \text{constant} \tag{9.12}$$

Equation 9.12 is a conservation law! If $(p_x)_1 + (p_x)_2$ is a constant, then the sum of the momenta *after* the collision equals the sum of the momenta *before* the collision. That is,

$$(p_{fx})_1 + (p_{fx})_2 = (p_{ix})_1 + (p_{ix})_2 \tag{9.13}$$

Furthermore, this equality is independent of the interaction force. We don't need to know anything about $\vec{F}_{1\text{ on }2}$ and $\vec{F}_{2\text{ on }1}$ to make use of Equation 9.13.

As an example, Figure 9.12 is a before-and-after pictorial representation of two equal-mass train cars colliding and coupling. Equation 9.13 relates the momenta of the cars after the collision to their momenta before the collision:

$$m_1(v_{fx})_1 + m_2(v_{fx})_2 = m_1(v_{ix})_1 + m_2(v_{ix})_2$$

Initially, car 1 is moving with velocity $(v_{ix})_1 = v_i$ while car 2 is at rest. Afterward, they roll together with the common final velocity v_f. Furthermore, $m_1 = m_2 = m$. With this information, the sum of the momenta is

$$mv_f + mv_f = 2mv_f = mv_i + 0$$

The mass cancels, and we find that the train cars' final velocity is $v_f = \frac{1}{2}v_i$. We were able to make this prediction of a simple outcome without knowing anything at all about the very complex interaction between the two cars as they collide.

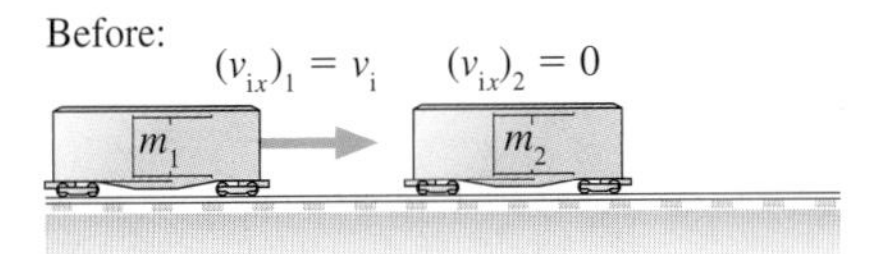

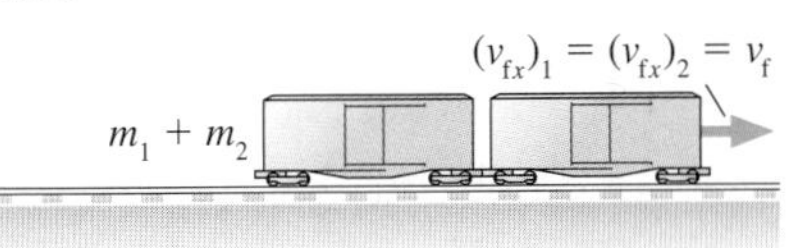

FIGURE 9.12 Two colliding train cars.

Law of Conservation of Momentum

Equation 9.13 illustrates the idea of a conservation law for momentum, but it was derived for the specific case of two particles colliding in one dimension. Our goal is to develop a more general law of conservation of momentum, a law that will be valid in three dimensions and that will work for any type of interaction. The next few paragraphs are fairly mathematical, so you might want to begin by looking ahead to Equation 9.20 and the statement of the law of conservation of momentum to see where we're heading.

Consider a *system* consisting of *N* particles. Figure 9.13 shows a simple case where $N = 3$. The particles might be large entities (cars, baseballs, etc.), or they might be the microscopic atoms in a gas. We can identify each particle by an identification number *k*. Every particle in the system *interacts* with every other particle via action/reaction pairs of forces $\vec{F}_{j\text{ on }k}$ and $\vec{F}_{k\text{ on }j}$. In addition, every particle is subjected to possible *external forces* $\vec{F}_{\text{ext on }k}$ from agents outside the system.

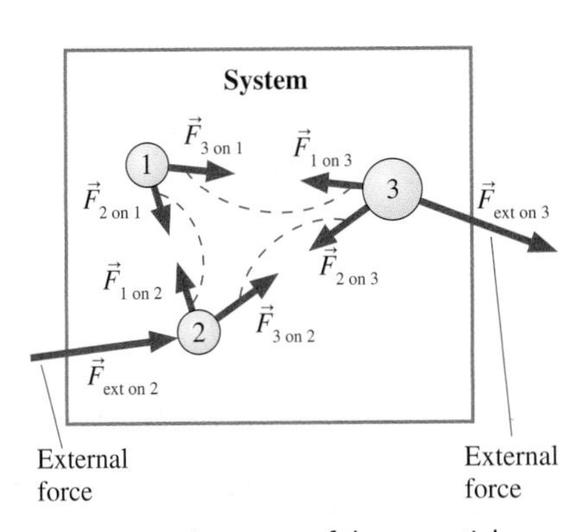

FIGURE 9.13 A system of three particles.

NOTE ▶ This definition of "the system" differs from the one we used in Chapter 8. There, where we were interested in how objects interact with each other, we identified each particle-like object as a separate system. When we use conservation laws, it is more useful to think of all the interacting objects as a single entity that we'll call "the system." ◀

If particle k has velocity $\vec{v}_k$, its momentum is $\vec{p}_k = m_k\vec{v}_k$. Define the **total momentum** $\vec{P}$ of the system as the vector sum

$$\vec{P} = \text{total momentum} = \vec{p}_1 + \vec{p}_2 + \vec{p}_3 + \cdots + \vec{p}_N = \sum_{k=1}^{N} \vec{p}_k \tag{9.14}$$

In other words, the total momentum *of the system* is the sum of all the individual momenta.

The time derivative of $\vec{P}$ tells us how the total momentum of the system changes with time:

$$\frac{d\vec{P}}{dt} = \sum_k \frac{d\vec{p}_k}{dt} = \sum_k \vec{F}_k \tag{9.15}$$

where we used Newton's second law from Equation 9.4 for each particle in the form $\vec{F}_k = d\vec{p}_k/dt$.

The net force acting on particle k can be divided into *external forces*, from outside the system, and *interaction forces* due to the other particles in the system:

$$\vec{F}_k = \sum_{j \neq k} \vec{F}_{j \text{ on } k} + \vec{F}_{\text{ext on } k} \tag{9.16}$$

The restriction $j \neq k$ expresses the fact that particle k does not exert a force on itself. Using this in Equation 9.15 gives the rate of change of the total momentum P of the system:

$$\frac{d\vec{P}}{dt} = \sum_k \sum_{j \neq k} \vec{F}_{j \text{ on } k} + \sum_k \vec{F}_{\text{ext on } k} \tag{9.17}$$

The double sum on $\vec{F}_{j \text{ on } k}$ adds *every* interaction force within the system. But the interaction forces come in action/reaction pairs, with $\vec{F}_{k \text{ on } j} = -\vec{F}_{j \text{ on } k}$, so $\vec{F}_{k \text{ on } j} + \vec{F}_{j \text{ on } k} = \vec{0}$. Consequently, **the sum of all the interaction forces is zero.** As a result, Equation 9.17 becomes

$$\frac{d\vec{P}}{dt} = \sum_k \vec{F}_{\text{ext on } k} = \vec{F}_{\text{net}} \tag{9.18}$$

where $\vec{F}_{\text{net}}$ is the net force exerted on the system by agents outside the system. But this is just Newton's second law written for the system as a whole! That is, the rate of change of the total momentum of the whole system is equal to the net force applied to the whole system.

Equation 9.18 has two very important implications. First, it tells us that we can analyze the motion of the system as a whole without needing to consider interaction forces between the particles that make up the system. In fact, we have been using this idea all along as an *assumption* of the particle model. When we treat cars and rocks and baseballs as particles, we assume that the internal forces between the atoms—the forces that hold the object together—do not affect the motion of the object as a whole. Now we have *justified* that assumption.

The second implication of Equation 9.18, and the more important one from the perspective of this chapter, applies to what we call an *isolated system*. An **isolated system** is a system for which the *net* external force is zero: $\vec{F}_{\text{net}} = \vec{0}$.

That is, an isolated system is one on which there are *no* external forces or for which the external forces are balanced and add to zero.

For an isolated system, Equation 9.18 is simply

$$\frac{d\vec{P}}{dt} = \vec{0} \quad \text{(isolated system)} \tag{9.19}$$

In other words, **the *total* momentum of an isolated system does not change.** The total momentum $\vec{P}$ remains constant, *regardless* of whatever interactions are going on *inside* the system. The importance of this result is sufficient to elevate it to a law of nature, alongside Newton's laws.

Law of conservation of momentum The total momentum $\vec{P}$ of an isolated system is a constant. Interactions within the system do not change the system's total momentum.

NOTE ► It is worth emphasizing the critical role of Newton's third law in the derivation of Equation 9.19. The law of conservation of momentum is a direct consequence of the fact that interactions within an isolated system are action/reaction pairs. ◄

Mathematically, the law of conservation of momentum for an isolated system is

$$\vec{P}_f = \vec{P}_i \tag{9.20}$$

The total momentum after an interaction is equal to the total momentum before the interaction. Because Equation 9.20 is a vector equation, the equality is true for each of the components of the momentum vector. That is,

$$\begin{aligned} (p_{fx})_1 + (p_{fx})_2 + (p_{fx})_3 + \cdots &= (p_{ix})_1 + (p_{ix})_2 + (p_{ix})_3 + \cdots \\ (p_{fy})_1 + (p_{fy})_2 + (p_{fy})_3 + \cdots &= (p_{iy})_1 + (p_{iy})_2 + (p_{iy})_3 + \cdots \end{aligned} \tag{9.21}$$

The x-equation is an extension of Equation 9.13 to N interacting particles.

EXAMPLE 9.3 Two balls shot from a tube

A 10 g ball and a 30 g ball are placed in a tube with a massless compressed spring between them. When the spring is released, the 10 g ball flies out of the tube at a speed of 6.0 m/s. With what speed does the 30 g ball emerge from the other end?

MODEL The two balls are the system. The balls interact with each other, but they form an isolated system because, for each ball, the upward normal force of the tube balances the downward weight force to make $\vec{F}_{net} = \vec{0}$. Thus the total momentum of the system is conserved.

VISUALIZE Figure 9.14 shows a before-and-after pictorial representation for the two balls. The total momentum before the spring is released is $\vec{P}_i = \vec{0}$ because both balls are at rest. Consequently, the *total* momentum will be $\vec{0}$ after the spring is released. The mathematical statement of momentum conservation, Equation 9.21, is

$$m_1(v_{fx})_1 + m_2(v_{fx})_2 = m_1(v_{ix})_1 + m_2(v_{ix})_2 = 0$$

where we've written the x-component of the momenta in terms of v_x and made use of the fact that the initial velocities are both zero.

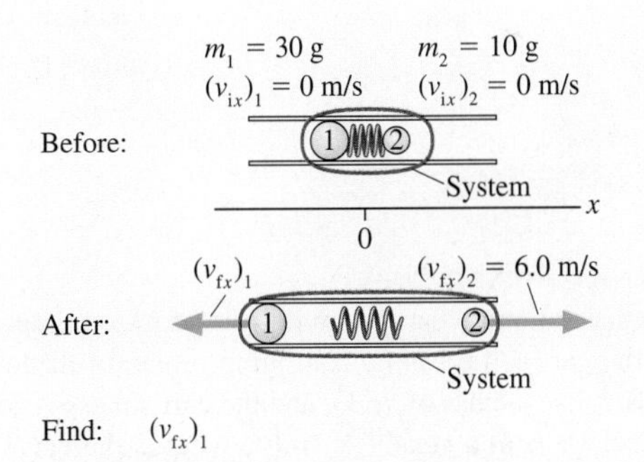

FIGURE 9.14 Before-and-after pictorial representation for two balls shot out of a tube.

Solving for $(v_{fx})_1$, we find

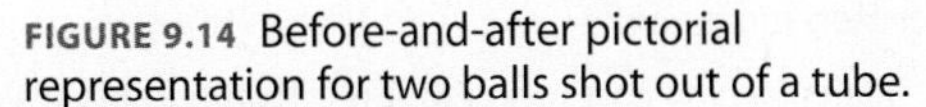

$$(v_{fx})_1 = -\frac{m_2}{m_1}(v_{fx})_2 = -\frac{1}{3}(v_{fx})_2 = -2.0 \text{ m/s}$$

The 30 g ball emerges with a *speed* of 2.0 m/s, one-third the speed of the 10 g ball.

ASSESS The *total* momentum of the system is zero, but the individual momenta are not. Because $(p_{fx})_2$ is positive (the 10 g ball moves to the right), $(p_{fx})_1$ must have the same magnitude but the opposite sign (the 30 g ball moves to the left). Notice that we didn't need to know any details about the spring to conclude that the 30 g ball has one-third the speed of the 10 g ball. Conservation of momentum *mandates* this result.

A Strategy for Conservation of Momentum Problems

6.3, 6.4, 6.6, 6.7, 6.10 ActivPhysics ONLINE

Our derivation of the law of conservation of momentum, and the conditions under which it holds, suggests a problem-solving strategy.

MP PROBLEM-SOLVING STRATEGY 9.1 Conservation of momentum

MODEL Clearly define *the system.*

- If possible, choose a system that is isolated ($\vec{F}_{net} = \vec{0}$) or within which the interactions are sufficiently short and intense that you can ignore external forces for the duration of the interaction (the impulse approximation). Momentum is conserved.
- If it's not possible to choose an isolated system, try to divide the problem into parts such that momentum is conserved during one segment of the motion. Other segments of the motion can be analyzed using Newton's laws or, as you'll learn in Chapters 10 and 11, conservation of energy.

VISUALIZE Draw a before-and-after pictorial representation. Define symbols that will be used in the problem, list known values, and identify what you're trying to find.

SOLVE The mathematical representation is based on the law of conservation of momentum: $\vec{P}_f = \vec{P}_i$. In component form, this is

$$(p_{fx})_1 + (p_{fx})_2 + (p_{fx})_3 + \cdots = (p_{ix})_1 + (p_{ix})_2 + (p_{ix})_3 + \cdots$$
$$(p_{fy})_1 + (p_{fy})_2 + (p_{fy})_3 + \cdots = (p_{iy})_1 + (p_{iy})_2 + (p_{iy})_3 + \cdots$$

ASSESS Check that your result has the correct units, is reasonable, and answers the question.

EXAMPLE 9.4 Rolling away

Bob sees a stationary cart 8.0 m in front of him. He decides to run to the cart as fast as he can, jump on, and roll down the street. Bob has a mass of 75 kg and the cart's mass is 25 kg. If Bob accelerates at a steady 1.0 m/s^2, what is the cart's speed just after Bob jumps on?

MODEL This is a two-part problem. First Bob accelerates across the ground. Then Bob lands on and sticks to the cart, a "collision" between Bob and the cart. The interaction forces between Bob and the cart (i.e., friction) act only over the fraction of a second it takes Bob's feet to become stuck to the cart. Using the impulse approximation allows the system Bob + cart to be treated as an isolated system during the brief interval of the "collision," and thus the total momentum of Bob + cart is conserved during this interaction. But the system Bob + cart is *not* an isolated system for the entire problem because Bob's initial acceleration has nothing to do with the cart.

VISUALIZE Our strategy is to divide the problem into an *acceleration* part, which we can analyze using kinematics, and a *collision* part that we can analyze with momentum conservation. The pictorial representation of Figure 9.15 includes information about both parts. Notice two important points. First, Bob's velocity $(v_{1x})_B$ at the end of his run is his "before" velocity for the collision. Second, Bob and the cart move together at the end, so v_{2x} is their common final velocity.

SOLVE The first part of the mathematical representation is kinematics. We don't know how long Bob accelerates, but we do know his acceleration and the distance. Thus

$$(v_{1x})_B^2 = (v_{0x})_B^2 + 2a_x(x_1 - x_0) = 2a_x x_1$$

Kinematic eq

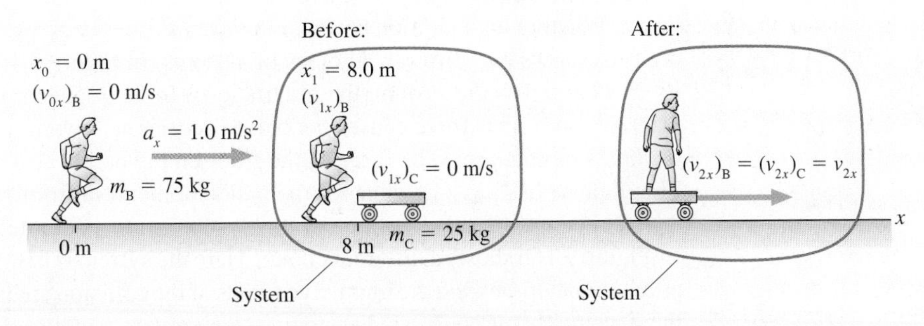

FIGURE 9.15 Pictorial representation of Bob and the cart.

His velocity after accelerating for 8.0 m is

$$(v_{1x})_B = \sqrt{2a_x x_1} = 4.0 \text{ m/s}$$

The second part of the problem, the collision, uses conservation of momentum: $P_{2x} = P_{1x}$. Written in terms of the individual momenta, this is

$$\begin{aligned} m_B(v_{2x})_B + m_C(v_{2x})_C &= (m_B + m_C)v_{2x} \\ &= m_B(v_{1x})_B + m_C(v_{1x})_C \\ &= m_B(v_{1x})_B \end{aligned}$$

where we've used $(v_{1x})_C = 0$ m/s because the cart starts at rest. Solving for v_{2x}, we find

$$v_{2x} = \frac{m_B}{m_B + m_C}(v_{1x})_B = \frac{75 \text{ kg}}{100 \text{ kg}} \times 4.0 \text{ m/s} = 3.0 \text{ m/s}$$

The cart's speed is 3.0 m/s immediately after Bob jumps on.

Notice how easy this was! No forces, no acceleration constraints, no simultaneous equations. Why didn't we think of this before? While conservation laws are indeed powerful, they can only answer certain questions. Had we wanted to know how far Bob slid across the cart before sticking to it, how long the slide took, or what the cart's acceleration was during the collision, we would not have been able to answer such questions on the basis of the conservation law. There is a price to pay for finding a simple connection between before and after, and that price is the loss of information about the details of the interaction. If we are satisfied with knowing only about before and after, then conservation laws are a simple and straightforward way to proceed. But many problems *do* require us to understand the interaction, and for these there is no avoiding Newton's laws and all they entail.

It Depends on the System

The first step in the problem-solving strategy asks you to clearly define *the system*. This is worth emphasizing, because many problem-solving errors arise from trying to apply momentum conservation to an inappropriate system. **The goal is to choose a system whose momentum will be conserved.** Even then, it is the *total* momentum of the system that is conserved, not the momenta of the individual particles within the system.

As an example, consider what happens if you drop a rubber ball and let it bounce off a hard floor. Is momentum conserved during the collision of the ball with the floor? You might be tempted to answer yes because the ball's rebound speed is very nearly equal to its impact speed. But there are two errors in this reasoning.

First, momentum depends on *velocity*, not speed. The ball's velocity and momentum just before the collision are negative. They are positive after the collision. Even if their magnitudes are equal, the ball's momentum after the collision is *not* equal to its momentum before the collision.

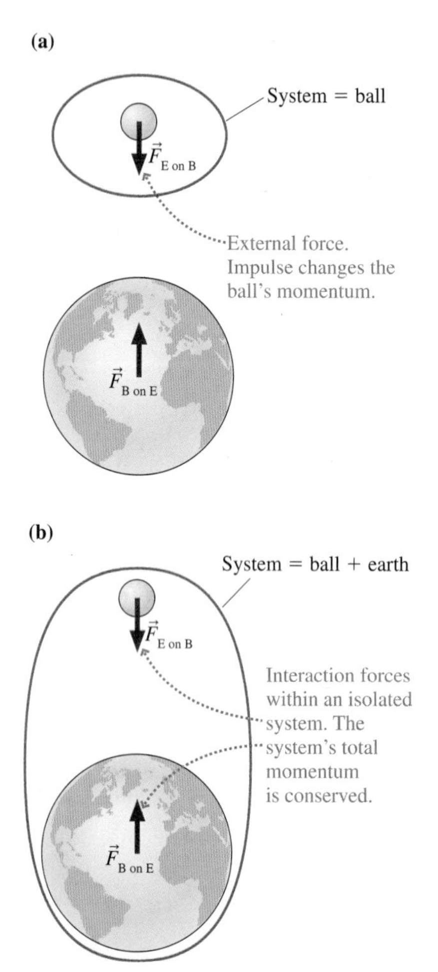

FIGURE 9.16 Whether or not momentum is conserved as a ball falls to earth depends on your choice of the system.

But more importantly, we haven't defined the system. The momentum of what? Whether or not momentum is conserved depends on the system. Figure 9.16 shows two different choices of systems. In Figure 9.16a, where the ball itself is chosen as the system, the gravitational force of the earth on the ball is an external force. This force causes the ball to accelerate toward the earth, changing the ball's momentum. The force of the floor on the ball is also an external force. The impulse of $\vec{F}_{\text{floor on ball}}$ changes the ball's momentum from "down" to "up" as the ball bounces. The momentum of this system is most definitely *not* conserved.

Figure 9.16b shows a different choice. Here the system is ball + earth. Now the gravitational forces and the impulsive forces of the collision are interactions *within* the system. This is an isolated system, so the *total* momentum $\vec{P} = \vec{p}_{\text{ball}} + \vec{p}_{\text{earth}}$ is conserved.

In fact, the total momentum is $\vec{P} = \vec{0}$. Before you release the ball, both the ball and the earth are at rest (in the earth's reference frame). The total momentum is zero before you release the ball, so it will *always* be zero. Consider the situation just before the ball hits the floor. If the ball's velocity is $v_{\text{B}y}$, it must be the case that

$$m_{\text{B}}v_{\text{B}y} + m_{\text{E}}v_{\text{E}y} = 0$$

and thus

$$v_{\text{E}y} = -\frac{m_{\text{B}}}{m_{\text{E}}}v_{\text{B}y}$$

In other words, as the ball is pulled down toward the earth, the ball pulls up on the earth (action/reaction pair of forces) until the entire earth reaches velocity $v_{\text{E}y}$. The earth's momentum is equal and opposite to the ball's momentum.

Why don't we notice the earth "leaping up" toward us each time we drop something? Because of the earth's enormous mass relative to everyday objects. A typical rubber ball has a mass of 60 g and hits the ground with a velocity of about -5 m/s. The earth's upward velocity is thus

$$v_{\text{E}y} \approx -\frac{6 \times 10^{-2}\ \text{kg}}{6 \times 10^{24}\ \text{kg}}(-5\ \text{m/s}) = 5 \times 10^{-26}\ \text{m/s}$$

The earth does, indeed, have a momentum equal and opposite to that of the ball, but the earth is so massive that it needs only an infinitesimal velocity to match the ball's momentum. At this speed, it would take the earth 300 million years to move the diameter of an atom!

STOP TO THINK 9.3 Objects A and C are made of different materials, with different "springiness," but they have the same mass and are initially at rest. When ball B collides with object A, the ball ends up at rest. When ball B is thrown with the same speed and collides with object C, the ball rebounds to the left. Compare the velocities of A and C after the collisions. Is v_{A} greater than, equal to, or less than v_{C}?

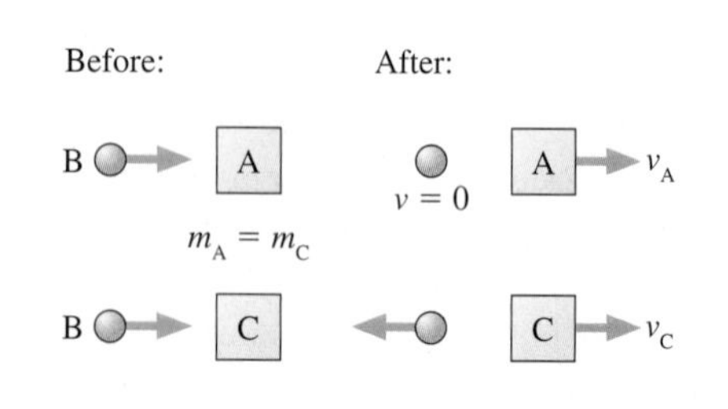

9.4 Explosions

An **explosion,** where the particles of the system move apart from each other after a brief, intense interaction, is the opposite of a collision. The explosive forces, which could be from an expanding spring or from expanding hot gases, are *internal* forces. If the system is isolated, its total momentum during the explosion will be conserved.

EXAMPLE 9.5 Recoil

A 10 g bullet is fired from a 3.0 kg rifle with a speed of 500 m/s. What is the recoil speed of the rifle?

MODEL A simple analysis would say that the rifle exerts a force on the bullet and the bullet, by Newton's third law, exerts a force on the rifle, causing the rifle to recoil. However, this is a little *too* simple. After all, the rifle has no means by which to exert a force on the bullet. Instead, the rifle causes a small mass of gunpowder to explode, releasing energy stored in the gunpowder. The expanding gas exerts forces on *both* the bullet and the rifle.

Let's define the system to be bullet + gas + rifle. The forces due to the expanding gas during the explosion are internal forces, within the system. Any friction forces between the bullet and the rifle as the bullet travels down the barrel are also internal forces. The only external forces, the weights, are balanced by the normal forces of the barrel on the bullet and the person holding the rifle, so $\vec{F}_{net} = \vec{0}$. This is an isolated system and the law of conservation of momentum applies.

VISUALIZE Figure 9.17 shows a pictorial representation before and after the bullet is fired. We'll assume the bullet is fired in the positive x-direction.

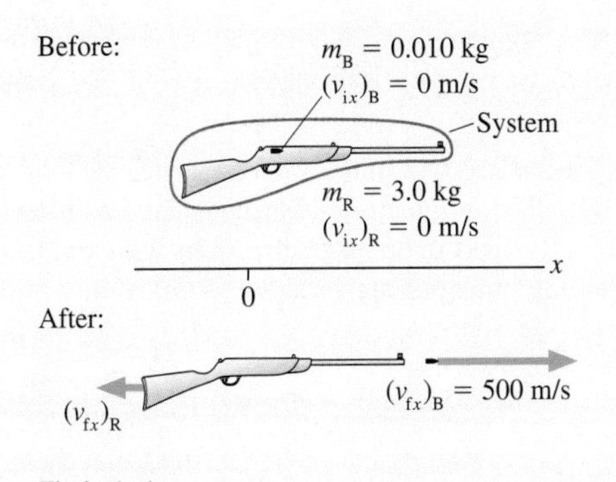
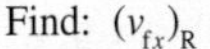

FIGURE 9.17 Before-and-after pictorial representation of a rifle firing a bullet.

SOLVE The x-component of the total momentum is $P_x = (p_x)_{bullet} + (p_x)_{rifle} + (p_x)_{gas}$. Everything is at rest before the trigger is pulled, so the initial momentum is zero. After the trigger is pulled, the momentum of the expanding gas is actually the sum of the momenta of all the molecules in the gas. For every molecule moving in the forward direction with velocity v and momentum mv there is, on average, another molecule moving in the opposite direction with velocity $-v$ and thus momentum $-mv$. When summed over the enormous number of molecules in the gas, we will be left with $p_{gas} \approx 0$. Thus the final momentum is that of the rifle and bullet. The law of conservation of momentum is

$$P_{fx} = m_B(v_{fx})_B + m_R(v_{fx})_R = P_{ix} = 0$$

Solving for the rifle's velocity, we find

$$(v_{fx})_R = -\frac{m_B}{m_R}(v_{fx})_B = -\frac{0.010 \text{ kg}}{3.0 \text{ kg}} \times 500 \text{ m/s} = -1.67 \text{ m/s}$$

The minus sign indicates that the rifle's recoil is to the left. The recoil *speed* is 1.67 m/s.

We would not know where to begin to solve a problem such as this using Newton's laws. But Example 9.5 is a simple problem when approached from the before-and-after perspective of a conservation law. The selection of bullet + gas + rifle as "the system" was the critical step. For momentum conservation to be a useful principle, we had to select a system in which the complicated forces due to expanding gas and friction were all internal forces. The rifle by itself is *not* an isolated system, so its momentum is *not* conserved.

EXAMPLE 9.6 Radioactivity

A ^{238}U uranium nucleus is radioactive. It spontaneously disintegrates into a small fragment that is ejected with a measured speed of 1.50×10^7 m/s and a "daughter nucleus" that recoils with a measured speed of 2.56×10^5 m/s. What are the atomic masses of the ejected fragment and the daughter nucleus?

MODEL The notation ^{238}U indicates the isotope of uranium with an atomic mass of 238 u, where u is the abbreviation for the *unified atomic mass unit*. The nucleus contains 92 protons (uranium is atomic number 92) and 146 neutrons. The disintegration of a nucleus is, in essence, an explosion. Only *internal* nuclear forces are involved, so the total momentum is conserved in the radioactive decay.

VISUALIZE Figure 9.18 shows the before-and-after pictorial representation. The mass of the daughter nucleus is m_1 and that of

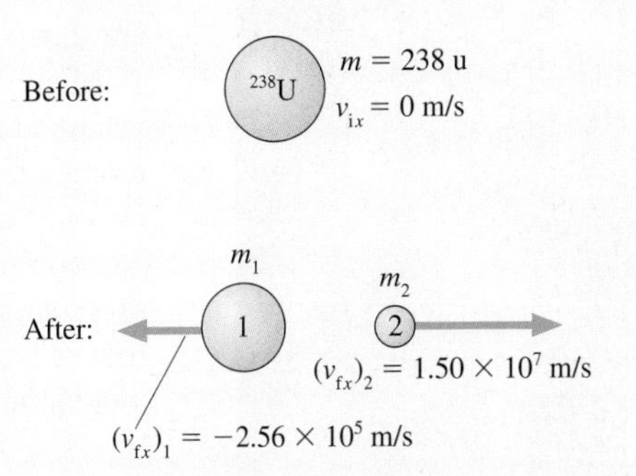
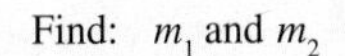

FIGURE 9.18 Before-and-after pictorial representation of the decay of a ^{238}U nucleus.

the ejected fragment is m_2. Notice that we converted the speed information to velocity information, giving $(v_{fx})_1$ and $(v_{fx})_2$ opposite signs.

SOLVE The nucleus was initially at rest, hence the total momentum is zero. The momentum after the decay is still zero if the two pieces fly apart in opposite directions with momenta equal in magnitude but opposite in sign. Momentum conservation requires

$$P_{fx} = m_1(v_{fx})_1 + m_2(v_{fx})_2 = P_{ix} = 0$$

Although we know both final velocities, this is not enough information to find the two unknown masses. However, we also have another conservation law, conservation of mass, that requires

$$m_1 + m_2 = 238 \text{ u}$$

Combining these two conservation laws gives

$$m_1(v_{fx})_1 + (238 \text{ u} - m_1)(v_{fx})_2 = 0$$

The mass of the daughter nucleus is

$$m_1 = \frac{(v_{fx})_2}{(v_{fx})_2 - (v_{fx})_1} \times 238 \text{ u}$$

$$= \frac{1.50 \times 10^7 \text{ m/s}}{(1.50 \times 10^7 - (-2.56 \times 10^5)) \text{ m/s}} \times 238 \text{ u} = 234 \text{ u}$$

With m_1 known, the mass of the ejected fragment is $m_2 = 238 - m_1 = 4$ u.

ASSESS All we learn from a momentum analysis is the masses. Chemical analysis of the daughter nucleus shows that it is the element thorium, atomic number 90, with two fewer protons than the uranium nucleus. The ejected fragment carried away two protons as part of its mass of 4 u, so it must be a particle consisting of two protons and two neutrons. This is the nucleus of a helium atom, ^{4}He, which in nuclear physics is called an *alpha particle* α. Thus the radioactive decay of ^{238}U can be written as $^{238}\text{U} \rightarrow {}^{234}\text{Th} + \alpha$.

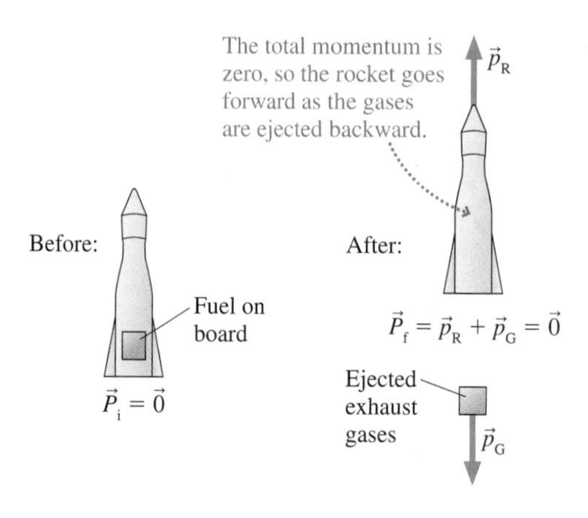
The total momentum is zero, so the rocket goes forward as the gases are ejected backward.
$\vec{p}_R$
Before:
Fuel on board
$\vec{P}_i = \vec{0}$
After:
$\vec{P}_f = \vec{p}_R + \vec{p}_G = \vec{0}$
Ejected exhaust gases
$\vec{p}_G$

FIGURE 9.19 Rocket propulsion is an example of conservation of momentum.

The total momentum of the rocket + gases system must be conserved, so the rocket accelerates forward as the gases are expelled backward.

Much the same reasoning explains how a rocket or jet aircraft accelerates. Figure 9.19 shows a rocket with a parcel of fuel on board. Burning converts the fuel to hot gases that are expelled from the rocket motor. If we choose rocket + gases to be the system, the burning and expulsion are all internal forces. There are no other forces, so the total momentum of the rocket + gases system must be conserved. The rocket gains forward velocity and momentum as the exhaust gases are shot out the back, but the *total* momentum of the system remains zero.

Many people find it hard to understand how a rocket can accelerate in the vacuum of space because there is nothing to "push against." Thinking in terms of momentum, you can see that the rocket does not push against anything *external*, but only against the gases that it pushes out the back. In return, in accordance with Newton's third law, the gases push forward on the rocket. The details of rocket propulsion are more complex than we want to handle, because the mass of the rocket is changing, but you should be able to use the law of conservation of momentum to understand the basic principle by which rocket propulsion occurs.

STOP TO THINK 9.4 An explosion in a rigid pipe shoots out three pieces. A 6 g piece comes out the right end. A 4 g piece comes out the left end with twice the speed of the 6 g piece. From which end, left or right, does the third piece emerge?

9.5 Inelastic Collisions

Collisions can have different possible outcomes. A rubber ball dropped on the floor bounces, but a ball of clay sticks to the floor without bouncing. A golf club hitting a golf ball causes the ball to rebound away from the club, but a bullet striking a block of wood embeds itself in the block.

A collision in which the two objects stick together and move with a common final velocity is called a **perfectly inelastic collision.** The clay hitting the floor and the bullet embedding itself in the wood are examples of perfectly inelastic collisions. Other examples include railroad cars coupling together upon impact and darts hitting a dart board. Figure 9.20 emphasizes the fact that the two objects have a common final velocity after they collide.

In an *elastic collision*, by contrast, the two objects bounce apart. We've looked at some examples of elastic collisions, but a full analysis requires some ideas about energy. We will return to elastic collisions in Chapter 10.

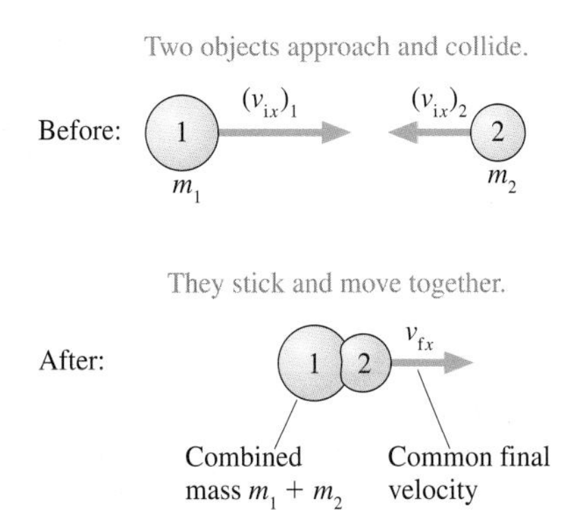

FIGURE 9.20 An inelastic collision.

EXAMPLE 9.7 An inelastic glider collision

In a laboratory experiment, a 200 g air-track glider and a 400 g air-track glider are pushed toward each other from opposite ends of the track. The gliders have Velcro tabs on the front so that they will stick together when they collide. The 200 g glider is pushed with an initial speed of 3.0 m/s. The collision causes it to reverse direction at 0.50 m/s. What was the initial speed of the 400 g glider?

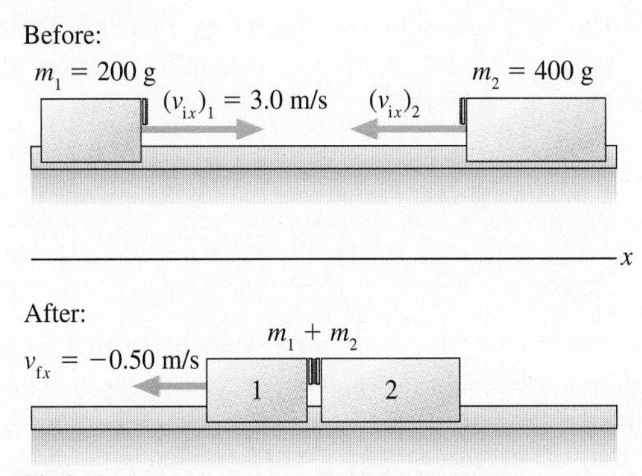

FIGURE 9.21 The before-and-after pictorial representation of two gliders colliding on an air track.

MODEL Model the gliders as particles. Define the two gliders together as the system. This is an isolated system, so its total momentum is conserved in the collision. The gliders stick together, so this is a perfectly inelastic collision.

VISUALIZE Figure 9.21 shows a pictorial representation. We've chosen to let the 200 g glider (glider 1) start out moving to the right, so $(v_{ix})_1$ is a positive 3.0 m/s. The gliders move to the left after the collision, so their common final velocity is $v_{fx} = -0.50$ m/s. Velocity $(v_{ix})_2$ will be negative.

SOLVE The law of conservation of momentum, $P_{fx} = P_{ix}$, is

$$(m_1 + m_2)v_{fx} = m_1(v_{ix})_1 + m_2(v_{ix})_2$$

where we made use of the fact that the combined mass $m_1 + m_2$ moves together after the collision. We can easily solve for the initial velocity of the 400 g glider:

$$(v_{ix})_2 = \frac{(m_1 + m_2)v_{fx} - m_1(v_{ix})_1}{m_2}$$

$$= \frac{(0.60\text{ kg})(-0.50\text{ m/s}) - (0.20\text{ kg})(3.0\text{ m/s})}{0.40\text{ kg}}$$

$$= -2.25\text{ m/s}$$

The negative sign, which we anticipated, indicates that the 400 g glider started out moving to the left. The initial *speed* of the glider, which we were asked to find, is 2.25 m/s.

EXAMPLE 9.8 Momentum in a car crash

A 2000 kg Cadillac had just started forward from a stop sign when it was struck from behind by a 1000 kg Volkswagen. The bumpers became entangled, and the two cars skidded forward together until they come to rest. Fortunately, both cars were equipped with airbags and the drivers were using seat belts, so no one was injured. Officer Tom, responding to the accident, measured the skid marks to be 3.0 m long. He also took testimony from the driver that the Cadillac's speed just before the impact was 5.0 m/s. Officer Tom charged the Volkswagen driver with reckless driving. Should the Volkswagen driver also be charged with exceeding the 50 km/hr speed limit? The judge calls you as an "expert witness" to analyze the evidence. What is your conclusion?

MODEL This is really *two* problems. First, there is an inelastic collision. The two cars are not an isolated system, because of external friction forces, but friction is not going to be significant during the brief collision. Within the impulse approximation, the momentum of the Volkswagen + Cadillac system will be conserved in the collision. Then we have a second problem, a dynamics problem of the two cars sliding.

VISUALIZE Figure 9.22a on the next page is a pictorial representation showing both the before and after of the collision and the

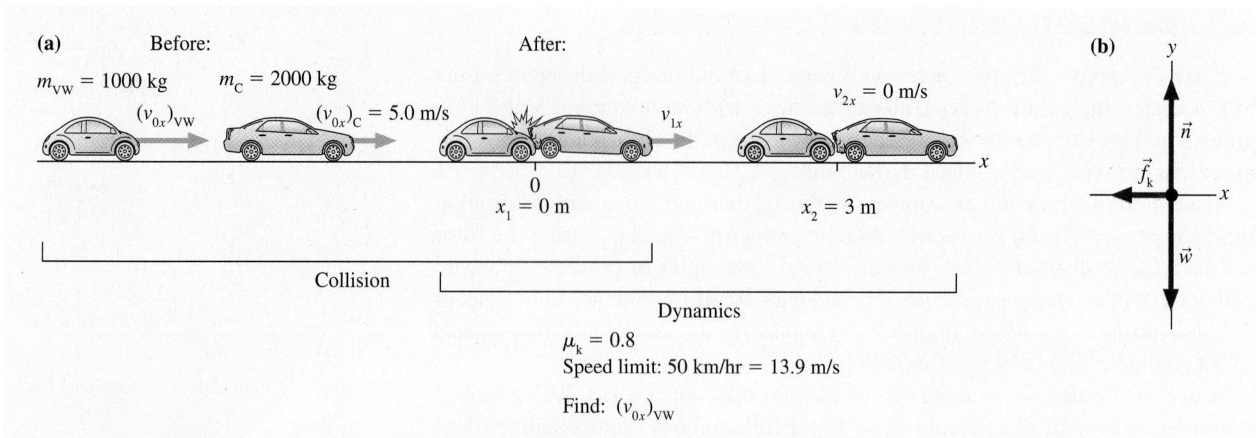

FIGURE 9.22 Pictorial representation and a free-body diagram of the cars as they skid.

more familiar picture for the dynamics of the skidding. We do not need to consider forces during the collision, because we will use the law of conservation of momentum, but we do need a free-body diagram of the cars during the subsequent skid. This is shown in Figure 9.22b.

The cars have a common velocity v_{1x} just after the collision. This is the *initial* velocity for the dynamics problem. Our goal is to find $(v_{0x})_{VW}$, the Volkswagen's velocity at the moment of impact. The 50 km/hr speed limit has been converted to 13.9 m/s.

SOLVE First, the inelastic collision. The law of conservation of momentum is

$$(m_{VW} + m_C)v_{1x} = m_{VW}(v_{0x})_{VW} + m_C(v_{0x})_C$$

Solving for the initial velocity of the Volkswagen, we find

$$(v_{0x})_{VW} = \frac{(m_{VW} + m_C)v_{1x} - m_C(v_{0x})_C}{m_{VW}}$$

To evaluate $(v_{0x})_{VW}$, we need to know v_{1x}, the velocity *immediately* after the collision as the cars begin to skid. This information will come out of the dynamics of the skid. Newton's second law, based on the free-body diagram, and the model of kinetic friction are

$$\sum F_x = -f_k = (m_{VW} + m_C)a_x$$

$$\sum F_y = n - (m_{VW} + m_C)g = 0$$

$$f_k = \mu_k n$$

where we have noted that $\vec{f}_k$ points to the left (negative x-component) and that the total mass is $m_{VW} + m_C$. From the y-equation and the friction equation,

$$f_k = \mu_k(m_{VW} + m_C)g$$

Using this in the x-equation gives us the acceleration during the skid,

$$a_x = \frac{-f_k}{m_{VW} + m_C} = -\mu_k g = -7.84 \text{ m/s}^2$$

where the coefficient of kinetic friction for rubber on concrete is taken from Table 5.1. With the acceleration determined, we can move on to the kinematics. This is constant acceleration, so

$$v_{2x}^2 = 0 = v_{1x}^2 + 2a_x(x_2 - x_1) = v_{1x}^2 + 2a_x x_2$$

Hence the skid starts with velocity

$$v_{1x} = \sqrt{-2a_x x_1} = \sqrt{-2(-7.84 \text{ m/s}^2)(3.0 \text{ m})} = 6.86 \text{ m/s}$$

As we have noted, this is the final velocity of the collision. Inserting v_{1x} back into the momentum conservation equation, we finally determine that

$$(v_{0x})_{VW} = \frac{(3000 \text{ kg})(6.86 \text{ m/s}) - (2000 \text{ kg})(5.0 \text{ m/s})}{1000 \text{ kg}}$$

$$= 10.6 \text{ m/s}$$

On the basis of your testimony, the Volkswagen driver is *not* charged with speeding!

NOTE ► Momentum is *not always* conserved. In this example, momentum was conserved during the collision but *not* during the skid. Momentum is conserved only for an isolated system. In practice, it is not unusual for momentum to be conserved in one part or one aspect of a problem but not in others. ◄

STOP TO THINK 9.5 The two particles are both moving to the right. Particle 1 catches up with particle 2 and collides with it. The particles stick together and continue on with velocity v_f. Which of these statements is true?

Before: (1) v_1 → (2) v_2 → After: (1)(2) v_f →

a. v_f is greater than v_1.
b. $v_f = v_1$.
c. v_f is greater than v_2 but less than v_1.
d. $v_f = v_2$.
e. v_f is less than v_2.
f. Can't tell without knowing the masses.

9.6 Momentum in Two Dimensions

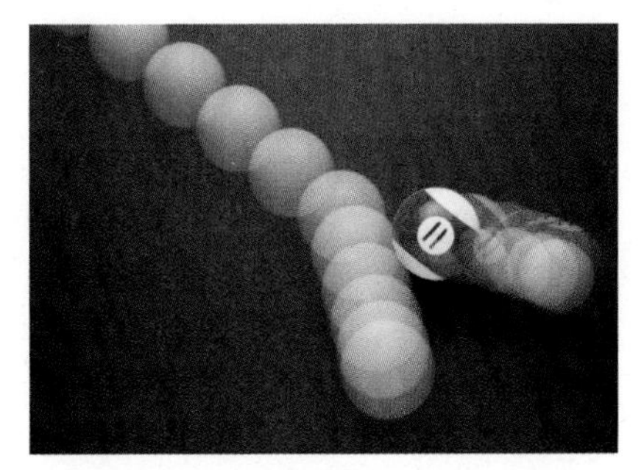

Collisions and explosions often involve motion in two dimensions.

Our examples thus far have been confined to motion along a one-dimensional axis. Many practical examples of momentum conservation involve motion in a plane. The total momentum $\vec{P}$ is a *vector* sum of the momenta $\vec{p} = m\vec{v}$ of the individual particles. Consequently, as we found in Section 9.3, momentum is conserved only if each component of $\vec{P}$ is conserved:

$$(p_{fx})_1 + (p_{fx})_2 + (p_{fx})_3 + \cdots = (p_{ix})_1 + (p_{ix})_2 + (p_{ix})_3 + \cdots$$
$$(p_{fy})_1 + (p_{fy})_2 + (p_{fy})_3 + \cdots = (p_{iy})_1 + (p_{iy})_2 + (p_{iy})_3 + \cdots \quad (9.22)$$

In this section we'll apply momentum conservation to motion in two dimensions.

EXAMPLE 9.9 Momentum in a 2D car crash

The 2000 kg Cadillac and the 1000 kg Volkswagen of Example 9.8 meet again the following week, just after leaving the auto body shop where they had been repaired. The stoplight has just turned green, and the Cadillac, heading north, drives forward into the intersection. The Volkswagen, traveling east, fails to stop. The Volkswagen crashes into the left front fender of the Cadillac, then the cars stick together and slide to a halt. Officer Tom, responding to the accident, sees that the skid marks go 35° northeast from the point of impact. The Cadillac driver, who keeps a close eye on the speedometer, reports that he was traveling at 3.0 m/s when the accident occurred. How fast was the Volkswagen going just before the impact?

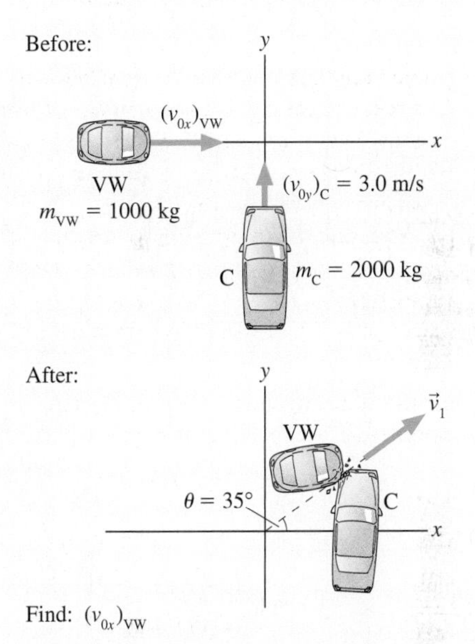

FIGURE 9.23 Pictorial representation of the collision between the Cadillac and the Volkswagen.

MODEL This is another inelastic collision. The total momentum of the Volkswagen + Cadillac system is conserved.

VISUALIZE Figure 9.23 is a before-and-after pictorial representation. The Volkswagen travels on the x-axis and the Cadillac on the y-axis, hence $(v_{0y})_{VW} = 0$ and $(v_{0x})_C = 0$.

SOLVE After the collision, the two cars move with the common velocity $\vec{v}_1$. The velocity components, as in a projectile motion problem, are $v_{1x} = v_1\cos\theta$ and $v_{1y} = v_1\sin\theta$. Thus the simultaneous x- and y-momentum equations are

$$\begin{aligned}(m_C + m_{VW})v_{1x} &= (m_C + m_{VW})v_1\cos\theta \\ &= m_C(v_{0x})_C + m_{VW}(v_{0x})_{VW} = m_{VW}(v_{0x})_{VW} \\ (m_C + m_{VW})v_{1y} &= (m_C + m_{VW})v_1\sin\theta \\ &= m_C(v_{0y})_C + m_{VW}(v_{0y})_{VW} = m_C(v_{0y})_C\end{aligned}$$

We can use the y-equation to find:

$$v_1 = \frac{m_C(v_{0y})_C}{(m_C + m_{VW})\sin\theta} = \frac{(2000\text{ kg})(3.0\text{ m/s})}{(3000\text{ kg})\sin 35°} = 3.49\text{ m/s}$$

Using this value for v_1 in the x-equation, we find that the Volkswagen's velocity was

$$(v_{0x})_{VW} = \frac{(m_C + m_{VW})v_1\cos\theta}{m_{VW}} = 8.6\text{ m/s}$$

It's instructive to examine this collision with a picture of the momentum vectors. Before the collision, $\vec{p}_{VW} = (1000\text{ kg})(8.6\text{ m/s})\hat{\imath} = 8600\hat{\imath}$ kg m/s and $\vec{p}_C = (2000\text{ kg})(3.0\text{ m/s})\hat{\jmath} = 6000\hat{\jmath}$ kg m/s. These vectors, and their sum $\vec{P} = \vec{p}_{VW} + \vec{p}_C$ are shown in Figure 9.24. You can see that the total momentum vector makes a 35° angle with the x-axis. The individual momenta change in the collision, *but the total momentum does not*. That is why the skid marks are 35° north of east.

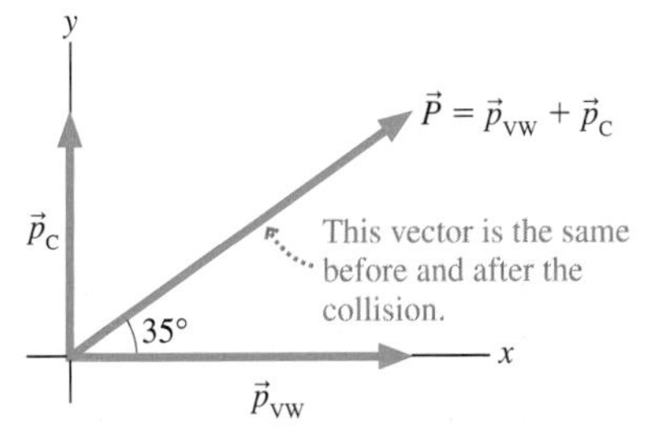

FIGURE 9.24 The momentum vectors of the car crash.

EXAMPLE 9.10 A three-piece explosion

A 10 g projectile is traveling east at 2.0 m/s when it suddenly explodes into three pieces. A 3.0 g fragment is shot due west at 10 m/s while another 3.0 g fragment travels 40° north of east at 12 m/s. What are the speed and direction of the third fragment?

MODEL Although many complex forces are involved in the explosion, they are all internal to the system. There are no external forces, so this is an isolated system and its total momentum is conserved.

VISUALIZE Figure 9.25 shows a before-and-after pictorial representation. We'll use uppercase M and V to distinguish the initial object from the three pieces into which it explodes.

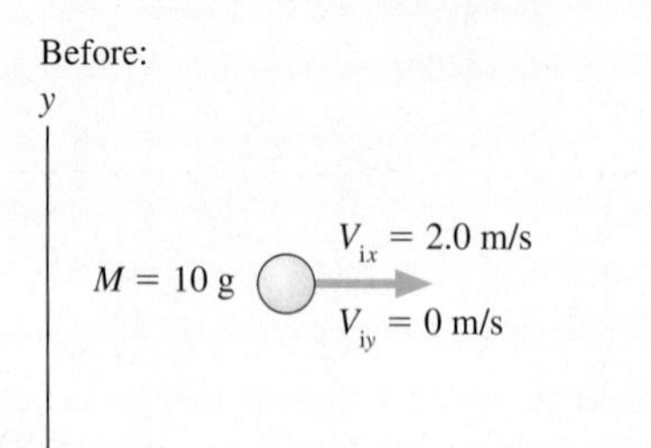

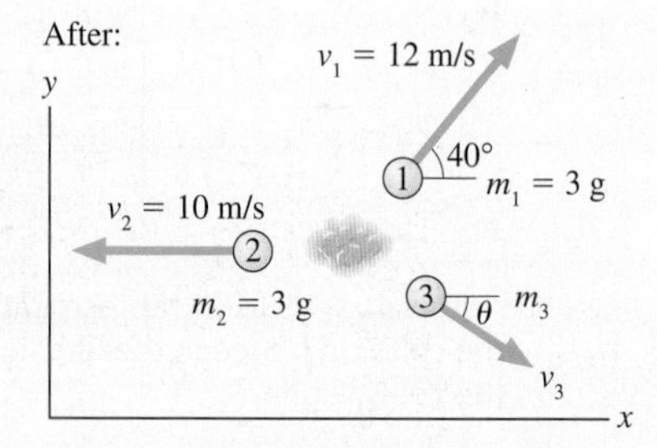

FIGURE 9.25 Before-and-after pictorial representation of the three-piece explosion.

SOLVE The system is the initial object and the subsequent three pieces. Conservation of momentum requires

$$m_1(v_{fx})_1 + m_2(v_{fx})_2 + m_3(v_{fx})_3 = MV_{ix}$$
$$m_1(v_{fy})_1 + m_2(v_{fy})_2 + m_3(v_{fy})_3 = MV_{iy}$$

Conservation of mass implies that

$$m_3 = M - m_1 - m_2 = 4.0 \text{ g}$$

Neither the original object nor m_2 have any momentum along the y-axis. We can use Figure 9.24 to write out the x- and y-components of $\vec{v}_1$ and $\vec{v}_3$, leading to

$$m_1 v_1 \cos 40° - m_2 v_2 + m_3 v_3 \cos\theta = MV$$
$$m_1 v_1 \sin 40° - m_3 v_3 \sin\theta = 0$$

where we used $(v_{fx})_2 = -v_2$ because m_2 is moving in the negative x-direction. Inserting known values in these equations gives us

$$-2.42 + 4v_3 \cos\theta = 20$$
$$23.14 - 4v_3 \sin\theta = 0$$

We can leave the masses in grams in this situation because the conversion factor to kilograms appears on both sides of the equation and thus cancels out. To solve, first use the second equation to write $v_3 = 5.79/\sin\theta$. Substitute this result into the first equation, noting that $\cos\theta/\sin\theta = 1/\tan\theta$, to get

$$-2.42 + 4\left(\frac{5.79}{\sin\theta}\right)\cos\theta = -2.42 + \frac{23.14}{\tan\theta} = 20$$

Now solve for θ:

$$\tan\theta = \frac{23.14}{20 + 2.42} = 1.032$$
$$\theta = \tan^{-1}(1.032) = 45.9°$$

Finally, use this result in the earlier expression for v_3 to find

$$v_3 = \frac{5.79}{\sin 45.9°} = 8.06 \text{ m/s.}$$

The third fragment, with a mass of 4.0 g, is shot 45.9° south of east at a speed of 8.06 m/s.

9.7 Angular Momentum

The momentum $\vec{p}$ describes linear motion, such as motion along a line or motion in a plane. For a single particle, the law of conservation of momentum is an alternative way of stating Newton's first law. Rather than saying that a particle will continue to move in a straight line at constant velocity unless acted on by a net force, we can say that the momentum of an isolated particle is conserved. Both express the idea that a particle in linear motion tends to "keep going" unless something acts on it to change its motion.

Another important class of motion that you've studied is motion in a circle. The momentum $\vec{p}$ is *not* a conserved quantity for a particle in circular motion. Momentum is a vector, and the momentum of a particle in circular motion changes as the direction of motion changes.

Nonetheless, a spinning bicycle wheel would keep turning if it were not for friction. A ball moving in a circle at the end of string tends to "keep going," like the bicycle wheel, as long as the conditions for circular motion haven't changed. The quantity that expresses this idea for circular motion is called *angular momentum*.

This section is a first introduction to the concept of angular momentum. We will restrict our attention to particles in circular motion, and we will not present any proofs or derivations. A more rigorous discussion is deferred to Chapter 13, where we will find that angular momentum is an important idea for understanding the rotational motion of solid objects.

Figure 9.26a shows a particle moving in a circle of radius r with velocity $\vec{v}$. We define the particle's **angular momentum** as

$$L = mrv_t \tag{9.23}$$

where v_t is the tangential component of $\vec{v}$. The angular momentum L has the same sign as v_t, which you will recall from Chapter 7 is positive for counterclockwise rotation, negative for clockwise rotation. The units of angular momentum are $\text{kg m}^2/\text{s}$.

Figure 9.26b shows a net force $\vec{F}_{net}$ acting on the particle. As in Chapter 7, we can decompose the vector $\vec{F}_{net}$ into a tangential component F_t and a radial component F_r. The radial component of the force causes the centripetal acceleration of circular motion. The component F_t parallel to the trajectory causes the particle to speed up or slow down.

If there's a tangential force, the particle's angular momentum $L = mrv_t$ changes as v_t changes. If there is no tangential force, $F_t = 0$, the angular momentum remains constant. Without formal proof, we'll state this idea as the *law of conservation of angular momentum*:

Law of conservation of angular momentum The angular momentum of a particle (or system of particles) in circular motion does not change unless there is a net tangential force on the particle. That is, $L_f = L_i$ if $F_t = 0$.

As stated, the law of conservation of angular momentum seems rather limited in scope. The significance of angular momentum will become more apparent in later chapters, where you'll see how it can be applied to noncircular orbits and to the rotation of solid objects. But we can give the law of conservation of momentum one interesting application in its current form: circular trajectories that change diameter.

Suppose you tie a ball to a string and swing it in a horizontal circle over your head. The tension in the rope is the radial force F_r that causes the ball's centripetal acceleration. There is no tangential force (assuming air resistance is negligible) and, once you get it going, the ball will continue to turn in a circle with constant tangential velocity v_t and constant angular momentum L. What if you now begin to pull the rope in, decreasing the radius of the ball's trajectory? Because there's no tangential force, *the ball's angular momentum is conserved* as it slowly spirals inward into smaller and smaller circles. Using the definition $L = mrv_t$, you can see that the ball's speed has to increase as r decreases in order to keep L constant.

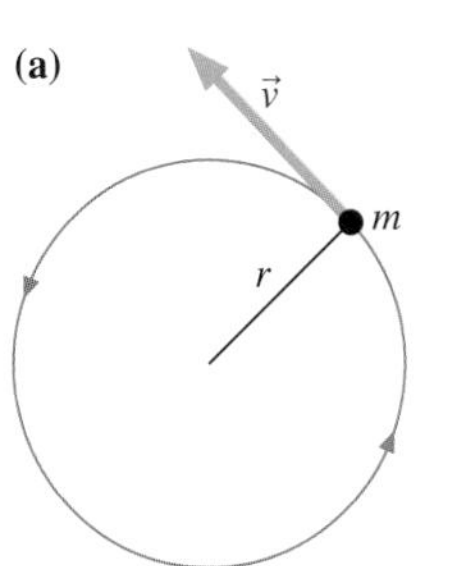
(a)
$\vec{v}$
m
r

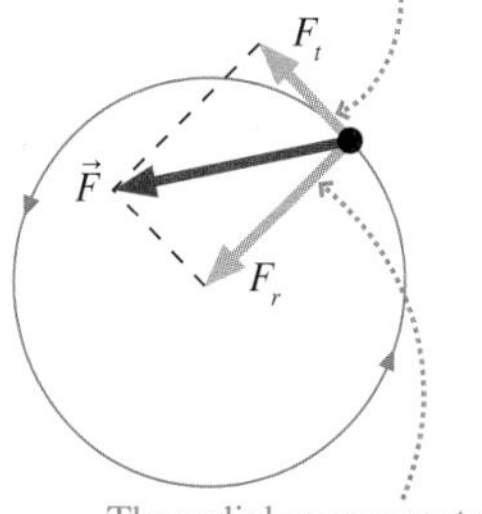
(b) The tangential component of $\vec{F}$ causes the particle's speed and angular momentum to change.
F_t
$\vec{F}$
F_r
The radial component of $\vec{F}$ causes the centripetal acceleration of circular motion.

FIGURE 9.26 A particle in circular motion has angular momentum.

The angular momentum of a spinning skater is conserved. This allows the skater to use her arms to control her speed.

EXAMPLE 9.11 A spinning ice skater

An ice skater spins around on the tips of his blades while holding a 5.0 kg weight in each hand. He begins with his arms straight out from his body and his hands 140 cm apart. While spinning at 2.0 rev/s, he pulls the weights in and holds them 50 cm apart against his shoulders. If we neglect the mass of the skater, how fast is he spinning after pulling the weights in?

MODEL Although the mass of the skater is larger than the mass of the weights, neglecting the skater's mass is not a bad approximation. Angular momentum depends on the radius of the circle. The skater's mass is concentrated in his torso, which has an effective radius (i.e., where most of the mass is concentrated) of only 9 or 10 cm. The weights move in much larger circles and have a disproportionate influence on his motion. The skater's arms exert radial forces on the weights just to keep them moving in circles, and even larger radial forces as he pulls them in. But there is no tangential force on the weights, so their total angular momentum is conserved.

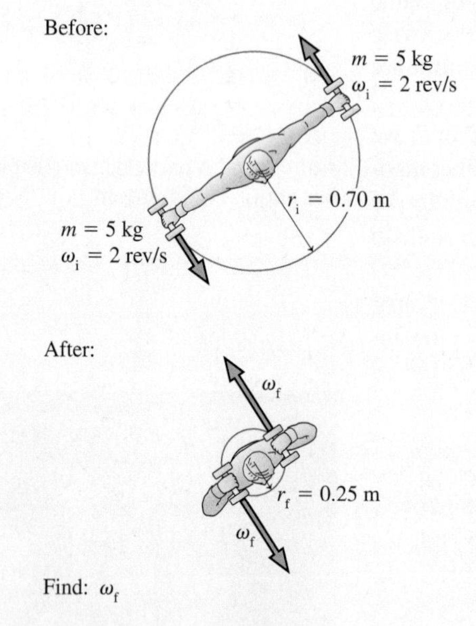

FIGURE 9.27 Top view pictorial representation of the spinning ice skater.

VISUALIZE Figure 9.27 shows a before-and-after pictorial representation, as seen from above.

SOLVE The two weights have the same mass, move in circles with the same radius, and have the same tangential velocity. Thus the total angular momentum is twice that of one weight. The mathematical statement of angular momentum conservation, $L_f = L_i$, is

$$2mr_f v_{ft} = 2mr_i v_{it}$$

You learned in Chapter 7 that a particle's tangential velocity is related to its angular velocity by $v_t = r\omega$. Using this result, and canceling the $2m$, we find

$$r_f^2 \omega_f = r_i^2 \omega_i$$

$$\omega_f = \left(\frac{r_i}{r_f}\right)^2 \omega_i$$

Although we would need ω in rad/s to calculate v_t from $v_t = r\omega$, we don't need to convert ω because the conversion factor would be the same on both sides of the equation and would cancel. We can leave the skater's initial angular velocity as $\omega_i = 2.0$ rev/s. When he pulls the weights in, his angular velocity increases to

$$\omega_f = \left(\frac{0.70\text{ m}}{0.25\text{ m}}\right)^2 \times 2.0\text{ rev/s} = 15.7\text{ rev/s}$$

ASSESS Pulling in the weights increases the skater's spin from 2 rev/s to nearly 16 rev/s. This is actually somewhat high, because we neglected the mass of the skater. Nonetheless, you've probably seen how figure skaters start a slow spin with arms outstretched, then dramatically speed up by raising their arms over their heads. They use the mass of their arms, not extra weights, to control their speed. While this example has been over simplified, it does illustrate that a skater's spin is governed by conservation of angular momentum.

STOP TO THINK 9.6 A dry ice (solid carbon dioxide) puck revolves in a circle on the end of a lightweight rigid rod that turns on frictionless bearings. A cushion of CO_2 gas allows the puck to glide across the surface without friction. As the puck sublimates (changes from a solid to a gas), does its speed increase, decrease, or stay the same?

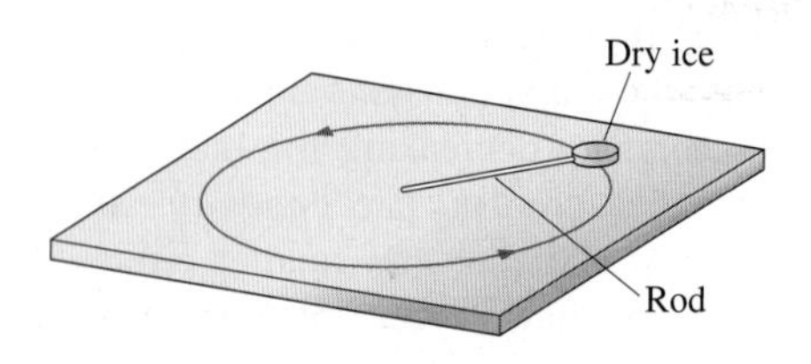

SUMMARY

The goal of Chapter 9 has been to introduce the ideas of impulse, momentum, and angular momentum and to learn a new problem-solving strategy based on conservation laws.

GENERAL PRINCIPLES

Law of Conservation of Momentum

The total momentum $\vec{P} = \vec{p}_1 + \vec{p}_2 + \cdots$ of an isolated system is a constant. Thus

$$\vec{P}_f = \vec{P}_i$$

Law of Conservation of Angular Momentum

The angular momentum L of a particle or system of particles in circular motion does not change unless there is a net tangential force. Thus

$$L_f = L_i$$

Solving Momentum Conservation Problems

MODEL Choose an isolated system or a system that is isolated during at least part of the problem.

VISUALIZE Draw a pictorial representation of the system before and after the interaction.

SOLVE Write the law of conservation of momentum in terms of vector components

$$(p_{fx})_1 + (p_{fx})_2 + \cdots = (p_{ix})_1 + (p_{ix})_2 + \cdots$$

$$(p_{fy})_1 + (p_{fy})_2 + \cdots = (p_{iy})_1 + (p_{iy})_2 + \cdots$$

ASSESS Is the result reasonable?

IMPORTANT CONCEPTS

Momentum $\vec{p} = m\vec{v}$

Impulse $J_x = \int_{t_i}^{t_f} F_x(t)\,dt$ = area under force curve

Impulse and momentum are related by the impulse-momentum theorem

$$\Delta p_x = J_x$$

This is an alternative statement of Newton's second law.

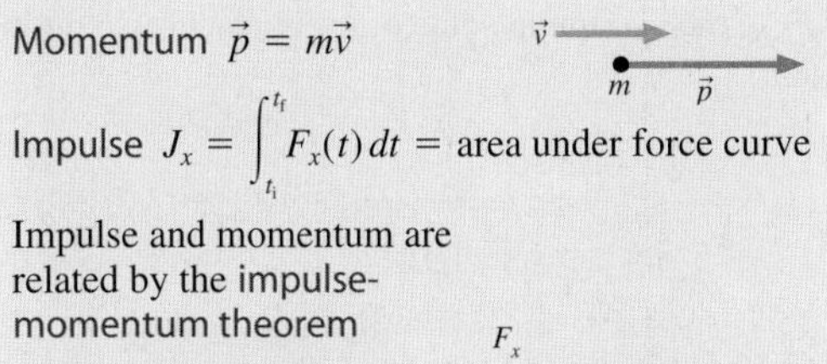

Angular momentum $L = mrv_t$

System A group of interacting particles.

Isolated system A system on which there are no external forces or the net external force is zero.

Before-and-after pictorial representation

- Define the system.
- Use two drawings to show the system *before* and *after* the interaction.
- List known information and identify what you are trying to find.

Before: m_1 (1) $(v_{ix})_1$ → ← $(v_{ix})_2$ (2) m_2

After: ← $(v_{fx})_1$ (1) (2) $(v_{fx})_2$ →

APPLICATIONS

Collisions Two or more particles come together. In a perfectly inelastic collision, they stick together and move with a common final velocity.

Explosions Two or more particles move away from each other.

Two dimensions No new ideas, but both the x- and y-components of P must be conserved, giving two simultaneous equations.

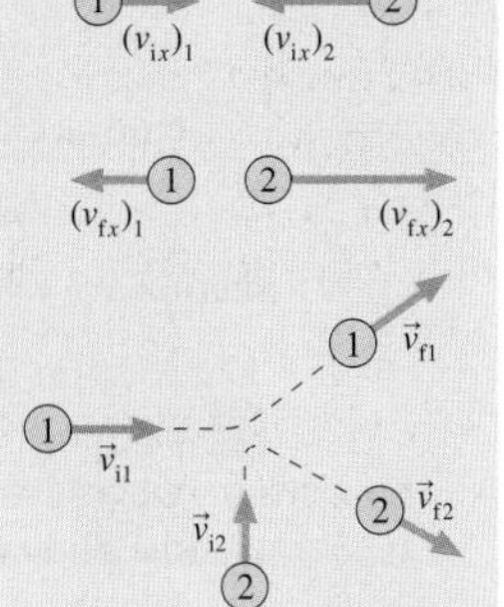

Momentum bar charts display the impulse-momentum theorem $p_{fx} = p_{ix} + J_x$ in graphical form.

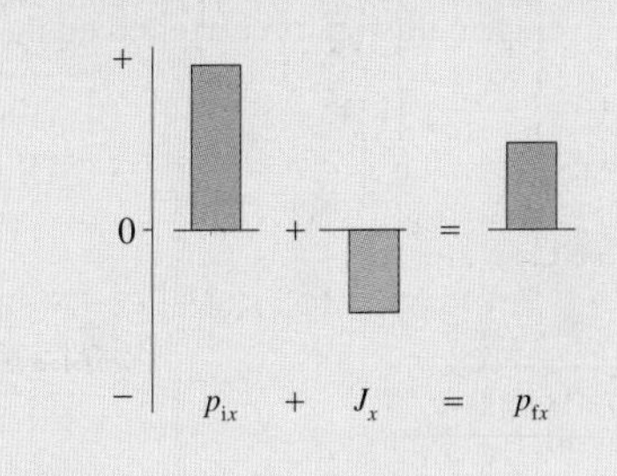

TERMS AND NOTATION

collision
impulsive force
momentum, $\vec{p}$
impulse, J
impulse-momentum theorem
impulse approximation
momentum bar chart
total momentum, $\vec{P}$
isolated system
law of conservation of momentum
explosion
perfectly inelastic collision
angular momentum, L
law of conservation of angular momentum

EXERCISES AND PROBLEMS

The icon indicates that the problem can be done on a Momentum Worksheet.

Exercises

Section 9.1 Momentum and Impulse

1. What is the magnitude of the momentum of
   a. A 1500 kg car traveling at 10 m/s?
   b. A 200 g baseball thrown at 40 m/s?
2. At what speed do a bicycle and its rider, with a combined mass of 100 kg, have the same momentum as a 1500 kg car traveling at 5.0 m/s?
3. What value of F_{max} gives an impulse of 6.0 Ns?

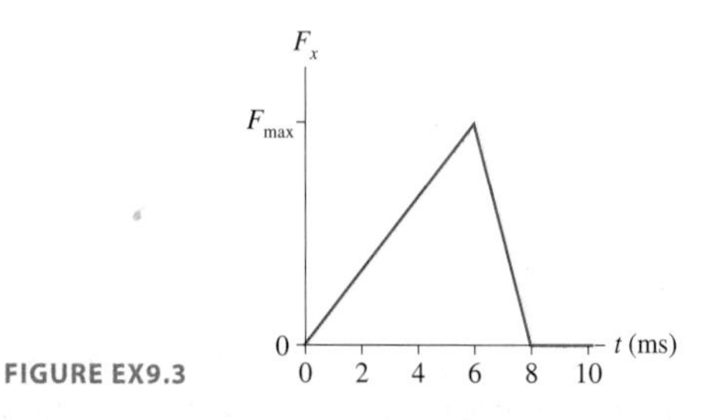

FIGURE EX9.3

4. Suppose a rubber ball and a steel ball collide. Which, if either, receives the larger impulse? Explain.
5. What is the impulse on a 3.0 kg particle that experiences this force?

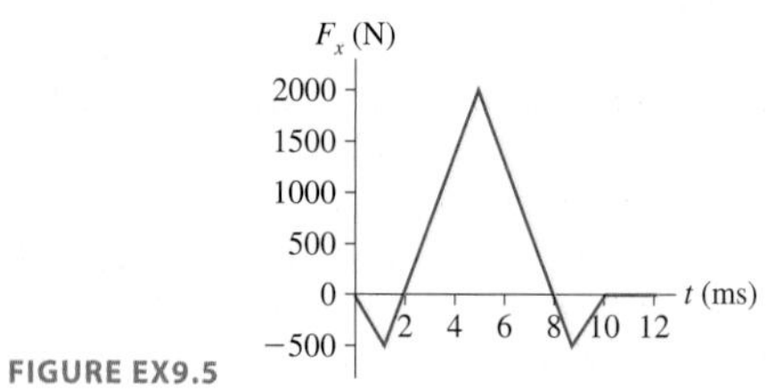

FIGURE EX9.5

Section 9.2 Solving Impulse and Momentum Problems

6. Figure Ex9.6 is an incomplete momentum bar chart for a collision that lasts 10 ms. What are the magnitude and direction of the average collision force exerted on the object?

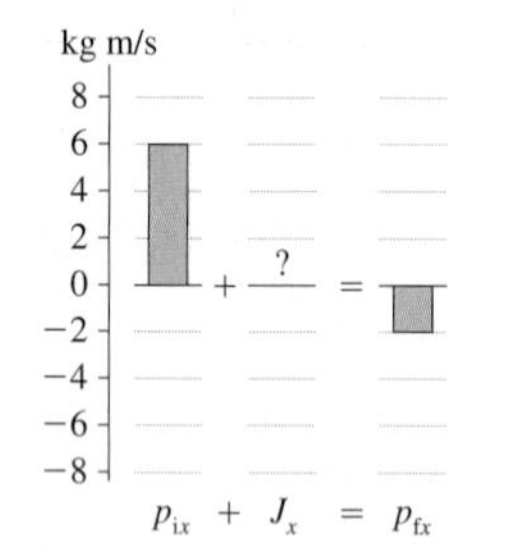

FIGURE EX9.6

7. a. A 2.0 kg object is moving to the right with a speed of 1.0 m/s when it experiences the force shown in Figure Ex9.7a. What are the object's speed and direction after the force ends?
   b. Answer this question for the force shown in Figure Ex9.7b.

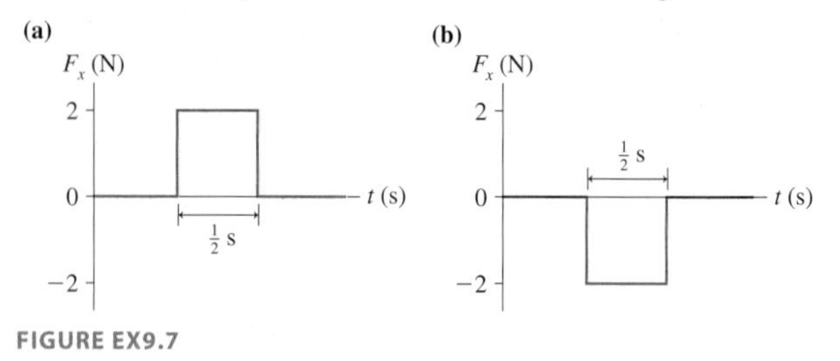

FIGURE EX9.7

8. A sled slides along a horizontal surface on which the coefficient of kinetic friction is 0.25. Its velocity at point A is 8.0 m/s and at point B is 5.0 m/s. Use the impulse-momentum theorem to find how long the sled takes to travel from A to B.
9. Use the impulse-momentum theorem to find how long a falling object takes to increase its speed from 5.5 m/s to 10.4 m/s.
10. A 60 g tennis ball with an initial speed of 32 m/s hits a wall and rebounds with the same speed. Figure Ex9.10 shows the force of the wall on the ball during the collision. What is the value of F_{max}, the maximum value of the contact force during the collision?

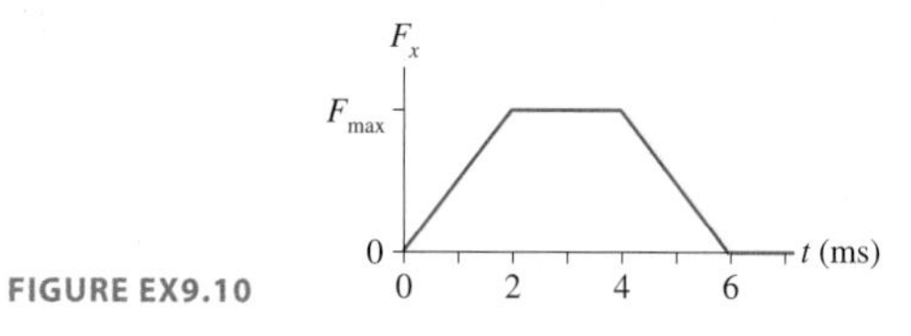

FIGURE EX9.10

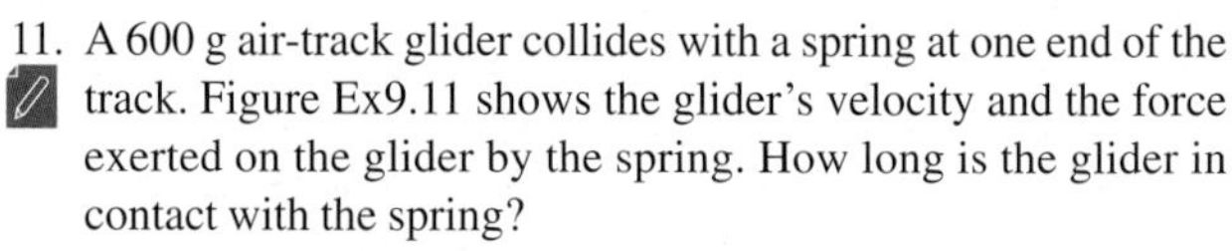

11. A 600 g air-track glider collides with a spring at one end of the track. Figure Ex9.11 shows the glider's velocity and the force exerted on the glider by the spring. How long is the glider in contact with the spring?

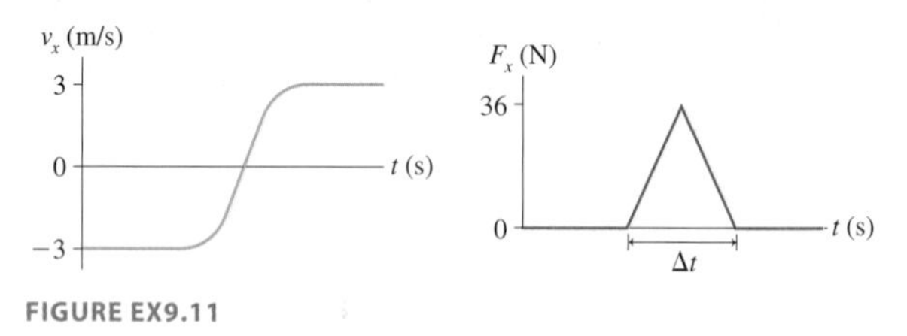

FIGURE EX9.11

Section 9.3 Conservation of Momentum

12. A 10-m-long glider with a mass of 680 kg (including the passengers) is gliding horizontally through the air at 30 m/s when a 60 kg skydiver drops out by releasing his grip on the glider. What is the glider's velocity just after the skydiver lets go?
13. A 10,000 kg railroad car is rolling at 2.0 m/s when a 4000 kg load of gravel is suddenly dropped in. What is the car's speed just after the gravel is loaded?
14. A 5000 kg open train car is rolling on frictionless rails at 22 m/s when it starts pouring rain. A few minutes later, the car's speed is 20 m/s. What mass of water has collected in the car?

Section 9.4 Explosions

15. A 50 kg archer, standing on frictionless ice, shoots a 100 g arrow at a speed of 100 m/s. What is the recoil speed of the archer?
16. In Problem 24 of Chapter 8 you found the recoil speed of Bob as he throws a rock while standing on frictionless ice. Bob has a mass of 75 kg and can throw a 500 g rock with a speed of 30 m/s. Find Bob's recoil speed again, this time using momentum.
17. Dan is gliding on his skateboard at 4.0 m/s. He suddenly jumps backward off the skateboard, kicking the skateboard forward at 8.0 m/s. How fast is Dan going as his feet hit the ground? Dan's mass is 50 kg and the skateboard's mass is 5.0 kg.

Section 9.5 Inelastic Collisions

18. A 300 g bird flying along at 6.0 m/s sees a 10 g insect heading straight toward it with a speed of 30 m/s. The bird opens its mouth wide and enjoys a nice lunch. What is the bird's speed immediately after swallowing?
19. The parking brake on a 2000 kg Cadillac has failed, and it is rolling slowly, at 1 mph, toward a group of small children. Seeing the situation, you realize you have just enough time to drive your 1000 kg Volkswagen head-on into the Cadillac and save the children. With what speed should you impact the Cadillac to bring it to a halt?
20. A 1500 kg car is rolling at 2.0 m/s. You would like to stop the car by firing a 10 kg blob of sticky clay at it. How fast should you fire the clay?

Section 9.6 Momentum in Two Dimensions

21. A 20 g ball of clay traveling east at 3.0 m/s collides with a 30 g ball of clay traveling north at 2.0 m/s. What are the speed and the direction of the resulting 50 g ball of clay?
22. Two particles collide and bounce apart. Figure Ex9.22 shows the initial momenta of both and the final momentum of particle 2. What is the final momentum of particle 1? Show the momentum vector on the diagram *and* write it in component form.

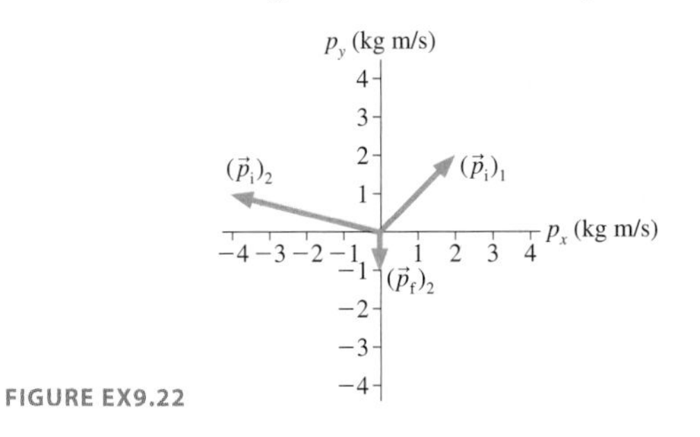

FIGURE EX9.22

Section 9.7 Angular Momentum

23. What is the angular momentum of the moon around the earth? Relevant astronomical data are found inside the cover of the book.
24. A 200 g puck revolves in a circle on a frictionless table at the end of a 50-cm-long string. The puck's angular momentum is 3.0 kg m^2/s. What is the tension in the string?

Problems

25. A 50 g ball is launched from ground level at an angle 30° above the horizon. Its initial speed is 25 m/s.
    a. What are the values of p_x and p_y an instant after the ball is thrown, at the point of maximum altitude, and an instant before the ball hits the ground?
    b. Why is one component of $\vec{p}$ constant? Explain.
    c. For the component of $\vec{p}$ that changes, show that the change in momentum is equal to the weight of the ball multiplied by the time of flight. Explain why this is so.
26. Far in space, where gravity is negligible, a 425 kg rocket traveling at 75 m/s fires its engines. Figure P9.26 shows the thrust force as a function of time. The mass lost by the rocket during these 30 s is negligible.
    a. What impulse does the engine impart to the rocket?
    b. At what time does the rocket reach its maximum speed? What is the maximum speed?

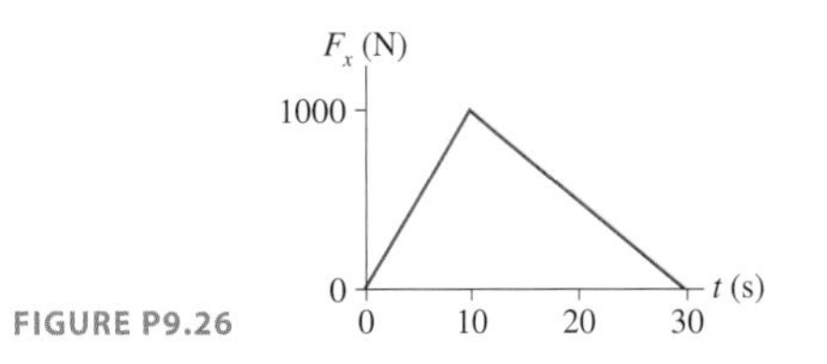

FIGURE P9.26

27. A tennis player swings her 1000 g racket with a speed of 10 m/s. She hits a 60 g tennis ball that was approaching her at a speed of 20 m/s. The ball rebounds at 40 m/s.
    a. How fast is her racket moving immediately after the impact? You can ignore the interaction of the racket with her hand for the brief duration of the collision.
    b. If the tennis ball and racket are in contact for 10 ms, what is the average force that the racket exerts on the ball? How does this compare to the ball's weight?
28. A 200 g ball is dropped from a height of 2.0 m, bounces on a hard floor, and rebounds to a height of 1.5 m. Figure P9.28 shows the impulse received from the floor. What maximum force does the floor exert on the ball?

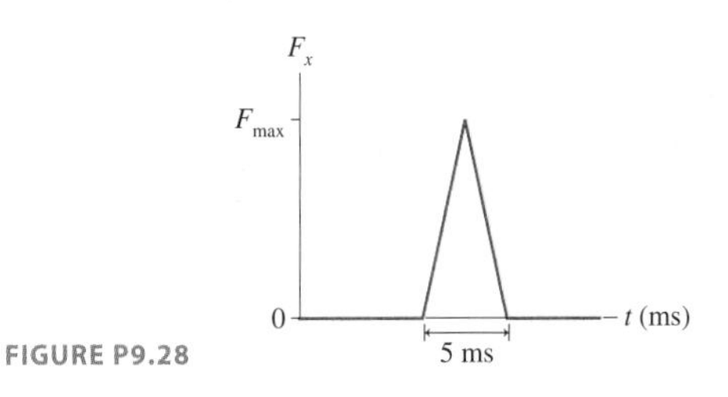

FIGURE P9.28

29. A 40 g rubber ball is dropped from a height of 1.8 m and rebounds to two-thirds of its initial height.
    a. Find the magnitude and direction of the impulse that the floor exerts on the ball.
    b. Using simple observations of an ordinary rubber ball, sketch a physically plausible graph of the force of the floor on the ball as a function of time.
    c. Make a plausible estimate of how long the ball is in contact with the floor, then use this quantity to calculate the approximate average force of the floor on the ball.

30. A 500 g cart is released from rest 1.0 m from the bottom of a frictionless, 30° ramp. The cart rolls down the ramp and bounces off a rubber block at the bottom. Figure P9.30 shows the force during the collision. After the cart bounces, how far does it roll back up the ramp?

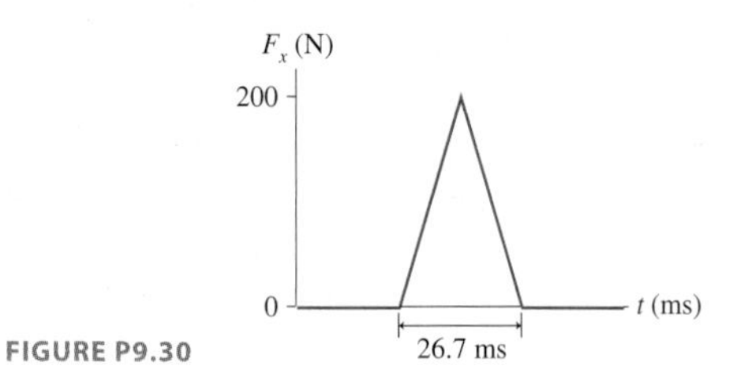

FIGURE P9.30

31. A 20 g ball of clay is thrown horizontally at 30 m/s toward a 1.0 kg block sitting at rest on a frictionless surface. The clay hits and sticks to the block.
    a. What impulse does the clay exert on the block?
    b. What impulse does the block exert on the clay?
    c. Does $J_{\text{block on clay}} = -J_{\text{clay on block}}$?
32. Three identical train cars, coupled together, are rolling east at 2.0 m/s. A fourth car traveling east at 4.0 m/s catches up with the three and couples to make a four-car train. A moment later, the train cars hit a fifth car that was at rest on the tracks, and it couples to make a five-car train. What is the speed of the five-car train?
33. Most geologists believe that the dinosaurs became extinct 65 million years ago when a large comet or asteroid struck the earth, throwing up so much dust that the sun was blocked out for a period of many months. Suppose an asteroid with a diameter of 2.0 km and a mass of 1.0×10^{13} kg hits the earth with an impact speed of 4.0×10^4 m/s.
    a. What is the earth's recoil speed after such a collision? (Use a reference frame in which the earth was initially at rest.)
    b. What percentage is this of the earth's speed around the sun? (Use the astronomical data inside the cover.)
34. At the center of a 50-m-diameter circular ice rink, a 75 kg skater traveling north at 2.5 m/s collides with and holds onto a 60 kg skater who had been heading west at 3.5 m/s.
    a. How long will it take them to glide to the edge of the rink?
    b. Where will they reach it? Give your answer as an angle north of west.
35. Two ice skaters, with masses of 50 kg and 75 kg, are at the center of a 60-m-diameter circular rink. The skaters push off against each other and glide to opposite edges of the rink. If the heavier skater reaches the edge in 20 s, how long does the lighter skater take to reach the edge?
36. A firecracker in a coconut blows the coconut into three pieces. Two pieces of equal mass fly off south and west, perpendicular to each other, at 20 m/s. The third piece has twice the mass as the other two. What are the speed and direction of the third piece?
37. One billiard ball is shot east at 2.0 m/s. A second, identical billiard ball is shot west at 1.0 m/s. The balls have a glancing collision, not a head-on collision, deflecting the second ball by 90° and sending it north at 1.41 m/s. What are the speed and direction of the first ball after the collision?
38. A 10 g bullet is fired into a 10 kg wood block that is at rest on a wood table. The block, with the bullet embedded, slides 5.0 cm across the table. What was the speed of the bullet?
39. Fred (mass 60 kg) is running with the football at a speed of 6.0 m/s when he is met head-on by Brutus (mass 120 kg), who is moving at 4.0 m/s. Brutus grabs Fred in a tight grip, and they fall to the ground. Which way do they slide, and how far? The coefficient of kinetic friction between football uniforms and Astroturf is 0.30.
40. You are part of a search-and-rescue mission that has been called out to look for a lost explorer. You've found the missing explorer, but you're separated from him by a 200-m-high cliff and a 30-m-wide raging river. To save his life, you need to get a 5.0 kg package of emergency supplies across the river. Unfortunately, you can't throw the package hard enough to make it across. Fortunately, you happen to have a 1.0 kg rocket intended for launching flares. Improvising quickly, you attach a sharpened stick to the front of the rocket, so that it will impale itself into the package of supplies, then fire the rocket at ground level toward the supplies. What minimum speed must the rocket have just before impact in order to save the explorer's life?

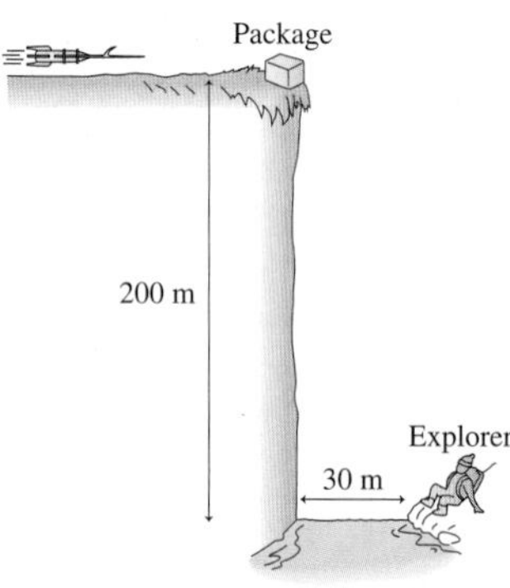

FIGURE P9.40

41. An object at rest on a flat, horizontal surface explodes into two fragments, one seven times as massive as the other. The heavier fragment slides 8.2 m before stopping. How far does the lighter fragment slide? Assume that both fragments have the same coefficient of kinetic friction.
42. A 1500 kg weather rocket accelerates upward at 10 m/s^2. It explodes 2.0 s after liftoff and breaks into two fragments, one twice as massive as the other. Photos reveal that the lighter fragment traveled straight up and reached a maximum height of 530 m. What were the speed and direction of the heavier fragment just after the explosion?
43. In a ballistics test, a 25 g bullet traveling horizontally at 1200 m/s goes through a 30-cm-thick 350 kg stationary target and emerges with a speed of 900 m/s. The target is free to slide on a smooth horizontal surface.
    a. How long is the bullet in the target? What average force does it exert on the target?
    b. What is the target's speed just after the bullet emerges?
44. Two 500 g blocks of wood are 2.0 m apart on a frictionless table. A 10 g bullet is fired at 400 m/s toward the blocks. It passes all the way through the first block, then embeds itself in the second block. The speed of the first block immediately afterward is 6.0 m/s. What is the speed of the second block after the bullet stops?
45. The skiing duo of Brian (80 kg) and Ashley (50 kg) is always a crowd pleaser. In one routine, Brian, wearing wood skis, starts at the top of a 200-m-long, 20° slope. Ashley waits for

him halfway down. As he skis past, she leaps into his arms and he carries her the rest of the way down. What is their speed at the bottom of the slope?

46. In a military test, a 575 kg unmanned spy plane is traveling north at an altitude of 2700 m and a speed of 450 m/s. It is intercepted by a 1280 kg rocket traveling east at 725 m/s. If the rocket and the spy plane become enmeshed in a tangled mess, where, relative to the point of impact, do they hit the ground?
47. The Army of the Nation of Whynot has a plan to propel small vehicles across the battlefield by shooting them, from behind, with a machine gun. They've hired you as a consultant to help with an upcoming test. A 100 kg test vehicle will roll along frictionless rails. The vehicle has a tall "sail" made of ultra-hard steel. Previous tests have shown that a 20 g bullet traveling at 400 m/s rebounds from the sail at 200 m/s. The design objective is for the cart to reach a speed of 12 m/s in 20 s. You need to tell them how many bullets to fire per second from the machine gun. A five-star general will watch the test, so your ability to get future consulting jobs will depend on the outcome.
48. A 500 kg cannon fires a 10 kg cannonball with a speed of 200 m/s relative to the muzzle. The cannon is on wheels that roll without friction. When the cannon fires, what is the speed of the cannonball relative to the earth?
49. A spaceship of mass 2.0×10^6 kg is cruising at a speed of 5.0×10^6 m/s when the antimatter reactor fails, blowing the ship into three pieces. One section, having a mass of 5.0×10^5 kg, is blown straight backward with a speed of 2.0×10^6 m/s. A second piece, with mass 8.0×10^5 kg, continues forward at 1.0×10^6 m/s. What are the direction and speed of the third piece?
50. A proton (mass 1 u) is shot at a speed of 5.0×10^7 m/s toward a gold target. The nucleus of a gold atom (mass 197 u) repels the proton and deflects it straight back toward the source with 90% of its initial speed. What is the recoil velocity of the gold nucleus?
51. A proton (mass 1 u) is shot toward an unknown target nucleus at a speed of 2.5×10^6 m/s. The proton rebounds with its speed reduced by 25% while the target nucleus acquires a speed of 3.12×10^5 m/s. What is the mass, in atomic mass units, of the target nucleus?
52. The nucleus of the polonium isotope ^{214}Po (mass 214 u) is radioactive and decays by emitting an alpha particle (a helium nucleus with mass 4 u). Laboratory experiments measure the speed of the alpha particle to be 1.92×10^7 m/s. Assuming the polonium nucleus was initially at rest, what is the recoil velocity of the nucleus that remains after the decay?
53. A neutron is an electrically neutral subatomic particle with a mass just slightly greater than that of a proton. A free neutron is radioactive and decays after a few minutes into other subatomic particles. In one experiment, a neutron at rest was observed to decay into a proton (mass 1.67×10^{-27} kg) and an electron (mass 9.11×10^{-31} kg). The proton and electron were shot out back-to-back. The proton speed was measured to be 1.0×10^5 m/s and the electron speed was 3.0×10^7 m/s. No other decay products were detected.
    a. Was momentum conserved in the decay of this neutron?

    NOTE ▶ Experiments such as this were first performed in the 1930s and seemed to indicate a failure of the law of conservation of momentum. However, physicists were reluctant to give up a conservation law that had worked so well in every other circumstance. In 1933, Wolfgang Pauli postulated that the neutron might have a *third* decay product that is virtually impossible to detect. Even so, it can carry away just enough momentum to keep the total momentum conserved. This proposed particle was named the *neutrino*, meaning "little neutral one." Neutrinos were, indeed, discovered nearly 20 years later. ◀

    b. If a neutrino was emitted in the above neutron decay, in which direction did it travel? Explain your reasoning.
    c. How much momentum did this neutrino "carry away" with it?
54. A 20 g ball of clay traveling east at 2.0 m/s collides with a 30 g ball of clay traveling 30° south of west at 1.0 m/s. What are the speed and direction of the resulting 50 g blob of clay?
55. Figure P9.55 shows a collision between three balls of clay. The three hit simultaneously and stick together. What are the speed and direction of the resulting blob of clay?

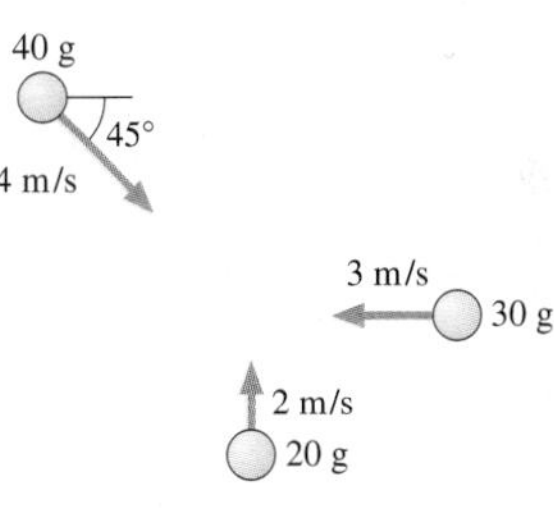

FIGURE P9.55

56. A 2100 kg truck is traveling east through an intersection at 2.0 m/s when it is hit simultaneously from the side and the rear. (Some people have all the luck!) One car is a 1200 kg compact traveling north at 5.0 m/s. The other is a 1500 kg midsize traveling east at 10 m/s. The three vehicles become entangled and slide as one body. What are their speed and direction just after the collision?
57. The carbon isotope ^{14}C is used for carbon dating of archeological artifacts. ^{14}C (mass 2.34×10^{-26} kg) decays by the process known as *beta decay* in which the nucleus emits an electron (the beta particle) and a subatomic particle called a neutrino. In one such decay, the electron and the neutrino are emitted at right angles to each other. The electron (mass 9.11×10^{-31} kg) has a speed of 5.0×10^7 m/s and the neutrino has a momentum of 8.0×10^{-24} kg m/s. What is the recoil speed of the nucleus?
58. A block of dry ice (solid carbon dioxide) tied to a 50-cm-long string swings in a circle with an initial speed of 2.0 m/s. As the CO_2 sublimates (that is, goes from a solid directly to a gas without melting), the gas vapor provides a "cushion" on which the block slides without friction. What is the speed of the block after half its mass has sublimated?
59. A 1.0-m-long massless rod is pivoted at one end and swings around in a circle on a frictionless table. A block with a hole through the center can slide in and out along the rod. Initially, a small piece of wax holds the block 30 cm from the pivot. The block is spun at 50 rpm, then the temperature of the rod is slowly increased. When the wax melts, the block slides out to the end of the rod. What is the final angular velocity? Give your answer in rpm.

60. Two people are standing on a very light board that is balanced on a fulcrum. The lighter person suddenly jumps straight up at 1.5 m/s. Just after he jumps, how fast, and in which direction, will the heavier person be moving?

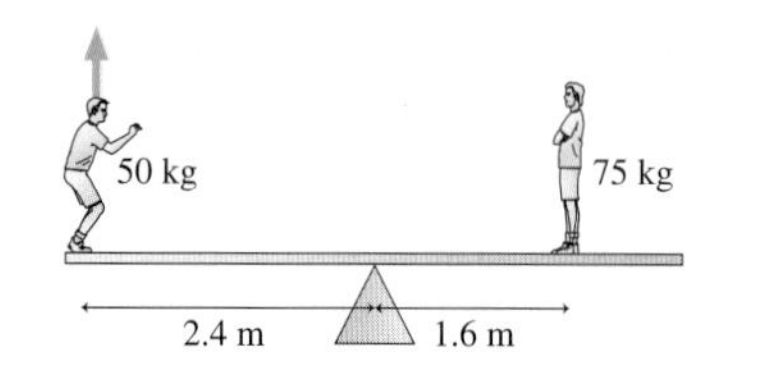

FIGURE P9.60

61. Figure P9.61 shows a 100 g puck revolving in a 20-cm-radius circle on a frictionless table. The string passes through a hole in the center of the table and is tied to two 200 g weights.
    a. What speed does the puck need to support the two weights?
    b. Suppose a flame burns through the string and causes the lower weight to fall off while the puck is revolving. What will be the puck's speed and the radius of its trajectory after the weight drops?

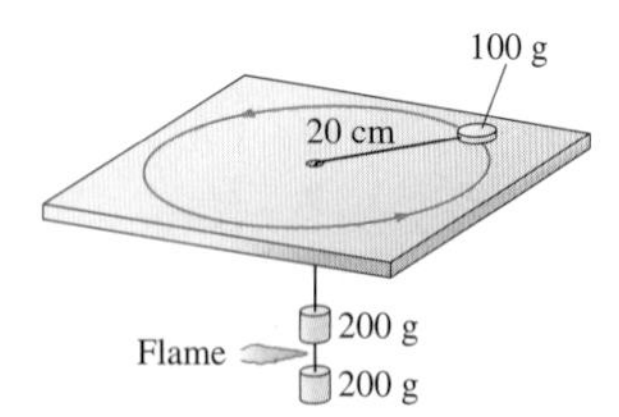

FIGURE P9.61

In Problems 62 through 65 you are given the equation used to solve a problem. For each of these, you are to
   a. Write a realistic problem for which this is the correct equation.
   b. Draw the before-and-after pictorial representation.
   c. Finish the solution of the problem.

62. $(0.10\text{ kg})(40\text{ m/s}) - (0.10\text{ kg})(-30\text{ m/s}) = \frac{1}{2}(1400\text{ N})\Delta t$
63. $(600\text{ g})(4.0\text{ m/s}) = (400\text{ g})(3.0\text{ m/s}) + (200\text{ g})(v_{ix})_2$
64. $(3000\text{ kg})v_{fx} = (2000\text{ kg})(5.0\text{ m/s}) + (1000\text{ kg})(-4.0\text{ m/s})$
65. $(50\text{ g})(v_{fx})_1 + (100\text{ g})(7.5\text{ m/s}) = (150\text{ g})(1.0\text{ m/s})$

Challenge Problems

66. A 75 kg shell is fired with an initial speed of 125 m/s at an angle 55° above horizontal. Air resistance is negligible. At its highest point, the shell explodes into two fragments, one four times more massive than the other. The heavier fragment lands directly below the point of the explosion. If the explosion exerts forces only in the horizontal direction, how far from the launch point does the lighter fragment land?
67. A 1000 kg cart is rolling to the right at 5.0 m/s. A 70 kg man is standing on the right end of the cart. What is the speed of the cart if the man suddenly starts running to the left with a speed of 10 m/s relative to the cart?
68. Ann (mass 50 kg) is standing at the left end of a 15-m-long, 500 kg cart that has frictionless wheels and rolls on a frictionless track. Initially both Ann and the cart are at rest. Suddenly, Ann starts running along the cart at a speed of 5.0 m/s relative to the cart. How far will Ann have run *relative to the ground* when she reaches the right end of the cart?
69. A 20 kg wood ball hangs from a 2.0-m-long wire. The maximum tension the wire can withstand without breaking is 400 N. A 1.0 kg projectile traveling horizontally hits and embeds itself in the wood ball. What is the largest speed this projectile can have without causing the cable to break?
70. A two-stage rocket is traveling at 1200 m/s with respect to the earth when the first stage runs out of fuel. Explosive bolts release the first stage and push it backward with a speed of 35 m/s relative to the second stage. The first stage is three times as massive as the second stage. What is the speed of the second stage after the separation?
71. You are the ground-control commander of a 2000 kg scientific rocket that is approaching Mars at a speed of 25,000 km/hr. It needs to quickly slow to 15,000 km/hr to begin a controlled descent to the surface. If the rocket enters the Martian atmosphere too fast it will burn up, and if it enters too slowly, it will use up its maneuvering fuel before reaching the surface and will crash. The rocket has a new braking system: Several 5.0 kg "bullets" on the front of the rocket can be fired straight ahead. Each has a high-explosive charge that fires it at a speed of 139,000 m/s relative to the rocket. You need to send the rocket an instruction to tell it how many bullets to fire. Success will bring you fame and glory, but failure of this $500,000,000 mission will ruin your career.
72. You are a world-famous physicist-lawyer defending a client who has been charged with murder. It is alleged that your client, Mr. Smith, shot the victim, Mr. Wesson. The detective who investigated the scene of the crime found a second bullet, from a shot that missed Mr. Wesson, that had embedded itself into a chair. You arise to cross-examine the detective.

    You: In what type of chair did you find the bullet?
    Det: A wooden chair.
    You: How massive was this chair?
    Det: It had a mass of 20 kg.
    You: How did the chair respond to being struck with a bullet?
    Det: It slid across the floor.
    You: How far?
    Det: Three centimeters. The slide marks on the dusty floor are quite distinct.
    You: What kind of floor was it?
    Det: A wood floor, very nice oak planks.
    You: What was the mass of the bullet you retrieved from the chair?
    Det: Its mass was 10 g.
    You: And how far had it penetrated into the chair?
    Det: A distance of 4 cm.
    You: Have you tested the gun you found in Mr. Smith's possession?
    Det: I have.
    You: What is the muzzle velocity of bullets fired from that gun?
    Det: The muzzle velocity is 450 m/s.
    You: And the barrel length?
    Det: The gun has a barrel length of 62 cm.

    With only a slight hesitation, you turn confidently to the jury and proclaim, "My client's gun did not fire these shots!" How are you going to convince the jury and the judge?

STOP TO THINK ANSWERS

Stop to Think 9.1: f. The cart is initially moving in the negative x-direction, so $p_{ix} = -20$ kg m/s. After it bounces, $p_{fx} = 10$ kg m/s. Thus $\Delta p = (10 \text{ kg m/s}) - (-20 \text{ kg m/s}) = 30$ kg m/s.

Stop to Think 9.2: b. The clay ball goes from $v_{ix} = v$ to $v_{fx} = 0$, so $J_{clay} = \Delta p_x = -mv$. The rubber ball rebounds, going from $v_{ix} = v$ to $v_{fx} = -v$ (same speed, opposite direction). Thus $J_{rubber} = \Delta p_x = -2mv$. The rubber ball has a larger momentum change, and this requires a larger impulse.

Stop to Think 9.3: Less than. The ball's momentum $m_B v_B$ is the same in both cases. Momentum is conserved, so the *total* momentum is the same after both collisions. The ball that rebounds from C has *negative* momentum, so C must have a larger momentum than A.

Stop to Think 9.4: Right end. The pieces started at rest, so the total momentum of the system is zero. It's an isolated system, so the total momentum after the explosion is still zero. The 6 g piece has momentum $6v$. The 4 g piece, with velocity $-2v$, has momentum $-8v$. The combined momentum of these two pieces is $-2v$. In order for P to be zero, the third piece must have a *positive* momentum $(+2v)$ and thus a positive velocity.

Stop to Think 9.5: c. Momentum conservation requires $(m_1 + m_2) \times v_f = m_1 v_1 + m_2 v_2$. Because $v_1 > v_2$, it must be that $(m_1 + m_2) \times v_f = m_1 v_1 + m_2 v_2 > m_1 v_2 + m_2 v_2 = (m_1 + m_2)v_2$. Thus $v_f > v_2$. Similarly, $v_2 < v_1$ so $(m_1 + m_2)v_f = m_1 v_1 + m_2 v_2 < m_1 v_1 + m_2 v_1 = (m_1 + m_2)v_1$. Thus $v_f < v_1$. The collision causes m_1 to slow down and m_2 to speed up.

Stop to Think 9.6: Increase. There are no tangential forces as the puck sublimates, so its angular momentum $L = mrv_t$ is conserved. The tangential velocity has to increase as m decreases to keep L constant.

10 Energy

A pole vaulter can lift himself nearly 6 m (20 ft) off the ground by transforming the kinetic energy of his run into gravitational potential energy.

▶ Looking Ahead

The goal of Chapter 10 is to introduce the ideas of kinetic and potential energy and to learn a new problem-solving strategy based on conservation of energy. In this chapter you will learn to:

- Understand and use the concepts of kinetic and potential energy.
- Use energy bar graphs.
- Use and interpret energy diagrams.
- Solve problems using the law of conservation of mechanical energy.
- Apply these ideas to elastic collisions.

◀ Looking Back

Our introduction to energy will be based on free fall. We will also use the before-and-after pictorial representation developed for impulse and momentum problems. Please review

- Section 2.6 Free-fall kinematics.
- Sections 9.2–9.3 Before-and-after pictorial representations and conservation of momentum.

Energy. It's a word you hear all the time. We use chemical energy to heat our homes and bodies, electrical energy to power our lights and computers, and solar energy to grow our crops and forests. We're told to use energy wisely and not to waste it. Athletes and weary students consume "energy bars" and "energy drinks" to gain quick energy.

But just what is energy? The concept of energy has grown and changed with time, and it is not easy to define in a general way just what energy is. Rather than starting with a formal definition, we're going to let the concept of energy expand slowly over the course of several chapters. The purpose of this chapter is to introduce the two most fundamental forms of energy, kinetic energy and potential energy. Our goal is to understand the characteristics of energy, how energy is used, and, especially important, how energy is transformed from one form to another. For example, this pole vaulter, after years of training, has become extraordinarily proficient at transforming kinetic energy into gravitational potential energy.

Ultimately we will discover a very powerful conservation law for energy. Some scientists consider the law of conservation of energy to be the most important of all the laws of nature. But all that in due time; first we have to start with the basic ideas.

10.1 A "Natural Money" Called Energy

We will start by discussing what seems to be a completely unrelated topic: money. As you will discover, monetary systems have much in common with energy. Let's begin with a short story.

The Parable of the Lost Penny

John was a hard worker. His only source of income was the paycheck he received each month. Even though most of each paycheck had to be spent on basic necessities, John managed to keep a respectable balance in his checking account. He even saved enough to occasionally buy a few stocks and bonds, his investment in the future.

John never cared much for pennies, so he kept a jar by the door and dropped all his pennies into it at the end of each day. Eventually, he reasoned, his saved pennies would be worth taking to the bank and converting into crisp new dollar bills.

John found it fascinating to keep track of these various forms of money. He noticed, to his dismay, that the amount of money in his checking account did not spontaneously increase overnight. Furthermore, there seemed to be a definite correlation between the size of his paycheck and the amount of money he had in the bank. So John decided to embark on a systematic study of money.

He began, as would any good scientist, by using his initial observations to formulate a hypothesis. John called this hypothesis a *model* of the monetary system. He found that he could represent his monetary model with the flowchart shown in Figure 10.1.

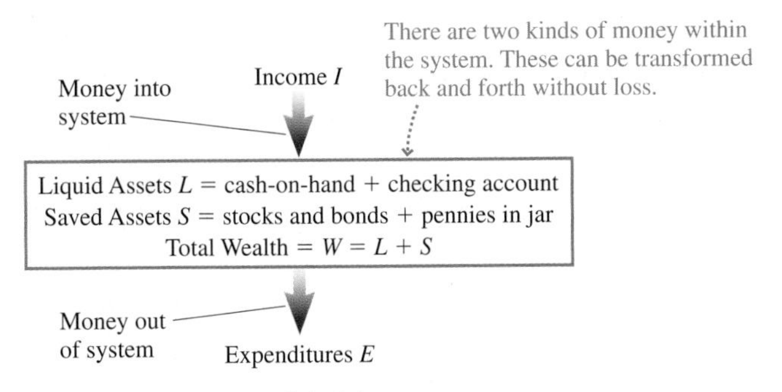

FIGURE 10.1 John's model of the monetary system.

As the chart shows, John divided his money into two basic types, liquid assets and saved assets. The *liquid assets L*, which included his checking account and the cash in his pockets, were moneys available for immediate use. His *saved assets S*, which included his stocks and bonds as well as the jar of pennies, had the *potential* to be converted into liquid assets, but they were not available for immediate use. John decided to call the sum total of assets his *wealth*: $W = L + S$.

John's assets were, more or less, simply definitions. The more interesting question, he thought, was how his wealth depended on his *income I* and *expenditures E*. These represented money transferred *to* him by his employer and money transferred *by* him to stores and bill collectors. After painstakingly collecting and analyzing his data, John finally determined that the relationship between monetary transfers and wealth is

$$\Delta W = I - E$$

John interpreted this equation to mean that the *change* in his wealth, ΔW, was numerically equal to the *net* monetary transfer $I - E$.

During a week-long period when John stayed home sick, isolated from the rest of the world, he had neither income nor expenses. In grand confirmation of his hypothesis, he found that his wealth W_f at the end of the week was identical to his wealth W_i at the week's beginning. That is, $W_f = W_i$. This occurred despite the fact that he had moved pennies from his pocket to the jar and also, by telephone, had sold some stocks and transferred the money to his checking account. In other words, John found that he could make all of the *internal* conversions of assets from one form to another that he wanted, but his total wealth remained constant (W = constant) as long as he was isolated from the world. This seemed such a remarkable rule that John named it the *law of conservation of wealth.*

One day, however, John added up his income and expenditures for the week, and the changes in his various assets, and he was 1¢ off! Inexplicably, some money seemed to have vanished. He was devastated. All those years of careful research, and now it seemed that his monetary hypothesis might not be true. Under some circumstances, yet to be discovered, it looked like $\Delta W \neq I - E$. Off by a measly penny. A wasted scientific life. . . .

But wait! In a flash of inspiration, John realized that perhaps there were other types of assets, yet to be discovered, and that his monetary hypothesis would still be valid if *all* assets were included. Weeks went by as John, in frantic activity, searched fruitlessly for previously *hidden* assets. Then one day, as John lifted the cushion off the sofa to vacuum out the potato chip crumbs—lo and behold, there it was!—the missing penny!

John raced to complete his theory, now including the sofa (as well as the washing machine) as a previously unknown form of saved assets that needed to be included in S. Other researchers soon discovered other types of assets, such as the remarkable find of the "cash in the mattress." To this day, when *all* known assets are included, monetary scientists have never found a violation of John's simple hypothesis that $\Delta W = I - E$. John was last seen sailing for Stockholm to collect the Nobel Prize for his Theory of Wealth.

This photovoltaic panel is transforming solar energy into electrical energy.

Energy

John, despite his diligent efforts, did not discover a law of nature. The monetary system is a human construction that, by design, obeys John's "laws." Monetary system laws, such that you cannot print money in your basement, are enforced by society, not by nature. But suppose that physical objects possessed a "natural money" that was governed by a theory, or model, similar to John's. An object might have several different forms of natural money that could be converted back and forth, but the total amount of an object's natural money would *change* only if natural money were *transferred* to or from the object. Two key words here, as in John's model, are *transfer* and *change.*

One of the greatest and most significant discoveries of science is that there is such a "natural money" called **energy.** You have heard of some of the many forms of energy, such as solar energy or nuclear energy, but others may be new to you. These forms of energy can differ as much as a checking account differs from loose change in the sofa. Much of our study is going to be focused on the *transformation* of energy from one form to another. Much of modern technology is concerned with transforming energy, such as changing the chemical energy of oil molecules to electrical energy or to the kinetic energy of your car.

As we use energy concepts, we will be "accounting" for energy that is transferred in or out of a system or that is transformed from one form to another within a system. Figure 10.2 shows a simple model of energy that is based on John's model of the monetary system. There are many details that must be added to this model, but it's a good starting point. The fact that nature "balances the books" for energy is one of the most profound discoveries of science.

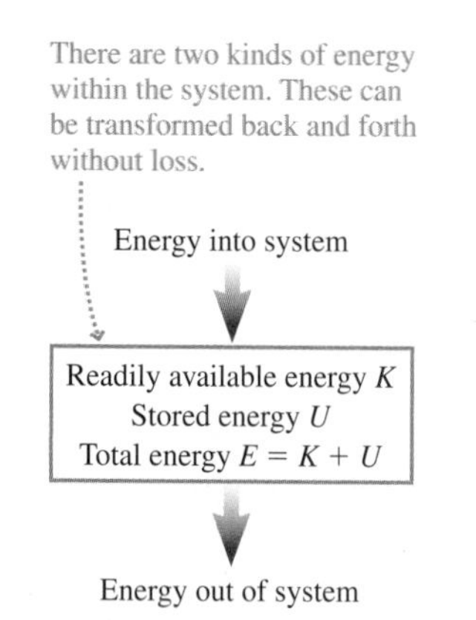

FIGURE 10.2 An initial model of energy.

A major goal is to discover the conditions under which energy is conserved. Surprisingly, the *law of conservation of energy* was not recognized until the mid-nineteenth century, long after Newton. The reason, similar to John's lost penny, was that it took scientists a long time to realize how many types of energy there are and the various ways that energy can be converted from one form to another. As you'll soon learn, energy ideas go well beyond Newtonian mechanics to include new concepts about heat, about chemical energy, and about the energy of the individual atoms and molecules that comprise a system. All of these forms of energy will ultimately have to be included in our accounting scheme for energy.

There's a lot to say about energy, and energy is an abstract idea, so we'll take it one step at a time. Much of the "theory" will be postponed until Chapter 11, after you've had some practice using the basic concepts of energy introduced in this chapter. We will extend these ideas in Part IV when we reach the study of thermodynamics.

10.2 Kinetic Energy and Gravitational Potential Energy

To begin, consider an object in vertical free fall. It can only move up or down, and the only force acting on it is the gravitational weight force $\vec{w}$. The object's position and velocity are given by the free-fall kinematics of Chapter 2, with $a_y = -g$.

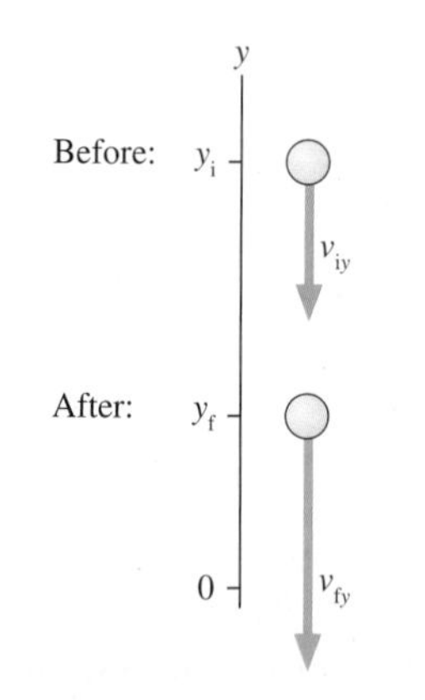

FIGURE 10.3 The before-and-after representation of an object in free fall.

Figure 10.3 is a before-and-after pictorial representation of an object in free fall, as you learned to draw in Chapter 9. We didn't call attention to it in Chapter 2, but one of the free-fall equations also relates "before" and "after." In particular, the kinematic equation

$$v_{fy}^2 = v_{iy}^2 + 2a_y\Delta y = v_{iy}^2 - 2g(y_f - y_i) \tag{10.1}$$

can easily be rewritten as

$$v_{fy}^2 + 2gy_f = v_{iy}^2 + 2gy_i \tag{10.2}$$

Equation 10.2 is a conservation law for free-fall motion. It tells us that the quantity $v_y^2 + 2gy$ has the same value *after* free fall (regardless of whether the motion is upward or downward) that it had *before* free fall. But free fall is a very specific type of motion, so it's not clear if this "law" has any wider validity. Let's introduce a more general technique to arrive at the same result, but a technique that can be extended to other types of motion.

Newton's second law for one-dimensional motion along the y-axis is

$$(F_{\text{net}})_y = ma_y = m\frac{dv_y}{dt} \tag{10.3}$$

The net force on an object in free fall is $(F_{\text{net}})_y = -mg$, so Equation 10.3 is

$$m\frac{dv_y}{dt} = -mg \tag{10.4}$$

Recall, from calculus, that we can use the chain rule to write

$$\frac{dv_y}{dt} = \frac{dv_y}{dy}\frac{dy}{dt} = v_y\frac{dv_y}{dy} \tag{10.5}$$

where we used $v_y = dy/dt$. Substituting this into Equation 10.4 gives

$$mv_y\frac{dv_y}{dy} = -mg \tag{10.6}$$

The chain rule has allowed us to change from a derivative of v_y with respect to time to a derivative of v_y with respect to position. This simple change will be the key step on the road to energy.

Rewrite Equation 10.6 as

$$mv_y\,dv_y = -mg\,dy \tag{10.7}$$

Now we can integrate both sides of the equation. However, we have to be careful to make sure the limits of integration match. We want to integrate from "before," when the object is at position y_i and has velocity v_{iy}, to "after," when the object is at position y_f and has velocity v_{fy}. Figure 10.3 shows these points in the motion. With these limits, the integrals are

$$\int_{v_{iy}}^{v_{fy}} mv_y\,dv_y = -\int_{y_i}^{y_f} mg\,dy \tag{10.8}$$

Carrying out the integrations, with m and g as constants, we find

$$\frac{1}{2}mv_y^2\bigg|_{v_{iy}}^{v_{fy}} = \frac{1}{2}mv_{fy}^2 - \frac{1}{2}mv_{iy}^2 = -mgy\bigg|_{y_i}^{y_f} = -mgy_f + mgy_i \tag{10.9}$$

The quantity v_y is the y-component of the vector $\vec{v}$. The sign of v_y indicates the direction of motion. But because v_y is squared wherever it appears in Equation 10.9, the sign is not relevant. All we need to know are the initial and final *speeds* v_i and v_f. With this, Equation 10.9 can be written

$$\frac{1}{2}mv_f^2 + mgy_f = \frac{1}{2}mv_i^2 + mgy_i \tag{10.10}$$

You should recognize that Equation 10.10, other than a constant factor of $\frac{1}{2}m$, is the same as Equation 10.2. This seems like a lot of effort to get to a result we already knew. However, our purpose was not to get the answer but to introduce a *procedure* that will turn out to have other valuable applications.

Kinetic and Potential Energy

6.1 ActivPhysics ONLINE

The quantity

$$K = \frac{1}{2}mv^2 \quad \text{(kinetic energy)} \tag{10.11}$$

is called the **kinetic energy** of the object. The quantity

$$U_g = mgy \quad \text{(gravitational potential energy)} \tag{10.12}$$

is the object's **gravitational potential energy.** These are the two basic forms of energy. **Kinetic energy is an energy of motion.** It depends on the object's speed but not its location. **Potential energy is an energy of position.** It depends on the object's position but not its speed.

NOTE ▶ Gravitational potential energy is called simply *potential energy* if there's no ambiguity. ◀

One of the most important characteristics of energy is that it is a scalar, not a vector. Kinetic energy depends on an object's speed v but *not* on the direction of motion. The kinetic energy of a particle will be the same regardless of whether it is moving up or down or left or right. Consequently, the mathematics of using energy is often much easier than the vector mathematics required by force and acceleration.

NOTE ▶ By its definition, kinetic energy can never be a negative number. If you find, in the course of solving a problem, that K is negative—stop! You have made an error somewhere. Don't just "lose" the minus sign and hope that everything turns out OK. ◀

The unit of kinetic energy is that of mass multiplied by velocity squared. In the SI system of units, this is kg m²/s². The unit of energy is so important that is has been given its own name, the **joule.** We define:

$$1 \text{ joule} = 1 \text{ J} \equiv 1 \text{ kg m}^2/\text{s}^2$$

The unit of potential energy, kg × m/s² × m = kg m²/s², is also the joule.

To give you an idea about the size of a joule, consider a 0.5 kg mass (weight on earth ≈ 1 lb) moving at 4 m/s (≈10 mph). Its kinetic energy is

$$K = \frac{1}{2}mv^2 = \frac{1}{2}(0.5 \text{ kg})(4 \text{ m/s})^2 = 4 \text{ J}$$

Its gravitational potential energy at a height of 1 m (≈3 ft) is

$$U_g = mgy = (0.5 \text{ kg})(9.8 \text{ m/s}^2)(1 \text{ m}) \approx 5 \text{ J}$$

This suggests that ordinary-sized objects moving at ordinary speeds will have energies of a fraction of a joule up to, perhaps, a few thousand joules (a running person has $K \approx 1000$ J). A high-speed truck might have $K \approx 10^6$ J.

NOTE ▶ You *must* have masses in kg and velocities in m/s before doing energy calculations. ◀

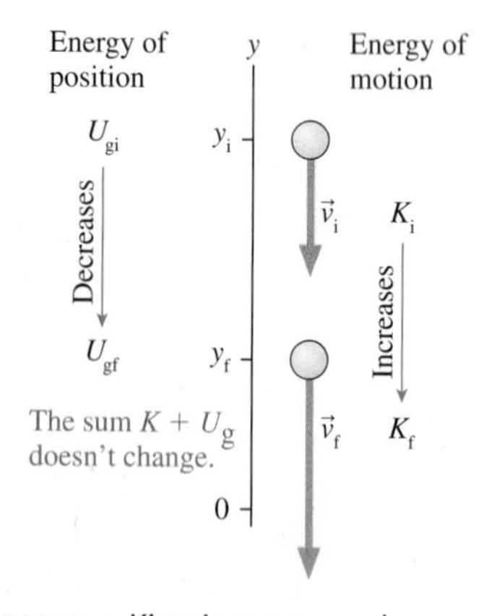

FIGURE 10.4 Kinetic energy and gravitational potential energy.

In terms of energy, Equation 10.10 says that for an object in free fall,

$$K_f + U_{gf} = K_i + U_{gi} \tag{10.13}$$

In other words, the sum $K + U_g$ of kinetic energy and gravitational potential energy is not changed by free fall. Its value *after* free fall (regardless of whether the motion is upward or downward) is the same as *before* free fall. Figure 10.4 illustrates this important idea.

EXAMPLE 10.1 Launching a pebble

Bob uses a slingshot to shoot a 20 g pebble straight up with a speed of 25 m/s. How high does the pebble go?

MODEL This is free-fall motion, so the sum of the kinetic and gravitational potential energy does not change as the pebble rises.

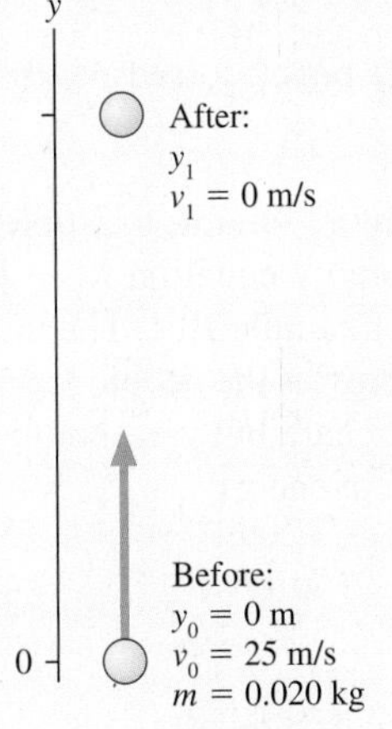

FIGURE 10.5 The before-and-after pictorial representation of a pebble shot upward from a slingshot.

VISUALIZE Figure 10.5 shows a before-and-after pictorial representation. The pictorial representation for energy problems is essentially the same as the pictorial representation you learned in Chapter 9 for momentum problems. We'll use numerical subscripts 0 and 1 for the initial and final points.

SOLVE Equation 10.13,

$$K_1 + U_{g1} = K_0 + U_{g0}$$

tells us that the sum $K + U_g$ is not changed by the motion. Using the definitions of K and U_g gives us this statement about energy:

$$\frac{1}{2}mv_1^2 + mgy_1 = \frac{1}{2}mv_0^2 + mgy_0$$

Here $y_0 = 0$ m and $v_1 = 0$ m/s, so the energy equation simplifies to

$$mgy_1 = \frac{1}{2}mv_0^2$$

This is easily solved for the height y_1:

$$y_1 = \frac{v_0^2}{2g} = \frac{(25\text{ m/s})^2}{2(9.8\text{ m/s}^2)} = 31.9\text{ m}$$

ASSESS Notice that the mass canceled and wasn't needed, a fact about free fall that you should remember from Chapter 2.

STOP TO THINK 10.1 Rank in order, from largest to smallest, the gravitational potential energies of balls 1 to 4.

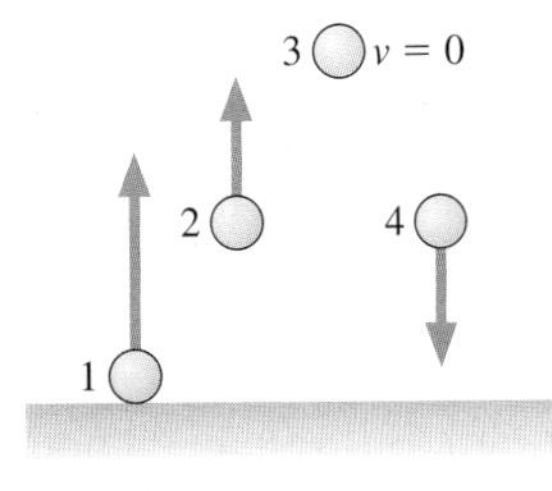

Energy Bar Charts

The pebble of Example 10.1 starts out with all kinetic energy, an energy of motion. As the pebble ascends, kinetic energy is converted into gravitational potential energy, *but the sum of the two doesn't change.* At the top, the pebble's energy is entirely potential energy. The simple bar chart in Figure 10.6 on the next page shows graphically how kinetic energy is transformed into gravitational potential energy as a pebble rises. The potential energy is then transformed back into kinetic energy as the pebble falls. The sum $K + U_g$ remains constant throughout the motion.

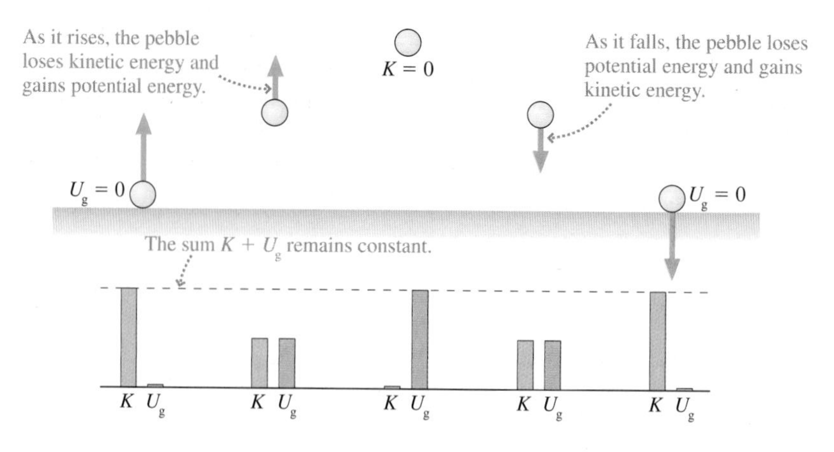

FIGURE 10.6 Simple energy bar charts for a pebble tossed into the air.

Figure 10.7a is an energy bar chart more suitable to problem solving. The chart is a graphical representation of the energy equation $K_f + U_{gf} = K_i + U_{gi}$. Figure 10.7b applies this to the pebble of Example 10.1. The initial kinetic energy is transformed entirely into potential energy as the pebble reaches its highest point. There are no numerical scales on a bar chart, but you should draw the bar heights proportional to the amount of each type of energy.

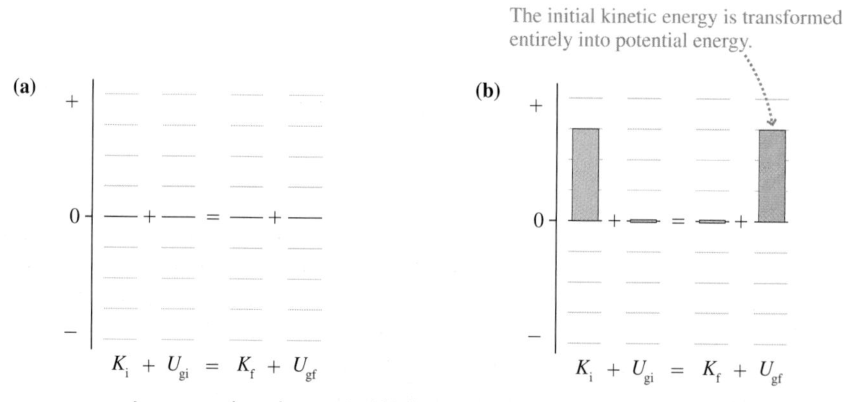

FIGURE 10.7 An energy bar chart suitable for problem solving.

The Zero of Potential Energy

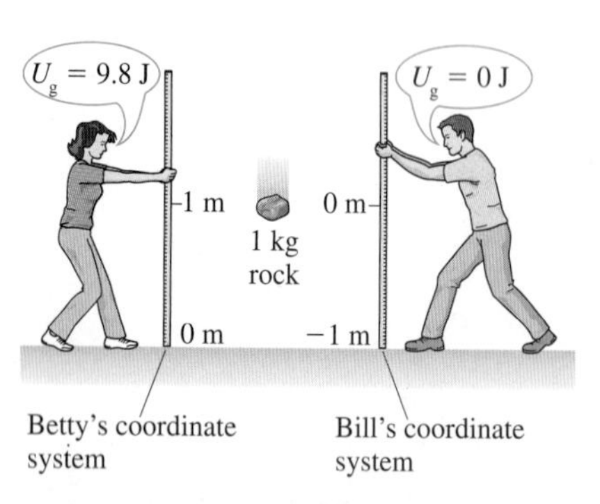

FIGURE 10.8 Betty and Bill use coordinate systems with different origins to determine the potential energy of a rock.

Our expression for the gravitational potential energy $U_g = mgy$ seems straightforward. But you might notice, on further reflection, that the value of U_g depends on where you choose to put the origin of your coordinate system. Consider Figure 10.8, where Betty and Bill are attempting to determine the potential energy of a 1 kg rock that is 1 m above the ground. Betty chooses to put the origin of her coordinate system on the ground, measures $y_{rock} = 1$ m, and quickly computes $U_g = mgy = 9.8$ J. Bill, on the other hand, read Chapter 1 very carefully and recalls that it is entirely up to him where to locate the origin of his coordinate system. So he places his origin next to the rock, measures $y_{rock} = 0$ m, and declares that $U_g = mgy = 0$ J!

How can the potential energy of one rock at one position in space have two different values? The source of this apparent difficulty comes from our interpretation of Equation 10.9. The integral of $-mgdy$ resulted in the expression $-mg(y_f - y_i)$, and this led us to propose that $U_g = mgy$. But all we are *really* justified in concluding is that the potential energy *changes* by $\Delta U = -mg(y_f - y_i)$. To go beyond this and claim $U_g = mgy$ is consistent with $\Delta U = -mg(y_f - y_i)$, but so also would be a claim that $U_g = mgy + C$ where C is any constant.

No matter where the rock is located, Betty's value of y will always equal Bill's value plus 1 m. Consequently, her value of the potential energy will always equal Bill's value plus 9.8 J. That is, their values of U_g differ by a constant. Nonetheless, both will calculate exactly the *same* value for ΔU if the rock changes position.

EXAMPLE 10.2 The speed of a falling rock

The rock shown in Figure 10.8 is released from rest. Use both Betty's and Bill's perspectives to calculate its speed just before it hits the ground.

MODEL This is free-fall motion, so the sum of the kinetic and gravitational potential energy does not change as the rock falls.

VISUALIZE Figure 10.9 shows a before-and-after pictorial representation using both Betty's and Bill's coordinate systems.

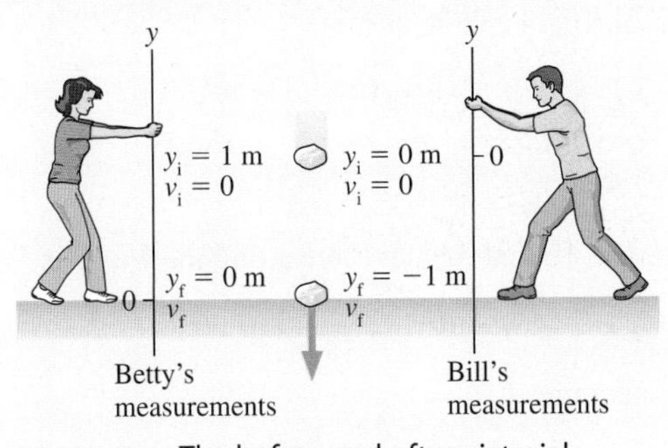

FIGURE 10.9 The before-and-after pictorial representation of a falling rock.

SOLVE The energy equation is $K_f + U_{gf} = K_i + U_{gi}$. Bill and Betty both agree that $K_i = 0$, because the rock was released from rest, so we have

$$K_f = \frac{1}{2}mv_f^2 = -(U_{gf} - U_{gi}) = -\Delta U$$

According to Betty, $U_{gi} = mgy_i = 9.8$ J and $U_{gf} = mgy_f = 0$ J. Thus

$$\Delta U_{Betty} = U_{gf} - U_{gi} = -9.8 \text{ J}$$

The rock *loses* potential energy as it falls. According to Bill, $U_{gi} = mgy_i = 0$ J and $U_{gf} = mgy_f = -9.8$ J. Thus

$$\Delta U_{Bill} = U_{gf} - U_{gi} = -9.8 \text{ J}$$

Bill has different values for U_{gi} and U_{gf} but the *same* value for ΔU. Thus they both agree that the rock hits the ground with speed

$$v_f = \sqrt{\frac{-2\Delta U}{m}} = \sqrt{\frac{-2(-9.8 \text{ J})}{1.0 \text{ kg}}} = 4.43 \text{ m/s}$$

Figure 10.10 shows energy bar charts for Betty and Bill. Despite their disagreement over the value of U_g, Betty and Bill arrive at the same value for v_f and their K_f bars are the same height. The reason is that only ΔU has physical significance, not U_g itself, and Betty and Bill found the same value for ΔU. You can place the origin of your coordinate system, and thus the "zero of potential energy," wherever you choose and be assured of getting the correct answer to a problem.

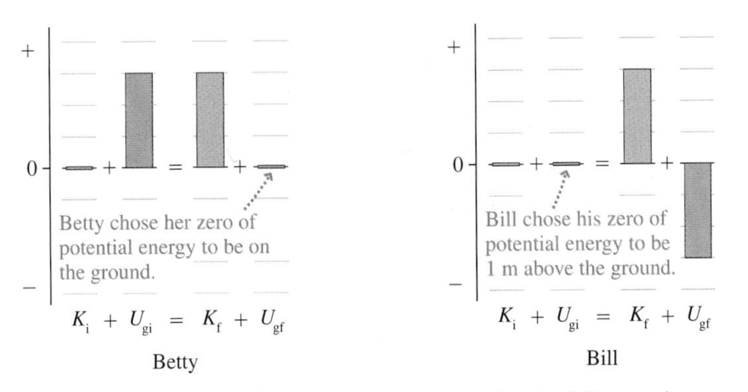

FIGURE 10.10 Betty's and Bill's energy bar charts for the falling rock.

NOTE ► Gravitational potential energy can be negative, as U_{gf} is for Bill. **A negative value for U_g means that the particle has *less* potential for motion at that point than it does at $y = 0$.** But there's nothing wrong with that. Contrast this with kinetic energy, which *cannot* be negative. ◄

10.3 A Closer Look at Gravitational Potential Energy

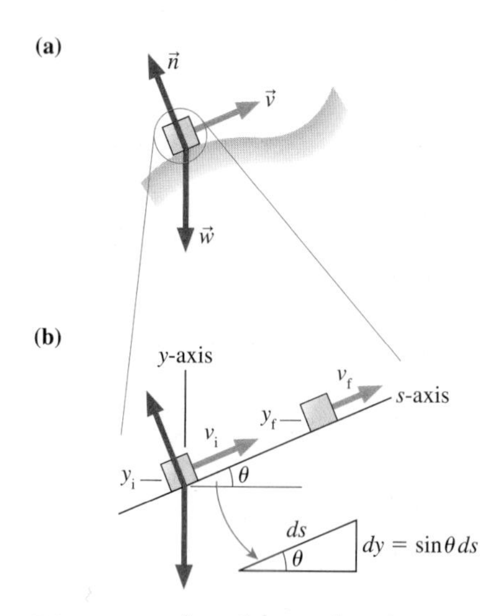

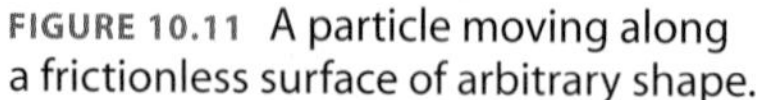
FIGURE 10.11 A particle moving along a frictionless surface of arbitrary shape.

The concept of energy would be of little interest or use if it applied only to free fall. Let's begin to expand the idea. Figure 10.11a shows an object of mass m sliding along a frictionless surface. The only forces acting on the object are its weight $\vec{w}$ and the normal force $\vec{n}$ from the surface. If the surface is curved, you know from calculus that we can subdivide the surface into many small (perhaps infinitesimal) straight-line segments. Figure 10.11b shows a magnified segment of the surface that, over some small distance, is a straight line at angle θ.

We can analyze the motion along this small segment using the procedure of Equations 10.3 through 10.10. Define an s-axis parallel to the direction of motion. Newton's second law along this axis is

$$(F_{\text{net}})_s = ma_s = m\frac{dv_s}{dt} \tag{10.14}$$

Using the chain rule, Equation 10.14 can be written

$$(F_{\text{net}})_s = m\frac{dv_s}{dt} = m\frac{dv_s}{ds}\frac{ds}{dt} = mv_s\frac{dv_s}{ds} \tag{10.15}$$

where, in the last step, we used $ds/dt = v_s$.

You can see from Figure 10.11b that the net force along the s-axis is

$$(F_{\text{net}})_s = -w\sin\theta = -mg\sin\theta \tag{10.16}$$

Thus Newton's second law becomes

$$-mg\sin\theta = mv_s\frac{dv_s}{ds} \tag{10.17}$$

Multiplying both sides by ds gives

$$mv_s dv_s = -mg\sin\theta\, ds \tag{10.18}$$

But you can see from the figure that $\sin\theta\, ds$ is simply dy, so Equation 10.18 becomes

$$mv_s dv_s = -mg\, dy \tag{10.19}$$

The roller coaster's potential energy is converted to kinetic energy as it speeds toward the bottom of the hill.

This is *identical* to Equation 10.7, which we found for free fall. Consequently, integrating this equation from "before" to "after" leads again to Equation 10.10:

$$\frac{1}{2}mv_f^{\,2} + mgy_f = \frac{1}{2}mv_i^{\,2} + mgy_i \tag{10.20}$$

where v_i^2 and v_f^2 are the squares of the *speeds* at the beginning and end of this segment of the motion.

We previously defined the kinetic energy $K = \frac{1}{2}mv^2$ and the gravitational potential energy $U_g = mgy$. Equation 10.20 shows that

$$K_f + U_{gf} = K_i + U_{gi} \tag{10.21}$$

for a particle moving along *any* frictionless surface, regardless of the shape.

NOTE ► For energy calculations, the y-axis is specifically a *vertical* axis. Gravitational potential energy depends on the *height* above the earth's surface. A tilted coordinate system, such as we often used in dynamics problems, doesn't work for problems with gravitational potential energy. ◄

STOP TO THINK 10.2 A small child slides down the four frictionless slides A–D. Each has the same height. Rank in order, from largest to smallest, her speeds v_A to v_D at the bottom.

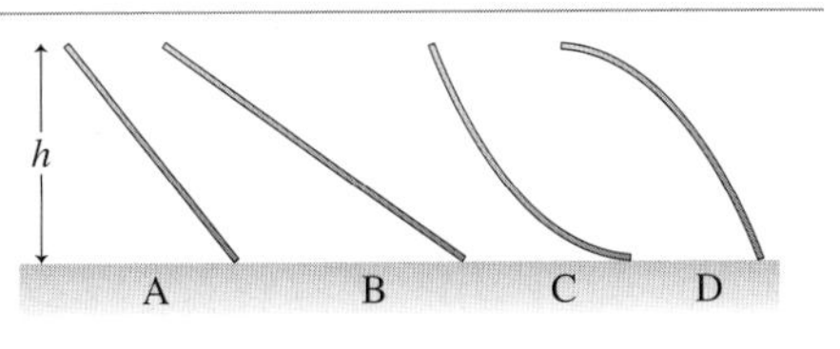

EXAMPLE 10.3 The speed of a sled

Christine runs forward with her sled at 2.0 m/s. She hops onto the sled at the top of a 5.0-m-high, very slippery slope. What is her speed at the bottom?

MODEL Model Christine and the sled as a particle. Assume the slope is frictionless. In that case, the sum of her kinetic and gravitational potential energy does not change as she slides down.

VISUALIZE Figure 10.12a shows a before-and-after pictorial representation. We are not told the angle of the slope, or even if it is a straight slope, but the *change* in potential energy depends only on the height Christine descends and *not* on the shape of the hill. Figure 10.12b shows an energy bar chart in which we see an initial kinetic *and* potential energy being transformed entirely into kinetic energy as she goes down the slope. The purpose of the bar chart is to visualize how the energy changes, and we can show that without yet knowing any numerical values of the energy.

SOLVE The quantity $K + U_g$ is the same at the bottom of the hill as it was at the top. Thus

$$\frac{1}{2}mv_1^2 + mgy_1 = \frac{1}{2}mv_0^2 + mgy_0$$

This is easily solved for Christine's speed at the bottom:

$$v_1 = \sqrt{v_0^2 + 2g(y_0 - y_1)} = \sqrt{v_0^2 + 2gh} = 10.1 \text{ m/s}$$

ASSESS We did not need to know the mass of either Christine or the sled.

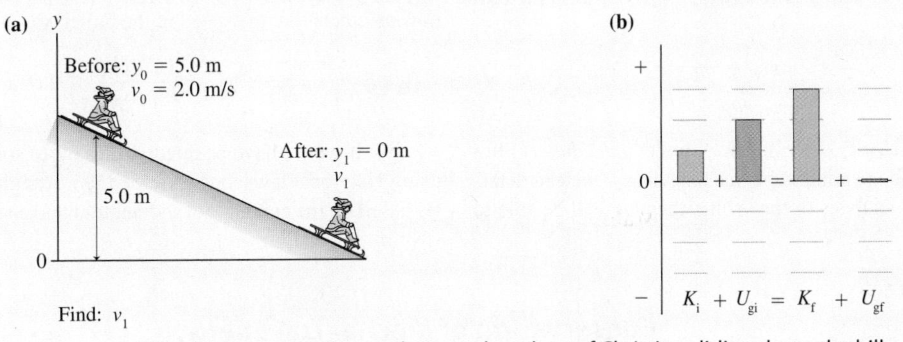

FIGURE 10.12 Pictorial representation and energy bar chart of Christine sliding down the hill.

Notice that the normal force $\vec{n}$ in Figure 10.12 didn't enter our analysis. The equation $K_f + U_{gf} = K_i + U_{gi}$ is a statement about how the particle's speed changes as it changes position. $\vec{n}$ does not have a component in the direction of motion, so it cannot change the particle's speed. Thus the normal force doesn't play a role in energy calculations.

The same is true for an object tied to a string and moving in a circle. The tension in the string causes the direction to change, but $\vec{T}$ does not have a component in the direction of motion and does not change the speed of the object. Hence Equation 10.21 also applies to a *pendulum*.

EXAMPLE 10.4 A ballistic pendulum

A 10 g bullet is fired into a 1200 g wood block that hangs from a 150-cm-long string. The bullet embeds itself into the block, and the block then swings out to an angle of 40°. What was the speed of the bullet? (This is called a *ballistic pendulum.*)

MODEL This is a two-part problem. The impact of the bullet with the block is an inelastic collision. We haven't done any analysis to let us know what happens to energy during a collision, but you learned in Chapter 9 that *momentum* is conserved in an inelastic collision. After the collision is over, the block swings out as a pendulum. The sum of the kinetic and gravitational potential energy does not change as the block swings to its largest angle.

VISUALIZE Figure 10.13 on the next page is a pictorial representation in which we've identified before-and-after quantities for both the collision and the swing.

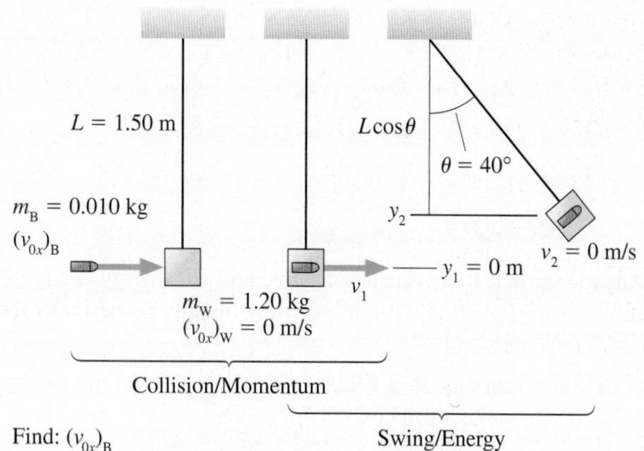

FIGURE 10.13 A ballistic pendulum is used to measure the speed of a bullet.

SOLVE The momentum conservation equation $P_f = P_i$ applied to the inelastic collision gives

$$(m_W + m_B)v_{1x} = m_W(v_{0x})_W + m_B(v_{0x})_B$$

The wood block is initially at rest, with $(v_{0x})_W = 0$, so the bullet's velocity is

$$(v_{0x})_B = \frac{m_W + m_B}{m_B} v_{1x}$$

where v_{1x} is the velocity of the block + bullet *immediately* after the collision, as the pendulum begins to swing. If we can determine v_{1x} from an analysis of the swing, then we will be able to calculate the speed of the bullet. Turning our attention to the swing, the energy equation $K_f + U_{gf} = K_i + U_{gi}$ is

$$\frac{1}{2}(m_W + m_B)v_2^2 + (m_W + m_B)gy_2$$

$$= \frac{1}{2}(m_W + m_B)v_1^2 + (m_W + m_B)gy_1$$

We used the *total* mass $(m_W + m_B)$ of the block and embedded bullet, but notice that it cancels out. We also dropped the x-subscript on v_1 because for energy calculations we need only speed, not velocity. The speed is zero at the top of the swing $(v_2 = 0)$, and we've defined the y-axis such that $y_1 = 0$ m thus

$$v_1 = \sqrt{2gy_2}$$

The initial speed is found simply from the maximum height of the swing. You can see from the geometry of Figure 10.13 that

$$y_2 = L - L\cos\theta = L(1 - \cos\theta) = 0.351 \text{ m}$$

With this, the initial velocity of the pendulum, immediately after the collision, is

$$v_{1x} = v_1 = \sqrt{2gy_2} = \sqrt{2(9.80 \text{ m/s}^2)(0.351 \text{ m})} = 2.62 \text{ m/s}$$

Having found v_{1x} from an energy analysis of the swing, we can now calculate that the speed of the bullet was

$$(v_{0x})_B = \frac{m_W + m_B}{m_B} v_{1x} = \frac{1.210 \text{ kg}}{0.010 \text{ kg}} \times 2.62 \text{ m/s} = 317 \text{ m/s}$$

ASSESS It would have been very difficult to solve this problem using Newton's laws, but it yielded to a straightforward analysis based on the concepts of momentum and energy.

Conservation of Mechanical Energy

The sum of the kinetic energy and the potential energy of a system is called the **mechanical energy**

$$E_{mech} = K + U \tag{10.22}$$

Here K is the total kinetic energy of all the particles in the system and U is the potential energy stored in the system. Our examples thus far suggest that a particle's mechanical energy does not change as it moves under the influence of gravity. The kinetic energy and the potential energy can change, as they are transformed back and forth into each other, but their sum remains constant. We can express the unchanging value of E_{mech} as

$$\Delta E_{mech} = \Delta K + \Delta U = 0 \tag{10.23}$$

This statement is called the **law of conservation of mechanical energy.**

NOTE ▶ The law of conservation of mechanical energy does *not* say $\Delta K = \Delta U$. One of the changes has to be positive while the other is negative if energy is to be conserved. ◀

But is this really a law of nature? Consider a box that is given a shove and then slides along the floor until it stops. The box loses kinetic energy as it slows down, but $\Delta U_g = 0$ because y doesn't change. Thus ΔE_{mech} is *not* zero for the box; its mechanical energy is not conserved.

In Chapter 9 you learned that momentum is conserved only for an isolated system. One of the important goals of this chapter and the next is to learn the conditions under which mechanical energy is conserved. We've seen thus far that mechanical energy *is* conserved for a particle that moves along a frictionless trajectory under the influence of gravity, but mechanical energy is *not* conserved when there is friction.

You know, of course, that after the box slides across the floor, both the box and the floor are slightly warmer than before. The kinetic energy has not been transformed into potential energy, but *something* has happened. We've already noted that there are different kinds of energy. Perhaps friction causes the kinetic energy to be transformed into a form of energy other than potential energy. This is a very important issue, one that we'll begin to explore in the next chapter.

The Basic Energy Model

We're beginning to develop what we'll call the *basic energy model.* This is a model of energy as a form of "natural money." It is based on three hypotheses:

1. There are two kinds of energy, a kinetic energy associated with the motion of a particle and a potential energy associated with the position of a particle.
2. Kinetic energy can be transformed into potential energy, and potential energy can be transformed into kinetic energy.
3. Under some circumstances the mechanical energy $E_{mech} = K + U$ is conserved. Its value at the end of a process equals its value at the beginning.

Our task, if the basic energy model is to be useful, is to answer three crucial questions:

1. Under what conditions is E_{mech} conserved?
2. What happens to the energy when E_{mech} isn't conserved?
3. How do you calculate the potential energy U for forces other than gravity?

There are many parts to the energy puzzle, and we must put them together piece by piece. This chapter is focused on how to use energy in situations were E_{mech} is conserved. We will turn our attention to answering these three important questions in the next chapter. For now, we can begin to develop a strategy for using energy to solve problems.

PROBLEM-SOLVING STRATEGY 10.1 Conservation of mechanical energy

MODEL Choose a system without friction or other losses of mechanical energy.

VISUALIZE Draw a before-and-after pictorial representation. Define symbols that will be used in the problem, list known values, and identify what you're trying to find.

SOLVE The mathematical representation is based on the law of conservation of mechanical energy:

$$K_f + U_f = K_i + U_i$$

ASSESS Check that your result has the correct units, is reasonable, and answers the question.

STOP TO THINK 10.3 A box slides along the frictionless surface shown in the figure. It is released from rest at the position shown. Is the highest point the box reaches on the other side at level a, level b, or level c?

Springs and rubber bands store energy—potential energy—that can be transformed into kinetic energy.

10.4 Restoring Forces and Hooke's Law

If you stretch a rubber band, a force appears that tries to pull the rubber band back to its equilibrium, or unstretched, length. A force that restores a system to an equilibrium position is called a **restoring force.** Systems that exhibit restoring forces are called **elastic.** The most basic examples of elasticity are things like springs and rubber bands. If you stretch a spring, a tension-like force pulls back. Similarly, a compressed spring tries to re-expand to its equilibrium length. Other examples of elasticity and restoring forces abound. The steel beams bend slightly as you drive your car over a bridge, but they are restored to equilibrium after your car passes by. Nearly everything that stretches, compresses, flexes, bends, or twists exhibits a restoring force and can be called elastic.

Elasticity has many important applications, but we have another motive for introducing elasticity at this point. One of the goals of this textbook is to understand the atomic structure of matter. We have already introduced a simple model of a solid in which particle-like atoms are held together by spring-like molecular bonds. Figure 10.14 reminds you of this model. We devoted the first nine chapters to understanding the dynamics of particles. Now it's time to look more closely at the elastic behavior of the bonds, the "glue" that holds matter together.

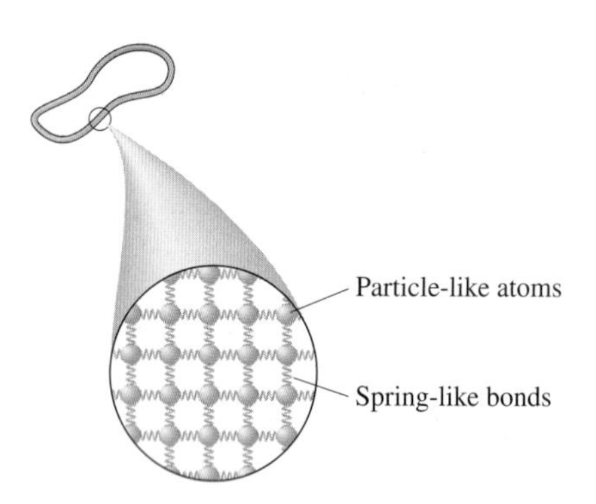

FIGURE 10.14 An elastic solid can be modeled as particle-like atoms held together by spring-like molecular bonds.

We're going to use a simple spring as a prototype of elasticity. Suppose you have a spring whose **equilibrium length** is L_0. This is the length of the spring when it is neither pushing nor pulling. If you now stretch the spring to length L, how hard does it pull back? One way to find out is to attach the spring to a bar, as shown in Figure 10.15, then to hang a mass m from the spring. The mass stretches the spring to length L. Lengths L_0 and L are easily measured with a meter stick.

The mass hangs in static equilibrium, so the upward spring force $\vec{F}_{sp}$ exactly balances the downward weight force $\vec{w}$ to give $\vec{F}_{net} = \vec{0}$. That is,

$$F_{sp} = w = mg \tag{10.24}$$

By using different masses to stretch the spring to different lengths, we can determine how F_{sp}, the magnitude of the spring's restoring force, depends on the length L.

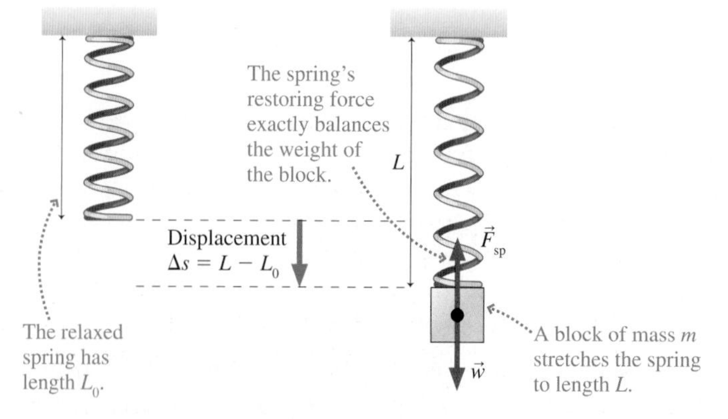

FIGURE 10.15 A spring of equilibrium length L_0 is stretched to length L by a hanging mass.

Figure 10.16 shows measured data for the restoring force of a real spring. Notice that the quantity graphed along the horizontal axis is $\Delta s = L - L_0$. This is the distance that the end of the spring has moved, which we call the **displacement from equilibrium.** The graph shows that the restoring force is proportional to the displacement. That is, the data fall along the straight line

$$F_{sp} = k\Delta s \qquad (10.25)$$

The proportionality constant k, which is the slope of the force-versus-displacement graph, is called the **spring constant.** The units of the spring constant are N/m.

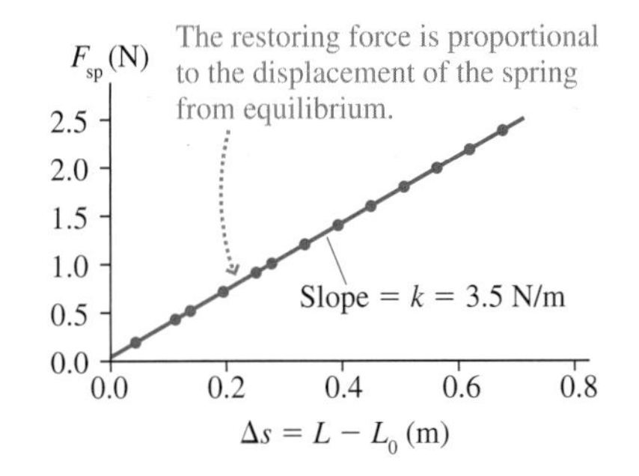

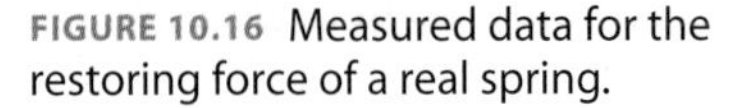
FIGURE 10.16 Measured data for the restoring force of a real spring.

NOTE ▶ The force does not depend on the spring's physical length L but, instead, on the *displacement* Δs of the end of the spring. ◀

The spring constant k is a property that characterizes a spring, just as mass m characterizes a particle. If k is large, it takes a large pull to cause a significant stretch, and we call the spring a "stiff" spring. If k is small, the spring can be stretched with very little force, and we call it a "soft" spring. Every spring has its own, unique value of k that remains constant for that spring. The spring constant for the spring in Figure 10.16 can be determined from the slope of the straight line to be $k = 3.5$ N/m.

NOTE ▶ Just as we used massless strings, we will adopt the idealization of a *massless spring.* While not a perfect description, it is a good approximation if the mass attached to a spring is much larger than the mass of the spring itself. ◀

Hooke's Law

Experiments show that Equation 10.25 is valid whether the spring is stretched or compressed. The only difference is the *direction* of the force—pulling in one case, pushing in the other. But because the spring force really is a vector, we do need to write Equation 10.25 in a form that gives the correct sign to the vector component of $\vec{F}_{sp}$.

Figure 10.17 shows a spring along a generic s-axis. The equilibrium position of the end of the spring is denoted s_e. This is the *position,* or coordinate, of the free end of the spring, *not* the spring's equilibrium length L_0.

When the spring is stretched, the displacement from equilibrium $\Delta s = s - s_e$ is *positive* while $(F_{sp})_s$, the s-component of the restoring force pointing to the left, is *negative.* If the spring is compressed, the displacement from equilibrium Δs is negative while the s-component of $\vec{F}_{sp}$, which now points to the right, is positive. No matter which way the end of the spring is displaced from equilibrium, the sign of the force component $(F_{sp})_s$ is always opposite to the sign of the displacement Δs. We can write this mathematically as

$$(F_{sp})_s = -k\Delta s \quad \text{(Hooke's law)} \qquad (10.26)$$

where $\Delta s = s - s_e$ is the displacement of the end of the spring from equilibrium. The minus sign is the mathematical indication of a *restoring* force.

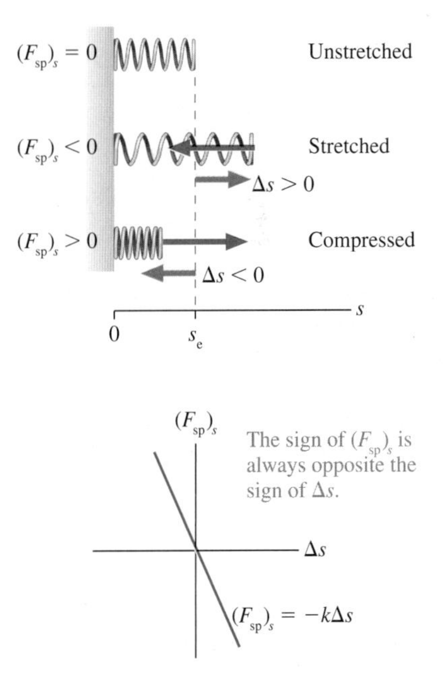

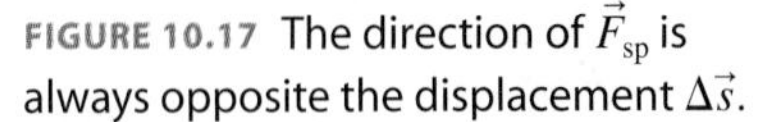
FIGURE 10.17 The direction of $\vec{F}_{sp}$ is always opposite the displacement $\Delta\vec{s}$.

Equation 10.26 for the restoring force of a spring is called **Hooke's law.** This "law" was first suggested by Robert Hooke, a contemporary (and sometimes bitter rival) of Newton. Hooke's law is not a true "law of nature," in the sense that Newton's laws are, but is actually just a *model* of a restoring force. It works extremely well for some springs, as in Figure 10.16, but less well for others. Hooke's law will fail for any spring if it is compressed or stretched too far.

NOTE ▶ Some of you, in an earlier physics course, may have learned to write Hooke's law as $F_{sp} = -kx$ (assuming the spring is along the x-axis), rather than as $-k\Delta x$. This can be misleading, and it is a common source of errors.

The restoring force will be $-kx$ *only* if the coordinate system in the problem is chosen such that the origin is at the equilibrium position of the free end of the spring. That is, $\Delta x = x$ only if $x_e = 0$. This is often done, but in complex problems it will sometimes be more convenient to locate the origin of the coordinate system elsewhere. If you try to use $F_{sp} = -kx$ after the origin has moved elsewhere—big trouble! So make sure you learn Hooke's law as $(F_{sp})_s = -k\Delta s$. ◄

EXAMPLE 10.5 Pull until it slips

Figure 10.18a shows a spring attached to a 2.0 kg block. The other end of the spring is pulled by a motorized toy train that moves forward at 5.0 cm/s. The spring constant is 50 N/m and the coefficient of static friction between the block and the surface is 0.60. The spring is at its equilibrium length at $t = 0$ s when the train starts to move. When does the block slip?

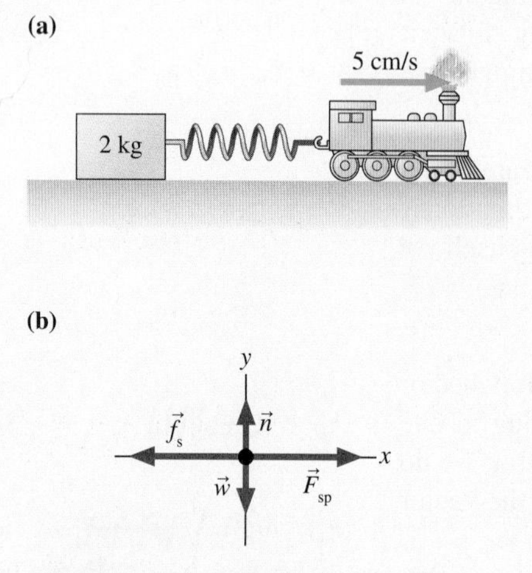

FIGURE 10.18 A toy train stretches the spring until the block slips.

MODEL Model the block as a particle and the spring as an ideal spring obeying Hooke's law.

VISUALIZE Figure 10.18b is a free-body diagram for the block.

SOLVE Recall that the tension in a massless string pulls equally at *both* ends of the string. The same is true for the spring force: It pulls (or pushes) equally at *both* ends. Imagine holding a rubber band with your left hand and stretching it with your right hand. Your left hand feels the pulling force, even though it was the right end of the rubber band that moved.

This is the key to solving the problem. As the right end of the spring moves, stretching the spring, the spring pulls backward on the train *and* forward on the block with equal strength. As the spring stretches, the static friction force on the block increases in magnitude to keep the block at rest. The block is in static equilibrium, so

$$\sum (F_{net})_x = (F_{sp})_x + (f_s)_x = F_{sp} - f_s = 0$$

where F_{sp} is the *magnitude* of the spring force. The magnitude is $F_{sp} = k\Delta x$, where $\Delta x = v_x t$ is the distance the train has moved. Thus

$$f_s = F_{sp} = k\Delta x$$

The block slips when the static friction force reaches its maximum value $f_{s\,max} = \mu_s n = \mu_s mg$. This occurs when the train has moved

$$\Delta x = \frac{f_{s\,max}}{k} = \frac{\mu_s mg}{k} = \frac{(0.60)(2.0\text{ kg})(9.8\text{ m/s}^2)}{50\text{ N/m}}$$
$$= 0.235\text{ m} = 23.5\text{ cm}$$

The time at which the block slips is

$$t = \frac{\Delta x}{v} = \frac{23.5\text{ cm}}{5.0\text{ cm/s}} = 4.7\text{ s}$$

This example illustrates an important class of motion called *stick-slip motion.* Once the block slips, it will shoot forward some distance, then stop and stick again. If the train continues to move forward, there will be a recurring sequence of stick, slip, stick, slip, stick. . . . Calculating the period of this stick-slip motion is not hard, although it's a bit beyond where we are right now.

The slip can range from a few centimeters in a relatively small earthquake to several meters in a very large earthquake.

Earthquakes are an important example of stick-slip motion. The large tectonic plates that make up the earth's crust are attempting to slide past each other, but friction forces cause the edges of the plates to stick together. The continued motion of the plates bends and deforms the rocks along the boundary. You may think of rocks as rigid and brittle, but large masses of rock, especially under the immense pressures within the earth, are somewhat elastic and can be "stretched." Eventually the elastic force of the deformed rocks exceeds the static friction force between the plates. An earthquake occurs as the plates slip and lurch forward. Once the tension is released, the plates stick together again and the process starts all over.

STOP TO THINK 10.4 The graph shows force versus displacement for three springs. Rank in order, from largest to smallest, the spring constants k_1, k_2, and k_3.

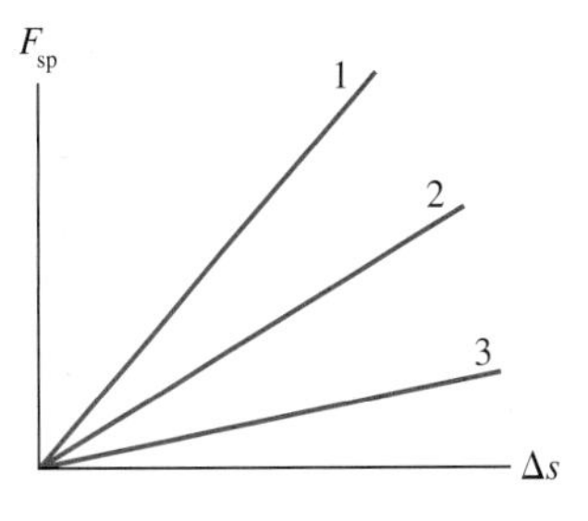

10.5 Elastic Potential Energy

The forces we have worked with thus far—gravity, friction, tension—have been constant forces. That is, their magnitudes do not change as an object moves. That feature has been important, because the kinematic equations we developed in Chapter 2 are for motion with constant acceleration. But a spring force exerts a *variable* force. The force is zero if $L = L_0$ (no displacement), and it steadily increases as the stretching length increases. The "natural motion" of a mass on a spring—think of pulling down on a spring and then releasing it—is an *oscillation*. This is *not* constant-acceleration motion, and we haven't yet developed the kinematics to handle oscillatory motion. That is why we haven't considered spring forces before now.

But suppose we're interested not in the time dependence of motion but only in before-and-after situations. For example, Figure 10.19 shows a before-and-after situation in which a spring launches a ball. Asking how the compression of the spring (the "before") affects the speed of the ball (the "after") is very different from wanting to know the ball's position as a function of time as the spring expands.

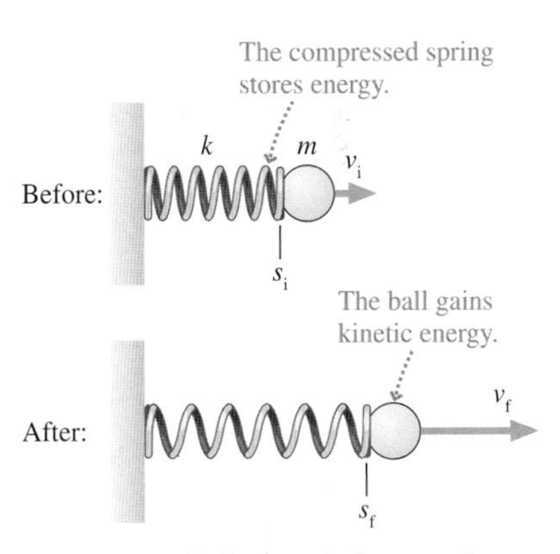

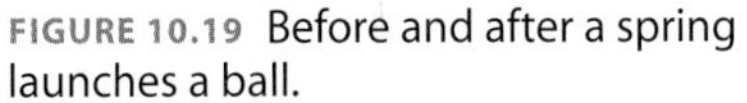

FIGURE 10.19 Before and after a spring launches a ball.

You certainly have a sense that a compressed spring has "stored energy," and Figure 10.19 shows clearly that the stored energy is transferred to the kinetic energy of the ball. Let's analyze this process with the same method we developed for motion under the influence of gravity. Newton's second law for the ball is

$$(F_{net})_s = ma_s = m\frac{dv_s}{dt} \tag{10.27}$$

The net force on the ball is given by Hooke's law, $(F_{net})_s = -k(s - s_e)$. Thus

$$m\frac{dv_s}{dt} = -k(s - s_e) \tag{10.28}$$

We'll use a generic s-axis, although it is better in actual problem solving to use x or y, depending on whether the motion is horizontal or vertical.

As we did before, use the chain rule to write

$$\frac{dv_s}{dt} = \frac{dv_s}{ds}\frac{ds}{dt} = v_s\frac{dv_s}{ds} \tag{10.29}$$

Substitute this into Equation 10.28, then multiply both sides by ds to get

$$mv_s\,dv_s = -k(s - s_e)\,ds \tag{10.30}$$

We can integrate both sides of the equation from the initial conditions i to the final conditions f—that is, integrate "from before to after," giving

$$\int_{v_i}^{v_f} mv_s\,dv_s = \frac{1}{2}mv_f^2 - \frac{1}{2}mv_i^2 = -k\int_{s_i}^{s_f}(s - s_e)\,ds \tag{10.31}$$

The integral on the right is not difficult, but many of you are new to calculus so we'll proceed step by step. The easiest way to get the answer in the most useful

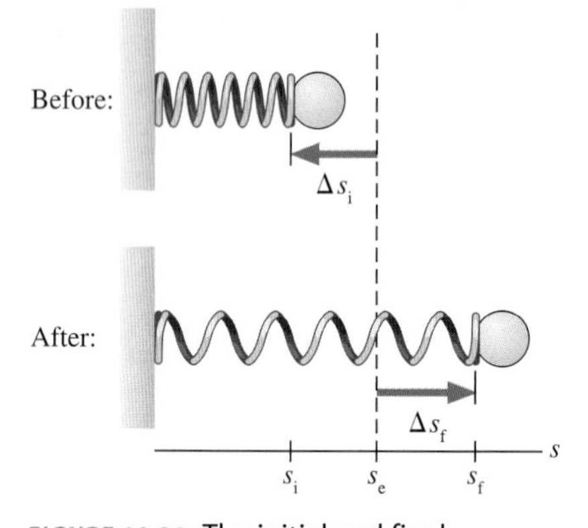

FIGURE 10.20 The initial and final displacements of the spring.

form is to make a change of variables. Define $u = (s - s_e)$, in which case $ds = du$. This changes the integrand from $(s - s_e)ds$ to $u\,du$.

When we change variables, we also must change the limits of integration. In particular, $s = s_i$ at the lower integration limit makes $u = s_i - s_e = \Delta s_i$, where Δs_i is the initial displacement of the spring from equilibrium. Likewise, $s = s_f$ makes $u = s_f - s_e = \Delta s_f$ at the upper limit. Figure 10.20 clarifies the meaning of Δs_i and Δs_f.

With this change of variables, the integral is

$$-k\int_{s_i}^{s_f} (s - s_e)\,ds = -k\int_{\Delta s_i}^{\Delta s_f} u\,du = -\frac{1}{2}ku^2\Big|_{\Delta s_i}^{\Delta s_f}$$
$$= -\frac{1}{2}k(\Delta s_f)^2 + \frac{1}{2}k(\Delta s_i)^2 \qquad (10.32)$$

Using this result makes Equation 10.31 become

$$\frac{1}{2}mv_f^2 - \frac{1}{2}mv_i^2 = -\frac{1}{2}k(\Delta s_f)^2 + \frac{1}{2}k(\Delta s_i)^2 \qquad (10.33)$$

which can be rewritten as

$$\frac{1}{2}mv_f^2 + \frac{1}{2}k(\Delta s_f)^2 = \frac{1}{2}mv_i^2 + \frac{1}{2}k(\Delta s_i)^2 \qquad (10.34)$$

We've succeeded in our goal of relating before and after. In particular, the quantity

$$\frac{1}{2}mv^2 + \frac{1}{2}k(\Delta s)^2 \qquad (10.35)$$

does not change as the spring compresses or expands. You recognize $\frac{1}{2}mv^2$ as the kinetic energy K. Let's define the **elastic potential energy** U_s of a spring to be

$$U_s = \frac{1}{2}k(\Delta s)^2 \quad \text{(elastic potential energy)} \qquad (10.36)$$

Then Equation 10.34 tells us that an object moving on a spring obeys

$$K_f + U_{sf} = K_i + U_{si} \qquad (10.37)$$

In other words, the mechanical energy $E_{mech} = K + U_s$ is conserved for an object moving *without friction* on an ideal spring.

NOTE ► Because Δs is squared, the elastic potential energy is positive for a spring that is either stretched or compressed. U_s is zero when the spring is at its equilibrium length L_0 and $\Delta s = 0$. ◄

EXAMPLE 10.6 A spring-launched plastic ball

A spring-loaded toy gun is used to launch a 10 g plastic ball. The spring, which has a spring constant of 10 N/m, is compressed by 10 cm as the ball is pushed into the barrel. When the trigger is pulled, the spring is released and shoots the ball back out. What is the ball's speed as it leaves the barrel? Assume that friction is negligible.

MODEL Assume an ideal spring that obeys Hooke's law. Also assume that the gun is held firmly enough to prevent recoil. There's no friction, hence the mechanical energy $K + U_s$ is conserved.

VISUALIZE Figure 10.21a shows a before-and-after pictorial representation. The compressed spring will push on the ball until the spring has returned to its equilibrium length. We have chosen to put the origin of the coordinate system at the equilibrium position of the free end of the spring, making $x_1 = -10$ cm and $x_2 = x_e = 0$ cm. It's also useful to look at an energy bar chart. The bar chart of Figure 10.21b shows the potential energy stored in the compressed spring being entirely transformed into the kinetic energy of the ball.

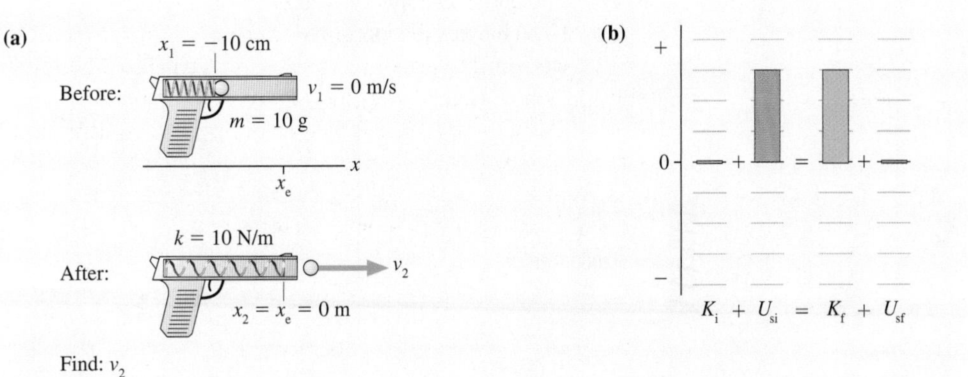

FIGURE 10.21 The before-and-after pictorial representation and energy bar chart of a ball being shot out of a spring-loaded toy gun. The gun is held and does not recoil.

SOLVE The energy conservation equation is $K_2 + U_{s2} = K_1 + U_{s1}$. We can use the elastic potential energy of the spring, Equation 10.36, to write this as

$$\frac{1}{2}mv_2^2 + \frac{1}{2}k(x_2 - x_e)^2 = \frac{1}{2}mv_1^2 + \frac{1}{2}k(x_1 - x_e)^2$$

Notice that we used x, rather than the generic s, and that we explicitly wrote out the meaning of Δx_1 and Δx_2. Using $x_2 = x_e = 0$ m and $v_1 = 0$ m/s simplifies this to

$$\frac{1}{2}mv_2^2 = \frac{1}{2}kx_1^2$$

It is now straightforward to solve for the ball's speed:

$$v_2 = \sqrt{\frac{kx_1^2}{m}} = \sqrt{\frac{(10\text{ N/m})(-0.10\text{ m})^2}{0.010\text{ kg}}} = 3.16\text{ m/s}$$

ASSESS This is a problem that we could *not* have solved with Newton's laws. The acceleration is not constant, and we have not learned how to handle the kinematics of nonconstant acceleration. But with conservation of energy—it's easy!

If an object attached to a spring moves vertically, the system has both elastic *and* gravitational potential energy. The mechanical energy then contains *two* potential energy terms:

$$E_{mech} = K + U_g + U_s \tag{10.38}$$

In other words, there are now two distinct ways of storing energy inside the system. The following example demonstrates this idea.

EXAMPLE 10.7 A spring-launched satellite

Prince Harry the Horrible wanted to be the first to launch a satellite. He placed a 2.0 kg payload on top of a very stiff 2.0-m-long spring with a spring constant of 50,000 N/m. Then the prince had his strongest men use a winch to crank the spring down to a length of 80 cm. When released, the spring shot the payload straight up. How high did it go?

MODEL Assume an ideal spring that obeys Hooke's law. There's no friction, hence the mechanical energy $K + U_g + U_s$ is conserved.

VISUALIZE Figure 10.22a on the next page shows the satellite ready for launch. We have chosen to place the origin of the coordinate system on the ground, which means that the equilibrium position of the end of the unstretched spring is *not* $y_e = 0$ m but, instead, $y_e = 2.0$ m. The payload reaches height y_2, where $v_2 = 0$ m/s.

SOLVE The energy conservation equation $K_2 + U_{s2} + U_{g2} = K_1 + U_{s1} + U_{g1}$ is

$$\frac{1}{2}mv_2^2 + \frac{1}{2}k(y_e - y_e)^2 + mgy_2 = \frac{1}{2}mv_1^2 + \frac{1}{2}k(y_1 - y_e)^2 + mgy_1$$

Notice that the elastic potential energy term on the left has $(y_e - y_e)^2$, not $(y_2 - y_e)^2$. The payload moves to position y_2, *but the spring does not!* The end of the spring stops at y_e. Both the initial and final speeds v_1 and v_2 are zero. Solving for the height:

$$y_2 = y_1 + \frac{k(y_1 - y_e)^2}{2mg} = 1840\text{ m}$$

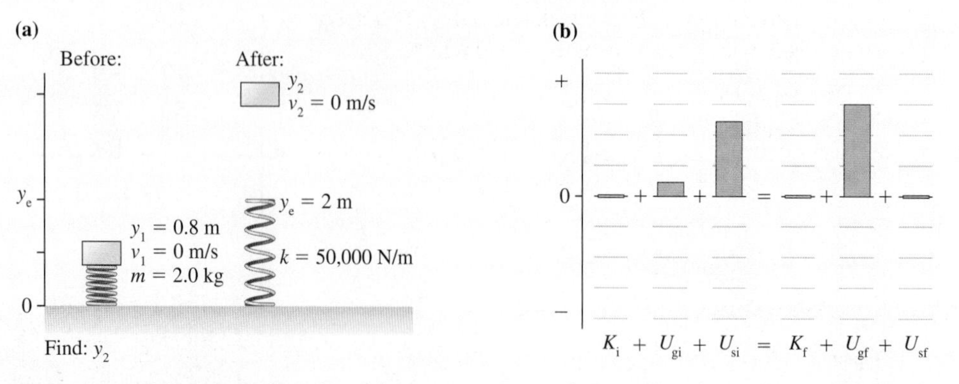

FIGURE 10.22 Before-and-after pictorial representation and energy bar chart of Prince Harry's spring-launched payload.

ASSESS Figure 10.22b shows an energy bar chart. The net effect of the launch is to transform the potential energy stored in the spring entirely into gravitational potential energy. The kinetic energy is zero at the beginning and zero again at the highest point. The payload does have kinetic energy as it comes off the spring; we did not need to know this energy to solve the problem.

EXAMPLE 10.8 Pushing apart

A spring with spring constant 2000 N/m is sandwiched between a 1.0 kg block and a 2.0 kg block on a frictionless table. The blocks are pushed together to compress the spring by 10 cm, then released. What are the velocities of the blocks as they fly apart?

MODEL Assume an ideal spring that obeys Hooke's law. There's no friction, hence the mechanical energy $K + U_s$ is conserved. In addition, because the blocks and spring form an isolated system, their total momentum is conserved.

VISUALIZE Figure 10.23 is a before-and-after pictorial representation.

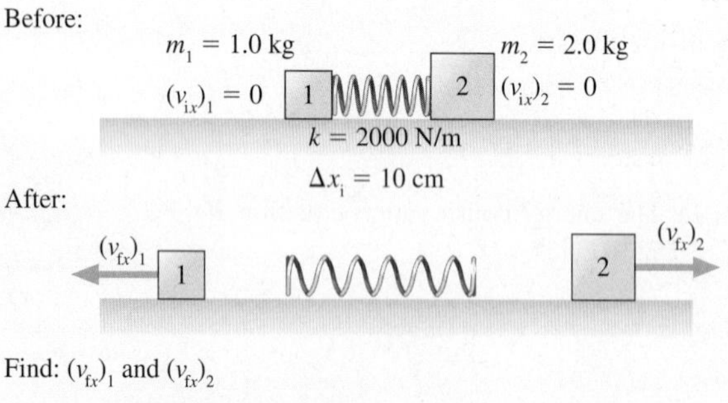

FIGURE 10.23 Before-and-after pictorial representation of the blocks and spring.

SOLVE The initial energy, with the spring compressed, is entirely potential. The final energy is entirely kinetic. The energy conservation equation $K_f + U_{sf} = K_i + U_{si}$ is

$$\frac{1}{2}m_1(v_f)_1^2 + \frac{1}{2}m_2(v_f)_2^2 + 0 = 0 + 0 + \frac{1}{2}k(\Delta x_i)^2$$

Notice that *both* blocks contribute to the kinetic energy. The energy equation has two unknowns, $(v_f)_1$ and $(v_f)_2$, and one equation is not enough to solve the problem. Fortunately, momentum is also conserved. The initial momentum is zero, because both blocks are at rest, so the momentum equation is

$$m_1(v_{fx})_1 + m_2(v_{fx})_2 = 0$$

which can be solved to give

$$(v_{fx})_1 = -\frac{m_2}{m_1}(v_{fx})_2$$

The minus sign indicates that the blocks move in opposite directions. The speed $(v_f)_1 = (m_2/m_1)(v_f)_2$ is all we need to calculate the kinetic energy. Substituting $(v_f)_1$ into the energy equation gives

$$\frac{1}{2}m_1\left(\frac{m_2}{m_1}(v_f)_2\right)^2 + \frac{1}{2}m_2(v_f)_2^2 = \frac{1}{2}k(\Delta x_i)^2$$

which simplifies to

$$m_2\left(1 + \frac{m_2}{m_1}\right)(v_f)_2^2 = k(\Delta x_i)^2$$

Solving for $(v_f)_2$, we find

$$(v_f)_2 = \sqrt{\frac{k(\Delta x_i)^2}{m_2(1 + m_2/m_1)}} = 1.826 \text{ m/s}$$

Finally, we can go back to find

$$(v_{fx})_1 = -\frac{m_2}{m_1}(v_{fx})_2 = -3.65 \text{ m/s}$$

The 2.0 kg block moves to the right at 1.826 m/s while the 1.0 kg block goes left at 3.65 m/s.

ASSESS This example shows just how powerful a problem-solving tool the conservation laws are.

STOP TO THINK 10.5 A spring-loaded gun shoots a plastic ball with a speed of 4 m/s. If the spring is compressed twice as far, the ball's speed will be

a. 2 m/s. b. 4 m/s.
c. 8 m/s. d. 16 m/s.

10.6 Elastic Collisions

Figure 9.1 showed a molecular-level view of a collision. Billions of spring-like molecular bonds are compressed as two objects collide, then the bonds expand and push the objects apart. In the language of energy, the kinetic energy of the objects is transformed into the elastic potential energy of molecular bonds, then back into kinetic energy as the two objects spring apart.

In some cases, such as the inelastic collisions of Chapter 9, some of the mechanical energy is dissipated inside the objects and not all of the kinetic energy is recovered. That is, $K_f < K_i$. (A homework problem will let you show this explicitly.) We're now interested in collisions in which *all* of the kinetic energy is stored as elastic potential energy in the bonds, and then *all* of the stored energy is transformed back into the post-collision kinetic energy of the objects. Such a collision, in which the mechanical energy is conserved, is called a **perfectly elastic collision.**

A perfectly elastic collision conserves both momentum and mechanical energy.

Needless to say, most real collisions fall somewhere between perfectly elastic and perfectly inelastic. A rubber ball bouncing on the floor might "lose" 20% of its kinetic energy on each bounce and return to only 80% of the height of the previous bounce. Perfectly elastic and perfectly inelastic collisions are limiting cases rarely seen in the real world, but they are nonetheless instructive for demonstrating the major ideas without making the mathematics too complex. Collisions between two very hard objects, such as two billiard balls or two steel balls, come close to being perfectly elastic.

ActivPhysics Online 6.2

Figure 10.24 shows a head-on, perfectly elastic collision of a ball of mass m_1, having initial velocity $(v_{ix})_1$, with a ball of mass m_2 that is initially at rest. The balls' velocities after the collision are $(v_{fx})_1$ and $(v_{fx})_2$. These are velocities, not speeds, and have signs. Ball 1, in particular, might bounce backward and have a negative value for $(v_{fx})_1$.

The collision must obey two conservation laws: conservation of momentum (obeyed in any collision) and conservation of mechanical energy (because the collision is perfectly elastic). Although the energy is transformed into potential energy during the collision, the mechanical energy before and after the collision is purely kinetic energy. Thus

Before: ① $(v_{ix})_1$ → ② K_i

During: ①② Energy is stored in compressed molecular bonds, then released as the bonds re-expand.

After: ① $(v_{fx})_1$ → ② $(v_{fx})_2$ → $K_f = K_i$

FIGURE 10.24 A perfectly elastic collision.

$$\text{momentum conservation:} \quad m_1(v_{fx})_1 + m_2(v_{fx})_2 = m_1(v_{ix})_1 \tag{10.39}$$

$$\text{energy conservation:} \quad \frac{1}{2}m_1(v_{fx})_1^2 + \frac{1}{2}m_2(v_{fx})_2^2 = \frac{1}{2}m_1(v_{ix})_1^2 \tag{10.40}$$

Momentum conservation alone is not sufficient to analyze the collision because there are two unknowns: the two final velocities. That is why we did not consider perfectly elastic collisions in Chapter 9. Energy conservation gives us another condition. Isolating $(v_{fx})_1$ in Equation 10.39 gives

$$(v_{fx})_1 = (v_{ix})_1 - \frac{m_2}{m_1}(v_{fx})_2 \tag{10.41}$$

Substitute this into Equation 10.40:

$$\frac{1}{2}m_1\left((v_{ix})_1 - \frac{m_2}{m_1}(v_{fx})_2\right)^2 + \frac{1}{2}m_2(v_{fx})_2^2$$

$$= \frac{1}{2}\left(m_1(v_{ix})_1^2 - 2m_2(v_{ix})_1(v_{fx})_2 + \frac{m_2^2}{m_1}(v_{fx})_2^2 + m_2(v_{fx})_2^2\right)$$

$$= \frac{1}{2}m_1(v_{ix})_1^2$$

This looks rather gruesome, but the first two terms on each side cancel and the resulting equation can be rearranged to give

$$(v_{fx})_2\left[\left(1 + \frac{m_2}{m_1}\right)(v_{fx})_2 - 2(v_{ix})_1\right] = 0 \tag{10.42}$$

One possible solution to this equation is seen to be $(v_{fx})_2 = 0$. However, this solution is of no interest; it is the case where ball 1 misses ball 2. The other solution is

$$(v_{fx})_2 = \frac{2}{1 + m_2/m_1}(v_{ix})_1 = \frac{2m_1}{m_1 + m_2}(v_{ix})_1$$

which, finally, can be substituted back into to Equation 10.41 to yield $(v_{fx})_1$. The complete solution is

$$(v_{fx})_1 = \frac{m_1 - m_2}{m_1 + m_2}(v_{ix})_1 \qquad (v_{fx})_2 = \frac{2m_1}{m_1 + m_2}(v_{ix})_1 \qquad \text{(perfectly elastic collision with ball 2 initially at rest)} \tag{10.43}$$

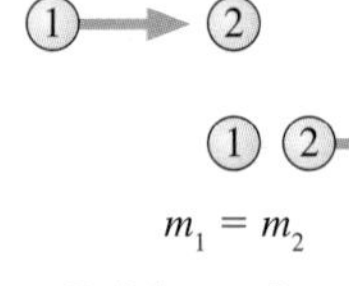

Ball 1 stops. Ball 2 goes forward with $v_2 = v_1$.

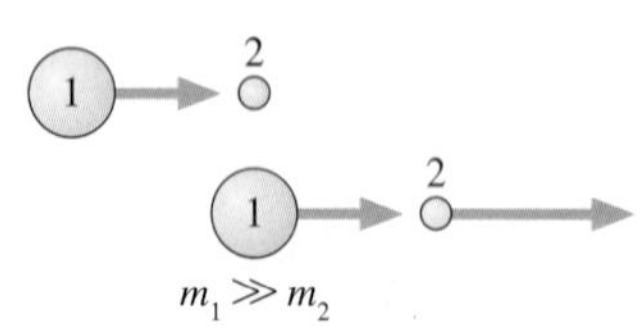

Ball 1 hardly slows down. Ball 2 is knocked forward at $v_2 \approx 2v_1$.

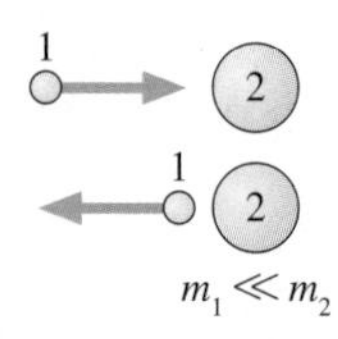

Ball 1 bounces off ball 2 with almost no loss of speed. Ball 2 hardly moves.

FIGURE 10.25 Three special elastic collisions.

Equations 10.43 allow us to compute the final velocity of each ball. These equations are a little difficult to interpret, so let us look at the three special cases shown in Figure 10.25.

Case 1: $m_1 = m_2$. This is the case of one billiard ball striking another of equal mass. For this case, Equations 10.43 give

$$v_{1f} = 0$$
$$v_{2f} = v_{1i}$$

Case 2: $m_1 \gg m_2$. This is the case of a bowling ball running into a Ping-Pong ball. We do not want an exact solution here, but an approximate solution for the limiting case that $m_1 \to \infty$. Equations 10.43 in this limit give

$$v_{1f} \approx v_{1i}$$
$$v_{2f} \approx 2v_{1i}$$

Case 3: $m_1 \ll m_2$. Now we have the reverse case of a Ping-Pong ball colliding with a bowling ball. Here we are interested in the limit $m_1 \to 0$, in which case Equations 10.43 become

$$v_{1f} \approx -v_{1i}$$
$$v_{2f} \approx 0$$

These cases agree well with our expectations and give us confidence that Equations 10.43 accurately describe a perfectly elastic collision.

EXAMPLE 10.9 A rebounding pendulum

A 200 g steel ball hangs on a 1.0-m-long string. The ball is pulled sideways so that the string is at a 45° angle, then released. At the very bottom of its swing the ball strikes a 500 g steel block that is resting on a frictionless table. To what angle does the ball rebound?

MODEL This is a challenging problem. We can divide it into three parts. First the ball swings down as a pendulum. Second, the ball and block have a collision. Steel balls bounce off each other very well, so we will assume that the collision is perfectly elastic. Third, the ball, after it bounces off the block, swings back up as a pendulum.

VISUALIZE Figure 10.26 shows four distinct moments of time: as the ball is released, an instant before the collision, an instant after the collision but before the ball and block have had time to move, and as the ball reaches its highest point on the rebound. Call the ball A and the block B, so $m_A = 0.20$ kg and $m_B = 0.50$ kg.

SOLVE Part 1: The first part involves the ball only. Its initial height is

$$(y_0)_A = L - L\cos\theta_0 = L(1 - \cos\theta_0) = 0.293 \text{ m}$$

We can use conservation of mechanical energy to find the ball's velocity at the bottom, just before impact on the block:

$$\frac{1}{2}m_A(v_1)_A^2 + m_A g(y_1)_A = \frac{1}{2}m_A(v_0)_A^2 + m_A g(y_0)_A$$

Solving for the velocity at the bottom, where $(v_0)_A = 0$ and $(y_1)_A = 0$, gives

$$(v_1)_A = \sqrt{2g(y_0)_A} = 2.40 \text{ m/s}$$

Part 2: The ball and block undergo a perfectly elastic collision in which the block is initially at rest. These are the conditions for which Equations 10.43 were derived. The velocities *immediately* after the collision, prior to any further motion, are

$$(v_{2x})_A = \frac{m_A - m_B}{m_A + m_B}(v_{1x})_A = -1.03 \text{ m/s}$$

$$(v_{2x})_B = \frac{2m_A}{m_A + m_B}(v_{1x})_A = +1.37 \text{ m/s}$$

The ball rebounds toward the left with a speed of 1.03 m/s while the block moves to the right at 1.37 m/s. Kinetic energy has been conserved (you might want to check this), but it is now shared between the ball and the block.

Part 3: Now the ball is a pendulum with an initial speed of 1.03 m/s. Mechanical energy is again conserved, so we can find its maximum height at the point where $(v_3)_A = 0$:

$$\frac{1}{2}m_A(v_3)_A^2 + m_A g(y_3)_A = \frac{1}{2}m_A(v_2)_A^2 + m_A g(y_2)_A$$

Solving for the maximum height gives

$$(y_3)_A = \frac{(v_2)_A^2}{2g} = 0.0541 \text{ m}$$

The height $(y_3)_A$ is related to angle θ_3 by $(y_3)_A = L(1 - \cos\theta_3)$. This can be solved to find the angle of rebound:

$$\theta_3 = \cos^{-1}\left(1 - \frac{(y_3)_A}{L}\right) = 18.9°$$

The block speeds away at 1.37 m/s and the ball rebounds to an angle of 18.9°.

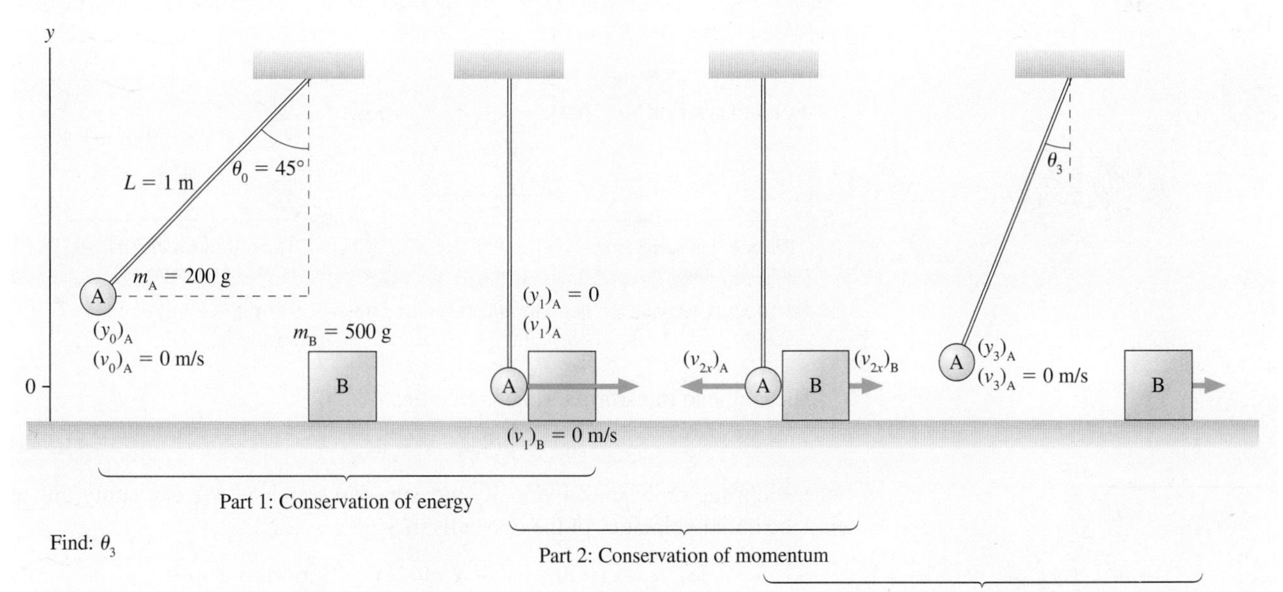

FIGURE 10.26 Four moments in the collision of a pendulum with a block.

FIGURE 10.27 A perfectly elastic collision in which both balls have an initial velocity.

Using Reference Frames

Equations 10.43 assumed that ball 2 was at rest prior to the collision. Suppose, however, you need to analyze the perfectly elastic collision that is just about to take place in Figure 10.27. What are the direction and speed of each ball after the collision? You could solve the simultaneous momentum and energy equations, but the mathematics becomes quite messy when both balls have an initial velocity. Fortunately, there's an easier way.

You already know the answer—Equations 10.43—when ball 2 is initially at rest. And in Chapter 6 you learned the Galilean transformation of velocity. This transformation relates an object's velocity v as measured in reference frame S to its velocity v' in a different reference frame S′ that moves with velocity V relative to S. The Galilean transformation provides an elegant and straightforward way to analyze the collision of Figure 10.27.

TACTICS BOX 10.1 Analyzing elastic collisions

1. Use the Galilean transformation to transform the initial velocities of balls 1 and 2 from the "lab frame" S to a reference frame S′ in which ball 2 is at rest;
2. Use Equations 10.43 to determine the outcome of the collision in frame S′; then
3. Transform the final velocities back to the "lab frame" S.

Figure 10.28a shows the "before" situation in reference frame S, which we can think of as the lab frame. Notice, compared to Figure 10.27, that we've given $(v_{ix})_2$ as a *velocity* with an appropriate sign. The frame S′ in which ball 2 is at rest is a frame that is traveling alongside ball 2 with the same velocity: $V = -3$ m/s.

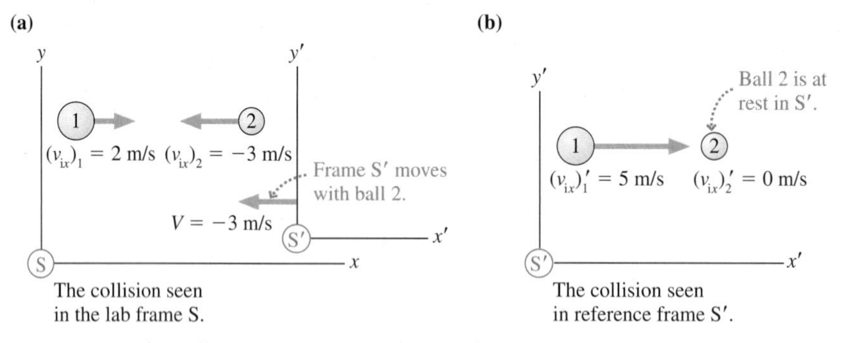

FIGURE 10.28 The collision seen in two reference frames, S and S′.

The Galilean transformation of velocities is

$$v' = v - V \tag{10.44}$$

where a prime represents a velocity measured in frame S′. We can apply this to find the initial velocities of the two balls in S′:

$$\begin{aligned}(v_{ix})'_1 &= (v_{ix})_1 - V = 2\text{ m/s} - (-3\text{ m/s}) = 5\text{ m/s}\\ (v_{ix})'_2 &= (v_{ix})_2 - V = -3\text{ m/s} - (-3\text{ m/s}) = 0\text{ m/s}\end{aligned} \tag{10.45}$$

Figure 10.28b shows the "before" situation in reference frame S′, where ball 2 is at rest.

Now we can use Equations 10.43 to find the post-collision velocities in frame S′:

$$(v_{fx})_1' = \frac{m_1 - m_2}{m_1 + m_2}(v_{ix})_1' = 1.67 \text{ m/s}$$
$$(v_{fx})_2' = \frac{2m_1}{m_1 + m_2}(v_{ix})_1' = 6.67 \text{ m/s} \qquad (10.46)$$

Frame S′ hasn't changed—it is still moving at $V = -3$ m/s—but the collision has caused both balls to have a velocity in S′.

Finally, we need to apply the reverse transformation $v = v' + V$, with the same V, to transform the post-collision velocities back to the lab frame:

$$(v_{fx})_1 = (v_{fx})_1' + V = 1.67 \text{ m/s} + (-3 \text{ m/s}) = -1.33 \text{ m/s}$$
$$(v_{fx})_2 = (v_{fx})_2' + V = 6.67 \text{ m/s} + (-3 \text{ m/s}) = 3.67 \text{ m/s} \qquad (10.47)$$

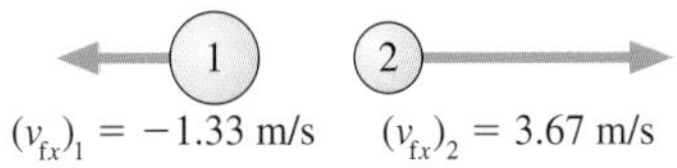

Figure 10.29 shows the situation after the collision. It's not hard to check that these final velocities do, indeed, conserve both momentum and energy.

$(v_{fx})_1 = -1.33$ m/s $\quad (v_{fx})_2 = 3.67$ m/s

FIGURE 10.29 The post-collision velocities in the lab frame.

10.7 Energy Diagrams

Potential energy is an energy of position. The gravitational potential energy depends on the height of an object and the elastic potential energy depends on a spring's displacement. Other potential energies you will meet in the future will depend in some way on position. Functions of position are easy to represent as graphs. A graph that shows a system's potential energy and total energy as a function of position is called an **energy diagram.** Energy diagrams allow you to visualize motion based on energy considerations. They can also be useful problem-solving tools, and they will play an important role when we get to quantum physics in Part VII.

Figure 10.30 is the energy diagram of a particle in free fall. The gravitational potential energy $U_g = mgy$ is graphed as a line through the origin with slope mg. The *potential-energy curve* is labeled PE. The line labeled TE is the *total energy line*, $E = K + U_g$. It is horizontal because the mechanical energy is conserved, meaning that the object's total mechanical energy E has the same value at every position.

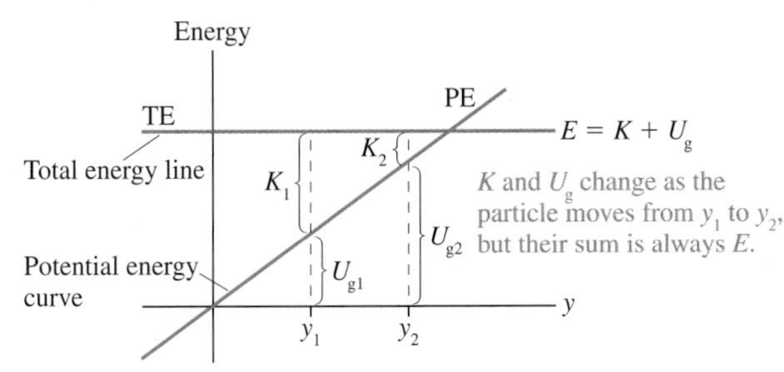

FIGURE 10.30 The energy diagram of a particle in free fall.

Suppose the particle is at position y_1. By definition, the distance from the axis to the potential-energy curve is the particle's potential energy U_{g1} at that position. Because $K_1 = E - U_{g1}$, the distance between the potential-energy curve and the total energy line is the particle's kinetic energy.

The four-frame "movie" of Figure 10.31 on the next page illustrates how an energy diagram is used to visualize motion. The first frame shows a particle projected upward from $y_a = 0$ with kinetic energy K_a. Initially the energy is entirely kinetic, with $U_{ga} = 0$. A pictorial representation and an energy bar chart help to illustrate what the energy diagram is showing.

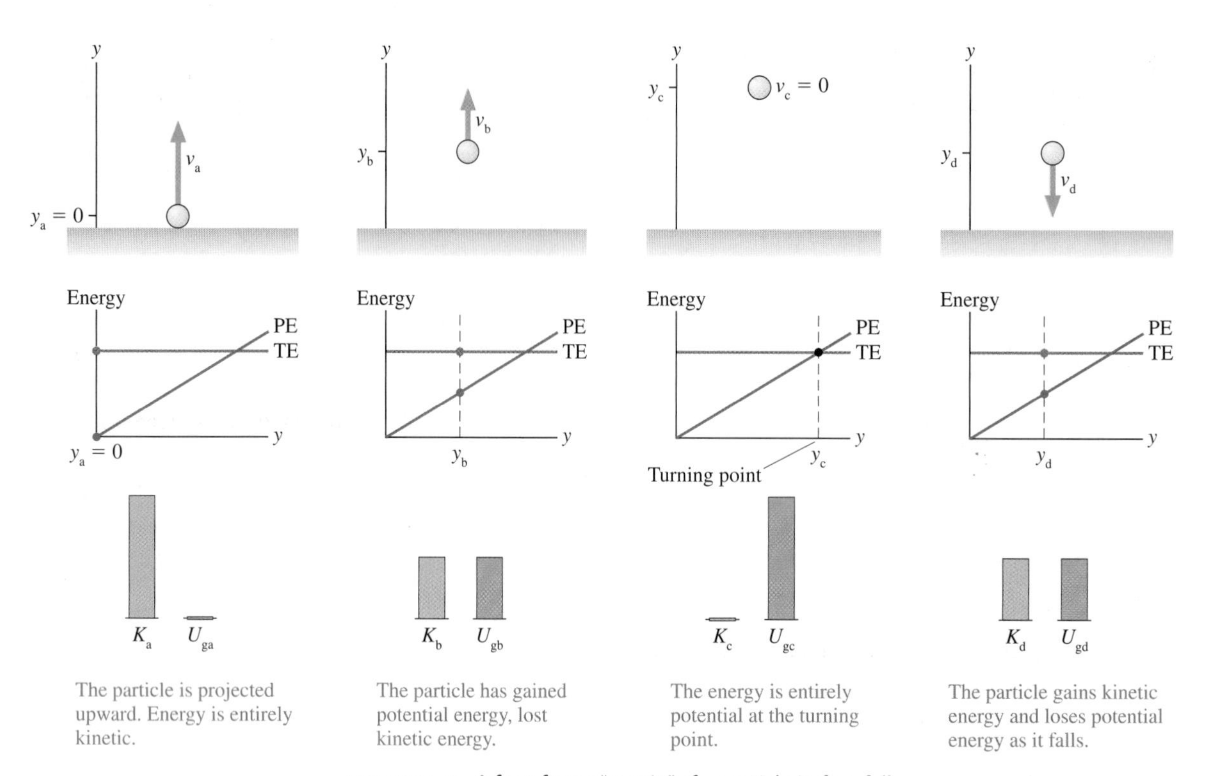

y
$y_a = 0$
v_a
Energy
PE
TE
$y_a = 0$
K_a U_{ga}
The particle is projected upward. Energy is entirely kinetic.
y_b
v_b
K_b U_{gb}
The particle has gained potential energy, lost kinetic energy.
y_c
$v_c = 0$
Turning point
K_c U_{gc}
The energy is entirely potential at the turning point.
y_d
v_d
K_d U_{gd}
The particle gains kinetic energy and loses potential energy as it falls.

FIGURE 10.31 A four-frame "movie" of a particle in free fall.

In the second frame, the particle has gained height but lost speed. The potential-energy curve U_{gb} is higher, and the distance K_b between the potential-energy curve and the total energy line is less. The particle continues rising and slowing until, in the third frame, it reaches the y-value where the total energy line crosses the potential-energy curve. This point, where $K = 0$ and the energy is entirely potential, is a *turning point* where the particle reverses direction. Finally, we see the particle speeding up as it falls.

A particle with this amount of total energy would need negative kinetic energy to be to the right of the point, at y_c, where the total energy line crosses the potential-energy curve. Negative K is not physically possible, so **the particle cannot be at positions with $U > E$.** Now, it's certainly true that you could make the particle reach a larger value of y simply by throwing it harder. But that would increase E and move the total energy line higher.

NOTE ▶ The TE line is under your control. You can move the TE line as far up or down as you wish by changing the initial conditions, such as projecting the particle upward with a different speed or dropping it from a different height. Once you've determined the initial conditions, you can use the energy diagram to analyze the motion for that amount of total energy. ◀

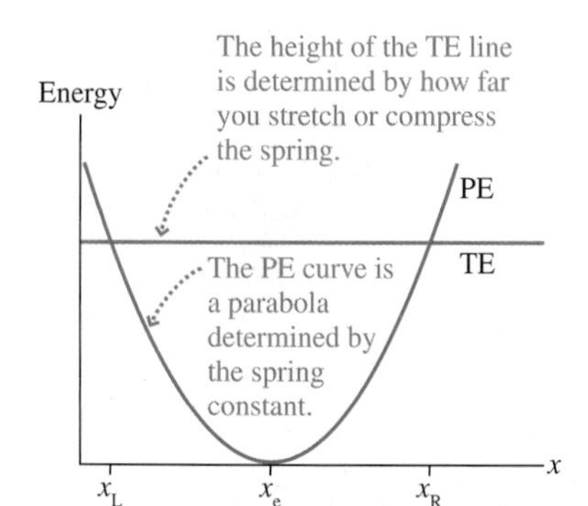

Energy
The height of the TE line is determined by how far you stretch or compress the spring.
PE
TE
The PE curve is a parabola determined by the spring constant.
x_L
x_e
x_R
x

FIGURE 10.32 The energy diagram of a mass on a horizontal spring.

Figure 10.32 shows the energy diagram of a mass on a horizontal spring. The potential-energy curve $U_s = \frac{1}{2}k(x - x_e)^2$ is a parabola centered at the equilibrium position x_e. The PE curve is determined by the spring constant; you can't change it. But you can set the TE to any height you wish simply by stretching the spring to the proper length. The figure shows one possible TE line.

Suppose you pull the mass out to position x_R and release it. Figure 10.33 is a four-frame movie of the subsequent motion. Initially, the energy is entirely potential. The restoring force of the spring pulls the mass toward x_e, increasing the kinetic energy as the potential energy decreases. The mass has maximum speed at position x_e, where $U_s = 0$, and then it slows down as the spring starts to compress.

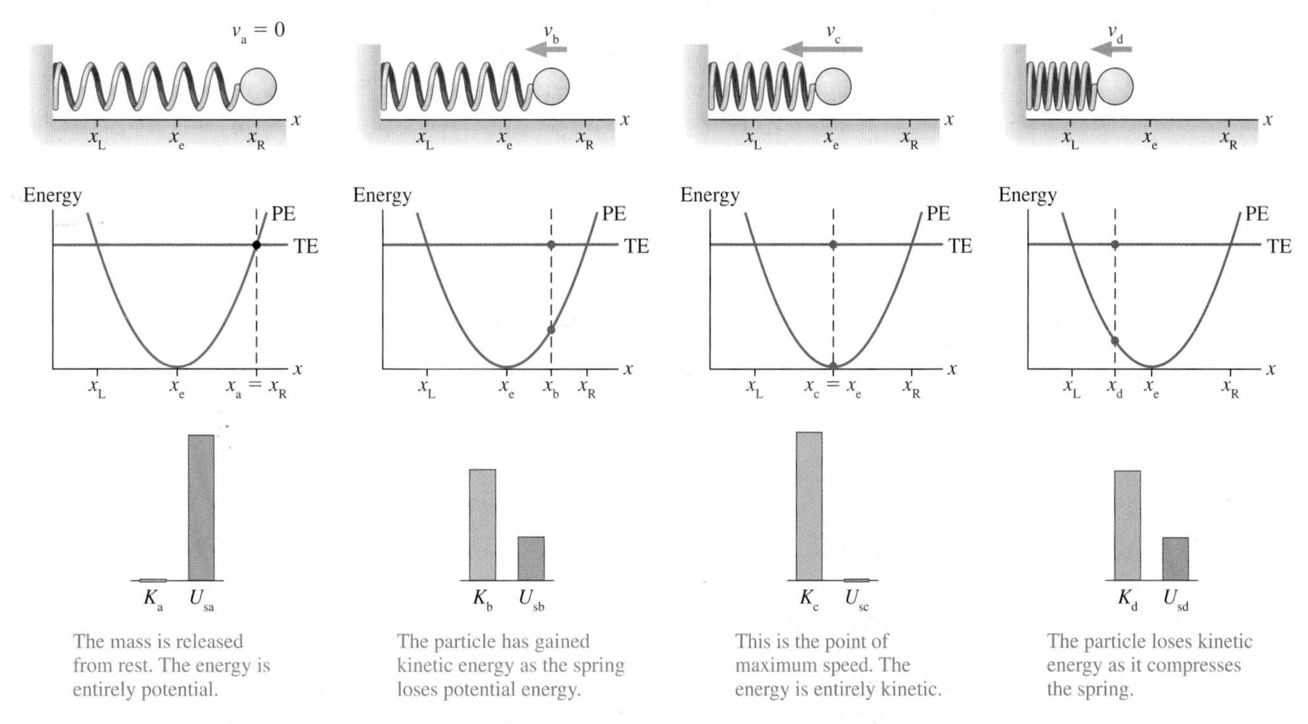

FIGURE 10.33 A four-frame movie of a mass oscillating on a spring.

If the movie were to continue, you should be able to visualize that position x_L is a turning point. The mass will instantaneously have $v_L = 0$ and $K_L = 0$, then reverse direction as the spring starts to expand. The mass will speed up until x_e, then slow down until reaching x_R, where it started. This is another turning point. It will reverse direction again and start the process over. In other words, the mass will *oscillate* back and forth between the left and right turning points at x_L and x_R where the TE line crosses the PE curve.

Figure 10.34 applies these ideas to a more general energy diagram. We don't know how this potential energy was created, but we can visualize the motion of a particle that has this potential energy. Suppose the particle is released from rest at position x_1. How will it then move?

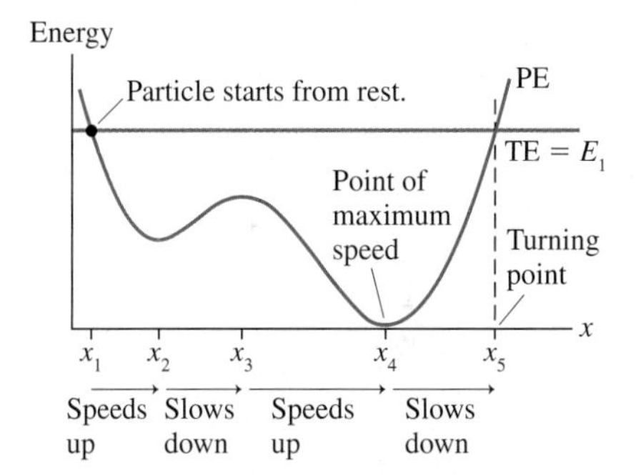

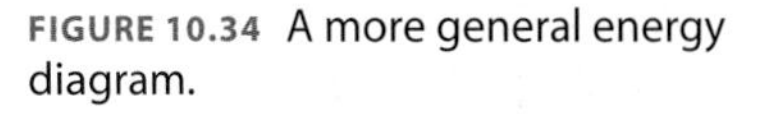

FIGURE 10.34 A more general energy diagram.

The particle's kinetic energy at x_1 is zero, hence the TE line must cross the PE curve at this point. The particle cannot move to the left, because $U > E$, so it begins to move toward the right. The particle speeds up from x_1 to x_2 as U decreases and K increases, then slows down from x_2 to x_3 as it goes up the "potential energy hill." The particle doesn't stop at x_3 because it still has kinetic energy. It speeds up from x_3 to x_4, reaching its maximum speed at x_4, then slows down between x_4 and x_5. Position x_5 is a turning point, a point where the TE line crosses the PE curve. The particle is instantaneously at rest, then reverses direction. The particle will oscillate back and forth between x_1 and x_5, following the pattern of slowing down and speeding up that we've outlined.

Equilibrium Positions

Positions x_2, x_3, and x_4, where the potential energy has a local minimum or maximum, are special positions. Consider a particle with the total energy E_2 shown in Figure 10.35. The particle can be at rest at x_2, with $K = 0$, but it cannot move away from x_2. In other words, a particle with energy E_2 is in *static equilibrium* at x_2. If you disturb the particle, giving it a small kinetic energy and a total energy just *slightly* larger than E_2, the particle will undergo a very small oscillation centered on x_2, like a marble in the bottom of a bowl. An equilibrium for which small

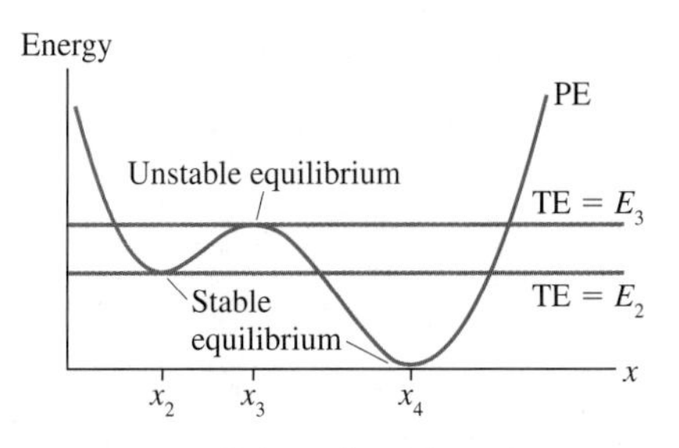

FIGURE 10.35 Points of stable and unstable equilibrium.

disturbances cause small oscillations is called a point of **stable equilibrium.** You should recognize that *any* minimum in the PE curve is a point of stable equilibrium. Position x_4 is also a point of stable equilibrium, in this case for a particle with $E = 0$.

Figure 10.35 also shows a particle with energy E_3 that is tangent to the curve at x_3. If a particle is placed *exactly* at x_3, it will stay there at rest ($K = 0$). But if you disturb the particle at x_3, giving it an energy only slightly more than E_3, it will speed up as it moves away from x_3. This is like trying to balance a marble on top of a hill. The slightest displacement will cause the marble to roll down the hill. A point of equilibrium for which a small disturbance causes the particle to move away is called a point of **unstable equilibrium.** Any maximum in the PE curve, such as x_3, is a point of unstable equilibrium.

We can summarize these lessons as follows:

TACTICS BOX 10.2 Interpreting an energy diagram

1. The distance from the axis to the PE curve is the particle's potential energy. The distance from the PE curve to the TE line is its kinetic energy. These are transformed as the position changes, causing the particle to speed up or slow down, but the sum $K + U$ doesn't change.
2. A point where the TE line crosses the PE curve is a turning point. The particle reverses direction.
3. The particle cannot be at a point where the PE curve is above the TE line.
4. The PE curve is determined by the properties of the system—mass, spring constant, and the like. You cannot change the PE curve. However, you can raise or lower the TE line simply by changing the initial conditions to give the particle more or less total energy.
5. A minimum in the PE curve is a point of stable equilibrium. A maximum in the PE curve is a point of unstable equilibrium.

EXAMPLE 10.10 Balancing a mass on a spring

A spring of length L_0 and spring constant k is standing on one end. A block of mass m is placed on the spring, compressing it. What is the length of the compressed spring?

MODEL Assume an ideal spring obeying Hooke's law. The block + spring system has both gravitational potential energy U_g *and* elastic potential energy U_s. The block sitting on top of the spring is at a point of stable equilibrium (small disturbances cause the block to oscillate slightly around the equilibrium position), so we can solve this problem by looking at the energy diagram.

VISUALIZE Figure 10.36a is a pictorial representation. We've used a coordinate system with the origin at ground level, so the equilibrium position of the uncompressed spring is $y_e = L_0$.

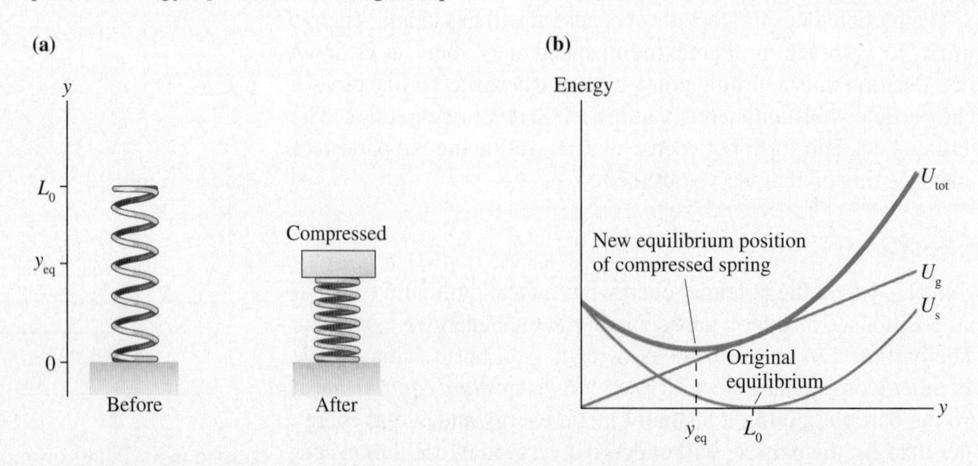

FIGURE 10.36 The block + spring system has both gravitational and elastic potential energy.

SOLVE Figure 10.36b shows the two potential energies separately and also shows the total potential energy

$$U_{tot} = U_g + U_s = mgy + \frac{1}{2}k(y - L_0)^2$$

The equilibrium position (the minimum of U_{tot}) has shifted from L_0 to a smaller value of y, closer to the ground. We can find the equilibrium by locating the position of the minimum in the PE curve. You know from calculus that the minimum of a function is at the point where the derivative (or slope) is zero. The derivative of U_{tot} is

$$\frac{dU_{tot}}{dy} = mg + k(y - L_0)$$

The derivative is zero at the point y_{eq}, so we can easily find

$$mg + k(y_{eq} - L_0) = 0$$

$$y_{eq} = L_0 - \frac{mg}{k}$$

The block compresses the spring by the length mg/k from its original length L_0, giving it a new equilibrium length $L_0 - mg/k$.

STOP TO THINK 10.6 A particle with the potential energy shown in the graph is moving to the right. It has 1 J of kinetic energy at $x = 1$ m. Where is the particle's turning point?

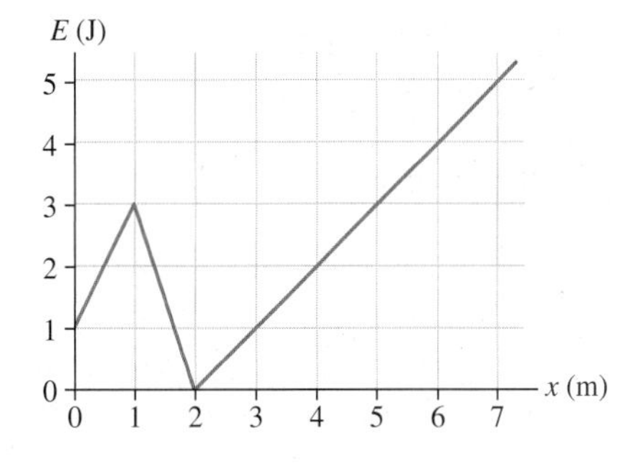

Molecular Bonds

Let's end this chapter by seeing how energy diagrams can allow us to understand something about molecular bonds. A *molecular bond* that holds two atoms together is an electric interaction between the charged electrons and nuclei. Figure 10.37 shows the potential-energy diagram for the diatomic molecule HCl (hydrogen chloride) as it has been experimentally determined. Distance x is the *atomic separation,* the distance between the hydrogen and the chlorine atoms. Note the very tiny distances: 1 nm $= 10^{-9}$ m.

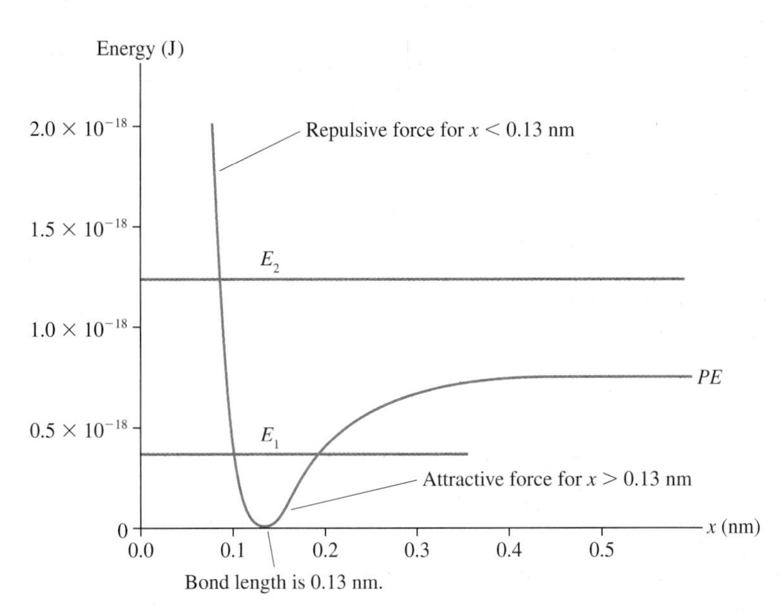

FIGURE 10.37 The energy diagram of the diatomic molecule HCl.

Although the potential energy is an electric energy, we can *interpret* the diagram using the steps in Tactics Box 10.2. The first thing you might notice is this potential-energy diagram has some similarities to a spring, with a deep potential-energy valley, but also some significant differences.

The molecule has a point of stable equilibrium at an atomic separation of $x_{eq} = 0.13$ nm. This is called the *bond length* of HCl, and you can find this value listed in chemistry books. If we try to push the atoms closer together (smaller x), the potential energy rises very rapidly. Physically, this is the repulsive electric force between the electrons orbiting each atom, preventing the atoms from getting too close.

There is also an attractive force between the atoms, called the *polarization force.* It is similar to the static electricity force by which a comb that has been brushed through your hair attracts small pieces of paper. If you try to pull the atoms apart (larger x), the attractive polarization force resists and is responsible for the increasing potential energy for $x > x_{eq}$. The equilibrium position is where the repulsive force between the electrons and the attractive polarization force are exactly balanced.

The repulsive force keeps getting stronger as you push the atoms together, and thus the potential-energy curve keeps getting steeper on the left. But the attractive polarization force gets *weaker* as the atoms get further apart. This is why the potential-energy curve becomes *less* steep as the atomic separation increases. Ultimately, at very large x, the potential energy no longer changes. This is not surprising, because two distant atoms do not interact with each other. This difference between the repulsive and attractive forces leads to an *asymmetric* curve.

It turns out that, for quantum physics reasons, a molecule cannot have $E = 0$ and thus cannot simply rest at the equilibrium position. By requiring the molecule to have some energy, such as E_1, we see that the atoms oscillate back and forth between two turning points. This is a *molecular vibration,* and atoms held together by molecular bonds are constantly vibrating. For a molecule having an energy $E_1 = 0.35 \times 10^{-18}$ J, as illustrated in Figure 10.37, the bond oscillates in length between roughly 0.10 nm and 0.18 nm.

Suppose we increase the molecule's energy to $E_2 = 1.25 \times 10^{-18}$ J. This could happen if the molecule absorbs some light. You can see from the energy diagram that atoms with this energy are not bound together at large values of x. There is no turning point on the right, so the atoms will keep moving apart. By raising the molecule's energy to E_2 we have *broken the molecular bond.* If the atoms happen to be moving together at the time the energy changes, they will "bounce" one last time (there is still a *left* turning point), then move away from each other and not return. The breaking of molecular bonds through the absorption of light is called *photodissociation.* It is an important process in the making of integrated circuits.

SUMMARY

The goal of Chapter 10 has been to introduce the ideas of kinetic and potential energy and to learn a new problem-solving strategy based on conservation of energy.

GENERAL PRINCIPLES

Law of Conservation of Mechanical Energy

If there are no friction or other energy-loss processes (to be explored more thoroughly in Chapter 11), then the mechanical energy $E_{mech} = K + U$ of a system is conserved. Thus

$$K_f + U_f = K_i + U_i$$

- K is the sum of the kinetic energies of all particles.
- U is the sum of all potential energies.

Solving Energy Conservation Problems

MODEL Choose a system without friction or other losses of mechanical energy.

VISUALIZE Draw a before-and-after pictorial representation.

SOLVE Use the law of conservation of energy

$$K_f + U_f = K_i + U_i$$

ASSESS Is the result reasonable?

IMPORTANT CONCEPTS

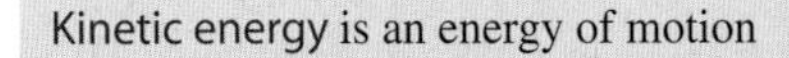

Kinetic energy is an energy of motion

$$K = \frac{1}{2}mv^2$$

Potential energy is an energy of position

- **Gravitational:** $U_g = mgy$
- **Elastic:** $U_s = \frac{1}{2}k(\Delta s)^2$

Basic Energy Model

Energy into system

Readily available energy K
Stored energy U
Mechanical energy $E_{mech} = K + U$

Energy can be transformed within the system without loss.

Energy out of system

Energy diagrams

These diagrams show the potential energy curve PE and the total mechanical energy line TE.

Energy
PE
K
TE
U
x

- The distance from the axis to the curve is PE.
- The distance from the curve to the TE line is KE.
- A point where the TE line crosses the PE curve is a **turning point.**
- Minima in the PE curve are points of **stable equilibrium.** Maxima are points of **unstable equilibrium.**

APPLICATIONS

Hooke's law

The restoring force of an ideal spring is

$$(F_{sp})_s = -k\Delta s$$

where k is the spring constant and $\Delta s = s - s_e$ is the displacement from equilibrium.

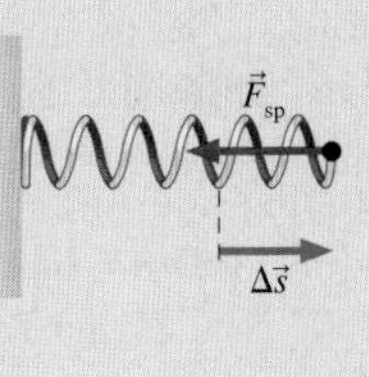

Perfectly elastic collisions

Both mechanical energy and momentum are conserved.

① $(v_{ix})_1$ → ② At rest

$$(v_{fx})_1 = \frac{m_1 - m_2}{m_1 + m_2}(v_{ix})_1 \qquad (v_{fx})_2 = \frac{2m_1}{m_1 + m_2}(v_{ix})_1$$

If ball 2 is moving, transform to a reference frame in which ball 2 is at rest.

TERMS AND NOTATION

energy
kinetic energy, K
gravitational potential energy, U_g
joule, J
mechanical energy
law of conservation of mechanical energy
restoring force
elastic
equilibrium length, L_0
displacement from equilibrium, Δs
spring constant, k
Hooke's law
elastic potential energy, U_s
perfectly elastic collision
energy diagram
stable equilibrium
unstable equilibrium

EXERCISES AND PROBLEMS

The icon indicates that the problem can be done on an Energy Worksheet.

Exercises

Section 10.2 Kinetic Energy and Gravitational Potential Energy

1. Which has the larger kinetic energy, a 10 g bullet fired at 500 m/s or a 10 kg bowling ball rolled at 10 m/s?
2. The lowest point in Death Valley is 85 m below sea level. The summit of nearby Mt. Whitney has an elevation of 4420 m. What is the change in potential energy of an energetic 65 kg hiker who makes it from the floor of Death Valley to the top of Mt. Whitney?
3. At what speed does a 1000 kg compact car have the same kinetic energy as a 20,000 kg truck going 25 km/hr?
4. An oxygen atom is four times as massive as a helium atom. In an experiment, a helium atom and an oxygen atom have the same kinetic energy. What is the ratio v_{He}/v_O of their speeds?
5. a. What is the kinetic energy of a 1500 kg car traveling at a speed of 30 m/s ($\approx$65 mph)?
   b. From what height would the car have to be dropped to have this same amount of kinetic energy just before impact?
   c. Does your answer to part b depend on the car's mass?
6. A boy reaches out of a window and tosses a ball straight up with a speed of 10 m/s. The ball is 20 m above the ground as he releases it. Use energy to find
   a. The ball's maximum height above the ground.
   b. The ball's speed as it passes the window on its way down.
   c. The speed of impact on the ground.
7. a. With what minimum speed must you toss a 100 g ball straight up to hit the 10-m-high roof of the gymnasium if you release the ball 1.5 m above the ground? Solve this problem using energy.
   b. With what speed does the ball hit the ground?

Section 10.3 A Closer Look at Gravitational Potential Energy

8. What minimum speed does a 100 g puck need to make it to the top of a 3.0-m-long, 20° frictionless ramp?
9. A 55 kg skateboarder wants to just make it to the upper edge of a "quarter pipe," a track that is one-quarter of a circle with a radius of 3.0 m. What speed does he need at the bottom?

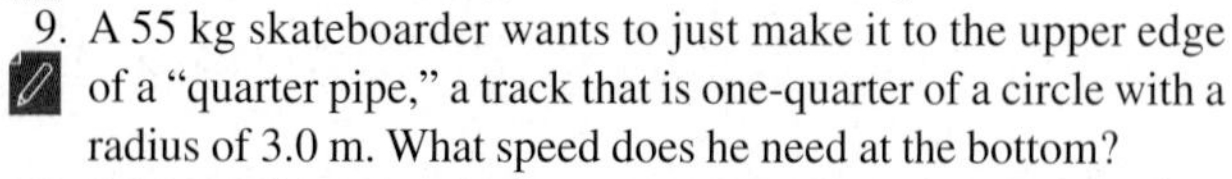

10. A 50 g ball is released from rest 1.0 m above the bottom of the track shown in Figure Ex10.10. It rolls down a straight 30° segment, then back up a parabolic segment whose shape is given by $y = \frac{1}{4}x^2$, where x and y are in m. How high will the ball go on the right before reversing direction and rolling back down?

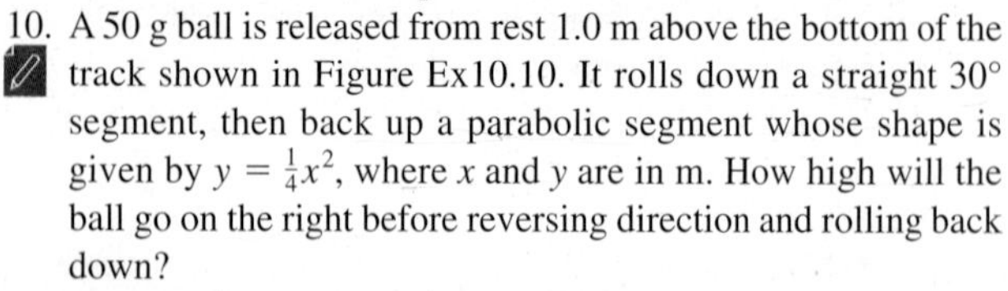

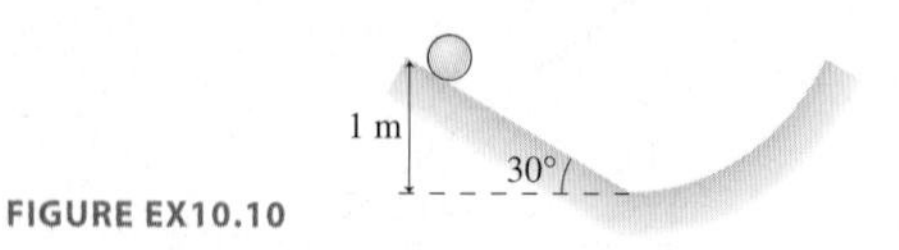

FIGURE EX10.10

11. A pendulum is made by tying a 500 g ball to a 75-cm-long string. The pendulum is pulled 30° to one side, then released.
    a. What is the ball's speed at the lowest point of its trajectory?
    b. To what angle does the pendulum swing on the other side?
12. A 20 kg child is on a swing that hangs from 3.0-m-long chains. What is her maximum speed if she swings out to a 45° angle?
13. A 1500 kg car is approaching the hill shown in Figure Ex10.13 at 10.0 m/s when it suddenly runs out of gas.
    a. Can the car make it to the top of the hill by coasting?
    b. If your answer to (a) is yes, what is the car's speed after coasting down the other side?

FIGURE EX10.13

Section 10.4 Restoring Forces and Hooke's Law

14. A 10-cm-long spring is attached to the ceiling. When a 2.0 kg mass is hung from it, the spring stretches to a length of 15 cm.
    a. What is the spring constant k?
    b. How long is the spring when a 3.0 kg mass is suspended from it?
15. A 5.0 kg mass hanging from a spring scale is slowly lowered onto a vertical spring, as shown in Figure Ex10.15.
    a. What does the spring scale read just before the mass touches the lower spring?
    b. The scale reads 20 N when the lower spring has been compressed by 2.0 cm. What is the value of the spring constant for the lower spring?
    c. At what compression length will the scale read zero?

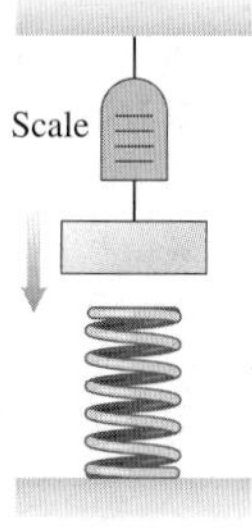

FIGURE EX10.15

16. A runner wearing spiked shoes pulls a 20 kg sled across frictionless ice using a horizontal spring with spring constant 150 N/m. The spring is stretched 20 cm from its equilibrium length. What is the acceleration of the sled?
17. You need to make a spring scale for measuring mass. You want each 1.0 cm length along the scale to correspond to a mass difference of 100 g. What should be the value of the spring constant?
18. A 60 kg student is standing atop a spring in an elevator that is accelerating upward at 3.0 m/s^2. The spring constant is 2500 N/m. By how much is the spring compressed?

Section 10.5 Elastic Potential Energy

19. How much energy can be stored in a spring with $k = 500$ N/m if the maximum possible stretch is 20 cm?
20. How far must you stretch a spring with $k = 1000$ N/m to store 200 J of energy?
21. A student places her 500 g physics book on a frictionless table. She pushes the book against a spring, compressing the spring by 4.0 cm, then releases the book. What is the book's speed as it slides away? The spring constant is 1250 N/m.

22. A block sliding along a horizontal frictionless surface with speed v collides with a spring and compresses it by 2.0 cm. What will be the compression if the same block collides with the spring at a speed of $2v$?
23. A 10 kg runaway grocery cart runs into a spring with spring constant 250 N/m and compresses it by 60 cm. What was the speed of the cart just before it hit the spring?
24. As a 15,000 kg jet plane lands on an aircraft carrier, its tail hook snags a cable to slow it down. The cable is attached to a spring with spring constant 60,000 N/m. If the spring stretches 30 m to stop the plane, what was the plane's landing speed?

Section 10.6 Elastic Collisions

25. A 50 g marble moving at 2.0 m/s strikes a 20 g marble at rest. What is the speed of each marble immediately after the collision?
26. A 50 g ball of clay traveling at speed v_0 hits and sticks to a 1.0 kg block sitting at rest on a frictionless surface.
    a. What is the speed of the block after the collision?
    b. Show that the mechanical energy is *not* conserved in this collision. What percentage of the ball's initial energy is "lost"?
27. Ball 1, with a mass of 100 g and traveling at 10 m/s, collides head on with ball 2, which has a mass of 300 g and is initially at rest. What are the final velocities of each ball if the collision is (a) perfectly elastic? (b) perfectly inelastic?
28. A proton is traveling to the right at 2.0×10^7 m/s. It has a head-on perfectly elastic collision with a carbon atom. The mass of the carbon atom is 12 times the mass of the proton. What are the speed and direction of each after the collision?

Section 10.7 Energy Diagrams

29. Figure Ex10.29 is the potential-energy diagram for a 20 g particle that is released from rest at $x = 1.0$ m.
    a. Will the particle move to the right or to the left? How can you tell?
    b. What is the particle's maximum speed? At what position does it have this speed?
    c. Where are the turning points of the motion?

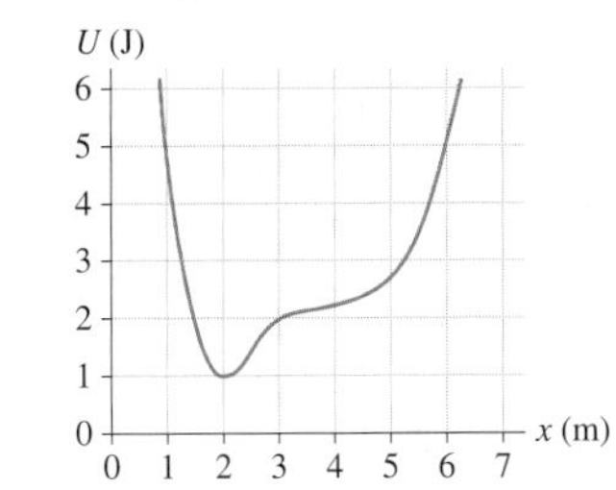

FIGURE EX10.29

30. Figure Ex10.30 is the potential-energy diagram for a 500 g particle that is released from rest at A. What are the particle's speeds at B, C, and D?

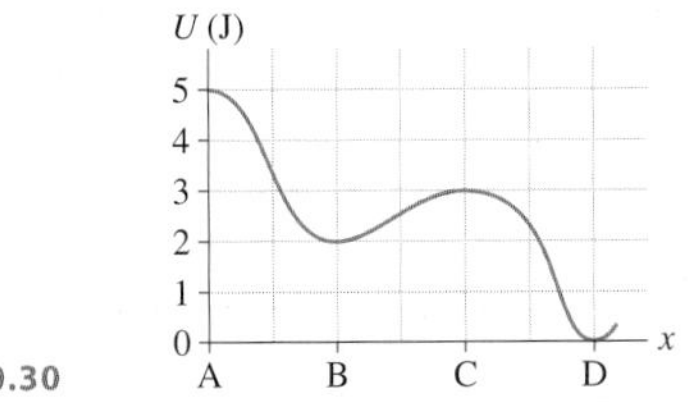

FIGURE EX10.30

31. What is the maximum speed of a 2.0 g particle that oscillates between $x = 2.0$ mm and $x = 8.0$ mm in Figure Ex10.31?

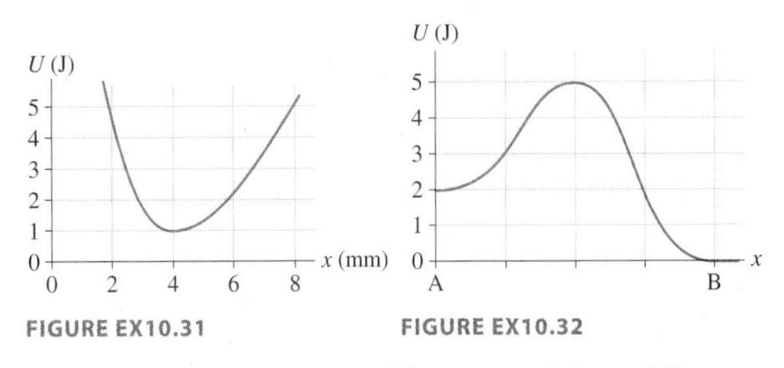

FIGURE EX10.31 FIGURE EX10.32

32. a. In Figure Ex10.32, what minimum speed does a 100 g particle need at point A to reach point B?
    b. What minimum speed does a 100 g particle need at point B to reach point A?

Problems

33. You're driving at 35 km/hr when the road suddenly descends 15 m into a valley. You take your foot off the accelerator and coast down the hill. Just as you reach the bottom you see the policeman hiding behind the speed limit sign that reads "70 km/hr." Are you going to get a speeding ticket?
34. A cannon tilted up at a 30° angle fires a cannon ball at 80 m/s from atop a 10-m-high fortress wall. What is the ball's impact speed on the ground below?
35. Your friend's Frisbee has become stuck 16 m above the ground in a tree. You want to dislodge the Frisbee by throwing a rock at it. The Frisbee is stuck pretty tight, so you figure the rock needs to be traveling at least 5.0 m/s when it hits the Frisbee.
    a. Does the speed with which you throw the rock depend on the angle at which you throw it? Explain.
    b. If you release the rock 2.0 m above the ground, with what minimum speed must you throw it?
36. A marble spins in a *vertical* plane around the inside of a smooth, 20-cm-diameter horizontal pipe. The marble's speed at the bottom of the circle is 3.0 m/s.
    a. What is the marble's speed at the top?
    b. Find an algebraic expression for the marble's speed when it is at angle θ, where the angle is measured from the bottom of the circle.
    c. Make a graph of v-versus-θ for one complete revolution.
37. A 50 g rock is placed in a slingshot and the rubber band is stretched. The force of the rubber band on the rock is shown by the graph in Figure P10.37.
    a. Is the rubber band stretched to the right or to the left? How can you tell?
    b. Does this rubber band obey Hooke's law? Explain.
    c. What is the rubber band's spring constant k?
    d. The rubber band is stretched 30 cm and then released. What is the speed of the rock?

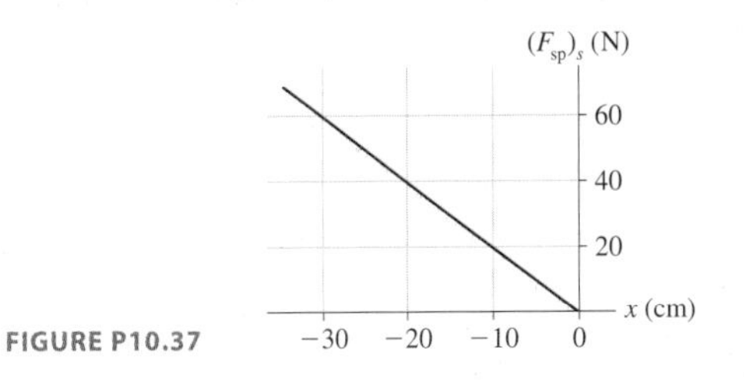

FIGURE P10.37

38. The spring in Figure P10.38a is compressed by length Δx. It launches the block across a frictionless surface with speed v_0. The two springs in Figure P10.38b are identical to the spring of Figure P10.38a. They are compressed by the same length Δx and used to launch the same block. What is the block's speed?

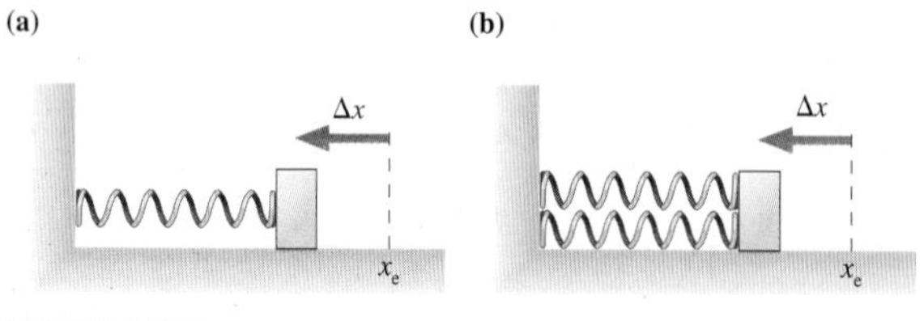

FIGURE P10.38

39. The spring in Figure P10.39a is compressed by length Δx. It launches the block across a frictionless surface with speed v_0. The two springs in Figure P10.39b are identical to the spring of Figure P10.39a. They are compressed the same *total* length Δx and used to launch the same block. What is the block's speed?

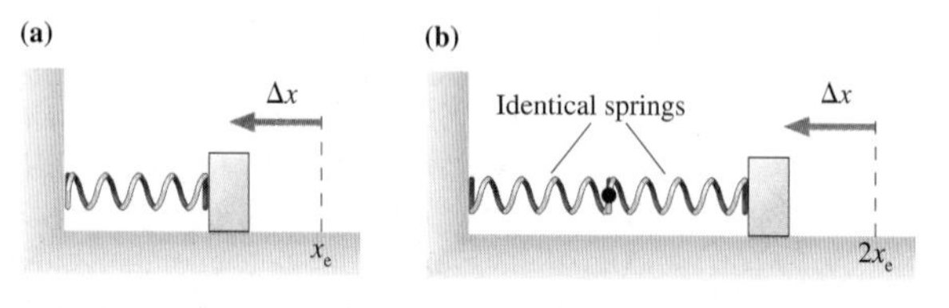

FIGURE P10.39

40. A 500 g rubber ball is dropped from a height of 10 m and undergoes a perfectly elastic collision with the earth.
    a. What is the earth's velocity after the collision? Assume the earth was at rest just before the collision.
    b. How many years would it take the earth to move 1.0 mm at this speed?
41. A 100 g marble rolls down a 40° incline. At the bottom, just after it exits onto a horizontal table, it collides with a 200 g steel ball at rest. How high above the table should the marble be released to give the steel ball a speed of 150 cm/s?
42. A package of mass m is released from rest at a warehouse loading dock and slides down a 3.0-m-high frictionless chute to a waiting truck. Unfortunately, the truck driver went on a break without having removed the previous package, of mass $2m$, from the bottom of the chute.
    a. Suppose the packages stick together. What is their common speed after the collision?
    b. Suppose the collision between the packages is elastic. To what height does the package of mass m rebound?

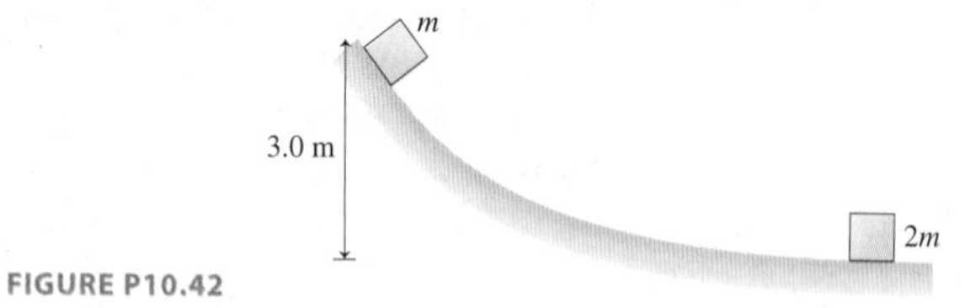

FIGURE P10.42

43. A 50 g ice cube can slide without friction up and down a 30° slope. The ice cube is pressed against a spring at the bottom of the slope, compressing the spring 10 cm. The spring constant is 25 N/m. When the ice cube is released, what distance will it travel up the slope before reversing direction?
44. A 1000 kg safe is 2.0 m above a heavy-duty spring when the rope holding the safe breaks. The safe hits the spring and compresses it 50 cm. What is the spring constant of the spring?
45. In a physics lab experiment, a spring clamped to the table is used to shoot a 20 g ball at a 30° angle. When the spring is compressed 20 cm, the ball travels horizontally 5.0 m and lands 1.5 m below the point at which it left the spring. What is the spring constant?
46. A vertical spring with $k = 490$ N/m is standing on the ground. You are holding a 5.0 kg block just above the spring, not quite touching it.
    a. How far does the spring compress if you let go of the block suddenly?
    b. How far does the spring compress if you slowly lower the block to the point where you can remove your hand without disturbing it?
    c. Why are your two answers different?
47. The desperate contestants on a TV survival show are very hungry. The only food they can see is some fruit hanging on a branch high in a tree. Fortunately, they have a spring they can use to launch a rock. The spring constant is 1000 N/m, and they can compress the spring a maximum of 30 cm. All the rocks on the island seem to have a mass of 400 g.
    a. With what speed does the rock leave the spring?
    b. If the fruit hangs 15 m above the ground, will they feast or go hungry?
48. A massless pan hangs from a spring that is suspended from the ceiling. When empty, the pan is 50 cm below the ceiling. If a 100 g clay ball is placed gently on the pan, the pan hangs 60 cm below the ceiling. Suppose the clay ball is dropped from the ceiling onto an empty pan. What is the pan's distance from the ceiling when the spring reaches its maximum length?
49. You have been hired to design a spring-launched roller coaster that will carry two passengers per car. The car goes up a 10-m-high hill, then descends 15 m to the track's lowest point. You've determined that the spring can be compressed a maximum of 2.0 m and that a loaded car will have a maximum mass of 400 kg. For safety reasons, the spring constant should be 10% larger than the minimum needed for the car to just make it over the top.
    a. What spring constant should you specify?
    b. What is the maximum speed of a 350 kg car if the spring is compressed the full amount?
50. It's been a great day of new, frictionless snow. Julie starts at the top of the 60° slope shown in Figure P10.50. At the bottom, a circular arc carries her through a 90° turn, and she then launches off a 3.0-m-high ramp. How far is her touchdown point from the base of the ramp?

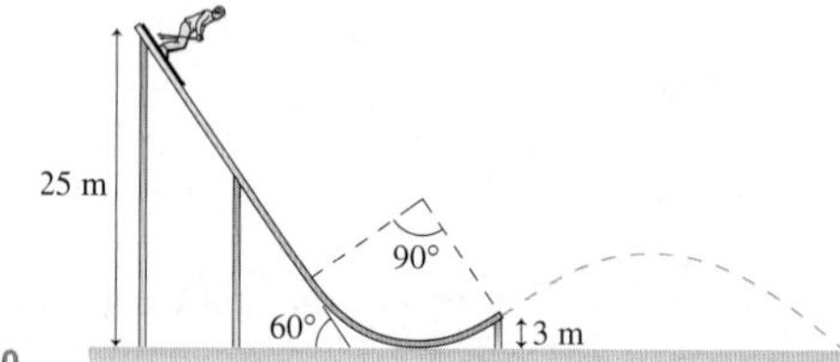

FIGURE P10.50

51. A 100 g block on a frictionless table is firmly attached to one end of a spring with $k = 20$ N/m. The other end of the spring is anchored to the wall. A 20 g ball is thrown horizontally toward the block with a speed of 5.0 m/s.
    a. If the collision is perfectly elastic, what is the ball's speed immediately after the collision?
    b. What is the maximum compression of the spring?
    c. Repeat parts a and b for the case of a perfectly inelastic collision.

52. You have been asked to design a "ballistic spring system" to measure the speed of bullets. A bullet of mass m is fired into a block of mass M. The block, with the embedded bullet, then slides across a frictionless table and collides with a horizontal spring whose spring constant is k. The opposite end of the spring is anchored to a wall. The spring's maximum compression d is measured.
    a. Find an expression for the bullet's initial speed v_B in terms of m, M, k, and d.
    b. What was the speed of a 5.0 g bullet if the block's mass is 2.0 kg and if the spring, with $k = 50$ N/m, was compressed by 10 cm?
    c. What fraction of the bullet's energy is "lost"? Where did it go?
53. A roller coaster car on the frictionless track shown in Figure P10.53 starts from rest at height h. The track is straight until point A. Between points A and D, the track consists of circle-shaped segments of radius R.
    a. What is the *maximum* height h_{max} from which the car can start so as not to fly off the track when going over the hill at point C? Give your answer in terms of the radius R. **Hint:** This is a two-part problem. First find v_{max} at C.
    b. Evaluate h_{max} for a roller coaster that has $R = 10$ m.

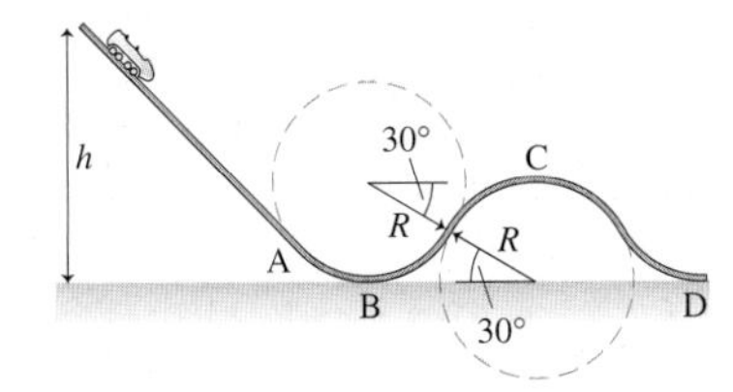

FIGURE P10.53

54. A pendulum is formed from a small ball of mass m on a string of length L. As Figure P10.54 shows, a peg is height $h = L/3$ above the pendulum's lowest point. From what minimum angle θ must the pendulum be released in order for the ball to go over the top of the peg without the string going slack?

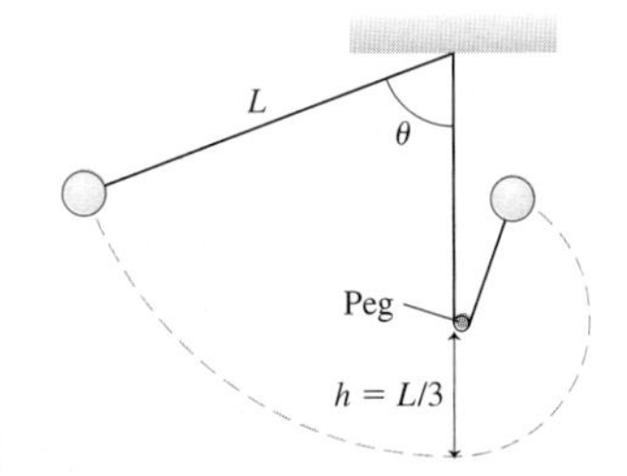

FIGURE P10.54

55. A block of mass m slides down a frictionless track, then around the inside of a circular loop-the-loop of radius R. From what minimum height h must the block start to make it around the loop without falling off? Give your answer as a multiple of R.
56. A new event has been proposed for the Winter Olympics. An athlete will sprint 100 m, starting from rest, then leap onto a 20 kg bobsled. The person and bobsled will then slide down a 50-m-long ice-covered ramp, sloped at 20°, and into a spring with a carefully calibrated spring constant of 2000 N/m. The athlete who compresses the spring the farthest wins the gold medal.

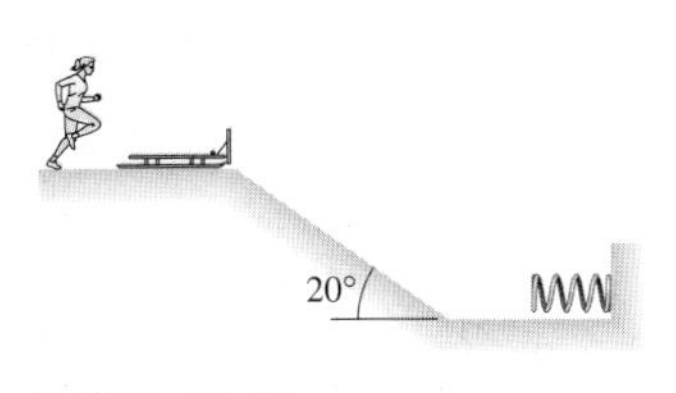

FIGURE P10.56

Lisa, whose mass is 40 kg, has been training for this event. She can reach a maximum speed of 12 m/s in the 100 m dash.
    a. How far will Lisa compress the spring?
    b. The Olympic committee has very exact specifications about the shape and angle of the ramp. Is this necessary? If the committee asks your opinion, what factors about the ramp will you tell them are important?
57. A 20 g ball is fired horizontally with initial speed v_0 toward a 100 g ball that is hanging motionless from a 1.0-m-long string. The balls undergo a head-on, perfectly elastic collision, after which the 100 g ball swings out to a maximum angle $\theta_{max} = 50°$. What was v_0?
58. A 100 g ball moving to the right at 4.0 m/s collides head-on with a 200 g ball that is moving to the left at 3.0 m/s.
    a. If the collision is perfectly elastic, what are the speed and direction of each ball after the collision?
    b. If the collision is perfectly inelastic, what are the speed and direction of the combined balls after the collision?
59. A 100 g ball moving to the right at 4.0 m/s catches up and collides with a 400 g ball that is moving to the right at 1.0 m/s. If the collision is perfectly elastic, what are the speed and direction of each ball after the collision?
60. Figure P10.60 shows the potential energy of a 500 g particle as it moves along the x-axis. Suppose the particle's mechanical energy is 12 J.
    a. Where are the particle's turning points?
    b. What is the particle's speed when it is at $x = 2.0$ m?
    c. What is the particle's maximum speed? At what position or positions does this occur?
    d. Give a written description of the motion of the particle as it moves from the left turning point to the right turning point.
    e. Suppose the particle's energy is lowered to 4.0 J. Describe the possible motions.

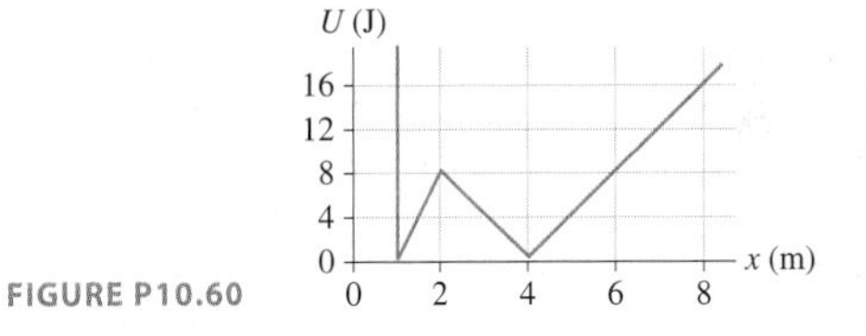

FIGURE P10.60

61. The ammonia molecule NH_3 has the tetrahedral structure shown in Figure P10.61a. The three hydrogen atoms form a triangle in the xy-plane at $z = 0$. The nitrogen atom is the apex of the pyramid. Figure P10.61b shows the potential energy of the nitrogen atom along a z-axis that is perpendicular to the H_3 triangle.
    a. At room temperature, the nitrogen atom has $\approx 0.4 \times 10^{-20}$ J of mechanical energy. Describe the position and motion of the nitrogen atom.
    b. The nitrogen atom can gain energy if the molecule absorbs light energy. Describe the motion of the nitrogen atom if its energy is 4×10^{-20} J.

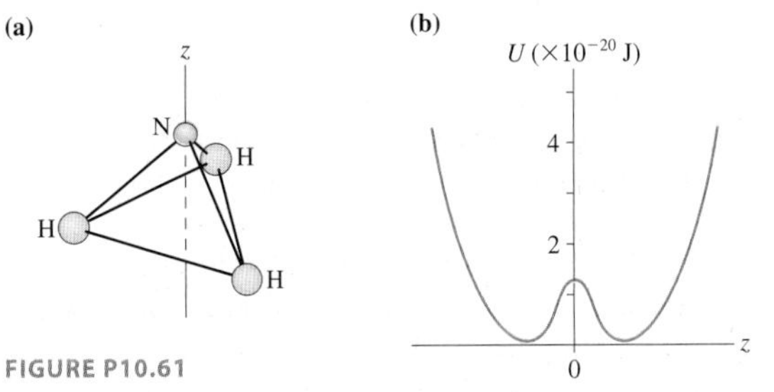

FIGURE P10.61

62. Protons and neutrons (together called *nucleons*) are held together in the nucleus of an atom by a force called the *strong force*. At very small separations, the strong force between two nucleons is larger than the repulsive electrical force between two protons—hence its name. But the strong force quickly weakens as the distance between the protons increases. A well-established model for the potential energy of two nucleons interacting via the strong force is

$$U = U_0\left[1 - e^{-x/x_0}\right]$$

where x is the distance between the centers of the two nucleons, x_0 is a constant having the value $x_0 = 2.0 \times 10^{-15}$ m, and $U_0 = 6.0 \times 10^{-11}$ J.

a. Calculate and draw an accurate potential-energy curve from $x = 0$ m to $x = 10 \times 10^{-15}$ m. Either calculate about 10 points by hand or use computer software.

b. Quantum effects are essential for a proper understanding of how nucleons behave. Nonetheless, let us innocently consider two neutrons *as if* they were small, hard, electrically neutral spheres of mass 1.67×10^{-27} kg and diameter 1.0×10^{-15} m. (We will consider neutrons rather than protons so as to avoid complications from the electric forces between protons.) You are going to hold two neutrons 5.0×10^{-15} m apart, measured between their centers, then release them. Draw the total energy line for this situation on your diagram of part a.

c. What speed do *each* of these neutrons have as they crash together? Keep in mind that *both* neutrons are moving.

63. Write a realistic problem for which the energy bar chart shown in Figure P10.63 correctly shows the energy at the beginning and end of the problem.

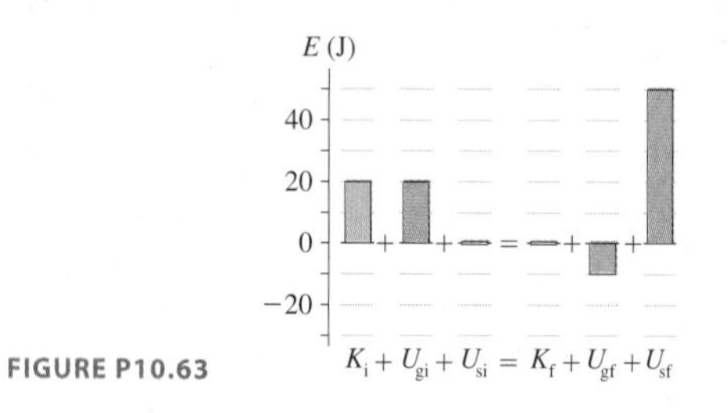

FIGURE P10.63

In Problems 64 through 67 you are given the equation used to solve a problem. For each of these, you are to

a. Write a realistic problem for which this is the correct equation.
b. Draw the before-and-after pictorial representation.
c. Finish the solution of the problem.

64. $\frac{1}{2}(1500\text{ kg})(5.0\text{ m/s})^2 + (1500\text{ kg})(9.80\text{ m/s}^2)(10\text{ m})$

$= \frac{1}{2}(1500\text{ kg})(v_i)^2 + (1500\text{ kg})(9.80\text{ m/s}^2)(0\text{ m})$

65. $\frac{1}{2}(0.20\text{ kg})(2.0\text{ m/s})^2 + \frac{1}{2}k(0\text{ m})^2$

$= \frac{1}{2}(0.20\text{ kg})(0\text{ m/s})^2 + \frac{1}{2}k(-0.15\text{ m})^2$

66. $(0.10\text{ kg} + 0.20\text{ kg})v_{1x} = (0.10\text{ kg})(3.0\text{ m/s})$

$\frac{1}{2}(0.30\text{ kg})(0\text{ m/s})^2 + \frac{1}{2}(3.0\text{ N/m})(\Delta x_2)^2$

$= \frac{1}{2}(0.30\text{ kg})(v_{1x})^2 + \frac{1}{2}(3.0\text{ N/m})(0\text{ m})^2$

67. $\frac{1}{2}(0.50\text{ kg})(v_f)^2 + (0.50\text{ kg})(9.80\text{ m/s}^2)(0\text{ m})$

$+ \frac{1}{2}(400\text{ N/m})(0\text{ m})^2$

$= \frac{1}{2}(0.50\text{ kg})(0\text{ m/s})^2$

$+ (0.50\text{ kg})(9.80\text{ m/s}^2)((-0.10\text{ m})\sin 30°)$

$+ \frac{1}{2}(400\text{ N/m})(-0.10\text{ m})^2$

Challenge Problems

68. It's your birthday, and to celebrate you're going to make your first bungee jump. You stand on a bridge that is 100 m above a river and attach a 30-m-long bungee cord to your harness. A bungee cord, for practical purposes, is just a long spring, and this cord has a spring constant of 40 N/m. Assume that your mass is 80 kg. After a long hesitation, you dive off the bridge. How far are you above the water when the cord reaches its maximum elongation?

69. A 10 kg box slides 4.0 m down the frictionless ramp shown in Figure CP10.69, then collides with a spring whose spring constant is 250 N/m.

a. What is the maximum compression of the spring?
b. At what compression of the spring does the box have its maximum velocity?

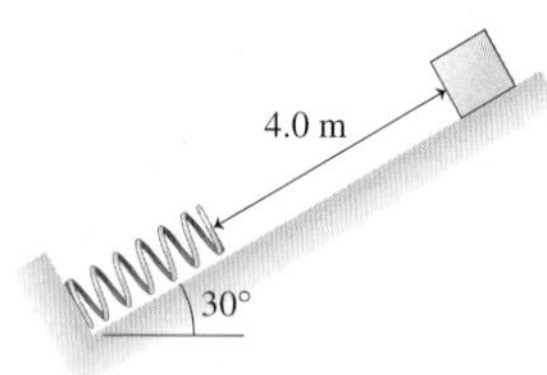

FIGURE CP10.69

70. Old naval ships fired 10 kg cannon balls from a 200 kg cannon. It was very important to stop the recoil of the cannon, since otherwise the heavy cannon would go careening across the deck of the ship. In one design, a large spring with spring constant 20,000 N/m was placed behind the cannon. The other end of the spring braced against a post that was firmly anchored to the ship's frame. What was the speed of the cannon ball if the spring compressed 50 cm when the cannon was fired?

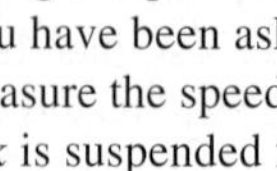

71. You have been asked to design a "ballistic spring system" to measure the speed of bullets. A spring whose spring constant is k is suspended from the ceiling. A block of mass M hangs from the spring. A bullet of mass m is fired vertically upward

into the bottom of the block. The spring's maximum compression d is measured.

a. Find an expression for the bullet's initial speed v_B in terms of m, M, k, and d.

b. What was the speed of a 10 g bullet if the block's mass is 2.0 kg and if the spring, with $k = 50$ N/m, was compressed by 45 cm?

72. A 2.0 kg cart has a spring with $k = 5000$ N/m attached to its front, parallel to the ground. This cart rolls at 4.0 m/s toward a stationary 1 kg cart.

a. What is the maximum compression of the spring during the collision?

b. What is the speed of each cart after the collision?

73. A 100 g steel ball and a 200 g steel ball each hang from 1-m-long strings. At rest, the balls hang side by side, barely touching. The 100 g ball is pulled to the left until its string is at a 45° angle. The 200 g ball is pulled to a 45° angle on the right. The balls are released so as to collide at the very bottom of their swings. To what angle does each ball rebound?

74. A sled starts from rest at the top of the frictionless, hemispherical, snow-covered hill shown in Figure CP10.74.

a. Find an expression for the sled's speed when it is at angle ϕ.

b. Use Newton's laws to find the maximum speed the sled can have at angle ϕ without leaving the surface.

c. At what angle ϕ_{max} does the sled "fly off" the hill?

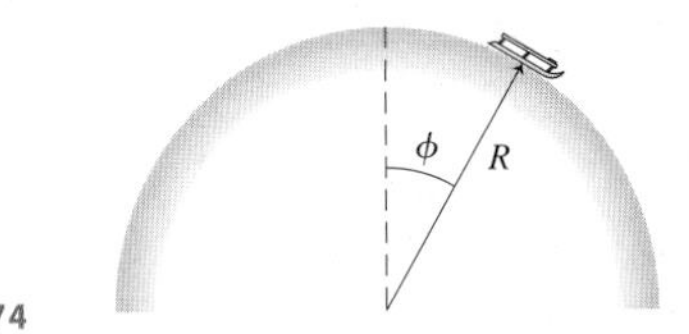

FIGURE CP10.74

STOP TO THINK ANSWERS

Stop to Think 10.1: $(U_g)_3 > (U_g)_2 = (U_g)_4 > (U_g)_1$. Gravitational potential energy depends only on height, not speed.

Stop to Think 10.2: $v_A = v_B = v_C = v_D$. Her increase in kinetic energy depends only on the vertical height through which she falls, not the shape of the slide.

Stop to Think 10.3: b. Mechanical energy is conserved on a frictionless surface. Because $K_i = 0$ and $K_f = 0$, it must be true that $U_f = U_i$ and thus $y_f = y_i$. The final height matches the initial height.

Stop to Think 10.4: $k_1 > k_2 > k_3$. The spring constant is the slope of the force-versus-displacement graph.

Stop to Think 10.5: c. U_s depends on Δs^2, so doubling the compression increases U_s by a factor of 4. All the potential energy is converted to kinetic energy, so K increases by a factor of 4. But K depends on v^2, so v increases by only a factor of $(4)^{1/2} = 2$.

Stop to Think 10.6: $x = 6$ m. From the graph, the particle's potential energy at $x = 1$ m is $U = 3$ J. Its total energy is thus $E = K + U = 4$ J. A TE line at 4 J crosses the PE curve at $x = 6$ m.

11 Work

This bobsled team is increasing the sled's kinetic energy by pushing it forward. In the language of physics, they are doing *work* on the sled.

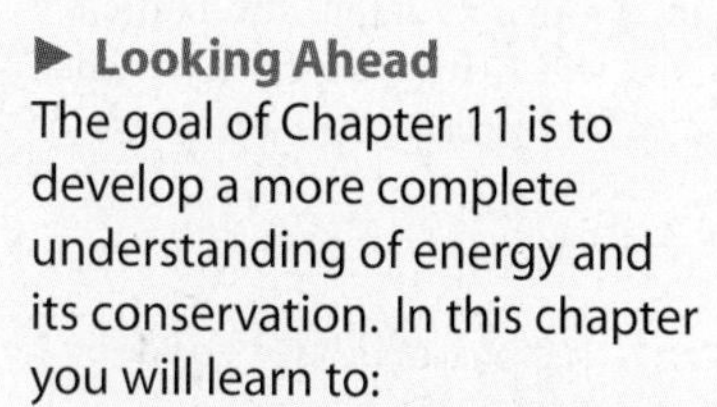

► Looking Ahead

The goal of Chapter 11 is to develop a more complete understanding of energy and its conservation. In this chapter you will learn to:

- Understand and apply the basic energy model.
- Calculate the work done on a system.
- Understand and use a more complete statement of conservation of energy.
- Use a general strategy for solving energy problems.
- Calculate the power supplied to or dissipated by a system.

◄ Looking Back

This chapter continues to develop energy ideas that were introduced in Chapter 10. Please review:

- Sections 10.2–10.3 Gravitational potential energy.
- Sections 10.4–10.5 Hooke's law and elastic potential energy.

Chapter 10 introduced the concept of energy. Although energy appears to be a useful idea, three major questions remain unanswered:

- How many kinds of energy are there?
- Under what conditions is energy conserved?
- How does a system gain or lose energy?

For example, this bobsled is gaining kinetic energy, but it's not doing so by losing potential energy. Instead, the runners are giving it kinetic energy by pushing it faster and faster. One of our goals in this chapter is to relate the energy gained by the sled to the strength of their push. Energy transferred by pushes and pulls is called *work*.

We will also explore how energy is *dissipated*. Because of friction, a bobsled sliding across a horizontal surface gradually slows and stops. What happens to its kinetic energy? By addressing these issues, we will put the concept of energy on a firmer foundation.

11.1 The Basic Energy Model

We will begin with an overview of where this chapter will take us, then come back to fill in the details. Consider a *system* of particles. The system can be characterized by two quantities that we call the *kinetic energy* and the *potential energy.* Kinetic energy is an energy due to the *motion* of the particles. The potential energy, which is often thought of as "stored energy," is due to *interactions* between the particles. For example, two particles connected by a stretched spring have a potential energy.

The sum of kinetic and potential energy is the *mechanical energy* $E_{mech} = K + U$. The term *mechanical* designates this form of energy as being due to motion and mechanical effects, such as stretching springs, rather than chemical effects or heat effects.

Mechanical energy is an energy of the object as a whole. That is, it is an energy *of the ball* or *of the rocket.* Suppose a ball is at rest, with zero mechanical energy. If you peered inside with a very powerful microscope, you would see the atoms inside the ball vibrating back and forth on their spring-like molecular bonds. This microscopic motion of the atoms and molecules within an object is a form of energy that is distinct from the object's mechanical energy. The total energy of the moving atoms is called the **thermal energy E_{th}.**

Thermal energy is associated with the system's *temperature.* A higher temperature means more microscopic motion and thus more thermal energy. Friction raises the temperature—think of rubbing your hands together briskly—so a system with friction "runs down" as its mechanical energy is transformed into thermal energy. We'll say more about thermal energy in Section 11.7.

We can define the **system energy E_{sys}** as the sum of the mechanical energy *of* the objects plus the thermal energy of the atoms *inside* the objects. That is,

$$E_{sys} = E_{mech} + E_{th} = K + U + E_{th} \tag{11.1}$$

As Figure 11.1 shows, kinetic and potential energy can be changed back and forth into each other. You studied this process in Chapter 10. Kinetic and potential energy can also be changed into thermal energy, but, as we'll discuss later, thermal energy is not normally changed into kinetic or potential energy. These energy exchanges within the system are called **energy transformations.** Energy transformations within the system do not change the value of E_{sys}.

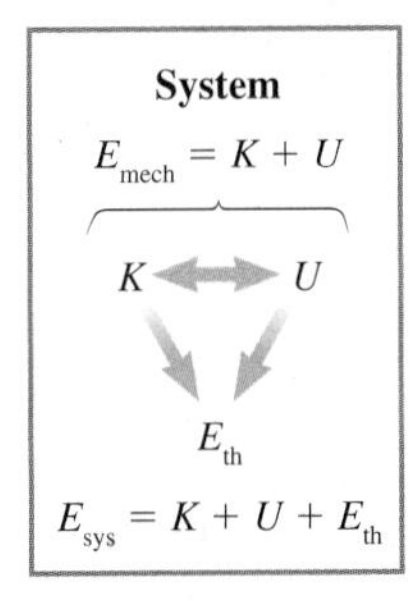

FIGURE 11.1 Energy can be transformed within the system.

NOTE ▶ We will use an arrow → as a shorthand way to indicate an energy transformation. If the kinetic energy of a box is transformed into thermal energy as it slides across a table, we will indicate this by $K \rightarrow E_{th}$. ◀

A system of particles is always surrounded by a larger *environment.* Unless the system is completely isolated, it has the possibility of exchanging energy with the environment. An energy exchange between the system and the environment is called an **energy transfer.** There are two primary energy transfer processes. The first, and the only one we will be concerned with for now, is due to forces—pushes and pulls—exerted on the system by the environment. For example, you give a ball kinetic energy by pushing on it. This *mechanical* transfer of energy to or from the system is called **work.** The symbol for work is W.

The second means of transferring energy between the system and its environment is a *nonmechanical* process called *heat.* Heat is a crucial idea that we will add to the energy model when we study thermodynamics, but for now we want to concentrate on the mechanical transfer of energy via work.

Figure 11.2 shows a **basic energy model** in which energy can be *transferred* to or from the system and energy can be *transformed* within the system. Notice how similar Figure 11.2 is to John's model of the monetary system in Chapter 10. As a *basic* model, Figure 11.2 is certainly not complete, and we will add new features to the model as needed. Nonetheless, it is a good starting point.

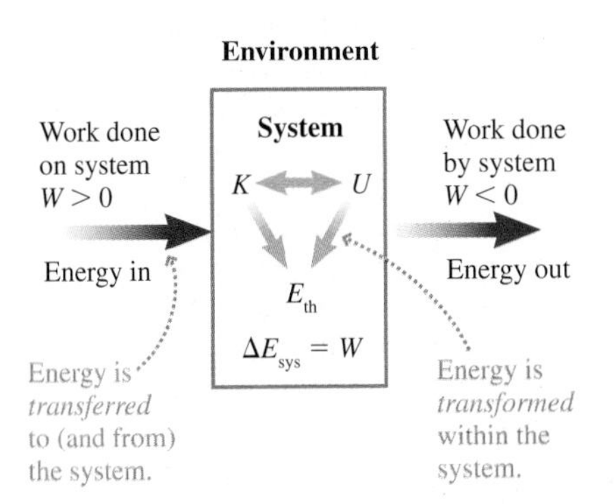

FIGURE 11.2 The basic energy model of a system interacting with its environment.

As the arrows in Figure 11.2 show, energy can both enter and leave the system. We'll distinguish between the two directions of energy flow by allowing the work W to be either positive or negative.

The sign of W is interpreted as follows:

$W > 0$ **The environment does work on the system and the system's energy increases,**

$W < 0$ **The system does work on the environment and the system's energy decreases.**

This is equivalent to considering expenditures (i.e., money out) to be negative income. In fact, this is how accountants really do handle expenditures.

Having established the quantities of the basic energy model, we can ask, what is the relationship among them? Our hypothesis, which is confirmed by experiment, is that

$$\Delta E_{sys} = \Delta K + \Delta U + \Delta E_{th} = W \tag{11.2}$$

The two essential ideas of the basic energy model and Equation 11.2 are:

1. Energy can be *transferred* to a system by doing work on the system. This process changes the total energy of the system: $\Delta E_{sys} = W$.
2. If no work is done, energy can be *transformed* within the system between K, U, and E_{th} as long as the total energy of the system doesn't change: $\Delta E_{sys} = 0$.

This is the essence of the basic energy model. The rest of Chapter 11 will substantiate Equation 11.2 and look at its many implications.

STOP TO THINK 11.1 A child slides down a playground slide at constant speed. The energy transformation is

a. $U \rightarrow K$ b. $K \rightarrow U$
c. There is no transformation because energy is conserved.
d. $U \rightarrow E_{th}$ e. $K \rightarrow E_{th}$

11.2 Work and Kinetic Energy

One dictionary defines *work* as:

1. Physical or mental effort; labor.
2. The activity by which one makes a living.
3. A task or duty.
4. Something produced as a result of effort, such as a *work of art*.
5. Plural *works:* A factory or plant where industry is carried on, such as *steel works*.
6. Plural *works:* The essential or operating parts of a mechanism.
7. The transfer of energy to a body by application of a force.

Work is a common word in the English language, with many meanings. When you first think of work, you probably think of the first two definitions in this list. After all, we talk about "working out," or we say, "I just got home from work." But that is *not* what work means in physics.

The basic energy model uses *work* in the sense of definition 7: Energy transferred to or from a body or system by the application of force. The critical question we must answer is: *How much energy* does a force transfer?

We can answer this question by following the procedure we used in Chapter 10 to find the potential energy of gravity and of a spring. Figure 11.3 shows a force $\vec{F}$ acting on a particle of mass m as the particle moves along an s-axis from an initial position s_i, with kinetic energy K_i, to a final position s_f where the kinetic energy is K_f. As the figure suggests, the force may vary in magnitude and/or direction as the particle moves.

NOTE ▶ $\vec{F}$ may not be the only force acting on the particle. For the particle to move along a straight line, some other force must cancel any component of $\vec{F}$ perpendicular to the s-axis. However, for now we'll assume that $\vec{F}$ is the only force with a component parallel to the s-axis and hence is the only force capable of changing the particle's speed. ◀

The force component F_s parallel to the s-axis causes the particle to speed up or slow down, thus transferring energy to or from the particle. We say that force $\vec{F}$

does work on the particle. Our goal is to find a relationship between F_s and ΔK. The s-component of Newton's second law is

$$F_s = ma_s = m\frac{dv_s}{dt} \tag{11.3}$$

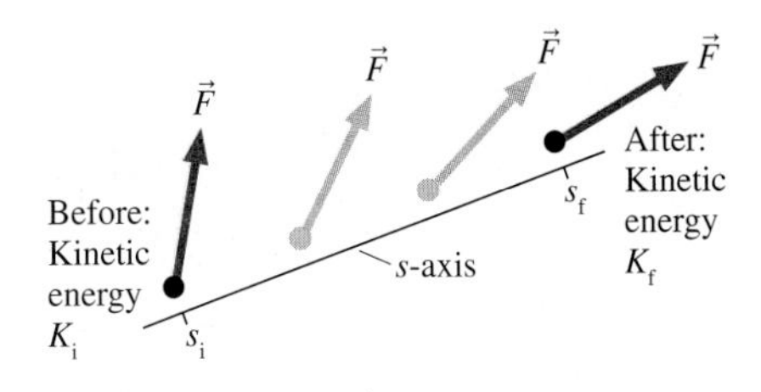

FIGURE 11.3 Force $\vec{F}$ does work as the particle moves from s_i to s_f.

where the v_s is the s-component of $\vec{v}$. As we did in Chapter 10, we can use the chain rule to write

$$m\frac{dv_s}{dt} = m\frac{dv_s}{ds}\frac{ds}{dt} = mv_s\frac{dv_s}{ds} \tag{11.4}$$

where $ds/dt = v_s$. Substituting Equation 11.4 into Equation 11.3 gives

$$F_s = mv_s\frac{dv_s}{ds} \tag{11.5}$$

The crucial step here, as it was in Chapter 10, was changing from a derivative with respect to time to a derivative with respect to position. We're going to want to integrate, so first multiply through by ds to get

$$mv_s\,dv_s = F_s\,ds \tag{11.6}$$

Now we can integrate both sides from "before," where the position is s_i and the speed is v_i, to "after," giving

$$\int_{v_i}^{v_f} mv_s\,dv_s = \frac{1}{2}mv_s^2\Big|_{v_i}^{v_f} = \frac{1}{2}mv_f^2 - \frac{1}{2}mv_i^2 = \int_{s_i}^{s_f} F_s\,ds \tag{11.7}$$

The left side of Equation 11.7 is ΔK, the change in the particle's kinetic energy as it moves from s_i to s_f. The integral on the right apparently specifies the extent to which the applied force changes the particle's kinetic energy. We define the *work* done by force $\vec{F}$ as the particle moves from s_i to s_f as

$$W = \int_{s_i}^{s_f} F_s\,ds \tag{11.8}$$

The unit of work, that of force multiplied by distance, is the N m. Using the definition of the newton gives

$$1\text{ N m} = 1\ (\text{kg m/s}^2)\text{ m} = 1\text{ kg m}^2/\text{s}^2 = 1\text{ J}$$

Thus the unit of work is really the unit of energy. This is consistent with the idea that work is a transfer of energy. Rather than use N m, we will measure work, just as we do energy, in joules.

Using Equation 11.8 as the definition of work, we can write Equation 11.7 as

$$\Delta K = W \tag{11.9}$$

Equation 11.9 is the quantitative statement that a force transfers kinetic energy to a particle, and thus *changes* the particle's kinetic energy, by pushing or pulling on it. Furthermore, **Equation 11.8 gives us a specific method to calculate *how much* energy is transferred by the push or pull.** This energy transfer, by mechanical means, is what we mean by the term *work*.

Notice that *no* work is done if $s_f = s_i$ because an integral that spans no interval is zero. **To change a particle's energy, a force must be applied as the particle undergoes a displacement.** If you were to hold a 200 lb weight over your head, you might break out in a sweat and your arms would tire. You might "feel" that you had done a lot of work, but you would have done *zero* work in the physics sense because the weight was not displaced while you were holding it and thus you transferred no energy to it.

This pitcher is increasing the ball's kinetic energy by doing work on it.

The Work-Kinetic Energy Theorem

Equation 11.8 is the work done by one force. Because $\vec{F}_{net} = \sum \vec{F}_i$, it's easy to see that the net work done on a particle by several forces is $W_{net} = \sum W_i$, where W_i is the work done by force $\vec{F}_i$. In that case, Equation 11.9 becomes

$$\Delta K = W_{net} \tag{11.10}$$

This basic idea—that the net work done on a particle causes the particle's kinetic energy to change—is a general principle, one worth giving a name:

> **The work-kinetic energy theorem** When one or more forces act on a particle as it is displaced from an initial position to a final position, the net work done on the particle by these forces causes the particle's kinetic energy to *change* by $\Delta K = W_{net}$.

One of the questions that opened this chapter was "How does a system gain or lose energy?" The work-kinetic energy theorem begins to answer that question by saying that a system gains or loses kinetic energy when work (done by forces originating in the environment) transfers energy between the environment and the system.

An Analogy with the Impulse-Momentum Theorem

You might have noticed that there is a similarity between the work-kinetic energy theorem and the impulse-momentum theorem of Chapter 9:

$$\begin{aligned} &\text{Work-kinetic energy theorem:} \quad \Delta K = W = \int_{s_i}^{s_f} F_s\, ds \\ &\text{Impulse-momentum theorem:} \quad \Delta p_s = J_s = \int_{t_i}^{t_f} F_s\, dt \end{aligned} \tag{11.11}$$

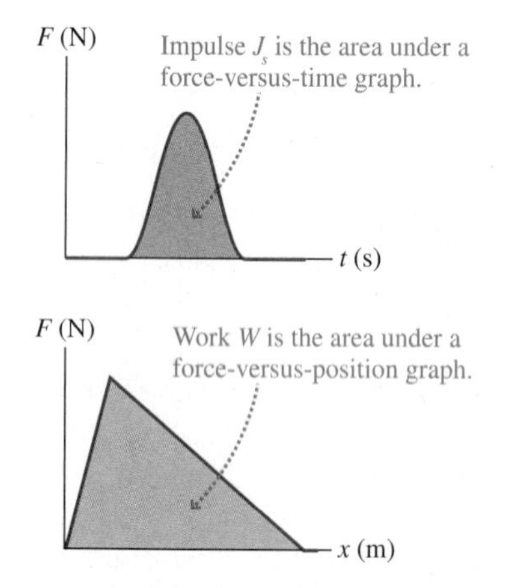

FIGURE 11.4 Impulse and work are both the area under a force graph, but it's very important to know what the horizontal axis is.

In both cases, a force acting on a particle changes the state of the system. If the force acts over a time interval from t_i to t_f, it creates an *impulse* that changes the particle's momentum. If the force acts over the spatial interval from s_i to s_f, it does *work* that changes the particle's kinetic energy. Figure 11.4 shows that the geometric interpretation of impulse as the area under the F-versus-t graph applies equally well to an interpretation of work as the area under the F-versus-s graph.

This does not mean that a force *either* creates an impulse *or* does work but does not do both. Quite the contrary. A force acting on a particle *both* creates an impulse *and* does work, changing both the momentum and the kinetic energy of the particle. Whether you use the work-kinetic energy theorem or the impulse-momentum theorem depends on the question you are trying to answer.

We can, in fact, express the kinetic energy in terms of the momentum as

$$K = \frac{1}{2}mv^2 = \frac{(mv)^2}{2m} = \frac{p^2}{2m} \tag{11.12}$$

You cannot change a particle's kinetic energy without also changing its momentum.

STOP TO THINK 11.2 A particle moving along the x-axis experiences the force shown in the graph. If the particle has 2.0 J of kinetic energy as it passes $x = 0$ m, what is its kinetic energy when it reaches $x = 4$ m?

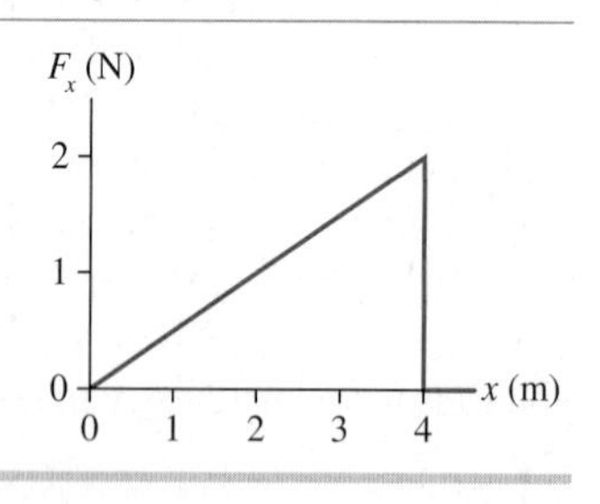

11.3 Calculating and Using Work

The work-kinetic energy theorem is a formal statement about the energy transferred to or from a particle by pushes and pulls. For this to be useful, we must be able to calculate the work. In this section we'll practice calculating work and using the work-kinetic energy theorem. We'll also introduce a new mathematical idea, the *dot product* of two vectors, that will allow us to write the work in a compact notation.

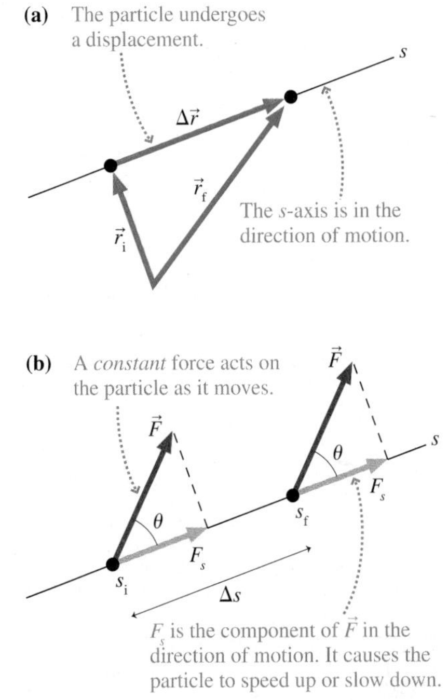
(a) The particle undergoes a displacement.
$\Delta\vec{r}$
$\vec{r}_i$
$\vec{r}_f$
s
The s-axis is in the direction of motion.
(b) A *constant* force acts on the particle as it moves.
$\vec{F}$
θ
F_s
s_i
s_f
Δs
F_s is the component of $\vec{F}$ in the direction of motion. It causes the particle to speed up or slow down.

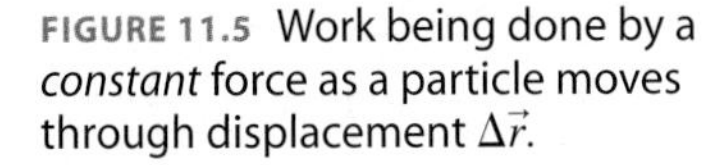

FIGURE 11.5 Work being done by a *constant* force as a particle moves through displacement $\Delta\vec{r}$.

Constant Force

We'll begin by calculating the work done by a force $\vec{F}$ that acts with a *constant* strength and in a *constant* direction as a particle moves along a straight line through a displacement $\Delta\vec{r}$. As Figure 11.5a shows, we'll define the s-axis to point in the direction of motion. In that case, Δs is equal to Δr, the magnitude of the displacement vector $\Delta\vec{r}$.

Figure 11.5b shows the force acting on the particle as it moves along the line. The force vector $\vec{F}$ makes an angle θ with respect to the displacement $\Delta\vec{r}$, so the component of the force vector along the direction of motion is $F_s = F\cos\theta$. According to Equation 11.8, the work done on the particle by this force is

$$W = \int_{s_i}^{s_f} F_s\, ds = \int_{s_i}^{s_f} F\cos\theta\, ds$$

Both F and θ are constant, so they can be taken outside the integral. Thus

$$W = F\cos\theta \int_{s_i}^{s_f} ds = F\cos\theta(s_f - s_i) = F\cos\theta(\Delta s) = F(\Delta r)\cos\theta \qquad (11.13)$$

We can use Equation 11.13 to calculate the work done by a constant force if we know the magnitude F of the force, the angle θ of the force from the line of motion, and the distance Δr through which the particle is displaced.

Activ Online Physics

5.1

EXAMPLE 11.1 Pulling a suitcase

A rope inclined upward at a 45° angle pulls a suitcase through the airport. The tension in the rope is 20 N. How much work does the tension do if the suitcase is pulled 100 m?

MODEL Model the suitcase as a particle.

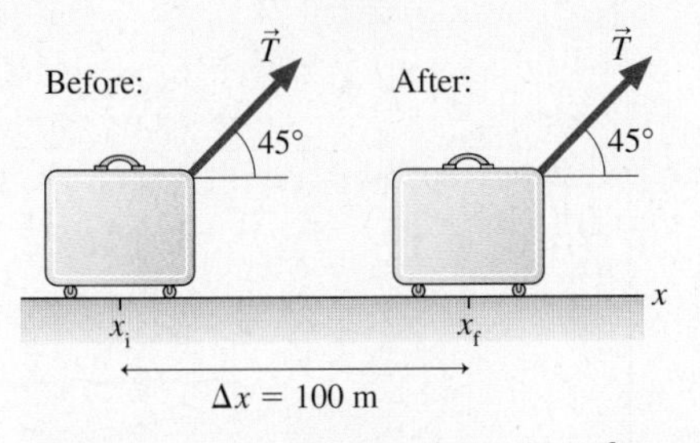
Before:
After:
$\vec{T}$
$\vec{T}$
45°
45°
x_i
x_f
x
$\Delta x = 100$ m

FIGURE 11.6 Pictorial representation of a suitcase pulled by a rope.

VISUALIZE Figure 11.6 shows a before-and-after pictorial representation.

SOLVE The motion is along the x-axis, so in this case $\Delta r = \Delta x$. We can use Equation 11.13 to find that the tension does work

$$W = T(\Delta x)\cos\theta = (20\text{ N})(100\text{ m})\cos 45° = 1410\text{ J}$$

ASSESS Because a person is pulling on the other end of the rope, we would say informally that the person does 1410 J of work on the suitcase.

According to the basic energy model, work can be either positive or negative to indicate energy transfer into or out of the system. The quantities F and Δr are always positive, so the sign of W is determined entirely by the angle θ between the force $\vec{F}$ and the displacement $\Delta\vec{r}$.

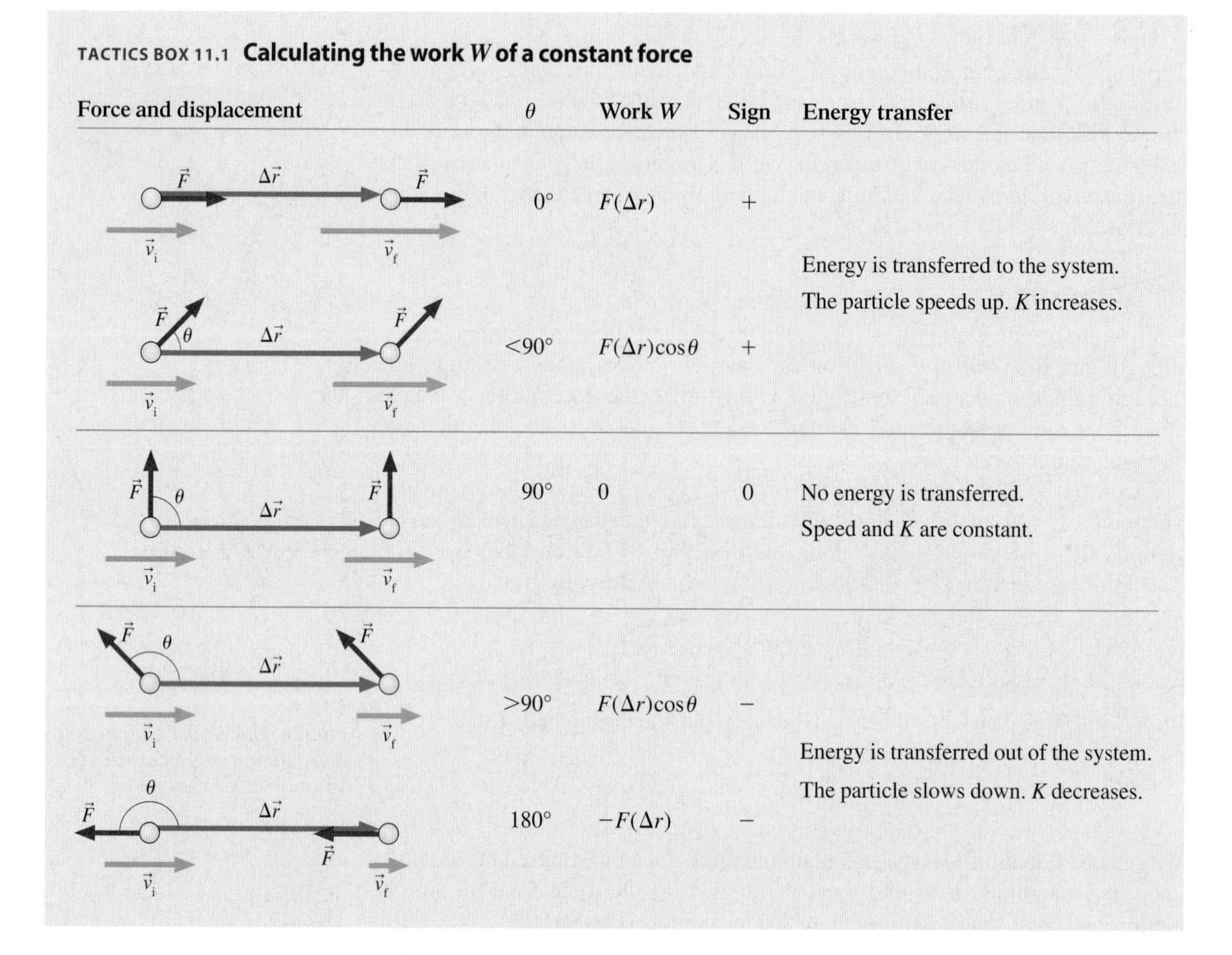

TACTICS BOX 11.1 Calculating the work W of a constant force

Force and displacement	θ	Work W	Sign	Energy transfer
	0°	$F(\Delta r)$	+	Energy is transferred to the system. The particle speeds up. K increases.
	<90°	$F(\Delta r)\cos\theta$	+	
	90°	0	0	No energy is transferred. Speed and K are constant.
	>90°	$F(\Delta r)\cos\theta$	−	Energy is transferred out of the system. The particle slows down. K decreases.
	180°	$-F(\Delta r)$	−	

NOTE ► You may have learned in an earlier physics course that work is "force times distance." This is *not* the definition of work, merely a special case. Work is "force times distance" only if the force is constant *and* parallel to the displacement (i.e., $\theta = 0°$). ◄

EXAMPLE 11.2 Work and kinetic energy during a rocket launch

A 150,000 kg rocket is launched straight up. The rocket motor generates a thrust of 4.0×10^6 N. What is the rocket's speed at a height of 500 m? Ignore air resistance and the slight mass loss due to burned fuel.

MODEL Model the rocket as a particle. Thrust and gravity are constant forces that do work on the rocket and change its energy.

VISUALIZE Figure 11.7 shows a pictorial representation and a free-body diagram.

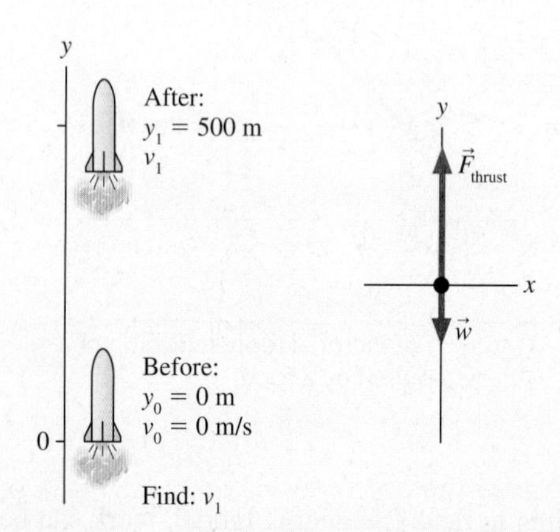

FIGURE 11.7 Pictorial representation and free-body diagram of a rocket launch.

SOLVE We can solve this problem with the work-kinetic energy theorem $\Delta K = W_{net}$. There are two forces, and both do work on the rocket. The thrust is in the direction of motion, with $\theta = 0°$, and thus

$$W_{thrust} = F_{thrust}(\Delta r) = (4.0 \times 10^6 \text{ N})(500 \text{ m}) = 2.00 \times 10^9 \text{ J}$$

The weight force points downward, opposite the displacement $\Delta\vec{r}$, so $\theta = 180°$. Thus the work done by gravity is

$$W_{grav} = -w(\Delta r) = -mg(\Delta r)$$
$$= -(1.5 \times 10^5\ \text{kg})(9.8\ \text{m/s}^2)(500\ \text{m}) = -0.74 \times 10^9\ \text{J}$$

The work done by the thrust is positive. By itself, the thrust would cause the rocket to speed up. The work done by gravity is negative. By itself, gravity would cause the rocket to slow down. The work-kinetic energy theorem, using $v_0 = 0$ m/s, is

$$\Delta K = \frac{1}{2}mv_1^2 - 0 = W_{net} = W_{thrust} + W_{grav} = 1.26 \times 10^9\ \text{J}$$

This is easily solved for the speed:

$$v_1 = \sqrt{\frac{2W_{net}}{m}} = 130\ \text{m/s}$$

ASSESS The net work is positive, meaning that energy is transferred *to* the rocket. In response, the rocket speeds up.

NOTE ► The work done by a force depends on the angle θ between the force $\vec{F}$ and the displacement $\Delta\vec{r}$, *not* on the direction the particle is moving. The work done on all four particles in Figure 11.8 is the same, despite the fact that they are moving in four different directions. ◄

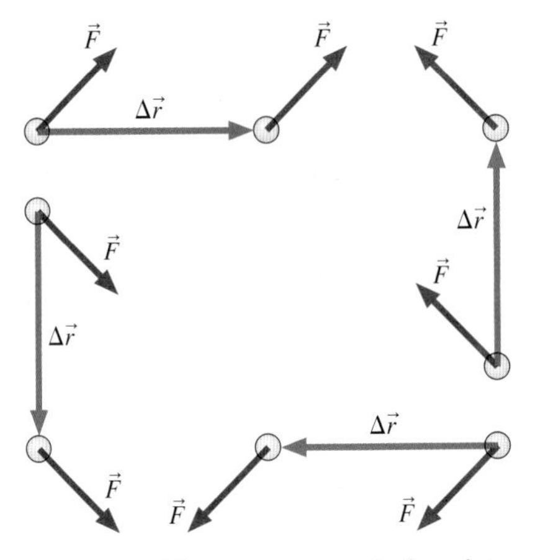

FIGURE 11.8 The same amount of work is done on each of these particles.

Force Perpendicular to the Direction of Motion

Figure 11.9a shows a bird's-eye view of a car turning a corner. As you learned in Chapter 7, a friction force points toward the center of the circle. How much work does friction do on the car?

Zero! In Figure 11.9b we've "bent" the s-axis to follow the curve. You can see that the friction force is everywhere perpendicular to the small displacement ds. F_s, the component of the force parallel to the displacement, is everywhere zero. Thus static friction does *no* work on the car. This shouldn't be surprising. You know that the car's speed, and hence its kinetic energy, doesn't change as it rounds the curve. Thus, according to the work-kinetic energy theorem, $W = \Delta K = 0$.

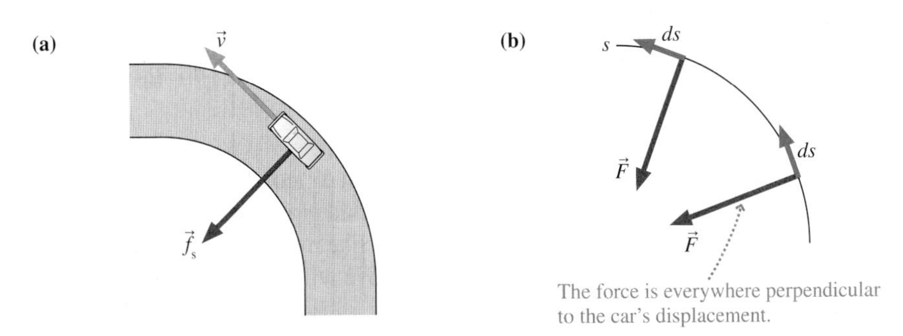

FIGURE 11.9 The force is everywhere perpendicular to the car's displacement. It does no work.

We used a car rounding a curve as a concrete example, but this is a general result: **A force that is everywhere perpendicular to the motion does no work.** A force perpendicular to the motion changes the *direction* of motion but not the particle's speed.

EXAMPLE 11.3 Pushing a puck

A 500 g ice hockey puck slides across frictionless ice with an initial speed of 2.0 m/s. A compressed-air gun can be used to exert a 1.0 N force on the puck. The air gun is aimed at the front edge of the puck with the compressed-air flow 30° below the horizontal. This force is applied continuously as the puck moves 50 cm. What is the puck's final speed?

EXAMPLE 11.6 Calculating work using the dot product
A 70 kg skier is gliding at 2.0 m/s when he starts down a very slippery 50-m-long, 10° slope. What is his speed at the bottom?

MODEL Model the skier as a particle and interpret "very slippery" to mean frictionless. Use the work-kinetic energy theorem to find his final speed.

VISUALIZE Figure 11.16 shows a pictorial representation.

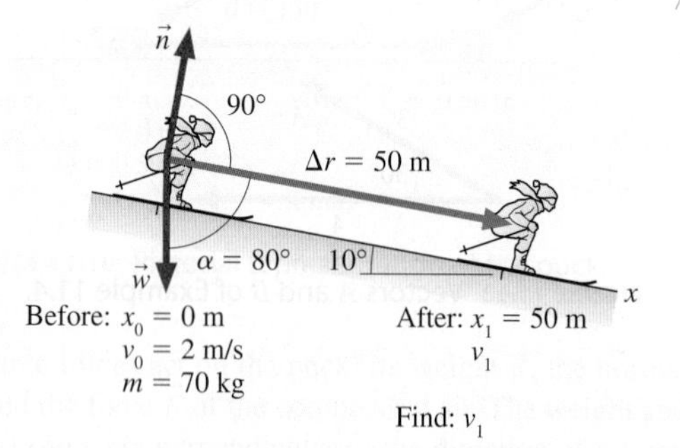

FIGURE 11.16 Pictorial representation of the skier.

SOLVE The only forces on the skier are $\vec{w}$ and $\vec{n}$. The normal force is perpendicular to the motion and thus does no work. The work done by gravity is easily calculated as a dot product:

$$W = \vec{w}\cdot\Delta\vec{r} = w(\Delta r)\cos\alpha$$
$$= (70\text{ kg})(9.8\text{ m/s}^2)(50\text{ m})\cos 80° = 5960\text{ J}$$

Notice that the angle *between* the vectors is 80°, not 10°. Then, from the work-kinetic energy theorem, we find

$$\Delta K = \frac{1}{2}mv_1^2 - \frac{1}{2}mv_0^2 = W$$

$$v_1 = \sqrt{v_0^2 + \frac{2W}{m}} = \sqrt{(2.0\text{ m/s})^2 + \frac{2(5960\text{ J})}{70\text{ kg}}} = 13.2\text{ m/s}$$

NOTE ► While in the midst of the mathematics of calculating work, do not lose sight of what the work-kinetic energy theorem is all about. It is a statement about *energy transfer,* saying that work is the energy transferred to or from the system due to forces exerted on the system. Work causes the system's kinetic energy to either increase or decrease. ◄

STOP TO THINK 11.4 Which force does the most work?

a. The 10 N force.
b. The 8 N force.
c. The 6 N force.
d. They all do the same amount of work.

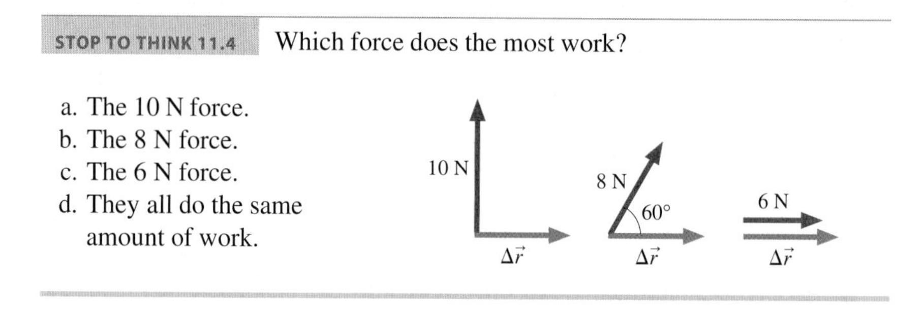

11.4 The Work Done by a Variable Force

We've learned how to calculate the work done on a particle by a force that has a constant magnitude and direction throughout the particle's displacement. But what about a force that changes in either magnitude or direction as the particle moves? Equation 11.8, the definition of work, is all we need:

$$W = \int_{s_i}^{s_f} F_s\,ds = \text{area under the force-versus-position graph} \qquad (11.17)$$

The integral sums up the small amounts of work $F_s\,ds$ done in each step along the trajectory. The only new feature, since F_s varies with position, is that we cannot take F_s outside the integral. We must evaluate the integral either geometrically, by finding the area under the curve, or by actually doing the integration. We'll restrict our applications to motion along a straight line in order to avoid unnecessarily complex mathematics.

EXAMPLE 11.7 Using work to find the speed of a car

A 1500 kg car accelerates from rest. Figure 11.17 shows the net force on the car (propulsion force minus any drag forces) as a function of the car's position. What is the car's speed after traveling 200 m?

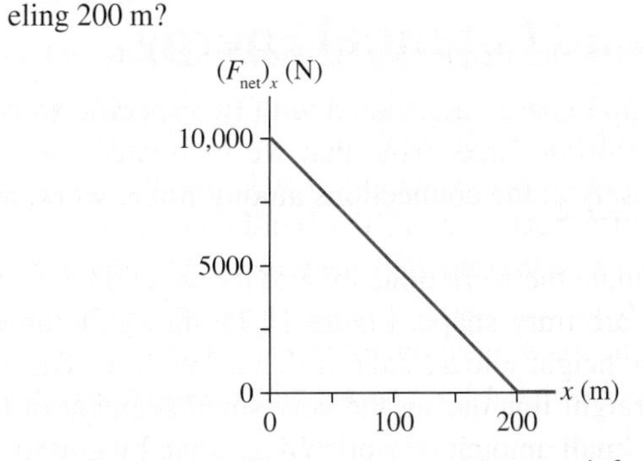

FIGURE 11.17 Force-versus-position graph for a car.

SOLVE The acceleration $a_x = (F_{net})_x/m$ is high as the car starts but decreases as the car picks up speed because of increasing drag. Figure 11.17 is a more realistic portrayal of the net force on a car than was our earlier model of a constant force. But a variable force means that we cannot use the familiar constant-acceleration kinematics. Instead, we can use the work-kinetic energy theorem. Because $v_0 = 0$ m/s, we have

$$\Delta K = \frac{1}{2}mv_1^2 - 0 = W_{net}$$

Starting from $x_0 = 0$ m, the work is

$$W_{net} = \int_{0\text{ m}}^{x_1} (F_{net})_x\, dx$$

$$= \text{area under the } (F_{net})_x\text{-versus-}x \text{ graph from 0 m to } x_1$$

The area under the curve of Figure 11.17 is that of a triangle of width 200 m. Thus

$$W_{net} = \text{area} = \frac{1}{2}(10{,}000\text{ N})(200\text{ m}) = 1{,}000{,}000\text{ J}$$

The work-kinetic energy theorem then gives

$$v_1 = \sqrt{\frac{2W_{net}}{m}} = \sqrt{\frac{2(1{,}000{,}000\text{ J})}{1500\text{ kg}}} = 36.5\text{ m/s}$$

ASSESS Because $1\text{ J} = 1\text{ kg m}^2/\text{s}^2$, the quantity W/m has units m^2/s^2. Thus the units of v_1 are m/s, as expected.

EXAMPLE 11.8 Using the work-kinetic energy theorem for a spring

The "pincube machine" was an ill-fated predecessor of the pinball machine. A 100 g cube is launched by pulling a spring back 20 cm and releasing it. What is the cube's launch speed, as it leaves the spring, if the spring constant is 20 N/m and the coefficient of kinetic friction is 0.10?

MODEL Model the cube as a particle and the spring as an ideal spring obeying Hooke's law. Use the work-kinetic energy theorem to find the launch speed.

VISUALIZE Figure 11.18 shows a before-and-after pictorial representation and a free-body diagram. We've placed the origin of the x-axis at the equilibrium position of the spring.

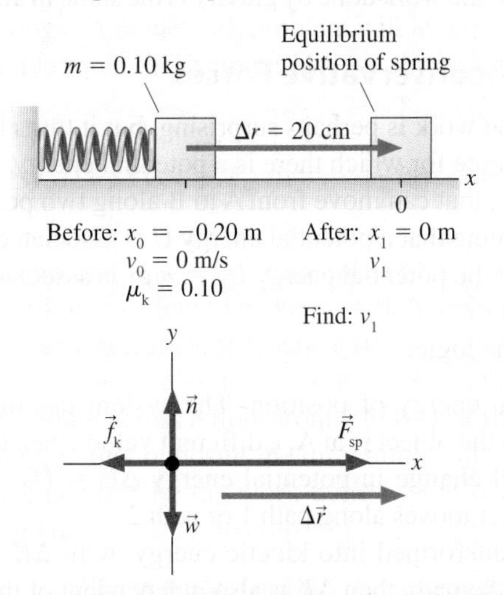

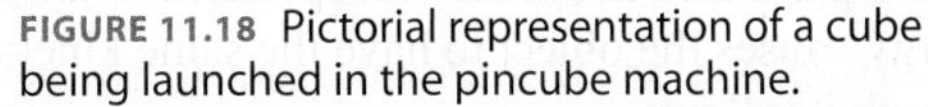

FIGURE 11.18 Pictorial representation of a cube being launched in the pincube machine.

SOLVE The normal force and weight are perpendicular to the motion and do no work. We can use the work-kinetic energy theorem, with $v_0 = 0$ m/s, to find the launch speed:

$$\Delta K = \frac{1}{2}mv_1^2 - 0 = W_{net} = W_{fric} + W_{sp}$$

Friction is a constant force $f_k = \mu_k mg$ in the direction *opposite* the motion, with $\theta = 180°$, so the work done by friction is

$$W_{fric} = \vec{f}_k \cdot \Delta\vec{r} = f_k(\Delta r)\cos 180° = -\mu_k mg\Delta x = -0.020\text{ J}$$

The negative work of friction would, by itself, slow the block down.

The spring force is a variable force: $(F_{sp})_x = -k\Delta x = -kx$, where $\Delta x = x - x_e = x$ because we chose a coordinate system with $x_e = 0$ m. Despite the minus sign, $(F_{sp})_x$ is a positive quantity (force pointing to the right) because x is negative throughout the motion. The spring force points in the direction of motion, so W_{sp} is positive. We can use Equation 11.17 to evaluate W_{sp}:

$$W_{sp} = \int_{x_0}^{x_1} (F_{sp})_x\, dx = -k\int_{x_0}^{x_1} x\, dx = -\frac{1}{2}kx^2\Big|_{x_0}^{x_1}$$

$$= -\left(\frac{1}{2}kx_1^2 - \frac{1}{2}kx_0^2\right)$$

Evaluating W_{sp} for $x_0 = -0.20$ m and $x_1 = 0$ m gives

$$W_{sp} = \frac{1}{2}(20\text{ N/m})(-0.20\text{ m})^2 = 0.400\text{ J}$$

The net work is $W_{net} = W_{fric} + W_{sp} = 0.380$ J, with which we can now find

$$v_1 = \sqrt{\frac{2W_{net}}{m}} = \sqrt{\frac{2(0.380\text{ J})}{0.100\text{ kg}}} = 2.76\text{ m/s}$$

$(\Delta s)_{PQR} = 2L$. More work (negative) is done on the longer path around the outside of the square. The work done by friction is *not* independent of the path followed.

A force for which the work is *not* independent of the path is called a **nonconservative force.** It is not possible to define a potential energy for a nonconservative force. Friction is a nonconservative force, so we cannot define a potential energy of friction.

This makes sense. If you toss a ball straight up, kinetic energy is transformed into gravitational potential energy. The ball has the potential to transform this energy back into kinetic energy, and it does so as the ball falls. But you cannot recover the kinetic energy lost to friction as a box slides to a halt. There's no "potential" that can be transformed back into kinetic energy.

Mechanical Energy

Consider a system of particles that interact via both conservative forces and nonconservative forces. The conservative forces do work W_c as the particles move from initial positions i to final positions f. The nonconservative forces do work W_{nc}. The total work done by *all* forces is $W_{net} = W_c + W_{nc}$. The change in the system's kinetic energy ΔK, as determined by the work-kinetic energy theorem, is

$$\Delta K = W_{net} = W_c(i \rightarrow f) + W_{nc}(i \rightarrow f) \tag{11.23}$$

The work done by the conservative forces can now be associated with a potential energy U. According to Equation 11.21, $W_c(i \rightarrow f) = -\Delta U$. With this definition, Equation 11.23 becomes

$$\Delta K + \Delta U = \Delta E_{mech} = W_{nc} \tag{11.24}$$

where, as in Chapter 10, the *mechanical energy* is $E_{mech} = K + U$.

Now we can see that **mechanical energy is conserved if there are no nonconservative forces.** That is

$$\Delta E_{mech} = 0 \text{ if } W_{nc} = 0 \tag{11.25}$$

This important conclusion is what we called the law of conservation of mechanical energy in Chapter 10. There we saw that friction prevents E_{mech} from being conserved, but we really didn't know why. Equation 11.24 tells us that any nonconservative force causes the mechanical energy to change. Friction and other "dissipative forces" lead to a loss of mechanical energy. Other outside forces, such as the pull of a rope, might increase the mechanical energy.

Equally important, Equation 11.24 tells us what to do if the mechanical energy isn't conserved. You can still use energy concepts to analyze the motion if you compute the work done by the nonconservative forces.

EXAMPLE 11.9 Using work and potential energy together

Use potential energy to find the launch speed of the "pincube machine" of Example 11.8. Recall that the 100 g cube is launched by pulling a spring back 20 cm and releasing it. The spring constant is 20 N/m and the coefficient of kinetic friction is 0.10.

MODEL This time let the system be the cube and the spring.

VISUALIZE Figure 11.18 shows the pictorial representation and free-body diagram..

SOLVE In solving this problem with the work-kinetic energy theorem, we had to explicitly calculate the work done by the spring. Now let's use Equation 11.24. The spring force is a conservative force that we can associate with the elastic potential energy U_s. The friction force is nonconservative. Thus

$$\Delta K + \Delta U_s = W_{nc} = W_{fric}$$

The spring's potential energy with $x_e = 0$ m is $U_s = \frac{1}{2}kx^2$. In Equation 11.22 we found that the work done by friction on an object sliding across a horizontal surface is $W_{fric} = -\mu_k mg\Delta s$. Thus the energy equation becomes

$$\frac{1}{2}mv_1^2 - \frac{1}{2}mv_0^2 + \frac{1}{2}kx_1^2 - \frac{1}{2}kx_0^2 = -\mu_k mg(x_1 - x_0)$$

Using $v_0 = 0$ m/s, $x_1 = 0$ m, and $x_0 = -0.20$ m, we find

$$v_1 = \sqrt{\frac{kx_0^2}{m} + 2\mu_k g x_0} = 2.76 \text{ m/s}$$

ASSESS What appeared to be a difficult problem, with both friction and a spring, turned out to be straightforward when analyzed with energy and work.

Example 11.9 illustrates an important idea. When we associate a potential energy with a conservative force we

- Enlarge the system to include all objects that interact via conservative forces.
- "Precompute" the work. We can do this because we don't need to know what paths the objects are going to follow. This precomputed work becomes a potential energy and moves from the right side of $\Delta K = W$ to the left side of Equation 11.24.

In Example 11.8, the system consisted of just the cube. We treated the spring force as a force from the environment doing work on the system. In Example 11.9, where we revisited the same problem, we brought the spring into the system and represented the conservative spring-cube interaction with a potential energy.

NOTE ▶ When you use a potential energy, you've already taken the work of that force into account. Don't compute the work explicitly, or you'll be double counting it! ◀

To analyze a problem using work and energy, you can either

1. Use the work-kinetic energy theorem $\Delta K = W$ and explicitly compute the work done by *every* force. This was the method of Example 11.8. Or
2. Represent the work done by conservative forces as potential energies, then use $\Delta K + \Delta U = W_{nc}$. The only work that must be computed is the work of any nonconservative forces. This was the method of Example 11.9.

It's important to recognize that **these two methods yield the same result!** It's simply that method 2 "precomputes" some of the work and represents it as a potential energy. In practice, **method 2 is always easier and is the preferred method.**

11.6 Finding Force from Potential Energy

We know how to find the potential energy due to a conservative force. Now we need to learn how to go in reverse. That is, if we know a particle's potential energy, how do we find the force acting on it?

Figure 11.23a shows a particle moving through a *small* displacement Δs while being acted on by a conservative force $\vec{F}$. If Δs is sufficiently small, the force component F_s in the direction of motion is essentially constant during the displacement. The work done on the particle as it moves from s to $s + \Delta s$ is

$$W(s \rightarrow s + \Delta s) = F_s \Delta s \qquad (11.26)$$

This work is shown in Figure 11.23b as the area under the force curve in the narrow rectangle of width Δs.

Because $\vec{F}$ is a conservative force, the change in the particle's potential energy over this interval is defined in Equation 11.21 as

$$\Delta U = -W(s \rightarrow s + \Delta s) = -F_s \Delta s$$

which we can rewrite as

$$F_s = -\frac{\Delta U}{\Delta s} \qquad (11.27)$$

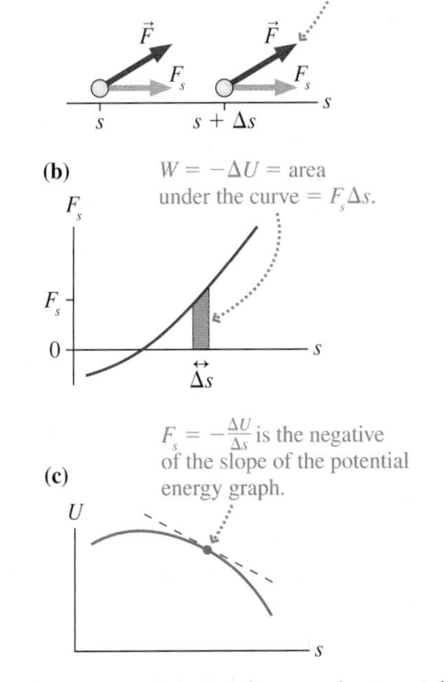
(a)
$\vec{F}$ is essentially constant if Δs is sufficiently small.
$\vec{F}$
F_s
s
$s + \Delta s$
(b)
$W = -\Delta U$ = area under the curve = $F_s \Delta s$.
F_s
0
Δs
(c)
$F_s = -\frac{\Delta U}{\Delta s}$ is the negative of the slope of the potential energy graph.
U
s

FIGURE 11.23 Relating force and potential energy.

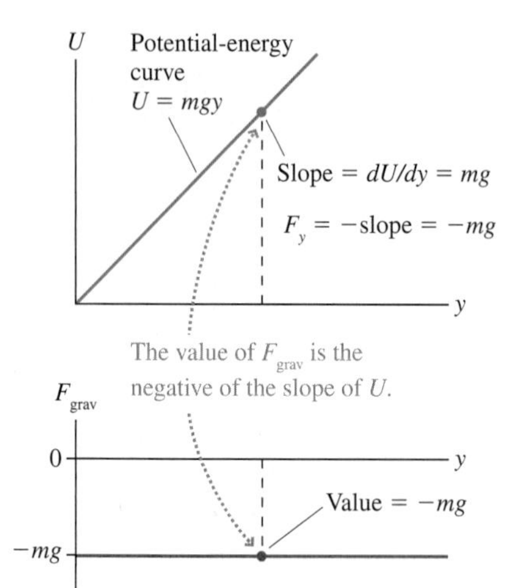

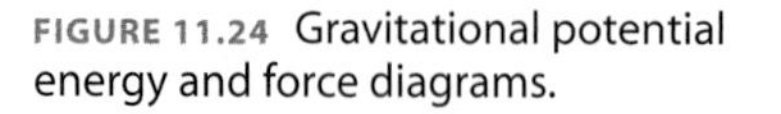

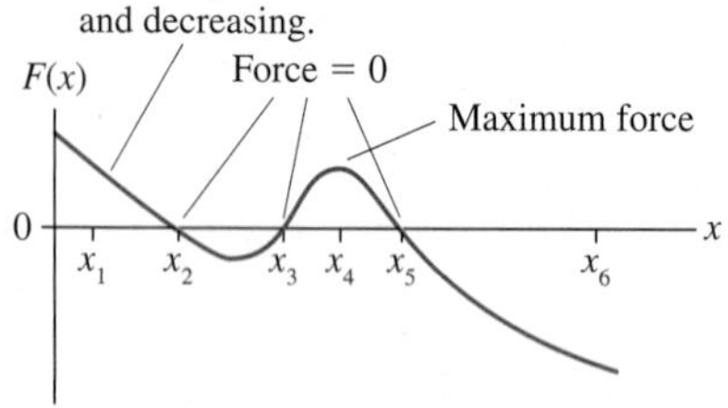

FIGURE 11.24 Gravitational potential energy and force diagrams.

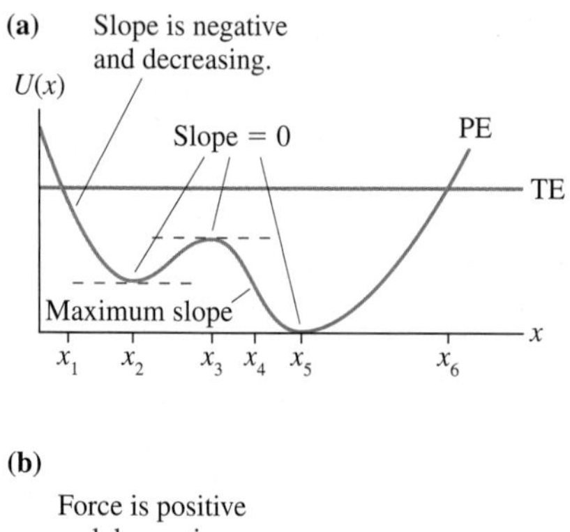

FIGURE 11.25 A potential-energy diagram and the corresponding force diagram.

In the limit $\Delta s \rightarrow 0$, we find that the force at position s is

$$F_s = \lim_{\Delta s \to 0}\left(-\frac{\Delta U}{\Delta s}\right) = -\frac{dU}{ds} \tag{11.28}$$

We see that the force on the particle is the *negative* of the derivative of the potential energy with respect to position. Figure 11.23b shows that we can interpret this result graphically by saying

$$F_s = \text{the negative of the slope of the } U\text{-versus-}s \text{ graph at } s. \tag{11.29}$$

In practice, of course, we will usually use either $F_x = -dU/dx$ or $F_y = -dU/dy$.

As an example, consider the gravitational potential energy $U_g = mgy$. Figure 11.24a shows the potential energy diagram U_g-versus-y. It is simply a straight-line graph passing through the origin. The force on the particle at position y, according to Equations 11.28 and 11.29, is simply

$$F_{grav} = -\frac{dU_g}{dy} = -(\text{slope of } U_g) = -mg$$

The negative sign, as always, indicates that the force points in the negative y-direction. Figure 11.24b shows the corresponding F-versus-y graph. At each point, the *value* of F is equal to the negative of the *slope* of the U-versus-y graph. This is similar to position and velocity graphs, where the value of v_x at any time t is equal to the slope of the x-versus-t graph.

We already knew that $F_{grav} = -mg$, of course, so the point of this particular example was to illustrate the meaning of Equation 11.29 rather than to find out anything new. Had we *not* known the force of gravity, we see is that it is possible to find it from the potential energy diagram.

Figure 11.25 is a more interesting example. You learned in Chapter 10 that a particle with the total energy TE will oscillate back and forth between turning points at x_1 and x_6 where the total energy line crosses the potential-energy curve. Now we can be more specific. The slope of the potential-energy graph is negative between x_1 and x_2. This means that the force on the particle, which is the negative of the slope of U, is *positive.* A particle between x_1 and x_2 experiences a force toward the right. The force decreases as the slope decreases until, at x_2, $F_x = 0$. This is consistent with our prior identification of x_2 as a point of stable *equilibrium.* The slope is positive (force negative and thus to the left) between x_2 and x_3, zero (zero force) at the unstable equilibrium point x_3, and so on. Point x_4, where the slope is most negative, is the point between x_1 and x_6 of maximum force.

Figure 11.25b is a plausible graph of F-versus-x. We don't know the exact shape, because we don't have an exact expression for U, but the force graph must look very much like this.

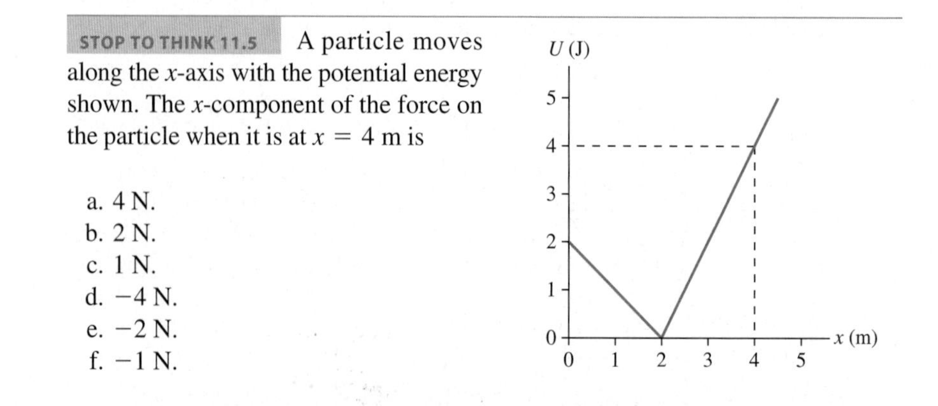

STOP TO THINK 11.5 A particle moves along the x-axis with the potential energy shown. The x-component of the force on the particle when it is at $x = 4$ m is

a. 4 N.
b. 2 N.
c. 1 N.
d. −4 N.
e. −2 N.
f. −1 N.

11.7 Thermal Energy

All of the objects we handle and use every day are extraordinarily large compared to the size of an atom. Each object is really a system consisting of vast numbers of particle-like atoms. We want to distinguish between the motion of the object as a whole and the motion of the atoms inside it. We will use the terms **macrophysics** to refer to the motion and dynamics of the object as a whole and **microphysics** to refer to the motion of atoms. You recognize the prefix *micro,* meaning "small." You may not be familiar with *macro,* which means "large" or "large-scale."

The connection between microscopic and macroscopic behavior is a topic we will explore in depth when we reach thermodynamics, in order to understand the bulk properties of materials, and again in electricity, to explain the electrical properties of conductors and insulators. For now, let's take a closer look at the microscopic energy in a system of atoms.

Kinetic and Potential Energy at the Microscopic Level

Figure 11.26 shows two different perspectives of an object. In the macrophysics perspective of Figure 11.26a you see an object of mass M moving as a whole with velocity v_{obj}. As a consequence of its motion, the object has macroscopic kinetic energy $K_{\text{macro}} = \frac{1}{2}Mv_{\text{obj}}^2$.

Figure 11.26b is a microphysics view of the same object, where now we see a *system of particles.* Each of these atoms is moving about, and in doing so they stretch and compress the spring-like molecular bonds between them. Consequently, there is a *microscopic potential energy* U_{micro} associated with molecular bonds. The energy stored in any one bond is very small, but there are incredibly many bonds.

Each moving atom has a kinetic energy $K_j = \frac{1}{2}mv_j^2$, where v_j is the speed of atom j. In this section we will use lowercase m to represent the mass of atoms within the system (which, for simplicity, we'll assume to all be the same) and uppercase M for the mass of the entire system. The kinetic energy of one atom is exceedingly small, but there are enormous numbers of atoms in a macroscopic object. The combined kinetic energy of all the atoms is what we call the *microscopic kinetic energy* K_{micro}. These microscopic energies associated with molecular bonds and moving atoms are quite distinct from the energies K_{macro} and U_{macro}.

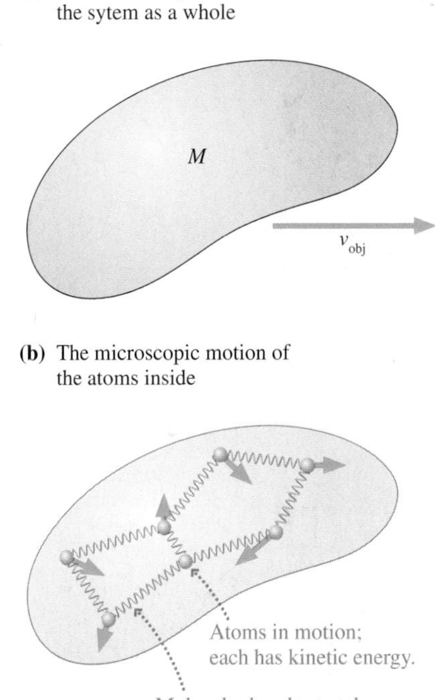
(a) The macroscopic motion of the sytem as a whole
M
v_{obj}
(b) The microscopic motion of the atoms inside
Atoms in motion; each has kinetic energy.
Molecular bonds stretch and compress; each has potential energy.

FIGURE 11.26 Two perspectives of motion and energy.

Is the microscopic energy worth worrying about? To see, consider a 500 g ($\approx$1 lb) iron ball moving at the respectable speed of 20 m/s ($\approx$45 mph). Its macroscopic kinetic energy is

$$K_{\text{macro}} = \frac{1}{2}Mv_{\text{obj}}^2 = 100 \text{ J}$$

A periodic table of the elements shows that iron has atomic mass 56. Recall from chemistry that 56 g of iron is 1 gram-molecular weight and has Avogadro's number ($N_A = 6.02 \times 10^{23}$) atoms. Thus 500 g of iron is $\approx$9 gram-molecular weights and contains $N \approx 9N_A \approx 5.4 \times 10^{24}$ iron atoms. Thus the mass of each atom is

$$m = \frac{M}{N} \approx \frac{0.50 \text{ kg}}{5.4 \times 10^{24}} \approx 9 \times 10^{-26} \text{ kg}$$

How fast do atoms move? In Part IV you'll learn that the speed of sound in air at room temperature is $\approx$340 m/s. Sound travels by atoms bumping into each other, so the atoms in the air must have a speed of *at least* 340 m/s. The speed of sound in solids is even higher, usually $>$1000 m/s, but that's partially influenced by the spring-like molecular bonds. As a rough estimate, $v \approx 500$ m/s is a reasonable guess. The kinetic energy of one iron atom at this speed is

$$K_{\text{atom}} = \frac{1}{2}mv^2 \approx 1.1 \times 10^{-20} \text{ J}$$

This is very tiny, but there are a great many atoms. If we assume, for our estimate, that all atoms move at this speed, the microscopic kinetic energy is

$$K_{\text{micro}} \approx NK_{\text{atom}} \approx 70{,}000 \text{ J}$$

The microscopic kinetic energy of the atoms moving inside the iron is much larger than the macroscopic kinetic energy of the object as a whole! We'll later see that, on average, U_{micro} for a solid is equal to K_{micro}, so the total microscopic energy is $\approx$140,000 J.

The combined microscopic kinetic and potential energy of the atoms is called the *thermal energy* of the system:

$$E_{\text{th}} = K_{\text{micro}} + U_{\text{micro}} \tag{11.30}$$

This energy is usually hidden from view in our macrophysics perspective, but it is quite real. We will discover later, when we reach thermodynamics, that the thermal energy is proportional to the *temperature* of the system. Raising the temperature causes the atoms to move faster and the bonds to stretch more, giving the system more thermal energy.

NOTE ► The microscopic energy of the atoms in a system is *not* called "heat." The word *heat,* like the word *work,* has a narrow and precise meaning in physics that is much more restricted than its use in everyday language. We will introduce the concept of heat later, when we need it. For the time being we want to use the correct term *thermal energy* to describe the random, thermal motion of the particles in a system. If the temperature of a system goes up (i.e., it gets hotter), it is because the system's thermal energy has increased. ◄

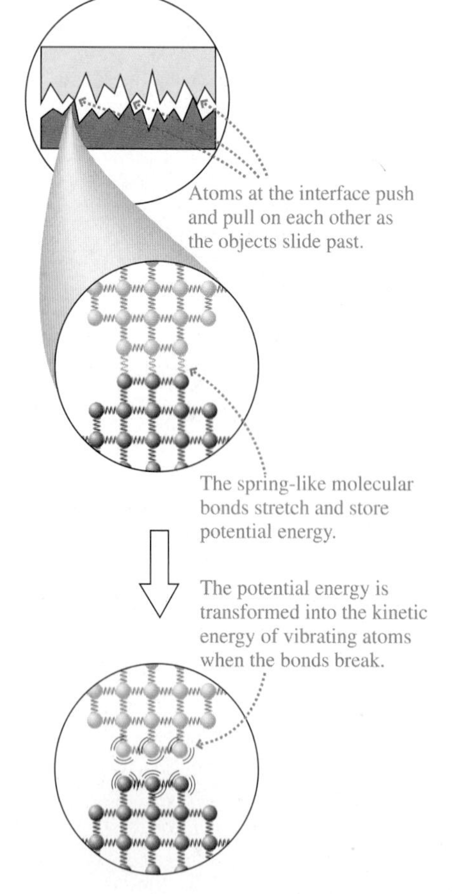

Atoms at the interface push and pull on each other as the objects slide past.
The spring-like molecular bonds stretch and store potential energy.
The potential energy is transformed into the kinetic energy of vibrating atoms when the bonds break.

FIGURE 11.27 The atomic-level view of friction.

Dissipative Forces

At the beginning of the chapter we asked "What happens to the energy?" when a system "runs down" because of friction. If you shove a book across the table, it gradually slows down and stops. Where did the energy go? The common answer "It went into heat" isn't quite right.

Figure 11.27 reminds you of the atomic-level model of friction that we introduced in Chapter 5. There we were interested in how the friction affects the macroscopic object. Now, think what friction does to the microscopic energy of the atoms. If two atoms temporarily stick together, the molecular bond gets stretched and U_{micro} increases. When the bond breaks and the atoms suddenly snap back into place, that potential energy gets transformed into microscopic kinetic energy K_{micro} of atoms bouncing around. Imagine having several balls connected by springs. If you pull one ball and then suddenly release it, you cause the whole system to jiggle and vibrate.

The forces of friction increase the microscopic kinetic and potential energy of the atoms. That is, they increase the object's thermal energy, and we perceive this as a higher temperature. That's why rubbing things together makes them warmer. Thus the correct answer to "What happens to the energy?" is "It is transformed into thermal energy."

Forces such as friction and drag that always oppose the motion are called **dissipative forces.** They cause the macroscopic kinetic energy of the system as a whole to be "dissipated" as thermal energy. Dissipative forces are always nonconservative forces.

Consider the book sliding to a halt. Call the work done by friction W_{diss} to indicate that it is a particular type of nonconservative work. The energy equation, Equation 11.24, is

$$\Delta K + \Delta U = \Delta K = W_{\text{diss}}$$

where we've used $\Delta U = 0$ for horizontal motion. If the kinetic energy is transformed entirely into thermal energy, then

$$\Delta E_{th} = -\Delta K$$

ΔK is a negative number (loss of kinetic energy), so the explicit negative sign makes ΔE_{th} a positive number (gain of thermal energy).

Comparing these two equations, you can see that the relationship between the work done by dissipative forces and the change in thermal energy is

$$\Delta E_{th} = -W_{diss} \tag{11.31}$$

W_{diss} is *always* a negative number because the force is always opposite the direction of motion. Consequently, ΔE_{th} is *always* a positive number. **Dissipative forces always increase the thermal energy, they never decrease it.**

The ballplayer's kinetic energy is being transformed into thermal energy.

EXAMPLE 11.10 Calculating the increase in thermal energy

A rope pulls a 10 kg wooden crate 3.0 m across a wood floor. What is the change in thermal energy?

SOLVE The change in thermal energy is $\Delta E_{th} = -W_{diss} = -W_{fric}$. From Equation 11.22,

$$W_{fric} = f_k(\Delta r)\cos 180° = -\mu_k mg\Delta x$$
$$= -(0.2)(10\text{ kg})(9.8\text{ m/s}^2)(3.0\text{ m}) = -58.8\text{ J}$$

Thus

$$\Delta E_{th} = -(-58.8\text{ J}) = 58.8\text{ J}$$

11.8 Conservation of Energy

Let's return to the basic energy model and start pulling together the many ideas introduced in this chapter. Figure 11.28 shows a general system consisting of several macroscopic objects. These objects interact with each other, and they may be acted on by external forces from the environment. Both the interaction forces and the external forces do work on the objects. The change in the system's kinetic energy is given by the work-kinetic energy theorem, $\Delta K = W_{net}$.

We previously divided W_{net} into the work W_c done by conservative forces and the work W_{nc} done by nonconservative forces. The work done by the conservative forces can be represented by a potential energy U. Let's now make a further distinction by dividing the nonconservative forces into *dissipative forces* and *external forces.* That is,

$$W_{nc} = W_{diss} + W_{ext} \tag{11.32}$$

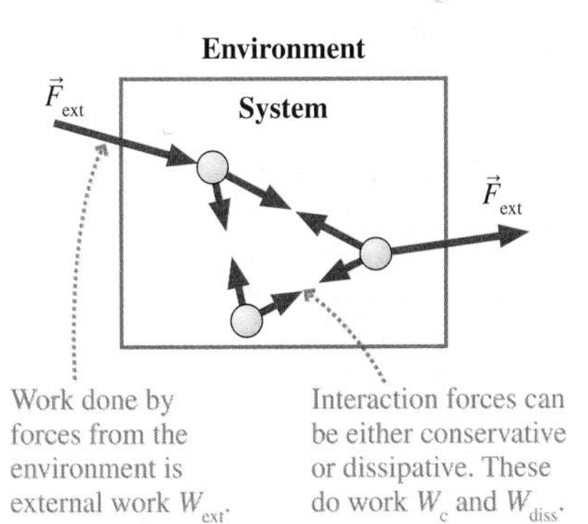
Environment
System
$\vec{F}_{ext}$
$\vec{F}_{ext}$
Work done by forces from the environment is external work W_{ext}.
Interaction forces can be either conservative or dissipative. These do work W_c and W_{diss}.

FIGURE 11.28 A system with both internal interaction forces and external forces.

To illustrate what we mean by an external force, suppose you pick up a box at rest on the floor and place it at rest on a table. The box gains gravitational potential energy, but $\Delta K = 0$. Or consider pulling the box across the table with a string. The box gains kinetic energy, but not by transforming potential energy. The force of your hand and the tension of the string are forces that "reach in" from the environment to change the system. Thus they are *external forces.*

We have to be careful choosing the system if we want this distinction to be valid. As you can imagine, we're going to associate W_{diss} with ΔE_{th}. We want the thermal energy E_{th} to be an energy *of the system.* Otherwise, it wouldn't make sense to talk about transforming kinetic energy into thermal energy. But for E_{th} to be an energy of the system, *both* objects involved in a dissipative interaction must be part of the system. The book sliding across the table raises the temperature of both the book *and the table*. Consequently, we must include both the book *and the table* in the system. The dissipative forces, like the conservative forces, are interaction forces *inside* the system.

With this distinction, the work-kinetic energy theorem is

$$\Delta K = W_c + W_{diss} + W_{ext} \tag{11.33}$$

As before, define the potential energy U such that $\Delta U = -W_c$. Remember that potential energy is really just the precomputed work of a conservative force. We've also seen that the work done by dissipative forces increases the system's thermal energy: $\Delta E_{th} = -W_{diss}$. With these substitutions, the work-kinetic energy theorem becomes

$$\Delta K = -\Delta U + -\Delta E_{th} + W_{ext}$$

We can write this more profitably as

$$\Delta K + \Delta U + \Delta E_{th} = \Delta E_{mech} + \Delta E_{th} = \Delta E_{sys} = W_{ext} \tag{11.34}$$

where $E_{sys} = E_{mech} + E_{th}$ is the total energy of the system. Equation 11.34 is the **energy equation** of the system.

Equation 11.34 is our most general statement about how the energy of a system changes, but we still need to give a clear interpretation as to what it says. In Chapter 9 we defined an *isolated system* as a system for which the *net* external force is zero. It follows that no external work is done on an isolated system: $W_{ext} = 0$. Thus one conclusion from Equation 11.34 is that **the total energy E_{sys} of an isolated system is conserved.** That is, $\Delta E_{sys} = 0$ for an isolated system. If, in addition, the system is also nondissipative (i.e., no friction forces), then $\Delta E_{th} = 0$. In that case, the mechanical energy E_{mech} is conserved.

These conclusions about energy can be summarized as the *law of conservation of energy:*

Law of conservation of energy The total energy $E_{sys} = E_{mech} + E_{th}$ of an isolated system is a constant. The kinetic, potential, and thermal energy within the system can be transformed into each other, but their sum cannot change. Further, the mechanical energy $E_{mech} = K + U$ is conserved if the system is both isolated and nondissipative.

The law of conservation of energy is one of the most powerful statements in physics.

Figure 11.29a redraws the basic energy model of Figure 11.2. Now you can see that this picture is a pictorial representation of Equation 11.34. E_{sys}, the total energy of the system, changes only if external forces transfer energy in or out of the system by doing work on the system. The kinetic, potential, and thermal energy within the system can be transformed into each other by interaction forces within the system. As Figure 11.29b shows, $E_{sys} = K + U + E_{th}$ remains constant if the system is isolated. The *transfer* and *transformation* of energy are what the basic energy model is all about.

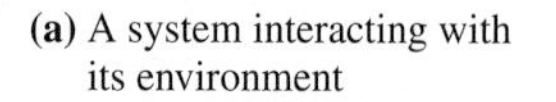

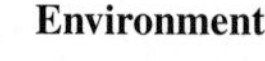

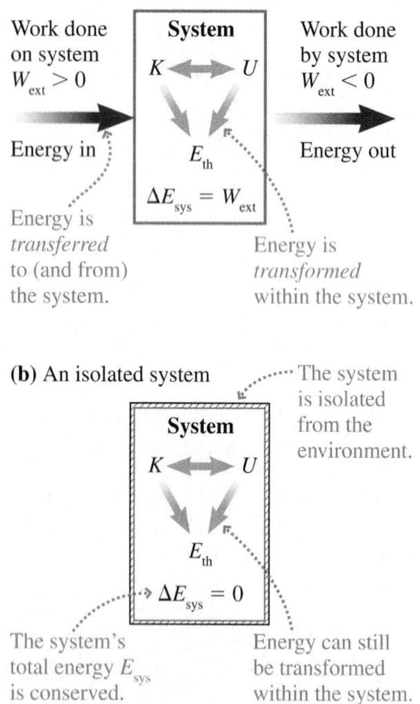

FIGURE 11.29 The basic energy model is a pictorial representation of the energy equation.

Energy Bar Charts

The energy bar charts of Chapter 10 can now be expanded to include the thermal energy and the work done by external forces. The energy equation, Equation 11.34, can be written

$$K_i + U_i + W_{ext} = K_f + U_f + \Delta E_{th} \tag{11.35}$$

The left side is the "before" condition ($K_i + U_i$) plus any energy that is added to or removed from the system. The right side is the "after" situation. The "energy accounting" of Equation 11.35 can be represented by the bar chart of Figure 11.30.

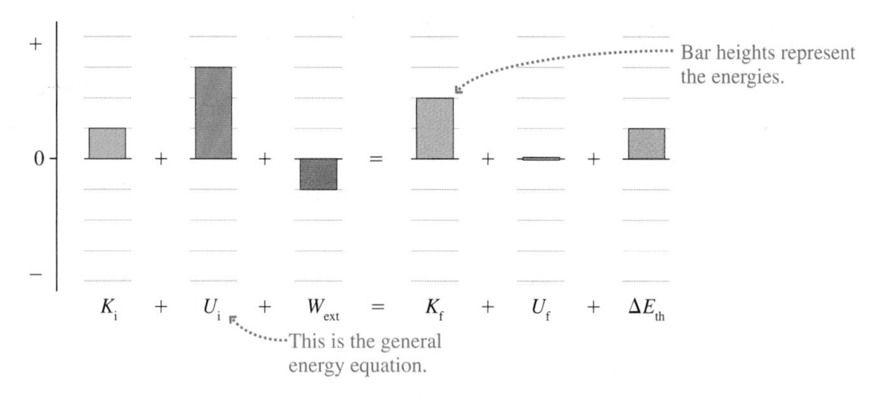

FIGURE 11.30 An energy bar chart shows how all the energy is accounted for.

NOTE ► We don't have any way to determine $(E_{th})_i$ or $(E_{th})_f$, but ΔE_{th} is always positive whenever the system contains dissipative forces. ◄

Let's look at a few examples.

EXAMPLE 11.11 Energy bar chart I

A box slides across a rough floor until it stops. Show the energy transfers and transformations on an energy bar chart.

SOLVE The box has an initial kinetic energy K_i. That energy is transformed into the thermal energy of the box and the floor. The potential energy doesn't change and no work is done by external forces, so the process is an energy transformation $K_i \rightarrow E_{th}$. This is shown in Figure 11.31. E_{sys} is conserved but E_{mech} is not.

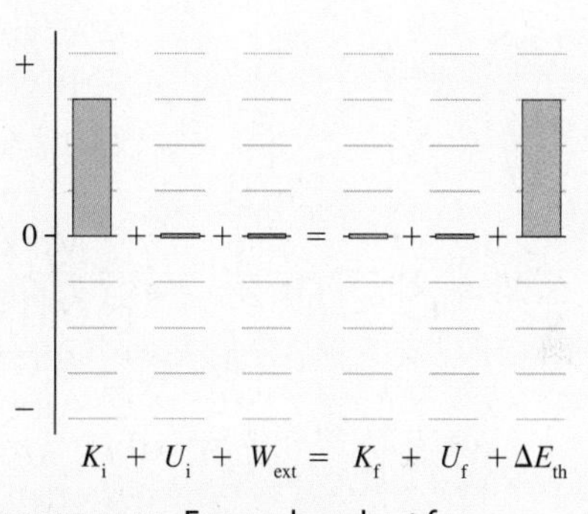

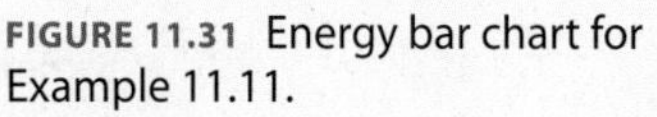

FIGURE 11.31 Energy bar chart for Example 11.11.

EXAMPLE 11.12 Energy bar chart II

A rope lifts a box at constant speed. Show the energy transfers and transformations on an energy bar chart.

SOLVE The tension in the rope is an external force that does work on the box, increasing the potential energy of the box. The kinetic energy is unchanged because the speed is constant. The process is an energy transfer $W_{ext} \rightarrow U_f$, as Figure 11.32 shows. This is not an isolated system, so E_{sys} is not conserved.

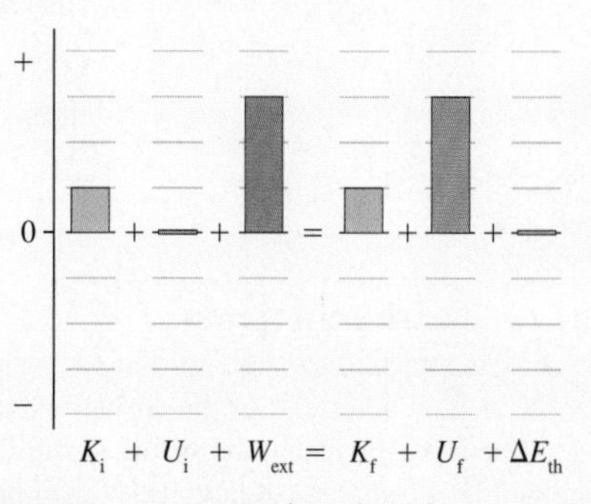

FIGURE 11.32 Energy bar chart for Example 11.12.

EXAMPLE 11.13 Energy bar chart III

The box that was lifted in Example 11.12 falls at a steady speed as the rope spins a generator and causes a light bulb to glow. Air resistance is negligible. Show the energy transfers and transformations on an energy bar chart.

SOLVE The initial potential energy decreases, but K does not change and $\Delta E_{th} = 0$. The tension in the rope is an external force that does work, but W_{ext} is negative in this case because $\vec{T}$ points up while the displacement $\Delta\vec{r}$ is down. Negative work means that energy is transferred from the system to the environment or, in more informal terms, that the *system does work on the environment.* The falling box does work on the generator to spin it and light the bulb. Energy is transferred out of the system and eventually ends up in the light bulb as electrical energy. The process is $U_i \rightarrow W_{ext}$. This is shown in Figure 11.33.

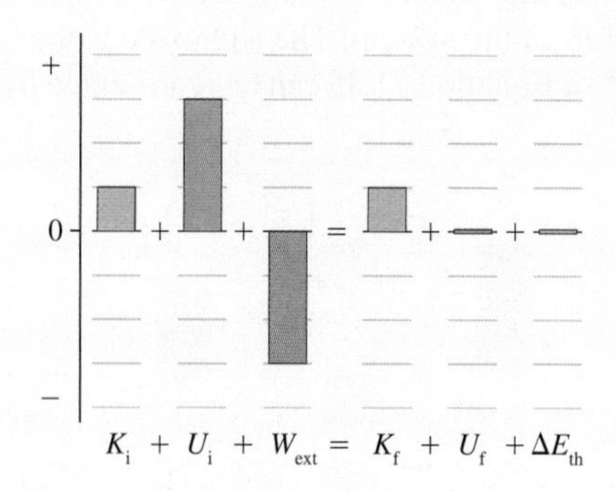

FIGURE 11.33 Energy bar chart for Example 11.13.

Strategy for Energy Problems

5.2–5.7, 6.5, 6.8, 6.9 ActivPhysics ONLINE

This is a good point at which to summarize the strategy we have been developing for using the concept of energy.

MP PROBLEM-SOLVING STRATEGY 11.1 Solving energy problems

MODEL Identify which objects are part of the system and which are in the environment. If possible, choose a system without friction or other dissipative forces. Some problems may need to be subdivided into two or more parts.

VISUALIZE Draw a before-and-after pictorial representation and an energy bar chart. A free-body diagram can be helpful if you're going to calculate work, although often the forces are simple enough to show on the pictorial representation.

SOLVE If the system is both isolated and nondissipative, then the mechanical energy is conserved:

$$K_f + U_f = K_i + U_i$$

If there are external or dissipative forces, calculate W_{ext} and W_{diss}. Then use the more general energy equation

$$K_f + U_f + \Delta E_{th} = K_i + U_i + W_{ext}$$

Kinematics and/or other conservation laws may be needed for some problems.

ASSESS Check that your result has the correct units, is reasonable, and answers the question.

EXAMPLE 11.14 Stretching a spring

The 5.0 kg box is attached to one end of a spring with spring constant 80 N/m. The other end of the spring is anchored to a wall. Initially the box is at rest at the spring's equilibrium position. A rope with a constant tension of 100 N then pulls the box away from the wall. What is the speed of the box after it has moved 50 cm? The coefficient of friction between the box and the floor is 0.30.

MODEL This is a complex situation, but one that we can analyze. First, identify the box, the spring, and the floor as the system. We need the floor inside the system because friction increases the temperature of the box *and* the floor. Friction is a dissipative force that does work W_{diss}. The tension in the rope is an external force. The work W_{ext} done by the rope's tension transfers energy into the system, causing K, U_s, and E_{th} all to increase.

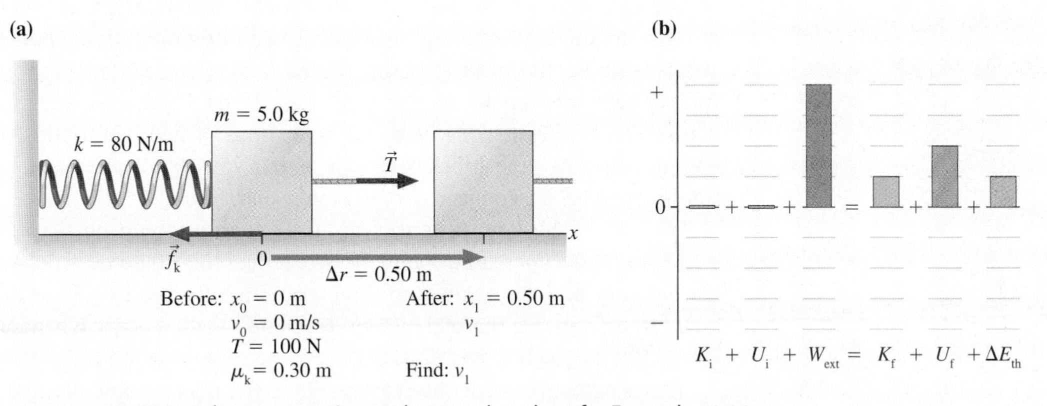

FIGURE 11.34 Pictorial representation and energy bar chart for Example 11.14.

VISUALIZE Figure 11.34a is a before-and-after pictorial representation. The energy transfers and transformations are shown in the energy bar chart of Figure 11.34b.

SOLVE We have an external force and a dissipative force for which we need to calculate the work. The work done by the rope's tension is

$$W_{ext} = \vec{T} \cdot \Delta\vec{r} = T(\Delta x)\cos 0° = (100\ \text{N})(0.50\ \text{m}) = 50\ \text{J}$$

The work done by friction is

$$\begin{aligned} W_{diss} &= f_k(\Delta r)\cos 180° = -\mu_k mg\Delta x \\ &= -(0.3)(5.0\ \text{kg})(9.8\ \text{m/s}^2)(0.50\ \text{m}) = -7.35\ \text{J} \end{aligned}$$

and thus $\Delta E_{th} = +7.35$ J. The energy equation $K_f + U_f + \Delta E_{th} = K_i + U_i + W_{ext}$ is

$$\frac{1}{2}mv_1^2 + \frac{1}{2}kx_1^2 + \Delta E_{th} = \frac{1}{2}mv_0^2 + \frac{1}{2}kx_0^2 + W_{ext}$$

We know that $x_0 = 0$ m and $v_0 = 0$ m/s, so the energy equation simplifies to

$$\frac{1}{2}mv_1^2 = W_{ext} - \Delta E_{th} - \frac{1}{2}kx_1^2$$

Solving for the final speed v_1 gives

$$v_1 = \sqrt{\frac{2(W_{ext} - \Delta E_{th} - \frac{1}{2}kx_1^2)}{m}} = 3.61\ \text{m/s}$$

ASSESS We had to bring all the energy ideas together to solve this problem.

A Good Start, But . . .

We've made a good start at understanding energy, but there's still much to do. First, work is not the only way to transfer energy to a system. Suppose you put a pan of water on the stove and turn on the burner. The water temperature goes up, meaning that E_{th} is increasing, but no work is done. Instead, energy is transferred to the water by a *nonmechanical* means that we will call *heat.* Heat, which has the symbol Q, is the transfer of energy when there is a temperature difference between the system and the environment.

If we include heat as a second means of transferring energy, Equation 11.34, the energy equation, becomes

$$\Delta E_{sys} = W_{ext} + Q \qquad (11.36)$$

This more general statement about energy is called the *first law of thermodynamics.* We will expand our basic energy model to include heat when we reach the study of thermodynamics in Part IV. Equation 11.36 will become essential for understanding thermal processes. The important point to notice for now is that we have developed a quite general model of energy that can later be expanded to include additional types of energy and energy transfers. Thermodynamics will be an extension of the ideas that we've established in Chapters 10 and 11, not an entirely new subject.

Second, there's something unusual about thermal energy. If you toss a block straight up, the kinetic energy is transformed into gravitational potential energy. After reaching the turning point, where $K = 0$, the potential energy is transformed back into kinetic energy as the block falls. If, instead, you push the block across a table, the block's kinetic energy is transformed (via friction) into thermal energy. But once the block stops ($K = 0$), you *never* see the thermal energy transformed back into macroscopic kinetic energy.

Why not? The law of conservation of energy would not be violated if E_{th} decreased and K increased. Think how practical this could be. The brakes of your car get very hot when you stop. You would hardly need your car engine if you could transform that thermal energy back into kinetic energy when the light turns green. But no one has ever done so.

There appears to be a one-way nature to the microscopic thermal energy that isn't true for an object's macroscopic kinetic or potential energy. It's easy to transform kinetic energy into thermal energy, difficult or impossible to transform it back. It seems as if some other law of physics is acting to prevent this. And indeed there is, a very important statement about energy transformation called the *second law of thermodynamics.*

So our basic energy model is a good start, but there's still much to do in Part IV as we expand these ideas into the full science of thermodynamics.

STOP TO THINK 11.6 A child at the playground slides down a pole at constant speed. This is a situation in which

a. $U \rightarrow K$. E_{mech} is not conserved but E_{sys} is.
b. $U \rightarrow E_{th}$. E_{mech} is conserved.
c. $U \rightarrow E_{th}$. E_{mech} is not conserved but E_{sys} is.
d. $K \rightarrow E_{th}$. E_{mech} is not conserved but E_{sys} is.
e. $U \rightarrow W_{ext}$. Neither E_{mech} nor E_{sys} are conserved.

11.9 Power

Work is a transfer of energy between the environment and a system. In many situations we would like to know *how fast* the energy is transferred. Does the force act quickly and transfer the energy very rapidly, or is it a slow and lazy transfer of energy? If you need to buy a motor to lift 2000 lb of bricks up 50 ft, it makes a *big* difference whether the motor has to do this in 30 s or 30 min!

The question "How fast?" implies that we are talking about a *rate.* For example, the velocity of an object—how fast it is going—is the *rate of change* of position. So when we raise the issue of how fast the energy is transferred, we are talking about the *rate of transfer* of energy. The rate at which energy is transferred or transformed is called the **power** P, and it is defined as

$$P \equiv \frac{dE_{sys}}{dt} \tag{11.37}$$

The English unit of power is the *horsepower.* The conversion factor to watts is

$$1 \text{ horsepower} = 1 \text{ hp} = 746 \text{ W}$$

Many common appliances, such as motors, are rated in hp.

The unit of power is the **watt,** which is defined as 1 watt $= 1 \text{ W} \equiv 1 \text{ J/s}$.

A force that is doing work (i.e., transferring energy) at a rate of 3 J/s has an "output power" of 3 W. The system gaining energy at the rate of 3 J/s is said to "consume" 3 W of power. Common prefixes used with power are mW (milliwatts), kW (kilowatts), and MW (megawatts).

EXAMPLE 11.15 Choosing a motor

What power motor is needed to lift a 2000 kg elevator at a steady 3.0 m/s?

SOLVE The tension in the cable does work on the elevator to lift it. Because the cable is pulled by the motor, we say that the motor does the work of lifting the elevator. The net force is zero, because the elevator moves at constant velocity, so the tension is simply $T = mg = 19{,}600$ N. The energy gained by the elevator is

$$\Delta E_{sys} = W_{ext} = T(\Delta y)$$

The power required to give the system this much energy in a time interval Δt is

$$P = \frac{\Delta E_{sys}}{\Delta t} = \frac{T(\Delta y)}{\Delta t}$$

But $\Delta y = v\Delta t$, so $P = Tv = (19{,}600\text{ N})(3.0\text{ m/s}) = 58{,}800\text{ W} = 79\text{ hp}$.

The idea of power as a *rate* of energy transfer applies no matter what the form of energy. Figure 11.35 shows several examples of the idea of power. For now, we primarily want to focus on *work* as the source of energy transfer. Within this more limited scope, power is simply the **rate of doing work:** $P = dW/dt$. If a particle moves through a small displacement $d\vec{r}$ while acted on by force $\vec{F}$, the force does a small amount of work dW given by

$$dW = \vec{F}\cdot d\vec{r}$$

Dividing both sides by dt, to give a rate of change, yields

$$\frac{dW}{dt} = \vec{F}\cdot\frac{d\vec{r}}{dt}$$

But $d\vec{r}/dt$ is the velocity $\vec{v}$, so we can write the power as

$$P = \vec{F}\cdot\vec{v} = Fv\cos\theta \tag{11.38}$$

In other words, the power delivered to a particle by a force acting on it is the dot product of the force with the particle's velocity. These ideas will become clearer with some examples.

Highly trained athletes have a tremendous power output.

Electrical energy ⟶ light and heat at 100 J/s.

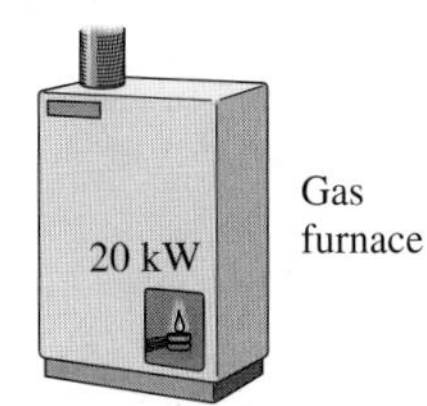

Chemical energy of glucose and fat ⟶ mechanical energy at ≈350 J/s ≈ $\frac{1}{2}$ hp.

20 kW

Gas furnace

Chemical energy of gas ⟶ thermal energy at 20,000 J/s.

FIGURE 11.35 Examples of power.

EXAMPLE 11.16 Power output of a motor

A factory uses a motor and a cable to drag a 300 kg machine into the proper place on the factory floor. What power must the motor supply to drag the machine at a speed of 0.50 m/s? The coefficient of friction between the machine and the floor is 0.60.

SOLVE The force applied by the motor, through the cable, is the tension force $\vec{T}$. This force does work on the machine with power $P = Tv$. The machine is in dynamic equilibrium, because the motion is at constant velocity, hence the tension in the rope balances the friction and is

$$T = f_k = \mu_k mg$$

The motor's power output is

$$P = Tv = \mu_k mgv = 882\text{ W}$$

EXAMPLE 11.17 Power output of a car engine

A 1500 kg car has a front profile that is 1.6 m wide and 1.4 m high. The coefficient of rolling friction is 0.02. What power must the engine provide to drive at a steady 30 m/s ($\approx$65 mph) if 25% of the power is "lost" before reaching the drive wheels?

SOLVE The net force on a car moving at a steady speed is zero. The motion is opposed both by rolling friction and by air resistance. The forward force on the car $\vec{F}_{car}$ (recall that this is really $\vec{F}_{car\ on\ ground}$, a reaction to the drive wheels pushing backward on the ground with $\vec{F}_{car\ on\ ground}$) exactly balances the two opposing forces:

$$F_{car} = f_r + D$$

where $\vec{D}$ is the drag due to the air. Using the results of Chapter 5, where both rolling friction and drag were introduced, this becomes

$$F_{car} = \mu_r mg + \frac{1}{4}Av^2 = 294\ \text{N} + 504\ \text{N} = 798\ \text{N}$$

$A = (1.6\ \text{m}) \times (1.4\ \text{m})$ is the front cross-section area of the car. The power required to push the car forward at this speed is

$$P_{car} = F_{car}v = (798\ \text{N})(30\ \text{m/s}) = 23{,}900\ \text{W} = 32\ \text{hp}$$

This is the power *needed* at the drive wheels to push the car against the dissipative forces of friction and air resistance. The power output of the engine is larger because some energy is used to run the water pump, the power steering, and other accessories. In addition, energy is lost to friction in the drive train. If 25% of the power is lost (a typical value), leading to $P_{car} = 0.75P_{engine}$, the engine's power output is

$$P_{engine} = \frac{P_{car}}{0.75} = 31{,}900\ \text{W} = 43\ \text{hp}$$

ASSESS Automobile engines are typically rated at $\approx$200 hp. Most of that power is reserved for fast acceleration and climbing hills.

STOP TO THINK 11.7 Four students run up the stairs in the time shown. Rank in order, from largest to smallest, their power outputs P_a to P_d.

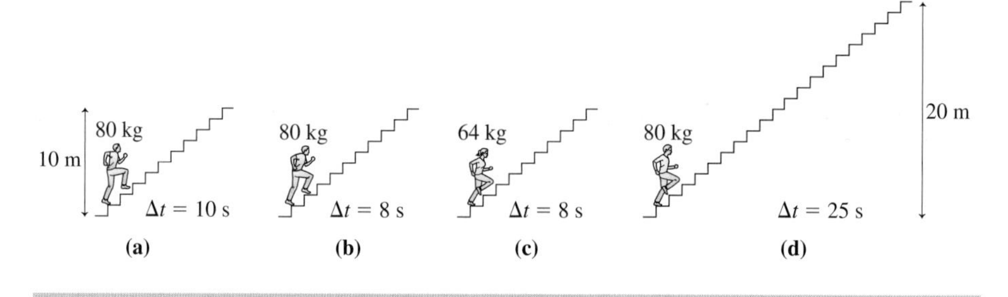

SUMMARY

The goal of Chapter 11 has been to develop a more complete understanding of energy and its conservation.

GENERAL PRINCIPLES

Basic Energy Model

- Energy is *transferred* to or from the system by work.
- Energy is *transformed* within the system.

Two versions of the energy equation are

$$\Delta E_{sys} = \Delta K + \Delta U + \Delta E_{th} = W_{ext}$$

$$K_f + U_f + \Delta E_{th} = K_i + U_i + W_{ext}$$

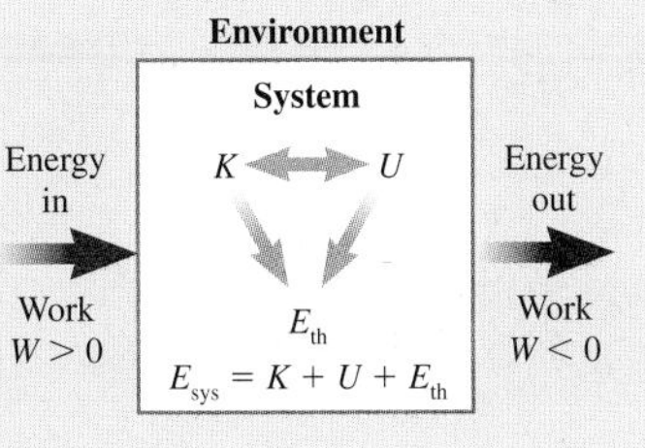

Solving Energy Problems

MODEL Identify objects in the system.

VISUALIZE Draw a before-and-after pictorial representation and an energy bar chart.

SOLVE Use the energy equation

$$K_f + U_f + \Delta E_{th} = K_i + U_i + W_{ext}$$

ASSESS Is the result reasonable?

Law of Conservation of Energy

- **Isolated system:** $W_{ext} = 0$. The total energy $E_{sys} = E_{mech} + E_{th}$ is conserved. $\Delta E_{sys} = 0$
- **Isolated, nondissipative system:** $W_{ext} = 0$ and $W_{diss} = 0$. The mechanical energy E_{mech} is conserved.

$$\Delta E_{mech} = 0 \text{ or } K_f + U_f = K_i + U_i$$

IMPORTANT CONCEPTS

The work-kinetic energy theorem is

$$\Delta K = W_{net} = W_c + W_{diss} + W_{ext}$$

Using $W_c = -\Delta U$ for conservative forces and $W_{diss} = -\Delta E_{th}$ for dissipative forces, this becomes the energy equation.

The work done by a force on a particle as it moves from s_i to s_f is

$$W = \int_{s_i}^{s_f} F_s\,ds = \text{area under the force curve}$$

$$= \vec{F} \cdot \Delta\vec{r} \text{ if } \vec{F} \text{ is a constant force}$$

Conservative forces are forces for which the work is independent of the path followed. The work done by a conservative force can be represented as a **potential energy**

$$\Delta U = U_f - U_i = -W_c(\mathrm{i} \rightarrow \mathrm{f})$$

A conservative force is found from the potential energy by

$$F = -dU/ds = \text{negative of the slope of the PE curve}$$

Dissipative forces transform **macroscopic energy** into thermal energy, which is the **microscopic energy** of the atoms and molecules.

$$\Delta E_{th} = -W_{diss}$$

APPLICATIONS

Power is the rate at which energy is transferred or transformed:

$$P = \frac{dE_{sys}}{dt}$$

For a particle moving with velocity $\vec{v}$, the power delivered to the particle by force $\vec{F}$ is $P = \vec{F} \cdot \vec{v} = Fv\cos\theta$.

Dot product

$$\vec{A} \cdot \vec{B} = AB\cos\alpha = A_xB_x + A_yB_y$$

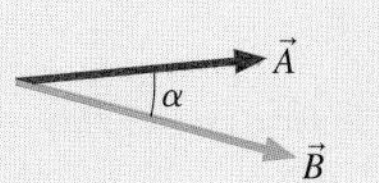

Energy bar charts display the energy equation $K_f + U_f + \Delta E_{th} = K_i + U_i + W_{ext}$ in graphical form.

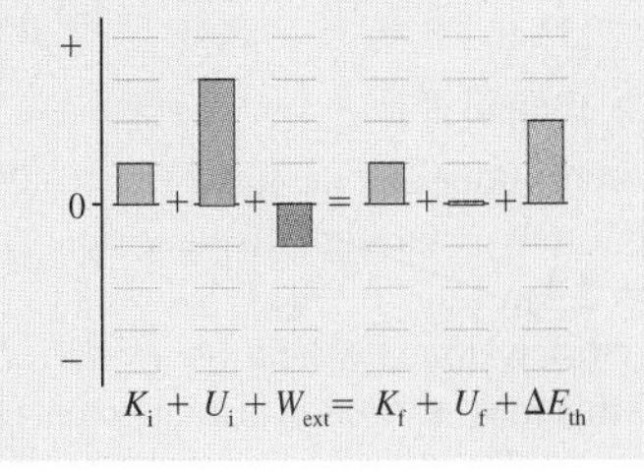

TERMS AND NOTATION

thermal energy, E_{th}
system energy, E_{sys}
energy transformation
energy transfer
work, W
basic energy model
work-kinetic energy theorem
dot product
scalar product
conservative force
nonconservative force
macrophysics
microphysics
dissipative force
energy equation
law of conservation of energy
power, P
watt, W

EXERCISES AND PROBLEMS

The icon indicates that the problem can be done on an Energy Worksheet.

Exercises

Section 11.2 Work and Kinetic Energy

Section 11.3 Calculating and Using Work

1. Evaluate the dot product of the three pairs of vectors in Figure Ex11.1.

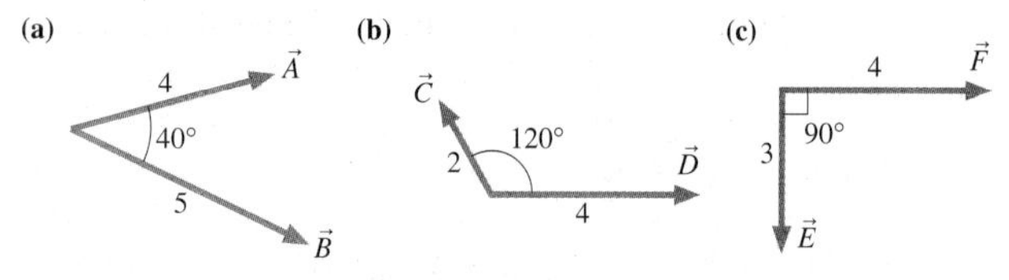

FIGURE EX11.1

2. Evaluate the dot product of the three pairs of vectors in Figure Ex11.2.

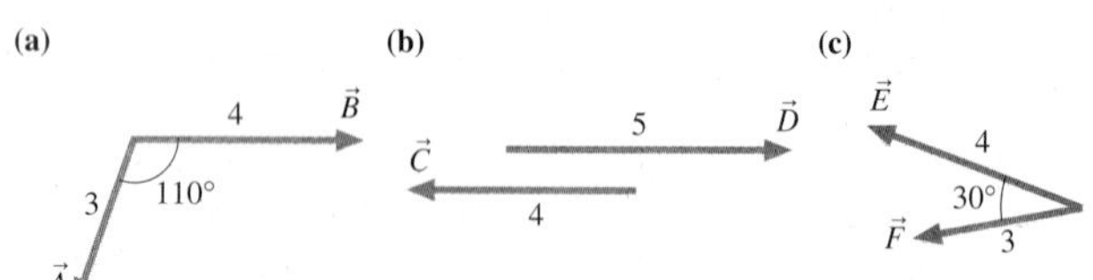

FIGURE EX11.2

3. Evaluate the dot product $\vec{A} \cdot \vec{B}$ if
   a. $\vec{A} = 3\hat{\imath} - 4\hat{\jmath}$ and $\vec{B} = -2\hat{\imath} + 6\hat{\jmath}$.
   b. $\vec{A} = 2\hat{\imath} + 3\hat{\jmath}$ and $\vec{B} = 6\hat{\imath} - 4\hat{\jmath}$.
4. Evaluate the dot product $\vec{A} \cdot \vec{B}$ if
   a. $\vec{A} = 5\hat{\imath} - 3\hat{\jmath}$ and $\vec{B} = 2\hat{\imath} - 2\hat{\jmath}$.
   b. $\vec{A} = 4\hat{\imath} - 2\hat{\jmath}$ and $\vec{B} = -3\hat{\imath} + 6\hat{\jmath}$.
5. How much work is done by the force $\vec{F} = (6.0\hat{\imath} - 3.0\hat{\jmath})$ N on a particle that moves through displacement (a) $\Delta\vec{r} = 2.0\hat{\imath}$ m and (b) $\Delta\vec{r} = 2.0\hat{\jmath}$ m?
6. How much work is done by the force $\vec{F} = (-5.0\hat{\imath} + 4.0\hat{\jmath})$ N on a particle that moves through displacement (a) $\Delta\vec{r} = 3.0\hat{\imath}$ m and (b) $\Delta\vec{r} = -3.0\hat{\jmath}$ m?
7. A 20 g particle is moving to the left at 30 m/s. How much work must be done on the particle to cause it to move to the right at 30 m/s?
8. A 2.0 kg book is lying on a 0.75-m-high table. You pick it up and place it on a bookshelf 2.25 m above the floor.
   a. How much work does gravity do on the book?
   b. How much work does your hand do on the book?
9. The two ropes seen in Figure Ex11.9 are used to lower a 255 kg piano 5.0 m from a second-story window to the ground. How much work is done by each of the three forces?

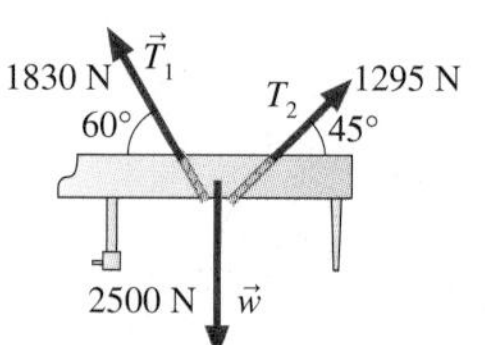

FIGURE EX11.9

10. The two ropes shown in the bird's-eye view of Figure Ex11.10 are used to drag a crate 3.0 m across the floor. How much work is done by each of the three forces?

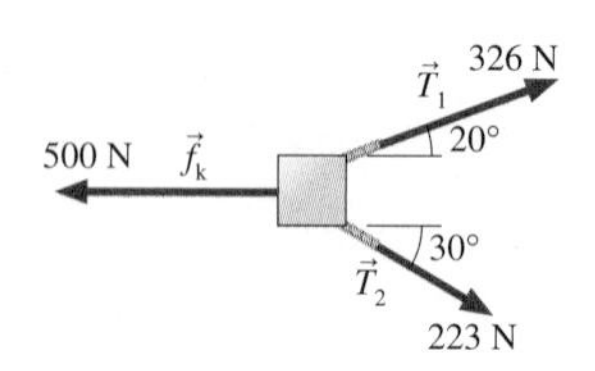

FIGURE EX11.10

11. Figure Ex11.11 is the velocity-versus-time graph for a 2.0 kg object moving along the x-axis. Determine the work done on the object during each of the five intervals AB, BC, CD, DE, and EF.

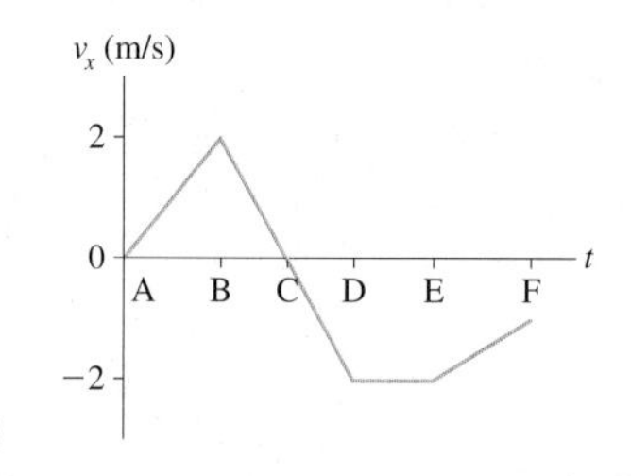

FIGURE EX11.11

Section 11.4 The Work Done by a Variable Force

12. Figure Ex11.12 is the force-versus-position graph for a particle moving along the x-axis. Determine the work done on the particle during each of the three intervals 0–1 m, 1–2 m, and 2–3 m.

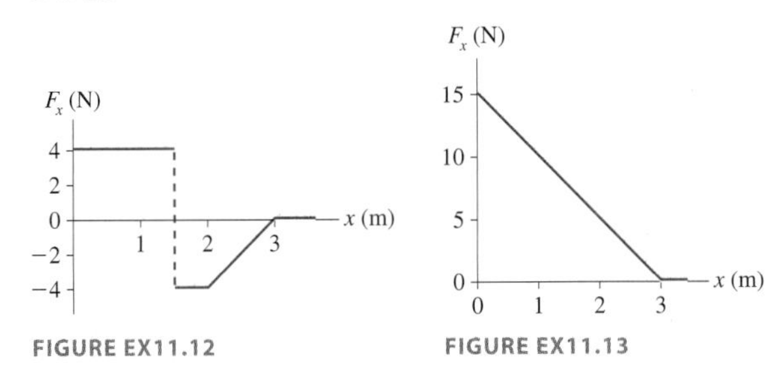

FIGURE EX11.12

FIGURE EX11.13

13. A 500 g particle moving along the x-axis experiences the force shown in Figure Ex11.13. The particle's velocity is 2.0 m/s at $x = 0$ m. What is its velocity at $x = 1$ m, 2 m, and 3 m?
14. A 2.0 kg particle moving along the x-axis experiences the force shown in Figure Ex11.14. The particle's velocity is 4.0 m/s at $x = 0$ m. What is its velocity at $x = 2$ m and 4 m?

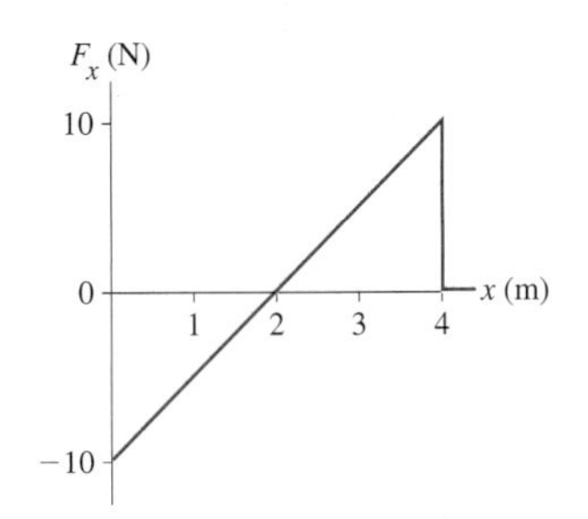

FIGURE EX11.14

15. A 500 g particle moving along the x-axis experiences the force shown in Figure Ex11.15. The particle goes from $v_x = 2.0$ m/s at $x = 0$ m to $v_x = 6.0$ m/s at $x = 2$ m. What is F_{max}?

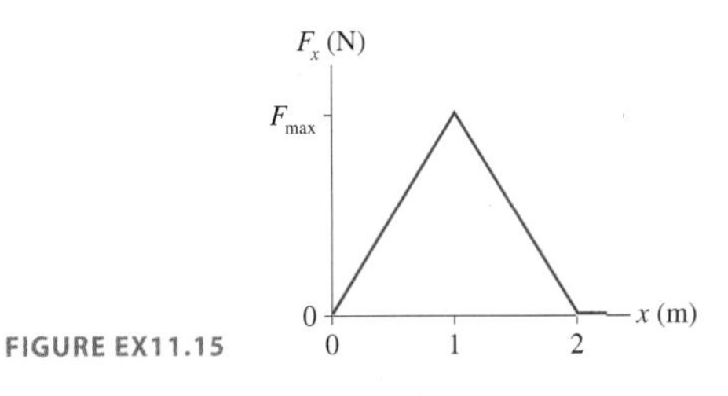

FIGURE EX11.15

Section 11.5 Force, Work, and Potential Energy

Section 11.6 Finding Force from Potential Energy

16. A particle has the potential energy shown in Figure Ex11.16. What is the x-component of the force on the particle at $x = 5$, 15, 25, and 35 cm?

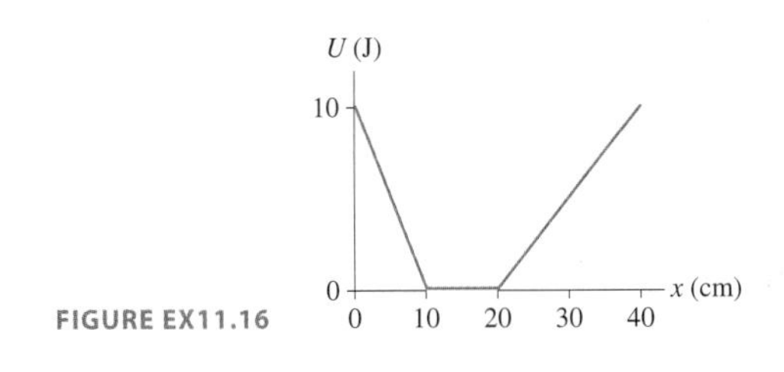

FIGURE EX11.16

17. A particle has the potential energy shown in Figure Ex11.17. What is the x-component of the force on the particle at $x = 1$ m and 3 m?

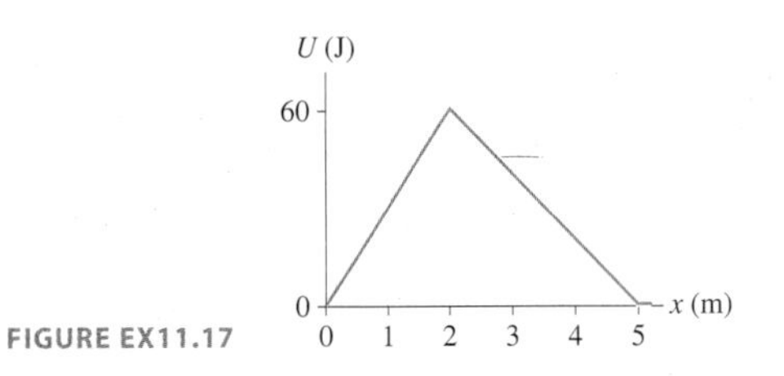

FIGURE EX11.17

18. A particle moving along the y-axis has the potential energy $U = 4y^3$ J, where y is in m.
    a. Graph the potential energy from $y = 0$ m to $y = 2$ m.
    b. What is the y-component of the force on the particle at $y = 0$ m, 1 m, and 2 m?
19. A particle moving along the x-axis has the potential energy $U = 10/x$ J, where x is in m.
    a. Graph the potential energy from $x = 1$ m to $x = 10$ m.
    b. What is the x-component of the force on the particle at $x = 2$ m, 5 m, and 8 m?

Section 11.7 Thermal Energy

20. The mass of a carbon atom is 2.00×10^{-26} kg.
    a. What is the kinetic energy of a carbon atom moving with a speed of 500 m/s?
    b. Two carbon atoms are joined by a spring-like carbon-carbon bond. The potential energy stored in the bond has the value you calculated in part a if the bond is stretched 0.050 nm. What is the value of the bond's spring constant?
21. In Part IV you'll learn to calculate that 1 mole (6.02×10^{23} atoms) of helium atoms in the gas phase has 3700 J of microscopic kinetic energy at room temperature. If we assume that all atoms move with the same speed, what is that speed? The mass of a helium atom is 6.68×10^{-27} kg.
22. A 1500 kg car traveling at 20 m/s skids to a halt.
    a. What energy transfers and transformations occur during the skid?
    b. What is the change in the thermal energy of the car and the road surface?
23. A 20 kg child slides down a 3.0-m-high playground slide. She starts from rest, and her speed at the bottom is 2.0 m/s.
    a. What energy transfers and transformations occur during the slide?
    b. What is the change in the thermal energy of the slide and the seat of her pants?

Section 11.8 Conservation of Energy

24. A system loses 1000 J of potential energy. In the process, it does 500 J of work on the environment and the thermal energy increases by 100 J. Show this process on an energy bar chart.
25. A system gains 500 J of kinetic energy while losing 200 J of potential energy. The thermal energy increases 100 J. Show this process on an energy bar chart.

26. How much work is done by the environment in the process shown in Figure Ex11.26? Is energy transferred from the environment to the system or from the system to the environment?

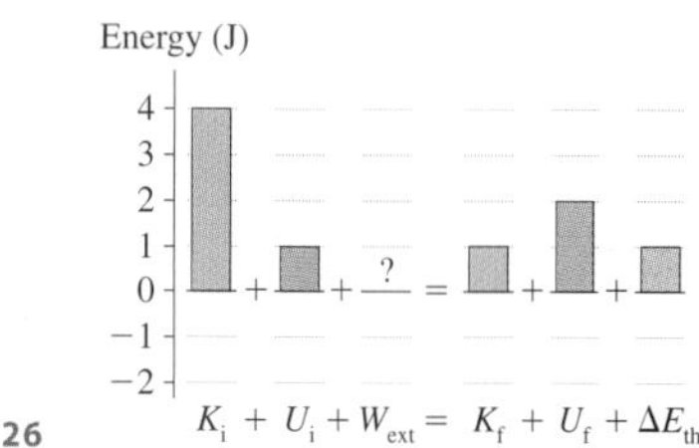

FIGURE EX11.26

27. A cable with 20 N of tension pulls straight up on a 1.02 kg block that is initially at rest. What is the block's speed after being lifted 2.0 m?

Section 11.9 Power

28. a. How much work does an elevator motor do to lift a 1000 kg elevator a height of 100 m?
    b. How much power must the motor supply to do this in 50 s at constant speed?
29. a. How much work must you do to push a 10 kg block of steel across a steel table at a steady speed of 1.0 m/s for 3.0 s?
    b. What is your power output while doing so?
30. At midday, solar energy strikes the earth with an intensity of about 1 kW/m^2. What is the area of a solar collector that could collect 150 MJ of energy in 1 hr? This is roughly the energy content of 1 gallon of gasoline.
31. Which consumes more energy, a 1.2 kW hair dryer used for 10 min or a 10 W night light left on for 24 hr?
32. The electric company bills you in "kilowatt hours," abbreviated kwh.
    a. Is this energy, power, or force? Explain.
    b. Monthly electric use for a typical household is 500 kwh. What is this in basic SI units?
33. A 50 kg sprinter, starting from rest, runs 50 m in 7.0 s at constant acceleration.
    a. What is the magnitude of the horizontal force acting on the sprinter?
    b. What is the sprinter's power output at 2.0 s, 4.0 s, and 6.0 s?

Problems

34. A particle moves from A to D in Figure P11.34 while experiencing force $\vec{F} = (6\hat{\imath} + 8\hat{\jmath})$ N. How much work does the force do if the particle follows path (a) ABD, (b) ACD, and (c) AD? Is this a conservative force? Explain.

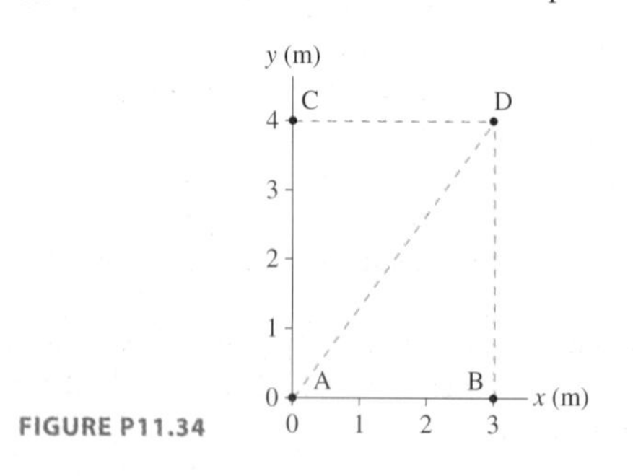

FIGURE P11.34

35. A 100 g particle experiences the one-dimensional, conservative force F_x shown in Figure P11.35.
    a. Draw a graph of the potential energy U from $x = 0$ m to $x = 5$ m. Let the zero of the potential energy be at $x = 0$ m. **Hint:** Think about the definition of potential energy *and* the geometric interpretation of the work done by a varying force.
    b. The particle is shot toward the right from $x = 1$ m with a speed of 25 m/s. What is the particle's mechanical energy?
    c. Draw the total energy line on your graph of part a.
    d. Where is the particle's turning point?

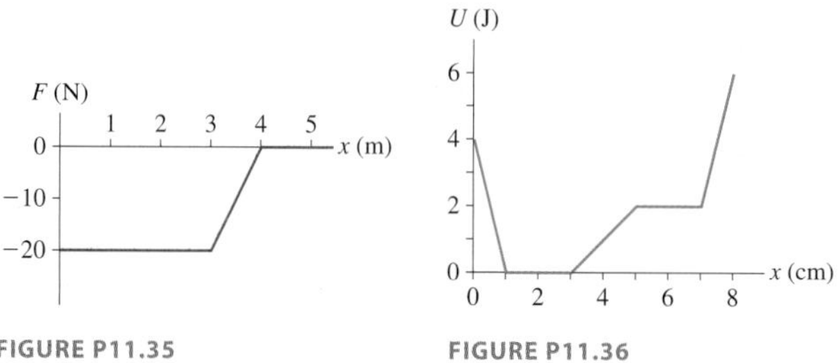

FIGURE P11.35 FIGURE P11.36

36. A 10 g particle has the potential energy shown in Figure P11.36.
    a. Draw a force-versus-position graph from $x = 0$ cm to $x = 8$ cm.
    b. How much work does the force do as the particle moves from $x = 2$ cm to $x = 6$ cm?
    c. What speed does the particle need at $x = 2$ cm to arrive at $x = 6$ cm with a speed of 10 m/s?
37. a. Figure P11.37a shows the force F_x exerted on a particle that moves along the x-axis. Draw a graph of the particle's potential energy as a function of position x. Let U be zero at $x = 0$ m.
    b. Figure P11.37b shows the potential energy U of a particle that moves along the x-axis. Draw a graph of the force F_x as a function of position x.

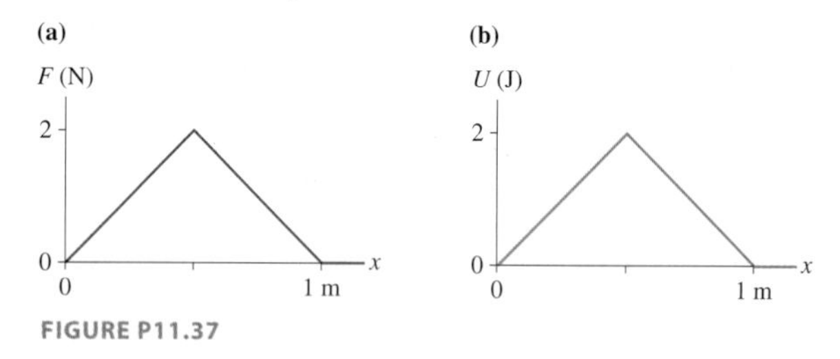

FIGURE P11.37

38. Figure P11.38 is the velocity-versus-time graph of a 500 g particle that starts at $x = 0$ m and moves along the x-axis. Draw graphs of the following by calculating and plotting numerical values at $t = 0, 1, 2, 3,$ and 4 s. Then sketch lines or curves of the appropriate shape between the points. Make sure you include appropriate scales on both axes of each graph.
    a. Acceleration versus time.
    b. Position versus time.
    c. Kinetic energy versus time.
    d. Force versus time.
    e. Use your F_x-versus-t graph to determine the *impulse* delivered to the particle during the time interval 0–2 s and also the interval 2–4 s.

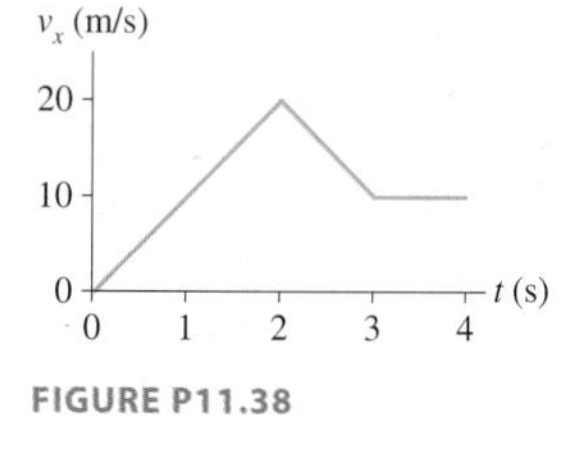

FIGURE P11.38

f. Use the impulse-momentum theorem to determine the particle's velocity at $t = 2$ s and at $t = 4$ s. Do your results agree with the velocity graph?
g. Now draw a graph of force versus *position*. This requires no calculations; just think carefully about what you learned in parts a to d.
h. Use your F_x-versus-x graph to determine the *work* done on the particle during the time interval 0–2 s and also the interval 2–4 s.
i. Use the work-kinetic energy theorem to determine the particle's velocity at $t = 2$ s and at $t = 4$ s. Do your results agree with the velocity graph?

39. A 1000 kg elevator accelerates upward at 1.0 m/s^2 for 10 m, starting from rest.
   a. How much work does gravity do on the elevator?
   b. How much work does the tension in the elevator cable do on the elevator?
   c. Use the work-kinetic energy theorem to find the kinetic energy of the elevator as it reaches 10 m.
   d. What is the speed of the elevator as it reaches 10 m?
40. Bob can throw a 500 g rock with a speed of 30 m/s. He moves his hand forward 1.0 m while doing so.
   a. How much work does Bob do on the rock?
   b. How much force, assumed to be constant, does Bob apply to the rock?
   c. What is Bob's maximum power output as he throws the rock?
41. Doug pushes a 5.0 kg crate up a 2.0-m-high 20° frictionless slope by pushing it with a constant *horizontal* force of 25 N. What is the speed of the crate as it reaches the top of the slope?
   a. Solve this problem using work and energy.
   b. Solve this problem using Newton's laws.
42. Sam, whose mass is 75 kg, straps on his skis and starts down a 50-m-high, 20° frictionless slope. A strong headwind exerts a *horizontal* force of 200 N on him as he skies. Find Sam's speed at the bottom (a) using work and energy, (b) using Newton's laws.
43. Susan's 10 kg baby brother Paul sits on a mat. Susan pulls the mat across the floor using a rope that is angled 30° above the floor. The tension is a constant 30 N and the coefficient of friction is 0.20. Use work and energy to find Paul's speed after being pulled 3.0 m.
44. A horizontal spring with spring constant 100 N/m is compressed 20 cm and used to launch a 2.5 kg box across a frictionless, horizontal surface. After the box travels some distance, the surface becomes rough. The coefficient of kinetic friction of the box on the surface is 0.15. Use work and energy to find how far the box slides across the rough surface before stopping.
45. A baggage handler throws a 15 kg suitcase horizontally along the floor of an airplane luggage compartment with an initial speed of 1.2 m/s. The suitcase slides 2.0 m before stopping. Use work and energy to find the suitcase's coefficient of kinetic friction on the floor.
46. Truck brakes can fail if they get too hot. In some mountainous areas, ramps of loose gravel are constructed to stop runaway trucks that have lost their brakes. The combination of a slight upward slope and a large coefficient of friction in the gravel brings the truck safely to a halt. Suppose a gravel ramp slopes upward at 6.0° and the coefficient of friction is 0.40. Use work and energy to find the length of a ramp that will stop a 15,000 kg truck that enters the ramp at 35 m/s (≈75 mph).
47. A freight company uses a compressed spring to shoot 2.0 kg packages up a 1.0-m-high frictionless ramp into a truck, as Figure P11.47 shows. The spring constant is 500 N/m and the spring is compressed 30 cm.
   a. What is the speed of the package when it reaches the truck?
   b. A careless worker spills his soda on the ramp. This creates a 50-cm-long sticky spot with a coefficient of kinetic friction 0.30. Will the next package make it into the truck?

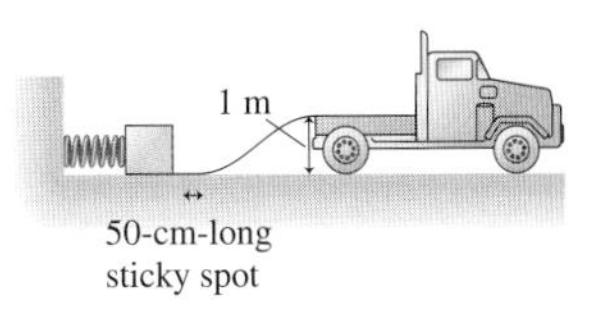

FIGURE P11.47

48. Use work and energy to find the speed of the 2.0 kg block just before it hits the floor if (a) the table is frictionless and if (b) the coefficient of kinetic friction of the 3.0 kg block is 0.15.

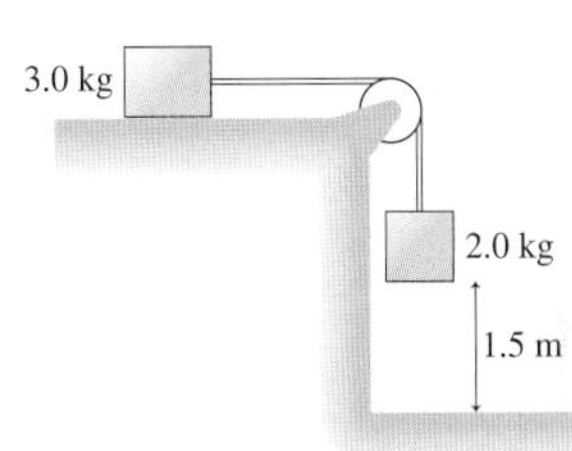

FIGURE P11.48

49. An 8.0 kg crate is pulled up a 30° incline by a person pulling on a rope that makes an 18° angle with the incline. The tension in the rope is 120 N and the crate's coefficient of kinetic friction on the incline is 0.25. If the crate moves 5.0 m, what is the work done by (a) the tension in the rope, (b) gravity, (c) friction, and (d) the normal force?
50. A 10.2 kg weather rocket generates a thrust of 200 N. The rocket, pointing upward, is clamped to the top of a vertical spring. The bottom of the spring, whose spring constant is 500 N/m, is anchored to the ground.
   a. Initially, before the engine is ignited, the rocket sits at rest on top of the spring. How much is the spring compressed?
   b. After the engine is ignited, what is the rocket's speed when the spring has stretched 40 cm? For comparison, what would be the rocket's speed after traveling this distance if it weren't attached to the spring?
51. A 50 kg ice skater is gliding along the ice, heading due north at 4.0 m/s. The ice has a small coefficient of static friction, to prevent the skater from slipping sideways, but $\mu_k = 0$. Suddenly, a wind from the northeast exerts a force of 4.0 N on the skater.
   a. Use work and energy to find the skater's speed after gliding 100 m in this wind.
   b. What is the minimum value of μ_s that allows her to continue moving straight north?
52. a. A 50 g ice cube can slide without friction up and down a 30° slope. The ice cube is pressed against a spring at the bottom of the slope, compressing the spring 10 cm. The spring constant is 25 N/m. When the ice cube is released, what total distance will it travel up the slope before reversing direction?
   b. The ice cube is replaced by a 50 g plastic cube whose coefficient of kinetic friction is 0.20. How far will the plastic cube travel up the slope?

53. A 5.0 kg box slides down a 5.0-m-high frictionless hill, starting from rest, across a 2.0-m-long horizontal surface, then hits a horizontal spring with spring constant 500 N/m. The other end of the spring is anchored against a wall. The ground under the spring is frictionless, but the 2.0-m-long horizontal surface is rough. The coefficient of kinetic friction of the box on this surface is 0.25.
   a. What is the speed of the box just before reaching the rough surface?
   b. What is the speed of the box just before hitting the spring?
   c. How far is the spring compressed?
   d. Including the first crossing, how many *complete* trips will the box make across the rough surface before coming to rest?
54. The spring shown in Figure P11.54 is compressed 50 cm and used to launch a 100 kg physics student. The track is frictionless until it starts up the incline. The student's coefficient of kinetic friction on the 30° incline is 0.15.
   a. What is the student's speed just after losing contact with the spring?
   b. How far up the incline does the student go?

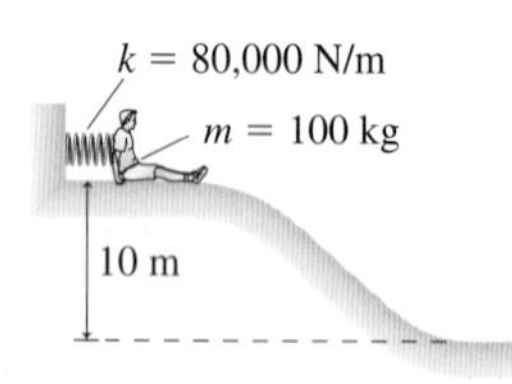

FIGURE P11.54

55. A block of mass m starts from rest at height h. It slides down a frictionless incline, across a rough horizontal surface of length L, then up a frictionless incline. The coefficient of kinetic friction on the rough surface is μ_k.
   a. What is the block's speed at the bottom of the first incline?
   b. How high does the block go on the second incline?
   Give your answers in terms of m, h, L, μ_k, and g.
56. Show that Hooke's law for an ideal spring is a conservative force. To do so, first calculate the work done by the spring as it expands from A to B. Then calculate the work done by the spring as it expands from A to point C, which is beyond B, then returns from C to B.
57. A clever engineer designs a "sprong" that obeys the force law $F_x = -q(x - x_e)^3$, where x_e is the equilibrium position of the end of the sprong and q is the sprong constant. For simplicity, we'll let $x_e = 0$ m. Then $F_x = -qx^3$.
   a. What are the units of q?
   b. Draw a graph of F_x versus x.
   c. Find an expression for the potential energy of a stretched or compressed sprong.
   d. A sprong-loaded toy gun shoots a 20 g plastic ball. What is the launch speed if the sprong constant is 40,000, with the units you found in part a, and the sprong is compressed 10 cm? Assume the barrel is frictionless.
58. A particle of mass m has initial conditions $x_0 = 0$ m and $v_0 > 0$ m/s. The particle experiences the variable force $F_x = F_0\sin(cx)$ as it moves to the right along the x-axis, where F_0 and c are constants.
   a. What are the units of F_0?
   b. What are the units of c?
   c. What is the value of F_x when the particle is at x_0?
   d. At what position x_{max} does the force first reach a maximum value? Your answer will be in terms of the constants F_0 and c and perhaps other numerical constants.
   e. Sketch a graph of F-versus-x from x_0 to x_{max}.
   f. What is the particle's velocity as it reaches x_{max}? Give your answer in terms of m, v_0, F_0, and c.
59. a. Estimate the height in meters of the two flights of stairs that go from the first to the third floor of a building.
   b. Estimate how long it takes you to *run* up these two flights of stairs.
   c. Estimate your power output in both watts and horsepower while running up the stairs.
60. A gardener pushes a 12 kg lawnmower whose handle is tilted up 37° above horizontal. The lawnmower's coefficient of rolling friction is 0.15. How much power does the gardener have to supply to push the lawnmower at a constant speed of 1.2 m/s?
61. A 5.0 kg cat leaps from the floor to the top of a 95-cm-high table. If the cat pushes against the floor for 0.20 s to accomplish this feat, what was her average power output during the pushoff period?
62. A 2.0 hp electric motor on a water well pumps water from 10 m below the surface. The density of water is 1.0 kg per liter. How many liters of water does the motor pump in 1 hr?
63. In a hydroelectric dam, water falls 25 m and then spins a turbine to generate electricity.
   a. What is ΔU of 1.0 kg of water?
   b. Suppose the dam is 80% efficient at converting the water's potential energy to electrical energy. How many kilograms of water must pass through the turbines each second to generate 50 MW of electricity? This is a typical value for a small hydroelectric dam.
64. The force required to tow a water skier at speed v is proportional to the speed. That is, $F_{tow} = Av$, where A is a proportionality constant. If a speed of 2.5 mph requires 2 hp, how much power is required to tow a water skier at 7.5 mph?
65. Estimate the maximum speed of a horse. Assume that a horse is 1.8 m tall and 0.5 m wide.
66. The engine in a 1500 kg car has a maximum power output of 200 hp, but 25% of the power is lost before reaching the drive wheels. The car has a front profile that is 1.6 m wide and 1.4 high. The coefficient of rolling friction is 0.02. What is the car's top speed? Is this answer reasonable?
67. A Porsche 944 Turbo has a rated power of 217 hp and a mass, with the driver, of 1480 kg. Two-thirds of the car's weight is over the drive wheels.
   a. What is the maximum acceleration of the Porsche on a concrete surface where $\mu_s = 1.0$?
   Hint: What force pushes the car forward?
   b. What is the speed of the Porsche at maximum power output?
   c. If the Porsche accelerates at a_{max}, how long does it take until it reaches the maximum power output?

In Problems 68 through 71 you are given the equation(s) used to solve a problem. For each of these, you are to
   a. Write a realistic problem for which this is the correct equation(s).
   b. Draw a pictorial representation.
   c. Finish the solution of the problem.

68. $\frac{1}{2}(2.0\text{ kg})(4.0\text{ m/s})^2 + 0$
   $+ (0.15)(2.0\text{ kg})(9.8\text{ m/s}^2)(2.0\text{ m}) = 0 + 0 + T(2.0\text{ m})$
69. $\frac{1}{2}(20\text{ kg})v_1^2 + 0$
   $+ (0.15)(20\text{ kg})(9.8\text{ m/s}^2)\cos 40°((2.5\text{ m})/\sin 40°)$
   $= 0 + (20\text{ kg})(9.8\text{ m/s}^2)(2.5\text{ m}) + 0$

70. $F_{push} - (0.20)(30\text{ kg})(9.8\text{ m/s}^2) = 0$

$75\text{ W} = F_{push}v$

71. $T - (1500\text{ kg})(9.8\text{ m/s}^2) = (1500\text{ kg})(1.0\text{ m/s}^2)$

$P = T(2.0\text{ m/s})$

Challenge Problems

72. You've taken a summer job at a water park. In one stunt, a water skier is going to glide up the 2.0-m-high frictionless ramp shown in Figure CP11.72, then sail over a 5.0-m-wide tank filled with hungry sharks. You will be driving the boat that pulls her to the ramp. She'll drop the tow rope at the base of the ramp just as you veer away. What minimum speed must you have as you reach the ramp in order for her to live to do this again tomorrow?

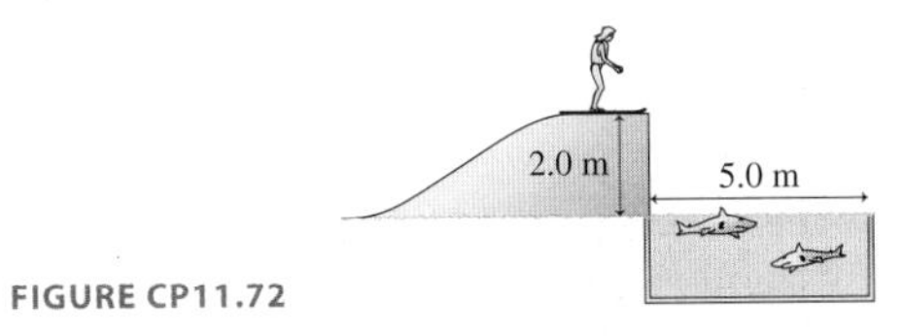

FIGURE CP11.72

73. The spring in Figure CP11.73 has a spring constant of 1000 N/m. It is compressed 15 cm, then launches a 200 g block. The horizontal surface is frictionless, but the block's coefficient of kinetic friction on the incline is 0.20. What distance d does the block sail through the air?

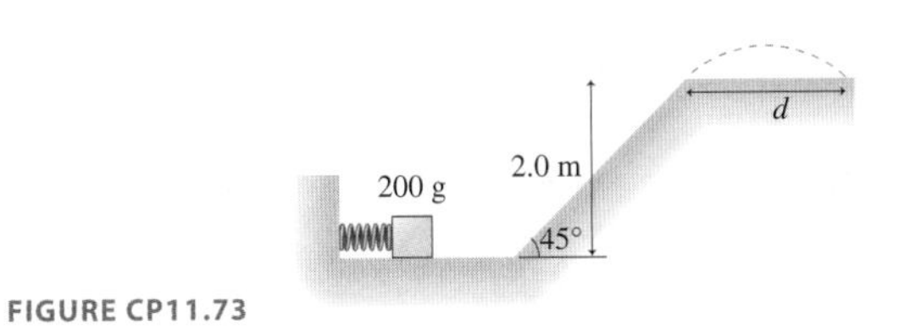

FIGURE CP11.73

74. As a hobby, you like to participate in reenactments of Civil War battles. Civil War cannons were "muzzle loaded," meaning that the gunpowder and the cannonball were inserted into the output end of the muzzle, then tamped into place with a long plunger. To recreate the authenticity of muzzle-loaded cannons, but without the danger of real cannons, Civil War buffs have invented a spring-powered cannon that fires a 1.0 kg plastic ball. A spring, with spring constant 3000 N/m, is mounted at the back of the barrel. You place a ball in the barrel, then use a long plunger to press the ball against the spring and lock the spring into place, ready for firing. In order for the latch to catch, the ball has to be moving at a speed of at least 2.0 m/s when the spring has been compressed 30 cm. The coefficient of friction of the ball in the barrel is 0.30. The plunger doesn't touch the sides of the barrel.
   a. If you push the plunger with a constant force, what is the minimum force that you must use to compress and latch the spring? You can assume that no effort was required to push the ball down the barrel to where it first contacts the spring.
   b. What is the cannon's muzzle velocity if the ball travels a total distance of 1.5 m to the end of the barrel?

75. The equation mgy for gravitational potential energy is valid only for objects near the surface of a planet. Consider two very large objects of mass m_1 and m_2, such as stars or planets, whose centers are separated by the large distance r. These two large objects exert gravitational forces on each other. You'll learn in Chapter 12 that the gravitational potential energy is

$$U = -\frac{Gm_1m_2}{r}$$

where $G = 6.67 \times 10^{-11}\text{ Nm}^2/\text{kg}^2$ is called the *gravitational constant.*
   a. Sketch a graph of U versus r. The mathematical difficulty at $r = 0$ is not a physically significant problem because the masses will collide before they get that close together.
   b. What separation r has been chosen as the point of zero potential energy? Does this make sense? Explain.
   c. Two stars are at rest 1.0×10^{14} m apart. This is about 10 times the diameter of the solar system. The first star is the size of our sun, with a mass of 2.0×10^{30} kg and a radius of 7.0×10^8 m. The second star has mass 8.0×10^{30} kg and radius of 11.0×10^8 m. Gravitational forces pull the two stars together. What is the speed of each star at the moment of impact?

STOP TO THINK ANSWERS

Stop to Think 11.1: d. Constant speed means $\Delta K = 0$. Gravitational potential energy is lost and friction heats up the slide and the child's pants.

Stop to Think 11.2: 6.0 J. $K_f = K_i + W$. W is the area under the curve, which is 4 J.

Stop to Think 11.3: b. The weight force $\vec{w}$ is in the same direction as the displacement. It does positive work. The tension force $\vec{T}$ is opposite the displacement. It does negative work.

Stop to Think 11.4: c. $W = F(\Delta r)\cos\theta$. The 10 N force at 90° does no work at all. $\cos 60° = \frac{1}{2}$, so the 8 N force does less work than the 6 N force.

Stop to Think 11.5: e. Force is the negative of the slope of the potential energy diagram. At $x = 4$ m the potential energy has risen by 4 J over a distance of 2 m, so the slope is 2 J/m = 2 N.

Stop to Think 11.6: c. Constant speed means $\Delta K = 0$. Gravitational potential energy is lost, and friction heats up the pole and the child's hands.

Stop to Think 11.7: $P_b > P_a = P_c > P_d$. The work done is $mg\Delta y$, so the power is $mg\Delta y/\Delta t$. Runner b does the same work as a but in less time. The ratio $m/\Delta t$ is the same for a and c. Runner d does twice the work as a but takes more than twice as long.

PART

II

Conservation Laws

SUMMARY

In Part II we have discovered that we don't need to know all the details of an interaction to relate the properties of a system "before" the interaction to the system's properties "after" the interaction. Along the way, we found two important quantities, momentum and energy, that are often conserved. Momentum and energy are specific characteristics of a system of particles.

Momentum and energy have specific conditions under which they are conserved. The total momentum $\vec{P}$ and the total energy E_{sys} are conserved for an *isolated system,* one on which the net external force is zero. Further, the system's mechanical energy is conserved if the system is both isolated and nondissipative (i.e., no friction forces). These ideas are captured in the two most important conservation laws, the law of conservation of momentum and the law of conservation of energy.

Of course, not all systems are isolated. For both momentum and energy, it was useful to develop a *model* of a system interacting with its environment. Interactions within the system do not change $\vec{P}$ or E_{sys}. The kinetic, potential, and thermal energy within the system can be transformed without changing E_{sys}. Interactions between the system and the environment *do* change the system's momentum and energy. In particular,

- Impulse is the transfer of momentum to or from the system: $\Delta p_s = J_s$.
- Work is the transfer of energy to or from the system: $\Delta E_{sys} = W_{ext}$.

The basic energy model is built around the twin ideas of the transfer and the transformation of energy.

The table below is a knowledge structure of conservation laws. You should compare this with the knowledge structure of Newtonian mechanics in the Part I Summary. Add the problem-solving strategies, and you now have a very powerful set of tools for understanding motion.

KNOWLEDGE STRUCTURE II Conservation Laws

ESSENTIAL CONCEPTS	Impulse, momentum, work, energy
BASIC GOALS	How is the system "after" an interaction related to the system "before"? What quantities are conserved, and under what conditions?
GENERAL PRINCIPLES	**Impulse-momentum theorem** $\Delta p_s = J_s$ **Work-kinetic energy theorem** $\Delta K = W = W_c + W_{diss} + W_{ext}$ **Energy equation** $\Delta E_{sys} = \Delta K + \Delta U + \Delta E_{th} = W_{ext}$
CONSERVATION LAWS	For an isolated system, with $\vec{F}_{net} = \vec{0}$ and $W_{net} = 0$ • The total momentum $\vec{P}$ is conserved. • The total energy $E_{sys} = E_{mech} + E_{th}$ is conserved. For an isolated and nondissipative system, with $W_{diss} = 0$ • The mechanical energy $E_{mech} = K + U$ is conserved.

BASIC PROBLEM-SOLVING STRATEGY Draw a before-and-after pictorial representation, then use the momentum or energy equations to relate "before" to "after." Where possible, choose a system for which momentum and/or energy are conserved. If necessary, calculate impulse and/or work.

Basic model of momentum and energy

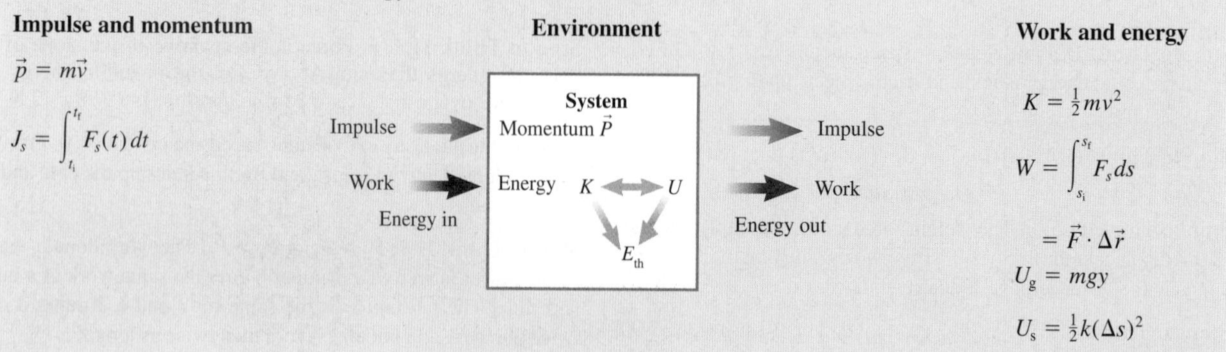

Energy Conservation

You hear it all the time. Turn off lights. Buy a more fuel-efficient car. Conserve energy. But why conserve energy if energy is already conserved? Consider the earth as a whole. No work is done on the earth. And while heat energy flows from the sun to the earth, the earth radiates an equal amount of heat back into space. With no work and no net heat flow, the earth's total energy E_{earth} is conserved.

Pumping oil, driving your car, running a nuclear reactor, or turning on the lights are all interactions *within* the earth system. They transform energy from one type to another, but they don't affect the value of E_{earth}. Consider some examples.

- Crude oil, stored in the earth, has chemical energy E_{chem}. Chemical energy, a form of microscopic potential energy, is released when chemical reactions rearrange the bonds. As you burn gasoline in your car engine, the chemical energy is transformed into the kinetic energy of the moving pistons. This kinetic energy, in turn, is transformed into the car's kinetic energy. The car's kinetic energy is ultimately dissipated as thermal energy in the brakes, air, tires, and road because of friction and drag. Overall, the energy process of driving looks like

$$E_{chem} \rightarrow K_{piston} \rightarrow K_{car} \rightarrow E_{th}$$

- Water stored behind a dam has gravitational potential energy U_g. Potential energy is transformed into kinetic energy as the water falls, then into the spinning turbine's kinetic energy. The turbine converts mechanical energy into electric energy E_{elec}. The electric energy reaches a light bulb where it is transformed partly into thermal energy (light bulbs are hot!) and partly into light energy. The light is absorbed by surfaces, heating them slightly and thus transforming the light energy into thermal energy. The overall energy process is

$$U_g \rightarrow K_{water} \rightarrow K_{turbine} \rightarrow E_{elec} \rightarrow E_{light} \rightarrow E_{th}.$$

Do you notice a trend? Stored energy (fossil fuel, water behind a dam) is transformed through a series of steps, some of which are considered "useful," until the energy is ultimately dissipated as thermal energy. **The total energy has not changed, but its "usefulness" has.**

Thermal energy is rarely "useful" energy. A room full of moving air molecules has a huge thermal energy, but you can't run your lights or your air conditioner with it. You can't turn the thermal energy of your hot brakes back into the kinetic energy of the car. Energy may be conserved, but there's a one-way characteristic of the transformations.

The energy stored in fuels and the energy of the sun is "high-quality energy" because of its potential to be transformed into such useful forms of energy as moving your car or heating your house. But as Figure II.1 shows, high-quality energy becomes "degraded" into thermal energy, where it is no longer useful. Thus the phrase "conserve energy" isn't used literally. Instead, it means to conserve or preserve the earth's sources of high-quality energy.

Conserving high-quality energy is important because fossil fuels are a finite resource. Experts may disagree as to how long fossil fuels will last, but all agree that it won't be forever. Oil and natural gas will likely become scarce during your lifetime. In addition, burning fossil fuel generates carbon dioxide, a major contributor to global warming. Energy conservation helps fuels last longer and minimizes their side effects.

There are two paths to conserving energy. One is to use less high-quality energy. Turning off lights and bicycling rather than driving are actions that preserve high-quality energy. A second path is to use energy more efficiently. That is, get more of the useful activity (miles driven, rooms lit) for the same amount of high-quality energy.

Light bulbs offer a good example. A 100 W incandescent light bulb actually produces only about 10 W of light energy. Ninety watts of the high-quality electric energy is immediately degraded as thermal energy without doing anything useful. By contrast, a 25 W compact fluorescent bulb generates the same 10 W of light but only 15 W of thermal energy. The same amount of high-quality energy can light four times as many rooms if 100 W incandescent bulbs are replaced by 25 W compact fluorescent bulbs.

So why conserve energy if energy is already conserved? Because technological society needs a dependable and sustainable supply of high-quality energy. Both technology improvement and lifestyle choices will help us achieve a sustainable energy future.

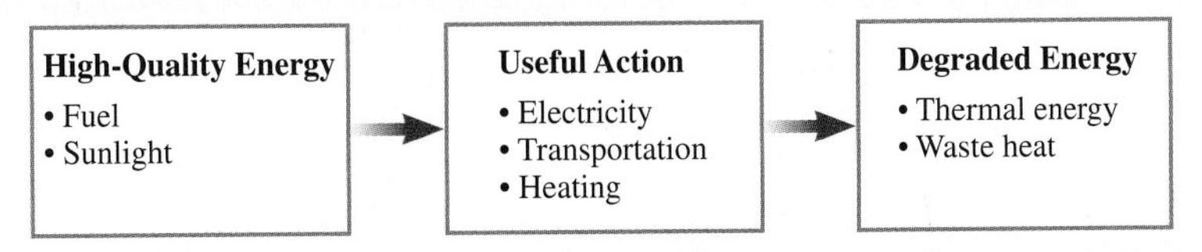

FIGURE II.1 "Using" energy transforms high-quality energy into thermal energy.

PART III

At what altitude above the surface of Alderaan does the Death Star orbit with a period of 10 hours? To find out, what properties of the planet do you need to estimate?

Applications of Newtonian Mechanics

From Nanostructures to the Stars

Engineering arose as a distinct discipline, separate from science, during the industrial revolution. Initially, the engineer's goal was to design new and better machines. Science and engineering in the 21st century may bear little resemblance to the practices of those early days, but the underlying principles remain the same. Whether your dream is to build nanomachines or to pilot spaceships to the stars, much of engineering and applied science is still built on a foundation of Newtonian mechanics.

Parts I and II of this textbook have established the *theory* of Newtonian mechanics. It is a theory of motion based on the concept of *forces* applied to *particles*. Our focus has been to develop the principal ideas of the theory, so you haven't seen too many examples of how this theory is applied to the "real world." In fact, most of the applications of Newtonian mechanics will be developed in other science and engineering courses. Nonetheless, we're now in a good position to examine some of the more practical aspects of Newtonian mechanics.

Our goal for Part III is to apply our newfound theory to four important topics:

- **Gravity.** By adding one more law, Newton's law of gravity, we'll be able to understand much about the physics of the space shuttle, communication satellites, the solar system, and interplanetary travel.
- **Rotation.** The primary limitation of the particle model is that we can't represent a spinning or rotating object as a particle. To understand rotation, a very important form of motion, we'll need to learn how to extend the particle model. We'll then be able to study the motion of objects such as rolling wheels and spinning space stations. Our study of rotation will also lead to another important conservation law, the law of conservation of angular momentum.
- **Oscillations.** Oscillations and vibrations are another important form of motion with many practical applications. Oscillations are seen in systems ranging from the pendulum in a grandfather's clock and the pistons in your car's engine to pieces of machinery that vibrate and the quartz crystal oscillator that provides the timing signals in sophisticated electronic circuits. The physics and mathematics of oscillations will also be the starting point for two other important topics later in this book: waves and AC circuits.
- **Fluids.** Liquids and gases are classified as fluids because they *flow*. Surprisingly, it takes no new physics to understand the basic mechanical properties of fluids. (The thermal properties of fluids, such as the ideal-gas law, will be studied in Part IV.) However, we will have to go well beyond the particle model to see how Newton's laws are applied to liquids and gases. By doing so we'll be able to understand what pressure is, how a steel ship can float, and how fluids flow through pipes.

Relatively few new concepts will be introduced in Part III. Instead, we will focus on *applying* our two primary sets of tools:

- Newton's laws of motion, and
- The conservation laws, especially conservation of energy.

These tools will help us analyze and understand a variety of interesting and practical applications.

Nanostructures

Nanostructures are motors, machines, and materials smaller than the width of a human hair. Some day, nanomachines introduced into a blood vessel may be able to perform microscopic surgery to repair a damaged heart. Nanostructures may become the basis for new materials that are stronger than steel but lighter than silk. Although nanostructures are mostly still in the research stage, they may come to dominate 21st-century technology.

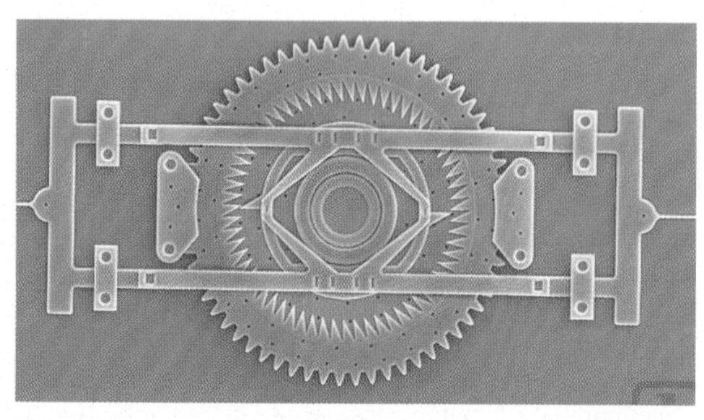

This microscopic machine is about the size of a human hair.

One interesting possibility is to develop self-assembling structures, not unlike mechanical DNA, wherein one type of nanomachine can build other nanomachines, including copies of itself, from raw materials. It's safe to say that nanostructures have megapotential.

But what do you need to know to design and use nanostructures? Mostly Newtonian mechanics. The forces and torques may be at the microscopic level, but you still have to analyze how the device will respond to those forces and torques. Friction, rather than being a minor nuisance, plays a crucial role that must be understood in detail. And many nanostructures will operate in fluids, so you must know the properties of both static and moving fluids.

Space

Humans have already visited the moon. It is entirely possible, perhaps even likely, that we'll begin to visit other planets during your lifetime. And it's undoubtedly true that unmanned spacecraft will continue to explore the solar system and beyond. Astronauts may get the glory, but countless engineers and scientists make space flight possible. They design the rockets, the instruments, and the life-support systems. They also determine the trajectory that a spacecraft must follow to reach its destination. These are all applications of Newtonian mechanics.

How do you get a satellite into orbit? How do you move a spacecraft from earth orbit onto a trajectory toward Mars? Questions that have to be answered include

- What is the optimum trajectory through space?
- How long will it take?
- How much energy will be required?
- How will you stop the spacecraft when it gets to its destination?
- How will you get it back?

These are complex problems, but you should recognize that they are not fundamentally different from the types of problems you have learned to solve in Parts I and II.

Power over Our Environment

Early humans, for hundreds of thousands of years, had to endure whatever nature provided. Only within the last few thousand years have agriculture, fire, and simple machines provided some level of control over the environment. And it has been a mere couple of centuries since machines, and later electronics, began to do much of our work and provide us with "creature comforts."

Is it a coincidence that machines began to appear about a hundred years after Newton? Of course not. Galileo, Newton, and others in the 17th and 18th centuries ignited what we now call the *scientific revolution*. The machines and other devices we take for granted today are direct consequences of scientific knowledge and the scientific method.

But science and engineering are a two-edged sword. Science has given us the power to control our environment. At the same time, much of the progress of the last two hundred years has come at the expense of the environment. We humans have deforested much of the world, polluted our air and water, and driven many of our fellow travelers on Spaceship Earth to extinction. Now, at the beginning of the 21st century, the evidence is increasingly clear that humans are altering the earth's climate and causing other global changes.

Fortunately, science also gives us the ability to understand the consequences of our actions and to develop better techniques and procedures. It is more important than ever that scientists and engineers in the 21st century distinguish control that is beneficial from control that is harmful. We'll return to some of these ideas in the Summary to Part III.

12 Newton's Theory of Gravity

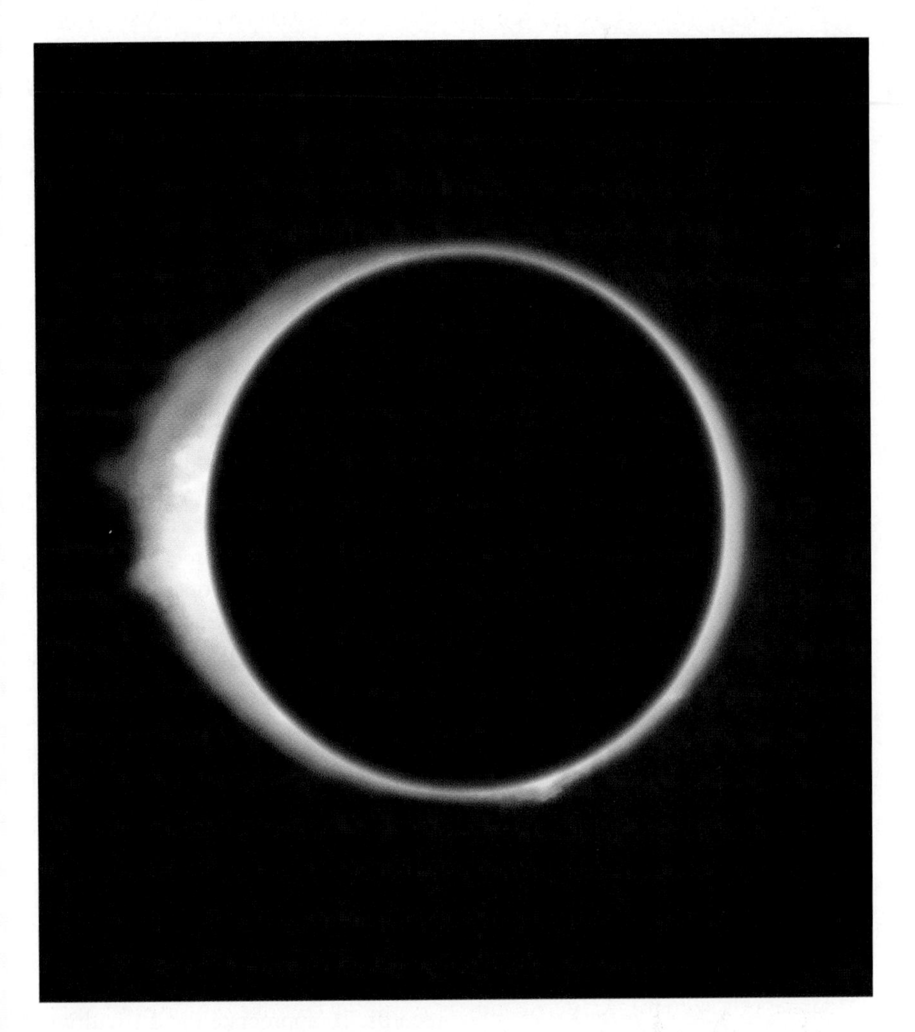

A solar eclipse occurs when the moon passes between the earth and the sun. The ancient Babylonians learned to predict eclipses, a remarkable feat of early science.

► Looking Ahead
The goal of Chapter 12 is to use Newton's theory of gravity to understand the motion of satellites and planets. In this chapter you will learn to:

- Place Newton's discovery of the law of gravity in historical context.
- Use Newton's theory of gravity to solve problems about orbital motion.
- Understand Kepler's laws of planetary orbits.
- Understand and use gravitational potential energy.

◄ Looking Back
Newton's theory of gravity depends on the properties of uniform circular motion. Please review:

- Sections 7.1–7.4 Uniform circular motion and circular orbits.
- Section 9.7 Angular momentum.
- Section 10.2 Gravitational potential energy.

Every ancient culture was fascinated with the motion of the heavens above. Without city lights or urban haze, the nighttime sky and the daytime sun were ever present, powerful experiences. The unknown people that built Stonehenge clearly used it as a solar observatory, and the ancient Babylonians learned to predict the occurrence of solar eclipses.

Our fascination with sky and the stars has not diminished in the 21st century. Today our interest may be in galaxies, black holes, and the Big Bang, but we're still exploring the heavens. One of the most important discoveries of science is that one pervasive force is dominant throughout the universe. This force is responsible for phenomena ranging from the orbiting space shuttle and solar eclipses to the dynamics of galaxies and the expansion of the universe. It is the force of gravity.

Newton's formulation of his theory of gravity was a pivotal event in the history of science. It was the first scientific theory to have broad explanatory and predictive power. Although Newton's theory is now over three centuries old, its importance has not diminished with time.

12.1 A Little History

FIGURE 12.1 The earth-centered cosmology of the ancient Greek and medieval periods.

The study of the structure of the universe is called **cosmology.** The ancient Greeks developed a cosmological model, illustrated in Figure 12.1, that placed the earth at the center of the universe while the moon, the sun, the planets, and the stars were points of light that turned about the earth on large "celestial spheres." This viewpoint was further expanded by the second-century Egyptian astronomer Ptolemy (the P is silent). He developed an elaborate mathematical model of the solar system that quite accurately predicted the complex planetary motions. Ptolemy's earth-centered cosmology was accepted without question for more than a thousand years in medieval Europe.

Then, in the year 1543, the medieval world was turned on its head with the publication of Nicholas Copernicus's *De Revolutionibus.* Copernicus argued that it is not the earth at rest in the center of the universe—it is the sun! Furthermore, Copernicus asserted that all of the planets, including the earth, revolve about the sun (hence his title) in circular orbits.

Copernicus suspected that his ideas would not sit well with the authorities, and he tactfully waited until he was on his deathbed to have his book published. The Catholic Church did, indeed, denounce his views and for more than a hundred years persecuted astronomers who used Copernican ideas. After all, Copernicus's attempt to remove the earth from the center of the universe undermined the very foundations of medieval theology. Not until many decades later, when Galileo used a telescope to study the heavens, did the Copernican view become widely accepted.

The greatest astronomer of this period was Tycho Brahe, a Dane born in 1546, just three years after Copernicus's death. Tycho began observing the heavens as a teenager, and for 30 years, from 1570 to 1600, compiled the most accurate astronomical observations the world had known. The invention of the telescope was still to come, so Tycho's observations were all with the naked eye. However, Tycho did develop ingenious mechanical sighting devices that allowed him to determine the positions of stars and planets in the sky with unprecedented accuracy. Despite his careful observations, Tycho never accepted Copernicus's assertion that the earth was in motion.

Kepler

Tycho willed his records to a young mathematical assistant named Johannes Kepler. Kepler had become one of the first outspoken defenders of Copernicus, and one of his goals was to find evidence for circular planetary orbits in Tycho's records. To appreciate the difficulty of this task, keep in mind that Kepler was working before the development of graphs or of calculus—and certainly before calculators! His mathematical tools were algebra, geometry, and trigonometry, and he was faced with thousands upon thousands of individual observations of planetary positions measured as angles above the horizon.

Ten years of effort without success eventually led Kepler to think that Copernicus's insistence on circular orbits was unnecessary. Perhaps orbits could have other shapes. This was the mental barrier that had to be broken, and once beyond it Kepler was quickly able to deduce that the orbit of Mars is, indeed, not a circle but an *ellipse!* Furthermore, the speed of the planet is not constant but varies as it moves around the ellipse.

Kepler made three major discoveries that today we call **Kepler's laws** of planetary orbits:

1. The planets move in elliptical orbits, with the sun at one focus of the ellipse.
2. A line drawn between the sun and a planet sweeps out equal areas during equal intervals of time.

3. The square of a planet's orbital period is proportional to the cube of the semimajor-axis length.

Figure 12.2a shows that an ellipse has two *foci* (plural of *focus*), and the sun occupies one of these. The long axis of the ellipse is the *major axis,* and half the length of this axis is called the *semimajor-axis length.* As the planet moves, a line drawn from the sun to the planet "sweeps out" an area. Figure 12.2b shows two such areas. Kepler's discovery that the areas are equal for equal Δt implies that the planet moves faster when near the sun, slower when farther away.

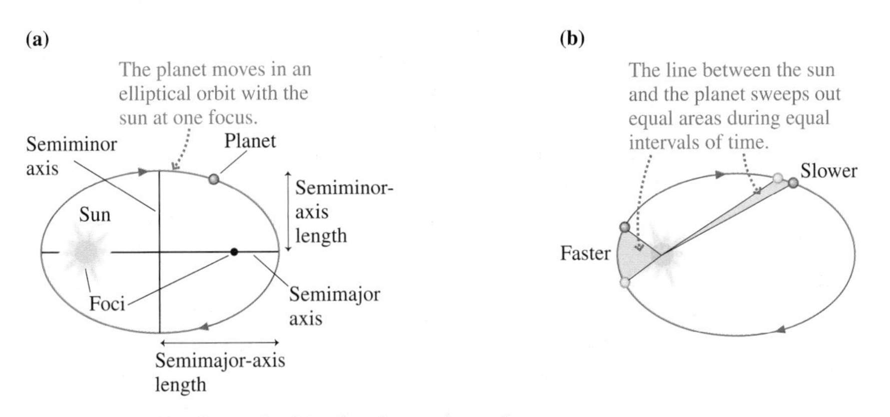
(a)
The planet moves in an elliptical orbit with the sun at one focus.
Semiminor axis
Planet
Semiminor-axis length
Sun
Foci
Semimajor axis
Semimajor-axis length
(b)
The line between the sun and the planet sweeps out equal areas during equal intervals of time.
Slower
Faster

FIGURE 12.2 The elliptical orbit of a planet about the sun.

All the planets except Mercury and Pluto have elliptical orbits that are only very slightly distorted circles. As Figure 12.3 shows, a circle is an ellipse in which the two foci move to the center, effectively making one focus, and the semimajor-axis length becomes the radius. Because the mathematics of ellipses is difficult, this chapter will focus on circular orbits.

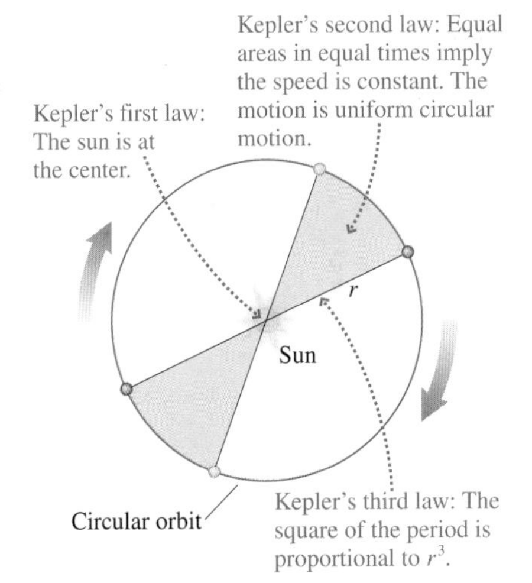
Kepler's first law: The sun is at the center.
Kepler's second law: Equal areas in equal times imply the speed is constant. The motion is uniform circular motion.
r
Sun
Circular orbit
Kepler's third law: The square of the period is proportional to r^3.

FIGURE 12.3 A circular orbit is a special case of an elliptical orbit.

Kepler made an additional contribution that is less widely recognized but that was essential to prepare the way for Newton. For Ptolemy and, later, Copernicus, the role of the sun was merely to light and warm the earth and planets. The sun had no role in their movement. Kepler was the first to suggest that the sun was a center of force that somehow *caused* the planetary motions. Now, Kepler was working before Galileo and Newton, so he did not speak in terms of forces and centripetal accelerations. He thought that some type of rays or spirit emanated from the sun and pushed the planets around their orbits. The value of his contribution was not the specific mechanism he proposed but his introduction of the idea that the sun somehow exerts forces on the planets to determine their motion.

Kepler published the first two of his laws in 1609, the same year in which Galileo first turned a telescope to the heavens. Through his telescope Galileo could *see* moons orbiting Jupiter, just as Copernicus had suggested the planets orbit the sun. He could *see* that Venus has phases, like the moon, which implied its orbital motion about the sun. And he could *see* spots moving across the face of the sun, indicating that it was not a perfect and unchanging symbol of God. Galileo was tried and convicted of heresy (this was still the time of the Inquisition), forced to publicly recant his views, and left a broken man. But his evidence was too persuasive, and by the time of his death in 1642 the Copernican revolution was complete.

In hindsight, we can see that Kepler's analysis and Galileo's observations had set the stage for a major theoretical leap. All that was needed was a great intellect to recognize and pull together these ideas. Enter Isaac Newton, one of the most brilliant scientists ever to live.

12.2 Isaac Newton

Isaac Newton, 1642–1727.

Isaac Newton was born to a poor farming family in 1642, the year of Galileo's death. He entered Trinity College at Cambridge University at age 19 as a "sub-sizar," a poor student who had to work his way through school. Newton graduated in 1665, at age 23, just as an outbreak of the plague in England forced the universities to close for two years. He returned to his family farm for that period, during which he made important experimental discoveries in optics, laid the foundations for his theories of mechanics and gravitation, and made major progress toward his invention of calculus as a whole new branch of mathematics.

A popular image has Newton thinking of the idea of gravity after an apple fell on his head. This amusing story is at least close to the truth. Newton himself said that the "notion of gravitation" came to him as he "sat in a contemplative mood" and "was occasioned by the fall of an apple." It occurred to him that, perhaps, the apple was attracted to the *center* of the earth but was prevented from getting there by the earth's surface. And if the apple was so attracted, why not the moon?

Robert Hooke, discoverer of Hooke's law, had already suggested that the planets might be attracted to the sun with a strength proportional to the inverse square of the distance between the sun and the planet. This seems to have been a hunch rather than being based on any particular evidence, and Hooke failed to follow up on the idea. Newton's genius was not just his successful application of Hooke's suggestion, but his sudden realization that **the force of the sun on the planets was identical to the force of the earth on the apple.** In other words, gravitation is a *universal* force between all objects in the universe! This is not shocking today, but no one before Newton had ever thought that the mundane motion of objects on earth had any connection at all with the stately motion of the planets around the sun.

Newton reasoned along the following lines. Suppose the moon's circular motion around the earth is due to the pull of the earth's gravity. Then, as Figure 12.4a shows, the force on the moon has magnitude $F_m = m_m g_{at\ moon}$, where $g_{at\ moon}$ is the acceleration due to the earth's gravity at the distance of the moon. We cannot assume that $g_{at\ moon}$ is the familiar 9.80 m/s^2. If the strength of gravity diminishes with distance, as Hooke had suggested, then the acceleration due to gravity will also decrease.

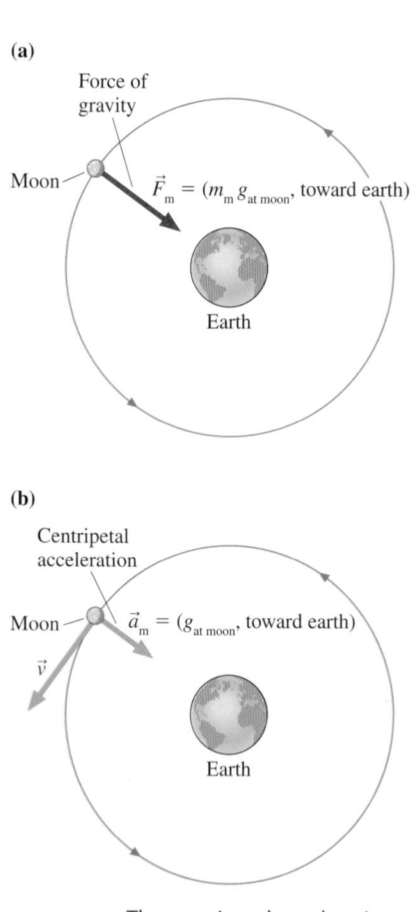

FIGURE 12.4 The centripetal acceleration of the moon is due to the pull of the earth's gravity.

NOTE ► We need to be careful with notation. The symbol g_{moon} would be the acceleration caused by the *moon's* gravity—that is, the acceleration of a falling object on the moon. Here we're interested in the acceleration *of* the moon by the earth's gravity, which we'll call $g_{at\ moon}$. ◄

The moon's acceleration, which is found from Newton's second law, is $\vec{a}_m = \vec{F}_m/m_m = (g_{at\ moon}$, toward the earth). This centripetal acceleration, shown in Figure 12.4b, has only the radial component $a_r = g_{at\ moon}$. As you learned in Chapter 7, the centripetal acceleration of an object in uniform circular motion is

$$a_r = g_{at\ moon} = \frac{v_m^2}{r_m} \tag{12.1}$$

The moon's speed is related to the radius r_m and period T_m of its orbit by v_m = circumference/period $= 2\pi r_m/T_m$. Combining these, Newton found

$$g_{at\ moon} = \frac{4\pi^2 r_m}{T_m^2} = \frac{4\pi^2(3.84 \times 10^8\ \text{m})}{(2.36 \times 10^6\ \text{s})^2} = 0.00272\ \text{m/s}^2$$

Astronomical measurements had established a reasonably good value for r_{moon} by the time of Newton, and the period $T_m = 27.3$ days was quite well known.

The moon's centripetal acceleration is significantly less than the acceleration due to gravity on the earth's surface. In fact,

$$\frac{g_{\text{at moon}}}{g_{\text{on earth}}} = \frac{0.00272 \text{ m/s}^2}{9.80 \text{ m/s}^2} = \frac{1}{3600}$$

This is an interesting result, but it was Newton's next step that was critical. He compared the radius of the moon's orbit to the radius of the earth:

$$\frac{r_m}{R_e} = \frac{3.84 \times 10^8 \text{ m}}{6.37 \times 10^6 \text{ m}} = 60.2$$

NOTE ▶ We'll use a lowercase r, as in r_m, to indicate the radius of an orbit. We'll use an uppercase R, as in R_e, to indicate the radius of a star or planet. ◀

Newton recognized that 60.2^2 is almost exactly 3600. Thus, he reasoned:

- If g has the value 9.80 at the earth's surface, and
- If the force of gravity and g decrease in size depending inversely on the square of the distance from the center of the earth,
- Then g will have exactly the value it needs at the distance of the moon to cause the moon to orbit the earth with a period of 27.3 days.

I deduced that the forces which keep the planets in their orbs must be reciprocally as the squares of their distances from the centers about which they revolve; and thereby compared the force requisite to keep the Moon in her orb with the force of gravity at the surface of the Earth; and found them answer pretty nearly.

Isaac Newton

His two ratios were not identical (because the earth isn't a perfect sphere and the moon's orbit isn't a perfect circle), but he found them to "answer pretty nearly" and knew that he had to be on the right track.

This flash of insight changed our most basic understanding of the universe. Copernicus displaced the earth from the center of the universe, and now Newton had shown that the laws of the heavens and the laws of earth are the same. Nonetheless, Newton did not publish his results for a long 22 years. The issue that troubled him was treating the sun, the earth, and the other planets as if they were single particles with all their mass at the center. If his idea about a universal force was correct, then *every atom* in the earth exerts a force on *every atom* in the moon. Newton had to show that all of these forces add up to give a result that is identical with treating the bodies as single particles. This is a problem in integral calculus, and Newton had first to develop the necessary mathematics. He did eventually succeed, and his theory of gravitation was published in 1687 along with his theory of mechanics (which we know as Newton's laws) in his great work *Philosophia Naturalis Principia Mathematica* (Mathematical Principles of Natural Philosophy). The rest is history.

STOP TO THINK 12.1 A satellite orbits the earth with constant speed at a height above the surface equal to the earth's radius. The magnitude of the satellite's acceleration is

a. $4g_{\text{on earth}}$ b. $2g_{\text{on earth}}$ c. $g_{\text{on earth}}$
d. $\frac{1}{2}g_{\text{on earth}}$ e. $\frac{1}{4}g_{\text{on earth}}$ f. 0

12.3 Newton's Law of Gravity

Newton proposed that *every* object in the universe attracts *every other* object with a force that has the following properties:

1. The force is inversely proportional to the square of the distance between the objects.
2. The force is directly proportional to the product of the masses of the two objects.

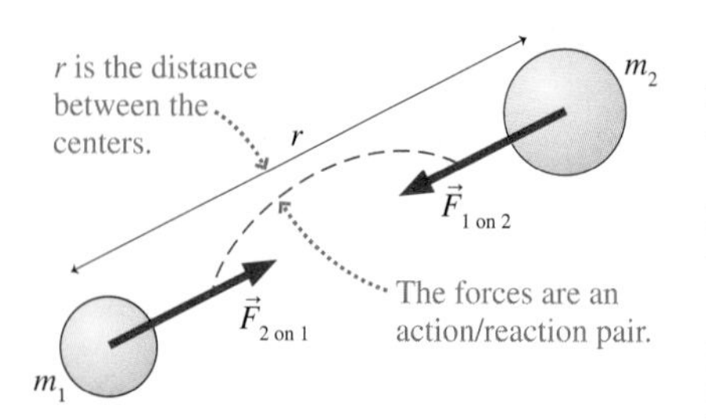

FIGURE 12.5 The gravitational forces on masses m_1 and m_2.

To make these ideas more specific, Figure 12.5 shows masses m_1 and m_2 separated by distance r. Each mass exerts an attractive force on the other, a force that we call the **gravitational force.** These two forces form an action/reaction pair, so $\vec{F}_{1\text{ on }2}$ is equal and opposite to $\vec{F}_{2\text{ on }1}$. The magnitude of the forces is given by Newton's law of gravity.

Newton's law of gravity If two objects with masses m_1 and m_2 are a distance r apart, the objects exert attractive forces on each other of magnitude

$$F_{1\text{ on }2} = F_{2\text{ on }1} = \frac{Gm_1m_2}{r^2} \tag{12.2}$$

The forces are directed along the straight line joining the two objects.

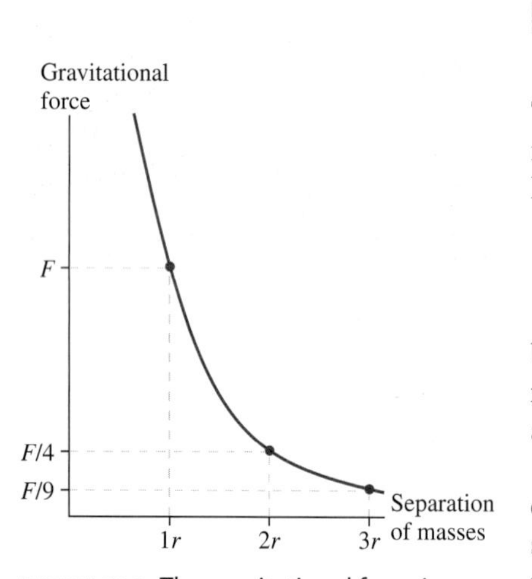

FIGURE 12.6 The gravitational force is an inverse-square force.

The constant G, called the **gravitational constant,** is a proportionality constant necessary to relate the masses, measured in kilograms, to the force, measured in Newtons. In the SI system of units, G has the value

$$G = 6.67 \times 10^{-11}\ \text{N m}^2/\text{kg}^2$$

Figure 12.6 is a graph of the gravitational force as a function of the distance between the two masses. As you can see, an inverse-square force decreases rapidly. Doubling the distance between the masses causes the force to decrease by a factor of 4.

Strictly speaking, Equation 12.2 is valid only for particles. As we noted, however, Newton was able to show that this equation also applies to spherical objects, such as planets, if r is the distance between their centers. Our intuition and common sense suggest this to us, as they did to Newton, but the proof is not essential, so we will omit it.

Gravitational Force and Weight

Knowing G, we can calculate the size of the gravitational force. Consider two 1.0 kg masses that are 1.0 m apart. According to Newton's law of gravity, these two masses exert an attractive gravitational force on each other. The magnitude of the force is

$$F_{1\text{ on }2} = F_{2\text{ on }1} = \frac{Gm_1m_2}{r^2}$$

$$= \frac{(6.67 \times 10^{-11}\ \text{N m}^2/\text{kg}^2)(1.0\ \text{kg})(1.0\ \text{kg})}{(1.0\ \text{m})^2} = 6.67 \times 10^{-11}\ \text{N}$$

This is an exceptionally tiny force, especially when compared to the weight of one of the masses: $w = mg = 9.8$ N.

The fact that the gravitational force between two ordinary-size objects is so small is the reason that we are not aware of it. As you sit there reading, you are being attracted to this book, to the person sitting next to you, and to every object around you, but the forces are so tiny in comparison to your weight and to the normal forces and friction forces acting on you that they are completely undetectable. Only when one (or both) of the masses is exceptionally large—planet-size—does the force of gravity become important.

We find a more respectable result if we calculate the force *of the earth* on a 1.0 kg mass at the earth's surface:

$$F_{\text{earth on 1 kg}} = \frac{GM_e m_{1\text{ kg}}}{R_e^2}$$

$$= \frac{(6.67 \times 10^{-11}\text{ N m}^2/\text{kg}^2)(5.98 \times 10^{24}\text{ kg})(1.0\text{ kg})}{(6.37 \times 10^6\text{ m})^2} = 9.8\text{ N}$$

where the distance between the mass and the center of the earth is the earth's radius. The earth's mass M_e and radius R_e were taken from Table 12.2 in Section 12.6. This table, which is also printed inside the cover of the book, contains astronomical data that will be used for examples and homework.

The force $F_{\text{earth on 1 kg}} = 9.8$ N is exactly the weight of a 1.0 kg mass: $w = mg = 9.8$ N. Is this a coincidence? Of course not. The weight $w = mg$ of an object *is* the "force of gravity" acting on it. This suggests that we should be able to use Newton's law of gravity to *predict* the value of g. We will return to this idea in Section 12.4.

Although weak, gravity is a *long-range* force. No matter how far apart two objects may be, there is a gravitational attraction between them given by Equation 12.2. Consequently, gravity is the most ubiquitous force in the universe. It not only keeps your feet on the ground, it also keeps the earth orbiting the sun, the solar system orbiting the center of the Milky Way galaxy, and the entire Milky Way galaxy performing an intricate orbital dance with other galaxies making up what is called the "local cluster" of galaxies.

The dynamics of stellar motions, spanning distances of hundreds of thousands of light years, are governed by Newton's law of gravity.

A galaxy of $\approx 10^{11}$ stars spanning a distance greater than 100,000 light years.

The Principle of Equivalence

Newton's law of gravity depends on a rather curious assumption. The concept of *mass* was introduced in Chapter 4 by considering the relationship between force and acceleration. The *inertial mass* of an object, which is the mass that appears in Newton's second law, is found by measuring the object's acceleration a in response to force F:

$$m_{\text{inert}} = \text{inertial mass} = \frac{F}{a} \tag{12.3}$$

Gravity plays no role in this definition of mass.

The quantities m_1 and m_2 in Newton's law of gravity are being used in a very different way. Masses m_1 and m_2 govern the strength of the gravitational attraction between two objects. The mass used in Newton's law of gravity is called the **gravitational mass.** The gravitational mass of an object can be determined by measuring the attractive force exerted on it by another mass M a distance r away:

$$m_{\text{grav}} = \text{gravitational mass} = \frac{r^2 F_{M\text{ on }m}}{GM} \tag{12.4}$$

Acceleration does not enter into the definition of the gravitational mass.

These are two very different concepts of mass. Yet Newton, in his theory of gravity, asserts that the inertial mass in his second law is the very same mass that governs the strength of the gravitational attraction between two objects. The assertion that $m_{\text{grav}} = m_{\text{inert}}$ is called the **principle of equivalence.** It says that inertial mass is *equivalent to* gravitational mass.

As a hypothesis about nature, the principle of equivalence is subject to experimental verification or disproof. Many exceptionally clever experiments have looked for any difference between the gravitational mass and the inertial mass, and they have shown that any difference, if it exists at all, is less than 10 parts in a trillion! As far as we know today, the gravitational mass and the inertial mass are exactly the same thing.

But why should a quantity associated with the dynamics of motion, relating force to acceleration, have anything at all to do with the gravitational attraction? This is a question that intrigued Einstein and eventually led to his general theory of relativity, the theory about curved space-time and black holes. General relativity is beyond the scope of this textbook, but it explains the principle of equivalence as a property of the space itself.

Newton's Theory of Gravity

Newton's theory of gravity is more than just Equation 12.2. The *theory* of gravity consists of:

1. A specific force law for gravity, given by Equation 12.2, *and*
2. The principle of equivalence, *and*
3. An assertion that Newton's three laws of motion are universally applicable. These laws are as valid for heavenly bodies, the planets and stars, as for earthly objects.

Consequently, everything we have learned about forces, motion, and energy is relevant to the dynamics of satellites, planets, and galaxies.

STOP TO THINK 12.2 The figure shows a binary star system. The mass of star 2 is twice the mass of star 1. Compared to $\vec{F}_{1\text{ on }2}$, the magnitude of the force $\vec{F}_{2\text{ on }1}$ is

a. four times as big.
b. twice as big.
c. the same size.
d. half as big.
e. one-quarter as big.

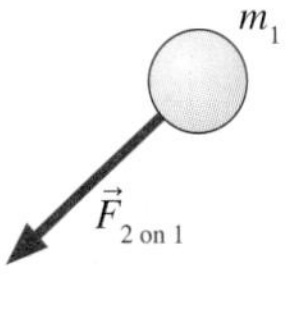

m_2

12.4 Little g and Big G

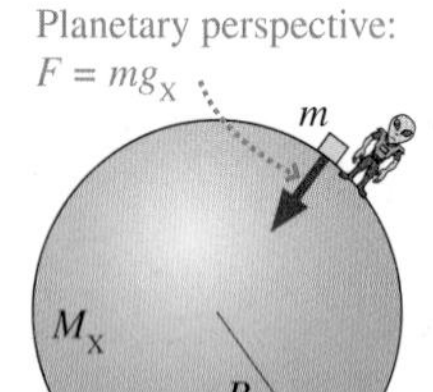

M_X R_X

Planet X

Universal perspective:
$F = \dfrac{GM_X m}{R_X^2}$
m M_X R_X

Planet X

FIGURE 12.7 Weighing an object of mass m on Planet X.

The familiar equation $w = mg$ works well when an object is on the surface of a planet, but mg will not help us find the force exerted on the same object if it is in orbit around the planet. Neither can we use mg to find the force of attraction between the earth and the moon. Newton's law of gravity provides a more fundamental starting point because it describes a *universal* force that exists between all objects.

To illustrate the connection between Newton's law of gravity and the familiar $w = mg$, Figure 12.7 shows an object of mass m on the surface of Planet X. Planet X inhabitant Mr. Xhzt places the object on a scale to weigh it and exclaims (translated), "Aha! This object has weight $w = mg_X$. The constant g_X is the acceleration due to gravity on my planet."

We, taking a more cosmic perspective, reply, "Yes, it has that weight *because* of a universal force of attraction between your planet and the object. The size of the force is determined by Newton's law of gravity."

We and Mr. Xhzt are both correct. Whether you choose to call it "weight" or "the force of gravitational attraction," we and Mr. Xhzt must arrive at the *same numerical value* for the magnitude of the force. Suppose an object of mass m is on the surface of a planet of mass M and radius R. The object's weight is

$$w = mg_{\text{surface}} \tag{12.5}$$

where $g_{surface}$ is the acceleration due to gravity at the planet's surface. The force of gravitational attraction, as given by Newton's law of gravity, is

$$F_{M \text{ on } m} = \frac{GMm}{R^2} \tag{12.6}$$

Because these are two names and two expressions for the same force, we can equate the right-hand sides to find that

$$g_{surface} = \frac{GM}{R^2} \tag{12.7}$$

We have used Newton's law of gravity to *predict* the value of g at the surface of a planet. The value depends on the mass and radius of the planet as well as on the value of G, which establishes the overall strength of the gravitational force.

The expression for $g_{surface}$ in Equation 12.7 is valid for any planet or star. Using the mass and radius of the earth, from Table 12.2, we can predict the earthly value of g:

$$g_{earth} = \frac{GM_e}{R_e^2} = \frac{(6.67 \times 10^{-11}\ \text{N m}^2/\text{kg}^2)(5.98 \times 10^{24}\ \text{kg})}{(6.37 \times 10^6\ \text{m})^2} = 9.83\ \text{m/s}^2$$

Galileo could measure g_{earth}, but he did not know *why* it was 9.8 m/s^2 rather than some other number. Now, with a more fundamental theory, we can see that the value for g_{earth} is a direct consequence of the size and mass of the earth. Similarly, you can show for homework that $g_{moon} = 1.62\ \text{m/s}^2$ and $g_{jupiter} = 25.9\ \text{m/s}^2$. A falling object on the moon would accelerate only about one-sixth as rapidly as on earth. On Jupiter, it would accelerate about 2.6 times more rapidly.

You probably noticed that the calculated value of g_{earth} is slightly larger than the "accepted" value of 9.80 m/s^2. Is there something wrong with the theory? No, the difference arises because we have not considered the earth's rotation. Figure 12.8 shows an object on the earth's equator as seen by an observer above the north pole. There are two forces on the object, the gravitational force $\vec{F}_{M \text{ on } m}$ and the normal force $\vec{n}$ of the ground. The *apparent weight* of the object, what a set of scales would read, is the magnitude n of the normal force. If the earth were not rotating, the object would be in static equilibrium and the two forces would exactly balance. Equation 12.7 gives the value of g on a nonrotating earth.

But the earth *is* rotating, and the object moves in a circle. An object in circular motion must have a *net* force acting on it, directed toward the center of the circle. Thus $n < F_{grav}$ for an object on a rotating planet. (This is the same reasoning we applied earlier to roller coasters doing loop-the-loops and buckets of water swinging overhead.) Because n is the apparent weight $w_{app} = mg_{app}$, the apparent value of g that we actually experience is slightly less than the value of g calculated from Newton's law of gravity. At midlatitudes, where most of us live, the reduction is just about 0.03 m/s^2, and hence the measured value for g on a rotating earth is 9.80 m/s^2.

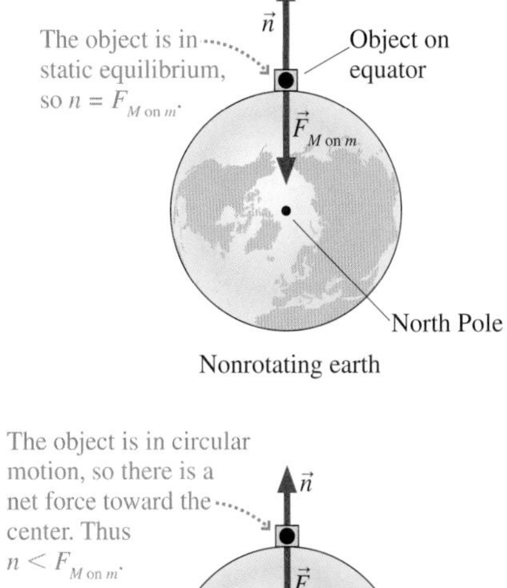

The object is in static equilibrium, so $n = F_{M \text{ on } m}$.
$\vec{n}$
Object on equator
$\vec{F}_{M \text{ on } m}$
North Pole
Nonrotating earth

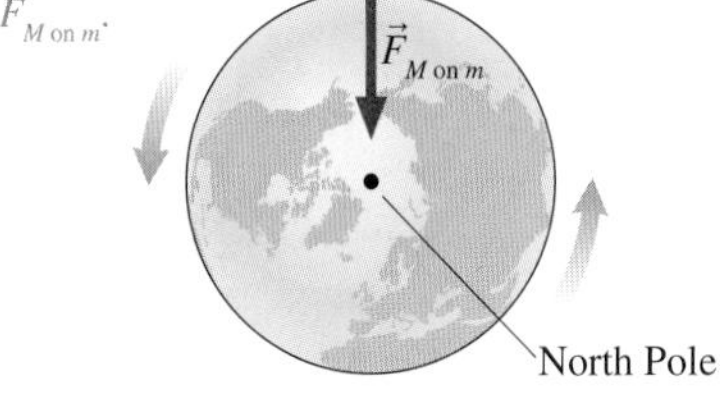

The object is in circular motion, so there is a net force toward the center. Thus $n < F_{M \text{ on } m}$.
$\vec{n}$
$\vec{F}_{M \text{ on } m}$
North Pole

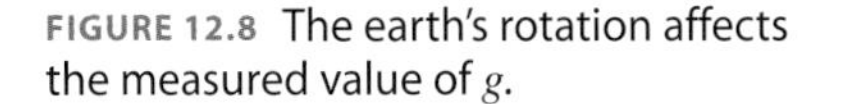

Rotating earth

FIGURE 12.8 The earth's rotation affects the measured value of g.

Decrease of g with Distance

Equation 12.7 gives $g_{surface}$ at the surface of a planet. More generally, imagine an object of mass m at distance $r > R$ from the center of a planet. Further, suppose that gravity from the planet is the only force acting on the object. Then its acceleration, the acceleration due to gravity, is given by Newton's second law:

$$g = \frac{F_{M \text{ on } m}}{m} = \frac{GM}{r^2} \tag{12.8}$$

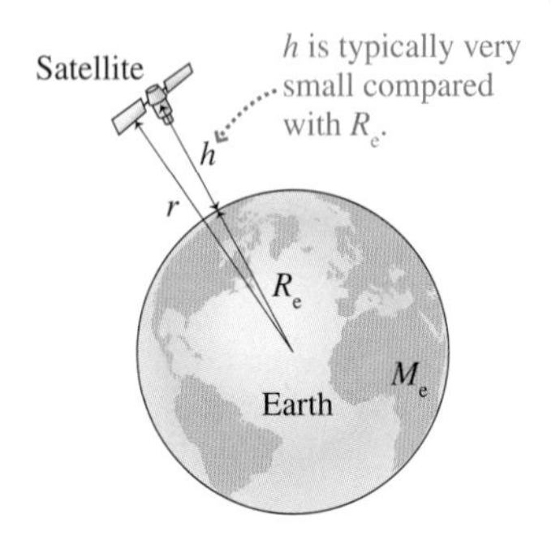

FIGURE 12.9 A satellite orbits the earth at height h.

This more general result agrees with Equation 12.7 if $r = R$, but it allows us to determine the "local" acceleration due to gravity at distances $r > R$. Equation 12.8 expresses Newton's discovery, with regard to the moon, that g decreases inversely with the square of the distance.

Figure 12.9 shows a satellite orbiting at height h above the earth's surface. Its distance from the center of the earth is $r = R_e + h$. Most people have a mental image that satellites orbit "far" from the earth, but in reality h is typically 200 miles $\approx 3 \times 10^5$ m while $R_e = 6.37 \times 10^6$ m. Thus the satellite is barely "skimming" the earth at a height only about 5% of the earth's radius!

The value of g at height h above the earth is

$$g = \frac{GM_e}{(R_e + h)^2} = \frac{GM_e}{R_e^2(1 + h/R_e)^2} = \frac{g_{earth}}{(1 + h/R_e)^2} \tag{12.9}$$

where $g_{earth} = 9.83$ m/s^2 is the value calculated in Equation 12.7 for $h = 0$ on a nonrotating earth. Table 12.1 shows the value of g evaluated at several values of h.

TABLE 12.1 Variation of the earth's acceleration due to gravity with height above the ground

Height h	Example	g (m/s^2)
0 m	ground	9.83
4500 m	Mt. Whitney	9.82
10,000 m	jet airplane	9.80
300,000 m	space shuttle	8.90
3,590,000 m	communications satellite	0.22

NOTE ▶ The acceleration due to gravity on a satellite such as the space shuttle is only slightly less than the ground-level value. An object in orbit is not "weightless" because there is no gravity in space but because it is in free fall, as you learned in Sections 5.3 and 7.4. ◀

Weighing the Earth

We can predict g if we know the earth's mass. But how do we know the value of M_e? We cannot place the earth on a giant pan balance, so how is its mass known? Furthermore, how do we know the value of G? These are interesting and important questions.

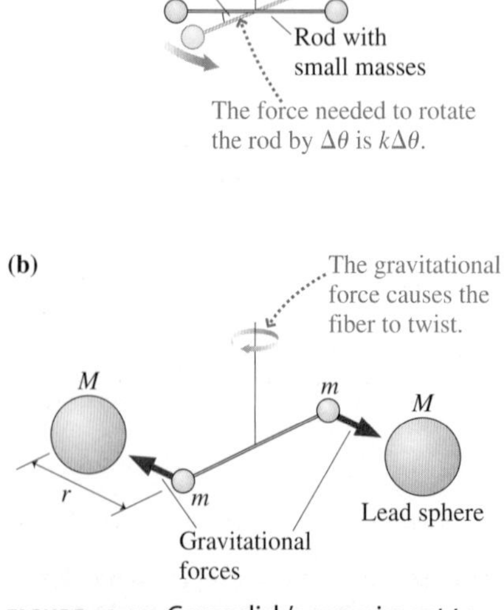

FIGURE 12.10 Cavendish's experiment to measure G.

Newton did not know the value of G. He could say that the gravitational force is proportional to the product m_1m_2 and inversely proportional to r^2, but he had no means of knowing the value of the proportionality constant.

Determining G requires a *direct* measurement of the gravitational force between two known masses at a known separation. The small size of the gravitational force between ordinary-size objects makes this quite a feat. Yet the English scientist Henry Cavendish came up with an ingenious way of doing so with a device called a *torsion balance.* Two fairly small masses m, typically about 10 g, are placed on the ends of a lightweight rod. The rod is hung from a thin fiber, as shown in Figure12.10a, and allowed to reach equilibrium.

If the rod is then rotated slightly and released, a *restoring force* will return it to equilibrium. This is analogous to displacing a spring from equilibrium, and in fact the restoring force and the angle of displacement obey a version of Hooke's law: $F_{restore} = k\Delta\theta$. The "torsion constant" k can be determined by timing the period of oscillations. Once k is known, a force that twists the rod slightly away from equilibrium can be measured by the product $k\Delta\theta$. It is possible to measure very small angular deflections, so this device can be used to determine very small forces.

Two larger masses M (typically lead spheres with $M \approx 10$ kg) are then brought close to the torsion balance, as shown in Figure 12.10b. The gravitational attraction that they exert on the smaller hanging masses causes a small but measurable twisting of the balance, enough to measure $F_{M \text{ on } m}$. Because m, M, and r are all known, Cavendish was able to determine G from

$$G = \frac{F_{M \text{ on } m} r^2}{Mm} \tag{12.10}$$

His first results were not highly accurate, but improvements over the years in this and similar experiments have produced the value of G accepted today.

With an independently determined value of G, we can return to Equation 12.7 to find

$$M_e = \frac{g_{\text{earth}} R_e^2}{G} \tag{12.11}$$

We have weighed the earth! The value of g_{earth} at the earth's surface is known with great accuracy from kinematics experiments. The earth's radius R_e is determined by surveying techniques. Combining our knowledge from these very different measurements has given us a way to determine the mass of the earth.

The acceleration due to gravity g is constant on the surface of any given planet, but is different for each planet. The gravitational constant G is a constant of a different nature. It is what we call a *universal constant.* Its value establishes the strength of one of the fundamental forces of nature. As far as we know, the gravitational force between two masses would be the same anywhere in the universe. Universal constants tell us something about the most basic and fundamental properties of nature. You will soon meet other universal constants.

STOP TO THINK 12.3 A planet has four times the mass of the earth, but the acceleration due to gravity on the planet's surface is the same as on the earth's surface. The planet's radius is

a. $4R_e$. b. $2R_e$. c. R_e. d. $\frac{1}{2}R_e$. e. $\frac{1}{4}R_e$.

12.5 Gravitational Potential Energy

Gravitational problems are ideal for the conservation-law tools we developed in Chapters 9 through 11. Because gravity is the only force, and it is a conservative force, both the momentum and the mechanical energy of the system $m_1 + m_2$ are conserved. To employ conservation of energy, however, we need to determine an appropriate form for the gravitational potential energy for two particles interacting via Newton's law of gravity.

The definition of potential energy that we developed in Chapter 11 is

$$\Delta U = U_f - U_i = -W_c(i \rightarrow f) \tag{12.12}$$

where $W_c(i \rightarrow f)$ is the work done by a conservative force as a particle moves from position i to position f. Strictly speaking, this defines only ΔU, the *change* in potential energy. To find an explicit expression for U, we must choose a zero point of the potential energy.

For a flat earth, we used $F = -mg$ and the choice that $U = 0$ at the surface ($y = 0$) to arrive at the now-familiar $U_g = mgy$. This result for U_g is valid only for $y \ll R_e$, when the earth's curvature and size are not apparent. We now need to find an expression for the gravitational potential energy of masses that interact over *large* distances.

Figure 12.11 shows two particles of mass m_1 and m_2. Let's calculate the work done on mass m_2 by the conservative force $\vec{F}_{1 \text{ on } 2}$ as m_2 moves from an initial position at distance r to a final position very far away. The force, which points to

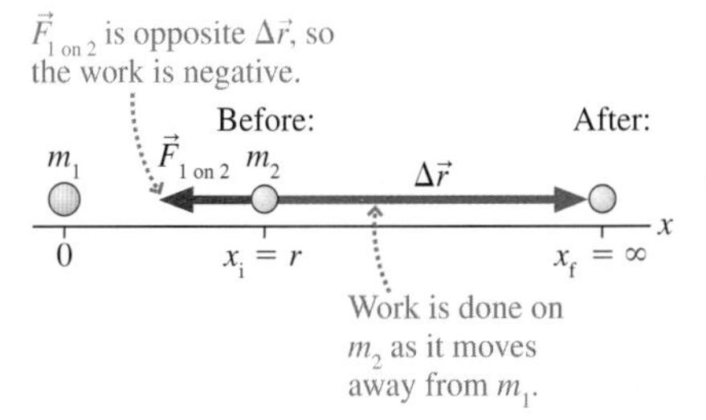

FIGURE 12.11 Calculating the work done by the gravitational force as mass m_2 moves from r to ∞.

the left, is opposite the displacement, hence this force does *negative* work. Consequently, due to the minus sign in Equation 12.12, ΔU is *positive.* A pair of masses *gains* potential energy as the masses move farther apart, just as a particle near the earth's surface gains potential energy as it moves to higher elevations.

Establish a coordinate system with m_1 at the origin and m_2 moving along the x-axis. The gravitational force is a variable force, so we need the full definition of work:

$$W(\mathrm{i} \rightarrow \mathrm{f}) = \int_{x_i}^{x_f} F_x\,dx \tag{12.13}$$

The $\vec{F}_{1 \text{ on } 2}$ points toward the left, so its x-component is $(F_{1 \text{ on } 2})_x = -Gm_1m_2/x^2$. As mass m_2 moves from $x_i = r$ to $x_f = \infty$, the potential energy changes by

$$\begin{aligned}\Delta U = U_{\text{at } \infty} - U_{\text{at } r} &= -\int_r^\infty (F_{1 \text{ on } 2})_x\,dx = -\int_r^\infty \left(\frac{-Gm_1m_2}{x^2}\right)dx \\ &= +Gm_1m_2\int_r^\infty \frac{dx}{x^2} = -\frac{Gm_1m_2}{x}\bigg|_r^\infty \\ &= \frac{Gm_1m_2}{r}\end{aligned} \tag{12.14}$$

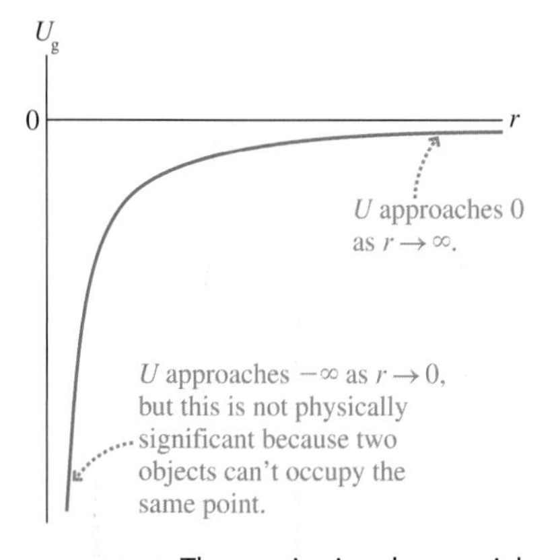

FIGURE 12.12 The gravitational potential-energy curve.

NOTE ▶ We chose to integrate along the x-axis, but the fact that gravity is a conservative force means that ΔU will have this value if m_2 moves from r to ∞ along *any* path. ◀

To proceed further, we need to choose the point where $U = 0$. We would like our choice to be valid for any star or planet, regardless of its mass and radius. This will be the case if we set $U = 0$ at the point where the interaction between the masses vanishes. According to Newton's law of gravity, the strength of the interaction is zero only when $r = \infty$. Two masses infinitely far apart will have no tendency, or potential, to move together, so we will *choose* to place the zero point of potential energy at $r = \infty$. That is, $U_{\text{at } \infty} = 0$.

This choice gives us the gravitational potential energy of masses m_1 and m_2:

$$U_g = -\frac{Gm_1m_2}{r} \tag{12.15}$$

This is the potential energy of masses m_1 and m_2 when their *centers* are separated by distance r. Figure 12.12 is a graph of U_g as a function of the distance r between the masses. Notice that it asymptotically approaches 0 as $r \rightarrow \infty$.

NOTE ▶ Although Equation 12.15 looks rather similar to Newton's law of gravity, it depends only on $1/r$, *not* on $1/r^2$. ◀

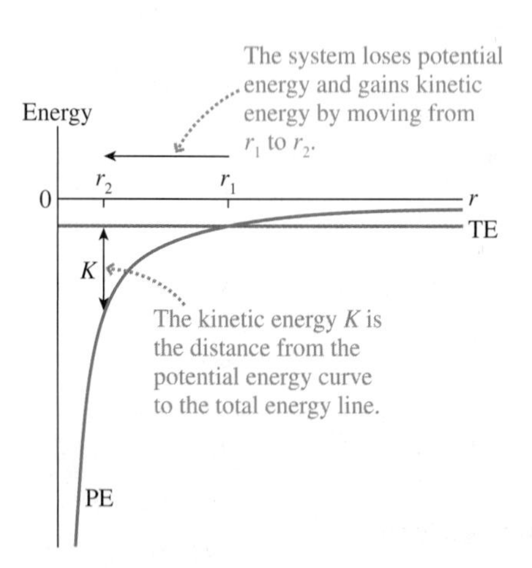

FIGURE 12.13 Two masses gain kinetic energy as the separation between them changes from r_1 to r_2.

It may seem disturbing that the potential energy is negative, but we encountered similar situations in Chapter 10. All a negative potential energy means is that the potential energy of the two masses at separation r is *less* than their potential energy at infinite separation. It is only the *change* in U that has physical significance, and the change will be the same no matter where we place the zero of potential energy.

Suppose two masses distance r_1 apart are released from rest. How will they move? From a force perspective, you would note that each mass experiences an attractive force and accelerates toward the other. The energy perspective of Figure 12.13 tells us the same thing. By moving toward smaller r (that is, $r_1 \rightarrow r_2$), the system *loses* potential energy and *gains* kinetic energy while conserving E_{mech}. The system is "falling downhill," although in a more general sense than we think about on a flat earth.

EXAMPLE 12.1 Crashing into the sun

Suppose the earth were suddenly to cease revolving around the sun. The gravitational force would then pull it directly into the sun. What would be the earth's speed as it crashed?

MODEL Model the earth and the sun as spherical masses. This is an isolated system, so its mechanical energy is conserved.

VISUALIZE Figure 12.14 is a before-and-after pictorial representation for this gruesome cosmic event. The "crash" occurs as the earth touches the sun, at which point the distance between their centers is $r_2 = R_s + R_e$. The initial separation r_1 is the radius of the earth's *orbit* about the sun, not the radius of the earth.

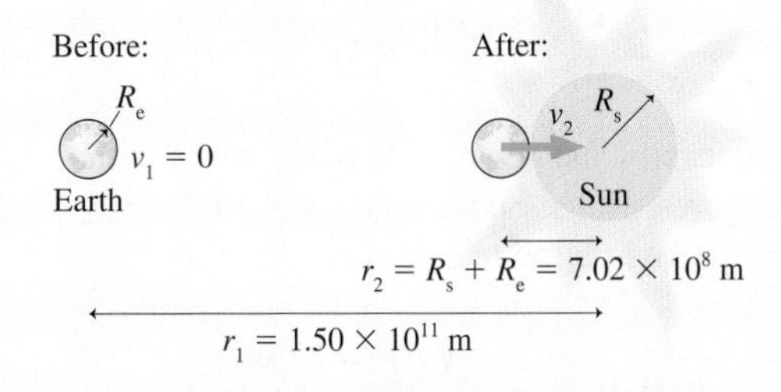

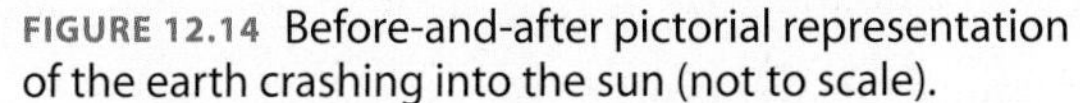

FIGURE 12.14 Before-and-after pictorial representation of the earth crashing into the sun (not to scale).

SOLVE Strictly speaking, the kinetic energy is the sum $K = K_{earth} + K_{sun}$. However, the sun is so much more massive than the earth that the lightweight earth does almost all of the moving. It is a reasonable approximation to consider the sun as remaining at rest. In that case, the energy conservation equation $K_2 + U_2 = K_1 + U_1$ is

$$\frac{1}{2}M_e v_2^2 - \frac{GM_sM_e}{(R_s + R_e)} = 0 - \frac{GM_sM_e}{r_1}$$

This is easily solved for the earth's speed at impact. Using data from Table 12.2, we find

$$v_2 = \sqrt{2GM_s\left(\frac{1}{R_s + R_e} - \frac{1}{r_1}\right)} = 6.13 \times 10^5 \text{ m/s}$$

ASSESS The earth is really flying along at over 1 million miles per hour! It is worth noting that we do not have the mathematical tools to solve this problem using Newton's second law because the acceleration is not constant. But the solution is straightforward when we use energy conservation.

EXAMPLE 12.2 Escape speed

A 1000 kg rocket is fired straight away from the surface of the earth. What speed does the rocket need to "escape" from the gravitational pull of the earth and never return? Assume a nonrotating earth.

MODEL In a simple universe, consisting of only the earth and the rocket, an insufficient launch speed will cause the rocket eventually to fall back to earth. Once the rocket finally slows to a halt, gravity will ever so slowly pull it back. The only way the rocket can escape is to never stop ($v = 0$) and thus never have a turning point! That is, the rocket must continue moving away from the earth forever. The *minimum* launch speed for escape, which is called the **escape speed,** will cause the rocket to stop ($v = 0$) only as it reaches $r = \infty$. Now ∞, of course, is not a "place," so a statement like this means that we want the rocket's speed to approach $v = 0$ asymptotically as $r \to \infty$.

VISUALIZE Figure 12.15 is a before-and-after pictorial representation.

SOLVE The energy conservation equation $K_2 + U_2 = K_1 + U_1$ is

$$0 + 0 = \frac{1}{2}mv_1^2 - \frac{GM_e m}{R_e}$$

where we used the fact that both the kinetic and potential energy are zero at $r = \infty$. Thus the escape speed is

$$v_{escape} = v_1 = \sqrt{\frac{2GM_e}{R_e}} = 11{,}200 \text{ m/s} \approx 25{,}000 \text{ mph}$$

ASSESS The problem was mathematically easy; the difficulty was deciding how to interpret it. That is why—as you have now seen many times—the "physics" of a problem consists of thinking, interpreting, and modeling. We will see variations on this problem in the future, both with gravity and electricity, so you might want to review the *reasoning* involved. Notice that the answer does *not* depend on the rocket's mass, so this is the escape speed for any object.

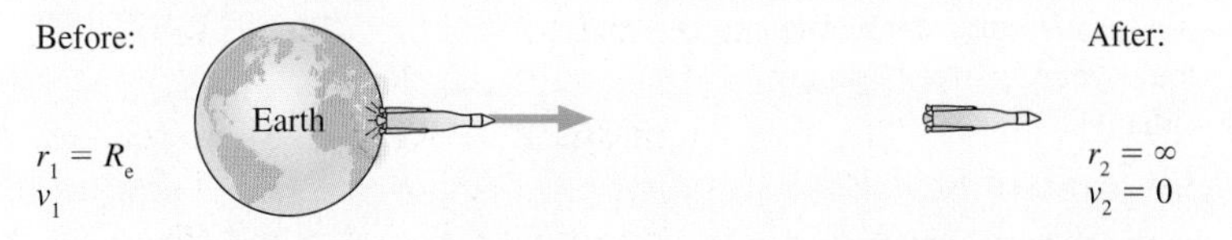

FIGURE 12.15 Pictorial representation of a rocket launched with sufficient speed to escape the earth's gravity.

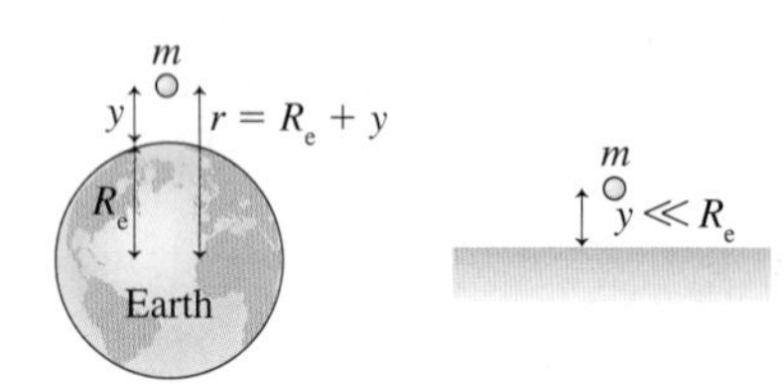

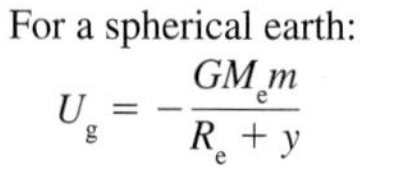

For a spherical earth:
$U_g = -\dfrac{GM_e m}{R_e + y}$
We can treat the earth as flat if $y \ll R_e$:
$U_g = mgy$

FIGURE 12.16 We can treat the earth as flat if $y \ll R_e$.

The Flat-Earth Approximation

Equation 12.15 is the general form of the gravitational potential energy, but how is it related to our previous use of $U_g = mgy$ on a flat earth? Figure 12.16 shows an object of mass m located at height y above the surface of the earth. The object's distance from the earth's center is $r = R_e + y$ and its gravitational potential energy is

$$U_g = -\frac{GM_e m}{r} = -\frac{GM_e m}{R_e + y} = -\frac{GM_e m}{R_e(1 + y/R_e)} \tag{12.16}$$

where, in the last step, we factored R_e out of the denominator.

Suppose the object is very close to the earth's surface ($y \ll R_e$). In that case, the ratio $y/R_e \ll 1$. There is an approximation you will learn about in calculus, called the *binomial approximation,* that says

$$\frac{1}{1 + x} \approx 1 - x \quad \text{if } x \ll 1 \tag{12.17}$$

As an illustration, you can easily use your calculator to find that $1/1.01 = 0.9901$, to four significant figures. But suppose you wrote $1.01 = 1 + 0.01$. You could then use the binomial approximation to calculate

$$\frac{1}{1.01} = \frac{1}{1 + 0.01} \approx 1 - 0.01 = 0.9900$$

You can see that the approximate answer is off by only 0.01%.

If we call $y/R_e = x$ in Equation 12.16 and use the binomial approximation, we find

$$U_g(\text{if } y \ll R_e) \approx -\frac{GM_e m}{R_e}\left(1 - \frac{y}{R_e}\right) = -\frac{GM_e m}{R_e} + m\left(\frac{GM_e}{R_e^2}\right)y \tag{12.18}$$

Now the first term is just the gravitational potential energy U_0 when the object is at ground level ($y = 0$). In the second term, you can recognize $GM_e/R_e^2 = g_{earth}$ from the definition of g in Equation 12.7. Thus we can write Equation 12.18 as

$$U_g(\text{if } y \ll R_e) = U_0 + mg_{earth}\,y \tag{12.19}$$

Although we chose U_g to be zero when $r = \infty$, we are always free to change our minds. If we change the zero point of potential energy to be $U_0 = 0$ at the surface, which is the choice we made in Chapter 10, then Equation 12.19 becomes

$$U_g(\text{if } y \ll R_e) = mg_{earth}\,y \tag{12.20}$$

We can sleep easier knowing that Equation 12.15 for the gravitational potential energy is consistent with our earlier "flat-earth" expression for the potential energy when $y \ll R_e$.

EXAMPLE 12.3 The speed of a projectile

A projectile is launched straight up from the earth.

a. With what speed should you launch it if it is to have a speed of 500 m/s at a height of 400 km? Ignore air resistance.
b. By what percentage would your answer be in error if you used a flat-earth approximation?

MODEL Mechanical energy is conserved.

VISUALIZE Figure 12.17 shows a pictorial representation.

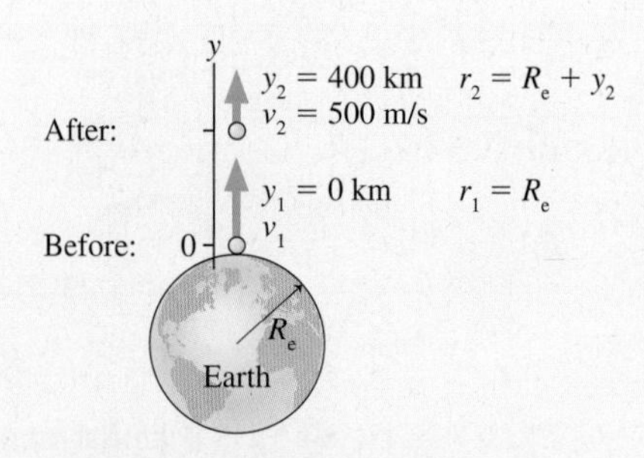

FIGURE 12.17 Pictorial representation of a projectile launched straight up.

SOLVE

a. Although the height is exaggerated in the figure, 400 km = 400,000 m is high enough that we cannot ignore the earth's spherical shape. The energy conservation equation $K_2 + U_2 = K_1 + U_1$ is

$$\frac{1}{2}mv_2^2 - \frac{GM_e m}{R_e + y_2} = \frac{1}{2}mv_1^2 - \frac{GM_e m}{R_e + y_1}$$

where we've written the distance between the projectile and the earth's center as $r = R_e + y$. The initial height is $y_1 = 0$. Notice that the projectile mass m cancels and is not needed. Solving for the launch speed,

$$v_1 = \sqrt{v_2^2 + 2GM_e\left(\frac{1}{R_e} - \frac{1}{R_e + y_2}\right)} = 2770 \text{ m/s}$$

This is about 6000 mph, much less than the escape speed.

b. The calculation is the same in the flat-earth approximation except that we use $U_g = mgy$ for the gravitational potential energy. Thus

$$\frac{1}{2}mv_2^2 + mgy_2 = \frac{1}{2}mv_1^2 + mgy_1$$

$$v_1 = \sqrt{v_2^2 + 2gy_2} = 2840 \text{ m/s}$$

The flat-earth value of 2840 m/s is 70 m/s too big. The error, as a percentage of the correct 2770 m/s, is

$$\text{error} = \frac{70}{2770} \times 100 = 2.5\%$$

ASSESS The true speed is less than the flat-earth approximation because the force of gravity decreases with height. Launching a rocket against a decreasing force takes less effort than it would with the flat-earth force of mg at all heights.

STOP TO THINK 12.4 Rank in order, from largest to smallest, the absolute values $|U_g|$ of the gravitational potential energies of these pairs of masses. The numbers give the relative masses and distances.

(a) $m_1 = 2$, $r = 4$, $m_2 = 2$

(b) $m_1 = 1$, $r = 1$, $m_2 = 1$

(c) $m_1 = 1$, $r = 2$, $m_2 = 1$

(d) $m_1 = 1$, $r = 4$, $m_2 = 4$

(e) $m_1 = 4$, $r = 8$, $m_2 = 4$

12.6 Satellite Orbits and Energies

Solving Newton's second law to find the trajectory of a mass moving under the influence of gravity is mathematically beyond this textbook. It turns out that the solution is a set of elliptical orbits. This is Kepler's first law, which he discovered empirically by analyzing Tycho Brahe's observations. Kepler had no *reason* to think that orbits should be ellipses rather than some other shape. Newton was able to show that ellipses are a *consequence* of his theory of gravity.

The mathematics of ellipses is rather difficult, so we will restrict most of our analysis to the limiting case in which an ellipse becomes a circle. Most planetary orbits differ only very slightly from being circular. The earth's orbit, for example has a (semiminor axis/semimajor axis) ratio of 0.99986—very close to a true circle!

Figure 12.18 shows a massive body M, such as the earth or the sun, with a lighter body m orbiting it. The lighter body is called a **satellite,** even though it may be a planet orbiting the sun. Newton's second law for the satellite is

$$F_{M \text{ on } m} = \frac{GMm}{r^2} = ma_r = \frac{mv^2}{r} \tag{12.21}$$

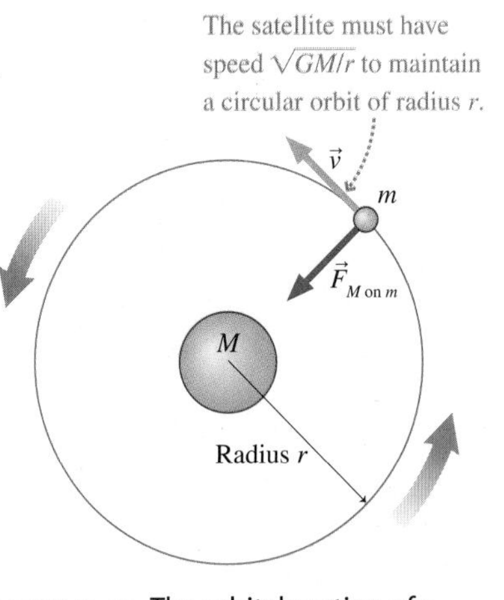

The satellite must have speed $\sqrt{GM/r}$ to maintain a circular orbit of radius r.
$\vec{v}$
m
$\vec{F}_{M \text{ on } m}$
M
Radius r

FIGURE 12.18 The orbital motion of a satellite due to the force of gravity.

The International Space Station appears to be floating, but it's actually traveling at nearly 8000 m/s as it orbits the earth.

Thus the speed of a satellite in a circular orbit is

$$v = \sqrt{\frac{GM}{r}} \tag{12.22}$$

A satellite must have this specific speed in order to have a circular orbit of radius r about the larger mass M. If the velocity differs from this value, the orbit will become elliptical rather than circular. Notice that the orbital speed does *not* depend on the satellite's mass m. This is consistent with our previous discovery, for motion on a flat earth, that motion due to gravity is independent of the mass.

EXAMPLE 12.4 The speed of the space shuttle

The space shuttle in a 300-km-high orbit ($\approx$180 mi) wants to capture a smaller satellite for repairs. What are the speeds of the shuttle and the satellite in this orbit?

SOLVE Despite their different masses, the shuttle, the satellite, and the astronaut working in space to make the repairs all travel side-by-side with the same speed. They are simply in free-fall together. Using $r = R_e + h$ with $h = 300 \text{ km} = 3.00 \times 10^5$ m, the speed is

$$v = \sqrt{\frac{(6.67 \times 10^{-11} \text{ N m}^2/\text{kg}^2)(5.98 \times 10^{24} \text{ kg})}{6.67 \times 10^6 \text{ m}}}$$

$$= 7730 \text{ m/s} \approx 16{,}000 \text{ mph}$$

ASSESS The answer depends on the mass of the earth but *not* on the mass of the satellite.

Kepler's Third Law

4.6 ActivPhysics Online

An important parameter of circular motion is the *period*. Recall that the period T is the time to complete one full orbit. The relationship among speed, radius, and period is

$$v = \frac{1 \text{ circumference}}{1 \text{ period}} = \frac{2\pi r}{T} \tag{12.23}$$

We can find a relationship between a satellite's period and the radius of its orbit by using Equation 12.22 for v:

$$v = \frac{2\pi r}{T} = \sqrt{\frac{GM}{r}} \tag{12.24}$$

Squaring both sides and solving for T gives

$$T^2 = \left(\frac{4\pi^2}{GM}\right) r^3 \tag{12.25}$$

In other words, the *square* of the period is proportional to the *cube* of the radius. This is Kepler's third law. You can see that Kepler's third law is a direct consequence of Newton's law of gravity.

Table 12.2 contains astronomical information about the sun, the earth, the moon, and other planets of the solar system. We can use these data to check the validity of Equation 12.25. Figure 12.19 is a graph of $\log T$ versus $\log r$ for all the planets in Table 12.2 except Mercury. Notice that the scales on each axis are increasing logarithmically—by *factors* of 10—rather than linearly. (Also, the vertical axis has converted T to the SI units of s.) As you can see, the graph is a very straight line with a statistical "best fit" equation

$$\log T = 1.500 \log r - 9.264$$

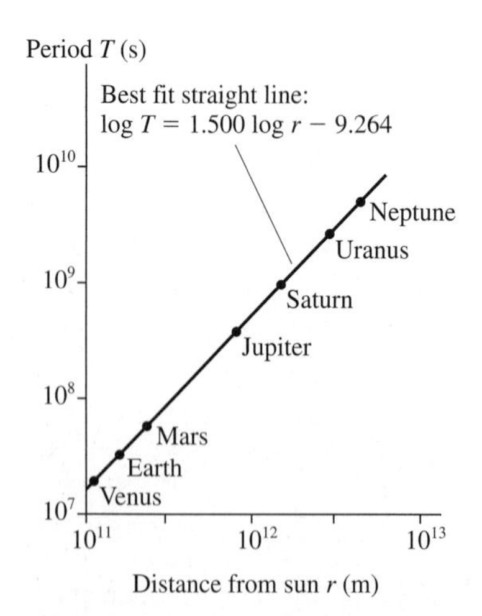

FIGURE 12.19 The graph of $\log T$ versus $\log r$ for the planetary data of Table 12.2.

As a homework problem, you can show that the slope of 1.500 for this "log-log graph" confirms the prediction of Equation 12.25. The y-intercept value of this

TABLE 12.2 Useful astronomical data

Planetary body	Mean distance from sun (m)	Period (years)	Mass (kg)	Mean radius (m)
Sun	–	–	1.99×10^{30}	6.96×10^{8}
Moon	3.84×10^{8}*	27.3 days	7.36×10^{22}	1.74×10^{6}
Mercury	5.79×10^{10}	0.241	3.18×10^{23}	2.43×10^{6}
Venus	1.08×10^{11}	0.615	4.88×10^{24}	6.06×10^{6}
Earth	1.50×10^{11}	1.00	5.98×10^{24}	6.37×10^{6}
Mars	2.28×10^{11}	1.88	6.42×10^{23}	3.37×10^{6}
Jupiter	7.78×10^{11}	11.9	1.90×10^{27}	6.99×10^{7}
Saturn	1.43×10^{12}	29.5	5.68×10^{26}	5.85×10^{7}
Uranus	2.87×10^{12}	84.0	8.68×10^{25}	2.33×10^{7}
Neptune	4.50×10^{12}	165	1.03×10^{26}	2.21×10^{7}

*Distance from earth.

line also contains useful information, and you can use it to determine the mass of the sun.

A particularly interesting application of Equation 12.25 is to communication satellites that are in **geosynchronous orbits** above the earth. These satellites have a period of 24 hours, making their orbital motion synchronous with the earth's rotation. As a result, a satellite in such an orbit appears to remain stationary over one point on the earth's equator. Equation 12.25 allows us to compute the radius of an orbit with this period:

$$r_{geo} = R_e + h_{geo} = \left[\left(\frac{GM}{4\pi^2}\right)T^2\right]^{1/3}$$

$$= \left[\left(\frac{(6.67 \times 10^{-11}\ \text{N m}^2/\text{kg}^2)(5.98 \times 10^{24}\ \text{kg})}{4\pi^2}\right)(86{,}400\ \text{s})^2\right]^{1/3}$$

$$= 4.225 \times 10^{7}\ \text{m}$$

The height of the orbit is

$$h_{geo} = r_{geo} - R_e = 3.59 \times 10^{7}\ \text{m} = 35{,}900\ \text{km} \approx 22{,}300\ \text{mi}$$

NOTE ► When using Equation 12.25, the period *must* be in SI units of s. ◄

Geosynchronous orbits are much higher than the low-earth orbits used by the space shuttle and remote-sensing satellites, where $h \approx 300$ km. Communications satellites in geosynchronous orbits were first proposed in 1948 by science fiction writer Arthur C. Clarke, 10 years before the first artificial satellite of any type!

EXAMPLE 12.5 Extrasolar planets

Astronomers using the most advanced telescopes have only recently seen evidence of planets orbiting nearby stars. These are called *extrasolar planets.* Suppose a planet is observed to have a 1200 day period as it orbits a star at the same distance that Jupiter is from the sun. What is the mass of the star in solar masses? (1 *solar mass* is defined to be the mass of the sun.)

SOLVE Here "day" means earth days, as used by astronomers to measure the period. Thus the planet's period in SI units is $T = 1200$ days $= 1.037 \times 10^{8}$ s. The orbital radius is that of Jupiter, which we can find in Table 12.2 to be $r = 7.78 \times 10^{11}$ m. Solving Equation 12.25 for the mass of the star gives

$$M = \frac{4\pi^2 r^3}{GT^2} = 2.59 \times 10^{31}\ \text{kg} \times \frac{1\ \text{solar mass}}{1.99 \times 10^{30}\ \text{kg}}$$

$$= 13\ \text{solar masses}$$

ASSESS This is a large, but not extraordinary, star.

STOP TO THINK 12.5 Two planets orbit a star. Planet 1 has orbital radius r_1 and planet 2 has $r_2 = 4r_1$. Planet 1 orbits with period T_1. Planet 2 orbits with period

a. $T_2 = 8T_1$. b. $T_2 = 4T_1$. c. $T_2 = 2T_1$.
d. $T_2 = \frac{1}{2}T_1$. e. $T_2 = \frac{1}{4}T_1$. f. $T_2 = \frac{1}{8}T_1$.

Kepler's Second Law

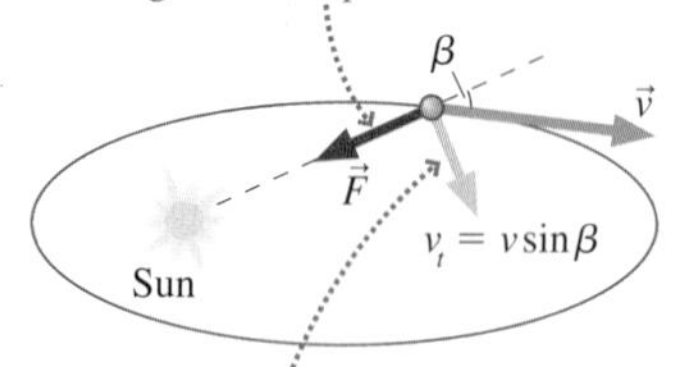
(a)
The gravitational force has no tangential component.
β
$\vec{v}$
$\vec{F}$
$v_t = v\sin\beta$
Sun
This is the component of $\vec{v}$ that would be tangent to a *circle* around the sun.

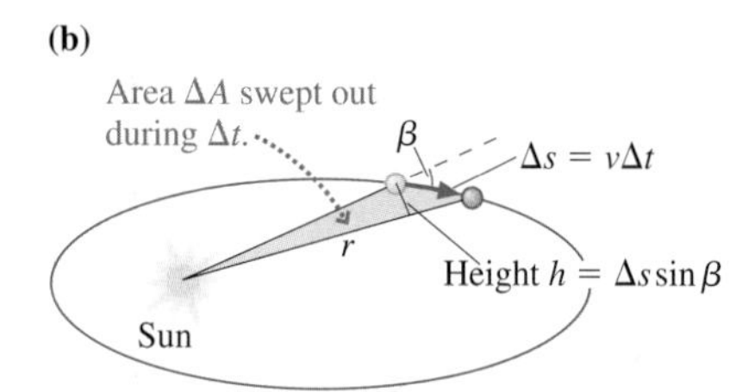
(b)
Area ΔA swept out during Δt.
β
Δs = vΔt
r
Height h = Δs sin β
Sun

FIGURE 12.20 Angular momentum is conserved for a planet in an elliptical orbit.

In Chapter 9 we defined a particle's *angular momentum* to be $L = mrv_t$, where v_t is the tangential component of the velocity vector. For a particle in circular motion, where v_t is simply the speed v, we showed that angular momentum is conserved if there is no tangential force on the particle. Although we won't prove it until Chapter 13, it turns out that angular momentum is conserved for a particle following a trajectory of *any* shape if the net tangential force is zero.

Figure 12.20a shows a satellite moving in an elliptical orbit around a star or planet at one focus. The tangential component of $\vec{v}$ is $v_t = v\sin\beta$, where β is the angle between $\vec{r}$ and $\vec{v}$. Consequently, the satellite's angular momentum is

$$L = mrv_t = mrv\sin\beta \tag{12.26}$$

For a circular orbit, where β is always 90°, this reduces to simply $L = mrv$.

The only force on a satellite, the gravitational force, points directly toward the star or planet that the satellite is orbiting. The gravitational force has no tangential component, thus **the satellite's angular momentum is conserved as it orbits.**

The satellite moves forward a small distance $\Delta s = v\Delta t$ during the small interval of time Δt. This motion defines the triangle of area ΔA shown in Figure 12.20b. ΔA is the area "swept out" by the satellite during Δt. You can see that the height of the triangle is $h = \Delta s\sin\beta$, so the triangle's area is

$$\Delta A = \frac{1}{2} \times \text{base} \times \text{height} = \frac{1}{2} \times r \times \Delta s\sin\beta = \frac{1}{2}rv\sin\beta\Delta t \tag{12.27}$$

The *rate* at which the area is swept out by the satellite as it moves is

$$\frac{\Delta A}{\Delta t} = \frac{1}{2}rv\sin\beta = \frac{mrv\sin\beta}{2m} = \frac{L}{2m} \tag{12.28}$$

The angular momentum L is conserved, so it has the same value at every point in the orbit. Consequently, the rate at which the area is swept out by the satellite is constant. This is Kepler's second law, which says that a line drawn between the sun and a planet sweeps out equal areas during equal intervals of time. We see that Kepler's second law is really a consequence of the conservation of angular momentum.

Kepler and Newton

Kepler's laws summarize observational data about the motions of the planets. They were an outstanding achievement, but they did not form a theory. Newton put forward a *theory,* a specific set of relationships between force and motion that allows *any* motion to be understood and calculated. Newton's theory of gravity has allowed us to *deduce* Kepler's laws and, thus, to understand them at a more fundamental level.

Furthermore, Kepler's laws are not perfectly accurate. The planets, in addition to being attracted to the sun, are also attracted toward each other and toward their orbiting moons. The consequences of these additional forces are small, but over time they provide measurable effects not contained in Kepler's laws. Newton's theory makes it possible to use the inverse-square law to calculate *all* of the forces, add them to get the net force acting on each planet, then proceed to solve Newton's second law to determine the dynamics. The mathematics of the solution can be exceedingly difficult, and today is all done with computers, but even with hand calculations this procedure in the mid-19th century predicted the existence of an undiscovered planet that was having minor effects on the orbital motion of Uranus. The planet Neptune was discovered in 1846, just where the calculations predicted.

Orbital Energetics

Let us conclude this chapter by thinking about the energetics of orbital motion. We found, with Equation 12.24, that a satellite in a circular orbit must have $v^2 = GM/r$. A satellite's speed is determined entirely by the size of its orbit. The satellite's kinetic energy is thus

$$K = \frac{1}{2}mv^2 = \frac{GMm}{2r} \tag{12.29}$$

But $-GMm/r$ is the potential energy, U_g, so

$$K = -\frac{1}{2}U_g \tag{12.30}$$

This is an interesting result. In all our earlier examples, the kinetic and potential energy were two independent parameters. In contrast, a satellite can move in a circular orbit *only* if there is a very specific relationship between K and U. It is not that K and U *have* to have this relationship, but if they do not, the trajectory will be elliptical rather than circular.

Equation 12.30 gives us the mechanical energy of a satellite in a circular orbit:

$$E_{mech} = K + U_g = \frac{1}{2}U_g \tag{12.31}$$

The gravitational potential energy is negative, hence the *total* mechanical energy is also negative. Negative total energy is characteristic of a **bound system,** a system in which the satellite is bound to the central mass by the gravitational force and cannot get away. In an unbound system, the satellite can move infinitely far away to where $U = 0$. Because the kinetic energy K must be ≥ 0, the total energy of an unbound system must be ≥ 0. A negative value of E_{mech} tells us that the satellite is unable to escape the central mass.

Figure 12.21 shows the kinetic, potential, and total energy of a satellite in a circular orbit as a function of the orbit's radius. Notice how $E_{mech} = \frac{1}{2}U_g$. This figure can help us understand the energetics of transferring a satellite from one orbit to another. Suppose a satellite is in an orbit of radius r_1 and that we'd like it to be in a larger orbit of radius r_2. The kinetic energy at r_2 is less than at r_1 (the satellite moves more slowly in the larger orbit), but you can see that the total energy *increases* as r

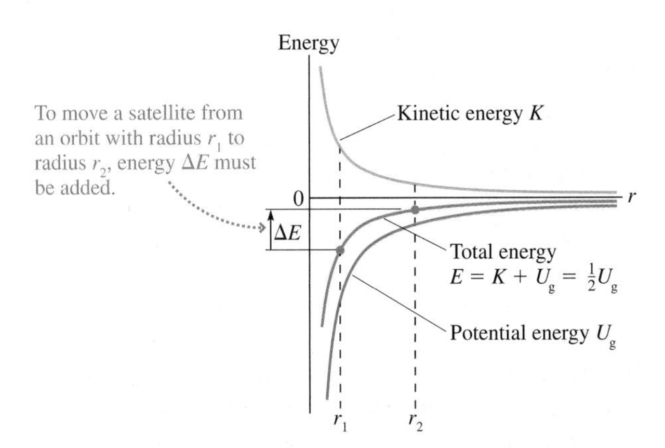

FIGURE 12.21 The kinetic, potential, and total energy of a satellite in a circular orbit.

increases. Consequently, transferring a satellite to a larger orbit requires a net energy increase $\Delta E > 0$. Where does this increase of energy come from?

Artificial satellites are raised to higher orbits by firing their rocket motors to create a forward thrust. This force does work on the satellite, and the energy equation of Chapter 11 tells us that this work increases the satellite's energy by $\Delta E_{mech} = W_{ext}$. Thus the energy to "lift" a satellite into a higher orbit comes from the chemical energy stored in the rocket fuel.

EXAMPLE 12.6 Raising a satellite

How much work must be done to boost a 1000 kg communications satellite from a low earth orbit with $h = 300$ km, where it is released by the space shuttle, to a geosynchronous orbit?

SOLVE The required work is $W_{ext} = \Delta E_{mech}$, and from Equation 12.31 we see that $\Delta E_{mech} = \frac{1}{2}\Delta U_g$. The initial orbit has radius $r_{shuttle} = R_e + h = 6.67 \times 10^6$ m. We earlier found the radius of a geosynchronous orbit to be 4.22×10^7 m. Thus

$$W_{ext} = \Delta E_{mech} = \frac{1}{2}\Delta U_g = \frac{1}{2}(-GM_e m)\left(\frac{1}{r_{geo}} - \frac{1}{r_{shuttle}}\right)$$

$$= 2.52 \times 10^{10} \text{ J}$$

ASSESS It takes a lot of energy to boost satellites to high orbits!

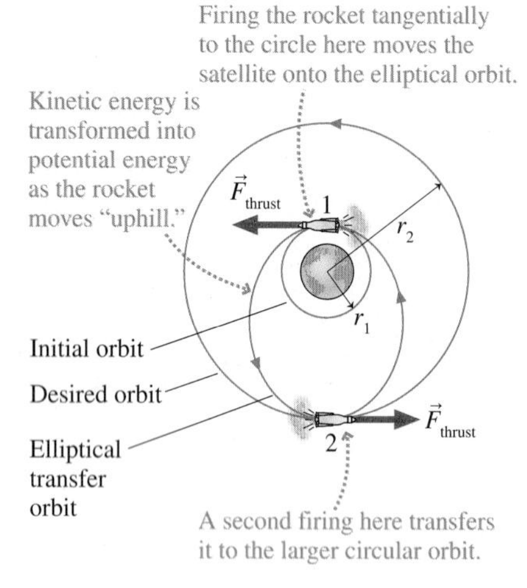
Firing the rocket tangentially to the circle here moves the satellite onto the elliptical orbit.
Kinetic energy is transformed into potential energy as the rocket moves "uphill."
$\vec{F}_{thrust}$
1
r_2
r_1
Initial orbit
Desired orbit
Elliptical transfer orbit
2
$\vec{F}_{thrust}$
A second firing here transfers it to the larger circular orbit.

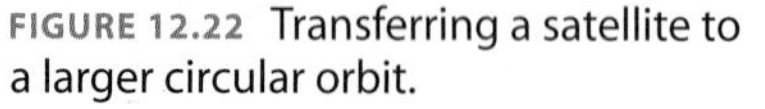
FIGURE 12.22 Transferring a satellite to a larger circular orbit.

You might think that the way to get a satellite into a larger orbit would be to point the thrusters toward the earth and blast outward. That would work fine *if* the satellite were initially at rest and moved straight out along a linear trajectory. But an orbiting satellite is already moving and has significant inertia. A force directed straight outward would *change* the satellite's velocity vector in that direction but would not cause it to *move* along that line. (Remember all those earlier motion diagrams for motion along curved trajectories.) In addition, a force directed outward would be almost at right angles to the motion and would do essentially zero work. Navigating in space is not as easy as it appears in *Star Wars!*

To move the satellite in Figure 12.22 from the orbit with radius r_1 to the larger circular orbit of radius r_2, the thrusters are turned on at point 1 to apply a brief *forward* thrust force in the direction of motion, *tangent* to the circle. This force does a significant amount of work because the force is parallel to the displacement, so the satellite quickly gains kinetic energy ($\Delta K > 0$). But $\Delta U_g = 0$ because the satellite does not have time to change its distance from the earth during a thrust of short duration. With the kinetic energy increased, but not the potential energy, the satellite no longer meets the requirement $K = -\frac{1}{2}U_g$ for a circular orbit. Instead, it goes into an elliptical orbit.

In the elliptical orbit, the satellite moves "uphill" toward point 2 by transforming kinetic energy into potential energy. At point 2, the satellite has arrived at the desired distance from earth and has the "right" value of the potential energy. But it turns out that its kinetic energy is now *less* than that needed for a circular orbit. (The analysis is more complex than we want to pursue here. It will be left for a homework Challenge Problem.) If no action is taken, the satellite will continue on its elliptical orbit and "fall" back to point 1. But another *forward* thrust at point 2 increases its kinetic energy, without changing U_g, until the kinetic energy reaches the value $K = -\frac{1}{2}U_g$ required for a circular orbit. Presto! The second burn kicks the satellite into the desired circular orbit of radius r_2. The work $W_{ext} = \Delta E_{mech}$ is the *total* work done in both burns. It takes a more extended analysis to see how the work has to be divided between the two burns, but even without those details you now have enough knowledge about orbits and energy to understand the ideas that are involved.

SUMMARY

The goal of Chapter 12 has been to use Newton's theory of gravity to understand the motion of satellites and planets.

GENERAL PRINCIPLES

Newton's Theory of Gravity

1. Two objects with masses M and m a distance r apart exert attractive **gravitational forces** on each other of magnitude

$$F_{M \text{ on } m} = F_{m \text{ on } M} = \frac{GMm}{r^2}$$

where the **gravitational constant** is $G = 6.67 \times 10^{-11}\,\text{N m}^2/\text{kg}^2$.

2. Gravitational mass and inertial mass are equivalent.
3. Newton's three laws of motion apply to satellites, planets, and stars.

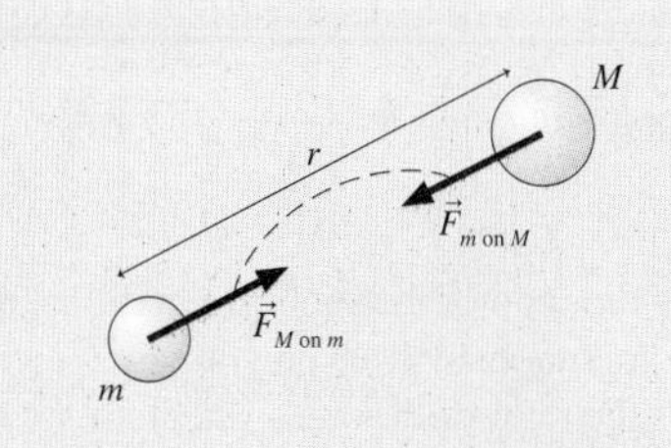

IMPORTANT CONCEPTS

Orbital motion of a planet (or satellite) is described by **Kepler's laws:**

1. Orbits are ellipses with the sun (or planet) at one focus.
2. A line between the sun and the planet sweeps out equal areas during equal intervals of time.
3. The square of the planet's period T is proportional to the cube of the orbit's semimajor axis.

Circular orbits are a special case of an ellipse. For a circular orbit around a mass M,

$$v = \sqrt{\frac{GM}{r}} \quad \text{and} \quad T^2 = \left(\frac{4\pi^2}{GM}\right)r^3$$

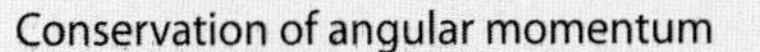

Conservation of angular momentum

The angular momentum $L = mrv\sin\beta$ remains constant throughout the orbit. Kepler's second law is a consequence of this law.

Orbital energetics

A satellite's mechanical energy $E_{mech} = K + U_g$ is conserved, where the gravitational potential energy is

$$U_g = -\frac{GMm}{r}$$

For circular orbits, $K = -\frac{1}{2}U_g$ and $E_{mech} = \frac{1}{2}U_g$. Negative total energy is characteristic of a **bound system.**

APPLICATIONS

For a planet of mass M and radius R,

- The acceleration due to gravity on the surface is $g_{surface} = \dfrac{GM}{R^2}$
- The escape speed is $v_{escape} = \sqrt{\dfrac{2GM}{R}}$
- The radius of a geosynchronous orbit is $r_{geo} = \left(\dfrac{GM}{4\pi^2}T^2\right)^{1/3}$

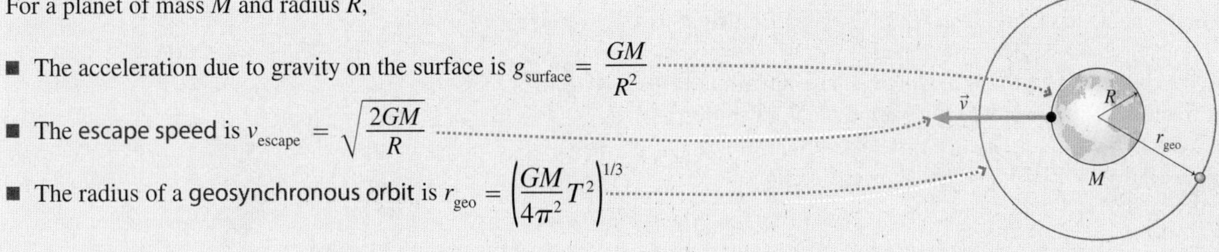

TERMS AND NOTATION

cosmology
Kepler's laws
gravitational force
Newton's law of gravity
gravitational constant, G
gravitational mass
principle of equivalence
Newton's theory of gravity
escape speed
satellite
geosynchronous orbit
bound system

EXERCISES AND PROBLEMS

Exercises

Section 12.3 Newton's Law of Gravity

1. a. What is the gravitational force of the sun on the earth?
   b. What is the gravitational force of the moon on the earth?
   c. The moon's force is what percent of the sun's force?
2. The centers of a 10 kg lead ball and a 100 g lead ball are separated by 10 cm.
   a. What gravitational force does each exert on the other?
   b. What is the ratio of this gravitational force to the weight of the 100 g ball?
3. What is the ratio of the sun's gravitational force on you to the earth's gravitational force on you?
4. What is the ratio of the sun's gravitational force on the moon to the earth's gravitational force on the moon?
5. A 1.0-m-diameter lead sphere has a mass of 5900 kg. A dust particle rests on the surface. What is the ratio of the gravitational force of the sphere on the dust particle to the weight of the dust particle?
6. Estimate the force of attraction between a 50 kg woman and a 70 kg man sitting 1.0 m apart.
7. The space shuttle orbits 300 km above the surface of the earth.
   a. What is the force of gravity on a 1.0 kg sphere inside the space shuttle?
   b. The sphere floats around inside the space shuttle, apparently "weightless." How is this possible?

Section 12.4 Little g and Big G

8. a. What is the acceleration due to gravity at the surface of the sun?
   b. What is the sun's acceleration due to gravity at the distance of the earth?
9. What is the acceleration due to gravity at the surface of (a) the moon and (b) Jupiter?
10. A starship is circling a distant planet of radius R. The astronauts find that the acceleration due to gravity at their altitude is half the value at the planet's surface. How far above the surface are they orbiting? Your answer will be a multiple of R.
11. A sensitive gravimeter at a mountain observatory finds that the acceleration due to gravity is $0.0075\ \text{m/s}^2$ less than that at sea level. What is the observatory's altitude?
12. Suppose we could shrink the earth without changing its mass. At what fraction of its current radius would the acceleration due to gravity at the surface be three times its present value?
13. Planet Z is 10,000 km in diameter. The acceleration due to gravity on Planet Z is $8.0\ \text{m/s}^2$.
    a. What is the mass of Planet Z?
    b. What is the acceleration due to gravity 10,000 km above Planet Z's north pole?

Section 12.5 Gravitational Potential Energy

14. a. What is the acceleration due to gravity on Mars?
    b. An astronaut on earth can throw a ball straight up to a height of 15 m. How high can he throw the ball on Mars?
15. A projectile is shot straight up from the earth's surface at a speed of 10,000 km/hr. How high does it go?
16. A rocket is launched straight up from the earth's surface at a speed of 15,000 m/s. What is its speed when it is very far away from the earth?
17. What is the escape speed from Jupiter?
18. A space station orbits the sun at the same distance as the earth but on the opposite side of the sun. A small probe is fired away from the station. What minimum speed does the probe need to escape the solar system?
19. You have been visiting a distant planet. Your measurements have determined that the planet's mass is twice that of earth but the acceleration due to gravity at the surface is only one-fourth as large.
    a. What is the planet's radius?
    b. To get back to earth, you need to escape the planet. What minimum speed does your rocket need?

Section 12.6 Satellite Orbits and Energies

20. A satellite orbits the sun with a period of 1.0 day. What is the radius of its orbit?
21. The space shuttle is in a 250-mile-high orbit. What are the shuttle's orbital period, in minutes, and its speed?
22. The *asteroid belt* circles the sun between the orbits of Mars and Jupiter. One asteroid has a period of 5.0 earth years. What are the asteroid's orbital radius and speed?
23. An earth satellite moves in a circular orbit at a speed of 5500 m/s. What is its orbital period?
24. What are the speed and altitude of a geosynchronous satellite orbiting Mars? Mars rotates on its axis once every 24.8 hours.
25. Use information about the earth and its orbit to determine the mass of the sun.
26. Three satellites orbit a planet of radius R, as shown in Figure Ex12.26. Satellites S_1 and S_3 have mass m. Satellite S_2 has mass $2m$. Satellite S_1 orbits in 250 minutes and the force on S_1 is 10,000 N.
    a. What are the periods of S_2 and S_3?
    b. What are the forces on S_2 and S_3?
    c. What is the kinetic-energy ratio K_1/K_3 for S_1 and S_3?

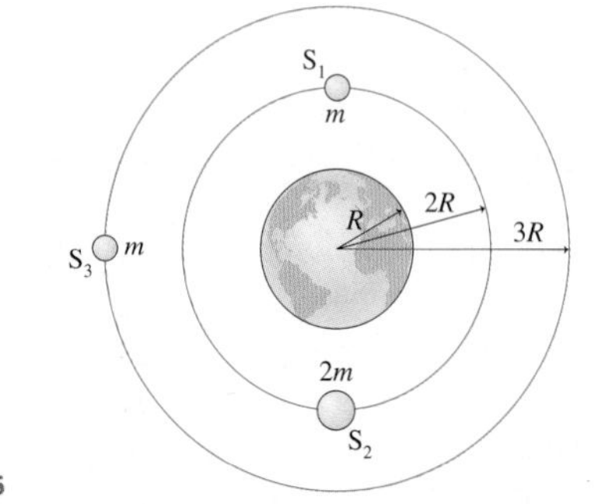

FIGURE EX12.26

27. A 4000 kg lunar lander is in orbit 50 km above the surface of the moon. It needs to move out to a 300-km-high orbit in order to link up with the mother ship that will take the astronauts home. How much work must the thrusters do?

28. The space shuttle is in a 250-km-high circular orbit. It needs to reach a 610-km-high circular orbit to catch the Hubble Space Telescope for repairs. The shuttle's mass is 75,000 kg. How much energy is required to boost it to the new orbit?

Problems

29. Two spherical objects have a combined mass of 150 kg. The gravitational attraction between them is 8.00×10^{-6} N when their centers are 20 cm apart. What is the mass of each?
30. Two 100 kg lead spheres are suspended from 100-m-long massless cables. The tops of the cables have been carefully anchored *exactly* 1 m apart. What is the distance between the centers of the spheres?
31. In a Cavendish balance experiment, 200 g masses are attached to the ends of a 40-cm-long rod. The rod is then suspended by a fiber whose torsion constant is 1.0×10^{-5} N/rad. That is, applying force F to *both* ends of the rod, in opposite directions, causes the rod to turn through angle $\Delta\theta$ such that $F = k\Delta\theta$. Once the rod is stationary, 20 kg lead spheres are brought close to each end of the rod. Through what angle does the rod turn if the distance between the center of each sphere and the corresponding 200 g mass is 15.0 cm? Give your answer in both radians and degrees. The angle is small, but it is measurable.
32. A new moon is almost exactly in line between the earth and the sun. A full moon is on the opposite side of the earth from the sun. What is the ratio of the *net* gravitational force on the moon when it is new to when it is full?
33. A 20 kg sphere is at the origin and a 10 kg sphere is at $(x, y) = (20 \text{ cm}, 0 \text{ cm})$. At what point or points could you place a small mass such that the net gravitational force on it due to the spheres is zero?
34. Figure P12.34 shows three masses. What are the magnitude and the direction of the net gravitational force on (a) the 20.0 kg mass and (b) the 5.0 kg mass? Give the direction as an angle cw or ccw of the y-axis.

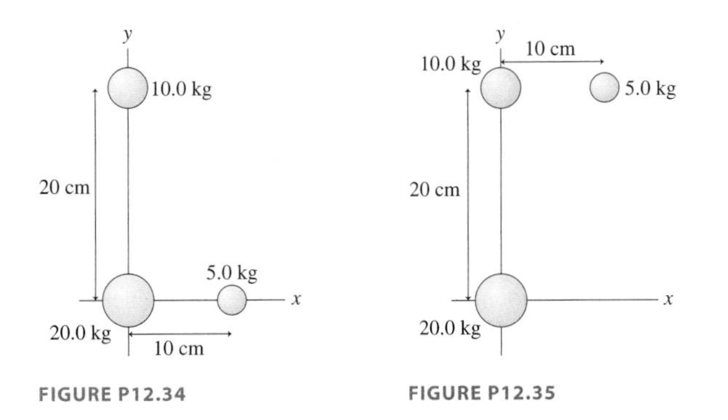

FIGURE P12.34

FIGURE P12.35

35. Figure P12.35 shows three masses. What are the magnitude and the direction of the net gravitational force on (a) the 10.0 kg mass and (b) the 20.0 kg mass? Give the direction as an angle cw or ccw of the y-axis.
36. What are the gravitational potential energies of (a) the 20.0 kg mass and (b) the 5.0 kg mass in Figure P12.34?
37. What are the gravitational potential energies of (a) the 10.0 kg mass and (b) the 20.0 kg mass in Figure P12.35?
38. Consider an object of mass m on the equator of a planet with mass M, radius R, and period of rotation T. If the planet did not rotate, the object would be in static equilibrium and its apparent weight would be $w_{app} = mg_{true}$, where $g_{true} = MG/R^2$. Because of the planet's rotation, the object is rather like a roller-coaster car going over the top of a hill. It has an effective weight $w_{app} = n = mg_{app}$, but $g_{app} \neq g_{true}$. Any experiment performed by a physics student to measure the acceleration due to gravity will determine the value of g_{app}, not g_{true}.
    a. Find an algebraic expression for g_{app} in terms of M, R, G, and T.
    b. Evaluate both g_{true} and g_{app} for the earth.
    c. Which of these g's have we been using in this text?
    d. How will the value of g measured at the North Pole compare to the value measured at the equator? Will it be larger, smaller, or the same? Explain.
39. a. At what height above the earth is the acceleration due to gravity 10% of its value at the surface?
    b. What is the speed of a satellite orbiting at that height?
40. A 1.0 kg object is released from rest 500 km ($\approx$300 miles) above the earth.
    a. What is its impact speed as it hits the ground? Ignore air resistance.
    b. What would the impact speed be if the earth were flat?
    c. By what percent is the flat-earth calculation in error?
    d. Suppose a space shuttle astronaut, while out for a stroll, "drops" a 1.0 kg hammer. Will it fall to earth? Explain why or why not.
41. A huge cannon is assembled on an airless planet. The planet has a radius of 5.0×10^6 m and a mass of 2.6×10^{24} kg. The cannon fires a projectile straight up at 5000 m/s.
    a. What height does the projectile reach above the surface?
    b. An observation satellite orbits the planet at a height of 1000 km. What is the projectile's speed as it passes the satellite?
42. Two meteoroids are heading for earth. Their speeds as they cross the moon's orbit are 2.0 km/s.
    a. The first meteoroid is heading straight for earth. What is its speed of impact?
    b. The second misses the earth by 5000 km. What is its speed at its closest point?
43. A binary star system has two stars, each with the same mass as our sun, separated by 1.0×10^{12} m. A comet is very far away and essentially at rest. Slowly but surely, gravity pulls the comet toward the stars. Suppose the comet travels along a straight line that passes through the midpoint between the two stars. What is the comet's speed at the midpoint?
44. Suppose that on earth you can jump straight up a distance of 50 cm. Can you escape from a 4.0-km-diameter asteroid with a mass of 1.0×10^{14} kg?
45. Figure P12.45 shows two identical planets of mass M and radius R spaced $6R$ apart. Ignore the rotation of the planets about each other. What is the escape speed of a rocket launched from (a) point A and (b) point B?

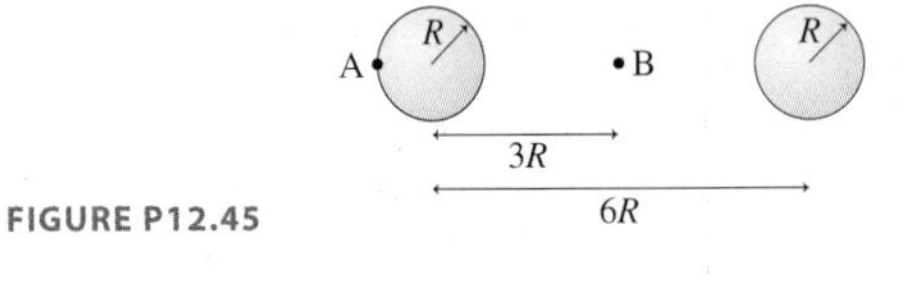

FIGURE P12.45

46. Two stars with the mass and radius of the sun are separated by distance $10R_s$, measured between their centers. A 10,000 kg space capsule moves along a line between the stars.
    a. Sketch a reasonably accurate graph of the space capsule's potential energy along a line passing through the two stars. Let one star be at $x = 0$ and the other at $x = 10R_s$.
    b. Suppose the space capsule is at rest exactly midway between the stars. Is this a point of equilibrium? If so, is it a stable or an unstable equilibrium? Explain.
    c. Suppose the space capsule is at rest 1.0 m closer to one star than the other. What will be the speed of the space capsule when it meets its ultimate fate?
47. a. At what distance from the center of the earth, along a line from the earth to the moon, is there no net force on a 10,000 kg spacecraft? This is the *crossover point* at which the attraction to the moon becomes larger than the attraction to the earth.
    b. The gravitational potential energy is the sum of the potential energies with respect to the earth and the moon. Calculate the spacecraft's potential energy at several points between the surface of the earth and the surface of the moon, including the crossover point. Then draw a reasonably accurate graph showing the potential energy from the earth to the moon.
    c. What is the mechanical energy of the spacecraft as it orbits 300 km above the moon? Don't forget to include a term due to the earth's gravitational potential energy.
    d. What is the minimum work the spacecraft engines must do to reach the crossover point and head home?
    e. If the spacecraft crosses the crossover point with the minimum possible energy, what will its speed be as it reaches the edge of the earth's 100-km-high atmosphere?
48. Two spherical asteroids have the same radius R. Asteroid 1 has mass M and asteroid 2 has mass $2M$. The two asteroids are released from rest with distance $10R$ between their centers. What is the speed of each asteroid just before they collide?
    Hint: You will need to use two conservation laws.
49. Two Jupiter-size planets are released from rest 1.0×10^{11} m apart. What are their speeds as they crash together?
50. a. How much energy must a 50,000 kg space shuttle lose to descend from a 500-km-high circular orbit to a 300-km-high orbit?
    b. Give a *qualitative* description, including a sketch, of how the shuttle would do this.
51. Suppose a cosmic accident increases the earth's rotation on its axis. At what rotational period, in hours, will objects at the equator begin to "fly off" the earth?
52. Mars has a small moon, Phobos, that orbits with a period of 7 h 39 min. The radius of Phobos' orbit is 9.4×10^6 m. What is the mass of Mars?
53. In 2000, NASA placed a satellite in orbit around an asteroid. Consider a spherical asteroid with a mass of 1.0×10^{16} kg and a radius of 8.8 km.
    a. What is the speed of a satellite orbiting 5.0 km above the surface?
    b. What is the escape speed from the asteroid?
54. You are the science officer on a visit to a distant solar system. Prior to landing on a planet you measure its diameter to be 1.8×10^7 m and its rotation period to be 22.3 hours. You have previously determined that the planet orbits 2.2×10^{11} m from its star with a period of 402 earth days. Once on the surface you find that the acceleration due to gravity is 12.2 m/s^2. What are the mass of (a) the planet and (b) the star?
55. NASA would like to place a satellite in orbit around the moon such that the satellite always remains in the same position over the lunar surface. What is the satellite's altitude?
56. A satellite orbiting the earth is directly over a point on the equator at 12:00 midnight every two days. It is not over that point at any time in between. What is the radius of the satellite's orbit?
57. Figure 12.19 showed a graph of log T versus log r for the planetary data given in Table 12.2. Such a graph is called a *log-log graph*. The scales in Figure 12.19 are logarithmic, not linear, meaning that each division along the axis corresponds to a *factor* of 10 increase in the value. Strictly speaking, the "correct" labels on the y-axis should be 7, 8, 9, and 10 because these are the logarithms of $10^7, \ldots, 10^{10}$.
    a. Consider two quantities u and v that are related by the expression $v^p = Cu^q$, where C is a constant. The exponents p and q are not necessarily integers. Define $x = \log u$ and $y = \log v$. Find an expression for y in terms of x.
    b. What *shape* will a graph of y versus x have? Explain.
    c. What *slope* will a graph of y versus x have? Explain.
    d. Figure 12.19 showed that the "best fit" line passing through all the planetary data points has the equation $\log T = 1.500 \log r - 9.264$. This is an *experimentally* determined relationship between log T and log r, using measured data. Is this experimental result consistent with what you would expect from Newton's theory of gravity? Explain.
    e. Use the experimentally determined "best fit" line to find the mass of the sun.
58. Figure P12.58 shows two planets of mass m orbiting a star of mass M. The planets are in the same orbit, with radius r, but are always at opposite ends of a diameter. Find an expression for the orbital period T.

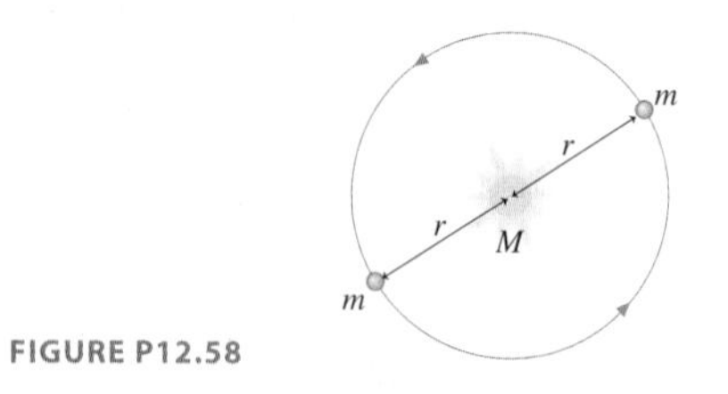

FIGURE P12.58

59. Humans currently use energy at the rate of about 10^{13} W. Suppose a future inventor finds a way to extract usable energy from the moon's orbital motion and beam that energy, without loss, to earth. If our energy use remains constant, by how much will the radius and period of the moon's orbit decrease after its energy has been tapped for 100 years? Give your answers as numerical values and as a percentage decrease.
60. Large stars can explode as they finish burning their nuclear fuel, causing a *supernova*. The explosion blows away the outer layers of the star. According to Newton's third law, the forces that push the outer layers away have *reaction forces* that are inwardly directed on the core of the star. These forces compress the core and can cause the core to undergo a *gravitational collapse*. The gravitational forces keep pulling all the matter together tighter and tighter, crushing atoms out

of existence. Under these extreme conditions, the protons and electrons can be squeezed together to form a neutron. If the collapse is halted when the neutrons all come into contact with each other, the result is an object called a *neutron star,* an entire star consisting of solid nuclear matter. Many neutron stars rotate about their axis with a period of ≈ 1 s and, as they do so, send out a pulse of electromagnetic waves once a second. These stars were discovered in the 1960s and are called *pulsars.*

a. Consider a neutron star with a mass equal to the sun, a radius of 10 km, and a rotation period of 1.0 s. What is the speed of a point on the equator of the star?
b. What is g at the surface of this neutron star?
c. A 1.0 kg mass has a weight on earth of 9.8 N. What would be its weight on the star?
d. How many revolutions per minute are made by a satellite orbiting 1.0 km above the surface?
e. What is the radius of a geosynchronous orbit about the neutron star?

61. Astronomers discover a binary star system that has a period of 90 days. The binary star system consists of two equal-mass stars, each with a mass twice that of the sun, that rotate like a dumbbell about the *center of mass* at the midpoint between them. How far apart are the two stars?
62. Three stars, each with the mass of our sun, form an equilateral triangle with sides 1.0×10^{12} m long. (This triangle would just about fit within the orbit of Jupiter.) The triangle has to rotate, because otherwise the stars would crash together in the center. What is the period of rotation? Give your answer in years.
63. Pluto moves in a fairly elliptical orbit around the sun. Pluto's speed at its closest approach of 4.43×10^9 km is 6.12 km/s. What is Pluto's speed at the most distant point in its orbit, where it is 7.30×10^9 km from the sun?
64. Mercury moves in a fairly elliptical orbit around the sun. Mercury's speed is 38.8 km/s when it is at its most distant point, 6.99×10^9 m from the sun. How far is Mercury from the sun at its closest point, where its speed is 59.0 km/s?
65. Comets move around the sun in very elliptical orbits. At its closet approach, in 1986, Comet Halley was 8.79×10^7 km from the sun and moving with a speed of 54.6 km/s. What will the comet's speed be when it crosses Neptune's orbit in 2006?
66. A spaceship is in a circular orbit of radius r_0 about a planet of mass M. A brief but intense firing of its engine in the forward direction decreases the spaceship's speed by 50%. This causes the spaceship to move into an elliptical orbit.
    a. What is the spaceship's new speed, just after the rocket burn is complete, in terms of M, G, and r_0?
    b. In terms of r_0, what are the spaceship's maximum and minimum distance from the planet in its new orbit?
67. A planet is orbiting a star when, for no apparent reason, the star's gravity suddenly vanishes. As Figure P12.67 shows, the planet then obeys Newton's first law and heads outward along a straight line. Is Kepler's second law still obeyed? That is, are equal areas swept out in equal intervals of time as the planet moves away?

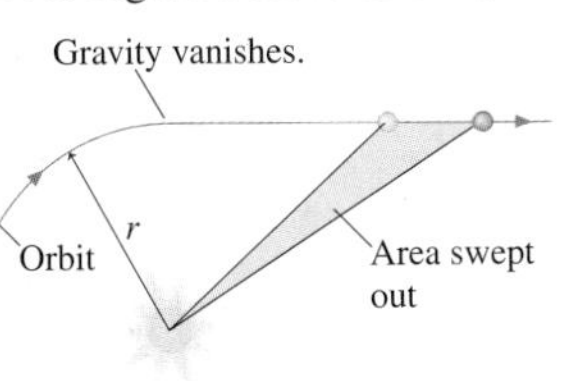

FIGURE P12.67

In Problems 68 through 71 you are given the equation(s) used to solve a problem. For each of these, you are to

a. Write a realistic problem for which this is the correct equation(s).
b. Draw a pictorial representation.
c. Finish the solution of the problem.

68. $$\frac{(6.67 \times 10^{-11}\text{ N m}^2/\text{kg}^2)(5.68 \times 10^{26}\text{ kg})}{r^2} = \frac{(6.67 \times 10^{-11}\text{ N m}^2/\text{kg}^2)(5.98 \times 10^{24}\text{ kg})}{(6.37 \times 10^6\text{ m})^2}$$

69. $$\frac{(6.67 \times 10^{-11}\text{ N m}^2/\text{kg}^2)(5.98 \times 10^{24}\text{ kg})(1000\text{ kg})}{r^2} = \frac{(1000\text{ kg})(1997\text{ m/s})^2}{r}$$

70. $$\frac{1}{2}(100\text{ kg})v_2^2 - \frac{(6.67 \times 10^{-11}\text{ N m}^2/\text{kg}^2)(7.36 \times 10^{22}\text{ kg})(100\text{ kg})}{(1.74 \times 10^6\text{ m})} = 0 - \frac{(6.67 \times 10^{-11}\text{ N m}^2/\text{kg}^2)(7.36 \times 10^{22}\text{ kg})(100\text{ kg})}{(3.48 \times 10^6\text{ m})}$$

71. $$(2.0 \times 10^{30}\text{ kg})v_{f1} + (4.0 \times 10^{30}\text{ kg})v_{f2} = 0$$
$$\frac{1}{2}(2.0 \times 10^{30}\text{ kg})v_{f1}{}^2 + \frac{1}{2}(4.0 \times 10^{30}\text{ kg})v_{f2}{}^2 - \frac{(6.67 \times 10^{-11}\text{ N m}^2/\text{kg}^2)(2.0 \times 10^{30}\text{ kg})(4.0 \times 10^{30}\text{ kg})}{(1.0 \times 10^9\text{ m})} = 0 + 0 - \frac{(6.67 \times 10^{-11}\text{ N m}^2/\text{kg}^2)(2.0 \times 10^{30}\text{ kg})(4.0 \times 10^{30}\text{ kg})}{(1.0 \times 10^{12}\text{ m})}$$

Challenge Problems

72. In 1996, the Solar and Heliospheric Observatory (SOHO) was "parked" in an orbit slightly inside the earth's orbit, as shown in Figure CP12.72. The satellite's period in this orbit is exactly one year, so it remains fixed relative to the earth. At this point, called a *Lagrange point,* the light from the sun is never blocked by the earth, yet the satellite remains "nearby" so that data are easily transmitted to earth. What is SOHO's distance from the earth?
    Hint: Use the binomial approximation. SOHO's distance from the earth is much less than the earth's distance from the sun.

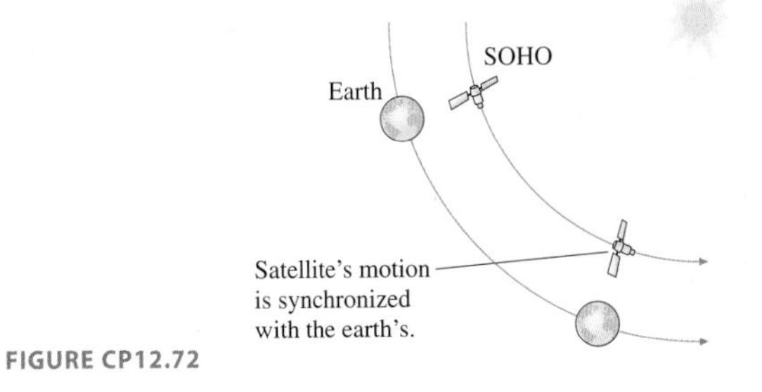

FIGURE CP12.72

73. The solar system is 25,000 light years from the center of our Milky Way galaxy. One *light year* is the distance light travels in one year at a speed of 3.0×10^8 m/s. Astronomers have determined that the solar system is orbiting the center of the galaxy at a speed of 230 km/s.
    a. Assuming the orbit is circular, what is the period of the solar system's orbit? Give your answer in years.
    b. Our solar system was formed roughly 5 billion years ago. How many orbits has it completed?
    c. The gravitational force on the solar system is the net force due to all the matter inside our orbit. Most of that matter is concentrated near the center of the galaxy. Assume that the matter has a spherical distribution, like a giant star. What is the approximate mass of the galactic center?
    d. Assume that the sun is a typical star with a typical mass. If galactic matter is made up of stars, approximately how many stars are in the center of the galaxy?

    Astronomers have spent many years trying to determine how many stars there are in the Milky Way. The number of stars seems to be only about 10% what you found in part (d). In other words, about 90% of the mass of the galaxy appears to be in some form other than stars. This is called the *dark matter* of the universe. No one knows what the dark matter is. This is one of the outstanding scientific questions of our day.
74. The space shuttle, in a 300-km-high orbit, needs to perform an experiment that has to take place well away from the spacecraft. To do so, a 100 kg payload is "lowered" toward the earth on a 10-km-long massless rope. (We'll overlook the details of how they do this and simply assume they can.) The payload can be hauled back on board the shuttle after the experiment. Assume that any initial motions associated with lowering the payload have damped out and that the shuttle and payload are flying in steady-state conditions.
    a. What is the angle of the rope as measured from a line drawn from the center of the earth through the payload? Explain.
    b. What is the tension in the rope?
75. Your job with NASA is to monitor satellite orbits. One day, during a routine survey, you find that a 400 kg satellite in a 1000-km-high circular orbit is going to collide with a smaller 100 kg satellite traveling in the same orbit but in the opposite direction. Knowing the construction of the two satellites, you expect they will become enmeshed into a single piece of space debris. When you notify your boss of this impending collision, he asks you to quickly determine whether the space debris will continue to orbit or crash into the earth. What will the outcome be?
76. Let's look in more detail at how a satellite is moved from one circular orbit to another. Figure CP12.76 shows two circular orbits, of radii r_1 and r_2, and an elliptical orbit that connects them. Points 1 and 2 are at the ends of the semimajor axis of the ellipse.
    a. A satellite moving along the elliptical orbit has to satisfy two conservation laws. Use these two laws to prove that the velocities at points 1 and 2 are

    $$v_1' = \sqrt{\frac{2GM(r_2/r_1)}{r_1 + r_2}} \text{ and } v_2' = \sqrt{\frac{2GM(r_1/r_2)}{r_1 + r_2}}$$

    The prime indicates that these are the velocities on the elliptical orbit. Both reduce to Equation 12.22 if $r_1 = r_2 = r$.
    b. Consider a 1000 kg communication satellite that needs to be boosted from an orbit 300 km above the earth to a geosynchronous orbit 35,900 km above the earth. Find the velocity v_1 on the lower circular orbit and the velocity v_1' at the low point on the elliptical orbit that spans the two circular orbits.
    c. How much work must the rocket motor do to transfer the satellite from the circular orbit to the elliptical orbit?
    d. Now find the velocity v_2' at the top of the elliptical orbit and the velocity v_2 of the upper circular orbit.
    e. How much work must the rocket motor do to transfer the satellite from the elliptical orbit to the upper circular orbit?
    f. Compute the total work done and compare your answer to the result of Example 12.6.

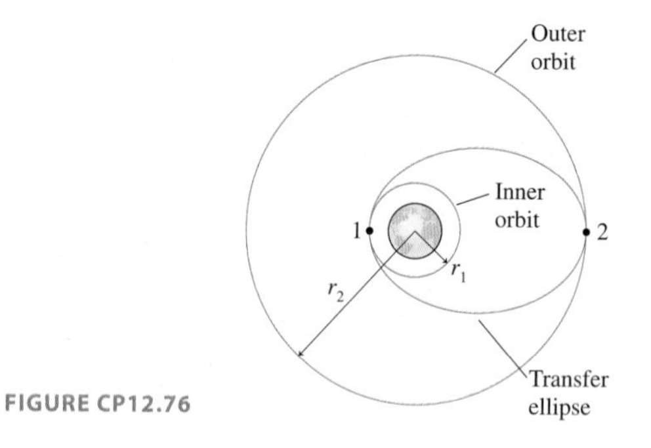

FIGURE CP12.76

STOP TO THINK ANSWERS

Stop to Think 12.1: e. The acceleration decreases inversely with the square of the distance. At height R_e, the distance from the center of the earth is $2R_e$.

Stop to Think 12.2: c. Newton's third law requires $F_{1 \text{ on } 2} = F_{2 \text{ on } 1}$.

Stop to Think 12.3: b. $g_{surface} = GM/R^2$. Because of the square, a radius twice as large balances a mass four times as large.

Stop to Think 12.4: In absolute value, $U_e > U_a = U_b = U_d > U_c$. $|U_g|$ is proportional to $m_1 m_2/r$.

Stop to Think 12.5: a. T^2 is proportional to r^3, or T is proportional to $r^{3/2}$. $4^{3/2} = 8$.

13 Rotation of a Rigid Body

Not all motion can be described as that of a particle. Rotation requires the idea of an extended object.

This diver is moving toward the water along a parabolic trajectory, much like a cannon ball. At the same time, she's rotating rapidly around her center of mass. This combination of two types of motion is what makes a great dive both interesting to watch and difficult to perform.

▶ Looking Ahead

The goal of Chapter 13 is to understand the physics of rotating objects. In this chapter you will learn to:

- Apply the rigid-body model to extended objects.
- Calculate torques and moments of inertia.
- Understand the rotation of a rigid body around a fixed axis.
- Understand rolling motion.
- Apply conservation of energy and angular momentum to rotational problems.
- Use vector mathematics to describe rotational motion.

◀ Looking Back

Rotational motion will revisit many of the major themes introduced in Parts I and II, especially the properties of circular motion. Please review:

- Section 4.5 Newton's second law.
- Sections 7.1 and 7.2 The mathematics of circular motion.
- Section 9.6 Angular momentum.
- Section 10.2 Kinetic and gravitational potential energy.

Our goal in this chapter is to understand rotational motion. We will focus our attention on what are called *rigid bodies*. Wheels, axles, and gyroscopes are examples of rigid bodies that rotate. Divers, gymnasts, and ice skaters also rotate, although the fact that they are *not* rigid bodies makes their motions more complex. Even so, we will be able to understand many aspects of their motion by modeling them as rigid bodies.

You will quickly discover that the physics of rotational motion is analogous to the physics of linear motion that you studied in Parts I and II. For example, the new concepts of torque and angular acceleration are the rotational analogs of force and acceleration, and we'll find a new version of Newton's second law that is the rotational equivalent of $\vec{F} = m\vec{a}$. Similarly, energy and conservation laws will continue to be important tools. Rotational motion is an important application of Newtonian mechanics to extended objects.

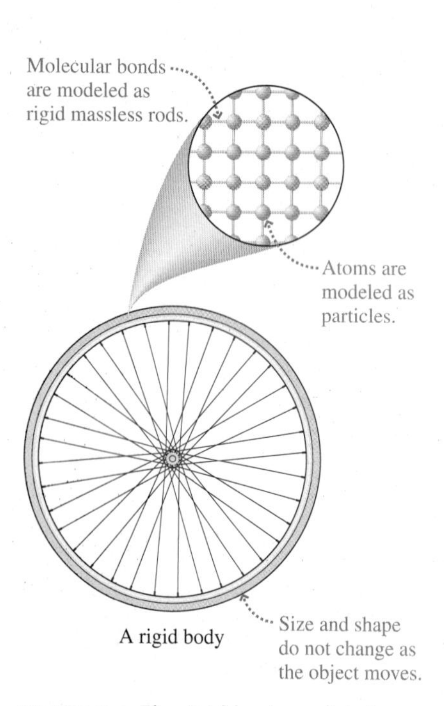
Molecular bonds are modeled as rigid massless rods.
Atoms are modeled as particles.
A rigid body
Size and shape do not change as the object moves.

FIGURE 13.1 The rigid-body model of an extended object.

13.1 Rotational Kinematics

Thus far, our study of physics has focused almost exclusively on the *particle model* in which an object is represented as a mass at a single point in space. The particle model is a perfectly good description of the physics in a vast number of situations, but there are other situations for which we need to consider the motion of an *extended object*—a system of particles for which the size and shape *do* make a difference and cannot be neglected.

A **rigid body** is an extended object whose size and shape do not change as it moves. For example, a bicycle wheel can be thought of as a rigid body. Figure 13.1 shows a rigid body as a collection of atoms held together by the rigid "massless rods" of molecular bonds.

Real molecular bonds are, of course, not perfectly rigid. That's why an object seemingly as rigid as a bicycle wheel can flex and bend. Thus Figure 13.1 is really a simplified *model* of an extended object, the **rigid-body model.** The rigid-body model is a very good approximation of many real objects of practical interest, such as wheels and axles. Even nonrigid objects can often be modeled as a rigid body during parts of their motion. For example, the diver in the opening photograph is well described as a rotating rigid body while she's in the tuck position.

Figure 13.2 illustrates the three basic types of motion of a rigid body: **translational motion, rotational motion,** and **combination motion.**

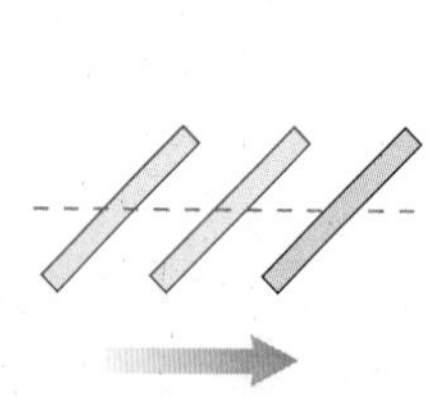

Translational motion: The object as a whole moves along a trajectory but does not rotate.

Rotational motion: The object rotates about a fixed point. Every point on the object moves in a circle.

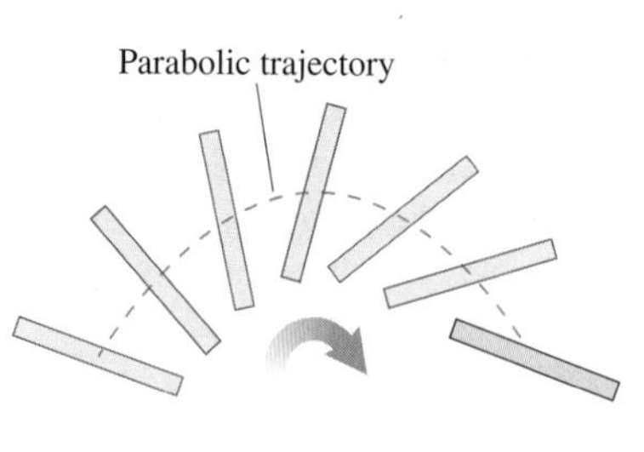
Parabolic trajectory

Combination motion: An object rotates as it moves along a trajectory.

FIGURE 13.2 Three basic types of motion of a rigid body.

Circular Motion

Rotation is an extension of circular motion, so we begin with a brief summary of some of the main results of Chapter 7. Figure 13.3 shows a particle of mass m rotating in a circle of radius r. Angle θ is the particle's angular position. The angular velocity

$$\omega = \frac{d\theta}{dt} \tag{13.1}$$

is the rate at which the particle moves around the circle. The SI units of ω are radians per second (rad/s). Revolutions per second (s^{-1}) and revolutions per minute (rpm) are frequently used units, but these need to be converted to rad/s for most calculations.

NOTE ▶ The sign convention is that ω is positive for counterclockwise (ccw) rotation, negative for clockwise (cw) rotation. ◀

A vector associated with the particle can be described in terms of its radial component, toward or away from the center, and its tangential component. Figure 13.3 reminds you of the radial and tangential components of $\vec{a}$ and $\vec{v}$ for circular motion.

You learned in Chapter 7 that the velocity, acceleration, and angular velocity of circular motion are related by

$$
\begin{aligned}
v_r &= 0 \qquad & a_r &= \frac{v_t^2}{r} = \omega^2 r \\
v_t &= \frac{ds}{dt} = r\omega \qquad & a_t &= \frac{dv_t}{dt}
\end{aligned}
\tag{13.2}
$$

Here s is the arc length around the circle ($s = r\theta$). The sign convention for ω implies that v_t and a_t are positive if they point in the ccw direction, negative if they point in the cw direction.

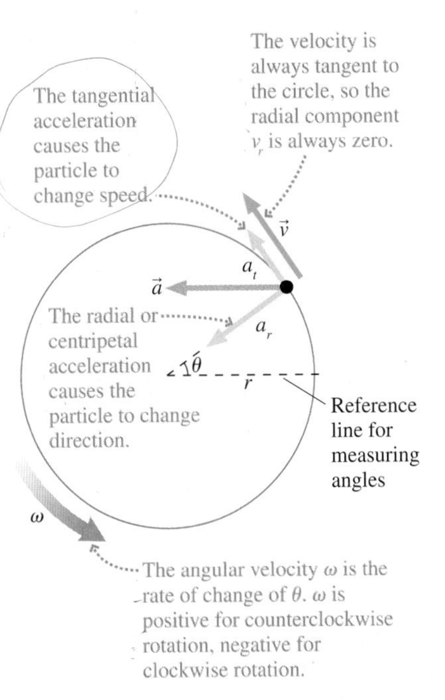

The tangential acceleration causes the particle to change speed.
The velocity is always tangent to the circle, so the radial component v_r is always zero.
$\vec{v}$
a_t
$\vec{a}$
a_r
The radial or centripetal acceleration causes the particle to change direction.
θ
r
Reference line for measuring angles
ω
The angular velocity ω is the rate of change of θ. ω is positive for counterclockwise rotation, negative for clockwise rotation.

FIGURE 13.3 A particle in circular motion.

Angular Velocity and Angular Acceleration

Figure 13.4 shows a wheel rotating on an axle. Notice that two points on the wheel, marked with dots, turn through the *same angle* as the wheel rotates, even through their radii may be different. That is, $\Delta\theta_1 = \Delta\theta_2$ during some time interval Δt. As a consequence, the two points have equal angular velocities: $\omega_1 = \omega_2$. Thus we can refer to the angular velocity ω *of the wheel*.

NOTE ▶ Two points in a rotating object have different tangential velocities v_t if they have different distances from the point of rotation, but *all* points have the *same* angular velocity ω. Thus angular velocity is one of the most important parameters of a rotating object. ◀

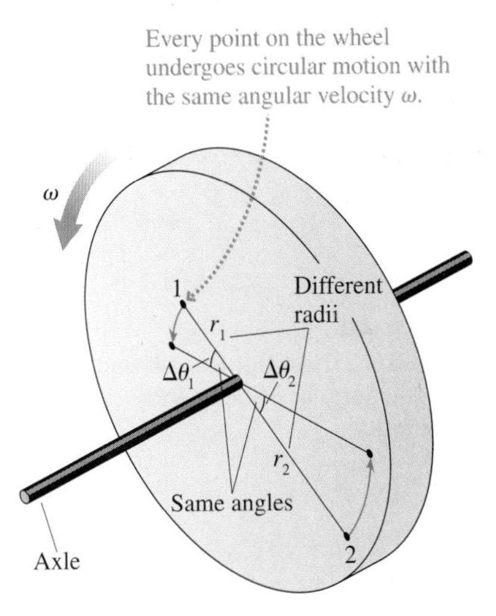

Every point on the wheel undergoes circular motion with the same angular velocity ω.
ω
1
Different radii
r_1
$\Delta\theta_1$
$\Delta\theta_2$
r_2
Same angles
Axle
2

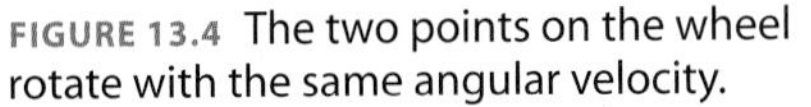
FIGURE 13.4 The two points on the wheel rotate with the same angular velocity.

We will want to consider situations in which an object's rotation speeds up or slows down; that is, situations in which the points on the object have a tangential acceleration a_t. You saw in Equation 13.2 that $a_t = dv_t/dt$ and that $v_t = r\omega$. Combining these two equations, we find

$$a_t = r\frac{d\omega}{dt} \tag{13.3}$$

In taking the derivative, we used the fact that r is a constant for circular motion.

We originally defined acceleration as $a = dv/dt$, the rate of change of the velocity. The derivative in Equation 13.3 is the rate of change of the *angular* velocity. By analogy, let's define the **angular acceleration** α (Greek alpha) to be

$$\alpha \equiv \frac{d\omega}{dt} \tag{13.4}$$

The units of angular acceleration are rad/s^2.

Angular acceleration is the *rate* at which the angular velocity ω changes, just as the linear acceleration is the rate at which the linear velocity v changes. If a wheel starts from rest with $\alpha = 2$ rad/s^2, its angular velocity will increase to 2 rad/s at $t = 1$ s, increase another 2 rad/s to 4 rad/s at $t = 2$ s, increase another 2 rad/s to 6 rad/s at $t = 3$ s, and so on. That is, the angular velocity increases by 2 rad/s per second.

NOTE ▶ Be careful with the sign of α. You learned in Chapter 2 that positive and negative values of the acceleration can't be interpreted as simply "speeding up" and "slowing down." ◀

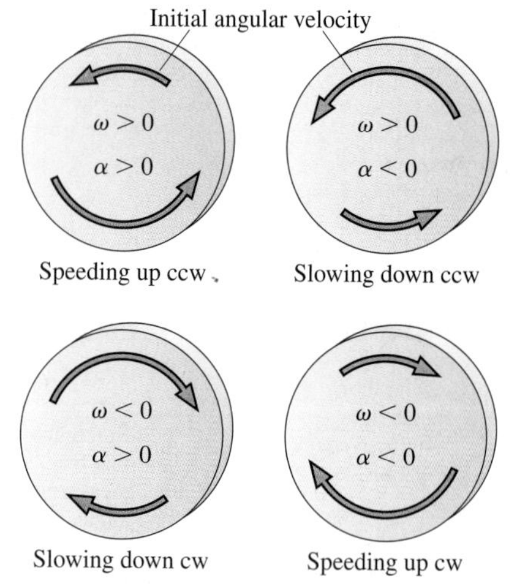

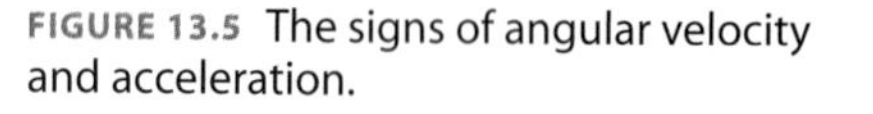

FIGURE 13.5 The signs of angular velocity and acceleration.

Because $\vec{a}$ is a vector, positive a_x means that v_x is increasing to the right or decreasing to the left. Negative a_x means that v_x is increasing to the left or decreasing to the right. For rotational motion, α is positive if ω is increasing ccw or decreasing cw, negative if ω is increasing cw or decreasing ccw. These are illustrated in Figure 13.5.

Comparing Equations 13.3 and 13.4, we see that the tangential and angular accelerations are related by

$$a_t = r\alpha \tag{13.5}$$

Two points on a rotating object have the *same* angular acceleration α, but in general they have *different* tangential accelerations because they are moving in circles of different radii. Notice the analogy between Equation 13.5 and the similar equation $v_t = r\omega$ for tangential and angular velocity.

Graphically, the angular acceleration α is the slope of the ω-versus-t graph. We can integrate Equation 13.4 to find

$$\omega_f = \omega_i + \text{area under the angular acceleration } \alpha \text{ curve between } t_i \text{ and } t_f \tag{13.6}$$

These relationships involving slopes and areas, illustrated in Figure 13.6, are exactly the same for rotational motion as they were for linear motion.

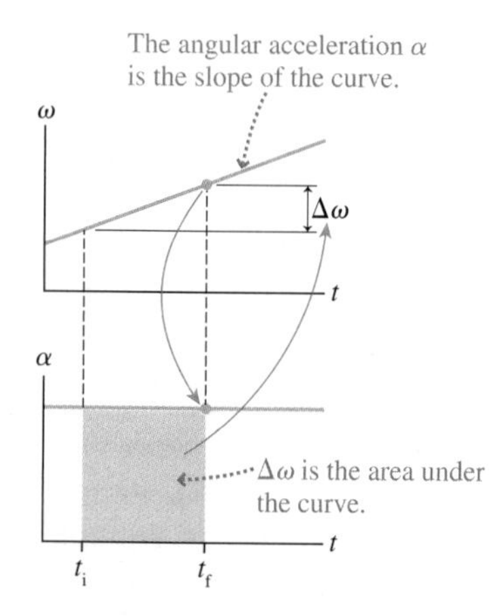

FIGURE 13.6 The graphical relationships between angular velocity and acceleration.

EXAMPLE 13.1 A rotating wheel

Figure 13.7a is a graph of angular velocity versus time for a rotating wheel. Describe the motion and draw a graph of angular acceleration versus time.

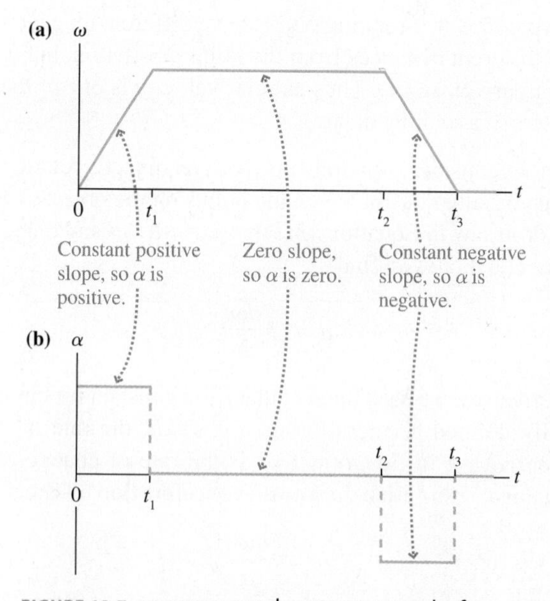

FIGURE 13.7 ω-versus-t and α-versus-t graphs for a rotating wheel.

MODEL The wheel is a rotating rigid body.

SOLVE This is a wheel that starts from rest, gradually speeds up *counterclockwise* until reaching top speed at t_1, maintains a constant angular velocity until t_2, then gradually slows down until stopping at t_3. The motion is always ccw because ω is always positive. The angular acceleration graph of Figure 13.7b is based on the fact that α is the slope of the ω-versus-t graph.

In Chapter 7 we developed kinematic equations for θ and ω. We can now replace the a_t/r that appeared in those equations with the angular acceleration α. Table 13.1 shows the resulting kinematic equations for constant angular acceleration. These equations apply to a particle in circular motion or to any rigid-body rotation. Notice that the rotational kinematic equations are exactly analogous to the linear kinematic equations.

7.7

TABLE 13.1 Rotational and linear kinematics for constant acceleration

Rotational kinematics	Linear kinematics
$\omega_f = \omega_i + \alpha\Delta t$	$v_f = v_i + a\Delta t$
$\theta_f = \theta_i + \omega_i\Delta t + \frac{1}{2}\alpha(\Delta t)^2$	$x_f = x_i + v_i\Delta t + \frac{1}{2}a(\Delta t)^2$
$\omega_f^2 = \omega_i^2 + 2\alpha\Delta\theta$	$v_f^2 = v_i^2 + 2a\Delta x$

EXAMPLE 13.2 A rotating crankshaft

A car's tachometer records the rotation frequency of the crankshaft. A car engine is idling at 500 rpm. When the light turns green, the crankshaft rotation speeds up at a constant rate to 2500 rpm over an interval of 3.0 s. How many revolutions does the crankshaft make during these 3 s?

MODEL The crankshaft is a rotating rigid body with constant angular acceleration.

SOLVE Imagine painting a dot on the crankshaft. Let the dot be at $\theta_i = 0$ rad at $t = 0$ s. Three seconds later the dot will have turned to angle

$$\theta_f = \omega_i\Delta t + \frac{1}{2}\alpha(\Delta t)^2$$

where $\Delta t = 3.0$ s. We can find the angular acceleration from the initial and final angular velocities, but first they must be converted to SI units:

$$\omega_i = 500\frac{\text{rev}}{\text{min}} \times \frac{1\text{ min}}{60\text{ s}} \times \frac{2\pi\text{ rad}}{1\text{ rev}} = 52.4\text{ rad/s}$$

$$\omega_f = 2500\frac{\text{rev}}{\text{min}} = 5\omega_i = 262.0\text{ rad/s}$$

The angular acceleration α is

$$\alpha = \frac{\Delta\omega}{\Delta t} = \frac{(262.0\text{ rad/s} - 52.4\text{ rad/s})}{3.0\text{ s}}$$

$$= \frac{209.6\text{ rad/s}}{3.0\text{ s}} = 69.9\text{ rad/s}^2$$

During these 3.0 s, the dot turns through an angle

$$\Delta\theta = \omega_i\Delta t + \frac{1}{2}\alpha(\Delta t)^2$$

$$= (52.4\text{ rad/s})(3.0\text{ s}) + \frac{1}{2}(69.9\text{ rad/s}^2)(3.0\text{ s})^2 = 472\text{ rad}$$

Because $472/2\pi = 75$, the crankshaft completes 75 revolutions as it spins up to 2500 rpm.

ASSESS This problem is solved just like the linear kinematics problems you learned to solve in Chapter 2.

STOP TO THINK 13.1 The fan blade is slowing down. What are the signs of ω and α?

a. ω is positive and α is positive.
b. ω is positive and α is negative.
c. ω is negative and α is positive.
d. ω is negative and α is negative.

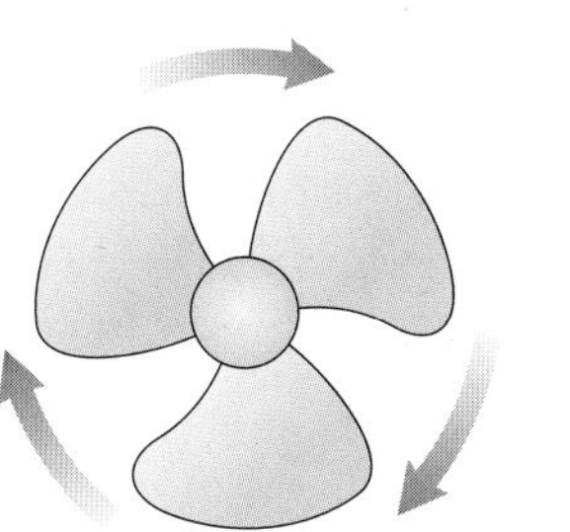

13.2 Rotation About the Center of Mass

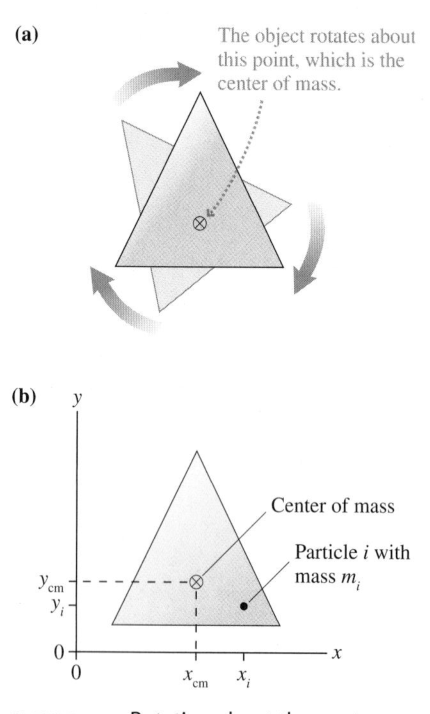

FIGURE 13.8 Rotation about the center of mass.

Imagine yourself floating in a space capsule deep in space. Suppose you take an object like that shown in Figure 13.8a, push two corners in opposite directions to spin it, then let go. The object will rotate, but it will have no translational motion as it floats beside you. *About what point does it rotate*? That is the question we need to answer.

An unconstrained object (i.e., one not on an axle or a pivot) on which there is no net force rotates about a point called the **center of mass.** The center of mass remains motionless while every other point in the object undergoes circular motion around it. You need not go deep into space to demonstrate rotation about the center of mass. If you have an air table, a flat object rotating on the air table rotates about its center of mass.

To locate the center of mass, Figure 13.8b models the object as if it were constructed from particles numbered $i = 1, 2, 3, \ldots$ Particle i has mass m_i and is located at position (x_i, y_i). We'll prove later in this section that the center of mass is located at position

$$x_{cm} = \frac{1}{M}\sum_i m_i x_i = \frac{m_1x_1 + m_2x_2 + m_3x_3 + \cdots}{m_1 + m_2 + m_3 + \cdots}$$
$$y_{cm} = \frac{1}{M}\sum_i m_i y_i = \frac{m_1y_1 + m_2y_2 + m_3y_3 + \cdots}{m_1 + m_2 + m_3 + \cdots} \qquad (13.7)$$

where $M = m_1 + m_2 + m_3 + \cdots$ is the object's total mass.

NOTE ► A three-dimensional object would need a similar equation for z_{cm}. For simplicity, we'll restrict ourselves to objects for which only the x- and y-coordinates are relevant. ◄

Let's see if Equations 13.7 make sense. Suppose you have an object consisting of N particles, all with the same mass m. That is, $m_1 = m_2 = \cdots = m_N = m$. We can factor the m out of the numerator and the denominator becomes simply Nm. The m cancels and x-coordinate of the center of mass is

$$x_{cm} = \frac{x_1 + x_2 + \cdots + x_N}{N} = x_{average}$$

In this case, x_{cm} is simply the *average* x-coordinate of all the particles. Likewise, y_{cm} will be the average of all the y-coordinates.

This *does* make sense! If the particle masses are all the same, the center of mass should be at the center of the object. And the "center of the object" is the average of the positions of all the particles. To allow for *unequal* masses, Equations 13.7 are called a *weighted average*. Particles of higher mass count more than particles of lower mass, but the basic idea remains the same. **The center of mass is the mass-weighted center of the object.**

EXAMPLE 13.3 The center of mass

A 500 g ball and a 2.0 kg ball are connected by a massless 50-cm-long rod.

a. Where is the center of mass?
b. What is the speed of each ball if they rotate about the center of mass at 40 rpm?

MODEL Model each ball as a particle.

VISUALIZE Figure 13.9 shows the two masses. We've chosen a coordinate system in which the masses are on the x-axis with the 2.0 kg mass at the origin.

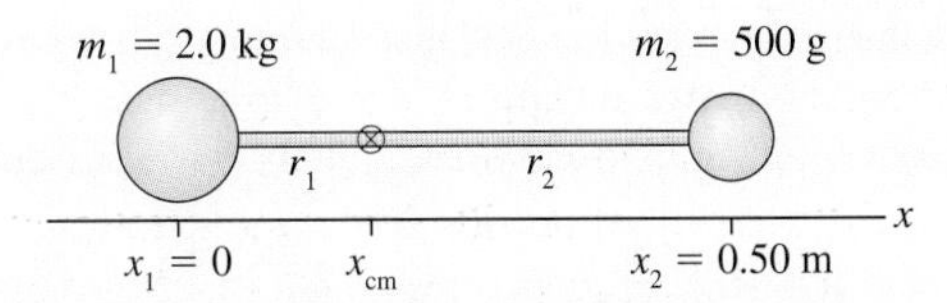

FIGURE 13.9 Finding the center of mass.

SOLVE

a. We can use Equation 13.7 to calculate that the center of mass is

$$x_{cm} = \frac{m_1x_1 + m_2x_2}{m_1 + m_2} = \frac{(2.0\text{ kg})(0.0\text{ m}) + (0.50\text{ kg})(0.50\text{ m})}{2.0\text{ kg} + 0.50\text{ kg}} = 0.10\text{ m}$$

$y_{cm} = 0$ because all the masses are on the x-axis. The center of mass is 20% of the way from the 2.0 g ball to the 0.50 kg ball.

b. Each ball rotates about the center of mass. The radii of the circles are $r_1 = 0.10$ m and $r_2 = 0.40$ m. The tangential velocities are $(v_i)_t = r_i\omega$, but this equation requires ω to be in rad/s. The conversion is

$$\omega = 40\frac{\text{rev}}{\text{min}} \times \frac{1\text{ min}}{60\text{ s}} \times \frac{2\pi\text{ rad}}{1\text{ rev}} = 4.19\text{ rad/s}$$

Consequently,

$$(v_1)_t = r_1\omega = (0.10\text{ m})(4.19\text{ rad/s}) = 0.419\text{ m/s}$$

$$(v_2)_t = r_2\omega = (0.40\text{ m})(4.19\text{ rad/s}) = 1.68\text{ m/s}$$

ASSESS The center of mass is closer to the heavier ball than to the lighter ball. We expected this because x_{cm} is a mass-weighted average of the positions.

For any realistic object, carrying out the summations of Equations 13.7 over all the atoms in the object is not practical. Instead, as Figure 13.10 shows, we can divide an extended object into many small cells or boxes, each with the very small mass Δm. We will number the cells 1, 2, 3, . . . , just as we did the particles. Cell i has coordinates (x_i, y_i) and mass $m_i = \Delta m$. The center-of-mass coordinates are then

$$x_{cm} = \frac{1}{M}\sum_i x_i\Delta m \quad \text{and} \quad y_{cm} = \frac{1}{M}\sum_i y_i\Delta m$$

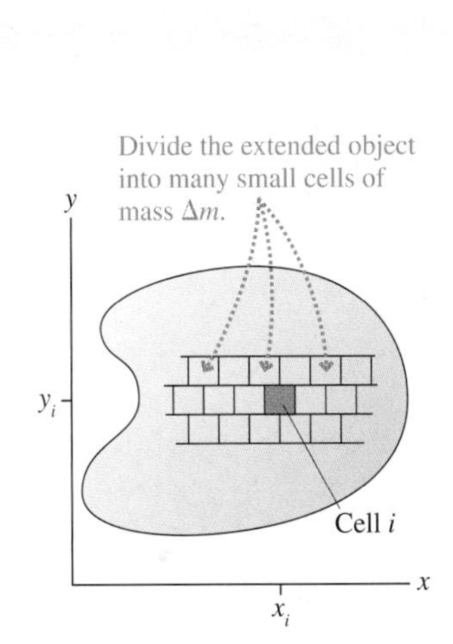

FIGURE 13.10 Calculating the center of mass of an extended object.

Now, as you might expect, we'll let the cells become smaller and smaller, with the total number increasing. As each cell becomes infinitesimally small, we can replace Δm with dm and the sum by an integral. Then

$$x_{cm} = \frac{1}{M}\int x\,dm \quad \text{and} \quad y_{cm} = \frac{1}{M}\int y\,dm \tag{13.8}$$

Equation 13.8 is a formal definition of the center of mass, but it is *not* ready to integrate in this form. First, integrals are carried out over *coordinates*, not over masses. Before we can integrate, we must replace dm by an equivalent expression involving a coordinate differential such as dx or dy. Second, no limits of integration have been specified. The procedure for using Equation 13.8 is best shown with an example.

EXAMPLE 13.4 The center of mass of a rod

Find the center of mass of a thin, uniform rod of length L and mass M. Use this result to find the tangential acceleration of one tip of a 1.60-m-long rod that rotates about its center of mass with an angular acceleration of 6.0 rad/s.

VISUALIZE Figure 13.11 shows the rod. We've chosen a coordinate system such that the rod lies along the x-axis from 0 to L. Because the rod is "thin," we'll assume that $y_{cm} = 0$.

SOLVE Our first task is to find x_{cm}, which lies somewhere on the x-axis. To do this, divide the rod into many small cells of

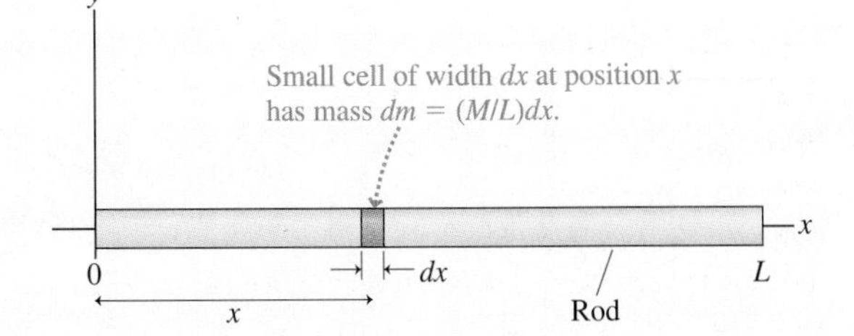

FIGURE 13.11 Finding the center of mass of a long, thin rod.

mass dm. One such cell, at position x, is shown. The cell's width is dx. Because the rod is *uniform*, the mass of this little cell is the *same fraction* of the total mass M that dx is of the total length L. That is,

$$\frac{dm}{M} = \frac{dx}{L}$$

Consequently, we can express dm in terms of the coordinate differential dx as

$$dm = \frac{M}{L}dx$$

NOTE ▶ The change of variables from dm to the differential of a coordinate is *the* key step in calculating the center of mass. ◀

With this expression for dm, Equation 13.8 for x_{cm} becomes

$$x_{cm} = \frac{1}{M}\left(\frac{M}{L}\int x\,dx\right) = \frac{1}{L}\int_0^L x\,dx$$

where in the last step we've noted that summing "all the mass in the rod" means integrating from $x = 0$ to $x = L$. This is a straightforward integral to carry out, giving

$$x_{cm} = \frac{1}{L}\left[\frac{x^2}{2}\right]_0^L = \frac{1}{L}\left[\frac{L^2}{2} - 0\right] = \frac{1}{2}L$$

The center of mass is at the center of the rod. For a 1.60-m-long rod, each tip of the rod rotates in a circle with $r = \frac{1}{2}L = 0.80$ m. The tangential acceleration, the rate at which the tip is speeding up, is

$$a_t = r\alpha = (0.80\text{ m})(6.0\text{ rad/s}^2) = 4.80\text{ m/s}^2$$

ASSESS You could have guessed that the center of mass is at the center of the rod, but now we've shown it rigorously.

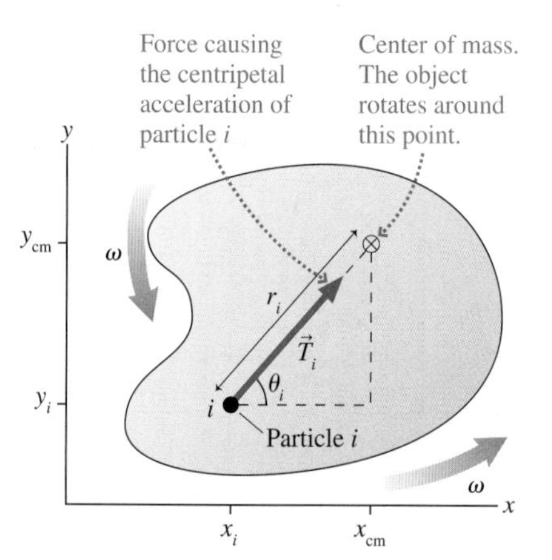

FIGURE 13.12 Finding the center of mass.

NOTE ▶ For any symmetrical object of uniform density, the center of mass is at the physical center of the object. ◀

To see where the center-of-mass equation comes from, Figure 13.12 shows an object rotating about its center of mass. Particle i is moving in a circle, so it *must* have a centripetal acceleration. Acceleration requires a force, and this force is due to tension in the molecular bonds that hold the object together. Force $\vec{T}_i$ on particle i has magnitude

$$T_i = m_i(a_i)_r = m_i r_i \omega^2 \tag{13.9}$$

where r_i is the distance of particle i from the center of mass and we used Equation 13.2 for a_r. All points in a rotating object have the *same* angular velocity, so ω doesn't need a subscript.

The internal tension forces are all paired as action/reaction forces, equal in magnitude but opposite in direction, so the sum of all the tension forces must be zero. That is, $\sum \vec{T}_i = \vec{0}$. The x-component of this sum is

$$\sum_i (T_i)_x = \sum_i T_i \cos\theta_i = \sum_i (m_i r_i \omega^2)\cos\theta_i = 0 \tag{13.10}$$

You can see from Figure 13.12 that $\cos\theta_i = (x_{cm} - x_i)/r_i$. Thus

$$\sum_i (T_i)_x = \sum_i (m_i r_i \omega^2)\frac{x_{cm} - x_i}{r_i} = \left(\sum_i m_i x_{cm} - \sum_i m_i x_i\right)\omega^2 = 0 \tag{13.11}$$

This equation will be true if the term in parentheses is zero. x_{cm} is a constant, so we can bring it outside the summation to write

$$\sum_i m_i x_{cm} - \sum_i m_i x_i = \left(\sum_i m_i\right)x_{cm} - \sum_i m_i x_i = Mx_{cm} - \sum_i m_i x_i = 0 \tag{13.12}$$

where we used the fact that $\sum m_i$ is simply the object's total mass M. Solving for x_{cm}, the x-coordinate of the object's center of mass is

$$x_{cm} = \frac{1}{M}\sum_i m_i x_i = \frac{m_1x_1 + m_2x_2 + m_3x_3 + \cdots}{m_1 + m_2 + m_3 + \cdots} \tag{13.13}$$

This was Equation 13.7. The y-equation is found similarly.

13.3 Torque

Torque is to rotational motion as force is to linear motion.

Newton's genius, summarized in his second law of motion, was to recognize force as the cause of acceleration. But what about *angular* acceleration? What do Newton's laws have to tell us about rotational motion? We need to find a rotational equivalent of force.

Consider the common experience of pushing open a door. Figure 13.13 is a top view of a door that is hinged on the left. Four pushing forces are shown, all of equal strength. Which of these will be most effective at opening the door?

Force $\vec{F}_1$ will open the door, but force $\vec{F}_2$, which pushes straight at the hinge, will not. Force $\vec{F}_3$ will open the door, but not as easily as $\vec{F}_1$. What about $\vec{F}_4$? It is perpendicular to the door, it has the same magnitude as $\vec{F}_1$, but you know from experience that pushing close to the hinge is not as effective as pushing at the outer edge of the door.

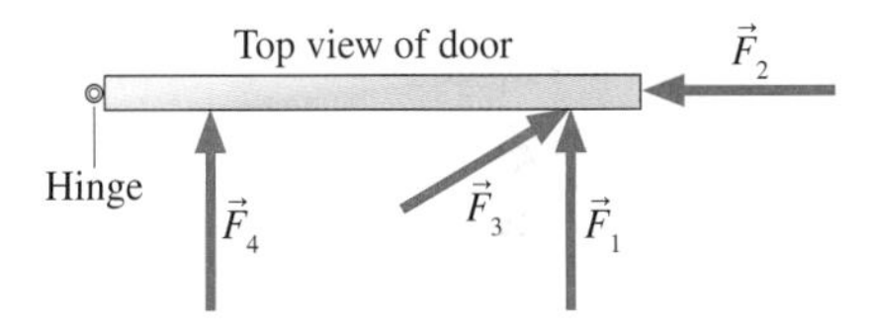

FIGURE 13.13 The four forces are the same strength, but they have different effects on the swinging door.

The ability of a force to cause a rotation or a twisting motion depends on three factors:

1. The magnitude F of the force.
2. The distance r from the point of application to the pivot.
3. The angle at which the force is applied.

To make these ideas specific, Figure 13.14 shows a force $\vec{F}$ applied at one point on a rigid body. For example, a string might be pulling on the object at that point, in which case the force would be a tension force. Figure 13.14 defines the distance r from the pivot to the point of application and the angle ϕ (Greek phi).

NOTE ▶ Angle ϕ is measured *counterclockwise* from the dotted line that extends outward along the radial line. This is consistent with our sign convention for the angular position θ. ◀

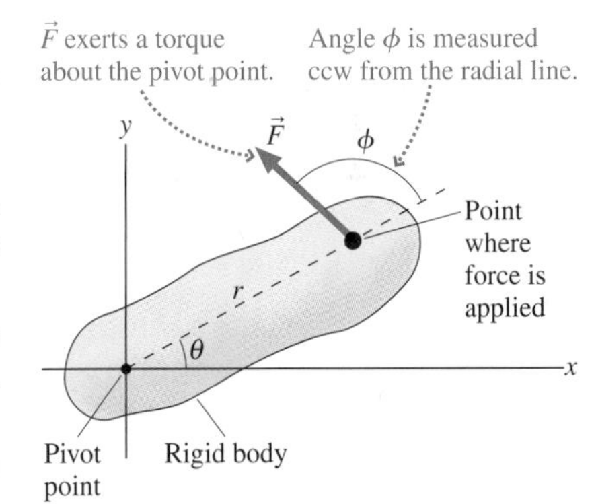

FIGURE 13.14 Force $\vec{F}$ exerts a torque about the pivot point.

Let's define a new quantity called the **torque** τ (Greek tau) as

$$\tau \equiv rF\sin\phi \qquad (13.14)$$

Torque depends on the three properties we just listed: the magnitude of the force, its distance from the pivot, and its angle. Loosely speaking, τ measures the "effectiveness" of the force at causing an object to rotate about a pivot. **Torque is the rotational equivalent of force.**

The SI units of torque are newton-meters, abbreviated N m. Although we defined 1 N m = 1 J during our study of energy, torque is not an energy-related quantity and so we do *not* use joules as a measure of torque.

Torque, like force, has a sign. A torque that rotates the object in a ccw direction is positive while a negative torque gives a cw rotation. Figure 13.15 summarizes the signs. Notice that a force pushing straight toward the pivot or pulling straight out from the pivot exerts *no* torque.

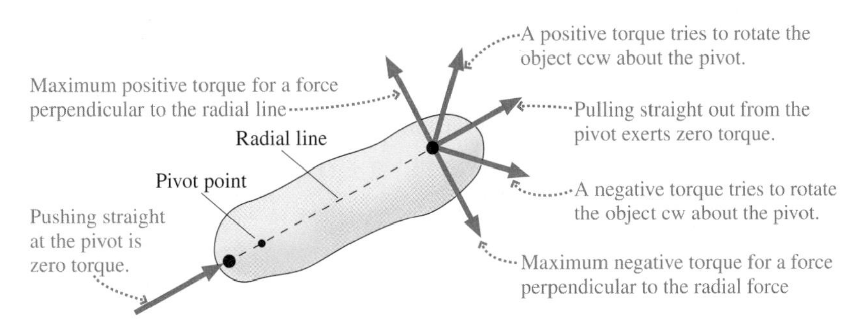

FIGURE 13.15 Signs and strengths of the torque.

Activ Online Physics 7.1

NOTE ▶ Torque differs from force in a very important way. Torque is calculated or measured *about a pivot point*. To say that a torque is 20 N m is meaningless. You need to say that the torque is 20 N m about a particular point. Torque can be calculated about any pivot point, but its value depends on the point chosen. In practice, we measure or calculate torques about the same point from which we measure an object's angular position θ (and thus its angular velocity ω and angular acceleration α). This assumption is built into the equations of rotational dynamics. ◀

Returning to the door of Figure 13.13, you can see that $\vec{F}_1$ is most effective at opening the door because $\vec{F}_1$ exerts the largest torque *about the pivot point*. $\vec{F}_3$ has equal magnitude, but it is applied at an angle less than 90° and thus exerts less torque. $\vec{F}_2$, pushing straight at the hinge with $\phi = 0°$, exerts no torque at all. And $\vec{F}_4$, with a smaller value for r, exerts less torque than $\vec{F}_1$.

(a)

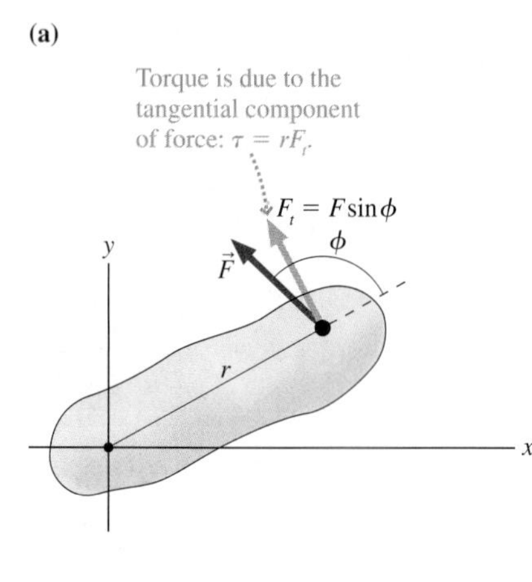

FIGURE 13.16 Two useful interpretations of the torque.

Interpreting Torque

Torque can be interpreted from two perspectives. First, Figure 13.16a shows that the quantity $F\sin\phi$ is the tangential force component F_t. Consequently, the torque can be written

$$\tau = rF_t \tag{13.15}$$

In other words, torque is the product of r with the force component F_t that is *perpendicular* to the radial line. This interpretation makes sense because the radial component of $\vec{F}$ points straight at the pivot point and cannot exert a torque.

Alternatively, Figure 13.16b shows that $d = r\sin\phi$ is the distance from the pivot to the **line of action,** the line along which force $\vec{F}$ acts. Thus the torque can also be written

$$|\tau| = dF \tag{13.16}$$

The distance d from the pivot to the line of action is called the **moment arm** (or the *lever arm*), so we can say that the torque is the product of the force with the moment arm. This second perspective on torque is widely used in applications.

NOTE ▶ Equation 13.16 gives only $|\tau|$, the magnitude of the torque; the sign has to be supplied by observing the direction in which the torque acts. ◀

EXAMPLE 13.5 Applying a torque

Luis uses a 20-cm-long wrench to turn a nut. The wrench handle is tilted 30° above the horizontal, and Luis pulls straight down on the end with a force of 100 N. How much torque does Luis exert on the nut?

VISUALIZE Figure 13.17 shows the situation. The angle is a negative $\phi = -120°$ because it is *clockwise* from the radial line.

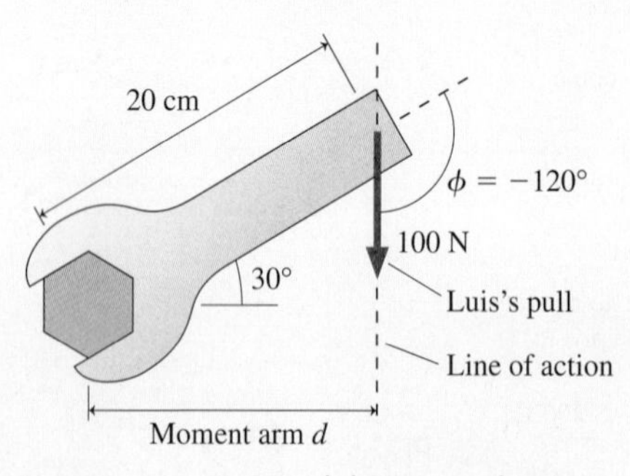

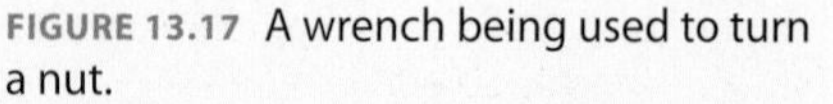

FIGURE 13.17 A wrench being used to turn a nut.

SOLVE The tangential component of the force is

$$F_t = F\sin\phi = -86.6\text{ N}$$

According to our sign convention, F_t is negative because it points in a cw direction. The torque, from Equation 13.15, is

$$\tau = rF_t = (0.20\text{ m})(-86.6\text{ N}) = -17.3\text{ N m}$$

The torque is negative because it tries to rotate the nut in a cw direction.

Alternatively, Figure 13.17 shows that the moment arm from the pivot to the line of action is

$$d = r\sin(60°) = 0.173\text{ m}$$

Inserting the moment arm in Equation 13.16 gives

$$|\tau| = dF = (0.173\text{ m})(100\text{ N}) = 17.3\text{ N m}$$

The torque acts to give a cw rotation, so we insert a minus sign to end up with

$$\tau = -17.3\text{ N m}$$

ASSESS Notice that Luis could increase the torque by changing the angle so that his pull is perpendicular to the wrench ($\phi = 90°$).

STOP TO THINK 13.2 Rank in order, from largest to smallest, the five torques τ_a to τ_e. The rods all have the same length and are pivoted at the dot.

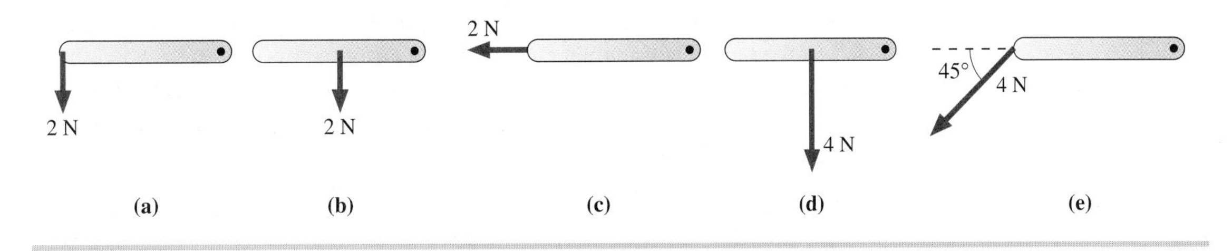

Net Torque

Figure 13.18 shows forces $\vec{F}_1, \vec{F}_2, \vec{F}_3, \ldots$ applied to an extended object. The object is free to rotate about the axle, but the axle prevents the object from having any translational motion. It does so by exerting force $\vec{F}_{axle}$ on the object to balance the other forces and keep $\vec{F}_{net} = \vec{0}$.

Forces $\vec{F}_1, \vec{F}_2, \vec{F}_3, \ldots$ exert torques $\tau_1, \tau_2, \tau_3, \ldots$ on the object, but $\vec{F}_{axle}$ does *not* exert a torque because it is applied at the pivot point and has zero moment arm. Thus the *net* torque about the axle is the sum of the torques due to the applied forces:

$$\tau_{net} = \tau_1 + \tau_2 + \tau_3 + \cdots = \sum_i \tau_i \tag{13.17}$$

In practice, usually only a small number of forces exert torques. For example, the only torque we considered in Example 13.5 was due to Luis's pull.

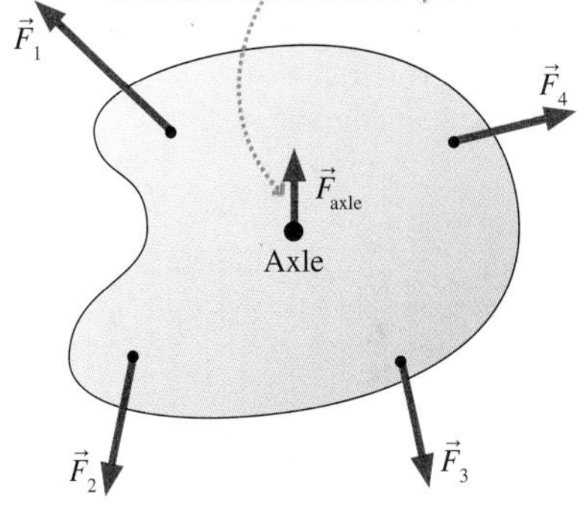

FIGURE 13.18 The forces exert a net torque about the pivot point.

Gravitational Torque

Gravity exerts a torque on many objects. If the object in Figure 13.19 is released, a torque due to the force of gravity will cause it to rotate around the axle. To calculate the torque about the axle, we start with the fact that gravity acts on *every* particle in the object, exerting a downward force of magnitude $F_i = m_i g$ on particle i. The *magnitude* of the gravitational torque on particle i is $|\tau_i| = d_i m_i g$, where d_i is the moment arm. But we need to be careful with signs.

A moment arm must be a positive number because it's a distance. If we establish a coordinate system with the origin at the axle, then you can see from Figure 13.19a that the moment arm d_i of particle i is $|x_i|$. A particle to the right of the axle (positive x_i) experiences a *negative* torque because gravity tries to rotate this particle in a clockwise direction. Similarly, a particle to the left of the axle (negative x_i) has a positive torque. The torque is opposite in sign to x_i, so we can get the sign of the torque right by writing

$$\tau_i = -x_i m_i g = -(m_i x_i)g \tag{13.18}$$

The net torque due to gravity is found by summing Equation 13.18 over all particles:

$$\tau_{grav} = \sum_i \tau_i = \sum_i (-m_i x_i g) = -\left(\sum_i m_i x_i\right) g \tag{13.19}$$

But according to the definition of center of mass, Equations 13.7, $\sum m_i x_i = M x_{cm}$. Thus the torque due to gravity is

$$\tau_{grav} = -Mgx_{cm} \tag{13.20}$$

where x_{cm} is the position of the center of mass *relative to the axis of rotation.*

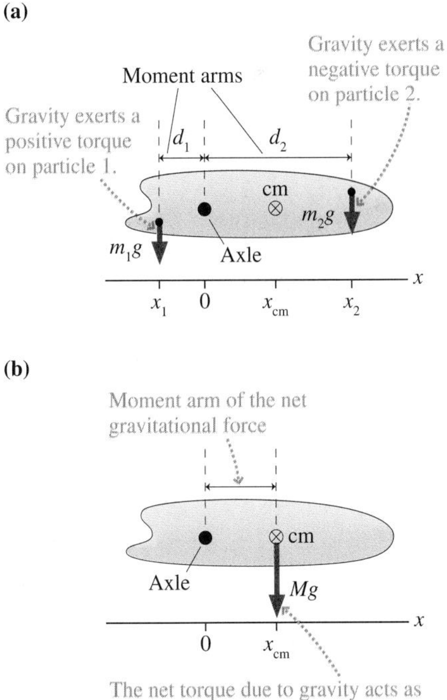

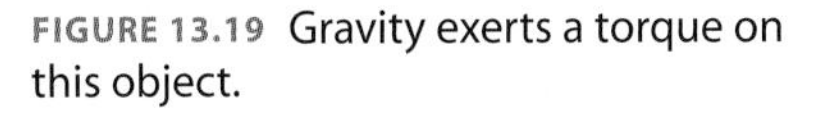

FIGURE 13.19 Gravity exerts a torque on this object.

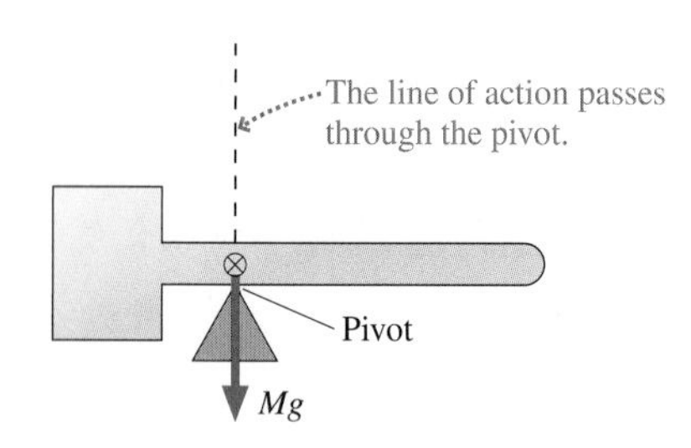

FIGURE 13.20 An object balances on a pivot that is directly under the center of mass.

Equation 13.20 has the simple interpretation shown in Figure 13.19b. Mg is the net gravitational force on the entire object, and x_{cm} is the moment arm between the rotation axis and the center of mass. The gravitational torque on an extended object of mass M is equivalent to the torque of a *single* force vector $\vec{F}_{grav} = -Mg\hat{j}$ acting at the object's center of mass.

In other words, **the gravitational torque is found by treating the object as if all its mass were concentrated at the center of mass.** This is the basis for the well-known technique of finding an object's center of mass by balancing it. An object will balance on a pivot, as shown in Figure 13.20, only if the center of mass is directly above the pivot point. If the pivot is *not* under the center of mass, the gravitational torque will cause the object to rotate.

EXAMPLE 13.6 The gravitational torque on a beam

The 4.00-m-long 500 kg steel beam shown in Figure 13.21 is supported 1.20 m from the right end. What is the gravitational torque about the support?

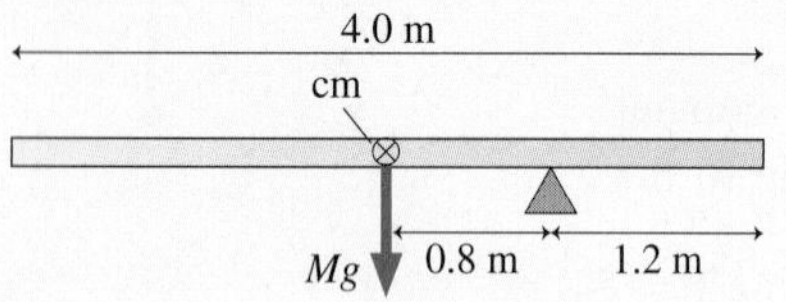

FIGURE 13.21 A steel beam supported at one point.

MODEL The center of mass of the beam is at the midpoint. $x_{cm} = -0.80$ m is measured from the pivot point.

SOLVE This is a straightforward application of Equation 13.20. The gravitational torque is

$$\tau_{grav} = -Mgx_{cm} = -(500\text{ kg})(9.80\text{ m/s}^2)(-0.80\text{ m})$$
$$= 3920\text{ Nm}$$

ASSESS The torque is positive because gravity tries to rotate the beam ccw around the point of support. Notice that the beam in Figure 13.21 is *not* in equilibrium. It will fall over unless other forces, not shown, are supporting it.

Couples

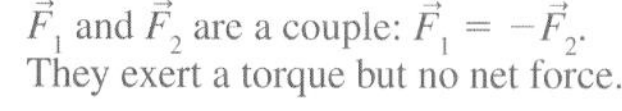

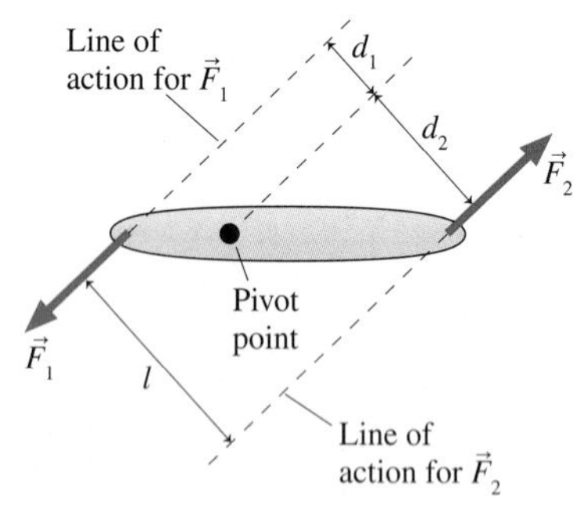

FIGURE 13.22 Two equal-but-opposite forces form a *couple*.

One way to rotate an extended object without translating it is to apply a pair of equal-but-opposite forces at two different points, as shown in Figure 13.22. This pair of forces is called a **couple.** The forces are in opposite directions, but the torques exerted by these forces act in the *same* direction. In this example, both torques are positive. Thus the two forces of a couple exert a *net torque*.

The forces are equal in magnitude, so we can write simply $F_1 = F_2 = F$. The moment arm of $\vec{F}_1$ is d_1, so the torque exerted by force $\vec{F}_1$ has magnitude $|\tau_1| = d_1F$. Similarly, force $\vec{F}_2$ exerts a torque of magnitude $|\tau_2| = d_2F$. The *net* torque is

$$|\tau_{net}| = d_1F + d_2F = (d_1 + d_2)F = lF \tag{13.21}$$

where l is the distance between the lines of action of the two forces. The sign of τ_{net} has to be supplied by observing the direction in which the torque acts.

There was nothing special about the point chosen as the pivot point in Figure 13.22. It could have been any point on the object. Hence a couple exerts the same net torque lF about *any* point on the object. This is not true of torques in general, just the torque exerted by a couple.

NOTE ► If an object is unconstrained (i.e., not on an axle), a couple causes the object to rotate about its center of mass. ◄

13.4 Rotational Dynamics

We've found that torque is the rotational equivalent of force. Now we need to learn what torque does. Figure 13.23 shows a model rocket engine attached to one end of a lightweight, rigid rod. When the engine is ignited, the thrust, which is *not* perpendicular to the rod, exerts force $\vec{F}_{\text{thrust}}$ on the rocket. The rod, which rotates around a pivot at the other end, exerts a tension force $\vec{T}$.

The tangential component of the thrust $F_t = F_{\text{thrust}} \sin\phi$ causes the rocket to speed up. In Chapter 7, we found that Newton's second law in the tangential direction is $F_t = ma_t$. This was perfectly adequate for single-particle motion, but our goal in this chapter is to understand the rotation of a rigid body. All points in a rotating object have the same angular acceleration so it will be useful to express Newton's second law in terms of angular acceleration rather than tangential acceleration.

The tangential and angular accelerations are related by $a_t = r\alpha$, allowing us to write Newton's second law as

$$F_t = ma_t = mr\alpha \tag{13.22}$$

Multiplying both sides by r gives us

$$rF_t = mr^2 a$$

But rF_t is the torque τ on the particle, hence Newton's second law for a *single particle* is

$$\tau = mr^2\alpha \tag{13.23}$$

At the beginning of Section 13.3 we asked the question, "What do Newton's laws have to tell us about rotational motion?" We now have the first part of an answer: **A torque causes an angular acceleration.** This is the rotational equivalent of our prior discovery, for motion along a line, that a force causes an acceleration. Now all that remains is to expand this idea from a single particle to an extended object.

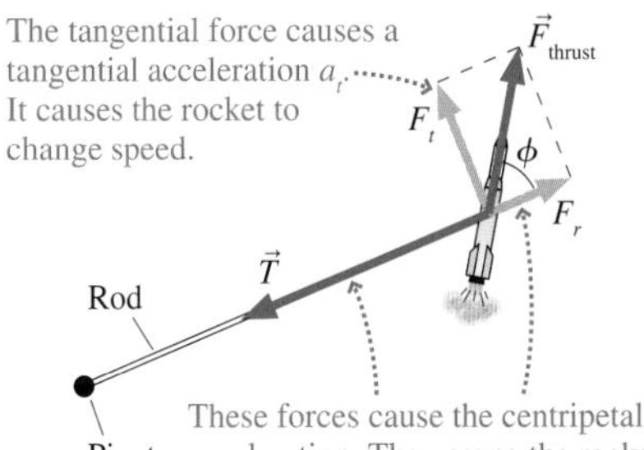
The tangential force causes a tangential acceleration a_t. It causes the rocket to change speed.
$\vec{F}_{\text{thrust}}$
F_t
ϕ
F_r
$\vec{T}$
Rod
Pivot point
These forces cause the centripetal acceleration. They cause the rocket to change direction, not speed.

FIGURE 13.23 The thrust force exerts a torque on the rocket and causes an angular acceleration.

Newton's Second Law

Figure 13.24 shows a rigid body that undergoes *pure rotational motion* about a fixed and unmoving axis. This might be an unconstrained rotation about the object's center of mass, such as we considered in Section 13.2. Or it might be an object, such as a pulley or a turbine, that is rotating on an axle. We'll assume that axles turn on frictionless bearings unless otherwise noted.

The forces $\vec{F}_1, \vec{F}_2, \vec{F}_3, \ldots$ in Figure 13.24 act on particles of masses $m_1, m_2, m_3 \ldots$ and exert torques $\tau_1, \tau_2, \tau_3, \ldots$ about the rotation axis. These torques cause the object to have an angular acceleration. The *net* torque on the object is the sum of the torques on all the individual particles. This is

$$\tau_{\text{net}} = \sum_i \tau_i = \sum_i (m_i r_i^2 \alpha) = \left(\sum_i m_i r_i^2\right)\alpha \tag{13.24}$$

where we used Equation 13.23, $\tau_i = m_i r_i^2 \alpha$, to relate the individual torques to the angular acceleration. We're making explicit use of the fact that every particle in a rotating rigid body has the *same* angular acceleration α.

The quantity $\Sigma m_i r_i^2$, which is the proportionality constant between angular acceleration and net torque, is called the object's **moment of inertia** I:

$$I = m_1 r_1^2 + m_2 r_2^2 + m_3 r_3^2 + \cdots = \sum_i m_i r_i^2 \tag{13.25}$$

The units of moment of inertia are kg m^2. An object's moment of inertia, like torque, *depends on the axis of rotation*. Once the axis is specified, allowing the values of r_i to be determined, the moment of inertia *about that axis* can be calculated from Equation 13.25.

NOTE ▶ The *moment* in *moment of inertia* and *moment arm* has nothing to do with time. These terms stem from the Latin *momentum*, meaning "motion." ◀

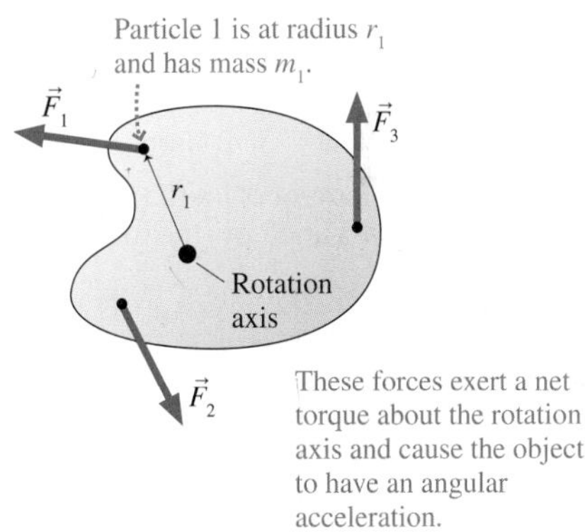
Particle 1 is at radius r_1 and has mass m_1.
$\vec{F}_1$
$\vec{F}_3$
r_1
Rotation axis
$\vec{F}_2$
These forces exert a net torque about the rotation axis and cause the object to have an angular acceleration.

FIGURE 13.24 The forces on a rigid body exert a torque about the rotation axis.

Substituting the moment of inertia into Equation 13.24 puts the final piece of the puzzle into place. An object that experiences a net torque τ_{net} about the axis of rotation undergoes an angular acceleration

$$\alpha = \frac{\tau_{net}}{I} \quad \text{(Newton's second law for rotational motion)} \qquad (13.26)$$

where I is the moment of inertia of the object *about the rotation axis*. This result, which is Newton's second law for rotation, is the fundamental equation of rigid-body dynamics.

In practice we often write $\tau_{net} = I\alpha$, but Equation 13.26 better conveys the idea that **torque is the cause of angular acceleration.** In the absence of a net torque ($\tau_{net} = 0$), the object either does not rotate ($\omega = 0$) or rotates with *constant* angular velocity ($\omega =$ constant).

Before rushing to calculate moments of inertia, let's get a better understanding of the meaning. First, notice that **moment of inertia is the rotational equivalent of mass.** It plays the same role in Equation 13.26 as mass m in the now-familiar $\vec{a} = \vec{F}_{net}/m$. Recall that the quantity we call *mass* was actually defined as the *inertial mass*. Objects with larger mass have a larger *inertia*, meaning that they're harder to accelerate. Similarly, an object with a larger moment of inertia is harder to rotate. Loosely speaking, it takes a larger torque to spin an object with a larger moment of inertia than an object with a smaller moment of inertia. The fact that *moment of inertia* retains the word *inertia* reminds us of this.

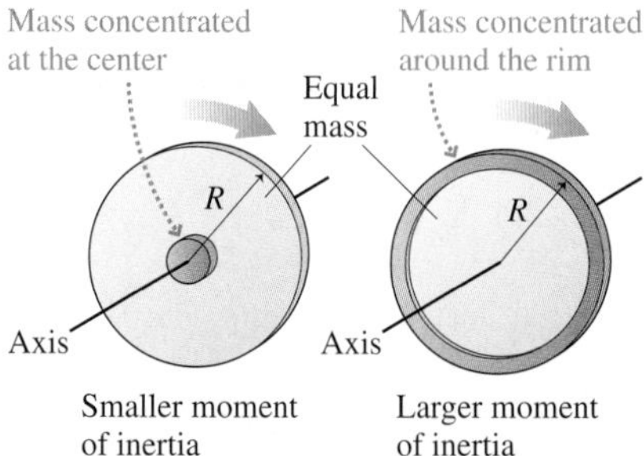

FIGURE 13.25 Moment of inertia depends both on the mass and how the mass is distributed.

But why does the moment of inertia depend on the distances r_i from the rotation axis? Think about the two wheels shown in Figure 13.25. They have the same total mass M and the same radius R. As you probably know from experience, it's much easier to spin the wheel whose mass is concentrated at the center than to spin the one whose mass is concentrated around the rim. This is because having the mass near the center (smaller values of r_i) lowers the moment of inertia.

Thus an object's moment of inertia depends not only on the object's mass but on *how the mass is distributed* around the rotation axis. This is well known to bicycle racers. Every time a cyclist accelerates, she has to "spin up" the wheels and tires. The larger the moment of inertia, the more effort it takes and the slower her acceleration. For this reason, racers use the lightest possible tires, and they put those tires on wheels that have been designed to keep the mass as close as possible to the center without sacrificing the necessary strength and rigidity.

Table 13.2 summarizes the analogies between linear and rotational dynamics.

TABLE 13.2 Rotational and linear dynamics

Rotational dynamics		Linear dynamics	
torque	τ_{net}	force	$\vec{F}_{net}$
moment of inertia	I	mass	m
angular acceleration	α	acceleration	$\vec{a}$
second law	$\alpha = \tau_{net}/I$	second law	$\vec{a} = \vec{F}_{net}/m$

EXAMPLE 13.7 Rotating rockets

Far out in space, a 100,000 kg rocket and a 200,000 kg rocket are docked at opposite ends of a motionless 90-m-long connecting tunnel. The tunnel is rigid and its mass is much less than that of either rocket. The rockets start their engines simultaneously, each generating 50,000 N of thrust in opposite directions. What is the structure's angular velocity after 30 s?

MODEL The entire structure can be modeled as two masses at the ends of a massless, rigid rod. The two thrust forces are a couple that exerts a torque on the structure. We'll assume the thrust forces are perpendicular to the connecting tunnel. This is an unconstrained rotation, so the structure will rotate about its center of mass.

VISUALIZE Figure 13.26 shows the rockets and defines distances r_1 and r_2 from the center of mass.

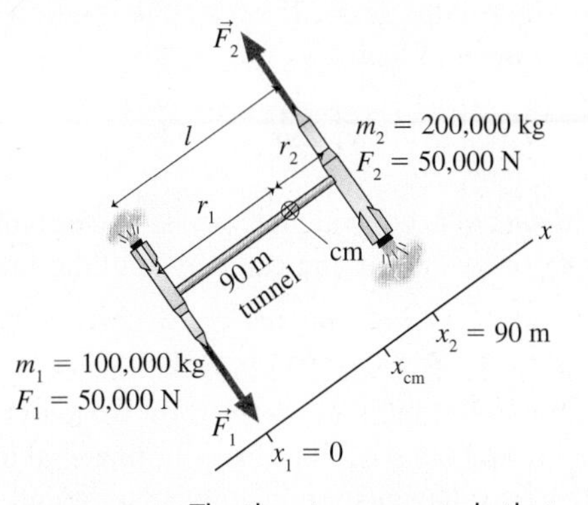

FIGURE 13.26 The thrusts are a couple that exerts a torque on the structure.

SOLVE Our strategy will be to use Newton's second law to find the angular acceleration, followed by rotational kinematics to find ω. We'll need to determine the moment of inertia, and that requires knowing the distances of the two rockets from the rotation axis. As we did in Example 13.3, choose a coordinate system in which the masses are on the x-axis and in which m_1 is at the origin. Then

$$x_{cm} = \frac{m_1 x_1 + m_2 x_2}{m_1 + m_2}$$

$$= \frac{(100{,}000\text{ kg})(0\text{ m}) + (200{,}000\text{ kg})(90\text{ m})}{100{,}000\text{ kg} + 200{,}000\text{ kg}} = 60\text{ m}$$

The structure's center of mass is $r_1 = 60$ m from the 100,000 kg rocket and $r_2 = 30$ m from the 200,000 kg rocket. The moment of inertia about the center of mass is

$$I = m_1 r_1^2 + m_2 r_2^2 = 540{,}000{,}000\text{ kg m}^2$$

The two rocket thrusts form a couple that exerts torque

$$\tau_{net} = lF = (90\text{ m})(50{,}000\text{ N}) = 4{,}500{,}000\text{ N m}$$

With I and τ_{net} now known, we can use Newton's second law to find the angular acceleration:

$$\alpha = \frac{\tau}{I} = \frac{4{,}500{,}000\text{ N m}}{540{,}000{,}000\text{ kg m}^2} = 0.00833\text{ rad/s}^2$$

After 30 seconds, the structure's angular velocity is

$$\omega = \alpha\Delta t = 0.250\text{ rad/s}$$

ASSESS Few of us have the experience to judge whether or not 0.25 rad/s is a reasonable answer to this problem. The significance of the example is to demonstrate the approach to a rotational dynamics problem.

STOP TO THINK 13.3 Rank in order, from largest to smallest, the angular accelerations α_a to α_e.

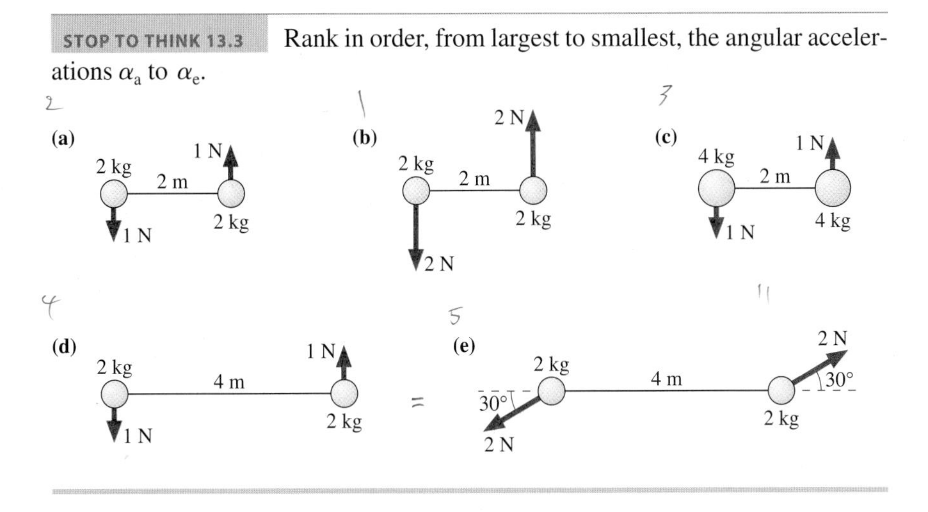

Moment of Inertia

Newton's second law for rotational motion is easy to write, but we can't make use of it without knowing an object's moment of inertia. Unlike mass, we can't measure moment of inertia by putting an object on a scale. And while we can guess that the center of mass of a symmetrical object is at the physical center of the object, we can *not* guess the moment of inertia of even a simple object. To find I, we really must carry through the calculation.

ActivPhysics ONLINE 7.6

Equation 13.25 defines the moment of inertia as a sum over all the particles in the system. As we did for the center of mass, we can replace the individual particles with cells 1, 2, 3, . . . of mass Δm. Then the moment of inertia summation can be converted to an integration:

$$I = \sum_i r_i^2 \Delta m \xrightarrow[\Delta m \to 0]{} I = \int r^2\, dm \tag{13.27}$$

where r is the distance from the rotation axis. If we let the rotation axis be the z-axis, then we can write the moment of inertia as

$$I = \int (x^2 + y^2)\, dm \tag{13.28}$$

NOTE ▶ You *must* replace dm by an equivalent expression involving a coordinate differential such as dx or dy before you can carry out the integration of Equation 13.28. ◀

You can use any coordinate system to calculate the coordinates x_{cm} and y_{cm} of the center of mass. But the moment of inertia is defined for rotation about a particular axis, and r is measured from that axis. Thus the coordinate system used for moment of inertia calculations *must* have its origin at the pivot point. Two examples will illustrate these ideas.

EXAMPLE 13.8 Moment of inertia of a rod about a pivot at one end

Find the moment of inertia of a thin, uniform rod of length L and mass M that rotates about a pivot at one end.

MODEL An object's moment of inertia depends on the axis of rotation. In this case, the rotation axis is at the end of the rod.

VISUALIZE Figure 13.27 defines an x-axis with the origin at the pivot point.

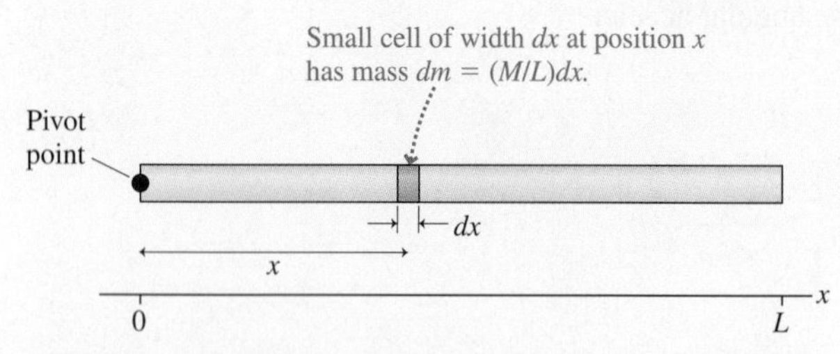
Small cell of width dx at position x
has mass dm = (M/L)dx.
Pivot
point
dx
x
x
0
L

FIGURE 13.27 Finding the moment of inertia about one end of a long, thin rod.

SOLVE Because the rod is thin, we can assume that $y \approx 0$ for all points on the rod. Thus

$$I = \int x^2\, dm$$

The small amount of mass dm in the small length dx is $dm = (M/L)\,dx$, as we found in Example 13.4. The rod extends from $x = 0$ to $x = L$, so the moment of inertia for a rod about one end is

$$I_{\text{end}} = \frac{M}{L}\int_0^L x^2\, dx = \frac{M}{L}\left[\frac{x^3}{3}\right]_0^L = \frac{1}{3}ML^2$$

ASSESS The moment of inertia involves a product of the total mass M with the *square* of a length, in this case L. All moments of inertia have a similar form, although the fraction in front will vary.

EXAMPLE 13.9 Moment of inertia of a circular disk about an axis through the center

Find the moment of inertia of a circular disk of radius R and mass M that rotates on an axis passing through its center.

VISUALIZE Figure 13.28 shows the disk and defines distance r from the axis.

SOLVE This is a situation of great practical importance. To solve this problem, we need to use a two-dimensional integration scheme that you learned in calculus. Rather than dividing the disk into little boxes, let's divide it into narrow *rings* of mass dm. Figure 13.28 shows one such ring, of radius r and width dr. Let dA represent the area of this ring. The mass dm in this ring is the same fraction of the total mass M as dA is of the total area A. That is,

$$\frac{dm}{M} = \frac{dA}{A}$$

Thus the mass in the small area dA is

$$dm = \frac{M}{A}dA$$

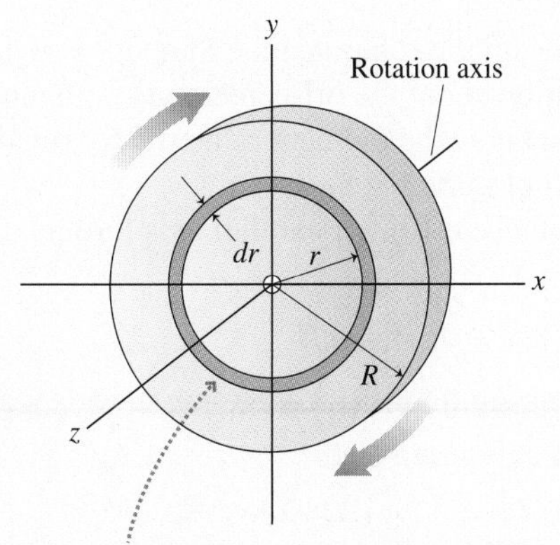

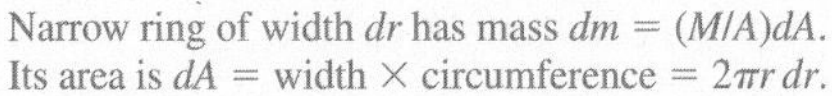

Narrow ring of width dr has mass $dm = (M/A)dA$.
Its area is dA = width × circumference = $2\pi r\,dr$.

FIGURE 13.28 Finding the moment of inertia of a disk about an axis through the center.

This is the reasoning we used to find the center of mass of the rod in Example 13.4, only now we're using it in two dimensions.

The total area of the disk is $A = \pi R^2$, but what is dA? If we imagine unrolling the little ring, it would form a long, thin rectangle of width $2\pi r$ and height dr. Thus the *area* of this little ring is $dA = 2\pi r dr$. With this information we can write

$$dm = \frac{M}{\pi R^2}(2\pi r dr) = \frac{2M}{R^2} r dr$$

Now we have an expression for dm in terms of a coordinate differential dr, so we can proceed to carry out the integration for I. Using Equation 13.27, we find

$$I_{\text{disk}} = \int r^2 dm = \int r^2\left(\frac{2M}{R^2} r dr\right) = \frac{2M}{R^2}\int_0^R r^3 dr$$

where in the last step we have used the fact that the disk extends from $r = 0$ to $r = R$. Performing the integration gives

$$I_{\text{disk}} = \frac{2M}{R^2}\left[\frac{r^4}{4}\right]_0^R = \frac{1}{2}MR^2$$

ASSESS Once again, the moment of inertia involves a product of the total mass M with the *square* of a length, in this case R.

Moments of inertia for most practical situations are tabulated and found in various science and engineering handbooks. You would need to compute I yourself only for an object of unusual shape. Table 13.3 is a short list of common moments of inertia.

TABLE 13.3 Moments of inertia of objects with uniform density

Object and axis	Picture	I	Object and axis	Picture	I
Thin rod, about center	L	$\frac{1}{12}ML^2$	Cylinder or disk, about center	R	$\frac{1}{2}MR^2$
Thin rod, about end	L	$\frac{1}{3}ML^2$	Cylindrical hoop, about center	R	MR^2
Plane or slab, about center	a, b	$\frac{1}{12}Ma^2$	Solid sphere, about diameter	R	$\frac{2}{5}MR^2$
Plane or slab, about edge	a, b	$\frac{1}{3}Ma^2$	Spherical shell, about diameter	R	$\frac{2}{3}MR^2$

The Parallel-Axis Theorem

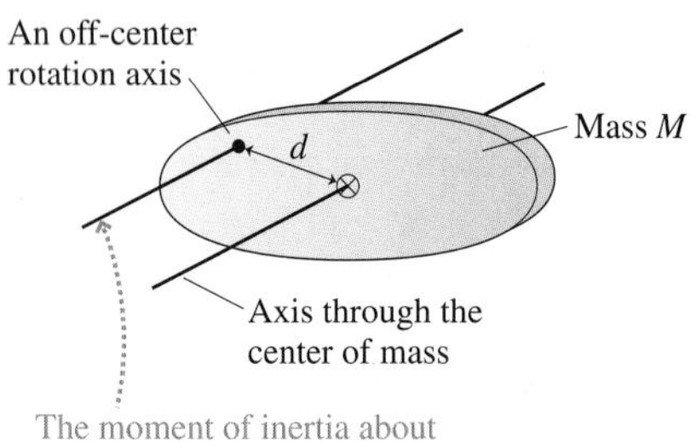
An off-center rotation axis
d
Mass M
Axis through the center of mass
The moment of inertia about this axis is $I = I_{cm} + Md^2$.

FIGURE 13.29 Rotation about an off-center axis.

The moment of inertia depends on the rotation axis. Suppose you need to know the moment of inertia for rotation about the off-center axis in Figure 13.29. You can find this quite easily if you know the moment of inertia for rotation around a *parallel axis* through the center of mass.

If the axis of interest is distance d from a parallel axis through the center of mass, the moment of inertia is

$$I = I_{cm} + Md^2 \tag{13.29}$$

Equation 13.29 is called the **parallel-axis theorem.** We'll give a proof for the one-dimensional object shown in Figure 13.30.

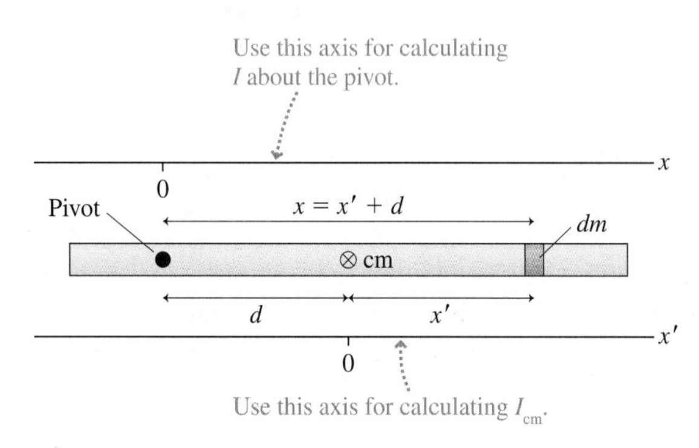
Use this axis for calculating I about the pivot.
x
0
Pivot
$x = x' + d$
dm
cm
d
x'
x'
0
Use this axis for calculating I_{cm}.

FIGURE 13.30 Proving the parallel-axis theorem.

The x-axis has its origin at the rotation axis and the x'-axis has its origin at the center of mass. You can see that the coordinates of dm along these two axes are related by $x = x' + d$. By definition, the moment of inertia about the rotation axis is

$$I = \int x^2 dm = \int (x' + d)^2 dm = \int (x')^2 dm + 2d\int x' dm + d^2 \int dm \tag{13.30}$$

The first of the three integrals on the right, by definition, is the moment of inertia I_{cm} about the center of mass. The third is simply Md^2 because adding up (integrating) all the dm gives the total mass M.

If you refer back to Equation 13.8, the definition of the center of mass, you'll see that the middle integral on the right is equal to Mx'_{cm}. But $x'_{cm} = 0$ because we specifically chose the x'-axis to have its origin at the center of mass. Thus the second integral is zero and we end up with Equation 13.29. The proof in two dimensions is similar.

EXAMPLE 13.10 The moment of inertia of a thin rod

Find the moment of inertia of a thin rod with mass M and length L about an axis one-third of the length from one end.

SOLVE From Table 13.3 we know the moment of inertia about the center of mass is $\frac{1}{12}ML^2$. The center of mass is at the center of the rod. An axis $\frac{1}{3}L$ from one end is $d = \frac{1}{6}L$ from the center of mass. Using the parallel-axis theorem,

$$I = I_{cm} + Md^2 = \frac{1}{12}ML^2 + M\left(\frac{1}{6}L\right)^2 = \frac{1}{9}ML^2$$

STOP TO THINK 13.4 Four Ts are made from two identical rods of equal mass and length. Rank in order, from largest to smallest, the moments of inertia I_a to I_d for rotation about the dotted line.

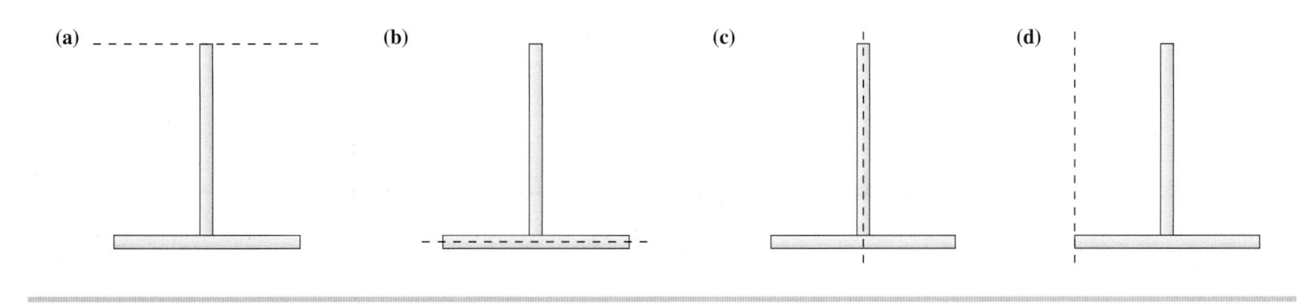

13.5 Rotation About a Fixed Axis

In this section we'll look at several examples of rotational dynamics for rigid bodies that rotate about a fixed axis. The restriction to a fixed axis avoids complications that arise for an object undergoing a combination of rotational and translational motion. The problem-solving strategy for rotational dynamics is very similar to that of linear dynamics.

ActivPhysics ONLINE 7.8–7.10

PROBLEM-SOLVING STRATEGY 13.1 Rotational dynamics problems

MODEL Model the object as a simple shape.

VISUALIZE Draw a pictorial representation to clarify the situation, define coordinates and symbols, and list known information.

- Identify the axis about which the object rotates.
- Identify forces and determine their distance from the axis. For most problems it will be useful to draw a free-body diagram.
- Identify any torques caused by the forces and the signs of the torques.

SOLVE The mathematical representation is based on Newton's second law for rotational motion

$$\tau_{net} = I\alpha \quad \text{or} \quad \alpha = \frac{\tau_{net}}{I}$$

- Find the moment of inertia in Table 13.3 or, if needed, calculate it as an integral or by using the parallel-axis theorem.
- Use rotational kinematics to find angles and angular velocities.

ASSESS Check that your result has the correct units, is reasonable, and answers the question.

EXAMPLE 13.11 Starting an airplane engine

The engine in a small airplane is specified to have a torque of 60 N m. This engine drives a 2.0-m-long, 40 kg propeller. On start-up, how long does it take the propeller to reach 200 rpm?

MODEL The propeller can be modeled as a rod that rotates about its center. The engine exerts a torque on the propeller.

VISUALIZE Figure 13.31 on the next page shows the propeller and the rotation axis.

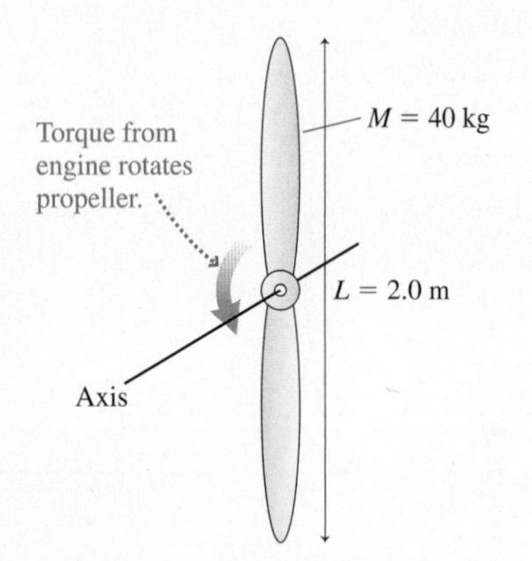

FIGURE 13.31 A rotating airplane propeller.

SOLVE The moment of inertia of a rod rotating about its center is found from Table 13.3:

$$I = \frac{1}{12}ML^2 = \frac{1}{12}(40\ \text{kg})(2.0\ \text{m})^2 = 13.33\ \text{kg m}^2$$

The 60 N m torque of the engine causes an angular acceleration

$$\alpha = \frac{\tau}{I} = \frac{60\ \text{N m}}{13.33\ \text{kg m}^2} = 4.50\ \text{rad/s}^2$$

The time needed to reach $\omega_f = 200$ rpm $= 3.33$ rev/s $=$ 20.9 rad/s is

$$\Delta t = \frac{\Delta\omega}{\alpha} = \frac{\omega_f - \omega_i}{\alpha} = \frac{20.9\ \text{rad/s} - 0\ \text{rad/s}}{4.5\ \text{rad/s}^2} = 4.6\ \text{s}$$

ASSESS We've assumed a constant angular acceleration, which is reasonable for the first few seconds while the propeller is still turning slowly. Eventually, air resistance and friction will cause opposing torques and the angular acceleration will decrease. At full speed, the negative torque due to air resistance and friction cancels the torque of the engine. Then $\tau_{net} = 0$ and the propeller turns at *constant* angular velocity with no angular acceleration.

EXAMPLE 13.12 An off-center disk

Figure 13.32 shows a piece of a large machine. A 10.0-cm-diameter, 5.0 kg disk turns on an axle. A vertical cable attached to the edge of the disk exerts a 100 N force but, initially, a pin keeps the disk from rotating. What is the initial angular acceleration of the disk when the pin is removed?

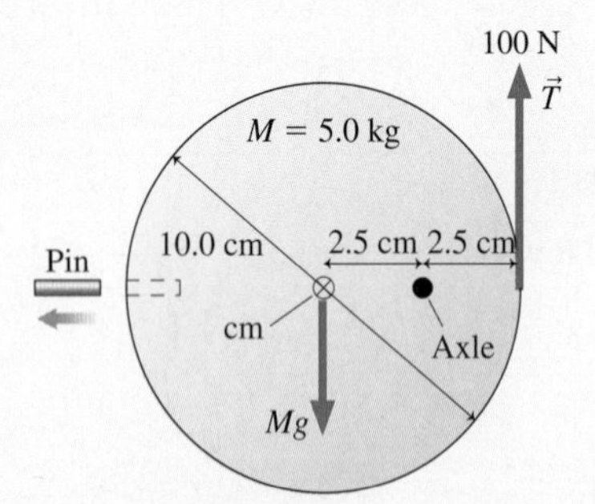

FIGURE 13.32 A disk rotates on an off-center axle after the pin is removed.

MODEL The disk has an off-center axle. Gravity exerts a torque about the axle.

VISUALIZE Both the cable and gravity rotate the disk ccw, so their torques are positive.

SOLVE After the pin is removed, the forces on the disk are a downward weight force, an upward force from the cable, and a force exerted by the axle. The axle force, which is exerted at the pivot, does not contribute to the torque and doesn't affect the rotation. The center of mass is to the *left* of the axle, at $x_{cm} = -\frac{1}{2}R$; thus the gravitational torque is

$$\tau_{grav} = -Mgx_{cm} = \frac{1}{2}MgR$$

This is a positive torque, as expected. The net torque, including the cable, is

$$\tau_{net} = \tau_{grav} + \tau_{cable} = \frac{1}{2}MgR + \frac{1}{2}RT = 3.73\ \text{N m}$$

Before finding the angular acceleration, we need to know the moment of inertia about the axle. This is where the parallel-axis theorem is useful. We know the moment of inertia about an axis through the center, from Table 13.3. The axle is offset by $d = \frac{1}{2}R$. Thus

$$I = I_{cm} + Md^2 = \frac{1}{2}MR^2 + M\left(\frac{1}{2}R\right)^2 = \frac{3}{4}MR^2$$
$$= 9.38 \times 10^{-3}\ \text{kg m}^2$$

The torque causes an angular acceleration

$$\alpha = \frac{\tau_{net}}{I} = \frac{3.73\ \text{N m}}{9.38 \times 10^{-3}\ \text{kg m}^2} = 397\ \text{rad/s}^2$$

The angular acceleration is positive, indicating that the disk begins rotating in a ccw direction.

ASSESS As the disk rotates, τ_{net} will change as the moment arms change. Consequently, the disk will *not* have constant angular acceleration. This is simply the *initial* value of α.

Constraints Due to Ropes and Pulleys

Many important applications of rotational dynamics involve objects, such as pulleys, that are connected via ropes or belts to other objects. Figure 13.33 shows a rope passing over a pulley and connected to an object in linear motion. If the rope turns on the pulley *without slipping*, then the rope's speed v_{rope} must exactly match the speed of the rim of the pulley, which is $v_{rim} = |\omega|R$. If the pulley has an angular acceleration, the rope's acceleration a_{rope} must match the *tangential* acceleration of the rim of the pulley, $a_t = |\alpha|R$.

The object attached to the other end of the rope has the same speed and acceleration as the rope. Consequently, an object connected to a pulley of radius R by a rope that does not slip must obey the constraints

$$\begin{aligned} v_{obj} &= |\omega|R \\ a_{obj} &= |\alpha|R \end{aligned} \quad \text{(motion constraints for a nonslipping rope)} \quad (13.31)$$

These constraints are very similar to the acceleration constraints introduced in Chapter 8 for two objects connected by a string or rope.

NOTE ▶ The constraints are given as magnitudes. Specific problems will need to introduce signs that depend on the direction of motion and on the choice of coordinate system. ◀

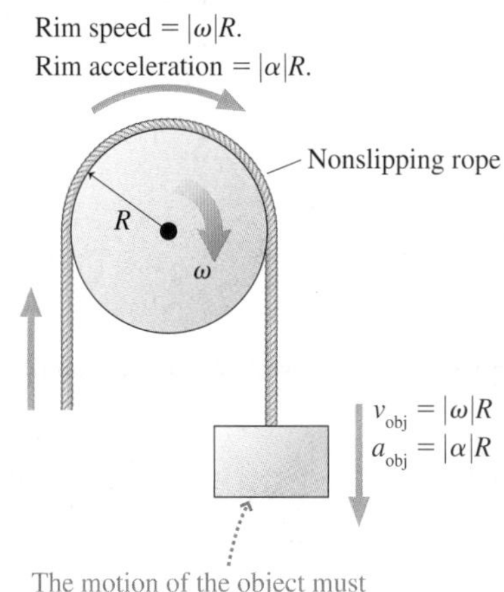

FIGURE 13.33 The rope's motion must match the motion of the rim of the pulley.

EXAMPLE 13.13 Lowering a block

A 2.0 kg block is attached to a massless string that is wrapped around a 1.0 kg, 4.0-cm-diameter cylinder, as shown in Figure 13.34a. The cylinder rotates on an axle through the center. The block is released from rest 1.0 m above the floor. How long does it take to reach the floor?

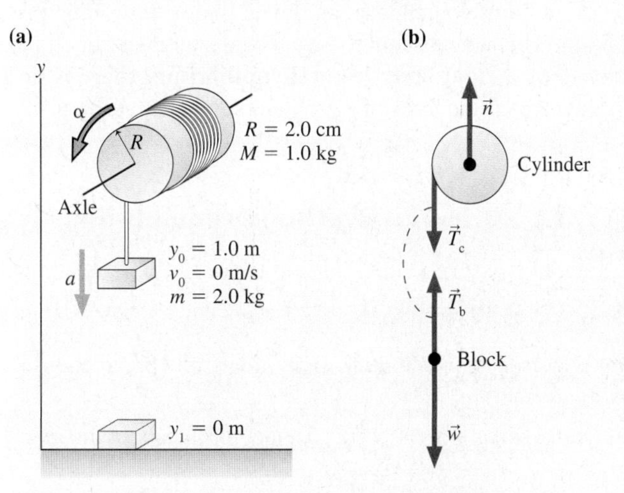

FIGURE 13.34 The falling block turns the cylinder.

MODEL Assume the string does not slip.

VISUALIZE Figure 13.34b shows the free-body diagram for the cylinder and the block. The string tension exerts an upward force on the block and a downward force on the outer edge of the cylinder. The string is massless, so these two tension forces act as if they are an action/reaction pair: $T_b = T_c = T$.

SOLVE Newton's second law applied to the linear motion of the block is

$$ma_y = T - mg$$

where, as usual, the y-axis points upward. What about the cylinder? There's no net force on it, because its center of mass is stationary, so the axle must exert an upward normal force $\vec{n}$ to balance $\vec{T}_c$. However, $\vec{n}$ does not exert a torque because it passes through the rotation axis. The only torque comes from the string tension. The moment arm for the tension is $d = R$, and the torque is positive because the string turns the cylinder ccw. Thus $\tau_{string} = TR$ and Newton's second law for the rotational motion is

$$\alpha = \frac{\tau_{net}}{I} = \frac{TR}{\frac{1}{2}MR^2} = \frac{2T}{MR}$$

The moment of inertia of a cylinder rotating about a center axis was taken from Table 13.3.

The last piece of information we need is the constraint due to the fact that the string doesn't slip. Equation 13.31 relates only the absolute values, but in this problem α is positive (ccw acceleration) while a_y is negative (downward acceleration). Hence

$$a_y = -\alpha R$$

Using α from the cylinder's equation in the constraint, we find

$$a_y = -\alpha R = -\frac{2T}{MR}R = -\frac{2T}{M}$$

Thus the tension is $T = -\frac{1}{2}Ma_y$. If we use this value of the tension in the block's equation, we can solve for the acceleration:

$$ma_y = -\frac{1}{2}Ma_y - mg$$

$$a_y = -\frac{g}{(1 + M/2m)} = -7.84 \text{ m/s}^2$$

The time to fall through $\Delta y = -1.0$ m is found from kinematics:

$$\Delta y = \frac{1}{2}a_y(\Delta t)^2$$

$$\Delta t = \sqrt{\frac{2\Delta y}{a_y}} = \sqrt{\frac{2(-1.0\text{ m})}{-7.84\text{ m/s}^2}} = 0.505\text{ s}$$

ASSESS The expression for the acceleration gives $a_y = -g$ if $M = 0$. This makes sense because the block would be in free fall if there were no cylinder. When the cylinder has mass, the downward force of gravity on the block has to accelerate the block *and* spin the cylinder. Consequently, the acceleration is reduced and the block takes longer to fall.

Structures such as bridges are analyzed in engineering statics.

13.6 Rigid-Body Equilibrium

We now have two versions of Newton's second law: $\vec{F}_{net} = M\vec{a}$ for translational motion and $\tau_{net} = I\alpha$ for rotational motion. Our goal in this short section is to look at how Newton's laws apply to objects in equilibrium. There are three cases:

- If $\vec{F}_{net} = \vec{0}$, then the object is in *translational equilibrium* and the center of mass is either at rest or moving in a straight line with constant speed. However, the object may be rotating.
- If $\tau_{net} = 0$, then the object is in *rotational equilibrium* and it is either not rotating or is rotating with a constant angular velocity. However, the object may have a translational motion.
- The condition for a rigid body to be in *total equilibrium* is both $\vec{F}_{net} = \vec{0}$ *and* $\tau_{net} = 0$. That is, no net force *and* no net torque.

A motionless rigid body is in total static equilibrium. An important branch of engineering called *statics* analyzes buildings, dams, bridges, and other structures in total static equilibrium.

No matter which pivot point you choose, an object that is not rotating is not rotating about that point. This would seem to be a trivial statement, but it has an important implication: **For a rigid body in total equilibrium, there is no net torque about any point.** This is the basis of a problem-solving strategy.

MP PROBLEM-SOLVING STRATEGY 13.2 Rigid-body equilibrium problems

MODEL Model the object as a simple shape.

VISUALIZE Draw a pictorial representation that shows all forces and distances. List known information.

- Pick any point you wish as a pivot point. The net torque about this point is zero.
- Determine the moment arms of all forces about this pivot point.
- Determine the sign of each torque about this pivot point.

SOLVE The mathematical representation is based on the fact that an object in total equilibrium has no net force and no net torque.

$$\vec{F}_{net} = \vec{0} \quad \text{and} \quad \tau_{net} = 0$$

- Write equations for $\Sigma F_x = 0$, $\Sigma F_y = 0$ and $\Sigma\tau = 0$.
- Solve the three simultaneous equations.

ASSESS Check that your result is reasonable and answers the question.

Although you can pick any point you wish as a pivot point, some choices make the problem easier than others. Often the best choice is a point at which several forces act because the torques exerted by those forces will be zero.

ActivPhysics ONLINE 7.2–7.5

EXAMPLE 13.14 Will the ladder slip?

A 3.0-m-long ladder leans against a frictionless wall at an angle of 60°. What is the minimum value of μ_s, the coefficient of static friction with the ground, that prevents the ladder from slipping?

MODEL The ladder is a rigid rod of length L. To not slip, it must be in both translational equilibrium ($\vec{F}_{net} = \vec{0}$) and rotational equilibrium ($\tau_{net} = 0$).

VISUALIZE Figure 13.35 shows the ladder and the forces acting on it.

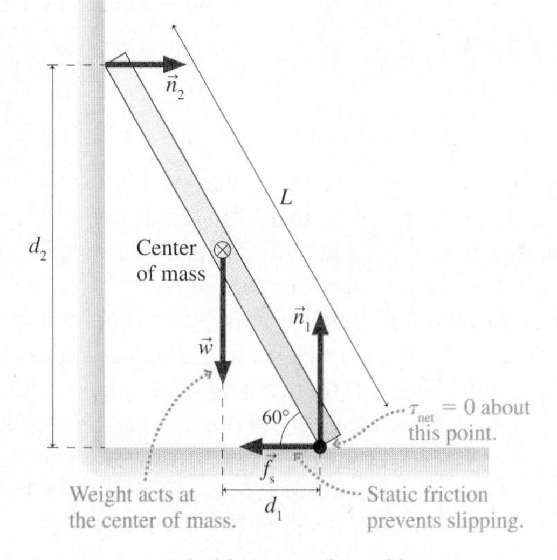

FIGURE 13.35 A ladder in total equilibrium.

SOLVE The x- and y-components of $\vec{F}_{net} = \vec{0}$ are

$$\sum F_x = n_2 - f_s = 0$$

$$\sum F_y = n_1 - w = n_1 - Mg = 0$$

The net torque is zero about *any* point, so which should we choose? The bottom corner of the ladder is a good choice because two forces pass through this point and have no torque about it. The torque about the bottom corner is

$$\tau_{net} = d_1 w - d_2 n_2 = \frac{1}{2}(L\cos 60°)Mg - (L\sin 60°)n_2 = 0$$

The signs are based on the observation that $\vec{w}$ would cause the ladder to rotate ccw while $\vec{n}_2$ would cause it to rotate cw. All together, we have three equations in the three unknowns n_1, n_2, and f_s. If we solve the third for n_2,

$$n_2 = \frac{\frac{1}{2}(L\cos 60°)Mg}{L\sin 60°} = \frac{Mg}{2\tan 60°}$$

we can then substitute this into the first to find

$$f_s = \frac{Mg}{2\tan 60°}$$

Our model of friction is $f_s \le f_{s\,max} = \mu_s n_1$. We can find n_1 from the second equation: $n_1 = Mg$. Using this, the model of static friction tells us that

$$f_s \le \mu_s Mg$$

Comparing these two expressions for f_s, we see that μ_s must obey

$$\mu_s \ge \frac{1}{2\tan 60°} = 0.29$$

Thus the minimum value of the coefficient of static friction is 0.29.

ASSESS You know from experience that you can lean a ladder or other object against a wall if the ground is "rough," but it slips if the surface is too smooth. 0.29 is a "medium" value for the coefficient of static friction, which is reasonable.

STOP TO THINK 13.5 A student holds a meter stick straight out with one or more masses dangling from it. Rank in order, from most difficult to least difficult, how hard it will be for the student to keep the meter stick from rotating.

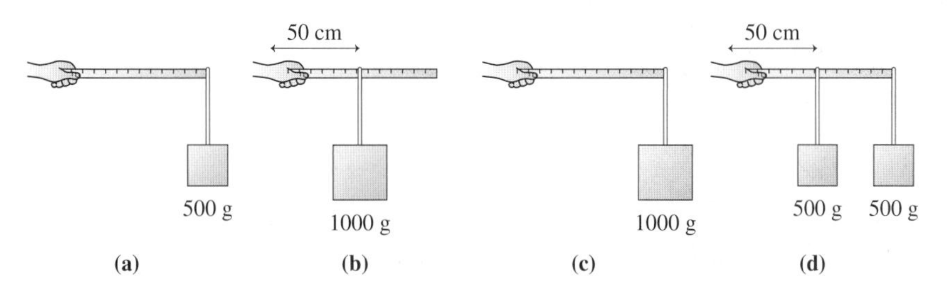

13.7 Rotational Energy

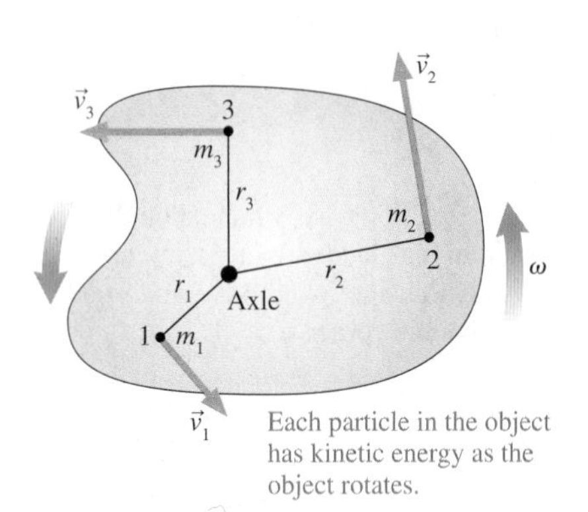

FIGURE 13.36 Rotational kinetic energy is due to the motion of the particles.

A rotating rigid body has kinetic energy because all atoms in the object are in motion. The kinetic energy due to rotation is called **rotational kinetic energy.** Rotational kinetic energy is *in addition* to any translational kinetic energy the object has if its center of mass is moving.

Figure 13.36 shows a few of the particles making up a solid object that rotates with angular velocity ω. Particle i, which rotates in a circle of radius r_i, moves with speed $v_i = r_i\omega$. The object's rotational kinetic energy is the sum of the kinetic energies of each of the particles:

$$\begin{aligned} K_{rot} &= \frac{1}{2}m_1 v_1^2 + \frac{1}{2}m_2 v_2^2 + \cdots \\ &= \frac{1}{2}m_1 r_1^2\omega^2 + \frac{1}{2}m_2 r_2^2\omega^2 + \cdots = \frac{1}{2}\left(\sum_i m_i r_i^2\right)\omega^2 \end{aligned} \tag{13.32}$$

You will recognize that the term in parentheses is our old friend, the moment of inertia I. Thus the rotational kinetic energy is

$$K_{rot} = \frac{1}{2}I\omega^2 \tag{13.33}$$

Rotational kinetic energy is *not* a new form of energy. This is the familiar kinetic energy of motion, only now expressed in a form that is especially convenient for rotational motion. Comparison to the familiar $\frac{1}{2}mv^2$ shows again that the moment of inertia I is the rotational equivalent of mass.

7.12, 7.13 ActivPhysics ONLINE

If the rotation axis is not through the center of mass, then rotation may cause the center of mass to move up or down. In that case, the object's gravitational potential energy $U_g = Mgy_{cm}$ will change. If there are no dissipative forces (i.e., if the axle is frictionless) and if no work is done by external forces, then the object's mechanical energy

$$E_{mech} = K_{rot} + U_g = \frac{1}{2}I\omega^2 + Mgy_{cm} \tag{13.34}$$

is a conserved quantity.

EXAMPLE 13.15 The speed of a rotating rod

A 1.0-m-long, 200 g rod is hinged at one end and connected to a wall. It is held out horizontally, then released. What is the speed of the tip of the rod as it hits the wall?

MODEL The mechanical energy is conserved if we assume the hinge is frictionless. The rod's gravitational potential energy is transformed into rotational kinetic energy as it "falls."

VISUALIZE Figure 13.37 is a familiar before-and-after pictorial representation of the rod. We've placed the origin of the coordinate system at the pivot point.

SOLVE Mechanical energy is conserved, so we can equate the rod's final mechanical energy to its initial mechanical energy:

$$\frac{1}{2}I\omega_1^2 + Mgy_{cm1} = \frac{1}{2}I\omega_0^2 + Mgy_{cm0}$$

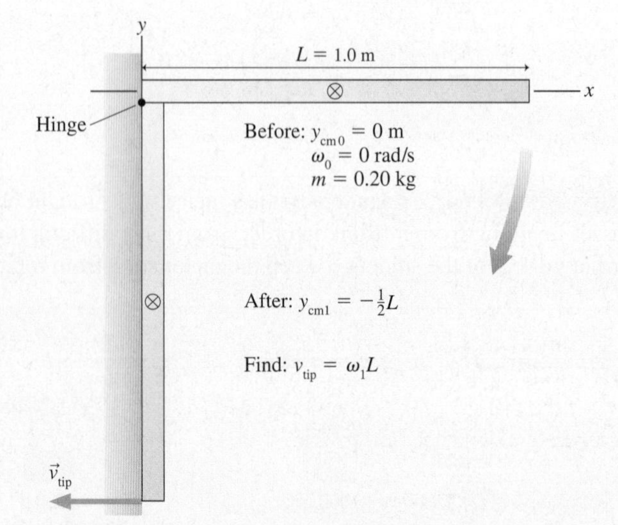

FIGURE 13.37 A before-and-after pictorial representation of the rod.

The initial conditions are $\omega_0 = 0$ and $y_{cm0} = 0$. The center of mass moves to $y_{cm1} = -\frac{1}{2}L$ as the rod hits the wall. Using $I = \frac{1}{3}ML^2$ for a rod rotating about one end, we have

$$\frac{1}{2}I\omega_1^2 + Mgy_{cm1} = \frac{1}{6}ML^2\omega_1^2 - \frac{1}{2}MgL = 0$$

We can solve this for the rod's angular velocity as it hits the wall:

$$\omega_1 = \sqrt{\frac{3g}{L}}$$

The tip of the rod is moving in a circle with radius $r = L$. Its final speed is

$$v_{tip} = \omega_1 L = \sqrt{3gL} = 5.42 \text{ m/s}$$

ASSESS Energy conservation is a powerful tool for rotational motion, just as it was for translational motion.

13.8 Rolling Motion

Rolling is a *combination motion* in which an object rotates about an axis that is moving along a straight-line trajectory. For example, Figure 13.38 is a time-exposure photo of a rolling wheel with one light bulb on the axis and a second light bulb at the edge. The axis light moves straight ahead, but the edge light follows a curve called a *cycloid*. Let's see if we can understand this interesting motion. We'll consider only objects that roll without slipping.

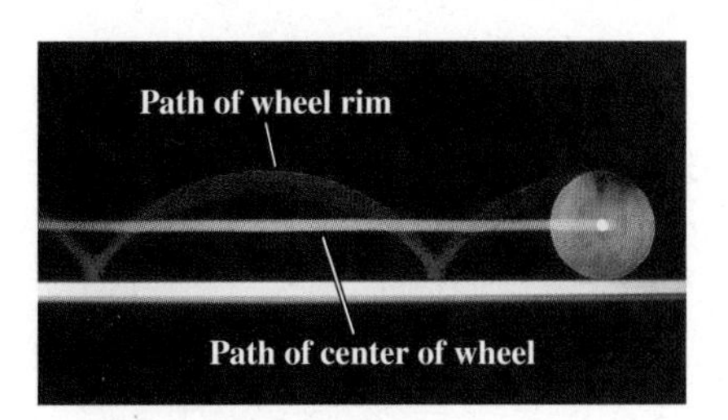

FIGURE 13.38 The trajectories of the center of a wheel and of a point on the rim are seen in a time-exposure photograph.

Figure 13.39 shows a round object—a wheel or a sphere—that rolls forward exactly one revolution. The point that had been on the bottom follows the cycloid, the curve you saw in Figure 13.38, to the top and back to the bottom. *Because the object doesn't slip*, the center of mass moves forward exactly one circumference: $\Delta x_{cm} = 2\pi R$.

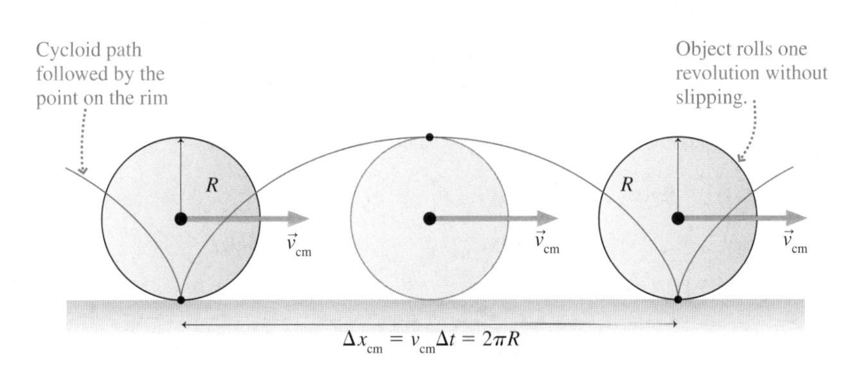

FIGURE 13.39 An object rolling through one revolution.

We can also write the distance traveled in terms of the velocity of the center of mass: $\Delta x_{cm} = v_{cm}\Delta t$. But Δt, the time it takes the object to make one complete revolution, is nothing other than the rotation period T. In other words, $\Delta x_{cm} = v_{cm}T$.

These two expressions for Δx_{cm} come from two perspectives on the motion, one looking at the rotation and the other at the translation of the center of mass. But it's the same distance no matter how you look at it, so these two expressions must be equal. Consequently,

$$\Delta x_{cm} = 2\pi R = v_{cm}T \qquad (13.35)$$

If we divide by T, we can write the center of mass velocity as

$$v_{cm} = \frac{2\pi}{T}R \qquad (13.36)$$

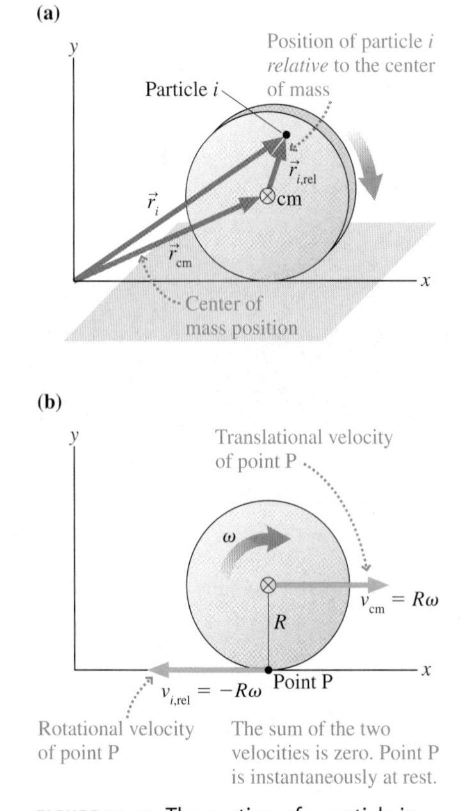

FIGURE 13.40 The motion of a particle in the rolling object.

But $2\pi/T$ is the angular velocity ω, as you learned in Chapter 7, leading to

$$v_{\mathrm{cm}} = R\omega \tag{13.37}$$

Equation 13.37 is the **rolling constraint,** the basic link between translation and rotation for objects that roll without slipping.

NOTE ▶ The rolling constraint is equivalent to Equation 13.31 for the speed of a rope that doesn't slip as it passes over a pulley. ◀

Let's look carefully at a particle in the rolling object. As Figure 13.40a shows, the position vector $\vec{r}_i$ for particle i is the vector sum $\vec{r}_i = \vec{r}_{\mathrm{cm}} + \vec{r}_{i,\mathrm{rel}}$. Taking the time derivative of this equation, we can write the velocity of particle i as

$$\vec{v}_i = \vec{v}_{\mathrm{cm}} + \vec{v}_{i,\mathrm{rel}} \tag{13.38}$$

In other words, the velocity of particle i can be divided into two parts: the velocity $\vec{v}_{\mathrm{cm}}$ of the object as a whole plus the velocity $\vec{v}_{i,\mathrm{rel}}$ of particle i relative to the center of mass (i.e., the velocity that particle i would have if the object were only rotating and had no translational motion).

Figure 13.40b applies this idea to point P at the very bottom of the rolling object, the point of contact between the object and the surface. This point is moving around the center of the object at angular velocity ω, so $v_{i,\mathrm{rel}} = -R\omega$. The negative sign indicates that the motion is cw. At the same time, the center-of-mass velocity, Equation 13.37, is $v_{\mathrm{cm}} = R\omega$. Adding these, we find that the velocity of point P, the lowest point, is $v_i = 0$. In other words, **the point on the bottom of a rolling object is instantaneously at rest.**

Although this seems surprising, it is really what we mean by "rolling without slipping." If the bottom point had a velocity, it would be moving horizontally relative to the surface. In other words, it would be slipping or sliding across the surface. To roll without slipping, the bottom point, the point touching the surface, must be at rest.

Figure 13.41 shows how the velocity vectors at the top, center, and bottom of a rotating wheel are found by adding the rotational velocity vectors to the center of mass velocity. You can see that $v_{\mathrm{bottom}} = 0$ and that $v_{\mathrm{top}} = 2R\omega = 2v_{\mathrm{cm}}$.

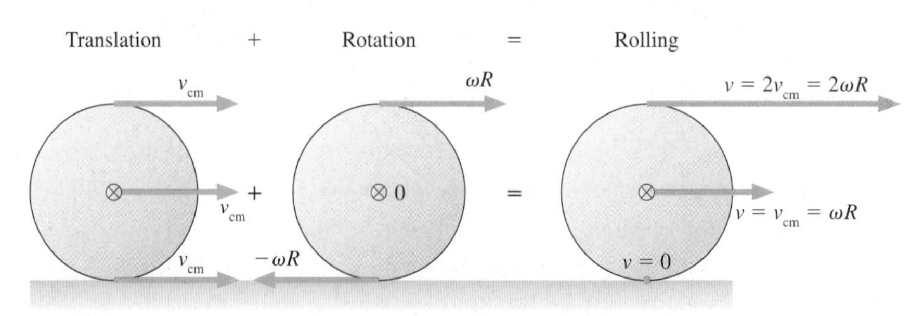

FIGURE 13.41 Rolling is a combination of translation and rotation.

Kinetic Energy of a Rolling Object

In the last section we found that the rotational kinetic energy of a rigid body in pure rotational motion is $K_{\mathrm{rot}} = \frac{1}{2}I\omega^2$. Now we would like to find the kinetic energy of an object that rolls, a combination of rotational and translation motion.

We begin with the observation that the bottom point in Figure 13.42 is instantaneously at rest. Consequently, we can think of an axis through P as an *instan-*

taneous axis of rotation. In other words, as Figure 13.42 shows, P is the pivot point for the entire object at this one instant of time.

The idea of an instantaneous axis of rotation seems a little far-fetched, but it is confirmed by looking at the instantaneous velocities of the center point and the top point. We found these in Figure 13.41 and they are shown again in Figure 13.42. They are exactly what you would expect as the tangential velocity $v_t = r\omega$ for rotation about P at distances R and $2R$.

From this perspective, the object's motion is pure rotation about point P. Thus the kinetic energy is that of pure rotation,

$$K = K_{\text{rotation about P}} = \frac{1}{2}I_P\omega^2 \tag{13.39}$$

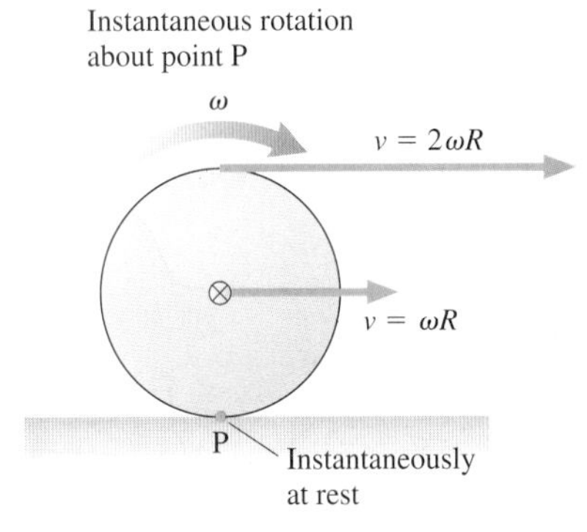

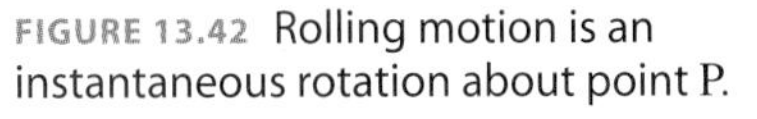
FIGURE 13.42 Rolling motion is an instantaneous rotation about point P.

I_P is the moment of inertia for rotation about point P. We can use the parallel-axis theorem to write I_P in terms of the moment of inertia I_{cm} about the center of mass. Point P is displaced by distance $d = R$, thus

$$I_P = I_{cm} + MR^2$$

Using this expression in Equation 13.39 gives us the kinetic energy:

$$K = \frac{1}{2}I_{cm}\omega^2 + \frac{1}{2}M(R\omega)^2 \tag{13.40}$$

We know from the rolling constraint that $R\omega$ is the center-of-mass velocity v_{cm}. Thus the kinetic energy of a rolling object is

$$K = \frac{1}{2}I_{cm}\omega^2 + \frac{1}{2}Mv_{cm}^2 = K_{rot} + K_{cm} \tag{13.41}$$

In other words, **the rolling motion of a rigid body can be described as a translation of the center of mass (with kinetic energy K_{cm}) plus a rotation about the center of mass (with kinetic energy K_{rot}).**

The Great Downhill Race

Figure 13.43 shows a contest in which a sphere, a cylinder, and a circular hoop, all of mass M and radius R, are placed at height h on a slope of angle θ. All three are released from rest at the same instant of time and roll down the ramp without slipping. To make things more interesting, they are joined by a particle of mass M that slides down the ramp without friction. Which one will win the race to the bottom of the hill?

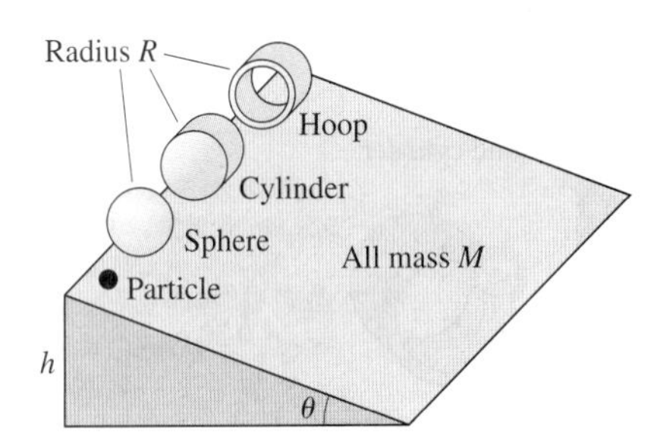

FIGURE 13.43 Which will win the downhill race?

If the particle model were valid, all four would arrive at the bottom at the same time and with the same speed. Does the fact that three of them rotate change this conclusion? We can use conservation of energy to answer this question.

An object's initial gravitational potential energy is transformed into kinetic energy as it rolls (or slides, in the case of the particle). The kinetic energy, as we just discovered, is a combination of translational and rotational kinetic energy. If we choose the bottom of the ramp as the zero point of potential energy, the statement of energy conservation $K_f = U_i$ can be written

$$\frac{1}{2}I_{cm}\omega^2 + \frac{1}{2}Mv_{cm}^2 = Mgh \tag{13.42}$$

The translational and rotational velocities are related by $\omega = v_{cm}/R$. In addition, notice from Table 13.3 that the moments of inertia of all the objects can be written in the form

$$I_{cm} = cMR^2 \tag{13.43}$$

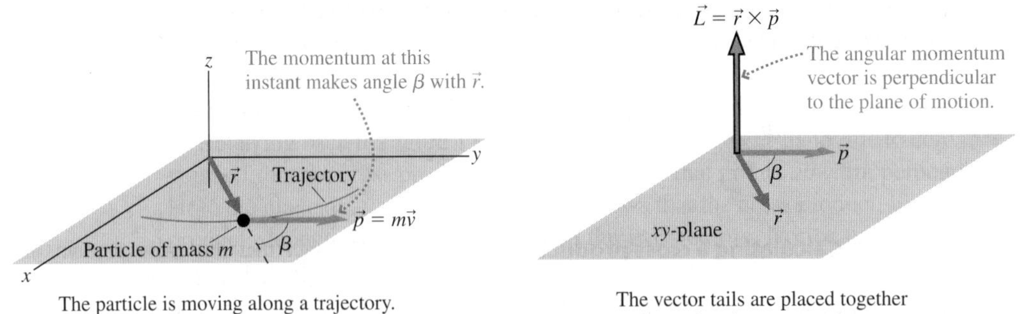

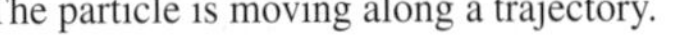
The particle is moving along a trajectory.

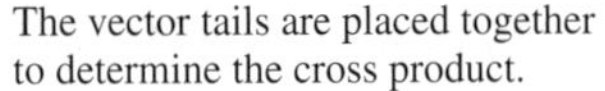
The vector tails are placed together to determine the cross product.

FIGURE 13.52 The angular momentum vector $\vec{L}$.

momentum vector $\vec{p}$ makes an angle β with the position vector $\vec{r}$. The particle's **angular momentum** $\vec{L}$ relative to the origin is a vector defined as

$$\vec{L} \equiv \vec{r} \times \vec{p} = (mrv\sin\beta, \text{ direction of right-hand rule}) \tag{13.50}$$

The **angular momentum vector is perpendicular to the plane of motion.** For circular motion, where $\beta = 90°$, the magnitude $L = rp = rmv_t$ agrees with our earlier definition. The units of angular momentum are kg m^2/s.

NOTE ▶ Angular momentum is the rotational equivalent of linear momentum in much the same way that torque is the rotational equivalent of force. Notice that the vector definitions are parallel: $\vec{\tau} \equiv \vec{r} \times \vec{F}$ and $\vec{L} \equiv \vec{r} \times \vec{p}$. ◀

The primary value of this more general definition of angular momentum lies in its connection to torque. To show this, take the time derivative of $\vec{L}$:

$$\begin{aligned} \frac{d\vec{L}}{dt} &= \frac{d}{dt}(\vec{r} \times \vec{p}) = \left(\frac{d\vec{r}}{dt} \times \vec{p} + \vec{r} \times \frac{d\vec{p}}{dt}\right) \\ &= \vec{v} \times \vec{p} + \vec{r} \times \vec{F}_{\text{net}} \end{aligned} \tag{13.51}$$

where we used Equation 13.48 for the derivative of a cross product. We also used the definition $\vec{v} = d\vec{r}/dt$ and, from Chapter 9, $\vec{F}_{\text{net}} = d\vec{p}/dt$.

Vectors $\vec{v}$ and $\vec{p}$ are parallel, and the cross product of two parallel vectors is $\vec{0}$. Thus the first term in Equation 13.51 vanishes. The second term $\vec{r} \times \vec{F}_{\text{net}}$ is the net torque, $\vec{\tau}_{\text{net}} = \vec{\tau}_1 + \vec{\tau}_2 + \cdots$, so we arrive at

$$\frac{d\vec{L}}{dt} = \vec{\tau}_{\text{net}} \tag{13.52}$$

Equation 13.52, which says **a net torque causes the particle's angular momentum to change,** is the rotational equivalent of $d\vec{p}/dt = \vec{F}_{\text{net}}$.

13.10 Angular Momentum of a Rigid Body

Equation 13.52 is the angular momentum of a single particle. The angular momentum of a system consisting of particles with individual angular momenta $\vec{L}_1, \vec{L}_2, \vec{L}_3, \ldots$ is the vector sum

$$\vec{L} = \vec{L}_1 + \vec{L}_2 + \vec{L}_3 + \cdots = \sum_i \vec{L}_i \tag{13.53}$$

We can combine Equations 13.52 and 13.53 to find the rate of change of the system's angular momentum:

$$\frac{d\vec{L}}{dt} = \sum_i \frac{d\vec{L}_i}{dt} = \sum_i \vec{\tau}_i = \vec{\tau}_{\text{net}} \tag{13.54}$$

The net torque includes the torque due to internal forces, acting within the system, and the torque due to external forces. Because the internal forces are action/reaction pairs of forces, acting with the same strength in opposite directions, the net torque due to internal forces is zero. That is, the forces within a system of particles do not exert a net torque on the system. Thus the only forces that contribute to the net torque are external forces exerted on the system by the environment.

For a system of particles, **the rate of change of the system's angular momentum is the net torque on the system.** Equation 13.54 is analogous to the Chapter 9 result $d\vec{P}/dt = \vec{F}_{\text{net}}$, which says that the rate of change of a system's total linear momentum is the net force on the system. Table 13.4 summarizes the analogies between linear and angular momentum and energy.

TABLE 13.4 Angular and linear momentum and energy

Angular momentum	Linear momentum
$K_{\text{rot}} = \frac{1}{2}I\omega^2$	$K_{\text{cm}} = \frac{1}{2}Mv_{\text{cm}}{}^2$
$\vec{L} = I\vec{\omega}$ *	$\vec{P} = M\vec{v}_{\text{cm}}$
$d\vec{L}/dt = \vec{\tau}_{\text{net}}$	$d\vec{P}/dt = \vec{F}_{\text{net}}$
The angular momentum of a system is conserved if there is no net torque.	The linear momentum of a system is conserved if there is no net force.

*Rotation about an axis of symmetry.

Conservation of Angular Momentum

A net torque on a rigid body causes its angular momentum to change. Conversely, the angular momentum does *not* change—it is *conserved*—for a system with no net torque. This is the basis of the law of conservation of angular momentum.

ActivPhysics ONLINE 7.14

Law of conservation of angular momentum The angular momentum $\vec{L}$ of an isolated system ($\vec{\tau}_{\text{net}} = \vec{0}$) is conserved. The final angular momentum $\vec{L}_{\text{f}}$ is equal to the initial angular momentum $\vec{L}_{\text{i}}$.

If the angular momentum is conserved, both the magnitude *and* the direction of $\vec{L}$ are unchanged.

EXAMPLE 13.18 An expanding rod

Two equal masses are at the ends of a massless 50-cm-long rod. The rod spins at 2.0 rev/s about an axis through its midpoint. Suddenly, a compressed gas expands the rod out to a length of 160 cm. What is the rotation frequency after the expansion?

MODEL The forces that expand the rod push outward from the pivot and exert no torques. Thus the system's angular momentum is conserved.

VISUALIZE Figure 13.53 on the next page is a before-and-after pictorial representation. The angular momentum vectors $\vec{L}_1$ and $\vec{L}_2$ are perpendicular to the plane of motion.

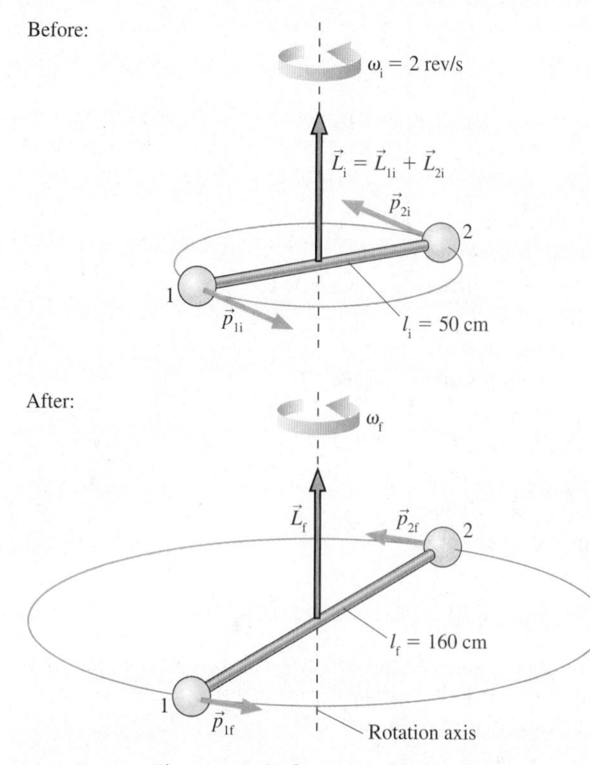

FIGURE 13.53 The system before and after the rod expands.

SOLVE The particles are moving in circles, so the magnitude of each angular momentum is $L = rp = mrv_t = mr^2\omega = \frac{1}{4}ml^2\omega$, where we used $r = \frac{1}{2}l$. Because $\vec{L}_1$ and $\vec{L}_2$ point in the same direction, the magnitude of their sum is simply the sum of their magnitudes. Thus the initial angular momentum of the system is

$$L_i = \frac{1}{4}ml_i^2\omega_i + \frac{1}{4}ml_i^2\omega_i = \frac{1}{2}ml_i^2\omega_i$$

Similarly, the angular momentum after the expansion is $L_f = \frac{1}{2}ml_f^2\omega_f$. Angular momentum is conserved as the rod expands, thus

$$\frac{1}{2}ml_f^2\omega_f = \frac{1}{2}ml_i^2\omega_i$$

Solving for ω_f, we find

$$\omega_f = \left(\frac{l_i}{l_f}\right)^2\omega_i = \left(\frac{50\text{ cm}}{160\text{ cm}}\right)^2(2.0\text{ rev/s}) = 0.195\text{ rev/s}$$

ASSESS The values of the masses weren't needed. All that matters is the ratio of the lengths.

The expansion of the rod in Example 13.18 causes a dramatic slowing of the rotation. Similarly, the rotation would speed up if the weights were pulled in. This is how an ice skater controls her speed as she does a spin. Pulling in her arms decreases her moment of inertia and causes her angular velocity to increase. Similarly, extending her arms increases her moment of inertia and her angular velocity drops until she can skate out of the spin. It's all a matter of conserving angular momentum.

Angular Momentum and Angular Velocity

The analogy between linear and rotational motion has been so consistent that you might expect one more. The Chapter 9 result $\vec{P} = M\vec{v}_{cm}$ might give us reason to anticipate that angular momentum and angular velocity are related by $\vec{L} = I\vec{\omega}$. Unfortunately, the analogy breaks down at this point. For an arbitrarily shaped object, the angular momentum vector and the angular velocity vector don't necessarily point in the same direction. The general relationship between $\vec{L}$ and $\vec{\omega}$ is beyond the scope of this text.

The good news is that the analogy *does* continue to hold for the rotation of a *symmetrical* object about the symmetry axis. For example, the axis of a cylinder or disk is a symmetry axis, as is any diameter through a sphere. For the rotation of a symmetrical object about the symmetry axis, the angular momentum and angular velocity are related by

$$\vec{L} = I\vec{\omega} \quad \text{(rotation about an axis of symmetry)} \tag{13.55}$$

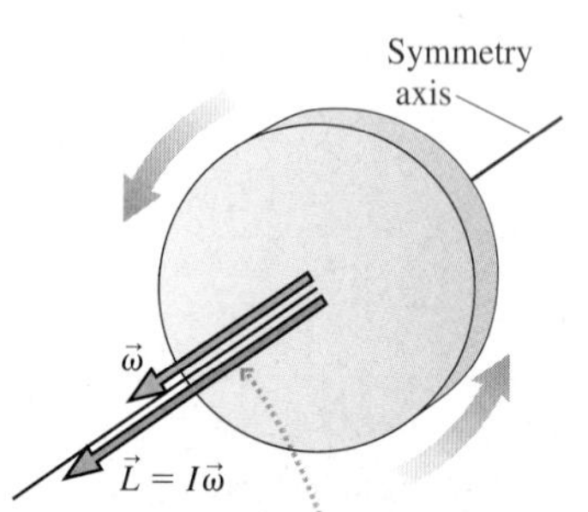

FIGURE 13.54 The angular momentum vector of a rigid body rotating about an axis of symmetry.

This relationship is shown for a spinning disk in Figure 13.54. Equation 13.55 is particularly important for applying the law of conservation of momentum to symmetrical rigid bodies.

EXAMPLE 13.19 Angular momentum of the earth

What is the angular momentum of the earth as it rotates on its axis?

MODEL The earth is a sphere that rotates about an axis of symmetry.

SOLVE This is a straightforward application of Equation 13.55. The earth's angular frequency is

$$\omega = \frac{1 \text{ rev}}{24 \text{ hr}} \times \frac{1 \text{ hr}}{3600 \text{ s}} \times \frac{2\pi \text{ rad}}{1 \text{ rev}} = 7.27 \times 10^{-5} \text{ rad/s}$$

The earth's moment of inertia is

$$I = \frac{2}{5}M_e R_e^2 = \frac{2}{5}(5.98 \times 10^{24} \text{ kg})(6.37 \times 10^6 \text{ m})^2$$
$$= 9.71 \times 10^{37} \text{ kg m}^2$$

Thus the magnitude of the earth's angular momentum is

$$L = I\omega = 7.06 \times 10^{33} \text{ kg m}^2\text{/s}$$

If you wrap your right fingers around a globe in the direction of the earth's rotation, from west to east, your thumb will point in the direction of the north pole. Consequently, the earth's angular momentum vector is

$$\vec{L} = (7.06 \times 10^{33} \text{ kg m}^2\text{/s, along the axis through the North Pole})$$

ASSESS The earth's angular momentum is very large not because the earth is spinning rapidly, but because the earth has such a large moment of inertia.

EXAMPLE 13.20 Two interacting disks

A 20-cm-diameter, 2.0 kg solid disk is rotating at 200 rpm. A 20-cm-diameter, 1.0 kg circular loop is dropped straight down onto the rotating disk. Friction causes the loop to accelerate until it is "riding" on the disk. What is the final angular velocity of the combined system?

MODEL The friction between the two objects creates torques that speed up the loop and slow down the disk. But these torques are internal to the combined disk + loop system, so $\tau_{net} = 0$ and the *total* angular momentum of the disk + loop system is conserved.

VISUALIZE Figure 13.55 is a before-and-after pictorial representation. Initially only the disk is rotating, at angular velocity $\vec{\omega}_i$. The rotation is about an axis of symmetry, so the angular momentum $\vec{L} = I\vec{\omega}$ is parallel to $\vec{\omega}$. At the end of the problem, $\vec{\omega}_{disk} = \vec{\omega}_{loop} = \vec{\omega}_f$.

SOLVE Both angular momentum vectors point along the rotation axis. Conservation of angular momentum tells us that the magnitude of $\vec{L}$ is unchanged. Thus

$$L_f = I_{disk}\omega_f + I_{loop}\omega_f = L_i = I_{disk}\omega_i$$

Solving for ω_f gives

$$\omega_f = \frac{I_{disk}}{I_{disk} + I_{loop}}\omega_i$$

The moments of inertia for a disk and a loop can be found in Table 13.3, leading to

$$\omega_f = \frac{\frac{1}{2}M_{disk}R^2}{\frac{1}{2}M_{disk}R^2 + M_{loop}R^2}\omega_i = 100 \text{ rpm}$$

ASSESS What appeared to be a difficult problem turns out to be fairly easy once you recognize that the total angular momentum is conserved.

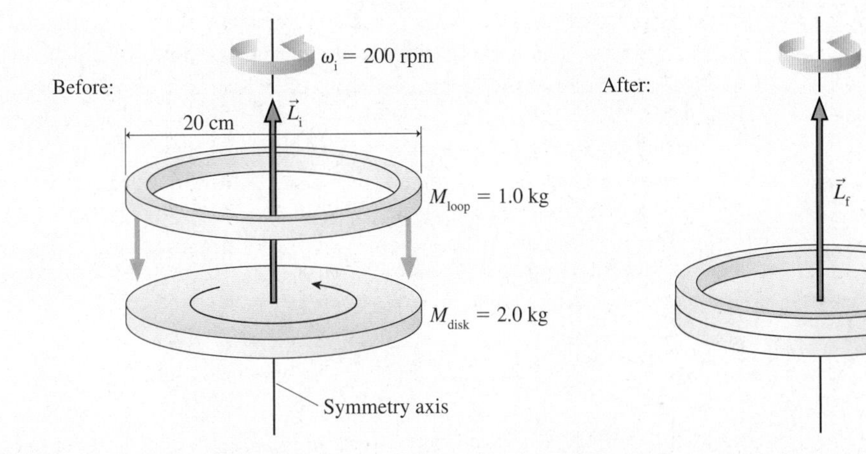

FIGURE 13.55 The circular hoop drops onto the rotating disk.

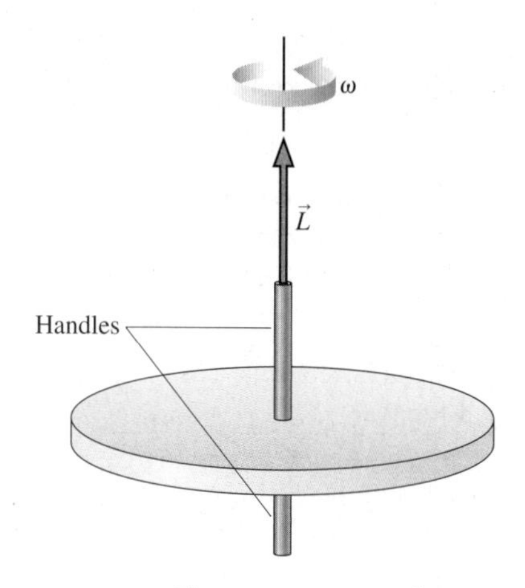

FIGURE 13.56 The vector nature of angular momentum makes it difficult to turn a rapidly spinning wheel.

When angular momentum is conserved, not only is its magnitude conserved, its direction—the direction of the rotation axis—must remain unchanged. This is often shown with the lecture demonstration illustrated in Figure 13.56. A bicycle wheel with two handles is given a spin, then handed to an unsuspecting student. The student is asked to turn the wheel 90°. Surprisingly, this is *very hard to do.*

The reason is that the wheel's angular momentum vector, which points straight up, is highly resistant to change. If the wheel is spinning fast, a *large* torque must be supplied to change $\vec{L}$. This directional stability of a rapidly spinning object is why gyroscopes are used as navigational devices on ships and planes. Once the axis of a spinning gyroscope is pointed north, it will maintain that direction as the ship or plane moves.

STOP TO THINK 13.6 Two buckets spin around in a horizontal circle on frictionless bearings. Suddenly, it starts to rain. As a result,

a. The buckets continue to rotate at constant angular velocity because the rain is falling vertically while the buckets move in a horizontal plane.
b. The buckets continue to rotate at constant angular velocity because the total mechanical energy of the bucket + rain system is conserved.
c. The buckets speed up because the potential energy of the rain is transformed into kinetic energy.
d. The buckets slow down because the angular momentum of the bucket + rain system is conserved.
e. Both a and b.
f. None of the above.

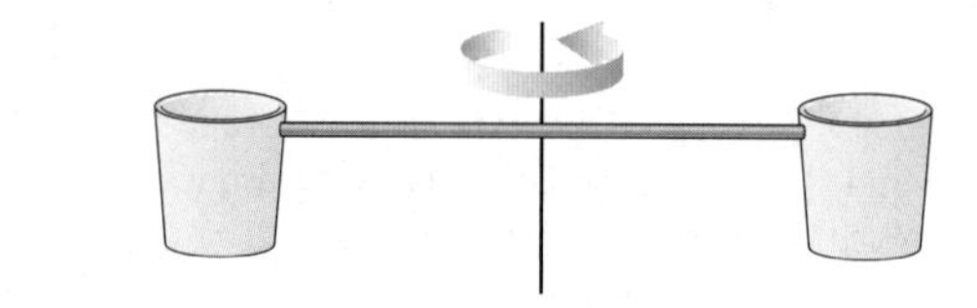

SUMMARY

The goal of Chapter 13 has been to understand the physics of rotating objects.

GENERAL PRINCIPLES

Rotational Dynamics

Every point on a **rigid body** rotating about a fixed axis has the same angular velocity ω and angular acceleration α.

Newton's second law for rotational motion is

$$\alpha = \frac{\tau_{net}}{I}$$

Use rotational kinematics to find angles and angular velocities.

Conservation Laws

Energy is conserved for an isolated system.

- Pure rotation $E = K_{rot} + U_g = \frac{1}{2}I\omega^2 + Mgy_{cm}$
- Rolling $E = K_{rot} + K_{cm} + U_g = \frac{1}{2}I\omega^2 + \frac{1}{2}Mv_{cm}{}^2 + Mgy_{cm}$

Angular momentum is conserved if $\vec{\tau}_{net} = \vec{0}$.

- Particle $\vec{L} = m\vec{r} \times \vec{p}$
- Rigid body rotating about axis of symmetry $\vec{L} = I\vec{\omega}$

IMPORTANT CONCEPTS

Angular velocity

$$\omega = \frac{d\theta}{dt}$$

Angular acceleration is the rotational equivalent of acceleration

$$\alpha = \frac{d\omega}{dt}$$

Torque is the rotational equivalent of force

$$\tau = rF\sin\phi = rF_t = dF$$

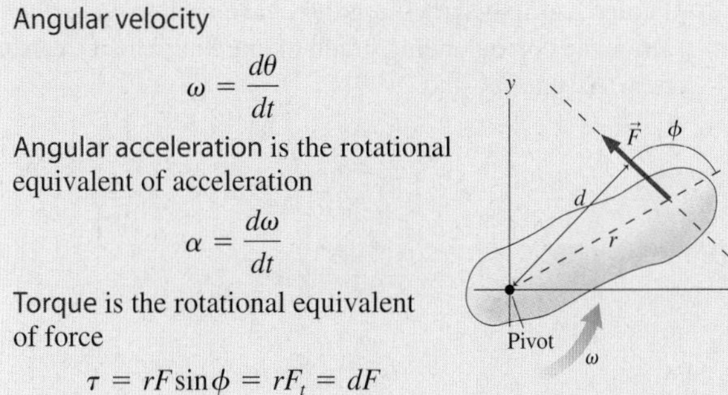

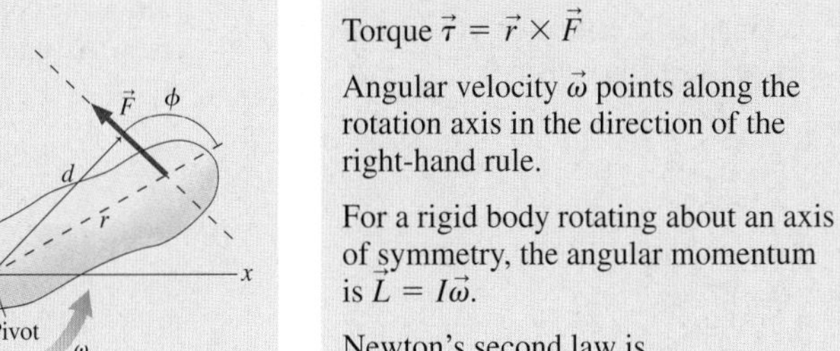

Vector description of rotation

Torque $\vec{\tau} = \vec{r} \times \vec{F}$

Angular velocity $\vec{\omega}$ points along the rotation axis in the direction of the right-hand rule.

For a rigid body rotating about an axis of symmetry, the angular momentum is $\vec{L} = I\vec{\omega}$.

Newton's second law is

$$\frac{d\vec{L}}{dt} = \vec{\tau}_{net}$$

A system of particles on which there is no net force undergoes unconstrained rotation about the center of mass

$$x_{cm} = \frac{1}{M}\int x\,dm \qquad y_{cm} = \frac{1}{M}\int y\,dm$$

The gravitational torque on a body can be found by treating the body as a particle with all the mass M concentrated at the center of mass.

The moment of inertia

$$I = \int r^2\,dm$$

is the rotational equivalent of mass. The moment of inertia depends on how the mass is distributed around the axis. If I_{cm} is known, the I about a parallel axis distance d away is given by the **parallel-axis theorem** $I = I_{cm} + Md^2$.

APPLICATIONS

Rotational kinematics

$$\omega_f = \omega_i + \alpha\Delta t$$

$$\theta_f = \theta_i + \omega_i\Delta t + \tfrac{1}{2}\alpha(\Delta t)^2$$

$$\omega_f^2 = \omega_i^2 + 2\alpha\Delta\theta$$

$$v_t = r\omega \qquad a_t = r\alpha$$

Rigid-body equilibrium

An object is in total equilibrium only if both $\vec{F}_{net} = \vec{0}$ and $\vec{\tau}_{net} = \vec{0}$.

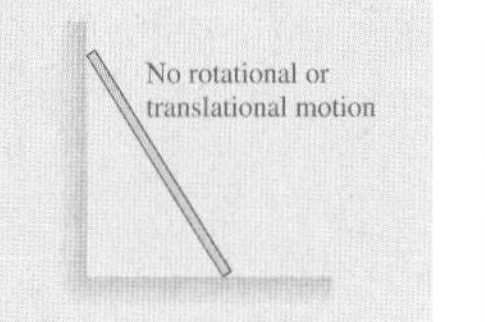

Rolling motion

For an object that rolls without slipping

$$v_{cm} = R\omega$$

$$K = K_{rot} + K_{cm}$$

TERMS AND NOTATION

rigid body
rigid-body model
translational motion
rotational motion
combination motion
angular acceleration, α
center of mass
torque, τ
line of action
moment arm, d
couple
moment of inertia, I
parallel-axis theorem
rotational kinetic energy, K_{rot}
rolling constraint
cross product
vector product
right-hand rule
angular momentum, $\vec{L}$
law of conservation of angular momentum

EXERCISES AND PROBLEMS

Exercises

Section 13.1 Rotational Kinematics

1. The graph shows the angular velocity of the crankshaft in a car. Draw a graph of the angular acceleration versus time. Include appropriate numerical scales on both axes.

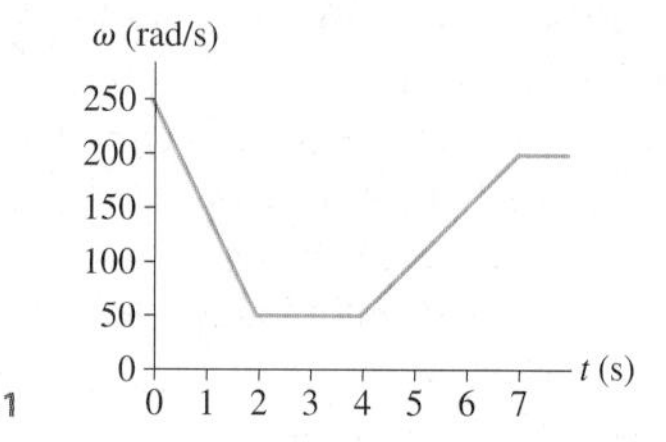

FIGURE EX13.1

2. The graph shows the angular acceleration of a turntable that starts from rest. Draw a graph of the angular velocity versus time. Include appropriate numerical scales on both axes.

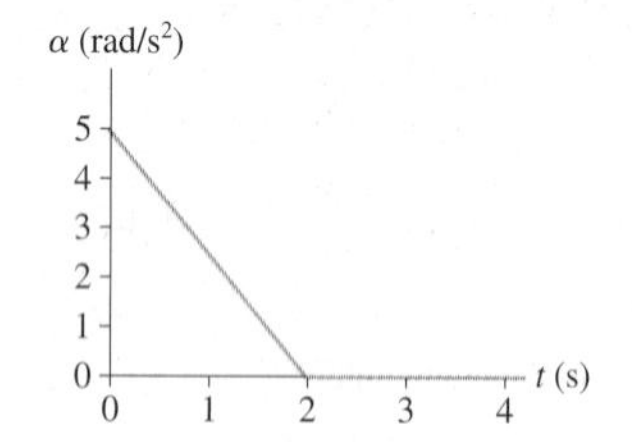

FIGURE EX13.2

3. The graph shows the angular velocity of one wheel on a car.
   a. Draw a graph of the angular acceleration versus time. Include appropriate numerical scales on both axes.
   b. Describe the car's motion.

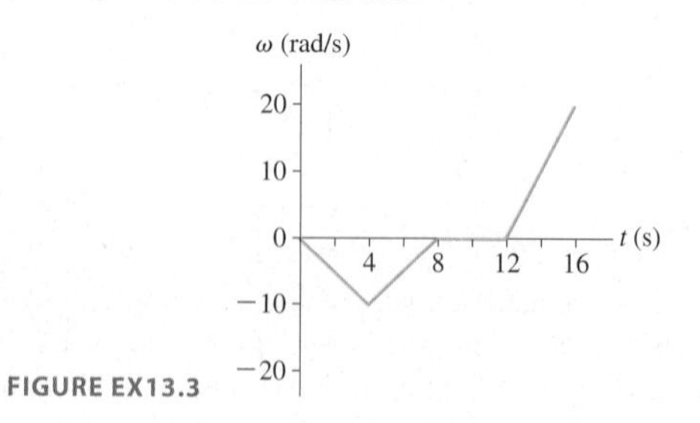

FIGURE EX13.3

4. a. Figure Ex13.4a shows angular velocity versus time. Draw the corresponding graph of angular acceleration versus time.
   b. Figure Ex13.4b shows angular acceleration versus time. Draw the corresponding graph of angular velocity versus time. Assume $\omega_0 = 0$.

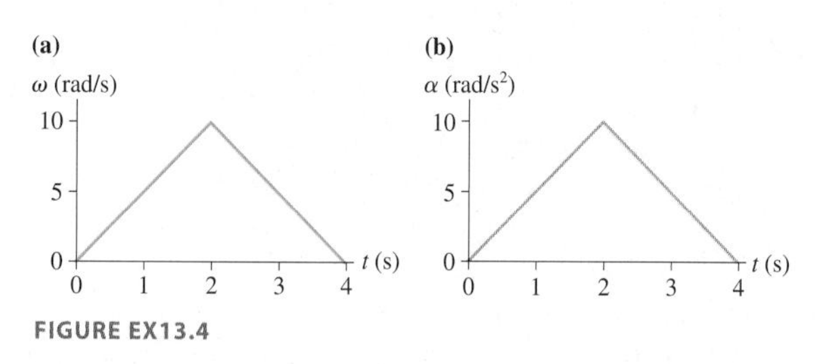

FIGURE EX13.4

5. A skater holds her arms outstretched as she spins at 180 rpm. What is the speed of her hands if they are 140 cm apart?
6. A magnetic computer disk 8.0 cm in diameter is initially at rest. A small dot is painted on the edge of the disk. The disk accelerates at 600 rad/s^2 for $\frac{1}{2}$ s, then coasts at a steady angular velocity for another $\frac{1}{2}$ s. What is the speed of the dot at $t = 1.0$ s? Through how many revolutions has it turned?
7. A high-speed drill rotating ccw at 2400 rpm comes to a halt in 2.5 s.
   a. What is the drill's angular acceleration?
   b. How many revolutions does it make as it stops?
8. An electric-generator turbine spins at 3600 rpm. Friction is so small that it takes the turbine 10 min to coast to a stop. How many revolutions does it make while stopping?

Section 13.2 Rotation About the Center of Mass

9. An equilateral triangle 5.0 cm on a side rotates about its center of mass at 120 rpm. What is the speed of one tip of the triangle?

10. The three masses shown in Figure Ex13.10 are connected by massless, rigid rods. What are the coordinates of the center of mass?

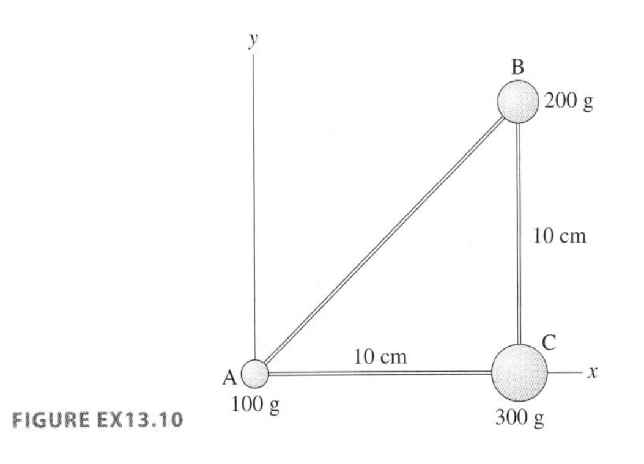

FIGURE EX13.10

Section 13.3 Torque

11. What is the net torque on the pulley about the axle?

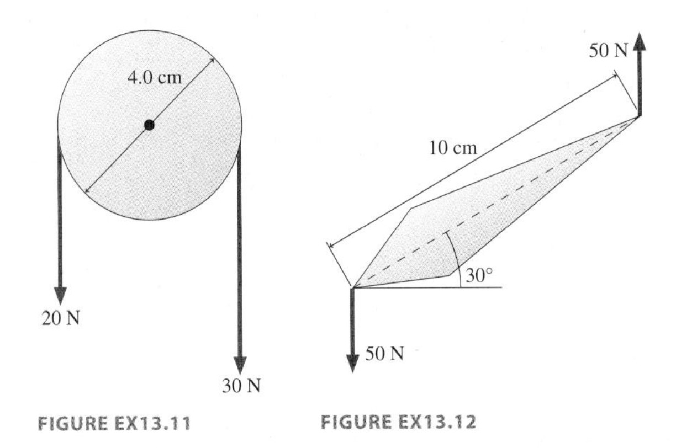

FIGURE EX13.11 FIGURE EX13.12

12. What is the net torque about the center of mass?
13. The tune-up specifications of a car call for the spark plugs to be tightened to a torque of 38 Nm. You plan to tighten the plugs by pulling on the end of a 25-cm-long wrench. Because of the cramped space under the hood, you'll need to pull at an angle of 120° with respect to the wrench shaft. With what force must you pull?
14. The 20-cm-diameter disk in Figure Ex13.14 can rotate on an axle through its center. What is the net torque about the axle?

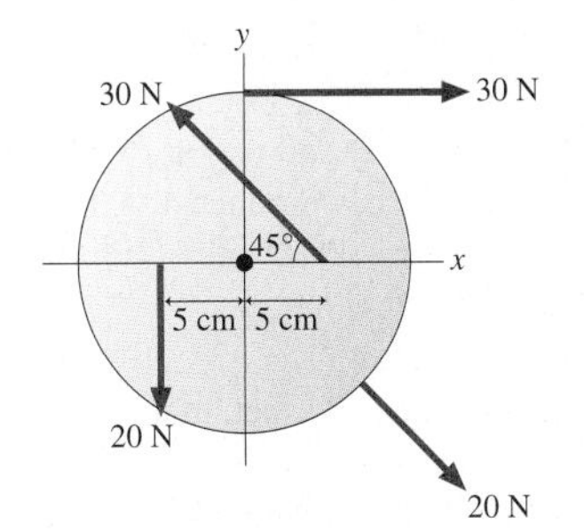

FIGURE EX13.14

15. A 4.0-m-long, 500 kg steel beam extends horizontally from the point where it has been bolted to the framework of a new building under construction. A 70 kg construction worker stands at the far end of the beam. What is the magnitude of the torque about the point where the beam is bolted into place?

16. An athlete at the gym holds a 3.0 kg steel ball in his hand. His arm is 70 cm long and has a mass of 4.0 kg. What is the magnitude of the torque about his shoulder if he holds his arm
    a. Straight out to his side, parallel to the floor?
    b. Straight, but 45° below horizontal?

Section 13.4 Rotational Dynamics

Section 13.5 Rotation About a Fixed Axis

17. The four masses shown in Figure Ex13.17 are connected by massless, rigid rods.
    a. Find the coordinates of the center of mass.
    b. Find the moment of inertia about an axis that passes through mass A and is perpendicular to the page.

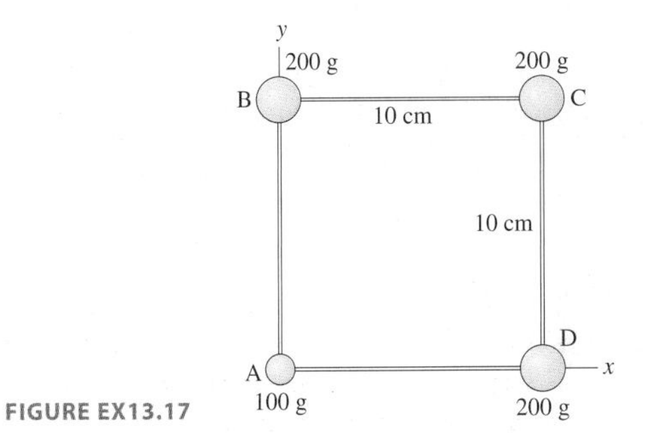

FIGURE EX13.17

18. The four masses shown in Figure Ex13.17 are connected by massless, rigid rods.
    a. Find the coordinates of the center of mass.
    b. Find the moment of inertia about a diagonal axis that passes through masses B and D.
19. The three masses shown in Figure Ex13.19 are connected by massless, rigid rods.
    a. Find the coordinates of the center of mass.
    b. Find the moment of inertia about an axis that passes through mass A and is perpendicular to the page.
    c. Find the moment of inertia about an axis that passes through masses B and C.

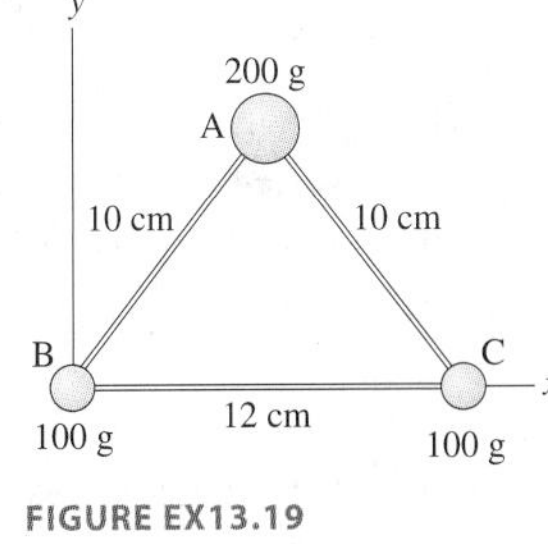

FIGURE EX13.19

20. An object's moment of inertia is 2.0 kg m^2. Its angular velocity is increasing at the rate of 4.0 rad/s per second. What is the torque on the object?
21. An object whose moment of inertia is 4.0 kg m^2 experiences the torque shown in Figure Ex13.21. What is the object's angular velocity at $t = 3.0$ s? Assume it starts from rest.

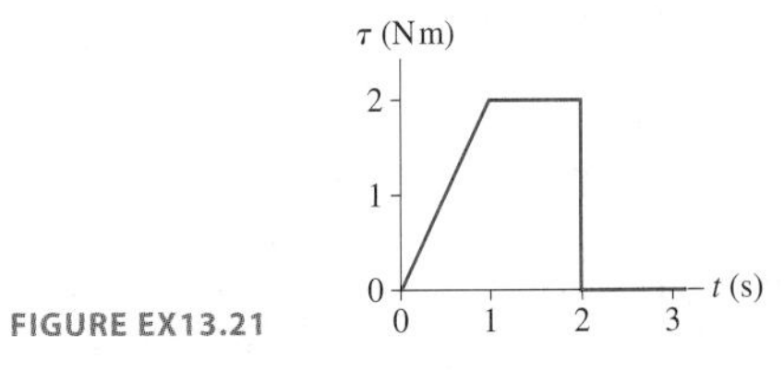

FIGURE EX13.21

22. A 1.0 kg ball and a 2.0 kg ball are connected by a 1.0-m-long rigid, massless rod. The rod is rotating cw about its center of mass at 20 rpm. What torque will bring the balls to a halt in 5.0 s?

23. A 200 g, 20-cm-diameter plastic disk is spun on an axle through its center by an electric motor. What torque must the motor supply to take the disk from 0 to 1800 rpm in 4.0 s?
24. The 200 g model rocket shown in Figure Ex13.24 generates 4.0 N of thrust. It spins in a horizontal circle at the end of a 100 g rigid rod. What is its angular acceleration?

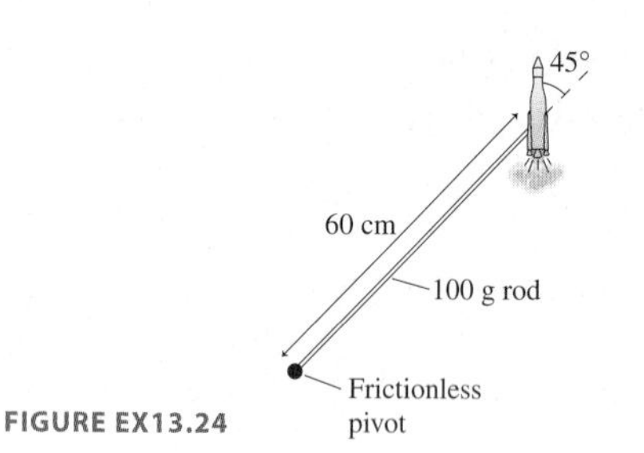

FIGURE EX13.24

Section 13.6 Rigid-Body Equilibrium

25. How much torque must the pin exert to keep the rod from rotating?

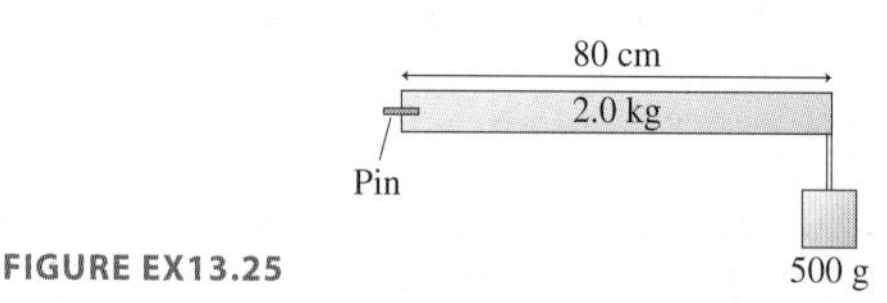

FIGURE EX13.25

26. Is the object in Figure Ex13.26 in equilibrium? Explain.

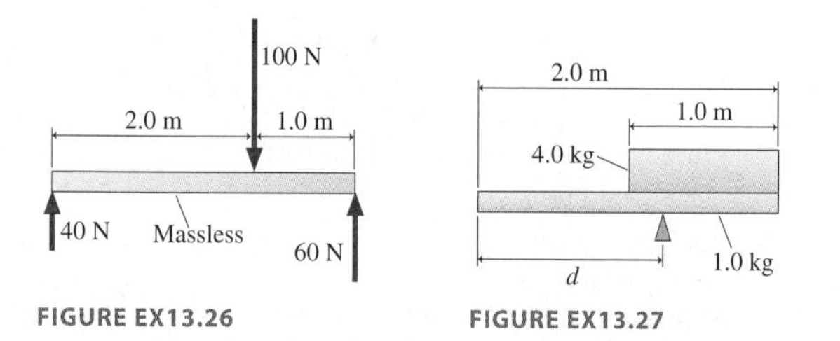

FIGURE EX13.26 FIGURE EX13.27

27. The two objects in Figure Ex13.27 are balanced on the pivot. What is distance d?

Section 13.7 Rotational Energy

28. What is the rotational kinetic energy of the earth? Assume the earth is a uniform sphere.
29. The three 200 g masses in Figure Ex13.29 are connected by massless, rigid rods to form a triangle. What is the triangle's rotational kinetic energy if it rotates at 5.0 rev/s about an axis through the center?

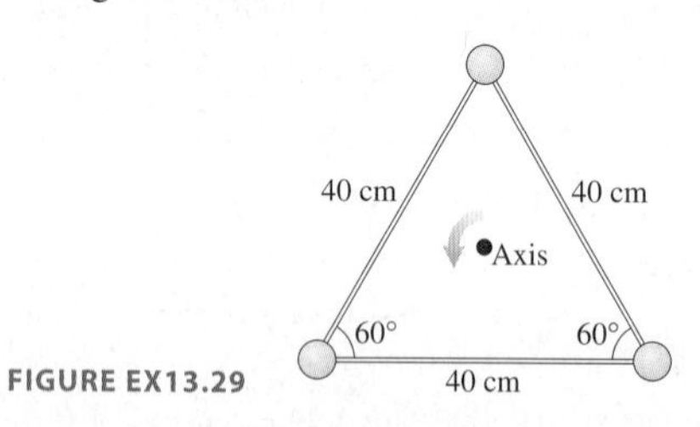

FIGURE EX13.29

30. A thin, 100 g disk with a diameter of 8.0 cm rotates about an axis through its center with 0.15 J of kinetic energy. What is the speed of a point on the rim?
31. A 300 g ball and a 600 g ball are connected by a 40-cm-long massless, rigid rod. The structure rotates about its center of mass at 100 rpm. What is its rotational kinetic energy?

Section 13.8 Rolling Motion

32. A car tire is 60 cm in diameter. The car is traveling at a speed of 20 m/s.
    a. What is the tire's rotation frequency, in rpm?
    b. What is the speed of a point at the top edge of the tire?
    c. What is the speed of a point at the bottom edge of the tire?
33. A 500 g, 8.0-cm-diameter can rolls across the floor at 1.0 m/s. What is the can's kinetic energy?
34. An 8.0-cm-diameter, 400 g sphere is released from rest at the top of a 2.1-m-long, 25° incline. It rolls, without slipping, to the bottom.
    a. What is the sphere's angular velocity at the bottom of the incline?
    b. What fraction of its kinetic energy is rotational?

Section 13.9 The Vector Description of Rotational Motion

35. Evaluate the cross products $\vec{A} \times \vec{B}$ and $\vec{C} \times \vec{D}$.

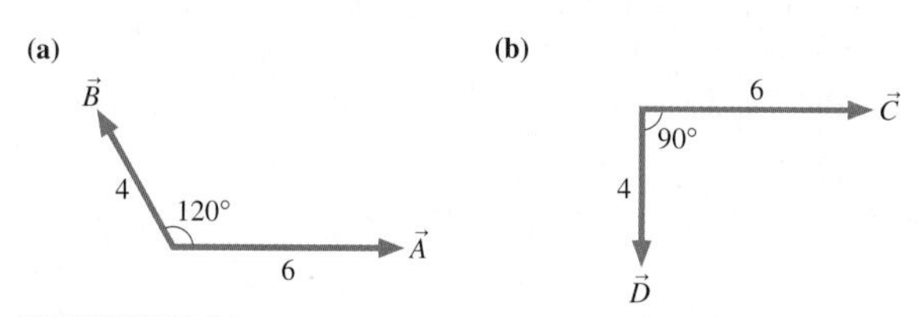

FIGURE EX13.35

36. Evaluate the cross products $\vec{A} \times \vec{B}$ and $\vec{C} \times \vec{D}$.

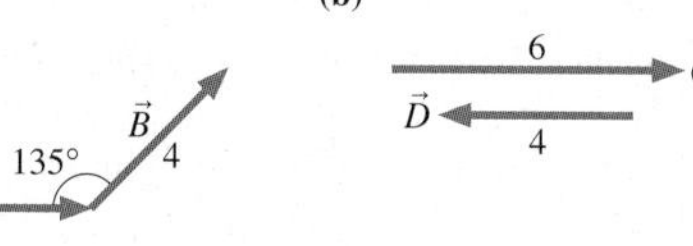

FIGURE EX13.36

37. a. What is $(\hat{\imath} \times \hat{\jmath}) \times \hat{\imath}$?
    b. What is $\hat{\imath} \times (\hat{\jmath} \times \hat{\imath})$?
38. a. What is $\hat{\imath} \times (\hat{\imath} \times \hat{\jmath})$?
    b. What is $(\hat{\imath} \times \hat{\jmath}) \times \hat{k}$?
39. Vector $\vec{A} = 3\hat{\imath} - \hat{\jmath}$ and vector $\vec{B} = 2\hat{\imath} + 3\hat{\jmath} - \hat{k}$.
    a. What is the cross product $\vec{A} \times \vec{B}$?
    b. Show vectors $\vec{A}$, $\vec{B}$, and $\vec{A} \times \vec{B}$ on a three-dimensional coordinate system.
40. Consider the vector $\vec{C} = 3\hat{\imath}$.
    a. What is a vector $\vec{D}$ such that $\vec{C} \times \vec{D} = \vec{0}$?
    b. What is a vector $\vec{E}$ such that $\vec{C} \times \vec{E} = 6\hat{k}$?
    c. What is a vector $\vec{F}$ such that $\vec{C} \times \vec{F} = -3\hat{\jmath}$?
41. Force $\vec{F} = -10\hat{\jmath}$ N is exerted on a particle at $\vec{r} = (5\hat{\imath} + 5\hat{\jmath})$ m. What is the torque on the particle about the origin?
42. Force $\vec{F} = (-10\hat{\imath} + 10\hat{\jmath})$N is exerted on a particle at $\vec{r} = 5\hat{\jmath}$ m. What is the torque on the particle about the origin?

Section 13.10 Angular Momentum of a Rigid Body

43. A 200 g block is attached to one end of a 40-cm-long massless rod. The rod rotates about a frictionless pivot at the other end. The block starts from rest and experiences a constant torque of 0.050 N m.
    a. How long does it take the block to complete 10 revolutions?
    b. What is the block's angular momentum at the time of completion?
    c. Show that $\Delta L/\Delta t = \tau$.
44. What are the magnitude and direction of the angular momentum of the 200 g particle in Figure Ex13.44?

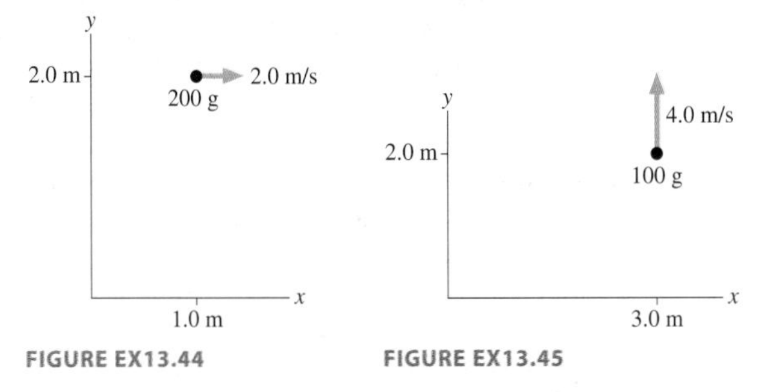

FIGURE EX13.44

FIGURE EX13.45

45. What are the magnitude and direction of the angular momentum of the 100 g particle in Figure Ex13.45?
46. What is the angular momentum of the 500 g rotating bar in Figure Ex13.46?

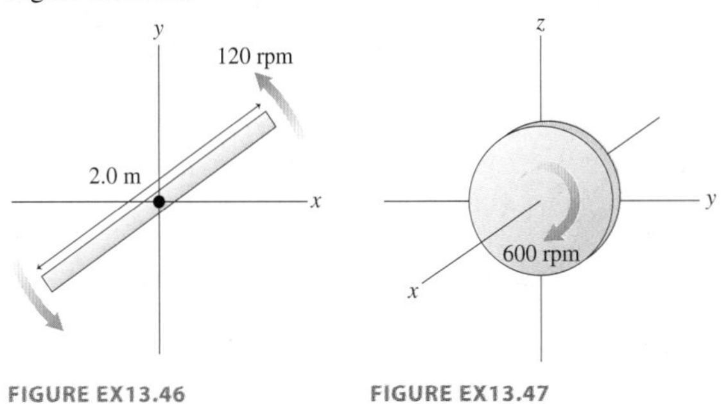

FIGURE EX13.46

FIGURE EX13.47

47. What is the angular momentum of the 2.0 kg, 4.0-cm-diameter rotating disk in Figure Ex13.47?

Problems

48. As the earth rotates, what is the speed of (a) a physics student in Miami, Florida, at latitude 26°, and (b) a physics student in Fairbanks, Alaska, at latitude 65°? Ignore the revolution of the earth around the sun.
49. A 60-cm-diameter wheel is rolling along at 20 m/s. What is the speed of a point at the front edge of the wheel?
50. An 800 g steel plate has the shape of the isosceles triangle shown in Figure P13.50. Locate the plate's center of mass.
    Hint: Divide the triangle into vertical strips of width dx, then relate the mass dm of a strip at position x to the values of x and dx.

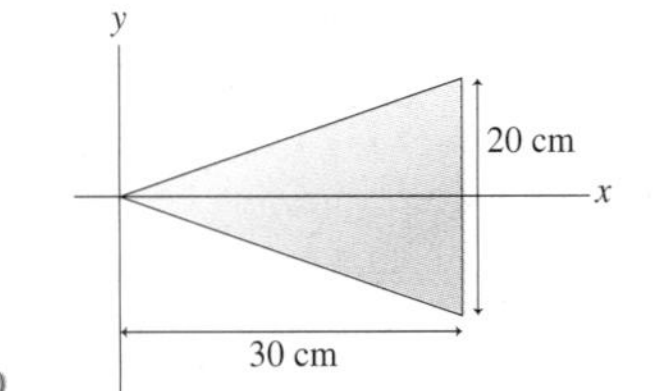

FIGURE P13.50

51. What is the moment of inertia of a 2.0 kg, 20-cm-diameter disk for rotation about an axis (a) through the center, and (b) through the edge of the disk.
52. Calculate by direct integration the moment of inertia for a thin rod of mass M and length L about an axis located distance d from one end. Confirm that your answer agrees with Table 13.3 when $d = 0$ and when $d = L/2$.
53. a. A disk of mass M and radius R has a hole of radius r centered on the axis. Calculate the moment of inertia of the disk.
    b. Confirm that your answer agrees with Table 13.3 when $r = 0$ and when $r = R$.
    c. A 4.0-cm-diameter disk with a 3.0-cm-diameter hole rolls down a 50-cm-long, 20° ramp. What is its speed at the bottom? What percent is this of the speed of a particle sliding down a frictionless ramp?
54. Determine the moment of inertia about the axis of the object shown in Figure P13.54.

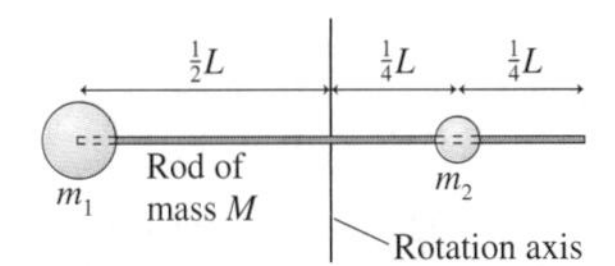

FIGURE P13.54

55. Calculate the moment of inertia about a center axis of a long rod with an $L \times L$ square cross section.

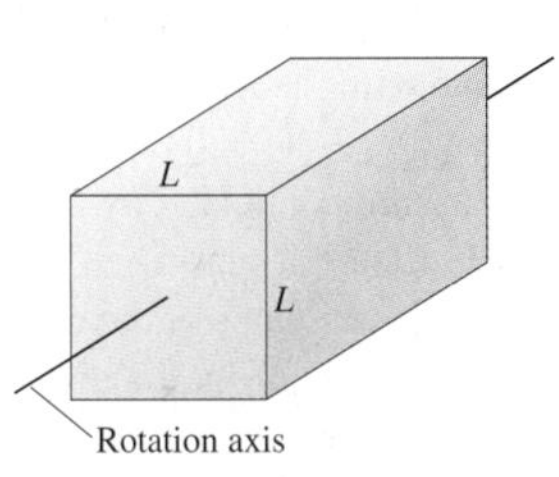

FIGURE P13.55

56. A 3.0-m-long rigid beam with a mass of 100 kg is supported at each end. An 80 kg student stands 2.0 m from support 1. How much upward force does each support exert on the beam?

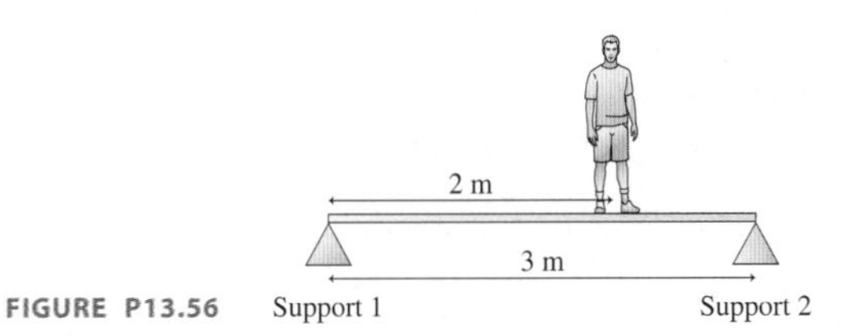

FIGURE P13.56

57. An 80 kg construction worker sits down 2.0 m from the end of a 1450 kg steel beam to eat his lunch. The cable supporting the beam is rated at 15,000 N. Should the worker be worried?

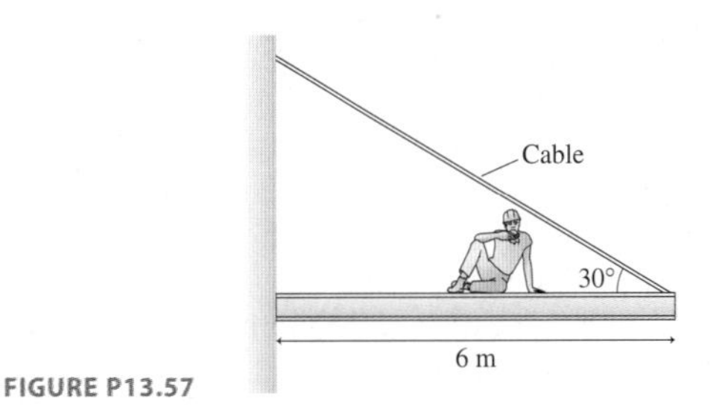

FIGURE P13.57

58. A forearm can be modeled as a 1.2 kg, 32-cm-long "beam" that pivots at the elbow and is supported by the biceps. How much force must the biceps exert to hold a 500 g ball with the forearm parallel to the floor?

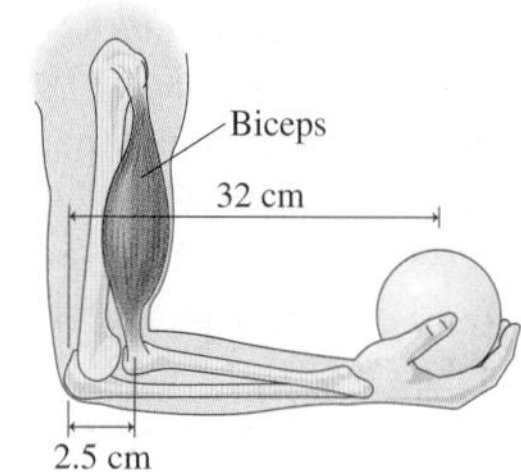

FIGURE P13.58

59. A 40 kg, 5.0-m-long beam is supported, but not attached to, the two posts in Figure P13.59. A 20 kg boy starts walking along the beam. How close can he get to the right end of the beam without it falling over?

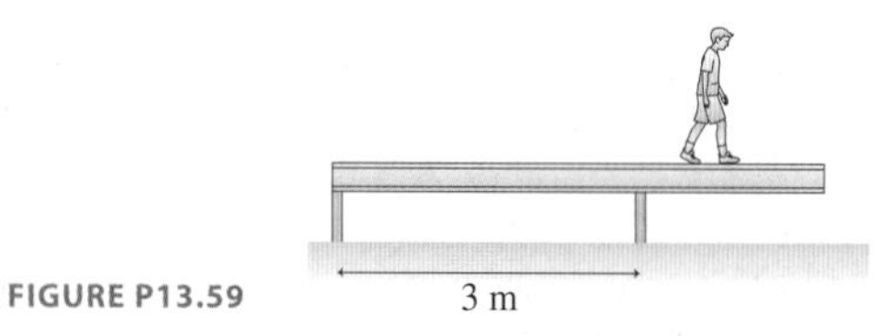

FIGURE P13.59

60. A 3.0-m-long ladder, as shown in Figure 13.35, leans against a frictionless wall. The coefficient of static friction between the ladder and the floor is 0.40. What is the minimum angle the ladder can make with the floor without slipping?
61. A 1.0 kg mass at $(x, y) = (20 \text{ cm}, 20 \text{ cm})$ and a 3.0 kg mass at (20 cm, 100 cm) are connected by a massless, rigid rod. They rotate about the center of mass.
    a. What are the coordinates of the center of mass?
    b. What is the moment of inertia about the center of mass?
    c. What constant torque will cause an angular velocity of 6.25 rad/s at the end of 3.0 s, starting from rest?
    d. At what angle, with respect to the axis of the rod, should 1.5 N forces be applied to each mass to give the torque you found in part c?
62. Starting from rest, a 12-cm-diameter compact disk takes 3.0 s to reach its operating angular velocity of 2000 rpm. Assume that the angular acceleration is constant. The disk's moment of inertia is 2.5×10^{-5} kg m^2.
    a. How much torque is applied to the disk?
    b. How many revolutions does it make before reaching full speed?
63. The two stars in a binary star system have masses 2.0×10^{30} kg and 6.0×10^{30} kg. They are separated by 2.0×10^{12} m. What are
    a. The system's rotation period, in years?
    b. The speed of each star?
64. A 60-cm-long, 500 g bar rotates in a horizontal plane on an axle that passes through the center of the bar. Compressed air is fed in through the axle, passes through a small hole down the length of the bar, and escapes as air jets from holes at the ends of the bar. The jets are perpendicular to the bar's axis. Starting from rest, the bar spins up to an angular velocity of 150 rpm at the end of 10 s.
    a. How much force does each jet of escaping air exert on the bar?
    b. If the axle is moved to one end of the bar while the air jets are unchanged, what will be the bar's angular velocity at the end of 10 seconds?
65. Flywheels are large, massive wheels used to store energy. They can be spun up slowly, then the wheel's energy can be released quickly to accomplish a task that demands high power. An industrial flywheel has a 1.5 m diameter and a mass of 250 kg. Its maximum angular velocity is 1200 rpm.
    a. A motor spins up the flywheel with a constant torque of 50 Nm. How long does it take the flywheel to reach top speed?
    b. How much energy is stored in the flywheel?
    c. The flywheel is disconnected from the motor and connected to a machine to which it will deliver energy. Half the energy stored in the flywheel is delivered in 2 seconds. What is the average power delivered to the machine?
    d. How much torque does the flywheel exert on the machine?
66. A 3.0 kg block is attached to a string that is wrapped around a 2.0 kg, 4.0-cm-diameter *hollow* cylinder that is free to rotate. (Use Figure 13.34 but treat the cylinder as hollow.) The block is released 1.0 m above the ground.
    a. Use Newton's second law to find the speed of the block as it hits the ground.
    b. Use conservation of energy to find the speed of the block as it hits the ground.
67. The two blocks in Figure P13.67 are connected by a massless rope that passes over a pulley. The pulley is 12 cm in diameter and has a mass of 2.0 kg. As the pulley turns, friction at the axle exerts a torque of magnitude 0.50 Nm. If the blocks are released from rest, how long does it take the 4.0 kg block to reach the floor?

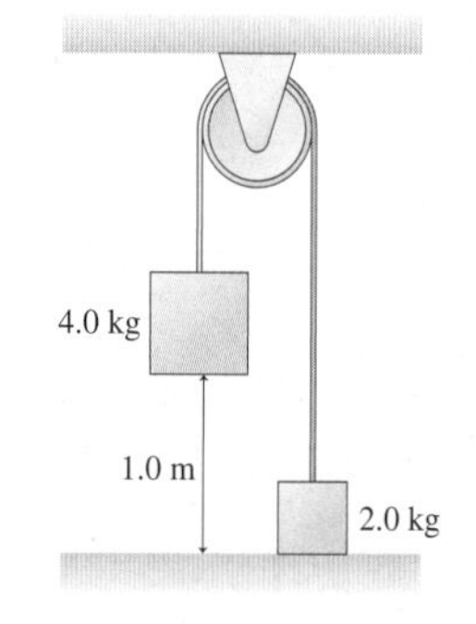

FIGURE P13.67

68. Blocks of mass m_1 and m_2 are connected by a massless string that passes over the frictionless pulley in Figure P13.68. Mass m_1 slides on a horizontal, frictionless surface. Mass m_2 is released while the blocks are at rest.
    a. Assume the pulley is massless. Find the acceleration of m_1 and the tension in the string. This is a Chapter 8 review problem.
    b. Suppose the pulley has mass m_p and radius R. Find the acceleration of m_1 and the tensions in the upper and lower portions of the string. Verify that your answers agree with part a if you set $m_p = 0$.

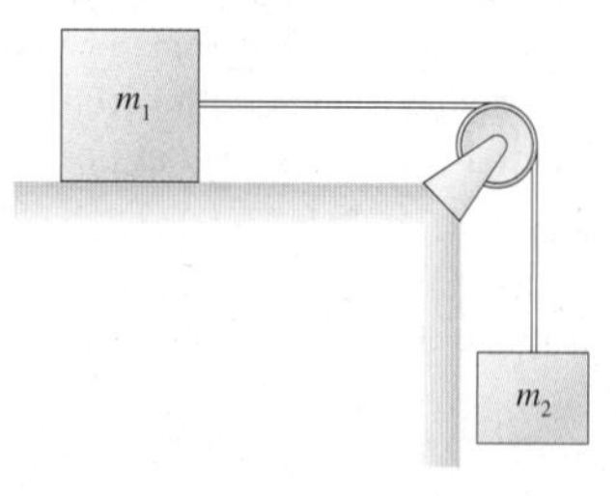

FIGURE P13.68

69. The 2.0 kg, 30-cm-diameter disk in Figure P13.69 is spinning at 300 rpm. How much friction force must the brake apply to the rim to bring the disk to a halt in 3.0 s?

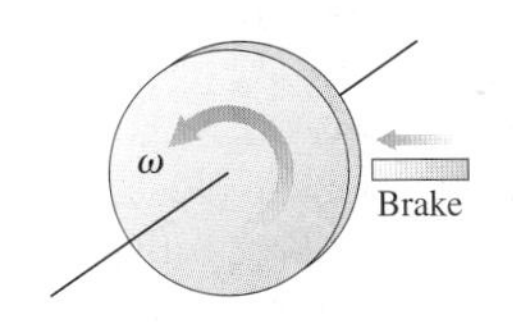

FIGURE P13.69

70. Suppose the connecting tunnel in Example 13.7 has a mass of 50,000 kg.
    a. How far from the 100,000 kg rocket is the center of mass of the entire structure?
    b. What is the structure's angular velocity after 30 s?
71. A hollow sphere is rolling along a horizontal floor at 5.0 m/s when it comes to a 30° incline. How far up the incline does it roll before reversing direction?
72. A 5.0 kg, 60-cm-diameter disk rotates on an axle passing through one edge. The axle is parallel to the floor. The cylinder is held with the center of mass at the same height as the axle, then released.
    a. What is the cylinder's initial angular acceleration?
    b. What is the cylinder's angular velocity when it is directly below the axle?

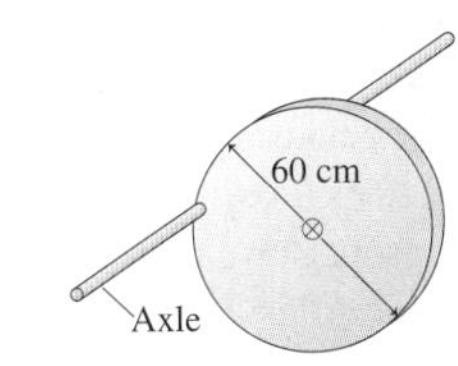

FIGURE P13.72

73. A hoop of mass M and radius R rotates about an axle at the edge of the hoop. The hoop starts at its highest position and is given a very small push to start it rotating. At its lowest position, what are (a) the angular velocity and (b) the speed of the lowest point on the hoop?

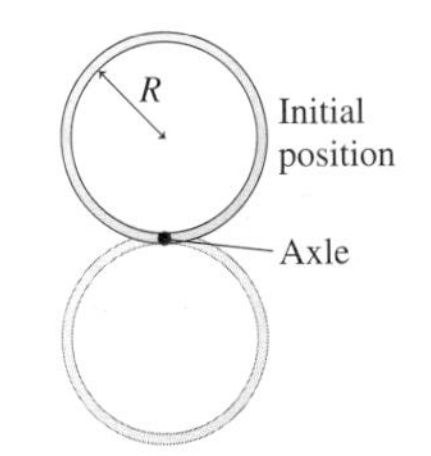

FIGURE P13.73

74. A long, thin rod of mass M and length L is standing straight up on a table. Its lower end rotates on a frictionless pivot. A very slight push causes the rod to fall over. As it hits the table, what are (a) the angular velocity and (b) the speed of the tip of the rod?
75. A sphere of mass M and radius R is rigidly attached to a thin rod that passes through the sphere at distance $\frac{1}{2}R$ from the center. A string wrapped around the rod exerts torque τ about the axis of the rod. Find an expression for the sphere's angular acceleration. The rod's moment of inertia is negligible.

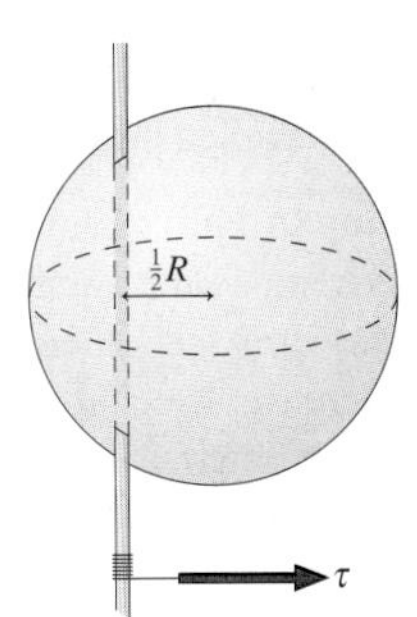

FIGURE P13.75

76. You've been given a pulley for your birthday. It's a fairly big pulley, 12 cm in diameter and with a mass of 2.0 kg. You get to wondering whether the pulley is uniform. That is, is the mass evenly distributed, or is it concentrated toward the center or near the rim? To find out, you hang the pulley on a hook, wrap a string around it several times, and suspend your 1.0 kg physics book 1.0 m above the floor. With your stopwatch, you find that it takes 0.71 s for your book to hit the floor. What can you conclude about the pulley?
77. A solid sphere of radius R is placed at a height of 30 cm on a 15° slope. It is released and rolls, without slipping, to the bottom.
    a. From what height should a circular hoop of radius R be released on the same slope in order to equal the sphere's speed at the bottom?
    b. Can a circular hoop of different diameter be released from a height of 30 cm and match the sphere's speed at the bottom? If so, what is the diameter? If not, why not?
78. A 2.0 kg, 20-cm-diameter turntable rotates at 100 rpm on frictionless bearings. Two 500 g blocks fall from above, hit the turntable simultaneously at opposite ends of a diagonal, and stick. What is the turntable's angular velocity, in rpm, just after this event?
79. A 200 g, 40-cm-diameter turntable rotates on frictionless bearings at 60 rpm. A 20 g block sits at the center of the turntable. A compressed spring shoots the block radially outward along a frictionless groove in the surface of the turntable. What is the turntable's rotation angular velocity when the block reaches the outer edge?
80. A merry-go-round is a common piece of playground equipment. A 3.0-m-diameter merry-go-round with a mass of 250 kg is spinning at 20 rpm. John runs tangent to the merry-go-round at 5.0 m/s, in the same direction that it is turning, and jumps onto the outer edge. John's mass is 30 kg. What is the merry-go-round's angular velocity, in rpm, after John jumps on?
81. A satellite follows the elliptical orbit shown. The only force on the satellite is the gravitational attraction of the planet. The satellite's speed at point a is 8000 m/s.
    a. Is there any torque on the satellite? Explain.
    b. What is the satellite's speed at point b?
    c. What is the satellite's speed at point c?

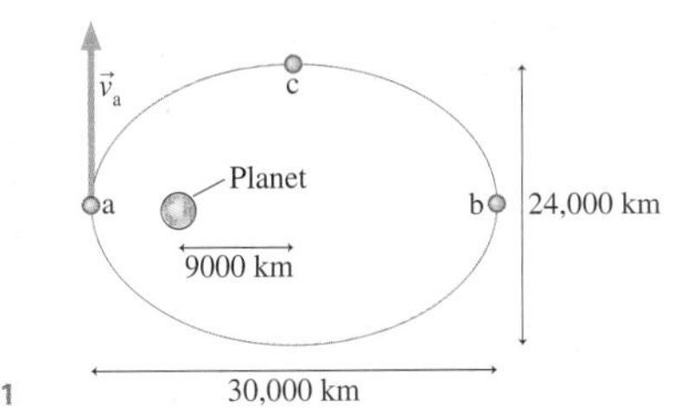

FIGURE P13.81

82. A 45 kg figure skater is spinning on the toes of her skates at 1.5 rpm. Her arms are outstretched as far as they will go. In this orientation, the skater can be modeled as a cylindrical torso (40 kg, 20 cm average diameter, 168 cm tall) plus two rod-like arms (2.5 kg each, 78 cm long) that each rotate about an axis through the inner end of the rod. The skater then raises her arms straight above her head, where she appears to be a 45 kg, 20-cm-diameter, 200-cm-tall cylinder. What is her new rotation frequency, in revolutions per second?

83. A billiard ball of mass m rolls without slipping across a table at speed v_0. It collides with a rod of length d and mass $M = 2m$. The rod is pivoted about a frictionless axle through its center, and it is initially hanging straight up and down at rest. After the collision, the billiard ball moves straight ahead with half its initial speed.
   a. What is the rod's angular velocity after the collision?
   b. Is the mechanical energy conserved?

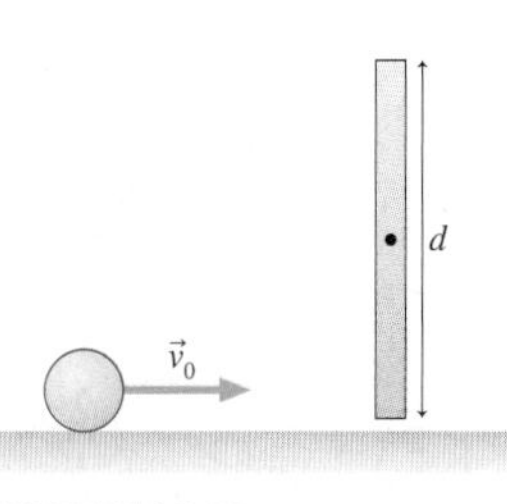

FIGURE P13.83

Challenge Problems

84. The marble rolls down a track and around a loop-the-loop of radius R. The marble has mass m and radius r. What minimum height h must the track have for the marble to make it around the loop-the-loop without falling off?

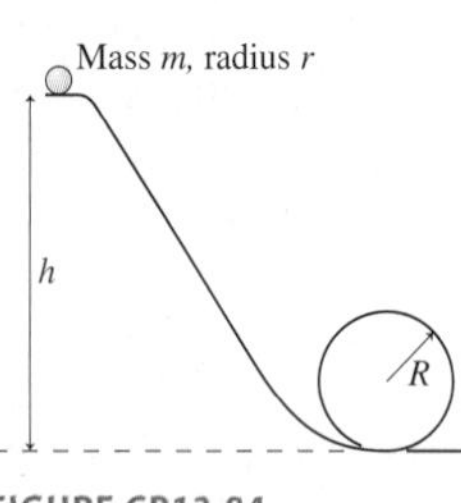

FIGURE CP13.84

85. A 2.0 kg wood block hangs from the bottom of a 1.0 kg, 1.0-m-long rod. The block and rod form a pendulum that swings on a frictionless pivot at the top end of the rod. A 10 g bullet is fired into the block, where it sticks, causing the pendulum to swing out to a 30° angle. What was the speed of the bullet? You can treat the wood block as a particle.
86. A 10 g bullet traveling at 400 m/s strikes a 10 kg, 1.0-m-wide door at the edge opposite the hinge. The bullet embeds itself in the door, causing the door to swing open. What is the angular velocity of the door just after impact?
87. During most of its lifetime, a star maintains an equilibrium size in which the inward force of gravity on each atom is balanced by an outward pressure force due to the heat of the nuclear reactions in the core. But after all the hydrogen "fuel" is consumed by nuclear fusion, the pressure force drops and the star undergoes a *gravitational collapse* until it becomes a *neutron star*. In a neutron star, the electrons and protons of the atoms are squeezed together by gravity until they fuse into neutrons. Neutron stars spin very rapidly and emit intense pulses of radio and light waves, one pulse per rotation. These "pulsing stars" were discovered in the 1960s and are called *pulsars*.
   a. A star with the mass ($M = 2.0 \times 10^{30}$ kg) and size ($R = 3.5 \times 10^8$ m) of our sun rotates once every 30 days. After undergoing gravitational collapse, the star forms a pulsar that is observed by astronomers to emit radio pulses every 0.1 s. By treating the neutron star as a solid sphere, deduce its radius.
   b. What is the speed of a point on the equator of the neutron star?

   Your answer will be somewhat too large because a star cannot be accurately modeled as a solid sphere. Even so, you will be able to show that a star, whose mass is 10^6 larger than the earth's, can be compressed by gravitational forces to a size smaller than a typical state in the United States!
88. A physics professor stands at rest on a 5.0 kg, 50-cm-diameter frictionless turntable. His assistant has a 64-cm-diameter bicycle wheel to which 4.0 kg of lead weights have been added around the rim. Handles extend outward from the axis so that the wheel can be held as it spins. The assistant spins the wheel to 180 rpm and holds it in a horizontal plane (the rotation axis is vertical) such that the rotation is ccw as seen from the ceiling. He then hands the spinning wheel to the professor.
   a. When the professor takes the wheel by the handles and the assistant lets go, does anything happen to the professor? If so, *describe* the professor's motion and *calculate* any relevant numerical quantities. If not, explain why not.
   b. Then the professor turns the spinning wheel over 180° so that the handle that had been pointing toward the ceiling now points toward the floor. Does anything happen to the professor? If so, *describe* the professor's motion and *calculate* any relevant numerical quantities. If not, explain why not.

   Hint: You'll need to *model* both the professor and the wheel. The professor has a total mass of 75 kg. His legs and torso are 70 kg. They have an average diameter of 25 cm and a height of 180 cm. His arms are 2.5 kg each, and he hold the handles of the wheel 45 cm from the center of his body. Don't forget that the wheel both spins *and* moves with the professor.

STOP TO THINK ANSWERS

Stop to Think 13.1: c. ω is negative because the rotation is cw. Because ω is negative but becoming *less* negative, the change $\Delta\omega$ is *positive*. So α is positive.

Stop to Think 13.2: $\tau_e > \tau_a = \tau_d > \tau_b > \tau_c$. The tangential component in e is larger than 2 N.

Stop to Think 13.3: $\alpha_b > \alpha_a > \alpha_c = \alpha_d = \alpha_e$. Angular acceleration is proportional to torque and inversely proportional to the moment of inertia. The moment of inertia depends on the *square* of the radius. The tangential force component in e is the same as in d.

Stop to Think 13.4: $I_a > I_d > I_b > I_c$. The moment of inertia is smaller when mass is more concentrated near the rotation axis.

Stop to Think 13.5: c > d > a = b. To keep the meter stick in equilibrium, the student must supply a torque equal and opposite to the torque due to the hanging masses. Torque depends on the mass *and* on how far the mass is from the pivot point.

Stop to Think 13.6: d. There is no net torque on the bucket + rain system, so the angular momentum is conserved. The addition of mass on the outer edge of the circle increases I, so ω must decrease. Mechanical energy is not conserved because the raindrop collisions are inelastic.

14 Oscillations

A playground swing is just one example of oscillatory motion.

► **Looking Ahead**
The goal of Chapter 14 is to understand systems that oscillate with simple harmonic motion. In this chapter you will learn to:

- Understand the kinematics of simple harmonic motion.
- Use graphical and mathematical representations of oscillatory motion.
- Understand energy in oscillating systems.
- Understand the dynamics of oscillating systems.
- Recognize the importance of resonance and damping in oscillating systems.

◄ **Looking Back**
Simple harmonic motion is closely related to circular motion. Much of our analysis of oscillating systems will be based on the law of conservation of energy. Please review:

- Section 7.1 Uniform circular motion.
- Sections 10.4 and 10.5 Restoring forces and elastic potential energy.
- Section 10.7 Energy diagrams.

This girl is having a good time on the swing. At the same time, she is demonstrating an important type of motion—*oscillatory motion*. Examples of oscillatory motion abound. A marble rolling back and forth in the bottom of a bowl and a car bouncing up and down on its springs are oscillating. So are a piece of vibrating machinery, a ringing bell, and the current in an electric circuit used to drive an antenna. A vibrating guitar string pushes the air molecules back and forth to send out a sound wave, showing that oscillations are closely related to waves.

Oscillatory motion is a repetitive motion back and forth about an equilibrium position. Swinging motions and vibrations of all kinds are oscillatory motions. All oscillatory motion is *periodic*.

Our goal in this chapter is to study the physics of oscillations. Much of our analysis will be focused on the most basic form of oscillatory motion, *simple harmonic motion*. We will start with the kinematics of simple harmonic motion—a mathematical description of the motion. Then we will examine oscillatory motion from the twin perspectives of energy and Newton's laws. Finally, we will look at how oscillations are built up by driving forces and how they decay over time.

14.1 Simple Harmonic Motion

Objects or systems of objects that undergo oscillatory motion are called **oscillators.** Figure 14.1 shows position-versus-time graphs for several different oscillating systems. Although the shapes of the graphs are different, all these oscillators have two things in common:

1. The oscillation takes place about an equilibrium position, and
2. The motion is periodic.

The time to complete one full cycle, or one oscillation, is called the **period** of the motion. Period is given the symbol T.

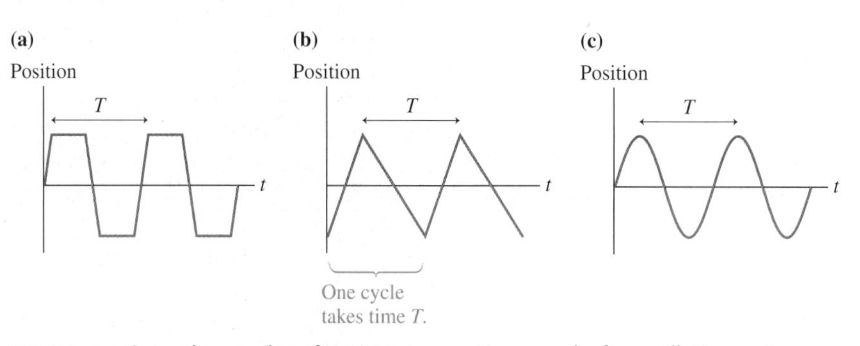

FIGURE 14.1 Several examples of position-versus-time graphs for oscillating systems.

A closely related piece of information is the number of cycles, or oscillations, completed per second. If the period is $\frac{1}{10}$ s, then the oscillator can complete 10 cycles in one second. Conversely, an oscillation period of 10 s allows only $\frac{1}{10}$ of a cycle to be completed per second. In general, T seconds per cycle implies that $1/T$ cycles will be completed each second. The number of cycles per second is called the **frequency** f of the oscillation. The relationship between frequency and period is

$$f = \frac{1}{T} \quad \text{or} \quad T = \frac{1}{f} \tag{14.1}$$

The units of frequency are **hertz,** abbreviated Hz, named in honor of the German physicist Heinrich Hertz who produced the first artificially generated radio waves in 1887. By definition,

$$1 \text{ Hz} \equiv 1 \text{ cycle per second} = 1 \text{ s}^{-1}$$

We will frequently deal with very rapid oscillations and make use of the units shown in Table 14.1.

TABLE 14.1 Units of frequency

Frequency	Period
10^3 Hz = 1 kilohertz = 1 kHz	1 ms
10^6 Hz = 1 megahertz = 1 MHz	1 μs
10^9 Hz = 1 gigahertz = 1 GHz	1 ns

NOTE ▶ Uppercase and lowercase letters *are* important. 1 MHz is 1 megahertz = 10^6 Hz, but 1 mHz would be 1 millihertz = 10^{-3} Hz! ◀

EXAMPLE 14.1 Frequency and period of a radio station

What is the oscillation period for the broadcast of a 100 MHz FM radio station?

SOLVE The frequency of current oscillations in the radio transmitter is 100 MHz = 1.0×10^8 Hz. The period is the inverse of the frequency, hence

$$T = \frac{1}{f} = \frac{1}{1.0 \times 10^8 \text{ Hz}} = 1.0 \times 10^{-8} \text{ s} = 10 \text{ ns}$$

A system can oscillate in many ways, but we will be especially interested in the smooth *sinusoidal* oscillation of Figure 14.1c. This sinusoidal oscillation, the most basic of all oscillatory motions, is called **simple harmonic motion,** often abbreviated SHM. Let's look at a graphical description before we dive into the mathematics of simple harmonic motion.

Figure 14.2a shows an air-track glider attached to a spring. If the glider is pulled out a few centimeters and released, it will oscillate back and forth on the nearly frictionless air track. Figure 14.2b shows actual results from an experiment in which a computer was used to measure the glider's position 20 times every second. This is a position-versus-time graph that has been rotated 90° from its usual orientation in order for the x-axis to match the motion of the glider.

The object's maximum displacement from equilibrium is called the **amplitude** A of the motion. The object's position oscillates between $x = -A$ and $x = +A$. When using a graph, notice that the amplitude is the distance from the *axis* to the maximum, *not* the distance from the minimum to the maximum.

Figure 14.3a shows the data with the graph axes in their "normal" positions. You can see that the amplitude in this experiment was $A = 0.17$ m, or 17 cm. You can also measure the period to be $T = 1.60$ s. Thus the oscillation frequency was $f = 1/T = 0.625$ Hz.

Figure 14.3b is a velocity-versus-time graph that the computer produced by using $\Delta x/\Delta t$ to find the slope of the position graph at each point. The velocity graph is also sinusoidal, oscillating between $-v_{max}$ (maximum speed to the left) and $+v_{max}$ (maximum speed to the right). As the figure shows,

- The instantaneous velocity is zero at the points where $x = \pm A$. These are the *turning points* in the motion.
- The maximum speed v_{max} is reached as the object passes through the equilibrium position at $x = 0$ m. The *velocity* is positive as the object moves to the right but *negative* as it moves to the left.

We can ask three important questions about this oscillating system:

1. How is the maximum speed v_{max} related to the amplitude A?
2. How are the period and frequency related to the object's mass m, the spring constant k, and the amplitude A?
3. Is the sinusoidal oscillation a consequence of Newton's laws?

A mass oscillating on a spring is the prototype of simple harmonic motion. Our analysis, in which we answer these questions, will be of a spring-mass system. Even so, most of what we learn will be applicable to other types of SHM.

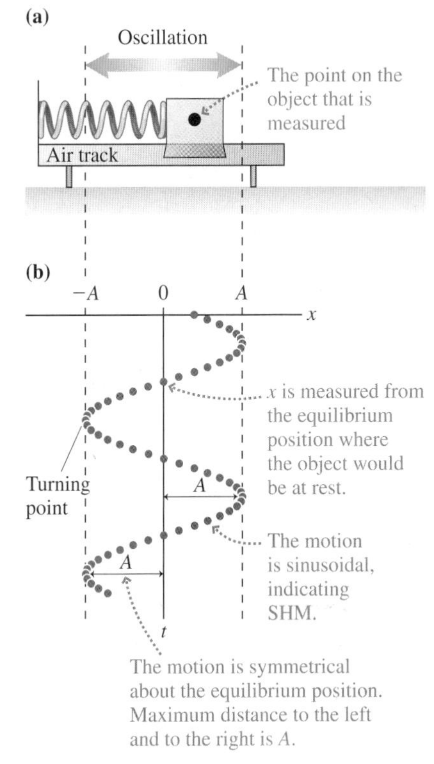

FIGURE 14.2 A prototype simple-harmonic-motion experiment.

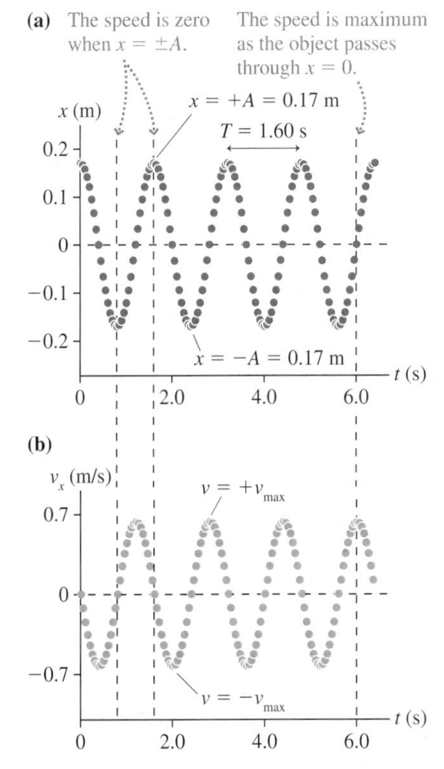

FIGURE 14.3 Position and velocity graphs of the experimental data.

Kinematics of Simple Harmonic Motion

Figure 14.4 on the next page redraws the position-versus-time graph of Figure 14.3 as a smooth curve. Although these are empirical data (we don't yet have any "theory" of oscillation) the position-versus-time graph is clearly a cosine function. We can write the object's position as

$$x(t) = A\cos\left(\frac{2\pi t}{T}\right) \tag{14.2}$$

where the notation $x(t)$ indicates that the position x is a *function* of time t. Because $\cos(2\pi) = \cos(0)$, it's easy to see that the position at time $t = T$ is the same as the position at $t = 0$. In other words, this is a cosine function with period T. Be sure to convince yourself that this function agrees with the five special points shown in Figure 14.4.

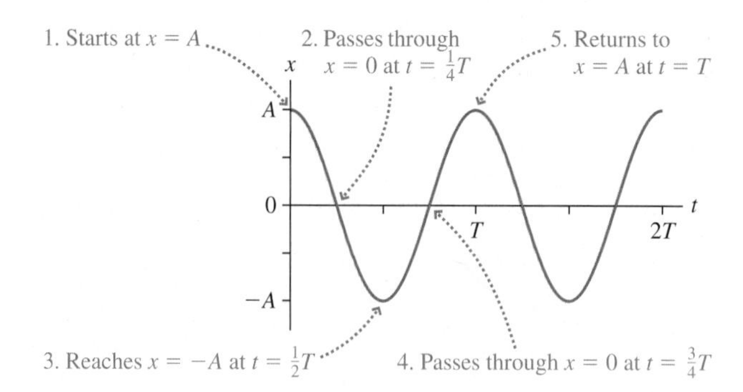

FIGURE 14.4 The position-versus-time graph for simple harmonic motion.

NOTE ▶ The argument of the cosine function is in *radians*. That will be true throughout this chapter. It's especially important to remember to set your calculator to radian mode before working oscillation problems. Leaving it in degree mode will lead to major errors. ◀

We can write Equation 14.2 in two alternative forms. Because the oscillation frequency is $f = 1/T$, we can write

$$x(t) = A\cos(2\pi f t) \tag{14.3}$$

Recall from Chapter 7 that a particle in circular motion has an *angular velocity* ω that is related to the period by $\omega = 2\pi/T$, where ω is in rad/s. Now that we've defined the frequency f, you can see that ω and f are related by

$$\omega \text{ (in rad/s)} = \frac{2\pi}{T} = 2\pi f \text{ (in Hz)} \tag{14.4}$$

In this context, ω is called the **angular frequency.** The position can be written in terms of ω as

$$x(t) = A\cos\omega t \tag{14.5}$$

Equations 14.2, 14.3, and 14.5 are equivalent ways to write the position of an object moving in simple harmonic motion.

Just as the position graph was clearly a cosine function, the velocity graph shown in Figure 14.5 is clearly an "upside-down" sine function with the same period T. The velocity v_x, which is a function of time, can be written

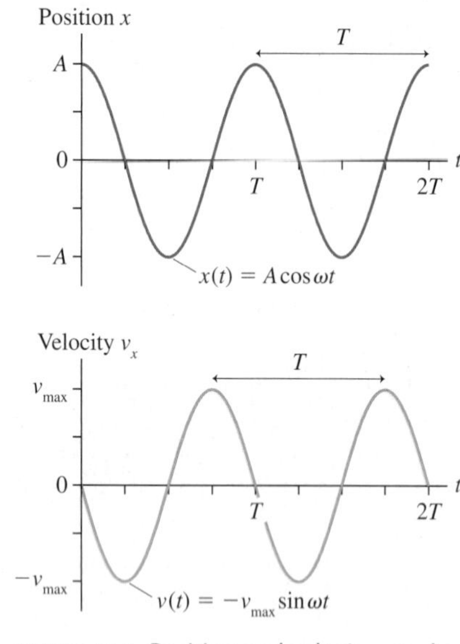

FIGURE 14.5 Position and velocity graphs for simple harmonic motion.

$$v_x(t) = -v_{max}\sin\left(\frac{2\pi t}{T}\right) = -v_{max}\sin(2\pi f t) = -v_{max}\sin\omega t \tag{14.6}$$

NOTE ▶ v_{max} is the maximum *speed* and thus is inherently a *positive* number. The minus sign in Equation 14.6 is needed to turn the sine function upside down. ◀

We deduced Equation 14.6 from the experimental results, but we could equally well find it from the position function of Equation 14.2. After all, velocity is the time derivative of position. Table 14.2 reminds you of the derivatives of sine and cosine functions. Using the derivative of the cosine function, we find

TABLE 14.2 Derivatives of sine and cosine functions

$\frac{d}{dt}(a\sin(bt + c)) = +ab\cos(bt + c)$
$\frac{d}{dt}(a\cos(bt + c)) = -ab\sin(bt + c)$

$$v_x(t) = \frac{dx}{dt} = -\frac{2\pi A}{T}\sin\left(\frac{2\pi t}{T}\right) = -2\pi f A\sin(2\pi f t) = -\omega A\sin\omega t \tag{14.7}$$

We can draw an important conclusion by comparing Equation 14.7, the mathematical definition of velocity, to Equation 14.6, the empirical description of the velocity. Namely, the maximum speed of an oscillation is related to the oscillation amplitude by

$$v_{max} = \frac{2\pi A}{T} = 2\pi f A = \omega A \quad (14.8)$$

Equation 14.8 answers the first question we posed above, which was how the maximum speed v_{max} is related to the amplitude A. Not surprisingly, the object moves faster if you stretch the spring further and give the oscillation a larger amplitude.

EXAMPLE 14.2 A system in simple harmonic motion

An air-track glider is attached to a spring, pulled 20 cm to the right, and released at $t = 0$. It makes 15 oscillations in 10 s.

a. What is the period of oscillation?
b. What is the object's maximum speed?
c. What are the position and velocity at $t = 0.80$ s?

MODEL An object oscillating on a spring is in simple harmonic motion.

SOLVE

a. The oscillation frequency is

$$f = \frac{15 \text{ oscillations}}{10 \text{ s}} = 1.5 \text{ oscillations/s} = 1.5 \text{ Hz}$$

Thus the period is $T = 1/f = 0.667$ s.

b. The oscillation amplitude is $A = 0.20$ m. Thus the maximum speed is

$$v_{max} = \frac{2\pi A}{T} = \frac{2\pi(0.20 \text{ m})}{0.667 \text{ s}} = 1.88 \text{ m/s}$$

c. The object starts at $x = +A$ at $t = 0$. This is exactly the oscillation described by Equations 14.2 and 14.6. The position at $t = 0.80$ s is

$$x = A\cos\left(\frac{2\pi t}{T}\right) = (0.20 \text{ m})\cos\left(\frac{2\pi(0.80 \text{ s})}{0.667 \text{ s}}\right)$$

$$= (0.20 \text{ m})\cos(7.54 \text{ rad}) = 0.062 \text{ m} = 6.2 \text{ cm}$$

The velocity at this instant of time is

$$v_x = -v_{max}\sin\left(\frac{2\pi t}{T}\right) = -(1.88 \text{ m/s})\sin\left(\frac{2\pi(0.80 \text{ s})}{0.667 \text{ s}}\right)$$

$$= -(1.88 \text{ m/s})\sin(7.54 \text{ rad}) = -1.79 \text{ m/s} = -179 \text{ cm/s}$$

At $t = 0.80$ s, which is slightly more than one period, the object is 6.2 cm to the right of equilibrium and moving to the *left* at 179 cm/s. Notice the use of radians in the calculations.

EXAMPLE 14.3 Finding the time

A mass oscillating in simple harmonic motion starts at $x = A$ and has period T. At what time, as a fraction of T, does the object first pass through $x = \frac{1}{2}A$?

SOLVE Figure 14.4 noted that the object passes through the equilibrium position $x = 0$ at $t = \frac{1}{4}T$. This is one-quarter of the total distance in one-quarter of a period. You might expect it to take $\frac{1}{8}T$ to reach $\frac{1}{2}A$, but this is *not* the case because the SHM graph is not linear between $x = A$ and $x = 0$. We need to use $x(t) = A\cos(2\pi t/T)$. First, write the equation with $x = \frac{1}{2}A$:

$$x = \frac{A}{2} = A\cos\left(\frac{2\pi t}{T}\right)$$

Then solve for the time at which this position is reached:

$$t = \frac{T}{2\pi}\cos^{-1}\left(\frac{1}{2}\right) = \frac{T}{2\pi}\frac{\pi}{3} = \frac{1}{6}T$$

ASSESS The motion is slow at the beginning and then speeds up, so it takes longer to move from $x = A$ to $x = \frac{1}{2}A$ than it does to move from $x = \frac{1}{2}A$ to $x = 0$. Notice that the answer is independent of the amplitude A.

STOP TO THINK 14.1 An object moves with simple harmonic motion. If the amplitude and the period are both doubled, the object's maximum speed is

a. quadrupled.
b. doubled.
c. unchanged.
d. halved.
e. quartered.

14.2 Simple Harmonic Motion and Circular Motion

The graphs of Figure 14.5 and the position function $x(t) = A\cos\omega t$ are for an oscillation in which the object just happened to be at $x_0 = A$ at $t = 0$. But you will recall that $t = 0$ is an arbitrary choice, the instant of time when you or someone else starts a stopwatch. What if you had started the stopwatch when the object was at $x_0 = -A$, or when the object was somewhere in the middle of an oscillation? In other words, what if the oscillator had different *initial conditions*. The position graph would still show an oscillation, but neither Figure 14.5 nor $x(t) = A\cos\omega t$ would describe the motion correctly.

To learn how to describe the oscillation for other initial conditions it will help to turn to a topic you studied in Chapter 7—circular motion. There's a very close connection between simple harmonic motion and circular motion.

Imagine you have a turntable with a small ball glued to the edge. Figure 14.6a shows how to make a "shadow movie" of the ball by projecting a light past the ball and onto a screen. The ball's shadow oscillates back and forth as the turntable rotates. This is certainly periodic motion, with the same period as the turntable, but is it simple harmonic motion?

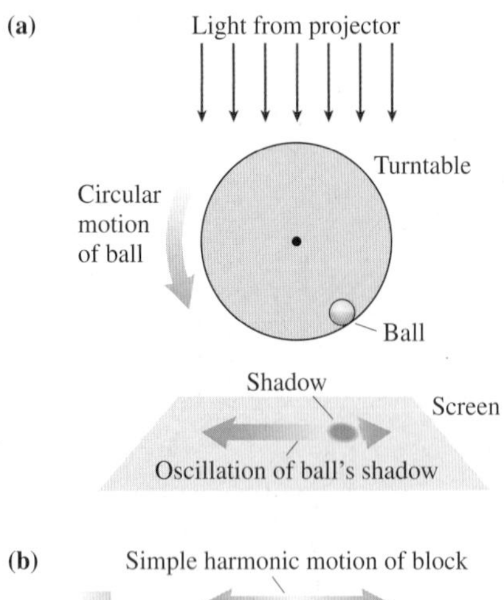

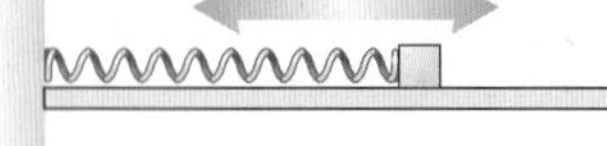

FIGURE 14.6 A projection of the circular motion of a rotating ball matches the simple harmonic motion of an object on a spring.

To find out, you could place a real object on a real spring directly below the shadow, as shown in Fig, 14.6b. If you did so, and if you adjusted the turntable to have the same period as the spring, you would find that the shadow's motion exactly matches the simple harmonic motion of the object on the spring. **Uniform circular motion projected onto one dimension is simple harmonic motion.**

To understand this, consider the particle in Figure 14.7. It is in uniform circular motion, moving *counterclockwise* in a circle with radius A. As in Chapter 7, we can locate the particle by the angle ϕ measured ccw from the x-axis. Projecting the ball's shadow onto a screen in Figure 14.6 is equivalent to observing just the x-component of the particle's motion. Figure 14.7 shows that the x-component, when the particle is at angle ϕ, is

$$x = A\cos\phi \tag{14.9}$$

Recall that the particle's *angular velocity*, in rad/s, is

$$\omega = \frac{d\phi}{dt} \tag{14.10}$$

This is the rate at which the angle ϕ is increasing. If the particle starts from $\phi_0 = 0$ at $t = 0$, its angle at a later time t is simply

$$\phi = \omega t \tag{14.11}$$

As ϕ increases, the particle's x-component is

$$x(t) = A\cos\omega t \tag{14.12}$$

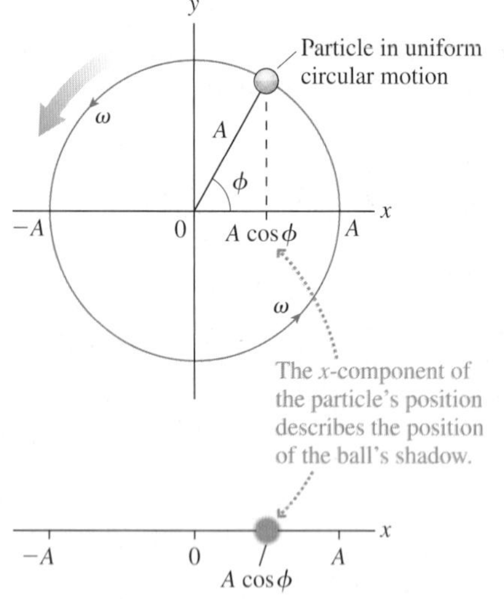

FIGURE 14.7 A particle in uniform circular motion with radius A and angular velocity ω.

This is identical to Equation 14.5 for the position of a mass on a spring! Thus the x-component of a particle in uniform circular motion is simple harmonic motion.

NOTE ► When used to describe oscillatory motion, ω is called the *angular frequency* rather than the angular velocity. The angular frequency of an oscillator has the same numerical value, in rad/s, as the angular velocity of the corresponding particle in circular motion. ◄

The names and units can be a bit confusing until you get used to them. It may help to notice that *cycle* and *oscillation* are not true units. Unlike the "standard meter" or the "standard kilogram," to which you could compare a length or a mass, there is no "standard cycle" to which you can compare an oscillation. Cycles and oscillations are simply counted events. Thus the frequency f has units of hertz, where 1 Hz = 1 s^{-1}. We may *say* "cycles per second" just to be clear, but the actual units are only "per second."

The radian is the SI unit of angle. However, the radian is a *defined* unit. Further, its definition as a ratio of two lengths ($\theta = s/r$) makes it a *pure number* without dimensions. As we noted in Chapter 7, the unit of angle, be it radians or degrees, is really just a *name* to remind us that we're dealing with an angle. The 2π in Equation 14.12 (and in many similar situations), which is stated without units, *means* 2π rad/cycle. When multiplied by the frequency f in cycles/s, it gives the frequency in rad/s. That is why, in this context, ω is called the angular *frequency*.

NOTE ▶ *Hertz* is specifically "cycles per second" or "oscillations per second." It is used for f but *not* for ω. We'll always be careful to use rad/s for ω, but you should be aware that many books give the units of ω as simply s^{-1}. ◀

The Phase Constant

Now we're ready to consider the issue of other initial conditions. The particle in Figure 14.7 started at $\phi_0 = 0$. This was equivalent to an oscillator starting at the far right edge, $x_0 = A$. Figure 14.8 shows a more general situation in which the initial angle ϕ_0 can have any value. The angle at a later time t is then

$$\phi = \omega t + \phi_0 \tag{14.13}$$

In this case, the particle's projection onto the x-axis at time t is

$$x(t) = A\cos(\omega t + \phi_0) \tag{14.14}$$

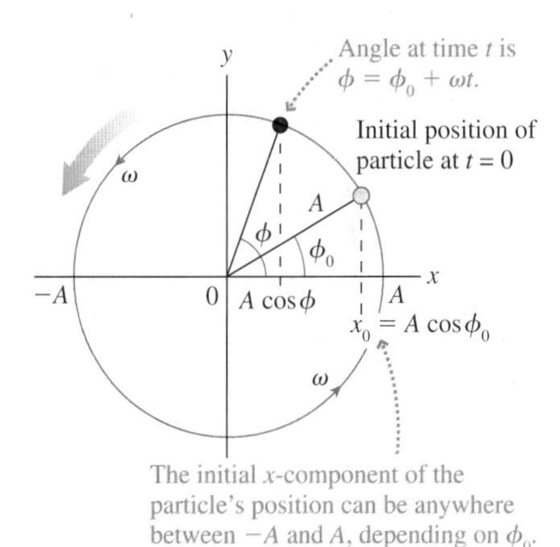

FIGURE 14.8 A particle in uniform circular motion with initial angle ϕ_0.

If Equation 14.14 describes the particle's projection, then it must also be the position of an oscillator in simple harmonic motion. The oscillator's velocity v_x is found by taking the derivative dx/dt. The resulting equations,

$$\begin{aligned} x(t) &= A\cos(\omega t + \phi_0) \\ v_x(t) &= -\omega A\sin(\omega t + \phi_0) = -v_{max}\sin(\omega t + \phi_0) \end{aligned} \tag{14.15}$$

are the two primary kinematic equations of simple harmonic motion.

The quantity $\phi = \omega t + \phi_0$, which steadily increases with time, is called the **phase** of the oscillation. The phase is simply the *angle* of the circular-motion particle whose shadow matches the oscillator. The constant ϕ_0 is called the **phase constant.** It specifies the *initial conditions* of the oscillator.

To see what the phase constant means, set $t = 0$ in Equations 14.15:

$$\begin{aligned} x_0 &= A\cos\phi_0 \\ v_{0x} &= -\omega A\sin\phi_0 \end{aligned} \tag{14.16}$$

The position x_0 and velocity v_{0x} at $t = 0$ are the initial conditions. **Different values of the phase constant correspond to different starting points on the circle and thus to different initial conditions.**

The perfect cosine function of Figure 14.5 and the equation $x(t) = A\cos\omega t$ are for an oscillation with $\phi_0 = 0$ rad. You can see from Equations 14.16 that $\phi_0 = 0$ rad implies $x_0 = A$ and $v_0 = 0$. That is, the particle starts from rest at the point of maximum displacement.

Figure 14.9 on the next page illustrates these ideas by looking at three values of the phase constant: $\phi_0 = \pi/3$ rad (60°), $-\pi/3$ rad (−60°), and π rad (180°). For each value of ϕ_0 you see the oscillator at its starting position, the starting position shown on a circle, and both position and velocity graphs. All the graphs have the same amplitude and the same period, but they are *shifted* relative to the graphs of Figure 14.5 (which were for $\phi_0 = 0$ rad) so that the maximum displacement $x = A$ occurs at a time other than $t = 0$.

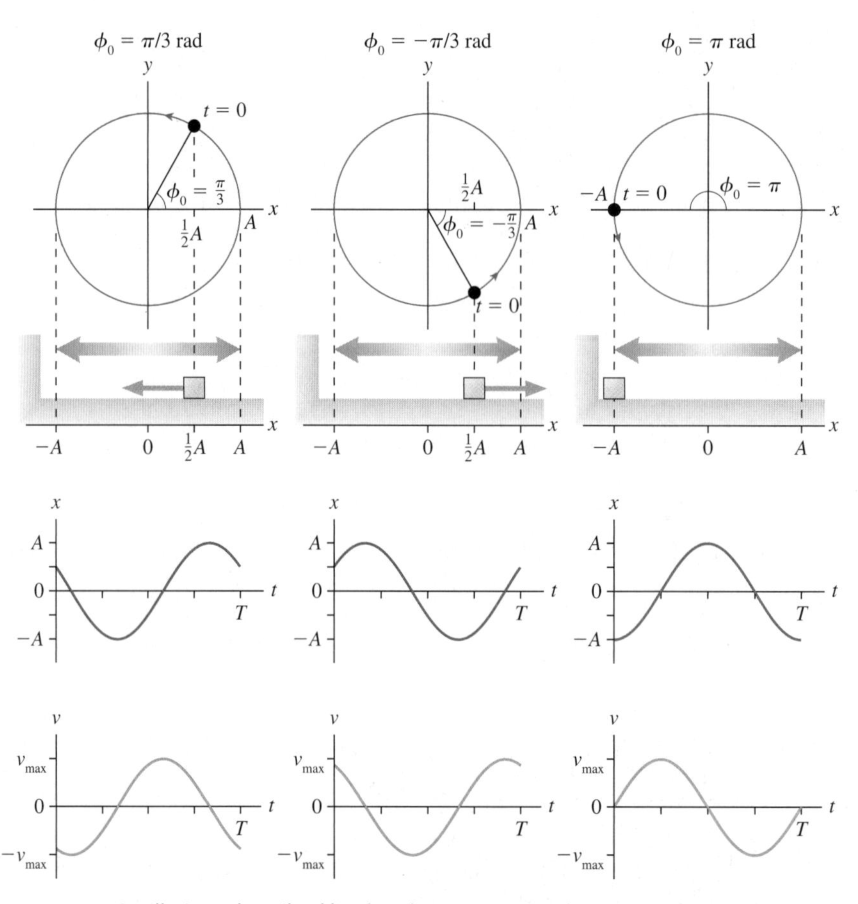

FIGURE 14.9 Oscillations described by the phase constants $\phi_0 = \pi/3$ rad, $-\pi/3$ rad, and π rad.

Notice that $\phi_0 = \pi/3$ rad and $\phi_0 = -\pi/3$ rad have the same starting position, $x_0 = \frac{1}{2}A$. This is a property of the cosine function in Equation 14.16. But these are *not* the same initial conditions. In one case the oscillator starts at $\frac{1}{2}A$ while moving to the right, in the other case it starts at $\frac{1}{2}A$ while moving to the left. You can distinguish between the two by visualizing the circular motion.

All values of the phase constant ϕ_0 between 0 and π rad correspond to a particle in the upper half of the circle and *moving to the left*. Thus v_{0x} is negative. All values of the phase constant ϕ_0 between π and 2π rad (or, as they are usually stated, between $-\pi$ and 0 rad) have the particle in the lower half of the circle and *moving to the right*. Thus v_{0x} is positive. If you're told that the oscillator is at $x = \frac{1}{2}A$ and moving to the right at $t = 0$, then the phase constant must be $\phi_0 = -\pi/3$ rad, not $+\pi/3$ rad.

EXAMPLE 14.4 Using the initial conditions

An object on a spring oscillates with a period of 0.80 s and an amplitude of 10 cm. At $t = 0$ s, it is 5.0 cm to the left of equilibrium and moving to the left. What are its position and direction of motion at $t = 2.0$ s?

MODEL An object oscillating on a spring is in simple harmonic motion.

SOLVE We can find the phase constant ϕ_0 from the initial condition $x_0 = -5.0 \text{ cm} = A\cos\phi_0$. This condition gives:

$$\phi_0 = \cos^{-1}\left(\frac{x_0}{A}\right) = \cos^{-1}\left(-\frac{1}{2}\right) = \pm\frac{2}{3}\pi \text{ rad} = \pm 120°$$

Because the oscillator is moving to the *left* at $t = 0$, it is in the upper half of the circular-motion diagram and must have a phase constant between 0 and π rad. Thus ϕ_0 is $\frac{2}{3}\pi$ rad. The angular frequency is

$$\omega = \frac{2\pi}{T} = \frac{2\pi}{0.80\ \text{s}} = 7.854\ \text{rad/s}$$

Thus the object's position at time $t = 2.0$ s is

$$\begin{aligned} x(t) &= A\cos(\omega t + \phi_0) \\ &= (10\ \text{cm})\cos\left((7.854\ \text{rad/s})(2.0\ \text{s}) + \frac{2}{3}\pi\right) \\ &= (10\ \text{cm})\cos(17.80\ \text{rad}) \\ &= 5.00\ \text{cm} \end{aligned}$$

The object is now 5.0 cm to the right of equilibrium. But which way is it moving? There are two ways to find out. The direct way is to calculate the velocity at $t = 2.0$ s:

$$v_x = -\omega A \sin(\omega t + \phi_0) = +68.1\ \text{cm/s}$$

The velocity is positive, so the motion is to the right. Alternatively, we could note that the phase at $t = 2.0$ s is $\phi = 17.80$ rad. Dividing by π, you can see that

$$\phi = 17.80\ \text{rad} = 5.67\pi\ \text{rad} = (4\pi + 1.67\pi)\ \text{rad}$$

The 4π rad represents complete revolutions. The "extra" phase of 1.67π rad falls between π and 2π rad, so the particle in the circular-motion diagram is in the lower half of the circle and moving to the right.

NOTE ▶ The inverse-cosine function $\cos^{-1}$ is a *two-valued* function. Your calculator returns a single value, an angle between 0 rad and π rad. But the negative of this angle is also a solution. As Example 14.4 demonstrates, you must use additional information to choose between them. ◀

STOP TO THINK 14.2 The figure shows four oscillators at $t = 0$. Which one has the phase constant $\phi_0 = \pi/4$ rad?

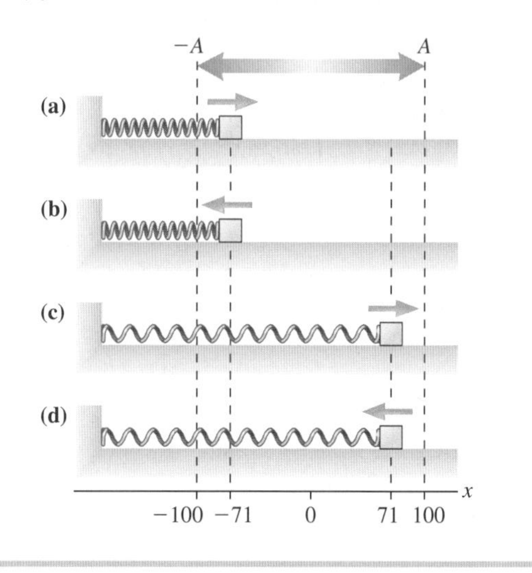

14.3 Energy in Simple Harmonic Motion

We've begun to develop the mathematical language of simple harmonic motion, but thus far we haven't included any physics. We've made no mention of the mass of the object or the spring constant of the spring. An energy analysis, using the tools of Chapters 10 and 11, is a good starting place. We already have all of the information we need because we found the elastic potential energy of a spring in Section 10.5.

9.3

Figure 14.10a on the next page shows an object oscillating on a spring, our prototype of simple harmonic motion. Now we'll specify that the object has mass m, the spring has spring constant k, and the motion takes place on a frictionless surface. You learned in Chapter 10 that the elastic potential energy when the object is at position x is $U_s = \frac{1}{2}k(\Delta x)^2$, where $\Delta x = x - x_e$ is the displacement

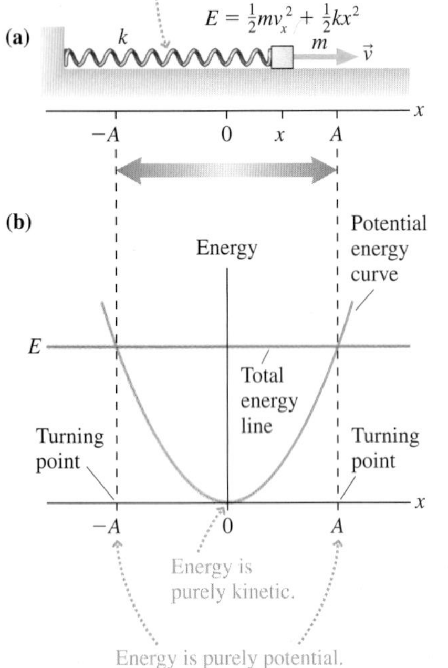

FIGURE 14.10 The energy is transformed between kinetic energy and potential energy as the object oscillates, but the mechanical energy $E = K + U$ doesn't change.

from the equilibrium position x_e. In this chapter we'll always use a coordinate system in which $x_e = 0$, making $\Delta x = x$. There's no chance for confusion with gravitational potential energy, so we can omit the subscript s and write the elastic potential energy as

$$U = \frac{1}{2}kx^2 \tag{14.17}$$

Thus the mechanical energy of an object oscillating on a spring is

$$E = K + U = \frac{1}{2}mv_x^2 + \frac{1}{2}kx^2 \tag{14.18}$$

Figure 14.10b is an energy diagram, showing the potential-energy curve $U = \frac{1}{2}kx^2$ as a parabola. Recall that a particle oscillates between the *turning points* where the total energy line E crosses the potential-energy curve. The left turning point is at $x = -A$ and the right turning point at $x = +A$. To go beyond these points would require a negative kinetic energy, which is physically impossible.

You can see that **the particle has purely potential energy at $x = \pm A$ and purely kinetic energy as it passes through the equilibrium point at $x = 0$.** At maximum displacement, with $x = \pm A$ and $v = 0$, the energy is

$$E(\text{at } x = \pm A) = U = \frac{1}{2}kA^2 \tag{14.19}$$

At $x = 0$, where $v_x = \pm v_{\text{max}}$, the energy is

$$E(\text{at } x = 0) = K = \frac{1}{2}mv_{\text{max}}{}^2 \tag{14.20}$$

These two cases are indicated on Figure 14.10.

The system's mechanical energy is conserved, because the surface is frictionless and there are no external forces, so the energy at maximum displacement and the energy at maximum speed, Equations 14.19 and 14.20, must be equal. That is

$$\frac{1}{2}m(v_{\text{max}})^2 = \frac{1}{2}kA^2 \tag{14.21}$$

From Equation 14.21 we can see that the maximum speed is related to the amplitude by

$$v_{\text{max}} = \sqrt{\frac{k}{m}}A \tag{14.22}$$

This is a relationship based on the physics of the situation.

Earlier, using kinematics, we found that

$$v_{\text{max}} = \frac{2\pi A}{T} = 2\pi f A = \omega A \tag{14.23}$$

Comparing Equations 14.22 and 14.23, we see that frequency and period of an oscillating spring are determined by the spring constant k and the object's mass m:

$$\omega = \sqrt{\frac{k}{m}} \qquad f = \frac{1}{2\pi}\sqrt{\frac{k}{m}} \qquad T = 2\pi\sqrt{\frac{m}{k}} \tag{14.24}$$

These three expressions are really only one equation. They say the same thing, but each expresses it in slightly different terms.

Equation 14.24 is the answer to the second question we posed at the beginning of the chapter, where we asked how the period and frequency are related to the object's mass m, the spring constant k, and the amplitude A. It is perhaps surprising, but **the period and frequency do not depend on the amplitude A.** A small oscillation and a large oscillation have the same period.

Because energy is conserved, we can combine Equations 14.18, 14.19, and 14.20 to write

$$E = \frac{1}{2}mv_x^2 + \frac{1}{2}kx^2 = \frac{1}{2}kA^2 = \frac{1}{2}m(v_{\text{max}})^2 \quad \text{(conservation of energy)} \quad (14.25)$$

Any pair of these may be useful, depending on the known information. For example, you can use the amplitude A to find the speed at any point x by combining the first and second expressions for E. The speed v at position x is

$$v = \sqrt{\frac{k}{m}(A^2 - x^2)} = \omega\sqrt{A^2 - x^2} \quad (14.26)$$

Similarly, you can use the first and second expressions to find the amplitude from the initial conditions x_0 and v_0:

$$A = \sqrt{x_0^2 + \frac{mv_0^2}{k}} = \sqrt{x_0^2 + \left(\frac{v_0}{\omega}\right)^2} \quad (14.27)$$

Figure 14.11 shows graphically how the kinetic and potential energy change with time. They both oscillate but remain *positive* because x and v_x are squared. Energy is continuously being transformed back and forth between the kinetic energy of the moving block and the stored potential energy of the spring, but their sum remains constant. Notice that K and U both oscillate *twice* each period; make sure you understand why.

FIGURE 14.11 Kinetic energy, potential energy, and the total mechanical energy for simple harmonic motion.

EXAMPLE 14.5 Using conservation of energy

A 500 g block on a spring is pulled a distance of 20 cm and released. The subsequent oscillations are measured to have a period of 0.80 s. At what position or positions is the block's speed 1.0 m/s?

MODEL The motion is simple harmonic motion. Energy is conserved.

SOLVE The block starts from the point of maximum displacement, where $E = U = \frac{1}{2}kA^2$. At a later time, when the position is x and the velocity is v_x, energy conservation requires

$$\frac{1}{2}mv_x^2 + \frac{1}{2}kx^2 = \frac{1}{2}kA^2$$

Solving for x, we find

$$x = \sqrt{A^2 - \frac{mv_x^2}{k}} = \sqrt{A^2 - \left(\frac{v}{\omega}\right)^2}$$

where we used $k/m = \omega^2$ from Equation 14.24. The angular frequency is easily found from the period: $\omega = 2\pi/T = 7.85$ rad/s. Thus

$$x = \sqrt{(0.20\text{ m})^2 - \left(\frac{1.0\text{ m/s}}{7.85\text{ rad/s}}\right)^2} = \pm 0.154\text{ m} = \pm 15.4\text{ cm}$$

There are two positions because the block has this speed on either side of equilibrium.

STOP TO THINK 14.3 Four springs have been compressed from their equilibrium position at $x = 0$ cm. When released, they will start to oscillate. Rank in order, from highest to lowest, the maximum speeds of the oscillators.

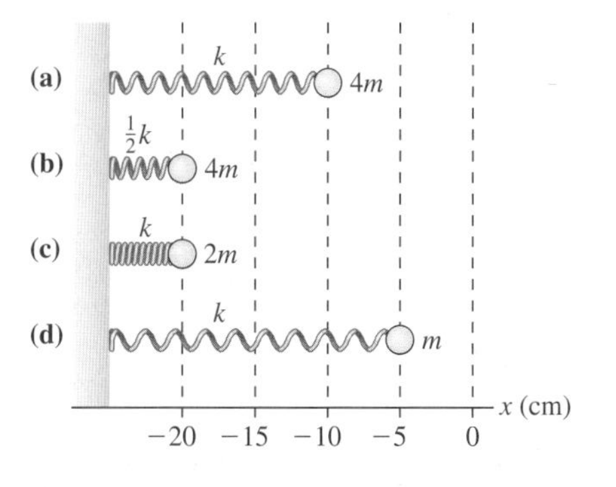

14.4 The Dynamics of Simple Harmonic Motion

9.1, 9.2 ActivPhysics Online

Our analysis thus far has been based on the experimental observation that the oscillation of a spring "looks" sinusoidal. It's time to show that Newton's second law *predicts* sinusoidal motion.

A motion diagram will help us visualize the object's acceleration. Figure 14.12 shows one cycle of the motion, separating motion to the left and motion to the right to make the diagram clear. As you can see, the object's velocity is large as it passes through the equilibrium point at $x = 0$, but $\vec{v}$ is *not changing* at that point. Acceleration measures the *change* of the velocity, hence $\vec{a} = \vec{0}$ at $x = 0$.

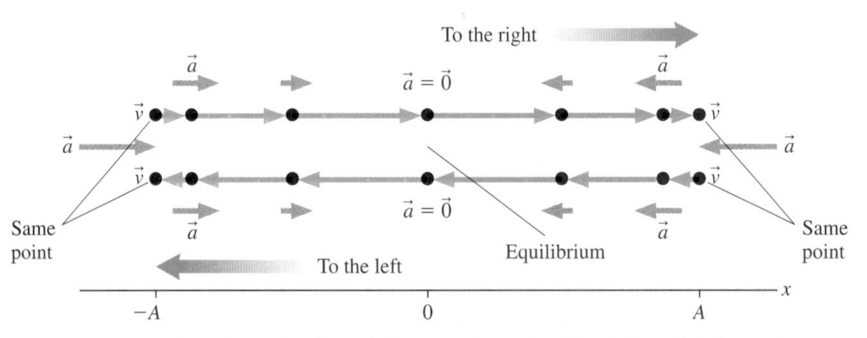

FIGURE 14.12 Motion diagram of simple harmonic motion. The left and right motions are separated for clarity but really occur along the same line.

On the other hand, the velocity is changing rapidly at the turning points. At the right turning point, $\vec{v}$ changes from a right-pointing vector to a left-pointing vector. Thus the acceleration $\vec{a}$ at the right turning point is large and *to the left*. In one-dimensional motion, the acceleration component a_x has a large *negative* value at the right turning point. Similarly, the acceleration $\vec{a}$ at the left turning point is large and *to the right*. Consequently, a_x has a large positive value at the left turning point.

NOTE ► This is exactly the same motion-diagram analysis we used in Chapters 1 and 2 to determine the acceleration at the turning point of a ball tossed straight up. ◄

Our motion diagram analysis suggests that the acceleration a_x is a maximum (most positive) when the displacement is most negative, a minimum (most negative) when the displacement is a maximum, and zero when $x = 0$. This is confirmed by taking the derivative of the velocity,

$$a_x = \frac{dv_x}{dt} = \frac{d}{dt}(-\omega A \sin\omega t) = -\omega^2 A \cos\omega t \tag{14.28}$$

then graphing it.

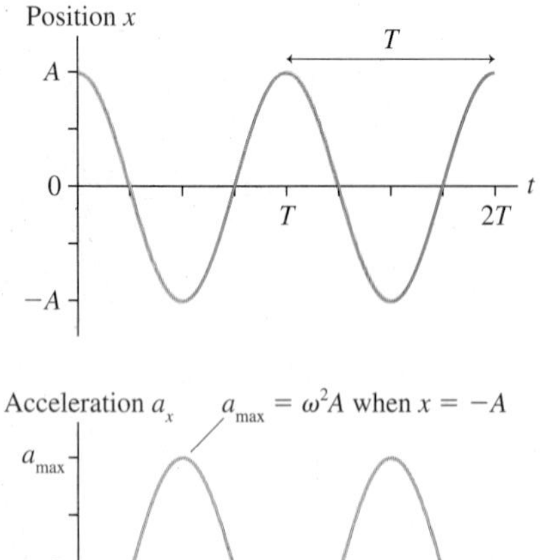

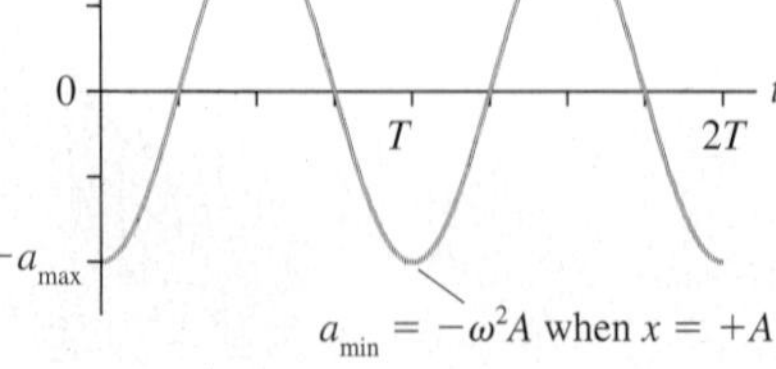

FIGURE 14.13 Position and acceleration graphs for an oscillating spring. We've chosen $\phi_0 = 0$.

Figure 14.13 shows the position graph that we started with in Figure 14.4 and the corresponding acceleration graph. Comparing the two, you can see that the acceleration graph looks like an upside-down position graph. In fact, because $x = A\cos\omega t$, Equation 14.28 for the acceleration can be written

$$a_x = -\omega^2 x \tag{14.29}$$

That is, **the acceleration is proportional to the *negative* of the displacement.** The acceleration is, indeed, maximum when the displacement is most negative and minimum when the displacement is most positive.

Our interest in the acceleration is that the acceleration is related to the net force by Newton's second law. Consider again our prototype mass on a spring, shown in Figure 14.14. This is the simplest possible oscillation, with no distractions due to friction or gravitational forces. We will assume the spring itself to be massless, as we have done previously. The only force acting on the object in the x-direction is the spring force $\vec{F}_{sp}$.

As you learned in Chapter 10, the spring force is given by Hooke's law:

$$(F_{sp})_x = -k\Delta x \tag{14.30}$$

The minus sign indicates that the spring force is a **restoring force,** a force that always points back toward the equilibrium position. If we place the origin of the coordinate system at the equilibrium position, as we've done throughout this chapter, then $\Delta x = x$ and Hooke's law is simply $(F_{sp})_x = -kx$.

The x-component of Newton's second law for the object attached to the spring is

$$(F_{net})_x = (F_{sp})_x = -kx = ma_x \tag{14.31}$$

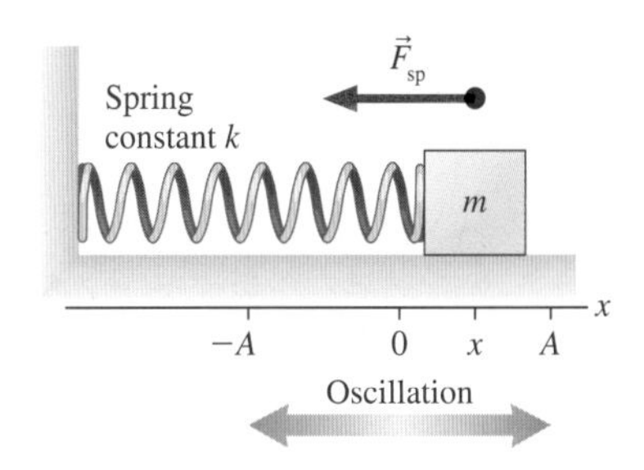

FIGURE 14.14 The prototype of simple harmonic motion: a mass oscillating on a horizontal spring without friction.

Equation 14.31 is easily rearranged to read

$$a_x = -\frac{k}{m}x \tag{14.32}$$

You can see that Equation 14.32 is identical to Equation 14.29 if the system oscillates with angular frequency $\omega = \sqrt{k/m}$. We previously found this expression for ω from an energy analysis. Our experimental observation that the acceleration is proportional to the *negative* of the displacement is exactly what Hooke's law would lead us to expect. That's the good news.

The bad news is that a_x is not a constant. As the object's position changes, so does the acceleration. Nearly all of our kinematic tools have been based on constant acceleration. We can't use those tools to analyze oscillations, so we must go back to the very definition of acceleration:

$$a_x = \frac{dv_x}{dt} = \frac{d^2x}{dt^2}$$

Acceleration is the second derivative of position with respect to time. If we use this definition in Equation 14.32, it becomes

$$\frac{d^2x}{dt^2} = -\frac{k}{m}x \quad \text{(equation of motion for a mass on a spring)} \tag{14.33}$$

Equation 14.33, which is called the **equation of motion,** is a second-order differential equation. Unlike other equations we've dealt with, Equation 14.33 cannot be solved by direct integration. We'll need to take a different approach.

Solving the Equation of Motion

The solution to an algebraic equation such as $x^2 = 4$ is a number. The solution to a differential equation is a *function*. The x in Equation 14.33 is really $x(t)$, the position as a function of time. The solution to this equation is a function $x(t)$ whose second derivative is the function itself multiplied by $(-k/m)$.

One important property of differential equations that you will learn about in math is that the solutions are *unique*. That is, there is only *one* solution to Equation 14.33. If we were able to *guess* a solution, the uniqueness property would tell us that we had found the *only* solution. That might seem a rather strange way to solve equations, but in fact differential equations are frequently solved by using your knowledge of what the solution needs to look like to guess an appropriate functional form. Let us give it a try!

We know from experimental evidence that the oscillatory motion of a spring appears to be sinusoidal. Let us *guess* that the solution to Equation 14.33 should have the functional form

$$x(t) = A\cos(\omega t + \phi_0) \tag{14.34}$$

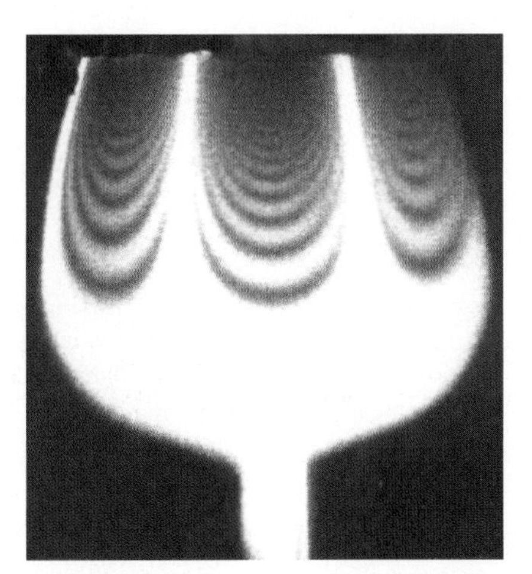

An optical technique called *interferometry* reveals the bell-like vibrations of a wine glass.

where A, ω, and ϕ_0 are unspecified constants that we can adjust to any values that might be necessary to satisfy the differential equation.

If you were to guess that a solution to the algebraic equation $x^2 = 4$ is $x = 2$, you would verify your guess by substituting it into the original equation to see if it works. We need to do the same thing here: substitute our guess for $x(t)$ into Equation 14.33 to see if, for an appropriate choice of the three constants, it works. To do so, we need the second derivative of $x(t)$. That is straightforward:

$$x(t) = A\cos(\omega t + \phi_0)$$
$$\frac{dx}{dt} = -\omega A\sin(\omega t + \phi_0) \tag{14.35}$$
$$\frac{d^2x}{dt^2} = -\omega^2 A\cos(\omega t + \phi_0)$$

If we now substitute the first and third of Equations 14.35 into Equation 14.33 we find

$$-\omega^2 A\cos(\omega t + \phi_0) = -\frac{k}{m}A\cos(\omega t + \phi_0) \tag{14.36}$$

Equation 14.36 will be true at all instants of time if and only if $\omega^2 = k/m$. There do not seem to be any restrictions on the two constants A and ϕ_0.

So we have found—by guessing!—that *the* solution to the equation of motion for a mass oscillating on a spring is

$$x(t) = A\cos(\omega t + \phi_0) \tag{14.37}$$

where the angular frequency

$$\omega = 2\pi f = \sqrt{\frac{k}{m}} \tag{14.38}$$

is determined by the mass and the spring constant.

NOTE ▶ Once again we see that the oscillation frequency is independent of the amplitude A. ◀

Equations 14.37 and 14.38 seem somewhat anticlimactic because we've been using these results for the last several pages. But keep in mind that we had been *assuming* $x = A\cos\omega t$ simply because the experimental observations "looked" like a cosine function. We've now justified that assumption by showing that Equation 14.37 really is the solution to Newton's second law for a mass on a spring. **The *theory* of oscillation, based on Hooke's law for a spring and Newton's second law, is in good agreement with the experimental observations.** This conclusion gives an affirmative answer to the last of the three questions that we asked early in the chapter, which was whether the sinusoidal oscillation of SHM is a consequence of Newton's laws.

EXAMPLE 14.6 Analyzing an oscillator

At $t = 0$ s, a 500 g block oscillating on a spring is observed moving to the right at $x = 15$ cm. It reaches a maximum displacement of 25 cm at $t = 0.300$ s.

a. Draw a position-versus-time graph for one cycle of the motion.
b. At what times during the first cycle does the mass pass through $x = 20$ cm?

MODEL The motion is simple harmonic motion.

SOLVE

a. The position equation of the block is $x(t) = A\cos(\omega t + \phi_0)$. We know that the amplitude is $A = 0.25$ m and that $x_0 = 0.15$ m. From these two pieces of information we obtain the phase constant:

$$\phi_0 = \cos^{-1}\left(\frac{x_0}{A}\right) = \cos^{-1}(.60) = \pm 0.927 \text{ rad}$$

The object is initially moving to the right, which tells us that the phase constant must be between $-\pi$ and 0 rad. Thus $\phi_0 = -0.927$ rad. The block reaches its maximum displacement $x_{max} = A$ at time $t = 0.30$ s. At that instant of time

$$x_{max} = A = A\cos(\omega t + \phi_0)$$

This can be true only if $\cos(\omega t + \phi_0) = 1$, which requires $\omega t + \phi_0 = 0$. Thus

$$\omega = \frac{-\phi_0}{t} = \frac{-(-0.927 \text{ rad})}{(0.300 \text{ s})} = 3.09 \text{ rad/s}$$

Now that we know ω, it is straightforward to compute the period

$$T = \frac{2\pi}{\omega} = 2.03 \text{ s}$$

Figure 14.15 graphs $x(t) = (25 \text{ cm})\cos(3.09t - 0.927)$, where t is in s, from $t = 0$ s to $t = 2.03$ s.

b. From $x = A\cos(\omega t + \phi_0)$, the time at which the mass reaches position $x = 20$ cm is

$$t = \frac{1}{\omega}\left(\cos^{-1}\left(\frac{x}{A}\right) - \phi_0\right)$$

$$= \frac{1}{3.09 \text{ rad/s}}\left(\cos^{-1}\left(\frac{20 \text{ cm}}{25 \text{ cm}}\right) + 0.927 \text{ rad}\right) = 0.508 \text{ s}$$

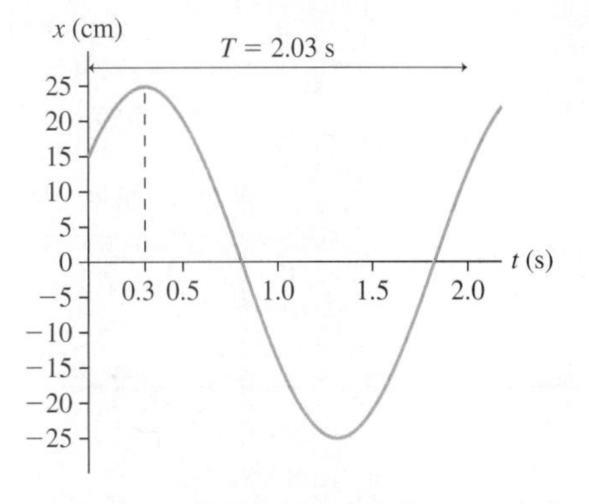

FIGURE 14.15 Position-versus-time graph for the oscillator of Example 14.6.

A calculator returns only one value of $\cos^{-1}$, in the range 0 to π rad, but we noted earlier that $\cos^{-1}$ actually has two values. Indeed, you can see in Figure 14.15 that there are two times at which the mass passes $x = 20$ cm. Because they are symmetrical on either side of $t = 0.300$ s, when $x = A$, the first point is $(0.508 \text{ s} - 0.300 \text{ s}) = 0.208$ s *before* the maximum. Thus the mass passes through $x = 20$ cm at $t = 0.092$ s and again at $t = 0.508$ s.

STOP TO THINK 14.4 This is the position graph of a mass on a spring. What can you say about the velocity and the force at the instant indicated by the dotted line?

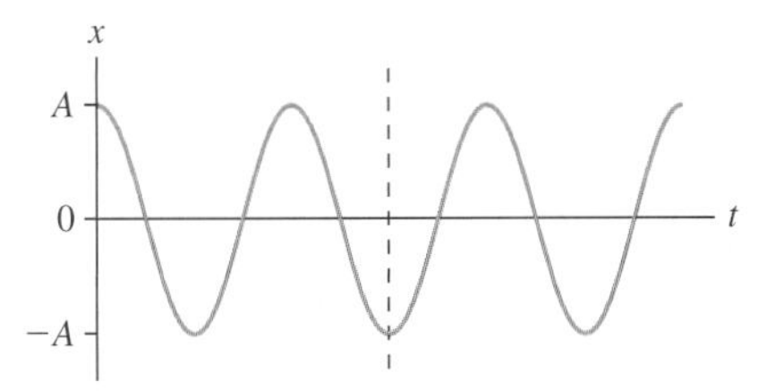

a. Velocity is positive; force is to the right.
b. Velocity is negative; force is to the right.
c. Velocity is zero; force is to the right.
d. Velocity is positive; force is to the left.
e. Velocity is negative; force is to the left.
f. Velocity is zero; force is to the left.
g. Velocity and force are both zero.

14.5 Vertical Oscillations

We have focused our analysis on a horizontally oscillating spring. But the typical demonstration you'll see in class is a mass bobbing up and down on a spring hung vertically from a support. Is it safe to assume that a vertical oscillation is the same as a horizontal oscillation? Or does the additional force of gravity change the motion? Let us look at this more carefully.

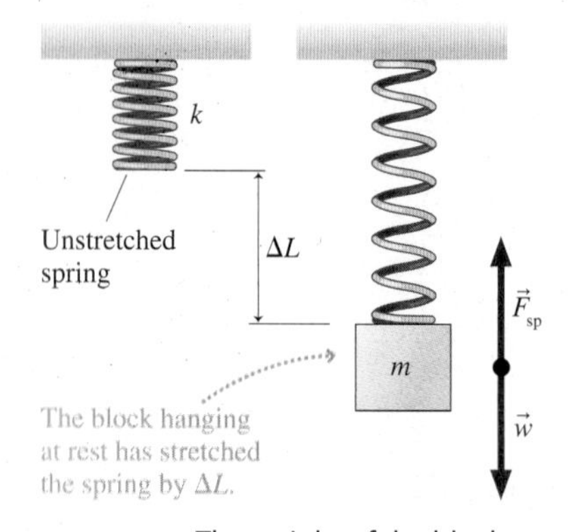

FIGURE 14.16 The weight of the block stretches the spring.

Figure 14.16 shows a block of mass m hanging from a spring of spring constant k. An important fact to notice is that the equilibrium position of the block is *not* where the spring is at its unstretched length. At the equilibrium position of the block, where it hangs motionless, the spring has stretched by ΔL.

Finding ΔL is a static-equilibrium problem in which the upward spring force balances the downward weight force of the block. The y-component of the spring force is given by Hooke's law,

$$(F_{sp})_y = -k\Delta y = +k\Delta L \tag{14.39}$$

Equation 14.39 makes a distinction between ΔL, which is simply a *distance* and is a positive number, and the displacement Δy. The block is displaced downward, so $\Delta y = -\Delta L$. Newton's first law for the block in equilibrium is

$$(F_{net})_y = (F_{sp})_y + w_y = k\Delta L - mg = 0 \tag{14.40}$$

from which we can find

$$\Delta L = \frac{mg}{k} \tag{14.41}$$

This is the distance the spring stretches when the block is attached to it.

Let the block oscillate around this equilibrium position, as shown in Figure 14.17. We've now placed the origin of the y-axis at the block's equilibrium position in order to be consistent with our analyses of oscillations throughout this chapter. If the block moves upward, as the figure shows, the spring gets shorter compared to its equilibrium length, but the spring is still *stretched* compared to its unstretched length in Figure 14.16. When the block is at position y, the spring is stretched by an amount $\Delta L - y$ and hence exerts an *upward* spring force $F_{sp} = k(\Delta L - y)$. The net force on the block at this point is

$$(F_{net})_y = (F_{sp})_y + w_y = k(\Delta L - y) - mg = (k\Delta L - mg) - ky \tag{14.42}$$

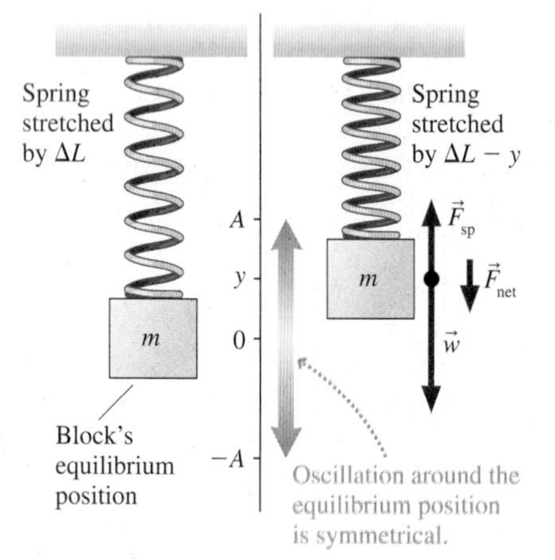

FIGURE 14.17 The block oscillates around the equilibrium position.

But $k\Delta L - mg$ is zero, from Equation 14.41, so the net force on the block is simply

$$(F_{net})_y = -ky \tag{14.43}$$

Equation 14.43 for vertical oscillations is *exactly* the same as Equation 14.31 for horizontal oscillations, where we found $(F_{net})_x = -kx$. That is, the restoring force for vertical oscillations is identical to the restoring force for horizontal oscillations. The role of gravity is to determine where the equilibrium position is, but it doesn't affect the motion around the equilibrium position.

Because the net force is the same, Newton's second law will have exactly the same oscillatory solution:

$$y(t) = A\cos(\omega t + \phi_0) \tag{14.44}$$

with, again, $\omega = \sqrt{k/m}$. The vertical oscillations of a mass on a spring are the same simple harmonic motion as that of a block on a horizontal spring. This is an important finding because it was not obvious that the motion would still be simple harmonic motion when gravity was included. Because the motions are the same, **everything we have learned about horizontal oscillations is equally valid for vertical oscillations.**

EXAMPLE 14.7 A vertical oscillation

A 200 g block hangs from a spring with spring constant 10 N/m. The block is pulled down to a point where the spring is 30 cm longer than its unstretched length, then released. Where is the block and what is its velocity 3.0 s later?

MODEL Vertical oscillations are simple harmonic motion.

VISUALIZE Figure 14.18 shows the situation.

SOLVE Although the spring begins by being stretched 30 cm, this is *not* the amplitude of the oscillation. Oscillations occur around the equilibrium position, so we have to begin by finding

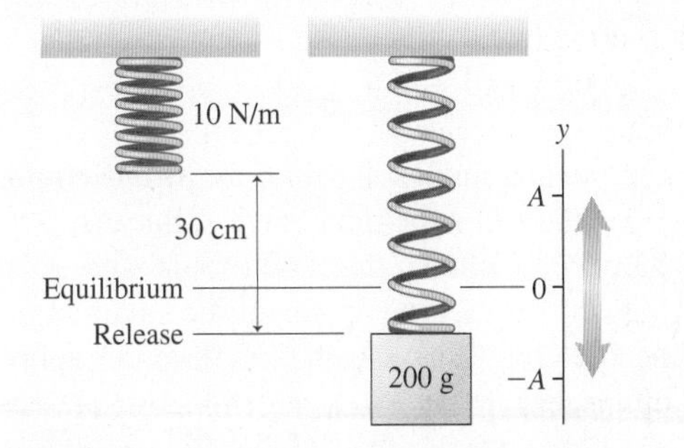

FIGURE 14.18 A block oscillates about its equilibrium position.

the equilibrium point where the block hangs motionless on the spring. The spring stretch at equilibrium is given by Equation 14.41,

$$\Delta L = \frac{mg}{k} = 0.196 \text{ m} = 19.6 \text{ cm}$$

Stretching the spring 30 cm pulls the block 10.4 cm below the equilibrium point, so $A = 10.4$ cm. That is, the block oscillates with an amplitude $A = 10.4$ cm about a point that is 19.6 cm beneath the spring's original end point. The block's position as a function of time, as measured from the equilibrium position, is

$$y(t) = (10.4 \text{ cm})\cos(\omega t + \phi_0)$$

where $\omega = \sqrt{k/m} = 7.07 \sqrt{\text{s}}$. The initial condition

$$y_0 = -A = A\cos\phi_0$$

requires the phase constant to be $\phi_0 = \pi$ rad. At $t = 3.0$ s the block's position is

$$y = (10.4 \text{ cm})\cos((7.07 \text{ rad/s})(3.0 \text{ s}) + \pi \text{ rad}) = 7.4 \text{ cm}$$

The block is 7.4 cm *above* the equilibrium position, or 12.2 cm *below* the original end of the spring. Its velocity at this instant is

$$v_x = -\omega A \sin(\omega t + \phi_0) = 52 \text{ cm/s}$$

14.6 The Pendulum

Now let's look at another very common oscillator: a pendulum. Figure 14.19a shows a mass m attached to a string of length L and free to swing back and forth. The pendulum's position can be described by the arc of length s, which is zero when the pendulum hangs straight down. Because angles are measured ccw, s and θ are positive when the pendulum is to the right of center, negative when it is to the left.

Two forces are acting on the mass: the string tension $\vec{T}$ and the weight $\vec{w}$. It will be convenient to do what we did in our study of circular motion: divide the forces into tangential components, parallel to the motion, and radial components parallel to the string. These are shown on the free-body diagram of Figure 14.19b.

Newton's second law for the tangential component, parallel to the motion, is

$$(F_{\text{net}})_t = \sum F_t = w_t = -mg\sin\theta = ma_t \tag{14.45}$$

Using $a_t = d^2s/dt^2$ for acceleration "around" the circle, and noting that the mass cancels, we can write Equation 14.45 as

$$\frac{d^2s}{dt^2} = -g\sin\theta \tag{14.46}$$

where angle θ is related to the arc length by $\theta = s/L$. This is the equation of motion for an oscillating pendulum. The sine function makes this equation more complicated than the equation of motion for an oscillating spring.

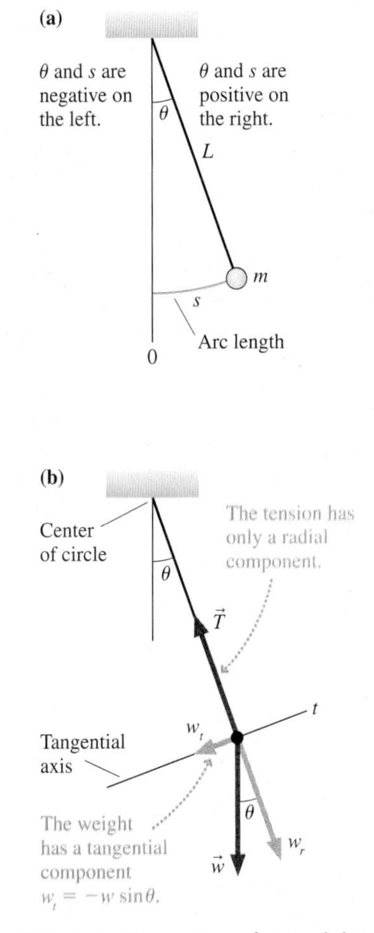

FIGURE 14.19 The motion of a pendulum.

The Small-Angle Approximation

Suppose we restrict the pendulum's oscillations to *small angles* of less than about 10°. This restriction allows us to make use of an interesting and important piece of geometry.

Figure 14.20 on the next page shows an angle θ and a circular arc of length $s = r\theta$. A right triangle has been constructed by dropping a perpendicular from the top of the arc to the axis. The height of the triangle is $h = r\sin\theta$. Suppose that

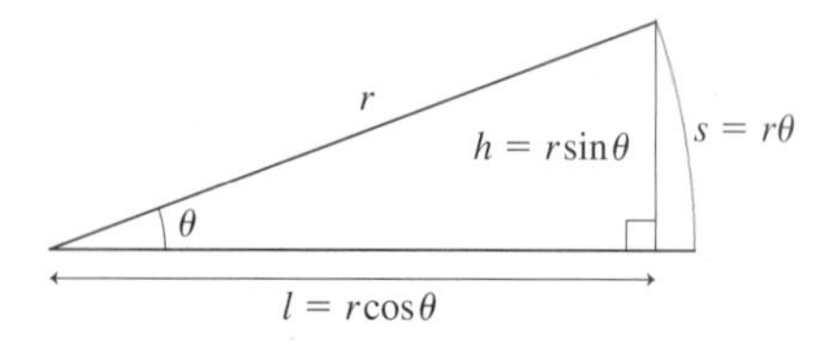

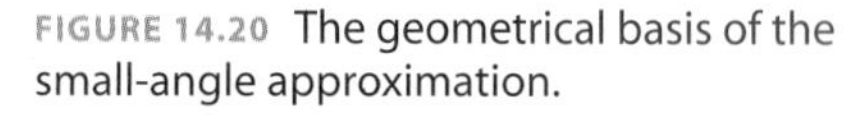

FIGURE 14.20 The geometrical basis of the small-angle approximation.

the angle θ is "small." In that case there is very little difference between h and s. If $h \approx s$, then $r\sin\theta \approx r\theta$. It follows that

$$\sin\theta \approx \theta \qquad (\theta \text{ in radians})$$

The result that $\sin\theta \approx \theta$ for small angles is called the **small-angle approximation.** We can similarly note that $l \approx r$ for small angles. Because $l = r\cos\theta$, it follows that $\cos\theta \approx 1$. Finally, we can take the ratio of sine and cosine to find $\tan\theta \approx \sin\theta \approx \theta$. Table 14.3 summarizes the results of the small-angle approximation. We will have other occasions to use the small-angle approximation throughout the remainder of this text.

TABLE 14.3 Small-angle approximations. θ must be in radians.

$\sin\theta \approx \theta$
$\cos\theta \approx 1$
$\tan\theta \approx \sin\theta \approx \theta$

NOTE ▶ The small-angle approximation is valid *only* if the angle θ is in radians! ◀

How small does θ have to be to justify using the small-angle approximation? It's easy to use your calculator to find that the small-angle approximation is good to three significant figures, an error of $\leq 0.1\%$, up to angles of ≈ 0.10 rad ($\approx 5°$). In practice, we will use the approximation up to about 10°, but for angles any larger it rapidly loses validity and produces unacceptable results.

9.10–9.12 ActivPhysics ONLINE

If we restrict the pendulum to $\theta < 10°$, we can use the small-angle approximation to write $\sin\theta \approx \theta = s/L$. In that case, Equation 14.45 for the net force on the mass is

$$(F_{\text{net}})_t = -(mg/L)s$$

and the equation of motion becomes

$$\frac{d^2s}{dt^2} = -\frac{g}{L}s \tag{14.47}$$

This is *exactly* the same equation as Equation 14.33 for a mass oscillating on a spring. The names are different, with x replaced by s and k/m by g/L, but that does not make it a different equation.

Because we know the solution to the spring problem, we can immediately write the solution to the pendulum problem just by changing variables and constants:

$$s(t) = A\cos(\omega t + \phi_0) \quad \text{or} \quad \theta(t) = \theta_{\max}\cos(\omega t + \phi_0) \tag{14.48}$$

The angular frequency

$$\omega = 2\pi f = \sqrt{\frac{g}{L}} \tag{14.49}$$

is determined by the length of the string. The pendulum is interesting in that **the frequency, and hence the period, is independent of the mass.** It depends only on the length of the pendulum. The amplitude A and the phase constant ϕ_0 are determined by the initial conditions, just as they were for an oscillating spring.

The pendulum clock has been used for hundreds of years.

EXAMPLE 14.8 **A pendulum clock**

What length pendulum has a period of exactly 1 s?

SOLVE The period is independent of the mass and depends only on the length. From Equation 14.49,

$$T = \frac{1}{f} = 2\pi\sqrt{\frac{L}{g}}$$

Solving for L, we find

$$L = g\left(\frac{T}{2\pi}\right)^2 = 0.248 \text{ m}$$

ASSESS This is a convenient length for a practical clock.

EXAMPLE 14.9 The maximum angle of a pendulum

A 300 g mass on a 30-cm-long string oscillates as a pendulum. It has a speed of 0.25 m/s as it passes through the lowest point. What maximum angle does the pendulum reach?

MODEL Assume that the angle remains small, in which case the motion is simple harmonic motion.

SOLVE The angular frequency of the pendulum is

$$\omega = \sqrt{\frac{g}{L}} = \sqrt{\frac{9.8\ \text{m/s}^2}{0.30\ \text{m}}} = 5.72\ \text{rad/s}$$

The speed at the lowest point is $v_{max} = \omega A$, so the amplitude is

$$A = s_{max} = \frac{v_0}{\omega} = \frac{0.25\ \text{m/s}}{5.72\ \text{rad/s}} = 0.0437\ \text{m}$$

The maximum angle, at the maximum arc length s_{max}, is

$$\theta_{max} = \frac{s_{max}}{L} = \frac{0.04347\ \text{m}}{0.30\ \text{m}} = 0.145\ \text{rad} = 8.3°$$

ASSESS Because the maximum angle is less than 10°, our analysis based on the small-angle approximation is valid.

The Conditions for Simple Harmonic Motion

You can begin to see how, in a sense, we have solved *all* simple-harmonic-motion problems once we have solved the problem of the horizontal spring. The restoring force of a spring, $F_{sp} = -kx$, is directly proportional to the displacement x from equilibrium. The pendulum's restoring force, in the small-angle approximation, is directly proportional to the displacement s. A restoring force that is directly proportional to the displacement from equilibrium is called a **linear restoring force.** For *any* linear restoring force, the equation of motion is identical to the spring equation (other than perhaps using different symbols). Consequently, **any system with a linear restoring force will undergo simple harmonic motion around the equilibrium position.**

This is why an oscillating spring is the prototype of SHM. Everything that we learn about an oscillating spring can be applied to the oscillations of any other linear restoring force, ranging from the vibration of airplane wings to the motion of electrons in electric circuits. Let's summarize this information with a Tactics Box.

TACTICS BOX 14.1 Identifying and analyzing simple harmonic motion

1. If the net force acting on a particle is a linear restoring force, the motion will be simple harmonic motion around the equilibrium position.
2. The position as a function of time is $x(t) = A\cos(\omega t + \phi_0)$. The velocity as a function of time is $v_x(t) = -\omega A \sin(\omega t + \phi_0)$. The maximum speed is $v_{max} = \omega A$. The equations are given here in terms of x, but they can be written in terms of y, θ, or some other parameter if the situation calls for it.
3. The amplitude A and the phase constant ϕ_0 are determined by the initial conditions through $x_0 = A\cos\phi_0$ and $v_{0x} = -\omega A \sin\phi_0$.
4. The angular frequency ω (and hence the period $T = 2\pi/\omega$) depends on the physics of the particular situation. But ω does *not* depend on A or ϕ_0.
5. Mechanical energy is conserved. Thus $\frac{1}{2}mv_x^2 + \frac{1}{2}kx^2 = \frac{1}{2}kA^2 = \frac{1}{2}m(v_{max})^2$. Energy conservation provides a relationship between position and velocity that is independent of time.

9.6–9.9

STOP TO THINK 14.5 One person swings on a swing and finds that the period is 3.0 s. A second person of equal mass joins him. With two people swinging, the period is

a. 6.0 s.
b. >3.0 s but not necessarily 6.0 s.
c. 3.0 s.
d. <3.0 s but not necessarily 1.5 s.
e. 1.5 s.
f. Can't tell without knowing the length.

The shock absorbers in cars and trucks are heavily damped springs. The vehicle's vertical motion, after hitting a rock or a pothole, is a damped oscillation.

14.7 Damped Oscillations

A pendulum left to itself gradually slows down and stops. The sound of a ringing bell gradually dies away. All real oscillators do run down—some very slowly but others quite quickly—as friction or other dissipative forces transform their mechanical energy into the thermal energy of the oscillator and its environment. An oscillation that runs down and stops is called a **damped oscillation.**

There are many possible reasons for the dissipation of energy: air resistance, friction, internal forces within the metal of the spring as it flexes, and so on. While it would be impractical to account for all of these, a reasonable model is to consider only air resistance because, in many cases, it will be the predominant dissipative force.

The drag force of air resistance is a complex force. There is no "law of air resistance" to tell us exactly how big air-resistance forces are. Chapter 5 introduced a *model* of air resistance in which the drag force was proportional to v^2. That's a good model when velocities are reasonably high, as they are for runners, baseballs, and cars, but a pendulum or an oscillating spring usually moves much more slowly. It is known from experiments that the drag force on *slowly* moving objects is *linearly* proportional to the velocity. Thus a reasonable model of the drag force for a slowly moving object is

$$\vec{D} = -b\vec{v} \quad \text{(model of the drag force)} \tag{14.50}$$

where the minus sign is the mathematical statement that the force is always opposite in direction to the velocity in order to slow the object.

The **damping constant** b depends in a complicated way on the shape of the object (long, narrow objects have less air resistance than wide, flat ones) *and* on the viscosity of the air or other medium in which the particle moves. The damping constant plays the same role in our model of air resistance that the coefficient of friction does in our model of friction.

The units of b need to be such that they will give units of force when multiplied by units of velocity. As you can confirm, these units are kg/s. A value $b = 0$ kg/s corresponds to the limiting case of no resistance, in which case the mechanical energy is conserved. A typical value of b for a spring or a pendulum in air is ≤ 0.10 kg/s. Poorly shaped objects or objects moving in a liquid (which is much more viscous than air) can have significantly larger values of b.

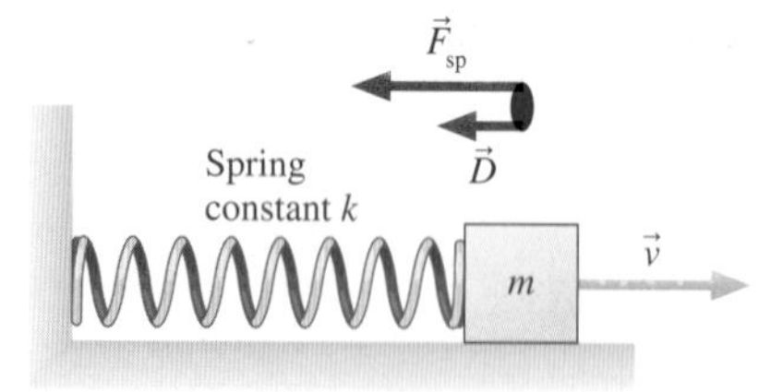

FIGURE 14.21 An oscillating mass in the presence of a drag force.

Figure 14.21 shows a mass oscillating on a spring in the presence of a drag force. With the drag included, Newton's second law is

$$(F_{net})_x = (F_{sp})_x + D_x = -kx - bv_x = ma_x \tag{14.51}$$

Using $v_x = dx/dt$ and $a_x = d^2x/dt^2$, we can write Equation 14.51 as

$$\frac{d^2x}{dt^2} + \frac{b}{m}\frac{dx}{dt} + \frac{k}{m}x = 0 \tag{14.52}$$

Equation 14.52 is the equation of motion of a damped oscillator. If you compare it to Equation 14.33, the equation of motion for a block on a frictionless surface, you'll see it differs by the inclusion of the term involving dx/dt.

Equation 14.52 is another second-order differential equation. We will simply assert (and, as a homework problem, you can confirm) that the solution is

$$x(t) = Ae^{-bt/2m}\cos(\omega t + \phi_0) \quad \text{(damped oscillator)} \tag{14.53}$$

where the angular frequency is given by

$$\omega = \sqrt{\frac{k}{m} - \frac{b^2}{4m^2}} = \sqrt{\omega_0^2 - \frac{b^2}{4m^2}} \tag{14.54}$$

Here $\omega_0 = \sqrt{k/m}$ is the angular frequency of an undamped oscillator ($b = 0$). The constant e is the base of natural logarithms, so $e^{-bt/2m}$ is an *exponential function*.

Notice that Equation 14.53 reduces to our previous solution, $x(t) = A\cos(\omega t + \phi_0)$, when $b = 0$. This makes sense and gives us confidence in Equation 14.53. A *lightly damped* system, which oscillates many times before stopping, is one for which $b/2m \ll \omega_0$. In that case, $\omega \approx \omega_0$ is a good approximation. That is, light damping does not affect the oscillation frequency. (*Heavy damping*, which stops the motion within a few oscillations, causes the oscillation frequency to be significantly lowered.) We will focus on lightly damped systems for the rest of this section.

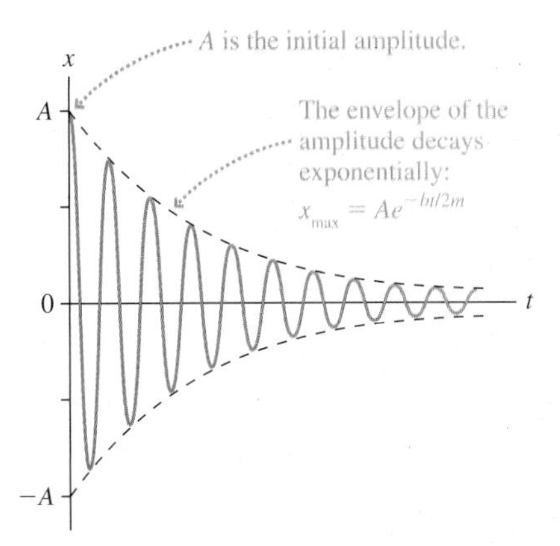

FIGURE 14.22 Position-versus-time graph for a damped oscillator.

Figure 14.22 is a graph of the position $x(t)$ for a lightly damped oscillator, as given by Equation 14.53. Notice that the term $Ae^{-bt/2m}$, which is shown by the dashed line, acts as a slowly varying amplitude:

$$x_{max}(t) = Ae^{-bt/2m} \tag{14.55}$$

where A is the *initial* amplitude, at $t = 0$. The oscillation keeps bumping up against this line, slowly dying out with time.

A slowly changing line that provides a border to a rapid oscillation is called the **envelope** of the oscillations. In this case, the oscillations have an *exponentially decaying envelope*. Figure 14.22 shows how real oscillations decay with time. Make sure you study Figure 14.22 long enough to see how both the oscillations and the decaying amplitude are related to Equation 14.53.

Changing the amount of damping, by changing the value of b, affects how quickly the oscillations decay. Figure 14.23 shows just the envelope $x_{max}(t)$ for several oscillators that are identical except for the value of the damping constant b. (You need to imagine a rapid oscillation within each envelope, as in Figure 14.22). Increasing b causes the oscillations to damp more quickly, while decreasing b makes them last longer.

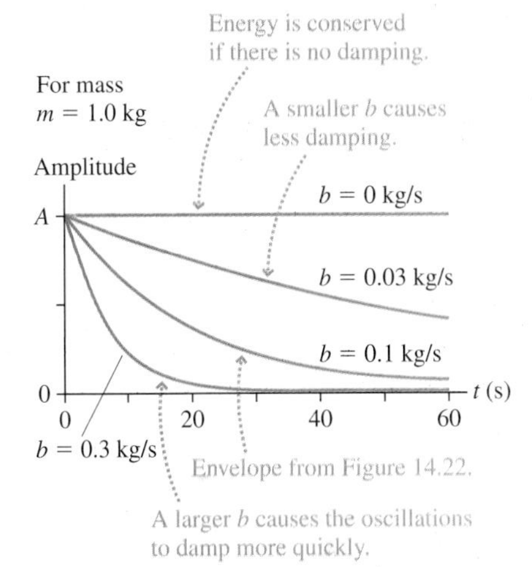

FIGURE 14.23 Several oscillation envelopes, corresponding to different values of the damping constant b.

Energy in Damped Systems

When considering the oscillator's mechanical energy, it is useful to define the **time constant** τ (Greek tau) to be

$$\tau = \frac{m}{b} \tag{14.56}$$

Because b has units of kg/s, τ has units of seconds. With this definition, we can write the oscillation amplitude as $x_{max}(t) = Ae^{-t/2\tau}$.

Because of the drag force, the mechanical energy is no longer conserved. At any particular time we can compute the mechanical energy from

$$E(t) = \frac{1}{2}kx_{max}^2 = \frac{1}{2}k(Ae^{-t/2\tau})^2 = \left(\frac{1}{2}kA^2\right)e^{-t/\tau} = E_0e^{-t/\tau} \tag{14.57}$$

where $E_0 = \frac{1}{2}kA^2$ is the initial energy at $t = 0$ and where we used $(z^m)^2 = z^{2m}$. In other words, **the oscillator's mechanical energy decays exponentially with time constant τ.**

As Figure 14.24 shows, the time constant is the amount of time needed for the energy to decay to e^{-1}, or 37%, of its initial value. By $t = 2\tau$, the energy has been reduced to $E(\text{at } 2\tau) = E_0e^{-2} = 0.13E_0$. We say that the time constant τ measures the "characteristic time" during which the energy of the oscillation is dissipated. Roughly two-thirds of the initial energy is gone after one time constant has elapsed and nearly 90% has dissipated after two time constants have gone by.

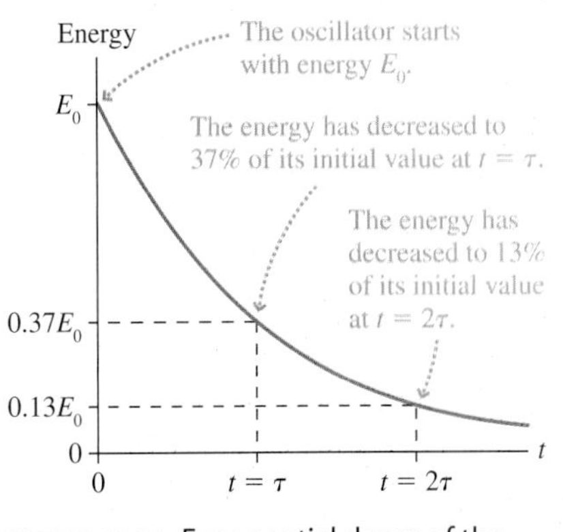

FIGURE 14.24 Exponential decay of the mechanical energy of an oscillator.

For practical purposes, we can speak of the time constant as the *lifetime* of the oscillation—about how long it lasts. An oscillator with $\tau = 3$ s will oscillate for roughly 3 s, while one with $\tau = 30$ s will continue for roughly 30 s. Mathematically, there is never a time when the oscillation is "over." Equation 14.57 approaches zero asymptotically, but it never gets there in any finite time. Thus there is not any specific time t at which we can say "The oscillation is over." The best we can do is define a characteristic time when the motion is "almost over," and that is what the time constant τ does.

There are many, many examples of exponential decay in science and engineering. You will see several more in this text, and you will find others in more advanced courses. Wherever exponential decay occurs, the idea of a time constant is used to characterize how long the decay lasts. No matter what is decaying, it will have only 37% of its original value (roughly one-third) after a time interval of one time constant has elapsed.

EXAMPLE 14.10 A damped pendulum

A 500 g mass swings on a 60-cm-string as a pendulum. The amplitude is observed to decay to half its initial value after 35 s.

a. What is the time constant for this oscillator?
b. At what time will the *energy* have decayed to half its initial value?

MODEL The motion is a damped oscillation.

SOLVE

a. The initial amplitude at $t = 0$ is $x_{max} = A$. At $t = 35$ s the amplitude is $x_{max} = \frac{1}{2}A$. The amplitude of oscillation at time t is given by Equation 14.55:

$$x_{max}(t) = Ae^{-bt/2m} = Ae^{-t/2\tau}$$

In this case,

$$\frac{1}{2}A = Ae^{-(35\text{ s})/2\tau}$$

Notice that we do not need to know A itself because it cancels out. To solve for τ, take the natural logarithm of both sides of the equation:

$$\ln\left(\frac{1}{2}\right) = -\ln 2 = \ln e^{-(35\text{ s})/2\tau} = -\frac{(35\text{ s})}{2\tau}$$

This is easily rearranged to give

$$\tau = \frac{35\text{ s}}{2\ln 2} = 25.2\text{ s}$$

If desired, we could now determine the damping constant to be $b = m/\tau = 0.020$ kg/s.

b. The energy at time t is given by

$$E(t) = E_0 e^{-t/\tau}$$

The time at which an exponential decay is reduced to $\frac{1}{2}E_0$, half its initial value, has a special name. It is called the **half-life** and given the symbol $t_{1/2}$. The concept of the half-life is widely used in applications such as radioactive decay. To relate $t_{1/2}$ to τ, first write

$$E(\text{at } t = t_{1/2}) = \frac{1}{2}E_0 = E_0 e^{-t_{1/2}/\tau}$$

The E_0 cancels, giving

$$\frac{1}{2} = e^{-t_{1/2}/\tau}$$

Again, take the natural logarithm of both sides:

$$\ln\left(\frac{1}{2}\right) = -\ln 2 = \ln e^{-t_{1/2}/\tau} = -t_{1/2}/\tau$$

Finally, solve for $t_{1/2}$:

$$t_{1/2} = \tau \ln 2 = 0.693\tau$$

This result that $t_{1/2}$ is 69% of τ is valid for any exponential decay. In this particular problem, half the energy is gone at

$$t_{1/2} = (.693)(25.2\text{ s}) = 17.5\text{ s}$$

ASSESS The oscillator loses energy faster than it loses amplitude. This is what we should expect because the energy depends on the *square* of the amplitude.

STOP TO THINK 14.6 Rank in order, from largest to smallest, the time constants τ_a to τ_d of the decays shown in the figure.

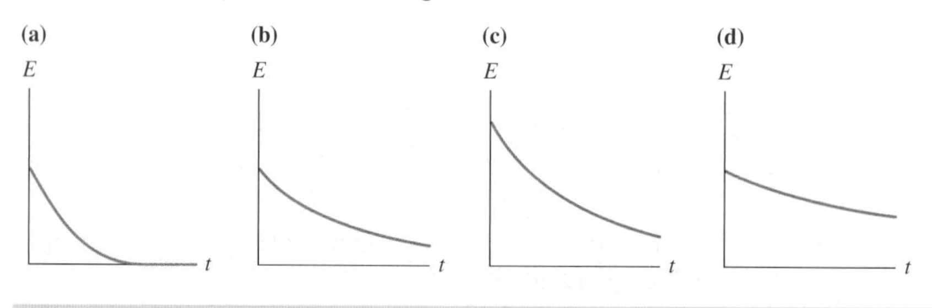

14.8 Driven Oscillations and Resonance

Thus far we have focused on the free oscillations of an isolated system. Some initial disturbance displaces the system from equilibrium, and it then oscillates freely until its energy is dissipated. These are very important situations, but they do not exhaust the possibilities. Another important situation is an oscillator that is subjected to a periodic external force. Its motion is called a **driven oscillation.**

A simple example of a driven oscillation is pushing a child on a swing, where your push is a periodic external force applied to the swing. A more complex example would be a car driving over a series of equally spaced bumps. Each bump causes a periodic upward force on the car's shock absorbers, which are big, heavily damped springs. The electromagnetic coil on the back of a loudspeaker cone provides a periodic magnetic force to drive the cone back and forth, causing it to send out sound waves. Air turbulence moving across the wings of an aircraft can exert periodic forces on the wings and other aerodynamic surfaces, causing them to vibrate if they are not properly designed.

As these examples suggest, driven oscillations have many important applications. However, driven oscillations are a mathematically complex subject. We will simply hint at some of the results, saving the details for more advanced classes.

Consider an oscillating system that, when left to itself, oscillates at a frequency f_0. We will call this the **natural frequency** of the oscillator. The natural frequency for a mass on a spring is $\sqrt{k/m}/2\pi$, but it might be given by some other expression for another type of oscillator. Regardless of the expression, f_0 is simply the frequency of the system if it is displaced from equilibrium and released.

Suppose that this system is subjected to a *periodic* external force of frequency f_{ext}. This frequency, which is called the **driving frequency,** is completely independent of the oscillator's natural frequency f_0. Somebody or something in the environment selects the frequency f_{ext} of the external force, causing the force to push on the system f_{ext} times every second.

Although it is possible to solve Newton's second law with an external driving force, we will be content to look at a graphical representation of the solution. The most important result is that the oscillation amplitude depends very sensitively on the frequency f_{ext} of the driving force. The response to the driving frequency is shown in Figure 14.25 for a system with $m = 1$ kg, a natural frequency $f_0 = 2$ Hz, and a damping constant $b = 0.20$ kg/s. This graph of amplitude versus driving frequency, called the **response curve,** occurs in many different applications.

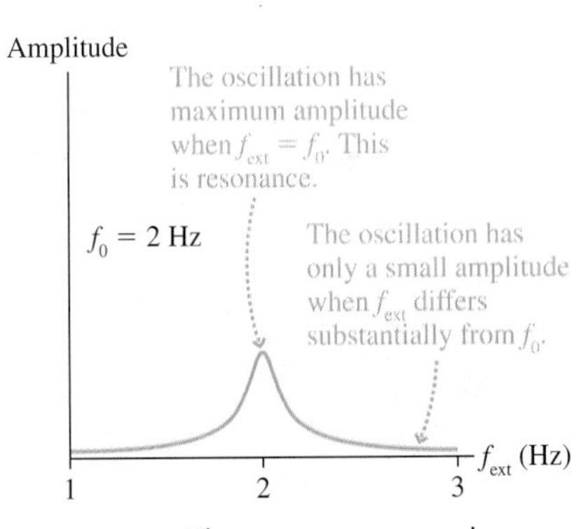
Amplitude
The oscillation has maximum amplitude when $f_{ext} = f_0$. This is resonance.
$f_0 = 2$ Hz
The oscillation has only a small amplitude when f_{ext} differs substantially from f_0.
f_{ext} (Hz)
1
2
3

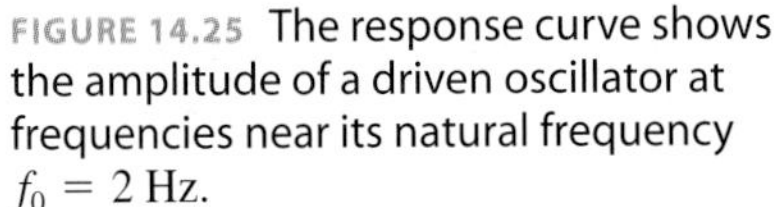
FIGURE 14.25 The response curve shows the amplitude of a driven oscillator at frequencies near its natural frequency $f_0 = 2$ Hz.

When the driving frequency is substantially different from the oscillator's natural frequency, at the right and left edges of Figure 14.25, the system oscillates but its amplitude is very small. The system simply does not respond well to a driving frequency that differs much from f_0. As the driving frequency gets closer and closer to the natural frequency, the amplitude of the oscillation rises dramatically. After all, f_0 is the frequency at which the system "wants" to oscillate, so it is quite happy to respond to a driving frequency near f_0. Hence the amplitude reaches a maximum exactly when the driving frequency matches the system's natural frequency: $f_{ext} = f_0$.

You can understand this if you think about the energy. When f_{ext} matches f_0, the external force always pushes the oscillator at the same point in its cycle. For example, you always push a child on a swing just as the swing reaches its highest point on your side. Such push always *adds energy* to the system, pushing the amplitude higher.

By contrast, suppose you try to push a swing at some frequency other than its natural oscillation frequency. Sometimes you would push it as it goes forward, thus adding energy to the system, but in other cycles you would be trying to push it forward as it comes back. This would decelerate the swing and remove energy from the system. The net result would be a small amplitude. Only the frequency-matching condition builds up the amplitude.

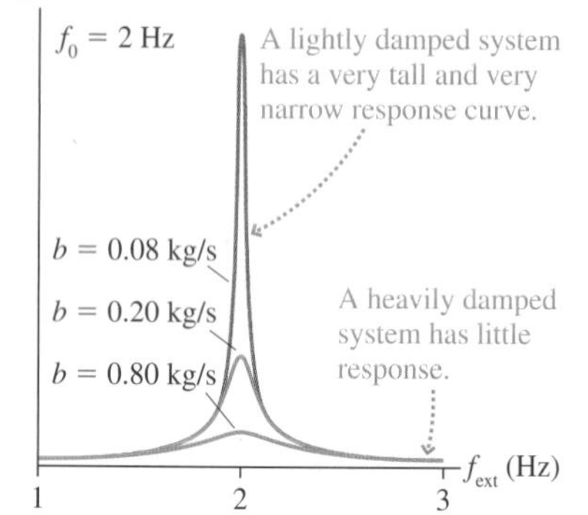
Amplitude
f_0 = 2 Hz
A lightly damped system has a very tall and very narrow response curve.
b = 0.08 kg/s
b = 0.20 kg/s
b = 0.80 kg/s
A heavily damped system has little response.
f_{ext} (Hz)
1
2
3

FIGURE 14.26 The resonance amplitude becomes higher and narrower as the damping constant decreases.

The amplitude can become exceedingly large when the frequencies match, especially if the damping constant is very small. Figure 14.26 shows the same oscillator with three different values of the damping constant. There's very little response if the damping constant is increased to 0.80 kg/s, but the amplitude for $f_{ext} = f_0$ becomes very large when the damping constant is reduced to 0.08 kg/s. This large-amplitude response to a driving force whose frequency matches the natural frequency of the system is a phenomenon called **resonance.** The condition for resonance is

$$f_{ext} = f_0 \quad \text{(resonance condition)} \tag{14.58}$$

Within the context of driven oscillations, the natural frequency f_0 is often called the **resonance frequency.**

There are many examples of resonance. We've seen that pushing a child on a swing is one. You may have heard that soldiers "break step" when crossing a footbridge, rather than marching in unison. The bridge has a natural frequency with which it bounces up and down. A group of soldiers marching across at exactly this frequency would cause a resonance, and potentially a disaster. A similar situation occurs when a singer is able to shatter a crystal goblet by singing the right note. The goblet has a natural frequency, which is the musical sound you hear if you lightly tap a wine glass. The singer causes a sound wave to impinge on the goblet, exerting a small driving force at the frequency of the note she is singing. If the singer's frequency matches the natural frequency of the goblet—resonance! The vibration amplitude may grow so large that the glass has a structural failure and shatters.

An important feature of Figure 14.26 is how the amplitude and width of the resonance depend on the damping constant. A heavily damped system responds fairly little, even at resonance, but it responds to a wide range of driving frequencies. Very lightly damped systems can reach exceptionally high amplitudes, but notice that the range of frequencies to which the system responds becomes narrower and narrower as b decreases.

A singer or musical instrument can shatter a crystal goblet by matching the goblet's natural oscillation frequency.

This allows us to understand why singers can break crystal goblets but not inexpensive, everyday glasses. An inexpensive glass gives a "thud" when tapped, but a fine crystal goblet "rings" for several seconds. In physics terms, the goblet has a much longer time constant than the glass. That, in turn, implies that the goblet is very lightly damped while the ordinary glass is heavily damped (because the internal forces within the glass are not those of a high-quality crystal structure). Only the lightly damped goblet, like the top curve in Figure 14.26, can reach amplitudes large enough to shatter. The restriction, though, is that its natural frequency has to be matched very precisely, which is why such demonstrations are done by professional singers or musicians.

It is worth noting that there are many mechanical systems, such as the wings on airplanes, for which it is essential that resonances be avoided! It is important to understand the conditions of resonance in order to design structures without them.

SUMMARY

The goal of Chapter 14 has been to understand systems that oscillate with simple harmonic motion.

GENERAL PRINCIPLES

Dynamics

SHM occurs when a linear restoring force acts to return a system to an equilibrium position.

Horizontal spring

$(F_{\text{net}})_x = -kx$

Vertical spring

The origin is at the equilibrium position $\Delta L = mg/k$.

$(F_{\text{net}})_y = -ky$

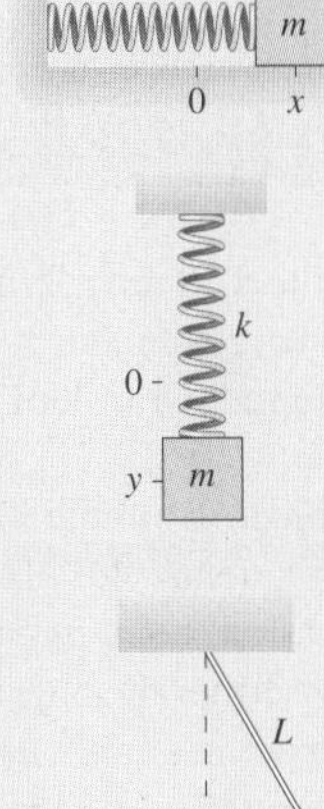

$$\omega = \sqrt{\frac{k}{m}} \qquad T = 2\pi\sqrt{\frac{m}{k}}$$

Pendulum

$$(F_{\text{net}})_t = -\left(\frac{mg}{L}\right)s$$

$$\omega = \sqrt{\frac{g}{L}} \qquad T = 2\pi\sqrt{\frac{L}{g}}$$

Energy

If there is **no friction** or dissipation, kinetic and potential energy are alternately transformed into each other, but the total mechanical energy $E = K + U$ is conserved.

$$E = \frac{1}{2}mv_x^{\,2} + \frac{1}{2}kx^2$$

$$= \frac{1}{2}m(v_{\text{max}})^2$$

$$= \frac{1}{2}kA^2$$

In a **damped system,** the energy decays exponentially

$$E = E_0 e^{-t/\tau}$$

where τ is the time constant.

IMPORTANT CONCEPTS

Simple harmonic motion (SHM) is a sinusoidal oscillation with period T and amplitude A.

Frequency $f = \dfrac{1}{T}$

Angular frequency

$$\omega = 2\pi f = \frac{2\pi}{T}$$

Position $x(t) = A\cos(\omega t + \phi_0)$

$$= A\cos\left(\frac{2\pi t}{T} + \phi_0\right)$$

Velocity $v_x(t) = -v_{\text{max}}\sin(\omega t + \phi_0)$ with maximum speed $v_{\text{max}} = \omega A$

Acceleration $a_x = -\omega^2 x$

SHM is the projection onto the x-axis of **uniform circular motion.**

$\phi = \omega t + \phi_0$ is the phase

The position at time t is

$$x(t) = A\cos\phi$$
$$= A\cos(\omega t + \phi_0)$$

The phase constant ϕ_0 determines the initial conditions:

$$x_0 = A\cos\phi_0 \qquad v_{0x} = -\omega A\sin\phi_0$$

APPLICATIONS

Resonance

When a system is driven by a periodic external force, it responds with a large-amplitude oscillation if $f_{\text{ext}} \approx f_0$ where f_0 is the system's natural oscillation frequency, or **resonant frequency.**

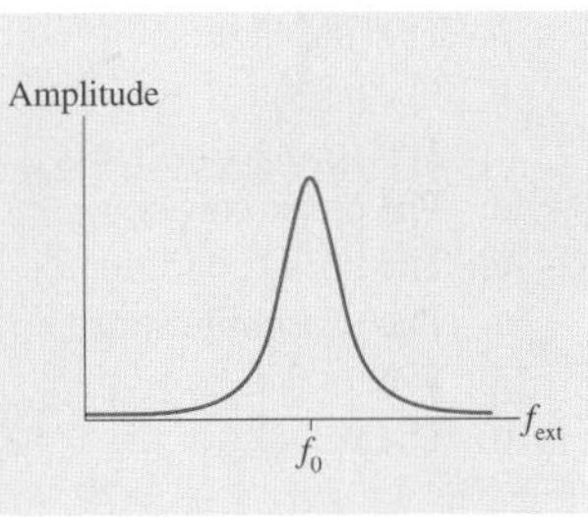

Damping

If there is a drag force $\vec{D} = -b\vec{v}$, where b is the damping constant, then (for lightly damped systems)

$$x(t) = Ae^{-bt/2m}\cos(\omega t + \phi_0)$$

The time constant for energy loss is $\tau = m/b$.

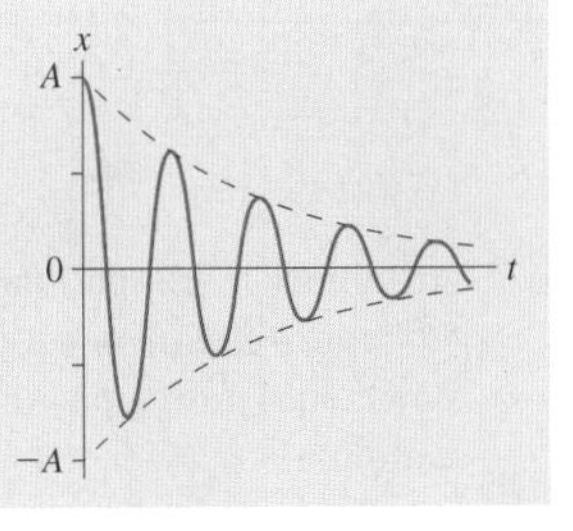

TERMS AND NOTATION

oscillatory motion
oscillator
period, T
frequency, f
hertz, Hz
simple harmonic motion, SHM
amplitude, A
angular frequency, ω
phase, ϕ
phase constant, ϕ_0
restoring force
equation of motion
small-angle approximation
linear restoring force
damped oscillation
damping constant, b
envelope
time constant, τ
half-life, $t_{1/2}$
driven oscillation
natural frequency, f_0
driving frequency, f_{ext}
response curve
resonance
resonance frequency, f_0

EXERCISES AND PROBLEMS

Exercises

Section 14.1 Simple Harmonic Motion

1. When a guitar string plays the note "A," the string vibrates at 440 Hz. What is the period of the vibration?
2. In taking your pulse, you count 75 heartbeats in 1 min. What are the period and frequency of your heart's oscillations?
3. An air-track glider attached to a spring oscillates between the 10 cm mark and the 60 cm mark on the track. The glider completes 10 oscillations in 33 s. What are the (a) period, (b) frequency, (c) angular frequency, (d) amplitude, and (e) maximum speed of the glider?
4. An air-track glider is attached to a spring. The glider is pulled to the right and released from rest at $t = 0$ s. It then oscillates with a period of 2.0 s and a maximum speed of 40 cm/s.
   a. What is the amplitude of the oscillation?
   b. What is the glider's position at $t = 0.25$ s?

Section 14.2 Simple Harmonic Motion and Circular Motion

5. What are the (a) amplitude, (b) frequency, and (c) phase constant of the oscillation shown in Figure Ex14.5?

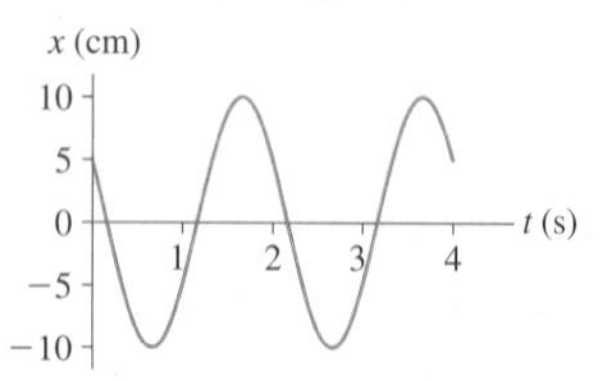

FIGURE EX14.5

6. What are the (a) amplitude, (b) frequency, and (c) phase constant of the oscillation shown in Figure Ex14.6?

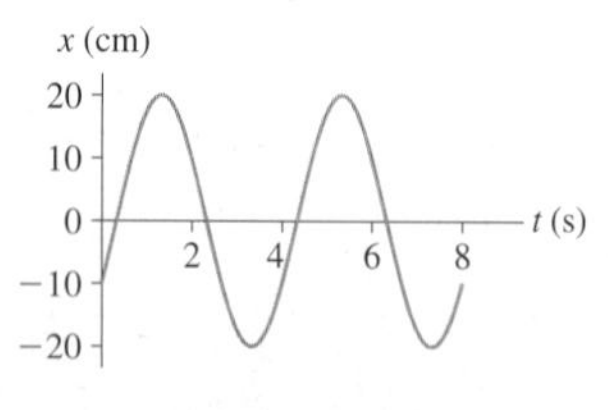

FIGURE EX14.6

7. An object in simple harmonic motion has an amplitude of 4.0 cm, a frequency of 2.0 Hz, and a phase constant of $2\pi/3$ rad. Draw a position graph showing two cycles of the motion.
8. An object in simple harmonic motion has an amplitude of 6.0 cm, a frequency of 0.50 Hz, and a phase constant of $-2\pi/3$ rad. Draw a position graph showing two cycles of the motion.
9. An object in simple harmonic motion has amplitude 4.0 cm and frequency 4.0 Hz, and at $t = 0$ s it passes through the equilibrium point moving to the right. Write the function $x(t)$ that describes the object's position.
10. An object in simple harmonic motion has amplitude 10.0 cm and frequency 0.50 Hz, and at $t = 0$ s it is at its largest distance to the left of the equilibrium position. Write the function $x(t)$ that describes the object's position.
11. An air-track glider attached to a spring oscillates with a period of 1.5 s. At $t = 0$ s the glider is 5.0 cm left of the equilibrium position and moving to the right at 36.3 cm/s.
    a. What is the phase constant?
    b. What is the phase at $t = 0$ s, 0.5 s, 1.0 s, and 1.5 s?

Section 14.3 Energy in Simple Harmonic Motion

Section 14.4 The Dynamics of Simple Harmonic Motion

12. A block attached to a spring with unknown spring constant oscillates with a period of 2.0 s. What is the period if
    a. The mass is doubled?
    b. The mass is halved?
    c. The amplitude is doubled?
    d. The spring constant is doubled?
    Parts a to d are independent questions, each referring to the initial situation.
13. A 200 g air-track glider is attached to a spring. The glider is pushed in 10 cm and released. A student with a stopwatch finds that 10 oscillations take 12.0 s. What is the spring constant?
14. The position of a 50 g oscillating mass is given by $x(t) = (2.0\text{ cm})\cos(10t - \pi/4)$, where t is in s. Determine:
    a. The amplitude.
    b. The period.
    c. The spring constant.
    d. The phase constant.
    e. The initial conditions.
    f. The maximum speed.
    g. The total energy.
    h. The velocity at $t = 0.40$ s.

15. A 200 g mass attached to a horizontal spring oscillates at a frequency of 2.0 Hz. At $t = 0$ s, the mass is at $x = 5.0$ cm and has $v_x = -30$ cm/s. Determine:
    a. The period.
    b. The angular frequency.
    c. The amplitude.
    d. The phase constant.
    e. The maximum speed.
    f. The maximum acceleration.
    g. The total energy.
    h. The position at $t = 0.40$ s.
16. A 507 g mass oscillates with an amplitude of 10 cm on a spring whose spring constant is 20 N/m. At $t = 0$ s the mass is 5.0 cm to the right of the equilibrium position and moving to the right. Determine:
    a. The period.
    b. The angular frequency.
    c. The phase constant.
    d. The initial velocity.
    e. The maximum speed.
    f. The total energy.
    g. The position at $t = 1.3$ s.
    h. The velocity at $t = 1.3$ s.
17. A 1.0 kg block is attached to a spring with spring constant 16 N/m. While the block is sitting at rest, a student hits it with a hammer and almost instantaneously gives it a speed of 40 cm/s. What are
    a. The amplitude of the subsequent oscillations?
    b. The block's speed at the point where $x = \frac{1}{2}A$?

Section 14.5 Vertical Oscillations

18. A spring is hanging from the ceiling. Attaching a 500 g physics book to the spring causes it to stretch 20 cm in order to come to equilibrium.
    a. What is the spring constant?
    b. From equilibrium, the book is pulled down 10 cm and released. What is the period of oscillation?
    c. What is the book's maximum speed? At what position or positions does it have this speed?
19. A spring with spring constant 15.0 N/m hangs from the ceiling. A ball is attached to the spring and allowed to come to rest. It is then pulled down 6.0 cm and released. If the ball makes 30 oscillations in 20 s, what are its (a) mass and (b) maximum speed?
20. A spring is hung from the ceiling. When a block is attached to its end, it stretches 2.0 cm before reaching its new equilibrium length. The block is then pulled down slightly and released. What is the frequency of oscillation?

Section 14.6 The Pendulum

21. Make a table with 4 columns and 8 rows. In row 1, label the columns θ (°), θ (rad), $\sin\theta$, and $\cos\theta$. In the left column, starting in row 2, write 0, 2, 4, 6, 8, 10, and 12.
    a. Convert each of these angles, in degrees, to radians. Put the results in column 2. Show four decimal places.
    b. Calculate the sines and cosines. Put the results, showing four decimal places, in columns 3 and 4.
    c. What is the first angle for which θ and $\sin\theta$ differ by more than 0.0010?
    d. What is the first angle for which $\cos\theta$ is less than 0.9900?
    e. Over what range of angles does the small-angle approximation appear to be valid?
22. A mass on a string of unknown length oscillates as a pendulum with a period of 4.0 s. What is the period if
    a. The mass is doubled?
    b. The string length is doubled?
    c. The string length is halved?
    d. The amplitude is doubled?

    Parts a to d are independent questions, each referring to the initial situation.
23. A 200 g ball is tied to a string. It is pulled to an angle of 8.0° and released to swing as a pendulum. A student with a stopwatch finds that 10 oscillations take 12.0 s. How long is the string?
24. What is the period of a 1.0-m-long pendulum on (a) the earth and (b) Venus?
25. What is the length of a pendulum whose period on the moon matches the period of a 2.0-m-long pendulum on the earth?
26. The angle of a pendulum is $\theta(t) = (0.10 \text{ rad})\cos(5t + \pi)$, where t is in s. Determine:
    a. The amplitude.
    b. The frequency.
    c. The phase constant.
    d. The length of the string.
    e. The initial conditions.
    f. The angle at $t = 2.0$ s.

Section 14.7 Damped Oscillations

27. The amplitude of an oscillator decreases to 36.8% of its initial value in 10.0 s. What is the value of the time constant?
28. Calculate and draw an accurate position graph from $t = 0$ s to $t = 10$ s of a damped oscillator having a frequency of 1.0 Hz and a time constant of 4.0 s.
29. A 250 g air-track glider is attached to a spring with spring constant 4.0 N/m. The damping constant due to air resistance is 0.015 kg/s. The glider is pulled out 20 cm from equilibrium and released. How many oscillations will it make during the time in which the amplitude decays to e^{-1} of its initial value?
30. A spring with spring constant 15.0 N/m hangs from the ceiling. A 500 g ball is attached to the spring and allowed to come to rest. It is then pulled down 6.0 cm and released. What is the time constant if the ball's amplitude has decreased to 3.0 cm after 30 oscillations?

Problems

31. An object oscillating on a spring has the position graph shown in Figure P14.31. Draw a position graph if the following changes are made.
    a. The amplitude is halved and the frequency is halved.
    b. The mass is quadrupled.

    Parts a and b are independent questions, each starting from the graph shown.

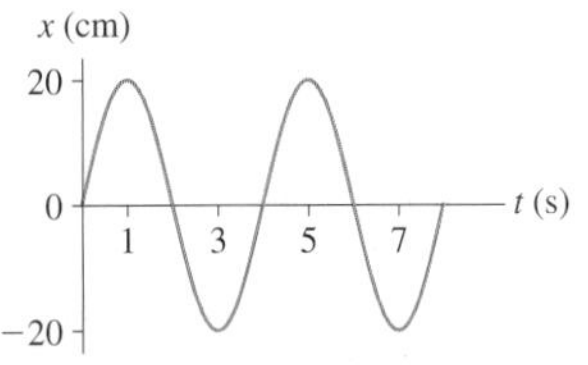

FIGURE P14.31

32. An object oscillating on a spring has the velocity graph shown in Figure P14.32. Draw a velocity graph if the following changes are made.
    a. The amplitude is doubled and the frequency is halved.
    b. The mass is quadrupled.

    Parts a and b are independent questions, each starting from the graph shown.

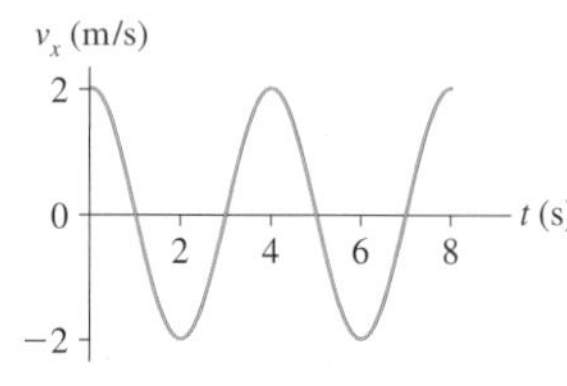

FIGURE P14.32

33. Figure P14.33 is the position-versus-time graph of a particle in simple harmonic motion.
    a. What is the phase constant?
    b. What is the velocity at $t = 0$ s?
    c. What is v_{max}?

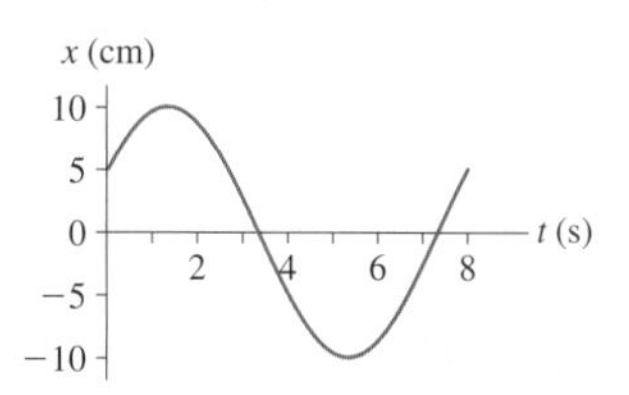

FIGURE P14.33

34. Figure P14.34 is the velocity-versus-time graph of a particle in simple harmonic motion.
    a. What is the amplitude of the oscillation?
    b. What is the phase constant?
    c. What is the position at $t = 0$ s?

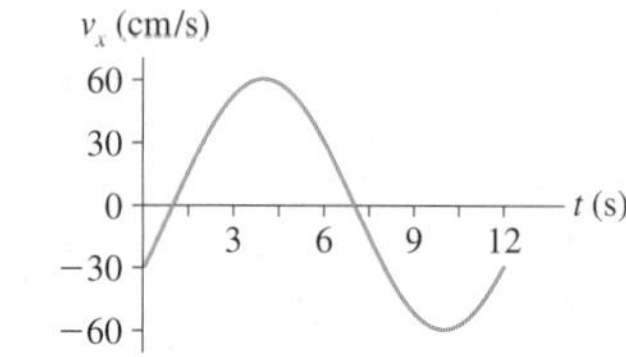

FIGURE P14.34

35. The two graphs in Figure P14.35 are for two different vertical mass/spring systems.
    a. What is the frequency of system A? What is the first time at which the mass has maximum speed while traveling in the upward direction?
    b. What is the period of system B? What is the first time at which the energy is all potential?
    c. If both systems have the same mass, what is the ratio k_A/k_B of their spring constants?

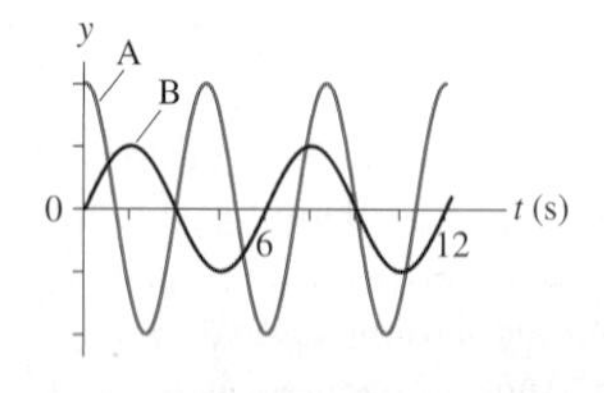

FIGURE P14.35

36. Astronauts in space cannot weigh themselves by standing on a bathroom scale. Instead, they determine their mass by oscillating on a large spring. Suppose an astronaut attaches one end of a large spring to her belt and the other end to a hook on the wall of the space capsule. A fellow astronaut then pulls her away from the wall and releases her. The spring's length as a function of time is shown in Figure P14.36.
    a. What is her mass if the spring constant is 240 N/m?
    b. What is her speed when the spring's length is 1.2 m?

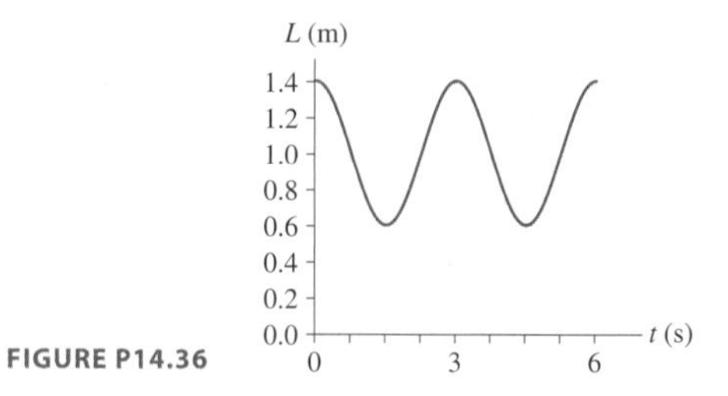

FIGURE P14.36

37. A 1.0 kg block oscillates on a spring with spring constant 20 N/m. At $t = 0$ s the block is 20 cm to the right of the equilibrium position and moving to the left at a speed of 100 cm/s. Determine the period of oscillation and draw a graph of position versus time.
38. An object in SHM oscillates with a period of 4.0 s and an amplitude of 10 cm. How long does the object take to move from $x = 0.0$ cm to $x = 6.0$ cm?
39. The motion of a particle is given by $x(t) = (25\text{ cm})\cos(10t)$, where t is in s. At what time is the kinetic energy twice the potential energy?
40. a. When the displacement of a mass on a spring is $\frac{1}{2}A$, what fraction of the energy is kinetic energy and what fraction is potential energy?
    b. At what displacement, as a fraction of A, is the energy half kinetic and half potential?
41. For a particle in simple harmonic motion, show that $v_{max} = (\pi/2)v_{avg}$ where v_{avg} is the average speed during one cycle of the motion.
42. A block on a spring is pulled to the right and released at $t = 0$ s. It passes $x = 3.0$ cm at $t = 0.685$ s, and it passes $x = -3.0$ cm at $t = 0.886$ s.
    a. What is the angular frequency?
    b. What is the amplitude?

    Hint: $\cos(\pi - \theta) = -\cos\theta$.
43. A 100 g ball attached to a spring with spring constant 2.5 N/m oscillates horizontally on a frictionless table. Its velocity is 20 cm/s when $x = -5.0$ cm.
    a. What is the amplitude of oscillation?
    b. What is the ball's maximum acceleration?
    c. What is the ball's position when the acceleration is maximum?
    d. What is the speed of the ball when $x = 3.0$ cm?
44. The velocity of an object in simple harmonic motion is given by $v_x(t) = -(0.35\text{ m/s})\sin(20t + \pi)$, where t is in s.
    a. What is the first time after $t = 0$ s at which the velocity is -0.25 m/s?
    b. What is the object's position at that time?
45. A 300 g oscillator has a speed of 95.4 cm/s when its displacement is 3.0 cm and 71.4 cm/s when its displacement is 6.0 cm. What is the oscillator's maximum speed?
46. An ultrasonic transducer, of the type used in medical ultrasound imaging, is a very thin disk ($m = 0.10$ g) driven back and forth in SHM at 1.0 MHz by an electromagnetic coil.
    a. The maximum restoring force that can be applied to the disk without breaking it is 40,000 N. What is the maximum oscillation amplitude that won't rupture the disk?
    b. What is the disk's maximum speed at this amplitude?

47. A 5.0 kg block hangs from a spring with spring constant 2000 N/m. The block is pulled down 5.0 cm from the equilibrium position and given an initial velocity of 1.0 m/s back toward equilibrium. What are the (a) frequency, (b) amplitude, and (c) total mechanical energy of the motion?
48. A 2.0 g spider is dangling at the end of a silk thread. You can make the spider bounce up and down on the thread by tapping lightly on his feet with a pencil. You soon discover that you can give the spider the largest amplitude on his little bungee cord if you tap exactly once every second. What is the spring constant of the silk thread?
49. A 200 g block hangs from a spring with spring constant 10 N/m. At $t = 0$ s the block is 20 cm below the equilibrium point and moving upward with a speed of 100 cm/s. What are the block's
    a. Oscillation frequency?
    b. Displacement from equilibrium when the speed is 50 cm/s?
    c. Position at $t = 1.0$ s?
50. A spring with spring constant k is suspended vertically from a support and a mass m is attached. The mass is held at the point where the spring is not stretched. Then the mass is released and begins to oscillate. The lowest point in the oscillation is 20 cm below the point where the mass was released. What is the oscillation frequency?
51. A compact car has a mass of 1200 kg. Assume that the car has one spring on each wheel, that the springs are identical, and that the mass is equally distributed over the four springs.
    a. What is the spring constant of each spring if the empty car bounces up and down 2.0 times each second?
    b. What will be the car's oscillation frequency while carrying four 70 kg passengers?
52. A block hangs in equilibrium from a vertical spring. When a second identical block is added, the original block sags by 5.0 cm. What is the oscillation frequency of the two-block system?
53. A 500 g block slides along a frictionless surface at a speed of 0.35 m/s. It runs into a horizontal massless spring with spring constant 50 N/m that extends outward from a wall. It compresses the spring, then is pushed back in the opposite direction by the spring, eventually losing contact with the spring.
    a. How long does the block remain in contact with the spring?
    b. How would your answer to part a change if the block's initial speed were doubled?
54. Figure P14.54 shows a 1.0 kg mass riding on top of a 5.0 kg mass as it oscillates on a frictionless surface. The spring constant is 50 N/m and the coefficient of static friction between the two blocks is 0.50. What is the maximum oscillation amplitude for which the upper block does not slip?

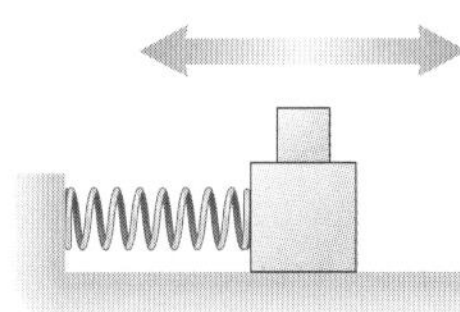

FIGURE P14.54

55. The two blocks in Figure P14.54 oscillate on a frictionless surface with a period of 1.5 s. The upper block just begins to slip when the amplitude is increased to 40 cm. What is the coefficient of static friction between the two blocks?
56. It is said that Galileo discovered a basic principle of the pendulum—that the period is independent of the amplitude—by using his pulse to time the period of swinging lamps in the cathedral as they swayed in the breeze. Suppose that one oscillation of a swinging lamp takes 5.5 s.
    a. How long is the lamp chain?
    b. What maximum speed does the lamp have if its maximum angle from vertical is 3.0°?
57. A 100 g mass on a 1.0-m-long string is pulled 8.0° to one side and released. How long does it take for the pendulum to reach 4.0° on the opposite side?
58. Astronauts on the first trip to Mars take along a pendulum that has a period on earth of 1.50 s. The period on Mars turns out to be 2.45 s. What is the Martian acceleration due to gravity?
59. The earth's acceleration due to gravity varies from 9.78 m/s^2 at the equator to 9.83 m/s^2 at the poles, because the earth is not a perfect sphere. A pendulum whose length is precisely 1.000 m can be used to measure g. Such a device is called a *gravimeter.*
    a. How long do 100 oscillations take at the equator?
    b. How long do 100 oscillations take at the north pole?
    c. Is the difference between your answers to parts a and b measurable? What kind of instrument could you use to measure the difference?
    d. Suppose you take your gravimeter to the top of a high mountain peak near the equator. There you find that 100 oscillations take 201.0 s. What is g on the mountain top?
60. In a science museum, a 110 kg brass pendulum bob swings at the end of a 15.0-m-long wire. The pendulum is started at exactly 8:00 A.M. every morning by pulling it 1.5 m to the side and releasing it. Because of its compact shape and smooth surface, the pendulum's damping constant is only 0.010 kg/s. At exactly 12:00 noon, how many oscillations will the pendulum have completed and what is its amplitude?
61. A 500 g air-track glider attached to a spring with spring constant 10 N/m is sitting at rest on a frictionless air track. A 250 g glider is pushed toward it from the far end of the track at a speed of 120 cm/s. It collides with and sticks to the 500 g glider. What are the amplitude and period of the subsequent oscillations?
62. A 200 g block attached to a horizontal spring is oscillating with an amplitude of 2.0 cm and a frequency of 2.0 Hz. Just as it passes through the equilibrium point, moving to the right, a sharp blow directed to the left exerts a 20 N force for 1.0 ms. What are the new (a) frequency and (b) amplitude?
63. A 1.00 kg block is attached to a horizontal spring with spring constant 2500 N/m. The block is at rest on a frictionless surface. A 10 g bullet is fired into the block, in the face opposite the spring, and sticks.
    a. What was the bullet's speed if the subsequent oscillations have an amplitude of 10.0 cm?
    b. Could you determine the bullet's speed by measuring the oscillation frequency? If so, how? If not, why not?
64. A pendulum consists of a massless, rigid rod with a mass at one end. The other end is pivoted on a frictionless pivot so that it can turn through a complete circle. The pendulum is inverted, so the mass is directly above the pivot point, then released. The speed of the mass as it passes through the lowest point is 5.0 m/s. If the pendulum undergoes small-amplitude oscillations at the bottom of the arc, what will the frequency be?

65. Figure P14.65 is a top view of an object of mass m connected between two stretched rubber bands of length L. The object rests on a frictionless surface. At equilibrium, the tension in each rubber band is T. Find an expression for the frequency of oscillations *perpendicular* to the rubber bands. Assume the amplitude is sufficiently small that the magnitude of the tension in the rubber bands is essentially unchanged as the mass oscillates.

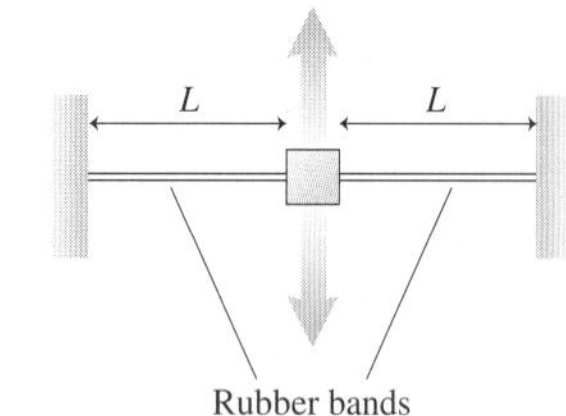

FIGURE P14.65

66. A molecular bond can be modeled as a spring between two atoms that vibrate with simple harmonic motion. Figure P14.66 shows an SHM approximation for the potential energy of an HCl molecule. For $E < 4 \times 10^{-19}$ J it is a good approximation to the more accurate HCl potential energy curve that was shown in Figure 10.37. Because the chlorine atom is so much more massive than the hydrogen atom, it is reasonable to assume that the hydrogen atom ($m = 1.67 \times 10^{-27}$ kg) vibrates back and forth while the chlorine atom remains at rest. Use the graph to estimate the vibrational frequency of the HCl molecule.
Hint: How can you use the graph to determine the spring constant of the molecular bond?

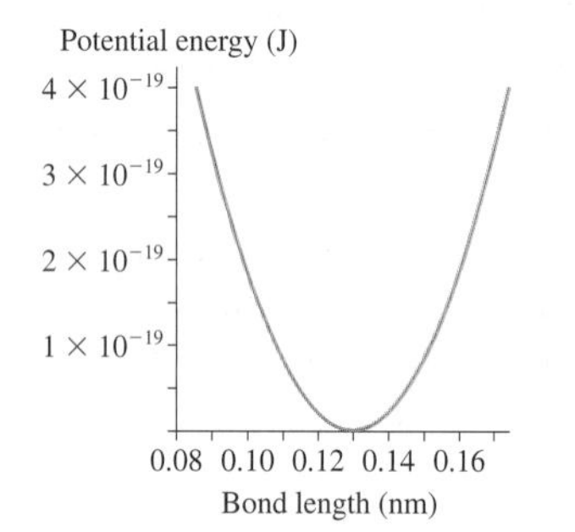

FIGURE P14.66

67. A marble can roll around the inside of a vertical circular hoop of radius R. It undergoes small-amplitude oscillations if displaced slightly from the equilibrium position at the lowest point. Find an expression for the period of these small-amplitude oscillations.
68. A penny rides on top of a piston as it undergoes vertical simple harmonic motion with an amplitude of 4.0 cm. If the frequency is low, the penny rides up and down without difficulty. If the frequency is steadily increased, there comes a point at which the penny leaves the surface.
    a. At what point in the cycle does the penny first lose contact with the piston?
    b. What is the maximum frequency for which the penny just barely remains in place for the full cycle?
69. On your first trip to Planet X you happen to take along a 200 g mass, a 40-cm-long spring, a meter stick, and a stopwatch. You're curious about the acceleration due to gravity on Planet X, where ordinary tasks seem easier than on earth, but you can't find this information in your Visitor's Guide. One night you suspend the spring from the ceiling in your room and hang the mass from it. You find that the mass stretches the spring by 31.2 cm. You then pull the mass down 10.0 cm and release it. With the stopwatch you find that 10 oscillations take 14.5 s. Can you now satisfy your curiosity?
70. The 15 g head of a bobble-head doll oscillates in SHM at a frequency of 4.0 Hz.
    a. What is the spring constant of the spring on which the head is mounted?
    b. Suppose the head is pushed 2.0 cm against the spring, then released. What is the head's maximum speed as it oscillates?
    c. The amplitude of the head's oscillations decreases to 0.5 cm in 4.0 s. What is the head's damping constant?
71. An oscillator with a mass of 500 g and a period of 0.50 s has an amplitude that decreases by 2.0% during each complete oscillation.
    a. If the initial amplitude is 10 cm, what will be the amplitude after 25 oscillations?
    b. At what time will the energy be reduced to 60% of its initial value?
72. A 200 g oscillator in a vacuum chamber has a frequency of 2.0 Hz. When air is admitted, the oscillation decreases to 60% of its initial amplitude in 50 s. How many oscillations will have been completed when the amplitude is 30% of its initial value?
73. You've been hired by the circus to help design a new act. Samson, the great trapeze artist, will swing back and forth on a trapeze that hangs from 6.0-m-long ropes. Delilah, his beautiful young assistant, will stand precariously on a small platform that is 2.0 m below the pivot point at which the trapeze ropes are tied. The goal is for Delilah to dive off the platform, fall 4.0 m, and then be caught by Samson as he swings through the lowest point of his arc. Obviously, timing is everything. Samson will be swinging in an arc with an amplitude of 8.0°. At what angle should Samson be when Delilah leaps? There's no safety net, and Delilah, who is also the circus treasurer, won't write your check until after the first show. Keep in mind that Samson and Delilah aren't scientists, so they'll need your answer in units they can understand.
74. Invent a device, based on what you have learned in this chapter, with which a "weightless" astronaut in space can measure her mass. To recognize your invention, the patent office will need:
    a. A sketch.
    b. A description of the operating principles.
    c. Values for all major components.
    d. Sample calculations showing that these component values are "reasonable."

Challenge Problems

75. A block on a frictionless table is connected as shown in Figure CP14.75 to two springs having spring constants k_1 and k_2. Show that the block's oscillation frequency is given by

$$f = \sqrt{f_1^2 + f_2^2}$$

where f_1 and f_2 are the frequencies at which it would oscillate if attached to spring 1 or spring 2 alone.

FIGURE CP14.75

76. A block on a frictionless table is connected as shown in Figure CP14.76 to two springs having spring constants k_1 and k_2. Find an expression for the block's oscillation frequency f in terms of the frequencies f_1 and f_2 at which it would oscillate if attached to spring 1 or spring 2 alone.

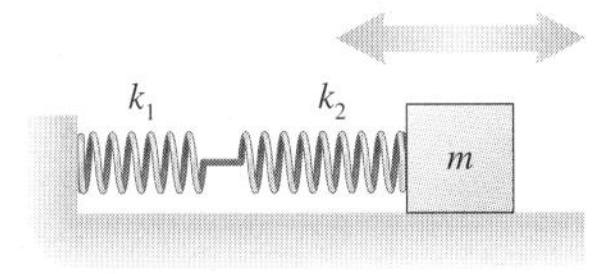

FIGURE CP14.76

77. A spring is standing upright on a table with its bottom end fastened to the table. A block is dropped from a height 3.0 cm above the top of the spring. The block sticks to the top end of the spring and then oscillates with an amplitude of 10 cm. What is the oscillation frequency?
78. Jose, whose mass is 75 kg, has just completed his first bungee jump and is now bouncing up and down at the end of the cord. His oscillations have an initial amplitude of 11.0 m and a period of 4.0 s.
    a. What is the spring constant of the bungee cord?
    b. What is Jose's maximum speed while oscillating?
    c. From what height above the lowest point did Jose jump?
    d. If the damping constant due to air resistance is 6.0 kg/s, how many oscillations will Jose make before his amplitude has decreased to 2.0 m?

    Hint: Although not entirely realistic, treat the bungee cord as an ideal spring that can be compressed to a shorter length as well as stretched to a longer length.
79. A 1000 kg car carrying two 100 kg football players travels over a bumpy "washboard" road with the bumps spaced 3.0 m apart. The driver finds that the car bounces up and down with maximum amplitude when he drives at a speed of 5.0 m/s ($\approx$11 mph). The car then stops and picks up three more 100 kg passengers. By how much does the car body sag on its suspension when these three additional passengers get in?
80. Prove that the expression for $x(t)$ in Equation 14.53 is a solution to the equation of motion for a damped oscillator, Equation 14.52, if and only if the angular frequency ω is given by the expression in Equation 14.54.

STOP TO THINK ANSWERS

Stop to Think 14.1: c. $v_{max} = 2\pi A/T$. Doubling A and T leaves v_{max} unchanged.

Stop to Think 14.2: d. Think of circular motion. At 45°, the particle is in the first quadrant (positive x) and moving to the left (negative v_x).

Stop to Think 14.3: c > b > a = d. Energy conservation $\frac{1}{2}kA^2 = \frac{1}{2}m(v_{max})^2$ gives $v_{max} = \sqrt{k/m}A$. k or m have to be increased or decreased by a factor of 4 to have the same effect as increasing or decreasing A by a factor of 2.

Stop to Think 14.4: c. $v_x = 0$ because the slope of the position graph is zero. The negative value of x shows that the particle is left of the equilibrium position, so the restoring force is to the right.

Stop to Think 14.5: c. The period of a pendulum does not depend on its mass.

Stop to Think 14.6: $\tau_d > \tau_b = \tau_c > \tau_a$. The time constant is the time to decay to 37% of the initial height. The time constant is independent of the initial height.

15 Fluids and Elasticity

Kayaking through the rapids requires an intuitive understanding of fluids.

▶ Looking Ahead

The goal of Chapter 15 is to understand macroscopic systems that flow or deform. In this chapter you will learn to:

- Understand and use the concept of mass density.
- Understand pressure in liquids and gases.
- Use a wide variety of units for measuring pressure.
- Use Archimedes' principle to understand buoyancy.
- Use an ideal-fluid model to investigate how fluids flow.
- Calculate the elastic deformation of solids and liquids.

◀ Looking Back

The material in this chapter depends on the conditions of equilibrium. Please review:

- Section 4.6 Equilibrium and Newton's first law.
- Section 10.4 Hooke's law and restoring forces.

This kayak is floating on water, a fluid. The water itself is in motion. Surprisingly, we need no new laws of physics to understand how fluids flow or why some objects float while others sink. The physics of fluids, often called *fluid mechanics,* is an important application of Newton's laws and the law of conservation of energy—physics that you learned in Parts I and II.

Fluids are macroscopic systems, and our study of fluids will take us well beyond the particle model. Two new concepts, *density* and *pressure,* will be introduced to describe macroscopic systems. We'll begin with *fluid statics,* situations in which the fluid remains at rest. Suction cups and floating aircraft carriers are just two of the applications we'll explore. Then we'll turn to fluids in motion. Bernoulli's equation, the governing principle of *fluid dynamics,* will explain how water flows through fire hoses, how airplanes stay aloft, and many things in between. We'll then end this chapter with a brief look at a different but related property of macroscopic systems, the *elasticity* of solids.

15.1 Fluids

Quite simply, a **fluid** is a substance that flows. Because they flow, fluids take the shape of their container rather than retaining a shape of their own. You may think that gases and liquids are quite different, but both are fluids, and their similarities are often more important than their differences.

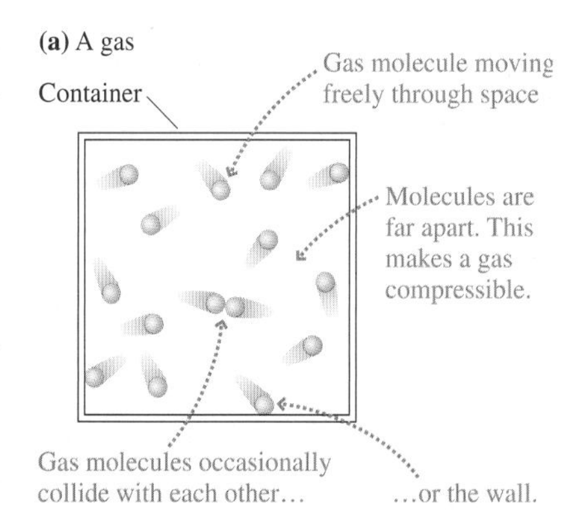

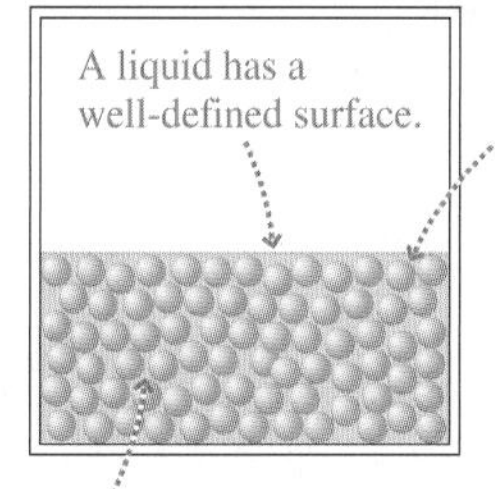

Gas molecules occasionally collide with each other...

(b) A liquid
A liquid has a well-defined surface.

FIGURE 15.1 Simple atomic-level models of gases and liquids.

Gases and Liquids

A **gas,** shown in Figure 15.1a, is a system in which each molecule moves through space as a free, noninteracting particle until, on occasion, it collides with another molecule or with the wall of the container. The gas you are most familiar with is air, a mixture of mostly nitrogen and oxygen molecules. Gases are fairly simple macroscopic systems, and Part IV of this textbook will delve into the thermal properties of gases. For now, two properties of gases interest us:

1. Gases are *fluids.* They flow, and they exert pressure on the walls of their container.
2. Gases are *compressible.* That is, the volume of a gas is easily increased or decreased, a consequence of the "empty space" between the molecules in a gas.

Liquids are more complicated than either gases or solids. Liquids, like solids, are nearly *incompressible.* This property tells us that the molecules in a liquid, as in a solid, are about as close together as they can get without coming into contact with each other. At the same time, a liquid flows and deforms to fit the shape of its container. The fluid nature of a liquid tells us that the molecules are free to move around.

Together, these observations suggest the model of a **liquid** shown in Figure 15.1b. Here you see a system in which the molecules are loosely held together by weak molecular bonds. The bonds are strong enough that the molecules never get far apart but not strong enough to prevent the molecules from sliding around each other.

Volume and Density

One important parameter that characterizes a macroscopic system is its volume V, the amount of space the system occupies. The SI unit of volume is m^3. Nonetheless, both cm^3 and, to some extent, liters (L) are widely used metric units of volume. In most cases, you *must* convert these to m^3 before doing calculations.

While it is true that $1\text{ m} = 100\text{ cm}$, it is *not* true that $1\text{ m}^3 = 100\text{ cm}^3$. Figure 15.2 shows that the volume conversion factor is $1\text{ m}^3 = 10^6\text{ cm}^3$. You can think of this process as cubing the linear conversion factor:

$$1\text{ m}^3 = 1\text{ m}^3 \times \left(\frac{100\text{ cm}}{1\text{ m}}\right)^3 = 10^6\text{ cm}^3$$

A liter is 1000 cm^3, so $1\text{ m}^3 = 10^3\text{ L}$. A milliliter (1 mL) is the same as 1 cm^3.

A system is also characterized by its *density.* Suppose you have several blocks of copper, each of different size. Each block has a different mass m and a different volume V. Nonetheless, all the blocks are copper, so there should be some quantity that has the *same* value for all the blocks, telling us, "This is copper, not some

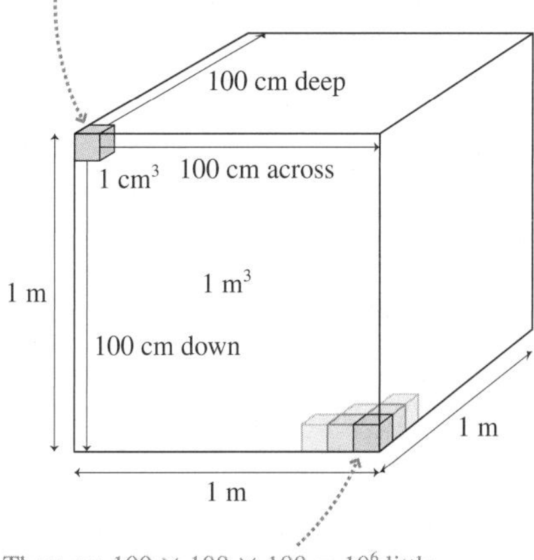

100 cm across

100 cm down

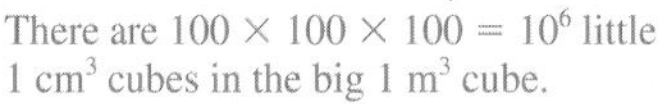

Subdivide the 1 m × 1 m × 1 m cube into little cubes 1 cm on a side. You will get 100 subdivisions along each edge.

There are 100 × 100 × 100 = 10⁶ little 1 cm³ cubes in the big 1 m³ cube.

FIGURE 15.2 There are 10^6 cm^3 in 1 m^3.

other material." The most important such parameter is the *ratio* of mass to volume, which we call the **mass density** ρ (lowercase Greek rho):

$$\rho = \frac{m}{V} \quad \text{(mass density)} \tag{15.1}$$

Conversely, an object of density ρ has mass

$$m = \rho V \tag{15.2}$$

The SI units of mass density are kg/m^3. Nonetheless, units of g/cm^3 are widely used. You need to convert these to SI units before doing most calculations. You must convert both the grams to kilograms and the cubic centimeters to cubic meters. The net result is the conversion factor

$$1\ g/cm^3 = 1000\ kg/m^3$$

TABLE 15.1 Densities of fluids at standard temperature (0°C) and pressure (1 atm)

Substance	ρ (kg/m^3)
Air	1.28
Ethyl alcohol	790
Gasoline	680
Glycerin	1260
Helium gas	0.18
Mercury	13,600
Oil (typical)	900
Seawater	1030
Water	1000

The mass density is usually called simply "the density" if there is no danger of confusion. However, we will meet other types of density as we go along, and sometimes it is important to be explicit about which density you are using. Table 15.1 provides a short list of mass densities of various fluids. Notice the enormous difference between the densities of gases and liquids. Gases have lower densities because the molecules in gases are farther apart than in liquids.

What does it *mean* to say that the density of gasoline is 680 kg/m^3 or, equivalently, 0.68 g/cm^3? Density is a mass-to-volume ratio. It is often described as the "mass per unit volume," but for this to make sense you have to know what is meant by "unit volume." Regardless of which system of length units you use, a **unit volume** is one of those units cubed. For example, if you measure lengths in meters, a unit volume is 1 m^3. But 1 cm^3 is a unit volume if you measure lengths in cm, and 1 mi^3 is a unit volume if you measure lengths in miles.

Density is the mass of one unit of volume, whatever the units happen to be. To say that the density of gasoline is 680 kg/m^3 is to say that the mass of 1 m^3 of gasoline is 680 kg. The mass of 1 cm^3 of gasoline is 0.68 g, so the density of gasoline in those units is 0.68 g/cm^3.

The mass density is independent of the object's size. That is, mass and volume are parameters that characterize a *specific piece* of some substance—say copper—whereas the mass density characterizes the substance itself. All pieces of copper have the same mass density, which differs from the mass density of any other substance. Thus mass density allows us to talk about the properties of copper in general without having to refer to any specific piece of copper.

EXAMPLE 15.1 Weighing the air

What is the mass of air in a living room having dimensions 4.0 m × 6.0 m × 2.5 m?

MODEL Table 15.1 gives air density at a temperature of 0°C. The air density doesn't vary significantly over a small range of temperatures (we'll study this issue in the next chapter), so we'll use this value even though most people keep their living room warmer than 0°C.

SOLVE The room's volume is

$$V = (4.0\ m) \times (6.0\ m) \times (2.5\ m) = 60\ m^3$$

The mass of the air is

$$m = \rho V = (1.28\ kg/m^3)(60\ m^3) = 77\ kg$$

ASSESS This is perhaps more mass than you might have expected from a substance that hardly seems to be there. For comparison, a swimming pool this size would contain 60,000 kg of water.

STOP TO THINK 15.1 A piece of glass is broken into two pieces of different size. Rank order, from largest to smallest, the mass densities of pieces 1, 2, and 3.

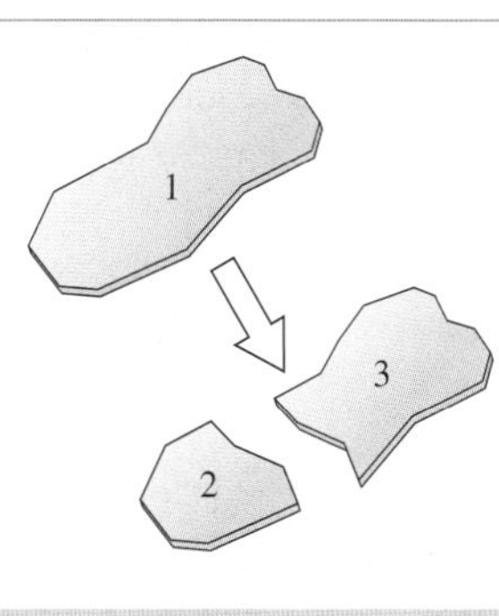

15.2 Pressure

Pressure is a word we all know and use. You probably have a commonsense idea of what pressure is. For example, you feel the effects of varying pressure against your eardrums when you swim underwater or take off in an airplane. Cans of whipped cream are "pressurized" to make the contents squirt out when you press the nozzle. It's hard to open a "vacuum sealed" jar of jelly the first time, but easy after the seal is broken.

You've undoubtedly seen water squirting out of a hole in the side of a container, as in Figure 15.3. Notice that the water emerges at greater speed from a hole at greater depth. And you've probably felt the air squirting out of a hole in a bicycle tire or inflatable air mattress. These observations suggest that

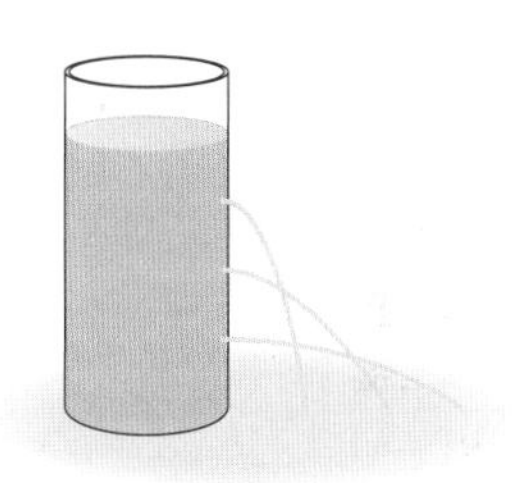

FIGURE 15.3 Water pressure pushes the water *sideways,* out of the holes.

- "Something" pushes the water or air *sideways,* out of the hole.
- In a liquid, the "something" is larger at greater depths. In a gas, the "something" appears to be the same everywhere.

Our goal is to turn these everyday observations into a precise definition of pressure.

Figure 15.4 shows a fluid—either a liquid or a gas—pressing against a small area A with force $\vec{F}$. This is the force that pushes the fluid out of a hole. In the absence of a hole, $\vec{F}$ pushes against the wall of the container. Let's define the **pressure** at this point in the fluid to be the ratio of the force to the area on which the force is exerted:

$$p = \frac{F}{A} \tag{15.3}$$

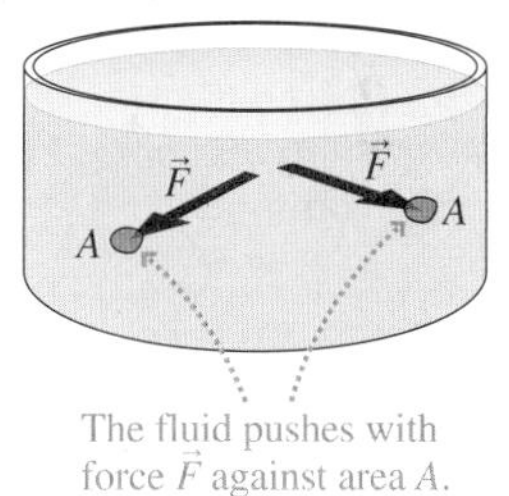

FIGURE 15.4 The fluid presses against area A with force $\vec{F}$.

Notice that pressure is a scalar, not a vector. You can see, from Equation 15.3, that a fluid exerts a force of magnitude

$$F = pA \tag{15.4}$$

on a surface of area A. The force is *perpendicular* to the surface.

NOTE ► Pressure itself is *not* a force, even though we sometimes talk informally about "the force exerted by the pressure." The correct statement is that the *fluid* exerts a force on a surface. ◄

From its definition, pressure has units of N/m^2. The SI unit of pressure is the **pascal,** defined as

$$1 \text{ pascal} = 1 \text{ Pa} \equiv 1 \frac{\text{N}}{\text{m}^2}$$

This unit is named for the 17th-century French scientist Blaise Pascal, who was one of the first to study fluids. Large pressures are often given in kilopascals, where 1 kPa = 1000 Pa.

Equation 15.3 is the basis for the simple pressure-measuring device shown in Figure 15.5a. Because the spring constant k and the area A are known, we can determine the pressure by measuring the compression of the spring. Once we've built such a device, we can place it in various liquids and gases to learn about pressure. Figure 15.5b shows what we can learn from a series of simple experiments.

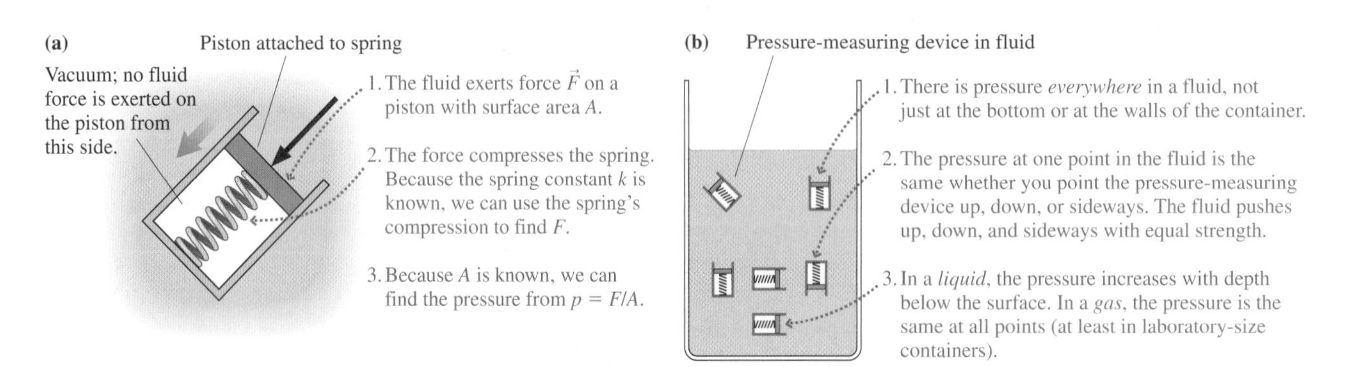

Piston attached to spring
Vacuum; no fluid force is exerted on the piston from this side.

3. Because A is known, we can find the pressure from $p = F/A$.

FIGURE 15.5 Learning about pressure.

The first statement in Figure 15.5b is especially important. Pressure exists at *all* points within a fluid, not just at the walls of the container. You may recall that tension exists at *all* points in a string, not only at its ends where it is tied to an object. We understood tension as the different parts of the string *pulling* against each other. Pressure is an analogous idea, except that the different parts of a fluid are *pushing* against each other.

Liquid

Nothing is touching the wall. There is no pressure.

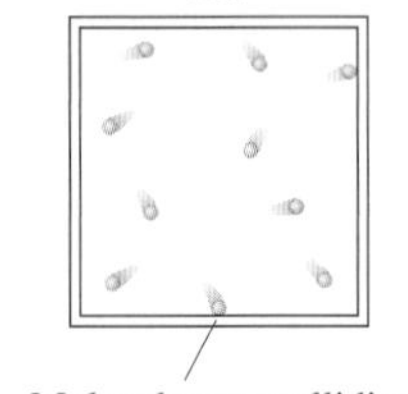

Gas
Molecules are colliding with the wall. There is pressure.

FIGURE 15.6 Liquids and gases in a "weightless" environment.

Causes of Pressure

Gases and liquids are both fluids, but they have some important differences. Liquids are nearly incompressible; gases are highly compressible. The molecules in a liquid attract each other via molecular bonds; the molecules in a gas do not interact other than through occasional collisions. These differences affect how we think about pressure in gases and liquids.

Imagine that you have two sealed jars, each containing a small amount of mercury and nothing else. All the air has been removed from the jars. Suppose you take the two jars into orbit on the space shuttle, where they are "weightless." One jar you keep cool, so that the mercury is a liquid. The other you heat until the mercury boils and becomes a gas. What can we say about the pressure in these two jars?

As Figure 15.6 shows, molecular bonds hold the liquid mercury together. It might quiver like Jello, but it remains a cohesive drop floating in the center of the jar. The liquid drop exerts no forces on the walls, so there's *no* pressure in the jar containing the liquid. (If we actually did this experiment, a very small fraction of the mercury would be in the vapor phase and create what is called *vapor pressure*. We can make the vapor pressure negligibly small by keeping the temperature low.)

The gas is different. Figure 15.1 introduced an atomic-level model of a gas in which a molecule moves freely until it collides with another molecule or with a wall of the container. Figure 15.7 shows a few of the gas molecules colliding with a wall. Recall, from our study of collisions in Chapter 9, that each molecule as it bounces exerts a tiny impulse on the wall. The impulse from any one collision is extremely small, but there are an extraordinarily large number of collisions every second. These collisions cause the gas to have a pressure.

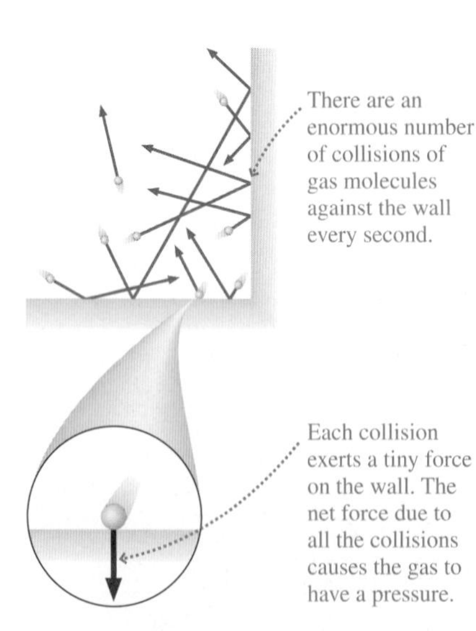

There are an enormous number of collisions of gas molecules against the wall every second.
Each collision exerts a tiny force on the wall. The net force due to all the collisions causes the gas to have a pressure.

FIGURE 15.7 The pressure in a gas is due to the net force of the molecules colliding with the walls.

The gas pressure can be calculated from the net force the molecules exert on the wall, divided by the area of the wall. We will do that calculation in Chapter 18. For now, we'll simply note that the pressure is proportional to the gas density in the container and to the absolute temperature.

Figure 15.8 shows the jars back on earth. Because of gravity, the liquid now fills the bottom of the jar and exerts a force on the bottom and the sides. Liquid mercury is incompressible, so the volume of liquid in Figure 15.8 is the same as in Figure 15.6. There is still no pressure on the top of the jar (other than the very small vapor pressure).

At first glance, the situation in the gas-filled jar seems unchanged from Figure 15.6. However, the earth's gravitational pull causes the gas density to be *slightly* more at the bottom of the jar than at the top. Because the pressure due to collisions is proportional to the density, the pressure is *slightly* larger at the bottom of the jar than at the top.

Thus there appear to be two contributions to the pressure in a container of fluid.

1. A *gravitational contribution* that arises from gravity pulling down on the fluid. Because a fluid can flow, forces are exerted on both the bottom and sides of the container. The gravitational contribution depends on the strength of the gravitational force.
2. A *thermal contribution* due to the collisions of freely moving gas molecules with the walls. The thermal contribution depends on the absolute temperature of the gas.

A detailed analysis finds that these two contributions are not entirely independent of each other, but the distinction is useful for a basic understanding of pressure. Let's see how these two contributions apply to different situations.

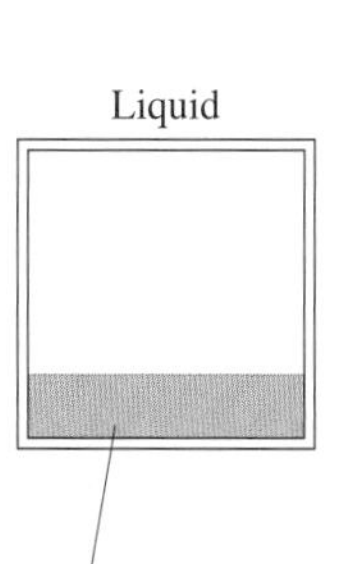

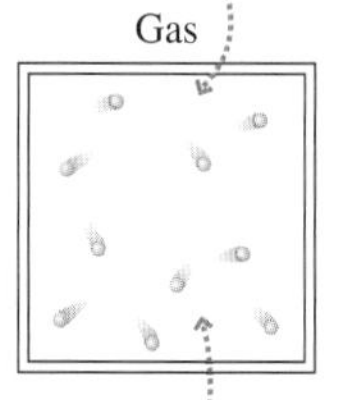

FIGURE 15.8 Gravity affects the pressure of the fluids.

Pressure in Gases

The pressure in a laboratory-size container of gas is due almost entirely to the thermal contribution. A container would have to be $\approx$100 m tall for gravity to cause the pressure at the top to be even 1% less than the pressure at the bottom. Laboratory-size containers are much less than 100 m tall, so we can quite reasonably assume that p has the *same* value at all points in a laboratory-size container of gas. Homework problems will let you verify that the gravitational contribution to the pressure in a container of gas is negligible.

Decreasing the number of molecules in a container decreases the gas pressure simply because there are fewer collisions with the walls. If a container is completely empty, with no atoms or molecules, then the pressure is $p = 0$ Pa. This is a *perfect vacuum.* No perfect vacuum exists in nature, not even in the most remote depths of outer space, because it is impossible to completely remove every atom from a region of space. In practice, a **vacuum** is an enclosed space in which $p \ll 1$ atm. Using $p = 0$ Pa is then a very good approximation.

Atmospheric Pressure

The earth's atmosphere is *not* a laboratory-size container. The height of the atmosphere is such that the gravitational contribution to pressure *is* important. As Figure 15.9 shows, the density of air slowly decreases with increasing height until reaching zero in the vacuum of space. Consequently, the pressure of the air, what we call the *atmospheric pressure* p_{atmos}, decreases with height. The air pressure is less in Denver than in Miami.

The atmospheric pressure *at sea level* varies slightly with the weather, but the global average sea-level pressure is 101,300 Pa. Consequently, we define the **standard atmosphere** as

$$1 \text{ standard atmosphere} = 1 \text{ atm} \equiv 101{,}300 \text{ Pa} = 101.3 \text{ kPa}$$

The standard atmosphere, usually referred to simply as "atmospheres," is a commonly used unit of pressure. But it is not an SI unit, so you must convert atmospheres to pascals before doing most calculations with pressure.

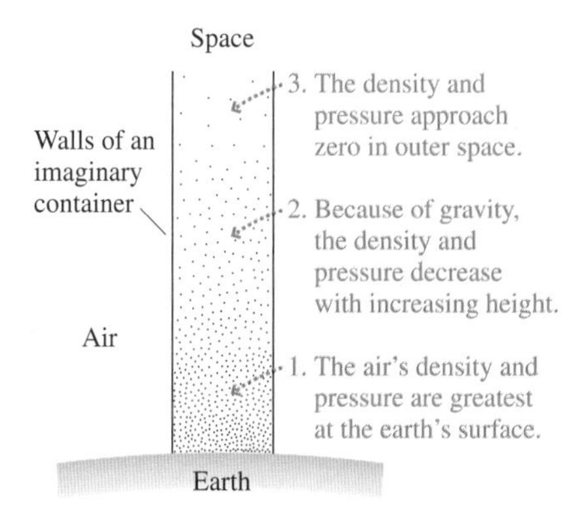

Space
Walls of an imaginary container
Air
Earth
3. The density and pressure approach zero in outer space.
2. Because of gravity, the density and pressure decrease with increasing height.
1. The air's density and pressure are greatest at the earth's surface.

FIGURE 15.9 The pressure and density decrease with increasing height in the atmosphere.

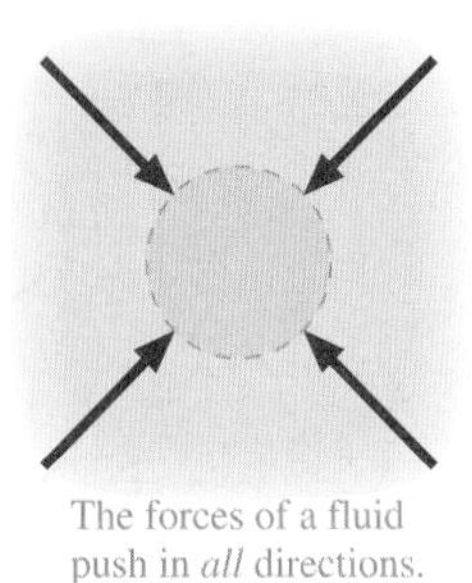

The forces of a fluid push in *all* directions.

FIGURE 15.10 Pressure forces in a fluid push with equal strength in all directions.

NOTE ▶ Unless you happen to live right at sea level, the atmospheric pressure around you is somewhat less than 1 atm. Pressure experiments use a barometer to determine the actual atmospheric pressure. For simplicity, this textbook will always assume that the pressure of the air is $p_{atmos} = 1$ atm unless stated otherwise. ◀

Given that the pressure of the air at sea level is 101.3 kPa, you might wonder why the weight of the air doesn't crush your forearm when you rest it on a table. Your forearm has a surface area of $\approx 200\ \text{cm}^2 = 0.02\ \text{m}^2$, so the force of the air pressing against it is ≈ 2000 N (≈ 450 pounds). How can you even lift your arm?

The reason, as Figure 15.10 shows, is that a fluid exerts pressure forces in *all* directions. There *is* a downward force of ≈ 2000 N on your forearm, but the air underneath your arm exerts an *upward* force of the same magnitude. The *net* force is very close to zero. (To be accurate, there is net *upward* force called the buoyant force. We'll study buoyancy in Section 15.4. For most objects, the buoyant force of the air is too small to notice.)

Removing the air from a container has very real consequences.

But, you say, there isn't any air under my arm if I rest it on a table. Actually, there is. There would be a *vacuum* under your arm if there were no air. Imagine placing your arm on the top of a large vacuum cleaner suction tube. What happens? You feel a downward force as the vacuum cleaner "tries to suck your arm in." However, the downward force you feel is not a *pulling* force from the vacuum cleaner. It is the *pushing* force of the air above your arm *when the air beneath your arm is removed and cannot push back.* Air molecules do not have hooks! They have no ability to "pull" on your arm. The air can only push.

Vacuum cleaners, suction cups, and other similar devices are powerful examples of how strong atmospheric pressure forces can be *if* the air is removed from one side of an object so as to produce an unbalanced force. The fact that we are *surrounded* by the fluid allows us to move around in the air, just as we swim underwater, oblivious of these strong forces.

EXAMPLE 15.2 A suction cup

A 10.0-cm-diameter suction cup is pushed against a smooth ceiling. What is the maximum weight of an object that can be suspended from the suction cup without pulling it off the ceiling? The weight of the suction cup is negligible.

MODEL Pushing the suction cup against the ceiling pushes the air out. We'll assume that the volume enclosed between the suction cup and the ceiling is a perfect vacuum with $p = 0$ Pa. We'll also assume that the atmospheric pressure in the room is 1 atm.

VISUALIZE Figure 15.11 shows a free-body diagram of the suction cup stuck to the ceiling. The downward normal force of the ceiling is distributed around the rim of the suction cup, but in the particle model we can show this as a single force vector.

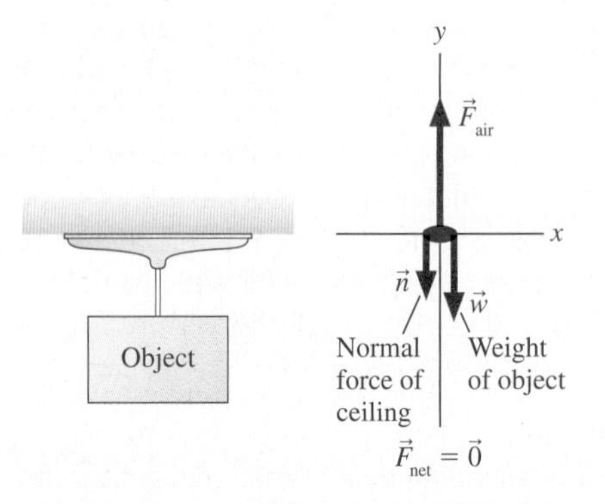

FIGURE 15.11 A suction cup is held to the ceiling by air pressure pushing upward on the bottom.

SOLVE The suction cup remains stuck to the ceiling, in static equilibrium, as long as $\vec{F}_{air} = \vec{n} + \vec{w}$. The magnitude of the upward force exerted by the air is

$$F_{air} = pA = p\pi r^2 = (101{,}300\ \text{Pa})\pi(0.050\ \text{m})^2 = 796\ \text{N}$$

There is no downward force from the air in this situation because there is no air inside the cup. Increasing the suspended weight decreases the normal force n by an equal amount. The maximum weight has been reached when n is reduced to zero. This happens when

$$w_{max} = F_{air} = 796\ \text{N}$$

Hence this suction cup can support a weight of up to 796 N.

Pressure in Liquids

Whereas a gas fills a container, gravity causes a liquid to fill the bottom of a container. Thus it's not surprising that the pressure in a liquid is due almost entirely to the gravitational contribution. We'd like to determine the pressure at depth d below the surface of the liquid. We will assume that the liquid is at rest; flowing liquids will be considered later in this chapter.

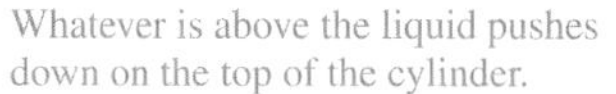

Whatever is above the liquid pushes
down on the top of the cylinder.

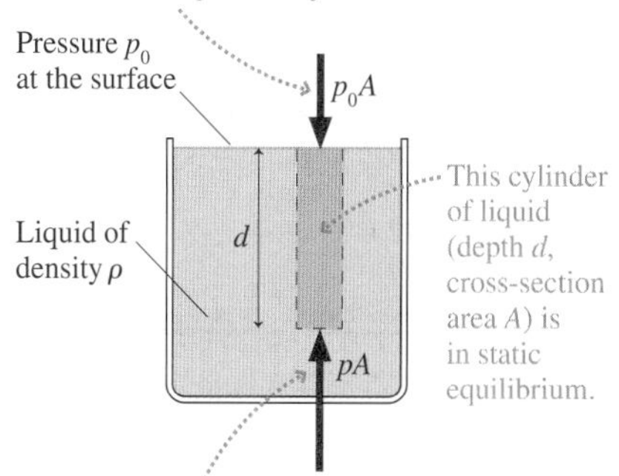

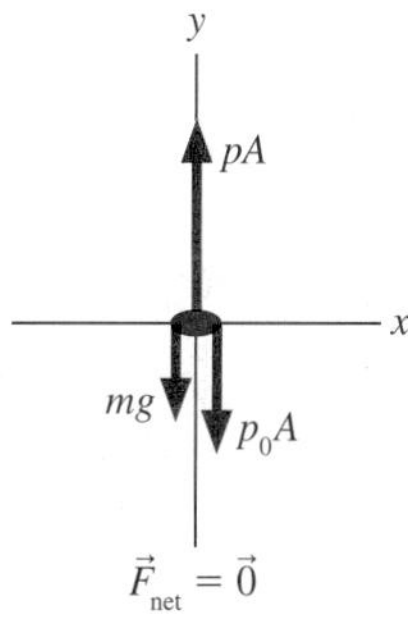

FIGURE 15.12 Measuring the pressure at depth d in a liquid.

The shaded cylinder of liquid in Figure 15.12 extends from the surface to depth d. This cylinder, like the rest of the liquid, is in static equilibrium with $\vec{F}_{net} = \vec{0}$. Three forces act on this cylinder: its weight mg, a downward force p_0A due to the pressure p_0 at the surface of the liquid, and an upward force pA due to the liquid beneath the cylinder pushing up on the bottom of the cylinder. This third force is a consequence of our earlier observation that different parts of a fluid push against each other. Pressure p, which is what we're trying to find, is the pressure at the bottom of the cylinder.

The upward force balances the two downward forces, so

$$pA = p_0A + mg \tag{15.5}$$

The liquid is a cylinder of cross-section area A and height d. Its volume is $V = Ad$ and its mass is $m = \rho V = \rho Ad$. Substituting this expression for the mass of the liquid into Equation 15.5, we find that the area A cancels from all terms. The pressure at depth d in a liquid is

$$p = p_0 + \rho gd \qquad \text{(hydrostatic pressure at depth } d\text{)} \tag{15.6}$$

where ρ is the liquid's density. Because of our assumption that the fluid is at rest, the pressure given by Equation 15.6 is called the **hydrostatic pressure.** The fact that g appears in Equation 15.6 reminds us that this a gravitational contribution to the pressure.

As expected, $p = p_0$ at the surface, where $d = 0$. Pressure p_0 is often due to the air or other gas above the liquid. $p_0 = 1$ atm $= 101.3$ kPa for a liquid that is open to the air. However, p_0 can also be the pressure due to a piston or a closed surface pushing down on the top of the liquid.

NOTE ► Equation 15.6 assumes that the liquid is *incompressible;* that is, its density ρ doesn't increase with depth. This is an excellent assumption for liquids, but not a good one for a gas, which *is* compressible. Even so, Equation 15.6 can be used with gases over fairly small distances, a few tens of meters or less, because the density is nearly constant over these distances. Equation 15.6 should not be used for calculating the pressure at different heights in the atmosphere. (A homework problem will let you derive a different equation for the pressure of the atmosphere.) ◄

EXAMPLE 15.3 The pressure on a submarine

A submarine cruises at a depth of 300 m. What is the pressure at this depth? Give the answer both in pascals and atmospheres.

SOLVE The density of seawater, from Table 15.1, is $\rho = 1030 \text{ kg/m}^3$. The pressure at depth $d = 300$ m is found from Equation 15.6 to be

$$\begin{aligned} p = p_0 + \rho gd &= 1.013 \times 10^5 \text{ Pa} \\ &\quad + (1030 \text{ kg/m}^3)(9.80 \text{ m/s}^2)(300 \text{ m}) \\ &= 3.13 \times 10^6 \text{ Pa} \end{aligned}$$

Converting the answer to atmospheres gives

$$p = 3.13 \times 10^6 \text{ Pa} \times \frac{1 \text{ atm}}{1.013 \times 10^5 \text{ Pa}} = 30.9 \text{ atm}$$

ASSESS The pressure deep in the ocean is very large. Windows on submersibles must be very thick to withstand the large forces.

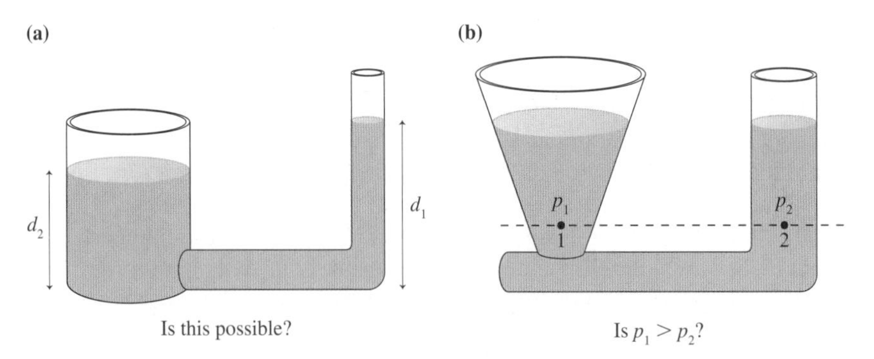

FIGURE 15.13 Some properties of a liquid in hydrostatic equilibrium are not what you might expect.

The hydrostatic pressure in a liquid depends only on the depth and the pressure at the surface. This observation has some important implications. Figure 15.13a shows two connected tubes. It's certainly true that the larger volume of liquid in the wide tube weighs more than the liquid in the narrow tube. You might think that this extra weight would push the liquid in the narrow tube higher than in the wide tube. But it doesn't. If d_1 were larger than d_2, then, according to the hydrostatic pressure equation, the pressure at the bottom of the narrow tube would be higher than the pressure at the bottom of the wide tube. This *pressure difference* would cause the liquid to *flow* from right to left until the heights were equal.

Thus a first conclusion: **A connected liquid in hydrostatic equilibrium rises to the same height in all open regions of the container.**

Figure 15.13b shows two connected tubes of different shape. The conical tube holds more liquid above the dotted line, so you might think that $p_1 > p_2$. But it isn't. Both points are at the same depth, thus $p_1 = p_2$. You can arrive at the same conclusion by thinking about the pressure at the bottom of the tubes. If p_1 were larger than p_2, the pressure at the bottom of the left tube would be larger than the pressure at the bottom of the right tube. This would cause the liquid to flow until the pressures were equal.

If $p_1 = p_2$, you might be wondering what's holding up the "extra" liquid in the conical tube. Figure 15.14 shows that the weight of this extra liquid is supported by the wall of the tube. Only the liquid that's *directly above* point 1 needs to be supported by the pressure at point 1.

Thus a second conclusion: **The pressure is the same at all points on a horizontal line through a connected liquid in hydrostatic equilibrium.**

NOTE ▶ Both of these conclusions are restricted to liquids in hydrostatic equilibrium. The situation is entirely different for flowing fluids, as we'll see later in the chapter. ◀

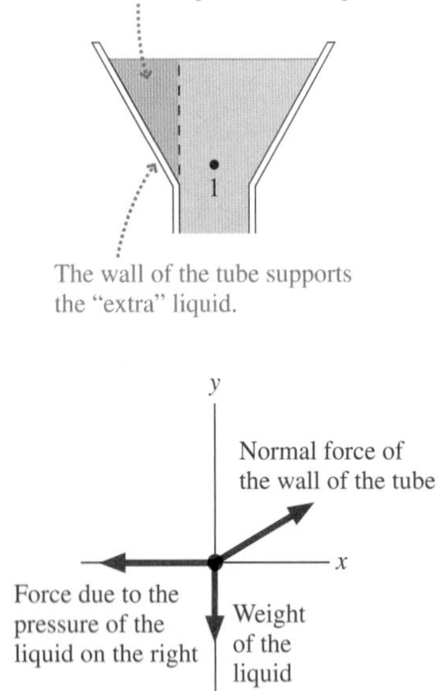

FIGURE 15.14 The weight of the liquid is supported by the wall of the tube.

EXAMPLE 15.4 Pressure in a closed tube

Water fills the tube shown in Figure 15.15. What is the pressure at the top of the closed tube?

MODEL This is a liquid in hydrostatic equilibrium. The closed tube is not an open region of the container, so the water cannot rise to an equal height. Nevertheless, the pressure is still the same at all points on a horizontal line. In particular, the pressure at the top of the closed tube equals the pressure in the open tube at the height of the dotted line. Assume $p_0 = 1$ atm.

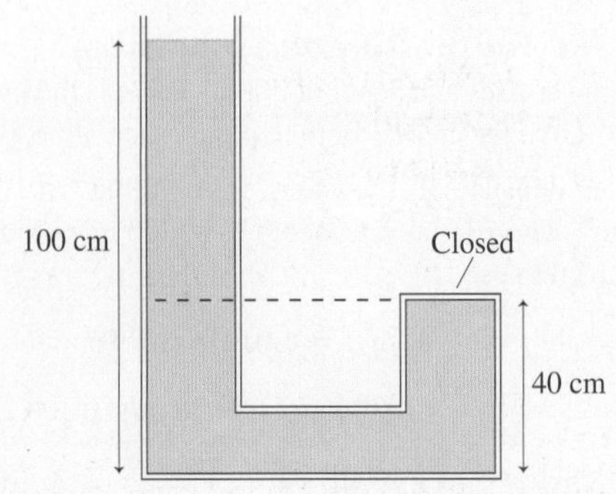

FIGURE 15.15 What is the pressure at the top of the closed tube?

SOLVE A point 40 cm above the bottom of the open tube is at a depth of 60 cm. The pressure at this depth is

$$p = p_0 + \rho g d = 1.013 \times 10^5 \text{ Pa} + (1000 \text{ kg/m}^3)(9.80 \text{ m/s}^2)(0.60 \text{ m})$$
$$= 1.072 \times 10^5 \text{ Pa}$$
$$= 1.06 \text{ atm}$$

This is the pressure at the top of the closed tube.

ASSESS The water in the open tube *pushes* the water in the closed tube up against the top of the tube. Consequently, in accordance with Newton's third law, the top of the tube *presses down on the liquid* with a force of magnitude $F = pA$.

We can draw one more conclusion from the hydrostatic pressure equation $p = p_0 + \rho g d$. If we change the pressure p_0 at the surface to p_1, the pressure at depth d becomes $p' = p_1 + \rho g d$. The *change* in pressure $\Delta p = p_1 - p_0$ is the same at all points in the fluid, independent of the size or shape of the container. This idea, that **a change in the pressure at one point in an incompressible fluid appears undiminished at all points in the fluid,** was first recognized by Blaise Pascal, and is called **Pascal's principle.**

For example, if we compressed the air above the open tube in Example 15.4 to a pressure of 1.5 atm, an increase of 0.5 atm, the pressure at the top of the closed tube would increase to 1.56 atm. Pascal's principle is the basis for hydraulic systems, as we'll see in the next section.

STOP TO THINK 15.2 Water is slowly poured into the container until the water level has risen into tubes A, B, and C. The water doesn't overflow from any of the tubes. How do the water depths in the three columns compare to each other?

a. $d_A > d_B > d_C$.
b. $d_A < d_B < d_C$.
c. $d_A = d_B = d_C$.
d. $d_A = d_C > d_B$.
e. $d_A = d_C < d_B$.

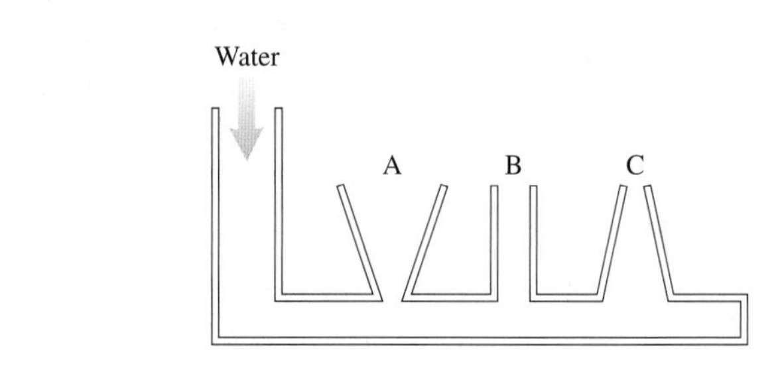

15.3 Measuring and Using Pressure

The pressure in a fluid is measured with a *pressure gauge,* which is often a device very similar to that in Figure 15.5. The fluid pushes against some sort of spring, usually a diaphragm, and the spring's displacement is registered by a pointer on a dial.

Many pressure gauges, such as tire gauges and the gauges on air tanks, measure not the actual or absolute pressure p but what is called **gauge pressure.** The gauge pressure, denoted p_g, is the pressure *in excess* of 1 atm. That is,

$$p_g = p - 1 \text{ atm} \tag{15.7}$$

You need to add 1 atm = 101.3 kPa to the reading of a pressure gauge to find the absolute pressure p that you need for doing most science or engineering calculations: $p = p_g + 1$ atm.

A tire-pressure gauge reads the gauge pressure p_g, not the absolute pressure p. The gauge reads zero when the tire is flat, but this doesn't mean there is a vacuum inside. Zero gauge pressure means the inside pressure is 1 atm.

EXAMPLE 15.5 An underwater pressure gauge

An underwater pressure gauge reads 60 kPa. What is its depth?

MODEL The gauge reads gauge pressure, not absolute pressure.

SOLVE The hydrostatic pressure at depth d, with $p_0 = 1$ atm, is $p = 1 \text{ atm} + \rho g d$. Thus the gauge pressure is

$$p_g = p - 1 \text{ atm} = (1 \text{ atm} + \rho g d) - 1 \text{ atm} = \rho g d$$

The term $\rho g d$ is the pressure *in excess* of atmospheric pressure and thus *is* the gauge pressure. Solving for d, we find

$$d = \frac{60{,}000 \text{ Pa}}{(1000 \text{ kg/m}^3)(9.80 \text{ m/s}^2)} = 6.1 \text{ m}$$

Solving Hydrostatic Problems

We now have enough information to formulate a set of rules for thinking about hydrostatic problems.

TACTICS BOX 15.1 Hydrostatics

1. **Draw a picture.** Show open surfaces, pistons, boundaries, and other features that affect the pressure. Include height and area measurements and fluid densities. Identify the points at which you need to find the pressure.
2. **Determine the pressure at surfaces.**
   - **Surface open to the air:** $p_0 = p_{atmos}$, usually 1 atm.
   - **Surface covered by a gas:** $p_0 = p_{gas}$.
   - **Closed surface:** $p = F/A$ where F is the force the surface, such as a piston, exerts on the fluid.
3. **Use horizontal lines.** Pressure in a connected fluid is the same at any point along a horizontal line.
4. **Allow for gauge pressure.** Pressure gauges read $p_g = p - 1$ atm.
5. **Use the hydrostatic pressure equation.** $p = p_0 + \rho g d$.

Manometers and Barometers

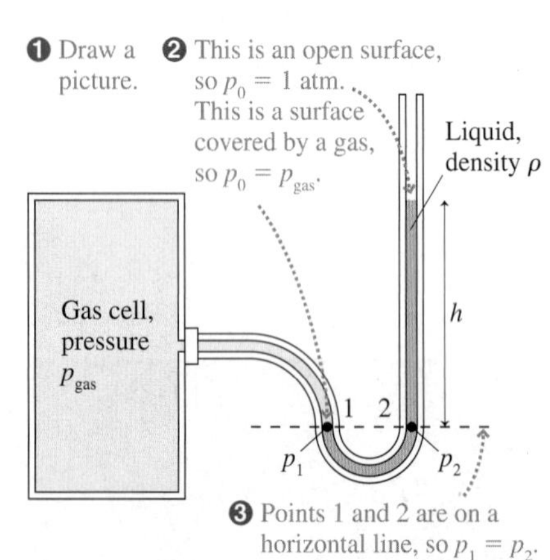

❶ Draw a picture.
❷ This is an open surface, so p_0 = 1 atm.
This is a surface covered by a gas, so $p_0 = p_{gas}$.
Liquid, density ρ
Gas cell, pressure p_{gas}
h
1 2
p_1
p_2
❸ Points 1 and 2 are on a horizontal line, so $p_1 = p_2$.

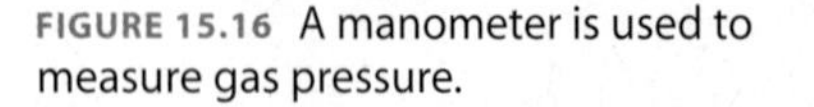
FIGURE 15.16 A manometer is used to measure gas pressure.

Gas pressure is sometimes measured with a device called a *manometer.* A manometer, shown in Figure 15.16, is a U-shaped tube connected to the gas at one end and open to the air at the other end. The tube is filled with a liquid—usually mercury—of density ρ. The liquid is in static equilibrium. A scale allows the user to measure the height h of the right side above the left side.

Steps 1–3 from Tactics Box 15.1 lead to the conclusion that the pressures p_1 and p_2 must be equal. Pressure p_1, at the surface on the left, is simply the gas pressure: $p_1 = p_{gas}$. Pressure p_2 is the hydrostatic pressure at depth $d = h$ in the liquid on the right: $p_2 = 1 \text{ atm} + \rho g h$. Equating these two pressures gives

$$p_{gas} = 1 \text{ atm} + \rho g h \tag{15.8}$$

Figure 15.16 assumed $p_{gas} > 1$ atm, so the right side of the liquid is higher than the left. Equation 15.8 is also valid for $p_{gas} < 1$ atm if the distance of the right side *below* the left side is considered to be a negative value of h.

EXAMPLE 15.6 Using a manometer

The pressure of a gas cell is measured with a mercury manometer. The mercury is 36.2 cm higher in the outside arm than in the arm connected to the gas cell.

a. What is the gas pressure?

b. What is the reading of a pressure gauge attached to the gas cell?

SOLVE

a. From Table 15.1, the density of mercury is $\rho = 13{,}600\ \text{kg/m}^3$. Equation 15.8 with $h = 0.362$ m gives

$$p_{\text{gas}} = 1\ \text{atm} + \rho g h = 149.5\ \text{kPa}$$

We had to change 1 atm to 101,300 Pa before adding. Converting the result to atmospheres, $p_{\text{gas}} = 1.476$ atm.

b. The pressure gauge reads gauge pressure: $p_g = p - 1\ \text{atm} = 0.476$ atm or 48.2 kPa.

ASSESS Manometers are useful over a pressure range from near vacuum up to ≈ 2 atm. For higher pressures, the mercury column would be too tall to be practical.

Another important pressure-measuring instrument is the *barometer,* which is used to measure the atmospheric pressure p_{atmos}. Figure 15.17a shows a glass tube, sealed at the bottom, that has been completely filled with a liquid. If we temporarily seal the top end, we can invert the tube and place it in a beaker of the same liquid. When the temporary seal is removed, some, but not all, of the liquid runs out, leaving a liquid column in the tube that is a height h above the surface of the liquid in the beaker. This device, shown in Figure 15.17b, is a barometer. What does it measure? And why doesn't *all* the liquid in the tube run out?

We can analyze the barometer much as we did the manometer. Points 1 and 2 in Figure 15.17b are on a horizontal line drawn even with the surface of the liquid. The liquid is in hydrostatic equilibrium, so the pressure at these two points must be equal. Liquid runs out of the tube only until a balance is reached between the pressure at the base of the tube and the pressure of the air.

You can think of a barometer as rather like a seesaw. If the pressure of the atmosphere increases, it presses down on the liquid in the beaker. This forces liquid up the tube until the pressures at points 1 and 2 are equal. If the atmospheric pressure falls, liquid has to flow out of the tube to keep the pressures equal at these two points.

The pressure at point 2 is the pressure due to the weight of the liquid in the tube plus the pressure of the gas above the liquid. But in this case there is no gas above the liquid! Because the tube had been completely full of liquid when it was inverted, the space left behind when the liquid ran out is a vacuum (ignoring a very slight *vapor pressure* of the liquid, negligible except in extremely precise measurements). Thus pressure p_2 is simply $p_2 = \rho g h$.

Equating p_1 and p_2 gives

$$p_{\text{atmos}} = \rho g h \tag{15.9}$$

Thus we can measure the atmosphere's pressure by measuring the height of the liquid column in a barometer.

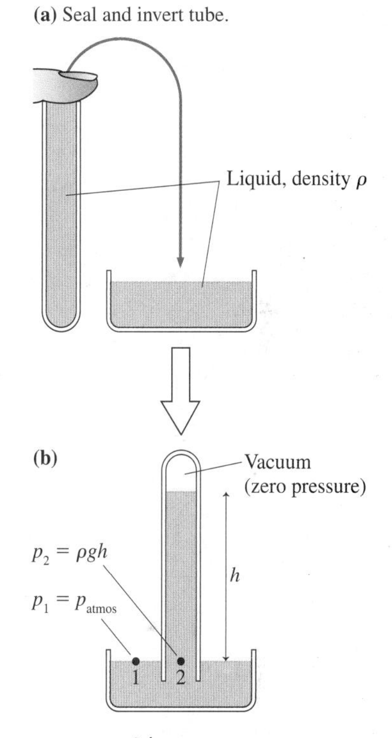

FIGURE 15.17 A barometer.

The average air pressure at sea level causes a column of mercury in a mercury barometer to stand 760 mm above the surface. Knowing that the density of mercury is $13{,}600\ \text{kg/m}^3$ (at 0°C), we can use Equation 15.9 to find that the average atmospheric pressure is

$$\begin{aligned} p_{\text{atmos}} &= \rho_{\text{Hg}} g h = (13{,}600\ \text{kg/m}^3)(9.80\ \text{m/s}^2)(0.760\ \text{m}) \\ &= 1.013 \times 10^5\ \text{Pa} = 101.3\ \text{kPa} \end{aligned}$$

This is the value given earlier as "one standard atmosphere." Now you can see that 1 atm = 101.3 kPa *because,* on average, a mercury barometer gives a reading of 760 mm.

The barometric pressure varies slightly from day to day as the weather changes. Weather systems are called *high-pressure systems* or *low-pressure systems,* depending upon whether the local sea-level pressure is higher or lower than one standard atmosphere. Higher pressure is usually associated with fair weather while lower pressure portends rain.

Pressure Units

In practice, pressure is measured in a number of different units. This plethora of units and abbreviations has arisen historically as scientists and engineers working on different subjects (liquids, high-pressure gases, low-pressure gases, weather, etc.) developed what seemed to them the most convenient units. These units continue in use through tradition, so it is necessary to become familiar with converting back and forth between them. Table 15.2 gives the basic conversions.

TABLE 15.2 Pressure units

Unit	Abbreviation	Conversion to 1 atm	Uses
pascal	Pa	101.3 kPa	SI unit: 1 Pa = 1 N/m^2 use in most calculations
atmosphere	atm	1 atm	general
millimeters of mercury	mm of Hg	760 mm of Hg	gases and barometric pressure
inches of mercury	in	29.92 in	barometric pressure in U.S. weather forecasting
pounds per square inch	psi	14.7 psi	engineering and industry

Blood Pressure

The last time you had a medical checkup, the doctor may have told you something like, "Your blood pressure is 120 over 80." What does that mean?

About every 0.8 s, assuming a pulse rate of 75 beats per minute, your heart "beats." The heart muscles contract and push blood out into your aorta. This contraction, like squeezing a balloon, raises the pressure in your heart. The pressure increase, in accordance with Pascal's principle, is transmitted through all your arteries.

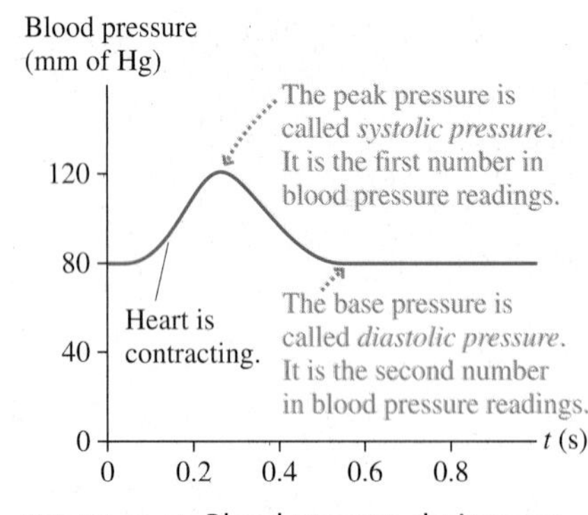

FIGURE 15.18 Blood pressure during one cycle of a heart beat.

Figure 15.18 is a pressure graph showing how blood pressure changes during one cycle of the heartbeat. The medical condition of *high blood pressure* usually means that your systolic pressure is higher than necessary for blood circulation. The high pressure causes undue stress and strain on your entire circulatory system, often leading to serious medical problems. Low blood pressure can cause you to get dizzy if you stand up quickly because the pressure isn't adequate to pump the blood up to your brain.

Blood pressure is measured with a cuff that goes around your arm. The doctor or nurse pressurizes the cuff, places a stethoscope over the artery in your arm, then slowly releases the pressure while watching a pressure gauge. Initially, the cuff squeezes the artery shut and cuts off the blood flow. When the cuff pressure drops below the systolic pressure, the pressure pulse during each beat of your heart forces the artery open briefly and a squirt of blood goes through. You can feel this, and the doctor or nurse records the pressure when she hears the blood start to flow. This is your systolic pressure.

This pulsing of the blood through your artery lasts until the cuff pressure reaches the diastolic pressure. Then the artery remains open continuously and the blood flows smoothly. This transition is easily heard in the stethoscope, and the doctor or nurse records your diastolic pressure.

Blood pressure is measured in millimeters of mercury. And it is a gauge pressure, the pressure in excess of 1 atm. A fairly typical blood pressure of a healthy young adult is 120/80, meaning that the systolic pressure is $p_g = 120$ mm of Hg (absolute pressure $p = 880$ mm of Hg) and the diastolic pressure is 80 mm of Hg.

Now that you know, be sure to ask your doctor questions next time you have your blood pressure measured!

The Hydraulic Lift

The use of pressurized liquids to do useful work is a technology known as **hydraulics.** Pascal's principle is the fundamental idea underlying hydraulic devices. If you increase the pressure at one point in a liquid by pushing a piston in, that pressure increase is transmitted to all points in the liquid. A second piston at some other point in the fluid can then push outward and do useful work.

The brake system in your car is a hydraulic system. Stepping on the brake pushes a piston into the *master brake cylinder* and increases the pressure in the *brake fluid.* The fluid itself hardly moves, but the pressure increase is transmitted to the four wheels where it pushes the brake pads against the spinning brake disk. You've used a pressurized liquid to achieve the useful goal of stopping your car.

One advantage of hydraulic systems over simple mechanical linkages is the possibility of *force multiplication.* To see how this works, we'll analyze a *hydraulic lift,* such as the one that lifts your car at the repair shop. Figure 15.19a shows force $\vec{F}_2$, perhaps due to a weight of mass m, pressing down on a liquid via a piston of area A_2. A much smaller force $\vec{F}_1$ presses down on a piston of area A_1. Can this system possibly be in equilibrium?

As you now know, the hydrostatic pressure is the same at all points along a horizontal line through a fluid. Consider the line passing through the liquid/piston interface on the left in Figure 15.19a. Pressures p_1 and p_2 must be equal, thus

$$p_0 + \frac{F_1}{A_1} = p_0 + \frac{F_2}{A_2} + \rho g h \tag{15.10}$$

The atmosphere presses equally on both sides, so p_0 cancels. The system is in static equilibrium if

$$F_2 = \frac{A_2}{A_1}F_1 - \rho g h A_2 \tag{15.11}$$

If the height h is very small, so that the term $\rho g h A_2$ is negligible, then F_2 (the weight of the heavy object) is larger than F_1 by the factor A_2/A_1. In other words, a small force applied to a small piston really can support a large car because both apply the *same pressure* to the fluid. The ratio A_2/A_1 is a force-multiplying factor.

NOTE ► Force $\vec{F}_2$ is the force of the heavy object pushing *down* on the liquid. According to Newton's third law, the liquid pushes *up* on the object with a force of equal magnitude. Thus F_2 in Equation 15.11 is the "lifting force." ◄

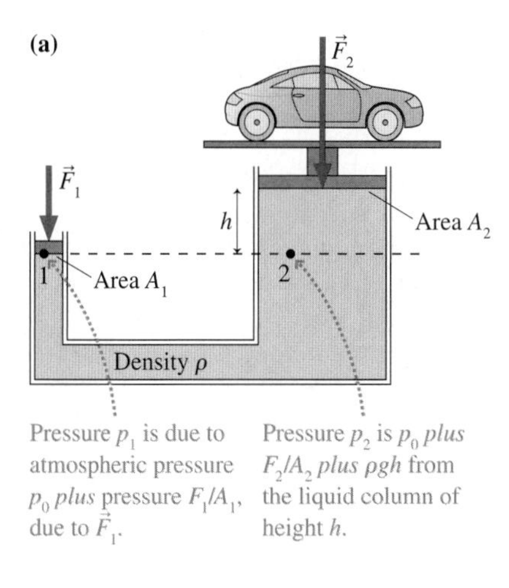
(a)
$\vec{F}_2$
$\vec{F}_1$
h
Area A_2
1
Area A_1
2
Density ρ
Pressure p_1 is due to atmospheric pressure p_0 *plus* pressure F_1/A_1, due to $\vec{F}_1$.
Pressure p_2 is p_0 *plus* F_2/A_2 *plus* $\rho g h$ from the liquid column of height h.

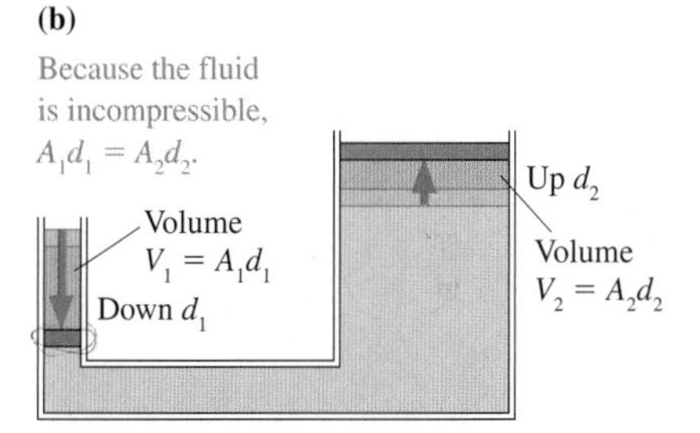
(b)
Because the fluid is incompressible, $A_1d_1 = A_2d_2$.
Up d_2
Volume $V_1 = A_1d_1$
Down d_1
Volume $V_2 = A_2d_2$

FIGURE 15.19 A hydraulic lift.

Suppose we need to lift the car higher. If piston 1 is pushed down distance d_1, as in Figure 15.19b, it displaces volume $V_1 = A_1d_1$ of liquid. Because the liquid is incompressible, V_1 must equal the volume $V_2 = A_2d_2$ added beneath piston 2 as it rises distance d_2. That is,

$$d_2 = \frac{d_1}{A_2/A_1} \tag{15.12}$$

The distance is *divided* by the same factor as that by which force is multiplied. A small force may be able to support a heavy weight, but you have to push the small piston a large distance to raise the heavy weight by a small amount.

This conclusion is really just a statement of energy conservation. Work is done *on* the liquid by a small force pushing the liquid through a large displacement. Work is done *by* the liquid when it lifts the heavy weight through a small distance. A full analysis must consider the fact that the gravitational potential energy of the liquid is also changing, so we can't simply equate the output work to the input work, but you can see that energy considerations require piston 1 to move further than piston 2.

Force $\vec{F}_1$ in Equation 15.11 is the force that balances the heavy object at height h. As a homework problem, you can show that the magnitude of force $\vec{F}_1$ must be increased by

$$\Delta F = \rho g (A_1 + A_2) d_2 \tag{15.13}$$

in order to lift the heavy object through distance d_2 to a new height $h + d_2$, where ρ is the density of the liquid. Surprisingly, ΔF is independent of the weight you're lifting.

EXAMPLE 15.7 Lifting a car

The hydraulic lift at a car repair shop is filled with oil. The car rests on a 25-cm-diameter piston. To lift the car, compressed air is used to push down on a 6.0-cm-diameter piston.

a. What air-pressure force will support a 1300 kg car level with the compressed-air piston?
b. By how much must the air-pressure force be increased to lift the car 2.0 m?

MODEL Assume that the oil is incompressible. Its density, from Table 15.1, is 900 kg/m^3.

SOLVE

a. The weight of the car pressing down on the piston is $F_2 = mg = 12{,}740$ N. The piston areas are $A_1 = \pi(0.030 \text{ m})^2 = 0.00283 \text{ m}^2$ and $A_2 = \pi(0.125 \text{ m})^2 = 0.0491 \text{ m}^2$. The force required to hold the car level with the compressed air piston, with $h = 0$ m, is

$$F_1 = \frac{F_2}{A_2/A_1} = \frac{12{,}740 \text{ N}}{(0.0491 \text{ m}^2)/(0.00283 \text{ m}^2)} = 734 \text{ N}$$

b. To raise the car $d_2 = 2.0$ m, the air-pressure force must be increased by

$$\Delta F = \rho g(A_1 + A_2)d_2 = 916 \text{ N}$$

ASSESS 734 N is roughly the weight of an average adult male. The multiplication factor $A_2/A_1 = (25 \text{ cm}/6 \text{ cm})^2 = 17$ makes it quite easy to hold up the car.

STOP TO THINK 15.3 Rank in order, from largest to smallest, the magnitudes of the forces $\vec{F}_1$, $\vec{F}_2$, and $\vec{F}_3$ required to balance the masses. The masses are in kilograms.

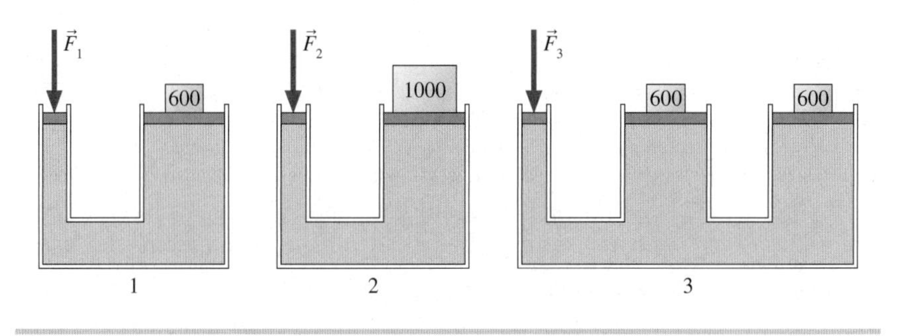
$\vec{F}_1$
600
1
$\vec{F}_2$
1000
2
$\vec{F}_3$
600
600
3

15.4 Buoyancy

A rock, as you know, sinks like a rock. Wood floats on the surface of a lake. A penny with a mass of a few grams sinks, but a massive steel aircraft carrier floats. How can we understand these diverse phenomena?

An air mattress floats effortlessly on the surface of a swimming pool. But if you've ever tried to push an air mattress underwater, you know it is nearly impossible. As you push down, the water pushes up. This net upward force of a fluid is called the **buoyant force.**

The basic reason for the buoyant force is easy to understand. Figure 15.20 shows a cylinder submerged in a liquid. The pressure in the liquid increases with depth, so the pressure at the bottom of the cylinder is larger than at the top. Both cylinder ends have equal area, so force $\vec{F}_{up}$ is larger than force $\vec{F}_{down}$. (Remember that pressure forces push in *all* directions.) Consequently, the pressure in the liquid exerts a *net upward force* on the cylinder of magnitude $F_{net} = F_{up} - F_{down}$. This is the buoyant force.

The submerged cylinder illustrates the idea in a simple way, but the result is not limited to cylinders or to liquids. Suppose we isolate a parcel of fluid of arbi-

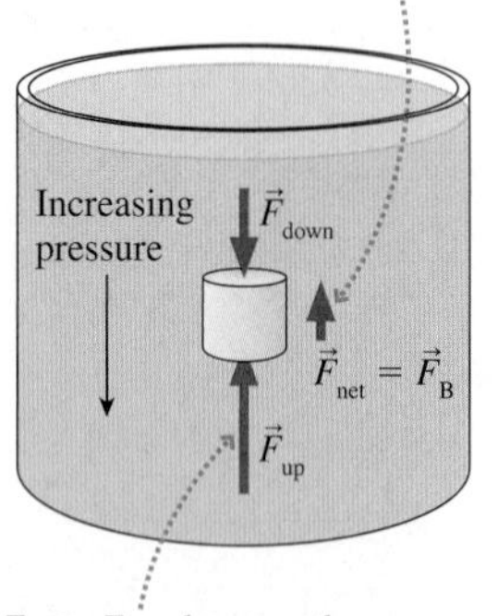
The net force of the fluid on the cylinder is the buoyant force $\vec{F}_B$.
Increasing pressure
$\vec{F}_{down}$
$\vec{F}_{net} = \vec{F}_B$
$\vec{F}_{up}$
$F_{up} > F_{down}$ because the pressure is greater at the bottom. Hence the fluid exerts a net upward force.

FIGURE 15.20 The buoyant force arises because the fluid pressure at the bottom of the cylinder is larger than at the top.

trary shape and volume by drawing an imaginary boundary around it, as shown in Figure 15.21a. This parcel is in static equilibrium. Consequently, the parcel's weight force pulling it down must be balanced by an upward force. The upward force, which is exerted on this parcel of fluid by the surrounding fluid, is the buoyant force $\vec{F}_B$. The buoyant force matches the weight of the fluid: $F_B = w$.

Now imagine that we could somehow remove this parcel of fluid and instantaneously replace it with an object having exactly the same shape and size, as shown in Figure 15.21b. Because the buoyant force is exerted by the *surrounding* fluid, and the surrounding fluid hasn't changed, the buoyant force on this new object is *exactly the same* as the buoyant force on the parcel of fluid that we removed.

When an object (or a portion of an object) is immersed in a fluid, it *displaces* fluid that would otherwise fill that region of space. This fluid is called the **displaced fluid.** The displaced fluid's volume is exactly the volume of the portion of the object that is immersed in the fluid. Figure 15.21 leads us to conclude that the magnitude of the upward buoyant force matches the weight of this displaced fluid.

This idea was first recognized by the ancient Greek mathematician and scientist Archimedes, perhaps the greatest scientist of antiquity, and today we know it as *Archimedes' principle.*

Archimedes' principle A fluid exerts an upward buoyant force $\vec{F}_B$ on an object immersed in or floating on the fluid. The magnitude of the buoyant force equals the weight of the fluid displaced by the object.

Suppose the fluid has density ρ_f and the object displaces volume V_f of fluid. The mass of the displaced fluid is $m_f = \rho_f V_f$ and so its weight is $w = \rho_f V_f g$. Thus Archimedes' principle in equation form is

$$F_B = \rho_f V_f g \tag{15.14}$$

NOTE ▶ It is important to distinguish the density and volume of the displaced fluid from the density and volume of the object. To do so, we'll use subscript f for the fluid and o for the object. ◀

(a)

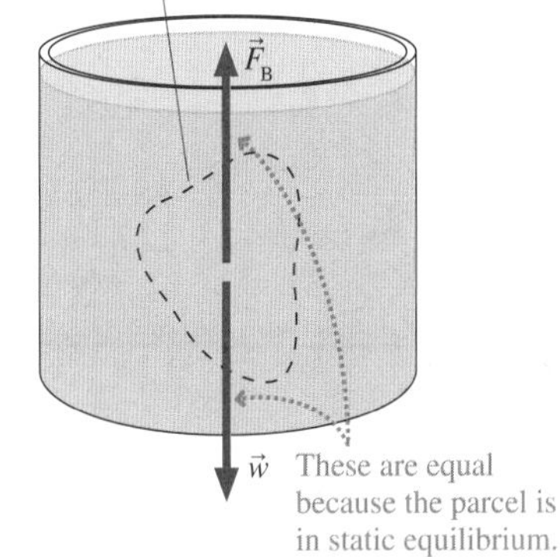
Imaginary boundary around a parcel of fluid
$\vec{F}_B$
$\vec{w}$
These are equal because the parcel is in static equilibrium.

(b)

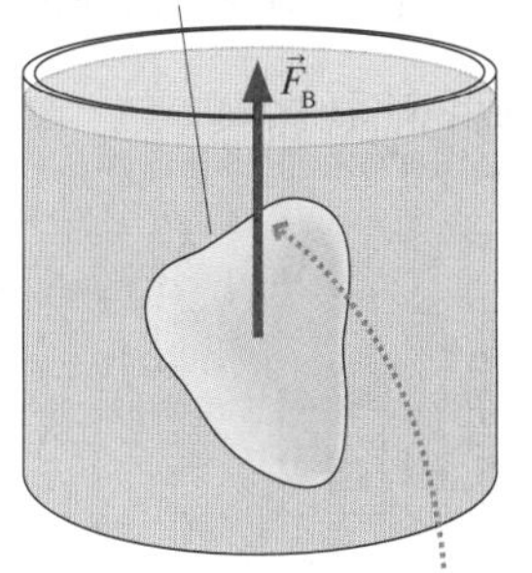
Real object with same size and shape as the parcel of fluid
$\vec{F}_B$
The buoyant force on the object is the same as on the parcel of fluid because the *surrounding* fluid has not changed.

FIGURE 15.21 The buoyant force on an object is the same as the buoyant force on the fluid it displaces.

EXAMPLE 15.8 Holding a block of wood underwater
A 10 cm × 10 cm × 10 cm block of wood with a density 700 kg/m^3 is held underwater by a string tied to the bottom of the container. What is the tension in the string?

MODEL The buoyant force on the wood is given by Archimedes' principle.

VISUALIZE Figure 15.22 shows the forces acting on the wood.

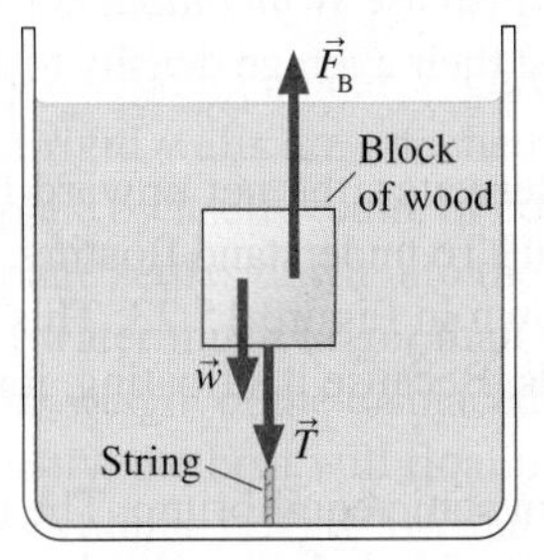
$\vec{F}_B$
Block of wood
$\vec{w}$
$\vec{T}$
String

FIGURE 15.22 The forces acting on the submerged wood.

SOLVE The block is in static equilibrium, so

$$\sum F_y = F_B - T - w = 0$$

Thus the tension is $T = F_B - w$. The mass of the block is $m_o = \rho_o V_o$ and its weight is $w = \rho_o V_o g$. The buoyant force, given by Equation 15.14, is $F_B = \rho_f V_f g$. Thus

$$T = \rho_f V_f g - \rho_o V_o g = (\rho_f - \rho_o) V_o g$$

where we've used the fact that $V_f = V_o$ for a completely submerged object. The volume is $V_o = 1000 \text{ cm}^3 = 1.0 \times 10^{-3} \text{ m}^3$, and hence the tension in the string is

$$T = ((1000 \text{ kg/m}^3) - (700 \text{ kg/m}^3)) \times (1.0 \times 10^{-3} \text{ m}^3)(9.8 \text{ m/s}^2) = 2.94 \text{ N}$$

ASSESS The tension depends on the *difference* in densities. The tension would vanish if the wood density matched the water density.

Thus the minimum height of the sides, a height that would allow the boat to float (in perfectly still water!) with water right up to the rails, is

$$h_{\text{min}} = \frac{m_o}{\rho_f A} \tag{15.18}$$

As a quick example, a 5 m × 10 m steel "barge" with a 2-cm-thick floor has an area of 50 m^2 and a mass of 7900 kg. The minimum height of the massless walls, as given by Equation 15.18, is 16 cm.

Real ships and boats are more complicated, but the same idea holds true. Whether it's made of concrete, steel, or lead, **a boat will float if its geometry allows it to displace enough water to equal the weight of the boat.**

STOP TO THINK 15.4 An ice cube is floating in a glass of water that is filled entirely to the brim. When the ice cube melts, the water level will

a. fall.
b. stay the same, right at the brim.
c. rise, causing the water to spill.

15.5 Fluid Dynamics

The wind blowing through your hair, a white-water river, and oil gushing from an oil well are examples of fluids in motion. We've focused thus far on fluid statics, but it's time to turn our attention to fluid dynamics.

Fluid flow is a complex subject. Many aspects, especially turbulence and the formation of eddies, are still not well understood and are areas of current science and engineering research. We will avoid these difficulties by using a simplified *model.* The **ideal-fluid model** provides a good, though not perfect, description of fluid flow in many situations. It captures the essence of fluid flow while eliminating unnecessary details.

The ideal-fluid model can be expressed in four assumptions about a fluid:

1. The fluid is *incompressible.* This is a good assumption for liquids but questionable for gases.
2. The fluid is *nonviscous.* Water flows much more easily than cold pancake syrup because the syrup is a very *viscous* fluid. **Viscosity** is a resistance to flow, and in a fluid is analogous to the kinetic friction of a solid object. Assuming that a fluid is nonviscous is equivalent to assuming that there's no friction. This is the weakest of the four assumptions for many liquids, but assuming a nonviscous liquid avoids major mathematical difficulties.
3. The flow is *steady.* That is, the fluid velocity at each point in the fluid is constant; it does not fluctuate or change with time. Flow under these conditions is called **laminar flow,** and it is distinguished from *turbulent flow.*
4. The flow is *irrotational.* The definition of *irrotational* is fairly technical, but there's a simple test. If a tiny paddle wheel anywhere in the fluid does not spin, then the flow is irrotational.

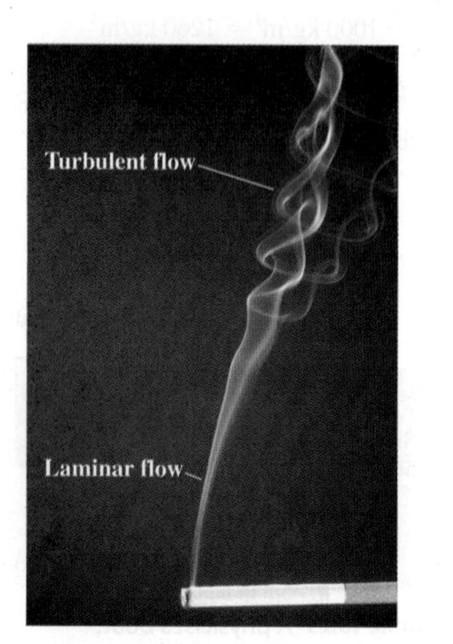

FIGURE 15.26 Rising smoke changes from laminar flow to turbulent flow.

The rising smoke in the photograph of Figure 15.26 begins as laminar flow, recognizable by the smooth contours, but at some point undergoes a transition to turbulent flow. A laminar-to-turbulent transition is not uncommon in fluid flow. The ideal-fluid model can be applied to the laminar flow, but not to the turbulent flow.

The Equation of Continuity

Figure 15.27 is another interesting photograph. Here smoke is being used to help engineers visualize the air flow around a car in a wind tunnel. The smoothness of the flow tells us this is laminar flow. But notice also how the individual smoke trails retain their identity. They don't cross or get mixed together. Each smoke trail represents a *streamline* in the fluid.

FIGURE 15.27 The laminar air flow around a car in a wind tunnel is made visible with smoke. Each smoke trail represents a streamline.

Imagine that we could inject a colored drop of water into a stream of water flowing as an ideal fluid. Because the flow is steady and frictionless, and the water is incompressible, this colored drop would maintain its identity as it flowed along. Its shape might change, becoming compressed or elongated, but it would not mix with the surrounding water.

The path or trajectory followed by this "particle of fluid" is called a **streamline.** Smoke particles mixed with the air allow you to see the streamlines in the wind-tunnel photograph of Figure 15.27. Notice also how the individual smoke trails retain their identity. Figure 15.28 illustrates three important properties of streamlines.

A bundle of neighboring streamlines, such as those shown in Figure 15.29a, form a **flow tube.** Because streamlines never cross, all the streamlines that cross plane 1 within area A_1 later cross plane 2 within area A_2. A flow tube is like an invisible pipe that keeps this portion of the flowing fluid distinct from other portions. Real pipes are also flow tubes.

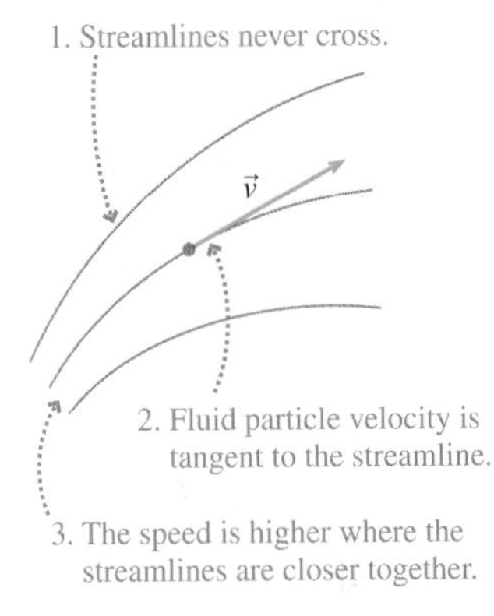

FIGURE 15.28 Particles in an ideal fluid move along streamlines.

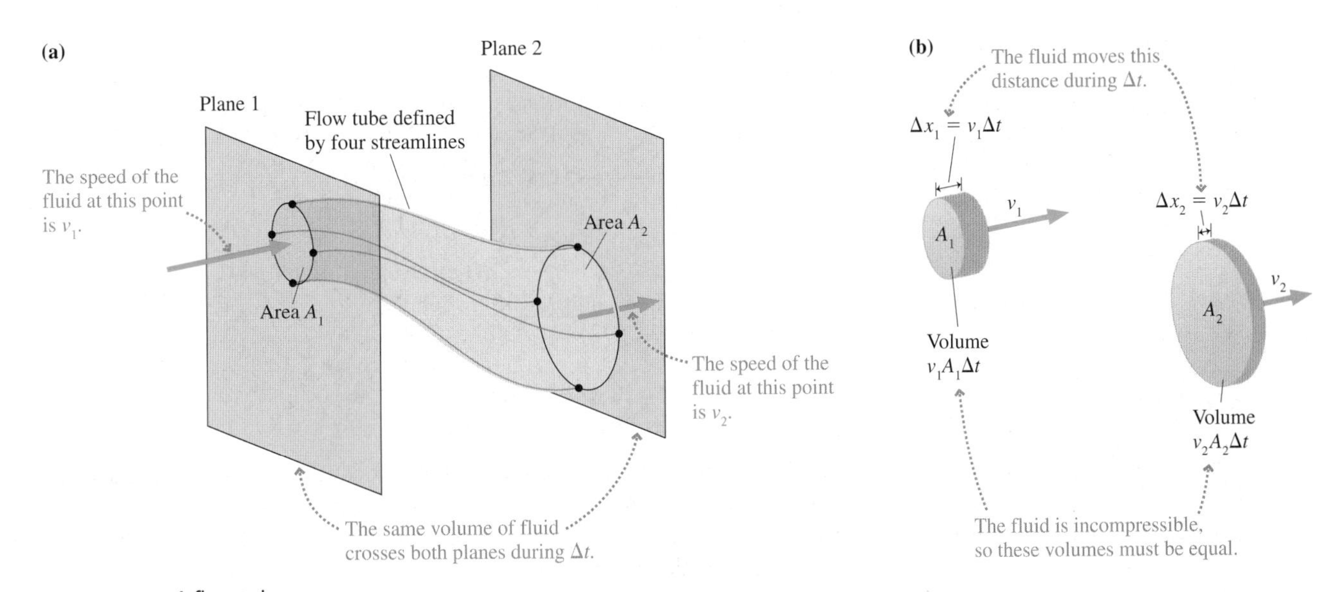

FIGURE 15.29 A flow tube.

(a)

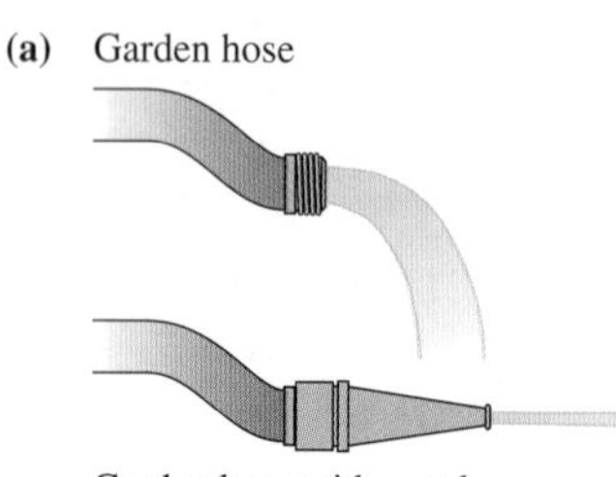

(b)

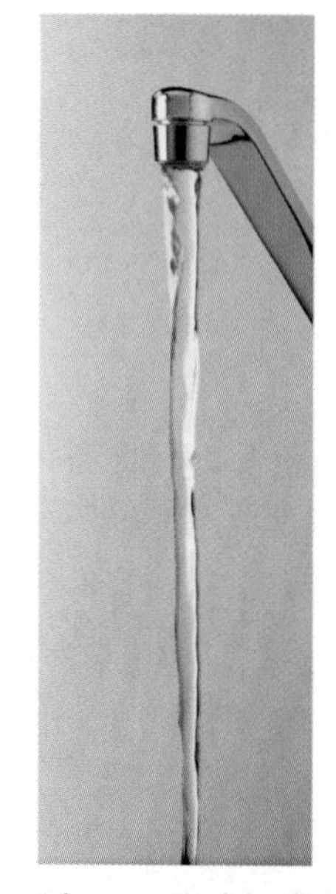

FIGURE 15.30 The speed of the fluid changes as the flow tube diameter changes. This is a consequence of the equation of continuity.

When you squeeze a toothpaste tube, the volume of toothpaste that emerges matches the amount by which you reduce the volume of the tube. An incompressible fluid in a flow tube acts the same way. Fluid is not created or destroyed within the flow tube, and it cannot be stored. If volume V enters the flow tube through area A_1 during some interval of time Δt, then an equal volume V must leave the flow tube through area A_2.

Figure 15.29b shows the flow crossing A_1 during a small interval of time Δt. If the fluid speed at this point is v_1, the fluid moves forward a small distance $\Delta x_1 = v_1 \Delta t$ and fills the volume $V_1 = A_1 \Delta x_1 = v_1 A_1 \Delta t$. The same analysis for the fluid crossing A_2 with fluid speed v_2 would find $V_2 = v_2 A_2 \Delta t$. These two volumes must be equal, leading to the conclusion that

$$v_1 A_1 = v_2 A_2 \tag{15.19}$$

Equation 15.19 is called the **equation of continuity,** and it is one of two important equations for the flow of an ideal fluid. The equation of continuity says that **the volume of an incompressible fluid entering one part of a flow tube must be matched by an equal volume leaving downstream.**

An important consequence of the equation of continuity is that **flow is faster in narrower parts of a flow tube, slower in wider parts.** You're familiar with this conclusion from many everyday observations. The garden hose shown in Figure 15.30a squirts farther after you put a nozzle on it. This is because the narrower opening of the nozzle gives the water a higher exit speed. Water flowing from the faucet shown in Figure 15.30b picks up speed as it falls. As a result, the flow tube "necks down" to a smaller diameter.

The quantity

$$Q = vA \tag{15.20}$$

is called the **volume flow rate.** The SI units of Q are m^3/s, although in practice Q may be measured in cm^3/s, liters per minute, or, in the United States, gallons per minute. Another way to express the meaning of the equation of continuity is to say that **the volume flow rate is constant at all points in a flow tube.**

EXAMPLE 15.10 Water through a garden hose

A garden hose has an inside diameter of 16 mm. The hose can fill a 10 L bucket in 20 s.

a. What is the speed of the water out of the end of the hose?
b. What diameter nozzle would increase the fluid speed by a factor of 4?

MODEL Treat the water as an ideal fluid. The hose itself is a flow tube, so the equation of continuity applies.

SOLVE

a. The volume flow rate is $Q = (10\text{ L})/(20\text{ s}) = 0.50\text{ L/s}$. To convert this to SI units you must recall that there are 1000 L in 1 m^3, or $1\text{ L} = 10^{-3}\text{ m}^3$. Thus $Q = 5.0 \times 10^{-4}\text{ m}^3/\text{s}$. We can find the speed of the water from Equation 15.20:

$$v = \frac{Q}{A} = \frac{Q}{\pi r^2} = \frac{5.0 \times 10^{-4}\text{ m}^3/\text{s}}{\pi(0.008\text{ m})^2} = 2.5\text{ m/s}$$

b. $Q = vA$ remains constant. To increase v by a factor of 4, A must be reduced by a factor of 4. The cross-section area depends on the square of the radius, so the area is reduced by a factor of 4 if the radius is reduced by a factor of 2. Thus the necessary nozzle diameter is 8 mm.

STOP TO THINK 15.5 The figure shows volume flow rates (in cm^3/s) for all but one tube. What is the volume flow rate through the unmarked tube? Is the flow direction in or out?

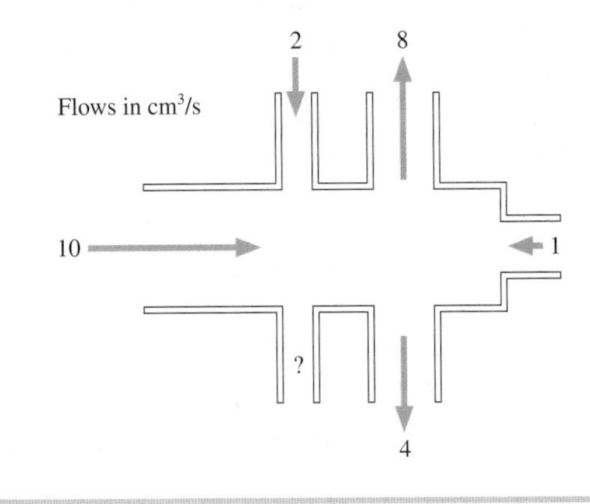

Flows in cm³/s
2
8
10
1
?
4

Bernoulli's Equation

The equation of continuity is one of two important relationships for ideal fluids. The other is a statement of energy conservation. The general statement of energy conservation that you learned in Chapter 11 is

$$\Delta K + \Delta U = W_{ext} \tag{15.21}$$

where W_{ext} is the work done by any external forces.

Let's see how this applies to the flow tube of Figure 15.31. Our system for analysis is the volume of fluid within the flow tube. Work is done on this volume of fluid by the pressure forces of the *surrounding* fluid. At point 1, the fluid to the left of the flow tube exerts force $\vec{F}_1$ on the system. This force points to the right. At the other end of the flow tube, at point 2, the fluid to the right of the flow tube exerts force $\vec{F}_2$ to the left. The pressure inside the flow tube is not relevant because those forces are internal to the system. Only external forces change the total energy.

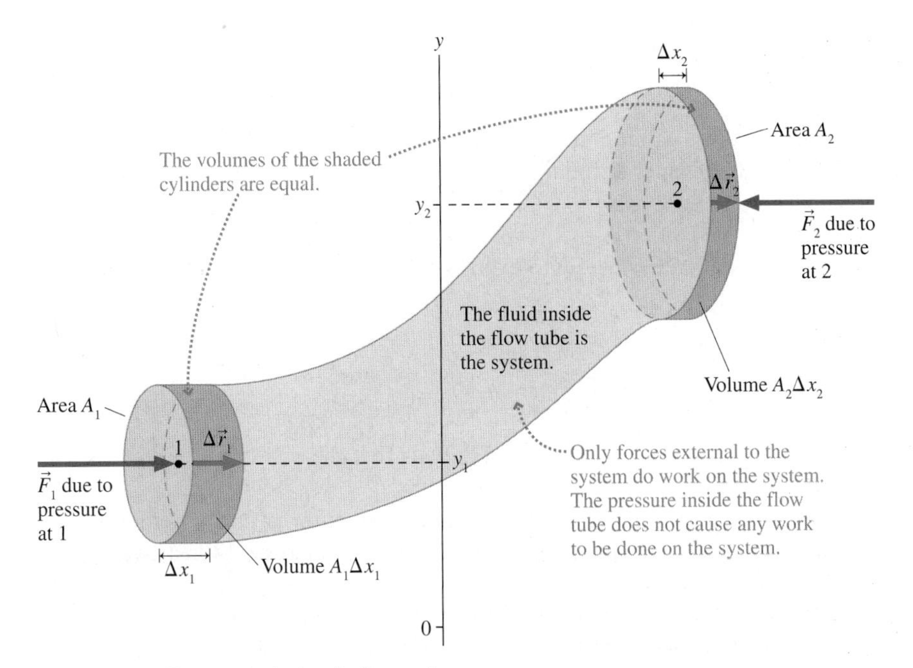

y
Δx₂
Area A₂
The volumes of the shaded cylinders are equal.
y₂
2
Δr₂
F₂ due to pressure at 2
The fluid inside the flow tube is the system.
Volume A₂Δx₂
Area A₁
1
Δr₁
y₁
F₁ due to pressure at 1
Only forces external to the system do work on the system. The pressure inside the flow tube does not cause any work to be done on the system.
Δx₁
Volume A₁Δx₁
0

FIGURE 15.31 Energy analysis of a flow tube.

At point 1, force $\vec{F}_1$ pushes the fluid through displacement $\Delta\vec{r}_1$. $\vec{F}_1$ and $\Delta\vec{r}_1$ are parallel, so the work done on the fluid at this point is

$$W_1 = \vec{F}_1 \cdot \Delta\vec{r}_1 = F_1 \Delta r_1 = (p_1 A_1)\Delta x_1 = p_1 V \tag{15.22}$$

The A_1 and Δx_1 enter the equation from different terms, but they conveniently combine to give the fluid volume V.

The situation is much the same at point 2 except that $\vec{F}_2$ points opposite the displacement $\Delta\vec{r}_2$. This introduces a $\cos(180°) = -1$ into the dot product for the work, giving

$$W_2 = \vec{F}_2 \cdot \Delta\vec{r}_2 = -F_1 \Delta r_1 = -(p_2 A_2)\Delta x_2 = -p_2 V \tag{15.23}$$

The pressure from the left at point 1 pushes the fluid ahead, a positive work. The pressure from the right at point 2 tries to slow the fluid down, a negative work. Together, the work by external forces is

$$W_{ext} = W_1 + W_2 = p_1 V - p_2 V \tag{15.24}$$

Now let's see how this work changes the kinetic and potential energy of the system. A small volume of fluid $V = A_1 \Delta x_1$ passes point 1 and, at some later time, arrives at point 2, where the unchanged volume is $V = A_2 \Delta x_2$. The change in gravitational potential energy for this volume of fluid is

$$\Delta U = mgy_2 - mgy_1 = \rho V g y_2 - \rho V g y_1 \tag{15.25}$$

where ρ is the fluid density. Similarly, the change in kinetic energy is

$$\Delta K = \frac{1}{2}mv_2^2 - \frac{1}{2}mv_1^2 = \frac{1}{2}\rho V v_2^2 - \frac{1}{2}\rho V v_1^2 \tag{15.26}$$

Combining Equations 15.24, 15.25, and 15.26 gives us the energy equation for the fluid in the flow tube:

$$\frac{1}{2}\rho V v_2^2 - \frac{1}{2}\rho V v_1^2 + \rho V g y_2 - \rho V g y_1 = p_1 V - p_2 V \tag{15.27}$$

The volume V cancels out of all the terms. If we regroup the terms, the energy equation becomes

$$p_1 + \frac{1}{2}\rho v_1^2 + \rho g y_1 = p_2 + \frac{1}{2}\rho v_2^2 + \rho g y_2 \tag{15.28}$$

Equation 15.28 is called **Bernoulli's equation.** It is named for the 18th-century Italian scientist Daniel Bernoulli, who made some of the earliest studies of fluid dynamics.

Bernoulli's equation is really nothing more than a statement about work and energy. It is sometimes useful to express Bernoulli's equation in the alternative form

$$p + \frac{1}{2}\rho v^2 + \rho g y = \text{constant} \tag{15.29}$$

This version of Bernoulli's equation tells us that the quantity $p + \frac{1}{2}\rho v^2 + \rho g y$ remains constant along a streamline.

One important implication of Bernoulli's equation is easily demonstrated. Before reading the next paragraph, try the simple experiment illustrated in Figure 15.32. Really, do try this!

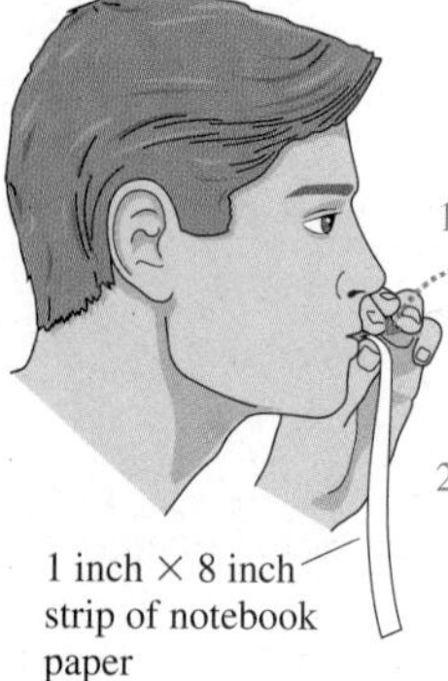
1. Hold strip at lower edge of bottom lip, just touching lip.
2. Pucker lips and blow hard straight out over the top of the strip.
1 inch × 8 inch strip of notebook paper

FIGURE 15.32 A simple demonstration of Bernoulli's equation.

What happened? You probably expected your breath to press the strip of paper down. Instead, the strip *rose.* In fact, the harder you blow, the more nearly the strip becomes parallel to the floor. This counterintuitive result is a consequence of Bernoulli's equation. As the air speed above the strip of paper increases, the pressure has to *decrease* to keep the quantity $p + \frac{1}{2}\rho v^2 + \rho g y$ constant. The air-pressure force pressing down on the top surface is then *less* than the air-pressure force pushing up on the bottom surface, resulting in a net upward force on the paper.

NOTE ▶ Using Bernoulli's equation is very much like using the law of conservation of energy. Rather than identifying a "before" and "after," you want to identify two points on a streamline. As the following examples show, Bernoulli's equation is often used in conjunction with the equation of continuity. ◀

EXAMPLE 15.11 An irrigation system

Water flows through the pipes shown in Figure 15.33. The water's speed through the lower pipe is 5.0 m/s and a pressure gauge reads 75 kPa. What is the reading of the pressure gauge on the upper pipe?

MODEL Treat the water as an ideal fluid obeying Bernoulli's equation. Consider a streamline connecting point 1 in the lower pipe with point 2 in the upper pipe.

SOLVE Bernoulli's equation, Equation 15.28, relates the pressure, fluid speed, and heights at points 1 and 2. It is easily solved for the pressure p_2 at point 2:

$$p_2 = p_1 + \frac{1}{2}\rho v_1^2 - \frac{1}{2}\rho v_2^2 + \rho g y_1 - \rho g y_2$$

$$= p_1 + \frac{1}{2}\rho(v_1^2 - v_2^2) + \rho g(y_1 - y_2)$$

All quantities on the right are known except v_2, and that is where the equation of continuity will be useful. The cross-section areas and water speeds at points 1 and 2 are related by

$$v_1 A_1 = v_2 A_2$$

from which we find

$$v_2 = \frac{A_1}{A_2}v_1 = \frac{r_1^2}{r_2^2}v_1 = \frac{(0.030\text{ m})^2}{(0.020\text{ m})^2}(5.0\text{ m/s}) = 11.25\text{ m/s}$$

The pressure at point 1 is $p_1 = 0.75\text{ kPa} + 1\text{ atm} = 176{,}300\text{ Pa}$. We can now use the above expression for p_2 to calculate $p_2 = 105{,}900\text{ Pa}$. This is the absolute pressure; the pressure gauge on the upper pipe will read

$$p_2 = 176{,}300\text{ Pa} - 1\text{ atm} = 4.6\text{ kPa}$$

ASSESS Reducing the pipe size decreases the pressure because it makes $v_2 > v_1$. Gaining elevation also reduces the pressure.

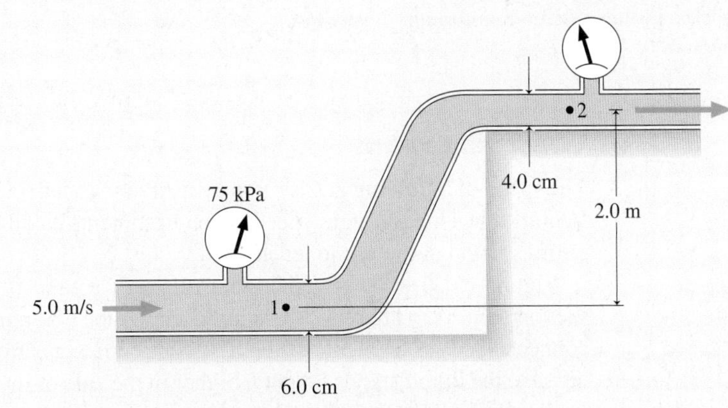

FIGURE 15.33 The water pipes of an irrigation system.

EXAMPLE 15.12 Hydroelectric power

Small hydroelectric plants in the mountains sometimes bring the water from a reservoir down to the power plant through enclosed tubes. In one such plant, the 100-cm-diameter intake tube in the base of the dam is 50 m below the reservoir surface. The water drops 200 m through the tube before flowing into the turbine through a 50-cm-diameter nozzle.

a. What is the water speed into the turbine?
b. By how much does the inlet pressure differ from the hydrostatic pressure at that depth?

MODEL Treat the water as an ideal fluid obeying Bernoulli's equation. Consider a streamline that begins at the surface of the reservoir and ends at the exit of the nozzle. The pressure at the surface is $p_1 = p_{atmos}$ and $v_1 \approx 0$ m/s. The water discharges into air, so $p_2 = p_{atm}$ at the exit.

VISUALIZE Figure 15.34 on the next page is a pictorial representation of the situation.

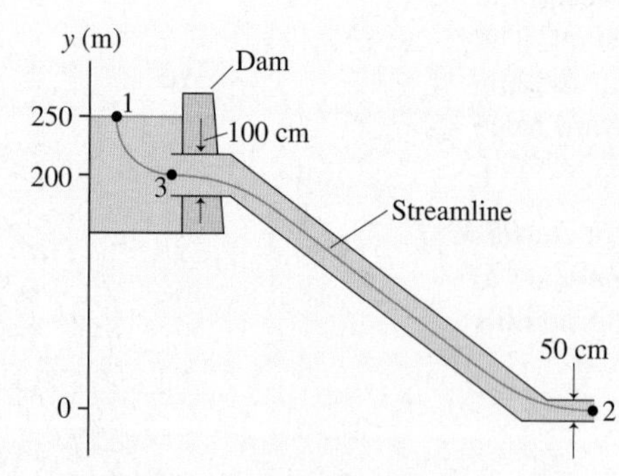

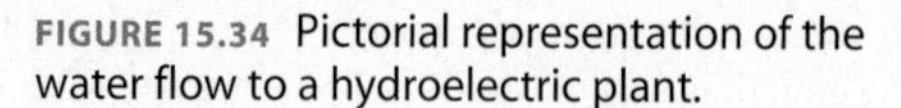

FIGURE 15.34 Pictorial representation of the water flow to a hydroelectric plant.

SOLVE

a. The power plant is in the mountains, where $p_{atmos} < 1$ atm, but p_{atmos} occurs on both sides of Bernoulli's equation and cancels. Bernoulli's equation, with $v_1 = 0$ m/s and $y_2 = 0$ m, is

$$p_{atmos} + \rho g y_1 = p_{atmos} + \frac{1}{2}\rho v_2^2$$

p_{atmos} cancels, as expected, as does the density ρ. Solving for v_2 gives

$$v_2 = \sqrt{2gy_1} = \sqrt{2(9.80 \text{ m/s}^2)(250 \text{ m})} = 70.0 \text{ m/s}$$

b. You might expect the pressure p_3 at the intake to be the hydrostatic pressure $p_{atmos} + \rho g d$ at depth d. But the water is *flowing* into the intake tube, so it's not in static equilibrium.

We can find the intake speed v_3 from the equation of continuity:

$$v_3 = \frac{A_2}{A_3}v_2 = \frac{r_2^2}{r_3^2}\sqrt{2gy_1}$$

The intake is along the streamline between points 1 and 2, so we can apply Bernoulli's equation to points 1 and 3:

$$p_{atmos} + \rho g y_1 = p_3 + \frac{1}{2}\rho v_3^2 + \rho g y_3$$

Solving this equation for p_3, and noting that $y_1 - y_3 = d$, we find

$$p_3 = p_{atmos} + \rho g(y_1 - y_3) - \frac{1}{2}\rho v_3^2$$

$$= p_{atmos} + \rho g d - \frac{1}{2}\rho\left(\frac{r_2}{r_3}\right)^4(2gy_1)$$

$$= p_{static} - \rho g y_1\left(\frac{r_2}{r_3}\right)^4$$

The intake pressure is *less* than hydrostatic pressure by the amount

$$\rho g y_1\left(\frac{r_2}{r_3}\right)^4 = 153{,}000 \text{ Pa} = 1.5 \text{ atm}$$

ASSESS The water's exit speed from the nozzle is the same as if it fell 250 m from the surface of the reservoir. This isn't surprising because we've assumed a nonviscous (i.e., frictionless) liquid. "Real" water would have less speed but still flow very fast.

Two Applications

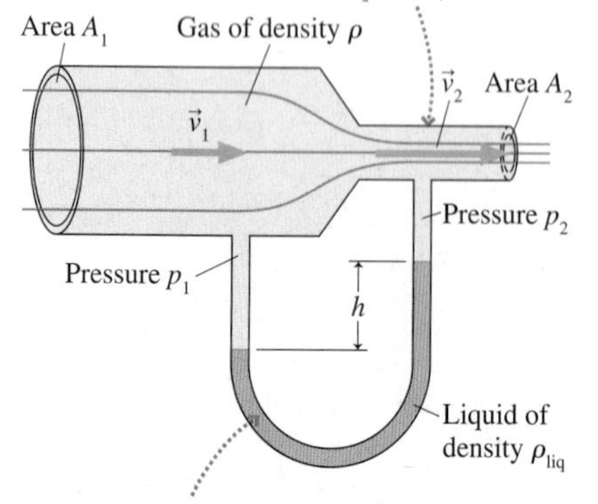

FIGURE 15.35 A Venturi tube measures gas-flow speeds.

The speed of a flowing gas is often measured with a device called a **Venturi tube.** Venturi tubes measure gas speeds in environments as different as chemistry laboratories, wind tunnels, and jet engines.

Figure 15.35 shows gas flowing through a tube that changes from cross-section area A_1 to area A_2. A U-shaped glass tube containing liquid of density ρ_{liq} connects the two segments of the flow tube. When gas flows through the horizontal tube, the liquid stands height h higher in the side of the U tube connected to the narrow segment of the flow tube.

Figure 15.35 shows how a Venturi tube works. We can make this analysis quantitative and determine the gas-flow speed from the liquid height h. Two pieces of information we have to work with are Bernoulli's equation

$$p_1 + \frac{1}{2}\rho v_1^2 + \rho g y_1 = p_2 + \frac{1}{2}\rho v_2^2 + \rho g y_2 \tag{15.30}$$

and the equation of continuity

$$v_2 A_2 = v_1 A_1 \tag{15.31}$$

In addition, the hydrostatic equation for the liquid tells us that the pressure p_2 above the right tube differs from the pressure p_1 above the left tube by $\rho_{liq} g h$. That is,

$$p_2 = p_1 - \rho_{liq} g h \tag{15.32}$$

First use Equations 15.31 and 15.32 to eliminate v_2 and p_2 in Bernoulli's equation:

$$p_1 + \frac{1}{2}\rho v_1^2 = (p_1 - \rho_{\text{liq}}gh) + \frac{1}{2}\rho\left(\frac{A_1}{A_2}\right)^2 v_1^2 \qquad (15.33)$$

The potential energy terms have disappeared because $y_1 = y_2$ for a horizontal tube. Equation 15.33 can now be solved for v_1, then v_2 is obtained from Equation 15.31. We'll skip a few algebraic steps and go right to the result:

$$v_1 = A_2\sqrt{\frac{2\rho_{\text{liq}}gh}{\rho(A_1^2 - A_2^2)}}$$
$$v_2 = A_1\sqrt{\frac{2\rho_{\text{liq}}gh}{\rho(A_1^2 - A_2^2)}} \qquad (15.34)$$

In practice, the equations for the gas-flow speeds have to be corrected for the fact that the gas, which is compressible, is not an ideal liquid. But Equation 15.34 is reasonably accurate even without corrections as long as the flow speeds are much less than the speed of sound, about 340 m/s. For us, the Venturi tube is an example of the power of Bernoulli's equation.

As a final example, we can use Bernoulli's equation to understand, at least qualitatively, how airplane wings generate *lift*. Figure 15.36 shows the cross section of an airplane wing. This shape is called an *airfoil.*

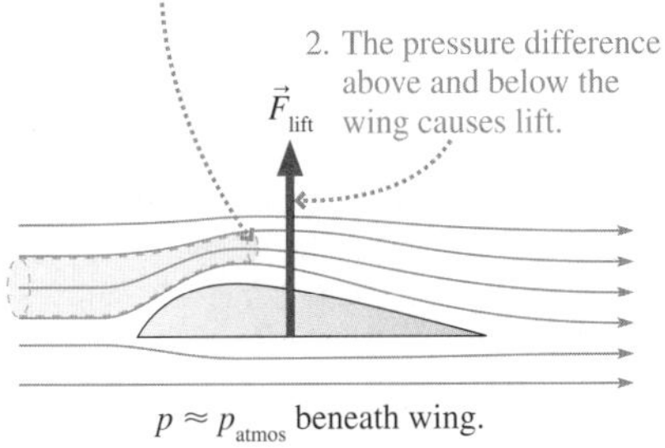
1. Flow tube decreases in size due to compression of streamlines. The higher speed lowers the pressure to $p < p_{\text{atmos}}$.
2. The pressure difference above and below the wing causes lift.
$\vec{F}_{\text{lift}}$
$p \approx p_{\text{atmos}}$ beneath wing.

FIGURE 15.36 Air flow over a wing generates lift by creating unequal pressures above and below.

Although you usually think of an airplane moving through the air, in the airplane's reference frame it is the air that flows across a stationary wing. As it does, the streamlines must separate. The flat bottom of the wing does not significantly alter the streamlines going under the wing. But the streamlines going over the wing get bunched together. This bunching reduces the cross-section area of a flow tube of streamlines. Consequently, in accordance with the equation of continuity, the air speed must increase as it flows across the top of the wing.

As you've seen several times, an increased air speed implies a decreased air pressure. This is the lesson of Bernoulli's equation. Because the air pressure above the wing is less than the air pressure below, the air exerts a net upward force on the wing, just as it did on the paper strip you blew across. The upward force of the air due to a pressure difference across the wing is called **lift.**

A complete analysis of the lift of a wing is quite complicated and involves many factors in addition to Bernoulli's equation. Nonetheless, you should now be able to understand one of the important physical principles that are involved.

STOP TO THINK 15.6 Rank in order, from highest to lowest, the liquid heights h_1 to h_4 in tubes 1 to 4. The air flow is from left to right.

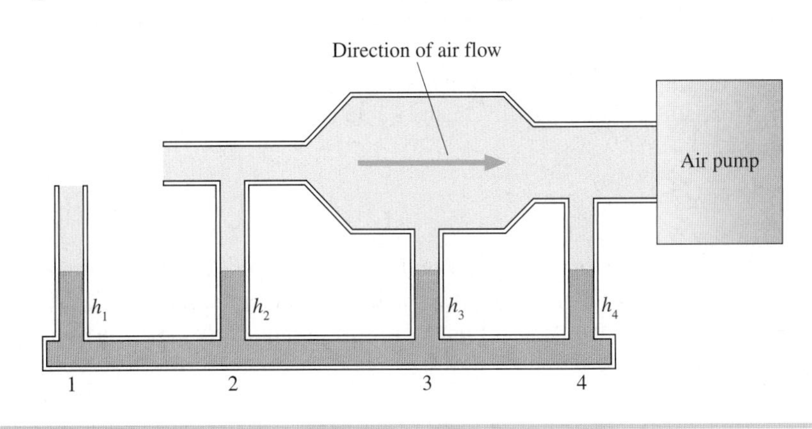
Direction of air flow
Air pump
h_1
h_2
h_3
h_4
1
2
3
4

15.6 Elasticity

The final subject to explore in this chapter is elasticity. Although elasticity applies primarily to solids rather than fluids, you will see that similar ideas come into play.

Tensile Stress and Young's Modulus

Suppose you clamp one end of a solid rod while using a strong machine to pull on the other with force $\vec{F}$. Figure 15.37a shows the experimental arrangement. We usually think of solids as being, well, solid. But any material, be it plastic, concrete, or steel, will stretch as the spring-like molecular bonds expand.

Figure 15.37b shows graphically the amount of force needed to stretch the rod by the amount ΔL. This graph contains several regions of interest. First is the *elastic region,* ending at the *elastic limit.* As long as ΔL is less than the elastic limit, the rod will return to its initial length L when the force is removed. Just such a reversible stretch is what we mean when we say a material is *elastic.* A stretch beyond the elastic limit will permanently deform the object; it will not return to its initial length when the force is removed. And, not surprisingly, there comes a point when the rod breaks.

For most materials, the graph begins with a *linear region,* which is where we will focus our attention. If ΔL is within the linear region, the force needed to stretch the rod is

$$F = k\Delta L \tag{15.35}$$

where k is the slope of the graph. You'll recognize Equation 15.35 as none other than Hooke's law.

(a)

(b)

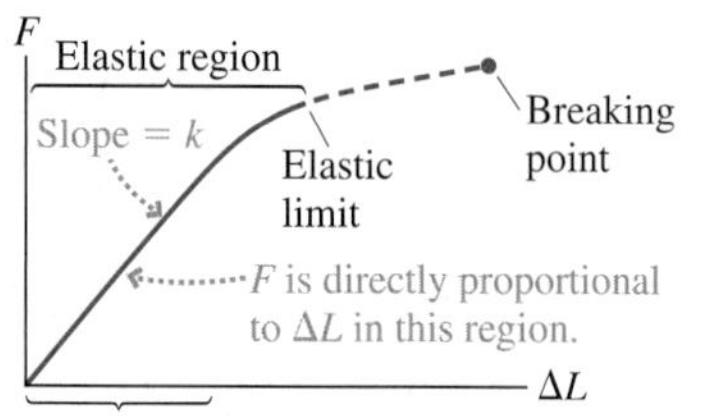

F
Elastic region
Slope = k
Elastic limit
Breaking point
F is directly proportional to ΔL in this region.
ΔL

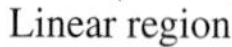

Linear region

FIGURE 15.37 Stretching a solid rod.

The difficulty with Equation 15.35 is that the proportionality constant k depends both on the composition of the rod—whether it is, say, plastic or aluminum—and on the rod's length and cross-section area. It would be useful to characterize the elastic properties of plastic in general, or aluminum in general, without needing to know the dimensions of a specific rod.

We can meet this goal by thinking about Hooke's law at the atomic scale. The elasticity of a material is directly related to the spring constant of the molecular bonds between neighboring atoms. As Figure 15.38 shows, the force pulling each bond is proportional to the quantity F/A. This force causes each bond to stretch by an amount proportional to $\Delta L/L$. We don't know what the proportionality constants are, but we don't need to. Hooke's law applied to a molecular bond tells us that the force pulling on a bond is proportional to the amount that the bond stretches. Thus F/A must be proportional to $\Delta L/L$. We can write their proportionality as

$$\frac{F}{A} = Y\frac{\Delta L}{L} \tag{15.36}$$

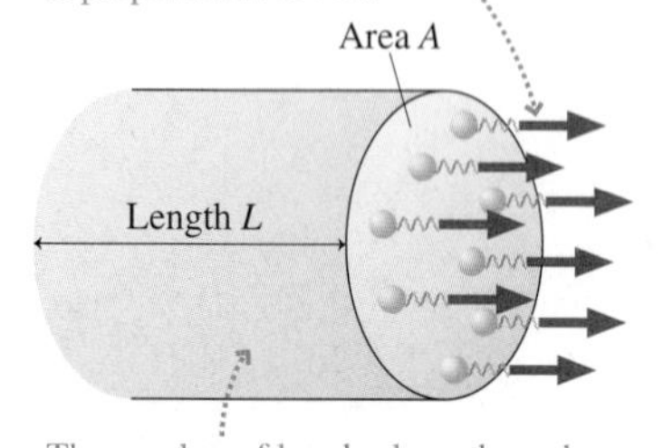

The number of bonds is proportional to area A. If the rod is pulled with force F, the force pulling on each bond is proportional to F/A.
Area A
Length L
The number of bonds along the rod is proportional to length L. If the rod stretches by ΔL, the stretch of each bond is proportional to ΔL/L.

FIGURE 15.38 A material's elasticity is directly related to the spring constant of the molecular bonds.

The proportionality constant Y is called **Young's modulus.** It is directly related to the spring constant of the molecular bonds, so it depends on the material from which the object is made but *not* on the object's geometry.

A comparison of Equations 15.35 and 15.36 shows that Young's modulus can be written

$$Y = \frac{kL}{A} \tag{15.37}$$

This is not a definition of Young's modulus but simply an expression for making an experimental determination of the value of Young's modulus. This k is the spring constant of the bar seen in Figure 15.37. It is a quantity easily measured in the laboratory.

The quantity F/A, where A is the cross-section area, is called **tensile stress.** Notice that it is essentially the same definition as pressure. Even so, tensile stress differs in that the stress is applied in a particular direction whereas pressure forces are exerted in all directions. Another difference is that stress is measured in N/m^2 rather than pascals. The quantity $\Delta L/L$, the fractional increase in the length, is called **strain.** Strain is dimensionless. The numerical values of strain are always very small because solids cannot be stretched very much before reaching the breaking point.

With these definitions, Equation 15.36 can be written

$$\text{stress} = Y \times \text{strain} \tag{15.38}$$

Because strain is dimensionless, Young's modulus Y has the same dimensions as stress, namely N/m^2. Table 15.3 gives values of Young's modulus for several common materials. Large values of Y characterize materials that are stiff and rigid. "Softer" materials, at least relatively speaking, have smaller values of Y. You can see that steel has a larger Young's modulus than aluminum.

TABLE 15.3 Elastic properties of various materials

Substance	Young's modulus (N/m^2)	Bulk modulus (N/m^2)
Aluminum	7×10^{10}	7×10^{10}
Concrete	3×10^{10}	–
Copper	11×10^{10}	14×10^{10}
Mercury	–	3×10^{10}
Plastic (polystyrene)	0.3×10^{10}	–
Steel	20×10^{10}	16×10^{10}
Water	–	0.2×10^{10}
Wood (Douglas fir)	1×10^{10}	–

We introduced Young's modulus by considering how materials stretch. But Equation 15.38 and Young's modulus also apply to the compression of materials. Compression is particularly important in engineering applications, where beams, columns, and support foundations are compressed by the load they bear. Concrete is often compressed, as in columns that support highway overpasses, but rarely stretched.

NOTE ▶ Whether the rod is stretched or compressed, Equation 15.38 is valid only in the linear region of the graph in Figure 15.37b. The breaking point is usually well outside the linear region, so you can't use Young's modulus to compute the maximum possible stretch or compression. ◀

EXAMPLE 15.13 Stretching a wire

A 2.0-m-long, 1.0-mm-diameter wire is suspended from the ceiling. Hanging a 4.5 kg mass from the wire stretches the wire's length by 1.0 mm. What is Young's modulus for this wire? Can you identify the material?

MODEL The hanging mass creates tensile stress in the wire.

SOLVE The force pulling on the wire, which is simply the weight of the hanging mass, produces tensile stress

$$\frac{F}{A} = \frac{mg}{\pi r^2} = \frac{(4.5\text{ kg})(9.80\text{ m/s}^2)}{\pi(0.0005\text{ m})^2} = 5.6 \times 10^7\text{ N/m}^2$$

The resulting stretch of 1.0 mm is a strain of $\Delta L/L = (1.0\text{ mm})/(2000\text{ mm}) = 5.0 \times 10^{-4}$. Thus Young's modulus for the wire is

$$Y = \frac{F/A}{\Delta L/L} = 11 \times 10^{10}\text{ N/m}^2$$

Referring to Table 15.3, we see that the wire is made of copper.

TERMS AND NOTATION

fluid	hydrostatic pressure	ideal-fluid model	lift
gas	Pascal's principle	viscosity	Young's modulus, Y
liquid	gauge pressure, p_g	laminar flow	tensile stress
mass density, ρ	hydraulics	streamline	strain
unit volume	buoyant force	flow tube	volume stress
pressure, p	displaced fluid	equation of continuity	volume strain
pascal, Pa	Archimedes' principle	volume flow rate, Q	bulk modulus, B
vacuum	average density, ρ_{avg}	Bernoulli's equation	
standard atmosphere, atm	neutral buoyancy	Venturi tube	

EXERCISES AND PROBLEMS

Exercises

Section 15.1 Fluids

1. A 100 mL beaker holds 120 g of liquid. What is the liquid's density in SI units?
2. Containers A and B have equal volumes. The mass of the liquid in B is 6300 times the mass of helium gas in container A. Identify the liquid in B.
3. A 6 m × 12 m swimming pool slopes linearly from a 1.0 m depth at one end to a 3.0 m depth at the other. What is the mass of water in the pool?
4. a. 50 g of gasoline are mixed with 50 g of water. What is the average density of the mixture?
   b. 50 cm^3 of gasoline are mixed with 50 cm^3 of water. What is the average density of the mixture?

Section 15.2 Pressure

5. The deepest point in the ocean is 11 km below sea level, deeper than Mt. Everest is tall. What is the pressure in atmospheres at this depth?
6. a. What volume of water has the same mass as 8.0 m^3 of ethyl alcohol?
   b. If this volume of water is in a cubic tank, what is the pressure at the bottom?
7. A 1.0-m-diameter vat of liquid is 2.0 m deep. The pressure at the bottom of the vat is 1.3 atm. What is the mass of the liquid in the vat?
8. A 50-cm-thick layer of oil floats on a 120-cm-thick layer of water. What is the pressure at the bottom of the water layer?
9. A research submarine has a 20-cm-diameter window 8.0 cm thick. The manufacturer says the window can withstand forces up to 1.0×10^6 N. What is the submarine's maximum safe depth? The pressure inside the submarine is maintained at 1.0 atm.
10. A 20-cm-diameter circular cover is placed over a 10-cm-diameter hole that leads into an evacuated chamber. The pressure in the chamber is 20 kPa. How much force is required to pull the cover off?

Section 15.3 Measuring and Using Pressure

11. What is the gas pressure inside the box?

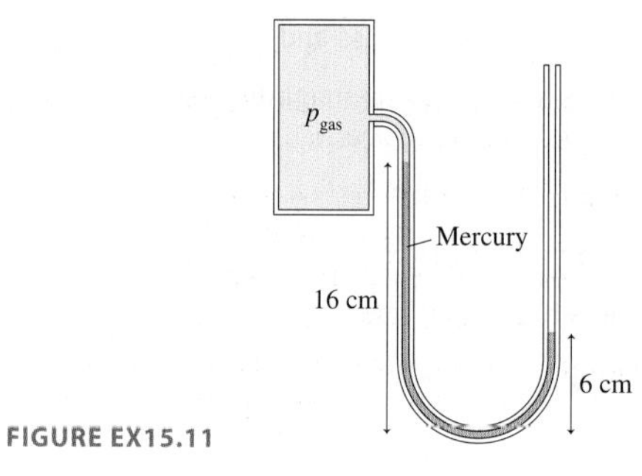

FIGURE EX15.11

12. What is the height of a water barometer at atmospheric pressure?
13. A 55 kg cheerleader uses an oil-filled hydraulic lift to hold four 110 kg football players at a height of 1.0 m. If her piston is 16 cm in diameter, what is the diameter of the football players' piston?
14. How far must a 2.0-cm-diameter piston be pushed down into one cylinder of a hydraulic lift to raise an 8.0-cm-diameter piston by 20 cm?

Section 15.4 Buoyancy

15. A 6.0-cm-diameter sphere with a mass of 89.3 g is neutrally buoyant in a liquid. Identify the liquid.
16. Two identical beakers are filled to the same height with water. Beaker B has a plastic sphere floating in it. Which beaker, with all its contents, weighs more? Or are they equal? Explain.

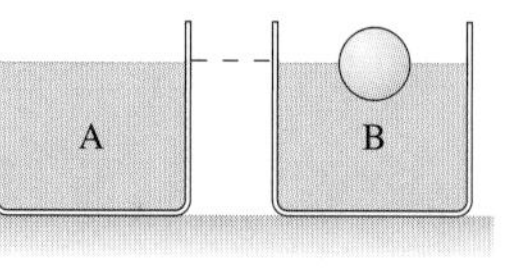

FIGURE EX15.16

17. Styrofoam has a density of 300 kg/m^3. What is the maximum mass that can hang without sinking from a 50-cm-diameter Styrofoam sphere in water? Assume the volume of the mass is negligible compared to that of the sphere.

18. A 10 cm × 10 cm × 10 cm wood block with a density of 700 kg/m^3 floats in water. What is the distance from the top of the block to the water if the water is fresh? If it's seawater?
19. What is the tension in the string?

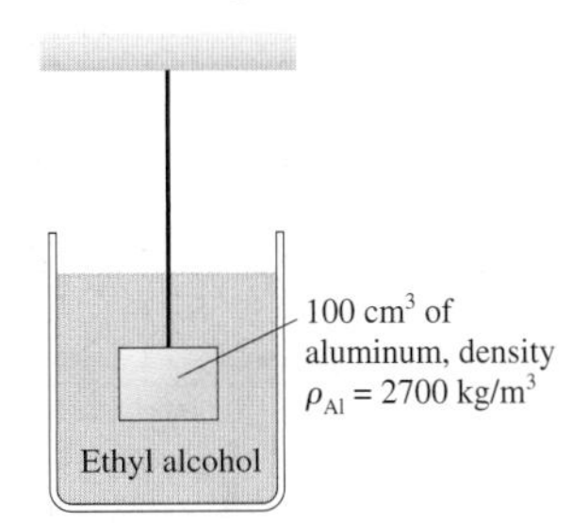

FIGURE EX15.19

Section 15.5 Fluid Dynamics

20. River Pascal with a volume flow rate of 5.0 × 10^5 L/s joins with River Archimedes, which carries 10.0 × 10^5 L/s, to form the Bernoulli River. The Bernoulli River is 150 m wide and 10 m deep. What is the speed of the water in the Bernoulli River?
21. Water flowing through a 2.0-cm-diameter pipe can fill a 300 L bathtub in 5.0 minutes. What is the speed of the water in the pipe?
22. A 1.0-cm-diameter pipe widens to 2.0 cm, then narrows to 0.5 cm. Liquid flows through the first segment at a speed of 4.0 m/s.
    a. What is the speed in the second and third segments?
    b. What is the volume flow rate through the pipe?
23. A long horizontal tube has a square cross section with sides of width L. A fluid moves through the tube with speed v_0. The tube then changes to a circular cross section with diameter L. What is the fluid's speed in the circular part of the tube?
24. What does the top pressure gauge read?

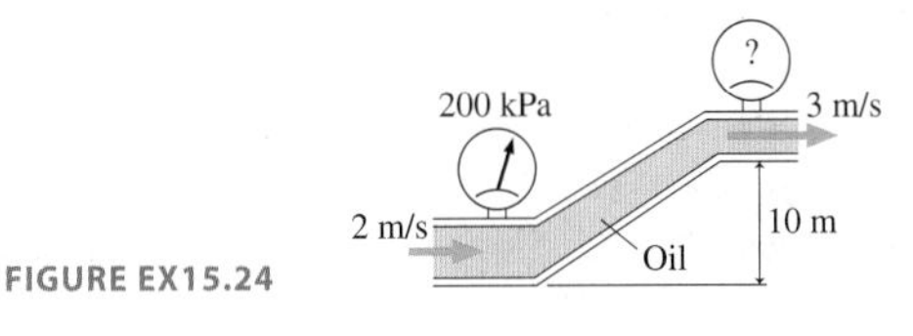

FIGURE EX15.24

Section 15.6 Elasticity

25. What hanging mass will stretch a 2.0-m-long, 0.50-mm-diameter steel wire by 1.0 mm?
26. An 80-cm-long, 1.0-mm-diameter steel guitar string must be tightened to a tension of 2000 N by turning the tuning screws. By how much is the string stretched?
27. A 3.0-m-tall, 50-cm-diameter concrete column supports a 200,000 kg load. By how much is the column compressed?
28. At what ocean depth would the volume of an aluminum sphere be reduced by 0.10%?

Problems

29. A gymnasium is 16 m high. By what percent is the air pressure at the floor greater than the air pressure at the ceiling?
30. The two 60-cm-diameter cylinders in Figure P15.30, closed at one end, open at the other, are joined to form a single cylinder, then the air inside is removed.
    a. How much force does the atmosphere exert on the flat end of each cylinder?
    b. Suppose one cylinder is bolted to a sturdy ceiling. How many 100 kg football players would need to hang from the lower cylinder to pull the two cylinders apart?

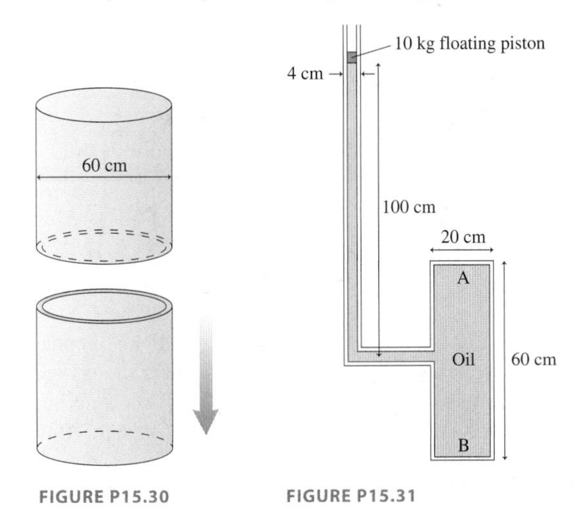

FIGURE P15.30 FIGURE P15.31

31. a. In Figure P15.31, how much force does the fluid exert on the end of the cylinder at A?
    b. How much force does the fluid exert on the end of the cylinder at B?
32. A friend asks you how much pressure is in your car tires. You know that the tire manufacturer recommends 30 psi, but it's been a while since you've checked. You can't find a tire gauge in the car, but you do find the owner's manual and a ruler. Fortunately, you've just finished taking physics, so you tell your friend, "I don't know, but I can figure it out." From the owner's manual you find that the car's mass is 1500 kg. It seems reasonable to assume that each tire supports one-fourth of the weight. With the ruler you find that the tires are 15 cm wide and the flattened segment of the tire in contact with the road is 13 cm long. What answer will you give your friend?
33. A 2.0 mL syringe has an inner diameter of 6.0 mm, a needle inner diameter of 0.25 mm, and a plunger pad diameter (where you place your finger) of 1.2 cm. A nurse uses the syringe to inject medicine into a patient whose blood pressure is 140/100.
    a. What is the minimum force the nurse needs to apply to the syringe?
    b. The nurse empties the syringe in 2.0 s. What is the flow speed of the medicine through the needle?
34. What is the longest vertical soda straw you could possibly drink from?
35. What is the total mass of the earth's atmosphere?
36. Suppose the density of the earth's atmosphere were a constant 1.3 kg/m^3, independent of height, until reaching the top. How thick would the atmosphere be?
37. Your science teacher has assigned you the task of building a water barometer. You've learned that the pressure of the atmosphere can vary by as much as 5% from 1 standard atmosphere as the weather changes.
    a. What minimum height must your barometer have?
    b. One stormy day the TV weather person says, "The barometric pressure this afternoon is a low 29.55 inches." What is the height of the water in your barometer?

38. The container shown in Figure P15.38 is filled with oil. It is open to the atmosphere on the left.
    a. What is the pressure at point A?
    b. What is the pressure difference between points A and B? Between points A and C?

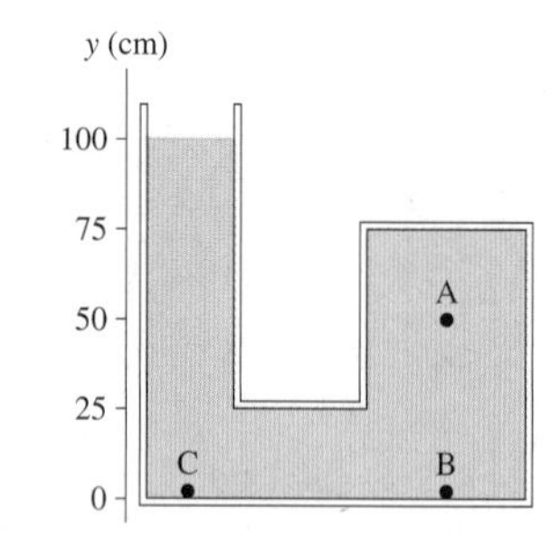

FIGURE P15.38

39. a. A 70 kg student balances a 1200 kg elephant on a hydraulic lift. What is the diameter of the piston the student is standing on?
    b. A second 70 kg student joins the first student. How high do they lift the elephant?

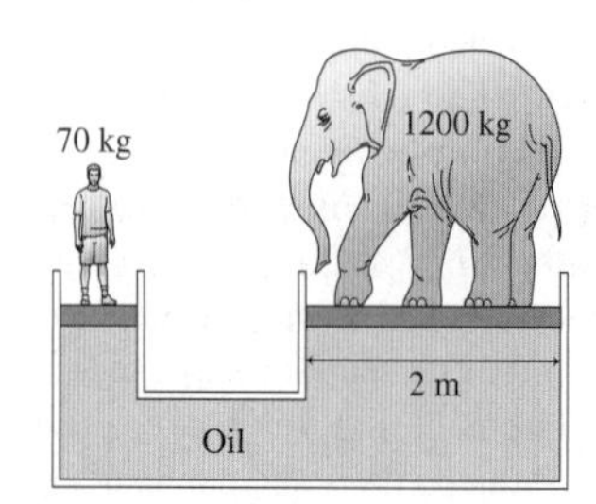

FIGURE P15.39

40. Figure 15.19 showed a hydraulic lift with force $\vec{F}_1$ balancing force $\vec{F}_2$. Assume that force $\vec{F}_2$ is the unchanging weight mg of an object of mass m. Derive Equation 15.13, which states that the force *increment* needed to lift the weight through distance d_2 is $\Delta F = \rho g(A_1 + A_2)d_2$, where ρ is the density of the liquid.
41. A U-shaped tube, open to the air on both ends, contains mercury. Water is poured into the left arm until the water column is 10.0 cm deep. How far upward from its initial position does the mercury in the right arm rise?
42. Glycerin is poured into an open U-shaped tube until the height in both sides is 20 cm. Ethyl alcohol is then poured into one arm until the height of the alcohol column is 20 cm. The two liquids do not mix. What is the difference in height between the top surface of the glycerin and the top surface of the alcohol?
43. Water stands at depth d behind a dam of width w.
    a. Find an expression for the net force of the water on the dam.
    b. Evaluate the net force on a 100-m-wide dam with a 60 m water depth.

    Hint: This problem requires an integration.
44. An aquarium tank is 100 cm long, 35 cm wide, and 40 cm deep. It is filled to the top.
    a. What is the force of the water on the bottom (100 cm × 35 cm) of the tank?
    b. What is the force of the water on the front window (100 cm × 40 cm) of the tank?

    Hint: This problem requires an integration.
45. It's possible to use the ideal gas law to show that the density of the earth's atmosphere decreases exponentially with height. That is, $\rho = \rho_0 \exp(-z/z_0)$, where z is the height above sea level, ρ_0 is the density at sea level (you can use the Table 15.1 value), and z_0 is called the *scale height* of the atmosphere. (See Challenge Problem 76.)
    a. Determine the value of z_0.
    b. What is the density of the air in Denver, at an elevation of 1600 m? What percent of sea-level density is this?

    Hint: This problem requires an integration. What is the weight of a column of air?
46. The average density of the body of a fish is 1080 kg/m^3. To keep from sinking, fish have an air bladder filled with air. What fraction of a fish's body must be filled with air to be neutrally buoyant? You can use the Table 15.1 value for the density of air.
47. A 6.0-cm-tall cylinder floats in water with its axis perpendicular to the surface. The length of the cylinder above water is 2.0 cm. What is the cylinder's mass density?
48. A sphere completely submerged in water is tethered to the bottom with a string. The tension in the string is one-third the weight of the sphere. What is the density of the sphere?
49. A 10 kg irregularly shaped rock with a density of 4200 kg/m^3 rests on the bottom of a swimming pool. What is the normal force of the pool bottom on the rock?
50. You need to determine the density of a ceramic statue. If you suspend it from a spring scale, the scale reads 28.4 N. If you then lower the statue into a tub of water, so that it is completely submerged, the scale reads 17.0 N. What is the density?
51. A 5.0 kg rock whose density is 4800 kg/m^3 is suspended by a string such that half of the rock's volume is under water. What is the tension in the string?
52. A 10 cm × 10 cm × 10 cm block of steel (ρ_{steel} = 7900 kg/m^3) is suspended from a spring scale. The scale is in newtons.
    a. What is the scale reading if the block is in air?
    b. What is the scale reading after the block has been lowered into a beaker of oil and is completely submerged?
53. A 10-cm-diameter, 20-cm-tall steel cylinder (ρ_{steel} = 7900 kg/m^3) floats in mercury. The axis of the cylinder is perpendicular to the surface. What length of steel is above the surface?
54. A cylinder with cross-section area A floats with its long axis vertical in a liquid of density ρ.
    a. Pressing down on the cylinder pushes it deeper into the liquid. Find an expression for the force needed to push the cylinder distance x deeper into the liquid and hold it there.
    b. A 4.0-cm-diameter cylinder floats in water. How much work must be done to push the cylinder 10 cm deeper into the water?

    Hint: An integration is required.
55. A less-dense liquid of density ρ_1 floats on top of a more-dense liquid of density ρ_2. A uniform cylinder of length l and density ρ, with $\rho_1 < \rho < \rho_2$, floats at the interface with its long axis vertical. What fraction of the length is in the more-dense liquid?
56. A 30-cm-tall, 4.0-cm-diameter plastic tube has a sealed bottom. 250 g of lead pellets are poured into the bottom of the tube, whose mass is 30 g, then the tube is lowered into a liquid. The tube floats with 5.0 cm extending above the surface. What is the density of the liquid?

57. A 355 mL soda can is 6.2 cm in diameter and has a mass of 20 g. Such a soda can half full of water is floating upright in water. What length of the can is above the water level?

58. The cylinder shown in the figure is completely submerged in a liquid of density ρ.
    a. Find expressions for the forces F_{top} and F_{bot} that the liquid exerts on the top and bottom of the cylinder.
    b. Write an expression for the magnitude of the buoyant force on the cylinder.
    c. Can you write your expression of part b in terms of the weight of the liquid that has been displaced by the cylinder?

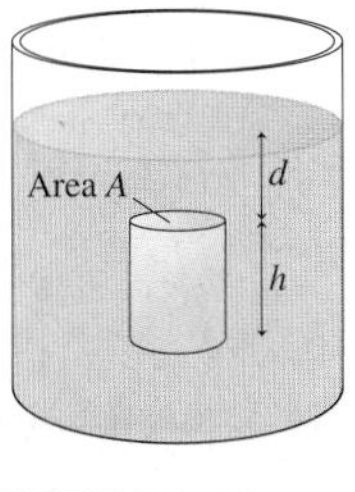

FIGURE P15.58

Congratulations! You've just rediscovered Archimedes' principle.

59. The bottom of a steel "boat" is a 5 m × 10 m × 2 cm piece of steel (ρ_{steel} = 7900 kg/m^3). The sides are made of 0.50-cm-thick steel. What minimum height must the sides have for this boat to float in perfectly calm water?

60. Water flows at 5.0 L/s through a horizontal pipe that narrows smoothly from 10 cm diameter to 5.0 cm diameter. A pressure gauge in the narrow section reads 50 kPa. What is the reading of a pressure gauge in the wide section?

61. Water flows from the pipe shown in the figure with a speed of 4.0 m/s.
    a. What is the water pressure as it exits into the air?
    b. What is the height h of the standing column of water?

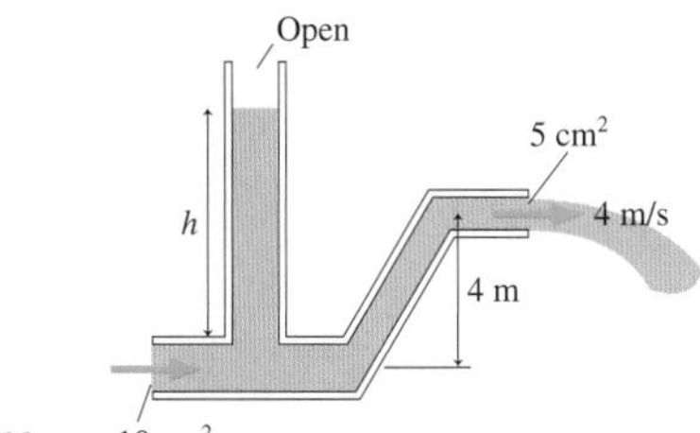

FIGURE P15.61

62. Water flows out of a 16-mm-diameter sink faucet at 1.0 m/s. At what distance below the faucet has the water stream narrowed to 10 mm diameter?

63. A hurricane wind blows across a 6.0 m × 15.0 m flat roof at a speed of 130 km/hr.
    a. Is the air pressure above the roof higher or lower than the pressure inside the house? Explain.
    b. What is the pressure difference?
    c. How much force is exerted on the roof? If the roof cannot withstand this much force, will it "blow in" or "blow out"?

64. Air flows through this tube at a rate of 1200 cm^3/s. Assume that air is an ideal fluid. What is the height h of mercury in the right side of the U-tube?

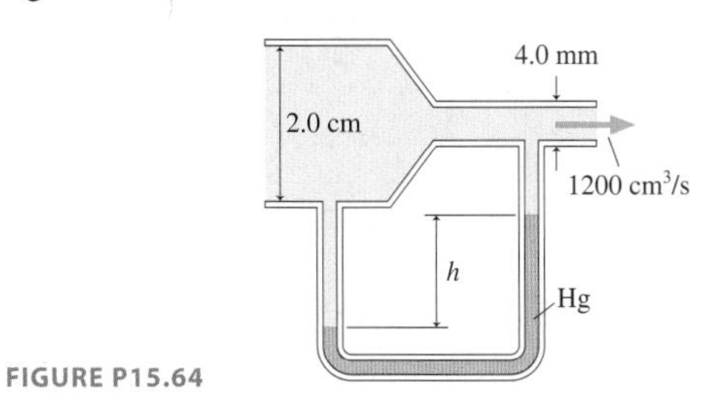

FIGURE P15.64

65. Air flows through the tube shown in the figure. Assume that air is an ideal fluid.
    a. What are the air speeds v_1 and v_2 at points 1 and 2?
    b. What is the volume flow rate?

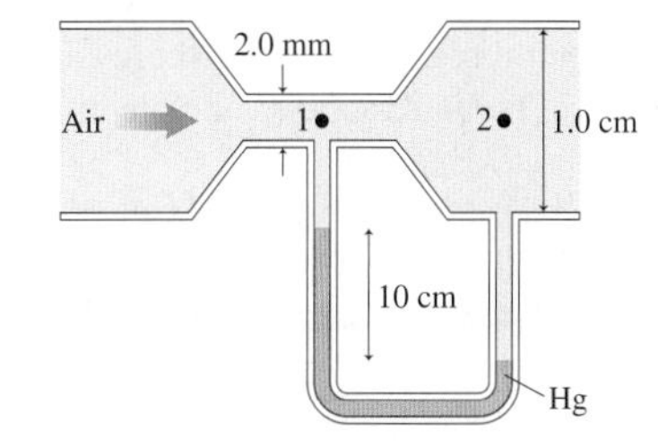

FIGURE P15.65

66. A water tank of height h has a small hole at height y. The water is replenished to keep h from changing. The water squirting from the hole has range x. The range approaches zero as $y \rightarrow 0$ because the water squirts right onto the table. The range also approaches zero as $y \rightarrow h$ because the horizontal velocity becomes zero. Thus there must be some height y between 0 and h for which the range is a maximum.
    a. Find an algebraic expression for the flow speed v with which the water exits the hole at height y.
    b. Find an algebraic expression for the range of a particle shot horizontally from height y with speed v.
    c. Combine your expressions from parts a and b. Then find the maximum range x_{max} and the height y of the hole. "Real" water won't achieve quite this range because of viscosity, but it will be close.

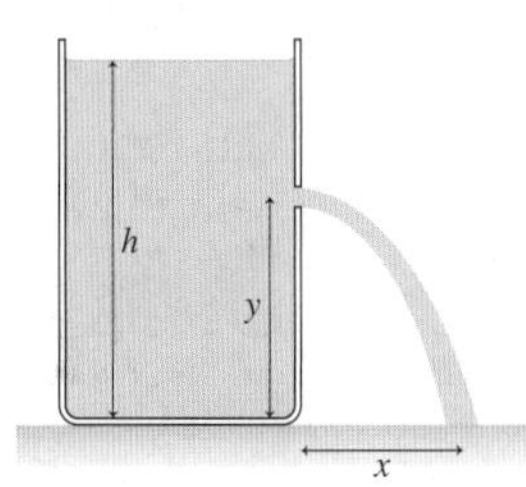

FIGURE P15.66

67. A 4.0-mm-diameter hole is 1.0 m below the surface of a 2.0-m-diameter tank of water.
    a. What is the volume flow rate through the hole, in L/min?
    b. What is the rate, in mm/min, at which the water level in the tank will drop if the water is not replenished?

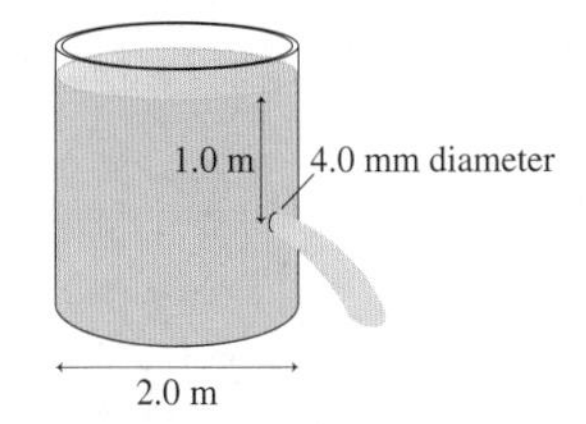

FIGURE P15.67

68. A 70 kg mountain climber dangling in a crevasse stretches a 50-m-long, 1.0-cm-diameter rope by 8.0 cm. What is Young's modulus for the rope?

69. A large 10,000 L aquarium is supported by four wood posts (Douglas fir) at the corners. Each post has a square 4.0 cm × 4.0 cm cross section and is 80 cm tall. By how much is each post compressed by the weight of the aquarium?

70. a. What is the pressure at a depth of 5000 m in the ocean?
    b. What is the fractional volume change $\Delta V/V$ of seawater at this pressure?

c. What is the density of seawater at this pressure?
d. The hydrostatic pressure equation that you used in part a assumes the liquid is incompressible. That's an excellent assumption for many applications, but the slight compressibility of water means that the hydrostatic pressure equation doesn't give perfect results at great depths. Is the actual pressure at 5000 m more or less than you computed in part a? Explain.

71. A cylindrical steel pressure vessel with volume 1.30 m^3 is to be tested. The vessel is entirely filled with water, then a piston at one end of the cylinder is pushed in until the pressure inside the vessel has increased by 2000 kPa. Suddenly, a safety plug on the top bursts. How many liters of water come out?

Challenge Problems

72. A 1.0-m-tall cylinder contains air at a pressure of 1 atm. A very thin, frictionless piston of negligible mass is placed at the top of the cylinder, to prevent any air from escaping, then mercury is slowly poured into the cylinder until no more can be added without the cylinder overflowing. What is the height h of the column of compressed air?
Hint: Boyle's law, which you learned in chemistry, says $p_1V_1 = p_2V_2$ for a gas compressed at constant temperature, which we will assume to be the case.

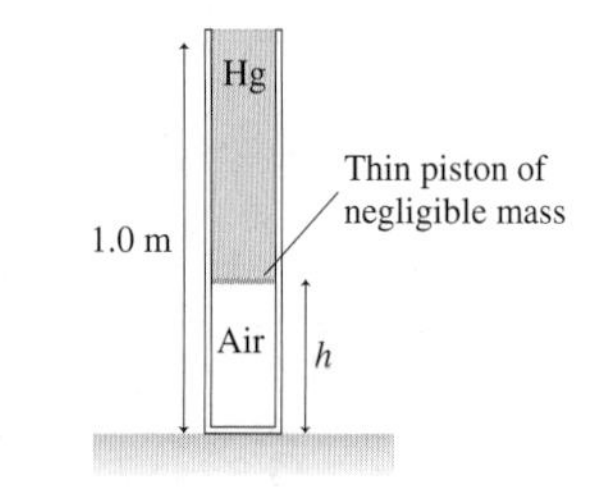

FIGURE CP15.72

73. A 1.0 g balloon is filled with helium gas until it becomes a 20-cm-diameter sphere. What maximum mass can be tied to the balloon (with a massless string) without the balloon sinking to the floor?

74. A cone of density ρ_o and total height l floats in a liquid of density ρ_f. The height of the cone above the liquid is h. What is the ratio h/l of the exposed height to the total height?

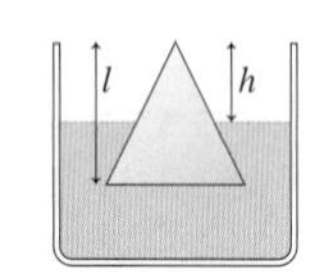

FIGURE CP15.74

75. A cylinder of density ρ_o, length l, and cross-section area A floats in a liquid of density ρ_f with its axis perpendicular to the surface. Length h of the cylinder is submerged when the cylinder floats at rest.
a. Write expressions for the object's weight w and for the buoyant force F_B. Use these to show that $h = (\rho_o/\rho_f)l$.
b. Suppose the cylinder is distance y *above* its equilibrium position. Find an expression for $(F_{net})_y$, the y-component of the net force. Use your expressions from part a to cancel some of the terms.
c. You should recognize your result of part b as a version of Hooke's law. What is the "spring constant" k?
d. If you push a floating object down and release it, it bobs up and down. So it is like a spring in the sense that it oscillates if displaced from equilibrium. Use your "spring constant" and what you know about simple harmonic motion to show that the cylinder's oscillation period is

$$T = 2\pi\sqrt{\frac{h}{g}}$$

e. What is the oscillation period for a 100-m-tall iceberg ($\rho_{ice} = 917 \text{ kg/m}^3$) in seawater?

76. The pressure of the atmosphere decreases with increasing elevation. Let's figure out how.
a. Establish a z-axis that points up, with $z = 0$ at sea level. Suppose the pressure at height z is known to be p and the air density is ρ. Use the hydrostatic pressure equation to write an expression for the pressure at height $z + dz$, where dz is so small that the density has not changed. Your expression will be in terms of p, ρ, dz, and perhaps some constants. The pressure *decreases* as you gain elevation, so be careful with signs.
b. Using your expression from part a, write an expression for dp, the amount by which the pressure *changes* in going from z to $z + dz$. Pressure is decreasing, so your expression should be negative.
c. You need to integrate your expression from part b, but you can't because the density ρ is not a constant. *If* the temperature remains constant, which we will assume, then the ideal gas law implies that pressure is directly proportional to density. That is, $p/\rho = p_0/\rho_0$, where p_0 and ρ_0 are the sea-level values of pressure and density. Use this to rewrite your expression for dp in terms of p, dz, and various constants.
d. Now you have an integrable expression, although you must first divide by p to get all the pressure terms on one side of the equation. Carry out the integration and use the fact that $p = p_0$ at $z = 0$ to determine the integration constant. Then solve for the pressure at height z. Your final result should be in the form $p = p_0\exp(-z/z_0)$.
e. z_0 is called the *scale height* of the atmosphere. It is the height at which $p = e^{-1}p_0$, or about 37% of the sea-level pressure. Determine the numerical value of z_0.
f. The lower layer of the atmosphere, called the troposphere, has a height of about 15,000 m. This is the region of the atmosphere where weather occurs. Above it is the stratosphere, where conditions are very different. Draw a graph of pressure versus height up to a height of 15,000 m.

Comment: We assumed a constant-temperature atmosphere. In the real atmosphere, the temperature in the troposphere decreases with increasing height. This alters how the pressure changes, but not enormously. Your result is a reasonably good approximation.

STOP TO THINK ANSWERS

Stop to Think 15.1: $\rho_1 = \rho_2 = \rho_3$. Density depends only on what the object is made of, not how big the pieces are.

Stop to Think 15.2: c. These are all open tubes, so the liquid rises to the same height in all three despite their different shapes.

Stop to Think 15.3: $F_2 > F_1 = F_3$. The masses in 3 do not add. The pressure underneath each of the two large pistons is mg/A_2, and the pressure under the small piston must be the same.

Stop to Think 15.4: b. The weight of the displaced water equals the weight of the ice cube. When the ice cube melts and turns into water, that amount of water will exactly fill the volume that the ice cube is now displacing.

Stop to Think 15.5: 1 cm^3/s out. The fluid is incompressible, so the sum of what flows in must match the sum of what flows out. 13 cm^3/s is known to be flowing in while 12 cm^3/s flows out. An additional 1 cm^3/s must flow out to achieve balance.

Stop to Think 15.6: $h_2 > h_4 > h_3 > h_1$. The liquid level is higher where the pressure is lower. The pressure is lower where the flow speed is higher. The flow speed is highest in the narrowest tube, zero in the open air.

PART III SUMMARY

Applications of Newtonian Mechanics

We have developed two parallel perspectives of motion, each with its own concepts and techniques. We focused on the first of these in Part I, where we dealt with the relationship between force and motion. Newton's second law is the principle most central to the force/motion perspective. Then, in Part II, we developed a before-and-after perspective based on the idea of conservation laws. Newton's laws were essential in the development of conservation laws, but they remain hidden in the background when the conservation laws are applied. Together, these two perspectives form the heart of Newtonian mechanics.

Our goal in Part III has been to see how Newtonian mechanics is applied to several diverse but important topics. We added only one new law of physics in Part III, Newton's law of gravity, and we introduced few completely new concepts. Instead, we've broadened our understanding of the force/motion perspective and the conservation-law perspective through our investigations of gravity, rotational motion, oscillations, and fluids. In reviewing Part III, pay close attention to the interplay between these two perspectives. Recognizing which is the best tool in a particular situation will help you improve your problem-solving ability.

Our knowledge of mechanics is now essentially complete. We will add a few additional ideas as we need them, but our journey into physics will be taking us in entirely new directions as we continue on. Hence this is an opportune moment to step back a bit to take a look at the "big picture." Newtonian mechanics may seem all very factual and straightforward to us today, but keep in mind that these ideas are all human inventions. There was a time when they did not exist and when our concepts of nature were quite different from what they are today.

KNOWLEDGE STRUCTURE III Applications of Newtonian Mechanics

Newton's Theory of Gravity

Any two masses exert attractive gravitational forces on each other.

Newton's law of gravity is

$$F_{m \text{ on } M} = F_{M \text{ on } m} = \frac{GMm}{r^2}$$

- Kepler's laws describe the elliptical orbits of satellites and planets.
- The gravitational potential energy is

$$U_g = -\frac{GMm}{r}$$

Rotation of a Rigid Body

A rigid body is a system of particles.
Rotational motion is analogous to linear motion.

Rotational motion	Linear motion
Angular acceleration α	Acceleration a
Torque τ	Force F
Moment of inertia I	Mass m
Angular momentum L	Momentum p

- **Newton's second law** $\tau_{net} = I\alpha$
- **Rotational kinetic energy** $K = \frac{1}{2}I\omega^2$

NEWTON'S LAWS
+
CONSERVATION LAWS

Oscillations

Systems with a linear restoring force exhibit simple harmonic oscillation.

- The **kinematic equations of SHM** are

$$x(t) = A\cos(\omega t + \phi_0)$$
$$v(t) = -v_{max}\sin(\omega t + \phi_0)$$

where $v_{max} = \omega A$ and the phase constant ϕ_0 describes the initial conditions.

- **Energy is transformed between kinetic and potential** as the system oscillates. In an undamped system, the total mechanical energy

$$E = \frac{1}{2}mv^2 + \frac{1}{2}kx^2 = \frac{1}{2}m(v_{max})^2 = \frac{1}{2}kA^2$$

is conserved.

Fluids and Elasticity

Fluids are systems that flow. Gases and liquids are fluids. Fluids are better characterized by density and pressure than by mass and force.

- **Liquids** Pressure is primarily gravitational. The hydrostatic pressure is

$$p = p_0 + \rho g d$$

- **Gases** Pressure is primarily thermal. Pressure in a container is constant.
- **Archimedes' principle** The buoyant force is equal to the weight of the displaced liquid.

For fluid flow, **Bernoulli's equation**

$$p_1 + \frac{1}{2}\rho v_1^2 + \rho g y_1 = p_2 + \frac{1}{2}\rho v_2^2 + \rho g y_2$$

is really a statement of energy conservation.

The Newtonian Synthesis

Newton's achievements, praised by no less than Einstein as "perhaps the greatest advance in thought that a single individual was ever privileged to make," are often called the *Newtonian synthesis*. "Synthesis" means "the uniting or combining of separate elements to form a coherent whole." It is often said of Newton that he "united the heavens and the earth." In doing so, he changed forever the way we view ourselves and our relationship to the universe.

As we noted in Chapter 12, medieval cosmology considered the heavenly bodies to be perfect, unchanging objects quite unrelated to imperfect and changeable earthly matter. Their perfection and immortality symbolized the perfection of God above, while the material bodies of humans were imperfect and mortal. This cosmology was mirrored in medieval feudal society. The king—ordained by God and whose symbol was the sun—was surrounded by a small circle of nobles and a larger circle of serfs and peasants. Taken together, the ideas and institutions of science, religion, and society of this time form what we call the medieval *worldview*. Their worldview, in its many facets, was hierarchical and authoritarian, reflecting their understanding of "natural order" in the universe.

Copernicus weakened medieval cosmology by questioning the position of the earth in the universe. Galileo, with his telescope, found that the heavens are not perfect and unchanging. Now, at the end of the 17th century, the success of Newton's theories implied that the sun and the planets were merely ordinary matter, obeying the same natural laws as earthly matter. This uniting of earthly motions and heavenly motions—the *synthesis* in the Newtonian synthesis—dealt the final blow to the medieval worldview.

Newton's success changed the way we see and think about the universe. Rather than seeing whirling celestial spheres, people began to think of the universe in terms of the motion of material particles following rigid laws. This Newtonian conception of the cosmos is often called a "clockwork universe." The technology of clocks was progressing rapidly in the 18th century, and people everywhere admired the consistency and predictability of these little machines. The Newtonian universe is a very large machine, but one that is consistent, predictable, and law-abiding. In other words, a perfect clock.

Major thinkers of the 17th and 18th centuries soon concluded that God had created the world by placing all the particles in their original positions, then giving them a push to get them going. God, in this role, was called the "prime mover." But once the universe was started, it went along perfectly well just by obeying Newton's laws. No divine intervention or guidance was needed. This is certainly a very different view of our relationship to God and the universe than was contained in the medieval worldview.

Newton also influenced the way people think about themselves and their society. His theories clearly demonstrated that the universe is not random or capricious but, instead, follows natural laws. Others soon began to apply the concept of natural law to human nature, human behavior, and human institutions. The main protagonist in this school of thought was the English philosopher and political scientist John Locke, a contemporary of Newton. Locke developed a theory of human behavior from the ideas of natural laws and empirical evidence. We cannot go into Locke's theories here, but Newton's success helped to propel Locke's ideas into the mainstream of 18th-century political thought.

Locke's writings had a great influence on a young American colonial named Thomas Jefferson. The concept of natural laws, as they apply to individuals, is very much behind Jefferson's enunciation of "unalienable rights" in the Declaration of Independence. In fact, the first sentence of the Declaration refers explicitly to "the Laws of Nature and of Nature's God." The idea of *checks and balances*, built into the Constitution of the United States, is very much a mechanical and clock-like model of how political institutions function.

Just as medieval feudalism mirrored the medieval understanding of the universe, contemporary constitutional democracy mirrors, in many ways, the Newtonian cosmology. Hierarchy and authority have been replaced by equality and law because they now seem to us the "natural order" of things. Having grown up with this modern worldview, we find it difficult to imagine any other. Nonetheless, it is important to realize that vastly different worldviews have existed at other times and in other cultures.

Science has changed dramatically in the last hundred-odd years. Newton's clockwork universe has been superseded by relativity and quantum physics. Entirely new theories and sciences, such as evolution, ecology, and psychology, have appeared. These new ideas are slowly working their way into other areas of thought and human activity, and bit by bit they are changing the ways in which we see ourselves, our society, and our relationship to nature. A future worldview is in the making.

Mathematics Review

APPENDIX A

Algebra

Using exponents: $a^{-x} = \dfrac{1}{a^x}$ $\quad a^x a^y = a^{(x+y)}$ $\quad \dfrac{a^x}{a^y} = a^{(x-y)}$ $\quad (a^x)^y = a^{xy}$

$a^0 = 1$ $\quad a^1 = a$ $\quad a^{1/n} = \sqrt[n]{a}$

Fractions: $\left(\dfrac{a}{b}\right)\left(\dfrac{c}{d}\right) = \dfrac{ac}{bd}$ $\quad \dfrac{a/b}{c/d} = \dfrac{ad}{bc}$ $\quad \dfrac{1}{1/a} = a$

Logarithms: If $a = e^x$, then $\ln(a) = x$ $\quad \ln(e^x) = x$ $\quad e^{\ln(x)} = x$

$\ln(ab) = \ln(a) + \ln(b)$ $\quad \ln\left(\dfrac{a}{b}\right) = \ln(a) - \ln(b)$ $\quad \ln(a^n) = n\ln(a)$

The expression $\ln(a + b)$ cannot be simplified.

Linear equations: The graph of the equation $y = ax + b$ is a straight line. a is the slope of the graph. b is the y-intercept.

Slope $a = \dfrac{\text{rise}}{\text{run}} = \dfrac{\Delta y}{\Delta x}$

y-intercept $= b$

Proportionality: To say that y is proportional to x, written $y \propto x$, means that $y = ax$, where a is a constant. Proportionality is a special case of linearity. A graph of a proportional relationship is a straight line that passes through the origin. If $y \propto x$, then

$$\frac{y_1}{y_2} = \frac{x_1}{x_2}$$

Quadratic equation: The quadratic equation $ax^2 + bx + c = 0$ has the two solutions $x = \dfrac{-b \pm \sqrt{b^2 - 4ac}}{2a}$.

Geometry and Trigonometry

Area and volume:

Rectangle
$A = ab$

Triangle
$A = \frac{1}{2}ab$

Circle
$C = 2\pi r$
$A = \pi r^2$

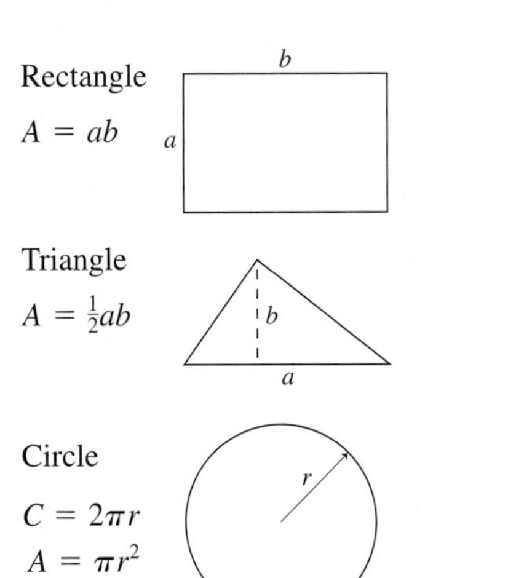

Rectangular box
$V = abc$

Right circular cylinder
$V = \pi r^2 l$

Sphere
$A = 4\pi r^2$
$V = \frac{4}{3}\pi r^3$

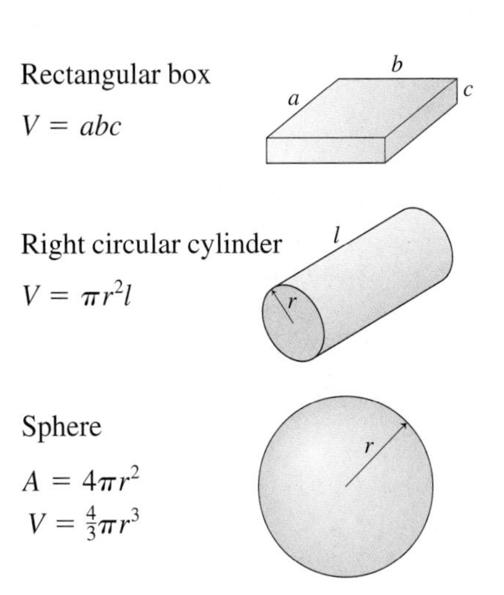

Arc length and angle: The angle θ in radians is defined as $\theta = s/r$.

The arc length that spans angle θ is $s = r\theta$.

$2\pi \text{ rad} = 360°$

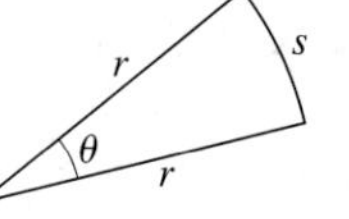

Right triangle: Pythagorean theorem $c = \sqrt{a^2 + b^2}$ or $a^2 + b^2 = c^2$

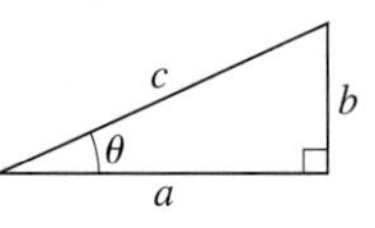

$$\sin\theta = \frac{b}{c} = \frac{\text{far side}}{\text{hypotenuse}} \qquad \theta = \sin^{-1}\left(\frac{b}{c}\right)$$

$$\cos\theta = \frac{a}{c} = \frac{\text{adjacent side}}{\text{hypotenuse}} \qquad \theta = \cos^{-1}\left(\frac{a}{c}\right)$$

$$\tan\theta = \frac{b}{a} = \frac{\text{far side}}{\text{adjacent side}} \qquad \theta = \tan^{-1}\left(\frac{b}{a}\right)$$

General triangle: $\alpha + \beta + \gamma = 180° = \pi \text{ rad}$

Law of cosines $c^2 = a^2 + b^2 - 2ab\cos\gamma$

Identities:

$$\tan\alpha = \frac{\sin\alpha}{\cos\alpha} \qquad \sin^2\alpha + \cos^2\alpha = 1$$

$$\sin(-\alpha) = -\sin\alpha \qquad \cos(-\alpha) = \cos\alpha$$

$$\sin(\alpha \pm \beta) = \sin\alpha\cos\beta \pm \cos\alpha\sin\beta \qquad \cos(\alpha \pm \beta) = \cos\alpha\cos\beta \mp \sin\alpha\sin\beta$$

$$\sin(2\alpha) = 2\sin\alpha\cos\alpha \qquad \cos(2\alpha) = \cos^2\alpha - \sin^2\alpha$$

$$\sin(\alpha \pm \pi/2) = \pm\cos\alpha \qquad \cos(\alpha \pm \pi/2) = \mp\sin\alpha$$

$$\sin(\alpha \pm \pi) = -\sin\alpha \qquad \cos(\alpha \pm \pi) = -\cos\alpha$$

Expansions and Approximations

Binomial expansion: $(1 + x)^n = 1 + nx + \dfrac{n(n-1)}{2}x^2 + \ldots$

Binomial approximation: $(1 + x)^n \approx 1 + nx$ if $x \ll 1$

Trigonometric expansions:

$$\sin\alpha = \alpha - \frac{\alpha^3}{3!} + \frac{\alpha^5}{5!} - \frac{\alpha^7}{7!} + \ldots \text{ for } \alpha \text{ in rad}$$

$$\cos\alpha = 1 - \frac{\alpha^2}{2!} + \frac{\alpha^4}{4!} - \frac{\alpha^6}{6!} + \ldots \text{ for } \alpha \text{ in rad}$$

Small-angle approximation: If $\alpha \ll 1$ rad, then $\sin\alpha \approx \tan\alpha \approx \alpha$ and $\cos\alpha \approx 1$.

The small-angle approximation is excellent for $\alpha < 5°$ (≈ 0.1 rad) and generally acceptable up to $\alpha \approx 10°$.

APPENDIX B

Periodic Table of Elements

Key: Atomic number 27, Symbol Co, Atomic mass 58.9

Period			Transition elements															
1	1 H 1.0																	2 He 4.0
2	3 Li 6.9	4 Be 9.0											5 B 10.8	6 C 12.0	7 N 14.0	8 O 16.0	9 F 19.0	10 Ne 20.2
3	11 Na 23.0	12 Mg 24.3											13 Al 27.0	14 Si 28.1	15 P 31.0	16 S 32.1	17 Cl 35.5	18 Ar 39.9
4	19 K 39.1	20 Ca 40.1	21 Sc 45.0	22 Ti 47.9	23 V 50.9	24 Cr 52.0	25 Mn 54.9	26 Fe 55.8	27 Co 58.9	28 Ni 58.7	29 Cu 63.5	30 Zn 65.4	31 Ga 69.7	32 Ge 72.6	33 As 74.9	34 Se 79.0	35 Br 79.9	36 Kr 83.8
5	37 Rb 85.5	38 Sr 87.6	39 Y 88.9	40 Zr 91.2	41 Nb 92.9	42 Mo 95.9	43 Tc 96.9	44 Ru 101.1	45 Rh 102.9	46 Pd 106.4	47 Ag 107.9	48 Cd 112.4	49 In 114.8	50 Sn 118.7	51 Sb 121.8	52 Te 127.6	53 I 126.9	54 Xe 131.3
6	55 Cs 132.9	56 Ba 137.3	57 La 138.9	72 Hf 178.5	73 Ta 180.9	74 W 183.9	75 Re 186.2	76 Os 190.2	77 Ir 192.2	78 Pt 195.1	79 Au 197.0	80 Hg 200.6	81 Tl 204.4	82 Pb 207.2	83 Bi 209.0	84 Po 209.0	85 At 210.0	86 Rn 222.0
7	87 Fr 223.0	88 Ra 226.0	89 Ac 227.0	104 Rf 261	105 Db 262	106 Sg 263	107 Bh 264	108 Hs 269	109 Mt 268	110 Ds 271	111 272	112 285						

	Inner transition elements													
Lanthanides 6	58 Ce 140.1	59 Pr 140.9	60 Nd 144.2	61 Pm 144.9	62 Sm 150.4	63 Eu 152.0	64 Gd 157.3	65 Tb 158.9	66 Dy 162.5	67 Ho 164.9	68 Er 167.3	69 Tm 168.9	70 Yb 173.0	71 Lu 175.0
Actinides 7	90 Th 232.0	91 Pa 231.0	92 U 238.0	93 Np 237.0	94 Pu 239.1	95 Am 241.1	96 Cm 244.1	97 Bk 249.1	98 Cf 252.1	99 Es 257.1	100 Fm 257.1	101 Md 258.1	102 No 259.1	103 Lr 262.1

Answers

Answers to Odd-Numbered Exercises and Problems

Solutions to questions posed in the Part Overview captions can be found at the end of this answer list.

Chapter 1

1. 7. 9. a. b. Greater 11. 13. 15.

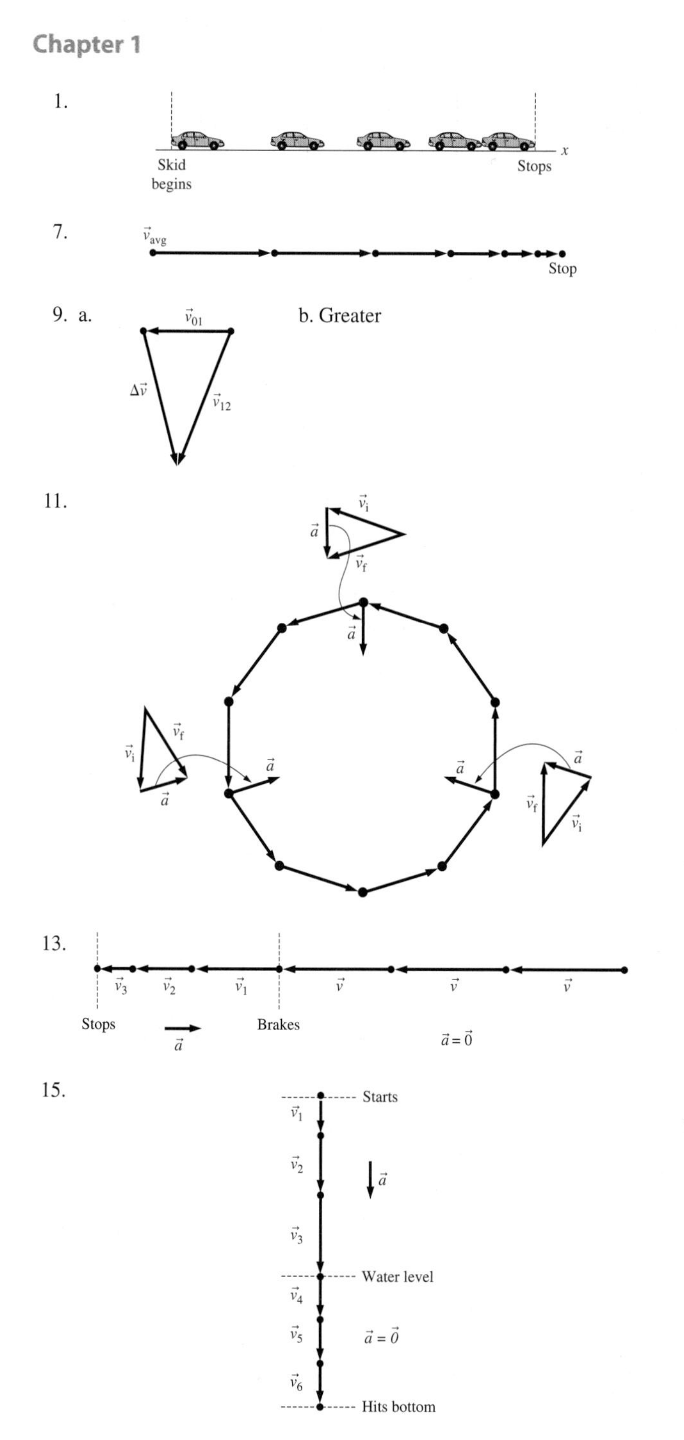

17.

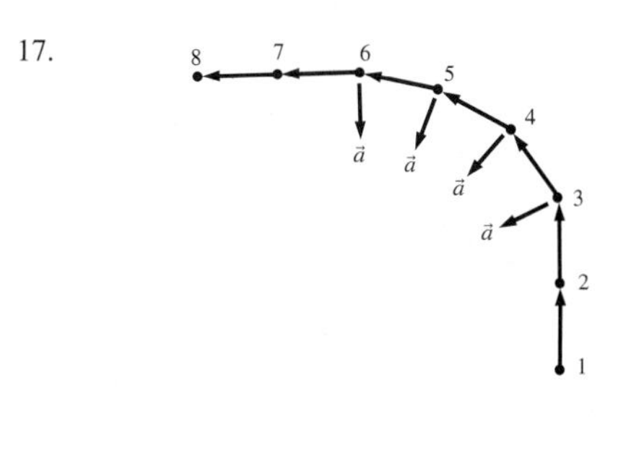

19.

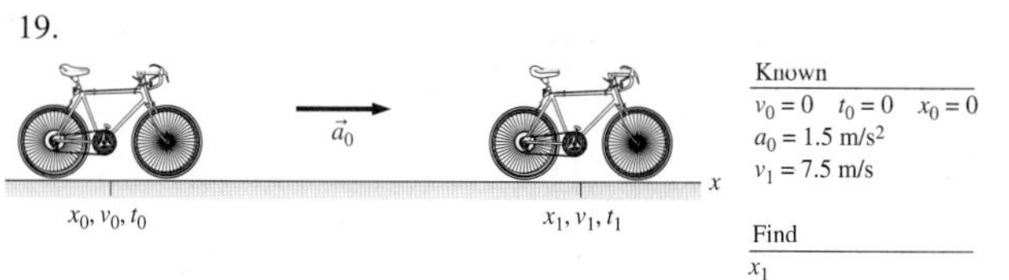

21. a. 9.12×10^{-6} s b. 3.42×10^{3} m c. 440 m/s d. 22.2 m/s
23. a. 3.60×10^{3} s b. 8.64×10^{4} s c. 3.16×10^{7} s d. 9.75 m/s²
25. 6.40×10^{3} m² and 8.25×10^{3} m²
27. a. 3 b. 3 c. 3 d. 2
29. a. 846 b. 7.9 c. 5.77 d. 13.1
35.

Pictorial representation

$\vec{a}_0$ $\vec{a}_1 = \vec{0}$

0 x_0, v_0, t_0 x_1, v_1, t_1 x_2, v_2, t_2

Known: $x_0 = 0$, $a_0 = 5$ m/s², $v_0 = 0$, $t_1 = 5$ s, $t_0 = 0$, $t_2 = 8$ s, $v_2 = v_1$, $a_1 = 0$

Find: x_2

Motion diagram

Accelerate Coast

Starts $\vec{a}$ $\vec{a} = \vec{0}$

37.

Pictorial representation

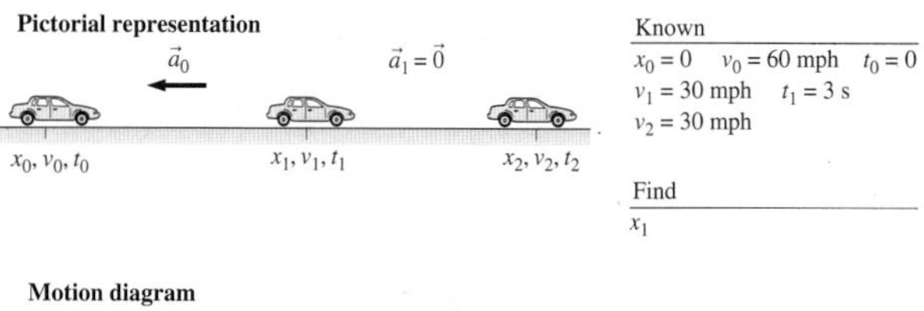

Motion diagram

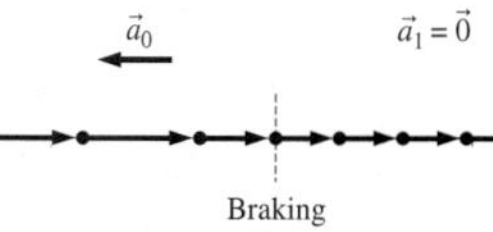

39.

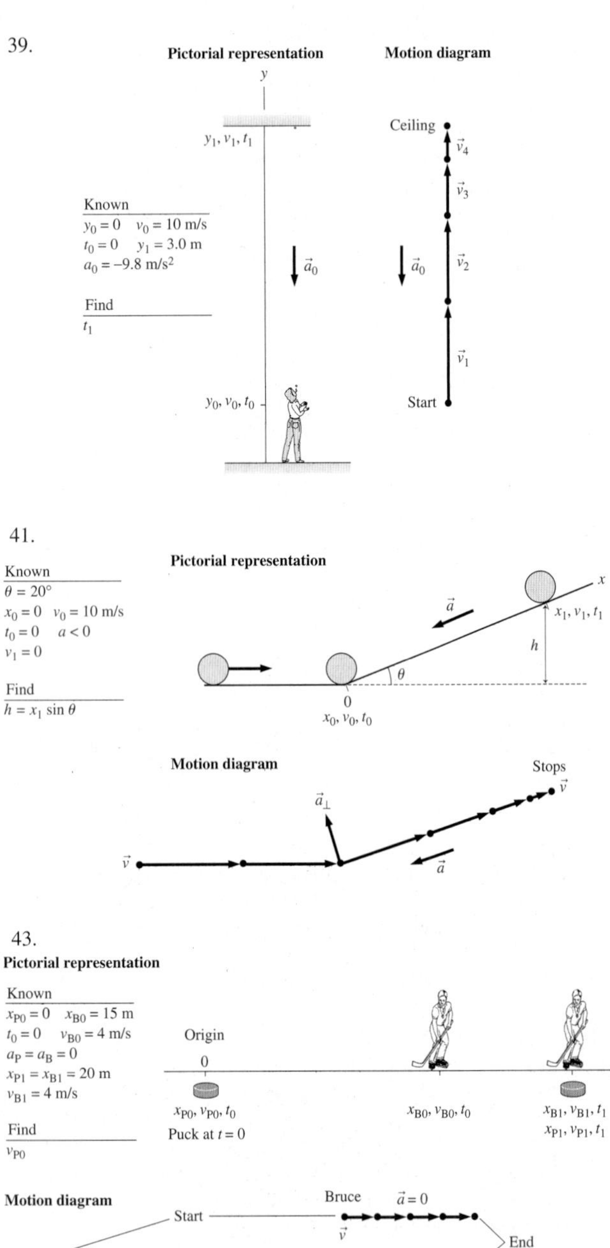

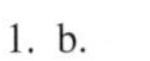

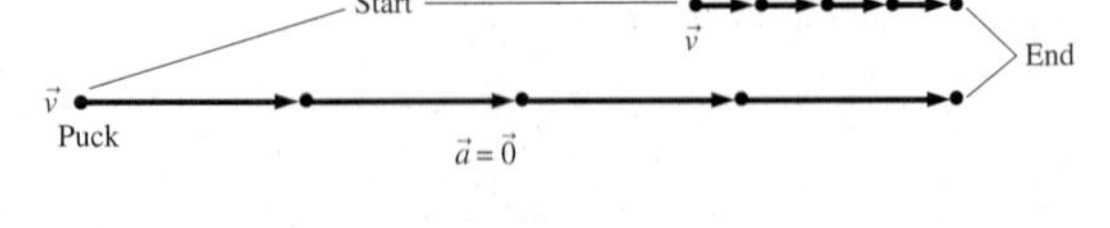

51. a.

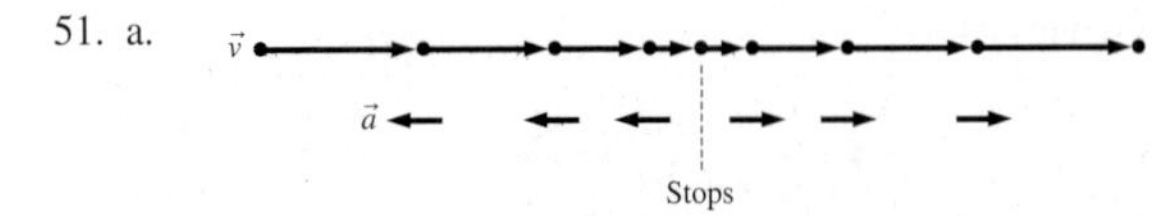

53. a.

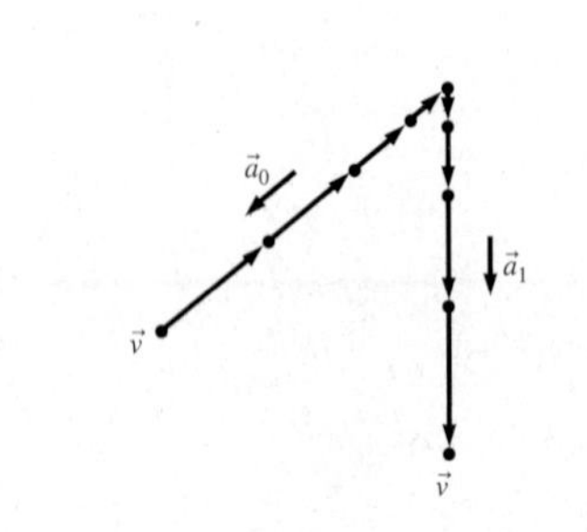

55. a.

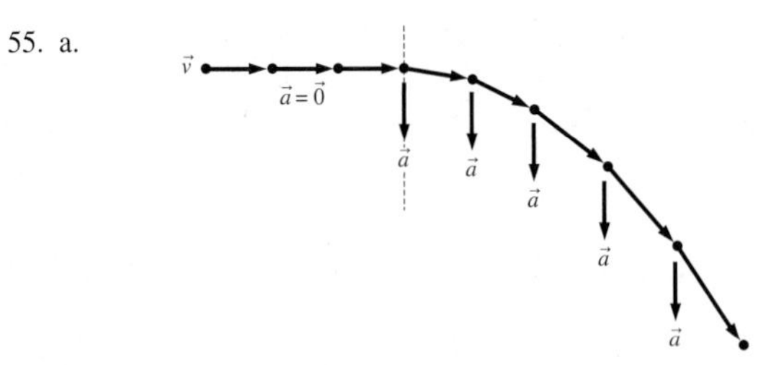

57. a. Neither is zero. b. Velocity is zero, acceleration is not zero.

Chapter 2

1. b.

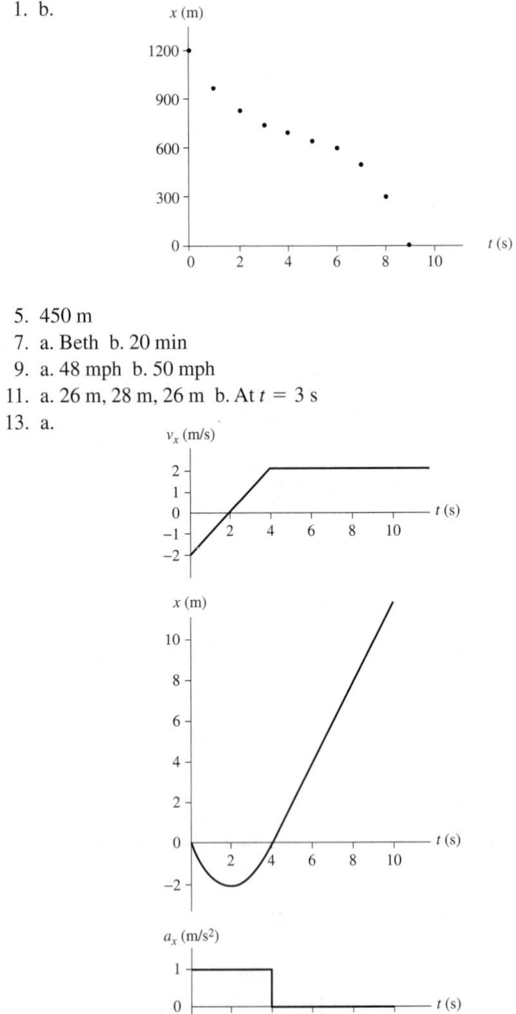

5. 450 m
7. a. Beth b. 20 min
9. a. 48 mph b. 50 mph
11. a. 26 m, 28 m, 26 m b. At $t = 3$ s
13. a.

b. 1 m/s²
15. a. 2.68 m/s² b. 27.3% c. 134 m or 440 ft
17. -2.8 m/s²
19. a. 78.4 m b. -39.2 m/s
21. 3.2 s
23. 134 m
25. a. 15 m b. 23 m/s c. 24 m/s²
27. 16 m/s

29. a.

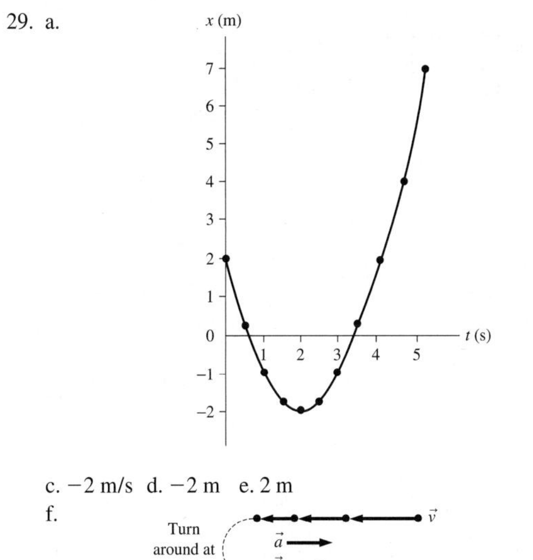

c. -2 m/s d. -2 m e. 2 m

f.

Turn around at $t = 2.0$ s

31. $v_A = -10$ m/s, $v_B = -20$ m/s $v_C = 75$ m/s
33. 0, 5, 20, 30, and 30 m/s
35. a.

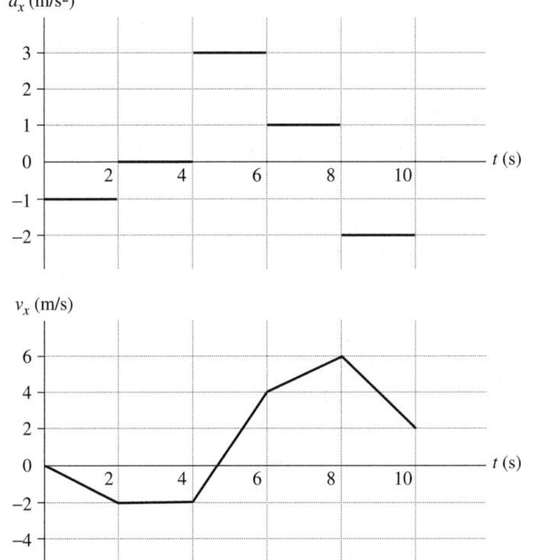

b. Displaced upward by 2.0 m/s

37. a. 0 s and 3 s b. 12 m and -18 m/s^2; -15 m and 18 m/s^2
39. 2.0 m/s^3
41.

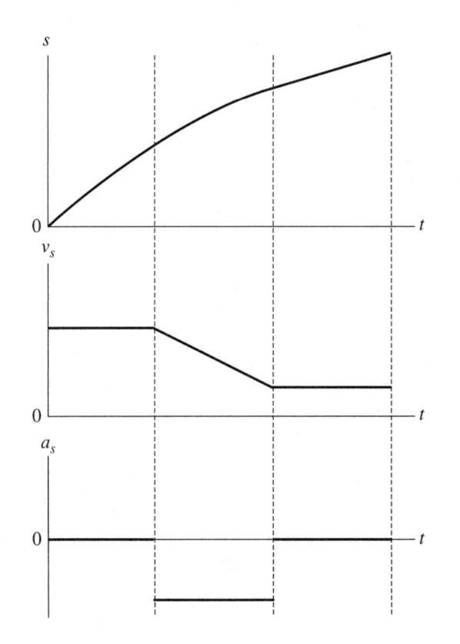

43.

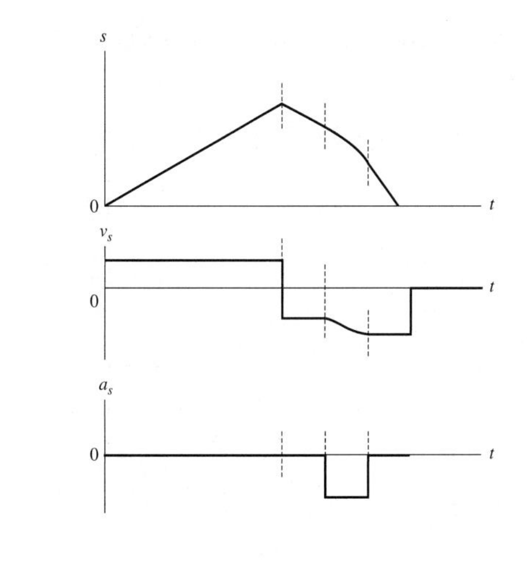

45.

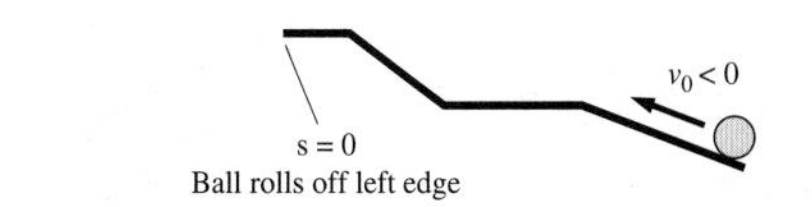

47. a. 179 mph b. Yes c. 35 s d. No
49. Yes
51. a. 100 m

b.

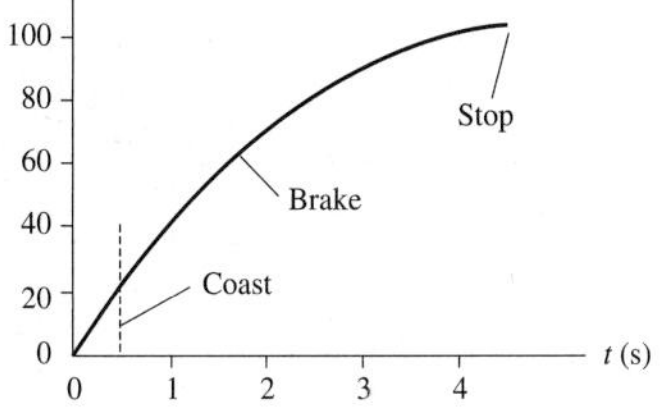

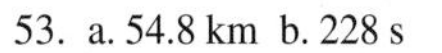

53. a. 54.8 km b. 228 s

c.

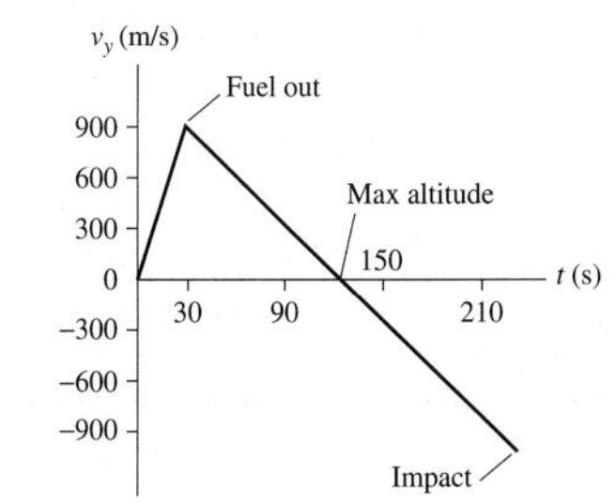

55. 19.7 m
57. 216 m
59. 9.9 m/s
61. a. 2.32 m/s b. 5.00 m/s c. 0%
63. Yes
65. a. 214 km/hr b. 16%
67. 14 m/s
69. a. 900 m b. 60 m/s
71. No
73. 5.5 m/s^2
75. c. 17.2 m/s
77. c. $x_1 = 250$ m, $x_2 = 750$ m
79. 70 m/s
81. a. 10.0 s b. 3.83 m/s^2 c. 6.4%
83. -4500 m/s^2

Chapter 3

1. a. Yes b. No
3. a. If $\vec{B}$ is in the same direction as $\vec{A}$. b. If $\vec{B}$ is opposite to $\vec{A}$.
7. 11.9 m/s
9. a. (70.7, −70.7) m b. (282, 103) m/s c. (0, −5.0) m/s^2 d. (−40, 30) N
11. $\vec{C}$: (−3.04, 0.815) m; $\vec{D}$: (12.8, −22.2)
13. a. 7.21, 56.3° below +x-axis b. 94.3 m, 58.0° above the +x-axis c. 44.7 m/s, 63.4° above the −x-axis d. 6.3 m/s^2, 18.4° right of the −y-axis
15.

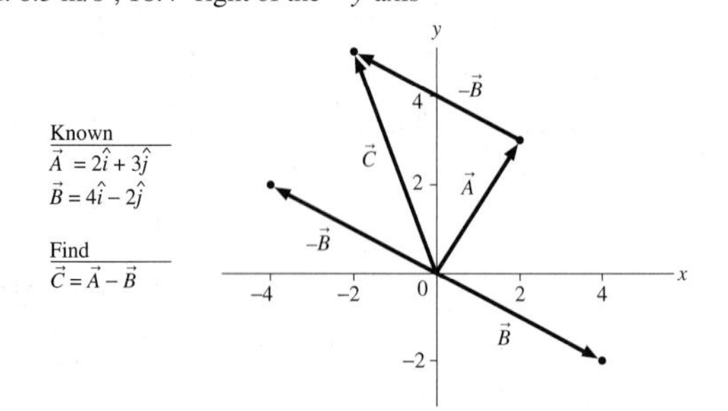

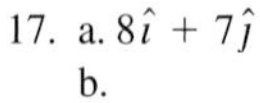

17. a. $8\hat{\imath} + 7\hat{\jmath}$
b.

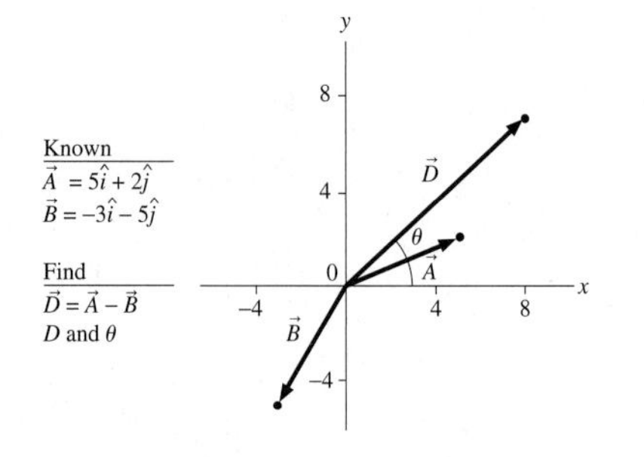

c. 10.6, 41.2° above the +x-axis

19. a. $17\hat{\imath} + 22\hat{\jmath}$
b.

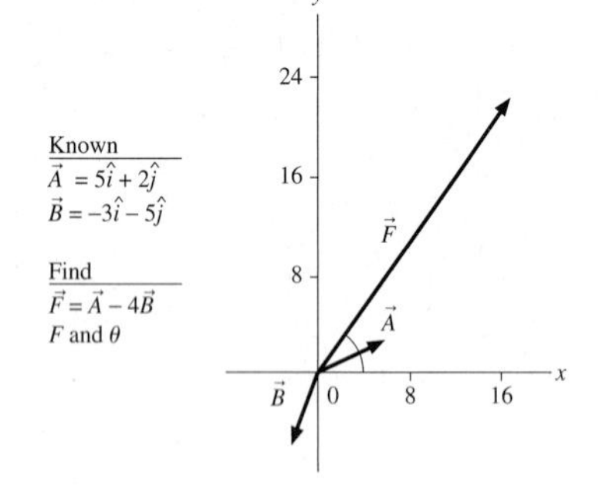

c. 27.8, 52.3° above the +x-axis

21. Coordinate system 1: $\vec{A} = -4\hat{\jmath}$ m, $\vec{B} = (-4.33\hat{\imath} + 2.50\hat{\jmath})$ m; Coordinate system 2: $\vec{A} = (-2.00\hat{\jmath} - 3.46\hat{\jmath})$ m, $\vec{B} = (-2.50\hat{\imath} + 4.33\hat{\jmath})$ m
23. a.

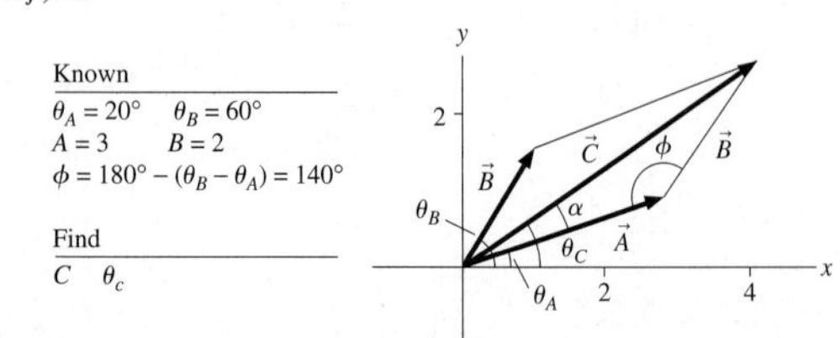

Known
$\theta_A = 20°$ $\theta_B = 60°$
$A = 3$ $B = 2$
$\phi = 180° - (\theta_B - \theta_A) = 140°$
Find
C θ_c

b. 4.71, 35.8° above the +x-axis c. 4.71, 35.8° above the +x-axis

25. a. $-6\hat{\imath} + 2\hat{\jmath}$ b. 6.32, 18.4° above the −x-axis
27. $(4.90\hat{\imath} + 2.83\hat{\jmath})$ m
29. a. b. c.
31. $0.707\hat{\imath} + 0.707\hat{\jmath}$
33. a. 100 m lower b. (500 m, east) + (5000 m, north) − (100 m, vertical)
35. a.

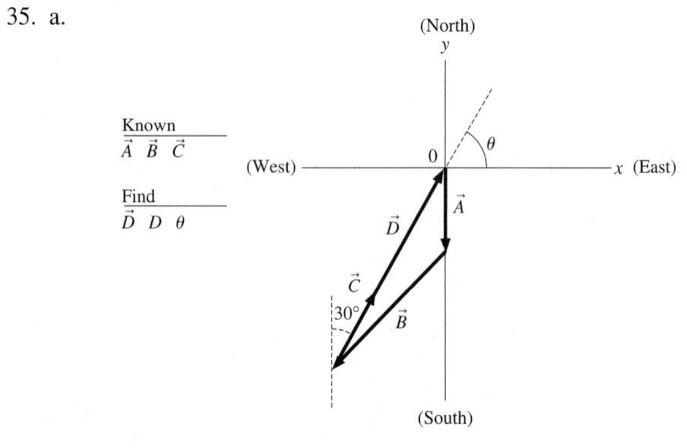

b. 360 m, 59.4° north of east

37. 7.5 m
39. 86.6 m/s
41. 385 paces, 24.6° west of north
43. −15.0 m/s
45. a. −3.4 m/s b. −9.4 m/s
47. 4.36 units, 83.4° below the −x-axis
49. 7.29 N, 79.2° below the −x-axis

Chapter 4

3.

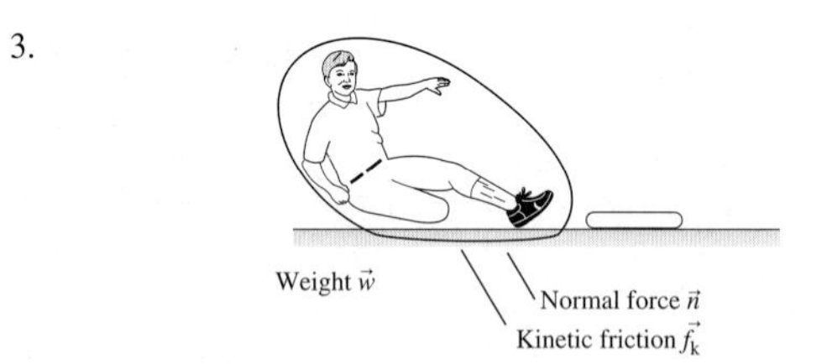

5. $m_1 = 0.08$ kg; $m_3 = 0.50$ kg
9. 0.25 kg
11. a. ≈0.05 N b. ≈100 N
13.

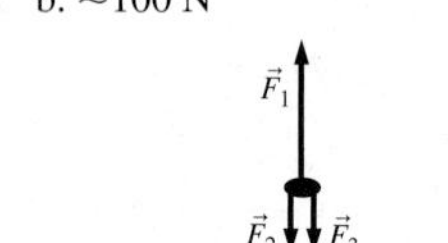

19. Force identification

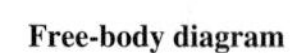

Normal force $\vec{n}$ Weight $\vec{w}$

21\. 23. 25. 27.

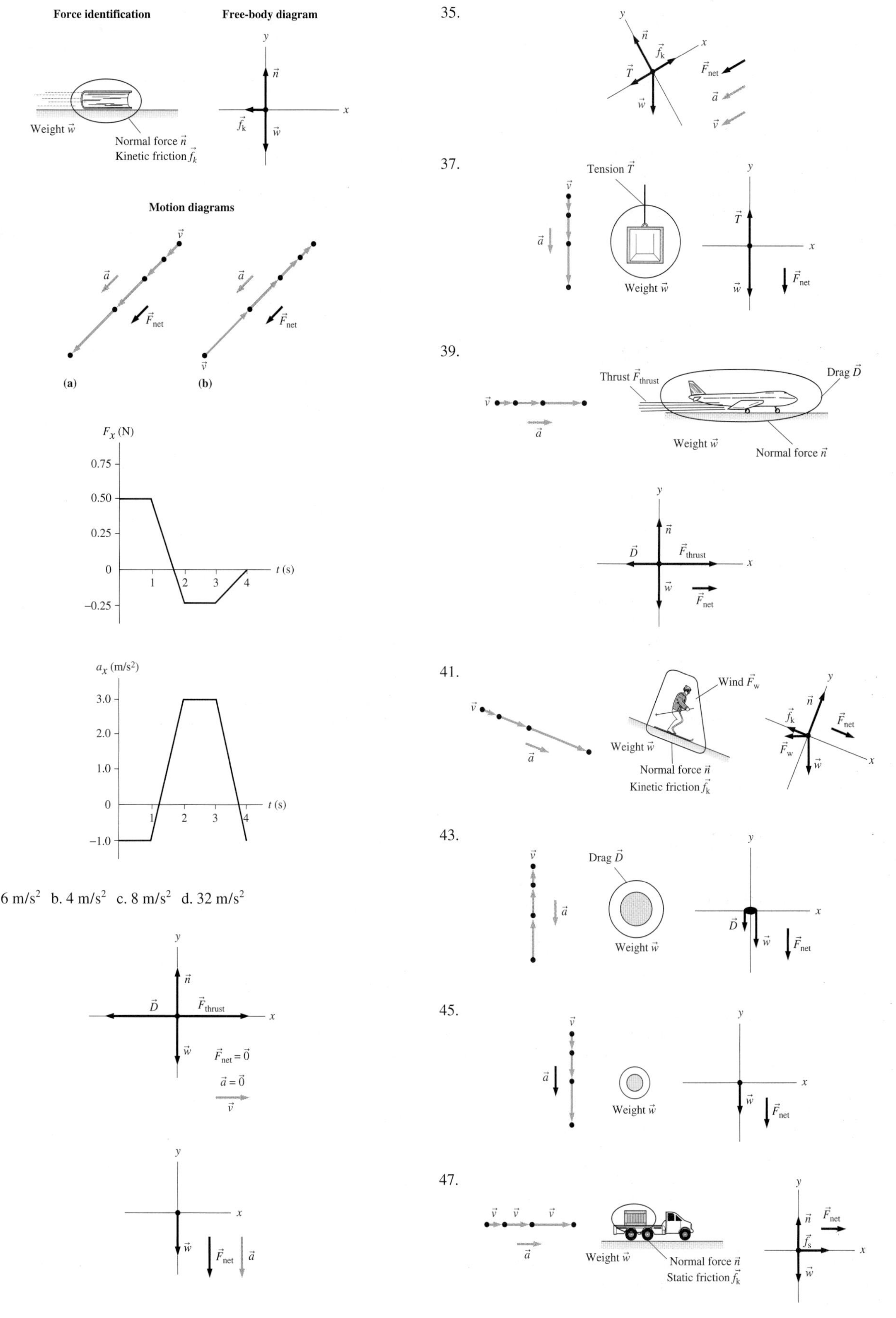

29\. a. 16 m/s^2 b. 4 m/s^2 c. 8 m/s^2 d. 32 m/s^2

31\. 33. 35. 37. 39. 41. 43. 45. 47.

49. a.

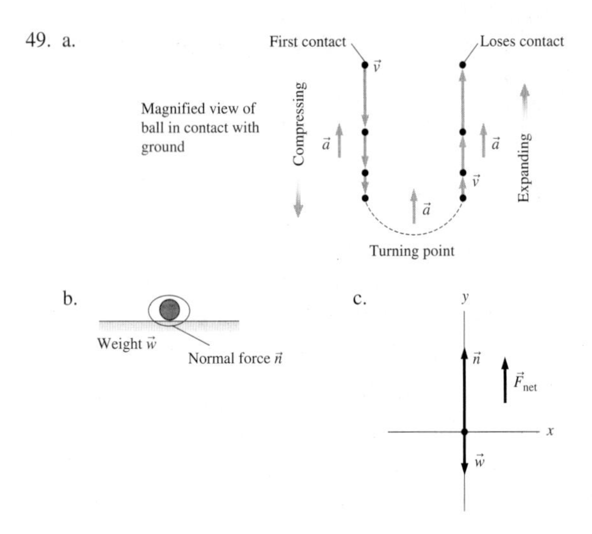

Chapter 5

1. $T_1 = 86.7$ N, $T_2 = 50.0$ N
3. 147 N
5. a. $a_x = 1.0$ m/s^2, $a_y = 0$ m/s^2 b. $a_x = 1.0$ m/s^2, $a_y = 0$ m/s^2
7. 8 N, 0 N, −12 N
9. a. 0 N b. 0 N c. 250 N
11. 307 N
13. a. 590 N b. 740 N c. 590 N
15. 0.25
17. 136 m
19. 2550 m
21. 192 m/s
23. ≈3 m/s^2
25. 4.0 m/s
27. Left first, then right.
29. a. 0.0036 N b. 0.0104 N
31. a. 784 N b. 1050 N
33. a. 58.8 N b. 67.8° c. 79.0 N
35. a. 6670 N b. 600 μs
    c.

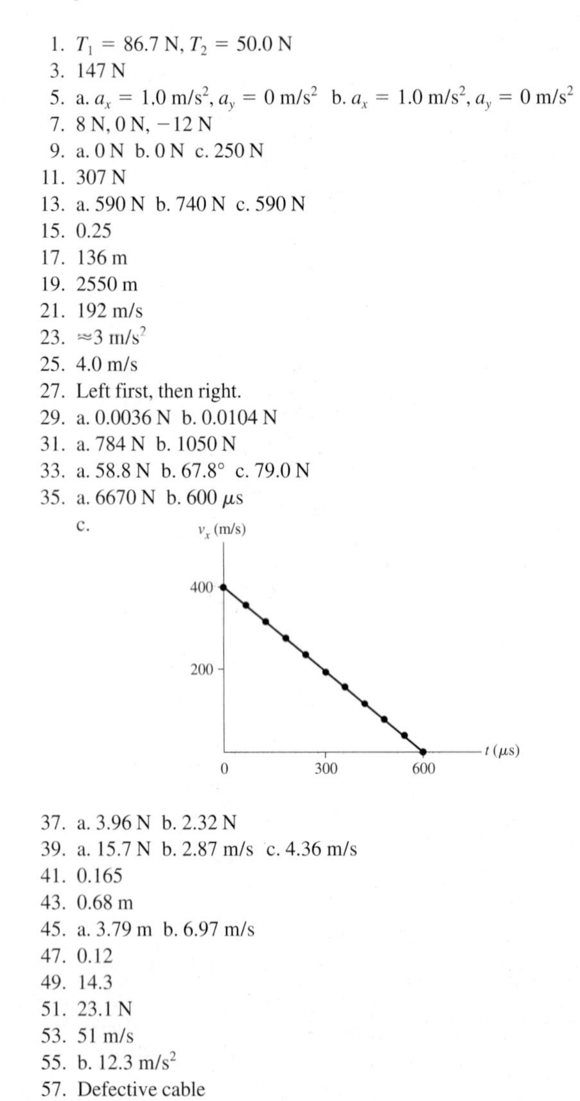

37. a. 3.96 N b. 2.32 N
39. a. 15.7 N b. 2.87 m/s c. 4.36 m/s
41. 0.165
43. 0.68 m
45. a. 3.79 m b. 6.97 m/s
47. 0.12
49. 14.3
51. 23.1 N
53. 51 m/s
55. b. 12.3 m/s^2
57. Defective cable
59. 13.0 m/s^2
67. $T = 144$ N
69. $\theta = 11.3°$
71. Green
73. b. 134 s and 402 s c. No

Chapter 6

1. a. $(2\hat{\imath} - 2\hat{\jmath})$ m/s^2 b. $(22\hat{\imath} - 16\hat{\jmath})$ m, $(9\hat{\imath} - 9\hat{\jmath})$ m/s, 12.7 m/s
3. a.

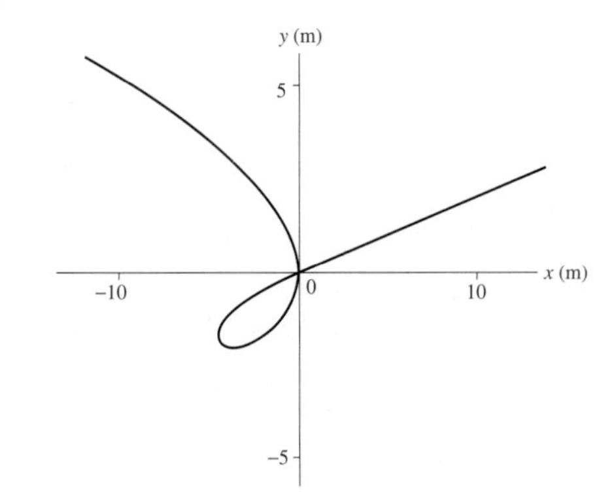

   b. $\vec{0}$ m, 2.0 m/s at $t = 0$ s; $\vec{0}$ m, 8.3 m/s at $t = 4$ s
   c. −90° at $t = 0$ s; 14° at $t = 4$ s
5. 38.8 m
9. a. At $t = 6$ s, $x = 240$ m, $y = 3.6$ m, $v_x = 40$ m/s, $v_y = -28.8$ m/s, $v = 49.3$ m/s
   b.

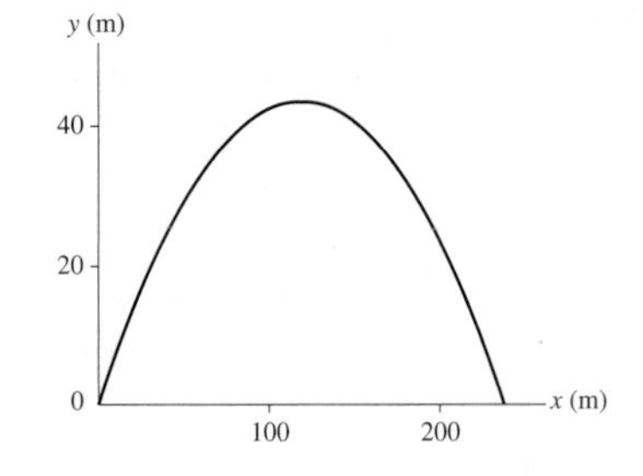

11. a. 0.0639 s b. 782 m/s
13. 2.0 km/hr
15. 0.40 m
17. a. 39.1 mi b. 19.5 mph
19. a. 55.6 hr b. 0.0917° c. Yes
21. 6.56×10^{12} m/s^2
23. a. $v_0^2 \sin^2\theta / 2g$ b. $h = 14.4$ m, 28.8 m, 43.2 m; $d = 99.8$ m, 115.2 m, 99.8 m
25. a. Launch point 80.8 m higher b. 34.4 m
    c. 49.8 m/s, 72.5° below horizontal
27. a. 276 m b. 12.75 s
29. Clears by 1.01 m
31. No
33. 34.3°
35. 678 m
37. a. 239 m b. 42.9 m
39. 106 m/s
41. 4.48 m/s
43. 105 m
45. Crocodile food.
47. 2.96 m
49. b. $\theta = 11.5°$
51. b. $x_1 = -29.2$ m

53. a. On the opposite bank 150 m east of where she started.
b.

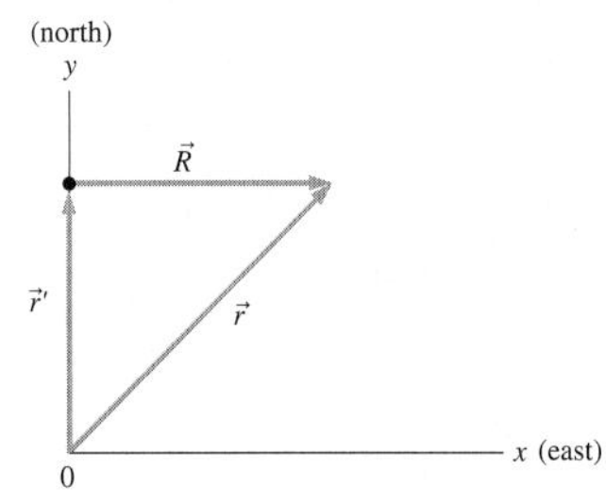

55. a. 44.4° above the $-x'$-axis
b.

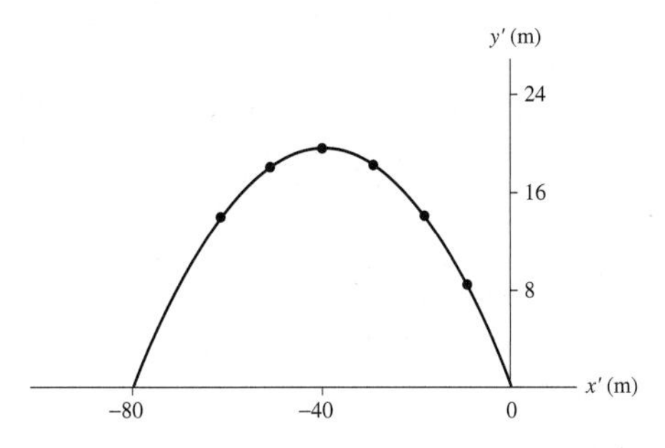

57. 10.1°
59. a. 30° toward the rear of the car b. 17.3 m/s
61. a. 7.18° south of east b. 2.48 hr
63. 3.0×10^8 m/s
65. 40.6° below horizontal
67. 4.78 m/s
69. a. Rotate the spacecraft 153.4° counterclockwise so that the exhaust is 26.6° below the positive x-axis. Fire with a thrust of 103,300 N for 433 s.
b.

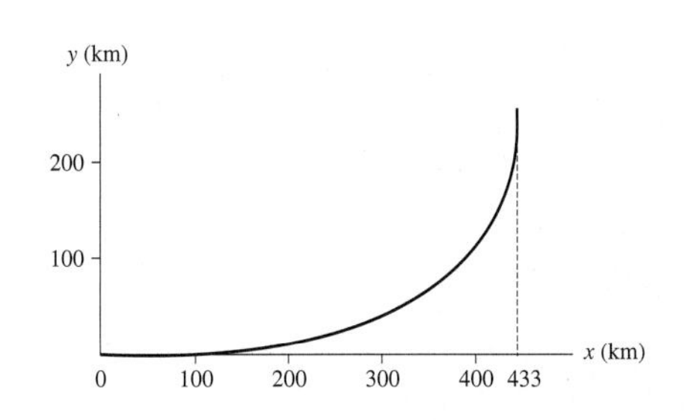

Chapter 7

1. b.

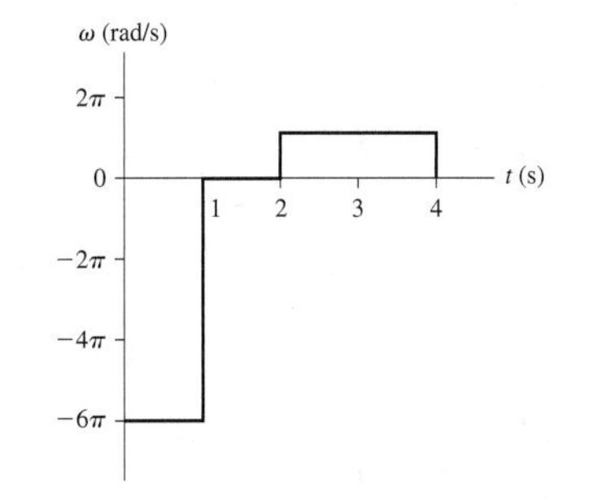

3. a. 1.5π rad/s b. 1.33 s
5. a. 3.0×10^4 m/s b. 2.0×10^{-7} rad/s c. 6.0×10^{-3} m/s^2
7. 5.65 m/s and 106 m/s^2
9. 34.3 m/s
11. a. 3.93 m/s b. 6.18 N
13. 7.27°
15. 2.0×10^{20} N
17. 1.58 m/s^2
19. 12.1 m/s
21. 19.8 m/s
23. $a_r = 2.72$ m/s^2; $a_t = 1.27$ m/s^2
25. a. -2.618 m/s^2 b. 31.25 rev
27. 49.5°
29. a. 0.967 m/s^2 b. 14.3g
31. 2.5 N higher at the north pole.
33. 172 N
35. 34.5 m/s
37. No
39. a. 5.00 N b. 30.2 rpm
41. 24.4 rpm
43. a. -9.80 m/s^2 b. -12.92 m/s^2 c. -6.68 m/s^2
45. a. 4.9 N b. 2.9 N c. 32.5 N
47. a. 319 N and 1397 N b. 5.68 s
49. 29.9 rpm
51. 2.63 m right of the point where the string was cut.
53. a. 1.90 m/s^2 at 20.6° from the r-axis b. 23.5 s
55. 3.75 rev
57. b. $\omega = 20$ rad/s
59. b. $\omega_f = 0.40$ rad/s
61. 14.19 N and 8.31 N
63. c. 94.5 rpm

Chapter 8

1.

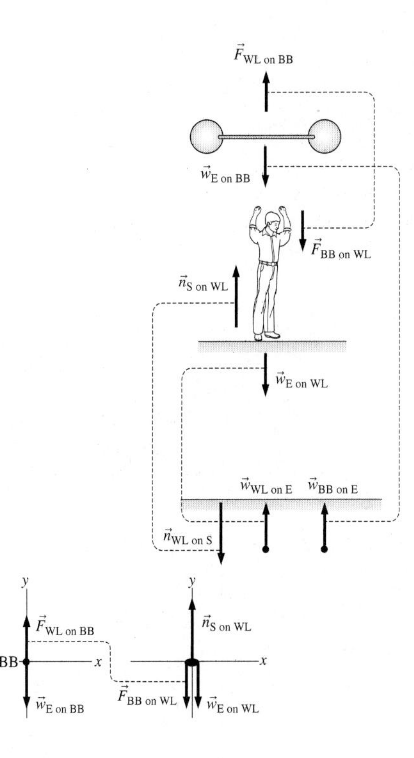

3.

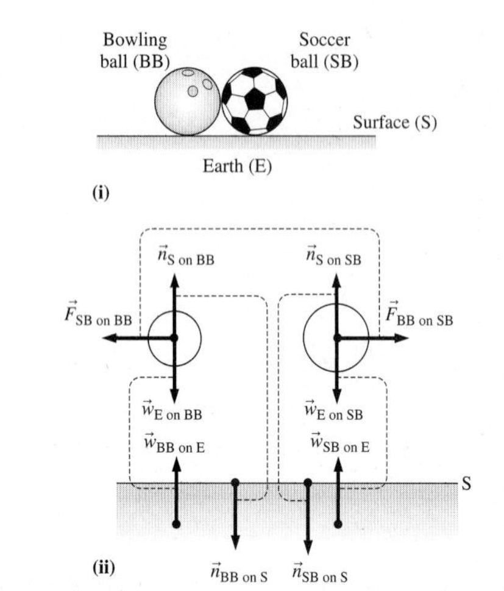

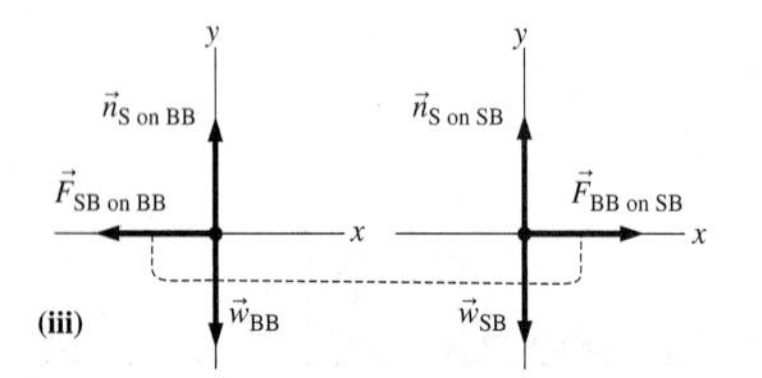

5. a.

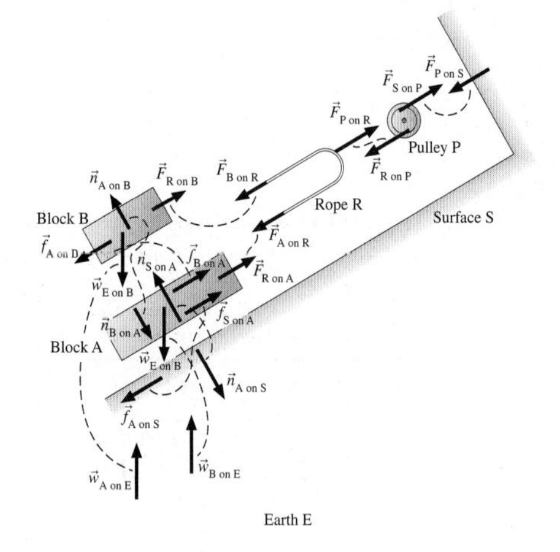

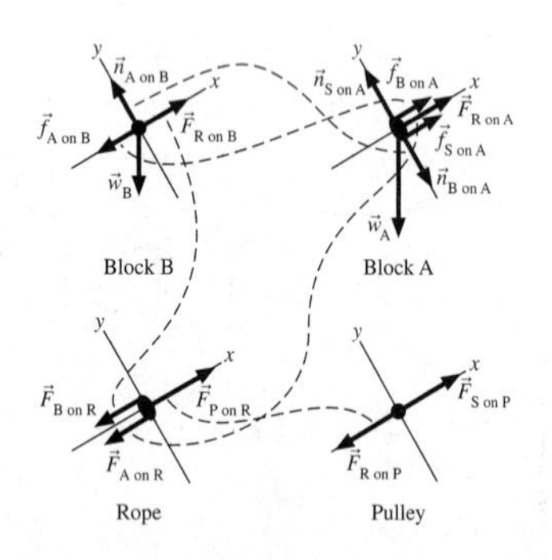

7. a. 784 N b. 1580 N
9. $F_{2\text{ on }3} = 6$ N; $F_{2\text{ on }1} = 10$ N

11. 588 N
13. a. 20 N b. 21 N
15. 66.6 N at 36°
17. 60 N
19.

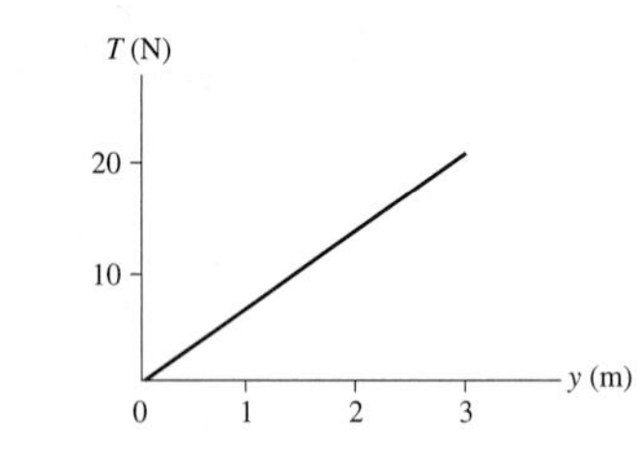

21. No
23. 99.0 m
25. 1.48 s
27. a. 3.92 N b. 2.16 m/s^2
29. 171.2 N
31. 200 kg
33. 98.9 kg
35. 2.29 m/s^2
37. a. 19.6 N b. Down c. 20.6 N
39. a. 3770 N b. 28.2 m/s
41. 3.27 m/s^2
43. 3590 N
47. a. 1.0 m/s b. 90 N
49. b. 8.99 N

Chapter 9

1. a. 1.5×10^4 kgm/s b. 8.0 kgm/s
3. 1500 N
5. 5.0 Ns
7. a. 1.5 m/s to the right b. 0.5 m/s to the right
9. 0.50 s
11. 0.20 s
13. 1.43 m/s
15. 0.20 m/s
17. 3.6 m/s
19. 2.0 mph
21. 1.7 m/s 45° north of east
23. 2.89×10^{34} kgm^2/s
25. a. (1.083, 0.625) kgm/s when thrown, (1.083, 0) kgm/s at the top
27. a. 6.4 m/s b. 360 N
29. a. 0.432 Ns upward c. 40 to 80 N is reasonable estimate
31. a. 0.588 Ns b. −0.588 Ns
33. a. 6.7×10^{-8} m/s b. 2×10^{-10}%
35. 13.3 s
37. 1.73 m/s at 54.7° south of east
39. 7.57 cm in the direction Brutus was running
41. 402 m
43. a. 286 μs, 26,200 N b. 0.0214 m/s
45. 27.8 m/s
47. 5 s^{-1}
49. 1.46×10^7 m/s in the forward direction
51. 14 u
53. b. and c. 1.40×10^{-22} kgm/s in the direction of the electron
55. 0.850 m/s, 72.5° below the x-axis
57. 1.97×10^3 m/s
59. 4.5 rpm

61. a. 2.80 m/s b. 2.22 m/s at a radius of 25.2 cm
63. c. $(v_{ix})_2 = 6.0$ m/s
65. c. $(v_{fx})_1 = -12$ m/s
67. 5.65 m/s
69. 90.3 m/s
71. 8

Chapter 10

1. The bullet
3. 112 km/hr
5. a. 6.75×10^5 J b. 45.9 m c. No
7. a. 12.9 m/s b. 14.0 m/s
9. 7.67 m/s
11. a. 1.403 m/s b. 30°
13. a. Yes b. 14.1 m/s
15. a. 49 N b. 1450 N/m c. 3.4 cm
17. 98 N/m
19. 10 J
21. 2.00 m/s
23. 3.00 m/s
25. 0.857 m/s and 2.86 m/s
27. a. −5.0 m/s and 5.0 m/s b. Both 2.5 m/s
29. a. Right b. 20.0 m/s at $x = 2.0$ m c. 1.0 and 6.0 m
31. 63.2 m/s
33. Yes
35. a. No b. 17.3 m/s
37. a. Left b. Yes c. 200 N/m d. 19.0 m/s
39. $v_0/\sqrt{2}$
41. 25.8 cm
43. 51.0 cm
45. 19.6 N/m
47. a. 14.8 m/s b. Go hungry.
49. a. 21,600 N/m b. 18.6 m/s
51. a. 3.33 m/s b. 11.8 cm c. 0.833 m/s and 6.45 cm
53. a. $\frac{3}{2}R$ b. 15 m
55. $2.5R$
57. 7.94 m/s
59. 100 g ball 0.80 m/s to the left; 400 g ball 2.2 m/s to the right
61. a. Vibrates about an equilibrium position on one side of the H_3 plane or the other.
    b. Oscillates from one side of the H_3 plane to the other side.
65. c. $k = 35.6$ N/m
67. $v_f = 2.65$ m/s
69. a. 1.46 m b. 19.6 cm
71. b. 453 m/s
73. 100 g ball to 79.3°, 200 g ball to 14.7°

Chapter 11

1. a. 15.3 b. −4.0 c. 0
3. a. −30 b. 0
5. a. 12.0 J b. −6.0 J
7. 0 J
9. 12,500 J by the weight, −7920 J by $\vec{T}_1$, −4580 J by $\vec{T}_2$
11. 4.0 J, −4.0 J, 4.0 J, 0 J, −3.0 J
13. 7.35 m/s, 9.17 m/s, 9.70 m/s
15. 8.0 N
17. −30 N at 1 m, 20 N at 3 m

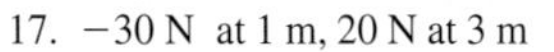

19. a.

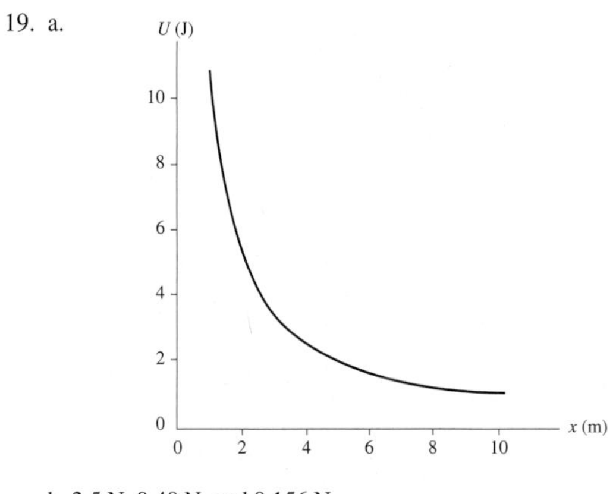

b. 2.5 N, 0.40 N, and 0.156 N

21. 1360 m/s
23. a. Potential energy is transformed to kinetic and thermal energy.
    b. 548 J
25.

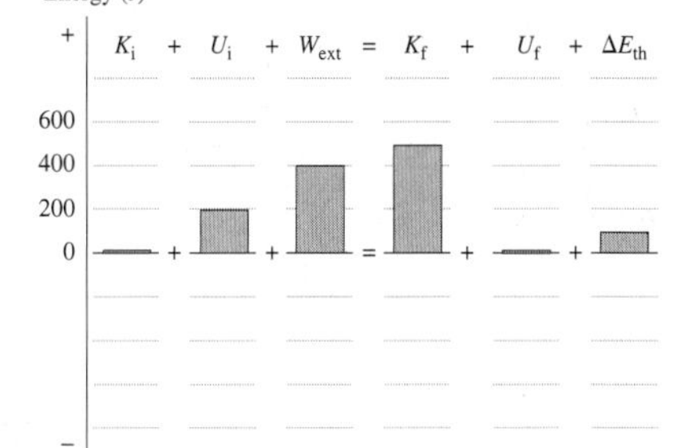

27. 6.26 m/s
29. a. 176.4 J b. 58.8 J
31. Night light
33. a. 102 N b. 416 W, 832 W, and 1248 W
35. a.

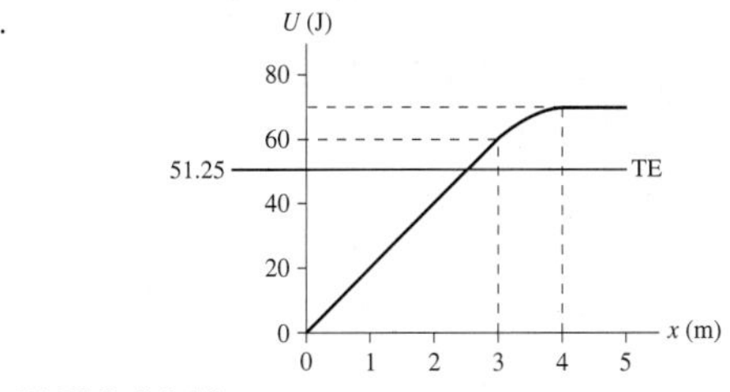

b. 51.25 J d. 2.56 m

37.

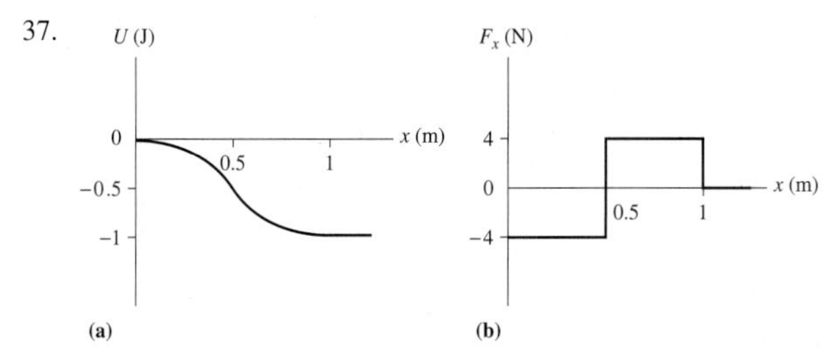

39. a. −98,000 J b. 108,000 J c. 10,000 J d. 4.47 m/s
41. a. and b. 3.97 m/s
43. 2.37 m/s
45. 0.037
47. a. 1.70 m/s b. No
49. a. 571 J b. −196 J c. −38.5 J d. 0 J
51. a. 2.16 m/s b. 0.0058
53. a. 9.90 m/s b. 9.39 m/s c. 93.9 cm d. 10
55. a. $\sqrt{2gh}$ b. $h - \mu_k L$

57. a. N/m^3
b.

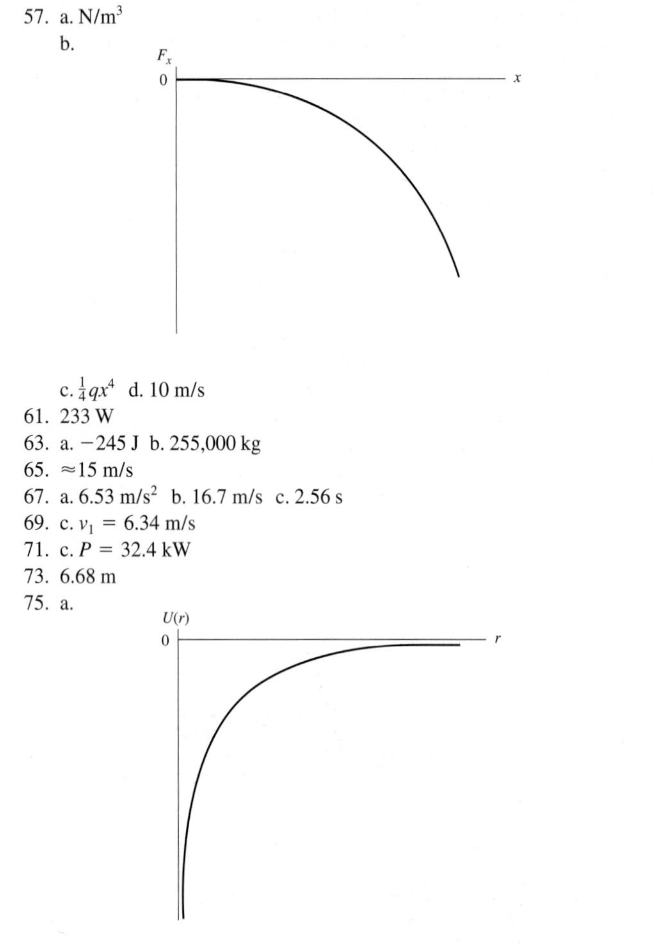

c. $\frac{1}{4}qx^4$ d. 10 m/s
61. 233 W
63. a. −245 J b. 255,000 kg
65. ≈15 m/s
67. a. 6.53 m/s^2 b. 16.7 m/s c. 2.56 s
69. c. $v_1 = 6.34$ m/s
71. c. $P = 32.4$ kW
73. 6.68 m
75. a.

b. Infinity c. 689,000 m/s and 172,000 m/s

Chapter 12

1. a. 3.53×10^{22} N b. 1.99×10^{20} N c. 0.56%
3. 6.00×10^{-4}
5. 1.60×10^{-7}
7. a. 8.97 N
9. a. 1.62 m/s^2 b. 25.9 m/s^2
11. 2430 m
13. a. 3.0×10^{24} kg b. 0.889 m/s^2
15. 418 km
17. 60.2 km/s
19. a. 1.80×10^7 m b. 9410 m/s
21. a. 7680 m/s b. 92.4 min
23. 4.2 hr
25. 2.01×10^{30} kg
27. 6.72×10^8 J
29. 46 kg and 104 kg
31. 1.19×10^{-3} rad or 0.0679°
33. (11.7 cm, 0 cm)
35. a. (4.72×10^{-7} N, 45° ccw from −y-axis)
b. (4.56×10^{-7} N, 7.6° cw from y-axis)
37. a. -10.0×10^{-8} J b. -9.65×10^{-8} J
39. a. 1.38×10^7 m b. 4450 m/s
41. a. 2.82×10^6 m b. 3670 m/s
43. 32,600 m/s
45. a. $\sqrt{16GM/7R}$ b. $\sqrt{4GM/3R}$
47. a. 3.46×10^8
b.

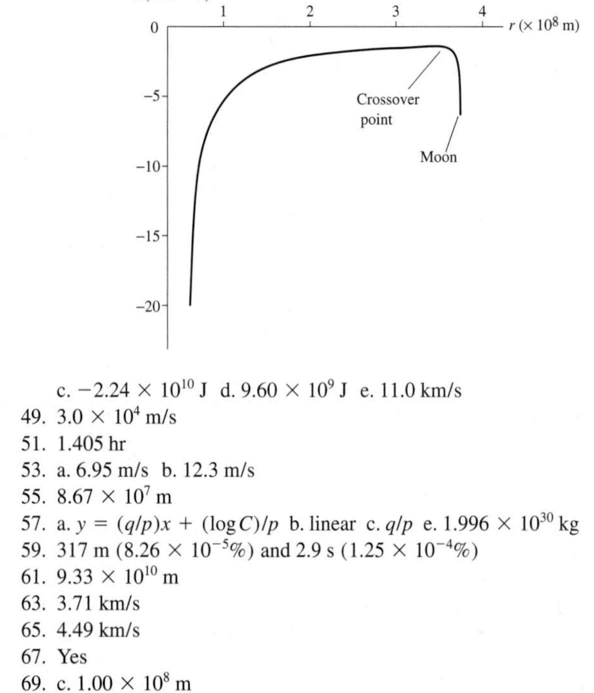

c. -2.24×10^{10} J d. 9.60×10^9 J e. 11.0 km/s
49. 3.0×10^4 m/s
51. 1.405 hr
53. a. 6.95 m/s b. 12.3 m/s
55. 8.67×10^7 m
57. a. $y = (q/p)x + (\log C)/p$ b. linear c. q/p e. 1.996×10^{30} kg
59. 317 m (8.26×10^{-5}%) and 2.9 s (1.25×10^{-4}%)
61. 9.33×10^{10} m
63. 3.71 km/s
65. 4.49 km/s
67. Yes
69. c. 1.00×10^8 m
71. $v_{f1} = 596$ m/s, $v_{f2} = 298$ m/s
73. a. 2.05×10^8 yr d. 9.4×10^{10}
75. Crash.

Chapter 13

1.

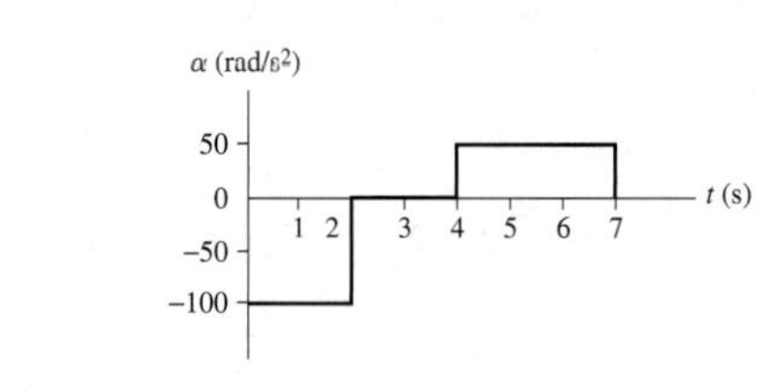

3. a.

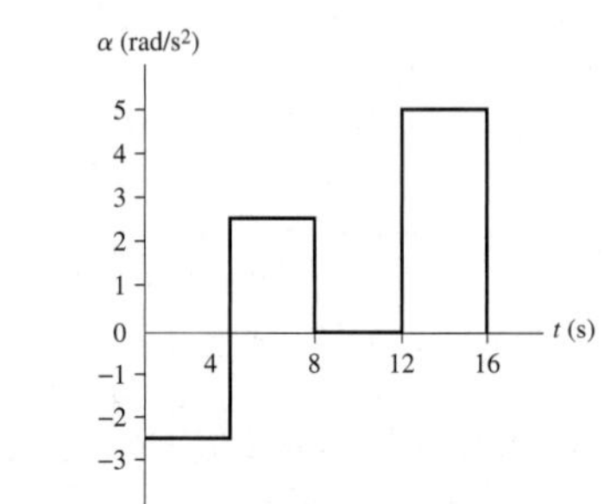

5. 13.2 m/s
7. a. −100.5 rad/s^2 b. 50.0
9. 36.3 cm/s
11. −0.20 Nm
13. 175.5 N
15. 12,500 Nm
17. a. (0.0571 m, 0.0571 m) b. 0.0080 kgm^2
19. a. (0.060 m, 0.040 m) b. 0.0020 kgm^2 c. 0.00128 kgm^2
21. 0.75 rad/s
23. 0.0471 Nm
25. 11.76 Nm

27. 1.40 m
29. 15.8 J
31. 1.75 J
33. 0.375 J
35. a. (20.78, out of page) b. (24, into page)
37. a. $\hat{j}$ b. $\hat{j}$
39. a. $\hat{\imath} + 3\hat{j} + 11\hat{k}$
b.

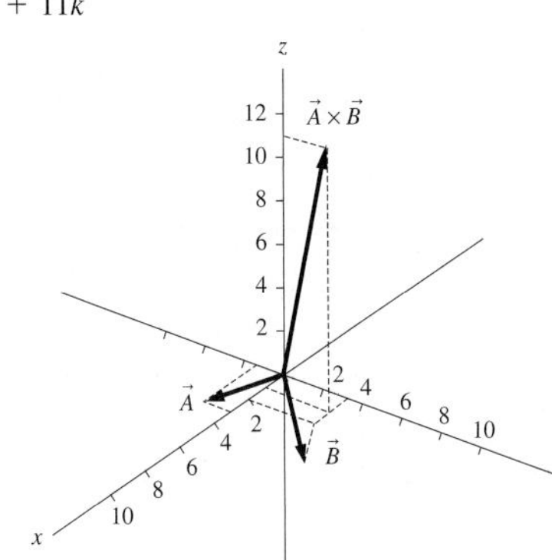

41. $-50\hat{k}$ N m
43. a. 8.97 s b. 0.448 kg m^2/s
45. $1.20\hat{k}$ kg m^2/s or (1.20 kg m^2/s, out of page)
47. $-0.0251\hat{\imath}$ kg m^2/s or (0.0251 kg m^2/s, into page)
49. 28.3 m/s
51. a. 0.010 kg m^2 b. 0.030 kg m^2
53. a. $\frac{1}{2}M(R^2 + r^2)$ c. 1.37 m/s
55. $\frac{1}{6}ML^2$
57. Yes
59. 1.00 m
61. a. (20 cm, 80 cm) b. 0.48 kg m^2 c. 1.0 N m d. 56.4°
63. a. 24.4 yr b. 4080 m/s and 12,250 m/s
65. a. 177 s b. 5.55×10^5 J c. 139 kW d. 1300 N m
67. 1.11 s
69. 1.57 N
71. 4.25 m
73. a. $\sqrt{2g/R}$ b. $\sqrt{8gR}$
75. $20\tau/13MR^2$
77. a. 42.9 cm b. No
79. 50 rpm
81. a. No b. 2000 m/s c. 4000 m/s
83. a. $3v_0/2d$ b. No
85. 393 m/s
87. a. 68,700 m b. 4.32×10^6 m/s

Chapter 14

1. 2.27 ms
3. a. 3.3 s b. 0.303 Hz c. 1.904 rad/s d. 0.25 m e. 0.476 m/s
5. a. 10 cm b. 0.50 Hz c. $\pi/3$ rad or 60°
7.

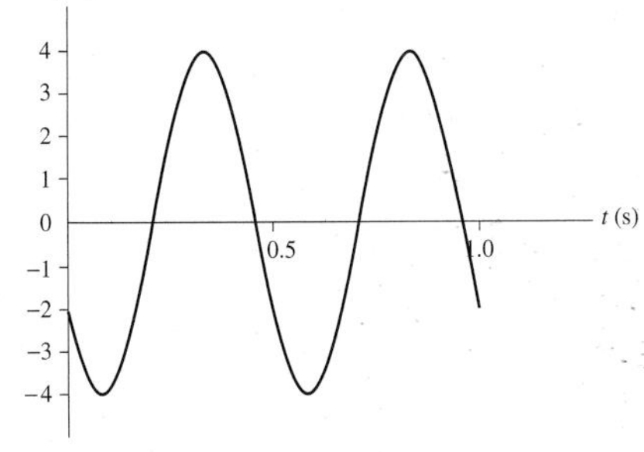

9. $x(t) = (4.0\text{ cm})\cos[(8.0\pi\text{ rad/s})t - \pi/2]$
11. a. $-2\pi/3$ rad or $-120°$ b. $-2\pi/3$ rad, 0 rad, $2\pi/3$ rad, $4\pi/3$ rad
13. 5.48 N/m
15. a. 0.50s b. 4π rad/s c. 5.54cm d. 0.445rad e. 69.6 cm/s f. 875 cm/s^2 g. 0.484 J h. 3.81 cm
17. a. 10.0 cm b. 34.6 cm/s
19. a. 0.169 kg b. 0.565 m/s
21. c. 12° d. 10° e. 0° to 10°
23. 35.7 cm
25. 0.330 m
27. 5.0 s
29. 21
31.

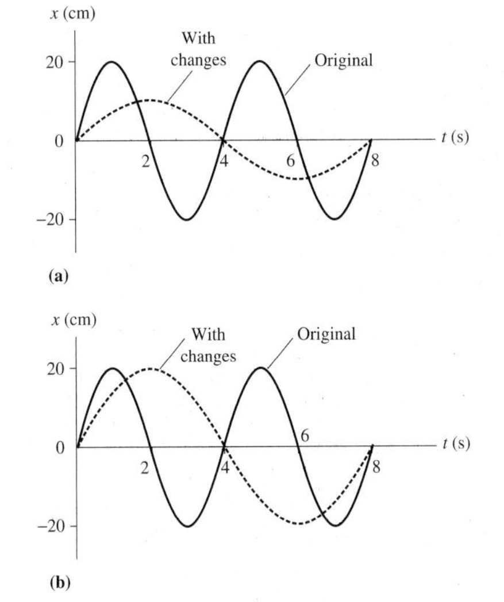

33. a. $-\pi/3$ rad or $-60°$ b. 6.80 cm/s b. 7.85 cm/s
35. a. 0.25 Hz, 3.0 s b. 6.0 s, 1.5 s c. 2.25
37. 1.405 s,

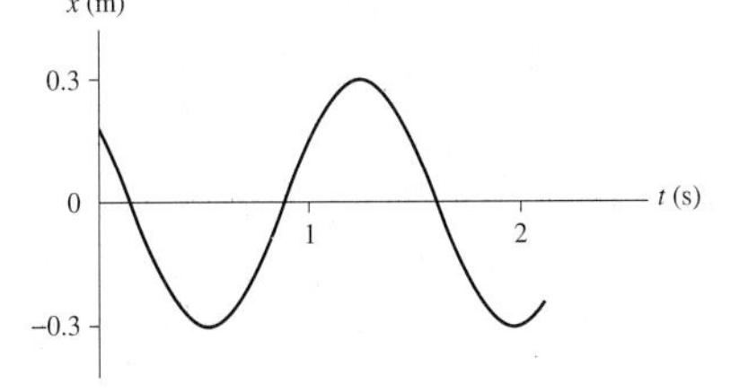

39. 0.0955 s
43. a. 6.40 cm b. 160 cm/s^2 c. -6.40 cm d. 0.283 m/s
45. 1.02 m/s
47. a. 3.18 Hz b. 0.0707 m c. 5.0 J
49. a. 1.125 Hz b. 23.5 cm c. -4.09 cm
51. a. 47,400 N/m b. 1.80 Hz
53. a. 0.314 s b. It would be unchanged.
55. 0.716
57. 0.669 s
59. a. 200.9 s b. 200.4 s c. Yes d. 9.77 m/s^2
61. 0.110 m at 1.72 s
63. a. 502 m/s b. No
65. $f = (1/2\pi)\sqrt{2T/mL}$
67. $T = 2\pi\sqrt{R/g}$
69. $g_X = 5.86$ m/s^2
71. a. 6.03 cm b. 6.32 s
73. 7.3°
77. 1.83 Hz
79. 2.23 cm

Chapter 15

1. $1200\ \text{kg/m}^3$
3. 1.44×10^5 kg
5. 1097 atm
7. 2440 kg
9. 3153 m
11. 88,000 Pa
13. 55.2 cm
15. Ethyl alcohol
17. 45.8 kg
19. 1.87 N
21. 3.18 m/s
23. $1.27v_0$
25. 2.0 kg
27. 1.0 mm
29. 0.2%
31. a. 5830 N b. 5990 N
33. a. 0.377 N b. 20.4 m/s
35. 5.27×10^{18} kg
37. a. 10.85 m b. 10.21 m
39. a. 0.483 m b. 2.34 cm
41. 3.7 mm
43. a. $\frac{1}{2}\rho g w d^2$ b. 1.76×10^9 N
45. a. 8080 m b. $1.05\ \text{kg/m}^3$, 82%
47. $667\ \text{kg/m}^3$
49. 74.7 N
51. 43.9 N
53. 8.38 cm
55. $(\rho - \rho_1)/(\rho_2 - \rho_1)$
57. 5.22 cm
59. 14.1 cm
61. a. p_{atmos} b. 4.61 m
63. a. Lower b. 835 Pa c. 75,100 N
65. a. 144 m/s and 5.78 m/s b. $4.54 \times 10^{-4}\ \text{m}^3/\text{s}$
67. a. 3.34 L/min b. 1.06 mm/min
69. 1.23 mm
71. 1.30 L
73. 3.6 g
75. e. 18.9 s

Part Overview Solutions

PART I Overview

If the acceleration of a Podracer can reach 50 m/s^2, what are the maximum tensions in the two large cables? To find out, what quantities do you need to estimate?

MODEL The cables from the two engines pull the passenger car, or Pod, forward. Treat the Pod as a particle. The size of the Pod suggests a mass of about 2000 kg. Because planets that have atmospheres and support life are likely to be similar in size to the earth, we'll estimate that the acceleration due to gravity is 10 m/s^2. The cables are curved and are angle θ above horizontal where they attach to the Pod. We'll neglect air resistance.

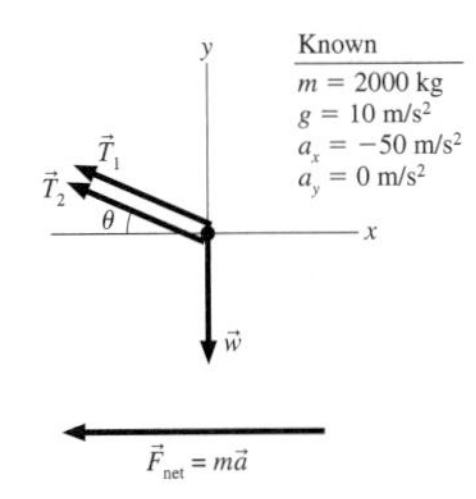

VISUALIZE The figure shows a free-body diagram of the Pod. Tensions $\vec{T}_1$ and $\vec{T}_2$ are due to the two cables. We'll assume equal magnitudes, so $T_1 = T_2 = T$. We want to find the tension when the acceleration reaches its maximum value $a_x = -50$ m/s^2, where the negative sign shows that the Podracer is accelerating to the left.

SOLVE Newton's second law for the Pod is

$$\sum F_x = -2T\cos\theta = ma_x$$

$$\sum F_y = 2T\sin\theta - mg = ma_y = 0$$

Thus $2T\sin\theta = mg$ and $2T\cos\theta = -ma_x$. Dividing the first of these equations by the second gives $\tan\theta = (-g/a_x)$, from which we find that

$$\theta = \tan^{-1}\left(-\frac{g}{a_x}\right) = \tan^{-1}\left(\frac{10 \text{ m/s}^2}{50 \text{ m/s}^2}\right) = 11.3°$$

We can see from the picture of the Podracer that this is a reasonable value for the angle. We can now use the angle to find the tension:

$$T = \frac{mg}{2\sin\theta} = \frac{(2000 \text{ kg})(10 \text{ m/s}^2)}{2\sin(11.3°)} = 51{,}000 \text{ N}$$

ASSESS The value of the tension would be slightly larger if we included air resistance, but 51,000 N is a reasonable estimate.

PART II Overview

The exploding Death Star releases 10^{33} J of energy. If the expanding shock wave exerts an impulse of 2.5×10^6 N s on the *Millennium Falcon*, by how much does the starship's velocity change? To find out, what property of the starship do you need to estimate?

MODEL The impulse-momentum theorem relates an object's change in momentum or change in velocity to the impulse exerted on it. Assume that the mass of the *Millennium Falcon* is 250,000 kg.

SOLVE The impulse momentum theorem is

$$\Delta p_x = m\Delta v_x = J_x$$

Thus the shock wave, when it hits, causes the starship's velocity to suddenly change by

$$\Delta v_x = \frac{J_x}{m} = \frac{2.5 \times 10^6 \text{ N s}}{2.5 \times 10^5 \text{ kg}} = 10 \text{ m/s}$$

ASSESS The starship experiences a sudden change of about 20 mph. This change will cause a noticeable jolt, but it is probably not sufficient to damage the starship.

PART III Overview

At what altitude above the surface of Alderaan does the Death Star orbit with a period of 10 hours? To find out, what properties of the planet do you need to estimate?

MODEL Assume that the Death Star is in a circular orbit around Alderaan. We need to know the mass and radius of Alderaan. Because planets that have atmospheres and support life are likely to be similar in size to the earth, we'll estimate that $M = 6.0 \times 10^{24}$ kg and $R = 6.1 \times 10^6$ m.

SOLVE The period T of a satellite in a circular orbit of radius r is given by

$$T^2 = \left(\frac{4\pi^2}{GM}\right)r^3$$

The radius of the orbit must be

$$r = \left[\left(\frac{GM}{4\pi^2}\right)T^2\right]^{1/3}$$

$$= \left[\frac{(6.67 \times 10^{-11} \text{ N m}^2/\text{kg}^2)(6.0 \times 10^{24} \text{ kg})(36{,}000 \text{ s})^2}{4\pi^2}\right]^{1/3}$$

$$= 23.6 \times 10^6 \text{ m}$$

Thus the Death Star's height above the surface is

$$h = r - R = 17.2 \times 10^6 \text{ m} = 17{,}200 \text{ km}$$

ASSESS The height is about 10,000 mi. This is much higher than the space shuttle orbits with a period of about 90 min, but less than the altitude of a geosynchronous satellite that orbits with a period of 24 hours. Thus the height is reasonable for a 10-hour period.

Credits

All Part Overview images are courtesy of Lucasfilm Ltd. Addison Wesley would like to give special thanks to Lucy Wilson, Christopher Holm, and the staff of Lucasfilm Ltd. for granting us permission to use these images and for their help in selecting them.

INTRODUCTION
Page **xxvi:** Courtesy of International Business Machines Corporation. Unauthorized use not permitted.

TITLE PAGE
Page **iii:** Rainbow/PictureQuest.

PART I
Part I Overview image: *Star Wars: Episode I – The Phantom Menace* © 1999 Lucasfilm Ltd. & ™. All rights reserved. Used under authorization. Unauthorized duplication is a violation of applicable law. Page **2:** Herman Eisenbeiss/Photo Researchers.

CHAPTER 1
Page **3:** Al Bello/Getty Images. Page **4** UL: David Woods/Corbis. Page **4** UR: Joseph Sohm/Corbis. Page **4** LL: Richard Megna/ Fundamental Photos. Page **4** LR: Fredrick M. Brown/Getty Images. Page **11:** Kevin Muggleton/Corbis. Page **13:** Wm. Sallaz/Corbis. Page **19:** Tony Freeman/PhotoEdit. Page **23:** Steve Smith/Getty Images. Page **26** T: U.S. Department of Commerce. Page **26** B: Bureau Int. des Poids et Mesures.

CHAPTER 2
Page **35:** Patrik Giardino/Corbis. Page **39:** Spencer Grant/PhotoEdit. Page **43** T: Tim Wright/Corbis. Page **43** B: Phil Boorman/Getty Images. Page **60:** James Sugar/Stockphoto.com. Page **63:** Scott Markewitz/Getty Images.

CHAPTER 3
Page **78:** Michael Yamashita/Corbis. Page **80:** Paul Chesley/Getty Images.

CHAPTER 4
Page **97:** George Lepp/Getty Images/Stone. Page **98** (a): Tony Freeman/ PhotoEdit. Page **98** (b): Brian Drake/Index Stock. Page **98** (c): Duomo/ Corbis. Page **98** (d): Dorling Kindersley Media Library. Page **98** (e): Chuck Savage/Corbis. Page **98** (f): Jeff Coolidge Photography. Page **101:** Jeff J. Daly/Fundamental Photographs. Page **112:** David Woods/Corbis.

CHAPTER 5
Page **122:** Joe McBride/Getty Images. Page **123:** PhotoDisc. Page **129:** Jonathan Nourak/PhotoEdit. Page **132:** Roger Ressmeyer/Corbis. Page **137** T: Corbis Digital Stock. Page **137** B: Jeff Coolidge Photography. Page **138:** Patrick Behar/Agence Vandystadt/Photo Researchers, Inc.

CHAPTER 6
Page **151:** Gerard Planchenault/Getty Images. Page **159:** Tony Freeman/ PhotoEdit. Page **161:** Richard Megna/Fundamental Photographs.

CHAPTER 7
Page **177:** Robin Smith/Getty Images. Page **185:** Robert Laberge/Getty Images. Page **192:** Sightseeing Archive/Getty Images. Page **194:** Tony Freeman/PhotoEdit.

CHAPTER 8
Page **207:** Chris Cole/Getty Images. Page **209:** Chuck Savage/Corbis. Page **212:** Pete Saloutos/Corbis.

PART II
Part II Overview image: *Star Wars: Episode VI – Return of the Jedi* © 1983 and 1997 Lucasfilm Ltd. & ™. All rights reserved. Used under authorization. Unauthorized duplication is a violation of applicable law. Page **238:** M.C. Escher Heirs/Cordon Art, Baarn, Holland.

CHAPTER 9
Page **239:** Russ Kinne/Comstock. Page **254:** Roger Ressmeyer/Corbis. Page **257:** Richard Megna/Fundamental Photos. Page **259:** Frederick M. Brown/Getty Images.

CHAPTER 10
Page **268:** Wally McNamee/Corbis. Page **270:** Roger Ressmeyer/Corbis. Page **276:** Lester Lefkowitz/Corbis. Page **280:** Gary Buss/Getty Images. Page **282:** Paul Harris/Getty Images. Page **287:** Dorling Kindersley Media Library.

CHAPTER 11
Page **304:** Shaun Botterill/Getty Images. Page **308:** Bettmann/Corbis. Page **323:** Al Behrman/AP Wide World. Page **329:** AFP/Corbis.

PART III
Part III Overview image: *Star Wars: Episode IV – A New Hope* © 1977 and 1997 Lucasfilm Ltd. & ™. All rights reserved. Used under authorization. Unauthorized duplication is a violation of applicable law. Page **342:** Courtesy Sandia National Laboratories/SUMMiT™ Technologies.

CHAPTER 12
Page **343:** PhotoDisc/Getty Images. Page **344:** Library of Congress. Page **346:** American Institute of Physics/Emilio Segre Visual Archives/ Regents of the University of California. Page **349:** Corbis Digital Stock. Page **358:** NASA.

CHAPTER 13
Page **369:** Glyn Kirk/Getty Images. Page **377:** Hart, G. K. & Vikki/Getty Images: Page **390:** Alain Choisnet/Getty Images. Page **393:** Richard Megna/Fundamental Photos.

CHAPTER 14
Page **413:** Michael Neveux/Corbis. Page **425:** Courtesy of Professor Thomas D. Rossing, Northern Illinois University. Page **430:** Richard Megna/Fundamental Photos. Page **432:** AP Photo/Toby Talbot. Page **436:** Martin Bough/Fundamental Photographs.

CHAPTER 15
Page **444:** Stu Forster/Getty Images. Page **450:** Richard Megna/ Fundamental Photos. Page **453:** Joseph Sinnot/Fundamental Photographs. Page **462:** Diane Hirsch/Fundamental Photos. Page **463:** Andy Sacks/ Getty Images. Page **464:** Don Farrall/Getty Images.

Index

For users of the five-volume edition: pages 1–481 are in Volume 1; pages 482–607 are in Volume 2; pages 608–779 are in Volume 3; pages 780–1194 are in Volume 4; and pages 1148–1383 are in Volume 5.
Pages 1195–1383 are not in the Standard Edition of this textbook.

D

E

F

G

H

N

O

S

T

U

V